MATRIX ANALYSIS AND APPLICATIONS

This balanced and comprehensive study presents the theory, methods and applications of matrix analysis in a new theoretical framework, allowing readers to understand second-order and higher-order matrix analysis in a completely new light.

Alongside the core subjects in matrix analysis, such as singular value analysis, the solution of matrix equations and eigenanalysis, the author introduces new applications and perspectives that are unique to this book. The very topical subjects of gradient analysis and optimization play a central role here. Also included are subspace analysis, projection analysis and tensor analysis, subjects which are often neglected in other books. Having provided a solid foundation to the subject, the author goes on to place particular emphasis on the many applications matrix analysis has in science and engineering, making this book suitable for scientists, engineers and graduate students alike.

XIAN-DA ZHANG is Professor Emeritus in the Department of Automation, at Tsinghua University, Beijing. He was a Distinguished Professor at Xidian University, Xi'an, China – a post awarded by the Ministry of Education of China, and funded by the Ministry of Education of China and the Cheung Kong Scholars Programme – from 1999 to 2002. His areas of research include signal processing, pattern recognition, machine learning and related applied mathematics. He has published over 120 international journal and conference papers, and 7 books in Chinese. He taught the graduate course "Matrix Analysis and Applications" at Tsinghua University from 2004 to 2011.

MATRIX ANALYSIS AND APPLICATIONS

XIAN-DA ZHANG
Tsinghua University, Beijing

CAMBRIDGE
UNIVERSITY PRESS

University Printing House, Cambridge CB2 8BS, United Kingdom

One Liberty Plaza, 20th Floor, New York, NY 10006, USA

477 Williamstown Road, Port Melbourne, VIC 3207, Australia

4843/24, 2nd Floor, Ansari Road, Daryaganj, Delhi – 110002, India

79 Anson Road, #06–04/06, Singapore 079906

Cambridge University Press is part of the University of Cambridge.

It furthers the University's mission by disseminating knowledge in the pursuit of education, learning, and research at the highest international levels of excellence.

www.cambridge.org
Information on this title: www.cambridge.org/9781108417419
DOI: 10.1017/9781108277587

First published 2017

Printed in the United Kingdom by Clays, St Ives plc

A catalogue record for this publication is available from the British Library.

ISBN 978-1-108-41741-9 Hardback

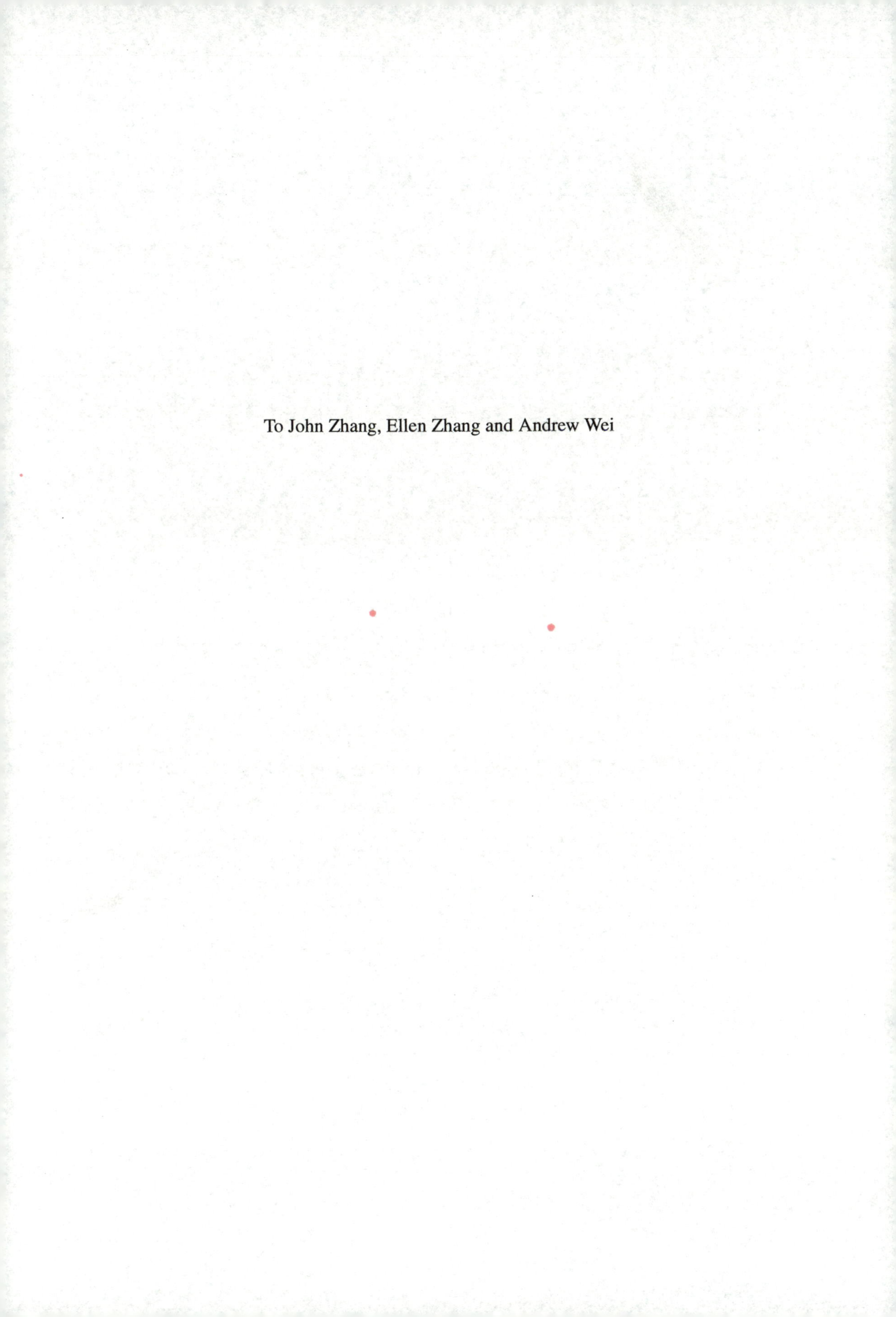

To John Zhang, Ellen Zhang and Andrew Wei

Contents

Preface

Linear algebra is a vast field of fundamental importance in most areas of pure (and applied) mathematics, while matrices are a key tool for the researchers, scientists, engineers and graduate students majoring in the science and engineering disciplines.

From the viewpoint of applications, matrix analysis provides a powerful mathematical modeling and computational framework for posing and solving important scientific and engineering problems. It is no exaggeration to say that matrix analysis is one of the most creative and flexible mathematical tools and that it plays an irreplaceable role in physics, mechanics, signal and information processing, wireless communications, machine learning, computer vision, automatic control, system engineering, aerospace, bioinformatics, medical image processing and many other disciplines, and it effectively supports research in them all. At the same time, novel applications in these disciplines have spawned a number of new results and methods of matrix analysis, such as quadratic eigenvalue problems, joint diagonalization, sparse representation and compressed sensing, matrix completion, nonnegative matrix factorization, tensor analysis and so on.

Goal of the Book

The main goal of this book is to help the reader develop the skills and background needed to recognize, formulate and solve linear algebraic problems by presenting systematically the theory, methods and applications of matrix analysis.

A secondary goal is to help the reader understand some recent applications, perspectives and developments in matrix analysis.

Structure of the Book

In order to provide a balanced and comprehensive account of the subject, this book covers the core theory and methods in matrix analysis, and places particular emphasis on its typical applications in various science and engineering disciplines. The book consists of ten chapters, spread over three parts.

Part I is on matrix algebra: it contains Chapters 1 through 3 and focuses on the necessary background material. Chapter 1 is an introduction to matrix algebra that is devoted to basic matrix operations. This is followed by a description of the vectorization of matrices, the representation of vectors as matrices, i.e. matricization, and the application of sparse matrices to face recognition. Chapter 2 presents some special matrices used commonly in matrix analysis. Chapter 3 presents the matrix differential, which is an important tool in optimization.

Part II is on matrix analysis: this is the heart of the book, and deals with the topics that are most frequently needed. It covers both theoretical and practical aspects and consists of six chapters, as follows.

Chapter 4 is devoted to the gradient analysis of matrices, with applications in smooth and nonsmooth convex optimization, constrained convex optimization, Newton's algorithm and the original–dual interior-point method.

In Chapter 5 we describe the singular value analysis of matrices, including singular value decomposition, generalized singular value decomposition, low-rank sparse matrix decomposition and matrix completion.

Researchers, scientists, engineers and graduate students from a wide variety of disciplines often have to use matrices for modeling purposes and to solve the resulting matrix equations. Chapter 6 focuses on ways to solve such equations and includes the Tikhonov regularization method, the total least squares method, the constrained total least squares method, nonnegative matrix factorization and the solution of sparse matrix equations.

Chapter 7 deals with eigenvalue decomposition, matrix reduction, generalized eigenvalue decomposition, the Rayleigh quotient, the generalized Rayleigh quotient, quadratic eigenvalue problems and joint diagonalization.

Chapter 8 is devoted to subspace analysis methods and subspace tracking algorithms in adaptive signal processing.

Chapter 9 focuses on orthogonal and oblique projections with their applications.

Part III is on higher-order matrix analysis and consists simply of Chapter 10. In it, matrix analysis is extended from the second-order case to higher orders via a presentation of the basic algebraic operations, representation as matrices, Tuckey decomposition, parallel factor decomposition, eigenvalue decomposition of tensors, nonnegative tensor decomposition and tensor completion, together with applications.

Features of the Book

The book introduces a novel theoretical framework for matrix analysis by dividing it into second-order matrix analysis (including gradient analysis, singular value analysis, eigenanalysis, subspace analysis and projection analysis) and higher-order matrix analysis (tensor analysis).

Gradient analysis and optimization play an important role in the book. This is a very topical subject and is central to many modern applications (such as communications, signal processing, pattern recognition, machine learning, radar, big data analysis, multimodal brain image fusion etc.) though quite classical in origin.

Some more contemporary topics of matrix analysis such as subspace analysis, projection analysis and tensor analysis, and which are often missing from other books, are included in our text.

Particular emphasis is placed on typical applications of matrix methods in science and engineering. The 80 algorithms for which summaries are given should help readers learn how to conduct computer experiments using related matrix analysis in their studies and research.

In order to make these methods easy to understand and master, this book adheres to the principle of both interpreting physics problems in terms of mathematics, and mathematical results in terms of physical ideas. Thus some typical or important matrix analysis problems are introduced by modeling a problem from physics, while some important mathematical results are explained and understood by revealing their physical meaning.

Reading the Book

The following diagram gives a schematic organization of this book to illustrate the chapter dependences.

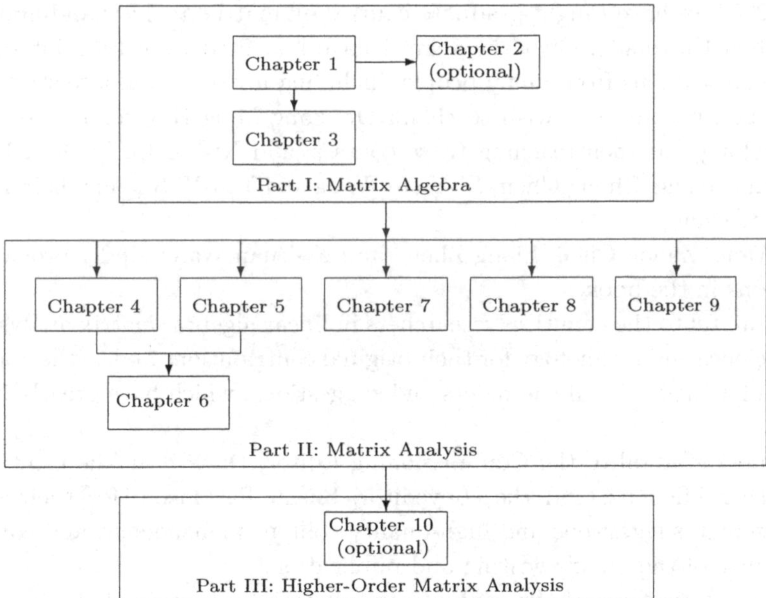

Chapters 2 and 10 are optional. In particular, Chapter 10 is specifically devoted to readers involved in multi-channel or multi-way data analysis and processing.

Intended Readership

Linear algebra and matrix analysis are used in a very wide range of subjects including physics, statistics, computer science, economics, information science and technology (including signal and image processing, communications, automation control, system engineering and pattern recognition), artificial intelligence, bioinformatics, biomedical engineering, to name just a selection. This book is dedicated to providing individuals in those disciplines with a solid foundation of the fundamental skills needed to develop and apply linear algebra and matrix analysis methods in their work.

The only background required of the reader is a good knowledge of advanced calculus, so the book will be suitable for graduate students in science and engineering.

Acknowledgments

The contents of this book reflect the author's collaboration with his own graduate students Jian Li, Zi-Zhe Ding, Yong-Tao Su, Xi-Lin Li, Heng Yang, Xi-Kai Zhao, Qi Lv, Qiu-Beng Gao, Li Zhang, Jian-Jiang Ding, Lu Wu, Feng Zhu, Ling Zhang, Dong-Xia Chang, De-Guang Xie, Chun-Yu Peng, Dao-Ming Zhang, Kun Wang, Xi-Yuan Wang, Zhong Chen, Tian-Xiang Luan, Liang Zheng, Yong Zhang, Yan-Yi Rao, all at Tsinghua University, and Shun-Tian Lou, Xiao-Long Zhu, Ji-Ming Ye, Fang-Ming Han, Xiao-Jun Li, Jian-Feng Chen at Xidian University.

Since 2004 we have taught graduate courses on matrix analysis and applications at Tsinghua University. Over the years I have benefited from keen interest, feedback and suggestions from many people, including my own graduate students, and students in our courses. I wish to thank Dr. Fang-Ming Han for his contribution to co-teaching and then teaching these courses, and Xi-Lin Li, Lu Wu, Dong-Xia Chang, Kun Wang, Zhong Chen, Xi-Yuan Wang and Yan-Yi Rao for their assistance with the teaching.

Kun Wang, Zhong Chen, Liang Zheng and Xi-Yuan Wang kindly provided some illustrations in the book.

I am grateful to the countless researchers in linear algebra, matrix analysis, information science and technology for their original contributions and to the anonymous reviewers for their critical comments and suggestions, which have greatly improved the text.

I am most grateful to the Commissioning Editor, David Liu, the Content Manager, Esther Miguéliz, and the copyeditor, Susan Parkinson, for their patience, understanding, suggestions and high-quality content management and copyediting in the course of the book's writing and publication.

This book uses some of the contents and materials of my book *Matrix Analysis and Applications* (Second Edition in Chinese, Tsinghua University Press, 2013).

Finally, I am grateful to my wife Xiao-Ying Tang, my son Yuan-Sheng Zhang, my daughter-in-law Lin Yan, my daughter Ye-Wei Zhang, my son-in-law Wei Wei for their support and encouragement in this project.

Notation

Sets

$\mathbb{R}$	real numbers
$\mathbb{R}^n$	real n-vectors ($n \times 1$ real matrices)
$\mathbb{R}^{m \times n}$	real $m \times n$ matrices
$\mathbb{R}[x]$	real polynomials
$\mathbb{R}[x]^{m \times n}$	real $m \times n$ polynomial matrices
$\mathbb{R}^{I \times J \times K}$	real third-order tensors
$\mathbb{R}^{I_1 \times \cdots \times I_N}$	real Nth-order tensor
$\mathbb{R}_+$	nonnegative real numbers, nonnegative orthant
$\mathbb{R}_{++}$	positive real numbers
$\mathbb{C}$	complex numbers
$\mathbb{C}^n$	complex n-vectors
$\mathbb{C}^{m \times n}$	complex $m \times n$ matrices
$\mathbb{C}[x]$	complex polynomials
$\mathbb{C}[x]^{m \times n}$	complex $m \times n$ polynomial matrices
$\mathbb{C}^{I \times J \times K}$	complex third-order tensors
$\mathbb{C}^{I_1 \times \cdots \times I_N}$	complex Nth-order tensors
$\mathbb{K}$	real or complex numbers
$\mathbb{K}^n$	real or complex n-vectors
$\mathbb{K}^{m \times n}$	real or complex $m \times n$ matrices
$\mathbb{K}^{I \times J \times K}$	real or complex third-order tensors
$\mathbb{K}^{I_1 \times \cdots \times I_N}$	real or complex Nth-order tensors
$\mathbb{Z}$	integers
$\mathbb{Z}_+$	nonnegative integers

Sets (*continued*)

$\mathbb{S}^{n\times n}$	symmetric $n \times n$ matrices
$\mathbb{S}_{+}^{n\times n}$	symmetric positive semi-definite $n \times n$ matrices
$\mathbb{S}_{++}^{n\times n}$	symmetric positive definite $n \times n$ matrices
$\mathbb{S}^{[m,n]}$	symmetric mth-order n-dimensional tensors $\mathcal{A}^{I_1\times\cdots\times I_m}$, $I_1 = \cdots = I_n$
$\mathbb{S}_{+}^{[m,n]}$	symmetric mth-order n-dimensional nonnegative tensors
$\forall$	for all
$x \in A$	x belongs to the set A, i.e. x is an element of A
$x \notin A$	x is not an element of the set A
$U \mapsto V$	U maps to V
$U \rightarrow W$	U transforms to W
$\ni$	such that
$\exists$	exists
$A \Rightarrow B$	A implies B
$A \subseteq B$	A is a subset of B
$A \subset B$	A is a proper subset of B
$A \cup B$	union of sets A and B
$A \cap B$	intersection of sets A and B
$A + B$	sum set of sets A and B
$A - B$	set-theoretic difference of sets A and B
$X \setminus A$	complement of the set A in the set X
$X_1 \times \cdots \times X_n$	Cartesian product of sets $X_1, \ldots, X_n$
$\mathcal{L}$	linear manifold
$\text{Gr}(n, r)$	Grassmann manifold
$\text{St}(n, r)$	Stiefel manifold
O_r	orthogonal group
$S^{\perp}$	orthogonal complement of the subspace S
$\mathcal{K}^m(\mathbf{A}, \mathbf{f})$	order-m Krylov subspace generated by $\mathbf{A}$ and $\mathbf{f}$
$\text{Col}(\mathbf{A})$	column space of the matrix $\mathbf{A}$
$\text{Ker}(\mathbf{A})$	kernel space of the matrix $\mathbf{A}$
$\text{Null}(\mathbf{A})$	null space of the matrix $\mathbf{A}$
$\text{nullity}(\mathbf{A})$	nullity of the matrix $\mathbf{A}$
$\text{Range}(\mathbf{A})$	range space of the matrix $\mathbf{A}$

Sets (*continued*)

Row($\mathbf{A}$)	row space of the matrix $\mathbf{A}$
Span($\mathbf{a}_1, \ldots, \mathbf{a}_m$)	span of vectors $\mathbf{a}_1, \ldots, \mathbf{a}_m$

Vectors

$\mathbf{x}^*$	conjugate of the vector $\mathbf{x}$
$\mathbf{x}^T$	transpose of the vector $\mathbf{x}$
$\mathbf{x}^H$	conjugate transpose (Hermitian conjugate) of the vector $\mathbf{x}$
$\mathcal{L}(\mathbf{u})$	linear transform of the vector $\mathbf{u}$
$\\|\mathbf{x}\\|_0$	ℓ_0-norm: the number of nonzero entries in the vector $\mathbf{x}$
$\\|\mathbf{x}\\|_1$	ℓ_1-norm of the vector $\mathbf{x}$
$\\|\mathbf{x}\\|_2$	Euclidean norm of the vector $\mathbf{x}$
$\\|\mathbf{x}\\|_p$	ℓ_p-norm or Hölder norm of the vector $\mathbf{x}$
$\\|\mathbf{x}\\|_*$	nuclear norm of the vector $\mathbf{x}$
$\\|\mathbf{x}\\|_\infty$	ℓ_∞-norm of the vector $\mathbf{x}$
$\langle \mathbf{x}, \mathbf{y} \rangle = \mathbf{x}^H \mathbf{y}$	inner product of vectors $\mathbf{x}$ and $\mathbf{y}$
$\mathbf{x} \circ \mathbf{y} = \mathbf{x}\mathbf{y}^H$	outer product of vectors $\mathbf{x}$ and $\mathbf{y}$
$\mathbf{x} \perp \mathbf{y}$	orthogonality of vectors $\mathbf{x}$ and $\mathbf{y}$
$\mathbf{x} > \mathbf{0}$	positive vector, with components $x_i > 0, \forall\, i$
$\mathbf{x} \geq \mathbf{0}$	nonnegative vector, with components $x_i \geq 0, \forall\, i$
$\mathbf{x} \geq \mathbf{y}$	vector elementwise inequality $x_i \geq y_i, \forall\, i$
unvec($\mathbf{x}$)	matricization of the column vector $\mathbf{x}$
unrvec($\mathbf{x}$)	row matricization of the column vector $\mathbf{x}$
$\theta_i^{(m)}, \mathbf{y}_i^{(m)}$	Rayleigh–Ritz (RR) values, RR vectors
$(\theta_i^{(m)}, \mathbf{y}_i^{(m)})$	Ritz pair

Matrices

$\mathbf{A} \in \mathbb{R}^{m \times n}$	real $m \times n$ matrix $\mathbf{A}$
$\mathbf{A} \in \mathbb{C}^{m \times n}$	complex $m \times n$ matrix $\mathbf{A}$
$\mathbf{A}[x] \in \mathbb{R}[x]^{m \times n}$	real $m \times n$ polynomial matrix $\mathbf{A}$
$\mathbf{A}[x] \in \mathbb{C}[x]^{m \times n}$	complex $m \times n$ polynomial matrix $\mathbf{A}$
$\mathbf{A}^*$	conjugate of $\mathbf{A}$
$\mathbf{A}^T$	transpose of $\mathbf{A}$

Matrices (*continued*)

$\mathbf{A}^H$	conjugate transpose (Hermitian conjugate) of $\mathbf{A}$
$(\mathbf{A}, \mathbf{B})$	matrix pencil
$\det(\mathbf{A}), \lvert\mathbf{A}\rvert$	determinant of $\mathbf{A}$
$\mathrm{tr}(\mathbf{A})$	trace of $\mathbf{A}$
$\mathrm{rank}(\mathbf{A})$	rank of $\mathbf{A}$
$\mathrm{eig}(\mathbf{A})$	eigenvalues of the Hermitian matrix $\mathbf{A}$
$\lambda_i(\mathbf{A})$	ith eigenvalue of the Hermitian matrix $\mathbf{A}$
$\lambda_{\max}(\mathbf{A})$	maximum eigenvalue(s) of the Hermitian matrix $\mathbf{A}$
$\lambda_{\min}(\mathbf{A})$	minimum eigenvalue(s) of the Hermitian matrix $\mathbf{A}$
$\lambda(\mathbf{A}, \mathbf{B})$	generalized eigenvalue of the matrix pencil $(\mathbf{A}, \mathbf{B})$
$\sigma_i(\mathbf{A})$	ith singular value of $\mathbf{A}$
$\sigma_{\max}(\mathbf{A})$	maximum singular value(s) of $\mathbf{A}$
$\sigma_{\min}(\mathbf{A})$	minimum singular value(s) of $\mathbf{A}$
$\rho(\mathbf{A})$	spectral radius of $\mathbf{A}$
$\mathbf{A}^{-1}$	inverse of the nonsingular matrix $\mathbf{A}$
$\mathbf{A}^\dagger$	Moore–Penrose inverse of $\mathbf{A}$
$\mathbf{A} \succ 0$	positive definite matrix $\mathbf{A}$
$\mathbf{A} \succeq 0$	positive semi-definite matrix $\mathbf{A}$
$\mathbf{A} \prec 0$	negative definite matrix $\mathbf{A}$
$\mathbf{A} \preceq 0$	negative semi-definite matrix $\mathbf{A}$
$\mathbf{A} > \mathbf{O}$	positive (or elementwise positive) matrix $\mathbf{A}$
$\mathbf{A} \geq \mathbf{O}$	nonnegative (or elementwise nonnegative) matrix $\mathbf{A}$
$\mathbf{A} \geq \mathbf{B}$	matrix elementwise inequality $a_{ij} \geq b_{ij}, \forall\, i, j$
$\lVert\mathbf{A}\rVert_1$	maximum absolute column-sum norm of $\mathbf{A}$
$\lVert\mathbf{A}\rVert_\infty$	maximum absolute row-sum norm of $\mathbf{A}$
$\lVert\mathbf{A}\rVert_{\mathrm{spec}}$	spectrum norm of $\mathbf{A}: \sigma_{\max}(\mathbf{A})$
$\lVert\mathbf{A}\rVert_F$	Frobenius norm of $\mathbf{A}$
$\lVert\mathbf{A}\rVert_\infty$	max norm of $\mathbf{A}$: the absolute maximum of all entries of $\mathbf{A}$
$\lVert\mathbf{A}\rVert_{\mathbf{G}}$	Mahalanobis norm of $\mathbf{A}$
$\mathrm{vec}\,\mathbf{A}$	column vectorization of $\mathbf{A}$
$\mathrm{rvec}\,\mathbf{A}$	row vectorization of $\mathbf{A}$
$\mathrm{off}(\mathbf{A})$	off function of $\mathbf{A} = [a_{ij}]$: $\sum_{i=1,i\neq j}^m \sum_{j=1}^n \lvert a_{ij}\rvert^2$
$\mathrm{diag}(\mathbf{A})$	diagonal function of $\mathbf{A} = [a_{ij}]$: $\sum_{i=1}^n \lvert a_{ii}\rvert^2$

Matrices (*continued*)

$\mathbf{diag}(\mathbf{A})$	diagonal vector of $\mathbf{A} = [a_{ij}]: [a_{11}, \ldots, a_{nn}]^T$
$\mathbf{Diag}(\mathbf{A})$	diagonal matrix of $\mathbf{A} = [a_{ij}]: \mathbf{Diag}(a_{11}, \ldots, a_{nn})$
$\langle \mathbf{A}, \mathbf{B} \rangle$	inner product of $\mathbf{A}$ and $\mathbf{B}$: $(\text{vec}\,\mathbf{A})^H \text{vec}\,\mathbf{B}$
$\mathbf{A} \otimes \mathbf{B}$	Kronecker product of matrices $\mathbf{A}$ and $\mathbf{B}$
$\mathbf{A} \odot \mathbf{B}$	Khatri–Rao product of matrices $\mathbf{A}$ and $\mathbf{B}$
$\mathbf{A} * \mathbf{B}$	Hadamard product of matrices $\mathbf{A}$ and $\mathbf{B}$
$\mathbf{A} \oplus \mathbf{B}$	direct sum of matrices $\mathbf{A}$ and $\mathbf{B}$
$\{\mathbf{A}\}_N$	matrix group consisting of matrices $\mathbf{A}_1, \ldots, \mathbf{A}_N$
$\{\mathbf{A}\}_N \otimes \mathbf{B}$	generalized Kronecker product of $\{\mathbf{A}\}_N$ and $\mathbf{B}$
$\delta\mathbf{x}, \delta\mathbf{X}$	perturbations of the vector $\mathbf{x}$ and the matrix $\mathbf{X}$
$\text{cond}(\mathbf{A})$	condition number of the matrix $\mathbf{A}$
$\text{In}(\mathbf{A})$	inertia of a symmetric matrix $\mathbf{A}$
$i_+(\mathbf{A})$	number of positive eigenvalues of $\mathbf{A}$
$i_-(\mathbf{A})$	number of negative eigenvalues of $\mathbf{A}$
$i_0(\mathbf{A})$	number of zero eigenvalues of $\mathbf{A}$
$\mathbf{A} \sim \mathbf{B}$	similarity transformation
$\mathbf{A}(\lambda) \cong \mathbf{B}(\lambda)$	balanced transformation
$\mathbf{A} \doteq \mathbf{B}$	essentially equal matrices
$\mathbf{J} = \mathbf{PAP}^{-1}$	Jordan canonical form of the matrix $\mathbf{A}$
$d_k(x)$	kth determinant divisor of a polynomial matrix $\mathbf{A}(x)$
$\sigma_k(x)$	kth invariant factor of a polynomial matrix $\mathbf{A}(x)$
$\mathbf{A}(\lambda)$	λ-matrix of the matrix $\mathbf{A}$
$\mathbf{S}(\lambda)$	Smith normal form of the λ-matrix $\mathbf{A}(\lambda)$

Special Vectors and Special Matrices

$\mathbf{P}_S$	projector onto the subspace S
$\mathbf{P}_S^\perp$	orthogonal projector onto the subspace S
$\mathbf{E}_{H\|S}$	oblique projector onto the subspace H along the subspace S
$\mathbf{E}_{S\|H}$	oblique projector onto the subspace S along the subspace H
$\mathbf{1}$	summing vector with all entries 1
$\mathbf{0}$	null or zero vector with all components 0
$\mathbf{e}_i$	basic vector with $e_i = 1$ and all other entries 0
$\boldsymbol{\pi}$	extracting vector with the last nonzero entry 1

Special Vectors and Special Matrices (*continued*)

$\mathbf{i}_N$	index vector: $[\langle 0\rangle, \langle 1\rangle, \ldots, \langle N-1\rangle]^T$
$\mathbf{i}_{N,\text{rev}}$	bit-reversed index vector of $\mathbf{i}_N$
$\mathbf{O}$	null or zero matrix, with all components zero
$\mathbf{I}$	identity matrix
$\mathbf{K}_{mn}$	$mn \times mn$ commutation matrix
$\mathbf{J}_n$	$n \times n$ exchange matrix: $\mathbf{J}_n = [\mathbf{e}_n, \ldots, \mathbf{e}_1]$
$\mathbf{P}$	$n \times n$ permutation matrix: $\mathbf{P} = [\mathbf{e}_{i_1}, \ldots, \mathbf{e}_{i_n}], i_1, \ldots, i_n \in \{1, \ldots, n\}$
$\mathbf{G}$	generalized permutation matrix or g-matrix: $\mathbf{G} = \mathbf{PD}$
$\mathbf{C}_n$	$n \times n$ centering matrix $= \mathbf{I}_n - n^{-1}\mathbf{1}_n\mathbf{1}_n^T$
$\mathbf{F}_N$	$N \times N$ Fourier matrix with entry $F(i,k) = (\mathrm{e}^{-\mathrm{j}2\pi/N})^{(i-1)(k-1)}$
$\mathbf{F}_{N,\text{rev}}$	$N \times N$ bit-reversed Fourier matrix
$\mathbf{H}_n$	Hadamard matrix: $\mathbf{H}_n\mathbf{H}_n^T = \mathbf{H}_n^T\mathbf{H}_n = n\mathbf{I}_n$
$\mathbf{A}$	symmetric Toeplitz matrix: $[a_{\|i-j\|}]_{i,j=1}^n$
$\mathbf{A}$	complex Toeplitz matrix: $\text{Toep}[a_0, a_1, \ldots, a_n]$ with $a_{-i} = a_i^*$
$\mathbf{Q}_n$	$n \times n$ real orthogonal matrix: $\mathbf{QQ}^T = \mathbf{Q}^T\mathbf{Q} = \mathbf{I}$
$\mathbf{U}_n$	$n \times n$ unitary matrix: $\mathbf{UU}^H = \mathbf{U}^H\mathbf{U} = \mathbf{I}$
$\mathbf{Q}_{m\times n}$	$m \times n$ semi-orthogonal matrix: $\mathbf{Q}^T\mathbf{Q} = \mathbf{I}_n$ or $\mathbf{QQ}^T = \mathbf{I}_m$
$\mathbf{U}_{m\times n}$	$m \times n$ para-unitary matrix: $\mathbf{U}^H\mathbf{U} = \mathbf{I}_n$, $m > n$, or $\mathbf{UU}^H = \mathbf{I}_m$, $m < n$
$\mathbf{S}_b$	between-class scatter matrix
$\mathbf{S}_w$	within-class scatter matrix

Tensors

$\mathcal{A} \in \mathbb{K}^{I_1\times\cdots\times I_N}$	Nth-order real or complex tensor
$\mathcal{I}, \mathcal{E}$	identity tensor
$\mathbf{A}_{i::}$	horizontal slice matrix of $\mathcal{A} \in \mathbb{K}^{I\times J\times K}$
$\mathbf{A}_{:j:}$	lateral slice matrix of $\mathcal{A} \in \mathbb{K}^{I\times J\times K}$
$\mathbf{A}_{::k}$	frontal slice matrix of $\mathcal{A} \in \mathbb{K}^{I\times J\times K}$
$\mathbf{a}_{:jk}, \mathbf{a}_{i:k}, \mathbf{a}_{ij:}$	mode-1, model-2, model-3 vectors of $\mathcal{A} \in \mathbb{K}^{I\times J\times K}$
$\text{vec}\,\mathcal{A}$	vectorization of tensor $\mathcal{A}$
$\text{unvec}\,\mathcal{A}$	matricization of tensor $\mathcal{A}$
$\mathbf{A}^{(JK\times I)}, \mathbf{A}^{(KI\times J)}, \mathbf{A}^{(IJ\times K)}$	matricization of tensor $\mathcal{A} \in \mathbb{K}^{I\times J\times K}$

Tensors (*continued*)

$\langle \mathcal{A}, \mathcal{B} \rangle$	inner product of tensors: $(\text{vec}\,\mathcal{A})^H \text{vec}\,\mathcal{B}$
$\|\mathcal{A}\|_F$	Frobenius norm of tensor $\mathcal{A}$
$\mathcal{A} = \mathbf{u} \circ \mathbf{v} \circ \mathbf{w}$	outer product of three vectors $\mathbf{u}, \mathbf{v}, \mathbf{w}$
$\mathcal{X} \times_n \mathbf{A}$	Tucker mode-n product of $\mathcal{X}$ and $\mathbf{A}$
$\text{rank}(\mathcal{A})$	rank of tensor $\mathcal{A}$
$[\![\mathcal{G}; \mathbf{U}^{(1)}, \ldots, \mathbf{U}^{(N)}]\!]$	Tucker operator of tensor $\mathcal{G}$
$\mathcal{G} \times_1 \mathbf{A} \times_2 \mathbf{B} \times_3 \mathbf{C}$	Tucker decomposition (third-order SVD)
$\mathcal{G} \times_1 \mathbf{U}^{(1)} \cdots \times_N \mathbf{U}^{(N)}$	higher-order SVD of Nth-order tensor
$x_{ijk} = \sum_{p=1}^{P} \sum_{q=1}^{Q} \sum_{r=1}^{R} a_{ip}\, b_{jq}\, c_{kr}$	CP decomposition of the third-order tensor $\mathcal{X}$
$\mathcal{A}\mathbf{x}^m, \mathcal{A}\mathbf{x}^{m-1}$	tensor–vector product of $\mathcal{A} \in \mathbb{S}^{[m,n]}, \mathbf{x} \in \mathbb{C}^{n \times 1}$
$\det(\mathcal{A})$	determinant of tensor $\mathcal{A}$
$\lambda_i(\mathcal{A})$	ith eigenvalue of tensor $\mathcal{A}$
$\sigma(\mathcal{A})$	spectrum of tensor $\mathcal{A}$

Functions and Derivatives

$\stackrel{\text{def}}{=}$	defined to be equal
$\sim$	asymptotically equal (in scaling sense)
$\approx$	approximately equal (in numerical value)
$f : \mathbb{R}^m \to \mathbb{R}$	real function $f(\mathbf{x}), \mathbf{x} \in \mathbb{R}^m, f \in \mathbb{R}$
$f : \mathbb{R}^{m \times n} \to \mathbb{R}$	real function $f(\mathbf{X}), \mathbf{X} \in \mathbb{R}^{m \times n}, f \in \mathbb{R}$
$f : \mathbb{C}^m \times \mathbb{C}^m \to \mathbb{R}$	real function $f(\mathbf{z}, \mathbf{z}^*), \mathbf{z} \in \mathbb{C}^m, f \in \mathbb{R}$
$f : \mathbb{C}^{m \times n} \times \mathbb{C}^{m \times n} \to \mathbb{R}$	real function $f(\mathbf{Z}, \mathbf{Z}^*), \mathbf{Z} \in \mathbb{C}^{m \times n}, f \in \mathbb{R}$
$\mathbf{dom}\, f, \mathcal{D}$	definition domain of function f
$\mathcal{E}$	domain of equality constraint function
$\mathcal{I}$	domain of inequality constraint function
$\mathcal{F}$	feasible set
$B_c(\mathbf{c}; r), B_o(\mathbf{c}; r)$	closed, open neighborhoods of $\mathbf{c}$ with radius r
$B_c(\mathbf{C}; r), B_o(\mathbf{C}; r)$	closed, open neighborhoods of $\mathbf{C}$ with radius r
$f(\mathbf{z}, \mathbf{z}^*), f(\mathbf{Z}, \mathbf{Z}^*)$	function of complex variables $\mathbf{z}$, or $\mathbf{Z}$
$\mathrm{d}f(\mathbf{z}, \mathbf{z}^*), \mathrm{d}f(\mathbf{Z}, \mathbf{Z}^*)$	complex differentials
$D(\mathbf{p}\|\mathbf{g})$	distance between vectors $\mathbf{p}$ and $\mathbf{g}$

Functions and Derivatives (*continued*)

$D(\mathbf{x},\mathbf{y})$	dissimilarity between vectors $\mathbf{x}$ and $\mathbf{y}$
$D_E(\mathbf{x},\mathbf{y})$	Euclidean distance between vectors $\mathbf{x}$ and $\mathbf{y}$
$D_M(\mathbf{x},\mathbf{y})$	Mahalanobis distance between vectors $\mathbf{x}$ and $\mathbf{y}$
$D_J^g(\mathbf{x},\mathbf{y})$	Bregman distance between vectors $\mathbf{x}$ and $\mathbf{y}$
$\mathrm{D}_{\mathbf{x}}, \mathrm{D}_{\mathrm{vec}\,\mathbf{X}}$	row partial derivative operators
$\mathrm{D}_{\mathbf{x}} f(\mathbf{x})$	row partial derivative vectors of $f(\mathbf{x})$
$\mathrm{D}_{\mathrm{vec}\,\mathbf{X}} f(\mathbf{X})$	row partial derivative vectors of $f(\mathbf{X})$
$\mathrm{D}_{\mathbf{X}}$	Jacobian operator
$\mathrm{D}_{\mathbf{X}} f(\mathbf{X})$	Jacobian matrix of the function $f(\mathbf{X})$
$\mathrm{D}_{\mathbf{z}}, \mathrm{D}_{\mathrm{vec}\,\mathbf{Z}}$	complex cogradient operator
$\mathrm{D}_{\mathbf{z}^*}, \mathrm{D}_{\mathrm{vec}\,\mathbf{Z}^*}$	complex conjugate cogradient operator
$\mathrm{D}_{\mathbf{z}} f(\mathbf{z},\mathbf{z}^*)$	cogradient vector of complex function $f(\mathbf{z},\mathbf{z}^*)$
$\mathrm{D}_{\mathrm{vec}\,\mathbf{Z}} f(\mathbf{Z},\mathbf{Z}^*)$	cogradient vector of complex function $f(\mathbf{Z},\mathbf{Z}^*)$
$\mathrm{D}_{\mathbf{z}^*} f(\mathbf{z},\mathbf{z}^*)$	conjugate cogradient vector of $f(\mathbf{z},\mathbf{z}^*)$
$\mathrm{D}_{\mathrm{vec}\,\mathbf{Z}^*} f(\mathbf{Z},\mathbf{Z}^*)$	conjugate cogradient vector of $f(\mathbf{Z},\mathbf{Z}^*)$
$\mathrm{D}_{\mathbf{Z}}, \nabla_{\mathbf{Z}^*}$	Jacobian, gradient matrix operator
$\mathrm{D}_{\mathbf{Z}} f(\mathbf{Z},\mathbf{Z}^*)$	Jacobian matrices of $f(\mathbf{Z},\mathbf{Z}^*)$
$\nabla_{\mathbf{Z}} f(\mathbf{Z},\mathbf{Z}^*)$	gradient matrices of $f(\mathbf{Z},\mathbf{Z}^*)$
$\mathrm{D}_{\mathbf{Z}^*} f(\mathbf{Z},\mathbf{Z}^*)$	conjugate Jacobian matrices of $f(\mathbf{Z},\mathbf{Z}^*)$
$\nabla_{\mathbf{Z}^*} f(\mathbf{Z},\mathbf{Z}^*)$	conjugate gradient matrices of $f(\mathbf{Z},\mathbf{Z}^*)$
$\nabla_{\mathbf{x}}, \nabla_{\mathrm{vec}\,\mathbf{X}}$	gradient vector operator
$\nabla_{\mathbf{x}} f(\mathbf{x})$	gradient vector of function $f(\mathbf{x})$
$\nabla_{\mathrm{vec}\,\mathbf{X}} f(\mathbf{X})$	gradient vector of function $f(\mathbf{X})$
$\nabla f(\mathbf{X})$	gradient matrix of function f
$\nabla^2 f$	Hessian matrix of function f
$\mathbf{H}_{\mathbf{x}} f(\mathbf{x})$	Hessian matrix of function f
$\mathbf{H} f(\mathbf{z},\mathbf{z}^*)$	full Hessian matrix of $f(\mathbf{z},\mathbf{z}^*)$
$\mathbf{H}_{\mathbf{z},\mathbf{z}}, \mathbf{H}_{\mathbf{z},\mathbf{z}^*}, \mathbf{H}_{\mathbf{z}^*,\mathbf{z}^*}$	part Hessian matrices of function $f(\mathbf{z},\mathbf{z}^*)$
$\mathrm{d}f, \partial f$	differential or subdifferential of function f
$\mathbf{g} \in \partial f$	subgradient of function f
$\Delta\mathbf{x}$	descent direction of function $f(\mathbf{x})$
$\Delta\mathbf{x}_{\mathrm{nt}}$	Newton step of function $f(\mathbf{x})$
$\max f, \min f$	maximize, minimize function f

Functions and Derivatives (*continued*)

$\max\{x, y\}$	maximum of x and y
$\min\{x, y\}$	minimum of x and y
inf	infimum
sup	supremum
Re, Im	real part, imaginary part of complex number
arg	argument of objective function or complex number
$\mathcal{P}_C(\mathbf{y}), \mathbf{P}_C\mathbf{y}$	projection operator of the vector $\mathbf{y}$ onto the subspace C
$\mathbf{x}^+$	nonnegative vector with entry $[\mathbf{x}^+]_i = \max\{x_i, 0\}$
$\mathbf{prox}_h(\mathbf{u})$	proximal operator of function $h(\mathbf{x})$ to point $\mathbf{u}$
$\mathbf{prox}_h(\mathbf{U})$	proximal operator of function $h(\mathbf{X})$ to point $\mathbf{U}$
$\text{soft}(x, \tau), \mathcal{S}_\tau[x]$	soft thresholding operator of real variable x

Probability

$\text{soft}(\mathbf{x}, \tau), \text{soft}(\mathbf{X}, \tau)$	soft thresholding operator of real variables $\mathbf{x}, \mathbf{X}$
$\mathcal{D}_\mu(\mathbf{\Sigma})$	singular value (matrix) thresholding (operation)
$I_C(\mathbf{x})$	indicator function
$\mathbf{x}_{\text{LS}}, \mathbf{X}_{\text{LS}}$	least squares solutions to $\mathbf{Ax} = \mathbf{b}, \mathbf{AX} = \mathbf{B}$
$\mathbf{x}_{\text{DLS}}, \mathbf{X}_{\text{DLS}}$	data least squares solutions to $\mathbf{Ax} = \mathbf{b}, \mathbf{AX} = \mathbf{B}$
$\mathbf{x}_{\text{WLS}}, \mathbf{X}_{\text{WLS}}$	weighted least squares solutions to $\mathbf{Ax} = \mathbf{b}, \mathbf{AX} = \mathbf{B}$
$\mathbf{x}_{\text{opt}}, \mathbf{X}_{\text{opt}}$	optimal solutions to $\mathbf{Ax} = \mathbf{b}, \mathbf{AX} = \mathbf{B}$
$\mathbf{x}_{\text{Tik}}, \mathbf{X}_{\text{Tik}}$	Tikhonov solutions to $\mathbf{Ax} = \mathbf{b}, \mathbf{AX} = \mathbf{B}$
$\mathbf{x}_{\text{TLS}}, \mathbf{X}_{\text{TLS}}$	total least squares (TLS) solutions to $\mathbf{Ax} = \mathbf{b}, \mathbf{AX} = \mathbf{B}$
$\mathbf{x}_{\text{GTLS}}, \mathbf{X}_{\text{GTLS}}$	generalized TLS solutions to $\mathbf{Ax} = \mathbf{b}, \mathbf{AX} = \mathbf{B}$
$\mathbf{x}_{\text{ML}}, \mathbf{X}_{\text{ML}}$	maximum likelihood solutions to $\mathbf{Ax} = \mathbf{b}, \mathbf{AX} = \mathbf{B}$
$D_{AB}^{(\alpha,\beta)}(\mathbf{P}\|\mathbf{G})$	alpha–beta (AB) divergence of matrices $\mathbf{P}$ and $\mathbf{G}$
$D_\alpha(\mathbf{P}\|\mathbf{G})$	alpha-divergence of matrices $\mathbf{P}$ and $\mathbf{G}$
$D_\beta(\mathbf{P}\|\mathbf{G})$	beta-divergence of matrices $\mathbf{P}$ and $\mathbf{G}$
$D_{\text{KL}}(\mathbf{P}\|\mathbf{G})$	Kullback–Leibler divergence of matrices $\mathbf{P}$ and $\mathbf{G}$
$\ln_q(x)$	Tsallis logarithm
$\exp_q(x)$	q-exponential
$\ln_{1-\alpha}(x)$	deformed logarithm
$\exp_{1-\alpha}(x)$	deformed exponential

Probability (*continued*)

$\text{sign}(x)$	signum function of real valued variable x
$\text{SGN}(x)$	signum multifunction of real valued variable x
$\text{shrink}(y, \alpha)$	shrink operator
$R(\mathbf{x})$	Rayleigh quotient, generalized Rayleigh quotient
$\mathbf{M}_{\text{off}}$	off-diagonal matrix corresponding to matrix $\mathbf{M}$
$z^{-j}\mathbf{x}(n)$	time-shifting operation on vector $\mathbf{x}(n)$
$E\{\mathbf{x}\} = \bar{\mathbf{x}}$	expectation (mean) of random vector $\mathbf{x}$
$E\{\mathbf{x}\mathbf{x}^H\}$	autocorrelation matrix of random vector $\mathbf{x}$
$E\{\mathbf{x}\mathbf{y}^H\}$	cross-correlation matrix of random vectors $\mathbf{x}$ and $\mathbf{y}$
ρ_{xy}	correlation coefficient of random vectors $\mathbf{x}$ and $\mathbf{y}$
$N(\mathbf{c}, \boldsymbol{\Sigma})$	Gaussian random vector with mean (vector) $\mathbf{c}$ and covariance (matrix) $\boldsymbol{\Sigma}$
$CN(\mathbf{c}, \boldsymbol{\Sigma})$	complex Gaussian random vector with mean (vector) $\mathbf{c}$ and covariance (matrix) $\boldsymbol{\Sigma}$
$f(x_1, \ldots, x_m)$	joint probability density function of random vector $\mathbf{x} = [x_1, \ldots, x_m]^T$
$\Phi_{\mathbf{x}}(\boldsymbol{\omega})$	characteristic function of random vector $\mathbf{x}$

Abbreviations

AB	alpha–beta
ADMM	alternating direction method of multipliers
ALS	alternating least squares
ANLS	alternating nonnegative least squares
APGL	accelerated proximal gradient line
ARNLS	alternating regularization nonnegative least squares
BBGP	Barzilai–Borwein gradient projection
BCQP	bound-constrained quadratic program
BFGS	Broyden–Fletcher–Goldfarb–Shanno
BP	basis pursuit
BPDN	basis pursuit denoising
BSS	blind source separation
CANDECOMP	canonical factor decomposition
CNMF	constrained nonnegative matrix factorization
CoSaMP	compression sampling matching pursuit
CP	CANDECOMP/PARAFAC
DCT	discrete cosine transform
DFT	discrete Fourier transform
DLS	data least squares
DOA	direction of arrival
EEG	electroencephalography
EM	expectation-maximization
EMML	expectation-maximization maximum likelihood
ESPRIT	estimating signal parameters via rotational invariance technique

EVD	eigenvalue decomposition
FAJD	fast approximate joint diagonalization
FAL	first-order augmented Lagrangian
FFT	fast Fourier transform
FISTA	fast iterative soft thresholding algorithm
GEAP	generalized eigenproblem adaptive power
GEVD	generalized eigenvalue decomposition
GSVD	generalized singular value decomposition
GTLS	generalized total least squares
HBM	heavy ball method
HOOI	higher-order orthogonal iteration
HOSVD	higher-order singular value decomposition
ICA	independent component analysis
IDFT	inverse discrete Fourier transform
iid	independent and identically distributed
inf	infimum
KKT	Karush–Kuhn–Tucker
KL	Kullback–Leibler
LARS	least angle regressive
Lasso	least absolute shrinkage and selection operator
LDA	linear discriminant analysis
LMV	Lathauwer–Moor–Vanderwalle
LP	linear programming
LS	least squares
LSI	latent semantic indexing
max	maximize, maximum
MCA	minor component analysis
MIMO	multiple-input–multiple-output
min	minimize, minimum
ML	maximum likelihood
MP	matching pursuit
MPCA	multilinear principal component analysis
MUSIC	multiple signal classification

NeNMF	Nesterov nonnegative matrix factorization
NMF	nonnegative matrix factorization
NTD	nonnegative tensor decomposition
OGM	optimal gradient method
OMP	orthogonal matching pursuit
PARAFAC	parallel factor decomposition
PAST	projection approximation subspace tracking
PASTd	projection approximation subspace tracking via deflation
PCA	principal component analysis
PCG	preconditioned conjugate gradient
PCP	principal component pursuit
pdf	positive definite
PMF	positive matrix factorization
psdf	positive semi-definite
PSF	point-spread function
PSVD	product singular value decomposition
RIC	restricted isometry constant
RIP	restricted isometry property
ROMP	regularization orthogonal matching pursuit
RR	Rayleigh–Ritz
QCLP	quadratically constrained linear programming
QEP	quadratic eigenvalue problem
QP	quadratic programming
QV	quadratic variation
sign	signum
SPC	smooth PARAFAC tensor completion
StOMP	stagewise orthogonal matching pursuit
sup	supremum
SVD	singular value decomposition
SVT	singular value thresholding
TLS	total least squares
TV	total variation
UPCA	unfold principal component analysis
VQ	vector quantization

Algorithms

PART I
MATRIX ALGEBRA

1

Introduction to Matrix Algebra

In science and engineering, we often encounter the problem of solving a system of linear equations. Matrices provide the most basic and useful mathematical tool for describing and solving such systems. Matrices not only have many basic mathematics operations (such as transposition, inner product, outer product, inverse, generalized inverse etc.) but also have a variety of important scalar functions (e.g., a norm, a quadratic form, a determinant, eigenvalues, rank and trace etc.). There are also special matrix operations, such as the direct sum, direct product, Hadamard product, Kronecker product, vectorization, etc.

In this chapter, we begin our introduction to matrix algebra by relating matrices to the problem of solving systems of linear equations.

1.1 Basic Concepts of Vectors and Matrices

First we introduce the basic concepts of and notation for vectors and matrices.

1.1.1 Vectors and Matrices

Let $\mathbb{R}$ (or $\mathbb{C}$) denote the set of real (or complex) numbers. An m-dimensional *column vector* is defined as

$$\mathbf{x} = \begin{bmatrix} x_1 \\ \vdots \\ x_m \end{bmatrix}. \tag{1.1.1}$$

If the ith component x_i is a real number, i.e., $x_i \in \mathbb{R}$, for all $i = 1, \ldots, m$, then $\mathbf{x}$ is an m-dimensional *real vector* and is denoted $\mathbf{x} \in \mathbb{R}^{m\times 1}$ or simply $\mathbf{x} \in \mathbb{R}^m$. Similarly, if $x_i \in \mathbb{C}$ for some i, then $\mathbf{x}$ is known as an m-dimensional *complex vector* and is denoted $\mathbf{x} \in \mathbb{C}^m$. Here, $\mathbb{R}^m$ and $\mathbb{C}^m$ represent the sets of all real and complex m-dimensional column vectors, respectively.

An m-dimensional *row vector* $\mathbf{x} = [x_1, \ldots, x_m]$ is represented as $\mathbf{x} \in \mathbb{R}^{1\times m}$ or

$\mathbf{x} \in \mathbb{C}^{1\times m}$. To save space, an m-dimensional column vector is usually written as the transposed form of a row vector, denoted $\mathbf{x} = [x_1, \ldots, x_m]^T$.

An $m \times n$ *matrix* is expressed as

$$\mathbf{A} = \begin{bmatrix} a_{11} & \cdots & a_{1n} \\ \vdots & \ddots & \vdots \\ a_{m1} & \cdots & a_{mn} \end{bmatrix} = [a_{ij}]_{i=1,j=1}^{m,n}. \tag{1.1.2}$$

The matrix $\mathbf{A}$ with (i, j)th real entry $a_{ij} \in \mathbb{R}$ is called an $m \times n$ *real matrix*, and denoted by $\mathbf{A} \in \mathbb{R}^{m\times n}$. Similarly, $\mathbf{A} \in \mathbb{C}^{m\times n}$ is an $m \times n$ *complex matrix*.

An $m\times n$ matrix can be represented as $\mathbf{A} = [\mathbf{a}_1, \ldots, \mathbf{a}_n]$ where its column vectors are $\mathbf{a}_j = [a_{1j}, \ldots, a_{mj}]^T$, $j = 1, \ldots, n$.

The *system of linear equations*

$$\left.\begin{aligned} a_{11}x_1 + a_{12}x_2 + \cdots + a_{1n}x_n &= b_1, \\ a_{21}x_1 + a_{22}x_2 + \cdots + a_{2n}x_n &= b_2, \\ &\;\vdots \\ a_{m1}x_1 + a_{m2}x_2 + \cdots + a_{mn}x_n &= b_m \end{aligned}\right\} \tag{1.1.3}$$

can be simply rewritten using vector and matrix symbols as a *matrix equation*

$$\mathbf{A}\mathbf{x} = \mathbf{b}, \tag{1.1.4}$$

where

$$\mathbf{A} = \begin{bmatrix} a_{11} & \cdots & a_{1n} \\ \vdots & \ddots & \vdots \\ a_{m1} & \cdots & a_{mn} \end{bmatrix}, \quad \mathbf{x} = \begin{bmatrix} x_1 \\ \vdots \\ x_n \end{bmatrix}, \quad \mathbf{b} = \begin{bmatrix} b_1 \\ \vdots \\ b_m \end{bmatrix}. \tag{1.1.5}$$

In modeling physical problems, the matrix $\mathbf{A}$ is usually the symbolic representation of a physical system (e.g., a linear system, a filter, or a wireless communication channel). There are three different types of vector in science and engineering [232]:

(1) *Physical vector* Its elements are physical quantities with magnitude and direction, such as a displacement vector, a velocity vector, an acceleration vector and so forth.

(2) *Geometric vector* A directed line segment or arrow is usually used to visualize a physical vector. Such a representation is called a geometric vector. For example, $\mathbf{v} = \overrightarrow{AB}$ represents the directed line segment with initial point A and the terminal point B.

(3) *Algebraic vector* A geometric vector can be represented in algebraic form. For a geometric vector $\mathbf{v} = \overrightarrow{AB}$ on a plane, if its initial point is $A = (a_1, a_2)$ and its terminal point is $B = (b_1, b_2)$, then the geometric vector $\mathbf{v} = \overrightarrow{AB}$ can

be represented in an algebraic form $\mathbf{v} = \begin{bmatrix} b_1 - a_1 \\ b_2 - a_2 \end{bmatrix}$. Such a geometric vector described in algebraic form is known as an algebraic vector.

Physical vectors are those often encountered in practical applications, while geometric vectors and algebraic vectors are respectively the visual representation and the algebraic form of physical vectors. Algebraic vectors provide a computational tool for physical vectors.

Depending on the different types of element value, algebraic vectors can be divided into the following three types:

(1) *Constant vector* Its entries are real constant numbers or complex constant numbers, e.g., $\mathbf{a} = [1, 5, 4]^T$.
(2) *Function vector* It uses functions as entries, e.g., $\mathbf{x} = [x^1, \ldots, x^n]^T$.
(3) *Random vector* Its entries are random variables or signals, e.g., $\mathbf{x}(n) = [x_1(n), \ldots, x_m(n)]^T$ where $x_1(n), \ldots, x_m(n)$ are m random variables or random signals.

Figure 1.1 summarizes the classification of vectors.

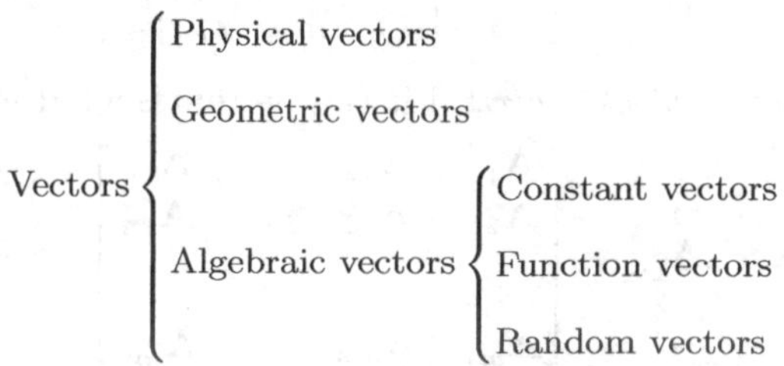

Figure 1.1 Classification of vectors.

Now we turn to matrices. An $m \times n$ matrix $\mathbf{A}$ is called a *square matrix* if $m = n$, a *broad matrix* for $m < n$, and a *tall matrix* for $m > n$.

The *main diagonal* of an $n \times n$ matrix $\mathbf{A} = [a_{ij}]$ is the segment connecting the top left to the bottom right corner. The entries located on the main diagonal, $a_{11}, a_{22}, \ldots, a_{nn}$, are known as the (main) *diagonal elements*.

An $n \times n$ matrix $\mathbf{A} = [a_{ij}]$ is called a *diagonal matrix* if all entries off the main diagonal are zero; it is then denoted by

$$\mathbf{D} = \mathbf{Diag}(d_{11}, \ldots, d_{nn}). \tag{1.1.6}$$

In particular, a diagonal matrix $\mathbf{I} = \mathbf{Diag}(1, \ldots, 1)$ is called an *identity matrix*, and $\mathbf{O} = \mathbf{Diag}(0, \ldots, 0)$ is known as a *zero matrix*.

A vector all of whose components are equal to zero is called a *zero vector* and is denoted as $\mathbf{0} = [0, \ldots, 0]^T$.

An $n \times 1$ vector $\mathbf{x} = [x_1, \ldots, x_n]^T$ with only one nonzero entry $x_i = 1$ constitutes a *basis vector*, denoted $\mathbf{e}_i$; e.g.,

$$\mathbf{e}_1 = \begin{bmatrix} 1 \\ 0 \\ 0 \\ \vdots \\ 0 \end{bmatrix}, \quad \mathbf{e}_2 = \begin{bmatrix} 0 \\ 1 \\ 0 \\ \vdots \\ 0 \end{bmatrix}, \quad \ldots, \quad \mathbf{e}_n = \begin{bmatrix} 0 \\ 0 \\ 0 \\ \vdots \\ 1 \end{bmatrix}. \tag{1.1.7}$$

Clearly, an $n \times n$ identity matrix $\mathbf{I}$ can be represented as $\mathbf{I} = [\mathbf{e}_1, \ldots, \mathbf{e}_n]$ using basis vectors.

In this book, we use often the following matrix symbols.

$\mathbf{A}(i, :)$ means the ith row of $\mathbf{A}$.

$\mathbf{A}(:, j)$ means the jth column of $\mathbf{A}$.

$\mathbf{A}(p : q, r : s)$ means the $(q - p + 1) \times (s - r + 1)$ *submatrix* consisting of the pth row to the qth row and the rth column to the sth column of $\mathbf{A}$. For example,

$$\mathbf{A}(3:6, 2:4) = \begin{bmatrix} a_{32} & a_{33} & a_{34} \\ a_{42} & a_{43} & a_{44} \\ a_{52} & a_{53} & a_{54} \\ a_{62} & a_{63} & a_{64} \end{bmatrix}.$$

A matrix $\mathbf{A}$ is an $m \times n$ *block matrix* if it can be represented in the form

$$\mathbf{A} = [\mathbf{A}_{ij}] = \begin{bmatrix} \mathbf{A}_{11} & \mathbf{A}_{12} & \cdots & \mathbf{A}_{1n} \\ \mathbf{A}_{21} & \mathbf{A}_{22} & \cdots & \mathbf{A}_{2n} \\ \vdots & \vdots & \ddots & \vdots \\ \mathbf{A}_{m1} & \mathbf{A}_{m2} & \cdots & \mathbf{A}_{mn} \end{bmatrix},$$

where the $\mathbf{A}_{ij}$ are matrices. The notation $[\mathbf{A}_{ij}]$ refers to a matrix consisting of block matrices.

1.1.2 Basic Vector Calculus

Basic vector calculus requires vector addition, vector multiplication by a scalar and vector products.

The *vector addition* of $\mathbf{u} = [u_1, \ldots, u_n]^T$ and $\mathbf{v} = [v_1, \ldots, v_n]^T$ is defined as

$$\mathbf{u} + \mathbf{v} = [u_1 + v_1, \ldots, u_n + v_n]^T. \tag{1.1.8}$$

Vector addition has the following two main properties:

- *Commutative law* $\mathbf{u} + \mathbf{v} = \mathbf{v} + \mathbf{u}$.
- *Associative law* $(\mathbf{u} + \mathbf{v}) \pm \mathbf{w} = \mathbf{u} + (\mathbf{v} \pm \mathbf{w}) = (\mathbf{u} \pm \mathbf{w}) + \mathbf{v}$.

The *vector multiplication* of an $n \times 1$ vector $\mathbf{u}$ by a scalar α is defined as

$$\alpha\mathbf{u} = [\alpha u_1, \ldots, \alpha u_n]^T. \tag{1.1.9}$$

The basic property of vector multiplication by a scalar is that it obeys the distributive law:

$$\alpha(\mathbf{u}+\mathbf{v}) = \alpha\mathbf{u}+\alpha\mathbf{v}. \tag{1.1.10}$$

The *inner product* (or *dot product* or *scalar product*) of two real or complex $n \times 1$ vectors $\mathbf{u} = [u_1, \ldots, u_n]^T$ and $\mathbf{v} = [v_1, \ldots, v_n]^T$, denoted $\langle \mathbf{u}, \mathbf{v} \rangle$, is defined as the real number

$$\langle \mathbf{u}, \mathbf{v} \rangle = \mathbf{u}^T\mathbf{v} = u_1v_1 + \cdots + u_nv_n = \sum_{i=1}^{n} u_iv_i, \tag{1.1.11}$$

or the complex number

$$\langle \mathbf{u}, \mathbf{v} \rangle = \mathbf{u}^H\mathbf{v} = u_1^*v_1 + \cdots + u_n^*v_n = \sum_{i=1}^{n} u_i^*v_i, \tag{1.1.12}$$

where $\mathbf{u}^H$ is the complex conjugate transpose, or Hermitian conjugate, of $\mathbf{u}$.

The inner product of two vectors $\langle \mathbf{u}, \mathbf{v} \rangle = \mathbf{u}^H\mathbf{v}$ has several important applications; for example, it may be used to measure the size (or length) of a vector, the distance between vectors, the neighborhood of a vector and so on. We will present these applications later.

The *outer product* (or *cross product*) of an $m \times 1$ real vector and an $n \times 1$ real vector, denoted $\mathbf{u} \circ \mathbf{v}$, is defined as the $m \times n$ real matrix

$$\mathbf{u} \circ \mathbf{v} = \mathbf{u}\mathbf{v}^T = \begin{bmatrix} u_1v_1 & \cdots & u_1v_n \\ \vdots & \vdots & \vdots \\ u_mv_1 & \cdots & u_mv_n \end{bmatrix}; \tag{1.1.13}$$

if $\mathbf{u}$ and $\mathbf{v}$ are complex then the outer product is the $m \times n$ complex matrix

$$\mathbf{u} \circ \mathbf{v} = \mathbf{u}\mathbf{v}^H = \begin{bmatrix} u_1v_1^* & \cdots & u_1v_n^* \\ \vdots & \vdots & \vdots \\ u_mv_1^* & \cdots & u_mv_n^* \end{bmatrix}. \tag{1.1.14}$$

In signal processing, wireless communications, pattern recognition, etc., for two $m \times 1$ data vectors $\mathbf{x}(t) = [x_1(t), \ldots, x_m(t)]^T$ and $\mathbf{y}(t) = [y_1(t), \ldots, y_m(t)]^T$, the $m \times m$ *autocorrelation matrix* is given by $\mathbf{R}_{xx} = E\{\mathbf{x}(t) \circ \mathbf{x}(t)\} = E\{\mathbf{x}(t)\mathbf{x}^H(t)\}$ and the $m \times m$ *cross-correlation matrix* is given by $\mathbf{R}_{xy} = E\{\mathbf{x}(t) \circ \mathbf{y}(t)\} = E\{\mathbf{x}(t)\mathbf{y}^H(t)\}$, where E is the expectation operator. Given the sample data $x_i(t), y_i(t), i = 1, \ldots, m$, $t = 1, \ldots, N$, the sample autocorrelation matrix $\hat{\mathbf{R}}_{xx}$ and the sample cross-correlation matrix $\hat{\mathbf{R}}_{xy}$ can be respectively estimated by

$$\hat{\mathbf{R}}_{xx} = \frac{1}{N}\sum_{t=1}^{N} \mathbf{x}(t)\mathbf{x}^H(t), \quad \hat{\mathbf{R}}_{xy} = \frac{1}{N}\sum_{t=1}^{N} \mathbf{x}(t)\mathbf{y}^H(t). \tag{1.1.15}$$

1.1.3 Basic Matrix Calculus

Basic matrix calculus requires the matrix transpose, conjugate, conjugate transpose, addition and multiplication.

Definition 1.1 If $\mathbf{A} = [a_{ij}]$ is an $m \times n$ matrix, then its *transpose* $\mathbf{A}^T$ is an $n \times m$ matrix with the (i, j)th entry $[\mathbf{A}^T]_{ij} = a_{ji}$. The *conjugate* of $\mathbf{A}$ is represented as $\mathbf{A}^*$ and is an $m \times n$ matrix with (i, j)th entry $[\mathbf{A}^*]_{ij} = a^*_{ij}$, while the *conjugate or Hermitian transpose* of $\mathbf{A}$, denoted $\mathbf{A}^H \in \mathbb{C}^{n\times m}$, is defined as

$$\mathbf{A}^H = (\mathbf{A}^*)^T = \begin{bmatrix} a^*_{11} & a^*_{21} & \cdots & a^*_{m1} \\ a^*_{12} & a^*_{22} & \cdots & a^*_{m2} \\ \vdots & \vdots & \ddots & \vdots \\ a^*_{1n} & a^*_{2n} & \cdots & a^*_{mn} \end{bmatrix}. \tag{1.1.16}$$

Definition 1.2 An $n \times n$ real (complex) matrix satisfying $\mathbf{A}^T = \mathbf{A}$ ($\mathbf{A}^H = \mathbf{A}$) is called a *symmetric matrix* (*Hermitian matrix*).

There are the following relationships between the transpose and conjugate transpose of a matrix:

$$\mathbf{A}^H = (\mathbf{A}^*)^T = (\mathbf{A}^T)^*. \tag{1.1.17}$$

For an $m \times n$ block matrix $\mathbf{A} = [\mathbf{A}_{ij}]$, its conjugate transpose $\mathbf{A}^H = [\mathbf{A}^H_{ji}]$ is an $n \times m$ block matrix:

$$\mathbf{A}^H = \begin{bmatrix} \mathbf{A}^H_{11} & \mathbf{A}^H_{21} & \cdots & \mathbf{A}^H_{m1} \\ \mathbf{A}^H_{12} & \mathbf{A}^H_{22} & \cdots & \mathbf{A}^H_{m2} \\ \vdots & \vdots & \ddots & \vdots \\ \mathbf{A}^H_{1n} & \mathbf{A}^H_{2n} & \cdots & \mathbf{A}^H_{mn} \end{bmatrix}.$$

The simplest algebraic operations with matrices are the addition of two matrices and the multiplication of a matrix by a scalar.

Definition 1.3 Given two $m \times n$ matrices $\mathbf{A} = [a_{ij}]$ and $\mathbf{B} = [b_{ij}]$, *matrix addition* $\mathbf{A} + \mathbf{B}$ is defined by $[\mathbf{A} + \mathbf{B}]_{ij} = a_{ij} + b_{ij}$. Similarly, *matrix subtraction* $\mathbf{A} - \mathbf{B}$ is defined as $[\mathbf{A} - \mathbf{B}]_{ij} = a_{ij} - b_{ij}$.

By using this definition, it is easy to verify that the addition and subtraction of two matrices obey the following rules:

- *Commutative law* $\mathbf{A} + \mathbf{B} = \mathbf{B} + \mathbf{A}$.
- *Associative law* $(\mathbf{A} + \mathbf{B}) \pm \mathbf{C} = \mathbf{A} + (\mathbf{B} \pm \mathbf{C}) = (\mathbf{A} \pm \mathbf{C}) + \mathbf{B}$.

Definition 1.4 Let $\mathbf{A} = [a_{ij}]$ be an $m \times n$ matrix, and α be a scalar. The product $\alpha\mathbf{A}$ is an $m \times n$ matrix and is defined as $[\alpha\mathbf{A}]_{ij} = \alpha a_{ij}$.

Definition 1.5 Consider an $m \times n$ matrix $\mathbf{A} = [a_{ij}]$ and an $r \times 1$ vector $\mathbf{x} = [x_1, \ldots, x_r]^T$. The product $\mathbf{Ax}$ exists only when $n = r$ and is an $m \times 1$ vector whose entries are given by

$$[\mathbf{Ax}]_i = \sum_{j=1}^{n} a_{ij} x_j, \quad i = 1, \ldots, m.$$

Definition 1.6 The *matrix product* of an $m \times n$ matrix $\mathbf{A} = [a_{ij}]$ and an $r \times s$ matrix $\mathbf{B} = [b_{ij}]$, denoted $\mathbf{AB}$, exists only when $n = r$ and is an $m \times s$ matrix with entries

$$[\mathbf{AB}]_{ij} = \sum_{k=1}^{n} a_{ik} b_{kj}, \quad i = 1, \ldots, m; \; j = 1, \ldots, s.$$

Theorem 1.1 *The matrix product obeys the following rules of operation:*

(a) *Associative law of multiplication* *If* $\mathbf{A} \in \mathbb{C}^{m\times n}, \mathbf{B} \in \mathbb{C}^{n\times p}$ *and* $\mathbf{C} \in \mathbb{C}^{p\times q}$, *then* $\mathbf{A}(\mathbf{BC}) = (\mathbf{AB})\mathbf{C}$.

(b) *Left distributive law of multiplication* *For two* $m \times n$ *matrices* $\mathbf{A}$ *and* $\mathbf{B}$, *if* $\mathbf{C}$ *is an* $n \times p$ *matrix then* $(\mathbf{A} \pm \mathbf{B})\mathbf{C} = \mathbf{AC} \pm \mathbf{BC}$.

(c) *Right distributive law of multiplication* *If* $\mathbf{A}$ *is an* $m \times n$ *matrix, while* $\mathbf{B}$ *and* $\mathbf{C}$ *are two* $n \times p$ *matrices, then* $\mathbf{A}(\mathbf{B} \pm \mathbf{C}) = \mathbf{AB} \pm \mathbf{AC}$.

(d) *If* α *is a scalar and* $\mathbf{A}$ *and* $\mathbf{B}$ *are two* $m\times n$ *matrices then* $\alpha(\mathbf{A}+\mathbf{B}) = \alpha\mathbf{A} + \alpha\mathbf{B}$.

Proof We will prove only (a) and (b) here, while the proofs of (c) and (d) are left to the reader as an exercise.

(a) Let $\mathbf{A}_{m\times n} = [a_{ij}], \mathbf{B}_{n\times p} = [b_{ij}], \mathbf{C}_{p\times q} = [c_{ij}]$, then

$$\begin{aligned}
[\mathbf{A}(\mathbf{BC})]_{ij} &= \sum_{k=1}^{n} a_{ik} (\mathbf{BC})_{kj} = \sum_{k=1}^{n} a_{ik} \left(\sum_{l=1}^{p} b_{kl} c_{lj} \right) \\
&= \sum_{l=1}^{p} \sum_{k=1}^{n} (a_{ik} b_{kl}) c_{lj} = \sum_{l=1}^{p} [\mathbf{AB}]_{il} c_{lj} = [(\mathbf{AB})\mathbf{C}]_{ij}
\end{aligned}$$

which means that $\mathbf{A}(\mathbf{BC}) = (\mathbf{AB})\mathbf{C}$.

(b) From the rule for matrix multiplication it is known that

$$[\mathbf{AC}]_{ij} = \sum_{k=1}^{n} a_{ik} c_{kj}, \quad [\mathbf{BC}]_{ij} = \sum_{k=1}^{n} b_{ik} c_{kj}.$$

Then, according to the matrix addition rule, we have

$$[\mathbf{AC} + \mathbf{BC}]_{ij} = [\mathbf{AC}]_{ij} + [\mathbf{BC}]_{ij} = \sum_{k=1}^{n} (a_{ik} + b_{ik}) c_{kj} = [(\mathbf{A} + \mathbf{B})\mathbf{C}]_{ij}.$$

This gives $(\mathbf{A} + \mathbf{B})\mathbf{C} = \mathbf{AC} + \mathbf{BC}$. □

Generally speaking, the product of two matrices does not satisfy the commutative law, namely $\mathbf{AB} \neq \mathbf{BA}$.

Another important operation on a square matrix is that of finding its inverse.

Put $\mathbf{x} = [x_1, \ldots, x_n]^T$ and $\mathbf{y} = [y_1, \ldots, y_n]^T$. The matrix–vector product $\mathbf{Ax} = \mathbf{y}$ can be regarded as a *linear transform* of the vector $\mathbf{x}$, where the $n \times n$ matrix $\mathbf{A}$ is called the *linear transform matrix*. Let $\mathbf{A}^{-1}$ denote the *linear inverse transform* of the vector $\mathbf{y}$ onto $\mathbf{x}$. If $\mathbf{A}^{-1}$ exists then one has

$$\mathbf{x} = \mathbf{A}^{-1}\mathbf{y}. \tag{1.1.18}$$

This equation can be viewed as the result of using $\mathbf{A}^{-1}$ to premultiply the original linear transform $\mathbf{Ax} = \mathbf{y}$, giving $\mathbf{A}^{-1}\mathbf{Ax} = \mathbf{A}^{-1}\mathbf{y} = \mathbf{x}$, which means that the *linear inverse transform matrix* $\mathbf{A}^{-1}$ must satisfy $\mathbf{A}^{-1}\mathbf{A} = \mathbf{I}$. Furthermore, $\mathbf{x} = \mathbf{A}^{-1}\mathbf{y}$ should be invertible as well. In other words, after premultiplying $\mathbf{x} = \mathbf{A}^{-1}\mathbf{y}$ by $\mathbf{A}$, we get $\mathbf{Ax} = \mathbf{AA}^{-1}\mathbf{y}$, which should be consistent with the original linear transform $\mathbf{Ax} = \mathbf{y}$. This means that $\mathbf{A}^{-1}$ must also satisfy $\mathbf{AA}^{-1} = \mathbf{I}$.

On the basis of the discussion above, the inverse matrix can be defined as follows.

Definition 1.7 Let $\mathbf{A}$ be an $n \times n$ matrix. The matrix $\mathbf{A}$ is said to be invertible if there is an $n \times n$ matrix $\mathbf{A}^{-1}$ such that $\mathbf{AA}^{-1} = \mathbf{A}^{-1}\mathbf{A} = \mathbf{I}$, and $\mathbf{A}^{-1}$ is referred to as the *inverse matrix* of $\mathbf{A}$.

The following are properties of the conjugate, transpose, conjugate transpose and inverse matrices.

1. The matrix conjugate, transpose and conjugate transpose satisfy the distributive law:

$$(\mathbf{A}+\mathbf{B})^* = \mathbf{A}^* + \mathbf{B}^*, \quad (\mathbf{A}+\mathbf{B})^T = \mathbf{A}^T + \mathbf{B}^T, \quad (\mathbf{A}+\mathbf{B})^H = \mathbf{A}^H + \mathbf{B}^H.$$

2. The transpose, conjugate transpose and inverse matrix of product of two matrices satisfy the following relationship:

$$(\mathbf{AB})^T = \mathbf{B}^T\mathbf{A}^T, \quad (\mathbf{AB})^H = \mathbf{B}^H\mathbf{A}^H, \quad (\mathbf{AB})^{-1} = \mathbf{B}^{-1}\mathbf{A}^{-1}$$

   in which both $\mathbf{A}$ and $\mathbf{B}$ are assumed to be invertible.

3. Each of the symbols for the conjugate, transpose and conjugate transpose can be exchanged with the symbol for the inverse:

$$(\mathbf{A}^*)^{-1} = (\mathbf{A}^{-1})^*, \quad (\mathbf{A}^T)^{-1} = (\mathbf{A}^{-1})^T, \quad (\mathbf{A}^H)^{-1} = (\mathbf{A}^{-1})^H.$$

   The notations $\mathbf{A}^{-*} = (\mathbf{A}^{-1})^*, \mathbf{A}^{-T} = (\mathbf{A}^{-1})^T$ and $\mathbf{A}^{-H} = (\mathbf{A}^{-1})^H$ are sometimes used.

4. For any $m \times n$ matrix $\mathbf{A}$, the $n \times n$ matrix $\mathbf{B} = \mathbf{A}^H\mathbf{A}$ and the $m \times m$ matrix $\mathbf{C} = \mathbf{AA}^H$ are Hermitian matrices.

1.1.4 Linear Independence of Vectors

Consider the system of linear equations (1.1.3). It can be written as the matrix equation $\mathbf{Ax} = \mathbf{b}$. Denoting $\mathbf{A} = [\mathbf{a}_1, \ldots, \mathbf{a}_n]$, the m equations of (1.1.3) can be written as

$$x_1\mathbf{a}_1 + \cdots + x_n\mathbf{a}_n = \mathbf{b}.$$

This is called a *linear combination* of the column vectors $\mathbf{a}_1, \ldots, \mathbf{a}_n$.

Definition 1.8 A set of n vectors, denoted $\{\mathbf{u}_1, \ldots, \mathbf{u}_n\}$, is said to be *linearly independent* if the matrix equation

$$c_1\mathbf{u}_1 + \cdots + c_n\mathbf{u}_n = \mathbf{0} \tag{1.1.19}$$

has only zero solutions $c_1 = \cdots = c_n = 0$. If the above equation may hold for a set of coefficients that are not all zero then the n vectors $\mathbf{u}_1, \ldots, \mathbf{u}_n$ are said to be *linearly dependent.*

An $n \times n$ matrix $\mathbf{A}$ is *nonsingular* if and only if the matrix equation $\mathbf{Ax} = \mathbf{0}$ has only the zero solution $\mathbf{x} = \mathbf{0}$. If $\mathbf{Ax} = \mathbf{0}$ exists for any nonzero solution $\mathbf{x} \neq \mathbf{0}$ then the matrix $\mathbf{A}$ is *singular.*

For an $n \times n$ matrix $\mathbf{A} = [\mathbf{a}_1, \ldots, \mathbf{a}_n]$, the matrix equation $\mathbf{Ax} = \mathbf{0}$ is equivalent to

$$\mathbf{a}_1x_1 + \cdots + \mathbf{a}_nx_n = \mathbf{0}. \tag{1.1.20}$$

From the above definition it follows that the matrix equation $\mathbf{Ax} = \mathbf{0}$ has a zero solution vector only, i.e., the matrix $\mathbf{A}$ is nonsingular, if and only if the column vectors $\mathbf{a}_1, \ldots, \mathbf{a}_n$ of $\mathbf{A}$ are linearly independent. Because of importance of this result, it is described in a theorem below.

Theorem 1.2 *An $n \times n$ matrix $\mathbf{A} = [\mathbf{a}_1, \ldots, \mathbf{a}_n]$ is nonsingular if and only if its n column vectors $\mathbf{a}_1, \ldots, \mathbf{a}_n$ are linearly independent.*

To summarize the above discussions, the nonsingularity of an $n \times n$ matrix $\mathbf{A}$ can be determined in any of the following three ways:

(1) its column vectors are linearly independent;

(2) for the matrix equation $\mathbf{Ax} = \mathbf{b}$ there exists a unique nonzero solution;

(3) the matrix equation $\mathbf{Ax} = \mathbf{0}$ has only a zero solution.

1.1.5 Matrix Functions

The following are five common *matrix functions*:

1. *Triangle matrix function*

$$\sin \mathbf{A} = \sum_{n=0}^{\infty} \frac{(-1)^n \mathbf{A}^{2n+1}}{(2n+1)!} = \mathbf{A} - \frac{1}{3!}\mathbf{A}^3 + \frac{1}{5!}\mathbf{A}^5 - \cdots \tag{1.1.21}$$

$$\cos \mathbf{A} = \sum_{n=0}^{\infty} \frac{(-1)^n \mathbf{A}^{2n}}{(2n)!} = \mathbf{I} - \frac{1}{2!}\mathbf{A}^2 + \frac{1}{4!}\mathbf{A}^4 - \cdots . \tag{1.1.22}$$

2. *Logarithm matrix function*

$$\ln(\mathbf{I} + \mathbf{A}) = \sum_{n=1}^{\infty} \frac{(-1)^{n-1}}{n} \mathbf{A}^n = \mathbf{A} - \frac{1}{2}\mathbf{A}^2 + \frac{1}{3}\mathbf{A}^3 - \cdots . \tag{1.1.23}$$

3. *Exponential matrix function* [311], [179]

$$\mathrm{e}^{\mathbf{A}} = \sum_{n=0}^{\infty} \frac{1}{n!} \mathbf{A}^n = \mathbf{I} + \mathbf{A} + \frac{1}{2}\mathbf{A}^2 + \frac{1}{3!}\mathbf{A}^3 + \cdots \tag{1.1.24}$$

$$\mathrm{e}^{-\mathbf{A}} = \sum_{n=0}^{\infty} \frac{1}{n!} (-1)^n \mathbf{A}^n = \mathbf{I} - \mathbf{A} + \frac{1}{2}\mathbf{A}^2 - \frac{1}{3!}\mathbf{A}^3 + \cdots \tag{1.1.25}$$

$$\mathrm{e}^{\mathbf{A}t} = \mathbf{I} + \mathbf{A}t + \frac{1}{2}\mathbf{A}^2 t^2 + \frac{1}{3!}\mathbf{A}^3 t^3 + \cdots . \tag{1.1.26}$$

4. *Matrix derivative* If the entries a_{ij} of the matrix $\mathbf{A}$ are the functions of a parameter t then the derivative of the matrix $\mathbf{A}$ is defined as follows:

$$\frac{\mathrm{d}\mathbf{A}}{\mathrm{d}t} = \dot{\mathbf{A}} = \begin{bmatrix} \frac{\mathrm{d}a_{11}}{\mathrm{d}t} & \frac{\mathrm{d}a_{12}}{\mathrm{d}t} & \cdots & \frac{\mathrm{d}a_{1n}}{\mathrm{d}t} \\ \frac{\mathrm{d}a_{21}}{\mathrm{d}t} & \frac{\mathrm{d}a_{22}}{\mathrm{d}t} & \cdots & \frac{\mathrm{d}a_{2n}}{\mathrm{d}t} \\ \vdots & \vdots & \ddots & \vdots \\ \frac{\mathrm{d}a_{m1}}{\mathrm{d}t} & \frac{\mathrm{d}a_{m2}}{\mathrm{d}t} & \cdots & \frac{\mathrm{d}a_{mn}}{\mathrm{d}t} \end{bmatrix} . \tag{1.1.27}$$

- The derivative of a matrix exponential function is given by

$$\frac{\mathrm{d}\mathrm{e}^{\mathbf{A}t}}{\mathrm{d}t} = \mathbf{A}\mathrm{e}^{\mathbf{A}t} = \mathrm{e}^{\mathbf{A}t}\mathbf{A}. \tag{1.1.28}$$

- The derivative of matrix product is given by

$$\frac{\mathrm{d}}{\mathrm{d}t}(\mathbf{A}\mathbf{B}) = \frac{\mathrm{d}\mathbf{A}}{\mathrm{d}t}\mathbf{B} + \mathbf{A}\frac{\mathrm{d}\mathbf{B}}{\mathrm{d}t}, \tag{1.1.29}$$

where $\mathbf{A}$ and $\mathbf{B}$ are matrix functions of the variable t.

5. *Matrix integral*

$$\int \mathbf{A}\mathrm{d}t = \begin{bmatrix} \int a_{11}\mathrm{d}t & \int a_{12}\mathrm{d}t & \cdots & \int a_{1n}\mathrm{d}t \\ \int a_{21}\mathrm{d}t & \int a_{22}\mathrm{d}t & \cdots & \int a_{2n}\mathrm{d}t \\ \vdots & \vdots & \ddots & \vdots \\ \int a_{m1}\mathrm{d}t & \int a_{m2}\mathrm{d}t & \cdots & \int a_{mn}\mathrm{d}t \end{bmatrix}. \tag{1.1.30}$$

1.2 Elementary Row Operations and Applications

Simple operations related to the rows of a matrix are referred to as elementary row operations; they can efficiently solve matrix equations, find the inverse of a matrix, construct the basis vectors of a vector space and so on.

1.2.1 Elementary Row Operations

When solving an $m \times n$ system of linear equations, it is useful to reduce the number of equations. The principle of reduction processes is to keep the solutions of the system of equations unchanged.

Definition 1.9 Two systems of linear equations in n unknowns are said to be *equivalent systems of equations* if they have the same sets of solutions.

To transform a given $m \times n$ matrix equation $\mathbf{A}_{m\times n}\mathbf{x}_n = \mathbf{b}_m$ into an equivalent matrix equation, a simple and efficient way is to apply successive elementary operations on the given matrix equation.

Definition 1.10 The following three types of operation on the rows of a system of linear equations are called *elementary row operations.*

Type I Interchange any two equations, say the pth and qth equations; this is denoted by $R_p \leftrightarrow R_q$.

Type II Multiply the pth equation by a nonzero number α; this is denoted by $\alpha R_p \to R_p$.

Type III Add β times the pth equation to the qth equation; this is denoted by $\beta R_p + R_q \to R_q$.

Clearly, any type of elementary row operation does not change the solution of a system of linear equations, so after elementary row operations, the reduced system of linear equations and the original system of linear equations are equivalent.

As a matter of fact, any elementary operation on an $m \times n$ system of equations $\mathbf{Ax} = \mathbf{b}$ is equivalent to the same type of elementary operation on the *augmented matrix* $\mathbf{B} = [\mathbf{A}, \mathbf{b}]$, where the column vector $\mathbf{b}$ is written alongside $\mathbf{A}$. Hence, performing elementary row operations on a system of linear equations $\mathbf{Ax} = \mathbf{b}$ is in practice implemented by using the same elementary row operations on the augmented matrix $\mathbf{B} = [\mathbf{A}, \mathbf{b}]$.

The discussions above show that if, after a sequence of elementary row operations, the augmented matrix $\mathbf{B}_{m\times(n+1)}$ becomes another simpler matrix $\mathbf{C}_{m\times(n+1)}$ then two matrices are *row equivalent*.

For the convenience of solving a system of linear equations, the final row equivalent matrix should be of echelon form (see below). The leftmost nonzero entry of a nonzero row is called the *leading entry* of the row. If the leading entry is equal to 1 then it is said to be the *leading-1 entry.*

DEFINITION 1.11 A matrix is said to be an *echelon matrix* if it has the following forms:

(1) all rows with all entries zero are located the bottom of the matrix;
(2) the leading entry of each nonzero row appears always to the right of the leading entry of the nonzero row above;
(3) all entries below the leading entry of the same column are equal to zero.

The following are some examples of echelon matrices:

$$\mathbf{A}=\begin{bmatrix}2 & * & *\\ 0 & 5 & *\\ 0 & 0 & 3\\ 0 & 0 & 0\end{bmatrix},\ \mathbf{A}=\begin{bmatrix}1 & * & *\\ 0 & 3 & *\\ 0 & 0 & 0\\ 0 & 0 & 0\end{bmatrix},\ \mathbf{A}=\begin{bmatrix}3 & * & *\\ 0 & 0 & 5\\ 0 & 0 & 0\\ 0 & 0 & 0\end{bmatrix},\ \mathbf{A}=\begin{bmatrix}0 & 1 & *\\ 0 & 0 & 9\\ 0 & 0 & 0\\ 0 & 0 & 0\end{bmatrix},$$

where $*$ indicates that the entry can be any value.

DEFINITION 1.12 [232] An echelon matrix **B** is said to be a *reduced row-echelon form (RREF) matrix* if the leading entry of each nonzero row is a leading-1 entry, and each leading-1 entry is the only nonzero entry in the column in which it is located.

THEOREM 1.3 *Any $m\times n$ matrix* **A** *is row equivalent to one and only one matrix in reduced row-echelon form.*

Proof See [285, Appendix A]. □

Given an $m \times n$ matrix **B**, Algorithm 1.1 transforms **B** to a reduced row-echelon form by performing suitable elementary operations.

DEFINITION 1.13 [285, p. 15] A *pivot position* of an $m \times n$ matrix **A** is the position of some leading entry of its echelon form. Each column containing a pivot position is called a *pivot column* of the matrix **A**.

The following examples show how to perform elementary operations for transforming a matrix to its row-echelon form and reduced row-echelon form, and how to determine the pivot columns of the original matrix.

Algorithm 1.1 Reduced row-echelon form [229]

input: $\mathbf{B} \in \mathbb{R}^{m \times n}$.

1. Select the row that has the first nonzero entry, say R_i. If this happens to be the first row, then go to the next step. Otherwise, perform a Type-I elementary operation $R_1 \leftrightarrow R_i$ such that the first entry of the new first row is nonzero.
2. Make a Type-II elementary operation $\alpha R_1 \to R_1$ to get a leading-1 entry for the first row.
3. Perform a Type-III elementary operation $\alpha R_i + \alpha R_1 \to R_1, i > 1$, to make all entries below the leading entry 1 of the first row equal to 0.
4. For the ith row, $i = 2, \ldots, m$, perform suitable elementary operations similar to the above steps to get a leading-1 entry for the ith row, and make all entries below the leading-1 entry of the ith row equal to 0.

output: The reduced row-echelon form of **B**.

Example 1.1 Consider the 3×5 matrix

$$\mathbf{A} = \begin{bmatrix} -3 & 6 & -1 & 1 & -7 \\ 1 & -2 & 2 & 3 & -1 \\ 2 & -4 & 5 & 8 & -4 \end{bmatrix}.$$

First, perform the elementary operations $(-2)R_2 + R_3 \to R_3$ and $3R_2 + R_1 \to R_1$:

$$\begin{bmatrix} 0 & 0 & 5 & 10 & -10 \\ 1 & -2 & 2 & 3 & -1 \\ 0 & 0 & 1 & 2 & -2 \end{bmatrix}.$$

Then, perform $(-\frac{2}{5})R_1 + R_2 \to R_2$ and $(-\frac{1}{5})R_1 + R_3 \to R_3$:

$$\begin{bmatrix} 0 & 0 & 5 & 10 & -10 \\ 1 & -2 & 0 & -1 & 3 \\ 0 & 0 & 0 & 0 & 0 \end{bmatrix}.$$

Finally, perform $R_1 \leftrightarrow R_2$:

$$\begin{bmatrix} \underline{1} & -2 & 0 & -1 & 3 \\ 0 & 0 & \underline{5} & 10 & -10 \\ 0 & 0 & 0 & 0 & 0 \end{bmatrix} \qquad \text{(row-echelon form)}.$$

The underlined positions give the pivot positions of the reworked matrix **A**. Hence, the first and third columns are the two pivot columns of the original version **A**. That is, the pivot columns are given by

$$\begin{bmatrix} -3 \\ 1 \\ 2 \end{bmatrix}, \quad \begin{bmatrix} -1 \\ 2 \\ 5 \end{bmatrix}.$$

After performing the row elementary operation $(-\frac{1}{5})R_2 \to R_2$, the above echelon matrix comes to the reduced row-echelon form:

$$\begin{bmatrix} 1 & -2 & 0 & -1 & 3 \\ 0 & 0 & 1 & 2 & -2 \\ 0 & 0 & 0 & 0 & 0 \end{bmatrix}.$$

1.2.2 Gauss Elimination Methods

Elementary row operations can be used to solve matrix equations and perform matrix inversion.

1. Gauss Elimination Method for Solving Matrix Equations

Consider how to solve an $n \times n$ matrix equation $\mathbf{Ax} = \mathbf{b}$, where the inverse matrix $\mathbf{A}^{-1}$ of the matrix $\mathbf{A}$ exists. It is to be hoped that the solution vector $\mathbf{x} = \mathbf{A}^{-1}\mathbf{b}$ can be obtained by using elementary row operations.

Form the $n \times (n+1)$ augmented matrix $\mathbf{B} = [\mathbf{A}, \mathbf{b}]$. Since the solution $\mathbf{x} = \mathbf{A}^{-1}\mathbf{b}$ can be written as a new matrix equation $\mathbf{Ix} = \mathbf{A}^{-1}\mathbf{b}$, we get a new augmented matrix $\mathbf{C} = [\mathbf{I}, \mathbf{A}^{-1}\mathbf{b}]$ associated with the solution $\mathbf{x} = \mathbf{A}^{-1}\mathbf{b}$. Hence, we can write the solution process for the two matrix equations $\mathbf{Ax} = \mathbf{b}$ and $\mathbf{x} = \mathbf{A}^{-1}\mathbf{b}$ respectively as follows:

$$\text{matrix equations} \quad \mathbf{Ax} = \mathbf{b} \xrightarrow[\text{operations}]{\text{Elementary row}} \mathbf{x} = \mathbf{A}^{-1}\mathbf{b},$$

$$\text{augmented matrices} \quad [\mathbf{A}, \mathbf{b}] \xrightarrow[\text{operations}]{\text{Elementary row}} [\mathbf{I}, \mathbf{A}^{-1}\mathbf{b}].$$

This implies that, after suitable elementary row operations on the augmented matrix $[\mathbf{A}, \mathbf{b}]$, if the left-hand part of the new augmented matrix is an $n \times n$ identity matrix $\mathbf{I}$ then the $(n+1)$th column gives the solution $\mathbf{x} = \mathbf{A}^{-1}\mathbf{b}$ of the original equation $\mathbf{Ax} = \mathbf{b}$ directly. This method is called the *Gauss* or *Gauss–Jordan elimination method*.

Example 1.2 Use the Gauss elimination method to solve

$$\begin{aligned} x_1 + x_2 + 2x_3 &= 6, \\ 3x_1 + 4x_2 - x_3 &= 5, \\ -x_1 + x_2 + x_3 &= 2. \end{aligned}$$

As before, perform elementary row operations on the augmented matrix of the given matrix equation to yield the following results:

$$\begin{bmatrix} 1 & 1 & 2 & 6 \\ 3 & 4 & -1 & 5 \\ -1 & 1 & 1 & 2 \end{bmatrix} \xrightarrow{(-3)R_1+R_2\to R_2} \begin{bmatrix} 1 & 1 & 2 & 6 \\ 0 & 1 & -7 & -13 \\ -1 & 1 & 1 & 2 \end{bmatrix} \xrightarrow{R_1+R_3\to R_3}$$

$$\begin{bmatrix} 1 & 1 & 2 & 6 \\ 0 & 1 & -7 & -13 \\ 0 & 2 & 3 & 8 \end{bmatrix} \xrightarrow{(-1)R_2+R_1\to R_1} \begin{bmatrix} 1 & 0 & 9 & 19 \\ 0 & 1 & -7 & -13 \\ 0 & 2 & 3 & 8 \end{bmatrix} \xrightarrow{(-2)R_2+R_3\to R_3}$$

$$\begin{bmatrix} 1 & 0 & 9 & 19 \\ 0 & 1 & -7 & -13 \\ 0 & 0 & 17 & 34 \end{bmatrix} \xrightarrow{\frac{1}{17}R_3\to R_3} \begin{bmatrix} 1 & 0 & 9 & 19 \\ 0 & 1 & -7 & -13 \\ 0 & 0 & 1 & 2 \end{bmatrix} \xrightarrow{(-9)R_3+R_1\to R_1}$$

$$\begin{bmatrix} 1 & 0 & 0 & 1 \\ 0 & 1 & -7 & -13 \\ 0 & 0 & 1 & 2 \end{bmatrix} \xrightarrow{7R_3+R_2\to R_2} \begin{bmatrix} 1 & 0 & 0 & 1 \\ 0 & 1 & 0 & 1 \\ 0 & 0 & 1 & 2 \end{bmatrix}.$$

Hence, the solution to the linear system of equations is given by $x_1 = 1$, $x_2 = 1$ and $x_3 = 2$.

Elementary row operations are also applicable for solving the $m \times n$ (where $m > n$) matrix equation $\mathbf{Ax} = \mathbf{b}$, as shown in Algorithm 1.2. To do this, it is necessary to transform the augmented matrix $\mathbf{B} = [\mathbf{A}, \mathbf{b}]$ to its reduced row-echelon form.

Algorithm 1.2 Solving the $m \times n$ matrix equations $\mathbf{Ax} = \mathbf{b}$ [232]

---

1. Form the augmented matrix $\mathbf{B} = [\mathbf{A}, \mathbf{b}]$.
2. Run Algorithm 1.1 to transform the augmented matrix $\mathbf{B}$ to its reduced row-echelon form, which is equivalent to the original augmented matrix.
3. From the reduced row-echelon form write down the corresponding system of linear equations equivalent to the original system of linear equations.
4. Solve the new system of linear equations to yield its general solution.

Example 1.3 Solve the system of linear equations

$$\begin{aligned} 2x_1 + 2x_2 - x_3 &= 1, \\ -2x_1 - 2x_2 + 4x_3 &= 1, \\ 2x_1 + 2x_2 + 5x_3 &= 5, \\ -2x_1 - 2x_2 - 2x_3 &= -3. \end{aligned}$$

Perform the Type-II operation $(\frac{1}{2})R_1 \to R_1$ on the augmented matrix:

$$\mathbf{B} = \begin{bmatrix} 2 & 2 & -1 & 1 \\ -2 & -2 & 4 & 1 \\ 2 & 2 & 5 & 5 \\ -2 & -2 & -2 & -3 \end{bmatrix} \to \begin{bmatrix} 1 & 1 & -\frac{1}{2} & \frac{1}{2} \\ -2 & -2 & 4 & 1 \\ 2 & 2 & 5 & 5 \\ -2 & -2 & -2 & -3 \end{bmatrix}.$$

Use elementary row operations to make the first entry zero for the ith row, with $i = 2, 3, 4$:

$$\begin{bmatrix} 1 & 1 & -\frac{1}{2} & \frac{1}{2} \\ -2 & -2 & 4 & 1 \\ 2 & 2 & 5 & 5 \\ -2 & -2 & -2 & -3 \end{bmatrix} \to \begin{bmatrix} 1 & 1 & -\frac{1}{2} & \frac{1}{2} \\ 0 & 0 & 3 & 2 \\ 0 & 0 & 6 & 4 \\ 0 & 0 & -3 & -2 \end{bmatrix}.$$

Perform the Type-II operation $(\frac{1}{3})R_2 \to R_2$:

$$\begin{bmatrix} 1 & 1 & -\frac{1}{2} & \frac{1}{2} \\ 0 & 0 & 3 & 2 \\ 0 & 0 & 6 & 4 \\ 0 & 0 & -3 & -2 \end{bmatrix} \to \begin{bmatrix} 1 & 1 & -\frac{1}{2} & \frac{1}{2} \\ 0 & 0 & 1 & \frac{2}{3} \\ 0 & 0 & 6 & 4 \\ 0 & 0 & -3 & -2 \end{bmatrix}.$$

Then, using elementary row operations, transform to zero all entries directly above and below the leading-1 entry of the second row:

$$\begin{bmatrix} 1 & 1 & -\frac{1}{2} & \frac{1}{2} \\ 0 & 0 & 1 & \frac{2}{3} \\ 0 & 0 & 6 & 4 \\ 0 & 0 & -3 & -2 \end{bmatrix} \to \begin{bmatrix} 1 & 1 & 0 & \frac{5}{6} \\ 0 & 0 & 1 & \frac{2}{3} \\ 0 & 0 & 0 & 0 \\ 0 & 0 & 0 & 0 \end{bmatrix},$$

from which one has two new linear equations, $x_1 + x_2 = \frac{5}{6}$ and $x_3 = \frac{2}{3}$. This linear system of equations has infinitely many solutions, its general solution is given by $x_1 = \frac{5}{6} - x_2$, $x_3 = \frac{2}{3}$. If $x_2 = 1$ then a set of particular solutions is given by $x_1 = -\frac{1}{6}$, $x_2 = 1$ and $x_3 = \frac{2}{3}$.

Any $m \times n$ complex matrix equation $\mathbf{Ax} = \mathbf{b}$ can be written as

$$(\mathbf{A}_\mathrm{r} + \mathrm{j}\,\mathbf{A}_\mathrm{i})(\mathbf{x}_\mathrm{r} + \mathrm{j}\,\mathbf{x}_\mathrm{i}) = \mathbf{b}_\mathrm{r} + \mathrm{j}\,\mathbf{b}_\mathrm{i}, \tag{1.2.1}$$

where $\mathbf{A}_\mathrm{r}, \mathbf{x}_\mathrm{r}, \mathbf{b}_\mathrm{r}$ and $\mathbf{A}_\mathrm{i}, \mathbf{x}_\mathrm{i}, \mathbf{b}_\mathrm{i}$ are the real and imaginary parts of $\mathbf{A}, \mathbf{x}, \mathbf{b}$, respectively. Expand the above equation to yield

$$\mathbf{A}_\mathrm{r}\mathbf{x}_\mathrm{r} - \mathbf{A}_\mathrm{i}\mathbf{x}_\mathrm{i} = \mathbf{b}_\mathrm{r}, \tag{1.2.2}$$

$$\mathbf{A}_\mathrm{i}\mathbf{x}_\mathrm{r} + \mathbf{A}_\mathrm{r}\mathbf{x}_\mathrm{i} = \mathbf{b}_\mathrm{i}. \tag{1.2.3}$$

The above equations can be combined into

$$\begin{bmatrix} \mathbf{A}_\mathrm{r} & -\mathbf{A}_\mathrm{i} \\ \mathbf{A}_\mathrm{i} & \mathbf{A}_\mathrm{r} \end{bmatrix} \begin{bmatrix} \mathbf{x}_\mathrm{r} \\ \mathbf{x}_\mathrm{i} \end{bmatrix} = \begin{bmatrix} \mathbf{b}_\mathrm{r} \\ \mathbf{b}_\mathrm{i} \end{bmatrix}. \tag{1.2.4}$$

Thus, m complex-valued equations with n complex unknowns become $2m$ real-valued equations with $2n$ real unknowns.

In particular, if $m = n$ then we have

$$\text{complex matrix equation} \quad \mathbf{Ax} = \mathbf{b} \quad \xrightarrow[\text{operations}]{\text{Elementary row}} \quad \mathbf{x} = \mathbf{A}^{-1}\mathbf{b},$$

$$\text{augmented matrix} \begin{bmatrix} \mathbf{A}_\mathrm{r} & -\mathbf{A}_\mathrm{i} & \mathbf{b}_\mathrm{r} \\ \mathbf{A}_\mathrm{i} & \mathbf{A}_\mathrm{r} & \mathbf{b}_\mathrm{i} \end{bmatrix} \xrightarrow[\text{operations}]{\text{Elementary row}} \begin{bmatrix} \mathbf{I}_n & \mathbf{O}_n & \mathbf{x}_\mathrm{r} \\ \mathbf{O}_n & \mathbf{I}_n & \mathbf{x}_\mathrm{i} \end{bmatrix}.$$

This shows that if we write the $n \times (n+1)$ complex augmented matrix $[\mathbf{A}, \mathbf{b}]$ as a $2n \times (2n+1)$ real augmented matrix and perform elementary row operations to make its left-hand side become an $2n \times 2n$ identity matrix then the upper and lower halves of the $(2n+1)$th column give respectively the real and imaginary parts of the complex solution vector $\mathbf{x}$ of the original complex matrix equation $\mathbf{Ax} = \mathbf{b}$.

2. Gauss Elimination Method for Matrix Inversion

Consider the inversion operation on an $n \times n$ nonsingular matrix $\mathbf{A}$. This problem can be modeled as an $n \times n$ matrix equation $\mathbf{AX} = \mathbf{I}$ whose solution $\mathbf{X}$ is the inverse matrix of $\mathbf{A}$. It is easily seen that the augmented matrix of the matrix equation $\mathbf{AX} = \mathbf{I}$ is $[\mathbf{A}, \mathbf{I}]$, whereas the augmented matrix of the solution equation $\mathbf{IX} = \mathbf{A}^{-1}$ is $[\mathbf{I}, \mathbf{A}^{-1}]$. Hence, we have the following relations:

$$\text{matrix equation} \quad \mathbf{AX} = \mathbf{I} \quad \xrightarrow[\text{operations}]{\text{Elementary row}} \mathbf{X} = \mathbf{A}^{-1},$$

$$\text{augmented matrix} \quad [\mathbf{A}, \mathbf{I}] \quad \xrightarrow[\text{operations}]{\text{Elementary row}} \quad [\mathbf{I}, \mathbf{A}^{-1}].$$

This result tells us that if we use elementary row operations on the $n \times 2n$ augmented matrix $[\mathbf{A}, \mathbf{I}]$ so that its left-hand part becomes an $n \times n$ identity matrix then the right-hand part yields the inverse $\mathbf{A}^{-1}$ of the given $n \times n$ matrix $\mathbf{A}$ directly. This is called the *Gauss elimination method for matrix inversion.*

Suppose that an $n \times n$ complex matrix $\mathbf{A}$ is nonsingular. Then, its inversion can be modeled as the complex matrix equation $(\mathbf{A}_\mathrm{r} + \mathrm{j}\mathbf{A}_\mathrm{i})(\mathbf{X}_\mathrm{r} + \mathrm{j}\mathbf{X}_\mathrm{i}) = \mathbf{I}$. This complex equation can be rewritten as

$$\begin{bmatrix} \mathbf{A}_\mathrm{r} & -\mathbf{A}_\mathrm{i} \\ \mathbf{A}_\mathrm{i} & \mathbf{A}_\mathrm{r} \end{bmatrix} \begin{bmatrix} \mathbf{X}_\mathrm{r} \\ \mathbf{X}_\mathrm{i} \end{bmatrix} = \begin{bmatrix} \mathbf{I}_n \\ \mathbf{O}_n \end{bmatrix}, \tag{1.2.5}$$

from which we get the following relation:

$$\text{complex matrix equation} \quad \mathbf{AX} = \mathbf{I} \quad \xrightarrow[\text{operations}]{\text{Elementary row}} \quad \mathbf{X} = \mathbf{A}^{-1},$$

$$\text{augmented matrix} \begin{bmatrix} \mathbf{A}_{\rm r} & -\mathbf{A}_{\rm i} & \mathbf{I}_n \\ \mathbf{A}_{\rm i} & \mathbf{A}_{\rm r} & \mathbf{O}_n \end{bmatrix} \xrightarrow[\text{operations}]{\text{Elementary row}} \begin{bmatrix} \mathbf{I}_n & \mathbf{O}_n & \mathbf{X}_{\rm r} \\ \mathbf{O}_n & \mathbf{I}_n & \mathbf{X}_{\rm i} \end{bmatrix}.$$

That is to say, if performing elementary row operations on the $2n \times 3n$ augmented matrix to transform its left-hand side to an $2n \times 2n$ identity matrix then the upper and lower halves of the $2n \times n$ matrix on the right give respectively the real and imaginary parts of the inverse matrix $\mathbf{A}^{-1}$ of the complex matrix $\mathbf{A}$.

1.3 Sets, Vector Subspaces and Linear Mapping

The set of all n-dimensional vectors with real (complex) components is called a real (complex) n-dimensional vector space, denoted $\mathbb{R}^n$ ($\mathbb{C}^n$). The n-dimensional vectors of real-world problems usually belong to subsets other than the whole set $\mathbb{R}^n$ or $\mathbb{C}^n$. These subsets are known as vector subspaces. In this section, we present the sets, the vector subspaces and the linear mapping of one vector subspace onto another.

1.3.1 Sets

Before introducing definitions of vector spaces and subspaces, it is necessary to present some set concepts. As the name implies, a *set* is a collection of elements.

A set is usually denoted by $S = \{\cdot\}$; inside the braces are the elements of the set S. If there are only a few elements in the set S, we write out these elements within the braces, e.g., $S = \{a, b, c, d\}$.

To describe the composition of a more complex set mathematically, we use the symbol "$|$" to mean "such that". For example, $S = \{x|P(x) = 0\}$ reads "the element x in set S such that $P(x) = 0$". A set with only one element α is called a *singleton*, denoted $\{\alpha\}$.

The following are several common notations for set operations:

$\forall$ denotes "for all $\cdots$";
$x \in A$ reads "x belongs to the set A", i.e., x is an element or member of A;
$x \notin A$ means that x is not an element of the set A;
$\ni$ denotes "such that";
$\exists$ denotes "there exists";
$A \Rightarrow B$ reads "condition A results in B" or "A implies B".

As an example, the trivial statement "there is a zero element θ in the set V such that the addition $x + \theta = x = \theta + x$ holds for all elements x in V" can be concisely expressed in the above notation as follows:

$$\exists\, \theta \in V \ \ni\ x + \theta = x = \theta + x,\ \forall x \in V.$$

Let A and B be two sets; then the sets have the following basic relations.

The notation $A \subseteq B$ reads "the set A is contained in the set B" or "A is a *subset* of B", which implies that each element in A is an element in B, namely $x \in A \Rightarrow x \in B$.

If $A \subset B$ then A is called a *proper subset* of B. The notation $B \supset A$ reads "B contains A" or "B is a *superset* of A". The set with no elements is denoted by $\emptyset$ and is called the *null set.*

The notation $A = B$ reads "the set A equals the set B", which means that $A \subseteq B$ and $B \subseteq A$, or $x \in A \Leftrightarrow x \in B$ (any element in A is an element in B, and vice versa). The negation of $A = B$ is written as $A \neq B$, implying that A does not belong to B, neither does B belong to A.

The union of A and B is denoted as $A \cup B$. If $X = A \cup B$ then X is called the *union set* of A and B. The union set is defined as follows:

$$X = A \cup B = \{x \in X |\, x \in A \text{ or } x \in B\}. \tag{1.3.1}$$

In other words, the elements of the union set $A \cup B$ consist of the elements of A and the elements of B.

The intersection of both sets A and B is represented by the notation $A \cap B$ and is defined as follows:

$$X = A \cap B = \{x \in X |\, x \in A \text{ and } x \in B\}. \tag{1.3.2}$$

The set $X = A \cap B$ is called the *intersection set* of A and B. Each element of the intersection set $A \cap B$ consists of elements common to both A and B.

The notation $Z = A + B$ means the sum set of sets A and B and is defined as follows:

$$Z = A + B = \{z = x + y \in Z |\, x \in A,\ y \in B\}, \tag{1.3.3}$$

namely, an element z of the *sum set* $Z = A + B$ consists of the sum of the element x in A and the element y in B.

The set-theoretic difference of the sets A and B, denoted "$A - B$", is also termed the *difference set* and is defined as follows:

$$X = A - B = \{x \in X |\, x \in A,\ \text{but } x \notin B\}. \tag{1.3.4}$$

That is to say, the difference set $A - B$ is the set of elements of A that are not in B. The difference set $A - B$ is also sometimes denoted by the notation $X = A \setminus B$. For example, $\{\mathbb{C}^n \setminus \mathbf{0}\}$ denotes the set of nonzero vectors in complex n-dimensional vector space.

The *relative complement* of A in X is defined as

$$A^{\mathrm{c}} = X - A = X \setminus A = \{x \in X |\, x \notin A\}. \tag{1.3.5}$$

Example 1.4 For the sets

$$A = \{1, 2, 3\}, \quad B = \{2, 3, 4\},$$

we have

$$A \cup B = \{1, 2, 3, 4\}, \quad A \cap B = \{2, 3\}, \quad A + B = \{3, 5, 7\},$$
$$A - B = A \setminus B = \{1\}, \quad B - A = B \setminus A = \{4\}.$$

If both X and Y are sets, and $x \in X$ and $y \in Y$, then the set of all ordered pairs (x, y) is denoted by $X \times Y$ and is termed the *Cartesian product* of the sets X and Y, and is defined as follows:

$$X \times Y = \big\{(x, y) \big| x \in X, \; y \in Y\big\}. \tag{1.3.6}$$

Similarly, $X_1 \times \cdots \times X_n$ denotes the Cartesian product of n sets $X_1, \ldots, X_n$, and its elements are the *ordered n-ples* $(x_1, \ldots, x_n)$:

$$X_1 \times \cdots \times X_n = \big\{(x_1, \ldots, x_n) \big| x_1 \in X_1, \ldots, x_n \in X_n\big\}. \tag{1.3.7}$$

If $f(\mathbf{X}, \mathbf{Y})$ is a scalar function with real matrices $\mathbf{X} \in \mathbb{R}^{n \times n}$ and $\mathbf{Y} \in \mathbb{R}^{n \times n}$ as variables then in linear mapping notation, the function can be denoted by the Cartesian product form $f : \mathbb{R}^{n \times n} \times \mathbb{R}^{n \times n} \to \mathbb{R}$.

1.3.2 Fields and Vector Spaces

The previous subsections uses the symbols $\mathbb{R}$, $\mathbb{R}^m$, $\mathbb{R}^{m \times n}$ and $\mathbb{C}$, $\mathbb{C}^m$, $\mathbb{C}^{m \times n}$. In this subsection we explain this notation from the viewpoint of vector spaces.

Definition 1.14 [229] A *field* is an algebraic structure $\mathbb{F}$ consisting of a nonempty set F together with two operations: for any two elements α, β in the set F, the addition operation $\alpha + \beta$ and the multiplication operation $\alpha\beta$ are uniquely determined such that, for any $\alpha, \beta, \gamma \in F$, the following conditions are satisfied:

(1) $\alpha + \beta = \beta + \alpha$ and $\alpha\beta = \beta\alpha$;
(2) $\alpha + (\beta + \gamma) = (\alpha + \beta) + \gamma$ and $\alpha(\beta\gamma) = (\alpha\beta)\gamma$;
(3) $\alpha(\beta + \gamma) = \alpha\beta + \alpha\gamma$;
(4) there is an element 0 in F such that $\alpha + 0 = \alpha$ for all $\alpha \in F$;
(5) there is an element $-\alpha$ such that $\alpha + (-\alpha) = 0$ for any $\alpha \in F$;
(6) there exists a nonzero element $1 \in F$ such that $\alpha 1 = \alpha$ for all $\alpha \in F$;
(7) there is an element $\alpha^{-1} \in F$ such that $\alpha\alpha^{-1} = 1$ for any nonzero element $\alpha \in F$.

The set of rational numbers, the set of real numbers and the set of complex numbers are fields, the *rational number field* $\mathbb{Q}$, the *real number field* $\mathbb{R}$ and the *complex number field* $\mathbb{C}$. However, the set of integers is not a field.

When considering a nonempty set V with vectors as elements, the concept of the field can be generalized to the concept of vector space.

Definition 1.15 A set V with vectors as elements is called a *vector space* if the addition operation is defined as the addition of two vectors, the multiplication operation is defined as the product of a vector and a scalar in the scalar field S and, for vectors $\mathbf{x}, \mathbf{y}, \mathbf{w}$ in the set V and scalars a_1, a_2 in the scalar field S, the following axioms, also termed postulates or laws, on addition and multiplication hold.

1. *Closure properties*

   (1) If $\mathbf{x} \in V$ and $\mathbf{y} \in V$ then $\mathbf{x} + \mathbf{y} \in V$, namely V is closed under addition. This is called the *closure for addition.*

   (2) If a_1 is a scalar and $\mathbf{y} \in V$ then $a_1\mathbf{y} \in V$, namely V is closed under scalar multiplication. This is called the *closure for scalar multiplication.*

2. *Axioms on addition*

   (1) *Commutative law for addition:* $\mathbf{x} + \mathbf{y} = \mathbf{y} + \mathbf{x}, \ \forall \mathbf{x}, \mathbf{y} \in V$.

   (2) *Associative law for addition:* $\mathbf{x} + (\mathbf{y} + \mathbf{w}) = (\mathbf{x} + \mathbf{y}) + \mathbf{w}, \ \forall \mathbf{x}, \mathbf{y}, \mathbf{w} \in V$.

   (3) *Existence of the null vector:* There exists a null vector $\mathbf{0}$ in V such that, for any vector $\mathbf{y} \in V$, $\mathbf{y} + \mathbf{0} = \mathbf{y}$ holds.

   (4) *Existence of negative vectors:* Given a vector $\mathbf{y} \in V$, there is another vector $-\mathbf{y} \in V$ such that $\mathbf{y} + (-\mathbf{y}) = \mathbf{0} = (-\mathbf{y}) + \mathbf{y}$.

3. *Axioms on scalar multiplication*

   (1) *Associative law for scalar multiplication:* $a(b\mathbf{y}) = (ab)\mathbf{y}$ holds for all vectors $\mathbf{y} \in V$ and all scalars $a, b \in S$.

   (2) *Right distributive law for scalar multiplication:* $a(\mathbf{x}+\mathbf{y}) = a\mathbf{x} + a\mathbf{y}$ holds for all vectors $\mathbf{x}, \mathbf{y} \in V$ and any scalar a in S.

   (3) *Left distributive law for scalar multiplication:* $(a + b)\mathbf{y} = a\mathbf{y} + b\mathbf{y}$ holds for all vectors $\mathbf{y}$ and all scalars $a, b \in S$.

   (4) *Unity law for scalar multiplication:* $1\mathbf{y} = \mathbf{y}$ holds for all vectors $\mathbf{y} \in V$.

As stated in the following theorem, vector spaces have other useful properties.

Theorem 1.4 *If V is a vector space, then*

- *The null vector $\mathbf{0}$ is unique.*
- *The inverse operation $-\mathbf{y}$ is unique for each $\mathbf{y} \in V$.*
- *For every vector $\mathbf{y} \in V$, $0\mathbf{y} = \mathbf{0}$ is true.*
- *For every scalar a, $a\mathbf{0} = \mathbf{0}$ holds.*
- *If $a\mathbf{y} = \mathbf{0}$ then $a = 0$ or $\mathbf{y} = \mathbf{0}$.*
- $(-1)\mathbf{y} = -\mathbf{y}$.

Proof See [232, pp. 365–366]. □

If the vectors in a vector space V are real valued, and the scalar field is a real field, then V is referred to as a *real vector space*. Similarly, if the vectors in V are complex valued, and the scalar field is a complex field, then V is called a *complex vector space.*

The two most typical vector spaces are the m-dimensional real vector space $\mathbb{R}^m$ and the m-dimensional complex vector space $\mathbb{C}^m$.

In many cases we are interested only in some given subset of a vector space V other than the whole space V.

Definition 1.16 Let W be a nonempty subset of the vector space V. Then W is called a *vector subspace* of V if all elements in W satisfy the following two conditions:

(1) $\mathbf{x} - \mathbf{y} \in W$ for all $\mathbf{x}, \mathbf{y} \in W$;
(2) $\alpha\mathbf{x} \in W$ for all $\alpha \in \mathbb{R}$ (or $\mathbb{C}$) and all $\mathbf{x} \in W$.

Vector subspaces are very useful in engineering applications such as automatic control, signal processing, pattern recognition, wireless communication and so forth. We will devote our attention to the theory and applications of vector subspaces in Chapter 8, "Subspace Analysis and Tracking".

1.3.3 Linear Mapping

In the previous subsections we discussed some simple operations of vectors, vector addition and the multiplication of a vector by a scalar, but we have not yet considered the transformation between vectors in two vector spaces.

By Wikipedia, mapping is the creation of maps, a graphic symbolic representation of the significant features of a part of the surface of the Earth. In mathematics, mapping is a synonym for mathematical function or for Morphism.

The notation for mappings follows the usual notation for functions. If V is a subspace in $\mathbb{R}^m$ and W is a subspace in $\mathbb{R}^n$ then the notation

$$T : V \mapsto W \tag{1.3.8}$$

denotes a general mapping, while

$$T : V \to W \tag{1.3.9}$$

represents a linear mapping or linear transformation, which is a mapping such that the following operations hold:

1. $T(\mathbf{v}_1 + \mathbf{v}_2) = T(\mathbf{v}_1) + T(\mathbf{v}_2)$ for any vectors $\mathbf{v}_1$ and $\mathbf{v}_2$ in V;
2. $T(\alpha\mathbf{v}) = \alpha T(\mathbf{v})$ for any scalar α.

The main example of a linear transformation is given by matrix multiplication.

Thus, a mapping represents a rule for transforming the vectors in V to corresponding vectors in W. The subspace V is said to be the *initial set* or *domain* of the mapping T and W its *final set* or *codomain.*

When $\mathbf{v}$ is some vector in the vector space V, $T(\mathbf{v})$ is referred to as the *image* of the vector $\mathbf{v}$ under the mapping T, or the value of the mapping T at the point $\mathbf{v}$, whereas $\mathbf{v}$ is called the *original image* of $T(\mathbf{v})$.

Given a subspace A of the vector space V, the mapping

$$T(A) = \{T(\mathbf{v}) \,|\, \mathbf{v} \in A\} \tag{1.3.10}$$

represents a collection of the vectors $\mathbf{v}$ under the mapping T.

In particular, if $T(\mathbf{v})$ represents a collection of transformed outputs of all vectors $\mathbf{v}$ in V, i.e.,

$$T(V) = \{T(\mathbf{v}) \,|\, \mathbf{v} \in V\}, \tag{1.3.11}$$

then we say that $T(V)$ is the *range* of the mapping T, denoted by

$$\mathrm{Im}(T) = T(V) = \{T(\mathbf{v}) \,|\, \mathbf{v} \in V\}. \tag{1.3.12}$$

Definition 1.17 A mapping $T : V \mapsto W$ is *one-to-one* if it is either *injective* or *surjective*, i.e., $T(\mathbf{x}) = T(\mathbf{y})$ implies $\mathbf{x} = \mathbf{y}$ or distinct elements have distinct images.

A one-to-one mapping $T : V \mapsto W$ has an inverse mapping $T^{-1}|W \mapsto V$. The inverse mapping T^{-1} restores what the mapping T has done. Hence, if $T(\mathbf{v}) = \mathbf{w}$ then $T^{-1}(\mathbf{w}) = \mathbf{v}$, resulting in $T^{-1}(T(\mathbf{v})) = \mathbf{v}, \forall\, \mathbf{v} \in V$ and $T(T^{-1}(\mathbf{w})) = \mathbf{w}, \forall\, \mathbf{w} \in W$.

Definition 1.18 A mapping T is called a *linear mapping* or *linear transformation* in the vector space V if it satisfies the linear relationship

$$T(c_1\mathbf{v}_1 + c_2\mathbf{v}_2) = c_1T(\mathbf{v}_1) + c_2T(\mathbf{v}_2) \tag{1.3.13}$$

for all $\mathbf{v}_1, \mathbf{v}_2 \in V$ and all scalars c_1, c_2.

If $\mathbf{u}_1, \ldots, \mathbf{u}_p$ are the input vectors of a system in engineering then $T(\mathbf{u}_1), \ldots,$ $T(\mathbf{u}_p)$ can be viewed as the output vectors of the system. The criterion for identifying whether a system is linear is this: if the system input is the linear expression $\mathbf{y} = c_1\mathbf{u}_1 + \cdots + c_p\mathbf{u}_p$ then we say that the system is linear only if its output satisfies the linear expression $T(\mathbf{y}) = T(c_1\mathbf{u}_1 + \cdots + c_p\mathbf{u}_p) = c_1T(\mathbf{u}_1) + \cdots + c_pT(\mathbf{u}_p)$. Otherwise, the system is nonlinear.

Example 1.5 Determine whether the following transformation T_1, $T_2 : \mathbb{R}^3 \mapsto \mathbb{R}^2$

are linear:

$$T_1(\mathbf{x}) = \begin{bmatrix} x_1 + x_2 \\ x_1^2 - x_2^2 \end{bmatrix}, \text{ where } \quad \mathbf{x} = [x_1, x_2, x_3]^T,$$

$$T_2(\mathbf{x}) = \begin{bmatrix} x_1 - x_2 \\ x_2 + x_3 \end{bmatrix}, \text{ where } \quad \mathbf{x} = [x_1, x_2, x_3]^T.$$

It is easily seen that the transformation $T_1 | \mathbb{R}^3 \mapsto \mathbb{R}^2$ does not satisfy a linear relationship and thus is not a linear transformation, while $T_2 | \mathbb{R}^3 \mapsto \mathbb{R}^2$ is a linear transformation since it satisfies a linear relationship.

An interesting special application of linear mappings is the electronic amplifier $\mathbf{A} \in \mathbb{C}^{n \times n}$ with high fidelity (Hi-Fi). By Hi-Fi, one means that there is the following linear relationship between any input signal vector $\mathbf{u}$ and the corresponding output signal vector $\mathbf{Au}$ of the amplifier:

$$\mathbf{Au} = \lambda \mathbf{u}, \tag{1.3.14}$$

where $\lambda > 1$ is the amplification factor or gain. The equation above is a typical characteristic equation of a matrix.

Another example is that the product of a matrix by a vector $\mathbf{A}_{m \times n} \mathbf{x}_{n \times 1}$ can also be viewed as a linear mapping, $T : \mathbf{x} \to \mathbf{Ax}$ or $T(\mathbf{x}) = \mathbf{Ax}$, that transforms the vector $\mathbf{x}$ in $\mathbb{C}^n$ to a vector $\mathbf{y} = \mathbf{Ax}$ in $\mathbb{C}^m$. In view of this linear mapping or transformation, the product of a matrix $\mathbf{A}$ and a vector is usually called the *matrix transformation* of the vector, where $\mathbf{A}$ is a *transformation matrix.*

The following are intrinsic relationships between a linear space and a linear mapping.

THEOREM 1.5 [35, p. 29] *Let V and W be two vector spaces, and let $T : V \to W$ be a linear mapping. Then the following relationships are true:*

- *if M is a linear subspace in V then $T(M)$ is a linear subspace in W;*
- *if N is a linear subspace in W then the linear inverse transform $T^{-1}(N)$ is a linear subspace in V.*

A linear mapping has the following basic properties: if $T : V \to W$ is a linear mapping then

$$T(\mathbf{0}) = \mathbf{0} \qquad \text{and} \qquad T(-\mathbf{x}) = -T(\mathbf{x}). \tag{1.3.15}$$

In particular, for a given linear transformation $\mathbf{y} = \mathbf{Ax}$ where the transformation matrix $\mathbf{A}$ is known, if our task is to obtain the output vector $\mathbf{y}$ from the input vector $\mathbf{x}$ then $\mathbf{Ax} = \mathbf{y}$ is said to be a *forward problem.* Conversely, the problem of finding the input vector $\mathbf{x}$ from the output vector $\mathbf{y}$ is known as an *inverse problem.* Clearly, the essence of the forward problem is a matrix–vector calculation, while the essence of the inverse problem is solving a matrix equation.

1.4 Inner Products and Vector Norms

In mathematics, physics and engineering, we often need to measure the size (or length) and neighborhood for a given vector, and to calculate the angle, distance and similarity between vectors. In these cases, inner products and vector norms are two essential mathematical tools.

1.4.1 Inner Products of Vectors

Let $\mathbb{K}$ represent a field that may be either the real field $\mathbb{R}$ or the complex field $\mathbb{C}$, and let V be an n-dimensional vector space $\mathbb{R}^n$ or $\mathbb{C}^n$.

Definition 1.19 The function $\langle \mathbf{x}, \mathbf{y} \rangle$ is called the *inner product* of two vectors $\mathbf{x}$ and $\mathbf{y}$, and V is known as the *inner product vector space*, if for all vectors $\mathbf{x}, \mathbf{y}, \mathbf{z} \in V$ and scalars $\alpha, \beta \in \mathbb{K}$, the mapping function $\langle \cdot, \cdot \rangle | V \times V \mapsto \mathbb{K}$ satisfies the following three axioms.

(1) *Conjugate symmetry* $\langle \mathbf{x}, \mathbf{y} \rangle = \langle \mathbf{y}, \mathbf{x} \rangle^*$.
(2) *Linearity in the first argument* $\langle \alpha\mathbf{x} + \beta\mathbf{y}, \mathbf{z} \rangle = \alpha\langle \mathbf{x}, \mathbf{z} \rangle + \beta\langle \mathbf{y}, \mathbf{z} \rangle$.
(3) *Nonnegativity* $\langle \mathbf{x}, \mathbf{x} \rangle \geq 0$ and $\langle \mathbf{x}, \mathbf{x} \rangle = 0 \Leftrightarrow \mathbf{x} = \mathbf{0}$ (strict positivity).

For a real inner product space, the conjugate symmetry becomes the real symmetry $\langle \mathbf{x}, \mathbf{y} \rangle = \langle \mathbf{y}, \mathbf{x} \rangle$.

Linearity in the first argument contains the homogeneity $\langle \alpha\mathbf{x}, \mathbf{y} \rangle = \alpha\langle \mathbf{x}, \mathbf{y} \rangle$ and the additivity $\langle \mathbf{x} + \mathbf{y}, \mathbf{z} \rangle = \langle \mathbf{x}, \mathbf{z} \rangle + \langle \mathbf{y}, \mathbf{z} \rangle$.

Conjugate symmetry and linearity in the first argument imply that

$$\langle \mathbf{x}, \alpha\mathbf{y} \rangle = \langle \alpha\mathbf{y}, \mathbf{x} \rangle^* = \alpha^* \langle \mathbf{y}, \mathbf{x} \rangle^* = \alpha^* \langle \mathbf{x}, \mathbf{y} \rangle, \tag{1.4.1}$$

$$\langle \mathbf{x}, \mathbf{y} + \mathbf{z} \rangle = \langle \mathbf{y} + \mathbf{z}, \mathbf{x} \rangle^* = \langle \mathbf{y}, \mathbf{x} \rangle^* + \langle \mathbf{z}, \mathbf{x} \rangle^* = \langle \mathbf{x}, \mathbf{y} \rangle + \langle \mathbf{x}, \mathbf{z} \rangle. \tag{1.4.2}$$

A vector space with an inner product is called an *inner product space.*

It is easy to verify that the function

$$\langle \mathbf{x}, \mathbf{y} \rangle = \mathbf{x}^H \mathbf{y} = \sum_{i=1}^{n} x_i^* y_i \tag{1.4.3}$$

is an inner product of two vectors, since the three axioms above are satisfied. The function $\langle \mathbf{x}, \mathbf{y} \rangle = \mathbf{x}^H \mathbf{y}$ is called the *canonical inner product* of $n \times 1$ vectors $\mathbf{x} = [x_1, \ldots, x_n]^T$ and $\mathbf{y} = [y_1, \ldots, y_n]^T$.

Note that in some literature the following canonical inner product formula is used:

$$\langle \mathbf{x}, \mathbf{y} \rangle = \mathbf{x}^T \mathbf{y}^* = \sum_{i=1}^{n} x_i y_i^* .$$

However, one sometimes adopts the weighted inner product

$$\langle \mathbf{x}, \mathbf{y} \rangle_{\mathbf{G}} = \mathbf{x}^H \mathbf{G} \mathbf{y}, \tag{1.4.4}$$

where $\mathbf{G}$ is a weighting matrix such that $\mathbf{x}^H\mathbf{G}\mathbf{x} > 0, \forall \mathbf{x} \in \mathbb{C}^n$.

EXAMPLE 1.6 Let $\{e^{j2\pi fn}\}_{n=0}^{N-1}$ be a sinusoidal sequence with frequency f, and $\mathbf{e}_n(f) = \left[1, e^{j(\frac{2\pi}{n+1})f}, \ldots, e^{j(\frac{2\pi}{n+1})nf}\right]^T$ be an $(n+1)\times 1$ complex sinusoid vector. Then the *discrete Fourier transform* (DFT) of N data samples $x(n), n = 0, 1, \ldots, N-1$ can be represented by the canonical inner product as follows:

$$X(f) = \sum_{n=0}^{N-1} x(n)e^{-j2\pi nf/N} = \mathbf{e}_{N-1}^H(f)\mathbf{x} = \langle \mathbf{e}_{N-1}(f), \mathbf{x}\rangle,$$

where $\mathbf{x} = [x(0), x(1), \ldots, x(N-1)]^T$ is the data vector.

1.4.2 Norms of Vectors

DEFINITION 1.20 Let V be a real or complex vector space. Given a vector $\mathbf{x} \in V$, the mapping function $p(\mathbf{x})|V \mapsto \mathbb{R}$ is called the *norm* of the vector $\mathbf{x} \in V$, if for all vectors $\mathbf{x}, \mathbf{y} \in V$ and any scalar $c \in \mathbb{K}$ (here $\mathbb{K}$ denotes $\mathbb{R}$ or $\mathbb{C}$), the following three *norm axioms* hold:

(1) *Nonnegativity:* $p(\mathbf{x}) \geq 0$ and $p(\mathbf{x}) = 0 \Leftrightarrow \mathbf{x} = \mathbf{0}$.
(2) *Homogeneity:* $p(c\mathbf{x}) = |c| \cdot p(\mathbf{x})$ is true for all complex constant c.
(3) *Triangle inequality:* $p(\mathbf{x} + \mathbf{y}) \leq p(\mathbf{x}) + p(\mathbf{y})$.

In a real or complex inner product space V, the vector norms have the following properties [35].

1. $\|\mathbf{0}\| = 0$ and $\|\mathbf{x}\| > 0$, $\forall \mathbf{x} \neq \mathbf{0}$.
2. $\|c\mathbf{x}\| = |c|\,\|\mathbf{x}\|$ holds for all vector $\mathbf{x} \in V$ and any scalar $c \in \mathbb{K}$.
3. *Polarization identity:* For real inner product spaces we have

$$\langle \mathbf{x}, \mathbf{y}\rangle = \frac{1}{4}\left(\|\mathbf{x}+\mathbf{y}\|^2 - \|\mathbf{x}-\mathbf{y}\|^2\right), \quad \forall \mathbf{x}, \mathbf{y}, \tag{1.4.5}$$

and for complex inner product spaces we have

$$\langle \mathbf{x}, \mathbf{y}\rangle = \frac{1}{4}\left(\|\mathbf{x}+\mathbf{y}\|^2 - \|\mathbf{x}-\mathbf{y}\|^2 - \mathrm{j}\,\|\mathbf{x}+\mathrm{j}\,\mathbf{y}\|^2 + \mathrm{j}\|\mathbf{x}-\mathrm{j}\,\mathbf{y}\|^2\right), \quad \forall \mathbf{x}, \mathbf{y}. \tag{1.4.6}$$

4. *Parallelogram law*

$$\|\mathbf{x}+\mathbf{y}\|^2 + \|\mathbf{x}-\mathbf{y}\|^2 = 2\left(\|\mathbf{x}\|^2 + \|\mathbf{y}\|^2\right), \quad \forall \mathbf{x}, \mathbf{y}. \tag{1.4.7}$$

5. *Triangle inequality*

$$\|\mathbf{x}+\mathbf{y}\| \leq \|\mathbf{x}\| + \|\mathbf{y}\|, \quad \forall \mathbf{x}, \mathbf{y} \in V. \tag{1.4.8}$$

6. *Cauchy–Schwartz inequality*

$$|\langle \mathbf{x}, \mathbf{y} \rangle| \leq \|\mathbf{x}\| \|\mathbf{y}\|. \tag{1.4.9}$$

The equality $|\langle \mathbf{x}, \mathbf{y} \rangle| = \|\mathbf{x}\| \|\mathbf{y}\|$ holds if and only if $\mathbf{y} = c\,\mathbf{x}$, where c is some nonzero complex constant.

1. Common Norms of Constant Vectors

The following are several common norms of constant vectors.

(1) *ℓ_0-norm*

$$\|\mathbf{x}\|_0 \stackrel{\text{def}}{=} \text{number of nonzero entries of } \mathbf{x}. \tag{1.4.10}$$

(2) *ℓ_1-norm*

$$\|\mathbf{x}\|_1 \stackrel{\text{def}}{=} \sum_{i=1}^{m} |x_i| = |x_1| + \cdots + |x_m|. \tag{1.4.11}$$

(3) *ℓ_2-norm* or the *Euclidean norm*

$$\|\mathbf{x}\|_2 \stackrel{\text{def}}{=} \|\mathbf{x}\|_E = \left(|x_1|^2 + \cdots + |x_m|^2\right)^{1/2}. \tag{1.4.12}$$

(4) *ℓ_∞-norm*

$$\|\mathbf{x}\|_\infty \stackrel{\text{def}}{=} \max\left\{|x_1|, \ldots, |x_m|\right\}. \tag{1.4.13}$$

(5) *ℓ_p-norm* or *Hölder norm* [275]

$$\|\mathbf{x}\|_p \stackrel{\text{def}}{=} \left(\sum_{i=1}^{m} |x_i|^p\right)^{1/p}, \quad p \geq 1. \tag{1.4.14}$$

The ℓ_0-norm does not satisfy the homogeneity $\|c\mathbf{x}\|_0 = |c|\,\|\mathbf{x}\|_0$, and thus is a *quasi-norm.* However, the ℓ_0-norm plays a key role in the representation and analysis of sparse vectors, as will be seen in Section 1.12.

The ℓ_p-norm is a quasi-norm if $0 < p < 1$ and a norm if $p \geq 1$.

Clearly, when $p = 1$ or $p = 2$, the ℓ_p-norm reduces to the ℓ_1-norm or the ℓ_2-norm, respectively.

The most commonly used vector norm is the Euclidean norm.

The following are several important applications of the Euclidean norm.

- Measuring the size or length of a vector:

$$\text{size}(\mathbf{x}) = \|\mathbf{x}\|_2 = \sqrt{x_1^2 + \cdots + x_m^2}, \tag{1.4.15}$$

which is called the *Euclidean length.*

- Defining the *ϵ-neighborhood* of a vector $\mathbf{x}$:

$$N_\epsilon(\mathbf{x}) = \left\{\mathbf{y} \,\middle|\, \|\mathbf{y} - \mathbf{x}\|_2 \leq \epsilon\right\}, \quad \epsilon > 0. \tag{1.4.16}$$

- Measuring the distance between vectors $\mathbf{x}$ and $\mathbf{y}$:

$$d(\mathbf{x},\mathbf{y}) = \|\mathbf{x}-\mathbf{y}\|_2 = \sqrt{(x_1-y_1)^2+\cdots+(x_m-y_m)^2}. \tag{1.4.17}$$

This is called the *Euclidean distance.*

- Defining the *angle* θ $(0 \leq \theta \leq 2\pi)$ between vectors $\mathbf{x}$ and $\mathbf{y}$:

$$\theta \stackrel{\text{def}}{=} \arccos\left(\frac{\langle \mathbf{x},\mathbf{y}\rangle}{\sqrt{\langle \mathbf{x},\mathbf{x}\rangle}\sqrt{\langle \mathbf{y},\mathbf{y}\rangle}}\right) = \arccos\left(\frac{\mathbf{x}^H\mathbf{y}}{\|\mathbf{x}\|\|\mathbf{y}\|}\right). \tag{1.4.18}$$

A vector with unit Euclidean length is referred to as a *normalized* (or *standardized*) *vector.* For any nonzero vector $\mathbf{x} \in \mathbb{C}^m$, $\mathbf{x}/\langle \mathbf{x},\mathbf{x}\rangle^{1/2}$ is the normalized version of the vector and has the same direction as $\mathbf{x}$.

The norm $\|\mathbf{x}\|$ is said to be a *unitary invariant norm* if $\|\mathbf{U}\mathbf{x}\| = \|\mathbf{x}\|$ holds for all vectors $\mathbf{x} \in \mathbb{C}^m$ and all unitary matrices $\mathbf{U} \in \mathbb{C}^{m\times m}$.

Proposition 1.1 [214] *The Euclidean norm* $\|\cdot\|_2$ *is unitary invariant.*

If the inner product $\langle \mathbf{x},\mathbf{y}\rangle = \mathbf{x}^H\mathbf{y} = 0$, this implies that the angle between the vectors $\theta = \pi/2$. In this case, the constant vectors $\mathbf{x}$ and $\mathbf{y}$ are said to be orthogonal, from which we have the following definition.

Definition 1.21 Two constant vectors $\mathbf{x}$ and $\mathbf{y}$ are said to be orthogonal, and this is denoted by $\mathbf{x}\perp\mathbf{y}$, if their inner product $\langle \mathbf{x},\mathbf{y}\rangle = \mathbf{x}^H\mathbf{y} = 0$.

2. *Inner Products and Norms of Function Vectors*

Definition 1.22 Let $\mathbf{x}(t), \mathbf{y}(t)$ be two function vectors in the complex vector space $\mathbb{C}^n$, and let the definition field of the function variable t be $[a,b]$ with $a < b$. Then the *inner product of the function vectors* $\mathbf{x}(t)$ and $\mathbf{y}(t)$ is defined as follows:

$$\langle \mathbf{x}(t),\mathbf{y}(t)\rangle \stackrel{\text{def}}{=} \int_a^b \mathbf{x}^H(t)\mathbf{y}(t)\mathrm{d}t. \tag{1.4.19}$$

It can be verified that the above function satisfies the three axioms of inner products (see Definition 1.19). Note that the real field $\mathbb{R}$ is an one-dimensional inner product space, but it is not a Euclidean space as the real field $\mathbb{R}$ is not finite dimensional.

The angle between function vectors is defined as follows:

$$\cos\theta \stackrel{\text{def}}{=} \frac{\langle \mathbf{x}(t),\mathbf{y}(t)\rangle}{\sqrt{\langle \mathbf{x}(t),\mathbf{x}(t)\rangle}\sqrt{\langle \mathbf{y}(t),\mathbf{y}(t)\rangle}} = \frac{\int_a^b \mathbf{x}^H(t)\mathbf{y}(t)\mathrm{d}t}{\|\mathbf{x}(t)\|\cdot\|\mathbf{y}(t)\|}, \tag{1.4.20}$$

where $\|\mathbf{x}(t)\|$ is the *norm of the function vector* $\mathbf{x}(t)$ and is defined as follows:

$$\|\mathbf{x}(t)\| \stackrel{\text{def}}{=} \left(\int_a^b \mathbf{x}^H(t)\mathbf{x}(t)\mathrm{d}t\right)^{1/2}. \tag{1.4.21}$$

Clearly, if the inner product of two function vectors is equal to zero, i.e.,

$$\int_a^b \mathbf{x}^H(t)\mathbf{y}(t)\mathrm{d}t = 0,$$

then the angle $\theta = \pi/2$. Hence, two function vectors are said to be orthogonal in $[a,b]$, denoted $\mathbf{x}(t) \perp \mathbf{y}(t)$, if their inner product is equal to zero.

3. Inner Products and Norms of Random Vectors

Definition 1.23 Let $\mathbf{x}(\xi)$ and $\mathbf{y}(\xi)$ be two $n \times 1$ random vectors of variable ξ. Then the *inner product of random vectors* is defined as follows:

$$\langle \mathbf{x}(\xi), \mathbf{y}(\xi) \rangle \stackrel{\text{def}}{=} E\{\mathbf{x}^H(\xi)\mathbf{y}(\xi)\}, \tag{1.4.22}$$

where E is the expectation operator $E\{\mathbf{x}(\xi)\} = [E\{x_1(\xi)\}, \ldots, E\{x_n(\xi)\}]^T$ and the function variable ξ may be time t, circular frequency f, angular frequency ω or space parameter s, and so on.

The square of the *norm of a random vector* $\mathbf{x}(\xi)$ is defined as

$$\|\mathbf{x}(\xi)\|^2 \stackrel{\text{def}}{=} E\{\mathbf{x}^H(\xi)\mathbf{x}(\xi)\}. \tag{1.4.23}$$

In contrast with the case for a constant vector or function vector, an $m \times 1$ random vector $\mathbf{x}(\xi)$ and an $n \times 1$ random vector $\mathbf{y}(\xi)$ are said to be orthogonal if any entry of $\mathbf{x}(\xi)$ is orthogonal to each entry of $\mathbf{y}(\xi)$. The following is the definition of *random vector orthogonality*.

Definition 1.24 An $m \times 1$ random vector $\mathbf{x}(\xi)$ and an $n \times 1$ random vector $\mathbf{y}(\xi)$ are orthogonal, denoted by $\mathbf{x}(\xi) \perp \mathbf{y}(\xi)$, if their cross-correlation matrix is equal to the $m \times n$ null matrix $\mathbf{O}$, i.e.,

$$E\{\mathbf{x}(\xi)\mathbf{y}^H(\xi)\} = \mathbf{O}. \tag{1.4.24}$$

The following proposition shows that, for any two orthogonal vectors, the square of the norm of their sum is equal to the sum of the squares of the respective vector norms.

Proposition 1.2 *If* $\mathbf{x} \perp \mathbf{y}$, *then* $\|\mathbf{x}+\mathbf{y}\|^2 = \|\mathbf{x}\|^2 + \|\mathbf{y}\|^2$.

Proof From the axioms of vector norms it is known that

$$\|\mathbf{x}+\mathbf{y}\|^2 = \langle \mathbf{x}+\mathbf{y}, \mathbf{x}+\mathbf{y} \rangle = \langle \mathbf{x}, \mathbf{x} \rangle + \langle \mathbf{x}, \mathbf{y} \rangle + \langle \mathbf{y}, \mathbf{x} \rangle + \langle \mathbf{y}, \mathbf{y} \rangle. \tag{1.4.25}$$

Since $\mathbf{x}$ and $\mathbf{y}$ are orthogonal, we have $\langle \mathbf{x}, \mathbf{y} \rangle = E\{\mathbf{x}^T\mathbf{y}\} = 0$. Moreover, from the axioms of inner products it is known that $\langle \mathbf{y}, \mathbf{x} \rangle = \langle \mathbf{x}, \mathbf{y} \rangle = 0$. Substituting this result into Equation (1.4.25), we immediately get

$$\|\mathbf{x}+\mathbf{y}\|^2 = \langle \mathbf{x}, \mathbf{x} \rangle + \langle \mathbf{y}, \mathbf{y} \rangle = \|\mathbf{x}\|^2 + \|\mathbf{y}\|^2.$$

This completes the proof of the proposition. □

This proposition is also referred to as the *Pythagorean theorem.*

On the *orthogonality* of two vectors, we have the following conclusions.

- *Mathematical definitions* Two vectors $\mathbf{x}$ and $\mathbf{y}$ are orthogonal if their inner product is equal to zero, i.e., $\langle \mathbf{x}, \mathbf{y} \rangle = 0$ (for constant vectors and function vectors), or the mathematical expectation of their outer product is equal to a null matrix $\mathbf{O}$, i.e., $E\{\mathbf{x}\mathbf{y}^H\} = \mathbf{O}$ (for random vectors).
- *Geometric interpretation* If two vectors are orthogonal then their angle is $\pi/2$, and the projection of one vector onto the other vector is equal to zero.
- *Physical significance* When two vectors are orthogonal, each vector contains no components of the other, that is, there exist no interactions or interference between these vectors.

The main points above are useful for adopting flexibly the orthogonality of two vectors in engineering applications.

1.4.3 Similarity Comparison Between Vectors

Clustering and pattern classification are two important techniques in statistical data analysis.

By *clustering*, we mean that a given big data set is divided into several data subsets (categories) in which the data in each subset have common or similar features.

Pattern classification involves classifying some unknown stimuli (activations) into a fixed number of categories (classes).

The main mathematical tool for clustering and classification is the distance measure.

The distance between two vectors $\mathbf{p}$ and $\mathbf{g}$, denoted $D(\mathbf{p}\|\mathbf{g})$, is a *measure* if it has following properties.

(1) *Nonnegativity and positiveness* $D(\mathbf{p}\|\mathbf{g}) \geq 0$, and equality holds if and only if $\mathbf{p} = \mathbf{g}$.
(2) *Symmetry* $D(\mathbf{p}\|\mathbf{g}) = D(\mathbf{g}\|\mathbf{p})$.
(3) *Triangle inequality* $D(\mathbf{p}\|\mathbf{z}) \leq D(\mathbf{p}\|\mathbf{g}) + D(\mathbf{g}\|\mathbf{z})$.

It is easily shown that the Euclidean distance is a measure.

In many applications, data vectors must become low-dimensional vectors via some transformation or processing method. These low-dimensional vectors are called the *pattern vectors* or feature vectors owing to the fact that they extract the features of the original data vectors, and are directly used for pattern clustering and classification. For example, the colors of clouds and the parameters of voice tones are pattern or feature vectors in weather forecasting and voice classification, respectively.

The basic rule of clustering and classification is to adopt some distance metric to measure the similarity of two feature vectors. As the name suggests, this *similarity* is a measure of the degree of similarity between vectors.

Consider a pattern classification problem. For simplicity, suppose that there are M classes of pattern vectors $\mathbf{s}_1, \ldots, \mathbf{s}_M$. Our problem is: given a feature vector $\mathbf{x}$ with an unknown pattern, we hope to recognize to which class it belongs. For this purpose, we need to compare the unknown pattern vector $\mathbf{x}$ with M known pattern vectors in order to recognize to which known pattern vector the given vector $\mathbf{x}$ is most similar. On the basis of a similarity comparison, we can obtain the pattern or signal classification.

A quantity known as the *dissimilarity* is used to make a reverse measurement of the similarity between vectors: two vectors with a small dissimilarity are similar.

Let $D(\mathbf{x}, \mathbf{s}_1), \ldots, D(\mathbf{x}, \mathbf{s}_M)$ be the dissimilarities between the unknown pattern vector $\mathbf{x}$ and the respective known pattern vectors $\mathbf{s}_1, \ldots, \mathbf{s}_M$. As an example, we will compare the dissimilarities between $\mathbf{x}$ and $\mathbf{s}_1, \mathbf{s}_2$. If

$$D(\mathbf{x}, \mathbf{s}_1) < D(\mathbf{x}, \mathbf{s}_2), \tag{1.4.26}$$

then we say the unknown pattern vector $\mathbf{x}$ is more similar to $\mathbf{s}_1$ than $\mathbf{s}_2$, where $\mathbf{s}_1$ and $\mathbf{s}_2$ are known pattern vectors.

The simplest and most intuitive dissimilarity parameter is the Euclidean distance between vectors. The Euclidean distance between the unknown pattern vector $\mathbf{x}$ and the ith known pattern vector $\mathbf{s}_i$, denoted $D_E(\mathbf{x}, \mathbf{s}_i)$, is given by

$$D_E(\mathbf{x}, \mathbf{s}_i) = \|\mathbf{x} - \mathbf{s}_i\|_2 = \sqrt{(\mathbf{x} - \mathbf{s}_i)^T(\mathbf{x} - \mathbf{s}_i)}. \tag{1.4.27}$$

The two vectors are exactly the same if their Euclidean distance is equal to zero:

$$D_E(\mathbf{x}, \mathbf{y}) = 0 \quad \Leftrightarrow \quad \mathbf{x} = \mathbf{y}.$$

If

$$D_E(\mathbf{x}, \mathbf{s}_i) = \min_k D_E(\mathbf{x}, \mathbf{s}_k), \quad k = 1, \ldots, M, \tag{1.4.28}$$

then $\mathbf{s}_i \in \{\mathbf{s}_1, \ldots, \mathbf{s}_M\}$ is said to be a *nearest neighbour* to $\mathbf{x}$.

A widely used classification method is *nearest neighbor classification*, which judges $\mathbf{x}$ to belong in the model type corresponding to its nearest neighbour.

Another frequently used distance function is the *Mahalanobis distance*, proposed by Mahalanobis in 1936 [312]. The Mahalanobis distance from the vector $\mathbf{x}$ to its mean $\boldsymbol{\mu}$ is given by

$$D_M(\mathbf{x}, \boldsymbol{\mu}) = \sqrt{(\mathbf{x} - \boldsymbol{\mu})^T \mathbf{C}_x^{-1} (\mathbf{x} - \boldsymbol{\mu})}, \tag{1.4.29}$$

where $\mathbf{C}_x = \mathrm{Cov}(\mathbf{x}, \mathbf{x}) = E\{(\mathbf{x} - \boldsymbol{\mu})(\mathbf{x} - \boldsymbol{\mu})^T\}$ is the autocovariance matrix of the vector $\mathbf{x}$.

The Mahalanobis distance between vectors $\mathbf{x} \in \mathbb{R}^n$ and $\mathbf{y} \in \mathbb{R}^n$ is denoted by $D_M(\mathbf{x}, \mathbf{y})$, and is defined as [312]

$$D_M(\mathbf{x}, \mathbf{y}) = \sqrt{(\mathbf{x} - \mathbf{y})^T \mathbf{C}_{xy}^{-1} (\mathbf{x} - \mathbf{y})}, \tag{1.4.30}$$

where $\mathbf{C}_{xy} = \text{Cov}(\mathbf{x}, \mathbf{y}) = E\{(\mathbf{x} - \boldsymbol{\mu}_x)(\mathbf{y} - \boldsymbol{\mu}_y)^T\}$ is the cross-covariance matrix of $\mathbf{x}$ and $\mathbf{y}$, while $\boldsymbol{\mu}_x$ and $\boldsymbol{\mu}_y$ are the means of $\mathbf{x}$ and $\mathbf{y}$, respectively.

Clearly, if the covariance matrix is the identity matrix, i.e., $\mathbf{C} = \mathbf{I}$, then the Mahalanobis distance reduces to the Euclidean distance. If the covariance matrix takes a diagonal form then the corresponding Mahalanobis distance is called the *normalized Euclidean distance* and is given by

$$D_M(\mathbf{x}, \mathbf{y}) = \sqrt{\sum_{i=1}^{n} \frac{(x_i - y_i)^2}{\sigma_i^2}}, \tag{1.4.31}$$

in which σ_i is the standard deviation of x_i and y_i in the whole sample set.

Let

$$\boldsymbol{\mu} = \frac{1}{M} \sum_{i=1}^{M} \mathbf{s}_i, \quad \mathbf{C} = \sum_{i=1}^{M} \sum_{j=1}^{M} (\mathbf{s}_i - \boldsymbol{\mu})(\mathbf{s}_j - \boldsymbol{\mu})^T \tag{1.4.32}$$

be the sample mean vector of M known pattern vectors $\mathbf{s}_i$ and the sample cross-covariance matrix. Then the Mahalanobis distance from the unknown pattern vector $\mathbf{x}$ to the ith known pattern vector $\mathbf{s}_i$ is defined as

$$D_M(\mathbf{x}, \mathbf{s}_i) = \sqrt{(\mathbf{x} - \mathbf{s}_i)^T \mathbf{C}^{-1} (\mathbf{x} - \mathbf{s}_i)}. \tag{1.4.33}$$

By the nearest neighbor classification method, if

$$D_M(\mathbf{x}, \mathbf{s}_i) = \min_k D_M(\mathbf{x}, \mathbf{s}_k), \quad k = 1, \ldots, M, \tag{1.4.34}$$

then the unknown pattern vector $\mathbf{x}$ is recognized as being in the pattern type to which $\mathbf{s}_i$ belongs.

The measure of dissimilarity between vectors is not necessarily limited to distance functions. The cosine function of the acute angle between two vectors,

$$D(\mathbf{x}, \mathbf{s}_i) = \cos \theta_i = \frac{\mathbf{x}^T \mathbf{s}_i}{\|\mathbf{x}\|_2 \|\mathbf{s}_i\|_2}, \tag{1.4.35}$$

is an effective measure of dissimilarities as well.

If $\cos \theta_i < \cos \theta_j, \forall j \neq i$, holds then the unknown pattern vector $\mathbf{x}$ is said to be most similar to the known pattern vector $\mathbf{s}_i$. The variant of Equation (1.4.35)

$$D(\mathbf{x}, \mathbf{s}_i) = \frac{\mathbf{x}^T \mathbf{s}_i}{\mathbf{x}^T \mathbf{x} + \mathbf{s}_i^T \mathbf{s}_i + \mathbf{x}^T \mathbf{s}_i} \tag{1.4.36}$$

is referred to as the *Tanimoto measure* [469] and is widely used in information retrieval, the classification of diseases, animal and plant classifications etc.

The signal to be classified as one of a set of objects is called the *target signal.* Signal classification is usually based on some physical or geometric attributes or concepts relating to the objects. Let X be a target signal and let A_i represent a classification attribute defining the ith object. Then we can adopt the *object–concept distance* $D(X, A_i)$ to describe the dissimilarity between the target signal and the ith object [448], and thus we have a relationship similar to Equation (1.4.26) as follows: if

$$D(X, A_i) < D(X, A_j), \quad \forall i, j, \tag{1.4.37}$$

then the target signal X is classified as the ith type of object, C_i, with the minimum object–concept distance $D(X, A_i)$.

1.4.4 Banach Space, Euclidean Space, Hilbert Space

A vector space equipped with a vector norm is called a *normed vector space*. Let X be a vector space. If, for every Cauchy sequence $\{\mathbf{x}_n\}$ in X, there exists an element $\mathbf{x}$ in X such that

$$\lim_{n\to\infty} \mathbf{x}_n = \mathbf{x}$$

then X is known as a *complete vector space*. The above convergence of a Cauchy sequence of vectors can be equivalently written as a convergence in the norm,

$$\lim_{n\to\infty} \|\mathbf{x}_n - \mathbf{x}\| = 0.$$

A *Banach space* is a complete normed vector space. Because the vector norm is a metric that allows the computation of vector length and the distance between vectors, a Banach space is a vector space with a metric that can compute vector length and the distance between vectors and is complete in the sense that a Cauchy sequence of vectors always converges in the norm.

A *Euclidean space* is an affine space with the standard Euclidean structure and encompasses the two-dimensional Euclidean plane, the three-dimensional space of Euclidean geometry, and other higher-dimensional spaces. Euclidean spaces have finite dimension [449]. An n-dimensional Euclidean space is usually denoted by $\mathbb{R}^n$, which is assumed to have the standard Euclidean structure.

The *standard Euclidean structure* includes the following:

(1) the standard inner product (also known as the dot product) on $\mathbb{R}^m$,

$$\langle \mathbf{x}, \mathbf{y} \rangle = \mathbf{x} \cdot \mathbf{y} = \mathbf{x}^T \mathbf{y} = x_1 y_1 + \cdots + x_m y_m;$$

(2) the *Euclidean length* of a vector $\mathbf{x}$ on $\mathbb{R}^m$,

$$\|\mathbf{x}\| = \sqrt{\langle \mathbf{x}, \mathbf{y} \rangle} = \sqrt{x_1^2 + \cdots + x_m^2};$$

(3) the *Euclidean distance* between $\mathbf{x}$ and $\mathbf{y}$ on $\mathbb{R}^m$,

$$d(\mathbf{x}, \mathbf{y}) = \|\mathbf{x} - \mathbf{y}\| = \sqrt{(x_1 - y_1)^2 + \cdots + (x_m - y_m)^2};$$

(4) the (nonreflex) angle $\theta\,(0^\circ \leq \theta \leq 180^\circ)$ between vectors $\mathbf{x}$ and $\mathbf{y}$ on $\mathbb{R}^m$,

$$\theta = \arccos\left(\frac{\langle \mathbf{x}, \mathbf{y}\rangle}{\|\mathbf{x}\|\|\mathbf{y}\|}\right) = \arccos\left(\frac{\mathbf{x}^T\mathbf{y}}{\|\mathbf{x}\|\|\mathbf{y}\|}\right).$$

A vector space equipped with an inner product is referred to as an *inner product space.*

Hilbert space, named after David Hilbert, is a generalization of Euclidean space: it is no longer limited to the finite-dimensional case. Similarly to Euclidean space, a Hilbert space is an inner product space that allows the length of a vector and the distance and angle between vectors to be measured, and thus it possesses the definition of orthogonality. Furthermore, a Hilbert space is a complete space, and all Cauchy sequences are equivalent to convergent sequences, so that most concepts in calculus can be extended to Hilbert spaces. Hilbert spaces provide an efficient representation for Fourier series and Fourier transforms based on any orthogonal series.

1.4.5 Inner Products and Norms of Matrices

The inner products and norms of vectors can be extended to the inner products and norms of matrices.

Consider $m \times n$ complex matrices $\mathbf{A} = [\mathbf{a}_1, \ldots, \mathbf{a}_n]$ and $\mathbf{B} = [\mathbf{b}_1, \ldots, \mathbf{b}_n]$. We stack $\mathbf{A}$ and $\mathbf{B}$ respectively into the following $mn \times 1$ vectors according to their columns:

$$\mathbf{a} = \mathrm{vec}(\mathbf{A}) = \begin{bmatrix} \mathbf{a}_1 \\ \vdots \\ \mathbf{a}_n \end{bmatrix}, \qquad \mathbf{b} = \mathrm{vec}(\mathbf{B}) = \begin{bmatrix} \mathbf{b}_1 \\ \vdots \\ \mathbf{b}_n \end{bmatrix},$$

where the elongated vector $\mathrm{vec}(\mathbf{A})$ is the vectorization of the matrix $\mathbf{A}$. We will discuss the vectorization of matrices in detail in Section 1.11.

Definition 1.25 The inner product of two $m \times n$ matrices $\mathbf{A}$ and $\mathbf{B}$, denoted $\langle \mathbf{A}, \mathbf{B}\rangle \,|\, \mathbb{C}^{m\times n} \times \mathbb{C}^{m\times n} \to \mathbb{C}$, is defined as the inner product of two elongated vectors:

$$\langle \mathbf{A}, \mathbf{B}\rangle = \langle \mathrm{vec}(\mathbf{A}), \mathrm{vec}(\mathbf{B})\rangle = \sum_{i=1}^{n} \mathbf{a}_i^H \mathbf{b}_i = \sum_{i=1}^{n} \langle \mathbf{a}_i, \mathbf{b}_i\rangle, \tag{1.4.38}$$

equivalently written as

$$\langle \mathbf{A}, \mathbf{B}\rangle = (\mathrm{vec}\,\mathbf{A})^H \mathrm{vec}(\mathbf{B}) = \mathrm{tr}(\mathbf{A}^H\mathbf{B}), \tag{1.4.39}$$

where $\mathrm{tr}(\mathbf{C})$ represents the trace function of a square matrix $\mathbf{C}$, defined as the sum of its diagonal entries.

Let $\mathbb{K}$ represent a real or complex field; thus $\mathbb{K}^{m\times n}$ denotes either $\mathbb{R}^{m\times n}$ or $\mathbb{C}^{m\times n}$.

DEFINITION 1.26 The norm of the matrix $\mathbf{A} \in \mathbb{K}^{m \times n}$, denoted $\|\mathbf{A}\|$, is defined as the real-valued function having the following properties:

(a) For any nonzero matrix $\mathbf{A} \neq \mathbf{O}$, its norm is larger than zero, i.e., $\|\mathbf{A}\| > 0$, if $\mathbf{A} \neq \mathbf{O}$, and $\|\mathbf{A}\| = 0$ if and only if $\mathbf{A} = \mathbf{O}$.
(b) $\|c\mathbf{A}\| = |c|\|\mathbf{A}\|$ for any $c \in \mathbb{K}$.
(c) $\|\mathbf{A} + \mathbf{B}\| \leq \|\mathbf{A}\| + \|\mathbf{B}\|$.
(d) The norm of the product of two matrices is less than or equal to the product of their norms, that is, $\|\mathbf{AB}\| \leq \|\mathbf{A}\|\,\|\mathbf{B}\|$.

EXAMPLE 1.7 For a real-valued function $f(\mathbf{A}) = \sum_{i=1}^n \sum_{j=1}^n |a_{ij}|$, it is easy to verify the following:

(a) $f(\mathbf{A}) \geq 0$; when $\mathbf{A} = 0$, i.e., when $a_{ij} \equiv 0$, $f(\mathbf{A}) = 0$.

(b) $f(c\mathbf{A}) = \displaystyle\sum_{i=1}^n \sum_{j=1}^n |ca_{ij}| = |c| \sum_{i=1}^n \sum_{j=1}^n |a_{ij}| = |c| f(\mathbf{A})$.

(c) $f(\mathbf{A} + \mathbf{B}) = \displaystyle\sum_{i=1}^n \sum_{j=1}^n (|a_{ij} + b_{ij}|) \leq \sum_{i=1}^n \sum_{j=1}^n (|a_{ij}| + |b_{ij}|) = f(\mathbf{A}) + f(\mathbf{B})$.

(d) For the product of two matrices, we have

$$\begin{aligned} f(\mathbf{AB}) &= \sum_{i=1}^n \sum_{j=1}^n \left| \sum_{k=1}^n a_{ik} b_{kj} \right| \leq \sum_{i=1}^n \sum_{j=1}^n \sum_{k=1}^n |a_{ik}||b_{kj}| \\ &\leq \sum_{i=1}^n \sum_{j=1}^n \left(\sum_{k=1}^n |a_{ik}| \sum_{l=1}^n |b_{kl}| \right) = f(\mathbf{A}) f(\mathbf{B}). \end{aligned}$$

Hence, by Definition 1.26, the real function $f(\mathbf{A}) = \sum_{i=1}^n \sum_{j=1}^n |a_{ij}|$ is a matrix norm.

There are three types of common matrix norms: induced norms, entrywise norms and Schatten norms.

1. Induced Norms

The *induced norm* of an $m \times n$ matrix $\mathbf{A} \in \mathbb{K}^{m \times n}$ is defined by means of the norm $\|\mathbf{x}\|$ of the vector $\mathbf{x} \in \mathbb{K}^n$ and the norm $\|\mathbf{Ax}\|$ of the vector $\mathbf{Ax} \in \mathbb{K}^m$ as

$$\begin{aligned} \|\mathbf{A}\| &= \sup \left\{ \frac{\|\mathbf{Ax}\|}{\|\mathbf{x}\|} \middle| \mathbf{x} \in \mathbb{K}^n \text{ with } \mathbf{x} \neq \mathbf{0} \right\} && (1.4.40) \\ &= \sup \left\{ \|\mathbf{Ax}\| \middle| \mathbf{x} \in \mathbb{K}^n \text{ with } \|\mathbf{x}\| = 1 \right\}. && (1.4.41) \end{aligned}$$

The induced norm is also called the operator norm. A common induced norm is the p-norm:

$$\|\mathbf{A}\|_p \stackrel{\text{def}}{=} \sup_{\mathbf{x} \neq \mathbf{0}} \frac{\|\mathbf{Ax}\|_p}{\|\mathbf{x}\|_p}, \tag{1.4.42}$$

where $p = 1, 2, \ldots$ The p-norm is also referred to as the *Minkowski p-norm* or *ℓ_p-norm*.

In particular, when $p = 1$ or ∞ the corresponding induced norms are respectively

$$\|\mathbf{A}\|_1 = \max_{1\leq j\leq n} \sum_{i=1}^{m} |a_{ij}|, \tag{1.4.43}$$

$$\|\mathbf{A}\|_\infty = \max_{1\leq i\leq m} \sum_{j=1}^{n} |a_{ij}|. \tag{1.4.44}$$

That is, $\|\mathbf{A}\|_1$ and $\|\mathbf{A}\|_\infty$ are simply the maximum absolute column sum and maximum absolute row sum of the matrix, respectively.

The induced norms $\|\mathbf{A}\|_1$ and $\|\mathbf{A}\|_\infty$ are also called the *absolute column sum norm* and the *absolute row sum norm*, respectively.

Example 1.8 For the matrix

$$\mathbf{A} = \begin{bmatrix} 1 & -2 & 3 \\ -4 & 5 & -6 \\ 7 & -8 & -9 \\ -10 & 11 & 12 \end{bmatrix},$$

its absolute column sum norm and row sum norm are respectively given by

$$\begin{aligned} \|\mathbf{A}\|_1 &= \max\big\{1 + |-4| + 7 + |-10|, |-2| + 5 + |-8| + 11, \\ &\quad 3 + |-6| + |-9| + 12\big\} = \max\{22, 26, 30\} = 30, \\ \|\mathbf{A}\|_\infty &= \max\big\{1 + |-2| + 3, |-4| + 5 + |-6|, 7 + |-8| + |-9|, \\ &\quad |-10| + 11 + 12\big\} = \max\{6, 15, 24, 33\} = 33. \end{aligned}$$

Another common induced matrix norm is the *spectral norm*, with $p = 2$, denoted $\|\mathbf{A}\|_2 = \|\mathbf{A}\|_{\text{spec}}$. This is defined as

$$\|\mathbf{A}\|_2 = \|\mathbf{A}\|_{\text{spec}} = \sqrt{\lambda_{\max}(\mathbf{A}^H\mathbf{A})} = \sigma_{\max}(\mathbf{A}), \tag{1.4.45}$$

i.e., the spectral norm is the largest singular value of $\mathbf{A}$ or the square root of the largest eigenvalue of the positive semi-definite matrix $\mathbf{A}^H\mathbf{A}$.

2. *Entrywise Norm*

Another type of matrix norm, the *entrywise norms*, treats an $m \times n$ matrix as a column vector of size mn and uses one of the familiar vector norms.

Let $\mathbf{a} = [a_{11}, \ldots, a_{m1}, a_{12}, \ldots, a_{m2}, \ldots, a_{1n}, \ldots, a_{mn}]^T = \text{vec}(\mathbf{A})$ be an $mn \times 1$ elongated vector of the $m \times n$ matrix $\mathbf{A}$. If we use the ℓ_p-norm definition of the elongated vector $\mathbf{a}$ then we obtain the ℓ_p-norm of the matrix $\mathbf{A}$ as follows:

$$\|\mathbf{A}\|_p \stackrel{\text{def}}{=} \|\mathbf{a}\|_p = \|\text{vec}(\mathbf{A})\|_p = \left(\sum_{i=1}^{m}\sum_{j=1}^{n} |a_{ij}|^p\right)^{1/p}. \tag{1.4.46}$$

Since this kind of matrix norm is represented by the matrix entries, it is named the *entrywise norm.*

The following are three typical entrywise matrix norms:

(1) ℓ_1-*norm* $(p = 1)$

$$\|\mathbf{A}\|_1 \stackrel{\text{def}}{=} \sum_{i=1}^{m}\sum_{j=1}^{n} |a_{ij}|; \tag{1.4.47}$$

(2) *Frobenius norm* $(p = 2)$

$$\|\mathbf{A}\|_F \stackrel{\text{def}}{=} \left(\sum_{i=1}^{m}\sum_{j=1}^{n} |a_{ij}|^2\right)^{1/2}; \tag{1.4.48}$$

(3) *max norm* or ℓ_∞-norm $(p = \infty)$

$$\|\mathbf{A}\|_\infty = \max_{i=1,\ldots,m;\, j=1,\ldots,n} \{|a_{ij}|\}. \tag{1.4.49}$$

The Frobenius norm is an extension of the Euclidean norm of the vector to the elongated vector $\mathbf{a} = [a_{11}, \ldots, a_{m1}, a_{12}, \ldots, a_{1n}, \ldots, a_{mn}]^T$.

The Frobenius norm can be also written in the form of the trace function as follows:

$$\|\mathbf{A}\|_F \stackrel{\text{def}}{=} \langle \mathbf{A}, \mathbf{A}\rangle^{1/2} = \sqrt{\text{tr}(\mathbf{A}^H\mathbf{A})} = \sqrt{\sum_{i=1}^{\min\{m,n\}} \sigma_i^2(\mathbf{A})}. \tag{1.4.50}$$

Given an $m \times n$ matrix $\mathbf{A}$, its Frobenius norm weighted by a positive definite matrix $\mathbf{\Omega}$, denoted $\|\mathbf{A}\|_{\mathbf{\Omega}}$, is defined by

$$\|\mathbf{A}\|_{\mathbf{\Omega}} = \sqrt{\text{tr}(\mathbf{A}^H\mathbf{\Omega}\mathbf{A})}. \tag{1.4.51}$$

This norm is usually called the *Mahalanobis norm.*

From $\|\mathbf{A}\|_2 = \sigma_{\max}(\mathbf{A})$ and $\|\mathbf{A}\|_F = \sqrt{\sum_{i=1}^{\min\{m,n\}} \sigma_i^2(\mathbf{A})}$ one gets a relationship between the induced norm and the entrywise norm with $p = 2$:

$$\|\mathbf{A}\|_2 \le \|\mathbf{A}\|_F, \tag{1.4.52}$$

and the equality holds if and only if the matrix $\mathbf{A}$ is a rank-1 matrix or a zero matrix, since $\sqrt{\sum_{i=1}^{\min\{m,n\}} \sigma_i^2(\mathbf{A})} = \sigma_{\max}(\mathbf{A})$ in this case.

The Frobenius norm $\|\mathbf{A}\|_F$ is very useful for numerical linear algebra and matrix analysis, since it is more easily calculated than the induced norm $\|\mathbf{A}\|_2$.

3. Schatten Norms

The common Schatten norms are the Schatten p-norms defined by the vector of the singular values of a matrix. If the singular values of a matrix $\mathbf{A} \in \mathbb{C}^{m\times n}$ are denoted by σ_i then the *Schatten p-norms* are defined by

$$\|\mathbf{A}\|_p = \left(\sum_{i=1}^{\min\{m,n\}} \sigma_i^p \right)^{1/p}, \quad p = 1, 2, \infty. \tag{1.4.53}$$

The Schatten p-norms share the same notation as the induced and entrywise p-norms, but they are different. The case $p = 2$ yields the Frobenius norm introduced before. The most familiar Schatten p-norm is the Schatten norm with $p = 1$. In order to avoid confusion, this norm will be denoted by $\|\mathbf{A}\|_*$ or $\|\mathbf{A}\|_{\text{tr}}$, rather than $\mathbf{A}_1$, and is called the *nuclear norm* (also known as the *trace norm*); it is defined as follows:

$$\|\mathbf{A}\|_* = \|\mathbf{A}\|_{\text{tr}} = \text{tr}(\sqrt{\mathbf{A}^H\mathbf{A}}) = \sum_{i=1}^{\min\{m,n\}} \sigma_i, \tag{1.4.54}$$

where $\text{tr}(\mathbf{A})$ is the trace of the square matrix $\mathbf{A}$.

If $\mathbf{A}$ and $\mathbf{B}$ are $m \times n$ matrices then their matrix norms have the following properties:

$$\|\mathbf{A}+\mathbf{B}\| + \|\mathbf{A}-\mathbf{B}\| = 2(\|\mathbf{A}\|^2 + \|\mathbf{B}\|^2) \quad \text{(parallelogram law)}, \tag{1.4.55}$$

$$\|\mathbf{A}+\mathbf{B}\|\,\|\mathbf{A}-\mathbf{B}\| \le \|\mathbf{A}\|^2 + \|\mathbf{B}\|^2. \tag{1.4.56}$$

The following are the relationships between the inner products and norms [214].

(1) *Cauchy–Schwartz inequlity*

$$|\langle \mathbf{A}, \mathbf{B}\rangle|^2 \le \|\mathbf{A}\|^2\|\mathbf{B}\|^2. \tag{1.4.57}$$

The equals sign holds if and only if $\mathbf{A} = c\mathbf{B}$, where c is a complex constant.

(2) *Pythagoras' theorem*

$$\langle \mathbf{A}, \mathbf{B}\rangle = 0 \quad \Rightarrow \quad \|\mathbf{A}+\mathbf{B}\|^2 = \|\mathbf{A}\|^2 + \|\mathbf{B}\|^2. \tag{1.4.58}$$

(3) *Polarization identity*

$$\text{Re}\left(\langle \mathbf{A}, \mathbf{B}\rangle\right) = \frac{1}{4}\left(\|\mathbf{A}+\mathbf{B}\|^2 - \|\mathbf{A}-\mathbf{B}\|^2\right), \tag{1.4.59}$$

$$\text{Re}\left(\langle \mathbf{A}, \mathbf{B}\rangle\right) = \frac{1}{2}\left(\|\mathbf{A}+\mathbf{B}\|^2 - \|\mathbf{A}\|^2 - \|\mathbf{B}\|^2\right), \tag{1.4.60}$$

where $\text{Re}\left(\langle \mathbf{A}, \mathbf{B}\rangle\right)$ represents the real part of the inner product $\langle \mathbf{A}, \mathbf{B}\rangle$.

1.5 Random Vectors

In engineering applications, the measured data are usually random variables. A vector with random variables as its entries is called a random vector.

In this section, we discuss the statistics and properties of random vectors by focusing on Gaussian random vectors.

1.5.1 Statistical Interpretation of Random Vectors

In the statistical interpretation of random vectors, the first-order and second-order statistics of random vectors are the most important.

Given a random vector $\mathbf{x}(\xi) = [x_1(\xi), \ldots, x_m(\xi)]^T$, the *mean vector* of $\mathbf{x}(\xi)$, denoted $\boldsymbol{\mu}_x$, is defined as follows:

$$\boldsymbol{\mu}_x = E\{\mathbf{x}(\xi)\} \stackrel{\text{def}}{=} \begin{bmatrix} E\{x_1(\xi)\} \\ \vdots \\ E\{x_m(\xi)\} \end{bmatrix} = \begin{bmatrix} \mu_1 \\ \vdots \\ \mu_m \end{bmatrix}, \tag{1.5.1}$$

where $E\{x_i(\xi)\} = \mu_i$ represents the mean of the random variable $x_i(\xi)$.

The *autocorrelation matrix* of the random vector $\mathbf{x}(\xi)$ is defined by

$$\mathbf{R}_x \stackrel{\text{def}}{=} E\{\mathbf{x}(\xi)\mathbf{x}^H(\xi)\} = \begin{bmatrix} r_{11} & \cdots & r_{1m} \\ \vdots & \ddots & \vdots \\ r_{m1} & \cdots & r_{mm} \end{bmatrix}, \tag{1.5.2}$$

where $r_{ii} (i = 1, \ldots, m)$ denotes the autocorrelation function of the random variable $x_i(\xi)$:

$$r_{ii} \stackrel{\text{def}}{=} E\{x_i(\xi)x_i^*(\xi)\} = E\{|x_i(\xi)|^2\}, \quad i = 1, \ldots, m, \tag{1.5.3}$$

whereas r_{ij} represents the cross-correlation function of $x_i(\xi)$ and $x_j(\xi)$:

$$r_{ij} \stackrel{\text{def}}{=} E\{x_i(\xi)x_j^*(\xi)\}, \quad i, j = 1, \ldots, m, \ i \neq j. \tag{1.5.4}$$

Clearly, the autocorrelation matrix is a complex-conjugate-symmetric matrix (i.e., a Hermitian matrix).

The *autocovariance matrix* of the random vector $\mathbf{x}(\xi)$, denoted $\mathbf{C}_x$, is defined as follows:

$$\begin{aligned} \mathbf{C}_x = \text{Cov}(\mathbf{x}, \mathbf{x}) &\stackrel{\text{def}}{=} E\{(\mathbf{x}(\xi) - \boldsymbol{\mu}_x)(\mathbf{x}(\xi) - \boldsymbol{\mu}_x)^H\} \\ &= \begin{bmatrix} c_{11} & \cdots & c_{1m} \\ \vdots & \ddots & \vdots \\ c_{m1} & \cdots & c_{mm} \end{bmatrix}, \end{aligned} \tag{1.5.5}$$

where the diagonal entries,

$$c_{ii} \stackrel{\text{def}}{=} E\{|x_i(\xi) - \mu_i|^2\}, \quad i = 1, \ldots, m, \tag{1.5.6}$$

represent the variance σ_i^2 of the random variable $x_i(\xi)$, i.e., $c_{ii} = \sigma_i^2$, whereas the other entries,

$$c_{ij} \stackrel{\text{def}}{=} E\{[x_i(\xi) - \mu_i][x_j(\xi) - \mu_j]^*\} = E\{x_i(\xi)x_j^*(\xi)\} - \mu_i\mu_j^* = c_{ji}^*, \tag{1.5.7}$$

express the covariance of the random variables $x_i(\xi)$ and $x_j(\xi)$. The autocovariance matrix is also Hermitian.

The relationship between the autocorrelation and autocovariance matrices is given by

$$\mathbf{C}_x = \mathbf{R}_x - \boldsymbol{\mu}_x \boldsymbol{\mu}_x^H. \tag{1.5.8}$$

By generalizing the autocorrelation and autocovariance matrices, one obtains the cross-correlation matrix of the random vectors $\mathbf{x}(\xi)$ and $\mathbf{y}(\xi)$,

$$\mathbf{R}_{xy} \stackrel{\text{def}}{=} E\{\mathbf{x}(\xi)\mathbf{y}^H(\xi)\} = \begin{bmatrix} r_{x_1,y_1} & \cdots & r_{x_1,y_m} \\ \vdots & \ddots & \vdots \\ r_{x_m,y_1} & \cdots & r_{x_m,y_m} \end{bmatrix}, \tag{1.5.9}$$

and the cross-covariance matrix

$$\mathbf{C}_{xy} \stackrel{\text{def}}{=} E\{[\mathbf{x}(\xi) - \boldsymbol{\mu}_x][\mathbf{y}(\xi) - \boldsymbol{\mu}_y]^H\} = \begin{bmatrix} c_{x_1,y_1} & \cdots & c_{x_1,y_m} \\ \vdots & \ddots & \vdots \\ c_{x_m,y_1} & \cdots & c_{x_m,y_m} \end{bmatrix}, \tag{1.5.10}$$

where $r_{x_i,y_j} \stackrel{\text{def}}{=} E\{x_i(\xi)y_j^*(\xi)\}$ is the cross-correlation of the random vectors $x_i(\xi)$ and $y_j(\xi)$ and $c_{x_i,y_j} \stackrel{\text{def}}{=} E\{[x_i(\xi)-\mu_{x_i}][y_j(\xi)-\mu_{y_j}]^*\}$ is the cross-covariance of $x_i(\xi)$ and $y_j(\xi)$.

It is easily seen that there exists the following relationship between the cross-covariance and cross-correlation matrices:

$$\mathbf{C}_{xy} = \mathbf{R}_{xy} - \boldsymbol{\mu}_x \boldsymbol{\mu}_y^H. \tag{1.5.11}$$

In real-world applications, a data vector $\mathbf{x} = [x(0), x(1), \ldots, x(N-1)]^T$ with nonzero mean μ_x usually needs to undergo a zero-mean normalization:

$$\mathbf{x} \leftarrow \mathbf{x} = [x(0) - \mu_x, x(1) - \mu_x, \ldots, x(N-1) - \mu_x]^T, \quad \mu_x = \frac{1}{N}\sum_{n=0}^{N-1} x(n).$$

After zero-mean normalization, the correlation matrices and covariance matrices are equal, i.e., $\mathbf{R}_x = \mathbf{C}_x$ and $\mathbf{R}_{xy} = \mathbf{C}_{xy}$.

Some properties of these matrices are as follows.

1. The autocorrelation matrix is Hermitian, i.e., $\mathbf{R}_x^H = \mathbf{R}_x$.
2. The autocorrelation matrix of the linear combination vector $\mathbf{y} = \mathbf{Ax}+\mathbf{b}$ satisfies $\mathbf{R}_y = \mathbf{A}\mathbf{R}_x\mathbf{A}^H$.
3. The cross-correlation matrix is not Hermitian but satisfies $\mathbf{R}_{xy}^H = \mathbf{R}_{yx}$.
4. $\mathbf{R}_{(x_1+x_2)y} = \mathbf{R}_{x_1y} + \mathbf{R}_{x_2y}$.
5. If $\mathbf{x}$ and $\mathbf{y}$ have the same dimension, then

$$\mathbf{R}_{x+y} = \mathbf{R}_x + \mathbf{R}_{xy} + \mathbf{R}_{yx} + \mathbf{R}_y.$$

6. $\mathbf{R}_{Ax,By} = \mathbf{A}\mathbf{R}_{xy}\mathbf{B}^H$.

The cross-correlation function describes the degree of correlation of two random variables $x_i(\xi)$ and $x_j(\xi)$. Generally speaking, the larger the autocorrelation function is, the greater is the degree of correlation of the two random vectors. The degree of correlation of two random variables $x(\xi)$ and $y(\xi)$ can be measured by their *correlation coefficient*,

$$\rho_{xy} \stackrel{\text{def}}{=} \frac{E\{(x(\xi)-\bar{x})(y(\xi)-\bar{y})^*\}}{\sqrt{E\{|x(\xi)-\bar{x}|^2\}E\{|y(\xi)-\bar{y}|^2\}}} = \frac{c_{xy}}{\sigma_x\sigma_y}. \tag{1.5.12}$$

Here $c_{xy} = E\{(x(\xi)-\bar{x})(y(\xi)-\bar{y})^*\}$ is the cross-covariance of the random variables $x(\xi)$ and $y(\xi)$, while σ_x^2 and σ_y^2 are respectively the variances of $x(\xi)$ and $y(\xi)$. Applying the Cauchy–Schwartz inequality to Equation (1.5.12), we have

$$0 \leq |\rho_{xy}| \leq 1. \tag{1.5.13}$$

The correlation coefficient ρ_{xy} measures the degree of similarity of two random variables $x(\xi)$ and $y(\xi)$. The closer ρ_{xy} is to zero, the weaker the degree of similarity of the random variables $x(\xi)$ and $y(\xi)$ is. On the other hand, the closer ρ_{xy} is to 1, the more similar $x(\xi)$ and $y(\xi)$ are. In particular, the two extreme values 0 and 1 of correlation coefficients have interesting physical meanings. The case $\rho_{xy} = 0$ means that the cross-covariance $c_{xy} = 0$, which implies there are no correlated components between the random variables $x(\xi)$ and $y(\xi)$. Thus, if $\rho_{xy} = 0$, the random variables $x(\xi)$ and $y(\xi)$ are said to be uncorrelated. Since this uncorrelation is defined in a statistical sense, it is usually said to be a *statistical uncorrelation.*

It is easy to verify that if $x(\xi) = c\,y(\xi)$, where c is a complex number, then $|\rho_{xy}| = 1$. Up to a fixed amplitude scaling factor $|c|$ and a phase $\phi(c)$, the random variables $x(\xi)$ and $y(\xi)$ are the same, so that $x(\xi) = cy(\xi) = |c|e^{j\phi(c)}y(\xi)$. Such a pair of random variables is said to be completely correlated or *coherent.*

Definition 1.27 Two random vectors $\mathbf{x}(\xi) = [x_1(\xi), \ldots, x_m(\xi)]^T$ and $\mathbf{y}(\xi) = [y_1(\xi), \ldots, y_n(\xi)]^T$ are said to be statistically uncorrelated if their cross-covariance matrix $\mathbf{C}_{xy} = \mathbf{O}_{m\times n}$ or, equivalently, $\rho_{x_i,y_j} = 0, \forall\, i, j$.

The random variables $x(\xi)$ and $y(\xi)$ are orthogonal if their cross-correlation is equal to zero, namely

$$r_{xy} = E\{x(\xi)y^*(\xi)\} = 0. \tag{1.5.14}$$

Similarly, two random vectors $\mathbf{x}(\xi) = [x_1(\xi), \ldots, x_m(\xi)]^T$ and $\mathbf{y}(\xi) = [y_1(\xi), \ldots, y_n(\xi)]^T$ are said to be orthogonal if any entry $x_i(\xi)$ of $\mathbf{x}(\xi)$ is orthogonal to any entry $y_j(\xi)$ of $\mathbf{y}(\xi)$, i.e., $r_{x_i,y_j} = E\{x_i(\xi)y_j(\xi)\} = 0$, $i = 1, \ldots, m, j = 1, \ldots, n$. Clearly, this implies that the cross-correlation matrix of the two random vectors is equal to the zero matrix, i.e., $\mathbf{R}_{xy} = \mathbf{O}_{m\times n}$.

Definition 1.28 The $m \times 1$ random vector $\mathbf{x}(\xi)$ is said to be orthogonal to the $n \times 1$ random vector $\mathbf{y}(\xi)$, if their cross-correlation matrix $\mathbf{R}_{xy} = \mathbf{O}_{m\times n}$.

Note that for the zero-mean normalized $m \times 1$ random vector $\mathbf{x}(\xi)$ and $n \times 1$ random vector $\mathbf{y}(\xi)$, their statistical uncorrelation and orthogonality are equivalent, as their cross-covariance and cross-correlation matrices are equal, i.e., $\mathbf{C}_{xy} = \mathbf{R}_{xy}$.

1.5.2 Gaussian Random Vectors

Definition 1.29 If each of its entries $x_i(\xi)$, $i = 1, \ldots, m$, is a Gaussian random variable then the random vector $\mathbf{x} = [x_1(\xi), \ldots, x_m(\xi)]^T$ is called a *Gaussian normal random vector*.

As stated below, the representations of the probability density functions of real and complex Gaussian random vectors are slightly different.

Let $\mathbf{x} \sim N(\bar{\mathbf{x}}, \mathbf{\Gamma}_x)$ denote a real Gaussian or normal random vector with the mean vector $\bar{\mathbf{x}} = [\bar{x}_1, \ldots, \bar{x}_m]^T$ and covariance matrix $\mathbf{\Gamma}_x = E\{(\mathbf{x} - \bar{\mathbf{x}})(\mathbf{x} - \bar{\mathbf{x}})^T\}$. If each entry of the Gaussian random vector is independent identically distributed (iid) then its covariance matrix $\mathbf{\Gamma}_x = E\{(\mathbf{x} - \bar{\mathbf{x}})(\mathbf{x} - \bar{\mathbf{x}})^T\} = \mathbf{Diag}(\sigma_1^2, \ldots, \sigma_m^2)$, where $\sigma_i^2 = E\{(x_i - \bar{x}_i)^2\}$ is the variance of the Gaussian random variable x_i.

Under the condition that all entries are statistically independent of each other, the probability density function of a Gaussian random vector $\mathbf{x} \sim N(\bar{\mathbf{x}}, \mathbf{\Gamma}_x)$ is the *joint probability density function* of its m random variables, i.e.,

$$\begin{aligned} f(\mathbf{x}) &= f(x_1, \ldots, x_m) \\ &= f(x_1) \cdots f(x_m) \\ &= \frac{1}{\sqrt{2\pi\sigma_1^2}} \exp\left(-\frac{(x_1 - \bar{x}_1)^2}{2\sigma_1^2}\right) \cdots \frac{1}{\sqrt{2\pi\sigma_m^2}} \exp\left(-\frac{(x_m - \bar{x}_m)^2}{2\sigma_m^2}\right) \\ &= \frac{1}{(2\pi)^{m/2}\sigma_1 \cdots \sigma_m} \exp\left(-\frac{(x_1 - \bar{x}_1)^2}{2\sigma_1^2} - \cdots - \frac{(x_m - \bar{x}_m)^2}{2\sigma_m^2}\right) \end{aligned}$$

or

$$f(\mathbf{x}) = \frac{1}{(2\pi)^{m/2}|\mathbf{\Gamma}_x|^{1/2}} \exp\left(-\frac{1}{2}(\mathbf{x} - \bar{\mathbf{x}})^T \mathbf{\Gamma}_x^{-1} (\mathbf{x} - \bar{\mathbf{x}})\right). \tag{1.5.15}$$

If the entries are not statistically independent of each other, then the probability density function of the Gaussian random vector $\mathbf{x} \sim N(\bar{\mathbf{x}}, \mathbf{\Gamma}_x)$ is also given by Equation (1.5.15), but the exponential term becomes [372], [397]

$$(\mathbf{x} - \bar{\mathbf{x}})^T \mathbf{\Gamma}_x^{-1} (\mathbf{x} - \bar{\mathbf{x}}) = \sum_{i=1}^{m} \sum_{j=1}^{m} [\mathbf{\Gamma}_x^{-1}]_{i,j} (x_i - \mu_i)(x_j - \mu_j), \tag{1.5.16}$$

where $[\mathbf{\Gamma}_x^{-1}]_{ij}$ represents the (i, j)th entry of the inverse matrix $\mathbf{\Gamma}_x^{-1}$ and $\mu_i = E\{x_i\}$ is the mean of the random variable x_i.

The *characteristic function* of a real Gaussian random vector is given by

$$\Phi_{\mathbf{x}}(\omega_1, \ldots, \omega_m) = \exp\left(\mathrm{j}\,\boldsymbol{\omega}^T \boldsymbol{\mu}_x - \frac{1}{2}\boldsymbol{\omega}^T \mathbf{\Gamma}_x \boldsymbol{\omega}\right), \tag{1.5.17}$$

where $\boldsymbol{\omega} = [\omega_1, \ldots, \omega_m]^T$.

If $x_i \sim CN(\mu_i, \sigma_i^2)$, then $\mathbf{x} = [x_1, \ldots, x_m]^T$ is called a *complex Gaussian random vector*, denoted $\mathbf{x} \sim CN(\boldsymbol{\mu}_x, \boldsymbol{\Gamma}_x)$, where $\boldsymbol{\mu}_x = [\mu_1, \ldots, \mu_m]^T$ and $\boldsymbol{\Gamma}$ are respectively the mean vector and the covariance matrix of the random vector $\mathbf{x}$. If $x_i = u_i + \mathrm{j}v_i$ and the random vectors $[u_1, v_1]^T, \ldots, [u_m, v_m]^T$ are statistically independent of each other then the probability density function of a complex Gaussian random vector $\mathbf{x}$ is given by [397, p. 35-5]

$$
\begin{aligned}
f(\mathbf{x}) &= \prod_{i=1}^{m} f(x_i) = \left(\pi^m \prod_{i=1}^{m} \sigma_i^2\right)^{-1} \exp\left(-\sum_{i=1}^{m} \frac{1}{\sigma_i^2}|x_i - \mu_i|^2\right) \\
&= \frac{1}{\pi^m |\boldsymbol{\Gamma}_x|} \exp\Big(-(\mathbf{x} - \boldsymbol{\mu}_x)^H \boldsymbol{\Gamma}_x^{-1} (\mathbf{x} - \boldsymbol{\mu}_x)\Big),
\end{aligned} \tag{1.5.18}
$$

where $\boldsymbol{\Gamma}_x = \mathbf{Diag}(\sigma_1^2, \ldots, \sigma_m^2)$. The characteristic function of the complex Gaussian random vector $\mathbf{x}$ is determined by

$$
\Phi_{\mathbf{x}}(\boldsymbol{\omega}) = \exp\left(\mathrm{j}\,\mathrm{Re}(\boldsymbol{\omega}^H \boldsymbol{\mu}_x) - \frac{1}{4}\boldsymbol{\omega}^H \boldsymbol{\Gamma}_x \boldsymbol{\omega}\right). \tag{1.5.19}
$$

A Gaussian random vector $\mathbf{x}$ has the following important properties.

(1) The probability density function of $\mathbf{x}$ is completely described by its mean vector and covariance matrix.
(2) If two Gaussian random vectors $\mathbf{x}$ and $\mathbf{y}$ are statistically uncorrelated then they are also statistically independent.
(3) Given a Gaussian random vector $\mathbf{x}$ with mean vector $\boldsymbol{\mu}_x$ and covariance matrix $\boldsymbol{\Gamma}_x$, the random vector $\mathbf{y}$ obtained by the linear transformation $\mathbf{y}(\xi) = \mathbf{A}\mathbf{x}(\xi)$ is also a Gaussian random vector, and its probability density function is given by

$$
f(\mathbf{y}) = \frac{1}{(2\pi)^{m/2}|\boldsymbol{\Gamma}_y|^{1/2}} \exp\left(-\frac{1}{2}(\mathbf{y} - \boldsymbol{\mu}_y)^T \boldsymbol{\Gamma}_y^{-1} (\mathbf{y} - \boldsymbol{\mu}_y)\right) \tag{1.5.20}
$$

for real Gaussian random vectors and

$$
f(\mathbf{y}) = \frac{1}{\pi^m |\boldsymbol{\Gamma}_y|} \exp\Big(-(\mathbf{y} - \boldsymbol{\mu}_y)^H \boldsymbol{\Gamma}_y^{-1} (\mathbf{y} - \boldsymbol{\mu}_y)\Big) \tag{1.5.21}
$$

for complex Gaussian random vectors.

In array processing, wireless communications, multiple-channel signal processing etc., one normally uses multiple sensors or array elements to receive multipath signals. In most cases it can be assumed that the additive noise in each sensor is a white Gaussian noise and that these white Gaussian noises are statistically uncorrelated.

Example 1.9 Consider a real Gaussian noise vector $\mathbf{x}(t) = [x_1(t), \ldots, x_m(t)]^T$ whose entries are all real Gaussian noise processes that are statistically uncorrelated. If these white Gaussian noises have the same variance σ^2 then

$$c_{x_i,x_j} = r_{x_i,x_j} = \begin{cases} \sigma^2, & i = j, \\ 0, & i \neq j, \end{cases} \tag{1.5.22}$$

and thus the autocovariance matrix of $\mathbf{x}(t)$ is

$$\mathbf{C}_x = \mathbf{R}_x = E\{\mathbf{x}(t)\mathbf{x}^T(t)\} = \begin{bmatrix} r_{x_1,x_1} & \cdots & r_{x_1,x_m} \\ \vdots & \ddots & \vdots \\ r_{x_m,x_1} & \cdots & r_{x_m,x_m} \end{bmatrix} = \sigma^2\mathbf{I}.$$

Hence the statistical expression of a real white Gaussian noise vector is given by

$$E\{\mathbf{x}(t)\} = \mathbf{0} \quad \text{and} \quad E\{\mathbf{x}(t)\mathbf{x}^T(t)\} = \sigma^2\mathbf{I}. \tag{1.5.23}$$

Example 1.10 Consider the complex Gaussian random vector $\mathbf{x}(t) = [x_1(t), \ldots, x_m(t)]^T$ whose components are complex white Gaussian noises and are statistically uncorrelated. If $x_i(t), i = 1, \ldots, m$, have zero mean and the same variance σ^2 then the real part $x_{\mathrm{R},i}(t)$ and the imaginary part $x_{\mathrm{I},i}(t)$ are two real white Gaussian noises that are statistically independent and have the same variance. This implies that

$$\begin{aligned} E\{x_{\mathrm{R},i}(t)\} = 0, \quad E\{x_{\mathrm{I},i}(t)\} = 0, \\ E\{x_{\mathrm{R},i}^2(t)\} = E\{x_{\mathrm{I},i}^2(t)\} = \frac{1}{2}\sigma^2, \\ E\{x_{\mathrm{R},i}(t)x_{\mathrm{I},i}(t)\} = 0, \\ E\{x_i(t)x_i^*(t)\} = E\{x_{\mathrm{R},i}^2(t)\} + E\{x_{\mathrm{I},i}^2(t)\} = \sigma^2. \end{aligned}$$

From the above conditions we know that

$$\begin{aligned} E\{x_i^2(t)\} &= E\{(x_{\mathrm{R},i}(t) + \mathrm{j}\,x_{\mathrm{I},i}(t))^2\} \\ &= E\{x_{\mathrm{R},i}^2(t)\} - E\{x_{\mathrm{I},i}^2(t)\} + \mathrm{j}\,2E\{x_{\mathrm{R},i}(t)x_{\mathrm{I},i}(t)\} \\ &= \frac{1}{2}\sigma^2 - \frac{1}{2}\sigma^2 + 0 = 0. \end{aligned}$$

Since $x_1(t), \ldots, x_m(t)$ are statistically uncorrelated, we have

$$E\{x_i(t)x_k(t)\} = 0, \quad E\{x_i(t)x_k^*(t)\} = 0, \quad i \neq k.$$

Summarizing the above conditions, we conclude that the statistical expression of the complex white Gaussian noise vector $\mathbf{x}(t)$ is given by

$$E\{\mathbf{x}(t)\} = \mathbf{0}, \tag{1.5.24}$$

$$E\{\mathbf{x}(t)\mathbf{x}^H(t)\} = \sigma^2\mathbf{I}, \tag{1.5.25}$$

$$E\{\mathbf{x}(t)\mathbf{x}^T(t)\} = \mathbf{O}. \tag{1.5.26}$$

Note that there is a difference between the statistical representations of the real and complex white Gaussian noise vectors. We will adopt the symbols $\mathbf{x}(t) \sim N(\mathbf{0}, \sigma^2\mathbf{I})$ and $\mathbf{x}(t) \sim CN(\mathbf{0}, \sigma^2\mathbf{I})$ to represent respectively the real and complex zero-mean white Gaussian noise vectors $\mathbf{x}(t)$.

1.6 Performance Indexes of Matrices

An $m \times n$ matrix is a multivariate representation having mn components. In mathematics, one often needs a multivariate representation to be described by a scalar. The *performance indexes* of a matrix are such mathematical tools.

In previous sections we have discussed two such performance indexes of matrices: the inner product and the norm. In this section, we present other several important performance indexes: the quadratic form, determinant, eigenvalues, trace and rank of a matrix.

1.6.1 Quadratic Forms

The *quadratic form* of an $n \times n$ matrix $\mathbf{A}$ is defined as $\mathbf{x}^H\mathbf{A}\mathbf{x}$, where $\mathbf{x}$ may be any $n \times 1$ nonzero vector. For example,

$$\begin{aligned}\mathbf{x}^T\mathbf{A}\mathbf{x} &= [x_1, x_2, x_3] \begin{bmatrix} 1 & 4 & 2 \\ -1 & 7 & 5 \\ -1 & 6 & 3 \end{bmatrix} \begin{bmatrix} x_1 \\ x_2 \\ x_3 \end{bmatrix} \\ &= x_1^2 + 7x_2^2 + 3x_3^2 + 3x_1x_2 + x_1x_3 + 11x_2x_3.\end{aligned}$$

If $\mathbf{x} = [x_1, \ldots, x_n]^T$, and the a_{ij} is the (i, j)th entry of the $n \times n$ matrix $\mathbf{A}$, the quadratic form of $\mathbf{A}$ can be expressed as

$$\begin{aligned}\mathbf{x}^T\mathbf{A}\mathbf{x} &= \sum_{i=1}^{n}\sum_{j=1}^{n} x_i x_j a_{ij} = \sum_{i=1}^{n} a_{ii}x_i^2 + \sum_{i=1,\, i\neq j}^{n}\sum_{j=1}^{n} a_{ij}x_i x_j \\ &= \sum_{i=1}^{n} a_{ii}x_i^2 + \sum_{i=1}^{n-1}\sum_{j=i+1}^{n} (a_{ij} + a_{ji})x_i x_j.\end{aligned}$$

From this formula, it is easy to know that the matrices

$$\mathbf{A} = \begin{bmatrix} 1 & 4 & 2 \\ -1 & 7 & 5 \\ -1 & 6 & 3 \end{bmatrix}, \qquad \mathbf{B} = \mathbf{A}^T = \begin{bmatrix} 1 & -1 & -1 \\ 4 & 7 & 6 \\ 2 & 5 & 3 \end{bmatrix},$$

$$\mathbf{C} = \begin{bmatrix} 1.0 & 1.5 & 0.5 \\ 1.5 & 7.0 & 5.5 \\ 0.5 & 5.5 & 3.0 \end{bmatrix}, \quad \mathbf{D} = \begin{bmatrix} 1 & 114 & 52 \\ -111 & 7 & 2 \\ -51 & 9 & 3 \end{bmatrix}, \quad \cdots$$

have the same quadratic form, i.e.,

$$\begin{aligned}\mathbf{x}^T\mathbf{A}\mathbf{x} &= \mathbf{x}^T\mathbf{B}\mathbf{x} = \mathbf{x}^T\mathbf{C}\mathbf{x} = \mathbf{x}^T\mathbf{D}\mathbf{x} \\ &= x_1^2 + 7x_2^2 + 3x_3^2 + 3x_1x_2 + x_1x_3 + 11x_2x_3.\end{aligned}$$

That is to say, for any quadratic form function

$$f(x_1,\ldots,x_n) = \sum_{i=1}^n \alpha_{ii}x_i^2 + \sum_{i=1,\,i\neq j}^n \sum_{j=1}^n \alpha_{ij}x_ix_j,$$

there exist many different matrices $\mathbf{A}$ such that their quadratic forms $\mathbf{x}^T\mathbf{A}\mathbf{x} = f(x_1,\ldots,x_n)$ are the same. However, there is only one symmetric matrix $\mathbf{A}^T = \mathbf{A}$ satisfying $\mathbf{x}^T\mathbf{A}\mathbf{x} = f(x_1,\ldots,x_n)$, whose entries are $a_{ij} = a_{ji} = \frac{1}{2}(\alpha_{ij} + \alpha_{ji})$, where $i = 1,\ldots,n$, $j = 1,\ldots,n$. Hence, in order to ensure the uniqueness of the definition, it is necessary to assume $\mathbf{A}$ to be a real symmetric matrix or a complex Hermitian matrix when discussing the quadratic forms of an $n \times n$ matrix $\mathbf{A}$. This assumption ensures that any quadratic form function is real-valued, since

$$(\mathbf{x}^H\mathbf{A}\mathbf{x})^* = (\mathbf{x}^H\mathbf{A}\mathbf{x})^H = \mathbf{x}^H\mathbf{A}^H\mathbf{x} = \mathbf{x}^H\mathbf{A}\mathbf{x}$$

holds for any Hermitian matrix $\mathbf{A}$ and any nonzero vector $\mathbf{x}$. One of the basic advantages of a real-valued function is its suitability for comparison with a zero value.

If the quadratic form $\mathbf{x}^H\mathbf{A}\mathbf{x} > 0$ is positive definite then the Hermitian matrix $\mathbf{A}$ is also positive definite. Similarly, one can define the positive semi-definiteness, negative definiteness and negative semi-definiteness of a Hermitian matrix.

Definition 1.30 A Hermitian matrix $\mathbf{A}$ is said to be:

(1) *positive definite*, denoted $\mathbf{A} \succ 0$, if the quadratic form $\mathbf{x}^H\mathbf{A}\mathbf{x} > 0, \forall\, \mathbf{x} \neq \mathbf{0}$;
(2) *positive semi-definite*, denoted $\mathbf{A} \succeq 0$, if the quadratic form $\mathbf{x}^H\mathbf{A}\mathbf{x} \geq 0, \forall\, \mathbf{x} \neq \mathbf{0}$;
(3) *negative definite*, denoted $\mathbf{A} \prec 0$, if the quadratic form $\mathbf{x}^H\mathbf{A}\mathbf{x} < 0, \forall\, \mathbf{x} \neq \mathbf{0}$;
(4) *negative semi-definite*, denoted $\mathbf{A} \preceq 0$, if the quadratic form $\mathbf{x}^H\mathbf{A}\mathbf{x} \leq 0, \forall\, \mathbf{x} \neq \mathbf{0}$;
(5) *indefinite* if $\mathbf{x}^H\mathbf{A}\mathbf{x} > 0$ for some nonzero vectors $\mathbf{x}$ and $\mathbf{x}^H\mathbf{A}\mathbf{x} < 0$ for other nonzero vectors $\mathbf{x}$.

For example, the real symmetric matrix $\mathbf{R} = \begin{bmatrix} 3 & -1 & 0 \\ -1 & 3 & -1 \\ 0 & -1 & 3 \end{bmatrix}$ is positive definite, as the quadratic form $\mathbf{x}^H\mathbf{R}\mathbf{x} = 2x_1^2 + x_2^2 + 2x_3^2 + (x_1 - x_2)^2 + (x_2 - x_3)^2 > 0$ unless $x_1 = x_2 = x_3 = 0$.

One sentence summary: *As a performance index, the quadratic form of a Hermitian matrix describes its positive definiteness.*

1.6.2 Determinants

The reader will recall that the determinant of a 2×2 matrix $[a_{ij}]$ is given by

$$\det[a_{ij}] = \begin{vmatrix} a_{11} & a_{12} \\ a_{21} & a_{22} \end{vmatrix} \stackrel{\text{def}}{=} a_{11}a_{22} - a_{12}a_{21}.$$

The determinant of an $n \times n$ matrix $\mathbf{A}$, denoted $\det(\mathbf{A})$ or $|\mathbf{A}|$, is written as follows:

$$\det(\mathbf{A}) = |\mathbf{A}| = \begin{vmatrix} a_{11} & a_{12} & \cdots & a_{1n} \\ a_{21} & a_{22} & \cdots & a_{2n} \\ \vdots & \vdots & \ddots & \vdots \\ a_{n1} & a_{n2} & \cdots & a_{nn} \end{vmatrix}. \tag{1.6.1}$$

After removing the ith row and the jth column from the matrix $\mathbf{A}$, the determinant A_{ij} of the remaining matrix is known as the *cofactor* of the entry a_{ij}. In particular, when $j = i$, A_{ii} is known as the principal minor of $\mathbf{A}$. Letting $\mathbf{A}_{ij}$ be the $(n-1) \times (n-1)$ submatrix obtained by removing the ith row and the jth column from the $n \times n$ matrix $\mathbf{A}$, the cofactor A_{ij} is related to the determinant of the submatrix $\mathbf{A}_{ij}$ as follows:

$$A_{ij} = (-1)^{i+j} \det(\mathbf{A}_{ij}). \tag{1.6.2}$$

The determinant of an $n \times n$ matrix $\mathbf{A}$ is $\forall n$ equal to the sum of the products of each entry of its any row (say the ith row) or any column (say the jth column) by its corresponding cofactor, namely

$$\det(\mathbf{A}) = a_{i1}A_{i1} + \cdots + a_{in}A_{in} = \sum_{j=1}^{n} a_{ij}(-1)^{i+j} \det(\mathbf{A}_{ij}), \tag{1.6.3}$$

$$\det(\mathbf{A}) = a_{1j}A_{1j} + \cdots + a_{nj}A_{nj} = \sum_{i=1}^{n} a_{ij}(-1)^{i+j} \det(\mathbf{A}_{ij}). \tag{1.6.4}$$

Hence the determinant of $\mathbf{A}$ can be recursively calculated: an nth-order determinant can be computed from the $(n-1)$th-order determinants, while each $(n-1)$th-order determinant can be calculated from the $(n-2)$th-order determinants, and so forth.

For a 3×3 matrix $\mathbf{A}$, its determinant is recursively given by

$$\det(\mathbf{A}) = \det \begin{bmatrix} a_{11} & a_{12} & a_{13} \\ a_{21} & a_{22} & a_{23} \\ a_{31} & a_{32} & a_{33} \end{bmatrix} = a_{11}A_{11} + a_{12}A_{12} + a_{13}A_{13}$$

$$= a_{11}(-1)^{1+1} \begin{vmatrix} a_{22} & a_{23} \\ a_{32} & a_{33} \end{vmatrix} + a_{12}(-1)^{1+2} \begin{vmatrix} a_{21} & a_{23} \\ a_{31} & a_{33} \end{vmatrix} + a_{13}(-1)^{1+3} \begin{vmatrix} a_{21} & a_{22} \\ a_{31} & a_{33} \end{vmatrix}$$

$$= a_{11}(a_{22}a_{33} - a_{23}a_{32}) - a_{12}(a_{21}a_{33} - a_{23}a_{31}) + a_{13}(a_{21}a_{33} - a_{22}a_{31}).$$

This is the *diagonal method* for a third-order determinant.

DEFINITION 1.31 A matrix with nonzero determinant is known as a *nonsingular matrix.*

1. Determinant Equalities [307]

1. If two rows (or columns) of a matrix $\mathbf{A}$ are exchanged then the value of $\det(\mathbf{A})$ remains unchanged, but the sign is changed.
2. If some row (or column) of a matrix $\mathbf{A}$ is a linear combination of other rows (or columns), then $\det(\mathbf{A}) = 0$. In particular, if some row (or column) is proportional or equal to another row (or column), or there is a zero row (or column), then $\det(\mathbf{A}) = 0$.
3. The determinant of an identity matrix is equal to 1, i.e., $\det(\mathbf{I}) = 1$.
4. Any square matrix $\mathbf{A}$ and its transposed matrix $\mathbf{A}^T$ have the same determinant, i.e., $\det(\mathbf{A}) = \det(\mathbf{A}^T)$; however, $\det(\mathbf{A}^H) = (\det(\mathbf{A}^T))^*$.
5. The determinant of a Hermitian matrix is real-valued, since
$$\det(\mathbf{A}) = \det(\mathbf{A}^H) = (\det(\mathbf{A}))^*. \tag{1.6.5}$$
6. The determinant of the product of two square matrices is equal to the product of their determinants, i.e.,
$$\det(\mathbf{AB}) = \det(\mathbf{A})\det(\mathbf{B}), \quad \mathbf{A}, \mathbf{B} \in \mathbb{C}^{n\times n}. \tag{1.6.6}$$
7. For any constant c and any $n \times n$ matrix $\mathbf{A}$, $\det(c\mathbf{A}) = c^n \det(\mathbf{A})$.
8. If $\mathbf{A}$ is nonsingular then $\det(\mathbf{A}^{-1}) = 1/\det(\mathbf{A})$.
9. For matrices $\mathbf{A}_{m\times m}, \mathbf{B}_{m\times n}, \mathbf{C}_{n\times m}, \mathbf{D}_{n\times n}$, the determinant of the block matrix for $\mathbf{A}$ nonsingular is
$$\det \begin{bmatrix} \mathbf{A} & \mathbf{B} \\ \mathbf{C} & \mathbf{D} \end{bmatrix} = \det(\mathbf{A})\det(\mathbf{D} - \mathbf{C}\mathbf{A}^{-1}\mathbf{B}) \tag{1.6.7}$$
and for $\mathbf{D}$ nonsingular is
$$\det \begin{bmatrix} \mathbf{A} & \mathbf{B} \\ \mathbf{C} & \mathbf{D} \end{bmatrix} = \det(\mathbf{D})\det(\mathbf{A} - \mathbf{B}\mathbf{D}^{-1}\mathbf{C}). \tag{1.6.8}$$
10. The determinant of a triangular (upper or lower triangular) matrix $\mathbf{A}$ is equal to the product of its main diagonal entries:
$$\det(\mathbf{A}) = \prod_{i=1}^{n} a_{ii}.$$
The determinant of a diagonal matrix $\mathbf{A} = \mathbf{Diag}(a_{11}, \ldots, a_{nn})$ is also equal to the product of its diagonal entries.

Here we give a proof of Equation (1.6.7):

$$\det\begin{bmatrix}\mathbf{A} & \mathbf{B}\\ \mathbf{C} & \mathbf{D}\end{bmatrix} = \det\left(\begin{bmatrix}\mathbf{A} & \mathbf{O}\\ \mathbf{C} & \mathbf{D}-\mathbf{C}\mathbf{A}^{-1}\mathbf{B}\end{bmatrix}\begin{bmatrix}\mathbf{I} & \mathbf{A}^{-1}\mathbf{B}\\ \mathbf{O} & \mathbf{I}\end{bmatrix}\right)$$
$$= \det(\mathbf{A})\det(\mathbf{D}-\mathbf{C}\mathbf{A}^{-1}\mathbf{B}).$$

We can prove Equation (1.6.8) in a similar way.

2. Determinant Inequalities [307]

1. *Cauchy–Schwartz inequality:* If $\mathbf{A}, \mathbf{B}$ are $m \times n$ matrices, then
$$|\det(\mathbf{A}^H\mathbf{B})|^2 \leq \det(\mathbf{A}^H\mathbf{A})\det(\mathbf{B}^H\mathbf{B}).$$
2. *Hadamard inequality:* For an $m \times m$ matrix $\mathbf{A}$, one has
$$\det(\mathbf{A}) \leq \prod_{i=1}^{m}\left(\sum_{j=1}^{m}|a_{ij}|^2\right)^{1/2}.$$
3. *Fisher inequality:* For $\mathbf{A}_{m\times m}, \mathbf{B}_{m\times n}, \mathbf{C}_{n\times n}$, one has
$$\det\left(\begin{bmatrix}\mathbf{A} & \mathbf{B}\\ \mathbf{B}^H & \mathbf{C}\end{bmatrix}\right) \leq \det(\mathbf{A})\det(\mathbf{C}).$$
4. *Minkowski inequality:* If $\mathbf{A}_{m\times m} \neq \mathbf{O}_{m\times m}$ and $\mathbf{B}_{m\times m} \neq \mathbf{O}_{m\times m}$ are positive semi-definite then
$$\sqrt[m]{\det(\mathbf{A}+\mathbf{B})} \geq \sqrt[m]{\det(\mathbf{A})} + \sqrt[m]{\det(\mathbf{B})}.$$
5. The determinant of a positive definite matrix $\mathbf{A}$ is larger than 0, i.e., $\det(\mathbf{A}) > 0$.
6. The determinant of a positive semi-definite matrix $\mathbf{A}$ is larger than or equal to 0, i.e., $\det(\mathbf{A}) \geq 0$.
7. If the $m\times m$ matrix $\mathbf{A}$ is positive semi-definite, then $(\det(\mathbf{A}))^{1/m} \leq m^{-1}\det(\mathbf{A})$.
8. If the matrices $\mathbf{A}_{m\times m}$ and $\mathbf{B}_{m\times m}$ are positive semi-definite then $\det(\mathbf{A}+\mathbf{B}) \geq \det(\mathbf{A}) + \det(\mathbf{B})$.
9. If $\mathbf{A}_{m\times m}$ is positive definite and $\mathbf{B}_{m\times m}$ is positive semi-definite then $\det(\mathbf{A}+\mathbf{B}) \geq \det(\mathbf{A})$.
10. If $\mathbf{A}_{m\times m}$ is positive definite and $\mathbf{B}_{m\times m}$ is negative semi-definite then $\det(\mathbf{A}+\mathbf{B}) \leq \det(\mathbf{A})$.

One sentence summary: *As a performance index, the value of the determinant of a matrix determines whether it is singular.*

1.6.3 Matrix Eigenvalues

Consider the output of a linear transformation $\mathcal{L}$ whose input is an $n \times 1$ nonzero vector $\mathbf{u}$. If the output is different from the input by a scale factor λ, i.e.,

$$\mathcal{L}\mathbf{u} = \lambda\mathbf{u}, \quad \mathbf{u} \neq \mathbf{0}, \tag{1.6.9}$$

then the scalar λ and the vector $\mathbf{u}$ are known as the eigenvalue and the corresponding eigenvector of the linear transformation $\mathcal{L}$. Since $\mathcal{L}\mathbf{u} = \lambda\mathbf{u}$ implies that the input vector keeps its "direction" unchanged, the vector $\mathbf{u}$ must depict an inherent feature of the linear transformation $\mathcal{L}$. This is the reason why the vector $\mathbf{u}$ is called the eigenvector of the linear transformation. In this sense, the eigenvalue λ can be regarded as the "gain" of the linear transformation $\mathcal{L}$ when $\mathbf{u}$ is inputted.

When the linear transformation $\mathcal{L}$ takes the form of an $n \times n$ matrix $\mathbf{A}$, the expression (1.6.9) can be extended to the definition of the eigenvalue and eigenvector of the matrix $\mathbf{A}$: if the linear algebraic equation

$$\mathbf{A}\mathbf{u} = \lambda\mathbf{u} \tag{1.6.10}$$

has a nonzero $n \times 1$ solution vector $\mathbf{u}$ then the scalar λ is called an eigenvalue of the matrix $\mathbf{A}$, and $\mathbf{u}$ is its eigenvector corresponding to λ.

The matrix equation (1.6.10) can be written equivalently as

$$(\mathbf{A} - \lambda\mathbf{I})\mathbf{u} = \mathbf{0}. \tag{1.6.11}$$

Since the above equation holds for a nonzero vector, $\mathbf{u}$, the only condition for it have a nonzero solution is that the determinant of the matrix $\mathbf{A} - \lambda\mathbf{I}$ is equal to zero, i.e.,

$$\det(\mathbf{A} - \lambda\mathbf{I}) = 0. \tag{1.6.12}$$

This equation is known as the *characteristic equation* of the matrix $\mathbf{A}$.

The characteristic equation (1.6.12) reflects the following facts:

- If (1.6.12) holds for $\lambda = 0$ then $\det(\mathbf{A}) = 0$. This implies that as long as the matrix $\mathbf{A}$ has a zero eigenvalue, this matrix must be a singular matrix.
- All the eigenvalues of a zero matrix are zero, and for any singular matrix there exists at least one zero eigenvalue. Clearly, if all n diagonal entries of an $n \times n$ singular matrix $\mathbf{A}$ contains a subtraction of the same scalar $x \neq 0$ that is not an eigenvalue of $\mathbf{A}$ then the matrix $\mathbf{A} - x\mathbf{I}$ must be nonsingular, since $|\mathbf{A} - x\mathbf{I}| \neq 0$.

Let $\text{eig}(\mathbf{A})$ represent the eigenvalues of the matrix $\mathbf{A}$. The basic properties of eigenvalues are listed below:

1. $\text{eig}(\mathbf{AB}) = \text{eig}(\mathbf{BA})$.
2. An $m \times n$ matrix $\mathbf{A}$ has at most $\min\{m, n\}$ different eigenvalues.
3. If $\text{rank}(\mathbf{A}) = r$, then the matrix $\mathbf{A}$ has at most r different eigenvalues.
4. The eigenvalues of the inverse matrix satisfy $\text{eig}(\mathbf{A}^{-1}) = 1/\text{eig}(\mathbf{A})$.

5. Let $\mathbf{I}$ be the identity matrix; then

$$\text{eig}(\mathbf{I} + c\mathbf{A}) = 1 + c\,\text{eig}(\mathbf{A}), \tag{1.6.13}$$

$$\text{eig}(\mathbf{A} - c\mathbf{I}) = \text{eig}(\mathbf{A}) - c. \tag{1.6.14}$$

PROPOSITION 1.3 *All eigenvalues of a positive definite matrix are positive real values.*

Proof Suppose that $\mathbf{A}$ is a positive definite matrix; then the quadratic form $\mathbf{x}^H\mathbf{A}\mathbf{x} > 0$ holds for any nonzero vector $\mathbf{x}$. If λ is any eigenvalue of $\mathbf{A}$, i.e., $\mathbf{A}\mathbf{u} = \lambda\mathbf{u}$, then $\mathbf{u}^H\mathbf{A}\mathbf{u} = \mathbf{u}^H\lambda\mathbf{u}$ and thus $\lambda = \mathbf{u}^H\mathbf{A}\mathbf{u}/(\mathbf{u}^H\mathbf{u})$ must be a positive real number, as it is the ratio of two positive real numbers. □

Since $\mathbf{u}^H\mathbf{u} > 0$ for any nonzero vector $\mathbf{u}$, from $\lambda = \mathbf{u}^H\mathbf{A}\mathbf{u}/(\mathbf{u}^H\mathbf{u})$ it is directly known that positive definite and nonpositive definite matrices can be described by their eigenvalues as follows.

(1) *Positive definite matrix* Its eigenvalues are positive real numbers.
(2) *Positive semi-definite matrix* Its eigenvalues are nonnegative.
(3) *Negative definite matrix* Its eigenvalues are negative.
(4) *Negative semi-definite matrix* Its eigenvalues are nonpositive.
(5) *Indefinite matrix* It has both positive and negative eigenvalues.

If $\mathbf{A}$ is a positive definite or positive semi-definite matrix then

$$\det(\mathbf{A}) \leq \prod_i a_{ii}. \tag{1.6.15}$$

This inequality is called the Hadamard inequality [214, p. 477].

The characteristic equation (1.6.12) suggests two methods for improving the numerical stability and the accuracy of solutions of the matrix equation $\mathbf{A}\mathbf{x} = \mathbf{b}$, as described below.

1. Method for Improving the Numerical Stability

Consider the matrix equation $\mathbf{A}\mathbf{x} = \mathbf{b}$, where $\mathbf{A}$ is usually positive definite or nonsingular. However, owing to noise or errors, $\mathbf{A}$ may sometimes be close to singular. We can alleviate this difficulty as follows. If λ is a small positive number then $-\lambda$ cannot be an eigenvalue of $\mathbf{A}$. This implies that the characteristic equation $|\mathbf{A} - x\mathbf{I}| = |\mathbf{A} - (-\lambda)\mathbf{I}| = |\mathbf{A} + \lambda\mathbf{I}| = 0$ cannot hold for any $\lambda > 0$, and thus the matrix $\mathbf{A} + \lambda\mathbf{I}$ must be nonsingular. Therefore, if we solve $(\mathbf{A} + \lambda\mathbf{I})\mathbf{x} = \mathbf{b}$ instead of the original matrix equation $\mathbf{A}\mathbf{x} = \mathbf{b}$, and λ takes a very small positive value, then one can overcome the singularity of $\mathbf{A}$ to improve greatly the numerical stability of solving $\mathbf{A}\mathbf{x} = \mathbf{b}$. This method of solving $(\mathbf{A} + \lambda\mathbf{I})\mathbf{x} = \mathbf{b}$, with $\lambda > 0$, instead of $\mathbf{A}\mathbf{x} = \mathbf{b}$ is the well-known Tikhonov regularization method for solving nearly singular matrix equations.

2. Method for Improving the Accuracy

For a matrix equation $\mathbf{Ax} = \mathbf{b}$, with the data matrix $\mathbf{A}$ nonsingular but containing additive interference or observation noise, if we choose a very small positive scalar λ to solve $(\mathbf{A} - \lambda\mathbf{I})\mathbf{x} = \mathbf{b}$ instead of $\mathbf{Ax} = \mathbf{b}$ then the influence of the noise of the data matrix $\mathbf{A}$ on the solution vector $\mathbf{x}$ will be greatly decreased. This is the basis of the well-known total least squares (TLS) method.

We will present the Tikhonov method and the TLS method in detail in Chapter 6. As is seen in the Tikhonov and TLS methods, the diagonal entries of a matrix play a more important role in matrix analysis than its off-diagonal entries.

One sentence summary: *As a performance index, the eigenvalues of a matrix describe its singularness, positive definiteness and the special structure of its diagonal entries.*

1.6.4 Matrix Trace

Definition 1.32 The sum of the diagonal entries of an $n \times n$ matrix $\mathbf{A}$ is known as its *trace*, denoted $\mathrm{tr}(\mathbf{A})$:

$$\mathrm{tr}(\mathbf{A}) = a_{11} + \cdots + a_{nn} = \sum_{i=1}^{n} a_{ii}. \tag{1.6.16}$$

The following are some properties of the matrix trace.

1. Trace Equality [307]

1. If both $\mathbf{A}$ and $\mathbf{B}$ are $n \times n$ matrices then $\mathrm{tr}(\mathbf{A} \pm \mathbf{B}) = \mathrm{tr}(\mathbf{A}) \pm \mathrm{tr}(\mathbf{B})$.
2. If both $\mathbf{A}$ and $\mathbf{B}$ are $n \times n$ matrices and c_1 and c_2 are constants then $\mathrm{tr}(c_1\mathbf{A} \pm c_2\mathbf{B}) = c_1\mathrm{tr}(\mathbf{A}) \pm c_2\mathrm{tr}(\mathbf{B})$. In particular, $\mathrm{tr}(c\mathbf{A}) = c\,\mathrm{tr}(\mathbf{A})$.
3. $\mathrm{tr}(\mathbf{A}^T) = \mathrm{tr}(\mathbf{A})$, $\mathrm{tr}(\mathbf{A}^*) = (\mathrm{tr}(\mathbf{A}))^*$ and $\mathrm{tr}(\mathbf{A}^H) = (\mathrm{tr}(\mathbf{A}))^*$.
4. If $\mathbf{A} \in \mathbb{C}^{m\times n}, \mathbf{B} \in \mathbb{C}^{n\times m}$ then $\mathrm{tr}(\mathbf{AB}) = \mathrm{tr}(\mathbf{BA})$.
5. If $\mathbf{A}$ is an $m \times n$ matrix then $\mathrm{tr}(\mathbf{A}^H\mathbf{A}) = 0$ implies that $\mathbf{A}$ is an $m \times n$ zero matrix.
6. $\mathbf{x}^H\mathbf{Ax} = \mathrm{tr}(\mathbf{Axx}^H)$ and $\mathbf{y}^H\mathbf{x} = \mathrm{tr}(\mathbf{xy}^H)$.
7. The trace of an $n \times n$ matrix is equal to the sum of its eigenvalues, namely $\mathrm{tr}(\mathbf{A}) = \lambda_1 + \cdots + \lambda_n$.
8. The trace of a block matrix satisfies

$$\mathrm{tr}\begin{bmatrix} \mathbf{A} & \mathbf{B} \\ \mathbf{C} & \mathbf{D} \end{bmatrix} = \mathrm{tr}(\mathbf{A}) + \mathrm{tr}(\mathbf{D}),$$

where $\mathbf{A} \in \mathbb{C}^{m\times m}, \mathbf{B} \in \mathbb{C}^{m\times n}, \mathbf{C} \in \mathbb{C}^{n\times m}$ and $\mathbf{D} \in \mathbb{C}^{n\times n}$.

9. For any positive integer k, we have

$$\mathrm{tr}(\mathbf{A}^k) = \sum_{i=1}^{n} \lambda_i^k. \tag{1.6.17}$$

By the trace equality $\mathrm{tr}(\mathbf{UV}) = \mathrm{tr}(\mathbf{VU})$, it is easy to see that

$$\mathrm{tr}(\mathbf{A}^H\mathbf{A}) = \mathrm{tr}(\mathbf{A}\mathbf{A}^H) = \sum_{i=1}^{n}\sum_{j=1}^{n} a_{ij}a_{ij}^* = \sum_{i=1}^{n}\sum_{j=1}^{n} |a_{ij}|^2. \tag{1.6.18}$$

However, if we substitute $\mathbf{U} = \mathbf{A}, \mathbf{V} = \mathbf{BC}$ and $\mathbf{U} = \mathbf{AB}, \mathbf{V} = \mathbf{C}$ into the trace equality $\mathrm{tr}(\mathbf{UV}) = \mathrm{tr}(\mathbf{VU})$, we obtain

$$\mathrm{tr}(\mathbf{ABC}) = \mathrm{tr}(\mathbf{BCA}) = \mathrm{tr}(\mathbf{CAB}). \tag{1.6.19}$$

Similarly, if we let $\mathbf{U} = \mathbf{A}, \mathbf{V} = \mathbf{BCD}$ or $\mathbf{U} = \mathbf{AB}, \mathbf{V} = \mathbf{CD}$ or $\mathbf{U} = \mathbf{ABC}, \mathbf{V} = \mathbf{D}$, respectively, we obtain

$$\mathrm{tr}(\mathbf{ABCD}) = \mathrm{tr}(\mathbf{BCDA}) = \mathrm{tr}(\mathbf{CDAB}) = \mathrm{tr}(\mathbf{DABC}). \tag{1.6.20}$$

Moreover, if $\mathbf{A}$ and $\mathbf{B}$ are $m \times m$ matrices, and $\mathbf{B}$ is nonsingular then

$$\mathrm{tr}(\mathbf{BAB}^{-1}) = \mathrm{tr}(\mathbf{B}^{-1}\mathbf{AB}) = \mathrm{tr}(\mathbf{ABB}^{-1}) = \mathrm{tr}(\mathbf{A}). \tag{1.6.21}$$

2. *Trace Inequality* [307]

1. If $\mathbf{A} \in \mathbb{C}^{m\times n}$ then $\mathrm{tr}(\mathbf{A}^H\mathbf{A}) = \mathrm{tr}(\mathbf{A}\mathbf{A}^H) \geq 0$.
2. *Schur inequality:* $\mathrm{tr}(\mathbf{A}^2) \leq \mathrm{tr}(\mathbf{A}^T\mathbf{A})$.
3. If $\mathbf{A}, \mathbf{B}$ are two $m \times n$ matrices then

$$\begin{aligned}
&\mathrm{tr}\big((\mathbf{A}^T\mathbf{B})^2\big) \leq \mathrm{tr}(\mathbf{A}^T\mathbf{A})\mathrm{tr}(\mathbf{B}^T\mathbf{B}) \quad \text{(Cauchy–Schwartz inequality)},\\
&\mathrm{tr}\big((\mathbf{A}^T\mathbf{B})^2\big) \leq \mathrm{tr}(\mathbf{A}^T\mathbf{A}\mathbf{B}^T\mathbf{B}),\\
&\mathrm{tr}\big((\mathbf{A}^T\mathbf{B})^2\big) \leq \mathrm{tr}(\mathbf{A}\mathbf{A}^T\mathbf{B}\mathbf{B}^T).
\end{aligned}$$

4. $\mathrm{tr}\big((\mathbf{A}+\mathbf{B})(\mathbf{A}+\mathbf{B})^T\big) \leq 2\big(\mathrm{tr}(\mathbf{A}\mathbf{A}^T) + \mathrm{tr}(\mathbf{B}\mathbf{B}^T)\big)$.
5. If $\mathbf{A}$ and $\mathbf{B}$ are two $m \times m$ symmetric matrices then $\mathrm{tr}(\mathbf{AB}) \leq \frac{1}{2}\mathrm{tr}(\mathbf{A}^2 + \mathbf{B}^2)$.

The Frobenius norm of an $m \times n$ matrix $\mathbf{A}$ can be also defined using the traces of the $m \times m$ matrix $\mathbf{A}^H\mathbf{A}$ or that of the $n \times n$ matrix $\mathbf{A}\mathbf{A}^H$, as follows [311, p. 10]:

$$\|\mathbf{A}\|_F = \sqrt{\mathrm{tr}(\mathbf{A}^H\mathbf{A})} = \sqrt{\mathrm{tr}(\mathbf{A}\mathbf{A}^H)}. \tag{1.6.22}$$

One sentence summary: *As a performance index, the trace of a matrix reflects the sum of its eigenvalues.*

1.6.5 Matrix Rank

Theorem 1.6 [433] *Among a set of p-dimensional (row or column) vectors, there are at most p linearly independent (row or column) vectors.*

Theorem 1.7 [433] *For an $m \times n$ matrix $\mathbf{A}$, the number of linearly independent rows and the number of linearly independent columns are the same.*

From this theorem we have the following definition of the rank of a matrix.

Definition 1.33 The *rank* of an $m \times n$ matrix $\mathbf{A}$ is defined as the number of its linearly independent rows or columns.

It needs to be pointed out that the matrix rank gives only the number of linearly independent rows or columns; it gives no information on the locations of these independent rows or columns.

The matrix equation $\mathbf{A}_{m\times n}\mathbf{x}_{n\times 1} = \mathbf{b}_{m\times 1}$ is said to be *consistent*, if it has at least one exact solution. A matrix equation with no exact solution is said to be *inconsistent*.

Matrix equations can be divided into three types.

(1) *Well-determined equation* If $m = n$ and $\text{rank}(\mathbf{A}) = n$, i.e., the matrix $\mathbf{A}$ is nonsingular, then the matrix equation $\mathbf{Ax} = \mathbf{b}$ is said to be well-determined.

(2) *Under-determined equation* The matrix equation $\mathbf{Ax} = \mathbf{b}$ is said to be under-determined if the number of linearly independent equations is less than the number of independent unknowns.

(3) *Over-determined equation* The matrix equation $\mathbf{Ax} = \mathbf{b}$ is said to be over-determined if the number of linearly independent equations is larger than the number of independent unknowns.

The terms "well-determined", "under-determined" and "over-determined" have the following meanings.

- *Meaning of well-determined equation* The number of independent equations and the number of independent unknowns are the same so that the solution of this system of equations is uniquely determined. The exact solution of a well-determined matrix equation $\mathbf{Ax} = \mathbf{b}$ is given by $\mathbf{x} = \mathbf{A}^{-1}\mathbf{b}$. A well-determined equation is a consistent equation.
- *Meaning of under-determined equation* The number of independent equations is less than the number of independent unknowns, which implies that the number of equations is not enough for determining a unique solution. As a matter of fact, such a system of linear equations has an infinitely many solutions. Hence, any under-determined matrix equation is a consistent equation.
- *Meaning of over-determined equation* Since the number of independent equations is larger than the number of independent unknowns, the number of independent equations appears surplus for determining the unique solution. An

over-determined matrix equation $\mathbf{Ax} = \mathbf{b}$ has no exact solution and thus is an inconsistent equation that may in some cases have an approximate solution, for example, a least squares solution.

A matrix $\mathbf{A}$ with $\text{rank}(\mathbf{A}) = r_A$ has r_A linearly independent column vectors. The linear combinations of the r_A linearly independent column vectors constitute a vector space, called the *column space* or the *range* or the *manifold* of $\mathbf{A}$.

The column space $\text{Col}(\mathbf{A})$ or the range $\text{Range}(\mathbf{A})$ is r_A-dimensional. Hence the rank of a matrix can be defined by using the dimension of its column space or range, as described below.

Definition 1.34 The dimension of the column space $\text{Col}(\mathbf{A})$ or the range $\text{Range}(\mathbf{A})$ of an $m \times n$ matrix $\mathbf{A}$ is defined as the rank of the matrix, namely

$$r_A = \dim(\text{Col}(\mathbf{A})) = \dim(\text{Range}(\mathbf{A})). \tag{1.6.23}$$

The following statements about the rank of the matrix $\mathbf{A}$ are equivalent:

(1) $\text{rank}(\mathbf{A}) = k$;
(2) there are k and not more than k columns of $\mathbf{A}$ that combine to give a linearly independent set;
(3) there are k and not more than k rows of $\mathbf{A}$ that combine to give a linearly independent set;
(4) there is a $k \times k$ submatrix of $\mathbf{A}$ with nonzero determinant, but all the $(k+1) \times (k+1)$ submatrices of $\mathbf{A}$ have zero determinant;
(5) the dimension of the column space $\text{Col}(\mathbf{A})$ or the range $\text{Range}(\mathbf{A})$ equals k;
(6) $k = n - \dim[\text{Null}(\mathbf{A})]$, where $\text{Null}(\mathbf{A})$ denotes the null space of the matrix $\mathbf{A}$.

Theorem 1.8 [433] *The rank of the product matrix* $\mathbf{AB}$ *satisfies the inequality*

$$\text{rank}(\mathbf{AB}) \leq \min\{\text{rank}(\mathbf{A}), \text{rank}(\mathbf{B})\}. \tag{1.6.24}$$

Lemma 1.1 *If premultiplying an* $m \times n$ *matrix* $\mathbf{A}$ *by an* $m \times m$ *nonsingular matrix* $\mathbf{P}$, *or postmultiplying it by an* $n \times n$ *nonsingular matrix* $\mathbf{Q}$, *then the rank of* $\mathbf{A}$ *is not changed, namely* $\text{rank}(\mathbf{PAQ}) = \text{rank}(\mathbf{A})$.

Lemma 1.2 $\text{rank}[\mathbf{A}, \mathbf{B}] \leq \text{rank}(\mathbf{A}) + \text{rank}(\mathbf{B})$.

Lemma 1.3 $\text{rank}(\mathbf{A} + \mathbf{B}) \leq \text{rank}[\mathbf{A}, \mathbf{B}] \leq \text{rank}(\mathbf{A}) + \text{rank}(\mathbf{B})$.

Lemma 1.4 *For an* $m \times n$ *matrix* $\mathbf{A}$ *and an* $n \times q$ *matrix* $\mathbf{B}$, *the rank inequality* $\text{rank}(\mathbf{AB}) \geq \text{rank}(\mathbf{A}) + \text{rank}(\mathbf{B}) - n$ *is true.*

1. *Properties of the Rank of a Matrix*

1. The rank is a positive integer.
2. The rank is equal to or less than the number of columns or rows of the matrix.

3. If the rank of an $n \times n$ matrix $\mathbf{A}$ is equal to n then $\mathbf{A}$ is nonsingular, or we say that $\mathbf{A}$ is a *full rank matrix*.
4. If $\text{rank}(\mathbf{A}_{m\times n}) < \min\{m,n\}$ then $\mathbf{A}$ is said to be a *rank-deficient matrix*.
5. If $\text{rank}(\mathbf{A}_{m\times n}) = m\,(< n)$ then the matrix $\mathbf{A}$ is a *full row rank matrix*.
6. If $\text{rank}(\mathbf{A}_{m\times n}) = n\,(< m)$ then the matrix $\mathbf{A}$ is a *full column rank matrix*.
7. Premultiplying any matrix $\mathbf{A}$ by a full column rank matrix or postmultiplying it by a full row rank matrix leaves the rank of the matrix $\mathbf{A}$ unchanged.

2. Rank Equalities

(1) If $\mathbf{A} \in \mathbb{C}^{m\times n}$, then $\text{rank}(\mathbf{A}^H) = \text{rank}(\mathbf{A}^T) = \text{rank}(\mathbf{A}^*) = \text{rank}(\mathbf{A})$.
(2) If $\mathbf{A} \in \mathbb{C}^{m\times n}$ and $c \neq 0$, then $\text{rank}(c\mathbf{A}) = \text{rank}(\mathbf{A})$.
(3) If $\mathbf{A} \in \mathbb{C}^{m\times m}$ and $\mathbf{C} \in \mathbb{C}^{n\times n}$ are nonsingular then $\text{rank}(\mathbf{AB}) = \text{rank}(\mathbf{B}) = \text{rank}(\mathbf{BC}) = \text{rank}(\mathbf{ABC})$ for $\mathbf{B} \in \mathbb{C}^{m\times n}$. That is, after premultiplying and/or postmultiplying by a nonsingular matrix, the rank of $\mathbf{B}$ remains unchanged.
(4) For $\mathbf{A}, \mathbf{B} \in \mathbb{C}^{m\times n}$, $\text{rank}(\mathbf{A}) = \text{rank}(\mathbf{B})$ if and only if there exist nonsingular matrices $\mathbf{X} \in \mathbb{C}^{m\times m}$ and $\mathbf{Y} \in \mathbb{C}^{n\times n}$ such that $\mathbf{B} = \mathbf{XAY}$.
(5) $\text{rank}(\mathbf{AA}^H) = \text{rank}(\mathbf{A}^H\mathbf{A}) = \text{rank}(\mathbf{A})$.
(6) If $\mathbf{A} \in \mathbb{C}^{m\times m}$ then $\text{rank}(\mathbf{A}) = m \Leftrightarrow \det(\mathbf{A}) \neq 0 \Leftrightarrow \mathbf{A}$ is nonsingular.

3. Rank Inequalities

- $\text{rank}(\mathbf{A}) \leq \min\{m,n\}$ for any $m \times n$ matrix $\mathbf{A}$.
- If $\mathbf{A}, \mathbf{B} \in \mathbb{C}^{m\times n}$ then $\text{rank}(\mathbf{A}+\mathbf{B}) \leq \text{rank}(\mathbf{A}) + \text{rank}(\mathbf{B})$.
- If $\mathbf{A} \in \mathbb{C}^{m\times k}$ and $\mathbf{B} \in \mathbb{C}^{k\times n}$ then

$$\text{rank}(\mathbf{A}) + \text{rank}(\mathbf{B}) - k \leq \text{rank}(\mathbf{AB}) \leq \min\{\text{rank}(\mathbf{A}), \text{rank}(\mathbf{B})\}.$$

One sentence summary: *As a performance index, the matrix rank describes the linear independence of the rows (or columns) of a matrix, which reflects the full rank or rank deficiency of the matrix.*

Table 1.1 summarizes the five important performance indexes discussed above and how they describe matrix performance.

Table 1.1 Performance indexes of matrices

Performance index	Matrix property determined by the performance index
quadratic form	positive definiteness and non-negative definiteness
determinant	singularity
eigenvalues	singularity and positive definiteness
trace	sum of diagonal entries, sum of eigenvalues
rank	linear independence of rows (or columns)

1.7 Inverse Matrices and Pseudo-Inverse Matrices

Matrix inversion is an important aspect of matrix calculus. In particular, the matrix inversion lemma is often used in signal processing, system sciences, automatic control, neural networks and so on. In this section we discuss the inverse of a full-rank square matrix and the pseudo-inverse of a non-square matrix with full row (or full column) rank. Regarding the inversion of a nonsquare or rank-deficient matrix, we will discuss this in the next section.

1.7.1 Definition and Properties of Inverse Matrices

An $n \times n$ matrix is called *nonsingular* if it has n linearly independent column vectors and n linearly independent row vectors. A nonsingular matrix can also be defined from the viewpoint of a linear system: a linear transformation or a square matrix $\mathbf{A}$ is said to be nonsingular if it produces a zero output only for zero input; otherwise it is singular. If a matrix is nonsingular then its inverse must exist. Conversely, a singular matrix has no inverse. The $n \times n$ matrix $\mathbf{B}$ such that $\mathbf{BA} = \mathbf{AB} = \mathbf{I}$ is called the inverse matrix of $\mathbf{A}$, denoted $\mathbf{B} = \mathbf{A}^{-1}$.

If $\mathbf{A}^{-1}$ exists then the matrix $\mathbf{A}$ is said to be nonsingular or invertible.

On the nonsingularity or invertibility of an $n \times n$ matrix $\mathbf{A}$, the following statements are equivalent [214].

(1) $\mathbf{A}$ is nonsingular.

(2) $\mathbf{A}^{-1}$ exists.

(3) $\text{rank}(\mathbf{A}) = n$.

(4) All rows of $\mathbf{A}$ are linearly independent.

(5) All columns of $\mathbf{A}$ are linearly independent.

(6) $\det(\mathbf{A}) \neq 0$.

(7) The dimension of the range of $\mathbf{A}$ is n.

(8) The dimension of the null space of $\mathbf{A}$ is equal to zero.

(9) $\mathbf{Ax} = \mathbf{b}$ is a consistent equation for every $\mathbf{b} \in \mathbb{C}^n$.

(10) $\mathbf{Ax} = \mathbf{b}$ has a unique solution for every $\mathbf{b}$.

(11) $\mathbf{Ax} = \mathbf{0}$ has only the trivial solution $\mathbf{x} = \mathbf{0}$.

1. Properties of the Inverse Matrix $\mathbf{A}^{-1}$ [25], [214]

1. $\mathbf{A}^{-1}\mathbf{A} = \mathbf{AA}^{-1} = \mathbf{I}$.
2. $\mathbf{A}^{-1}$ is unique.
3. The determinant of the inverse matrix is equal to the reciprocal of the determinant of the original matrix, i.e., $|\mathbf{A}^{-1}| = 1/|\mathbf{A}|$.
4. The inverse matrix $\mathbf{A}^{-1}$ is nonsingular.

5. The inverse matrix of an inverse matrix is the original matrix, i.e., $(\mathbf{A}^{-1})^{-1} = \mathbf{A}$.
6. The inverse matrix of a Hermitian matrix $\mathbf{A} = \mathbf{A}^H$ satisfies $(\mathbf{A}^H)^{-1} = (\mathbf{A}^{-1})^H = \mathbf{A}^{-H}$.
7. If $\mathbf{A}^H = \mathbf{A}$ then $(\mathbf{A}^{-1})^H = \mathbf{A}^{-1}$. That is to say, the inverse matrix of any Hermitian matrix is a Hermitian matrix as well.
8. $(\mathbf{A}^*)^{-1} = (\mathbf{A}^{-1})^*$.
9. If $\mathbf{A}$ and $\mathbf{B}$ are invertible then $(\mathbf{AB})^{-1} = \mathbf{B}^{-1}\mathbf{A}^{-1}$.
10. If $\mathbf{A} = \mathbf{Diag}(a_1, \ldots, a_m)$ is a diagonal matrix then its inverse matrix

$$\mathbf{A}^{-1} = \mathbf{Diag}(a_1^{-1}, \ldots, a_m^{-1}).$$

11. Let $\mathbf{A}$ be nonsingular. If $\mathbf{A}$ is an orthogonal matrix then $\mathbf{A}^{-1} = \mathbf{A}^T$, and if $\mathbf{A}$ is a unitary matrix then $\mathbf{A}^{-1} = \mathbf{A}^H$.

1.7.2 Matrix Inversion Lemma

Lemma 1.5 *Let $\mathbf{A}$ be an $n \times n$ invertible matrix, and $\mathbf{x}$ and $\mathbf{y}$ be two $n \times 1$ vectors such that $(\mathbf{A} + \mathbf{x}\mathbf{y}^H)$ is invertible; then*

$$(\mathbf{A} + \mathbf{x}\mathbf{y}^H)^{-1} = \mathbf{A}^{-1} - \frac{\mathbf{A}^{-1}\mathbf{x}\mathbf{y}^H\mathbf{A}^{-1}}{1 + \mathbf{y}^H\mathbf{A}^{-1}\mathbf{x}}. \tag{1.7.1}$$

Lemma 1.5 is called the *matrix inversion lemma*, and was presented by Sherman and Morrison [438], [439] in 1949 and 1950.

The matrix inversion lemma can be extended to an inversion formula for a sum of matrices:

$$\begin{aligned}(\mathbf{A} + \mathbf{UBV})^{-1} &= \mathbf{A}^{-1} - \mathbf{A}^{-1}\mathbf{UB}(\mathbf{B} + \mathbf{BVA}^{-1}\mathbf{UB})^{-1}\mathbf{BVA}^{-1} \\ &= \mathbf{A}^{-1} - \mathbf{A}^{-1}\mathbf{U}(\mathbf{I} + \mathbf{BVA}^{-1}\mathbf{U})^{-1}\mathbf{BVA}^{-1}\end{aligned} \tag{1.7.2}$$

or

$$(\mathbf{A} - \mathbf{UV})^{-1} = \mathbf{A}^{-1} + \mathbf{A}^{-1}\mathbf{U}(\mathbf{I} - \mathbf{VA}^{-1}\mathbf{U})^{-1}\mathbf{VA}^{-1}. \tag{1.7.3}$$

The above formula was obtained by Woodbury in 1950 [513] and is called the *Woodbury formula*.

Taking $\mathbf{U} = \mathbf{u}$, $\mathbf{B} = \beta$ and $\mathbf{V} = \mathbf{v}^H$, the Woodbury formula gives the result

$$(\mathbf{A} + \beta\mathbf{u}\mathbf{v}^H)^{-1} = \mathbf{A}^{-1} - \frac{\beta}{1 + \beta\mathbf{v}^H\mathbf{A}^{-1}\mathbf{u}}\mathbf{A}^{-1}\mathbf{u}\mathbf{v}^H\mathbf{A}^{-1}. \tag{1.7.4}$$

In particular, if we let $\beta = 1$ then Equation (1.7.4) reduces to formula (1.7.1), the matrix inversion lemma of Sherman and Morrison.

As a matter of fact, before Woodbury obtained the inversion formula (1.7.2),

Duncan [139] in 1944 and Guttman [192] in 1946 had obtained the following inversion formula:

$$(\mathbf{A}-\mathbf{U}\mathbf{D}^{-1}\mathbf{V})^{-1}=\mathbf{A}^{-1}+\mathbf{A}^{-1}\mathbf{U}(\mathbf{D}-\mathbf{V}\mathbf{A}^{-1}\mathbf{U})^{-1}\mathbf{V}\mathbf{A}^{-1}. \tag{1.7.5}$$

This formula is called the *Duncan–Guttman inversion formula* [391], [392].

In addition to the Woodbury formula, the inverse matrix of a sum of matrices also has the following forms [206]:

$$\begin{aligned}(\mathbf{A}+\mathbf{UBV})^{-1}&=\mathbf{A}^{-1}-\mathbf{A}^{-1}(\mathbf{I}+\mathbf{UBVA}^{-1})^{-1}\mathbf{UBVA}^{-1} && (1.7.6)\\ &=\mathbf{A}^{-1}-\mathbf{A}^{-1}\mathbf{UB}(\mathbf{I}+\mathbf{VA}^{-1}\mathbf{UB})^{-1}\mathbf{VA}^{-1} && (1.7.7)\\ &=\mathbf{A}^{-1}-\mathbf{A}^{-1}\mathbf{UBV}(\mathbf{I}+\mathbf{A}^{-1}\mathbf{UBV})^{-1}\mathbf{A}^{-1} && (1.7.8)\\ &=\mathbf{A}^{-1}-\mathbf{A}^{-1}\mathbf{UBVA}^{-1}(\mathbf{I}+\mathbf{UBVA}^{-1})^{-1}. && (1.7.9)\end{aligned}$$

The following are inversion formulas for block matrices.

When the matrix $\mathbf{A}$ is invertible, one has [22]:

$$\begin{aligned}&\begin{bmatrix}\mathbf{A} & \mathbf{U}\\ \mathbf{V} & \mathbf{D}\end{bmatrix}^{-1}\\ &=\begin{bmatrix}\mathbf{A}^{-1}+\mathbf{A}^{-1}\mathbf{U}(\mathbf{D}-\mathbf{VA}^{-1}\mathbf{U})^{-1}\mathbf{VA}^{-1} & -\mathbf{A}^{-1}\mathbf{U}(\mathbf{D}-\mathbf{VA}^{-1}\mathbf{U})^{-1}\\ -(\mathbf{D}-\mathbf{VA}^{-1}\mathbf{U})^{-1}\mathbf{VA}^{-1} & (\mathbf{D}-\mathbf{VA}^{-1}\mathbf{U})^{-1}\end{bmatrix}.\end{aligned} \tag{1.7.10}$$

If the matrices $\mathbf{A}$ and $\mathbf{D}$ are invertible, then [216], [217]

$$\begin{bmatrix}\mathbf{A} & \mathbf{U}\\ \mathbf{V} & \mathbf{D}\end{bmatrix}^{-1}=\begin{bmatrix}(\mathbf{A}-\mathbf{UD}^{-1}\mathbf{V})^{-1} & -\mathbf{A}^{-1}\mathbf{U}(\mathbf{D}-\mathbf{VA}^{-1}\mathbf{U})^{-1}\\ -\mathbf{D}^{-1}\mathbf{V}(\mathbf{A}-\mathbf{UD}^{-1}\mathbf{V})^{-1} & (\mathbf{D}-\mathbf{VA}^{-1}\mathbf{U})^{-1}\end{bmatrix} \tag{1.7.11}$$

or [139]

$$\begin{bmatrix}\mathbf{A} & \mathbf{U}\\ \mathbf{V} & \mathbf{D}\end{bmatrix}^{-1}=\begin{bmatrix}(\mathbf{A}-\mathbf{UD}^{-1}\mathbf{V})^{-1} & -(\mathbf{A}-\mathbf{UD}^{-1}\mathbf{V})^{-1}\mathbf{UD}^{-1}\\ -(\mathbf{D}-\mathbf{VA}^{-1}\mathbf{U})^{-1}\mathbf{VA}^{-1} & (\mathbf{D}-\mathbf{VA}^{-1}\mathbf{U})^{-1}\end{bmatrix}. \tag{1.7.12}$$

1.7.3 Inversion of Hermitian Matrices

Consider the inversion of a $(m+1)\times(m+1)$ nonsingular Hermitian matrix $\mathbf{R}_{m+1}$.

First, an $(m+1)\times(m+1)$ Hermitian matrix can always be written in a block form:

$$\mathbf{R}_{m+1}=\begin{bmatrix}\mathbf{R}_m & \mathbf{r}_m\\ \mathbf{r}_m^H & \rho_m\end{bmatrix}, \tag{1.7.13}$$

where ρ_m is the $(m+1,m+1)$th entry of $\mathbf{R}_{m+1}$, and $\mathbf{R}_m$ is an $m\times m$ Hermitian matrix.

Now, we consider how to compute the inverse matrix $\mathbf{R}_{m+1}^{-1}$ using the inverse matrix $\mathbf{R}_m^{-1}$. For this purpose, let

$$\mathbf{Q}_{m+1} = \begin{bmatrix} \mathbf{Q}_m & \mathbf{q}_m \\ \mathbf{q}_m^H & \alpha_m \end{bmatrix} \tag{1.7.14}$$

be the inverse matrix of $\mathbf{R}_{m+1}$. Then we have

$$\mathbf{R}_{m+1}\mathbf{Q}_{m+1} = \begin{bmatrix} \mathbf{R}_m & \mathbf{r}_m \\ \mathbf{r}_m^H & \rho_m \end{bmatrix} \begin{bmatrix} \mathbf{Q}_m & \mathbf{q}_m \\ \mathbf{q}_m^H & \alpha_m \end{bmatrix} = \begin{bmatrix} \mathbf{I}_m & \mathbf{0}_m \\ \mathbf{0}_m^H & 1 \end{bmatrix} \tag{1.7.15}$$

which gives the following four equations:

$$\mathbf{R}_m\mathbf{Q}_m + \mathbf{r}_m\mathbf{q}_m^H = \mathbf{I}_m, \tag{1.7.16}$$

$$\mathbf{r}_m^H\mathbf{Q}_m + \rho_m\mathbf{q}_m^H = \mathbf{0}_m^H, \tag{1.7.17}$$

$$\mathbf{R}_m\mathbf{q}_m + \mathbf{r}_m\alpha_m = \mathbf{0}_m, \tag{1.7.18}$$

$$\mathbf{r}_m^H\mathbf{q}_m + \rho_m\alpha_m = 1. \tag{1.7.19}$$

If $\mathbf{R}_m$ is invertible, then from Equation (1.7.18), it follows that

$$\mathbf{q}_m = -\alpha_m\mathbf{R}_m^{-1}\mathbf{r}_m. \tag{1.7.20}$$

Substitute this result into Equation (1.7.19) to yield

$$\alpha_m = \frac{1}{\rho_m - \mathbf{r}_m^H\mathbf{R}_m^{-1}\mathbf{r}_m}. \tag{1.7.21}$$

After substituting Equation (1.7.21) into Equation (1.7.20), we can obtain

$$\mathbf{q}_m = \frac{-\mathbf{R}_m^{-1}\mathbf{r}_m}{\rho_m - \mathbf{r}_b^H\mathbf{R}_m^{-1}\mathbf{r}_m}. \tag{1.7.22}$$

Then, substitute Equation (1.7.22) into Equation (1.7.16):

$$\mathbf{Q}_m = \mathbf{R}_m^{-1} - \mathbf{R}_m^{-1}\mathbf{r}_m\mathbf{q}_m^H = \mathbf{R}_m^{-1} + \frac{\mathbf{R}_m^{-1}\mathbf{r}_m(\mathbf{R}_m^{-1}\mathbf{r}_m)^H}{\rho_m - \mathbf{r}_m^H\mathbf{R}_m^{-1}\mathbf{r}_m}. \tag{1.7.23}$$

In order to simplify (1.7.21)–(1.7.23), put

$$\mathbf{b}_m \stackrel{\text{def}}{=} [b_0^{(m)}, b_1^{(m)}, \ldots, b_{m-1}^{(m)}]^T = -\mathbf{R}_m^{-1}\mathbf{r}_m, \tag{1.7.24}$$

$$\beta_m \stackrel{\text{def}}{=} \rho_m - \mathbf{r}_m^H\mathbf{R}_m^{-1}\mathbf{r}_m = \rho_m + \mathbf{r}_m^H\mathbf{b}_m. \tag{1.7.25}$$

Then (1.7.21)–(1.7.23) can be respectively simplified to

$$\alpha_m = \frac{1}{\beta_m}, \quad \mathbf{q}_m = \frac{1}{\beta_m}\mathbf{b}_m, \quad \mathbf{Q}_m = \mathbf{R}_m^{-1} + \frac{1}{\beta_m}\mathbf{b}_m\mathbf{b}_m^H.$$

Substituting the these results into (1.7.15), it is immediately seen that

$$\mathbf{R}_{m+1}^{-1} = \mathbf{Q}_{m+1} = \begin{bmatrix} \mathbf{R}_m^{-1} & \mathbf{0}_m \\ \mathbf{0}_m^H & 0 \end{bmatrix} + \frac{1}{\beta_m}\begin{bmatrix} \mathbf{b}_m\mathbf{b}_m^H & \mathbf{b}_m \\ \mathbf{b}_m^H & 1 \end{bmatrix}. \tag{1.7.26}$$

This formula for calculating the $(m+1)\times(m+1)$ inverse $\mathbf{R}_{m+1}$ from the $m\times m$ inverse $\mathbf{R}_m$ is called the *block inversion lemma* for Hermitian matrices [352].

Given an $M\times M$ nonsingular Hermitian matrix $\mathbf{H}$, using the block inversion lemma for $m=1,\ldots,M-1$ we can find the inverse matrix $\mathbf{H}^{-1}$ in a recursive way.

1.7.4 Left and Right Pseudo-Inverse Matrices

From a broader perspective, any $n\times m$ matrix $\mathbf{G}$ may be called the inverse of a given $m\times n$ matrix $\mathbf{A}$, if the product of $\mathbf{G}$ and $\mathbf{A}$ is equal to the identity matrix $\mathbf{I}$. According to the different possible structures of the $m\times n$ matrix $\mathbf{A}$, for the matrix $\mathbf{G}$ to be such that the product of it and $\mathbf{A}$ is equal to the identity matrix there exist three possibilities as shown in the next example.

EXAMPLE 1.11 Consider the inverses of the following three matrices,

$$\mathbf{A}_1=\begin{bmatrix}2 & -2 & -1\\ 1 & 1 & -2\\ 1 & 0 & -1\end{bmatrix},\quad \mathbf{A}_2=\begin{bmatrix}4 & 8\\ 5 & -7\\ -2 & 3\end{bmatrix},\quad \mathbf{A}_3=\begin{bmatrix}1 & 3 & 1\\ 2 & 5 & 1\end{bmatrix}.$$

For the matrix $\mathbf{A}_1$, there is a unique matrix

$$\mathbf{G}=\begin{bmatrix}-1 & -2 & 5\\ -1 & -1 & 3\\ -1 & -2 & 4\end{bmatrix}$$

such that $\mathbf{G}\mathbf{A}_1=\mathbf{A}_1\mathbf{G}=\mathbf{I}_{3\times 3}$. In this case, the matrix $\mathbf{G}$ is actually the inverse matrix of $\mathbf{A}_1$, i.e., $\mathbf{G}=\mathbf{A}_1^{-1}$.

In the case of the matrix $\mathbf{A}_2$, there exist an infinite number of 2×3 matrices $\mathbf{L}$ such that $\mathbf{L}\mathbf{A}_2=\mathbf{I}_{2\times 2}$, for example

$$\mathbf{L}=\begin{bmatrix}\frac{7}{68} & \frac{2}{17} & 0\\ 0 & 2 & 5\end{bmatrix},\quad \mathbf{L}=\begin{bmatrix}0 & 3 & 7\\ 0 & 2 & 5\end{bmatrix},\quad \cdots$$

For the matrix $\mathbf{A}_3$, there are no 3×2 matrices $\mathbf{G}_3$ such that $\mathbf{G}_3\mathbf{A}_3=\mathbf{I}_{3\times 3}$ but there exist more than one 3×2 matrix $\mathbf{R}$ such that $\mathbf{A}_3\mathbf{R}=\mathbf{I}_{2\times 2}$, for example

$$\mathbf{R}=\begin{bmatrix}1 & 1\\ -1 & 0\\ 3 & -1\end{bmatrix},\quad \mathbf{R}=\begin{bmatrix}-1 & 1\\ 0 & 0\\ 2 & -1\end{bmatrix},\quad \cdots$$

Summarizing the above discussion, we see that, in addition to the inverse matrix $\mathbf{A}^{-1}$ satisfying $\mathbf{A}\mathbf{A}^{-1}=\mathbf{A}^{-1}\mathbf{A}=\mathbf{I}$, there are two other forms of inverse matrix that satisfy only $\mathbf{L}\mathbf{A}=\mathbf{I}$ or $\mathbf{A}\mathbf{R}=\mathbf{I}$.

DEFINITION 1.35 [433] The matrix $\mathbf{L}$ satisfying $\mathbf{L}\mathbf{A}=\mathbf{I}$ but not $\mathbf{A}\mathbf{L}=\mathbf{I}$ is called the *left inverse* of the matrix $\mathbf{A}$. Similarly, the matrix $\mathbf{R}$ satisfying $\mathbf{A}\mathbf{R}=\mathbf{I}$ but not $\mathbf{R}\mathbf{A}=\mathbf{I}$ is said to be the *right inverse* of $\mathbf{A}$.

Remark A matrix $\mathbf{A} \in \mathbb{C}^{m\times n}$ has a left inverse only when $m \geq n$ and a right inverse only when $m \leq n$, respectively.

As shown in Example 1.11, for a given $m\times n$ matrix $\mathbf{A}$, when $m > n$ it is possible that other $n \times m$ matrices $\mathbf{L}$ to satisfy $\mathbf{LA} = \mathbf{I}_n$, while when $m < n$ it is possible that other $n \times m$ matrices $\mathbf{R}$ to satisfy $\mathbf{AR} = \mathbf{I}_m$. That is to say, the left or right inverse of a given matrix $\mathbf{A}$ is usually not unique. Let us consider the conditions for a unique solution of the left and right inverse matrices.

Let $m > n$ and let the $m \times n$ matrix $\mathbf{A}$ have full column rank, i.e., $\text{rank}(\mathbf{A}) = n$. In this case, the $n \times n$ matrix $\mathbf{A}^H\mathbf{A}$ is invertible. It is easy to verify that

$$\mathbf{L} = (\mathbf{A}^H\mathbf{A})^{-1}\mathbf{A}^H \tag{1.7.27}$$

satisfies $\mathbf{LA} = \mathbf{I}$ but not $\mathbf{AL} = \mathbf{I}$. This type of left inverse matrix is uniquely determined, and is usually called the *left pseudo-inverse matrix* of $\mathbf{A}$.

On the other hand, if $m < n$ and $\mathbf{A}$ has the full row rank, i.e., $\text{rank}(\mathbf{A}) = m$, then the $m \times m$ matrix $\mathbf{AA}^H$ is invertible. Define

$$\mathbf{R} = \mathbf{A}^H(\mathbf{AA}^H)^{-1}. \tag{1.7.28}$$

It is easy to see that $\mathbf{R}$ satisfies $\mathbf{AR} = \mathbf{I}$ rather than $\mathbf{RA} = \mathbf{I}$. The matrix $\mathbf{R}$ is also uniquely determined and is usually called the *right pseudo-inverse matrix* of $\mathbf{A}$.

The left pseudo-inverse matrix is closely related to the least squares solution of overdetermined equations, while the right pseudo-inverse matrix is closely related to the least squares minimum-norm solution of under-determined equations. A detailed discussion is given in Chapter 6.

The following are the dimensional recursions for the left and right pseudo-inverse matrices [542].

Consider an $n \times m$ matrix $\mathbf{F}_m$ (where $n > m$) and its left pseudo-inverse matrix $\mathbf{F}_m^\dagger = (\mathbf{F}_m^H\mathbf{F}_m)^{-1}\mathbf{F}_m^H$. Let $\mathbf{F}_m = [\mathbf{F}_{m-1}, \mathbf{f}_m]$, where $\mathbf{f}_m$ is the mth column of the matrix $\mathbf{F}_m$ and $\text{rank}(\mathbf{F}_m) = m$; then a recursive computation formula for $\mathbf{F}_m^\dagger$ is given by

$$\mathbf{F}_m^\dagger = \begin{bmatrix} \mathbf{F}_{m-1}^\dagger - \mathbf{F}_{m-1}^\dagger\mathbf{f}_m\mathbf{e}_m^H\Delta_m^{-1} \\ \mathbf{e}_m^H\Delta_m^{-1} \end{bmatrix}, \tag{1.7.29}$$

where $\mathbf{e}_m = (\mathbf{I}_n - \mathbf{F}_{m-1}\mathbf{F}_{m-1}^\dagger)\mathbf{f}_m$ and $\Delta_m^{-1} = (\mathbf{f}_m^H\mathbf{e}_m)^{-1}$; the initial recursion value is $\mathbf{F}_1^\dagger = \mathbf{f}_1^H/(\mathbf{f}_1^H\mathbf{f}_1)$.

Given a matrix $\mathbf{F}_m \in \mathbb{C}^{n\times m}$, where $n < m$, and again writing $\mathbf{F}_m = [\mathbf{F}_{m-1}, \mathbf{f}_m]$, the right pseudo-inverse matrix $\mathbf{F}_m^\dagger = \mathbf{F}_m^H(\mathbf{F}_m\mathbf{F}_m^H)^{-1}$ has the following recursive formula:

$$\mathbf{F}_m^\dagger = \begin{bmatrix} \mathbf{F}_{m-1}^\dagger - \Delta_m\mathbf{F}_{m-1}^\dagger\mathbf{f}_m\mathbf{c}_m \\ \Delta_m\mathbf{c}_m^H \end{bmatrix}. \tag{1.7.30}$$

Here $\mathbf{c}_m^H = \mathbf{f}_m^H(\mathbf{I}_n - \mathbf{F}_{m-1}\mathbf{F}_{m-1}^\dagger)$ and $\Delta_m = \mathbf{c}_m^H f_m$. The initial recursion value is $\mathbf{F}_1^\dagger = \mathbf{f}_1^H/(\mathbf{f}_1^H\mathbf{f}_1)$.

1.8 Moore–Penrose Inverse Matrices

In the above section we discussed the inverse matrix $\mathbf{A}^{-1}$ of a nonsingular square matrix $\mathbf{A}$, the left pseudo-inverse matrix $\mathbf{A}_L^{-1} = (\mathbf{A}^H\mathbf{A})^{-1}\mathbf{A}^H$ of an $m \times n$ $(m > n)$ matrix $\mathbf{A}$ with full column rank, and the right pseudo-inverse matrix $\mathbf{A}_R^{-1} = \mathbf{A}^H(\mathbf{A}\mathbf{A}^H)^{-1}$ of an $m \times n$ $(m < n)$ matrix $\mathbf{A}$ with full row rank. A natural question to ask is: does there exist an inverse matrix for an $m \times n$ rank-deficient matrix? If such a inverse matrix exists, what conditions should it meet?

1.8.1 Definition and Properties

Consider an $m \times n$ rank-deficient matrix $\mathbf{A}$, regardless of the size of m and n but with $\text{rank}(\mathbf{A}) = k < \min\{m, n\}$. The inverse of an $m \times n$ rank-deficient matrix is said to be its generalized inverse matrix and is an $n \times m$ matrix.

Let $\mathbf{A}^\dagger$ denote the generalized inverse of the matrix $\mathbf{A}$.

From the matrix rank property $\text{rank}(\mathbf{AB}) \leq \min\{\text{rank}(\mathbf{A}), \text{rank}(\mathbf{B})\}$, it follows that neither $\mathbf{A}\mathbf{A}^\dagger = \mathbf{I}_{m\times m}$ nor $\mathbf{A}^\dagger\mathbf{A} = \mathbf{I}_{n\times n}$ hold, because either the $m \times m$ matrix $\mathbf{A}\mathbf{A}^\dagger$ or the $n \times n$ matrix $\mathbf{A}^\dagger\mathbf{A}$ is rank-deficient, the maximum of their ranks is $\text{rank}(\mathbf{A}) = k$ and is less than $\min\{m, n\}$.

Since $\mathbf{A}\mathbf{A}^\dagger \neq \mathbf{I}_{m\times m}$ and $\mathbf{A}^\dagger\mathbf{A} \neq \mathbf{I}_{n\times n}$, it is necessary to consider using a product of three matrices to define the generalized inverse of a rank-deficient matrix $\mathbf{A}$.

For this purpose, let us consider solving the linear matrix equation $\mathbf{Ax} = \mathbf{y}$. If $\mathbf{A}^\dagger$ is the generalized inverse matrix of $\mathbf{A}$ then $\mathbf{Ax} = \mathbf{y} \Rightarrow \mathbf{x} = \mathbf{A}^\dagger\mathbf{y}$. Substituting $\mathbf{x} = \mathbf{A}^\dagger\mathbf{y}$ into $\mathbf{Ax} = \mathbf{y}$, we have $\mathbf{A}\mathbf{A}^\dagger\mathbf{y} = \mathbf{y}$, and thus $\mathbf{A}\mathbf{A}^\dagger\mathbf{A}\mathbf{x} = \mathbf{A}\mathbf{x}$. Since this equation should hold for any nonzero vector $\mathbf{x}$, the following condition should essentially be satisfied:

$$\mathbf{A}\mathbf{A}^\dagger\mathbf{A} = \mathbf{A}. \tag{1.8.1}$$

Unfortunately, the matrix $\mathbf{A}^\dagger$ meeting the condition (1.8.1) is not unique. To overcome this difficulty we must add other conditions.

The condition $\mathbf{A}\mathbf{A}^\dagger\mathbf{A} = \mathbf{A}$ in (1.8.1) can only guarantee that $\mathbf{A}^\dagger$ is a generalized inverse matrix of $\mathbf{A}$; it cannot conversely guarantee that $\mathbf{A}$ is also a generalized inverse matrix of $\mathbf{A}^\dagger$, whereas $\mathbf{A}$ and $\mathbf{A}^\dagger$ should essentially be mutually generalized inverses. This is one of the main reasons why the definition $\mathbf{A}\mathbf{A}^\dagger\mathbf{A} = \mathbf{A}$ does not uniquely determine $\mathbf{A}^\dagger$.

Now consider the solution equation of the original matrix equation $\mathbf{Ax} = \mathbf{y}$, i.e., $\mathbf{x} = \mathbf{A}^\dagger\mathbf{y}$. Our problem is: given any $\mathbf{y} \neq \mathbf{0}$, find $\mathbf{x}$. Clearly, $\mathbf{x} = \mathbf{A}^\dagger\mathbf{y}$ can be written as $\mathbf{x} = \mathbf{A}^\dagger\mathbf{A}\mathbf{x}$, yielding $\mathbf{A}^\dagger\mathbf{y} = \mathbf{A}^\dagger\mathbf{A}\mathbf{A}^\dagger\mathbf{y}$. Since $\mathbf{A}^\dagger\mathbf{y} = \mathbf{A}^\dagger\mathbf{A}\mathbf{A}^\dagger\mathbf{y}$ should hold for any nonzero vector $\mathbf{y}$, the condition

$$\mathbf{A}^\dagger\mathbf{A}\mathbf{A}^\dagger = \mathbf{A}^\dagger \tag{1.8.2}$$

must be satisfied as well.

This shows that if $\mathbf{A}^\dagger$ is the generalized inverse of the rank-deficient matrix $\mathbf{A}$, then the two conditions in Equations (1.8.1) and (1.8.2) must be satisfied at the same time. However, these two conditions are still not enough for a unique definition of $\mathbf{A}^\dagger$.

If an $m \times n$ matrix $\mathbf{A}$ is of full column rank or full row rank, we certainly hope that the generalized inverse matrix $\mathbf{A}^\dagger$ will include the left and right pseudo-inverse matrices as two special cases. Although the left pseudo-inverse matrix $\mathbf{L} = (\mathbf{A}^H\mathbf{A})^{-1}\mathbf{A}^H$ of the $m \times n$ full column rank matrix $\mathbf{A}$ is such that $\mathbf{LA} = \mathbf{I}_{n\times n}$ rather than $\mathbf{AL} = \mathbf{I}_{m\times m}$, it is clear that $\mathbf{AL} = \mathbf{A}(\mathbf{A}^H\mathbf{A})^{-1}\mathbf{A}^H = (\mathbf{AL})^H$ is an $m \times m$ Hermitian matrix. Coincidentally, for any right pseudo-inverse matrix $\mathbf{R} = \mathbf{A}^H(\mathbf{A}\mathbf{A}^H)^{-1}$, the matrix product $\mathbf{RA} = \mathbf{A}^H(\mathbf{A}\mathbf{A}^H)^{-1}\mathbf{A} = (\mathbf{RA})^H$ is an $n\times n$ Hermitian matrix. In other words, in order to guarantee that $\mathbf{A}^\dagger$ exists uniquely for any $m \times n$ matrix $\mathbf{A}$, the following two conditions must be added:

$$\mathbf{A}\mathbf{A}^\dagger = (\mathbf{A}\mathbf{A}^\dagger)^H, \tag{1.8.3}$$

$$\mathbf{A}^\dagger\mathbf{A} = (\mathbf{A}^\dagger\mathbf{A})^H. \tag{1.8.4}$$

On the basis of the conditions (1.8.1)–(1.8.4), one has the following definition of $\mathbf{A}^\dagger$.

Definition 1.36 [384] Let $\mathbf{A}$ be any $m \times n$ matrix; an $n \times m$ matrix $\mathbf{A}^\dagger$ is said to be the *Moore–Penrose inverse* of $\mathbf{A}$ if $\mathbf{A}^\dagger$ meets the following four conditions (usually called the *Moore–Penrose conditions*):

(a) $\mathbf{A}\mathbf{A}^\dagger\mathbf{A} = \mathbf{A}$;

(b) $\mathbf{A}^\dagger\mathbf{A}\mathbf{A}^\dagger = \mathbf{A}^\dagger$;

(c) $\mathbf{A}\mathbf{A}^\dagger$ is an $m \times m$ Hermitian matrix, i.e., $\mathbf{A}\mathbf{A}^\dagger = (\mathbf{A}\mathbf{A}^\dagger)^H$;

(d) $\mathbf{A}^\dagger\mathbf{A}$ is an $n \times n$ Hermitian matrix, i.e., $\mathbf{A}^\dagger\mathbf{A} = (\mathbf{A}^\dagger\mathbf{A})^H$.

Remark 1 From the projection viewpoint, Moore [330] showed in 1935 that the generalized inverse matrix $\mathbf{A}^\dagger$ of an $m\times n$ matrix $\mathbf{A}$ must meet two conditions, but these conditions are not convenient for practical use. After two decades, Penrose [384] in 1955 presented the four conditions (a)–(d) stated above. In 1956, Rado [407] showed that the four conditions of Penrose are equivalent to the two conditions of Moore. Therefore the conditions (a)–(d) are called the Moore–Penrose conditions, and the generalized inverse matrix satisfying the Moore–Penrose conditions is referred to as the *Moore–Penrose inverse* of $\mathbf{A}$.

Remark 2 In particular, the Moore–Penrose condition (a) is the condition that the generalized inverse matrix $\mathbf{A}^\dagger$ of $\mathbf{A}$ must meet, while (b) is the condition that the generalized inverse matrix $\mathbf{A} = (\mathbf{A}^\dagger)^\dagger$ of the matrix $\mathbf{A}^\dagger$ must meet.

Depending on to what extent the Moore–Penrose conditions are met, the generalized inverse matrix can be classified as follows [174].

(1) The generalized inverse matrix $\mathbf{A}^\dagger$ satisfying all four conditions is the Moore–Penrose inverse of $\mathbf{A}$.
(2) The matrix $\mathbf{G} = \mathbf{A}^\dagger$ satisfying conditions (a) and (b) is said to be the *self-reflexive generalized inverse* of $\mathbf{A}$.
(3) The matrix $\mathbf{A}^\dagger$ satisfying conditions (a), (b) and (c) is referred to as the *regularized generalized inverse* of $\mathbf{A}$.
(4) The matrix $\mathbf{A}^\dagger$ satisfying conditions (a), (b) and (d) is called the *weak generalized inverse* of $\mathbf{A}$.

It is easy to verify that the inverse matrices and the various generalized inverse matrices described early are special examples of the Moore–Penrose inverse matrices.

- The inverse matrix $\mathbf{A}^{-1}$ of an $n \times n$ nonsingular matrix $\mathbf{A}$ meets all four Moore–Penrose conditions.
- The left pseudo-inverse matrix $(\mathbf{A}^H\mathbf{A})^{-1}\mathbf{A}^H$ of an $m \times n$ $(m > n)$ matrix $\mathbf{A}$ meets all four Moore–Penrose conditions.
- The right pseudo-inverse matrix $\mathbf{A}^H(\mathbf{A}\mathbf{A}^H)^{-1}$ of an $m \times n$ $(m < n)$ matrix $\mathbf{A}$ meets all four Moore–Penrose conditions.
- The left inverse matrix $\mathbf{L}_{n\times m}$ such that $\mathbf{L}\mathbf{A}_{m\times n} = \mathbf{I}_n$ is a weak generalized inverse matrix satisfying the Moore–Penrose conditions (a), (b) and (d).
- The right inverse matrix $\mathbf{R}_{n\times m}$ such that $\mathbf{A}_{m\times n}\mathbf{R} = \mathbf{I}_m$ is a regularized generalized inverse matrix satisfying the Moore–Penrose conditions (a), (b) and (c).

The inverse matrix $\mathbf{A}^{-1}$, the left pseudo-inverse matrix $(\mathbf{A}^H\mathbf{A})^{-1}\mathbf{A}^H$ and the right pseudo-inverse matrix $\mathbf{A}^H(\mathbf{A}\mathbf{A}^H)^{-1}$ are uniquely determined, respectively. Similarly, any Moore–Penrose inverse matrix is uniquely determined as well.

The Moore–Penrose inverse of any $m \times n$ matrix $\mathbf{A}$ can be uniquely determined by [51]

$$\mathbf{A}^\dagger = (\mathbf{A}^H\mathbf{A})^\dagger\mathbf{A}^H \qquad (\text{if } m \geq n) \tag{1.8.5}$$

or [187]

$$\mathbf{A}^\dagger = \mathbf{A}^H(\mathbf{A}\mathbf{A}^H)^\dagger \qquad (\text{if } m \leq n). \tag{1.8.6}$$

From Definition 1.36 it is easily seen that both $\mathbf{A}^\dagger = (\mathbf{A}^H\mathbf{A})^\dagger\mathbf{A}^H$ and $\mathbf{A}^\dagger = \mathbf{A}^H(\mathbf{A}\mathbf{A}^H)^\dagger$ meet the four Moore–Penrose conditions.

From [307], [311], [400], [401] and [410], the Moore–Penrose inverse matrices $\mathbf{A}^\dagger$ have the following properties.

1. For an $m \times n$ matrix $\mathbf{A}$, its Moore–Penrose inverse $\mathbf{A}^\dagger$ is uniquely determined.
2. The Moore–Penrose inverse of the complex conjugate transpose matrix $\mathbf{A}^H$ is given by $(\mathbf{A}^H)^\dagger = (\mathbf{A}^\dagger)^H = \mathbf{A}^{\dagger \mathrm{H}} = \mathbf{A}^{H\dagger}$.
3. The generalized inverse of a Moore–Penrose inverse matrix is equal to the original matrix, namely $(\mathbf{A}^\dagger)^\dagger = \mathbf{A}$.
4. If $c \neq 0$ then $(c\mathbf{A})^\dagger = c^{-1}\mathbf{A}^\dagger$.
5. If $\mathbf{D} = \mathbf{Diag}(d_{11}, \ldots, d_{nn})$, then $\mathbf{D}^\dagger = \mathbf{Diag}(d_{11}^\dagger, \ldots, d_{nn}^\dagger)$, where $d_{ii}^\dagger = d_{ii}^{-1}$ (if $d_{ii} \neq 0$) or $d_{ii}^\dagger = 0$ (if $d_{ii} = 0$).
6. The Moore–Penrose inverse of an $m \times n$ zero matrix $\mathbf{O}_{m\times n}$ is an $n \times m$ zero matrix, i.e., $\mathbf{O}_{m\times n}^\dagger = \mathbf{O}_{n\times m}$.
7. The Moore–Penrose inverse of the $n \times 1$ vector $\mathbf{x}$ is an $1 \times n$ vector and is given by $\mathbf{x}^\dagger = (\mathbf{x}^H\mathbf{x})^{-1}\mathbf{x}^H$.
8. If $\mathbf{A}^H = \mathbf{A}$ and $\mathbf{A}^2 = \mathbf{A}$ then $\mathbf{A}^\dagger = \mathbf{A}$.
9. If $\mathbf{A} = \mathbf{BC}$, $\mathbf{B}$ is of full column rank, and $\mathbf{C}$ is of full row rank then
$$\mathbf{A}^\dagger = \mathbf{C}^\dagger\mathbf{B}^\dagger = \mathbf{C}^H(\mathbf{C}\mathbf{C}^H)^{-1}(\mathbf{B}^H\mathbf{B})^{-1}\mathbf{B}^H.$$
10. $(\mathbf{A}\mathbf{A}^H)^\dagger = (\mathbf{A}^\dagger)^H\mathbf{A}^\dagger$ and $(\mathbf{A}\mathbf{A}^H)^\dagger(\mathbf{A}\mathbf{A}^H) = \mathbf{A}\mathbf{A}^\dagger$.
11. If the matrices $\mathbf{A}_i$ are mutually orthogonal, i.e., $\mathbf{A}_i^H\mathbf{A}_j = \mathbf{O}$, $i \neq j$, then $(\mathbf{A}_1 + \cdots + \mathbf{A}_m)^\dagger = \mathbf{A}_1^\dagger + \cdots + \mathbf{A}_m^\dagger$.
12. Regarding the ranks of generalized inverse matrices, one has $\mathrm{rank}(\mathbf{A}^\dagger) = \mathrm{rank}(\mathbf{A}) = \mathrm{rank}(\mathbf{A}^H) = \mathrm{rank}(\mathbf{A}^\dagger\mathbf{A}) = \mathrm{rank}(\mathbf{A}\mathbf{A}^\dagger) = \mathrm{rank}(\mathbf{A}\mathbf{A}^\dagger\mathbf{A}) = \mathrm{rank}(\mathbf{A}^\dagger\mathbf{A}\mathbf{A}^\dagger)$.
13. The Moore–Penrose inverse of any matrix $\mathbf{A}_{m\times n}$ can be determined by $\mathbf{A}^\dagger = (\mathbf{A}^H\mathbf{A})^\dagger\mathbf{A}^H$ or $\mathbf{A}^\dagger = \mathbf{A}^H(\mathbf{A}\mathbf{A}^H)^\dagger$. In particular, the inverse matrices of the full rank matrices are as follows:
    - If $\mathbf{A}$ is of full column rank then $\mathbf{A}^\dagger = (\mathbf{A}^H\mathbf{A})^{-1}\mathbf{A}^H$, i.e., the Moore–Penrose inverse of a matrix $\mathbf{A}$ with full column rank reduces to the left pseudo-inverse matrix of $\mathbf{A}$.
    - If $\mathbf{A}$ is of full row rank then $\mathbf{A}^\dagger = \mathbf{A}^H(\mathbf{A}\mathbf{A}^H)^{-1}$, i.e., the Moore–Penrose inverse of a matrix $\mathbf{A}$ with full row rank reduces to the right pseudo-inverse matrix of $\mathbf{A}$.
    - If $\mathbf{A}$ is nonsingular then $\mathbf{A}^\dagger = \mathbf{A}^{-1}$, that is, the Moore–Penrose inverse of a nonsingular matrix $\mathbf{A}$ reduces to the inverse $\mathbf{A}^{-1}$ of $\mathbf{A}$.
14. For a matrix $\mathbf{A}_{m\times n}$, even if $\mathbf{A}\mathbf{A}^\dagger \neq \mathbf{I}_m, \mathbf{A}^\dagger\mathbf{A} \neq \mathbf{I}_n, \mathbf{A}^H(\mathbf{A}^H)^\dagger \neq \mathbf{I}_n$ and $(\mathbf{A}^H)^\dagger\mathbf{A}^H \neq \mathbf{I}_m$, the following results are true:
    - $\mathbf{A}^\dagger\mathbf{A}\mathbf{A}^H = \mathbf{A}^H$ and $\mathbf{A}^H\mathbf{A}\mathbf{A}^\dagger = \mathbf{A}^H$;
    - $\mathbf{A}\mathbf{A}^\dagger(\mathbf{A}^\dagger)^H = (\mathbf{A}^\dagger)^H$ and $(\mathbf{A}^H)^\dagger\mathbf{A}^\dagger\mathbf{A} = (\mathbf{A}^\dagger)^H$;
    - $(\mathbf{A}^H)^\dagger\mathbf{A}\mathbf{A} = \mathbf{A}$ and $\mathbf{A}\mathbf{A}^H(\mathbf{A}^H)^\dagger = \mathbf{A}$;
    - $\mathbf{A}^H(\mathbf{A}^\dagger)^H\mathbf{A}^\dagger = \mathbf{A}^\dagger$ and $\mathbf{A}^\dagger(\mathbf{A}^\dagger)^H\mathbf{A}^H = \mathbf{A}^\dagger$.

1.8.2 Computation of Moore–Penrose Inverse Matrix

Consider an $m \times n$ matrix $\mathbf{A}$ with rank r, where $r \leq \min(m, n)$. The following are four methods for computing the Moore–Penrose inverse matrix $\mathbf{A}^\dagger$.

1. Equation-Solving Method [384]

Step 1 Solve the matrix equations $\mathbf{A}\mathbf{A}^H\mathbf{X}^H = \mathbf{A}$ and $\mathbf{A}^H\mathbf{A}\mathbf{Y} = \mathbf{A}^H$ to yield the solutions $\mathbf{X}^H$ and $\mathbf{Y}$, respectively.

Step 2 Compute the generalized inverse matrix $\mathbf{A}^\dagger = \mathbf{X}\mathbf{A}\mathbf{Y}$.

The following are two equation-solving algorithms for Moore–Penrose inverse matrices.

Algorithm 1.3 Equation-solving method 1 [187]

1. Compute the matrix $\mathbf{B} = \mathbf{A}\mathbf{A}^H$.
2. Solve the matrix equation $\mathbf{B}^2\mathbf{X}^H = \mathbf{B}$ to get the matrix $\mathbf{X}^H$.
3. Calculate the Moore–Penrose inverse matrix $\mathbf{B}^\dagger = (\mathbf{A}\mathbf{A}^H)^\dagger = \mathbf{X}\mathbf{B}\mathbf{X}^H$.
4. Compute the Moore–Penrose inverse matrix $\mathbf{A}^\dagger = \mathbf{A}^H\mathbf{B}^\dagger$.

Algorithm 1.4 Equation-solving method 2 [187]

1. Calculate the matrix $\mathbf{B} = \mathbf{A}^H\mathbf{A}$.
2. Solve the matrix equation $\mathbf{B}^2\mathbf{X}^H = \mathbf{X}$ to obtain the matrix $\mathbf{X}^H$.
3. Compute the Moore–Penrose inverse matrix $\mathbf{B}^\dagger = (\mathbf{A}^H\mathbf{A})^\dagger = \mathbf{X}\mathbf{B}\mathbf{X}^H$.
4. Calculate the Moore–Penrose inverse matrix $\mathbf{A}^\dagger = \mathbf{B}^\dagger\mathbf{A}^H$.

Algorithm 1.3 computes $\mathbf{A}^\dagger = \mathbf{A}^H(\mathbf{A}\mathbf{A}^H)^\dagger$ and Algorithm 1.4 computes $\mathbf{A}^\dagger = (\mathbf{A}^H\mathbf{A})^\dagger\mathbf{A}^H$. If the number of columns of the matrix $\mathbf{A}_{m\times n}$ is larger than the number of its rows then the dimension of the matrix product $\mathbf{A}\mathbf{A}^H$ is less than the dimension of $\mathbf{A}^H\mathbf{A}$. In this case Algorithm 1.3 needs less computation. If, on the contrary, the number of rows is larger than the number of columns then one should select Algorithm 1.4.

2. Full-Rank Decomposition Method

Definition 1.37 Let a rank-deficient matrix $\mathbf{A}_{m\times n}$ have rank $r < \min\{m, n\}$. If $\mathbf{A} = \mathbf{F}\mathbf{G}$, where $\mathbf{F}_{m\times r}$ is of full column rank and $\mathbf{G}_{r\times n}$ is of full row rank, then $\mathbf{A} = \mathbf{F}\mathbf{G}$ is called the *full-rank decomposition* of the matrix $\mathbf{A}$.

Lemma 1.6 [433] *A matrix $\mathbf{A} \in \mathbb{C}^{m\times n}$ with $\operatorname{rank}(\mathbf{A}) = r$ can be decomposed into $\mathbf{A} = \mathbf{F}\mathbf{G}$, where $\mathbf{F} \in \mathbb{C}^{m\times r}$ and $\mathbf{G} \in \mathbb{C}^{r\times n}$ have full-column rank and full-row rank, respectively.*

If $\mathbf{A} = \mathbf{FG}$ is a full-rank decomposition of the $m \times n$ matrix $\mathbf{A}$, then $\mathbf{A}^\dagger = \mathbf{G}^\dagger\mathbf{F}^\dagger = \mathbf{G}^H(\mathbf{G}\mathbf{G}^H)^{-1}(\mathbf{F}^H\mathbf{F})^{-1}\mathbf{F}^H$ meets the four Moore–Penrose conditions, so the $n \times m$ matrix $\mathbf{A}^\dagger$ must be the Moore–Penrose inverse matrix of $\mathbf{A}$.

Elementary row operations easily implement the full-rank decomposition of a rank-deficient matrix $\mathbf{A} \in \mathbb{C}^{m\times n}$, as shown in Algorithm 1.5.

Algorithm 1.5 Full-rank decomposition

---

1. Use elementary row operations to get the reduced row-echelon form of $\mathbf{A}$.
2. Use the pivot columns of $\mathbf{A}$ as the column vectors of a matrix $\mathbf{F}$.
3. Use the nonzero rows in the reduced row-echelon to get the row vectors of a matrix $\mathbf{G}$.
4. The full-rank decomposition is given by $\mathbf{A} = \mathbf{FG}$.

Example 1.12 In Example 1.1, via elementary row operations on the matrix

$$\mathbf{A} = \begin{bmatrix} -3 & 6 & -1 & 1 & -7 \\ 1 & -2 & 2 & 3 & -1 \\ 2 & -4 & 5 & 8 & -4 \end{bmatrix},$$

we obtained its reduced row-echelon form,

$$\begin{bmatrix} 1 & -2 & 0 & -1 & 3 \\ 0 & 0 & 1 & 2 & -2 \\ 0 & 0 & 0 & 0 & 0 \end{bmatrix}.$$

The pivot columns of $\mathbf{A}$ are the first and third columns, so we have

$$\mathbf{F} = \begin{bmatrix} -3 & -1 \\ 1 & 2 \\ 2 & 5 \end{bmatrix}, \quad \mathbf{G} = \begin{bmatrix} 1 & -2 & 0 & -1 & 3 \\ 0 & 0 & 1 & 2 & -2 \end{bmatrix}.$$

Hence we get the full-rank decomposition $\mathbf{A} = \mathbf{FG}$:

$$\begin{bmatrix} -3 & 6 & -1 & 1 & -7 \\ 1 & -2 & 2 & 3 & -1 \\ 2 & -4 & 5 & 8 & -4 \end{bmatrix} = \begin{bmatrix} -3 & -1 \\ 1 & 2 \\ 2 & 5 \end{bmatrix} \begin{bmatrix} 1 & -2 & 0 & -1 & 3 \\ 0 & 0 & 1 & 2 & -2 \end{bmatrix}.$$

3. *Recursive Methods*

Block the matrix $\mathbf{A}_{m\times n}$ into $\mathbf{A}_k = [\mathbf{A}_{k-1}, \mathbf{a}_k]$, where $\mathbf{A}_{k-1}$ consists of the first $k-1$ columns of $\mathbf{A}$ and $\mathbf{a}_k$ is the kth column of $\mathbf{A}$. Then, the Moore–Penrose inverse $\mathbf{A}_k^\dagger$ of the block matrix $\mathbf{A}_k$ can be recursively calculated from $\mathbf{A}_{k-1}^\dagger$. When $k = n$, we get the Moore–Penrose inverse matrix $\mathbf{A}^\dagger$. Such a recursive algorithm was presented by Greville in 1960 [188].

Algorithm 1.6 below is available for all matrices. However, when the number of

Algorithm 1.6 Column recursive method

initialize: $\mathbf{A}_1^\dagger = \mathbf{a}_1^\dagger = (\mathbf{a}_1^H\mathbf{a}_1)^{-1}\mathbf{a}_1^H$. Put $k = 2$.

repeat

1. Compute $\mathbf{d}_k = \mathbf{A}_{k-1}^\dagger \mathbf{a}_k$.
2. Compute $\mathbf{b}_k = \begin{cases} (1 + \mathbf{d}_k^H\mathbf{d}_k)^{-1}\mathbf{d}_k^H\mathbf{A}_{k-1}^\dagger, & \text{if } \mathbf{d}_k - \mathbf{A}_{k-1}\mathbf{d}_k = \mathbf{0}; \\ (\mathbf{a}_k - \mathbf{A}_{k-1}\mathbf{d}_k)^\dagger, & \text{if } \mathbf{a}_k - \mathbf{A}_{k-1}\mathbf{d}_k \neq \mathbf{0}. \end{cases}$
3. Compute $\mathbf{A}_k^\dagger = \begin{bmatrix} \mathbf{A}_{k-1}^\dagger - \mathbf{d}_k\mathbf{b}_k \\ \mathbf{b}_k \end{bmatrix}$.
4. **exit if** $k = n$.

return $k \leftarrow k+1$.

output: $\mathbf{A}^\dagger$.

rows of $\mathbf{A}$ is small, in order to reduce the number of recursions it is recommended that one should first use the column recursive algorithm 1.6 for finding the Moore–Penrose matrix $(\mathbf{A}^H)^\dagger = \mathbf{A}^{H\dagger}$, and then compute $\mathbf{A}^\dagger = (\mathbf{A}^{H\dagger})^H$.

4. Trace Method

The trace method for computing the Moore–Penrose inverse matrix is shown in Algorithm 1.7.

Algorithm 1.7 Trace method [397]

input: $\mathbf{A} \in \mathbb{R}^{m\times n}$ and $\text{rank}(\mathbf{A}) = r$.

initialize: $\mathbf{B} = \mathbf{A}^T\mathbf{A}$, and set $\mathbf{C}_1 = \mathbf{I}$, $k = 1$.

repeat

1. Compute $\mathbf{C}_{k+1} = k^{-1}\text{tr}(\mathbf{C}_k\mathbf{B})\mathbf{I} - \mathbf{C}_k\mathbf{B}$.
2. **exit if** $k = r - 1$.

return $k \leftarrow k+1$.

output: $\mathbf{A}^\dagger = \dfrac{r}{\text{tr}(\mathbf{C}_k\mathbf{B})}\mathbf{C}_k\mathbf{A}^T$.

1.9 Direct Sum and Hadamard Product

1.9.1 Direct Sum of Matrices

Definition 1.38 [186] The *direct sum* of an $m \times m$ matrix $\mathbf{A}$ and an $n \times n$ matrix $\mathbf{B}$, denoted $\mathbf{A} \oplus \mathbf{B}$, is an $(m+n) \times (m+n)$ matrix and is defined as follows:

$$\mathbf{A} \oplus \mathbf{B} = \begin{bmatrix} \mathbf{A} & \mathbf{O}_{m\times n} \\ \mathbf{O}_{n\times m} & \mathbf{B} \end{bmatrix}. \tag{1.9.1}$$

From the above definition, it is easily shown that the direct sum of matrices has the following properties [214], [390]:

1. If c is a constant then $c(\mathbf{A} \oplus \mathbf{B}) = c\mathbf{A} \oplus c\mathbf{B}$.
2. The direct sum does not satisfy exchangeability, i.e., $\mathbf{A} \oplus \mathbf{B} \neq \mathbf{B} \oplus \mathbf{A}$ unless $\mathbf{A} = \mathbf{B}$.
3. If $\mathbf{A}, \mathbf{B}$ are two $m \times m$ matrices and $\mathbf{C}$ and $\mathbf{D}$ are two $n \times n$ matrices then
$$(\mathbf{A} \pm \mathbf{B}) \oplus (\mathbf{C} \pm \mathbf{D}) = (\mathbf{A} \oplus \mathbf{C}) \pm (\mathbf{B} \oplus \mathbf{D}),$$
$$(\mathbf{A} \oplus \mathbf{C})(\mathbf{B} \oplus \mathbf{D}) = \mathbf{AB} \oplus \mathbf{CD}.$$
4. If $\mathbf{A}, \mathbf{B}, \mathbf{C}$ are $m \times m$, $n \times n$, $p \times p$ matrices, respectively, then
$$\mathbf{A} \oplus (\mathbf{B} \oplus \mathbf{C}) = (\mathbf{A} \oplus \mathbf{B}) \oplus \mathbf{C} = \mathbf{A} \oplus \mathbf{B} \oplus \mathbf{C}.$$
5. If $\mathbf{A}_{m \times m}$ and $\mathbf{B}_{n \times n}$ are respectively orthogonal matrices then $\mathbf{A} \oplus \mathbf{B}$ is an $(m+n) \times (m+n)$ orthogonal matrix.
6. The complex conjugate, transpose, complex conjugate transpose and inverse matrices of the direct sum of two matrices are given by
$$\begin{aligned}(\mathbf{A} \oplus \mathbf{B})^* &= \mathbf{A}^* \oplus \mathbf{B}^*,\\ (\mathbf{A} \oplus \mathbf{B})^T &= \mathbf{A}^T \oplus \mathbf{B}^T,\\ (\mathbf{A} \oplus \mathbf{B})^H &= \mathbf{A}^H \oplus \mathbf{B}^H,\\ (\mathbf{A} \oplus \mathbf{B})^{-1} &= \mathbf{A}^{-1} \oplus \mathbf{B}^{-1} \quad (\text{if } \mathbf{A}^{-1} \text{ and } \mathbf{B}^{-1} \text{ exist}).\end{aligned}$$
7. The trace, rank and determinant of the direct sum of N matrices are as follows:
$$\begin{aligned}\mathrm{tr}\left(\bigoplus_{i=0}^{N-1} \mathbf{A}_i\right) &= \sum_{i=0}^{N-1} \mathrm{tr}(\mathbf{A}_i),\\ \mathrm{rank}\left(\bigoplus_{i=0}^{N-1} \mathbf{A}_i\right) &= \sum_{i=0}^{N-1} \mathrm{rank}(\mathbf{A}_i),\\ \det\left(\bigoplus_{i=0}^{N-1} \mathbf{A}_i\right) &= \prod_{i=0}^{N-1} \det(\mathbf{A}_i).\end{aligned}$$

1.9.2 Hadamard Product

Definition 1.39 The *Hadamard product* of two $m \times n$ matrices $\mathbf{A} = [a_{ij}]$ and $\mathbf{B} = [b_{ij}]$ is denoted $\mathbf{A} * \mathbf{B}$ and is also an $m \times n$ matrix, each entry of which is defined as the product of the corresponding entries of the two matrices:
$$[\mathbf{A} * \mathbf{B}]_{ij} = a_{ij} b_{ij}. \tag{1.9.2}$$
That is, the Hadamard product is a mapping $\mathbb{R}^{m \times n} \times \mathbb{R}^{m \times n} \mapsto \mathbb{R}^{m \times n}$.

The Hadamard product is also known as the *Schur product* or the *elementwise product.*

The following theorem describes the positive definiteness of the Hadamard product and is usually known as the *Hadamard product theorem* [214].

THEOREM 1.9 *If two $m \times m$ matrices $\mathbf{A}$ and $\mathbf{B}$ are positive definite (positive semi-definite) then their Hadamard product $\mathbf{A} * \mathbf{B}$ is positive definite (positive semi-definite) as well.*

COROLLARY 1.1 (Fejer theorem) [214] *An $m \times m$ matrix $\mathbf{A} = [a_{ij}]$ is positive semi-definite if and only if*

$$\sum_{i=1}^{m}\sum_{j=1}^{m} a_{ij}b_{ij} \geq 0$$

holds for all $m \times m$ positive semi-definite matrices $\mathbf{B} = [b_{ij}]$.

The following theorems describe the relationship between the Hadamard product and the matrix trace.

THEOREM 1.10 [311, p. 46] *Let $\mathbf{A}, \mathbf{B}, \mathbf{C}$ be $m \times n$ matrices, $\mathbf{1} = [1, \ldots, 1]^T$ be an $n \times 1$ summing vector and $\mathbf{D} = \mathbf{Diag}(d_1, \ldots, d_m)$, where $d_i = \sum_{j=1}^{n} a_{ij}$; then*

$$\operatorname{tr}\left(\mathbf{A}^T(\mathbf{B} * \mathbf{C})\right) = \operatorname{tr}\left((\mathbf{A}^T * \mathbf{B}^T)\mathbf{C}\right), \tag{1.9.3}$$

$$\mathbf{1}^T\mathbf{A}^T(\mathbf{B} * \mathbf{C})\mathbf{1} = \operatorname{tr}(\mathbf{B}^T\mathbf{D}\mathbf{C}). \tag{1.9.4}$$

THEOREM 1.11 [311, p. 46] *Let $\mathbf{A}, \mathbf{B}$ be two $n \times n$ positive definite square matrices, and $\mathbf{1} = [1, \ldots, 1]^T$ be an $n \times 1$ summing vector. Suppose that $\mathbf{M}$ is an $n \times n$ diagonal matrix, i.e., $\mathbf{M} = \mathbf{Diag}(\mu_1, \ldots, \mu_n)$, while $\mathbf{m} = \mathbf{M}\mathbf{1}$ is an $n \times 1$ vector. Then one has*

$$\operatorname{tr}(\mathbf{A}\mathbf{M}\mathbf{B}^T\mathbf{M}) = \mathbf{m}^T(\mathbf{A} * \mathbf{B})\mathbf{m}, \tag{1.9.5}$$

$$\operatorname{tr}(\mathbf{A}\mathbf{B}^T) = \mathbf{1}^T(\mathbf{A} * \mathbf{B})\mathbf{1}, \tag{1.9.6}$$

$$\mathbf{M}\mathbf{A} * \mathbf{B}^T\mathbf{M} = \mathbf{M}(\mathbf{A} * \mathbf{B}^T)\mathbf{M}. \tag{1.9.7}$$

From the above definition, it is known that Hadamard products obey the exchange law, the associative law and the distributive law of the addition:

$$\mathbf{A} * \mathbf{B} = \mathbf{B} * \mathbf{A}, \tag{1.9.8}$$

$$\mathbf{A} * (\mathbf{B} * \mathbf{C}) = (\mathbf{A} * \mathbf{B}) * \mathbf{C}, \tag{1.9.9}$$

$$\mathbf{A} * (\mathbf{B} \pm \mathbf{C}) = \mathbf{A} * \mathbf{B} \pm \mathbf{A} * \mathbf{C}. \tag{1.9.10}$$

The properties of Hadamard products are summarized below [311].

1. If $\mathbf{A}, \mathbf{B}$ are $m \times n$ matrices then

$$(\mathbf{A} * \mathbf{B})^T = \mathbf{A}^T * \mathbf{B}^T, \quad (\mathbf{A} * \mathbf{B})^H = \mathbf{A}^H * \mathbf{B}^H, \quad (\mathbf{A} * \mathbf{B})^* = \mathbf{A}^* * \mathbf{B}^*.$$

2. The Hadamard product of a matrix $\mathbf{A}_{m\times n}$ and a zero matrix $\mathbf{O}_{m\times n}$ is given by $\mathbf{A} * \mathbf{O}_{m\times n} = \mathbf{O}_{m\times n} * \mathbf{A} = \mathbf{O}_{m\times n}$.
3. If c is a constant then $c(\mathbf{A} * \mathbf{B}) = (c\mathbf{A}) * \mathbf{B} = \mathbf{A} * (c\mathbf{B})$.
4. The Hadamard product of two positive definite (positive semi-definite) matrices $\mathbf{A}, \mathbf{B}$ is positive definite (positive semi-definite) as well.
5. The Hadamard product of the matrix $\mathbf{A}_{m\times m} = [a_{ij}]$ and the identity matrix $\mathbf{I}_m$ is an $m \times m$ diagonal matrix, i.e.,

$$\mathbf{A} * \mathbf{I}_m = \mathbf{I}_m * \mathbf{A} = \mathbf{Diag}(\mathbf{A}) = \mathbf{Diag}(a_{11}, \ldots, a_{mm}).$$

6. If $\mathbf{A}, \mathbf{B}, \mathbf{D}$ are three $m \times m$ matrices and $\mathbf{D}$ is a diagonal matrix then

$$(\mathbf{DA}) * (\mathbf{BD}) = \mathbf{D}(\mathbf{A} * \mathbf{B})\mathbf{D}.$$

7. If $\mathbf{A}, \mathbf{C}$ are two $m \times m$ matrices and $\mathbf{B}, \mathbf{D}$ are two $n \times n$ matrices then

$$(\mathbf{A} \oplus \mathbf{B}) * (\mathbf{C} \oplus \mathbf{D}) = (\mathbf{A} * \mathbf{C}) \oplus (\mathbf{B} * \mathbf{D}).$$

8. If $\mathbf{A}, \mathbf{B}, \mathbf{C}, \mathbf{D}$ are all $m \times n$ matrices then

$$(\mathbf{A} + \mathbf{B}) * (\mathbf{C} + \mathbf{D}) = \mathbf{A} * \mathbf{C} + \mathbf{A} * \mathbf{D} + \mathbf{B} * \mathbf{C} + \mathbf{B} * \mathbf{D}.$$

9. If $\mathbf{A}, \mathbf{B}, \mathbf{C}$ are $m \times n$ matrices then

$$\mathrm{tr}\left(\mathbf{A}^T(\mathbf{B} * \mathbf{C})\right) = \mathrm{tr}\left((\mathbf{A}^T * \mathbf{B}^T)\mathbf{C}\right).$$

The Hadamard product of $n \times n$ matrices $\mathbf{A}$ and $\mathbf{B}$ obeys the following inequality:

(1) *Oppenheim inequality* [23, p. 144] If $\mathbf{A}$ and $\mathbf{B}$ are positive semi-definite, then

$$|\mathbf{A} * \mathbf{B}| \geq a_{11} \cdots a_{nn} |\mathbf{B}|. \tag{1.9.11}$$

(2) If $\mathbf{A}$ and $\mathbf{B}$ are positive semi-definite, then [327]

$$|\mathbf{A} * \mathbf{B}| \geq |\mathbf{AB}|. \tag{1.9.12}$$

(3) *Eigenvalue inequality* [23, p.144] If $\mathbf{A}$ and $\mathbf{B}$ are positive semi-definite and $\lambda_1, \ldots, \lambda_n$ are the eigenvalues of the Hadamard product $\mathbf{A} * \mathbf{B}$, while $\hat{\lambda}_1, \ldots, \hat{\lambda}_n$ are the eigenvalues of the matrix product $\mathbf{AB}$, then

$$\prod_{i=k}^{n} \lambda_i \geq \prod_{i=k}^{n} \hat{\lambda}_i, \quad k = 1, \ldots, n. \tag{1.9.13}$$

(4) *Rank inequality* [327]

$$\mathrm{rank}(\mathbf{A} * \mathbf{B}) \leq \mathrm{rank}(\mathbf{A})\,\mathrm{rank}(\mathbf{B}). \tag{1.9.14}$$

The Hadamard products of matrices are useful in lossy compression algorithms (such as JPEG).

1.10 Kronecker Products and Khatri–Rao Product

The Hadamard product described in the previous section is a special product of two matrices. In this section, we discuss two other special products of two matrices: the Kronecker products and the Khatri–Rao product.

1.10.1 Kronecker Products

Kronecker products are divided into right and left Kronecker products.

Definition 1.40 (*Right Kronecker product*) [32] Given an $m \times n$ matrix $\mathbf{A} = [\mathbf{a}_1, \ldots, \mathbf{a}_n]$ and another $p \times q$ matrix $\mathbf{B}$, their right Kronecker product $\mathbf{A} \otimes \mathbf{B}$ is an $mp \times nq$ matrix defined by

$$\mathbf{A} \otimes \mathbf{B} = [a_{ij}\mathbf{B}]_{i=1,j=1}^{m,n} = \begin{bmatrix} a_{11}\mathbf{B} & a_{12}\mathbf{B} & \cdots & a_{1n}\mathbf{B} \\ a_{21}\mathbf{B} & a_{22}\mathbf{B} & \cdots & a_{2n}\mathbf{B} \\ \vdots & \vdots & \ddots & \vdots \\ a_{m1}\mathbf{B} & a_{m2}\mathbf{B} & \cdots & a_{mn}\mathbf{B} \end{bmatrix}. \tag{1.10.1}$$

Definition 1.41 (*Left Kronecker product*) [186], [413] For an $m \times n$ matrix $\mathbf{A}$ and a $p \times q$ matrix $\mathbf{B} = [\mathbf{b}_1, \ldots, \mathbf{b}_q]$, their left Kronecker product $\mathbf{A} \otimes \mathbf{B}$ is an $mp \times nq$ matrix defined by

$$[\mathbf{A} \otimes \mathbf{B}]_{\text{left}} = [\mathbf{A}b_{ij}]_{i=1,j=1}^{p,q} = \begin{bmatrix} \mathbf{A}b_{11} & \mathbf{A}b_{12} & \cdots & \mathbf{A}b_{1q} \\ \mathbf{A}b_{21} & \mathbf{A}b_{22} & \cdots & \mathbf{A}b_{2q} \\ \vdots & \vdots & \ddots & \vdots \\ \mathbf{A}b_{p1} & \mathbf{A}b_{p2} & \cdots & \mathbf{A}b_{pq} \end{bmatrix}. \tag{1.10.2}$$

Clearly, the left or right Kronecker product is a mapping $\mathbb{R}^{m\times n} \times \mathbb{R}^{p\times q} \mapsto \mathbb{R}^{mp\times nq}$.

It is easily seen that if we adopt the right Kronecker product form then the left Kronecker product can be written as $[\mathbf{A} \otimes \mathbf{B}]_{\text{left}} = \mathbf{B} \otimes \mathbf{A}$. Since the right Kronecker product form is the one generally adopted, this book uses the right Kronecker product hereafter unless otherwise stated.

In particular, when $n = 1$ and $q = 1$, the Kronecker product of two matrices reduces to the Kronecker product of two column vectors $\mathbf{a} \in \mathbb{R}^m$ and $\mathbf{b} \in \mathbb{R}^p$:

$$\mathbf{a} \otimes \mathbf{b} = [a_i\mathbf{b}]_{i=1}^{m} = \begin{bmatrix} a_1\mathbf{b} \\ \vdots \\ a_m\mathbf{b} \end{bmatrix}. \tag{1.10.3}$$

The result is an $mp \times 1$ vector. Evidently, the outer product of two vectors $\mathbf{x} \circ \mathbf{y} = \mathbf{x}\mathbf{y}^T$ can be also represented using the Kronecker product as $\mathbf{x} \circ \mathbf{y} = \mathbf{x} \otimes \mathbf{y}^T$.

The Kronecker product is also known as the *direct product* or *tensor product* [307].

Summarizing results from [32], [60] and other literature, Kronecker products have the following properties.

1. The Kronecker product of any matrix and a zero matrix is equal to the zero matrix, i.e., $\mathbf{A} \otimes \mathbf{O} = \mathbf{O} \otimes \mathbf{A} = \mathbf{O}$.
2. If α and β are constants then $\alpha\mathbf{A} \otimes \beta\mathbf{B} = \alpha\beta(\mathbf{A} \otimes \mathbf{B})$.
3. The Kronecker product of an $m \times m$ identity matrix and an $n \times n$ identity matrix is equal to an $mn \times mn$ identity matrix, i.e., $\mathbf{I}_m \otimes \mathbf{I}_n = \mathbf{I}_{mn}$.
4. For matrices $\mathbf{A}_{m\times n}, \mathbf{B}_{n\times k}, \mathbf{C}_{l\times p}, \mathbf{D}_{p\times q}$, we have
$$(\mathbf{AB}) \otimes (\mathbf{CD}) = (\mathbf{A} \otimes \mathbf{C})(\mathbf{B} \otimes \mathbf{D}). \tag{1.10.4}$$
5. For matrices $\mathbf{A}_{m\times n}, \mathbf{B}_{p\times q}, \mathbf{C}_{p\times q}$, we have
$$\mathbf{A} \otimes (\mathbf{B} \pm \mathbf{C}) = \mathbf{A} \otimes \mathbf{B} \pm \mathbf{A} \otimes \mathbf{C}, \tag{1.10.5}$$
$$(\mathbf{B} \pm \mathbf{C}) \otimes \mathbf{A} = \mathbf{B} \otimes \mathbf{A} \pm \mathbf{C} \otimes \mathbf{A}. \tag{1.10.6}$$
6. The inverse and generalized inverse matrix of Kronecker products satisfy
$$(\mathbf{A} \otimes \mathbf{B})^{-1} = \mathbf{A}^{-1} \otimes \mathbf{B}^{-1}, \quad (\mathbf{A} \otimes \mathbf{B})^{\dagger} = \mathbf{A}^{\dagger} \otimes \mathbf{B}^{\dagger}. \tag{1.10.7}$$
7. The transpose and the complex conjugate transpose of Kronecker products are given by
$$(\mathbf{A} \otimes \mathbf{B})^T = \mathbf{A}^T \otimes \mathbf{B}^T, \quad (\mathbf{A} \otimes \mathbf{B})^H = \mathbf{A}^H \otimes \mathbf{B}^H. \tag{1.10.8}$$
8. The rank of the Kronecker product is
$$\text{rank}(\mathbf{A} \otimes \mathbf{B}) = \text{rank}(\mathbf{A})\text{rank}(\mathbf{B}). \tag{1.10.9}$$
9. The determinant of the Kronecker product
$$\det(\mathbf{A}_{n\times n} \otimes \mathbf{B}_{m\times m}) = (\det \mathbf{A})^m (\det \mathbf{B})^n. \tag{1.10.10}$$
10. The trace of the Kronecker product is
$$\text{tr}(\mathbf{A} \otimes \mathbf{B}) = \text{tr}(\mathbf{A})\text{tr}(\mathbf{B}). \tag{1.10.11}$$
11. For matrices $\mathbf{A}_{m\times n}, \mathbf{B}_{m\times n}, \mathbf{C}_{p\times q}, \mathbf{D}_{p\times q}$, we have
$$(\mathbf{A} + \mathbf{B}) \otimes (\mathbf{C} + \mathbf{D}) = \mathbf{A} \otimes \mathbf{C} + \mathbf{A} \otimes \mathbf{D} + \mathbf{B} \otimes \mathbf{C} + \mathbf{B} \otimes \mathbf{D}. \tag{1.10.12}$$
12. For matrices $\mathbf{A}_{m\times n}, \mathbf{B}_{p\times q}, \mathbf{C}_{k\times l}$, it is true that
$$(\mathbf{A} \otimes \mathbf{B}) \otimes \mathbf{C} = \mathbf{A} \otimes (\mathbf{B} \otimes \mathbf{C}). \tag{1.10.13}$$
13. For matrices $\mathbf{A}_{m\times n}, \mathbf{B}_{k\times l}, \mathbf{C}_{p\times q}, \mathbf{D}_{r\times s}$, there is the following relationship:
$$(\mathbf{A} \otimes \mathbf{B}) \otimes (\mathbf{C} \otimes \mathbf{D}) = \mathbf{A} \otimes \mathbf{B} \otimes \mathbf{C} \otimes \mathbf{D}. \tag{1.10.14}$$
14. For matrices $\mathbf{A}_{m\times n}, \mathbf{B}_{p\times q}, \mathbf{C}_{n\times r}, \mathbf{D}_{q\times s}, \mathbf{E}_{r\times k}, \mathbf{F}_{s\times l}$, we have
$$(\mathbf{A} \otimes \mathbf{B})(\mathbf{C} \otimes \mathbf{D})(\mathbf{E} \otimes \mathbf{F}) = (\mathbf{ACE}) \otimes (\mathbf{BDF}). \tag{1.10.15}$$

15. The following is a special example of Equation (1.10.15) (see [32], [186]):

$$\mathbf{A} \otimes \mathbf{D} = (\mathbf{A}\mathbf{I}_p) \otimes (\mathbf{I}_q\mathbf{D}) = (\mathbf{A} \otimes \mathbf{I}_q)(\mathbf{I}_p \otimes \mathbf{D}), \tag{1.10.16}$$

where $\mathbf{I}_p \otimes \mathbf{D}$ is a block diagonal matrix (for the right Kronecker product) or a sparse matrix (for the left Kronecker product), while $\mathbf{A} \otimes \mathbf{I}_q$ is a sparse matrix (for the right Kronecker product) or a block diagonal matrix (for the left Kronecker product).

16. For matrices $\mathbf{A}_{m\times n}, \mathbf{B}_{p\times q}$, we have $\exp(\mathbf{A} \otimes \mathbf{B}) = \exp(\mathbf{A}) \otimes \exp(\mathbf{B})$.
17. Let $\mathbf{A} \in \mathbb{C}^{m\times n}$ and $\mathbf{B} \in \mathbb{C}^{p\times q}$ then [311, p. 47]

$$\mathbf{K}_{pm}(\mathbf{A} \otimes \mathbf{B}) = (\mathbf{B} \otimes \mathbf{A})\mathbf{K}_{qn}, \tag{1.10.17}$$

$$\mathbf{K}_{pm}(\mathbf{A} \otimes \mathbf{B})\mathbf{K}_{nq} = \mathbf{B} \otimes \mathbf{A}, \tag{1.10.18}$$

$$\mathbf{K}_{pm}(\mathbf{A} \otimes \mathbf{B}) = \mathbf{B} \otimes \mathbf{A}, \tag{1.10.19}$$

$$\mathbf{K}_{mp}(\mathbf{B} \otimes \mathbf{A}) = \mathbf{A} \otimes \mathbf{B}, \tag{1.10.20}$$

where $\mathbf{K}$ is a commutation matrix (see Subsection 1.11.1).

1.10.2 Generalized Kronecker Products

The generalized Kronecker product is the Kronecker product of a matrix group consisting of more matrices and another matrix.

Definition 1.42 [390] Given N matrices $\mathbf{A}_i \in \mathbb{C}^{m\times n}, i = 1, \ldots, N$ that constitute a matrix group $\{\mathbf{A}\}_N$, the Kronecker product of the matrix group $\{\mathbf{A}\}_N$ and an $N \times l$ matrix $\mathbf{B}$, known as the *generalized Kronecker product*, is defined by

$$\{\mathbf{A}\}_N \otimes \mathbf{B} = \begin{bmatrix} \mathbf{A}_1 \otimes \mathbf{b}_1 \\ \vdots \\ \mathbf{A}_N \otimes \mathbf{b}_N \end{bmatrix}, \tag{1.10.21}$$

where $\mathbf{b}_i$ is the ith row vector of the matrix $\mathbf{B}$.

Example 1.13 Put

$$\{\mathbf{A}\}_2 = \left\{ \begin{bmatrix} 1 & 1 \\ 2 & -1 \end{bmatrix} \atop \begin{bmatrix} 2 & -\mathrm{j} \\ 1 & \mathrm{j} \end{bmatrix} \right\}, \quad \mathbf{B} = \begin{bmatrix} 1 & 2 \\ 1 & -1 \end{bmatrix}.$$

Then the generalized Kronecker product $\{\mathbf{A}\}_2 \otimes \mathbf{B}$ is given by

$$\{\mathbf{A}\}_2 \otimes \mathbf{B} = \begin{bmatrix} \begin{bmatrix} 1 & 1 \\ 2 & -1 \end{bmatrix} \otimes [1,\ 2] \\ \begin{bmatrix} 2 & -\mathrm{j} \\ 1 & \mathrm{j} \end{bmatrix} \otimes [1, -1] \end{bmatrix} = \begin{bmatrix} 1 & 1 & 2 & 2 \\ 2 & -1 & 4 & -2 \\ 2 & -\mathrm{j} & -2 & \mathrm{j} \\ 1 & \mathrm{j} & -1 & -\mathrm{j} \end{bmatrix}.$$

It should be noted that the generalized Kronecker product of two matrix groups $\{\mathbf{A}\}$ and $\{\mathbf{B}\}$ is still a matrix group rather than a single matrix.

Example 1.14 Let

$$\{\mathbf{A}\}_2 = \left\{ \begin{bmatrix} 1 & 1 \\ 1 & -1 \end{bmatrix} \atop \begin{bmatrix} 1 & -\mathrm{j} \\ 1 & \mathrm{j} \end{bmatrix} \right\}, \quad \{\mathbf{B}\} = \left\{ [1,1] \atop [1,-1] \right\}.$$

The generalized Kronecker product is given by

$$\{\mathbf{A}\}_2 \otimes \{\mathbf{B}\} = \left\{ \begin{bmatrix} 1 & 1 \\ 1 & -1 \end{bmatrix} \otimes [1,1] \atop \begin{bmatrix} 1 & -\mathrm{j} \\ 1 & \mathrm{j} \end{bmatrix} \otimes [1,-1] \right\} = \left\{ \begin{bmatrix} 1 & 1 & 1 & 1 \\ 1 & -1 & 1 & -1 \end{bmatrix} \atop \begin{bmatrix} 1 & -\mathrm{j} & -1 & \mathrm{j} \\ 1 & \mathrm{j} & -1 & -\mathrm{j} \end{bmatrix} \right\}.$$

Generalized Kronecker products have important applications in filter bank analysis and the derivation of fast algorithms in the Haar transform and the Hadamard transform [390]. Based on the generalized Kronecker product, we can derive the fast Fourier transform (FFT) algorithm, which we will discuss in Subsection 2.7.4 in Chapter 2.

1.10.3 Khatri–Rao Product

Definition 1.43 The *Khatri–Rao product* of two matrices with the same number of columns, $\mathbf{G} \in \mathbb{R}^{p\times n}$ and $\mathbf{F} \in \mathbb{R}^{q\times n}$, is denoted by $\mathbf{F} \odot \mathbf{G}$ and is defined as [244], [409]

$$\mathbf{F} \odot \mathbf{G} = [\mathbf{f}_1 \otimes \mathbf{g}_1, \mathbf{f}_2 \otimes \mathbf{g}_2, \ldots, \mathbf{f}_n \otimes \mathbf{g}_n] \in \mathbb{R}^{pq\times n}. \tag{1.10.22}$$

Thus the Khatri–Rao product consists of the Kronecker product of the corresponding column vectors of two matrices. Hence the Khatri–Rao product is called the *columnwise Kronecker product.*

More generally, if $\mathbf{A} = [\mathbf{A}_1, \ldots, \mathbf{A}_u]$ and $\mathbf{B} = [\mathbf{B}_1, \ldots, \mathbf{B}_u]$ are two block matrices, and the submatrices $\mathbf{A}_i$ and $\mathbf{B}_i$ have the same number of columns, then

$$\mathbf{A} \odot \mathbf{B} = [\mathbf{A}_1 \otimes \mathbf{B}_1, \mathbf{A}_2 \otimes \mathbf{B}_2, \ldots, \mathbf{A}_u \otimes \mathbf{B}_u]. \tag{1.10.23}$$

Khatri–Rao products have a number of properties [32], [301].

1. The basic properties of the Khatri–Rao product itself are follows.
   - *Distributive law* $(\mathbf{A} + \mathbf{B}) \odot \mathbf{D} = \mathbf{A} \odot \mathbf{D} + \mathbf{B} \odot \mathbf{D}$.
   - *Associative law* $\mathbf{A} \odot \mathbf{B} \odot \mathbf{C} = (\mathbf{A} \odot \mathbf{B}) \odot \mathbf{C} = \mathbf{A} \odot (\mathbf{B} \odot \mathbf{C})$.
   - *Commutative law* $\mathbf{A} \odot \mathbf{B} = \mathbf{K}_{nn}(\mathbf{B} \odot \mathbf{A})$, where $\mathbf{K}_{nn}$ is a commutation matrix; see the next section.

2. The relationship between the Khatri–Rao product and the Kronecker product is

$$(\mathbf{A} \otimes \mathbf{B})(\mathbf{F} \odot \mathbf{G}) = \mathbf{AF} \odot \mathbf{BG}. \tag{1.10.24}$$

3. The relationships between the Khatri–Rao product and the Hadamard product are

$$(\mathbf{A} \odot \mathbf{B}) * (\mathbf{C} \odot \mathbf{D}) = (\mathbf{A} * \mathbf{C}) \odot (\mathbf{B} * \mathbf{D}), \tag{1.10.25}$$

$$(\mathbf{A} \odot \mathbf{B})^T (\mathbf{A} \odot \mathbf{B}) = (\mathbf{A}^T \mathbf{A}) * (\mathbf{B}^T \mathbf{B}), \tag{1.10.26}$$

$$(\mathbf{A} \odot \mathbf{B})^\dagger = \big((\mathbf{A}^T \mathbf{A}) * (\mathbf{B}^T \mathbf{B})\big)^\dagger (\mathbf{A} \odot \mathbf{B})^T. \tag{1.10.27}$$

More generally,

$$(\mathbf{A} \odot \mathbf{B} \odot \mathbf{C})^T (\mathbf{A} \odot \mathbf{B} \odot \mathbf{C}) = (\mathbf{A}^T \mathbf{A}) * (\mathbf{B}^T \mathbf{B}) * (\mathbf{C}^T \mathbf{C}), \tag{1.10.28}$$

$$(\mathbf{A} \odot \mathbf{B} \odot \mathbf{C})^\dagger = \big((\mathbf{A}^T \mathbf{A}) * (\mathbf{B}^T \mathbf{B}) * (\mathbf{C}^T \mathbf{C})\big)^\dagger (\mathbf{A} \odot \mathbf{B} \odot \mathbf{C})^T. \tag{1.10.29}$$

1.11 Vectorization and Matricization

There exist functions or operators that transform a matrix into a vector or vice versa. These functions or operators are the vectorization of a matrix and the matricization of a vector.

1.11.1 Vectorization and Commutation Matrix

The *vectorization* of a matrix $\mathbf{A} \in \mathbb{R}^{m \times n}$, denoted $\text{vec}(\mathbf{A})$, is a linear transformation that arranges the entries of $\mathbf{A} = [a_{ij}]$ as an $mn \times 1$ vector via column stacking:

$$\text{vec}(\mathbf{A}) = [a_{11}, \ldots, a_{m1}, \ldots, a_{1n}, \ldots, a_{mn}]^T. \tag{1.11.1}$$

A matrix $\mathbf{A}$ can be also arranged as a row vector by stacking the rows; this is known as the *row vectorization* of the matrix, denoted $\text{rvec}(\mathbf{A})$, and is defined as

$$\text{rvec}(\mathbf{A}) = [a_{11}, \ldots, a_{1n}, \ldots, a_{m1}, \ldots, a_{mn}]. \tag{1.11.2}$$

For instance, given a matrix $\mathbf{A} = \begin{bmatrix} a_{11} & a_{12} \\ a_{21} & a_{22} \end{bmatrix}$, $\text{vec}(\mathbf{A}) = [a_{11}, a_{21}, a_{12}, a_{22}]^T$ and $\text{rvec}(\mathbf{A}) = [a_{11}, a_{12}, a_{21}, a_{22}]$.

Clearly, there exist the following relationships between the vectorization and the row vectorization of a matrix:

$$\text{rvec}(\mathbf{A}) = (\text{vec}(\mathbf{A}^T))^T, \quad \text{vec}(\mathbf{A}^T) = (\text{rvec}\,\mathbf{A})^T. \tag{1.11.3}$$

One obvious fact is that, for a given $m \times n$ matrix $\mathbf{A}$, the two vectors $\text{vec}(\mathbf{A})$ and $\text{vec}(\mathbf{A}^T)$ contain the same entries but the orders of their entries are different. Interestingly, there is a unique $mn \times mn$ permutation matrix that can transform

$\text{vec}(\mathbf{A})$ into $\text{vec}(\mathbf{A}^T)$. This permutation matrix is known as the *commutation matrix*; it is denoted $\mathbf{K}_{mn}$ and is defined by

$$\mathbf{K}_{mn}\text{vec}(\mathbf{A}) = \text{vec}(\mathbf{A}^T). \tag{1.11.4}$$

Similarly, there is an $nm \times nm$ permutation matrix transforming $\text{vec}(\mathbf{A}^T)$ into $\text{vec}(\mathbf{A})$. Such a commutation matrix, denoted $\mathbf{K}_{nm}$, is defined by

$$\mathbf{K}_{nm}\text{vec}(\mathbf{A}^T) = \text{vec}(\mathbf{A}). \tag{1.11.5}$$

From (1.11.4) and (1.11.5) it can be seen that $\mathbf{K}_{nm}\mathbf{K}_{mn}\text{vec}(\mathbf{A}) = \mathbf{K}_{nm}\text{vec}(\mathbf{A}^T) = \text{vec}(\mathbf{A})$. Since this formula holds for any $m \times n$ matrix $\mathbf{A}$, we have $\mathbf{K}_{nm}\mathbf{K}_{mn} = \mathbf{I}_{mn}$ or $\mathbf{K}_{mn}^{-1} = \mathbf{K}_{nm}$.

The $mn \times mn$ commutation matrix $\mathbf{K}_{mn}$ has the following properties [310].

1. $\mathbf{K}_{mn}\text{vec}(\mathbf{A}) = \text{vec}(\mathbf{A}^T)$ and $\mathbf{K}_{nm}\text{vec}(\mathbf{A}^T) = \text{vec}(\mathbf{A})$, where $\mathbf{A}$ is an $m \times n$ matrix.
2. $\mathbf{K}_{mn}^T\mathbf{K}_{mn} = \mathbf{K}_{mn}\mathbf{K}_{mn}^T = \mathbf{I}_{mn}$, or $\mathbf{K}_{mn}^{-1} = \mathbf{K}_{nm}$.
3. $\mathbf{K}_{mn}^T = \mathbf{K}_{nm}$.
4. $\mathbf{K}_{mn}$ can be represented as a Kronecker product of the essential vectors:
$$\mathbf{K}_{mn} = \sum_{j=1}^{n}(\mathbf{e}_j^T \otimes \mathbf{I}_m \otimes \mathbf{e}_j).$$
5. $\mathbf{K}_{1n} = \mathbf{K}_{n1} = \mathbf{I}_n$.
6. $\mathbf{K}_{nm}\mathbf{K}_{mn}\text{vec}(\mathbf{A}) = \mathbf{K}_{nm}\text{vec}(\mathbf{A}^T) = \text{vec}(\mathbf{A})$.
7. The eigenvalues of the commutation matrix $\mathbf{K}_{nn}$ are 1 and -1 and their multiplicities are respectively $\frac{1}{2}n(n+1)$ and $\frac{1}{2}n(n-1)$.
8. The rank of the commutation matrix is given by $\text{rank}(\mathbf{K}_{mn}) = 1 + d(m-1, n-1)$, where $d(m,n)$ is the greatest common divisor of m and n and $d(n,0) = d(0,n) = n$.
9. $\mathbf{K}_{mn}(\mathbf{A} \otimes \mathbf{B})\mathbf{K}_{pq} = \mathbf{B} \otimes \mathbf{A}$, and thus can be equivalently written as $\mathbf{K}_{mn}(\mathbf{A} \otimes \mathbf{B}) = (\mathbf{B} \otimes \mathbf{A})\mathbf{K}_{qp}$, where $\mathbf{A}$ is an $n \times p$ matrix and $\mathbf{B}$ is an $m \times q$ matrix. In particular, $\mathbf{K}_{mn}(\mathbf{A}_{n\times n} \otimes \mathbf{B}_{m\times m}) = (\mathbf{B} \otimes \mathbf{A})\mathbf{K}_{mn}$.
10. $\text{tr}(\mathbf{K}_{mn}(\mathbf{A}_{m\times n} \otimes \mathbf{B}_{m\times n})) = \text{tr}(\mathbf{A}^T\mathbf{B}) = \left(\text{vec}(\mathbf{B}^T)\right)^T \mathbf{K}_{mn}\text{vec}(\mathbf{A})$.

The construction of an $mn \times mn$ commutation matrix is as follows. First let

$$\mathbf{K}_{mn} = \begin{bmatrix} \mathbf{K}_1 \\ \vdots \\ \mathbf{K}_m \end{bmatrix}, \quad \mathbf{K}_i \in \mathbb{R}^{n\times mn}, i = 1, \ldots, m; \tag{1.11.6}$$

then the (i,j)th entry of the first submatrix $\mathbf{K}_1$ is given by

$$K_1(i,j) = \begin{cases} 1, & j = (i-1)m+1,\ i = 1,\ldots,n, \\ 0, & \text{otherwise.} \end{cases} \tag{1.11.7}$$

Next, the ith submatrix $\mathbf{K}_i$ ($i = 2,\ldots,m$) is constructed from the $(i-1)$th submatrix $\mathbf{K}_{i-1}$ as follows:

$$\mathbf{K}_i = [\mathbf{0}, \mathbf{K}_{i-1}(1:mn-1)], \qquad i = 2,\ldots,m, \tag{1.11.8}$$

where $\mathbf{K}_{i-1}(1:mn-1)$ denotes a submatrix consisting of the first $(mn-1)$ columns of the $n \times m$ submatrix $\mathbf{K}_{i-1}$.

Example 1.15 For $m = 2, n = 4$, we have

$$\mathbf{K}_{24} = \begin{bmatrix} 1 & 0 & 0 & 0 & 0 & 0 & 0 & 0 \\ 0 & 0 & 1 & 0 & 0 & 0 & 0 & 0 \\ 0 & 0 & 0 & 0 & 1 & 0 & 0 & 0 \\ 0 & 0 & 0 & 0 & 0 & 0 & 1 & 0 \\ 0 & 1 & 0 & 0 & 0 & 0 & 0 & 0 \\ 0 & 0 & 0 & 1 & 0 & 0 & 0 & 0 \\ 0 & 0 & 0 & 0 & 0 & 1 & 0 & 0 \\ 0 & 0 & 0 & 0 & 0 & 0 & 0 & 1 \end{bmatrix}.$$

If $m = 4, n = 2$ then

$$\mathbf{K}_{42} = \begin{bmatrix} 1 & 0 & 0 & 0 & 0 & 0 & 0 & 0 \\ 0 & 0 & 0 & 0 & 1 & 0 & 0 & 0 \\ 0 & 1 & 0 & 0 & 0 & 0 & 0 & 0 \\ 0 & 0 & 0 & 0 & 0 & 1 & 0 & 0 \\ 0 & 0 & 1 & 0 & 0 & 0 & 0 & 0 \\ 0 & 0 & 0 & 0 & 0 & 0 & 1 & 0 \\ 0 & 0 & 0 & 1 & 0 & 0 & 0 & 0 \\ 0 & 0 & 0 & 0 & 0 & 0 & 0 & 1 \end{bmatrix}.$$

Hence we get

$$\mathbf{K}_{42}\,\mathrm{vec}(\mathbf{A}) = \begin{bmatrix} 1 & 0 & 0 & 0 & 0 & 0 & 0 & 0 \\ 0 & 0 & 0 & 0 & 1 & 0 & 0 & 0 \\ 0 & 1 & 0 & 0 & 0 & 0 & 0 & 0 \\ 0 & 0 & 0 & 0 & 0 & 1 & 0 & 0 \\ 0 & 0 & 1 & 0 & 0 & 0 & 0 & 0 \\ 0 & 0 & 0 & 0 & 0 & 0 & 1 & 0 \\ 0 & 0 & 0 & 1 & 0 & 0 & 0 & 0 \\ 0 & 0 & 0 & 0 & 0 & 0 & 0 & 1 \end{bmatrix} \begin{bmatrix} a_{11} \\ a_{21} \\ a_{31} \\ a_{41} \\ a_{12} \\ a_{22} \\ a_{32} \\ a_{42} \end{bmatrix} = \begin{bmatrix} a_{11} \\ a_{12} \\ a_{21} \\ a_{22} \\ a_{31} \\ a_{32} \\ a_{41} \\ a_{42} \end{bmatrix} = \mathrm{vec}(\mathbf{A}^T).$$

1.11.2 Matricization of a Vector

The operation for transforming an $mn \times 1$ vector $\mathbf{a} = [a_1, \ldots, a_{mn}]^T$ into an $m \times n$ matrix $\mathbf{A}$ is known as the *matricization of the column vector* $\mathbf{a}$, denoted $\text{unvec}_{m,n}(\mathbf{a})$, and is defined as

$$\mathbf{A}_{m\times n} = \text{unvec}_{m,n}(\mathbf{a}) = \begin{bmatrix} a_1 & a_{m+1} & \cdots & a_{m(n-1)+1} \\ a_2 & a_{m+2} & \cdots & a_{m(n-1)+2} \\ \vdots & \vdots & \ddots & \vdots \\ a_m & a_{2m} & \cdots & a_{mn} \end{bmatrix}. \tag{1.11.9}$$

Clearly, the (i,j)th entry A_{ij} of the matrix $\mathbf{A}$ is given by the kth entry a_k of the vector $\mathbf{a}$ as follows:

$$A_{ij} = a_{i+(j-1)m}, \quad i = 1, \ldots, m,\ j = 1, \ldots, n. \tag{1.11.10}$$

Similarly, the operation for transforming a $1 \times mn$ row vector $\mathbf{b} = [b_1, \ldots, b_{mn}]$ into an $m \times n$ matrix $\mathbf{B}$ is called the *matricization of the row vector* $\mathbf{b}$. This matricization is denoted $\text{unrvec}_{m,n}(\mathbf{b})$, and is defined as follows:

$$\mathbf{B}_{m\times n} = \text{unrvec}_{m,n}(\mathbf{b}) = \begin{bmatrix} b_1 & b_2 & \cdots & b_n \\ b_{n+1} & b_{n+2} & \cdots & b_{2n} \\ \vdots & \vdots & \ddots & \vdots \\ b_{(m-1)n+1} & b_{(m-1)n+2} & \cdots & b_{mn} \end{bmatrix}. \tag{1.11.11}$$

This is equivalently represented in element form as

$$B_{ij} = b_{j+(i-1)n}, \quad i = 1, \ldots, m,\ j = 1, \ldots, n. \tag{1.11.12}$$

It can be seen from the above definitions that there are the following relationships between matricization (unvec) and column vectorization (vec) or row vectorization (rvec):

$$\begin{bmatrix} A_{11} & \cdots & A_{1n} \\ \vdots & \ddots & \vdots \\ A_{m1} & \cdots & A_{mn} \end{bmatrix} \underset{\text{unvec}}{\overset{\text{vec}}{\rightleftarrows}} [A_{11}, \ldots, A_{m1}, \ldots, A_{1n}, \ldots, A_{mn}]^T,$$

$$\begin{bmatrix} A_{11} & \cdots & A_{1n} \\ \vdots & \ddots & \vdots \\ A_{m1} & \cdots & A_{mn} \end{bmatrix} \underset{\text{unvec}}{\overset{\text{rvec}}{\rightleftarrows}} [A_{11}, \ldots, A_{1n}, \ldots, A_{m1}, \ldots, A_{mn}],$$

which can be written as

$$\text{unvec}_{m,n}(\mathbf{a}) = \mathbf{A}_{m\times n} \quad \Leftrightarrow \quad \text{vec}(\mathbf{A}_{m\times n}) = \mathbf{a}_{mn\times 1}, \tag{1.11.13}$$

$$\text{unrvec}_{m,n}(\mathbf{b}) = \mathbf{B}_{m\times n} \quad \Leftrightarrow \quad \text{rvec}(\mathbf{B}_{m\times n}) = \mathbf{b}_{1\times mn}. \tag{1.11.14}$$

1.11.3 Properties of Vectorization Operator

The vectorization operator has the following properties [68], [207], [311].

1. The vectorization of a transposed matrix is given by $\text{vec}(\mathbf{A}^T) = \mathbf{K}_{mn}\text{vec}(\mathbf{A})$ for $\mathbf{A} \in \mathbb{C}^{m\times n}$.
2. The vectorization of a matrix sum is given by $\text{vec}(\mathbf{A}+\mathbf{B}) = \text{vec}(\mathbf{A}) + \text{vec}(\mathbf{B})$.
3. The vectorization of a Kronecker product [311, p. 184] is given by
$$\text{vec}(\mathbf{X}\otimes\mathbf{Y}) = (\mathbf{I}_m \otimes \mathbf{K}_{qp} \otimes \mathbf{I}_n)(\text{vec}(\mathbf{X})\otimes\text{vec}(\mathbf{Y})). \tag{1.11.15}$$
4. The trace of a matrix product is given by
$$\text{tr}(\mathbf{A}^T\mathbf{B}) = (\text{vec}(\mathbf{A}))^T\,\text{vec}(\mathbf{B}), \tag{1.11.16}$$
$$\text{tr}(\mathbf{A}^H\mathbf{B}) = (\text{vec}(\mathbf{A}))^H\,\text{vec}(\mathbf{B}), \tag{1.11.17}$$
$$\text{tr}(\mathbf{ABC}) = (\text{vec}(\mathbf{A}))^T(\mathbf{I}_p\otimes\mathbf{B})\,\text{vec}(\mathbf{C}), \tag{1.11.18}$$
while the trace of the product of four matrices is determined by [311, p. 31]
$$\begin{aligned}\text{tr}(\mathbf{ABCD}) &= (\text{vec}(\mathbf{D}^T))^T(\mathbf{C}^T\otimes\mathbf{A})\,\text{vec}(\mathbf{B})\\ &= (\text{vec}(\mathbf{D}))^T(\mathbf{A}\otimes\mathbf{C}^T)\,\text{vec}(\mathbf{B}^T).\end{aligned}$$
5. The Kronecker product of two vectors $\mathbf{a}$ and $\mathbf{b}$ can be represented as the vectorization of their outer product $\mathbf{ba}^T$ as follows:
$$\mathbf{a}\otimes\mathbf{b} = \text{vec}(\mathbf{ba}^T) = \text{vec}(\mathbf{b}\circ\mathbf{a}). \tag{1.11.19}$$
6. The vectorization of the Hadamard product is given by
$$\text{vec}(\mathbf{A}*\mathbf{B}) = \text{vec}(\mathbf{A}) * \text{vec}(\mathbf{B}) = \mathbf{Diag}(\text{vec}(\mathbf{A}))\,\text{vec}(\mathbf{B}), \tag{1.11.20}$$
where $\mathbf{Diag}(\text{vec}(\mathbf{A}))$ is a diagonal matrix whose entries are the vectorization function $\text{vec}(\mathbf{A})$.
7. The relation of the vectorization function to the Khatri–Rao product [60] is as follows:
$$\text{vec}(\mathbf{U}_{m\times p}\mathbf{V}_{p\times p}\mathbf{W}_{p\times n}) = (\mathbf{W}^T\odot\mathbf{U})\,\mathbf{diag}(\mathbf{V}), \tag{1.11.21}$$
where $\mathbf{diag}(\mathbf{V}) = [v_{11},\ldots,v_{pp}]^T$ and v_{ii} are the diagonal entries of $\mathbf{V}$.
8. The relation of the vectorization of the matrix product $\mathbf{A}_{m\times p}\mathbf{B}_{p\times q}\mathbf{C}_{q\times n}$ to the Kronecker product [428, p. 263] is given by
$$\text{vec}(\mathbf{ABC}) = (\mathbf{C}^T\otimes\mathbf{A})\,\text{vec}(\mathbf{B}), \tag{1.11.22}$$
$$\text{vec}(\mathbf{ABC}) = (\mathbf{I}_q\otimes\mathbf{AB})\,\text{vec}(\mathbf{C}) = (\mathbf{C}^T\mathbf{B}^T\otimes\mathbf{I}_m)\,\text{vec}(\mathbf{A}), \tag{1.11.23}$$
$$\text{vec}(\mathbf{AC}) = (\mathbf{I}_p\otimes\mathbf{A})\,\text{vec}(\mathbf{C}) = (\mathbf{C}^T\otimes\mathbf{I}_m)\,\text{vec}(\mathbf{A}). \tag{1.11.24}$$

Example 1.16 Consider the matrix equation $\mathbf{AXB} = \mathbf{C}$, where $\mathbf{A} \in \mathbb{R}^{m\times n}, \mathbf{X} \in \mathbb{R}^{n\times p}$, $\mathbf{B} \in \mathbb{R}^{p\times q}$ and $\mathbf{C} \in \mathbb{R}^{m\times q}$. By using the vectorization function property $\text{vec}(\mathbf{AXB}) = (\mathbf{B}^T \otimes \mathbf{A})\,\text{vec}(\mathbf{X})$, the vectorization $\text{vec}(\mathbf{AXB}) = \text{vec}(\mathbf{C})$ of the original matrix equation can be rewritten as the Kronecker product form [410]

$$(\mathbf{B}^T \otimes \mathbf{A})\text{vec}(\mathbf{X}) = \text{vec}(\mathbf{C}),$$

and thus $\text{vec}(\mathbf{X}) = (\mathbf{B}^T \otimes \mathbf{A})^\dagger \text{vec}(\mathbf{C})$. Then by matricizing $\text{vec}(\mathbf{X})$, we get the solution matrix $\mathbf{X}$ of the original matrix equation $\mathbf{AXB} = \mathbf{C}$.

Example 1.17 Consider solving the matrix equation $\mathbf{AX} + \mathbf{XB} = \mathbf{Y}$, where all the matrices are $n \times n$. By the vectorization operator property $\text{vec}(\mathbf{ADB}) = (\mathbf{B}^T \otimes \mathbf{A})\,\text{vec}(\mathbf{D})$, we have $\text{vec}(\mathbf{AX}) = \text{vec}(\mathbf{AXI}) = (\mathbf{I}_n \otimes \mathbf{A})\text{vec}(\mathbf{X})$ and $\text{vec}(\mathbf{XB}) = \text{vec}(\mathbf{IXB}) = (\mathbf{B}^T \otimes \mathbf{I}_n)\text{vec}(\mathbf{X})$, and thus the original matrix equation $\mathbf{AX}+\mathbf{XB} = \mathbf{Y}$ can be rewritten as $(\mathbf{I}_n \otimes \mathbf{A} + \mathbf{B}^T \otimes \mathbf{I}_n)\,\text{vec}(\mathbf{X}) = \text{vec}(\mathbf{Y})$, from which we get

$$\text{vec}(\mathbf{X}) = (\mathbf{I}_n \otimes \mathbf{A} + \mathbf{B}^T \otimes \mathbf{I}_n)^\dagger \text{vec}(\mathbf{Y}).$$

Then, by matricizing $\text{vec}(\mathbf{X})$, we get the solution $\mathbf{X}$.

1.12 Sparse Representations

A sparse linear combination of prototype signal-atoms (see below) representing a target signal is known as a sparse representation.

1.12.1 Sparse Vectors and Sparse Representations

A vector or matrix most of whose elements are zero is known as a *sparse vector* or a *sparse matrix.* A signal vector $\mathbf{y} \in \mathbb{R}^m$ can be decomposed into at most m orthogonal basis (vectors) $\mathbf{g}_k \in \mathbb{R}^m, k = 1,\ldots,m$. Such an orthogonal basis is called a *complete orthogonal basis.* In the signal decomposition

$$\mathbf{y} = \mathbf{Gc} = \sum_{i=1}^{m} c_i \mathbf{g}_i, \tag{1.12.1}$$

the coefficient vector $\mathbf{c}$ must be nonsparse.

If the signal vector $\mathbf{y} \in \mathbb{R}^m$ is decomposed into a linear combination of n vectors $\mathbf{a}_i \in \mathbb{R}^m, i = 1,\ldots,n$ (where $n > m$), i.e.,

$$\mathbf{y} = \mathbf{Ax} = \sum_{i=1}^{n} x_i \mathbf{a}_i \quad (n > m), \tag{1.12.2}$$

then the $n\,(> m)$ vectors $\mathbf{a}_i \in \mathbb{R}^m$, $i = 1,\ldots,n$ cannot be an orthogonal basis set. In order to distinguish them from true basis vectors, nonorthogonal column vectors are called *atoms* or *frames.* Because the number n of atoms is larger than the dimension m of the vector space $\mathbb{R}^m$, such a set of atoms is said to be *overcomplete.*

A matrix consisting of overcomplete atoms, $\mathbf{A} = [\mathbf{a}_1, \ldots, \mathbf{a}_n] \in \mathbb{R}^{m\times n} (n > m)$, is referred to as a *dictionary*.

For a dictionary (matrix) $\mathbf{A} \in \mathbb{R}^{m\times n}$, it is usually assumed that

(1) the row number m of $\mathbf{A}$ is less than its column number n;
(2) the dictionary $\mathbf{A}$ has full-row rank, i.e., $\text{rank}(\mathbf{A}) = m$;
(3) all columns of $\mathbf{A}$ have unit Euclidean norms, i.e., $\|\mathbf{a}_j\|_2 = 1, j = 1, \ldots, n$.

The overcomplete signal decomposition formula (1.12.2) is a under-determined equation that has infinitely many solution vectors $\mathbf{x}$. There are two common methods for solving this type of under-determined equations.

1. *Classical method* (finding the minimum ℓ_2-norm solution)

$$\min \|\mathbf{x}\|_2 \quad \text{subject to} \quad \mathbf{A}\mathbf{x} = \mathbf{y}. \tag{1.12.3}$$

   The advantage of this method is that the solution is unique and is the minimum norm solution, also called the minimum energy (sum of squared amplitudes of each component) solution or the shortest distance (from the origin) solution. However, because each entry of this solution vector usually takes a nonzero value, it does not meet the requirements of sparse representation in many practical applications.
2. *Modern method* (finding the minimum ℓ_0-norm solution)

$$\min \|\mathbf{x}\|_0 \quad \text{subject to} \quad \mathbf{A}\mathbf{x} = \mathbf{y}, \tag{1.12.4}$$

   where the ℓ_0-norm $\|\mathbf{x}\|_0$ is defined as the number of nonzero elements of the vector $\mathbf{x}$. The advantage of this method is that it selects a sparse solution vector, which makes it available for many practical applications. The disadvantage of this method is that its computation is more complex.

If in the observation data there exist errors or background noise then the minimum ℓ_0-norm solution becomes

$$\min \|\mathbf{x}\|_0 \quad \text{subject to} \quad \|\mathbf{A}\mathbf{x} - \mathbf{y}\|_2 \leq \epsilon, \tag{1.12.5}$$

where $\epsilon > 0$ is very small.

When the coefficient vector $\mathbf{x}$ is sparse, the signal decomposition $\mathbf{y} = \mathbf{A}\mathbf{x}$ is known as a *sparse decomposition* of signals. The columns of the dictionary matrix $\mathbf{A}$ are called *explanatory variables*; the signal vector $\mathbf{y}$ is known as the *response variable* or *target signal*; $\mathbf{A}\mathbf{x}$ is said to be the linear response predictor, while $\mathbf{x}$ can be regarded as the *sparse representation* of the target signal $\mathbf{y}$ corresponding to the dictionary $\mathbf{A}$.

Equation (1.12.4) is a *sparse representation problem* of the target signal $\mathbf{y}$ corresponding to the dictionary $\mathbf{A}$, while Equation (1.12.5) is a *sparse approximation problem* of the target signal.

Given a positive integer K, if the ℓ_0-norm of the vector $\mathbf{x}$ is less than or equal to

K, i.e., $\|\mathbf{x}\|_0 \leq K$, then $\mathbf{x}$ is said to be *K-sparse*. For a given signal vector $\mathbf{y}$ and a dictionary $\mathbf{A}$, if the coefficient vector $\mathbf{x}$ satisfying $\mathbf{Ax} = \mathbf{y}$ has the minimum ℓ_0-norm then $\mathbf{x}$ is said to be the *sparsest representation* of the target signal $\mathbf{y}$ corresponding to the dictionary $\mathbf{A}$.

Sparse representation is a type of linear inverse problem. In communications and information theory, the matrix $\mathbf{A} \in \mathbb{R}^{m\times N}$ and the vector $\mathbf{x} \in \mathbb{R}^N$ represent respectively the *coding matrix* and the *plaintext* to be sent, and the observation vector $\mathbf{y} \in \mathbb{R}^m$ is the *ciphertext*. The linear inverse problem becomes a decoding problem: how to recover the plaintext $\mathbf{x}$ from the ciphertext $\mathbf{y}$.

1.12.2 Sparse Representation of Face Recognition

As a typical application of sparse representation, let us consider a face recognition problem. Suppose that there are c classes of target faces that the $R_1 \times R_2$ matrix representation results for each image of a given face have been vectorized to an $m \times 1$ vector $\mathbf{d}$ (where $m = R_1 \times R_2$ is the number of samples creating each facial image, e.g., $m = 512 \times 512$), and suppose that each column vector is normalized to the unit Euclidean norm. Hence N_i training images of the ith known face under different types of illumination can be represented as an $m \times N_i$ matrix $\mathbf{D}_i = [\mathbf{d}_{i,1}, \mathbf{d}_{i,2}, \ldots, \mathbf{d}_{i,N_i}] \in \mathbb{R}^{m\times N_i}$. Given a large enough training set $\mathbf{D}_i$, then a new image $\mathbf{y}$ of the ith known face, shot under another illumination, can be represented as a linear combination of the known training images, i.e., $\mathbf{y} \approx \mathbf{D}_i\boldsymbol{\alpha}_i$, where $\boldsymbol{\alpha}_i \in \mathbb{R}^m$ is a coefficient vector. The problem is that in practical applications we do not usually know to which target collection of known faces the new experimental samples belong, so we need to make face recognition: determine to which target class these samples belong.

If we know or make a rough guess that the new testing sample is a certain unknown target face in the c classes of known target faces, a dictionary consisting of training samples for c classes can be written as a training data matrix as follows:

$$\mathbf{D} = [\mathbf{D}_1, \ldots, \mathbf{D}_c] = [\mathbf{D}_{1,1}, \ldots, \mathbf{D}_{1,N_1}, \ldots, \mathbf{D}_{c,1}, \ldots, \mathbf{D}_{c,N_c}] \in \mathbb{R}^{m\times N}, \tag{1.12.6}$$

where $N = \sum_{i=1}^{c} N_i$ denotes the total number of the training images of the c target faces. Hence the face image $\mathbf{y}$ to be recognized can be represented as the linear combination

$$\mathbf{y} = \mathbf{D}\boldsymbol{\alpha}_0 = [\mathbf{D}_{1,1} \;\cdots\; \mathbf{D}_{1,N_1} \;\cdots\; \mathbf{D}_{c,1} \;\cdots\; \mathbf{D}_{c,N_c}] \begin{bmatrix} \mathbf{0}_{N_1} \\ \vdots \\ \mathbf{0}_{N_{i-1}} \\ \boldsymbol{\alpha}_i \\ \mathbf{0}_{N_{i+1}} \\ \vdots \\ \mathbf{0}_{N_c} \end{bmatrix}, \tag{1.12.7}$$

where $\mathbf{0}_{N_k}, k = 1, \ldots, i-1, i+1, \ldots, c$ is an $N_k \times 1$ zero vector.

The face recognition becomes a problem for solving a matrix equation or a linear inverse problem: given a data vector $\mathbf{y}$ and a data matrix $\mathbf{D}$, solve for the solution vector $\boldsymbol{\alpha}_0$ of the matrix equation $\mathbf{y} = \mathbf{D}\boldsymbol{\alpha}_0$.

It should be noted that, since $m < N$ in the general case, the matrix equation $\mathbf{y} = \mathbf{D}\boldsymbol{\alpha}_0$ is under-determined and has infinitely many solutions, among which the sparsest solution is the solution of interest.

Since the solution vector must be a sparse vector, the face recognition problem can be described as an optimization problem:

$$\min \|\boldsymbol{\alpha}_0\|_0 \quad \text{subject to} \quad \mathbf{y} = \mathbf{D}\boldsymbol{\alpha}_0. \tag{1.12.8}$$

This is a typical ℓ_0-norm minimization problem. We will discuss the solution of this problem in Chapter 6.

Exercises

1.1 Let $\mathbf{x} = [x_1, \ldots, x_m]^T, \mathbf{y} = [y_1, \ldots, y_n]^T, \mathbf{z} = [z_1, \ldots, z_k]^T$ be complex vectors. Use the matrix form to represent the outer products $\mathbf{x} \circ \mathbf{y}$ and $\mathbf{x} \circ \mathbf{y} \circ \mathbf{z}$.

1.2 Show the associative law of matrix addition $(\mathbf{A}+\mathbf{B})+\mathbf{C} = \mathbf{A}+(\mathbf{B}+\mathbf{C})$ and the right distributive law of matrix product $(\mathbf{A}+\mathbf{B})\mathbf{C} = \mathbf{AC}+\mathbf{BC}$.

1.3 Given the system of linear equations

$$\begin{cases} 2y_1 - y_2 = x_1, \\ y_1 + 2y_2 = x_2, \\ -2y_1 + 3y_2 = x_3, \end{cases} \quad \text{and} \quad \begin{cases} 3z_1 - z_2 = y_1, \\ 5z_1 + 2z_2 = y_2, \end{cases}$$

use z_1, z_2 to represent x_1, x_2, x_3.

1.4 By using elementary row operations, simplify the matrix

$$\mathbf{A} = \begin{bmatrix} 0 & 0 & 0 & 0 & 2 & 8 & 4 \\ 0 & 0 & 0 & 1 & 4 & 9 & 7 \\ 0 & 3 & -11 & -3 & -8 & -15 & -32 \\ 0 & -2 & -8 & 1 & 6 & 13 & 21 \end{bmatrix}$$

into its reduced-echelon form.

1.5 Use elementary row operations to solve the system of linear equations

$$\begin{aligned} 2x_1 - 4x_2 + 3x_3 - 4x_4 - 11x_5 &= 28, \\ -x_1 + 2x_2 - x_3 + 2x_4 + 5x_5 &= -13, \\ -3x_3 + 2x_4 + 5x_5 &= -10, \\ 3x_1 - 5x_2 + 10x_3 - 7x_4 + 12x_5 &= 31. \end{aligned}$$

1.6 Set

$$1^3 + 2^3 + \cdots + n^3 = a_1 n + a_2 n^2 + a_3 n^3 + a_4 n^4.$$

Find the constants a_1, a_2, a_3, a_4. (*Hint:* Let $n = 1, 2, 3, 4$, to get a system of linear equations.)

1.7 Suppose that $F : \mathbb{R}^3 \to \mathbb{R}^2$ is a transformation defined as

$$F(\mathbf{x}) = \begin{bmatrix} 2x_1 - x_2 \\ x_2 + 5x_3 \end{bmatrix}, \quad \mathbf{x} = \begin{bmatrix} x_1 \\ x_2 \\ x_3 \end{bmatrix}.$$

Determine whether F is a linear transformation.

1.8 Show that $\mathbf{x}^T\mathbf{A}\mathbf{x} = \mathrm{tr}(\mathbf{x}^T\mathbf{A}\mathbf{x})$ and $\mathbf{x}^T\mathbf{A}\mathbf{x} = \mathrm{tr}(\mathbf{A}\mathbf{x}\mathbf{x}^T)$.

1.9 Given $\mathbf{A} \in \mathbb{R}^{n\times n}$, demonstrate the Schur inequality $\mathrm{tr}(\mathbf{A}^2) \le \mathrm{tr}(\mathbf{A}^T\mathbf{A})$, where equality holds if and only if $\mathbf{A}$ is a symmetric matrix.

1.10 The roots λ satisfying $|\mathbf{A} - \lambda\mathbf{I}| = 0$ are known as the eigenvalues of the matrix $\mathbf{A}$. Show that if λ is a single eigenvalue of $\mathbf{A}$ then there is at least one determinant $|\mathbf{A}_k - \lambda\mathbf{I}| = 0$, where $\mathbf{A}_k$ is the matrix remaining after the kth row and kth column of $\mathbf{A}$ have been removed.

1.11 The equation of a straight line can be represented as $ax + by = -1$. Show that the equation of the line through two points (x_1, y_1) and (x_2, y_2) is given by

$$\begin{vmatrix} 1 & x & y \\ 1 & x_1 & y_1 \\ 1 & x_2 & y_2 \end{vmatrix} = 0.$$

1.12 The equation of a plane can be represented as $ax + by + cz = -1$. Show that the equation of the plane through the three points (x_i, y_i, z_i), $i = 1, 2, 3$, is determined by

$$\begin{vmatrix} 1 & x & y & z \\ 1 & x_1 & y_1 & z_1 \\ 1 & x_2 & y_2 & z_2 \\ 1 & x_3 & y_3 & z_3 \end{vmatrix} = 0.$$

1.13 Given a vector set $\{\mathbf{x}_1, \ldots, \mathbf{x}_p\}$ with $\mathbf{x}_p \neq \mathbf{0}$, let $a_1, a_2, \ldots, a_{p-1}$ be any constants, and let $\mathbf{y}_i = \mathbf{x}_i + a_i\mathbf{x}_p$, $i = 1, \ldots, p-1$. Show that the vectors in the set $\{\mathbf{y}_1, \ldots, \mathbf{y}_{p-1}\}$ are linearly independent, if and only if the vectors $\mathbf{x}_1, \ldots, \mathbf{x}_p$ are linearly independent.

1.14 Without expanding the determinants, prove the following results:

$$2\begin{vmatrix} a & b & c \\ d & e & f \\ x & y & z \end{vmatrix} = \begin{vmatrix} a+b & b+c & c+a \\ d+e & e+f & f+d \\ x+y & y+z & z+x \end{vmatrix}$$

and

$$2\begin{vmatrix} 0 & a & b \\ a & 0 & c \\ b & c & 0 \end{vmatrix} = \begin{vmatrix} b+a & c & c \\ b & a+c & b \\ a & a & c+b \end{vmatrix}.$$

1.15 Suppose that $\mathbf{A}_{n\times n}$ is positive definite and $\mathbf{B}_{n\times n}$ is positive semi-definite; show that $\det(\mathbf{A}+\mathbf{B}) \geq \det(\mathbf{A})$.

1.16 Let $\mathbf{A}_{12\times 12}$ satisfy $\mathbf{A}^5 = 3\mathbf{A}$. Find all the possible values of $|\mathbf{A}|$.

1.17 Given a block matrix $\mathbf{X} = [\mathbf{A}, \mathbf{B}]$, show that

$$|\mathbf{x}|^2 = |\mathbf{A}\mathbf{A}^T + \mathbf{B}\mathbf{B}^T| = \begin{vmatrix} \mathbf{A}^T\mathbf{A} & \mathbf{A}^T\mathbf{B} \\ \mathbf{B}^T\mathbf{A} & \mathbf{B}^T\mathbf{B} \end{vmatrix}.$$

1.18 Let $\mathbf{A}^2 = \mathbf{A}$. Use two methods to show that $\text{rank}(\mathbf{I}-\mathbf{A}) = n-\text{rank}(\mathbf{A})$: (a) use the property that the matrix trace and the matrix rank are equal; (b) consider linearly independent solutions of the matrix equation $(\mathbf{I}-\mathbf{A})\mathbf{x} = \mathbf{0}$.

1.19 Given the matrix

$$\mathbf{A} = \begin{bmatrix} 1 & 1 & 1 & 2 \\ -1 & 0 & 2 & -3 \\ 2 & 4 & 8 & 5 \end{bmatrix},$$

find the rank of $\mathbf{A}$. (*Hint:* Transform the matrix $\mathbf{A}$ into its echelon form.)

1.20 Let $\mathbf{A}$ be an $m \times n$ matrix. Show that $\text{rank}(\mathbf{A}) \leq m$ and $\text{rank}(\mathbf{A}) \leq n$.

1.21 Consider the system of linear equations

$$\begin{aligned} x_1 + 3x_2 - x_3 &= a_1, \\ x_1 + 2x_2 &= a_2, \\ 3x_1 + 7x_2 - x_3 &= a_3. \end{aligned}$$

(a) Determine the necessary and sufficient condition for the above system of linear equations to be consistent.

(b) Consider three cases:

(i) $a_1 = 2, a_2 = 2, a_3 = 6$;
(ii) $a_1 = 1, a_2 = 0, a_3 = -2$;
(iii) $a_1 = 0, a_2 = 1, a_3 = 2$.

Determine whether the above linear system of equations is consistent. If it is, then give the corresponding solution.

1.22 Determine that for what values of α, the system of linear equations

$$\begin{aligned} (\alpha+3)x_1 + x_2 + 2x_3 &= \alpha, \\ 3(\alpha+1)x_1 + \alpha x_2 + (\alpha+3)x_3 &= 3, \\ \alpha x_1 + (\alpha-1)x_2 + x_3 &= \alpha, \end{aligned}$$

has a unique solution, no solution or infinitely many solutions. In the case where the system of linear equations has infinitely many solutions, find a general solution.

1.23 Set $\mathbf{a}_1 = [1,1,1,3]^T$, $\mathbf{a}_2 = [-1,-3,5,1]^T$, $\mathbf{a}_3 = [3,2,-1,p+2]^T$ and $\mathbf{a}_4 = [-2,-6,10,p]^T$.

(a) For what values of p are the above four vectors linearly independent? Use a linear combination of $\mathbf{a}_1, \ldots, \mathbf{a}_4$ to express $\mathbf{a} = [4,1,6,10]^T$.

(b) For what values of p are the above four vectors linearly dependent? Find the rank of the matrix $[\mathbf{a}_1, \mathbf{a}_2, \mathbf{a}_3, \mathbf{a}_4]$ in these cases.

1.24 Given

$$\mathbf{a}_1 = \begin{bmatrix} a \\ 2 \\ 10 \end{bmatrix}, \quad \mathbf{a}_2 = \begin{bmatrix} -2 \\ 1 \\ 5 \end{bmatrix}, \quad \mathbf{a}_3 = \begin{bmatrix} -1 \\ 1 \\ 4 \end{bmatrix}, \quad \mathbf{b} = \begin{bmatrix} 1 \\ b \\ c \end{bmatrix},$$

find respectively the values of a, b, c such that the following conditions are satisfied:

(a) $\mathbf{b}$ can be linearly expressed by $\mathbf{a}_1, \mathbf{a}_2, \mathbf{a}_3$ and this expression is unique;

(b) $\mathbf{b}$ cannot be linearly expressed by $\mathbf{a}_1, \mathbf{a}_2, \mathbf{a}_3$;

(c) $\mathbf{b}$ can be linearly expressed by $\mathbf{a}_1, \mathbf{a}_2, \mathbf{a}_3$, but not uniquely. Find a general expression for $\mathbf{b}$ in this case.

1.25 Show that

$$\mathbf{A} \text{ nonsingular} \quad \Leftrightarrow \quad \det \begin{bmatrix} \mathbf{A} & \mathbf{B} \\ \mathbf{C} & \mathbf{D} \end{bmatrix} = \det(\mathbf{A}) \det(\mathbf{D} - \mathbf{C}\mathbf{A}^{-1}\mathbf{B}).$$

1.26 Verify that the vector group

$$\left\{ \begin{bmatrix} 1 \\ 0 \\ 1 \\ 2 \end{bmatrix}, \begin{bmatrix} -1 \\ 1 \\ 1 \\ 0 \end{bmatrix}, \begin{bmatrix} -1 \\ -2 \\ 1 \\ 0 \end{bmatrix} \right\}$$

is a set of orthogonal vectors.

1.27 Suppose that $\mathbf{A} = \mathbf{A}^2$ is an idempotent matrix. Show that all its eigenvalues are either 1 or 0.

1.28 If $\mathbf{A}$ is an idempotent matrix, show that each of $\mathbf{A}^H, \mathbf{IA}$ and $\mathbf{I} - \mathbf{A}^H$ is a idempotent matrix.

1.29 Given the 3×5 matrix

$$\mathbf{A} = \begin{bmatrix} 1 & 3 & 2 & 5 & 7 \\ 2 & 1 & 0 & 6 & 1 \\ 1 & 1 & 2 & 5 & 4 \end{bmatrix},$$

use the MATLAB functions orth(A) and null(A) to find an orthogonal basis for the column space Span($\mathbf{A}$) and for the null space Null($\mathbf{A}$).

1.30 Let $\mathbf{x}$ and $\mathbf{y}$ be any two vectors in the Euclidean space $\mathbb{R}^n$. Show the Cauchy–Schwartz inequality $|\mathbf{x}^T\mathbf{y}| \leq \|\mathbf{x}\|\|\mathbf{y}\|$. (*Hint:* Observe that $\|\mathbf{x} - c\mathbf{y}\|^2 \geq 0$ holds for all scalars c.)

1.31 Define a transformation $H : \mathbb{R}^2 \to \mathbb{R}^2$ as

$$H(\mathbf{x}) = \begin{bmatrix} x_1 + x_2 - 1 \\ 3x_1 \end{bmatrix}, \qquad \mathbf{x} = \begin{bmatrix} x_1 \\ x_2 \end{bmatrix}.$$

Determine whether H is a linear transformation.

1.32 Put

$$\mathbf{P} = \begin{bmatrix} 0.5 & 0.2 & 0.3 \\ 0.3 & 0.8 & 0.3 \\ 0.2 & 0 & 0.4 \end{bmatrix}, \qquad \mathbf{x}_0 = \begin{bmatrix} 1 \\ 0 \\ 0 \end{bmatrix}.$$

Assume that the state vector of a particular system can be described by the Markov chain $\mathbf{x}_{k+1} = \mathbf{P}\mathbf{x}_k$, $k = 0, 1, \ldots$ Compute the state vectors $\mathbf{x}_1, \ldots, \mathbf{x}_{15}$ and analyze the change in the system over time.

1.33 Given a matrix function

$$\mathbf{A}(x) = \begin{bmatrix} 2x & -1 & x & 2 \\ 4 & x & 1 & -1 \\ 3 & 2 & x & 5 \\ 1 & -2 & 3 & x \end{bmatrix},$$

find $\mathrm{d}^3|\mathbf{A}(x)|/\mathrm{d}x^3$. (*Hint:* Expand the determinant $|\mathbf{A}(x)|$ according to any row or column; only the terms x^4 and x^3 are involved.)

1.34 Show that if $\mathbf{A}_1$ is nonsingular then

$$\begin{vmatrix} \mathbf{A}_1 & \mathbf{A}_2 \\ \mathbf{A}_3 & \mathbf{A}_4 \end{vmatrix} = |\mathbf{A}_1||\mathbf{A}_4 - \mathbf{A}_3\mathbf{A}_1^{-1}\mathbf{A}_2|.$$

1.35 Determine the positive definite status of the following quadratics:

(a) $f = -2x_1^2 - 8x_2^2 - 6x_3^2 + 2x_1x_2 + 2x_1x_3$,
(b) $f = x_1^2 + 4x_2^2 + 9x_3^2 + 15x_4^2 - 2x_1x_2 + 4x_1x_3 + 2x_1x_4 - 6x_2x_4 - 12x_3x_4$.

1.36 Show that

(a) if $\mathbf{B}$ is a real-valued nonsingular matrix then $\mathbf{A} = \mathbf{B}\mathbf{B}^T$ is positive definite.
(b) if $|\mathbf{C}| \neq 0$ then $\mathbf{A} = \mathbf{C}\mathbf{C}^H$ is positive definite.

1.37 Suppose that $\mathbf{A}$ and $\mathbf{B}$ are $n \times n$ real matrices. Show the following Cauchy–Schwartz inequalities for the trace function:

(a) $(\mathrm{tr}(\mathbf{A}^T\mathbf{B}))^2 \leq \mathrm{tr}(\mathbf{A}^T\mathbf{A}\mathbf{B}^T\mathbf{B})$, where the equality holds if and only if $\mathbf{AB}$ is a symmetric matrix;
(b) $(\mathrm{tr}(\mathbf{A}^T\mathbf{B}))^2 \leq \mathrm{tr}(\mathbf{A}\mathbf{A}^T\mathbf{B}\mathbf{B}^T)$, where the equality holds if and only if $\mathbf{A}^T\mathbf{B}$ is a symmetric matrix.

1.38 Show that $\mathrm{tr}(\mathbf{ABC}) = \mathrm{tr}(\mathbf{BCA}) = \mathrm{tr}(\mathbf{CAB})$.

1.39 Assuming that the inverse matrices appearing below exist, show the following results:

(a) $(\mathbf{A}^{-1}+\mathbf{I})^{-1}=\mathbf{A}(\mathbf{A}+\mathbf{I})^{-1}$.
(b) $(\mathbf{A}^{-1}+\mathbf{B}^{-1})^{-1}=\mathbf{A}(\mathbf{A}+\mathbf{B})^{-1}\mathbf{B}=\mathbf{B}(\mathbf{A}+\mathbf{B})^{-1}\mathbf{A}$.
(c) $(\mathbf{I}+\mathbf{AB})^{-1}\mathbf{A}=\mathbf{A}(\mathbf{I}+\mathbf{BA})^{-1}$.
(d) $\mathbf{A}-\mathbf{A}(\mathbf{A}+\mathbf{B})^{-1}\mathbf{A}=\mathbf{B}-\mathbf{B}(\mathbf{A}+\mathbf{B})^{-1}\mathbf{B}$.

1.40 Show that an eigenvalue of the inverse of a matrix $\mathbf{A}$ is equal to the reciprocal of the corresponding eigenvalue of the original matrix, i.e., $\text{eig}(\mathbf{A}^{-1})=1/\text{eig}(\mathbf{A})$.

1.41 Verify that $\mathbf{A}^\dagger=(\mathbf{A}^H\mathbf{A})^\dagger\mathbf{A}^H$ and $\mathbf{A}^\dagger=\mathbf{A}^H(\mathbf{A}\mathbf{A}^H)^\dagger$ meet the four Moore–Penrose conditions.

1.42 Find the Moore–Penrose inverse of the matrix $\mathbf{A}=[1,5,7]^T$.

1.43 Prove that $\mathbf{A}(\mathbf{A}^T\mathbf{A})^{-2}\mathbf{A}^T$ is the Moore–Penrose inverse of the matrix $\mathbf{A}\mathbf{A}^T$.

1.44 Given the matrix

$$\mathbf{X}=\begin{bmatrix}1 & 0 & -2\\ 0 & 1 & -1\\ -1 & 1 & 1\\ 2 & -1 & 2\end{bmatrix},$$

use respectively the recursive method (Algorithm 1.6) and the trace method (Algorithm 1.7) find the Moore–Penrose inverse matrix $\mathbf{X}^\dagger$.

1.45 Consider the mapping $\mathbf{UV}=\mathbf{W}$, where $\mathbf{U}\in\mathbb{C}^{m\times n},\mathbf{V}\in\mathbb{C}^{n\times p},\mathbf{W}\in\mathbb{C}^{m\times p}$ and $\mathbf{U}$ is a rank-deficient matrix. Show that $\mathbf{V}=\mathbf{U}^\dagger\mathbf{W}$, where $\mathbf{U}^\dagger\in\mathbb{C}^{n\times m}$ is the Moore–Penrose inverse of $\mathbf{U}$.

1.46 Show that if $\mathbf{Ax}=\mathbf{b}$ is consistent then its general solution is $\mathbf{x}=\mathbf{A}^\dagger\mathbf{b}+(\mathbf{I}-\mathbf{A}^\dagger\mathbf{A})\mathbf{z}$, where $\mathbf{A}^\dagger$ is the Moore–Penrose inverse of $\mathbf{A}$ and $\mathbf{z}$ is any constant vector.

1.47 Show that the linear mapping $T:V\mapsto W$ has the properties $T(\mathbf{0})=\mathbf{0}$ and $T(-\mathbf{x})=-T(\mathbf{x})$.

1.48 Show that, for any $m\times n$ matrix $\mathbf{A}$, its vectorization function is given by

$$\text{vec}(\mathbf{A})=(\mathbf{I}_n\otimes\mathbf{A})\text{vec}(\mathbf{I}_n)=(\mathbf{A}^T\otimes\mathbf{I}_n)\text{vec}(\mathbf{I}_m).$$

1.49 Show that the necessary and sufficient condition for $\mathbf{A}\otimes\mathbf{B}$ to be nonsingular is that both the matrices $\mathbf{A},\mathbf{B}$ are nonsingular. Furthermore, show that $(\mathbf{A}\otimes\mathbf{B})^{-1}=\mathbf{A}^{-1}\otimes\mathbf{B}^{-1}$.

1.50 Show that $(\mathbf{A}\otimes\mathbf{B})^\dagger=\mathbf{A}^\dagger\otimes\mathbf{B}^\dagger$.

1.51 If $\mathbf{A},\mathbf{B},\mathbf{C}$ are $m\times m$ matrices and $\mathbf{C}^T=\mathbf{C}$, show that

$$(\text{vec}(\mathbf{C}))^T(\mathbf{A}\otimes\mathbf{B})\,\text{vec}(\mathbf{C})=(\text{vec}(\mathbf{C}))^T(\mathbf{B}\otimes\mathbf{A})\,\text{vec}(\mathbf{C}).$$

1.52 Show that

$$\text{tr}(\mathbf{ABCD})=(\text{vec}(\mathbf{D}^T))^T(\mathbf{C}^T\otimes\mathbf{A})\,\text{vec}(\mathbf{B})=(\text{vec}(\mathbf{D})^T)(\mathbf{A}\otimes\mathbf{C}^T)\,\text{vec}(\mathbf{B}^T).$$

1.53 Let $\mathbf{A}, \mathbf{B} \in \mathbb{R}^{m\times n}$, show $\mathrm{tr}(\mathbf{A}^T\mathbf{B}) = (\mathrm{vec}(\mathbf{A}))^T \mathrm{vec}(\mathbf{B}) = \sum \mathrm{vec}(\mathbf{A}*\mathbf{B})$, where $\sum \mathrm{vec}(\cdot)$ represents the sum of all elements of the column vector functions.

1.54 Let $\mathbf{x}_i$ and $\mathbf{x}_j$ be two column vectors of the matrix $\mathbf{X}$, their covariance matrix is $\mathrm{Cov}\left(\mathbf{x}_i, \mathbf{x}_j^T\right) = \mathbf{M}_{ij}$. Hence the variance–covariance matrix of the vectorization function $\mathrm{vec}(\mathbf{X})$, denoted $\mathrm{Var}(\mathrm{vec}(\mathbf{X}))$ is a block matrix with the submatrix $\mathbf{M}_{ij}$, i.e., $\mathrm{Var}(\mathrm{vec}(\mathbf{X})) = \{\mathbf{M}_{ij}\}$. For the special cases of $\mathbf{M}_{ij} = m_{ij}\mathbf{V}$, if m_{ij} is the entry of the matrix $\mathbf{M}$, show that:

(a) $\mathrm{Var}(\mathrm{vec}(\mathbf{X})) = \mathbf{M} \otimes \mathbf{V}$;
(b) $\mathrm{Var}(\mathrm{vec}(\mathbf{TX})) = \mathbf{M} \otimes \mathbf{TVT}^T$;
(c) $\mathrm{Var}(\mathrm{vec}(\mathbf{X}^T)) = \mathbf{V} \otimes \mathbf{M}$.

1.55 Consider two $n \times n$ matrices $\mathbf{A}$ and $\mathbf{B}$.

(a) Let $\mathbf{d} = [d_1, \ldots, d_n]^T$ and $\mathbf{D} = \mathbf{Diag}(d_1, \ldots, d_n)$. Show that $\mathbf{d}^T(\mathbf{A}*\mathbf{B})\mathbf{d} = \mathrm{tr}(\mathbf{ADB}^T\mathbf{D})$.
(b) If both $\mathbf{A}$ and $\mathbf{B}$ are the positive definite matrices, show that the Hadamard product $\mathbf{A} * \mathbf{B}$ is a positive definite matrix. This property is known as the positive definiteness of the Hadamard product.

1.56 Show that $\mathbf{A} \otimes \mathbf{B} = \mathrm{vec}(\mathbf{BA}^T)$.

1.57 Show that $\mathrm{vec}(\mathbf{PQ}) = (\mathbf{Q}^T \otimes \mathbf{P})\,\mathrm{vec}(\mathbf{I}) = (\mathbf{Q}^T \otimes \mathbf{I})\,\mathrm{vec}(\mathbf{P}) = (\mathbf{I} \otimes \mathbf{P})\,\mathrm{vec}(\mathbf{Q})$.

1.58 Verify the following result:

$$\mathrm{tr}(\mathbf{XA}) = (\mathrm{vec}(\mathbf{A}^T))^T\mathrm{vec}(\mathbf{X}) = (\mathrm{vec}(\mathbf{X}^T))^T\mathrm{vec}(\mathbf{A}).$$

1.59 Let $\mathbf{A}$ be an $m \times n$ matrix, and $\mathbf{B}$ be a $p \times q$ matrix. Show

$$(\mathbf{A} \otimes \mathbf{B})\mathbf{K}_{nq} = \mathbf{K}_{mp}(\mathbf{B} \otimes \mathbf{A}), \quad \mathbf{K}_{pm}(\mathbf{A} \otimes \mathbf{B})\mathbf{K}_{nq} = \mathbf{B} \otimes \mathbf{A},$$

where $\mathbf{K}_{ij}$ is the commutation matrix.

1.60 For any $m \times n$ matrix $\mathbf{A}$ and $p \times 1$ vector $\mathbf{b}$, show the following results.

(a) $\mathbf{K}_{pm}(\mathbf{A} \otimes \mathbf{b}) = \mathbf{b} \otimes \mathbf{A}$.
(b) $\mathbf{K}_{mp}(\mathbf{b} \otimes \mathbf{A}) = \mathbf{A} \otimes \mathbf{b}$.
(c) $(\mathbf{A} \otimes \mathbf{b})\mathbf{K}_{np} = \mathbf{b}^T \otimes \mathbf{A}$.
(d) $(\mathbf{b}^T \otimes \mathbf{A})\mathbf{K}_{pn} = \mathbf{A} \otimes \mathbf{b}$.

1.61 Show that the Kronecker product of the $m \times m$ identity matrix and the $n \times n$ identity matrix yields the $mn \times mn$ identity matrix, namely $\mathbf{I}_m \otimes \mathbf{I}_n = \mathbf{I}_{mn}$.

1.62 Set $\mathbf{x} \in \mathbb{R}^{I\times JK}, \mathbf{G} \in \mathbb{R}^{P\times QR}, \mathbf{A} \in \mathbb{R}^{I\times P}, \mathbf{B} \in \mathbb{R}^{J\times Q}, \mathbf{C} \in \mathbb{R}^{K\times R}$ and $\mathbf{A}^T\mathbf{A} = \mathbf{I}_P, \mathbf{B}^T\mathbf{B} = \mathbf{I}_Q, \mathbf{C}^T\mathbf{C} = \mathbf{I}_R$. Show that if $\mathbf{X} = \mathbf{AG}(\mathbf{C} \otimes \mathbf{B})^T$ and $\mathbf{X}$ and $\mathbf{A}, \mathbf{B}, \mathbf{C}$ are given then the matrix $\mathbf{G}$ can be recovered or reconstructed from $\mathbf{G} = \mathbf{A}^T\mathbf{X}(\mathbf{C} \otimes \mathbf{B})$.

1.63 Use $\mathrm{vec}(\mathbf{UVW}) = (\mathbf{W}^T \otimes \mathbf{U})\,\mathrm{vec}(\mathbf{V})$ to find the vectorization representation of $\mathbf{X} = \mathbf{AG}(\mathbf{C} \otimes \mathbf{B})^T$.

1.64 Apply the Kronecker product to solve for the unknown matrix $\mathbf{X}$ in the matrix equation $\mathbf{AXB} = \mathbf{C}$.

1.65 Apply the Kronecker product to solve for the unknown matrix $\mathbf{X}$ in the generalized continuous-time Lyapunov equation $\mathbf{LX} + \mathbf{XN} = \mathbf{Y}$, where the dimensions of all matrices are $n \times n$ and the matrices L and N are invertible.

1.66 Show the following results:

$$
\begin{aligned}
&(\mathbf{A} \odot \mathbf{B}) * (\mathbf{C} \odot \mathbf{D}) = (\mathbf{A} * \mathbf{C}) \odot (\mathbf{B} * \mathbf{D}), \\
&(\mathbf{A} \odot \mathbf{B} \odot \mathbf{C})^T (\mathbf{A} \odot \mathbf{B} \odot \mathbf{C}) = (\mathbf{A}^T \mathbf{A}) * (\mathbf{B}^T \mathbf{B}) * (\mathbf{C}^T \mathbf{C}), \\
&(\mathbf{A} \odot \mathbf{B})^\dagger = \left((\mathbf{A}^T \mathbf{A}) * (\mathbf{B}^T \mathbf{B}) \right)^\dagger (\mathbf{A} \odot \mathbf{B})^T.
\end{aligned}
$$

1.67 Try extending the ℓ_0-norm of a vector to the ℓ_0-norm of an $m \times n$ matrix. Use this to define whether a matrix is K-sparse.

2

Special Matrices

In real-world applications, we often meet matrices with special structures. Such matrices are collectively known as special matrices. Understanding the internal structure of these special matrices is helpful for using them and for simplifying their representation. This chapter will focus on some of the more common special matrices with applications.

2.1 Hermitian Matrices

In linear algebra, the *adjoint* of a matrix $\mathbf{A}$ refers to its conjugate transpose and is commonly denoted by $\mathbf{A}^H$. If $\mathbf{A}^H = \mathbf{A}$, then $\mathbf{A}$ is said to be *self-adjoint*. A self-adjoint matrix $\mathbf{A} = \mathbf{A}^H \in \mathbb{C}^{n\times n}$ is customarily known as a *Hermitian matrix*.

A matrix $\mathbf{A}$ is known as *anti-Hermitian* if $\mathbf{A} = -\mathbf{A}^H$.

The *centro-Hermitian matrix* $\mathbf{R}$ is an $n \times n$ matrix whose entries have the symmetry $r_{ij} = r^*_{n-j+1,n-i+1}$.

Hermitian matrices have the following properties.

1. $\mathbf{A}$ is a Hermitian matrix, if and only if $\mathbf{x}^H\mathbf{A}\mathbf{x}$ is a real number for all complex vectors $\mathbf{x}$.
2. For all $\mathbf{A} \in \mathbb{C}^{n\times n}$, the matrices $\mathbf{A} + \mathbf{A}^H, \mathbf{A}\mathbf{A}^H$ and $\mathbf{A}^H\mathbf{A}$ are Hermitian.
3. If $\mathbf{A}$ is a Hermitian matrix, then $\mathbf{A}^k$ are Hermitian matrices for all $k = 1, 2, 3, \ldots$ If $\mathbf{A}$ is a nonsingular Hermitian matrix then its inverse, $\mathbf{A}^{-1}$, is Hermitian as well.
4. If $\mathbf{A}$ and $\mathbf{B}$ are Hermitian matrices then $\alpha\mathbf{A} + \beta\mathbf{B}$ is Hermitian for all real numbers α and β.
5. If $\mathbf{A}$ and $\mathbf{B}$ are Hermitian then $\mathbf{AB} + \mathbf{BA}$ and $\mathrm{j}(\mathbf{AB} - \mathbf{BA})$ are also Hermitian.
6. If $\mathbf{A}$ and $\mathbf{B}$ are anti-Hermitian matrices then $\alpha\mathbf{A} + \beta\mathbf{B}$ is anti-Hermitian for all real numbers α and β.
7. For all $\mathbf{A} \in \mathbb{C}^{n\times n}$, the matrix $\mathbf{A} - \mathbf{A}^H$ is anti-Hermitian.
8. If $\mathbf{A}$ is a Hermitian matrix then $\mathrm{j}\,\mathbf{A}$ ($\mathrm{j} = \sqrt{-1}$) is anti-Hermitian.
9. If $\mathbf{A}$ is an anti-Hermitian matrix, then $\mathrm{j}\,\mathbf{A}$ is Hermitian.

Positive definition criterion for a Hermitian matrix An $n \times n$ Hermitian matrix $\mathbf{A}$ is positive definite, if and only if any of the following conditions is satisfied.

- The quadratic form $\mathbf{x}^H \mathbf{A}\mathbf{x} > 0, \forall \mathbf{x} \neq \mathbf{0}$.
- The eigenvalues of $\mathbf{A}$ are all larger than zero.
- There is an $n \times n$ nonsingular matrix $\mathbf{R}$ such that $\mathbf{A} = \mathbf{R}^H\mathbf{R}$.
- There exists an $n \times n$ nonsingular matrix $\mathbf{P}$ such that $\mathbf{P}^H\mathbf{A}\mathbf{P}$ is positive definite.

Let $\mathbf{z}$ be a Gaussian random vector with zero mean. Then its correlation matrix $\mathbf{R}_{zz} = E\{\mathbf{z}\mathbf{z}^H\}$ is Hermitian, and is always positive definite, because the quadratic form $\mathbf{x}^H\mathbf{R}_{zz}\mathbf{x} = E\{|\mathbf{x}^H\mathbf{z}|^2\} > 0$.

Positive definite and positive semi-definite matrices obey the following inequalities [214, Section 8.7]:

(1) *Hadamard inequality* If an $m \times m$ matrix $\mathbf{A} = [a_{ij}]$ is positive definite then

$$\det(\mathbf{A}) \leqslant \prod_{i=1}^{m} a_{ii}$$

and equality holds if and only if $\mathbf{A}$ is a diagonal matrix.

(2) *Fischer inequality* If a block matrix $\mathbf{P} = \begin{bmatrix} \mathbf{A} & \mathbf{B} \\ \mathbf{B}^H & \mathbf{C} \end{bmatrix}$ is position definite, where submatrices $\mathbf{A}$ and $\mathbf{C}$ are the square nonzero matrices, then

$$\det(\mathbf{P}) \leqslant \det(\mathbf{A}) \det(\mathbf{C}).$$

(3) *Oppenheim inequality* For two $m \times m$ positive semi-definite matrices $\mathbf{A}$ and $\mathbf{B}$,

$$\det(\mathbf{A}) \prod_{i=1}^{m} b_{ii} \leqslant \det(\mathbf{A} \odot \mathbf{B}),$$

where $\mathbf{A} \odot \mathbf{B}$ is the Hadamard product of $\mathbf{A}$ and $\mathbf{B}$.

(4) *Minkowski inequality* For two $m \times m$ positive definite matrices $\mathbf{A}$ and $\mathbf{B}$, one has

$$\sqrt[n]{\det(\mathbf{A} + \mathbf{B})} \geq \sqrt[n]{\det(\mathbf{A})} + \sqrt[n]{\det(\mathbf{B})}.$$

2.2 Idempotent Matrix

In some applications one uses a multiple product of an $n \times n$ matrix $\mathbf{A}$, resulting into three special matrices: an idempotent matrix, a unipotent matrix and a tripotent matrix.

Definition 2.1 A matrix $\mathbf{A}_{n \times n}$ is *idempotent*, if $\mathbf{A}^2 = \mathbf{A}\mathbf{A} = \mathbf{A}$.

The eigenvalues of any idempotent matrix except for the identity matrix **I** take only the values 1 and 0, but a matrix whose eigenvalues take only 1 and 0 is not necessarily an idempotent matrix. For example, the matrix

$$\mathbf{B} = \frac{1}{8}\begin{bmatrix} 11 & 3 & 3 \\ 1 & 1 & 1 \\ -12 & -4 & -4 \end{bmatrix}$$

has three eigenvalues $1, 0, 0$, but it is not an idempotent matrix, because

$$\mathbf{B}^2 = \frac{1}{8}\begin{bmatrix} 11 & 3 & 3 \\ 0 & 0 & 0 \\ -11 & -3 & -3 \end{bmatrix} \neq \mathbf{B}.$$

An idempotent matrix has the following useful properties [433], [386]:

1. The eigenvalues of an idempotent matrix take only the values 1 and 0.
2. All idempotent matrices $\mathbf{A} \neq \mathbf{I}$ are singular.
3. The rank and trace of any idempotent matrix are equal, i.e., $\text{rank}(\mathbf{A}) = \text{tr}(\mathbf{A})$.
4. If $\mathbf{A}$ is an idempotent matrix then $\mathbf{A}^H$ is also an idempotent matrix, i.e., $\mathbf{A}^H\mathbf{A}^H = \mathbf{A}^H$.
5. If $\mathbf{A}$ is an idempotent $n \times n$ matrix then $\mathbf{I}_n - \mathbf{A}$ is also an idempotent matrix, and $\text{rank}(\mathbf{I}_n - \mathbf{A}) = n - \text{rank}(\mathbf{A})$.
6. All symmetric idempotent matrices are positive semi-definite.
7. Let the $n \times n$ idempotent matrix $\mathbf{A}$ have the rank r_A; then $\mathbf{A}$ has r_A eigenvalues equal to 1 and $n - r_A$ eigenvalues equal to zero.
8. A symmetric idempotent matrix $\mathbf{A}$ can be expressed as $\mathbf{A} = \mathbf{L}\mathbf{L}^T$, where $\mathbf{L}$ satisfies $\mathbf{L}^T\mathbf{L} = \mathbf{I}_{r_A}$ and r_A is the rank of $\mathbf{A}$.
9. All the idempotent matrices $\mathbf{A}$ are diagonalizable, i.e.,

$$\mathbf{U}^{-1}\mathbf{A}\mathbf{U} = \mathbf{\Sigma} = \begin{bmatrix} \mathbf{I}_{r_A} & \mathbf{O} \\ \mathbf{O} & \mathbf{O} \end{bmatrix}, \tag{2.2.1}$$

where $r_A = \text{rank}(\mathbf{A})$ and $\mathbf{U}$ is a unitary matrix.

Definition 2.2 A matrix $\mathbf{A}_{n\times n}$ is called *unipotent* or *involutory*, if $\mathbf{A}^2 = \mathbf{A}\mathbf{A} = \mathbf{I}$.

If an $n\times n$ matrix $\mathbf{A}$ is unipotent, then the function $f(\cdot)$ has the following property [386]:

$$f(s\mathbf{I} + t\mathbf{A}) = \frac{1}{2}[(\mathbf{I} + \mathbf{A})f(s+t) + (\mathbf{I} - \mathbf{A})f(s-t)]. \tag{2.2.2}$$

There exists a relationship between an idempotent matrix and an unipotent matrix: the matrix $\mathbf{A}$ is an unipotent matrix if and only if $\frac{1}{2}(\mathbf{A} + \mathbf{I})$ is an idempotent matrix.

Definition 2.3 A matrix $\mathbf{A}_{n\times n}$ is called a *nilpotent matrix*, if $\mathbf{A}^2 = \mathbf{A}\mathbf{A} = \mathbf{O}$.

If $\mathbf{A}$ is a nilpotent matrix, then the function $f(\cdot)$ has the following property [386]:

$$f(s\mathbf{I} + t\mathbf{A}) = \mathbf{I}f(s) + t\mathbf{A}f'(s), \tag{2.2.3}$$

where $f'(s)$ is the first-order derivative of $f(s)$.

Definition 2.4 A matrix $\mathbf{A}_{n\times n}$ is called *tripotent*, if $\mathbf{A}^3 = \mathbf{A}$.

It is easily seen that if $\mathbf{A}$ is a tripotent matrix then $-\mathbf{A}$ is also a tripotent matrix.

It should be noted that a tripotent matrix is not necessarily an idempotent matrix, although an idempotent matrix must be a triponent matrix (because if $\mathbf{A}^2 = \mathbf{A}$ then $\mathbf{A}^3 = \mathbf{A}^2\mathbf{A} = \mathbf{A}\mathbf{A} = \mathbf{A}$). In order to show this point, consider the eigenvalues of a tripotent matrix. Let λ be an eigenvalue of the tripotent matrix $\mathbf{A}$, and $\mathbf{u}$ the eigenvector corresponding to λ, i.e., $\mathbf{A}\mathbf{u} = \lambda\mathbf{u}$. Premultiplying both sides of $\mathbf{A}\mathbf{u} = \lambda\mathbf{u}$ by $\mathbf{A}$, we have $\mathbf{A}^2\mathbf{u} = \lambda\mathbf{A}\mathbf{u} = \lambda^2\mathbf{u}$. Premultiplying $\mathbf{A}^2\mathbf{u} = \lambda^2\mathbf{u}$ by $\mathbf{A}$, we immediately have

$$\mathbf{A}^3\mathbf{u} = \lambda^2\mathbf{A}\mathbf{u} = \lambda^3\mathbf{u}.$$

Because $\mathbf{A}^3 = \mathbf{A}$ is a tripotent matrix, the above equation can be written as $\mathbf{A}\mathbf{u} = \lambda^3\mathbf{u}$, and thus the eigenvalues of a tripotent matrix satisfies the relation $\lambda = \lambda^3$. That is to say, the eigenvalues of a tripotent matrix have three possible values -1, 0, $+1$; which is different from the eigenvalues of an idempotent matrix, which take only the values 0 and $+1$. In this sense, an idempotent matrix is a special tripotent matrix with no eigenvalue equal to -1.

2.3 Permutation Matrix

The $n \times n$ identity matrix $\mathbf{I} = [\mathbf{e}_1, \ldots, \mathbf{e}_n]$ consists of the basis vectors in an order such that the matrix is diagonal. On changing the order of the basis vectors, we obtain four special matrices: the permutation matrix, the commutation matrix, the exchange matrix and the selection matrix.

2.3.1 Permutation Matrix and Exchange Matrix

Definition 2.5 A square matrix is known as a *permutation matrix*, if each of its rows and columns have one and only one nonzero entry equal to 1.

A permutation matrix $\mathbf{P}$ has the following properties [60].

(1) $\mathbf{P}^T\mathbf{P} = \mathbf{P}\mathbf{P}^T = \mathbf{I}$, i.e., a permutation matrix is orthogonal.
(2) $\mathbf{P}^T = \mathbf{P}^{-1}$.
(3) $\mathbf{P}^T\mathbf{A}\mathbf{P}$ and $\mathbf{A}$ have the same diagonal entries but their order may be different.

EXAMPLE 2.1 Given the 5×4 matrix

$$\mathbf{A} = \begin{bmatrix} a_{11} & a_{12} & a_{13} & a_{14} \\ a_{21} & a_{22} & a_{23} & a_{24} \\ a_{31} & a_{32} & a_{33} & a_{34} \\ a_{41} & a_{42} & a_{43} & a_{44} \\ a_{51} & a_{52} & a_{53} & a_{54} \end{bmatrix}.$$

For the permutation matrices

$$\mathbf{P}_4 = \begin{bmatrix} 0 & 0 & 0 & 1 \\ 0 & 1 & 0 & 0 \\ 1 & 0 & 0 & 0 \\ 0 & 0 & 1 & 0 \end{bmatrix}, \quad \mathbf{P}_5 = \begin{bmatrix} 0 & 0 & 0 & 0 & 1 \\ 0 & 0 & 1 & 0 & 0 \\ 0 & 1 & 0 & 0 & 0 \\ 0 & 0 & 0 & 1 & 0 \\ 1 & 0 & 0 & 0 & 0 \end{bmatrix},$$

we have

$$\mathbf{P}_5\mathbf{A} = \begin{bmatrix} a_{51} & a_{52} & a_{53} & a_{54} \\ a_{31} & a_{32} & a_{33} & a_{34} \\ a_{21} & a_{22} & a_{23} & a_{24} \\ a_{41} & a_{42} & a_{43} & a_{44} \\ a_{11} & a_{12} & a_{13} & a_{14} \end{bmatrix}, \quad \mathbf{AP}_4 = \begin{bmatrix} a_{13} & a_{12} & a_{14} & a_{11} \\ a_{23} & a_{22} & a_{24} & a_{21} \\ a_{33} & a_{32} & a_{34} & a_{31} \\ a_{43} & a_{42} & a_{44} & a_{41} \\ a_{53} & a_{52} & a_{54} & a_{51} \end{bmatrix}.$$

That is, premultiplying an $m \times n$ matrix $\mathbf{A}$ by an $m \times m$ permutation matrix is equivalent to rearranging the rows of $\mathbf{A}$, and postmultiplying the matrix $\mathbf{A}$ by an $n \times n$ permutation matrix is equivalent to rearranging the columns of $\mathbf{A}$. The new sequences of rows or columns depend on the structures of the permutation matrices.

A $p \times q$ permutation matrix can be a random arrangement of q basis vectors $\mathbf{e}_1, \ldots, \mathbf{e}_q \in \mathbb{R}^{p\times 1}$. If the basis vectors are arranged according to some rule, the permutation matrix may become one of three special permutation matrices: a commutation matrix, an exchange matrix or a shift matrix.

1. Commutation Matrix

As stated in Section 1.11, the commutation matrix $\mathbf{K}_{mn}$ is defined as the special permutation matrix such that $\mathbf{K}_{mn}\text{vec}\,\mathbf{A}_{m\times n} = \text{vec}(\mathbf{A}^T)$. The role of this matrix is to commute the entry positions of an $mn \times 1$ vector in such a way that the new vector $\mathbf{K}_{mn}\text{vec}(\mathbf{A})$ becomes $\text{vec}(\mathbf{A}^T)$, so this special permutation matrix is known as the commutation matrix.

2. Exchange Matrix

The *exchange matrix* is usually represented using the notation $\mathbf{J}$, and is defined as

$$\mathbf{J} = \begin{bmatrix} 0 & & 1 \\ & 1 & \\ & \iddots & \\ 1 & & 0 \end{bmatrix} \tag{2.3.1}$$

which has every entry on the cross-diagonal line equal to 1, whereas all other entries in the matrix are equal to zero.

The exchange matrix is called the *reflection matrix* or *backward identity matrix*, since it can be regarded as a backward arrangement of the basis vectors, namely $\mathbf{J} = [\mathbf{e}_n, \mathbf{e}_{n-1}, \ldots, \mathbf{e}_1]$.

Premultiplying an $m \times n$ matrix $\mathbf{A}$ by the exchange matrix $\mathbf{J}_m$, we can invert the order of the rows of $\mathbf{A}$, i.e., we exchange all row positions with respect to its central horizontal axis, obtaining

$$\mathbf{J}_m\mathbf{A} = \begin{bmatrix} a_{m1} & a_{m2} & \cdots & a_{mn} \\ \vdots & \vdots & \vdots & \vdots \\ a_{21} & a_{22} & \cdots & a_{2n} \\ a_{11} & a_{12} & \cdots & a_{1n} \end{bmatrix}. \tag{2.3.2}$$

Similarly, postmultiplying $\mathbf{A}$ by the exchange matrix $\mathbf{J}_n$, we invert the order of the columns of $\mathbf{A}$, i.e., we exchange all column positions with respect to its central vertical axis:

$$\mathbf{A}\mathbf{J}_n = \begin{bmatrix} a_{1n} & \cdots & a_{12} & a_{11} \\ a_{2n} & \cdots & a_{22} & a_{21} \\ \vdots & \vdots & \vdots & \vdots \\ a_{mn} & \cdots & a_{m2} & a_{m1} \end{bmatrix}. \tag{2.3.3}$$

It is easily seen that

$$\mathbf{J}^2 = \mathbf{J}\mathbf{J} = \mathbf{I}, \qquad \mathbf{J}^T = \mathbf{J}. \tag{2.3.4}$$

Here $\mathbf{J}^2 = \mathbf{I}$ is the *involutory property* of the exchange matrix, and $\mathbf{J}^T = \mathbf{J}$ expresses its symmetry. That is, an exchange matrix is both an *involutory matrix* ($\mathbf{J}^2 = \mathbf{I}$) and a symmetric matrix ($\mathbf{J}^T = \mathbf{J}$).

As stated in Chapter 1, the commutation matrix has the property $\mathbf{K}_{mn}^T = \mathbf{K}_{mn}^{-1} = \mathbf{K}_{nm}$. Clearly, if $m = n$ then $\mathbf{K}_{nn}^T = \mathbf{K}_{nn}^{-1} = \mathbf{K}_{nn}$. This implies that

$$\mathbf{K}_{nn}^2 = \mathbf{K}_{nn}\mathbf{K}_{nn} = \mathbf{I}_{nn}, \tag{2.3.5}$$

$$\mathbf{K}_{nn}^T = \mathbf{K}_{nn}. \tag{2.3.6}$$

That is to say, like the exchange matrix, a square commutation matrix is both an involutory matrix and a symmetric matrix.

In particular, when premultiplying an $m \times n$ matrix $\mathbf{A}$ by an $m \times m$ exchange matrix $\mathbf{J}_m$ and postmultiplying an $n \times n$ exchange matrix $\mathbf{J}_n$, we have

$$\mathbf{J}_m\mathbf{A}\mathbf{J}_n = \begin{bmatrix} a_{m,n} & a_{m,n-1} & \cdots & a_{m,1} \\ a_{m-1,n} & a_{m-1,n-1} & \cdots & a_{m-1,1} \\ \vdots & \vdots & \vdots & \vdots \\ a_{1,n} & a_{1,n-1} & \cdots & a_{1,1} \end{bmatrix}. \tag{2.3.7}$$

Similarly, the operation $\mathbf{Jc}$ inverts the element order of the column vector $\mathbf{c}$ and $\mathbf{c}^T\mathbf{J}$ inverts the element order of the row vector $\mathbf{c}^T$.

3. Shift Matrix

The $n \times n$ *shift matrix* is defined as

$$\mathbf{P} = \begin{bmatrix} 0 & 1 & 0 & \cdots & 0 \\ 0 & 0 & 1 & \cdots & 0 \\ \vdots & \vdots & \vdots & \ddots & \vdots \\ 0 & 0 & 0 & \cdots & 1 \\ 1 & 0 & 0 & \cdots & 0 \end{bmatrix}. \tag{2.3.8}$$

In other words, the entries of the shift matrix are $p_{i,i+1} = 1\,(1 \leqslant i \leqslant n-1)$ and $p_{n1} = 1$; all other entries are equal to zero. Evidently, a shift matrix can be represented using basis vectors as $\mathbf{P} = [\mathbf{e}_n, \mathbf{e}_1, \ldots, \mathbf{e}_{n-1}]$.

Given an $m \times n$ matrix $\mathbf{A}$, the operation $\mathbf{P}_m\mathbf{A}$ will make $\mathbf{A}$'s first row move to its lowermost row, namely

$$\mathbf{P}_m\mathbf{A} = \begin{bmatrix} a_{21} & a_{22} & \cdots & a_{2n} \\ \vdots & \vdots & \vdots & \vdots \\ a_{m1} & a_{m2} & \cdots & a_{mn} \\ a_{11} & a_{12} & \cdots & a_{1n} \end{bmatrix}.$$

Similarly, postmultiplying $\mathbf{A}$ by an $n \times n$ shift matrix $\mathbf{P}_n$, we have

$$\mathbf{AP}_n = \begin{bmatrix} a_{1n} & a_{11} & \cdots & a_{1,n-1} \\ a_{2n} & a_{21} & \cdots & a_{2,n-1} \\ \vdots & \vdots & \vdots & \vdots \\ a_{mn} & a_{m1} & \cdots & a_{m,n-1} \end{bmatrix}.$$

The role of $\mathbf{P}_n$ is to move the rightmost column of $\mathbf{A}$ to the leftmost position.

It is easy to see that, unlike a square exchange matrix $\mathbf{J}_n$ or a square commutation matrix $\mathbf{K}_{nn}$, a shift matrix is neither involutory nor symmetric.

2.3.2 Generalized Permutation Matrix

1. Generalized Permutation Matrix

Consider the observation-data model

$$\mathbf{x}(t) = \mathbf{As}(t) = \sum_{i=1}^{n} \mathbf{a}_i s_i(t), \tag{2.3.9}$$

where $\mathbf{s}(t) = [s_1(t), \ldots, s_n(t)]^T$ represents a source vector and $\mathbf{A} = [\mathbf{a}_1, \ldots, \mathbf{a}_n]$, with $\mathbf{a}_i \in \mathbb{R}^{m\times 1}$, $i = 1, \cdots, n$, is an $m \times n$ constant-coefficient matrix ($m \geq n$)

that represents the linear mixing process of n sources and is known as the *mixture matrix*. The mixture matrix $\mathbf{A}$ is assumed to be of full-column rank.

The question is how to recover the source vector $\mathbf{s}(t) = [s_1(t), \ldots, s_n(t)]^T$ using only the m-dimensional observation-data vector $\mathbf{x}(t)$. This is the well-known blind source separation (BSS) problem. Here, the terminology "blind" has two meanings: (a) the sources $s_1(t), \ldots, s_n(t)$ are unobservable; (b) how the n sources are mixed is unknown (i.e., the mixture matrix $\mathbf{A}$ is unknown).

The core issue of the BSS problem is to identify the generalized inverse of the mixture matrix, $\mathbf{A}^\dagger = (\mathbf{A}^T\mathbf{A})^{-1}\mathbf{A}^T$, because the source vector $\mathbf{s}(t)$ is then easily recovered using $\mathbf{s}(t) = \mathbf{A}^\dagger\mathbf{x}(t)$. Unfortunately, in mixture-matrix identification there exist two uncertainties or ambiguities.

(1) If the ith and jth source signals are exchanged in order, and the ith and the jth columns of the mixture matrix $\mathbf{A}$ are exchanged at the same time, then the observation data vector $\mathbf{x}(t)$ is unchanged. This shows that the source signal ordering cannot be identified from only the observation-data vector $\mathbf{x}(t)$. Such an ambiguity is known as the ordering uncertainty of separated signals.

(2) Using only the observation data vector $\mathbf{x}(t)$, it is impossible to identify accurately the amplitudes, $s_i(t), i = 1, \ldots, n$, of the source signals, because

$$\mathbf{x}(t) = \sum_{i=1}^{n} \frac{\mathbf{a}_i}{\alpha_i} \alpha_i s_i(t), \tag{2.3.10}$$

where the α_i are unknown scalars. This ambiguity is called the amplitude uncertainty of the separate signals.

The two uncertainties above can be described by a generalized permutation matrix.

Definition 2.6 An $m \times m$ matrix $\mathbf{G}$ is known as a *generalized permutation matrix* (or a *g-matrix*), if each row and each column has one and only one nonzero entry.

It is easy to show that a square matrix is a g-matrix if and only if it can be decomposed into the product $\mathbf{G}$ of a permutation matrix $\mathbf{P}$ and a nonsingular diagonal matrix $\mathbf{D}$:

$$\mathbf{G} = \mathbf{PD}. \tag{2.3.11}$$

For instance,

$$\mathbf{G} = \begin{bmatrix} 0 & 0 & 0 & 0 & \alpha \\ 0 & 0 & \beta & 0 & 0 \\ 0 & \gamma & 0 & 0 & 0 \\ 0 & 0 & 0 & \lambda & 0 \\ \rho & 0 & 0 & 0 & 0 \end{bmatrix} = \begin{bmatrix} 0 & 0 & 0 & 0 & 1 \\ 0 & 0 & 1 & 0 & 0 \\ 0 & 1 & 0 & 0 & 0 \\ 0 & 0 & 0 & 1 & 0 \\ 1 & 0 & 0 & 0 & 0 \end{bmatrix} \begin{bmatrix} \rho & & & & 0 \\ & \gamma & & & \\ & & \beta & & \\ & & & \lambda & \\ 0 & & & & \alpha \end{bmatrix}.$$

By the above definition, if premultiplying (or postmultiplying) a matrix $\mathbf{A}$ by a g-matrix, then the rows (or columns) of $\mathbf{A}$ will be rearranged, and all entries of each row (or column) of $\mathbf{A}$ will be multiplied by a scalar factor. For example,

$$\begin{bmatrix} 0 & 0 & 0 & 0 & \alpha \\ 0 & 0 & \beta & 0 & 0 \\ 0 & \gamma & 0 & 0 & 0 \\ 0 & 0 & 0 & \lambda & 0 \\ \rho & 0 & 0 & 0 & 0 \end{bmatrix} \begin{bmatrix} a_{11} & a_{12} & a_{13} & a_{14} \\ a_{21} & a_{22} & a_{23} & a_{24} \\ a_{31} & a_{32} & a_{33} & a_{34} \\ a_{41} & a_{42} & a_{43} & a_{44} \\ a_{51} & a_{52} & a_{53} & a_{54} \end{bmatrix} = \begin{bmatrix} \alpha a_{51} & \alpha a_{52} & \alpha a_{53} & \alpha a_{54} \\ \beta a_{31} & \beta a_{32} & \beta a_{33} & \beta a_{34} \\ \gamma a_{21} & \gamma a_{22} & \gamma a_{23} & \gamma a_{24} \\ \lambda a_{41} & \lambda a_{42} & \lambda a_{43} & \lambda a_{44} \\ \rho a_{11} & \rho a_{12} & \rho a_{13} & \rho a_{14} \end{bmatrix}.$$

Using the g-matrix, Equation (2.3.10) can be equivalently written as

$$\mathbf{x}(t) = \mathbf{A}\mathbf{G}^{-1}(\mathbf{G}\mathbf{s})(t) = \mathbf{A}\mathbf{G}^{-1}\tilde{\mathbf{s}}(t), \tag{2.3.12}$$

where $\tilde{\mathbf{s}}(t) = \mathbf{G}\mathbf{s}(t)$.

Therefore, the task of BSS is now to identify the product $\mathbf{A}\mathbf{G}^{-1}$ of the matrix $\mathbf{A}$ and the generalized inverse matrix $\mathbf{G}^{-1}$ rather than the generalized inverse $\mathbf{A}^{\dagger}$, which greatly eases the solution of the BSS problem. That is to say, the solution of the BSS is now given by $\tilde{\mathbf{s}}(t) = (\mathbf{A}\mathbf{G}^{-1})^{\dagger}\mathbf{x}(t) = \mathbf{G}\mathbf{A}^{\dagger}\mathbf{x}(t)$.

If we let $\mathbf{W} = \mathbf{G}\mathbf{A}^{\dagger}$ be a de-mixing or separation matrix, then adaptive BSS is used to update adaptively both the de-mixing matrix $\mathbf{W}(t)$ and the reconstructed source signal vector $\tilde{\mathbf{s}}(t) = \mathbf{W}(t)\mathbf{x}(t)$.

2. *Selection Matrix*

As the name implies, the *selection matrix* is a type of matrix that can select certain rows or columns of a matrix. Let $\mathbf{x}_i = [x_i(1), \ldots, x_i(N)]^T$ and, given an $m \times N$ matrix

$$\mathbf{X} = \begin{bmatrix} x_1(1) & x_1(2) & \cdots & x_1(N) \\ x_2(1) & x_2(2) & \cdots & x_2(N) \\ \vdots & \vdots & \vdots & \vdots \\ x_m(1) & x_m(2) & \cdots & x_m(N) \end{bmatrix},$$

we have

$$\mathbf{J}_1\mathbf{X} = \begin{bmatrix} x_1(1) & x_1(2) & \cdots & x_1(N) \\ x_2(1) & x_2(2) & \cdots & x_2(N) \\ \vdots & \vdots & \vdots & \vdots \\ x_{m-1}(1) & x_{m-1}(2) & \cdots & x_{m-1}(N) \end{bmatrix} \quad \text{if} \quad \mathbf{J}_1 = [\mathbf{I}_{m-1}, \mathbf{0}_{m-1}],$$

$$\mathbf{J}_2\mathbf{X} = \begin{bmatrix} x_2(1) & x_2(2) & \cdots & x_2(N) \\ x_3(1) & x_3(2) & \cdots & x_3(N) \\ \vdots & \vdots & \vdots & \vdots \\ x_m(1) & x_m(2) & \cdots & x_m(N) \end{bmatrix} \quad \text{if} \quad \mathbf{J}_2 = [\mathbf{0}_{m-1}, \mathbf{I}_{m-1}].$$

That is, the matrix $\mathbf{J}_1\mathbf{X}$ selects the uppermost $m-1$ rows of the original matrix $\mathbf{X}$, and the matrix $\mathbf{J}_2\mathbf{X}$ selects the lowermost $m-1$ rows of $\mathbf{X}$.

Similarly, we have

$$\mathbf{X}\mathbf{J}_1 = \begin{bmatrix} x_1(1) & x_1(2) & \cdots & x_1(N-1) \\ x_2(1) & x_2(2) & \cdots & x_2(N-1) \\ \vdots & \vdots & \vdots & \vdots \\ x_m(1) & x_m(2) & \cdots & x_m(N-1) \end{bmatrix} \quad \text{if} \quad \mathbf{J}_1 = \begin{bmatrix} \mathbf{I}_{N-1} \\ \mathbf{0}_{N-1} \end{bmatrix},$$

$$\mathbf{X}\mathbf{J}_2 = \begin{bmatrix} x_1(2) & x_1(3) & \cdots & x_1(N) \\ x_2(2) & x_2(3) & \cdots & x_2(N) \\ \vdots & \vdots & \vdots & \vdots \\ x_m(2) & x_m(3) & \cdots & x_m(N) \end{bmatrix} \quad \text{if} \quad \mathbf{J}_2 = \begin{bmatrix} \mathbf{0}_{N-1} \\ \mathbf{I}_{N-1} \end{bmatrix}.$$

In other words, the matrix $\mathbf{X}\mathbf{J}_1$ selects the leftmost $N-1$ columns of the matrix $\mathbf{X}$, and the matrix $\mathbf{X}\mathbf{J}_2$ selects the rightmost $N-1$ columns of $\mathbf{X}$.

2.4 Orthogonal Matrix and Unitary Matrix

The vectors $\mathbf{x}_1, \ldots, \mathbf{x}_k \in \mathbb{C}^n$ constitute an *orthogonal set* if $\mathbf{x}_i^H \mathbf{x}_j = 0,\ 1 \le i < j \le k$. Moreover, if the vectors are normalized, i.e., $\|\mathbf{x}\|_2^2 = \mathbf{x}_i^H \mathbf{x}_i = 1,\ i = 1, \ldots, k$, then the orthogonal set is known as an *orthonormal set.*

THEOREM 2.1 *A set of orthogonal nonzero vectors is linearly independent.*

Proof Suppose that $\{\mathbf{x}_1, \ldots, \mathbf{x}_k\}$ is an orthogonal set, and $\mathbf{0} = \alpha_1 \mathbf{x}_1 + \cdots + \alpha_k \mathbf{x}_k$. Then we have

$$0 = \mathbf{0}^H \mathbf{0} = \sum_{i=1}^{k} \sum_{j=1}^{k} \alpha_i^* \alpha_j \mathbf{x}_i^H \mathbf{x}_j = \sum_{i=1}^{k} |\alpha_i|^2 \mathbf{x}_i^H \mathbf{x}_i.$$

Since the vectors are orthogonal, and $\mathbf{x}_i^H \mathbf{x}_i > 0$, $\sum_{i=1}^{k} |\alpha_i|^2 \mathbf{x}_i^H \mathbf{x}_i = 0$ implies that all $|\alpha_i|^2 = 0$ and so all $\alpha_i = 0$, and thus $\mathbf{x}_1, \ldots, \mathbf{x}_k$ are linearly independent. □

DEFINITION 2.7 A real square matrix $\mathbf{Q} \in \mathbb{R}^{n \times n}$ is *orthogonal*, if

$$\mathbf{Q}\mathbf{Q}^{\mathrm{T}} = \mathbf{Q}^{\mathrm{T}}\mathbf{Q} = \mathbf{I}. \tag{2.4.1}$$

A complex square matrix $\mathbf{U} \in \mathbb{C}^{n \times n}$ is *unitary matrix* if

$$\mathbf{U}\mathbf{U}^H = \mathbf{U}^H\mathbf{U} = \mathbf{I}. \tag{2.4.2}$$

DEFINITION 2.8 A real $m \times n$ matrix $\mathbf{Q}_{m \times n}$ is called *semi-orthogonal* if it satisfies only $\mathbf{Q}\mathbf{Q}^T = \mathbf{I}_m$ or $\mathbf{Q}^T\mathbf{Q} = \mathbf{I}_n$. Similarly, a complex matrix $\mathbf{U}_{m \times n}$ is known as *para-unitary* if it satisfies only $\mathbf{U}\mathbf{U}^H = \mathbf{I}_m$ or $\mathbf{U}^H\mathbf{U} = \mathbf{I}_n$.

THEOREM 2.2 [214] *If* $\mathbf{U} \in \mathbb{C}^{n \times n}$ *then the following statements are equivalent:*

- *$\mathbf{U}$ is a unitary matrix;*
- *$\mathbf{U}$ is nonsingular and $\mathbf{U}^H = \mathbf{U}^{-1}$;*
- *$\mathbf{U}\mathbf{U}^H = \mathbf{U}^H\mathbf{U} = \mathbf{I}$;*
- *$\mathbf{U}^H$ is a unitary matrix;*
- *the columns of $\mathbf{U} = [\mathbf{u}_1, \ldots, \mathbf{u}_n]$ constitute an orthonormal set, namely*

$$\mathbf{u}_i^H \mathbf{u}_j = \delta(i-j) = \begin{cases} 1, & i = j, \\ 0, & i \neq j; \end{cases}$$

- *the rows of $\mathbf{U}$ constitute an orthonormal set;*
- *for all $\mathbf{x} \in \mathbb{C}^n$, the Euclidean lengths of $\mathbf{y} = \mathbf{U}\mathbf{x}$ and $\mathbf{x}$ are the same, namely $\|\mathbf{y}\|_2 = \|\mathbf{x}\|_2$.*

If a linear transformation matrix $\mathbf{A}$ is unitary, then the linear transformation $\mathbf{A}\mathbf{x}$ is known as a *unitary transformation.* A unitary transformation has the following properties.

(1) The vector inner product is invariant under unitary transformation: $\langle \mathbf{x}, \mathbf{y}\rangle = \langle \mathbf{A}\mathbf{x}, \mathbf{A}\mathbf{y}\rangle$, since $\langle \mathbf{A}\mathbf{x}, \mathbf{A}\mathbf{y}\rangle = (\mathbf{A}\mathbf{x})^H \mathbf{A}\mathbf{y} = \mathbf{x}^H\mathbf{A}^H\mathbf{A}\mathbf{y} = \mathbf{x}^H\mathbf{y} = \langle \mathbf{x}, \mathbf{y}\rangle$.

(2) The vector norm is invariant under unitary transformation, i.e., $\|\mathbf{A}\mathbf{x}\|^2 = \|\mathbf{x}\|^2$, since $\|\mathbf{A}\mathbf{x}\|^2 = \langle \mathbf{A}\mathbf{x}, \mathbf{A}\mathbf{x}\rangle = \langle \mathbf{x}, \mathbf{x}\rangle = \|\mathbf{x}\|^2$.

(3) The angle between the vectors $\mathbf{A}\mathbf{x}$ and $\mathbf{A}\mathbf{y}$ is also invariant under unitary transformation, namely

$$\cos\theta = \frac{\langle \mathbf{A}\mathbf{x}, \mathbf{A}\mathbf{y}\rangle}{\|\mathbf{A}\mathbf{x}\|\|\mathbf{A}\mathbf{y}\|} = \frac{\langle \mathbf{x}, \mathbf{y}\rangle}{\|\mathbf{x}\|\|\mathbf{y}\|}. \tag{2.4.3}$$

This is the result of combining the former two properties.

The determinant of a unitary matrix is equal to 1, i.e.,

$$|\det(\mathbf{A})| = 1, \quad \text{if } \mathbf{A} \text{ is unitary.} \tag{2.4.4}$$

Definition 2.9 The matrix $\mathbf{B} \in \mathbb{C}^{n\times n}$ such that $\mathbf{B} = \mathbf{U}^H\mathbf{A}\mathbf{U}$ is said to be *unitarily equivalent* to $\mathbf{A} \in \mathbb{C}^{n\times n}$. If $\mathbf{U}$ is a real orthogonal matrix then we say that $\mathbf{B}$ is orthogonally equivalent to $\mathbf{A}$.

Definition 2.10 The matrix $\mathbf{A} \in \mathbb{C}^{n\times n}$ is known as a *normal matrix* if $\mathbf{A}^H\mathbf{A} = \mathbf{A}\mathbf{A}^H$.

The following list summarizes the properties of a unitary matrix [307].

1. $\mathbf{A}_{m\times m}$ is a unitary matrix $\Leftrightarrow$ the columns of $\mathbf{A}$ are orthonormal vectors.
2. $\mathbf{A}_{m\times m}$ is a unitary matrix $\Leftrightarrow$ the rows of $\mathbf{A}$ are orthonormal vectors.
3. $\mathbf{A}_{m\times m}$ is real that $\mathbf{A}$ is a unitary matrix $\Leftrightarrow$ $\mathbf{A}$ is orthogonal.

4. $\mathbf{A}_{m\times m}$ is a unitary matrix $\Leftrightarrow$ $\mathbf{A}\mathbf{A}^H = \mathbf{A}^H\mathbf{A} = \mathbf{I}_m$
$\Leftrightarrow$ $\mathbf{A}^T$ is a unitary matrix
$\Leftrightarrow$ $\mathbf{A}^H$ is a unitary matrix
$\Leftrightarrow$ $\mathbf{A}^*$ is a unitary matrix
$\Leftrightarrow$ $\mathbf{A}^{-1}$ is a unitary matrix
$\Leftrightarrow$ $\mathbf{A}^i$ is a unitary matrix, $i = 1, 2, \ldots$
5. $\mathbf{A}_{m\times m}, \mathbf{B}_{m\times m}$ are unitary matrices $\Rightarrow$ $\mathbf{AB}$ is a unitary matrix.
6. If $\mathbf{A}_{m\times m}$ is a unitary matrix then
   - $|\det(\mathbf{A})| = 1$.
   - $\text{rank}(\mathbf{A}) = m$.
   - $\mathbf{A}$ is normal, i.e., $\mathbf{A}\mathbf{A}^H = \mathbf{A}^H\mathbf{A}$.
   - λ is an eigenvalue of $\mathbf{A}$ $\Rightarrow$ $|\lambda| = 1$.
   - For a matrix $\mathbf{B}_{m\times n}$, $\|\mathbf{AB}\|_F = \|\mathbf{B}\|_F$.
   - For a matrix $\mathbf{B}_{n\times m}$, $\|\mathbf{BA}\|_F = \|\mathbf{B}\|_F$.
   - For a vector $\mathbf{x}_{m\times 1}$, $\|\mathbf{Ax}\|_2 = \|\mathbf{x}\|_2$.
7. If $\mathbf{A}_{m\times m}$ and $\mathbf{B}_{n\times n}$ are two unitary matrices then
   - $\mathbf{A} \oplus \mathbf{B}$ is a unitary matrix.
   - $\mathbf{A} \otimes \mathbf{B}$ is a unitary matrix.

Definition 2.11 An $N \times N$ diagonal matrix whose diagonal entries take the values either $+1$ or -1 is known as a *signature matrix*.

Definition 2.12 Let $\mathbf{J}$ be an $N\times N$ signature matrix. Then the $N\times N$ matrix $\mathbf{Q}$ such that $\mathbf{QJQ}^T = \mathbf{J}$ is called the *J-orthogonal matrix* or the *hypernormal matrix*.

From the above definition we know that when a signature matrix equals the identity matrix, i.e., $\mathbf{J} = \mathbf{I}$, the J-orthogonal matrix $\mathbf{Q}$ reduces to an orthogonal matrix.

Clearly, a J-orthogonal matrix $\mathbf{Q}$ is nonsingular, and the absolute value of its determinant is equal to 1.

Any $N \times N$ J-orthogonal matrix $\mathbf{Q}$ can be equivalently defined by

$$\mathbf{Q}^T\mathbf{JQ} = \mathbf{J}. \tag{2.4.5}$$

Substituting the definition formula $\mathbf{QJQ}^T = \mathbf{J}$ into the right-hand side of Equation (2.4.5), it immediately follows that

$$\mathbf{Q}^T\mathbf{JQ} = \mathbf{QJQ}^T. \tag{2.4.6}$$

This symmetry is known as the *hyperbolic symmetry* of the signature matrix $\mathbf{J}$.

The matrix

$$\mathbf{Q} = \mathbf{J} - 2\frac{\mathbf{v}\mathbf{v}^T}{\mathbf{v}^T\mathbf{J}\mathbf{v}} \tag{2.4.7}$$

is called the *hyperbolic Householder matrix* [71]. In particular, if $\mathbf{J} = \mathbf{I}$ then the hyperbolic Householder matrix reduces to the *Householder matrix*.

2.5 Band Matrix and Triangular Matrix

A triangular matrix is one of the standard forms of matrix decomposition; the triangular matrix itself is a special example of the band matrix.

2.5.1 Band Matrix

A matrix $\mathbf{A} \in \mathbb{C}^{m\times n}$ such that $a_{ij} = 0, |i-j| > k$, is known as a *band matrix*. In particular, if $a_{ij} = 0, \forall\, i > j+p$, then $\mathbf{A}$ is said to have *under bandwidth* p, and if $a_{ij} = 0, \forall\, j > i+q$, then $\mathbf{A}$ is said to have *upper bandwidth* q. The following is an example of a 7×5 band matrix with under bandwidth 1 and upper bandwidth 2:

$$\begin{bmatrix} \times & \times & \times & 0 & 0 \\ \times & \times & \times & \times & 0 \\ 0 & \times & \times & \times & \times \\ 0 & 0 & \times & \times & \times \\ 0 & 0 & 0 & \times & \times \\ 0 & 0 & 0 & 0 & \times \\ 0 & 0 & 0 & 0 & 0 \end{bmatrix},$$

where $\times$ denotes any nonzero entry.

A special form of band matrix is of particular interest. This is the tridiagonal matrix. A matrix $\mathbf{A} \in \mathbb{C}^{n\times n}$ is *tridiagonal*, if its entries $a_{ij} = 0$ when $|i-j| > 1$. Clearly, the tridiagonal matrix is a square band matrix with upper bandwidth 1 and under bandwidth 1. However, the tridiagonal matrix is a special example of the Hessenberg matrix, which is defined below.

An $n\times n$ matrix $\mathbf{A}$ is known as an *upper Hessenberg matrix* if it has the form

$$\mathbf{A} = \begin{bmatrix} a_{11} & a_{12} & a_{13} & \cdots & a_{1n} \\ a_{21} & a_{22} & a_{23} & \cdots & a_{2n} \\ 0 & a_{32} & a_{33} & \cdots & a_{3n} \\ 0 & 0 & a_{43} & \cdots & a_{4n} \\ \vdots & \vdots & \vdots & \ddots & \vdots \\ 0 & 0 & 0 & \cdots & a_{nn} \end{bmatrix}.$$

A matrix $\mathbf{A}$ is known as a *lower Hessenberg matrix* if $\mathbf{A}^T$ is a upper Hessenberg matrix.

2.5.2 Triangular Matrix

Two common special band matrices are the upper triangular matrix and the lower triangular matrix. The triangular matrix is a canonical form in matrix decomposition.

A square matrix $\mathbf{U} = [u_{ij}]$ with $u_{ij} = 0$, $i > j$, is called an *upper triangular matrix*, and its general form is

$$\mathbf{U} = \begin{bmatrix} u_{11} & u_{12} & \cdots & u_{1n} \\ 0 & u_{22} & \cdots & u_{2n} \\ \vdots & \vdots & \ddots & \vdots \\ 0 & 0 & \cdots & u_{nn} \end{bmatrix} \quad \Rightarrow \quad |\mathbf{U}| = u_{11}u_{22}\cdots u_{nn}.$$

A square matrix $\mathbf{L} = [l_{ij}]$ such that $l_{ij} = 0$, $i < j$, is known as the *lower triangular matrix*, and its general form is

$$\mathbf{L} = \begin{bmatrix} l_{11} & & & 0 \\ l_{12} & l_{22} & & \\ \vdots & \vdots & \ddots & \\ l_{n1} & l_{n2} & \cdots & l_{nn} \end{bmatrix} \quad \Rightarrow \quad |\mathbf{L}| = l_{11}l_{22}\cdots l_{nn}.$$

In fact, a triangular matrix is not only an upper Hessenberg matrix but also a lower Hessenberg matrix.

Summarizing the definitions on triangular matrixes, a square matrix $\mathbf{A} = [a_{ij}]$ is said to be

(1) *lower triangular* if $a_{ij} = 0\,(i < j)$;
(2) *strict lower triangular* if $a_{ij} = 0\ (i \le j)$;
(3) *unit triangular* if $a_{ij} = 0\ (i < j)$, $a_{ii} = 1\ (\forall\ i)$;
(4) *upper triangular* if $a_{ij} = 0\ (i > j)$;
(5) *strict upper triangular* if $a_{ij} = 0\ (i \ge j)$;
(6) *unit upper triangular* if $a_{ij} = 0\ (i > j)$, $a_{ii} = 1\ (\forall\ i)$.

1. Properties of Upper Triangular Matrices

1. The product of upper triangular matrices is also an upper triangular matrix; i.e., if $\mathbf{U}_1, \ldots, \mathbf{U}_k$ are upper triangular matrices then $\mathbf{U} = \mathbf{U}_1 \cdots \mathbf{U}_k$ is also an upper triangular matrix.
2. The determinant of an upper triangular matrix $\mathbf{U} = [u_{ij}]$ is equal to the product of its diagonal entries:

$$\det(\mathbf{U}) = u_{11}\cdots u_{nn} = \prod_{i=1}^{n} u_{ii}.$$

3. The inverse of an upper triangular matrix is also an upper triangular matrix.
4. The kth power $\mathbf{U}^k$ of an upper triangular matrix $\mathbf{U}_{n\times n}$ is also an upper triangular matrix, and its ith diagonal entry is equal to u_{ii}^k.
5. The eigenvalues of an upper triangular matrix $\mathbf{U}_{n\times n}$ are $u_{11}, \ldots, u_{nn}$.
6. A positive definite Hermitian matrix $\mathbf{A}$ can be decomposed into $\mathbf{A} = \mathbf{T}^H\mathbf{D}\mathbf{T}$, where $\mathbf{T}$ is a unit upper triangular complex matrix, and $\mathbf{D}$ is a real diagonal matrix.

2. Properties of Lower Triangular Matrices

1. A product of lower triangular matrices is also a lower triangular matrix, i.e., if $\mathbf{L}_1, \ldots, \mathbf{L}_k$ are lower triangular matrices then $\mathbf{L} = \mathbf{L}_1 \cdots \mathbf{L}_k$ is also a lower triangular matrix.
2. The determinant of a lower triangular matrix is equal to the product of its diagonal entries, i.e.,
$$\det(\mathbf{L}) = l_{11} \cdots l_{nn} = \prod_{i=1}^{n} l_{ii}.$$
3. The inverse of a lower triangular matrix is also a lower triangular matrix.
4. The kth power $\mathbf{L}^k$ of a lower triangular matrix $\mathbf{L}_{n\times n}$ is also a lower triangular matrix, and its ith diagonal entry is l_{ii}^k.
5. The eigenvalues of a lower triangular matrix $\mathbf{L}_{n\times n}$ are $l_{11}, \ldots, l_{nn}$.
6. A positive definite matrix $\mathbf{A}_{n\times n}$ can be decomposed into the product of a lower triangular matrix $\mathbf{L}_{n\times n}$ and its transpose, i.e., $\mathbf{A} = \mathbf{L}\mathbf{L}^T$. This decomposition is called the Cholesky decomposition of $\mathbf{A}$.

The lower triangular matrix $\mathbf{L}$ such that $\mathbf{A} = \mathbf{L}\mathbf{L}^T$ is sometimes known as the square root of $\mathbf{A}$. More generally, any matrix $\mathbf{B}$ satisfying

$$\mathbf{B}^2 = \mathbf{A} \tag{2.5.1}$$

is known as the *square root matrix* of $\mathbf{A}$, denoted $\mathbf{A}^{1/2}$. It is should be noted that the square root matrix of a square matrix $\mathbf{A}$ is not necessarily unique.

3. Block Triangular Matrices

If, for a block triangular matrix, the matrix blocks on its diagonal line or cross-diagonal line are invertible, then its inverse is given respectively by

$$\begin{bmatrix} \mathbf{A} & \mathbf{O} \\ \mathbf{B} & \mathbf{C} \end{bmatrix}^{-1} = \begin{bmatrix} \mathbf{A}^{-1} & \mathbf{O} \\ -\mathbf{C}^{-1}\mathbf{B}\mathbf{A}^{-1} & \mathbf{C}^{-1} \end{bmatrix}, \tag{2.5.2}$$

$$\begin{bmatrix} \mathbf{A} & \mathbf{B} \\ \mathbf{C} & \mathbf{O} \end{bmatrix}^{-1} = \begin{bmatrix} \mathbf{O} & \mathbf{C}^{-1} \\ \mathbf{B}^{-1} & -\mathbf{B}^{-1}\mathbf{A}\mathbf{C}^{-1} \end{bmatrix}, \tag{2.5.3}$$

$$\begin{bmatrix} \mathbf{A} & \mathbf{B} \\ \mathbf{O} & \mathbf{C} \end{bmatrix}^{-1} = \begin{bmatrix} \mathbf{A}^{-1} & -\mathbf{A}^{-1}\mathbf{B}\mathbf{C}^{-1} \\ \mathbf{O} & \mathbf{C}^{-1} \end{bmatrix}. \tag{2.5.4}$$

2.6 Summing Vector and Centering Matrix

A special vector and a special matrix used frequently in statistics and data processing are the summing vector and centering matrix.

2.6.1 Summing Vector

A vector with all entries 1 is known as a *summing vector* and is denoted $\mathbf{1} = [1, \ldots, 1]^T$. It is called a summing vector because the sum of n scalars can be expressed as the inner product of a summing vector and another vector: given an m-vector $\mathbf{x} = [x_1, \ldots, x_m]^T$, the sum of its components can be expressed as $\sum_{i=1}^m x_i = \mathbf{1}^T\mathbf{x}$ or $\sum_{i=1}^m x_i = \mathbf{x}^T\mathbf{1}$.

Example 2.2 If we let $\mathbf{x} = [a, b, -c, d]^T$ then the sum $a + b - c + d$ can be expressed as

$$a + b - c + d = [1\ 1\ 1\ 1]\begin{bmatrix} a \\ b \\ -c \\ d \end{bmatrix} = \mathbf{1}^T\mathbf{x} = \langle \mathbf{1}, \mathbf{x} \rangle.$$

In some calculations we may meet summing vectors with different dimensions. In this case, we usually indicate by a subscript the dimension of each summing vector as a subscript in order to avoid confusion. For example, $\mathbf{1}_3 = [1, 1, 1]^T$. Consider this product of the summing vector and a matrix:

$$\mathbf{1}_3^T\mathbf{X}_{3\times 2} = [1\ 1\ 1]\begin{bmatrix} 4 & -1 \\ -4 & 3 \\ 1 & -1 \end{bmatrix} = [1\ 1] = \mathbf{1}_2^T.$$

The inner product of a summing vector and itself is an integer equal to its dimension, namely

$$\langle \mathbf{1}, \mathbf{1} \rangle = \mathbf{1}_n^T\mathbf{1}_n = n. \tag{2.6.1}$$

The outer product of two summing vectors is a matrix all of whose entries are equal to 1; for example,

$$\mathbf{1}_2\mathbf{1}_3^T = \begin{bmatrix} 1 \\ 1 \end{bmatrix} [1\ 1\ 1] = \begin{bmatrix} 1 & 1 & 1 \\ 1 & 1 & 1 \end{bmatrix} = \mathbf{J}_{2\times 3}.$$

More generally,

$$\mathbf{1}_p\mathbf{1}_q^T = \mathbf{J}_{p\times q} \quad \text{(a matrix with all entries equal to 1).} \tag{2.6.2}$$

It is easy to verify that

$$\mathbf{J}_{m\times p}\mathbf{J}_{p\times n} = p\mathbf{J}_{m\times n}, \quad \mathbf{J}_{p\times q}\mathbf{1}_q = q\mathbf{1}_p, \quad \mathbf{1}_p^T\mathbf{J}_{p\times q} = p\mathbf{1}_q^T. \tag{2.6.3}$$

In particular, for an $n \times n$ matrix $\mathbf{J}_n$, we have

$$\mathbf{J}_n = \mathbf{1}_n\mathbf{1}_n^T, \quad \mathbf{J}_n^2 = n\mathbf{J}_n. \tag{2.6.4}$$

Hence, selecting

$$\bar{\mathbf{J}}_n = \frac{1}{n}\mathbf{J}_n, \tag{2.6.5}$$

we have $\bar{\mathbf{J}}_n^2 = \bar{\mathbf{J}}_n$. That is, $\bar{\mathbf{J}}_n$ is idempotent.

2.6.2 Centering Matrix

The matrix

$$\mathbf{C}_n = \mathbf{I}_n - \bar{\mathbf{J}}_n = \mathbf{I}_n - \frac{1}{n}\mathbf{J}_n \tag{2.6.6}$$

is known as a *centering matrix*.

It is easily verified that a centering matrix is not only symmetric but also idempotent, namely

$$\mathbf{C}_n = \mathbf{C}_n^T = \mathbf{C}_n^2. \tag{2.6.7}$$

In addition, a centering matrix has the following properties:

$$\left.\begin{aligned} \mathbf{C}_n\mathbf{1} &= \mathbf{0}, \\ \mathbf{C}_n\mathbf{J}_n &= \mathbf{J}_n\mathbf{C}_n = \mathbf{0}. \end{aligned}\right\} \tag{2.6.8}$$

The summing vector $\mathbf{1}$ and the centering matrix $\mathbf{C}$ are very useful in mathematical statistics [433, p. 67].

First, the mean of a set of data $x_1, \ldots, x_n$ can be expressed using a summing vector as follows:

$$\bar{x} = \frac{1}{n}\sum_{i=1}^{n} x_i = \frac{1}{n}(x_1 + \cdots + x_n) = \frac{1}{n}\mathbf{x}^T\mathbf{1} = \frac{1}{n}\mathbf{1}^T\mathbf{x}, \tag{2.6.9}$$

where $\mathbf{x} = [x_1, \ldots, x_n]^T$ is the data vector.

Next, using the definition formula (2.6.6) of the centering matrix and its property shown in (2.6.8), it follows that

$$\begin{aligned} \mathbf{C}\mathbf{x} &= \mathbf{x} - \bar{\mathbf{J}}\mathbf{x} = \mathbf{x} - \frac{1}{n}\mathbf{1}\mathbf{1}^T\mathbf{x} = \mathbf{x} - \bar{x}\mathbf{1} \\ &= [x_1 - \bar{x}, \ldots, x_n - \bar{x}]^T. \end{aligned} \tag{2.6.10}$$

In other words, the role of the linear transformation matrix $\mathbf{C}$ is to subtract the mean of n data from the data vector $\mathbf{x}$. This is the mathematical meaning of the centering matrix.

Moreover, for the inner product of the vector $\mathbf{C}\mathbf{x}$, we have

$$\begin{aligned} \langle \mathbf{C}\mathbf{x}, \mathbf{C}\mathbf{x} \rangle &= (\mathbf{C}\mathbf{x})^T\mathbf{C}\mathbf{x} = [x_1 - \bar{x}, \ldots, x_n - \bar{x}][x_1 - \bar{x}, \ldots, x_n - \bar{x}]^T \\ &= \sum_{i=1}^{n}(x_i - \bar{x})^2. \end{aligned}$$

From Equation (2.6.8) it is known that $\mathbf{C}^T\mathbf{C} = \mathbf{C}\mathbf{C} = \mathbf{C}$, and thus the above equation can be simplified to

$$\mathbf{x}^T\mathbf{C}\mathbf{x} = \sum_{i=1}^{n}(x_i - \bar{x})^2. \tag{2.6.11}$$

The right-hand side of (2.6.11) is the well-known covariance of the data $x_1, \ldots, x_n$. That is to say, the covariance of a set of data can be reexpressed in the quadratic form of the centering matrix as the kernel $\mathbf{x}^T\mathbf{C}\mathbf{x}$.

2.7 Vandermonde Matrix and Fourier Matrix

Consider two special matrices whose entries in each row (or column) constitute a geometric series. These two special matrices are the Vandermonde matrix and the Fourier matrix, and they have wide applications in engineering. In fact, the Fourier matrix is a special example of the Vandermonde matrix.

2.7.1 Vandermonde Matrix

The $n \times n$ *Vandermonde matrix* is a matrix taking the special form

$$\mathbf{A} = \begin{bmatrix} 1 & x_1 & x_1^2 & \cdots & x_1^{n-1} \\ 1 & x_2 & x_2^2 & \cdots & x_2^{n-1} \\ \vdots & \vdots & \vdots & \vdots & \vdots \\ 1 & x_n & x_n^2 & \cdots & x_n^{n-1} \end{bmatrix} \tag{2.7.1}$$

or

$$\mathbf{A} = \begin{bmatrix} 1 & 1 & \cdots & 1 \\ x_1 & x_2 & \cdots & x_n \\ x_1^2 & x_2^2 & \cdots & x_n^2 \\ \vdots & \vdots & \vdots & \vdots \\ x_1^{n-1} & x_2^{n-1} & \cdots & x_n^{n-1} \end{bmatrix}. \tag{2.7.2}$$

That is to say, the entries of each row (or column) constitute a geometric series.

The Vandermonde matrix has a prominent property: if the n parameters $x_1, \ldots, x_n$ are different then the Vandermonde matrix is nonsingular, because its determinant is given by [32, p. 193]

$$\det(\mathbf{A}) = \prod_{i,j=1,\, i>j}^{n} (x_i - x_j). \tag{2.7.3}$$

Given an $n \times n$ complex Vandermonde matrix

$$\mathbf{A} = \begin{bmatrix} 1 & 1 & \cdots & 1 \\ a_1 & a_2 & \cdots & a_n \\ \vdots & \vdots & \vdots & \vdots \\ a_1^{n-1} & a_2^{n-1} & \cdots & a_n^{n-1} \end{bmatrix}, \quad a_k \in \mathbb{C}, \tag{2.7.4}$$

its *inverse* is given by [339]

$$\mathbf{A}^{-1} = \left[(-1)^{i+j} \frac{\sigma_{n-j}(a_1, \ldots, a_{j-1}, a_{j+1}, \ldots, a_n)}{\prod_{k=1}^{j-1}(a_j - a_k) \prod_{k=j+1}^{n}(a_k - a_j)} \right]^T_{\substack{i=1,\ldots,n \\ j=1,\ldots,n}}, \tag{2.7.5}$$

where

$$\begin{aligned} \sigma_0(a_1, \ldots, a_{j-1}, a_{j+1}, \ldots, a_n) &= 1, \\ \sigma_1(a_1, \ldots, a_{j-1}, a_{j+1}, \ldots, a_n) &= \sum_{k=1, k\neq j}^{n} a_k, \\ \sigma_i(a_1, \ldots, a_{j-1}, a_{j+1}, \ldots, a_n) &= \sum_{k=1, k\neq j}^{n} a_k \cdots a_{k+i-1}, \quad i = 2, \ldots, n, \end{aligned}$$

with $a_i = 0, i > n$.

Example 2.3 In the extended Prony method, the signal model is assumed to be the superposition of p exponential functions, namely

$$\hat{x}_n = \sum_{i=1}^{p} b_i z_i^n, \quad n = 0, 1, \ldots, N-1 \tag{2.7.6}$$

is used as the mathematical model for fitting the data $x_0, x_1, \ldots, x_{N-1}$. In general, b_i and z_i are assumed to be complex numbers, and

$$b_i = A_i \exp(\mathrm{j}\,\theta_i), \quad z_i = \exp[(\alpha_i + \mathrm{j}\,2\pi f_i)\Delta t],$$

where A_i is the amplitude of the ith exponential function, θ_i is its phase (in radians), α_i its damping factor, f_i its oscillation frequency (in Hz), and Δt denotes the sampling time interval (in seconds).

The matrix form of Equation (2.7.6) is

$$\mathbf{\Phi b} = \hat{\mathbf{x}}$$

in which $\mathbf{b} = [b_0, b_1, \ldots, b_p]^T$, $\hat{\mathbf{x}} = [\hat{x}_0, \hat{x}_1, \ldots, \hat{x}_{N-1}]^T$, and $\mathbf{\Phi}$ is a complex Vandermonde matrix given by

$$\mathbf{\Phi} = \begin{bmatrix} 1 & 1 & 1 & \cdots & 1 \\ z_1 & z_2 & z_3 & \cdots & z_p \\ z_1^2 & z_2^2 & z_3^2 & \cdots & z_p^2 \\ \vdots & \vdots & \vdots & \vdots & \vdots \\ z_1^{N-1} & z_2^{N-1} & z_3^{N-1} & \cdots & z_p^{N-1} \end{bmatrix}. \tag{2.7.7}$$

By minimizing the square error $\epsilon = \sum_{n=1}^{N-1} |x_n - \hat{x}_n|^2$ we get a least squares solution of the matrix equation $\mathbf{\Phi b} = \hat{\mathbf{x}}$, as follows:

$$\mathbf{b} = [\mathbf{\Phi}^H \mathbf{\Phi}]^{-1} \mathbf{\Phi}^H \mathbf{x}. \tag{2.7.8}$$

It is easily shown that the computation of $\mathbf{\Phi}^H \mathbf{\Phi}$ in Equation (2.7.8) can be greatly simplified so that, without doing the multiplication operation on the Vandermonde matrix, we can get directly

$$\mathbf{\Phi}^H \mathbf{\Phi} = \begin{bmatrix} \gamma_{11} & \gamma_{12} & \cdots & \gamma_{1p} \\ \gamma_{21} & \gamma_{22} & \cdots & \gamma_{2p} \\ \vdots & \vdots & \vdots & \vdots \\ \gamma_{p1} & \gamma_{p2} & \cdots & \gamma_{pp} \end{bmatrix}, \tag{2.7.9}$$

where

$$\gamma_{ij} = \frac{(z_i^* z_j)^N - 1}{(z_i^* z_j) - 1}. \tag{2.7.10}$$

The $N \times p$ matrix $\mathbf{\Phi}$ shown in Equation (2.7.7) is one of two Vandermonde matrices widely applied in signal processing; another common complex Vandermonde matrix is given by

$$\mathbf{\Phi} = \begin{bmatrix} 1 & 1 & \cdots & 1 \\ \mathrm{e}^{\lambda_1} & \mathrm{e}^{\lambda_2} & \cdots & \mathrm{e}^{\lambda_d} \\ \vdots & \vdots & \vdots & \vdots \\ \mathrm{e}^{\lambda_1 (N-1)} & \mathrm{e}^{\lambda_2 (N-1)} & \cdots & \mathrm{e}^{\lambda_d (N-1)} \end{bmatrix}. \tag{2.7.11}$$

2.7.2 Fourier Matrix

The Fourier transform of the discrete-time signals $x_0, x_1, \ldots, x_{N-1}$ is known as the *discrete Fourier transform* (DFT) or *spectrum* of these signals, and is defined as

$$X_k = \sum_{n=0}^{N-1} x_n \mathrm{e}^{-\mathrm{j}\, 2\pi nk/N} = \sum_{n=0}^{N-1} x_n w^{nk}, \quad k = 0, 1, \ldots, N-1. \tag{2.7.12}$$

This equation can be expressed in matrix form as

$$\begin{bmatrix} X_0 \\ X_1 \\ \vdots \\ X_{N-1} \end{bmatrix} = \begin{bmatrix} 1 & 1 & \cdots & 1 \\ 1 & w & \cdots & w^{N-1} \\ \vdots & \vdots & \vdots & \vdots \\ 1 & w^{N-1} & \cdots & w^{(N-1)(N-1)} \end{bmatrix} \begin{bmatrix} x_0 \\ x_1 \\ \vdots \\ x_{N-1} \end{bmatrix} \tag{2.7.13}$$

or simply as

$$\hat{\mathbf{x}} = \mathbf{F}\mathbf{x}, \tag{2.7.14}$$

where $\mathbf{x} = [x_0, x_1, \ldots, x_{N-1}]^T$ and $\hat{\mathbf{x}} = [X_0, X_1, \ldots, X_{N-1}]^T$ are respectively the discrete-time signal vector and the spectrum vector, whereas

$$\mathbf{F} = \begin{bmatrix} 1 & 1 & \cdots & 1 \\ 1 & w & \cdots & w^{N-1} \\ \vdots & \vdots & \vdots & \vdots \\ 1 & w^{N-1} & \cdots & w^{(N-1)(N-1)} \end{bmatrix}, \quad w = \mathrm{e}^{-\mathrm{j}\,2\pi/N} \tag{2.7.15}$$

is known as the *Fourier matrix* with (i,k)th entry $F(i,k) = w^{(i-1)(k-1)}$.

Clearly, each row and each column of the Fourier matrix constitute a respective geometric series, and thus the Fourier matrix can be regarded as an $N \times N$ special Vandermonde matrix.

From the definition it can be seen that the Fourier matrix is symmetric, i.e., $\mathbf{F}^T = \mathbf{F}$.

Equation (2.7.14) shows that the DFT of a discrete-time signal vector can be expressed by the matrix $\mathbf{F}$, which is the reason why $\mathbf{F}$ is termed the Fourier matrix.

Moreover, from the definition of the Fourier matrix it is easy to verify that

$$\mathbf{F}^H\mathbf{F} = \mathbf{F}\mathbf{F}^H = N\mathbf{I}. \tag{2.7.16}$$

Since the Fourier matrix is a special Vandermonde matrix that is nonsingular, it follows from $\mathbf{F}^H\mathbf{F} = N\mathbf{I}$ that the inverse of the Fourier matrix is given by

$$\mathbf{F}^{-1} = \frac{1}{N}\mathbf{F}^H = \frac{1}{N}\mathbf{F}^*. \tag{2.7.17}$$

Hence, from Equation (2.7.2) it can be seen immediately that

$$\mathbf{x} = \mathbf{F}^{-1}\hat{\mathbf{x}} = \frac{1}{N}\mathbf{F}^*\hat{\mathbf{x}}, \tag{2.7.18}$$

which can be written as

$$\begin{bmatrix} x_0 \\ x_1 \\ \vdots \\ x_{N-1} \end{bmatrix} = \frac{1}{N} \begin{bmatrix} 1 & 1 & \cdots & 1 \\ 1 & w^* & \cdots & (w^{N-1})^* \\ \vdots & \vdots & \vdots & \vdots \\ 1 & (w^{N-1})^* & \cdots & (w^{(N-1)(N-1)})^* \end{bmatrix} \begin{bmatrix} X_0 \\ X_1 \\ \vdots \\ X_{N-1} \end{bmatrix}, \tag{2.7.19}$$

thus

$$x_n = \frac{1}{N}\sum_{k=0}^{N-1} X_k \mathrm{e}^{\mathrm{j}2\pi nk/N}, \quad n = 0, 1, \ldots, N-1. \tag{2.7.20}$$

This is the formula for the *inverse discrete Fourier transform* (IDFT).

From the definition it is easy to see that the $n \times n$ Fourier matrix $\mathbf{F}$ has the following properties [307].

(1) The Fourier matrix is symmetric, i.e., $\mathbf{F}^T = \mathbf{F}$.
(2) The inverse of the Fourier matrix is given by $\mathbf{F}^{-1} = N^{-1}\mathbf{F}^*$.
(3) $\mathbf{F}^2 = \mathbf{P} = [\mathbf{e}_1, \mathbf{e}_N, \mathbf{e}_{N-1}, \ldots, \mathbf{e}_2]$ (the permutation matrix), where $\mathbf{e}_k$ is the basis vector whose kth entry equals 1 and whose other entries are zero.
(4) $\mathbf{F}^4 = \mathbf{I}$.

2.7.3 Index Vectors

Given an $N \times 1$ vector $\mathbf{x} = [x_0, x_1, \ldots, x_{N-1}]^T$, we will consider the subscript representation of its elements in binary code.

Define the *index vector* [390]

$$\mathbf{i}_N = \begin{bmatrix} \langle 0 \rangle \\ \langle 1 \rangle \\ \vdots \\ \langle N-1 \rangle \end{bmatrix}, \quad N = 2^n, \tag{2.7.21}$$

in which $\langle i \rangle$ denotes the binary representation of the integer i, with $i = 0, 1, \ldots, N-1$.

Example 2.4 For $N = 2^2 = 4$ and $N = 2^3 = 8$, we have respectively

$$\begin{bmatrix} 0 \\ 1 \\ 2 \\ 3 \end{bmatrix} \Leftrightarrow \mathbf{i}_4 = \begin{bmatrix} \langle 0 \rangle \\ \langle 1 \rangle \\ \langle 2 \rangle \\ \langle 3 \rangle \end{bmatrix} = \begin{bmatrix} 00 \\ 01 \\ 10 \\ 11 \end{bmatrix},$$

$$\begin{bmatrix} 0 \\ 1 \\ 2 \\ 3 \\ 4 \\ 5 \\ 6 \\ 7 \end{bmatrix} \Leftrightarrow \mathbf{i}_8 = \begin{bmatrix} \langle 0 \rangle \\ \langle 1 \rangle \\ \langle 2 \rangle \\ \langle 3 \rangle \\ \langle 4 \rangle \\ \langle 5 \rangle \\ \langle 6 \rangle \\ \langle 7 \rangle \end{bmatrix} = \begin{bmatrix} 000 \\ 001 \\ 010 \\ 011 \\ 100 \\ 101 \\ 110 \\ 111 \end{bmatrix}.$$

Importantly, an index vector can be repressed as the Kronecker product of binary codes 0 and 1. Let a, b, c, d be binary codes; then the *binary Kronecker product* of binary codes is defined as

$$\begin{bmatrix} a \\ b \end{bmatrix}_2 \otimes \begin{bmatrix} c \\ d \end{bmatrix}_2 = \begin{bmatrix} ac \\ ad \\ bc \\ bd \end{bmatrix}_2, \tag{2.7.22}$$

where xy denotes the arrangement in order of the binary code x and y rather than their product.

Example 2.5 The binary Kronecker product representations of the index vectors $\mathbf{i}_4$ and $\mathbf{i}_8$ are as follows:

$$\mathbf{i}_4 = \begin{bmatrix} 0_1 \\ 1_1 \end{bmatrix} \otimes \begin{bmatrix} 0_0 \\ 1_0 \end{bmatrix} = \begin{bmatrix} 0_1 0_0 \\ 0_1 1_0 \\ 1_1 0_0 \\ 1_1 1_0 \end{bmatrix} = \begin{bmatrix} 00 \\ 01 \\ 10 \\ 11 \end{bmatrix} = \begin{bmatrix} \langle 0 \rangle \\ \langle 1 \rangle \\ \langle 2 \rangle \\ \langle 3 \rangle \end{bmatrix}$$

and

$$\mathbf{i}_8 = \begin{bmatrix} 0_2 \\ 1_2 \end{bmatrix} \otimes \begin{bmatrix} 0_1 \\ 1_1 \end{bmatrix} \otimes \begin{bmatrix} 0_0 \\ 1_0 \end{bmatrix} = \begin{bmatrix} 0_2 \\ 1_2 \end{bmatrix} \otimes \begin{bmatrix} 0_1 0_0 \\ 0_1 1_0 \\ 1_1 0_0 \\ 1_1 1_0 \end{bmatrix} = \begin{bmatrix} 0_2 0_1 0_0 \\ 0_2 0_1 1_0 \\ 0_2 1_1 0_0 \\ 0_2 1_1 1_0 \\ 1_2 0_1 0_0 \\ 1_2 0_1 1_0 \\ 1_2 1_1 0_0 \\ 1_2 1_1 1_0 \end{bmatrix} = \begin{bmatrix} 000 \\ 001 \\ 010 \\ 011 \\ 100 \\ 101 \\ 110 \\ 111 \end{bmatrix} = \begin{bmatrix} \langle 0 \rangle \\ \langle 1 \rangle \\ \langle 2 \rangle \\ \langle 3 \rangle \\ \langle 4 \rangle \\ \langle 5 \rangle \\ \langle 6 \rangle \\ \langle 7 \rangle \end{bmatrix}.$$

More generally, the binary Kronecker product representation of the index vector $\mathbf{i}_N$ is given by

$$\begin{aligned} \mathbf{i}_N &= \begin{bmatrix} 0_{N-1} \\ 1_{N-1} \end{bmatrix} \otimes \begin{bmatrix} 0_{N-2} \\ 1_{N-2} \end{bmatrix} \otimes \cdots \otimes \begin{bmatrix} 0_1 \\ 1_1 \end{bmatrix} \otimes \begin{bmatrix} 0_0 \\ 1_0 \end{bmatrix} \\ &= \begin{bmatrix} 0_{N-1} \cdots 0_1 0_0 \\ 0_{N-1} \cdots 0_1 1_0 \\ \vdots \\ 1_{N-1} \cdots 1_1 0_0 \\ 1_{N-1} \cdots 1_1 1_0 \end{bmatrix} = \begin{bmatrix} \langle 0 \rangle \\ \langle 1 \rangle \\ \vdots \\ \langle N-2 \rangle \\ \langle N-1 \rangle \end{bmatrix}. \end{aligned} \tag{2.7.23}$$

On the other hand, the *bit-reversed index vector* of $\mathbf{i}_N$ is denoted $\mathbf{i}_{N,\text{rev}}$ and is defined as follows:

$$\mathbf{i}_{N,\text{rev}} = \begin{bmatrix} \langle 0\rangle_{\text{rev}} \\ \langle 1\rangle_{\text{rev}} \\ \vdots \\ \langle N-2\rangle_{\text{rev}} \\ \langle N-1\rangle_{\text{rev}} \end{bmatrix}, \tag{2.7.24}$$

where $\langle i\rangle_{\text{rev}}$ represents the bit-reversed result of the binary code of the integer i, $i = 0, 1, \ldots, N-1$. For instance, if $N = 8$, then the reversed results of the binary code 001 and 011 are respectively 100 and 110.

Interestingly, on reversing the order of the binary Kronecker product of the index vector $\mathbf{i}_N$, we obtain directly the bit-reversed index vector

$$\mathbf{i}_{N,\text{rev}} = \begin{bmatrix} 0_0 \\ 1_0 \end{bmatrix} \otimes \begin{bmatrix} 0_1 \\ 1_1 \end{bmatrix} \otimes \cdots \otimes \begin{bmatrix} 0_{N-1} \\ 1_{N-1} \end{bmatrix}. \tag{2.7.25}$$

Example 2.6 The bit-reversed index vectors of $\mathbf{i}_4$ and $\mathbf{i}_8$ are respectively given by

$$\mathbf{i}_{4,\text{rev}} = \begin{bmatrix} 0_0 \\ 1_0 \end{bmatrix} \otimes \begin{bmatrix} 0_1 \\ 1_1 \end{bmatrix} = \begin{bmatrix} 0_0 0_1 \\ 0_0 1_1 \\ 1_0 0_1 \\ 1_0 1_1 \end{bmatrix} = \begin{bmatrix} 00 \\ 10 \\ 01 \\ 11 \end{bmatrix} = \begin{bmatrix} \langle 0\rangle \\ \langle 2\rangle \\ \langle 1\rangle \\ \langle 3\rangle \end{bmatrix}$$

and

$$\mathbf{i}_{8,\text{rev}} = \begin{bmatrix} 0_0 \\ 1_0 \end{bmatrix} \otimes \begin{bmatrix} 0_1 \\ 1_1 \end{bmatrix} \otimes \begin{bmatrix} 0_2 \\ 1_2 \end{bmatrix} = \begin{bmatrix} 0_0 0_1 \\ 0_0 1_1 \\ 1_0 0_1 \\ 1_0 1_1 \end{bmatrix} \otimes \begin{bmatrix} 0_2 \\ 1_2 \end{bmatrix} = \begin{bmatrix} 0_0 0_1 0_2 \\ 0_0 0_1 1_2 \\ 0_0 1_1 0_2 \\ 0_0 1_1 1_2 \\ 1_0 0_1 0_2 \\ 1_0 0_1 1_2 \\ 1_0 1_1 0_2 \\ 1_0 1_1 1_2 \end{bmatrix} = \begin{bmatrix} 000 \\ 100 \\ 010 \\ 110 \\ 001 \\ 101 \\ 011 \\ 111 \end{bmatrix} = \begin{bmatrix} \langle 0\rangle \\ \langle 4\rangle \\ \langle 2\rangle \\ \langle 6\rangle \\ \langle 1\rangle \\ \langle 5\rangle \\ \langle 3\rangle \\ \langle 7\rangle \end{bmatrix}.$$

Notice that the intermediate result of a binary Kronecker product should be written in conventional binary code sequence; for example, $0_0 0_1 1_2$ should be written as 100.

2.7.4 FFT Algorithm

We now consider a fast algorithm for the DFT $\hat{\mathbf{x}} = \mathbf{F}\mathbf{x}$ shown in Equation (2.7.14). To this end, perform Type-I elementary row operation on the augmented matrix $[\mathbf{F}, \hat{\mathbf{x}}]$. Clearly, during the elementary row operations, the subscript indices of the row vectors of the Fourier matrix $\mathbf{F}$ and the subscript indices of the elements of the output vector $\hat{\mathbf{x}}$ are the same. Hence, we have

$$\hat{\mathbf{x}}_{\text{rev}} = \mathbf{F}_{N,\text{rev}}\mathbf{x}. \tag{2.7.26}$$

Therefore, the key step in deriving the N-point FFT algorithm is to construct the $N \times N$ bit-reversed Fourier matrix $\mathbf{F}_{N,\text{rev}}$.

Letting

$$\{\mathbf{A}\}_2 = \left\{ \begin{bmatrix} \begin{bmatrix} 1 & 1 \\ 1 & -1 \end{bmatrix} \\ \begin{bmatrix} 1 & -\mathrm{j} \\ 1 & \mathrm{j} \end{bmatrix} \end{bmatrix} \right\}, \quad \mathbf{B} = \begin{bmatrix} 1 & 1 \\ 1 & -1 \end{bmatrix}, \tag{2.7.27}$$

the $N \times N$ bit-reversed Fourier matrix can be recursively constructed by

$$\mathbf{F}_{N,\text{rev}} = \{\mathbf{A}\}_{N/2} \otimes \{\mathbf{A}\}_{N/4} \otimes \cdots \otimes \{\mathbf{A}\}_2 \otimes \mathbf{B}, \tag{2.7.28}$$

where

$$\{\mathbf{A}\}_{N/2} = \left\{ \begin{matrix} \{\mathbf{A}\}_{N/4} \\ \{\mathbf{R}\}_{N/4} \end{matrix} \right\}, \quad \{\mathbf{R}\}_{N/4} = \left\{ \begin{matrix} \mathbf{R}_1 \\ \vdots \\ \mathbf{R}_{N/4} \end{matrix} \right\} \tag{2.7.29}$$

together with

$$\mathbf{R}_k = \begin{bmatrix} 1 & \mathrm{e}^{-\mathrm{j}(2k-1)\pi/N} \\ 1 & -\mathrm{e}^{-\mathrm{j}(2k-1)\pi/N} \end{bmatrix}, \quad k = 1, \ldots, \frac{N}{2}. \tag{2.7.30}$$

EXAMPLE 2.7 When $N = 4$, we have the 4×4 bit-reversed Fourier matrix

$$\mathbf{F}_{4,\text{rev}} = \{\mathbf{A}\}_2 \otimes \mathbf{B} = \begin{bmatrix} \begin{bmatrix} 1 & 1 \\ 1 & -1 \end{bmatrix} \otimes [1, 1] \\ \begin{bmatrix} 1 & -\mathrm{j} \\ 1 & \mathrm{j} \end{bmatrix} \otimes [1, -1] \end{bmatrix} = \begin{bmatrix} 1 & 1 & 1 & 1 \\ 1 & -1 & 1 & -1 \\ 1 & -\mathrm{j} & -1 & \mathrm{j} \\ 1 & \mathrm{j} & -1 & -j \end{bmatrix}. \tag{2.7.31}$$

From Equation (2.7.26), we have

$$\begin{bmatrix} X_0 \\ X_2 \\ X_1 \\ X_3 \end{bmatrix} = \begin{bmatrix} 1 & 1 & 1 & 1 \\ 1 & -1 & 1 & -1 \\ 1 & -\mathrm{j} & -1 & \mathrm{j} \\ 1 & \mathrm{j} & -1 & -j \end{bmatrix} \begin{bmatrix} x_0 \\ x_1 \\ x_2 \\ x_3 \end{bmatrix}. \tag{2.7.32}$$

This is just the four-point *FFT algorithm*.

EXAMPLE 2.8 When $N = 8$, from (2.7.29) we have

$$\{\mathbf{A}\}_4 = \left\{ \begin{matrix} \{\mathbf{A}\}_2 \\ \{\mathbf{R}\}_2 \end{matrix} \right\} = \left\{ \begin{matrix} \begin{bmatrix} 1 & 1 \\ 1 & -1 \end{bmatrix} \\ \begin{bmatrix} 1 & -\mathrm{j} \\ 1 & \mathrm{j} \end{bmatrix} \\ \begin{bmatrix} 1 & \mathrm{e}^{-\mathrm{j}\pi/4} \\ 1 & -\mathrm{e}^{-\mathrm{j}\pi/4} \end{bmatrix} \\ \begin{bmatrix} 1 & \mathrm{e}^{-\mathrm{j}3\pi/4} \\ 1 & -\mathrm{e}^{-\mathrm{j}3\pi/4} \end{bmatrix} \end{matrix} \right\}, \tag{2.7.33}$$

and thus the 8×8 bit-reversed Fourier matrix is given by

$$\mathbf{F}_{8,\mathrm{rev}} = \{\mathbf{A}\}_4 \otimes (\{\mathbf{A}\}_2 \otimes \mathbf{B}) = \{\mathbf{A}\}_4 \otimes \mathbf{F}_{4,\mathrm{rev}}$$

$$= \left\{ \begin{matrix} \begin{bmatrix} 1 & 1 \\ 1 & -1 \end{bmatrix} \\ \begin{bmatrix} 1 & -\mathrm{j} \\ 1 & \mathrm{j} \end{bmatrix} \\ \begin{bmatrix} 1 & \mathrm{e}^{-\mathrm{j}\pi/4} \\ 1 & -\mathrm{e}^{-\mathrm{j}\pi/4} \end{bmatrix} \\ \begin{bmatrix} 1 & \mathrm{e}^{-\mathrm{j}3\pi/4} \\ 1 & -\mathrm{e}^{-\mathrm{j}3\pi/4} \end{bmatrix} \end{matrix} \right\} \otimes \begin{bmatrix} 1 & 1 & 1 & 1 \\ 1 & -1 & 1 & -1 \\ 1 & -\mathrm{j} & -1 & \mathrm{j} \\ 1 & \mathrm{j} & -1 & -j \end{bmatrix}$$

$$= \begin{bmatrix} 1 & 1 & 1 & 1 & 1 & 1 & 1 & 1 \\ 1 & -1 & 1 & -1 & 1 & -1 & 1 & -1 \\ 1 & -\mathrm{j} & -1 & \mathrm{j} & 1 & -\mathrm{j} & -1 & \mathrm{j} \\ 1 & \mathrm{j} & -1 & -\mathrm{j} & 1 & \mathrm{j} & -1 & -\mathrm{j} \\ 1 & \mathrm{e}^{-\mathrm{j}\pi/4} & -\mathrm{j} & \mathrm{e}^{-\mathrm{j}3\pi/4} & -1 & -\mathrm{e}^{-\mathrm{j}\pi/4} & \mathrm{j} & -\mathrm{e}^{-\mathrm{j}3\pi/4} \\ 1 & -\mathrm{e}^{-\mathrm{j}\pi/4} & -\mathrm{j} & -\mathrm{e}^{-\mathrm{j}3\pi/4} & -1 & \mathrm{e}^{-\mathrm{j}\pi/4} & \mathrm{j} & \mathrm{e}^{-\mathrm{j}3\pi/4} \\ 1 & \mathrm{e}^{-\mathrm{j}3\pi/4} & \mathrm{j} & \mathrm{e}^{-\mathrm{j}\pi/4} & -1 & -\mathrm{e}^{-\mathrm{j}3\pi/4} & -\mathrm{j} & -\mathrm{e}^{-\mathrm{j}\pi/4} \\ 1 & -\mathrm{e}^{-\mathrm{j}3\pi/4} & \mathrm{j} & -\mathrm{e}^{-\mathrm{j}\pi/4} & -1 & \mathrm{e}^{-\mathrm{j}3\pi/4} & -\mathrm{j} & \mathrm{e}^{-\mathrm{j}\pi/4} \end{bmatrix}. \tag{2.7.34}$$

From Equation (2.7.26) it immediately follows that

$$\begin{bmatrix} X_0 \\ X_4 \\ X_2 \\ X_6 \\ X_1 \\ X_5 \\ X_3 \\ X_7 \end{bmatrix} = \begin{bmatrix} 1 & 1 & 1 & 1 & 1 & 1 & 1 & 1 \\ 1 & -1 & 1 & -1 & 1 & -1 & 1 & -1 \\ 1 & -\mathrm{j} & -1 & \mathrm{j} & 1 & -\mathrm{j} & -1 & \mathrm{j} \\ 1 & \mathrm{j} & -1 & -\mathrm{j} & 1 & \mathrm{j} & -1 & -\mathrm{j} \\ 1 & \mathrm{e}^{-\mathrm{j}\pi/4} & -\mathrm{j} & \mathrm{e}^{-\mathrm{j}3\pi/4} & -1 & -\mathrm{e}^{-\mathrm{j}\pi/4} & \mathrm{j} & -\mathrm{e}^{-\mathrm{j}3\pi/4} \\ 1 & -\mathrm{e}^{-\mathrm{j}\pi/4} & -\mathrm{j} & -\mathrm{e}^{-\mathrm{j}3\pi/4} & -1 & \mathrm{e}^{-\mathrm{j}\pi/4} & \mathrm{j} & \mathrm{e}^{-\mathrm{j}3\pi/4} \\ 1 & \mathrm{e}^{-\mathrm{j}3\pi/4} & \mathrm{j} & \mathrm{e}^{-\mathrm{j}\pi/4} & -1 & -\mathrm{e}^{-\mathrm{j}3\pi/4} & -\mathrm{j} & -\mathrm{e}^{-\mathrm{j}\pi/4} \\ 1 & -\mathrm{e}^{-\mathrm{j}3\pi/4} & \mathrm{j} & -\mathrm{e}^{-\mathrm{j}\pi/4} & -1 & \mathrm{e}^{-\mathrm{j}3\pi/4} & -\mathrm{j} & \mathrm{e}^{-\mathrm{j}\pi/4} \end{bmatrix} \begin{bmatrix} x_0 \\ x_1 \\ x_2 \\ x_3 \\ x_4 \\ x_5 \\ x_6 \\ x_7 \end{bmatrix}.$$

$$(2.7.35)$$

This is the eight-point FFT algorithm.

Similarly, we can derive other $N(=16, 32, \ldots)$-point FFT algorithms.

2.8 Hadamard Matrix

The Hadamard matrix is an important special matrix in communication, information theory, signal processing and so on.

DEFINITION 2.13 $\mathbf{H}_n \in \mathbb{R}^{n\times n}$ is known as a *Hadamard matrix* if its entries take the values either $+1$ or -1 such that

$$\mathbf{H}_n\mathbf{H}_n^T = \mathbf{H}_n^T\mathbf{H}_n = n\mathbf{I}_n. \tag{2.8.1}$$

The properties of the Hadamard matrix are as follows.

(1) After multiplying the entries of any row or column of the Hadamard matrix by -1, the result is still a Hadamard matrix. Hence, we can obtain a special Hadamard matrix, with all entries of the first row and of the first column equal to $+1$. Such a special Hadamard matrix is known as a *standardized Hadamard matrix.*
(2) An $n \times n$ Hadamard matrix exists only when $n = 2^k$, where k is an integer.
(3) $(1/\sqrt{n})\mathbf{H}_n$ is an orthonormal matrix.
(4) The determinant of the $n \times n$ Hadamard matrix $\mathbf{H}_n$ is $\det(\mathbf{H}_n) = n^{n/2}$.

Standardizing a Hadamard matrix will greatly help to construct a Hadamard matrix with higher-dimension. The following theorem gives a general construction approach for the standardized orthonomal Hadamard matrix.

THEOREM 2.3 *Putting $n = 2^k$, where $k = 1, 2, \ldots$, the standardized orthonormal Hadamard matrix has the following general construction formula:*

$$\bar{\mathbf{H}}_n = \frac{1}{\sqrt{2}}\begin{bmatrix}\bar{\mathbf{H}}_{n/2} & \bar{\mathbf{H}}_{n/2}\\ \bar{\mathbf{H}}_{n/2} & -\bar{\mathbf{H}}_{n/2}\end{bmatrix}, \tag{2.8.2}$$

where

$$\bar{\mathbf{H}}_2 = \frac{1}{\sqrt{2}}\begin{bmatrix}1 & 1\\ 1 & -1\end{bmatrix}. \tag{2.8.3}$$

Proof First, $\bar{\mathbf{H}}_2$ is clearly a standardizing orthonormal Hadamard matrix, since $\bar{\mathbf{H}}_2^T\bar{\mathbf{H}}_2 = \bar{\mathbf{H}}_2\bar{\mathbf{H}}_2^T = \mathbf{I}_2$. Next, let $\bar{\mathbf{H}}_{2^k}$ be the standardizing orthonormal Hadamard matrix for $n = 2^k$, namely $\bar{\mathbf{H}}_{2^k}^T\bar{\mathbf{H}}_{2^k} = \bar{\mathbf{H}}_{2^k}\bar{\mathbf{H}}_{2^k}^T = \mathbf{I}_{2^k\times 2^k}$. Hence, for $n = 2^{k+1}$, it is easily seen that

$$\bar{\mathbf{H}}_{2^{k+1}} = \frac{1}{\sqrt{2}}\begin{bmatrix}\bar{\mathbf{H}}_{2^k} & \bar{\mathbf{H}}_{2^k}\\ \bar{\mathbf{H}}_{2^k} & -\bar{\mathbf{H}}_{2^k}\end{bmatrix}$$

satisfies the orthogonality condition, i.e.,

$$\bar{\mathbf{H}}_{2^{k+1}}^T \bar{\mathbf{H}}_{2^{k+1}} = \frac{1}{2}\begin{bmatrix} \bar{\mathbf{H}}_{2^k}^T & \bar{\mathbf{H}}_{2^k}^T \\ \bar{\mathbf{H}}_{2^k}^T & -\bar{\mathbf{H}}_{2^k}^T \end{bmatrix}\begin{bmatrix} \bar{\mathbf{H}}_{2^k} & \bar{\mathbf{H}}_{2^k} \\ \bar{\mathbf{H}}_{2^k} & -\bar{\mathbf{H}}_{2^k} \end{bmatrix} = \mathbf{I}_{2^{k+1}\times 2^{k+1}}.$$

Similarly, it is easy to show that $\bar{\mathbf{H}}_{2^{k+1}} \bar{\mathbf{H}}_{2^{k+1}}^T = \mathbf{I}_{2^{k+1}\times 2^{k+1}}$. Moreover, since $\bar{\mathbf{H}}_{2^k}$ is standardized, $\bar{\mathbf{H}}_{2^{k+1}}$ is standardized as well. Hence, the theorem holds also for $n = 2^{k+1}$. By mathematical induction, the theorem is proved. □

A nonstandardizing Hadamard matrix can be written as

$$\mathbf{H}_n = \mathbf{H}_{n/2} \otimes \mathbf{H}_2 = \mathbf{H}_2 \otimes \cdots \otimes \mathbf{H}_2 \quad (n = 2^k), \tag{2.8.4}$$

where $\mathbf{H}_2 = \begin{bmatrix} 1 & 1 \\ 1 & -1 \end{bmatrix}$.

Evidently, there is the following relationship between the standardizing and nonstandardizing Hadamard matrices:

$$\bar{\mathbf{H}}_n = \frac{1}{\sqrt{n}}\mathbf{H}_n. \tag{2.8.5}$$

Example 2.9 When $n = 2^3 = 8$, we have the Hadamard matrix

$$\mathbf{H}_8 = \begin{bmatrix} 1 & 1 \\ 1 & -1 \end{bmatrix} \otimes \begin{bmatrix} 1 & 1 \\ 1 & -1 \end{bmatrix} \otimes \begin{bmatrix} 1 & 1 \\ 1 & -1 \end{bmatrix}$$
$$= \begin{bmatrix} 1 & 1 & 1 & 1 & 1 & 1 & 1 & 1 \\ 1 & -1 & 1 & -1 & 1 & -1 & 1 & -1 \\ 1 & 1 & -1 & -1 & 1 & 1 & -1 & -1 \\ 1 & -1 & -1 & 1 & 1 & -1 & -1 & 1 \\ 1 & 1 & 1 & 1 & -1 & -1 & -1 & -1 \\ 1 & -1 & 1 & -1 & -1 & 1 & -1 & 1 \\ 1 & 1 & -1 & -1 & -1 & -1 & 1 & 1 \\ 1 & -1 & -1 & 1 & -1 & 1 & 1 & -1 \end{bmatrix}.$$

Figure 2.1 shows the waveforms of the eight rows of the Hadamard matrix $\phi_0(t), \phi_1(t), \ldots, \phi_7(t)$.

It is easily seen that the Hadamard waveforms $\phi_0(t), \phi_1(t), \ldots, \phi_7(t)$ are orthogonal to each other, namely $\int_0^1 \phi_i(t)\phi_j(t)\mathrm{d}t = \delta(i-j)$.

If $\mathbf{H}$ is a Hadamard matrix, the linear transformation $\mathbf{Y} = \mathbf{HX}$ is known as the *Hadamard transform* of the matrix $\mathbf{X}$. Since the Hadamard matrix is a standardizing orthonormal matrix, and its entries are only $+1$ or -1, the Hadamard matrix is the only orthonormal transformation using addition and subtraction alone.

The Hadamard matrix may be used for mobile communication coding. The resulting code is called the Hadamard (or Walsh–Hadamard) code. In addition, owing to the orthogonality between its row vectors, they can be used to simulate the spread waveform vector of each user in a code division multiple access (CDMA) system.

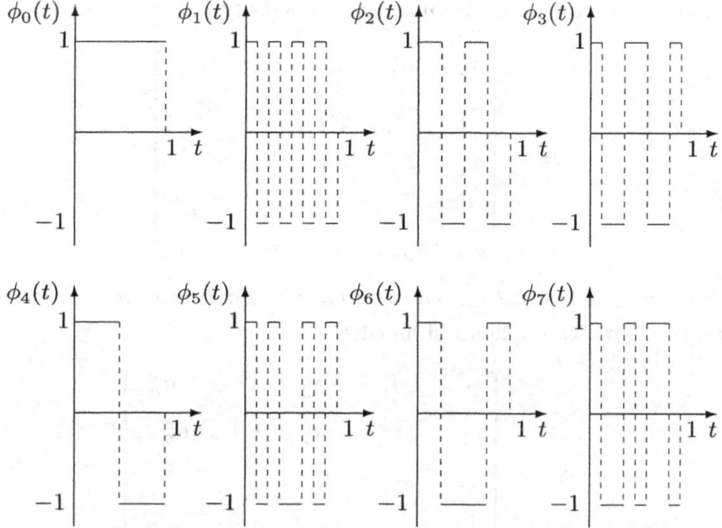

Figure 2.1 Waveform of each row of the Hadamard matrix $\mathbf{H}_8$.

2.9 Toeplitz Matrix

In the early twentieth century, Toeplitz [467] presented a special matrix in his paper on the bilinear function relating to the Laurent series. All entries on any diagonal of this special matrix take the same values, namely

$$\mathbf{A} = \begin{bmatrix} a_0 & a_{-1} & a_{-2} & \cdots & a_{-n} \\ a_1 & a_0 & a_{-1} & \cdots & a_{-n+1} \\ a_2 & a_1 & a_0 & \ddots & \vdots \\ \vdots & \vdots & \ddots & \ddots & a_{-1} \\ a_n & a_{n-1} & \cdots & a_1 & a_0 \end{bmatrix} = [a_{i-j}]_{i,j=0}^{n}. \tag{2.9.1}$$

This form of matrix with $\mathbf{A} = [a_{i-j}]_{i,j=0}^{n}$ is known as the *Toeplitz matrix*. Clearly, an $(n+1) \times (n+1)$ Toeplitz matrix is determined completely by the entries $a_0, a_{-1}, \ldots, a_{-n}$ of its first row and the entries $a_0, a_1, \ldots, a_n$ of its first column.

2.9.1 Symmetric Toeplitz Matrix

The entries of a *symmetric Toeplitz matrix* $\mathbf{A} = [a_{|i-j|}]_{i,j=0}^{n}$ satisfy the symmetry relations $a_{-i} = a_i, i = 1, 2, \ldots, n$. Thus, a symmetric Toeplitz matrix is completely described by the entries of its first row. An $(n+1) \times (n+1)$ symmetric Toeplitz matrix is usually denoted $\text{Toep}[a_0, a_1, \ldots, a_n]$.

If the entries of a complex Toeplitz matrix satisfy $a_{-i} = a_i^*$, i.e.,

$$\mathbf{A} = \begin{bmatrix} a_0 & a_1^* & a_2^* & \cdots & a_n^* \\ a_1 & a_0 & a_1^* & \cdots & a_{n-1}^* \\ a_2 & a_1 & a_0 & \ddots & \vdots \\ \vdots & \vdots & \ddots & \ddots & a_1^* \\ a_n & a_{n-1} & \cdots & a_1 & a_0 \end{bmatrix}, \tag{2.9.2}$$

then it is known as a *Hermitian Toeplitz matrix*. Furthermore, an $(n+1)\times(n+1)$ Toeplitz matrix with the special structure

$$\mathbf{A}_S = \begin{bmatrix} 0 & -a_1^* & -a_2^* & \cdots & -a_n^* \\ a_1 & 0 & -a_1^* & \cdots & -a_{n-1}^* \\ a_2 & a_1 & 0 & \ddots & \vdots \\ \vdots & \vdots & \ddots & \ddots & -a_1^* \\ a_n & a_{n-1} & \cdots & a_1 & 0 \end{bmatrix} \tag{2.9.3}$$

is called a *skew-Hermitian Toeplitz matrix*, whereas

$$\mathbf{A} = \begin{bmatrix} a_0 & -a_1^* & -a_2^* & \cdots & -a_n^* \\ a_1 & a_0 & -a_1^* & \cdots & -a_{n-1}^* \\ a_2 & a_1 & a_0 & \ddots & \vdots \\ \vdots & \vdots & \ddots & \ddots & -a_1^* \\ a_n & a_{n-1} & \cdots & a_1 & a_0 \end{bmatrix} \tag{2.9.4}$$

is known as a *skew-Hermitian-type Toeplitz matrix*.

Toeplitz matrices have the following properties [397].

(1) The linear combination of several Toeplitz matrices is also a Toeplitz matrix.
(2) If the entries of a Toeplitz matrix $\mathbf{A}$ satisfy $a_{ij} = a_{|i-j|}$, then $\mathbf{A}$ is a symmetric Toeplitz matrix.
(3) The transpose $\mathbf{A}^T$ of the Toeplitz matrix $\mathbf{A}$ is also a Toeplitz matrix.
(4) The entries of a Toeplitz matrix are symmetric relative to its cross-diagonal lines.

In statistical signal processing and other related fields, it is necessary to solve the matrix equation $\mathbf{Ax} = \mathbf{b}$ with symmetric Toeplitz matrix $\mathbf{A}$. Such a system of linear equations is called a *Toeplitz system*.

By using the special structure of the Toeplitz matrix, a class of Levinson recursive algorithms has been proposed for solving a Toeplitz system of linear equations. For a real positive-definite Toeplitz matrix, its classical Levinson recursion [291] involved a large redundancy in the calculation. In order to reduce the computational redundancy, Delsarte and Genin presented successively a split Levinson algorithm [120] and a split Schur algorithm [121]. Later, Krishna and Morgera [265] extended the split Levinson recursion to a Toeplitz system of complex linear equations.

Although these algorithms are recursive, their computational complexity is $O(n^2)$. In addition to the Levinson recursive algorithms, one can apply an FFT algorithm to solve the Toeplitz system of linear equations, and this class of algorithms only requires a computational complexity $O(n \log_2 n)$, which is faster than the Levinson recursive algorithms. In view of this, FFT algorithms for solving the Toeplitz system of linear equations are called fast Toeplitz algorithms in most of the literature. These fast Toeplitz algorithms include the Kumar algorithm [269], the Davis algorithm [117] and so on, while some publications refer to this class of FFT algorithms as superfast algorithms.

The computation complexity of the Kumar algorithm is $O(n \log^2 n)$. In particular, the conjugate gradient algorithm in [89] requires only a computational complexity $O(n \log n)$ for solving the Toeplitz system of complex linear equations.

2.9.2 Discrete Cosine Transform of Toeplitz Matrix

An $N \times N$ Toeplitz matrix can be computed fast using an Nth-order *discrete cosine transform* (DCT).

The Nth-order DCT can be expressed as an $N \times N$ matrix $\mathbf{T}$, where the (m, l)th entry of $\mathbf{T} = [t_{m,l}]_{m,l=0}^{N-1}$ is defined as

$$t_{m,l} = \tau_m \cos\left(\frac{\pi}{2N} m(2l+1)\right) \quad \text{for } m, l = 0, 1, \ldots, N-1, \tag{2.9.5}$$

$$\tau_m = \begin{cases} \sqrt{1/N}, & m = 0, \\ \sqrt{2/N}, & m = 1, 2, \ldots, N-1. \end{cases} \tag{2.9.6}$$

From the above definition, it is easy to see that $\mathbf{T}^{-1} = \mathbf{T}^T$, namely $\mathbf{T}$ is an orthogonal matrix. Hence, given any $N \times N$ matrix $\mathbf{A}$, its DCT matrix $\hat{\mathbf{A}} = \mathbf{T}\mathbf{A}\mathbf{T}^T$ and the original matrix $\mathbf{A}$ have the same eigenvalues. Algorithm 2.1 shows the *fast DCT algorithm* for a Toeplitz matrix.

The common way of computing Toeplitz matrix eigenvalues is to adopt some procedure (such as the Given rotation) to transform the nondiagonal entries of $\mathbf{A}$ to zero, namely to diagonalize $\mathbf{A}$. As mentioned above, since $\mathbf{T}$ is an orthogonal matrix, $\hat{\mathbf{A}} = \mathbf{T}\mathbf{A}\mathbf{T}^T$ and $\mathbf{A}$ have the same eigenvalues. Hence, we can diagonalize the DCT $\hat{\mathbf{A}}$ instead of $\mathbf{A}$ itself to find the eigenvalues of $\mathbf{A}$. In some cases, the diagonal entries of the DCT matrix $\hat{\mathbf{A}}$ are already exact estimates of the eigenvalues of $\mathbf{A}$; see e.g. [185].

Some examples show [357] that the fast DCT matrix $\hat{\mathbf{A}}$ can be used as an approximation of the Toeplitz matrix $\mathbf{A}$ because the main components of $\hat{\mathbf{A}}$ are focused on a much smaller matrix partition that corresponds to the steady part of $\mathbf{A}$. From

Algorithm 2.1 Fast DCT algorithm for a Toeplitz matrix [357]

input: Toeplitz matrix $\mathbf{A} = [a_{l,k}]_{l,k=0}^{N-1} = [a_{k-l}]_{l,k=0}^{N-1}$.

1. For $m = 0, 1, \ldots, N-1$, compute
$$x_{m,0} = \sum_{l=0}^{N-1} w^{m(2l+1)} a_{l,0}, \quad w = \exp\left(-\mathrm{j}\frac{\pi}{2N}\right).$$
(Complexity: one $2N$-point DFT.)
2. For $n = -(N-1), \ldots, (N-1)$, compute
$$v_1(n) = \sum_{k=0}^{N-1} w^{2nk} a_{k+1}, \qquad v_2(n) = \sum_{k=0}^{N-1} w^{2nk} a_{1-N+k}.$$
(Complexity: two $2N$-point DFTs.)
3. For $m = 0, 1, \ldots, N-1$, compute
$$x_{m,N} = (-1)^m x_{m,0} + w^{-m}((-1)^m v_1(n) - v_2(n)).$$
(Complexity: N-times multiplication.)
4. For $n = -(N-1), \ldots, (N-1)$, compute
$$u_1(n) = \sum_{k=0}^{N-1} k w^{2nk} a_{k+1}, \quad u_2(n) = \sum_{k=0}^{N-1} k w^{2nk} a_{1-N+k}.$$
(Complexity: two $2N$-point DFTs.)
5. For $m = 0, 1, \ldots, N-1$; $n = -(N-1), \ldots, (N-1)$, compute
$$y_{m,n} = \frac{1}{w^{-2n} - w^{2m}} \left(x_{m,0} w^{-2n} - x_{m,N} w^{-2n} (-1)^n + w^m \left(v_1(n) - (-1)^m v_2(n)\right)\right),$$
$$y_{m,-m} = x_{m,0} + (N-1)(-1)^m x_{m,N} - w^{-m} \left(u_1(-m) - (-1)^m u_2(-m)\right).$$
(Complexity: several multiplications are needed for each value.)
6. For $m = 0, 1, \ldots, N-1$; $n = -(N-1), \ldots, (N-1)$, compute
$$\hat{a}_{m,n} = \tau_m \tau_n \mathrm{Re}\left(w^n \frac{y_{m,n} + y^*_{m,-n}}{2}\right).$$
(Complexity: several multiplications are required for each value.)

the viewpoint of a decision about rank, that of the DCT matrix $\hat{\mathbf{A}}$ is easily found seen but the rank of $\mathbf{A}$ is difficult to ascertain. Notice that, since $\mathbf{A}$ and its DCT $\hat{\mathbf{A}}$ have the same eigenvalues, their ranks are the same.

Exercises

2.1 Let $\mathbf{A}$ be an idempotent and symmetric matrix. Show that $\mathbf{A}$ is semi-definite.

2.2 Let $\mathbf{A}$ be an $n \times n$ antisymmetric matrix such that $\mathbf{A} = -\mathbf{A}^T$. Show that if n is an odd number then the determinant of $\mathbf{A}$ must be equal to zero.

2.3 Show that if $\mathbf{A} = -\mathbf{A}^T$ then $\mathbf{A} + \mathbf{I}$ is nonsingular.

2.4 Assume that $\mathbf{A} = -\mathbf{A}^T$. Show that the Cayley transform $\mathbf{T} = (\mathbf{I} - \mathbf{A})(\mathbf{I} + \mathbf{A})^{-1}$ is an orthogonal matrix.

2.5 Prove the following properties of an idempotent matrix.

(a) The eigenvalues of any idempotent matrix take only the values 1 and 0.
(b) All idempotent matrices (except for the identity matrix) $\mathbf{A}$ are singular.
(c) If $\mathbf{A}$ is an idempotent matrix then $\text{rank}(\mathbf{A}) = \text{tr}(\mathbf{A})$.
(d) If $\mathbf{A}$ is an idempotent matrix then $\mathbf{A}^H$ is also an idempotent matrix.

2.6 If $\mathbf{A}$ is an idempotent matrix, show that

(a) the matrices $\mathbf{A}^k$ and $\mathbf{A}$ have the same eigenvalues.
(b) $\mathbf{A}^k$ and $\mathbf{A}$ have the same rank.

2.7 Let $\mathbf{A}$ be an orthogonal matrix, and let $\mathbf{A} + \mathbf{I}$ be nonsingular. Show that the matrix $\mathbf{A}$ can be expressed as the Cayley transform $\mathbf{A} = (\mathbf{I} - \mathbf{S})(\mathbf{I} + \mathbf{S})^{-1}$, where $\mathbf{S}$ is a real antisymmetric matrix such that $\mathbf{A} = -\mathbf{A}^T$.

2.8 Let $\mathbf{P}$ be an $n \times n$ permutation matrix. Show that there exists a positive integer k such that $\mathbf{P}^k = \mathbf{I}$. (*Hint:* Consider the matrix sequence $\mathbf{P}, \mathbf{P}^2, \mathbf{P}^3, \ldots$)

2.9 Assume that $\mathbf{P}$ and $\mathbf{Q}$ are two $n \times n$ permutation matrices. Show that $\mathbf{PQ}$ is also an $n \times n$ permutation matrix.

2.10 Show that, for any matrix $\mathbf{A}$, there is a triangular matrix $\mathbf{T}$ such that $\mathbf{TA}$ is unitary.

2.11 Show that, for a given matrix $\mathbf{A}$, there is a matrix $\mathbf{J}$, whose diagonal entries take the values $+1$ or -1, such that $\mathbf{JA} + \mathbf{I}$ is nonsingular.

2.12 Show that if $\mathbf{H} = \mathbf{A} + \mathrm{j}\,\mathbf{B}$ is a Hermitian matrix and $\mathbf{A}$ is nonsingular then $|\det(\mathbf{H})|^2 = |\mathbf{A}|^2 |\mathbf{I} + \mathbf{A}^{-1}\mathbf{B}\mathbf{A}^{-1}\mathbf{B}|$.

2.13 Let $\mathbf{A}, \mathbf{S}$ be $n \times n$ matrices and let $\mathbf{S}$ be nonsingular.

(a) Show that $(\mathbf{S}^{-1}\mathbf{A}\mathbf{S})^2 = \mathbf{S}^{-1}\mathbf{A}^2\mathbf{S}$ and $(\mathbf{S}^{-1}\mathbf{A}\mathbf{S})^3 = \mathbf{S}^{-1}\mathbf{A}^3\mathbf{S}$.
(b) Use mathematical induction to show that $(\mathbf{S}^{-1}\mathbf{A}\mathbf{S})^k = \mathbf{S}^{-1}\mathbf{A}^k\mathbf{S}$, where k is a positive integer.

2.14 Let $\mathbf{A}$ be an $n \times n$ real matrix. Show that $\mathbf{B} = (\mathbf{A} + \mathbf{A}^T)/2$ is a symmetric matrix, while $\mathbf{C} = (\mathbf{A} - \mathbf{A}^T)/2$ is an antisymmetric matrix.

2.15 Let $y_i = y_i(x_1, \ldots, x_n)$, $i = 1, \ldots, n$ be n functions of $x_1, \ldots, x_n$. The matrix $\mathbf{J} = \mathbf{J}(\mathbf{y}, \mathbf{x}) = [\partial y_i / \partial x_j]$ is called the Jacobian matrix of the function $y_i(x_1, \ldots, x_n)$, $i = 1, \ldots, n$, and its determinant is the Jacobian determinant; $\mathbf{y} = [y_1, \ldots, y_n]^T$ and $\mathbf{x} = [x_1, \ldots, x_n]^T$. Show that $\mathbf{J}(\mathbf{z}, \mathbf{y})\mathbf{J}(\mathbf{y}, \mathbf{x}) = \mathbf{J}(\mathbf{z}, \mathbf{x})$.

2.16 Let $n \times n$ matrices $\mathbf{X}$ and $\mathbf{Y}$ be symmetric, and let $\mathbf{Y} = \mathbf{A}\mathbf{X}\mathbf{A}^T$. Show that the Jacobian determinant $|\mathbf{J}(\mathbf{Y}, \mathbf{X})| = |\mathbf{A}|^{n+1}$.

2.17 [32, p. 265] An $n \times n$ matrix $\mathbf{M}$ is known as a Markov matrix if its entries satisfy the condition $m_{ij} \geq 0$, $\sum_{i=1}^{n} m_{ij} = 1$, $j = 1, 2, \ldots, n$. Assuming that $\mathbf{P}$ and $\mathbf{Q}$ are two Markov matrices, show that

(a) for a constant $0 \leq \lambda \leq 1$, the matrix $\lambda\mathbf{P} + (1 - \lambda)\mathbf{Q}$ is a Markov matrix.
(b) the matrix product $\mathbf{PQ}$ is a Markov matrix as well.

2.18 Show the following properties of the Fourier matrix $\mathbf{F}$:

(a) $\mathbf{F}^{-1} = N^{-1}\mathbf{F}^*$.
(b) $\mathbf{F}^2 = [\mathbf{e}_1, \mathbf{e}_N, \mathbf{e}_{N-1}, \ldots, \mathbf{e}_2]$.
(c) $\mathbf{F}^4 = \mathbf{I}$.

2.19 The square matrix $\mathbf{A}$ satisfying $\mathbf{A}\mathbf{A}^H = \mathbf{A}^H\mathbf{A}$ is known as a normal matrix. Show that if $\mathbf{A}$ is a normal matrix then the matrix $\mathbf{A} - \lambda\mathbf{I}$ is also a normal matrix.

2.20 Two matrices $\mathbf{A}$ and $\mathbf{B}$ such that their product satisfies the commutative law $\mathbf{AB} = \mathbf{BA}$ are called *exchangeable matrices*. Show that if $\mathbf{A}$ and $\mathbf{B}$ are exchangeable then the condition for $\mathbf{A}^H$ and $\mathbf{B}$ to be exchangeable is that $\mathbf{A}$ is a normal matrix.

2.21 Let $\mathbf{A}$ be an $n \times n$ complex matrix with eigenvalues $\lambda_1, \ldots, \lambda_n$. Show that $\mathbf{A}$ is a normal matrix if and only if one of the following conditions holds:

(a) $\mathbf{A} = \mathbf{B} + \mathrm{j}\mathbf{C}$ where $\mathbf{B}$ and $\mathbf{C}$ are Hermitian and exchangeable matrices.
(b) The eigenvalues of $\mathbf{A}\mathbf{A}^H$ are $|\lambda_1|^2, \ldots, |\lambda_n|^2$.
(c) The eigenvalues of $\mathbf{A} + \mathbf{A}^H$ are $\lambda_1 + \lambda_1^*, \ldots, \lambda_n + \lambda_n^*$.
(d) $\mathbf{A} = \mathbf{U}\boldsymbol{\Sigma}\mathbf{U}^H$, where $\mathbf{U}$ is a unitary matrix and $\boldsymbol{\Sigma}$ is a diagonal matrix.

3

Matrix Differential

The matrix differential is a generalization of the multivariate function differential. The matrix differential (including the matrix partial derivative and gradient) is an important operation tool in matrix analysis and calculation and has wide applications in statistics, optimization, manifold computation, geometric physics, differential geometry, econometrics and many engineering disciplines.

3.1 Jacobian Matrix and Gradient Matrix

In the first half of this chapter we discuss the partial derivatives of a real scalar function, real vector function and real matrix function with respect to a real vector or matrix variable. In order to facilitate understanding, the symbols for variables and functions are first introduced below.

- $\mathbf{x} = [x_1, \ldots, x_m]^T \in \mathbb{R}^m$ denotes a real vector variable.
- $\mathbf{X} = [\mathbf{x}_1, \ldots, \mathbf{x}_n] \in \mathbb{R}^{m\times n}$ denotes a real matrix variable.
- $f(\mathbf{x}) \in \mathbb{R}$ is a real scalar function with an $m \times 1$ real vector $\mathbf{x}$ as variable, denoted $f : \mathbb{R}^m \to \mathbb{R}$, such as $f(\mathbf{x}) = \text{tr}(\mathbf{x}\mathbf{x}^T)$.
- $f(\mathbf{X}) \in \mathbb{R}$ is a real scalar function whose variable is an $m \times n$ real matrix $\mathbf{X}$, denoted $f : \mathbb{R}^{m\times n} \to \mathbb{R}$; e.g., $f(\mathbf{X}) = \det(\mathbf{X}^T\mathbf{X})$.
- $\mathbf{f}(\mathbf{x}) \in \mathbb{R}^p$ is a p-dimensional column vector function with an $m \times 1$ real vector $\mathbf{x}$ as variable, denoted $\mathbf{f} : \mathbb{R}^m \to \mathbb{R}^p$; e.g., $\mathbf{f}(\mathbf{x}) = b\mathbf{x} + c$.
- $\mathbf{f}(\mathbf{X}) \in \mathbb{R}^p$ is a p-dimensional real column vector function whose variable is an $m \times n$ real matrix $\mathbf{X}$, denoted $\mathbf{f} : \mathbb{R}^{m\times n} \to \mathbb{R}^p$, such as $\mathbf{f}(\mathbf{X}) = \text{vec}(\mathbf{X}^T\mathbf{X})$.
- $\mathbf{F}(\mathbf{x}) \in \mathbb{R}^{p\times q}$ is a $p \times q$ real matrix function whose variable is an $m \times 1$ real vector $\mathbf{x}$, denoted $\mathbf{F} : \mathbb{R}^m \to \mathbb{R}^{p\times q}$; e.g., $\mathbf{F}(\mathbf{x}) = \mathbf{x}\mathbf{x}^T$.
- $\mathbf{F}(\mathbf{X}) \in \mathbb{R}^{p\times q}$ is a $p \times q$ real matrix function whose variable is an $m \times n$ real matrix $\mathbf{X}$, denoted $\mathbf{F} : \mathbb{R}^{m\times n} \to \mathbb{R}^{p\times q}$; e.g., $\mathbf{F}(\mathbf{X}) = \mathbf{X}^T\mathbf{X}$.

Table 3.1 summarizes the classification of the real functions above.

Table 3.1 Classification of real functions

Function type	Variable $\mathbf{x} \in \mathbb{R}^m$	Variable $\mathbf{X} \in \mathbb{R}^{m\times n}$
scalar function $f \in \mathbb{R}$	$f(\mathbf{x})$ $f:\ \mathbb{R}^m \to \mathbb{R}$	$f(\mathbf{X})$ $f:\ \mathbb{R}^{m\times n} \to \mathbb{R}$
vector function $\mathbf{f} \in \mathbb{R}^p$	$\mathbf{f}(\mathbf{x})$ $\mathbf{f}:\ \mathbb{R}^m \to \mathbb{R}^p$	$\mathbf{f}(\mathbf{X})$ $\mathbf{f}:\ \mathbb{R}^{m\times n} \to \mathbb{R}^p$
matrix function $\mathbf{F} \in \mathbb{R}^{p\times q}$	$\mathbf{F}(\mathbf{x})$ $\mathbf{F}:\ \mathbb{R}^m \to \mathbb{R}^{p\times q}$	$\mathbf{F}(\mathbf{X})$ $\mathbf{F}:\ \mathbb{R}^{m\times n} \to \mathbb{R}^{p\times q}$

In this section we discuss the partial derivatives of a real scalar function and a real matrix function.

3.1.1 Jacobian Matrix

Definition 3.1 The *row partial derivative operator* with respect to an $m \times 1$ vector or an $m \times n$ matrix $\mathbf{X}$ are respectively defined as

$$\mathrm{D}_{\mathbf{x}} \stackrel{\text{def}}{=} \frac{\partial}{\partial \mathbf{x}^T} = \left[\frac{\partial}{\partial x_1}, \ldots, \frac{\partial}{\partial x_m}\right] \tag{3.1.1}$$

and

$$\mathrm{D}_{\mathrm{vec}\,\mathbf{X}} \stackrel{\text{def}}{=} \frac{\partial}{\partial(\mathrm{vec}\,\mathbf{X})^T} = \left[\frac{\partial}{\partial x_{11}}, \ldots, \frac{\partial}{\partial x_{m1}}, \ldots, \frac{\partial}{\partial x_{1n}}, \ldots, \frac{\partial}{\partial x_{mn}}\right]. \tag{3.1.2}$$

The *Jacobian operator* with respect to an $m \times n$ matrix $\mathbf{X}$ is defined as

$$\mathrm{D}_{\mathbf{X}} \stackrel{\text{def}}{=} \frac{\partial}{\partial \mathbf{X}^T} = \begin{bmatrix} \frac{\partial}{\partial x_{11}} & \cdots & \frac{\partial}{\partial x_{m1}} \\ \vdots & \ddots & \vdots \\ \frac{\partial}{\partial x_{1n}} & \cdots & \frac{\partial}{\partial x_{mn}} \end{bmatrix}. \tag{3.1.3}$$

Definition 3.2 The *partial derivative vector* resulting from the operation of $\mathbf{D}_{\mathbf{x}}$ on a real scalar function $f(\mathbf{x})$ with $m \times 1$ vector variable $\mathbf{x}$ is an $1 \times m$ row vector and is given by

$$\mathrm{D}_{\mathbf{x}} f(\mathbf{x}) = \frac{\partial f(\mathbf{x})}{\partial \mathbf{x}^T} = \left[\frac{\partial f(\mathbf{x})}{\partial x_1}, \ldots, \frac{\partial f(\mathbf{x})}{\partial x_m}\right]. \tag{3.1.4}$$

Definition 3.3 The partial derivative matrix

$$\mathrm{D}_{\mathbf{X}} f(\mathbf{X}) = \frac{\partial f(\mathbf{X})}{\partial \mathbf{X}^T} = \begin{bmatrix} \frac{\partial f(\mathbf{X})}{\partial x_{11}} & \cdots & \frac{\partial f(\mathbf{X})}{\partial x_{m1}} \\ \vdots & \ddots & \vdots \\ \frac{\partial f(\mathbf{X})}{\partial x_{1n}} & \cdots & \frac{\partial f(\mathbf{X})}{\partial x_{mn}} \end{bmatrix} \in \mathbb{R}^{n\times m} \tag{3.1.5}$$

is known as the *Jacobian matrix* of the real scalar function $f(\mathbf{X})$, and the partial derivative vector

$$\begin{aligned} \mathrm{D}_{\mathrm{vec}\,\mathbf{X}} f(\mathbf{X}) &= \frac{\partial f(\mathbf{X})}{\partial (\mathrm{vec}\,\mathbf{X})^T} \\ &= \left[\frac{\partial f(\mathbf{X})}{\partial x_{11}}, \ldots, \frac{\partial f(\mathbf{X})}{\partial x_{m1}}, \ldots, \frac{\partial f(\mathbf{X})}{\partial x_{1n}}, \ldots, \frac{\partial f(\mathbf{X})}{\partial x_{mn}} \right] \end{aligned} \tag{3.1.6}$$

is the *row partial derivative vector* of the real scalar function $f(\mathbf{X})$ with matrix variable $\mathbf{X}$.

There is the following relationship between the Jacobian matrix and the row partial derivative vector:

$$\mathrm{D}_{\mathrm{vec}\,\mathbf{X}} f(\mathbf{X}) = \mathrm{rvec}\,\mathrm{D}_{\mathbf{X}} f(\mathbf{X}) = \left(\mathrm{vec}\,\mathrm{D}_{\mathbf{X}}^T f(\mathbf{X}) \right)^T. \tag{3.1.7}$$

That is to say, the row vector partial derivative $\mathrm{D}_{\mathrm{vec}\,\mathbf{X}} f(\mathbf{X})$ is equal to the transpose of the column vectorization $\mathrm{vec}(\mathrm{D}_{\mathbf{X}}^T f(\mathbf{X}))$ of the Jacobian matrix $\mathrm{D}_{\mathbf{X}}^T f(\mathbf{X})$. This important relation is the basis of the Jacobian matrix identification.

As a matter of fact, the Jacobian matrix is more useful than the row partial derivative vector.

The following theorem provides a specific expression for the Jacobian matrix of a $p \times q$ real-valued matrix function $\mathbf{F}(\mathbf{X})$ with $m \times n$ matrix variable $\mathbf{X}$.

Definition 3.4 [311] Let the vectorization of a $p \times q$ matrix function $\mathbf{F}(\mathbf{X})$ be given by

$$\mathrm{vec}\,\mathbf{F}(\mathbf{X}) \stackrel{\mathrm{def}}{=} [f_{11}(\mathbf{X}), \ldots, f_{p1}(\mathbf{X}), \ldots, f_{1q}(\mathbf{X}), \ldots, f_{pq}(\mathbf{X})]^T \in \mathbb{R}^{pq}. \tag{3.1.8}$$

Then the $pq \times mn$ Jacobian matrix of $\mathbf{F}(\mathbf{X})$ is defined as

$$\mathrm{D}_{\mathbf{X}} \mathbf{F}(\mathbf{X}) \stackrel{\mathrm{def}}{=} \frac{\partial\, \mathrm{vec}\,\mathbf{F}(\mathbf{X})}{\partial (\mathrm{vec}\,\mathbf{X})^T} \in \mathbb{R}^{pq \times mn} \tag{3.1.9}$$

whose specific expression $\mathrm{D}_{\mathbf{X}} \mathbf{F}(\mathbf{X})$ is given by

$$\begin{bmatrix} \frac{\partial f_{11}}{\partial (\mathrm{vec}\,\mathbf{X})^T} \\ \vdots \\ \frac{\partial f_{p1}}{\partial (\mathrm{vec}\,\mathbf{X})^T} \\ \vdots \\ \frac{\partial f_{1q}}{\partial (\mathrm{vec}\,\mathbf{X})^T} \\ \vdots \\ \frac{\partial f_{pq}}{\partial (\mathrm{vec}\,\mathbf{X})^T} \end{bmatrix} = \begin{bmatrix} \frac{\partial f_{11}}{\partial x_{11}} & \cdots & \frac{\partial f_{11}}{\partial x_{m1}} & \cdots & \frac{\partial f_{11}}{\partial x_{1n}} & \cdots & \frac{\partial f_{11}}{\partial x_{mn}} \\ \vdots & \vdots & \vdots & \vdots & \vdots & \vdots & \vdots \\ \frac{\partial f_{p1}}{\partial x_{11}} & \cdots & \frac{\partial f_{p1}}{\partial x_{m1}} & \cdots & \frac{\partial f_{p1}}{\partial x_{1n}} & \cdots & \frac{\partial f_{p1}}{\partial x_{mn}} \\ \vdots & \vdots & \vdots & \vdots & \vdots & \vdots & \vdots \\ \frac{\partial f_{1q}}{\partial x_{11}} & \cdots & \frac{\partial f_{1q}}{\partial x_{m1}} & \cdots & \frac{\partial f_{1q}}{\partial x_{1n}} & \cdots & \frac{\partial f_{1q}}{\partial x_{mn}} \\ \vdots & \vdots & \vdots & \vdots & \vdots & \vdots & \vdots \\ \frac{\partial f_{pq}}{\partial x_{11}} & \cdots & \frac{\partial f_{pq}}{\partial x_{m1}} & \cdots & \frac{\partial f_{pq}}{\partial x_{1n}} & \cdots & \frac{\partial f_{pq}}{\partial x_{mn}} \end{bmatrix}. \tag{3.1.10}$$

3.1.2 Gradient Matrix

The partial derivative operator in column form is known as the gradient vector operator.

Definition 3.5 The *gradient vector operators* with respect to an $m \times 1$ vector $\mathbf{x}$ and to an $m \times n$ matrix $\mathbf{X}$ are respectively defined as

$$\nabla_{\mathbf{x}} \stackrel{\text{def}}{=} \frac{\partial}{\partial \mathbf{x}} = \left[\frac{\partial}{\partial x_1}, \ldots, \frac{\partial}{\partial x_m} \right]^T, \tag{3.1.11}$$

and

$$\nabla_{\text{vec}\,\mathbf{X}} \stackrel{\text{def}}{=} \frac{\partial}{\partial\, \text{vec}\,\mathbf{X}} = \left[\frac{\partial}{\partial x_{11}}, \ldots, \frac{\partial}{\partial x_{1n}}, \ldots, \frac{\partial}{\partial x_{m1}}, \ldots, \frac{\partial}{\partial x_{mn}} \right]^T. \tag{3.1.12}$$

Definition 3.6 The *gradient matrix operator* with respect to an $m \times n$ matrix $\mathbf{X}$ is defined as

$$\nabla_{\mathbf{X}} \stackrel{\text{def}}{=} \frac{\partial}{\partial \mathbf{X}} = \begin{bmatrix} \frac{\partial}{\partial x_{11}} & \cdots & \frac{\partial}{\partial x_{1n}} \\ \vdots & \ddots & \vdots \\ \frac{\partial}{\partial x_{m1}} & \cdots & \frac{\partial}{\partial x_{mn}} \end{bmatrix}. \tag{3.1.13}$$

Definition 3.7 The *gradient vectors* of functions $f(\mathbf{x})$ and $f(\mathbf{X})$ are respectively defined as

$$\nabla_{\mathbf{x}} f(\mathbf{x}) \stackrel{\text{def}}{=} \left[\frac{\partial f(\mathbf{x})}{\partial x_1}, \ldots, \frac{\partial f(\mathbf{x})}{\partial x_m} \right]^T = \frac{\partial f(\mathbf{x})}{\partial \mathbf{x}}, \tag{3.1.14}$$

$$\nabla_{\text{vec}\,\mathbf{X}} f(\mathbf{X}) \stackrel{\text{def}}{=} \left[\frac{\partial f(\mathbf{X})}{\partial x_{11}}, \ldots, \frac{\partial f(\mathbf{X})}{\partial x_{m1}}, \ldots, \frac{\partial f(\mathbf{X})}{\partial x_{1n}}, \ldots, \frac{\partial f(\mathbf{X})}{\partial x_{mn}} \right]^T. \tag{3.1.15}$$

Definition 3.8 The *gradient matrix* of the function $f(\mathbf{X})$ is defined as

$$\nabla_{\mathbf{X}} f(\mathbf{X}) = \begin{bmatrix} \frac{\partial f(\mathbf{X})}{\partial x_{11}} & \cdots & \frac{\partial f(\mathbf{X})}{\partial x_{1n}} \\ \vdots & \ddots & \vdots \\ \frac{\partial f(\mathbf{X})}{\partial x_{m1}} & \cdots & \frac{\partial f(\mathbf{X})}{\partial x_{mn}} \end{bmatrix} = \frac{\partial f(\mathbf{X})}{\partial \mathbf{X}}. \tag{3.1.16}$$

Clearly, there is the following relationship between the gradient matrix and the gradient vector of a function $f(\mathbf{X})$:

$$\nabla_{\mathbf{X}} f(\mathbf{X}) = \text{unvec}\, \nabla_{\text{vec}\,\mathbf{X}} f(\mathbf{X}). \tag{3.1.17}$$

Comparing Equation (3.1.16) with Equation (3.1.5), we have

$$\nabla_{\mathbf{X}} f(\mathbf{X}) = \mathrm{D}_{\mathbf{X}}^T f(\mathbf{X}). \tag{3.1.18}$$

That is to say, the gradient matrix of a real scalar function $f(\mathbf{X})$ is equal to the transpose of its Jacobian matrix.

As pointed out by Kreutz-Delgado [262], in manifold computation [3], geometric physics [430], [162], differential geometry [452] and so forth, the row partial derivative vector and the Jacobian matrix are the "most natural" choices. As we will see later, the row partial derivative vector and the Jacobian matrix are also the most natural choices for matrix derivatives. However, in optimization and many engineering applications, the gradient vector and the gradient matrix are a more natural choice than the row partial derivative vector and the Jacobian matrix.

An obvious fact is that, given a real scalar function $f(\mathbf{x})$, its gradient vector is directly equal to the transpose of the partial derivative vector. In this sense, the partial derivative in row vector form is a covariant form of the gradient vector, so the row partial derivative vector is also known as the *cogradient vector*. Similarly, the Jacobian matrix is sometimes called the *cogradient matrix*. The cogradient is a *covariant operator* [162] that itself is not the gradient, but is related to the gradient.

For this reason, the partial derivative operator $\partial/\partial\mathbf{x}^T$ and the Jacobian operator $\partial/\partial\mathbf{X}^T$ are known as the (row) partial derivative operator, the covariant form of the gradient operator or the cogradient operator.

The direction of the negative gradient $-\nabla_{\mathbf{x}}f(\mathbf{x})$ is known as the *gradient flow direction* of the function $f(\mathbf{x})$ at the point $\mathbf{x}$, and is expressed as

$$\dot{\mathbf{x}} = -\nabla_{\mathbf{x}}f(\mathbf{x}) \quad \text{or} \quad \dot{\mathbf{X}} = -\nabla_{\mathrm{vec}\,\mathbf{X}}f(\mathbf{X}). \tag{3.1.19}$$

From the definition formula of the gradient vector the following can be deduced.

(1) In the gradient flow direction, the function $f(\mathbf{x})$ decreases at the *maximum descent rate*. On the contrary, in the oppositive direction (i.e., the positive gradient direction), the function increases at the *maximum ascent rate*.
(2) Each component of the gradient vector gives the rate of change of the scalar function $f(\mathbf{x})$ in the component direction.

Definition 3.9 For a real matrix function $\mathbf{F}(\mathbf{X}) \in \mathbb{R}^{p\times q}$ with matrix variable $\mathbf{X} \in \mathbb{R}^{m\times n}$, its gradient matrix is defined as

$$\nabla_{\mathbf{X}}\mathbf{F}(\mathbf{X}) = \frac{\partial\,(\mathrm{vec}\,\mathbf{F}(\mathbf{X}))^T}{\partial\,\mathrm{vec}\,\mathbf{X}} = \left(\frac{\partial\,\mathrm{vec}\,\mathbf{F}(\mathbf{X})}{\partial(\mathrm{vec}\,\mathbf{X})^T}\right)^T. \tag{3.1.20}$$

Obviously, one has

$$\nabla_{\mathbf{X}}\mathbf{F}(\mathbf{X}) = (\mathrm{D}_{\mathbf{X}}\mathbf{F}(\mathbf{X}))^T. \tag{3.1.21}$$

In other words, the gradient matrix of a real matrix function is a transpose of its Jacobian matrix.

3.1.3 Calculation of Partial Derivative and Gradient

The *gradient computation* of a real function with respect to its matrix variable has the following properties and rules [307].

1. If $f(\mathbf{X}) = c$, where c is a real constant and $\mathbf{X}$ is an $m \times n$ real matrix, then the gradient $\partial c/\partial \mathbf{X} = \mathbf{O}_{m\times n}$.
2. *Linear rule* If $f(\mathbf{X})$ and $g(\mathbf{X})$ are respectively real-valued functions of the matrix variable $\mathbf{X}$, and c_1 and c_2 are two real constants, then
$$\frac{\partial[c_1 f(\mathbf{X}) + c_2 g(\mathbf{X})]}{\partial \mathbf{X}} = c_1 \frac{\partial f(\mathbf{X})}{\partial \mathbf{X}} + c_2 \frac{\partial g(\mathbf{X})}{\partial \mathbf{X}}. \tag{3.1.22}$$
3. *Product rule* If $f(\mathbf{X}), g(\mathbf{X})$ and $h(\mathbf{X})$ are real-valued functions of the matrix variable $\mathbf{X}$ then
$$\frac{\partial (f(\mathbf{X})g(\mathbf{X}))}{\partial \mathbf{X}} = g(\mathbf{X})\frac{\partial f(\mathbf{X})}{\partial \mathbf{X}} + f(\mathbf{X})\frac{\partial g(\mathbf{X})}{\partial \mathbf{X}} \tag{3.1.23}$$
and
$$\begin{aligned}\frac{\partial (f(\mathbf{X})g(\mathbf{X})h(\mathbf{X}))}{\partial \mathbf{X}} &= g(\mathbf{X})h(\mathbf{X})\frac{\partial f(\mathbf{X})}{\partial \mathbf{X}} + f(\mathbf{X})h(\mathbf{X})\frac{\partial g(\mathbf{X})}{\partial \mathbf{X}} \\ &\quad + f(\mathbf{X})g(\mathbf{X})\frac{\partial h(\mathbf{X})}{\partial \mathbf{X}}.\end{aligned} \tag{3.1.24}$$
4. *Quotient rule* If $g(\mathbf{X}) \neq 0$ then
$$\frac{\partial (f(\mathbf{X})/g(\mathbf{X}))}{\partial \mathbf{X}} = \frac{1}{g^2(\mathbf{X})}\left(g(\mathbf{X})\frac{\partial f(\mathbf{X})}{\partial \mathbf{X}} - f(\mathbf{X})\frac{\partial g(\mathbf{X})}{\partial \mathbf{X}}\right). \tag{3.1.25}$$
5. *Chain rule* If $\mathbf{X}$ is an $m \times n$ matrix and $y = f(\mathbf{X})$ and $g(y)$ are respectively the real-valued functions of the matrix variable $\mathbf{X}$ and of the scalar variable y then
$$\frac{\partial g(f(\mathbf{X}))}{\partial \mathbf{X}} = \frac{\mathrm{d}g(y)}{\mathrm{d}y}\frac{\partial f(\mathbf{X})}{\partial \mathbf{X}}. \tag{3.1.26}$$

As an extension, if $g(\mathbf{F}(\mathbf{X})) = g(\mathbf{F})$, where $\mathbf{F} = [f_{kl}] \in \mathbb{R}^{p\times q}$ and $\mathbf{X} = [x_{ij}] \in \mathbb{R}^{m\times n}$ then the chain rule is given by [386]
$$\left[\frac{\partial g(\mathbf{F})}{\partial \mathbf{X}}\right]_{ij} = \frac{\partial g(\mathbf{F})}{\partial x_{ij}} = \sum_{k=1}^{p}\sum_{l=1}^{q}\frac{\partial g(\mathbf{F})}{\partial f_{kl}}\frac{\partial f_{kl}}{\partial x_{ij}}. \tag{3.1.27}$$

When computing the partial derivative of the functions $f(\mathbf{x})$ and $f(\mathbf{X})$, we have to make the following basic assumption.

Independence assumption Given a real-valued function f, we assume that the vector variable $\mathbf{x} = [x_i]_{i=1}^{m} \in \mathbb{R}^m$ and the matrix variable $\mathbf{X} = [x_{ij}]_{i=1,j=1}^{m,n} \in \mathbb{R}^{m\times n}$ do not themselves have any special structure; namely, the entries of $\mathbf{x}$ and $\mathbf{X}$ are independent.

The independence assumption can be expressed in mathematical form as follows:
$$\frac{\partial x_i}{\partial x_j} = \delta_{ij} = \begin{cases} 1, & i = j, \\ 0, & \text{otherwise}, \end{cases} \tag{3.1.28}$$

and

$$\frac{\partial x_{kl}}{\partial x_{ij}} = \delta_{ki}\delta_{lj} = \begin{cases} 1, & k = i \text{ and } l = j, \\ 0, & \text{otherwise.} \end{cases} \tag{3.1.29}$$

Equations (3.1.28) and (3.1.29) are the basic formulas for partial derivative computation. Below we will give a few examples.

Example 3.1 Find the Jacobian matrix of the real-valued function $f(\mathbf{x}) = \mathbf{x}^T\mathbf{A}\mathbf{x}$.

Solution Because $\mathbf{x}^T\mathbf{A}\mathbf{x} = \sum_{k=1}^{n}\sum_{l=1}^{n} a_{kl}x_kx_l$, by using Equation (3.1.28) we can find the ith component of the row partial derivative vector $\partial\,\mathbf{x}^T\mathbf{A}\mathbf{x}/\partial\mathbf{x}^T$, given by

$$\left[\frac{\partial \mathbf{x}^T\mathbf{A}\mathbf{x}}{\partial \mathbf{x}^T}\right]_i = \frac{\partial}{\partial x_i}\sum_{k=1}^{n}\sum_{l=1}^{n} a_{kl}x_kx_l = \sum_{k=1}^{n} x_k a_{ki} + \sum_{l=1}^{n} x_l a_{il},$$

from which we immediately get the row partial derivative vector $\mathrm{D}f(\mathbf{x}) = \mathbf{x}^T\mathbf{A} + \mathbf{x}^T\mathbf{A}^T = \mathbf{x}^T(\mathbf{A} + \mathbf{A}^T)$ and the gradient vector $\nabla_{\mathbf{x}} f(\mathbf{x}) = (\mathrm{D}f(\mathbf{x}))^T = (\mathbf{A} + \mathbf{A}^T)\mathbf{x}$.

Example 3.2 Find the Jacobian matrix of the real function $f(\mathbf{x}) = \mathbf{a}^T\mathbf{X}\mathbf{X}^T\mathbf{b}$, where $\mathbf{X} \in \mathbb{R}^{m\times n}$ and $\mathbf{a}, \mathbf{b} \in \mathbb{R}^{n\times 1}$.

Solution Noticing that

$$\mathbf{a}^T\mathbf{X}\mathbf{X}^T\mathbf{b} = \sum_{k=1}^{m}\sum_{l=1}^{m} a_k \left(\sum_{p=1}^{n} x_{kp}x_{lp}\right) b_l$$

and using Equation (3.1.29), we can see easily that

$$\begin{aligned}
\left[\frac{\partial f(\mathbf{X})}{\partial \mathbf{X}^T}\right]_{ij} = \frac{\partial f(\mathbf{X})}{\partial x_{ji}} &= \sum_{k=1}^{m}\sum_{l=1}^{m}\sum_{p=1}^{n} \frac{\partial a_k x_{kp} x_{lp} b_l}{\partial x_{ji}} \\
&= \sum_{k=1}^{m}\sum_{l=1}^{m}\sum_{p=1}^{n}\left[a_k x_{lp} b_l \frac{\partial x_{kp}}{\partial x_{ji}} + a_k x_{kp} b_l \frac{\partial x_{lp}}{\partial x_{ji}}\right] \\
&= \sum_{i=1}^{m}\sum_{l=1}^{m}\sum_{j=1}^{n} a_j x_{li} b_l + \sum_{k=1}^{m}\sum_{i=1}^{m}\sum_{j=1}^{n} a_k x_{ki} b_j \\
&= \sum_{i=1}^{m}\sum_{j=1}^{n} \left[\mathbf{X}^T\mathbf{b}\right]_i a_j + \left[\mathbf{X}^T\mathbf{a}\right]_i b_j,
\end{aligned}$$

which yields respectively the Jacobian matrix and the gradient matrix,

$$\mathrm{D}_{\mathbf{X}} f(\mathbf{X}) = \mathbf{X}^T(\mathbf{b}\mathbf{a}^T + \mathbf{a}\mathbf{b}^T) \quad \text{and} \quad \nabla_{\mathbf{X}} f(\mathbf{X}) = (\mathbf{b}\mathbf{a}^T + \mathbf{a}\mathbf{b}^T)\mathbf{X}.$$

Example 3.3 Consider the objective function $f(\mathbf{X}) = \mathrm{tr}(\mathbf{X}\mathbf{B})$, where $\mathbf{X}$ and $\mathbf{B}$

are $m \times n$ and $n \times m$ real matrices, respectively. Find the Jacobian matrix and the gradient matrix of $f(\mathbf{X}) = \mathrm{tr}(\mathbf{XB})$.

Solution First, the (k, l)th entry of the matrix product $\mathbf{XB}$ is given by $[\mathbf{XB}]_{kl} = \sum_{p=1}^{n} x_{kp} b_{pl}$, so the trace of the matrix product is $\mathrm{tr}(\mathbf{XB}) = \sum_{p=1}^{m} \sum_{l=1}^{n} x_{lp} b_{pl}$. Hence, using Equation (3.1.29), it is easily found that

$$\left[\frac{\partial\, \mathrm{tr}(\mathbf{XB})}{\partial\, \mathbf{X}^T}\right]_{ij} = \frac{\partial}{\partial x_{ji}} \left(\sum_{p=1}^{m} \sum_{l=1}^{n} x_{lp} b_{pl} \right) = \sum_{p=1}^{m} \sum_{l=1}^{n} \frac{\partial x_{lp}}{\partial x_{ji}} b_{pl} = b_{ij},$$

that is, $\partial\, \mathrm{tr}(\mathbf{XB})/\partial\, \mathbf{X}^T = \mathbf{B}$. Moreover, since $\mathrm{tr}(\mathbf{BX}) = \mathrm{tr}(\mathbf{XB})$, we see that the $n \times m$ Jacobian matrix and the $m \times n$ gradient matrix are respectively given by

$$\mathrm{D}_{\mathbf{X}} \mathrm{tr}(\mathbf{XB}) = \mathrm{D}_{\mathbf{X}} \mathrm{tr}(\mathbf{BX}) = \mathbf{B} \quad \text{and} \quad \nabla_{\mathbf{X}} \mathrm{tr}(\mathbf{XB}) = \nabla_{\mathbf{X}} \mathrm{tr}(\mathbf{BX}) = \mathbf{B}^T.$$

The following are examples of computing the Jacobian matrix and the gradient matrix of a real matrix function.

Example 3.4 Setting $\mathbf{F}(\mathbf{X}) = \mathbf{X} \in \mathbb{R}^{m \times n}$ and then computing the partial derivative directly yields

$$\frac{\partial f_{kl}}{\partial x_{ij}} = \frac{\partial x_{kl}}{\partial x_{ij}} = \delta_{lj} \delta_{ki},$$

which yields the Jacobian matrix $\mathrm{D}_{\mathbf{X}} \mathbf{X} = \mathbf{I}_n \otimes \mathbf{I}_m = \mathbf{I}_{mn} \in \mathbb{R}^{mn \times mn}$.

Example 3.5 Let $\mathbf{F}(\mathbf{X}) = \mathbf{AXB}$, where $\mathbf{A} \in \mathbb{R}^{p \times m}, \mathbf{X} \in \mathbb{R}^{m \times n}, \mathbf{B} \in \mathbb{R}^{n \times q}$. From the partial derivative

$$\frac{\partial f_{kl}}{\partial x_{ij}} = \frac{\partial [\mathbf{AXB}]_{kl}}{\partial x_{ij}} = \frac{\partial \left(\sum_{u=1}^{m} \sum_{v=1}^{n} a_{ku} x_{uv} b_{vl} \right)}{\partial x_{ij}} = b_{jl} a_{ki},$$

we get respectively the $pq \times mn$ Jacobian matrix and the $mn \times pq$ gradient matrix:

$$\mathrm{D}_{\mathbf{X}}(\mathbf{AXB}) = \mathbf{B}^T \otimes \mathbf{A} \quad \text{and} \quad \nabla_{\mathbf{X}}(\mathbf{AXB}) = \mathbf{B} \otimes \mathbf{A}^T.$$

Example 3.6 Let $\mathbf{F}(\mathbf{X}) = \mathbf{AX}^T\mathbf{B}$, where $\mathbf{A} \in \mathbb{R}^{p \times n}, \mathbf{X} \in \mathbb{R}^{m \times n}, \mathbf{B} \in \mathbb{R}^{m \times q}$. Compute the partial derivative

$$\frac{\partial f_{kl}}{\partial x_{ij}} = \frac{\partial [\mathbf{AX}^T\mathbf{B}]_{kl}}{\partial x_{ij}} = \frac{\partial \left(\sum_{u=1}^{m} \sum_{v=1}^{n} a_{ku} x_{vu} b_{vl} \right)}{\partial x_{ij}} = b_{il} a_{kj}$$

to obtain respectively the $pq \times mn$ Jacobian matrix and the $mn \times pq$ gradient matrix:

$$\mathrm{D}_{\mathbf{X}}(\mathbf{AX}^T\mathbf{B}) = (\mathbf{B}^T \otimes \mathbf{A})\mathbf{K}_{mn} \quad \text{and} \quad \nabla_{\mathbf{X}}(\mathbf{AX}^T\mathbf{B}) = \mathbf{K}_{nm}(\mathbf{B} \otimes \mathbf{A}^T),$$

where $\mathbf{K}_{mn}$ and $\mathbf{K}_{nm}$ are commutation matrices.

Example 3.7 Let $\mathbf{X} \in \mathbb{R}^{m\times n}$, then

$$\frac{\partial f_{kl}}{\partial x_{ij}} = \frac{\partial [\mathbf{X}\mathbf{X}^T]_{kl}}{\partial x_{ij}} = \frac{\partial \left(\sum_{u=1}^n x_{ku}x_{lu}\right)}{\partial x_{ij}} = \delta_{li}x_{kj} + x_{lj}\delta_{ki},$$

$$\frac{\partial f_{kl}}{\partial x_{ij}} = \frac{\partial [\mathbf{X}^T\mathbf{X}]_{kl}}{\partial x_{ij}} = \frac{\partial \left(\sum_{u=1}^n x_{uk}x_{ul}\right)}{\partial x_{ij}} = x_{il}\delta_{kj} + \delta_{lj}x_{ik}.$$

Hence the Jacobian matrices are given by

$$\begin{aligned}\mathrm{D}_{\mathbf{X}}(\mathbf{X}\mathbf{X}^T) &= (\mathbf{I}_m \otimes \mathbf{X})\mathbf{K}_{mn} + (\mathbf{X} \otimes \mathbf{I}_m) \\ &= (\mathbf{K}_{mm} + \mathbf{I}_{m^2})(\mathbf{X} \otimes \mathbf{I}_m) \ \in \mathbb{R}^{mm\times mn}, \\ \mathrm{D}_{\mathbf{X}}(\mathbf{X}^T\mathbf{X}) &= (\mathbf{X}^T \otimes \mathbf{I}_n)\mathbf{K}_{mn} + (\mathbf{I}_n \otimes \mathbf{X}^T) \\ &= (\mathbf{K}_{nn} + \mathbf{I}_{n^2})(\mathbf{I}_n \otimes \mathbf{X}^T) \ \in \mathbb{R}^{nn\times mn}\end{aligned}$$

and the gradient matrices are given by

$$\begin{aligned}\nabla_{\mathbf{X}}(\mathbf{X}\mathbf{X}^T) &= (\mathbf{X}^T \otimes \mathbf{I}_m)(\mathbf{K}_{mm} + \mathbf{I}_{m^2}) \ \in \mathbb{R}^{mn\times mm}, \\ \nabla_{\mathbf{X}}(\mathbf{X}^T\mathbf{X}) &= (\mathbf{I}_n \otimes \mathbf{X})(\mathbf{K}_{nn} + \mathbf{I}_{n^2}) \ \in \mathbb{R}^{mn\times nn}\end{aligned}$$

in which we have used $(\mathbf{A}_{p\times m}\otimes\mathbf{B}_{q\times n})\mathbf{K}_{mn} = \mathbf{K}_{pq}(\mathbf{B}_{q\times n}\otimes\mathbf{A}_{p\times m})$ and $\mathbf{K}_{mn}^T = \mathbf{K}_{nm}$.

Example 3.8 The partial derivatives of the products of three matrices are given by

$$\begin{aligned}\frac{\partial [\mathbf{X}^T\mathbf{B}\mathbf{X}]_{kl}}{\partial x_{ij}} &= \frac{\partial \sum_p \left(x_{pk}[\mathbf{B}\mathbf{X}]_{pl} + [\mathbf{X}^T\mathbf{B}]_{kp}x_{pl}\right)}{\partial x_{ij}} \\ &= \sum_p \left(\delta_{pi}\delta_{kj}[\mathbf{B}\mathbf{X}]_{pl} + \delta_{pi}\delta_{lj}[\mathbf{X}^T\mathbf{B}]_{kp}\right) \\ &= [\mathbf{B}\mathbf{X}]_{il}\delta_{kj} + \delta_{lj}[\mathbf{X}^T\mathbf{B}]_{ki} \\ \frac{\partial [\mathbf{X}\mathbf{B}\mathbf{X}^T]_{kl}}{\partial x_{ij}} &= \frac{\partial \sum_p \left(x_{kp}[\mathbf{B}\mathbf{X}^T]_{pl} + [\mathbf{X}\mathbf{B}]_{kp}x_{lp}\right)}{\partial x_{ij}} \\ &= \sum_p \left(\delta_{ki}\delta_{pj}[\mathbf{B}\mathbf{X}^T]_{pl} + \delta_{pj}\delta_{li}[\mathbf{X}\mathbf{B}]_{kp}\right) \\ &= [\mathbf{B}\mathbf{X}^T]_{jl}\delta_{ki} + \delta_{li}[\mathbf{X}\mathbf{B}]_{kj},\end{aligned}$$

and thus we obtain respectively the Jacobian matrix and the gradient matrix:

$$\begin{aligned}\mathrm{D}_{\mathbf{X}}(\mathbf{X}^T\mathbf{B}\mathbf{X}) &= \left((\mathbf{B}\mathbf{X})^T \otimes \mathbf{I}_n\right)\mathbf{K}_{mn} + \left(\mathbf{I}_n \otimes (\mathbf{X}^T\mathbf{B})\right) \ \in \mathbb{R}^{nn\times mn}, \\ \mathrm{D}_{\mathbf{X}}(\mathbf{X}\mathbf{B}\mathbf{X}^T) &= (\mathbf{X}\mathbf{B}^T) \otimes \mathbf{I}_m + (\mathbf{I}_m \otimes (\mathbf{X}\mathbf{B}))\,\mathbf{K}_{mn} \ \in \mathbb{R}^{mm\times mn}; \\ \nabla_{\mathbf{X}}(\mathbf{X}^T\mathbf{B}\mathbf{X}) &= \mathbf{K}_{nm}\left((\mathbf{B}\mathbf{X}) \otimes \mathbf{I}_n\right) + \left(\mathbf{I}_n \otimes (\mathbf{B}^T\mathbf{X})\right) \ \in \mathbb{R}^{mn\times nn}, \\ \nabla_{\mathbf{X}}(\mathbf{X}\mathbf{B}\mathbf{X}^T) &= (\mathbf{B}\mathbf{X}^T) \otimes \mathbf{I}_m + \mathbf{K}_{nm}\left(\mathbf{I}_m \otimes (\mathbf{X}\mathbf{B})^T\right) \ \in \mathbb{R}^{mn\times mm}.\end{aligned}$$

Table 3.2 summarizes the partial derivative, the Jacobian matrix and the gradient matrix for several typical matrix functions.

Table 3.2 Partial derivative, Jacobian matrix and gradient matrix of $\mathbf{F}(\mathbf{X})$

$\mathbf{F}(\mathbf{X})$	$\partial f_{kl}/\partial x_{ij}$	$\mathrm{D}_{\mathbf{X}}\mathbf{F}(\mathbf{X})$	$\nabla_{\mathbf{X}}\mathbf{F}(\mathbf{X})$
$\mathbf{AXB}$	$b_{jl}a_{ki}$	$\mathbf{B}^T \otimes \mathbf{A}$	$\mathbf{B} \otimes \mathbf{A}^T$
$\mathbf{A}^T\mathbf{XB}$	$b_{jl}a_{ik}$	$\mathbf{B}^T \otimes \mathbf{A}^T$	$\mathbf{B} \otimes \mathbf{A}$
$\mathbf{AXB}^T$	$b_{lj}a_{ki}$	$\mathbf{B} \otimes \mathbf{A}$	$\mathbf{B}^T \otimes \mathbf{A}^T$
$\mathbf{A}^T\mathbf{XB}^T$	$b_{lj}a_{ik}$	$\mathbf{B} \otimes \mathbf{A}^T$	$\mathbf{B}^T \otimes \mathbf{A}$
$\mathbf{AX}^T\mathbf{B}$	$b_{il}a_{kj}$	$(\mathbf{B}^T \otimes \mathbf{A})\mathbf{K}_{mn}$	$\mathbf{K}_{nm}(\mathbf{B} \otimes \mathbf{A}^T)$
$\mathbf{A}^T\mathbf{X}^T\mathbf{B}$	$b_{il}a_{jk}$	$(\mathbf{B}^T \otimes \mathbf{A}^T)\mathbf{K}_{mn}$	$\mathbf{K}_{nm}(\mathbf{B} \otimes \mathbf{A})$
$\mathbf{AX}^T\mathbf{B}^T$	$b_{li}a_{kj}$	$(\mathbf{B} \otimes \mathbf{A})\mathbf{K}_{mn}$	$\mathbf{K}_{nm}(\mathbf{B}^T \otimes \mathbf{A}^T)$
$\mathbf{A}^T\mathbf{X}^T\mathbf{B}^T$	$b_{li}a_{jk}$	$(\mathbf{B} \otimes \mathbf{A}^T)\mathbf{K}_{mn}$	$\mathbf{K}_{nm}(\mathbf{B}^T \otimes \mathbf{A})$
$\mathbf{XX}^T$	$\delta_{li}x_{kj} + x_{lj}\delta_{ki}$	$(\mathbf{K}_{mm} + \mathbf{I}_{m^2})(\mathbf{X} \otimes \mathbf{I}_m)$	$(\mathbf{X}^T \otimes \mathbf{I}_m)(\mathbf{K}_{mm} + \mathbf{I}_{m^2})$
$\mathbf{X}^T\mathbf{X}$	$x_{il}\delta_{kj} + \delta_{lj}x_{ik}$	$(\mathbf{K}_{nn} + \mathbf{I}_{n^2})(\mathbf{I}_n \otimes \mathbf{X}^T)$	$(\mathbf{I}_n \otimes \mathbf{X})(\mathbf{K}_{nn} + \mathbf{I}_{n^2})$
$\mathbf{X}^T\mathbf{BX}$	$[\mathbf{BX}]_{il}\delta_{kj}$ $+ \delta_{lj}[\mathbf{X}^T\mathbf{B}]_{ki}$	$((\mathbf{BX})^T \otimes \mathbf{I}_n)\,\mathbf{K}_{mn}$ $+ (\mathbf{I}_n \otimes (\mathbf{X}^T\mathbf{B}))$	$\mathbf{K}_{nm}\,((\mathbf{BX}) \otimes \mathbf{I}_n)$ $+ (\mathbf{I}_n \otimes (\mathbf{B}^T\mathbf{X}))$
$\mathbf{XBX}^T$	$[\mathbf{BX}^T]_{jl}\delta_{ki}$ $+ \delta_{li}[\mathbf{XB}]_{kj}$	$(\mathbf{XB}^T) \otimes \mathbf{I}_m$ $+ (\mathbf{I}_m \otimes (\mathbf{XB}))\,\mathbf{K}_{mn}$	$(\mathbf{BX}^T) \otimes \mathbf{I}_m$ $+\mathbf{K}_{nm}\,(\mathbf{I}_m \otimes (\mathbf{XB})^T)$

If $\mathbf{X} = \mathbf{x} \in \mathbb{R}^{m\times 1}$ then from Table 3.2 we obtain

$$\mathrm{D}_{\mathbf{x}}(\mathbf{xx}^T) = \mathbf{I}_{m^2}(\mathbf{x} \otimes \mathbf{I}_m) + \mathbf{K}_{mm}(\mathbf{x} \otimes \mathbf{I}_m) = (\mathbf{x} \otimes \mathbf{I}_m) + (\mathbf{I}_m \otimes \mathbf{x}), \tag{3.1.30}$$

since $\mathbf{K}_{mm}(\mathbf{x} \otimes \mathbf{I}_m) = (\mathbf{I}_m \otimes \mathbf{x})\mathbf{K}_{m1}$ and $\mathbf{K}_{m1} = \mathbf{I}_m$. Similarly, we can obtain

$$\mathrm{D}_{\mathbf{x}}(\mathbf{xx}^T) = (\mathbf{K}_{11} + \mathbf{I}_1)(\mathbf{I}_1 \otimes \mathbf{x}^T) = 2\mathbf{x}^T. \tag{3.1.31}$$

It should be noted that direct computation of partial derivatives $\partial f_{kl}/\partial x_{ij}$ can be used to find the Jacobian matrices and the gradient matrices of many matrix functions, but, for more complex functions (such as the inverse matrix, the Moore–Penrose inverse matrix and the exponential functions of a matrix), direct computation of their partial derivatives is more complicated and difficult. Hence, naturally, we want to have an easily remembered and effective mathematical tool for computing the Jacobian matrices and the gradient matrices of real scalar functions and real matrix functions. Such a mathematical tool is the matrix differential, which is the topic of the next section.

3.2 Real Matrix Differential

The matrix differential is an effective mathematical tool for computing the partial derivatives of a function with respect to its variables. This section introduces the first-order real matrix differential together with its theory, calculation methods and applications.

3.2.1 Calculation of Real Matrix Differential

The differential of an $m \times n$ matrix $\mathbf{X} = [x_{ij}]$ is known as the *matrix differential*, denoted $\mathrm{d}\mathbf{X}$, and is defined as $\mathrm{d}\mathbf{X} = [\mathrm{d}x_{ij}]_{i=1,j=1}^{m,n}$.

Example 3.9 Consider the differential of the scalar function $\mathrm{tr}(\mathbf{U})$. We have

$$\mathrm{d}(\mathrm{tr}\,\mathbf{U}) = \mathrm{d}\left(\sum_{i=1}^{n} u_{ii}\right) = \sum_{i=1}^{n} \mathrm{d}u_{ii} = \mathrm{tr}(\mathrm{d}\mathbf{U}),$$

namely $\mathrm{d}(\mathrm{tr}\,\mathbf{U}) = \mathrm{tr}(\mathrm{d}\mathbf{U})$.

Example 3.10 An entry of the matrix differential of the matrix product $\mathbf{UV}$ is given by

$$\begin{aligned}
[\mathrm{d}(\mathbf{UV})]_{ij} &= \mathrm{d}\left([\mathbf{UV}]_{ij}\right) = \mathrm{d}\left(\sum_k u_{ik}v_{kj}\right) = \sum_k \mathrm{d}(u_{ik}v_{kj}) \\
&= \sum_k \left((\mathrm{d}u_{ik})v_{kj} + u_{ik}\mathrm{d}v_{kj}\right) = \sum_k (\mathrm{d}u_{ik})v_{kj} + \sum_k u_{ik}\mathrm{d}v_{kj} \\
&= [(\mathrm{d}\mathbf{U})\mathbf{V}]_{ij} + [\mathbf{U}(\mathrm{d}\mathbf{V})]_{ij}.
\end{aligned}$$

Hence, we have for the matrix differential $\mathrm{d}(\mathbf{UV}) = (\mathrm{d}\mathbf{U})\mathbf{V} + \mathbf{U}(\mathrm{d}\mathbf{V})$.

The real matrix differential has the following two basic properties.

(1) *Transpose* The differential of a matrix transpose is equal to the transpose of the matrix differential, i.e., $\mathrm{d}(\mathbf{X}^T) = (\mathrm{d}\mathbf{X})^T = \mathrm{d}\mathbf{X}^T$.
(2) *Linearity* $\mathrm{d}(\alpha\mathbf{X} + \beta\mathbf{Y}) = \alpha\,\mathrm{d}\mathbf{X} + \beta\,\mathrm{d}\mathbf{Y}$.

The following summarizes common computation formulas for the matrix differential [311, pp. 148–154].

1. The differential of a constant matrix is a zero matrix, namely $\mathrm{d}\mathbf{A} = \mathbf{O}$.
2. The matrix differential of the product $\alpha\mathbf{X}$ is given by $\mathrm{d}(\alpha\mathbf{X}) = \alpha\,\mathrm{d}\mathbf{X}$.
3. The matrix differential of a transposed matrix is equal to the transpose of the original matrix differential, namely $\mathrm{d}(\mathbf{X}^T) = (\mathrm{d}\mathbf{X})^T = \mathrm{d}\mathbf{X}^T$.
4. The matrix differential of the sum (or difference) of two matrices is given by $\mathrm{d}(\mathbf{U} \pm \mathbf{V}) = \mathrm{d}\mathbf{U} \pm \mathrm{d}\mathbf{V}$.
5. The matrix differential of the matrix product is $\mathrm{d}(\mathbf{AXB}) = \mathbf{A}(\mathrm{d}\mathbf{X})\mathbf{B}$.

6. The matrix differentials of the functions $\mathbf{UV}$ and $\mathbf{UVW}$, where $\mathbf{U} = \mathbf{F}(\mathbf{X}), \mathbf{V} = \mathbf{G}(\mathbf{X}), \mathbf{W} = \mathbf{H}(\mathbf{X})$, are respectively given by

$$\mathrm{d}(\mathbf{UV}) = (\mathrm{d}\mathbf{U})\mathbf{V} + \mathbf{U}(\mathrm{d}\mathbf{V}) \tag{3.2.1}$$
$$\mathrm{d}(\mathbf{UVW}) = (\mathrm{d}\mathbf{U})\mathbf{VW} + \mathbf{U}(\mathrm{d}\mathbf{V})\mathbf{W} + \mathbf{UV}(\mathrm{d}\mathbf{W}). \tag{3.2.2}$$

7. The differential of the matrix trace $\mathrm{d}(\mathrm{tr}\,\mathbf{X})$ is equal to the trace of the matrix differential $\mathrm{d}\mathbf{X}$, namely

$$\mathrm{d}(\mathrm{tr}\,\mathbf{X}) = \mathrm{tr}(\mathrm{d}\mathbf{X}). \tag{3.2.3}$$

In particular, the differential of the trace of the matrix function $\mathbf{F}(\mathbf{X})$ is given by $\mathrm{d}(\mathrm{tr}\,\mathbf{F}(\mathbf{X})) = \mathrm{tr}(\mathrm{d}\mathbf{F}(\mathbf{X}))$.

8. The differential of the determinant of $\mathbf{X}$ is

$$\mathrm{d}|\mathbf{X}| = |\mathbf{X}|\,\mathrm{tr}(\mathbf{X}^{-1}\mathrm{d}\mathbf{X}). \tag{3.2.4}$$

In particular, the differential of the determinant of the matrix function $\mathbf{F}(\mathbf{X})$ is given by $\mathrm{d}|\mathbf{F}(\mathbf{X})| = |\mathbf{F}(\mathbf{X})|\,\mathrm{tr}(\mathbf{F}^{-1}(\mathbf{X})\mathrm{d}\mathbf{F}(\mathbf{X}))$.

9. The matrix differential of the Kronecker product is given by

$$\mathrm{d}(\mathbf{U} \otimes \mathbf{V}) = (\mathrm{d}\mathbf{U}) \otimes \mathbf{V} + \mathbf{U} \otimes \mathrm{d}\mathbf{V}. \tag{3.2.5}$$

10. The matrix differential of the Hadamard product is given by

$$\mathrm{d}(\mathbf{U} * \mathbf{V}) = (\mathrm{d}\mathbf{U}) * \mathbf{V} + \mathbf{U} * \mathrm{d}\mathbf{V}. \tag{3.2.6}$$

11. The matrix differential of the inverse matrix is

$$\mathrm{d}(\mathbf{X}^{-1}) = -\mathbf{X}^{-1}(\mathrm{d}\mathbf{X})\mathbf{X}^{-1}. \tag{3.2.7}$$

12. The differential of the vectorization function $\mathrm{vec}\,\mathbf{X}$ is equal to the vectorization of the matrix differential, i.e.,

$$\mathrm{d}\,\mathrm{vec}\,\mathbf{X} = \mathrm{vec}(\mathrm{d}\mathbf{X}). \tag{3.2.8}$$

13. The differential of the matrix logarithm is

$$\mathrm{d}\log\mathbf{X} = \mathbf{X}^{-1}\mathrm{d}\mathbf{X}. \tag{3.2.9}$$

In particular, $\mathrm{d}\log\mathbf{F}(\mathbf{X}) = \mathbf{F}^{-1}(\mathbf{X})\,\mathrm{d}\mathbf{F}(\mathbf{X})$.

14. The matrix differentials of $\mathbf{X}^\dagger, \mathbf{X}^\dagger\mathbf{X}$ and $\mathbf{X}\mathbf{X}^\dagger$, where $\mathbf{X}^\dagger$ is the Moore–Penrose inverses of $\mathbf{X}$, are

$$\begin{aligned}\mathrm{d}(\mathbf{X}^\dagger) = &-\mathbf{X}^\dagger(\mathrm{d}\mathbf{X})\mathbf{X}^\dagger + \mathbf{X}^\dagger(\mathbf{X}^\dagger)^T(\mathrm{d}\mathbf{X}^T)(\mathbf{I} - \mathbf{X}\mathbf{X}^\dagger)\\ &+ (\mathbf{I} - \mathbf{X}^\dagger\mathbf{X})(\mathrm{d}\mathbf{X}^T)(\mathbf{X}^\dagger)^T\mathbf{X}^\dagger,\end{aligned} \tag{3.2.10}$$
$$\mathrm{d}(\mathbf{X}^\dagger\mathbf{X}) = \mathbf{X}^\dagger(\mathrm{d}\mathbf{X})(\mathbf{I} - \mathbf{X}^\dagger\mathbf{X}) + \left(\mathbf{X}^\dagger(\mathrm{d}\mathbf{X})(\mathbf{I} - \mathbf{X}^\dagger\mathbf{X})\right)^T, \tag{3.2.11}$$
$$\mathrm{d}(\mathbf{X}\mathbf{X}^\dagger) = (\mathbf{I} - \mathbf{X}\mathbf{X}^\dagger)(\mathrm{d}\mathbf{X})\mathbf{X}^\dagger + \left((\mathbf{I} - \mathbf{X}\mathbf{X}^\dagger)(\mathrm{d}\mathbf{X})\mathbf{X}^\dagger\right)^T. \tag{3.2.12}$$

3.2.2 Jacobian Matrix Identification

In multivariate calculus, the multivariate function $f(x_1,\ldots,x_m)$ is said to be differentiable at the point $(x_1,\ldots,x_m)$, if a change in $f(x_1,\ldots,x_m)$ can be expressed as

$$\begin{aligned}\Delta f(x_1,\ldots,x_m) &= f(x_1+\Delta x_1,\ldots,x_m+\Delta x_m) - f(x_1,\ldots,x_m)\\ &= A_1\Delta x_1 + \cdots + A_m\Delta x_m + O(\Delta x_1,\ldots,\Delta x_m),\end{aligned}$$

where $A_1,\ldots,A_m$ are independent of $\Delta x_1,\ldots,\Delta x_m$, respectively, and $O(\Delta x_1,\ldots,\Delta x_m)$ denotes the second-order and the higher-order terms in $\Delta x_1,\ldots,\Delta x_m$. In this case, the partial derivative $\partial f/\partial x_1,\ldots,\partial f/\partial x_m$ must exist, and

$$\frac{\partial f}{\partial x_1} = A_1, \quad \ldots \quad , \quad \frac{\partial f}{\partial x_m} = A_m.$$

The linear part of the change $\Delta f(x_1,\ldots,x_m)$,

$$A_1\Delta x_1 + \cdots + A_m\Delta x_m = \frac{\partial f}{\partial x_1}\mathrm{d}x_1 + \cdots + \frac{\partial f}{\partial x_m}\mathrm{d}x_m,$$

is said to be the *differential* or *first-order differential* of the multivariate function $f(x_1,\ldots,x_m)$ and is denoted by

$$\mathrm{d}f(x_1,\ldots,x_m) = \frac{\partial f}{\partial x_1}\,\mathrm{d}x_1 + \cdots + \frac{\partial f}{\partial x_m}\,\mathrm{d}x_m. \tag{3.2.13}$$

The sufficient condition for a multivariate function $f(x_1,\ldots,x_m)$ to be differentiable at the point $(x_1,\ldots,x_m)$ is that the partial derivatives $\partial f/\partial x_1,\ldots,\partial f/\partial x_m$ exist and are continuous.

1. Jacobian Matrix Identification for a Real Function $f(\mathbf{x})$

Consider a scalar function $f(\mathbf{x})$ with variable $\mathbf{x} = [x_1,\ldots,x_m]^T \in \mathbb{R}^m$. By regarding the elements $x_1,\ldots,x_m$ as m variables, and using Equation (3.2.13), we can obtain the differential of the scalar function $f(\mathbf{x})$ directly, as follows:

$$\mathrm{d}f(\mathbf{x}) = \frac{\partial f(\mathbf{x})}{\partial x_1}\mathrm{d}x_1 + \cdots + \frac{\partial f(\mathbf{x})}{\partial x_m}\mathrm{d}x_m = \begin{bmatrix}\frac{\partial f(\mathbf{x})}{\partial x_1} & \cdots & \frac{\partial f(\mathbf{x})}{\partial x_m}\end{bmatrix}\begin{bmatrix}\mathrm{d}x_1\\ \vdots\\ \mathrm{d}x_m\end{bmatrix}$$

or

$$\mathrm{d}f(\mathbf{x}) = \frac{\partial f(\mathbf{x})}{\partial \mathbf{x}^T}\,\mathrm{d}\mathbf{x} = (\mathrm{d}\mathbf{x})^T\frac{\partial f(\mathbf{x})}{\partial \mathbf{x}}, \tag{3.2.14}$$

where $\frac{\partial f(\mathbf{x})}{\partial \mathbf{x}^T} = \left[\frac{\partial f(\mathbf{x})}{\partial x_1},\ldots,\frac{\partial f(\mathbf{x})}{\partial x_m}\right]$ and $\mathrm{d}\mathbf{x} = [\mathrm{d}x_1,\ldots,\mathrm{d}x_m]^T$.

Equation (3.2.14) is a vector form of the differential rule that implies an important application. If we let $\mathbf{A} = \partial f(\mathbf{x})/\partial \mathbf{x}^T$ then the first-order differential can be denoted as a trace: $\mathrm{d}f(\mathbf{x}) = (\partial f(\mathbf{x})/\partial \mathbf{x}^T)\mathrm{d}\mathbf{x} = \mathrm{tr}(\mathbf{A}\,\mathrm{d}\mathbf{x})$. This shows that there is an equivalence relationship between the Jacobian matrix of the scalar function $f(\mathbf{x})$ and its matrix differential, namely

$$\mathrm{d}f(\mathbf{x}) = \mathrm{tr}(\mathbf{A}\,\mathrm{d}\mathbf{x}) \quad \Leftrightarrow \quad \mathrm{D}_{\mathbf{x}} f(\mathbf{x}) = \frac{\partial f(\mathbf{x})}{\partial \mathbf{x}^T} = \mathbf{A}. \tag{3.2.15}$$

In other words, if the differential of the function $f(\mathbf{x})$ is denoted as $\mathrm{d}f(\mathbf{x}) = \mathrm{tr}(\mathbf{A}\,\mathrm{d}\mathbf{x})$ then the matrix $\mathbf{A}$ is just the Jacobian matrix of the function $f(\mathbf{x})$.

2. Jacobian Matrix Identification for a Real Function $f(\mathbf{X})$

Consider a scalar function $f(\mathbf{X})$ with variable $\mathbf{X} = [\mathbf{x}_1, \ldots, \mathbf{x}_n] \in \mathbb{R}^{m\times n}$. Denoting $\mathbf{x}_j = [x_{1j}, \ldots, x_{mj}]^T, j = 1, \ldots, n$, then from Equation (3.2.2) it is easily known that

$$\begin{aligned}
\mathrm{d}f(\mathbf{X}) &= \frac{\partial f(\mathbf{X})}{\partial \mathbf{x}_1}\mathrm{d}\mathbf{x}_1 + \cdots + \frac{\partial f(\mathbf{X})}{\partial \mathbf{x}_n}\mathrm{d}\mathbf{x}_n \\
&= \begin{bmatrix} \dfrac{\partial f(\mathbf{X})}{\partial x_{11}} & \cdots & \dfrac{\partial f(\mathbf{X})}{\partial x_{m1}} & \cdots & \dfrac{\partial f(\mathbf{X})}{\partial x_{1n}} & \cdots & \dfrac{\partial f(\mathbf{X})}{\partial x_{mn}} \end{bmatrix} \begin{bmatrix} \mathrm{d}x_{11} \\ \vdots \\ \mathrm{d}x_{m1} \\ \vdots \\ \mathrm{d}x_{1n} \\ \vdots \\ \mathrm{d}x_{mn} \end{bmatrix} \\
&= \frac{\partial f(\mathbf{X})}{\partial (\mathrm{vec}\,\mathbf{X})^T}\,\mathrm{d}\,\mathrm{vec}\,\mathbf{X} = \mathrm{D}_{\mathrm{vec}\,\mathbf{X}} f(\mathbf{X})\,\mathrm{d}\mathrm{vec}\,\mathbf{X}.
\end{aligned} \tag{3.2.16}$$

By the relationship between the row partial derivative vector and the Jacobian matrix, $\mathrm{D}_{\mathrm{vec}\,\mathbf{X}} f(\mathbf{X}) = \left(\mathrm{vec}\,\mathrm{D}_{\mathbf{X}}^T f(\mathbf{X})\right)^T$, Equation (3.2.16) can be written as

$$\mathrm{d}f(\mathbf{X}) = (\mathrm{vec}\,\mathbf{A}^T)^T \mathrm{d}(\mathrm{vec}\,\mathbf{X}), \tag{3.2.17}$$

where

$$\mathbf{A} = \mathrm{D}_{\mathbf{X}} f(\mathbf{X}) = \frac{\partial f(\mathbf{X})}{\partial \mathbf{X}^T} = \begin{bmatrix} \dfrac{\partial f(\mathbf{X})}{\partial x_{11}} & \cdots & \dfrac{\partial f(\mathbf{X})}{\partial x_{m1}} \\ \vdots & \ddots & \vdots \\ \dfrac{\partial f(\mathbf{X})}{\partial x_{1n}} & \cdots & \dfrac{\partial f(\mathbf{X})}{\partial x_{mn}} \end{bmatrix} \tag{3.2.18}$$

is the Jacobian matrix of the scalar function $f(\mathbf{X})$.

Using the relationship between the vectorization operator vec and the trace function $\mathrm{tr}(\mathbf{B}^T\mathbf{C}) = (\mathrm{vec}\,\mathbf{B})^T \mathrm{vec}\,\mathbf{C}$, and letting $\mathbf{B} = \mathbf{A}^T$ and $\mathbf{C} = \mathrm{d}\mathbf{X}$, then Equation (3.2.17) can be expressed in the trace form as

$$\mathrm{d}f(\mathbf{X}) = \mathrm{tr}(\mathbf{A}\mathrm{d}\mathbf{X}). \tag{3.2.19}$$

This can be regarded as the canonical form of the differential of a scalar function $f(\mathbf{X})$.

The above discussion shows that once the matrix differential of a scalar function $\mathrm{d}f(\mathbf{X})$ is expressed in its canonical form, we can identify the Jacobian matrix and/or the gradient matrix of the scalar function $f(\mathbf{X})$, as shown below.

Proposition 3.1 *If a scalar function $f(\mathbf{X})$ is differentiable at the point $\mathbf{X}$ then the Jacobian matrix $\mathbf{A}$ can be directly identified as follows* [311]*:*

$$\mathrm{d}f(\mathbf{x}) = \mathrm{tr}(\mathbf{A}\mathrm{d}\mathbf{x}) \quad \Leftrightarrow \quad \mathrm{D}_{\mathbf{x}}f(\mathbf{x}) = \mathbf{A}, \tag{3.2.20}$$

$$\mathrm{d}f(\mathbf{X}) = \mathrm{tr}(\mathbf{A}\mathrm{d}\mathbf{X}) \quad \Leftrightarrow \quad \mathrm{D}_{\mathbf{X}}f(\mathbf{X}) = \mathbf{A}. \tag{3.2.21}$$

Proposition 3.1 motivates the following effective approach to directly identifying the Jacobian matrix $\mathrm{D}_{\mathbf{X}}f(\mathbf{X})$ of the scalar function $f(\mathbf{X})$:

Step 1 Find the differential $\mathrm{d}f(\mathbf{X})$ of the real function $f(\mathbf{X})$, and denote it in the canonical form as $\mathrm{d}f(\mathbf{X}) = \mathrm{tr}(\mathbf{A}\,\mathrm{d}\mathbf{X})$.

Step 2 The Jacobian matrix is directly given by $\mathbf{A}$.

It has been shown [311] that the Jacobian matrix $\mathbf{A}$ is uniquely determined: if there are $\mathbf{A}_1$ and $\mathbf{A}_2$ such that $\mathrm{d}f(\mathbf{X}) = \mathbf{A}_i\mathrm{d}\mathbf{X}, i = 1, 2$, then $\mathbf{A}_1 = \mathbf{A}_2$.

Since the gradient matrix is the transpose of the Jacobian matrix for a given real function $f(\mathbf{X})$, Proposition 3.1 implies in addition that

$$\mathrm{d}f(\mathbf{X}) = \mathrm{tr}(\mathbf{A}\mathrm{d}\mathbf{X}) \Leftrightarrow \nabla_{\mathbf{X}}f(\mathbf{X}) = \mathbf{A}^T. \tag{3.2.22}$$

Because the Jacobian matrix $\mathbf{A}$ is uniquely determined, the gradient matrix is unique determined as well.

3. Jacobian Matrices of Trace Functions

Example 3.11 Consider the quadratic form $f(\mathbf{x}) = \mathbf{x}^T\mathbf{A}\mathbf{x}$, where $\mathbf{A}$ is a square constant matrix. Writing $f(\mathbf{x}) = \mathrm{tr}(f(\mathbf{x}))$ and then using the expression for the differential of a matrix product, it is easily obtained that

$$\begin{aligned}\mathrm{d}f(\mathbf{x}) &= \mathrm{d}\,\mathrm{tr}(\mathbf{x}^T\mathbf{A}\mathbf{x}) = \mathrm{tr}\big((\mathrm{d}\mathbf{x})^T\mathbf{A}\mathbf{x} + \mathbf{x}^T\mathbf{A}\mathrm{d}\mathbf{x}\big)\\ &= \mathrm{tr}\big((\mathrm{d}\mathbf{x}^T\mathbf{A}\mathbf{x})^T + \mathbf{x}^T\mathbf{A}\mathrm{d}\mathbf{x}\big) = \mathrm{tr}\left(\mathbf{x}^T\mathbf{A}^T\mathrm{d}\mathbf{x} + \mathbf{x}^T\mathbf{A}\mathrm{d}\mathbf{x}\right)\\ &= \mathrm{tr}\big(\mathbf{x}^T(\mathbf{A} + \mathbf{A}^T)\mathrm{d}\mathbf{x}\big)\,.\end{aligned}$$

By Proposition 3.1, we can obtain the gradient vector directly:

$$\nabla_{\mathbf{x}}(\mathbf{x}^T\mathbf{A}\mathbf{x}) = \frac{\partial\mathbf{x}^T\mathbf{A}\mathbf{x}}{\partial\mathbf{x}} = \big(\mathbf{x}^T(\mathbf{A} + \mathbf{A}^T)\big)^T = (\mathbf{A}^T + \mathbf{A})\mathbf{x}. \tag{3.2.23}$$

Clearly, if $\mathbf{A}$ is a symmetric matrix, then $\nabla_{\mathbf{x}}(\mathbf{x}^T\mathbf{A}\mathbf{x}) = \partial\mathbf{x}^T\mathbf{A}\mathbf{x}/\partial\mathbf{x} = 2\mathbf{A}\mathbf{x}$.

Example 3.12 For $\text{tr}(\mathbf{X}^T\mathbf{X})$, since $\text{tr}(\mathbf{A}^T\mathbf{B}) = \text{tr}(\mathbf{B}^T\mathbf{A})$, we have

$$\begin{aligned}\text{d}(\text{tr}(\mathbf{X}^T\mathbf{X})) &= \text{tr}\big(\text{d}(\mathbf{X}^T\mathbf{X})\big) = \text{tr}\big((\text{d}\mathbf{X})^T\mathbf{X} + \mathbf{X}^T\text{d}\mathbf{X}\big)\\ &= \text{tr}\big((\text{d}\mathbf{X})^T\mathbf{X}\big) + \text{tr}(\mathbf{X}^T\text{d}\mathbf{X}) = \text{tr}\big(2\mathbf{X}^T\text{d}\mathbf{X}\big).\end{aligned}$$

On the one hand, from Proposition 3.1 it is known that the gradient matrix is given by

$$\frac{\partial\, \text{tr}(\mathbf{X}^T\mathbf{X})}{\partial\, \mathbf{X}} = (2\mathbf{X}^T)^T = 2\mathbf{X}. \tag{3.2.24}$$

On the other hand, the differential of the trace of three-matrix product $\text{tr}(\mathbf{X}^T\mathbf{A}\mathbf{X})$ is given by

$$\begin{aligned}\text{d}\,\text{tr}(\mathbf{X}^T\mathbf{A}\mathbf{X}) &= \text{tr}\big(\text{d}(\mathbf{X}^T\mathbf{A}\mathbf{X})\big) = \text{tr}\big((\text{d}\mathbf{X})^T\mathbf{A}\mathbf{X} + \mathbf{X}^T\mathbf{A}\text{d}\mathbf{X}\big)\\ &= \text{tr}\big((\text{d}\mathbf{X})^T\mathbf{A}\mathbf{X}\big) + \text{tr}(\mathbf{X}^T\mathbf{A}\text{d}\mathbf{X})\\ &= \text{tr}\big((\mathbf{A}\mathbf{X})^T\text{d}\mathbf{X}\big) + \text{tr}(\mathbf{X}^T\mathbf{A}\text{d}\mathbf{X})\\ &= \text{tr}\big(\mathbf{X}^T(\mathbf{A}^T + \mathbf{A})\text{d}\mathbf{X}\big),\end{aligned}$$

which yields the gradient matrix

$$\frac{\partial\, \text{tr}(\mathbf{X}^T\mathbf{A}\mathbf{X})}{\partial\, \mathbf{X}} = \big(\mathbf{X}^T(\mathbf{A}^T + \mathbf{A})\big)^T = (\mathbf{A} + \mathbf{A}^T)\mathbf{X}. \tag{3.2.25}$$

Example 3.13 Given a trace including an inverse matrix, $\text{tr}(\mathbf{A}\mathbf{X}^{-1})$, since

$$\begin{aligned}\text{d}\,\text{tr}(\mathbf{A}\mathbf{X}^{-1}) &= \text{tr}\big(\text{d}(\mathbf{A}\mathbf{X}^{-1})\big) = \text{tr}\big(\mathbf{A}\text{d}\mathbf{X}^{-1}\big)\\ &= -\text{tr}\big(\mathbf{A}\mathbf{X}^{-1}(\text{d}\mathbf{X})\mathbf{X}^{-1}\big) = -\text{tr}\big(\mathbf{X}^{-1}\mathbf{A}\mathbf{X}^{-1}\text{d}\mathbf{X}\big),\end{aligned}$$

the gradient matrix is given by $\partial\, \text{tr}(\mathbf{A}\mathbf{X}^{-1})/\partial\, \mathbf{X} = -(\mathbf{X}^{-1}\mathbf{A}\mathbf{X}^{-1})^T$.

The following are the main points in applying Proposition 3.1:

(1) Any scalar function $f(\mathbf{X})$ can always be written in the form of a trace function, because $f(\mathbf{X}) = \text{tr}(f(\mathbf{X}))$.
(2) No matter where $\text{d}\mathbf{X}$ appears initially in the trace function, we can place it in the rightmost position via the trace property $\text{tr}(\mathbf{C}(\text{d}\mathbf{X})\mathbf{B}) = \text{tr}(\mathbf{B}\mathbf{C}\,\text{d}\mathbf{X})$, giving the canonical form $\text{d}f(\mathbf{X}) = \text{tr}(\mathbf{A}\,\text{d}\mathbf{X})$.

Table 3.3 summarizes the differential matrices and Jacobian matrices of several typical trace functions [311]; note that $\mathbf{A}^{-2} = \mathbf{A}^{-1}\mathbf{A}^{-1}$.

4. Jacobian Matrices of Determinant Functions

From the matrix differential $\text{d}|\mathbf{X}| = |\mathbf{X}|\,\text{tr}(\mathbf{X}^{-1}\text{d}\mathbf{X})$ and Proposition 3.1, it can immediately be seen that the gradient matrix of the determinant $|\mathbf{X}|$ is

$$\frac{\partial |\mathbf{X}|}{\partial\, \mathbf{X}} = |\mathbf{X}|(\mathbf{X}^{-1})^T = |\mathbf{X}|\,\mathbf{X}^{-T}. \tag{3.2.26}$$

Table 3.3 Differential matrices and Jacobian matrices of trace functions

Trace function $f(\mathbf{X})$	$\mathrm{d}f(\mathbf{X})$	$\partial f(\mathbf{X})/\partial \mathbf{X}^T$
$\mathrm{tr}(\mathbf{X})$	$\mathrm{tr}(\mathbf{I}\mathrm{d}\mathbf{X})$	$\mathbf{I}$
$\mathrm{tr}(\mathbf{X}^{-1})$	$-\mathrm{tr}(\mathbf{X}^{-2}\mathrm{d}\mathbf{X})$	$-\mathbf{X}^{-2}$
$\mathrm{tr}(\mathbf{AX})$	$\mathrm{tr}(\mathbf{A}\mathrm{d}\mathbf{X})$	$\mathbf{A}$
$\mathrm{tr}(\mathbf{X}^2)$	$2\mathrm{tr}(\mathbf{X}\mathrm{d}\mathbf{X})$	$2\mathbf{X}$
$\mathrm{tr}(\mathbf{X}^T\mathbf{X})$	$2\mathrm{tr}(\mathbf{X}^T\mathrm{d}\mathbf{X})$	$2\mathbf{X}^T$
$\mathrm{tr}(\mathbf{X}^T\mathbf{AX})$	$\mathrm{tr}\big(\mathbf{X}^T(\mathbf{A}+\mathbf{A}^T)\mathrm{d}\mathbf{X}\big)$	$\mathbf{X}^T(\mathbf{A}+\mathbf{A}^T)$
$\mathrm{tr}(\mathbf{XAX}^T)$	$\mathrm{tr}\big((\mathbf{A}+\mathbf{A}^T)\mathbf{X}^T\mathrm{d}\mathbf{X}\big)$	$(\mathbf{A}+\mathbf{A}^T)\mathbf{X}^T$
$\mathrm{tr}(\mathbf{XAX})$	$\mathrm{tr}\left((\mathbf{AX}+\mathbf{XA})\mathrm{d}\mathbf{X}\right)$	$\mathbf{AX}+\mathbf{XA}$
$\mathrm{tr}(\mathbf{AX}^{-1})$	$-\mathrm{tr}\left(\mathbf{X}^{-1}\mathbf{AX}^{-1}\mathrm{d}\mathbf{X}\right)$	$-\mathbf{X}^{-1}\mathbf{AX}^{-1}$
$\mathrm{tr}(\mathbf{AX}^{-1}\mathbf{B})$	$-\mathrm{tr}\big(\mathbf{X}^{-1}\mathbf{BAX}^{-1}\mathrm{d}\mathbf{X}\big)$	$-\mathbf{X}^{-1}\mathbf{BAX}^{-1}$
$\mathrm{tr}\big((\mathbf{X}+\mathbf{A})^{-1}\big)$	$-\mathrm{tr}\big((\mathbf{X}+\mathbf{A})^{-2}\mathrm{d}\mathbf{X}\big)$	$-(\mathbf{X}+\mathbf{A})^{-2}$
$\mathrm{tr}(\mathbf{XAXB})$	$\mathrm{tr}\big((\mathbf{AXB}+\mathbf{BXA})\mathrm{d}\mathbf{X}\big)$	$\mathbf{AXB}+\mathbf{BXA}$
$\mathrm{tr}(\mathbf{XAX}^T\mathbf{B})$	$\mathrm{tr}\big((\mathbf{AX}^T\mathbf{B}+\mathbf{A}^T\mathbf{X}^T\mathbf{B}^T)\mathrm{d}\mathbf{X}\big)$	$\mathbf{AX}^T\mathbf{B}+\mathbf{A}^T\mathbf{X}^T\mathbf{B}^T$
$\mathrm{tr}(\mathbf{AXX}^T\mathbf{B})$	$\mathrm{tr}\left(\mathbf{X}^T(\mathbf{BA}+\mathbf{A}^T\mathbf{B}^T)\mathrm{d}\mathbf{X}\right)$	$\mathbf{X}^T(\mathbf{BA}+\mathbf{A}^T\mathbf{B}^T)$
$\mathrm{tr}(\mathbf{AX}^T\mathbf{XB})$	$\mathrm{tr}\big((\mathbf{BA}+\mathbf{A}^T\mathbf{B}^T)\mathbf{X}^T\mathrm{d}\mathbf{X}\big)$	$(\mathbf{BA}+\mathbf{A}^T\mathbf{B}^T)\mathbf{X}^T$

For the logarithm of the determinant, $\log|\mathbf{X}|$, its matrix differential is

$$\mathrm{d}\log|\mathbf{X}| = |\mathbf{X}|^{-1}\mathrm{d}|\mathbf{X}| = |\mathbf{X}|^{-1}\mathrm{tr}(|\mathbf{X}|\mathbf{X}^{-1}\mathrm{d}\mathbf{X}) = \mathrm{tr}(\mathbf{X}^{-1}\mathrm{d}\mathbf{X}), \tag{3.2.27}$$

hence the gradient matrix of the determinant logarithm function $\log|\mathbf{X}|$ is determined by

$$\frac{\partial \log|\mathbf{X}|}{\partial \mathbf{X}} = \mathbf{X}^{-T}. \tag{3.2.28}$$

Consider the determinant of $\mathbf{X}^2$. From $\mathrm{d}|\mathbf{X}| = |\mathbf{X}|\mathrm{tr}(\mathbf{X}^{-1}\mathrm{d}\mathbf{X})$ it is known that $\mathrm{d}|\mathbf{X}^2| = \mathrm{d}|\mathbf{X}|^2 = 2|\mathbf{X}|\,\mathrm{d}|\mathbf{X}| = 2|\mathbf{X}|^2\,\mathrm{tr}\big(\mathbf{X}^{-1}\mathrm{d}\mathbf{X}\big)$. By applying Proposition 3.1, we have

$$\frac{\partial|\mathbf{X}|^2}{\partial \mathbf{X}} = 2|\mathbf{X}|^2(\mathbf{X}^{-1})^T = 2|\mathbf{X}|^2\mathbf{X}^{-T}. \tag{3.2.29}$$

More generally, the matrix differential of $|\mathbf{X}^k|$ is as follows:

$$\mathrm{d}|\mathbf{X}^k| = |\mathbf{X}^k|\,\mathrm{tr}(\mathbf{X}^{-k}\mathrm{d}\mathbf{X}^k) = |\mathbf{X}^k\,|\mathrm{tr}(\mathbf{X}^{-k}(k\mathbf{X}^{k-1})\mathrm{d}\mathbf{X}) = k|\mathbf{X}^k|\,\mathrm{tr}(\mathbf{X}^{-1}\mathrm{d}\mathbf{X}).$$

Hence

$$\frac{\partial|\mathbf{X}^k|}{\partial \mathbf{X}} = k|\mathbf{X}^k|\,\mathbf{X}^{-T}. \tag{3.2.30}$$

Letting $\mathbf{X} \in \mathbb{R}^{m\times n}$ and $\text{rank}(\mathbf{X}) = m$, i.e., $\mathbf{X}\mathbf{X}^T$ is invertible, then we have

$$\begin{aligned}\mathrm{d}|\mathbf{X}\mathbf{X}^T| &= |\mathbf{X}\mathbf{X}^T|\,\mathrm{tr}\Big((\mathbf{X}\mathbf{X}^T)^{-1}\mathrm{d}(\mathbf{X}\mathbf{X}^T)\Big)\\ &= |\mathbf{X}\mathbf{X}^T|\,\Big(\mathrm{tr}\Big((\mathbf{X}\mathbf{X}^T)^{-1}(\mathrm{d}\mathbf{X})\mathbf{X}^T\Big) + \mathrm{tr}\Big((\mathbf{X}\mathbf{X}^T)^{-1}\mathbf{X}(\mathrm{d}\mathbf{X})^T\Big)\Big)\\ &= |\mathbf{X}\mathbf{X}^T|\,\Big(\mathrm{tr}\Big(\mathbf{X}^T(\mathbf{X}\mathbf{X}^T)^{-1}\mathrm{d}\mathbf{X}\Big) + \mathrm{tr}\Big(\mathbf{X}^T(\mathbf{X}\mathbf{X}^T)^{-1}\mathrm{d}\mathbf{X}\Big)\Big)\\ &= \mathrm{tr}\Big(2\,|\mathbf{X}\mathbf{X}^T|\,\mathbf{X}^T(\mathbf{X}\mathbf{X}^T)^{-1}\mathrm{d}\mathbf{X}\Big).\end{aligned}$$

By Proposition 3.1, we get the gradient matrix

$$\frac{\partial|\mathbf{X}\mathbf{X}^T|}{\partial\mathbf{X}} = 2|\mathbf{X}\mathbf{X}^T|\,(\mathbf{X}\mathbf{X}^T)^{-1}\mathbf{X}. \tag{3.2.31}$$

Similarly, set $\mathbf{X} \in \mathbb{R}^{m\times n}$. If $\text{rank}(\mathbf{X}) = n$, i.e., $\mathbf{X}^T\mathbf{X}$ is invertible, then

$$\mathrm{d}|\mathbf{X}^T\mathbf{X}| = \mathrm{tr}\big(2|\mathbf{X}^T\mathbf{X}|(\mathbf{X}^T\mathbf{X})^{-1}\mathbf{X}^T\mathrm{d}\mathbf{X}\big), \tag{3.2.32}$$

and hence $\partial|\mathbf{X}^T\mathbf{X}|/\partial\mathbf{X} = 2|\mathbf{X}^T\mathbf{X}|\,\mathbf{X}(\mathbf{X}^T\mathbf{X})^{-1}$.

Table 3.4 summarizes the real matrix differentials and the Jacobian matrices of several typical determinant functions.

Table 3.4 Differentials and Jacobian matrices of determinant functions

$f(\mathbf{X})$	$\mathrm{d}f(\mathbf{X})$	$\partial f(\mathbf{X})/\partial\mathbf{X}$
$\lvert\mathbf{X}\rvert$	$\lvert\mathbf{X}\rvert\,\mathrm{tr}(\mathbf{X}^{-1}\mathrm{d}\mathbf{X})$	$\lvert\mathbf{X}\rvert\mathbf{X}^{-1}$
$\log\lvert\mathbf{X}\rvert$	$\mathrm{tr}(\mathbf{X}^{-1}\mathrm{d}\mathbf{X})$	$\mathbf{X}^{-1}$
$\lvert\mathbf{X}^{-1}\rvert$	$-\lvert\mathbf{X}^{-1}\rvert\,\mathrm{tr}(\mathbf{X}^{-1}\mathrm{d}\mathbf{X})$	$-\lvert\mathbf{X}^{-1}\rvert\mathbf{X}^{-1}$
$\lvert\mathbf{X}^2\rvert$	$2\lvert\mathbf{X}\rvert^2\,\mathrm{tr}(\mathbf{X}^{-1}\mathrm{d}\mathbf{X})$	$2\lvert\mathbf{X}\rvert^2\mathbf{X}^{-1}$
$\lvert\mathbf{X}^k\rvert$	$k\lvert\mathbf{X}\rvert^k\,\mathrm{tr}(\mathbf{X}^{-1}\mathrm{d}\mathbf{X})$	$k\lvert\mathbf{X}\rvert^k\mathbf{X}^{-1}$
$\lvert\mathbf{X}\mathbf{X}^T\rvert$	$2\lvert\mathbf{X}\mathbf{X}^T\rvert\,\mathrm{tr}(\mathbf{X}^T(\mathbf{X}\mathbf{X}^T)^{-1}\mathrm{d}\mathbf{X})$	$2\lvert\mathbf{X}\mathbf{X}^T\rvert\mathbf{X}^T(\mathbf{X}\mathbf{X}^T)^{-1}$
$\lvert\mathbf{X}^T\mathbf{X}\rvert$	$2\lvert\mathbf{X}^T\mathbf{X}\rvert\,\mathrm{tr}((\mathbf{X}^T\mathbf{X})^{-1}\mathbf{X}^T\mathrm{d}\mathbf{X})$	$2\lvert\mathbf{X}^T\mathbf{X}\rvert(\mathbf{X}^T\mathbf{X})^{-1}\mathbf{X}^T$
$\log\lvert\mathbf{X}^T\mathbf{X}\rvert$	$2\mathrm{tr}((\mathbf{X}^T\mathbf{X})^{-1}\mathbf{X}^T\mathrm{d}\mathbf{X})$	$2(\mathbf{X}^T\mathbf{X})^{-1}\mathbf{X}^T$
$\lvert\mathbf{A}\mathbf{X}\mathbf{B}\rvert$	$\lvert\mathbf{A}\mathbf{X}\mathbf{B}\rvert\,\mathrm{tr}(\mathbf{B}(\mathbf{A}\mathbf{X}\mathbf{B})^{-1}\mathbf{A}\mathrm{d}\mathbf{X})$	$\lvert\mathbf{A}\mathbf{X}\mathbf{B}\rvert\mathbf{B}(\mathbf{A}\mathbf{X}\mathbf{B})^{-1}\mathbf{A}$
$\lvert\mathbf{X}\mathbf{A}\mathbf{X}^T\rvert$	$\lvert\mathbf{X}\mathbf{A}\mathbf{X}^T\rvert\,\mathrm{tr}\Big((\mathbf{A}\mathbf{X}^T(\mathbf{X}\mathbf{A}\mathbf{X}^T)^{-1} + (\mathbf{X}\mathbf{A})^T(\mathbf{X}\mathbf{A}^T\mathbf{X}^T)^{-1})\mathrm{d}\mathbf{X}\Big)$	$\lvert\mathbf{X}\mathbf{A}\mathbf{X}^T\rvert\Big(\mathbf{A}\mathbf{X}^T(\mathbf{X}\mathbf{A}\mathbf{X}^T)^{-1} + (\mathbf{X}\mathbf{A})^T(\mathbf{X}\mathbf{A}^T\mathbf{X}^T)^{-1}\Big)$
$\lvert\mathbf{X}^T\mathbf{A}\mathbf{X}\rvert$	$\lvert\mathbf{X}^T\mathbf{A}\mathbf{X}\rvert\mathrm{tr}\Big(((\mathbf{X}^T\mathbf{A}\mathbf{X})^{-T}(\mathbf{A}\mathbf{X})^T + (\mathbf{X}^T\mathbf{A}\mathbf{X})^{-1}\mathbf{X}^T\mathbf{A})\mathrm{d}\mathbf{X}\Big)$	$\lvert\mathbf{X}^T\mathbf{A}\mathbf{X}\rvert\Big((\mathbf{X}^T\mathbf{A}\mathbf{X})^{-T}(\mathbf{A}\mathbf{X})^T + (\mathbf{X}^T\mathbf{A}\mathbf{X})^{-1}\mathbf{X}^T\mathbf{A}\Big)$

3.2.3 Jacobian Matrix of Real Matrix Functions

Let $f_{kl} = f_{kl}(\mathbf{X})$ be the entry of the kth row and lth column of the real matrix function $\mathbf{F}(\mathbf{X})$; then $\mathrm{d}f_{kl}(\mathbf{X}) = [\mathrm{d}\mathbf{F}(\mathbf{X})]_{kl}$ represents the differential of the scalar function $f_{kl}(\mathbf{X})$. From Equation (3.2.16) we have

$$\mathrm{d}f_{kl}(\mathbf{X}) = \left[\frac{\partial f_{kl}(\mathbf{X})}{\partial x_{11}} \cdots \frac{\partial f_{kl}(\mathbf{X})}{\partial x_{m1}} \cdots \frac{\partial f_{kl}(\mathbf{X})}{\partial x_{1n}} \cdots \frac{\partial f_{kl}(\mathbf{X})}{\partial x_{mn}}\right] \begin{bmatrix} \mathrm{d}x_{11} \\ \vdots \\ \mathrm{d}x_{m1} \\ \vdots \\ \mathrm{d}x_{1n} \\ \vdots \\ \mathrm{d}x_{mn} \end{bmatrix}.$$

The above result can be rewritten in vectorization form as follows:

$$\mathrm{d}\,\mathrm{vec}\,\mathbf{F}(\mathbf{X}) = \mathbf{A}\,\mathrm{d}\,\mathrm{vec}\,\mathbf{X}, \tag{3.2.33}$$

where

$$\mathrm{d}\,\mathrm{vec}\,\mathbf{F}(\mathbf{X}) = [\mathrm{d}f_{11}(\mathbf{X}), \ldots, \mathrm{d}f_{p1}(\mathbf{X}), \ldots, \mathrm{d}f_{1q}(\mathbf{X}), \ldots, \mathrm{d}f_{pq}(\mathbf{X})]^T, \tag{3.2.34}$$

$$\mathrm{d}\,\mathrm{vec}\,\mathbf{X} = [\mathrm{d}x_{11}, \ldots, \mathrm{d}x_{m1}, \ldots, \mathrm{d}x_{1n}, \ldots, \mathrm{d}x_{mn}]^T \tag{3.2.35}$$

and

$$\mathbf{A} = \begin{bmatrix} \frac{\partial f_{11}(\mathbf{X})}{\partial x_{11}} & \cdots & \frac{\partial f_{11}(\mathbf{X})}{\partial x_{m1}} & \cdots & \frac{\partial f_{11}(\mathbf{X})}{\partial x_{1n}} & \cdots & \frac{\partial f_{11}(\mathbf{X})}{\partial x_{mn}} \\ \vdots & \vdots & \vdots & \vdots & \vdots & \vdots & \vdots \\ \frac{\partial f_{p1}(\mathbf{X})}{\partial x_{11}} & \cdots & \frac{\partial f_{p1}(\mathbf{X})}{\partial x_{m1}} & \cdots & \frac{\partial f_{p1}(\mathbf{X})}{\partial x_{1n}} & \cdots & \frac{\partial f_{p1}(\mathbf{X})}{\partial x_{mn}} \\ \vdots & \vdots & \vdots & \vdots & \vdots & \vdots & \vdots \\ \frac{\partial f_{1q}(\mathbf{X})}{\partial x_{11}} & \cdots & \frac{\partial f_{1q}(\mathbf{X})}{\partial x_{m1}} & \cdots & \frac{\partial f_{1q}(\mathbf{X})}{\partial x_{1n}} & \cdots & \frac{\partial f_{1q}(\mathbf{X})}{\partial x_{mn}} \\ \vdots & \vdots & \vdots & \vdots & \vdots & \vdots & \vdots \\ \frac{\partial f_{pq}(\mathbf{X})}{\partial x_{11}} & \cdots & \frac{\partial f_{pq}(\mathbf{X})}{\partial x_{m1}} & \cdots & \frac{\partial f_{pq}(\mathbf{X})}{\partial x_{1n}} & \cdots & \frac{\partial f_{pq}(\mathbf{X})}{\partial x_{mn}} \end{bmatrix}$$

$$= \frac{\partial\,\mathrm{vec}\,\mathbf{F}(\mathbf{X})}{\partial(\mathrm{vec}\,\mathbf{X})^T}. \tag{3.2.36}$$

In other words, the matrix $\mathbf{A}$ is the Jacobian matrix $\mathrm{D}_{\mathbf{X}}\mathbf{F}(\mathbf{X})$ of the matrix function $\mathbf{F}(\mathbf{X})$.

Let $\mathbf{F}(\mathbf{X}) \in \mathbb{R}^{p\times q}$ be a matrix function including $\mathbf{X}$ and $\mathbf{X}^T$ as variables, where $\mathbf{X} \in \mathbb{R}^{m\times n}$. Then the first-order matrix differential is given by

$$\mathrm{d}\,\mathrm{vec}\,\mathbf{F}(\mathbf{X}) = \mathbf{A}\,\mathrm{d}(\mathrm{vec}\mathbf{X}) + \mathbf{B}\,\mathrm{d}(\mathrm{vec}\,\mathbf{X}^T), \quad \mathbf{A}, \mathbf{B} \in \mathbb{R}^{pq\times mn}.$$

Since $\mathrm{d}(\mathrm{vec}\mathbf{X}^T) = \mathbf{K}_{mn}\mathrm{d}\mathrm{vec}\mathbf{X}$, the above equation can be rewritten as

$$\mathrm{d}\,\mathrm{vec}\,\mathbf{F}(\mathbf{X}) = (\mathbf{A} + \mathbf{B}\mathbf{K}_{mn})\mathrm{d}\,\mathrm{vec}\,\mathbf{X}. \tag{3.2.37}$$

The above discussion can be summarized in the following proposition.

PROPOSITION 3.2 *Given a matrix function* $\mathbf{F}(\mathbf{X}) : \mathbb{R}^{m\times n} \to \mathbb{R}^{p\times q}$, *then its* $pq \times mn$ *Jacobian matrix can be determined from*

$$\begin{aligned} \mathrm{d}\,\mathrm{vec}\mathbf{F}(\mathbf{X}) &= \mathbf{A}\,\mathrm{d}\mathrm{vec}\mathbf{X} + \mathbf{B}\,\mathrm{d}(\mathrm{vec}\mathbf{X})^T \\ \Leftrightarrow \quad \mathrm{D}_{\mathbf{X}}\mathbf{F}(\mathbf{X}) &= \frac{\partial\,\mathrm{vec}\,\mathbf{F}(\mathbf{X})}{\partial\,(\mathrm{vec}\mathbf{X})^T} = \mathbf{A} + \mathbf{B}\mathbf{K}_{mn} \end{aligned} \tag{3.2.38}$$

and its $mn \times pq$ *gradient matrix can be identified using*

$$\nabla_{\mathbf{X}}\mathbf{F}(\mathbf{X}) = (\mathrm{D}_{\mathbf{X}}\mathbf{F}(\mathbf{X}))^T = \mathbf{A}^T + \mathbf{K}_{nm}\mathbf{B}^T. \tag{3.2.39}$$

Importantly, because

$$\begin{aligned} \mathrm{d}\mathbf{F}(\mathbf{X}) = \mathbf{A}(\mathrm{d}\mathbf{X})\mathbf{B} \quad &\Leftrightarrow \quad \mathrm{d}\,\mathrm{vec}\mathbf{F}(\mathbf{X}) = (\mathbf{B}^T \otimes \mathbf{A})\mathrm{d}\,\mathrm{vec}\mathbf{X}, \\ \mathrm{d}\mathbf{F}(\mathbf{X}) = \mathbf{C}(\mathrm{d}\mathbf{X}^T)\mathbf{D} \quad &\Leftrightarrow \quad \mathrm{d}\mathrm{vec}\mathbf{F}(\mathbf{X}) = (\mathbf{D}^T \otimes \mathbf{C})\mathbf{K}_{mn}\mathrm{d}\mathrm{vec}\mathbf{X}, \end{aligned}$$

identification given in Proposition 3.2 can be simplified to the differential of $\mathbf{F}(\mathbf{X})$.

THEOREM 3.1 *Given a matrix function* $\mathbf{F}(\mathbf{X}) : \mathbb{R}^{m\times n} \to \mathbb{R}^{p\times q}$, *then its* $pq \times mn$ *Jacobian matrix can be identified as follows:* [311]

$$\begin{aligned} \mathrm{d}\mathbf{F}(\mathbf{X}) &= \mathbf{A}(\mathrm{d}\mathbf{X})\mathbf{B} + \mathbf{C}(\mathrm{d}\mathbf{X}^T)\mathbf{D}, \\ \Leftrightarrow \quad \mathrm{D}_{\mathbf{X}}\mathbf{F}(\mathbf{X}) &= \frac{\partial\,\mathrm{vec}\,\mathbf{F}(\mathbf{X})}{\partial(\mathrm{vec}\,\mathbf{X})^T} = (\mathbf{B}^T \otimes \mathbf{A}) + (\mathbf{D}^T \otimes \mathbf{C})\mathbf{K}_{mn}, \end{aligned} \tag{3.2.40}$$

the $mn \times pq$ *gradient matrix can be determined from*

$$\nabla_{\mathbf{X}}\mathbf{F}(\mathbf{X}) = \frac{\partial(\mathrm{vec}\,\mathbf{F}(\mathbf{X}))^T}{\partial\,\mathrm{vec}\,\mathbf{X}} = (\mathbf{B} \otimes \mathbf{A}^T) + \mathbf{K}_{nm}(\mathbf{D} \otimes \mathbf{C}^T). \tag{3.2.41}$$

Table 3.5 summarizes the matrix differentials and Jacobian matrices of some real functions.

EXAMPLE 3.14 Given a matrix function $\mathbf{A}\mathbf{X}^T\mathbf{B}$ whose matrix differential is $\mathrm{d}(\mathbf{A}\mathbf{X}^T\mathbf{B}) = \mathbf{A}(\mathrm{d}\mathbf{X}^T)\mathbf{B}$, the Jacobian matrix of $\mathbf{A}\mathbf{X}^T\mathbf{B}$ is

$$\mathrm{D}_{\mathbf{X}}(\mathbf{A}\mathbf{X}^T\mathbf{B}) = (\mathbf{B}^T \otimes \mathbf{A})\mathbf{K}_{mn}, \tag{3.2.42}$$

whose transpose yields the gradient matrix of $\mathbf{A}\mathbf{X}^T\mathbf{B}$.

EXAMPLE 3.15 For the function $\mathbf{X}^T\mathbf{B}\mathbf{X}$, its matrix differential is $\mathrm{d}(\mathbf{X}^T\mathbf{B}\mathbf{X}) = \mathbf{X}^T\mathbf{B}\mathrm{d}\mathbf{X} + \mathrm{d}(\mathbf{X}^T)\mathbf{B}\mathbf{X}$, which yields the Jacobian matrix of $\mathbf{X}^T\mathbf{B}\mathbf{X}$ as follows:

$$\mathrm{D}_{\mathbf{X}}(\mathbf{X}^T\mathbf{B}\mathbf{X}) = \mathbf{I} \otimes (\mathbf{X}^T\mathbf{B}) + \left((\mathbf{B}\mathbf{X})^T \otimes \mathbf{I}\right)\mathbf{K}_{mn}; \tag{3.2.43}$$

the transpose gives the gradient matrix of $\mathbf{X}^T\mathbf{B}\mathbf{X}$.

Table 3.5 Matrix differentials and Jacobian matrices of real functions

Functions	Matrix differential	Jacobian matrix
$f(x):\ \mathbb{R} \to \mathbb{R}$	$\mathrm{d}f(x) = A\mathrm{d}x$	$A \in \mathbb{R}$
$f(\mathbf{x}):\ \mathbb{R}^m \to \mathbb{R}$	$\mathrm{d}f(\mathbf{x}) = \mathbf{A}\mathrm{d}\mathbf{x}$	$\mathbf{A} \in \mathbb{R}^{1\times m}$
$f(\mathbf{X}):\ \mathbb{R}^{m\times n} \to \mathbb{R}$	$\mathrm{d}f(\mathbf{X}) = \mathrm{tr}(\mathbf{A}\mathrm{d}\mathbf{X})$	$\mathbf{A} \in \mathbb{R}^{n\times m}$
$\mathbf{f}(\mathbf{x}):\ \mathbb{R}^m \to \mathbb{R}^p$	$\mathrm{d}\mathbf{f}(\mathbf{x}) = \mathbf{A}\mathrm{d}\mathbf{x}$	$\mathbf{A} \in \mathbb{R}^{p\times m}$
$\mathbf{f}(\mathbf{X}):\ \mathbb{R}^{m\times n} \to \mathbb{R}^p$	$\mathrm{d}\mathbf{f}(\mathbf{X}) = \mathbf{A}\mathrm{d}(\mathrm{vec}\mathbf{X})$	$\mathbf{A} \in \mathbb{R}^{p\times mn}$
$\mathbf{F}(\mathbf{x}):\ \mathbb{R}^m \to \mathbb{R}^{p\times q}$	$\mathrm{d}\,\mathrm{vec}\mathbf{F}(\mathbf{x}) = \mathbf{A}\mathrm{d}\mathbf{x}$	$\mathbf{A} \in \mathbb{R}^{pq\times m}$
$\mathbf{F}(\mathbf{X}):\ \mathbb{R}^{m\times n} \to \mathbb{R}^{p\times q}$	$\mathrm{d}\mathbf{F}(\mathbf{X}) = \mathbf{A}(\mathrm{d}\mathbf{X})\mathbf{B}$	$(\mathbf{B}^T \otimes \mathbf{A}) \in \mathbb{R}^{pq\times mn}$
$\mathbf{F}(\mathbf{X}):\ \mathbb{R}^{m\times n} \to \mathbb{R}^{p\times q}$	$\mathrm{d}\mathbf{F}(\mathbf{X}) = \mathbf{C}(\mathrm{d}\mathbf{X}^T)\mathbf{D}$	$(\mathbf{D}^T \otimes \mathbf{C})\mathbf{K}_{mn} \in \mathbb{R}^{pq\times mn}$

Table 3.6 Jacobian matrices of some matrix functions

$\mathbf{F}(\mathbf{X})$	$\mathrm{d}\mathbf{F}(\mathbf{X})$	Jacobian matrix
$\mathbf{X}^T\mathbf{X}$	$\mathbf{X}^T\mathrm{d}\mathbf{X} + (\mathrm{d}\mathbf{X}^T)\mathbf{X}$	$(\mathbf{I}_n \otimes \mathbf{X}^T) + (\mathbf{X}^T \otimes \mathbf{I}_n)\mathbf{K}_{mn}$
$\mathbf{X}\mathbf{X}^T$	$\mathbf{X}(\mathrm{d}\mathbf{X}^T) + (\mathrm{d}\mathbf{X})\mathbf{X}^T$	$(\mathbf{I}_m \otimes \mathbf{X})\mathbf{K}_{mn} + (\mathbf{X} \otimes \mathbf{I}_m)$
$\mathbf{A}\mathbf{X}^T\mathbf{B}$	$\mathbf{A}(\mathrm{d}\mathbf{X}^T)\mathbf{B}$	$(\mathbf{B}^T \otimes \mathbf{A})\mathbf{K}_{mn}$
$\mathbf{X}^T\mathbf{B}\mathbf{X}$	$\mathbf{X}^T\mathbf{B}\mathrm{d}\mathbf{X} + (\mathrm{d}\mathbf{X}^T)\mathbf{B}\mathbf{X}$	$\mathbf{I} \otimes (\mathbf{X}^T\mathbf{B}) + ((\mathbf{B}\mathbf{X})^T \otimes \mathbf{I})\mathbf{K}_{mn}$
$\mathbf{A}\mathbf{X}^T\mathbf{B}\mathbf{X}\mathbf{C}$	$\mathbf{A}(\mathrm{d}\mathbf{X}^T)\mathbf{B}\mathbf{X}\mathbf{C} + \mathbf{A}\mathbf{X}^T\mathbf{B}(\mathrm{d}\mathbf{X})\mathbf{C}$	$((\mathbf{B}\mathbf{X}\mathbf{C})^T \otimes \mathbf{A})\mathbf{K}_{mn}$ $+\mathbf{C}^T \otimes (\mathbf{A}\mathbf{X}^T\mathbf{B})$
$\mathbf{A}\mathbf{X}\mathbf{B}\mathbf{X}^T\mathbf{C}$	$\mathbf{A}(\mathrm{d}\mathbf{X})\mathbf{B}\mathbf{X}^T\mathbf{C} + \mathbf{A}\mathbf{X}\mathbf{B}(\mathrm{d}\mathbf{X}^T)\mathbf{C}$	$(\mathbf{B}\mathbf{X}^T\mathbf{C})^T \otimes \mathbf{A}$ $+(\mathbf{C}^T \otimes (\mathbf{A}\mathbf{X}\mathbf{B}))\mathbf{K}_{mn}$
$\mathbf{X}^{-1}$	$-\mathbf{X}^{-1}(\mathrm{d}\mathbf{X})\mathbf{X}^{-1}$	$-(\mathbf{X}^{-T} \otimes \mathbf{X}^{-1})$
$\mathbf{X}^k$	$\sum_{j=1}^{k}\mathbf{X}^{j-1}(\mathrm{d}\mathbf{X})\mathbf{X}^{k-j}$	$\sum_{j=1}^{k}(\mathbf{X}^T)^{k-j} \otimes \mathbf{X}^{j-1}$
$\log \mathbf{X}$	$\mathbf{X}^{-1}\mathrm{d}\mathbf{X}$	$\mathbf{I} \otimes \mathbf{X}^{-1}$
$\exp(\mathbf{X})$	$\sum_{k=0}^{\infty}\frac{1}{(k+1)!}\sum_{j=0}^{k}\mathbf{X}^j(\mathrm{d}\mathbf{X})\mathbf{X}^{k-j}$	$\sum_{k=0}^{\infty}\frac{1}{(k+1)!}\sum_{j=0}^{k}(\mathbf{X}^T)^{k-j} \otimes \mathbf{X}^j$

Table 3.6 lists some matrix functions and their Jacobian matrices.

It should be noted that the matrix differentials of some matrix functions may not be representable in the canonical form required by Theorem 3.1, but nevertheless can be expressed in the canonical form in Proposition 3.2. In such cases, we must use Proposition 3.2 to identify the Jacobian matrix.

Example 3.16 Let $\mathbf{F}(\mathbf{X},\mathbf{Y}) = \mathbf{X}\otimes\mathbf{Y}$ be the Kronecker product of two matrices $\mathbf{X}\in\mathbb{R}^{p\times m}$ and $\mathbf{Y}\in\mathbb{R}^{n\times q}$. Consider the matrix differential $\mathrm{d}\mathbf{F}(\mathbf{X},\mathbf{Y}) = (\mathrm{d}\mathbf{X})\otimes\mathbf{Y} + \mathbf{X}\otimes(\mathrm{d}\mathbf{Y})$. By the vectorization formula

$$\mathrm{vec}(\mathbf{X}\otimes\mathbf{Y}) = (\mathbf{I}_m\otimes\mathbf{K}_{qp}\otimes\mathbf{I}_n)(\mathrm{vec}\,\mathbf{X}\otimes\mathrm{vec}\,\mathbf{Y})$$

we have

$$\begin{aligned}\mathrm{vec}(\mathrm{d}\mathbf{X}\otimes\mathbf{Y}) &= (\mathbf{I}_m\otimes\mathbf{K}_{qp}\otimes\mathbf{I}_n)(\mathrm{d}\,\mathrm{vec}\mathbf{X}\otimes\mathrm{vec}\,\mathbf{Y})\\ &= (\mathbf{I}_m\otimes\mathbf{K}_{qp}\otimes\mathbf{I}_n)(\mathbf{I}_{pm}\otimes\mathrm{vec}\,\mathbf{Y})\mathrm{d}\,\mathrm{vec}\mathbf{X}, \qquad (3.2.44)\\ \mathrm{vec}(\mathbf{X}\otimes\mathrm{d}\mathbf{Y}) &= (\mathbf{I}_m\otimes\mathbf{K}_{qp}\otimes\mathbf{I}_n)(\mathrm{vec}\,\mathbf{X}\otimes\mathrm{d}\,\mathrm{vec}\,\mathbf{Y})\\ &= (\mathbf{I}_m\otimes\mathbf{K}_{qp}\otimes\mathbf{I}_n)(\mathrm{vec}\,\mathbf{X}\otimes\mathbf{I}_{nq})\mathrm{d}\,\mathrm{vec}\mathbf{Y}. \qquad (3.2.45)\end{aligned}$$

Hence, the Jacobian matrices are respectively as follows:

$$\mathrm{D}_{\mathbf{X}}(\mathbf{X}\otimes\mathbf{Y}) = (\mathbf{I}_m\otimes\mathbf{K}_{qp}\otimes\mathbf{I}_n)(\mathbf{I}_{pm}\otimes\mathrm{vec}\,\mathbf{Y}), \qquad (3.2.46)$$

$$\mathrm{D}_{\mathbf{Y}}(\mathbf{X}\otimes\mathbf{Y}) = (\mathbf{I}_m\otimes\mathbf{K}_{qp}\otimes\mathbf{I}_n)(\mathrm{vec}\,\mathbf{X}\otimes\mathbf{I}_{nq}). \qquad (3.2.47)$$

The analysis and examples in this section show that the first-order real matrix differential is indeed an effective mathematical tool for identifying the Jacobian matrix and the gradient matrix of a real function. The operation of this tool is simple and easy to master.

3.3 Real Hessian Matrix and Identification

In real-world problems, we have to employ not only the first-order derivative but also the second-order derivative of a given real function in order to obtain more information about it. In the above section we presented the Jacobian matrix and the gradient matrix, which are two useful representations of the first-order derivative of a real-valued function. Here we discuss the second-order derivative of a real function. The main problem with which we are concerned is how to obtain the Hessian matrix of a real function.

3.3.1 Real Hessian Matrix

Consider a real vector $\mathbf{x}\in\mathbb{R}^{m\times 1}$, the second-order derivative of a real function $f(\mathbf{x})$ is known as the *Hessian matrix*, denoted $\mathbf{H}[f(\mathbf{x})]$, and is defined as

$$\mathbf{H}[f(\mathbf{x})] = \frac{\partial^2 f(\mathbf{x})}{\partial\mathbf{x}\partial\mathbf{x}^T} = \frac{\partial}{\partial\mathbf{x}}\left(\frac{\partial f(\mathbf{x})}{\partial\mathbf{x}^T}\right)\in\mathbb{R}^{m\times m}, \qquad (3.3.1)$$

which is simply represented as

$$\mathbf{H}[f(\mathbf{x})] = \nabla^2_{\mathbf{x}} f(\mathbf{x}) = \nabla_{\mathbf{x}}(\mathrm{D}_{\mathbf{x}} f(\mathbf{x})), \qquad (3.3.2)$$

where $\mathrm{D}_{\mathbf{x}}$ is the cogradient operator. Hence, the (i,j)th entry of the Hessian matrix is defined as

$$[\mathbf{H}f(\mathbf{x})]_{ij} = \left[\frac{\partial^2 f(\mathbf{x})}{\partial \mathbf{x}\,\partial \mathbf{x}^T}\right]_{ij} = \frac{\partial}{\partial x_i}\left(\frac{\partial f(\mathbf{x})}{\partial x_j}\right), \tag{3.3.3}$$

thus for the matrix itself we have

$$\mathbf{H}[f(\mathbf{x})] = \frac{\partial^2 f(\mathbf{x})}{\partial \mathbf{x}\,\partial \mathbf{x}^T} = \begin{bmatrix} \frac{\partial^2 f}{\partial x_1 \partial x_1} & \cdots & \frac{\partial^2 f}{\partial x_1 \partial x_m} \\ \vdots & \ddots & \vdots \\ \frac{\partial^2 f}{\partial x_m \partial x_1} & \cdots & \frac{\partial^2 f}{\partial x_m \partial x_m} \end{bmatrix} \in \mathbb{R}^{m\times m}. \tag{3.3.4}$$

This shows that the Hessian matrix of a real scalar function $f(\mathbf{x})$ is an $m \times m$ matrix that consists of m^2 second-order partial derivatives of $f(\mathbf{x})$ with respect to the entries x_i of the vector variable $\mathbf{x}$.

By the definition, it can be seen that the Hessian matrix of a real scalar function $f(\mathbf{x})$ is a real symmetric matrix, i.e,

$$(\mathbf{H}[f(\mathbf{x})])^T = \mathbf{H}[f(\mathbf{x})]. \tag{3.3.5}$$

The reason is that, for a second-differentiable continuous function $f(\mathbf{x})$, its second-order derivative is independent of the order of differentiation, i.e., $\frac{\partial^2 f}{\partial x_i \partial x_j} = \frac{\partial^2 f}{\partial x_j \partial x_i}$.

By mimicking the definition of the Hessian matrix of a real scalar function $f(\mathbf{x})$, we can define the Hessian matrix of a real scalar function $f(\mathbf{X})$:

$$\mathbf{H}[f(\mathbf{X})] = \frac{\partial^2 f(\mathbf{X})}{\partial \mathrm{vec}\mathbf{X}\,\partial (\mathrm{vec}\mathbf{X})^T} = \nabla_{\mathbf{X}}(\mathrm{D}_{\mathbf{X}} f(\mathbf{X})) \in \mathbb{R}^{mn\times mn}, \tag{3.3.6}$$

which can be written in element form as

$$\begin{bmatrix} \frac{\partial^2 f}{\partial x_{11}\partial x_{11}} & \cdots & \frac{\partial^2 f}{\partial x_{11}\partial x_{m1}} & \cdots & \frac{\partial^2 f}{\partial x_{11}\partial x_{1n}} & \cdots & \frac{\partial^2 f}{\partial x_{11}\partial x_{mn}} \\ \vdots & \vdots & \vdots & \vdots & \vdots & \vdots & \vdots \\ \frac{\partial^2 f}{\partial x_{m1}\partial x_{11}} & \cdots & \frac{\partial^2 f}{\partial x_{m1}\partial x_{m1}} & \cdots & \frac{\partial^2 f}{\partial x_{m1}\partial x_{1n}} & \cdots & \frac{\partial^2 f}{\partial x_{m1}\partial x_{mn}} \\ \vdots & \vdots & \vdots & \vdots & \vdots & \vdots & \vdots \\ \frac{\partial^2 f}{\partial x_{1n}\partial x_{11}} & \cdots & \frac{\partial^2 f}{\partial x_{1n}\partial x_{m1}} & \cdots & \frac{\partial^2 f}{\partial x_{1n}\partial x_{1n}} & \cdots & \frac{\partial^2 f}{\partial x_{1n}\partial x_{mn}} \\ \vdots & \vdots & \vdots & \vdots & \vdots & \vdots & \vdots \\ \frac{\partial^2 f}{\partial x_{mn}\partial x_{11}} & \cdots & \frac{\partial^2 f}{\partial x_{mn}\partial x_{m1}} & \cdots & \frac{\partial^2 f}{\partial x_{mn}\partial x_{1n}} & \cdots & \frac{\partial^2 f}{\partial x_{mn}\partial x_{mn}} \end{bmatrix}. \tag{3.3.7}$$

From $\frac{\partial^2 f}{\partial x_{ij}\partial x_{kl}} = \frac{\partial^2 f}{\partial x_{kl}\partial x_{ij}}$ it is easily seen that the Hessian matrix of a real scalar function $f(\mathbf{X})$ is a real symmetric matrix:

$$(\mathbf{H}[f(\mathbf{X})])^T = \mathbf{H}[f(\mathbf{X})]. \tag{3.3.8}$$

3.3.2 Real Hessian Matrix Identification

Let us discuss how to identify the Hessian matrices for real scalar functions $f(\mathbf{x})$ and $f(\mathbf{X})$. First, consider the case of a real scalar function $f(\mathbf{x})$.

In many cases, it may involve considerable trouble to compute a Hessian matrix from its definition formula. A simpler method is based on the relationship between the second-order differential matrix and the Hessian matrix.

Noting that the differential $\mathrm{d}\mathbf{x}$ is not a function of the vector $\mathbf{x}$, we have

$$\mathrm{d}^2\mathbf{x} = \mathrm{d}(\mathrm{d}\mathbf{x}) = 0. \tag{3.3.9}$$

Keeping this point in mind, from Equation (3.2.14) it is easy to find the second-order differential $\mathrm{d}^2 f(\mathbf{x}) = \mathrm{d}(\mathrm{d}f(\mathbf{x}))$:

$$\mathrm{d}^2 f(\mathbf{x}) = \mathrm{d}\mathbf{x}^T \frac{\partial\, \mathrm{d}f(\mathbf{x})}{\partial \mathbf{x}} = \mathrm{d}\mathbf{x}^T \frac{\partial}{\partial \mathbf{x}} \left(\frac{\partial f(\mathbf{x})}{\partial \mathbf{x}^T} \right) \mathrm{d}\mathbf{x} = (\mathrm{d}\mathbf{x})^T \frac{\partial f^2(\mathbf{x})}{\partial \mathbf{x}\, \partial \mathbf{x}^T} \mathrm{d}\mathbf{x}$$

which can simply be written as

$$\mathrm{d}^2 f(\mathbf{x}) = (\mathrm{d}\mathbf{x})^T \mathbf{H}[f(\mathbf{x})] \mathrm{d}\mathbf{x}, \tag{3.3.10}$$

where

$$\mathbf{H}[f(\mathbf{x})] = \frac{\partial f^2(\mathbf{x})}{\partial \mathbf{x}\, \partial \mathbf{x}^T} \tag{3.3.11}$$

is the *Hessian matrix* of the function $f(\mathbf{x})$ and the (i,j)th entry of $\mathbf{H}[f(\mathbf{x})]$ is given by

$$h_{ij} = \frac{\partial}{\partial x_i} \left(\frac{\partial f(\mathbf{x})}{\partial x_j} \right) = \frac{\partial f^2(\mathbf{x})}{\partial x_i \partial x_j}. \tag{3.3.12}$$

Noting that the matrix $\mathbf{A} \in \mathbb{R}^{1\times m}$ in the first-order differential $\mathrm{d}f(\mathbf{x}) = \mathbf{A}\mathrm{d}\mathbf{x}$ is usually a real row vector and that its differential is still a real row vector, we have

$$\mathrm{d}\mathbf{A} = (\mathrm{d}\mathbf{x})^T \mathbf{B} \in \mathbb{R}^{1\times m},$$

where $\mathbf{B} \in \mathbb{R}^{m\times m}$. Hence, the second-order differential of the function $f(\mathbf{x})$ is a quadratic function,

$$\mathrm{d}^2 f(\mathbf{x}) = \mathrm{d}(\mathbf{A}\mathrm{d}\mathbf{x}) = (\mathrm{d}\mathbf{x})^T \mathbf{B} \mathrm{d}\mathbf{x}. \tag{3.3.13}$$

By comparing Equation (3.3.13) with Equation (3.3.10), it can be seen that the Hessian matrix $\mathbf{H}_{\mathbf{x}}[f(\mathbf{x})] = \mathbf{B}$. In order to ensure that the Hessian matrix is real symmetric, we take

$$\mathbf{H}[f(\mathbf{x})] = \frac{1}{2}(\mathbf{B}^T + \mathbf{B}). \tag{3.3.14}$$

Now consider the case of a real scalar function $f(\mathbf{X})$.

From Equation (3.2.16) it follows that the second-order differential of the scalar function $f(\mathbf{X})$ is given by

$$\begin{aligned} \mathrm{d}^2 f(\mathbf{X}) &= \mathrm{d}(\mathrm{vec}\,\mathbf{X})^T \frac{\partial\, \mathrm{d}f(\mathbf{X})}{\partial\, \mathrm{vec}\,\mathbf{X}} \\ &= \mathrm{d}(\mathrm{vec}\,\mathbf{X})^T \frac{\partial}{\partial\, \mathrm{vec}\,\mathbf{X}} \left(\frac{\partial f(\mathbf{X})}{\partial (\mathrm{vec}\,\mathbf{X})^T} \right) \mathrm{d}\, \mathrm{vec}\,\mathbf{X} \\ &= (\mathrm{d}\, \mathrm{vec}\mathbf{X})^T \frac{\partial f^2(\mathbf{X})}{\partial\, \mathrm{vec}\mathbf{X}\, \partial (\mathrm{vec}\,\mathbf{X})^T} \mathrm{d}\, \mathrm{vec}\mathbf{X}, \end{aligned}$$

namely

$$\mathrm{d}^2 f(\mathbf{X}) = \mathrm{d}(\mathrm{vec}\mathbf{X})^T \mathbf{H}[f(\mathbf{X})] \mathrm{d}\,\mathrm{vec}\mathbf{X}. \tag{3.3.15}$$

Here

$$\mathbf{H}[f(\mathbf{X})] = \frac{\partial^2 f(\mathbf{X})}{\partial \mathrm{vec}\mathbf{X}\, \partial (\mathrm{vec}\mathbf{X})^T} \tag{3.3.16}$$

is the Hessian matrix of $f(\mathbf{X})$.

Formula (3.3.15) is called the *second-order (matrix) differential rule* for $f(\mathbf{X})$.

For the real scalar function $f(\mathbf{X})$, the matrix $\mathbf{A}$ in the first-order differential $\mathrm{d}f(\mathbf{X}) = \mathbf{A}\mathrm{d}\,\mathrm{vec}\mathbf{X}$ is usually a real row vector of the variable matrix $\mathbf{X}$ and the differential of $\mathbf{A}$ is still a real row vector; thus we have

$$\mathrm{d}\mathbf{A} = \mathrm{d}(\mathrm{vec}\,\mathbf{X})^T \mathbf{B} \quad \in \mathbb{R}^{1\times mn},$$

where $\mathbf{B} \in \mathbb{R}^{mn\times mn}$. Hence, the second-order differential of $f(\mathbf{X})$ is a quadratic function

$$\mathrm{d}^2 f(\mathbf{X}) = \mathrm{d}(\mathrm{vec}\,\mathbf{X})^T \mathbf{B}\,\mathrm{d}\,\mathrm{vec}\mathbf{X}. \tag{3.3.17}$$

By comparing Equation (3.3.17) with Equation (3.3.15), it follows that the Hessian matrix of the function $f(\mathbf{X})$ is

$$\mathbf{H}[f(\mathbf{X})] = \frac{1}{2}(\mathbf{B}^T + \mathbf{B}), \tag{3.3.18}$$

because a real Hessian matrix must be symmetric.

The results above can be summarized in the following proposition.

Proposition 3.3 *For a scalar function $f(\cdot)$ with vector $\mathbf{x}$ or matrix $\mathbf{X}$ as function variable, there is the following second-order identification relation between the second-order differential and the Hessian matrix:*

$$\mathrm{d}^2 f(\mathbf{x}) = (\mathrm{d}\mathbf{x})^T \mathbf{B}\,\mathrm{d}\mathbf{x} \quad \Leftrightarrow \quad \mathbf{H}[f(\mathbf{x})] = \frac{1}{2}(\mathbf{B}^T + \mathbf{B}), \tag{3.3.19}$$

$$\mathrm{d}^2 f(\mathbf{X}) = \mathrm{d}(\mathrm{vec}\mathbf{X})^T \mathbf{B}\,\mathrm{d}\,\mathrm{vec}\,\mathbf{X} \quad \Leftrightarrow \quad \mathbf{H}[f(\mathbf{X})] = \frac{1}{2}(\mathbf{B}^T + \mathbf{B}). \tag{3.3.20}$$

More generally, we have the second-order identification theorem for matrix function as follows.

THEOREM 3.2 [311, p. 115] *Let $\mathbf{F} : S \to \mathbb{R}^{m\times p}$ be a matrix function defined on a set $S \in \mathbb{R}^{n\times q}$, and twice differentiable at an interior point $\mathbf{C}$ of S. Then*

$$\mathrm{d}^2\mathrm{vec}(\mathbf{F}(\mathbf{C};\mathbf{X})) = (\mathbf{I}_{mp} \otimes \mathrm{d}\,\mathrm{vec}\mathbf{X})^T\mathbf{B}(\mathbf{C})\mathrm{d}\mathrm{vec}\mathbf{X} \tag{3.3.21}$$

for all $\mathbf{X} \in \mathbb{R}^{n\times q}$, if and only if $\mathbf{H}[\mathbf{F}(\mathbf{C})] = \frac{1}{2}(\mathbf{B}(\mathbf{C}) + (\mathbf{B}(\mathbf{C}))'_v)$, where $\mathbf{B}, (\mathbf{B}')_v \in \mathbb{R}^{pmqn\times mn}$ are given by

$$\mathbf{B} = \begin{bmatrix} \mathbf{B}_{11} \\ \vdots \\ \mathbf{B}_{p1} \\ \vdots \\ \mathbf{B}_{1q} \\ \vdots \\ \mathbf{B}_{pq} \end{bmatrix}, \quad (\mathbf{B}')_v = \begin{bmatrix} \mathbf{B}_{11}^T \\ \vdots \\ \mathbf{B}_{p1}^T \\ \vdots \\ \mathbf{B}_{1q}^T \\ \vdots \\ \mathbf{B}_{pq}^T \end{bmatrix} \tag{3.3.22}$$

with $\mathbf{B}_{ij} \in \mathbb{R}^{mn\times mn}, \forall\, i = 1,\ldots,p,\ j = 1,\ldots,q$.

Theorem 3.2 gives the following second-order identification formula for a matrix function:

$$\begin{aligned} &\mathrm{d}^2\,\mathrm{vec}\mathbf{F}(\mathbf{X}) = (\mathbf{I}_{mp} \otimes \mathrm{d}\,\mathrm{vec}\mathbf{X})^T\mathbf{B}\,\mathrm{d}\,\mathrm{vec}\mathbf{X} \\ &\Leftrightarrow \mathbf{H}[\mathbf{F}(\mathbf{X})] = \frac{1}{2}[\mathbf{B} + (\mathbf{B}')_v]. \end{aligned} \tag{3.3.23}$$

In particular, if $\mathbf{F} \in \mathbb{R}^{m\times p}$ takes the form of a scalar function f or a vector function $\mathbf{f} \in \mathbb{R}^{m\times 1}$, and the variable matrix $\mathbf{X} \in \mathbb{R}^{n\times q}$ a scalar x or a vector $\mathbf{x} \in \mathbb{R}^{n\times 1}$, then the above second-order identification formula yields the following identification formulas:

$$\begin{aligned}
&f(x)\text{: } \mathrm{d}^2[f(x)] = \beta(\mathrm{d}x)^2 \;\Leftrightarrow\; \mathbf{H}[f(x)] = \beta, \\
&f(\mathbf{x})\text{: } \mathrm{d}^2[f(\mathbf{x})] = (\mathrm{d}\mathbf{x})^T\mathbf{B}\mathrm{d}\mathbf{x} \;\Leftrightarrow\; \mathbf{H}[f(\mathbf{x})] = \tfrac{1}{2}(\mathbf{B} + \mathbf{B}^T), \\
&f(\mathbf{X})\text{: } \mathrm{d}^2[f(\mathbf{X})] = \mathrm{d}(\mathrm{vec}\,\mathbf{X})^T\mathbf{B}\mathrm{d}\mathrm{vec}\mathbf{X} \;\Leftrightarrow\; \mathbf{H}[f(\mathbf{X})] = \tfrac{1}{2}(\mathbf{B} + \mathbf{B}^T); \\
&\mathbf{f}(x)\text{: } \mathrm{d}^2[\mathbf{f}(x)] = \mathbf{b}(\mathrm{d}x)^2 \;\Leftrightarrow\; \mathbf{H}[\mathbf{f}(x)] = \mathbf{b}, \\
&\mathbf{f}(\mathbf{x})\text{: } \mathrm{d}^2[\mathbf{f}(\mathbf{x})] = (\mathbf{I}_m \otimes \mathrm{d}\mathbf{x})^T\mathbf{B}\mathrm{d}\mathbf{x} \;\Leftrightarrow\; \mathbf{H}[\mathbf{f}(\mathbf{x})] = \tfrac{1}{2}[\mathbf{B} + (\mathbf{B}')_v], \\
&\mathbf{f}(\mathbf{X})\text{: } \mathrm{d}^2[\mathbf{f}(\mathbf{X})] = (\mathbf{I}_m \otimes \mathrm{d}(\mathrm{vec}\,\mathbf{X}))^T\mathbf{B}\mathrm{d}\mathrm{vec}\mathbf{X} \;\Leftrightarrow\; \mathbf{H}[\mathbf{f}(\mathbf{X})] = \tfrac{1}{2}[\mathbf{B} + (\mathbf{B}')_v]; \\
&\mathbf{F}(x)\text{: } \mathrm{d}^2[\mathbf{F}(x)] = \mathbf{B}(\mathrm{d}x)^2 \;\Leftrightarrow\; \mathbf{H}[\mathbf{F}(x)] = \mathrm{vec}\,\mathbf{B}, \\
&\mathbf{F}(\mathbf{x})\text{: } \mathrm{d}^2[\mathrm{vec}\mathbf{F}] = (\mathbf{I}_{mp} \otimes \mathrm{d}\mathbf{x})^T\mathbf{B}\mathrm{d}\mathbf{x} \;\Leftrightarrow\; \mathbf{H}[\mathbf{F}(\mathbf{x})] = \tfrac{1}{2}[\mathbf{B} + (\mathbf{B}')_v].
\end{aligned}$$

The Hessian matrix identification based on Theorem 3.2 requires a row vectorization operation on the second-order matrix differential. An interesting question is whether we can directly identify the Hessian matrix from the second-order matrix differential without such a row vectorization operation.

As mentioned earlier, the first-order differential of a scalar function $f(\mathbf{X})$ can be written in the canonical form of trace function as $\mathrm{d}f(\mathbf{X}) = \mathrm{tr}(\mathbf{A}\mathrm{d}\mathbf{X})$, where $\mathbf{A} = \mathbf{A}(\mathbf{X}) \in \mathbb{R}^{n\times m}$ is generally a matrix function of the variable matrix $\mathbf{X} \in \mathbb{R}^{m\times n}$. Without loss of generality, it can be assumed that the differential matrix of the matrix function $\mathbf{A} = \mathbf{A}(\mathbf{X})$ takes one of two forms:

$$\mathrm{d}\mathbf{A} = \mathbf{B}(\mathrm{d}\mathbf{X})\mathbf{C}, \quad \mathbf{B},\mathbf{C} \in \mathbb{R}^{n\times m} \tag{3.3.24}$$

or

$$\mathrm{d}\mathbf{A} = \mathbf{U}(\mathrm{d}\mathbf{X})^T\mathbf{V}, \quad \mathbf{U} \in \mathbb{R}^{n\times n}, \mathbf{V} \in \mathbb{R}^{m\times m}. \tag{3.3.25}$$

Notice that the above two differential matrices include $(\mathrm{d}\mathbf{X})\mathbf{C}$, $\mathbf{B}\mathrm{d}\mathbf{X}$ and $(\mathrm{d}\mathbf{X})^T\mathbf{V}$, $\mathbf{U}(\mathrm{d}\mathbf{X})^T$ as two special examples.

Substituting (3.3.24) and (3.3.25) into $\mathrm{d}f(\mathbf{X}) = \mathrm{tr}(\mathbf{A}\mathrm{d}\mathbf{X})$, respectively, it is known that the second-order differential $\mathrm{d}^2 f(\mathbf{X}) = \mathrm{tr}(\mathrm{d}\mathbf{A}\mathrm{d}\mathbf{X})$ of the real scalar function $f(\mathbf{X})$ takes one of two forms

$$\mathrm{d}^2 f(\mathbf{X}) = \mathrm{tr}(\mathbf{B}(\mathrm{d}\mathbf{X})\mathbf{C}\mathrm{d}\mathbf{X}) \tag{3.3.26}$$

or

$$\mathrm{d}^2 f(\mathbf{X}) = \mathrm{tr}\big(\mathbf{V}(\mathrm{d}\mathbf{X})\mathbf{U}(\mathrm{d}\mathbf{X})^T\big). \tag{3.3.27}$$

By the trace property $\mathrm{tr}(\mathbf{ABCD}) = (\mathrm{vec}\,\mathbf{D}^T)^T(\mathbf{C}^T \otimes \mathbf{A})\mathrm{vec}\,\mathbf{B}$, we have

$$\begin{aligned}
\mathrm{tr}(\mathbf{B}(\mathrm{d}\mathbf{X})\mathbf{C}\mathrm{d}\mathbf{X}) &= (\mathrm{vec}(\mathrm{d}\mathbf{X})^T)^T(\mathbf{C}^T \otimes \mathbf{B})\mathrm{vec}(\mathrm{d}\mathbf{X})\\
&= (\mathrm{d}(\mathbf{K}_{mn}\mathrm{vec}\,\mathbf{X}))^T(\mathbf{C}^T \otimes \mathbf{B})\mathrm{d}\mathrm{vec}\mathbf{X}\\
&= (\mathrm{d}(\mathrm{vec}\,\mathbf{X}))^T\mathbf{K}_{nm}(\mathbf{C}^T \otimes \mathbf{B})\mathrm{d}\mathrm{vec}\mathbf{X}, \qquad (3.3.28)\\
\mathrm{tr}\big(\mathbf{V}\mathrm{d}\mathbf{X}\mathbf{U}(\mathrm{d}\mathbf{X})^T\big) &= (\mathrm{vec}(\mathrm{d}\mathbf{X}))^T(\mathbf{U}^T \otimes \mathbf{V})\mathrm{vec}(\mathrm{d}\mathbf{X})\\
&= (\mathrm{d}(\mathrm{vec}\,\mathbf{X}))^T(\mathbf{U}^T \otimes \mathbf{V})\mathrm{d}\mathrm{vec}\mathbf{X}, \qquad (3.3.29)
\end{aligned}$$

in which we have used the results $\mathbf{K}_{mn}\mathrm{vec}\,\mathbf{A}_{m\times n} = \mathrm{vec}(\mathbf{A}_{m\times n}^T)$ and $\mathbf{K}_{mn}^T = \mathbf{K}_{nm}$.

The expressions $\mathrm{d}^2 f(\mathbf{X}) = \mathrm{tr}(\mathbf{B}(\mathrm{d}\mathbf{X})\mathbf{C}\mathrm{d}\mathbf{X})$ and $\mathrm{d}^2 f(\mathbf{X}) = \mathrm{tr}\big(\mathbf{V}(\mathrm{d}\mathbf{X})\mathbf{U}(\mathrm{d}\mathbf{X})^T\big)$ can be viewed as two canonical forms for the second-order matrix differentials of the real function $f(\mathbf{X})$.

Each of the second-order differentials $\mathrm{tr}\big(\mathbf{V}(\mathrm{d}\mathbf{X})\mathbf{U}(\mathrm{d}\mathbf{X})^T\big)$ and $\mathrm{tr}(\mathbf{B}(\mathrm{d}\mathbf{X})\mathbf{C}\mathrm{d}\mathbf{X})$ can be equivalently expressed as the the canonical quadratic form required by Proposition 3.3, so this can be equivalently stated as the following theorem for Hessian matrix identification.

THEOREM 3.3 [311, p. 192] *Let the real function $f(\mathbf{X})$ with the $m\times n$ real matrix $\mathbf{X}$ as variable be second-order differentiable. Then there is the following relation between the second-order differential matrix $\mathrm{d}^2 f(\mathbf{X})$ and the Hessian matrix $\mathbf{H}[f(\mathbf{X})]$:*

$$\begin{aligned}
&\mathrm{d}^2 f(\mathbf{X}) = \mathrm{tr}\big(\mathbf{V}(\mathrm{d}\mathbf{X})\mathbf{U}(\mathrm{d}\mathbf{X})^T\big)\\
\Leftrightarrow\quad &\mathbf{H}[f(\mathbf{X})] = \frac{1}{2}(\mathbf{U}^T \otimes \mathbf{V} + \mathbf{U} \otimes \mathbf{V}^T), \qquad (3.3.30)
\end{aligned}$$

where $\mathbf{U} \in \mathbb{R}^{n\times n}$, $\mathbf{V} \in \mathbb{R}^{m\times m}$, *or*

$$\begin{aligned} &\mathrm{d}^2 f(\mathbf{X}) = \mathrm{tr}(\mathbf{B}(\mathrm{d}\mathbf{X})\mathbf{C}\mathrm{d}\mathbf{X}) \\ \Leftrightarrow\ &\mathbf{H}[f(\mathbf{X})] = \frac{1}{2}\mathbf{K}_{nm}(\mathbf{C}^T \otimes \mathbf{B} + \mathbf{B}^T \otimes \mathbf{C}), \end{aligned} \tag{3.3.31}$$

where $\mathbf{K}_{nm}$ *is the commutation matrix.*

Theorem 3.3 shows that the basic problem in the Hessian matrix identification is how to express the second-order matrix differential of a given real scalar function as one of the two canonical forms required by Theorem 3.3.

The following three examples show how to find the Hessian matrix of a given real function using Theorem 3.3.

Example 3.17 Consider the real function $f(\mathbf{X}) = \mathrm{tr}(\mathbf{X}^{-1})$, where $\mathbf{X}$ is an $n \times n$ matrix. First, computing the first-order differential

$$\mathrm{d}f(\mathbf{X}) = -\mathrm{tr}\big(\mathbf{X}^{-1}(\mathrm{d}\mathbf{X})\mathbf{X}^{-1}\big)$$

and then using $\mathrm{d}(\mathrm{tr}\,\mathbf{U}) = \mathrm{tr}(\mathrm{d}\mathbf{U})$, we get the second-order differential

$$\begin{aligned} \mathrm{d}^2 f(\mathbf{X}) &= -\mathrm{tr}\big((\mathrm{d}\mathbf{X}^{-1})(\mathrm{d}\mathbf{X})\mathbf{X}^{-1}\big) - \mathrm{tr}\,\big(\mathbf{X}^{-1}(\mathrm{d}\mathbf{X})(\mathrm{d}\mathbf{X}^{-1})\big) \\ &= 2\,\mathrm{tr}\big(\mathbf{X}^{-1}(\mathrm{d}\mathbf{X})\mathbf{X}^{-1}(\mathrm{d}\mathbf{X})\mathbf{X}^{-1}\big) \\ &= 2\,\mathrm{tr}\big(\mathbf{X}^{-2}(\mathrm{d}\mathbf{X})\mathbf{X}^{-1}\mathrm{d}\mathbf{X}\big)\,. \end{aligned}$$

Now, by Theorem 3.3, we get the Hessian matrix

$$\mathbf{H}[f(\mathbf{X})] = \frac{\partial^2 \mathrm{tr}(\mathbf{X}^{-1})}{\partial \mathrm{vec}\,\mathbf{X}\,\partial(\mathrm{vec}\,\mathbf{X})^T} = \mathbf{K}_{nn}\left(\mathbf{X}^{-T} \otimes \mathbf{X}^{-2} + (\mathbf{X}^{-2})^T \otimes \mathbf{X}^{-1}\right).$$

Example 3.18 For the real function $f(\mathbf{X}) = \mathrm{tr}(\mathbf{X}^T\mathbf{A}\mathbf{X})$, its first-order differential is

$$\mathrm{d}f(\mathbf{X}) = \mathrm{tr}\big(\mathbf{X}^T(\mathbf{A} + \mathbf{A}^T)\mathrm{d}\mathbf{X}\big)\,.$$

Thus, the second-order differential is

$$\mathrm{d}^2 f(\mathbf{X}) = \mathrm{tr}\big((\mathbf{A} + \mathbf{A}^T)(\mathrm{d}\mathbf{X})(\mathrm{d}\mathbf{X})^T\big)\,.$$

By Theorem 3.3, the Hessian matrix is given by

$$\mathbf{H}[f(\mathbf{X})] = \frac{\partial^2 \mathrm{tr}(\mathbf{X}^T\mathbf{A}\mathbf{X})}{\partial \mathrm{vec}\,\mathbf{X}\,\partial(\mathrm{vec}\,\mathbf{X})^T} = \mathbf{I} \otimes (\mathbf{A} + \mathbf{A}^T).$$

Example 3.19 The first-order differential of the function $\log|\mathbf{X}_{n\times n}|$ is $\mathrm{d}\log|\mathbf{X}| = \mathrm{tr}(\mathbf{X}^{-1}\mathrm{d}\mathbf{x})$, and the second-order differential is $-\mathrm{tr}(\mathbf{X}^{-1}(\mathrm{d}\mathbf{x})\mathbf{X}^{-1}\mathrm{d}\mathbf{x})$. From Theorem 3.3, the Hessian matrix is obtained as follows:

$$\mathbf{H}[f(\mathbf{X})] = \frac{\partial^2 \log|\mathbf{X}|}{\partial \mathrm{vec}\,\mathbf{X}\,\partial(\mathrm{vec}\,\mathbf{X})^T} = -\mathbf{K}_{nn}(\mathbf{X}^{-T} \otimes \mathbf{X}^{-1}).$$

3.4 Complex Gradient Matrices

In array processing and wireless communications, a narrowband signal is usually expressed as a complex equivalent baseband signal, and thus the transmit and receive signals with system parameters are expressed as complex vectors. In these applications, the objective function of an optimization problem is a real-valued function of a complex vector or matrix. Hence, the gradient of the objective function is a complex vector or matrix. Obviously, this complex gradient has the following two forms:

(1) *complex gradient* the gradient of the objective function with respect to the complex vector or matrix variable itself;

(2) *conjugate gradient* the gradient of the objective function with respect to the complex conjugate vector or matrix variable.

3.4.1 Holomorphic Function and Complex Partial Derivative

Before discussing the complex gradient and conjugate gradient, it is necessary to recall the relevant facts about complex functions.

For convenience, we first list the standard symbols for complex variables and complex functions.

- $\mathbf{z} = [z_1, \ldots, z_m]^T \in \mathbb{C}^m$ is a complex variable vector whose complex conjugate is $\mathbf{z}^*$.
- $\mathbf{Z} = [\mathbf{z}_1, \ldots, \mathbf{z}_n] \in \mathbb{C}^{m\times n}$ is a complex variable matrix with complex conjugate $\mathbf{Z}^*$.
- $f(\mathbf{z}, \mathbf{z}^*) \in \mathbb{C}$ is a complex scalar function of $m \times 1$ complex variable vectors $\mathbf{z}$ and $\mathbf{z}^*$, denoted $f : \mathbb{C}^m \times \mathbb{C}^m \to \mathbb{C}$.
- $f(\mathbf{Z}, \mathbf{Z}^*) \in \mathbb{C}$ is a complex scalar function of $m \times n$ complex variable matrices $\mathbf{Z}$ and $\mathbf{Z}^*$, denoted $f : \mathbb{C}^{m\times n} \times \mathbb{C}^{m\times n} \to \mathbb{C}$.
- $\mathbf{f}(\mathbf{z}, \mathbf{z}^*) \in \mathbb{C}^p$ is a $p\times 1$ complex vector function of $m\times 1$ complex variable vectors $\mathbf{z}$ and $\mathbf{z}^*$, denoted $\mathbf{f} : \mathbb{C}^m \times \mathbb{C}^m \to \mathbb{C}^p$.
- $\mathbf{f}(\mathbf{Z}, \mathbf{Z}^*) \in \mathbb{C}^p$ is a $p \times 1$ complex vector function of $m \times n$ complex variable matrices $\mathbf{Z}$ and $\mathbf{Z}^*$, denoted $\mathbf{f} : \mathbb{C}^{m\times n} \times \mathbb{C}^{m\times n} \to \mathbb{C}^p$.
- $\mathbf{F}(\mathbf{z}, \mathbf{z}^*) \in \mathbb{C}^{p\times q}$ is a $p \times q$ complex matrix function of $m \times 1$ complex variable vectors $\mathbf{z}$ and $\mathbf{z}^*$, denoted $\mathbf{F} : \mathbb{C}^m \times \mathbb{C}^m \to \mathbb{C}^{p\times q}$.
- $\mathbf{F}(\mathbf{Z}, \mathbf{Z}^*) \in \mathbb{C}^{p\times q}$ is a $p \times q$ complex matrix function of $m \times n$ complex variable matrices $\mathbf{Z}$ and $\mathbf{Z}^*$, denoted $\mathbf{F} : \mathbb{C}^{m\times n} \times \mathbb{C}^{m\times n} \to \mathbb{C}^{p\times q}$.

Table 3.7 summarizes the classification of complex-valued functions.

Table 3.7 Classification of complex-valued functions

Function	Variables $z, z^* \in C$	Variables $\mathbf{z}, \mathbf{z}^* \in \mathbb{C}^m$	Variables $\mathbf{Z}, \mathbf{Z}^* \in \mathbb{C}^{m\times n}$
$f \in \mathbb{C}$	$f(z, z^*)$ $f: \mathbb{C}\times\mathbb{C} \to \mathbb{C}$	$f(\mathbf{z}, \mathbf{z}^*)$ $\mathbb{C}^m \times \mathbb{C}^m \to \mathbb{C}$	$f(\mathbf{Z}, \mathbf{Z}^*)$ $\mathbb{C}^{m\times n} \times \mathbb{C}^{m\times n} \to \mathbb{C}$
$\mathbf{f} \in \mathbb{C}^p$	$\mathbf{f}(z, z^*)$ $\mathbf{f}: \mathbb{C}\times\mathbb{C} \to \mathbb{C}^p$	$\mathbf{f}(\mathbf{z}, \mathbf{z}^*)$ $\mathbb{C}^m \times \mathbb{C}^m \to \mathbb{C}^p$	$\mathbf{f}(\mathbf{Z}, \mathbf{Z}^*)$ $\mathbb{C}^{m\times n} \times \mathbb{C}^{m\times n} \to \mathbb{C}^p$
$\mathbf{F} \in \mathbb{C}^{p\times q}$	$\mathbf{F}(z, z^*)$ $\mathbf{F}: \mathbb{C}\times\mathbb{C} \to \mathbb{C}^{p\times q}$	$\mathbf{F}(\mathbf{z}, \mathbf{z}^*)$ $\mathbb{C}^m \times \mathbb{C}^m \to \mathbb{C}^{p\times q}$	$\mathbf{F}(\mathbf{Z}, \mathbf{Z}^*)$ $\mathbb{C}^{m\times n} \times \mathbb{C}^{m\times n} \to \mathbb{C}^{p\times q}$

Definition 3.10 [264] Let $D \subseteq \mathbb{C}$ be the definition domain of the function $f : D \to \mathbb{C}$. The function $f(z)$ with complex variable z is said to be a *complex analytic function* in the domain D if $f(z)$ is *complex differentiable*, namely $\lim_{\Delta z\to 0} \frac{f(z+\Delta z) - f(z)}{\Delta z}$ exists for all $z \in D$.

The terminology "complex analytic" is commonly replaced by the completely synonymous terminology "*holomorphic*". Hence, a complex analytic function is usually called a holomorphic function. It is noted that a complex function is (real) analytic in the real-variable x-domain and y-domain, but it is not necessarily holomorphic in the complex variable domain $z = x + \mathrm{j}y$, i.e., it may be complex nonanalytic.

It is assumed that the complex function $f(z)$ can be expressed in terms of its real part $u(x, y)$ and imaginary part $v(x, y)$ as

$$f(z) = u(x, y) + \mathrm{j}v(x, y),$$

where $z = x + \mathrm{j}y$ and both $u(x, y)$ and $v(x, y)$ are real functions.

For a holomorphic scalar function, the following four statements are equivalent [155].

(1) The complex function $f(z)$ is a holomorphic (i.e., complex analytic) function.
(2) The derivative $f'(z)$ of the complex function exists and is continuous.
(3) The complex function $f(z)$ satisfies the *Cauchy–Riemann condition*

$$\frac{\partial u}{\partial x} = \frac{\partial v}{\partial y} \quad \text{and} \quad \frac{\partial v}{\partial x} = -\frac{\partial u}{\partial y}. \tag{3.4.1}$$

(4) All derivatives of the complex function $f(z)$ exist, and $f(z)$ has a convergent power series.

The Cauchy–Riemann condition is also known as the *Cauchy–Riemann equations*, and its direct result is that the function $f(z) = u(x, y) + \mathrm{j}v(x, y)$ is a holomorphic function only when both the real functions $u(x, y)$ and $v(x, y)$ satisfy the *Laplace equation* at the same time:

$$\frac{\partial^2 u(x,y)}{\partial x^2}+\frac{\partial^2 u(x,y)}{\partial y^2}=0 \quad\text{and}\quad \frac{\partial^2 v(x,y)}{\partial x^2}+\frac{\partial^2 v(x,y)}{\partial y^2}=0. \tag{3.4.2}$$

A real function $g(x,y)$ is called a *harmonic function*, if it satisfies the *Laplace equation*

$$\frac{\partial^2 g(x,y)}{\partial x^2}+\frac{\partial^2 g(x,y)}{\partial y^2}=0. \tag{3.4.3}$$

A complex function $f(z) = u(x,y) + \mathrm{j}v(x,y)$ is not a holomorphic function if either of the real functions $u(x,y)$ and $v(x,y)$ does not meet the Cauchy–Riemann condition or the Laplace equation.

Although the power function z^n, the exponential function e^z, the logarithmic function $\ln z$, the sine function $\sin z$ and the cosine function $\cos z$ are holomorphic functions, i.e., analytic functions in the complex plane, many commonly used functions are not holomorphic, as shown in the following examples.

- For the complex function $f(z) = z^* = x - \mathrm{j}y = u(x,y) + \mathrm{j}v(x,y)$, its real part $u(x,y) = x$ and imaginary part $v(x,y) = -y$ obviously do not meet the Cauchy–Riemann condition $\frac{\partial u}{\partial x} = \frac{\partial v}{\partial y}$.
- Any nonconstant real function $f(z) \in \mathbb{R}$ does not satisfy the Cauchy–Riemann conditions $\frac{\partial u}{\partial x} = \frac{\partial v}{\partial y}$ and $\frac{\partial u}{\partial y} = -\frac{\partial v}{\partial x}$, because the imaginary part $v(x,y)$ of $f(z) = u(x,y) + \mathrm{j}v(x,y)$ is zero. In particular, the real function $f(z) = |z| = \sqrt{x^2+y^2}$ is nondifferentiable, while for $f(z) = |z|^2 = x^2 + y^2 = u(x,y) + \mathrm{j}v(x,y)$, its real part $u(x,y) = x^2 + y^2$ is not a harmonic function because the Laplace condition is not satisfied, namely, $\frac{\partial^2 u(x,y)}{\partial x^2}+\frac{\partial^2 u(x,y)}{\partial y^2} \neq 0$.
- The complex functions $f(z) = \mathrm{Re}\, z = x$ and $f(z) = \mathrm{Im}\, z = y$ do not meet the Cauchy–Riemann condition.

Since many common complex functions $f(z)$ are not holomorphic, a natural question to ask is whether there is a general representation form that can ensure that any complex function is holomorphic. To answer this question, it is necessary to recall the definition on the derivative of a complex number z and its conjugate z^* in complex function theory.

The *formal partial derivatives* of complex numbers are defined as

$$\frac{\partial}{\partial z} = \frac{1}{2}\left(\frac{\partial}{\partial x} - \mathrm{j}\frac{\partial}{\partial y}\right), \tag{3.4.4}$$

$$\frac{\partial}{\partial z^*} = \frac{1}{2}\left(\frac{\partial}{\partial x} + \mathrm{j}\frac{\partial}{\partial y}\right). \tag{3.4.5}$$

The formal partial derivatives above were presented by Wirtinger [511] in 1927, so they are sometimes called *Wirtinger partial derivatives*.

Regarding the partial derivatives of the complex variable $z = x + \mathrm{j}y$, there is a basic assumption on the independence of the real and imaginary parts:

$$\frac{\partial x}{\partial y} = 0 \quad \text{and} \quad \frac{\partial y}{\partial x} = 0. \tag{3.4.6}$$

By the definition of the partial derivative and by the above independence assumption, it is easy to find that

$$\begin{aligned} \frac{\partial z}{\partial z^*} &= \frac{\partial x}{\partial z^*} + \mathrm{j}\frac{\partial y}{\partial z^*} = \frac{1}{2}\left(\frac{\partial x}{\partial x} + \mathrm{j}\frac{\partial x}{\partial y}\right) + \mathrm{j}\frac{1}{2}\left(\frac{\partial y}{\partial x} + \mathrm{j}\frac{\partial y}{\partial y}\right) \\ &= \frac{1}{2}(1+0) + \mathrm{j}\frac{1}{2}(0+\mathrm{j}), \\ \frac{\partial z^*}{\partial z} &= \frac{\partial x}{\partial z} - \mathrm{j}\frac{\partial y}{\partial z} = \frac{1}{2}\left(\frac{\partial x}{\partial x} - \mathrm{j}\frac{\partial x}{\partial y}\right) - \mathrm{j}\frac{1}{2}\left(\frac{\partial y}{\partial x} - \mathrm{j}\frac{\partial y}{\partial y}\right) \\ &= \frac{1}{2}(1-0) - \mathrm{j}\frac{1}{2}(0-\mathrm{j}). \end{aligned}$$

This implies that

$$\frac{\partial z}{\partial z^*} = 0 \qquad \text{and} \qquad \frac{\partial z^*}{\partial z} = 0. \tag{3.4.7}$$

Equation (3.4.7) reveals a basic result in the theory of complex variables: the complex variable z and the complex conjugate variable z^* are independent variables.

In the standard framework of complex functions, a complex function $f(z)$ (where $z = x + \mathrm{j}y$) is written in the real polar coordinates $r \stackrel{\text{def}}{=} (x, y)$ as $f(r) = f(x, y)$. However, in the framework of complex derivatives, under the independence assumption of the real and imaginary parts, the complex function $f(z)$ is written as $f(c) = f(z, z^*)$ in the conjugate coordinates $c \stackrel{\text{def}}{=} (z, z^*)$ instead of the polar coordinates $r = (x, y)$. Hence, when finding the *complex partial derivative* $\nabla_z f(z, z^*)$ and the *complex conjugate partial derivative* $\nabla_{z^*} f(z, z^*)$, the complex variable z and the complex conjugate variable z^* are regarded as two independent variables:

$$\begin{aligned} \nabla_z f(z, z^*) &= \left.\frac{\partial f(z, z^*)}{\partial z}\right|_{z^* = \text{const}}, \\ \nabla_{z^*} f(z, z^*) &= \left.\frac{\partial f(z, z^*)}{\partial z^*}\right|_{z = \text{const}}. \end{aligned} \tag{3.4.8}$$

This implies that when any nonholomorphic function $f(z)$ is written as $f(z, z^*)$, it becomes holomorphic, because, for a fixed z^* value, the complex function $f(z, z^*)$ is analytic on the whole complex plane $z = x + \mathrm{j}y$; and, for a fixed z value, the complex function $f(z, z^*)$ is analytic on the whole complex plane $z^* = x - \mathrm{j}y$, see, e.g., [155], [263].

Example 3.20 Given a real function $f(z, z^*) = |z|^2 = zz^*$, its first-order partial derivatives $\partial|z|^2/\partial z = z^*$ and $\partial|z|^2/\partial z^* = z$ exist and are continuous. This is to say, although $f(z) = |z|^2$ itself is not a holomorphic function with respect to z, the function $f(z, z^*) = |z|^2 = zz^*$ is analytic on the whole complex plane $z = x + \mathrm{j}y$ (when z^* is fixed as a constant) and is analytic on the whole complex plane $z^* = x - \mathrm{j}y$ (when z is fixed as a constant).

Table 3.8 gives for comparison the nonholomorphic and holomorphic representation forms of complex functions.

Table 3.8 Nonholomorphic and holomorphic functions

Functions	Nonholomorphic	Holomorphic
Coordinates	$\begin{cases} r \stackrel{\text{def}}{=} (x, y) \in \mathbb{R} \times \mathbb{R} \\ z = x + \mathrm{j}y \end{cases}$	$\begin{cases} c \stackrel{\text{def}}{=} (z, z^*) \in \mathbb{C} \times \mathbb{C} \\ z = x + \mathrm{j}y,\ z^* = x - \mathrm{j}y \end{cases}$
Representation	$f(r) = f(x, y)$	$f(c) = f(z, z^*)$

The following are common formulas and rules for the complex partial derivatives [263]:

(1) The conjugate partial derivative of the complex conjugate function $\dfrac{\partial f^*(z, z^*)}{\partial z^*}$:

$$\frac{\partial f^*(z, z^*)}{\partial z^*} = \left(\frac{\partial f(z, z^*)}{\partial z} \right)^*. \tag{3.4.9}$$

(2) The partial derivative of the conjugate of the complex function $\dfrac{\partial f^*(z, z^*)}{\partial z}$:

$$\frac{\partial f^*(z, z^*)}{\partial z} = \left(\frac{\partial f(z, z^*)}{\partial z^*} \right)^*. \tag{3.4.10}$$

(3) *Complex differential rule*

$$\mathrm{d}f(z, z^*) = \frac{\partial f(z, z^*)}{\partial z} \mathrm{d}z + \frac{\partial f(z, z^*)}{\partial z^*} \mathrm{d}z^*. \tag{3.4.11}$$

(4) *Complex chain rule*

$$\frac{\partial h(g(z, z^*))}{\partial z} = \frac{\partial h(g(z, z^*))}{\partial g(z, z^*)} \frac{\partial g(z, z^*)}{\partial z} + \frac{\partial h(g(z, z^*))}{\partial g^*(z, z^*)} \frac{\partial g^*(z, z^*)}{\partial z}, \tag{3.4.12}$$

$$\frac{\partial h(g(z, z^*))}{\partial z^*} = \frac{\partial h(g(z, z^*))}{\partial g(z, z^*)} \frac{\partial g(z, z^*)}{\partial z^*} + \frac{\partial h(g(z, z^*))}{\partial g^*(z, z^*)} \frac{\partial g^*(z, z^*)}{\partial z^*}. \tag{3.4.13}$$

3.4.2 Complex Matrix Differential

The concepts of the complex function $f(z)$ and the holomorphic function $f(z, z^*)$ of a complex scalar z can be easily extended to the complex matrix function $\mathbf{F}(\mathbf{Z})$ and the holomorphic complex matrix function $\mathbf{F}(\mathbf{Z}, \mathbf{Z}^*)$.

On the *holomorphic complex matrix functions*, the following statements are equivalent [68].

(1) The matrix function $\mathbf{F}(\mathbf{Z})$ is a holomorphic function of the complex matrix variable $\mathbf{Z}$.
(2) The complex matrix differential $\mathrm{d}\,\mathrm{vec}\,\mathbf{F}(\mathbf{Z}) = \dfrac{\partial\,\mathrm{vec}\,\mathbf{F}(\mathbf{Z})}{\partial(\mathrm{vec}\,\mathbf{Z})^T}\mathrm{d}\,\mathrm{vec}\,\mathbf{Z}$.
(3) For all $\mathbf{Z}$, $\dfrac{\partial\,\mathrm{vec}\,\mathbf{F}(\mathbf{Z})}{\partial(\mathrm{vec}\,\mathbf{Z}^*)^T} = \mathbf{O}$ (the zero matrix) holds.
(4) For all $\mathbf{Z}$, $\dfrac{\partial\,\mathrm{vec}\,\mathbf{F}(\mathbf{Z})}{\partial(\mathrm{vec}\,\mathrm{Re}\mathbf{Z})^T} + \mathrm{j}\dfrac{\partial\,\mathrm{vec}\,\mathbf{F}(\mathbf{Z})}{\partial(\mathrm{vec}\,\mathrm{Im}\,\mathbf{Z})^T} = \mathbf{O}$ holds.

The complex matrix function $\mathbf{F}(\mathbf{Z},\mathbf{Z}^*)$ is obviously a holomorphic function, and its matrix differential is

$$\mathrm{d}\,\mathrm{vec}\,\mathbf{F}(\mathbf{Z},\mathbf{Z}^*) = \frac{\partial\,\mathrm{vec}\,\mathbf{F}(\mathbf{Z},\mathbf{Z}^*)}{\partial(\mathrm{vec}\,\mathbf{Z})^T}\mathrm{d}\,\mathrm{vec}\,\mathbf{Z} + \frac{\partial\,\mathrm{vec}\,\mathbf{F}(\mathbf{Z},\mathbf{Z}^*)}{\partial(\mathrm{vec}\,\mathbf{Z}^*)^T}\mathrm{d}\,\mathrm{vec}\,\mathbf{Z}^*. \tag{3.4.14}$$

The partial derivative of the holomorphic function $\mathbf{F}(\mathbf{Z},\mathbf{Z}^*)$ with respect to the real part $\mathrm{Re}\,\mathbf{Z}$ of the matrix variable $\mathbf{Z}$ is given by

$$\frac{\partial\,\mathrm{vec}\,\mathbf{F}(\mathbf{Z},\mathbf{Z}^*)}{\partial(\mathrm{vec}\,\mathrm{Re}\,\mathbf{Z})^T} = \frac{\partial\,\mathrm{vec}\,\mathbf{F}(\mathbf{Z},\mathbf{Z}^*)}{\partial(\mathrm{vec}\,\mathbf{Z})^T} + \frac{\partial\,\mathrm{vec}\,\mathbf{F}(\mathbf{Z},\mathbf{Z}^*)}{\partial(\mathrm{vec}\,\mathbf{Z}^*)^T},$$

and the partial derivative of $\mathbf{F}(\mathbf{Z},\mathbf{Z}^*)$ with respect to the imaginary part $\mathrm{Im}\,\mathbf{Z}$ of the matrix variable $\mathbf{Z}$ is as follows:

$$\frac{\partial\,\mathrm{vec}\,\mathbf{F}(\mathbf{Z},\mathbf{Z}^*)}{\partial(\mathrm{vec}\,\mathrm{Im}\mathbf{Z})^T} = \mathrm{j}\left(\frac{\partial\,\mathrm{vec}\,\mathbf{F}(\mathbf{Z},\mathbf{Z}^*)}{\partial(\mathrm{vec}\,\mathbf{Z})^T} - \frac{\partial\,\mathrm{vec}\,\mathbf{F}(\mathbf{Z},\mathbf{Z}^*)}{\partial(\mathrm{vec}\,\mathbf{Z}^*)^T}\right).$$

The complex matrix differential $\mathrm{d}\,\mathbf{Z} = [\mathrm{d}Z_{ij}]_{i=1,j=1}^{m,n}$ has the following properties [68].

1. *Transpose* $\mathrm{d}\mathbf{Z}^T = \mathrm{d}(\mathbf{Z}^T) = (\mathrm{d}\mathbf{Z})^T$.
2. *Hermitian transpose* $\mathrm{d}\mathbf{Z}^H = \mathrm{d}(\mathbf{Z}^H) = (\mathrm{d}\mathbf{Z})^H$.
3. *Conjugate* $\mathrm{d}\mathbf{Z}^* = \mathrm{d}(\mathbf{Z}^*) = (\mathrm{d}\mathbf{Z})^*$.
4. *Linearity (additive rule)* $\mathrm{d}(\mathbf{Y}+\mathbf{Z}) = \mathrm{d}\mathbf{Y} + \mathrm{d}\mathbf{Z}$.
5. *Chain rule* If $\mathbf{F}$ is a function of $\mathbf{Y}$, while $\mathbf{Y}$ is a function of $\mathbf{Z}$, then

$$\mathrm{d}\,\mathrm{vec}\,\mathbf{F} = \frac{\partial\,\mathrm{vec}\,\mathbf{F}}{\partial(\mathrm{vec}\,\mathbf{Y})^T}\mathrm{d}\,\mathrm{vec}\,\mathbf{Y} = \frac{\partial\,\mathrm{vec}\,\mathbf{F}}{\partial(\mathrm{vec}\,\mathbf{Y})^T}\frac{\partial\,\mathrm{vec}\,\mathbf{Y}}{\partial(\mathrm{vec}\,\mathbf{Z})^T}\mathrm{d}\,\mathrm{vec}\,\mathbf{Z},$$

where $\dfrac{\partial\,\mathrm{vec}\,\mathbf{F}}{\partial(\mathrm{vec}\,\mathbf{Y})^T}$ and $\dfrac{\partial\,\mathrm{vec}\,\mathbf{Y}}{\partial(\mathrm{vec}\,\mathbf{Z})^T}$ are the *normal complex partial derivative* and the *generalized complex partial derivative*, respectively.
6. *Multiplication rule*

$$\mathrm{d}(\mathbf{UV}) = (\mathrm{d}\mathbf{U})\mathbf{V} + \mathbf{U}(\mathrm{d}\mathbf{V})$$
$$\mathrm{d}\,\mathrm{vec}(\mathbf{UV}) = (\mathbf{V}^T\otimes\mathbf{I})\,\mathrm{d}\,\mathrm{vec}\,\mathbf{U} + (\mathbf{I}\otimes\mathbf{U})\,\mathrm{d}\,\mathrm{vec}\,\mathbf{V}.$$

7. *Kronecker product* $\mathrm{d}(\mathbf{Y}\otimes\mathbf{Z}) = \mathrm{d}\mathbf{Y}\otimes\mathbf{Z} + \mathbf{Y}\otimes\mathrm{d}\mathbf{Z}$.

8. *Hadamard product* $\mathrm{d}(\mathbf{Y} * \mathbf{Z}) = \mathrm{d}\mathbf{Y} * \mathbf{Z} + \mathbf{Y} * \mathrm{d}\mathbf{Z}$.

In the following we derive the relationship between the complex matrix differential and the complex partial derivative.

First, the complex differential rule for scalar variables,

$$\mathrm{d}f(z, z^*) = \frac{\partial f(z, z^*)}{\partial z}\,\mathrm{d}z + \frac{\partial f(z, z^*)}{\partial z^*}\,\mathrm{d}z^* \tag{3.4.15}$$

is easily extended to a complex differential rule for the multivariate real scalar function $f(\cdot) = f((z_1, z_1^*), \ldots, (z_m, z_m^*))$:

$$\begin{aligned}\mathrm{d}f(\cdot) &= \frac{\partial f(\cdot)}{\partial z_1}\,\mathrm{d}z_1 + \cdots + \frac{\partial f(\cdot)}{\partial z_m}\,\mathrm{d}z_m + \frac{\partial f(\cdot)}{\partial z_1^*}\,\mathrm{d}z_1^* + \cdots + \frac{\partial f(\cdot)}{\partial z_m^*}\,\mathrm{d}z_m^* \\ &= \frac{\partial f(\cdot)}{\partial \mathbf{z}^T}\,\mathrm{d}\mathbf{z} + \frac{\partial f(\cdot)}{\partial \mathbf{z}^H}\,\mathrm{d}\mathbf{z}^*. \end{aligned} \tag{3.4.16}$$

This complex differential rule is the basis of the complex matrix differential.

In particular, if $f(\cdot) = f(\mathbf{z}, \mathbf{z}^*)$, then

$$\mathrm{d}f(\mathbf{z}, \mathbf{z}^*) = \frac{\partial f(\mathbf{z}, \mathbf{z}^*)}{\partial \mathbf{z}^T}\,\mathrm{d}\mathbf{z} + \frac{\partial f(\mathbf{z}, \mathbf{z}^*)}{\partial \mathbf{z}^H}\,\mathrm{d}\mathbf{z}^*$$

or, simply denoted,

$$\mathrm{d}f(\mathbf{z}, \mathbf{z}^*) = \mathrm{D}_{\mathbf{z}} f(\mathbf{z}, \mathbf{z}^*)\,\mathrm{d}\mathbf{z} + \mathrm{D}_{\mathbf{z}^*} f(\mathbf{z}, \mathbf{z}^*)\,\mathrm{d}\mathbf{z}^*. \tag{3.4.17}$$

Here $\mathrm{d}\mathbf{z} = [\mathrm{d}z_1, \ldots, \mathrm{d}z_m]^T$, $\mathrm{d}\mathbf{z}^* = [\mathrm{d}z_1^*, \ldots, \mathrm{d}z_m^*]^T$ and

$$\mathrm{D}_{\mathbf{z}} f(\mathbf{z}, \mathbf{z}^*) = \left.\frac{\partial f(\mathbf{z}, \mathbf{z}^*)}{\partial \mathbf{z}^T}\right|_{\mathbf{z}^* = \mathrm{const}} = \left[\frac{\partial f(\mathbf{z}, \mathbf{z}^*)}{\partial z_1}, \ldots, \frac{\partial f(\mathbf{z}, \mathbf{z}^*)}{\partial z_m}\right], \tag{3.4.18}$$

$$\mathrm{D}_{\mathbf{z}^*} f(\mathbf{z}, \mathbf{z}^*) = \left.\frac{\partial f(\mathbf{z}, \mathbf{z}^*)}{\partial \mathbf{z}^H}\right|_{\mathbf{z} = \mathrm{const}} = \left[\frac{\partial f(\mathbf{z}, \mathbf{z}^*)}{\partial z_1^*}, \ldots, \frac{\partial f(\mathbf{z}, \mathbf{z}^*)}{\partial z_m^*}\right] \tag{3.4.19}$$

are respectively the *cogradient vector* and the *conjugate cogradient vector* of the real scalar function $f(\mathbf{z}, \mathbf{z}^*)$, while

$$\mathrm{D}_{\mathbf{z}} = \frac{\partial}{\partial \mathbf{z}^T} \stackrel{\text{def}}{=} \left[\frac{\partial}{\partial z_1}, \ldots, \frac{\partial}{\partial z_m}\right], \tag{3.4.20}$$

$$\mathrm{D}_{\mathbf{z}^*} = \frac{\partial}{\partial \mathbf{z}^H} \stackrel{\text{def}}{=} \left[\frac{\partial}{\partial z_1^*}, \ldots, \frac{\partial}{\partial z_m^*}\right] \tag{3.4.21}$$

are termed the *cogradient operator* and the *conjugate cogradient operator* of complex vector variable $\mathbf{z} \in \mathbb{C}^m$, respectively.

Let $\mathbf{z} = \mathbf{x} + \mathrm{j}\mathbf{y} = [z_1, \ldots, z_m]^T \in \mathbb{C}^m$ with $\mathbf{x} = [x_1, \ldots, x_m]^T \in \mathbb{R}^m, \mathbf{y} = [y_1, \ldots, y_m]^T \in \mathbb{R}^m$, namely $z_i = x_i + \mathrm{j}y_i, i = 1, \ldots, m$; the real part x_i and the imaginary part y_i are two independent variables.

Applying the complex partial derivative operators

$$\mathrm{D}_{z_i} = \frac{\partial}{\partial z_i} = \frac{1}{2}\left(\frac{\partial}{\partial x_i} - \mathrm{j}\frac{\partial}{\partial y_i}\right), \tag{3.4.22}$$

$$\mathrm{D}_{z_i^*} = \frac{\partial}{\partial z_i^*} = \frac{1}{2}\left(\frac{\partial}{\partial x_i} + \mathrm{j}\frac{\partial}{\partial y_i}\right) \tag{3.4.23}$$

to each element of the row vector $\mathbf{z}^T = [z_1, \ldots, z_m]$, we obtain the following *complex cogradient operator*

$$\mathrm{D}_{\mathbf{z}} = \frac{\partial}{\partial \mathbf{z}^T} = \frac{1}{2}\left(\frac{\partial}{\partial \mathbf{x}^T} - \mathrm{j}\frac{\partial}{\partial \mathbf{y}^T}\right) \tag{3.4.24}$$

and the *complex conjugate cogradient operator*

$$\mathrm{D}_{\mathbf{z}^*} = \frac{\partial}{\partial \mathbf{z}^H} = \frac{1}{2}\left(\frac{\partial}{\partial \mathbf{x}^T} + \mathrm{j}\frac{\partial}{\partial \mathbf{y}^T}\right). \tag{3.4.25}$$

Similarly, the *complex gradient operator* and the *complex conjugate gradient operator* are respectively defined as

$$\nabla_{\mathbf{z}} = \frac{\partial}{\partial \mathbf{z}} \stackrel{\text{def}}{=} \left[\frac{\partial}{\partial z_1}, \ldots, \frac{\partial}{\partial z_m}\right]^T, \tag{3.4.26}$$

$$\nabla_{\mathbf{z}^*} = \frac{\partial}{\partial \mathbf{z}^*} \stackrel{\text{def}}{=} \left[\frac{\partial}{\partial z_1^*}, \ldots, \frac{\partial}{\partial z_m^*}\right]^T. \tag{3.4.27}$$

Hence, the *complex gradient vector* and the *complex conjugate gradient vector* of the real scalar function $f(\mathbf{z}, \mathbf{z}^*)$ are respectively defined as

$$\nabla_{\mathbf{z}} f(\mathbf{z}, \mathbf{z}^*) = \left.\frac{\partial f(\mathbf{z}, \mathbf{z}^*)}{\partial \mathbf{z}}\right|_{\mathbf{z}^* = \text{const vector}} = (\mathrm{D}_{\mathbf{z}} f(\mathbf{z}, \mathbf{z}^*))^T \tag{3.4.28}$$

$$\nabla_{\mathbf{z}^*} f(\mathbf{z}, \mathbf{z}^*) = \left.\frac{\partial f(\mathbf{z}, \mathbf{z}^*)}{\partial \mathbf{z}^*}\right|_{\mathbf{z} = \text{const vector}} = (\mathrm{D}_{\mathbf{z}^*} f(\mathbf{z}, \mathbf{z}^*))^T. \tag{3.4.29}$$

Applying the complex partial derivative operator to each element of the complex vector $\mathbf{z} = [z_1, \ldots, z_m]^T$, we get the complex gradient operator

$$\nabla_{\mathbf{z}} = \frac{\partial}{\partial \mathbf{z}} = \frac{1}{2}\left(\frac{\partial}{\partial \mathbf{x}} - \mathrm{j}\frac{\partial}{\partial \mathbf{y}}\right) \tag{3.4.30}$$

and the complex conjugate gradient operator

$$\nabla_{\mathbf{z}^*} = \frac{\partial}{\partial \mathbf{z}^*} = \frac{1}{2}\left(\frac{\partial}{\partial \mathbf{x}} + \mathrm{j}\frac{\partial}{\partial \mathbf{y}}\right). \tag{3.4.31}$$

By the definitions of the complex gradient operator and the complex conjugate gradient operator, it is not difficult to obtain

$$\frac{\partial \mathbf{z}^T}{\partial \mathbf{z}} = \frac{\partial \mathbf{x}^T}{\partial \mathbf{z}} + \mathrm{j}\frac{\partial \mathbf{y}^T}{\partial \mathbf{z}} = \frac{1}{2}\left(\frac{\partial \mathbf{x}^T}{\partial \mathbf{x}} - \mathrm{j}\frac{\partial \mathbf{x}^T}{\partial \mathbf{y}}\right) + \mathrm{j}\frac{1}{2}\left(\frac{\partial \mathbf{y}^T}{\partial \mathbf{x}} - \mathrm{j}\frac{\partial \mathbf{y}^T}{\partial \mathbf{y}}\right) = \mathbf{I}_{m\times m},$$
$$\frac{\partial \mathbf{z}^T}{\partial \mathbf{z}^*} = \frac{\partial \mathbf{x}^T}{\partial \mathbf{z}^*} + \mathrm{j}\frac{\partial \mathbf{y}^T}{\partial \mathbf{z}^*} = \frac{1}{2}\left(\frac{\partial \mathbf{x}^T}{\partial \mathbf{x}} + \mathrm{j}\frac{\partial \mathbf{x}^T}{\partial \mathbf{y}}\right) + \mathrm{j}\frac{1}{2}\left(\frac{\partial \mathbf{y}^T}{\partial \mathbf{x}} + \mathrm{j}\frac{\partial \mathbf{y}^T}{\partial \mathbf{y}}\right) = \mathbf{O}_{m\times m}.$$

When finding the above two equations, we used $\frac{\partial \mathbf{x}^T}{\partial \mathbf{x}} = \mathbf{I}_{m\times m}$, $\frac{\partial \mathbf{x}^T}{\partial \mathbf{y}} = \mathbf{O}_{m\times m}$ and $\frac{\partial \mathbf{y}^T}{\partial \mathbf{y}} = \mathbf{I}_{m\times m}$, $\frac{\partial \mathbf{y}^T}{\partial \mathbf{x}} = \mathbf{O}_{m\times m}$, because the real part $\mathbf{x}$ and the imaginary part $\mathbf{y}$ of the complex vector variable $\mathbf{z}$ are independent.

Summarizing the above results and their conjugate, transpose and complex conjugate transpose, we have the following important results:

$$\frac{\partial \mathbf{z}^T}{\partial \mathbf{z}} = \mathbf{I}, \quad \frac{\partial \mathbf{z}^H}{\partial \mathbf{z}^*} = \mathbf{I}, \quad \frac{\partial \mathbf{z}}{\partial \mathbf{z}^T} = \mathbf{I}, \quad \frac{\partial \mathbf{z}^*}{\partial \mathbf{z}^H} = \mathbf{I}, \tag{3.4.32}$$
$$\frac{\partial \mathbf{z}^T}{\partial \mathbf{z}^*} = \mathbf{O}, \quad \frac{\partial \mathbf{z}^H}{\partial \mathbf{z}} = \mathbf{O}, \quad \frac{\partial \mathbf{z}}{\partial \mathbf{z}^H} = \mathbf{O}, \quad \frac{\partial \mathbf{z}^*}{\partial \mathbf{z}^T} = \mathbf{O}. \tag{3.4.33}$$

The above results reveal an important fact of the complex matrix differential: under the basic assumption that the real part and imaginary part of a complex vector are independent, the complex vector variable $\mathbf{z}$ and its complex conjugate vector variable $\mathbf{z}^*$ can be viewed as two independent variables. This important fact is not surprising because the angle between $\mathbf{z}$ and $\mathbf{z}^*$ is $\pi/2$, i.e., they are orthogonal to each other. Hence, we can summarize the rules for using the cogradient operator and gradient operator as follows.

(1) When using the complex cogradient operator $\partial/\partial \mathbf{z}^T$ or the complex gradient operator $\partial/\partial \mathbf{z}$, the complex conjugate vector variable $\mathbf{z}^*$ is handled as a constant vector.
(2) When using the complex conjugate cogradient operator $\partial/\partial \mathbf{z}^H$ or the complex conjugate gradient operator $\partial/\partial \mathbf{z}^*$, the vector variable $\mathbf{z}$ is handled as a constant vector.

We might regard the above rules as embodying the independent rule of complex partial derivative operators: when applying the complex partial derivative operator, the complex cogradient operator, the complex conjugate cogradient operator, the complex gradient operator or the complex conjugate gradient operator, the complex vector variables $\mathbf{z}$ and $\mathbf{z}^*$ can be handled as independent vector variables, namely, when one vector is the variable of a given function, the other can be viewed as a constant vector.

At this point, let us consider the real scalar function $f(\mathbf{Z}, \mathbf{Z}^*)$ with variables $\mathbf{Z}, \mathbf{Z}^* \in \mathbb{C}^{m\times n}$. Performing the vectorization of $\mathbf{Z}$ and $\mathbf{Z}^*$, respectively, from Equation (3.4.17) we get the first-order complex differential rule for the real scalar function $f(\mathbf{Z},\mathbf{Z}^*)$:

$$\begin{aligned}
\mathrm{d}f(\mathbf{Z},\mathbf{Z}^*) &= \frac{\partial f(\mathbf{Z},\mathbf{Z}^*)}{\partial(\mathrm{vec}\,\mathbf{Z})^T}\mathrm{d}\,\mathrm{vec}\,\mathbf{Z} + \frac{\partial f(\mathbf{Z},\mathbf{Z}^*)}{\partial(\mathrm{vec}\,\mathbf{Z}^*)^T}\mathrm{d}\,\mathrm{vec}\,\mathbf{Z}^* \\
&= \frac{\partial f(\mathbf{Z},\mathbf{Z}^*)}{\partial(\mathrm{vec}\,\mathbf{Z})^T}\mathrm{d}\,\mathrm{vec}\,\mathbf{Z} + \frac{\partial f(\mathbf{Z},\mathbf{Z}^*)}{\partial(\mathrm{vec}\,\mathbf{Z}^*)^T}\mathrm{d}\,\mathrm{vec}\,\mathbf{Z}^*, \qquad (3.4.34)
\end{aligned}$$

where

$$\begin{aligned}
\frac{\partial f(\mathbf{Z},\mathbf{Z}^*)}{\partial(\mathrm{vec}\,\mathbf{Z})^T} &= \left[\frac{\partial f(\mathbf{Z},\mathbf{Z}^*)}{\partial Z_{11}},\ldots,\frac{\partial f(\mathbf{Z},\mathbf{Z}^*)}{\partial Z_{m1}},\ldots,\frac{\partial f(\mathbf{Z},\mathbf{Z}^*)}{\partial Z_{1n}},\ldots,\frac{\partial f(\mathbf{Z},\mathbf{Z}^*)}{\partial Z_{mn}}\right], \\
\frac{\partial f(\mathbf{Z},\mathbf{Z}^*)}{\partial(\mathrm{vec}\,\mathbf{Z}^*)^T} &= \left[\frac{\partial f(\mathbf{Z},\mathbf{Z}^*)}{\partial Z_{11}^*},\ldots,\frac{\partial f(\mathbf{Z},\mathbf{Z}^*)}{\partial Z_{m1}^*},\ldots,\frac{\partial f(\mathbf{Z},\mathbf{Z}^*)}{\partial Z_{1n}^*},\ldots,\frac{\partial f(\mathbf{Z},\mathbf{Z}^*)}{\partial Z_{mn}^*}\right].
\end{aligned}$$

Now define the complex cogradient vector and the complex conjugate cogradient vector:

$$\mathrm{D}_{\mathrm{vec}\,\mathbf{Z}} f(\mathbf{Z},\mathbf{Z}^*) = \frac{\partial f(\mathbf{Z},\mathbf{Z}^*)}{\partial(\mathrm{vec}\,\mathbf{Z})^T}, \qquad (3.4.35)$$

$$\mathrm{D}_{\mathrm{vec}\,\mathbf{Z}^*} f(\mathbf{Z},\mathbf{Z}^*) = \frac{\partial f(\mathbf{Z},\mathbf{Z}^*)}{\partial(\mathrm{vec}\,\mathbf{Z}^*)^T}. \qquad (3.4.36)$$

The complex gradient vector and the complex conjugate gradient vector of the function $f(\mathbf{Z},\mathbf{Z}^*)$ are respectively defined as

$$\nabla_{\mathrm{vec}\,\mathbf{Z}} f(\mathbf{Z},\mathbf{Z}^*) = \frac{\partial f(\mathbf{Z},\mathbf{Z}^*)}{\partial\,\mathrm{vec}\,\mathbf{Z}}, \qquad (3.4.37)$$

$$\nabla_{\mathrm{vec}\,\mathbf{Z}^*} f(\mathbf{Z},\mathbf{Z}^*) = \frac{\partial f(\mathbf{Z},\mathbf{Z}^*)}{\partial\,\mathrm{vec}\,\mathbf{Z}^*}. \qquad (3.4.38)$$

The conjugate gradient vector $\nabla_{\mathrm{vec}\,\mathbf{Z}^*} f(\mathbf{Z},\mathbf{Z}^*)$ has the following properties [68]:

(1) The conjugate gradient vector of the function $f(\mathbf{Z},\mathbf{Z}^*)$ at an extreme point is equal to the zero vector, i.e., $\nabla_{\mathrm{vec}\,\mathbf{Z}^*} f(\mathbf{Z},\mathbf{Z}^*) = \mathbf{0}$.
(2) The conjugate gradient vector $\nabla_{\mathrm{vec}\,\mathbf{Z}^*} f(\mathbf{Z},\mathbf{Z}^*)$ and the negative conjugate gradient vector $-\nabla_{\mathrm{vec}\,\mathbf{Z}^*} f(\mathbf{Z},\mathbf{Z}^*)$ point in the direction of the steepest ascent and steepest descent of the function $f(\mathbf{Z},\mathbf{Z}^*)$, respectively.
(3) The step length of the steepest increase slope is $\|\nabla_{\mathrm{vec}\,\mathbf{Z}^*} f(\mathbf{Z},\mathbf{Z}^*)\|_2$.
(4) The conjugate gradient vector $\nabla_{\mathrm{vec}\,\mathbf{Z}^*} f(\mathbf{Z},\mathbf{Z}^*)$ is the normal to the surface $f(\mathbf{Z},\mathbf{Z}^*) = \mathrm{const}$. Hence, the conjugate gradient vector $\nabla_{\mathrm{vec}\,\mathbf{Z}^*} f(\mathbf{Z},\mathbf{Z}^*)$ and the negative conjugate gradient vector $-\nabla_{\mathrm{vec}\,\mathbf{Z}^*} f(\mathbf{Z},\mathbf{Z}^*)$ can be used separately in gradient ascent algorithms and gradient descent algorithms.

Furthermore, the *complex Jacobian matrix* and the *complex conjugate Jacobian* *matrix* of the real scalar function $f(\mathbf{Z}, \mathbf{Z}^*)$ are respectively as follows:

$$\mathrm{D}_{\mathbf{Z}} f(\mathbf{Z}, \mathbf{Z}^*) \stackrel{\text{def}}{=} \left.\frac{\partial f(\mathbf{Z}, \mathbf{Z}^*)}{\partial \mathbf{Z}^T}\right|_{\mathbf{Z}^* = \text{const matrix}} = \begin{bmatrix} \frac{\partial f(\mathbf{Z}, \mathbf{Z}^*)}{\partial Z_{11}} & \cdots & \frac{\partial f(\mathbf{Z}, \mathbf{Z}^*)}{\partial Z_{m1}} \\ \vdots & \ddots & \vdots \\ \frac{\partial f(\mathbf{Z}, \mathbf{Z}^*)}{\partial Z_{1n}} & \cdots & \frac{\partial f(\mathbf{Z}, \mathbf{Z}^*)}{\partial Z_{mn}} \end{bmatrix}, \tag{3.4.39}$$

$$\mathrm{D}_{\mathbf{Z}^*} f(\mathbf{Z}, \mathbf{Z}^*) \stackrel{\text{def}}{=} \left.\frac{\partial f(\mathbf{Z}, \mathbf{Z}^*)}{\partial \mathbf{Z}^H}\right|_{\mathbf{Z} = \text{const matrix}} = \begin{bmatrix} \frac{\partial f(\mathbf{Z}, \mathbf{Z}^*)}{\partial Z_{11}^*} & \cdots & \frac{\partial f(\mathbf{Z}, \mathbf{Z}^*)}{\partial Z_{m1}^*} \\ \vdots & \ddots & \vdots \\ \frac{\partial f(\mathbf{Z}, \mathbf{Z}^*)}{\partial Z_{1n}^*} & \cdots & \frac{\partial f(\mathbf{Z}, \mathbf{Z}^*)}{\partial Z_{mn}^*} \end{bmatrix}. \tag{3.4.40}$$

Similarly, the complex gradient matrix and the complex conjugate gradient matrix of the real scalar function $f(\mathbf{Z}, \mathbf{Z}^*)$ are respectively given by

$$\nabla_{\mathbf{Z}} f(\mathbf{Z}, \mathbf{Z}^*) \stackrel{\text{def}}{=} \left.\frac{\partial f(\mathbf{Z}, \mathbf{Z}^*)}{\partial \mathbf{Z}}\right|_{\mathbf{Z}^* = \text{const matrix}} = \begin{bmatrix} \frac{\partial f(\mathbf{Z}, \mathbf{Z}^*)}{\partial Z_{11}} & \cdots & \frac{\partial f(\mathbf{Z}, \mathbf{Z}^*)}{\partial Z_{1n}} \\ \vdots & \ddots & \vdots \\ \frac{\partial f(\mathbf{Z}, \mathbf{Z}^*)}{\partial Z_{m1}} & \cdots & \frac{\partial f(\mathbf{Z}, \mathbf{Z}^*)}{\partial Z_{mn}} \end{bmatrix}, \tag{3.4.41}$$

$$\nabla_{\mathbf{Z}^*} f(\mathbf{Z}, \mathbf{Z}^*) \stackrel{\text{def}}{=} \left.\frac{\partial f(\mathbf{Z}, \mathbf{Z}^*)}{\partial \mathbf{Z}^*}\right|_{\mathbf{Z} = \text{const matrix}} = \begin{bmatrix} \frac{\partial f(\mathbf{Z}, \mathbf{Z}^*)}{\partial Z_{11}^*} & \cdots & \frac{\partial f(\mathbf{Z}, \mathbf{Z}^*)}{\partial Z_{1n}^*} \\ \vdots & \ddots & \vdots \\ \frac{\partial f(\mathbf{Z}, \mathbf{Z}^*)}{\partial Z_{m1}^*} & \cdots & \frac{\partial f(\mathbf{Z}, \mathbf{Z}^*)}{\partial Z_{mn}^*} \end{bmatrix}. \tag{3.4.42}$$

Summarizing the definitions above, there are the following relations among the various complex partial derivatives of the real scalar function $f(\mathbf{Z}, \mathbf{Z})$:

(1) The conjugate gradient (cogradient) vector is equal to the complex conjugate of the gradient (cogradient) vector; and the conjugate Jacobian (gradient) matrix is equal to the complex conjugate of the Jacobian (gradient) matrix.

(2) The gradient (conjugate gradient) vector is equal to the transpose of the co-

gradient (conjugate cogradient) vector, namely

$$\nabla_{\text{vec}\,\mathbf{Z}} f(\mathbf{Z},\mathbf{Z}^*) = \mathrm{D}^T_{\text{vec}\,\mathbf{Z}} f(\mathbf{Z},\mathbf{Z}^*), \tag{3.4.43}$$

$$\nabla_{\text{vec}\,\mathbf{Z}^*} f(\mathbf{Z},\mathbf{Z}^*) = \mathrm{D}^T_{\text{vec}\,\mathbf{Z}^*} f(\mathbf{Z},\mathbf{Z}^*). \tag{3.4.44}$$

(3) The cogradient (conjugate cogradient) vector is equal to the transpose of the vectorization of Jacobian (conjugate Jacobian) matrix:

$$\mathrm{D}_{\text{vec}\,\mathbf{Z}} f(\mathbf{Z},\mathbf{Z}) = \left(\text{vec}\,\mathrm{D}_{\mathbf{Z}} f(\mathbf{Z},\mathbf{Z}^*)\right)^T, \tag{3.4.45}$$

$$\mathrm{D}_{\text{vec}\,\mathbf{Z}^*} f(\mathbf{Z},\mathbf{Z}) = \left(\text{vec}\,\mathrm{D}_{\mathbf{Z}^*} f(\mathbf{Z},\mathbf{Z}^*)\right)^T. \tag{3.4.46}$$

(4) The gradient (conjugate gradient) matrix is equal to the transpose of the Jacobian (conjugate Jacobian) matrix:

$$\nabla_{\mathbf{Z}} f(\mathbf{Z},\mathbf{Z}^*) = \mathrm{D}^T_{\mathbf{Z}} f(\mathbf{Z},\mathbf{Z}^*), \tag{3.4.47}$$

$$\nabla_{\mathbf{Z}^*} f(\mathbf{Z},\mathbf{Z}^*) = \mathrm{D}^T_{\mathbf{Z}^*} f(\mathbf{Z},\mathbf{Z}^*). \tag{3.4.48}$$

The following are the rules of operation for the complex gradient.

1. If $f(\mathbf{Z},\mathbf{Z}^*) = c$ (a constant), then its gradient matrix and conjugate gradient matrix are equal to the zero matrix, namely $\partial c/\partial\,\mathbf{Z} = \mathbf{O}$ and $\partial c/\partial\,\mathbf{Z}^* = \mathbf{O}$.
2. *Linear rule* If $f(\mathbf{Z},\mathbf{Z}^*)$ and $g(\mathbf{Z},\mathbf{Z}^*)$ are scalar functions, and c_1 and c_2 are complex numbers, then

$$\frac{\partial(c_1 f(\mathbf{Z},\mathbf{Z}^*) + c_2 g(\mathbf{Z},\mathbf{Z}^*))}{\partial\,\mathbf{Z}^*} = c_1\frac{\partial f(\mathbf{Z},\mathbf{Z}^*)}{\partial\,\mathbf{Z}^*} + c_2\frac{\partial g(\mathbf{Z},\mathbf{Z}^*)}{\partial\,\mathbf{Z}^*}.$$

3. *Multiplication rule*

$$\frac{\partial f(\mathbf{Z},\mathbf{Z}^*) g(\mathbf{Z},\mathbf{Z}^*)}{\partial \mathbf{Z}^*} = g(\mathbf{Z},\mathbf{Z}^*)\frac{\partial f(\mathbf{Z},\mathbf{Z}^*)}{\partial\,\mathbf{Z}^*} + f(\mathbf{Z},\mathbf{Z}^*)\frac{\partial g(\mathbf{Z},\mathbf{Z}^*)}{\partial\,\mathbf{Z}^*}.$$

4. *Quotient rule* If $g(\mathbf{Z},\mathbf{Z}^*) \neq 0$ then

$$\frac{\partial f/g}{\partial\,\mathbf{Z}^*} = \frac{1}{g^2(\mathbf{Z},\mathbf{Z}^*)}\left[g(\mathbf{Z},\mathbf{Z}^*)\frac{\partial f(\mathbf{Z},\mathbf{Z}^*)}{\partial\,\mathbf{Z}^*} - f(\mathbf{Z},\mathbf{Z}^*)\frac{\partial g(\mathbf{Z},\mathbf{Z}^*)}{\partial\,\mathbf{Z}^*}\right].$$

If $h(\mathbf{Z},\mathbf{Z}^*) = g(\mathbf{F}(\mathbf{Z},\mathbf{Z}^*),\mathbf{F}^*(\mathbf{Z},\mathbf{Z}^*))$ then the quotient rule becomes

$$\begin{aligned}\frac{\partial h(\mathbf{Z},\mathbf{Z}^*)}{\partial\,\text{vec}\,\mathbf{Z}} &= \frac{\partial g(\mathbf{F}(\mathbf{Z},\mathbf{Z}^*),\mathbf{F}^*(\mathbf{Z},\mathbf{Z}^*))}{\partial\,(\text{vec}\,\mathbf{F}(\mathbf{Z},\mathbf{Z}^*))^T}\frac{\partial\,(\text{vec}\,\mathbf{F}(\mathbf{Z},\mathbf{Z}^*))^T}{\partial\,\text{vec}\,\mathbf{Z}} \\ &\quad + \frac{\partial g(\mathbf{F}(\mathbf{Z},\mathbf{Z}^*),\mathbf{F}^*(\mathbf{Z},\mathbf{Z}^*))}{\partial\,(\text{vec}\,\mathbf{F}^*(\mathbf{Z},\mathbf{Z}^*))^T}\frac{\partial\,(\text{vec}\,\mathbf{F}^*(\mathbf{Z},\mathbf{Z}^*))^T}{\partial\,\text{vec}\,\mathbf{Z}},\end{aligned} \tag{3.4.49}$$

$$\begin{aligned}\frac{\partial h(\mathbf{Z},\mathbf{Z}^*)}{\partial\,\text{vec}\,\mathbf{Z}^*} &= \frac{\partial g(\mathbf{F}(\mathbf{Z},\mathbf{Z}^*),\mathbf{F}^*(\mathbf{Z},\mathbf{Z}^*))}{\partial\,(\text{vec}\,\mathbf{F}(\mathbf{Z},\mathbf{Z}^*))^T}\frac{\partial\,(\text{vec}\mathbf{F}(\mathbf{Z},\mathbf{Z}^*))^T}{\partial\,\text{vec}\,\mathbf{Z}^*} \\ &\quad + \frac{\partial g(\mathbf{F}(\mathbf{Z},\mathbf{Z}^*),\mathbf{F}^*(\mathbf{Z},\mathbf{Z}^*))}{\partial\,(\text{vec}\mathbf{F}^*(\mathbf{Z},\mathbf{Z}^*))^T}\frac{\partial\,(\text{vec}\,\mathbf{F}^*(\mathbf{Z},\mathbf{Z}^*))^T}{\partial\,\text{vec}\,\mathbf{Z}^*}.\end{aligned} \tag{3.4.50}$$

3.4.3 Complex Gradient Matrix Identification

If we let

$$\mathbf{A} = \mathrm{D}_{\mathbf{Z}} f(\mathbf{Z}, \mathbf{Z}^*) \quad \text{and} \quad \mathbf{B} = \mathrm{D}_{\mathbf{Z}^*} f(\mathbf{Z}, \mathbf{Z}^*), \tag{3.4.51}$$

then

$$\frac{\partial f(\mathbf{Z}, \mathbf{Z}^*)}{\partial (\mathrm{vec}\, \mathbf{Z})^T} = \mathrm{rvec}\, \mathrm{D}_{\mathbf{Z}} f(\mathbf{Z}, \mathbf{Z}^*) = \mathrm{rvec}\, \mathbf{A} = (\mathrm{vec}(\mathbf{A}^T))^T, \tag{3.4.52}$$

$$\frac{\partial f(\mathbf{Z}, \mathbf{Z}^*)}{\partial (\mathrm{vec}\, \mathbf{Z})^H} = \mathrm{rvec}\, \mathrm{D}_{\mathbf{Z}^*} f(\mathbf{Z}, \mathbf{Z}^*) = \mathrm{rvec}\, \mathbf{B} = (\mathrm{vec}(\mathbf{B}^T))^T. \tag{3.4.53}$$

Hence, the first-order complex matrix differential formula (3.4.34) can be rewritten as

$$\mathrm{d}f(\mathbf{Z}, \mathbf{Z}^*) = (\mathrm{vec}(\mathbf{A}^T))^T \mathrm{d}\,\mathrm{vec}\, \mathbf{Z} + (\mathrm{vec}(\mathbf{B}^T))^T \mathrm{d}\,\mathrm{vec}\, \mathbf{Z}^*. \tag{3.4.54}$$

Using $\mathrm{tr}(\mathbf{C}^T \mathbf{D}) = (\mathrm{vec}\, \mathbf{C})^T \mathrm{vec}\, \mathbf{D}$, Equation (3.4.54) can be written as

$$\mathrm{d}f(\mathbf{Z}, \mathbf{Z}^*) = \mathrm{tr}(\mathbf{A}\mathrm{d}\mathbf{Z} + \mathbf{B}\mathrm{d}\mathbf{Z}^*). \tag{3.4.55}$$

From Equations (3.4.51)–(3.4.55) we can obtain the following proposition for identifying the complex Jacobian matrix and the complex gradient matrix.

Proposition 3.4 *Given a scalar function $f(\mathbf{Z}, \mathbf{Z}^*) : \mathbb{C}^{m\times n} \times \mathbb{C}^{m\times n} \to \mathbb{C}$, its complex Jacobian and gradient matrices can be respectively identified by*

$$\mathrm{d}f(\mathbf{Z}, \mathbf{Z}^*) = \mathrm{tr}(\mathbf{A}\mathrm{d}\mathbf{Z} + \mathbf{B}\mathrm{d}\mathbf{Z}^*) \quad \Leftrightarrow \quad \begin{cases} \mathrm{D}_{\mathbf{Z}} f(\mathbf{Z}, \mathbf{Z}^*) = \mathbf{A}, \\ \mathrm{D}_{\mathbf{Z}^*} f(\mathbf{Z}, \mathbf{Z}^*) = \mathbf{B} \end{cases} \tag{3.4.56}$$

$$\mathrm{d}f(\mathbf{Z}, \mathbf{Z}^*) = \mathrm{tr}(\mathbf{A}\mathrm{d}\mathbf{Z} + \mathbf{B}\mathrm{d}\mathbf{Z}^*) \quad \Leftrightarrow \quad \begin{cases} \nabla_{\mathbf{Z}} f(\mathbf{Z}, \mathbf{Z}^*) = \mathbf{A}^T, \\ \nabla_{\mathbf{Z}^*} f(\mathbf{Z}, \mathbf{Z}^*) = \mathbf{B}^T. \end{cases} \tag{3.4.57}$$

That is to say, the complex gradient matrix and the complex conjugate gradient matrix are respectively identified as the transposes of the matrices $\mathbf{A}$ *and* $\mathbf{B}$.

This proposition shows that the key to identifying a complex Jacobian or gradient matrix is to write the matrix differential of the function $f(\mathbf{Z}, \mathbf{Z}^*)$ in the canonical form $\mathrm{d}f(\mathbf{Z}, \mathbf{Z}^*) = \mathrm{tr}(\mathbf{A}\mathrm{d}\mathbf{Z} + \mathbf{B}\mathrm{d}\mathbf{Z}^*)$. In particular, if $f(\mathbf{Z}, \mathbf{Z}^*)$ is a real function then $\mathbf{B} = \mathbf{A}^*$.

Example 3.21 The matrix differential of $\mathrm{tr}(\mathbf{Z}\mathbf{A}\mathbf{Z}^*\mathbf{B})$ is given by

$$\begin{aligned} \mathrm{d}(\mathrm{tr}(\mathbf{Z}\mathbf{A}\mathbf{Z}^*\mathbf{B})) &= \mathrm{tr}((\mathrm{d}\mathbf{Z})\mathbf{A}\mathbf{Z}^*\mathbf{B}) + \mathrm{tr}(\mathbf{Z}\mathbf{A}(\mathrm{d}\mathbf{Z}^*)\mathbf{B}) \\ &= \mathrm{tr}(\mathbf{A}\mathbf{Z}^*\mathbf{B}\mathrm{d}\mathbf{Z}) + \mathrm{tr}(\mathbf{B}\mathbf{Z}\mathbf{A}\mathrm{d}\mathbf{Z}^*) \end{aligned}$$

from which it follows that the gradient matrix and the complex conjugate gradient matrix of $\text{tr}(\mathbf{ZAZ}^*\mathbf{B})$ are respectively given by

$$\nabla_{\mathbf{Z}}\text{tr}(\mathbf{ZAZ}^*\mathbf{B}) = (\mathbf{AZ}^*\mathbf{B})^T = \mathbf{B}^T\mathbf{Z}^H\mathbf{A}^T,$$
$$\nabla_{\mathbf{Z}^*}\text{tr}(\mathbf{ZAZ}^*\mathbf{B}) = (\mathbf{BZA})^T = \mathbf{A}^T\mathbf{Z}^T\mathbf{B}^T.$$

Example 3.22 From the complex matrix differentials

$$\begin{aligned} \text{d}|\mathbf{ZZ}^*| &= |\mathbf{ZZ}^*|\text{tr}((\mathbf{ZZ}^*)^{-1}\text{d}(\mathbf{ZZ}^*)) \\ &= |\mathbf{ZZ}^*|\text{tr}(\mathbf{Z}^*(\mathbf{ZZ}^*)^{-1}\text{d}\mathbf{Z}) + |\mathbf{ZZ}^*|\text{tr}((\mathbf{ZZ}^*)^{-1}\mathbf{Z}\text{d}\mathbf{Z}^*), \\ \text{d}|\mathbf{ZZ}^H| &= |\mathbf{ZZ}^H|\text{tr}((\mathbf{ZZ}^H)^{-1}\text{d}(\mathbf{ZZ}^H)) \\ &= |\mathbf{ZZ}^H|\text{tr}(\mathbf{Z}^H(\mathbf{ZZ}^H)^{-1}\text{d}\mathbf{Z}) + |\mathbf{ZZ}^H|\text{tr}(\mathbf{Z}^T(\mathbf{Z}^*\mathbf{Z}^T)^{-1}\text{d}\mathbf{Z}^*), \end{aligned}$$

we get the complex gradient matrix and the complex conjugate gradient matrix as follows:

$$\nabla_{\mathbf{Z}}|\mathbf{ZZ}^*| = |\mathbf{ZZ}^*|(\mathbf{Z}^H\mathbf{Z}^T)^{-1}\mathbf{Z}^H, \quad \nabla_{\mathbf{Z}^*}|\mathbf{ZZ}^*| = |\mathbf{ZZ}^*|\mathbf{Z}^T(\mathbf{Z}^H\mathbf{Z}^T)^{-1},$$
$$\nabla_{\mathbf{Z}}|\mathbf{ZZ}^H| = |\mathbf{ZZ}^H|(\mathbf{Z}^*\mathbf{Z}^T)^{-1}\mathbf{Z}^*, \quad \nabla_{\mathbf{Z}^*}|\mathbf{ZZ}^H| = |\mathbf{ZZ}^H|(\mathbf{ZZ}^H)^{-1}\mathbf{Z}.$$

Table 3.9 lists the complex gradient matrices of several trace functions.

Table 3.9 Complex gradient matrices of trace functions

$f(\mathbf{Z},\mathbf{Z}^*)$	$\text{d}f$	$\partial f/\partial \mathbf{Z}$	$\partial f/\partial \mathbf{Z}^*$
$\text{tr}(\mathbf{AZ})$	$\text{tr}(\mathbf{A}\text{d}\mathbf{Z})$	$\mathbf{A}^T$	$\mathbf{O}$
$\text{tr}(\mathbf{AZ}^H)$	$\text{tr}(\mathbf{A}^T\text{d}\mathbf{Z}^*)$	$\mathbf{O}$	$\mathbf{A}$
$\text{tr}(\mathbf{ZAZ}^T\mathbf{B})$	$\text{tr}((\mathbf{AZ}^T\mathbf{B}+\mathbf{A}^T\mathbf{Z}^T\mathbf{B}^T)\text{d}\mathbf{Z})$	$\mathbf{B}^T\mathbf{ZA}^T+\mathbf{BZA}$	$\mathbf{O}$
$\text{tr}(\mathbf{ZAZB})$	$\text{tr}((\mathbf{AZB}+\mathbf{BZA})\text{d}\mathbf{Z})$	$(\mathbf{AZB}+\mathbf{BZA})^T$	$\mathbf{O}$
$\text{tr}(\mathbf{ZAZ}^*\mathbf{B})$	$\text{tr}(\mathbf{AZ}^*\mathbf{B}\text{d}\text{d}\mathbf{Z}+\mathbf{BZA}\text{d}\text{d}\mathbf{Z}^*)$	$\mathbf{B}^T\mathbf{Z}^H\mathbf{A}^T$	$\mathbf{A}^T\mathbf{Z}^T\mathbf{B}^T$
$\text{tr}(\mathbf{ZAZ}^H\mathbf{B})$	$\text{tr}(\mathbf{AZ}^H\mathbf{B}\text{d}\mathbf{Z}+\mathbf{A}^T\mathbf{Z}^T\mathbf{B}^T\text{d}\mathbf{Z}^*)$	$\mathbf{B}^T\mathbf{Z}^*\mathbf{A}^T$	$\mathbf{BZA}$
$\text{tr}(\mathbf{AZ}^{-1})$	$-\text{tr}(\mathbf{Z}^{-1}\mathbf{AZ}^{-1}\text{d}\mathbf{Z})$	$-\mathbf{Z}^{-T}\mathbf{A}^T\mathbf{Z}^{-T}$	$\mathbf{O}$
$\text{tr}(\mathbf{Z}^k)$	$k\text{tr}(\mathbf{Z}^{k-1}\text{d}\mathbf{Z})$	$k(\mathbf{Z}^T)^{k-1}$	$\mathbf{O}$

Table 3.10 lists the complex gradient matrices of several determinant functions.

If $\mathbf{f}(\mathbf{z},\mathbf{z}^*) = [f_1(\mathbf{z},\mathbf{z}^*),\ldots,f_n(\mathbf{z},\mathbf{z}^*)]^T$ is an $n \times 1$ complex vector function with $m \times 1$ complex vector variable then

$$\begin{bmatrix} \text{d}f_1(\mathbf{z},\mathbf{z}^*) \\ \vdots \\ \text{d}f_n(\mathbf{z},\mathbf{z}^*) \end{bmatrix} = \begin{bmatrix} \text{D}_{\mathbf{z}}f_1(\mathbf{z},\mathbf{z}^*) \\ \vdots \\ \text{D}_{\mathbf{z}}f_n(\mathbf{z},\mathbf{z}^*) \end{bmatrix} \text{d}\mathbf{z} + \begin{bmatrix} \text{D}_{\mathbf{z}^*}f_1(\mathbf{z},\mathbf{z}^*) \\ \vdots \\ \text{D}_{\mathbf{z}^*}f_n(\mathbf{z},\mathbf{z}^*) \end{bmatrix} \text{d}\mathbf{z}^*,$$

which can simply be written as

$$\text{d}\mathbf{f}(\mathbf{z},\mathbf{z}^*) = \text{D}_{\mathbf{z}}\mathbf{f}(\mathbf{z},\mathbf{z}^*)\text{d}\mathbf{z} + \text{D}_{\mathbf{z}^*}\mathbf{f}(\mathbf{z},\mathbf{z}^*)\text{d}\mathbf{z}^*, \tag{3.4.58}$$

Table 3.10 Complex gradient matrices of determinant functions

$f(\mathbf{Z},\mathbf{Z}^*)$	$\mathrm{d}f$	$\partial f/\partial \mathbf{Z}$	$\partial f/\partial \mathbf{Z}^*$
$\lvert\mathbf{Z}\rvert$	$\lvert\mathbf{Z}\rvert\mathrm{tr}(\mathbf{Z}^{-1}\mathrm{d}\mathbf{Z})$	$\lvert\mathbf{Z}\rvert\mathbf{Z}^{-T}$	$\mathbf{O}$
$\lvert\mathbf{Z}\mathbf{Z}^T\rvert$	$2\lvert\mathbf{Z}\mathbf{Z}^T\rvert\mathrm{tr}(\mathbf{Z}^T(\mathbf{Z}\mathbf{Z}^T)^{-1}\mathrm{d}\mathbf{Z})$	$2\lvert\mathbf{Z}\mathbf{Z}^T\rvert(\mathbf{Z}\mathbf{Z}^T)^{-1}\mathbf{Z}$	$\mathbf{O}$
$\lvert\mathbf{Z}^T\mathbf{Z}\rvert$	$2\lvert\mathbf{Z}^T\mathbf{Z}\rvert\,\mathrm{tr}((\mathbf{Z}^T\mathbf{Z})^{-1}\mathbf{Z}^T\mathrm{d}\mathbf{Z})$	$2\lvert\mathbf{Z}^T\mathbf{Z}\rvert\mathbf{Z}(\mathbf{Z}^T\mathbf{Z})^{-1}$	$\mathbf{O}$
$\lvert\mathbf{Z}\mathbf{Z}^*\rvert$	$\mathbf{Z}\mathbf{Z}^*\rvert\mathrm{tr}(\mathbf{Z}^*(\mathbf{Z}\mathbf{Z}^*)^{-1}\mathrm{d}\mathbf{Z}$ $+(\mathbf{Z}\mathbf{Z}^*)^{-1}\mathbf{Z}\mathrm{d}\mathbf{Z}^*)$	$\lvert\mathbf{Z}\mathbf{Z}^*\rvert(\mathbf{Z}^H\mathbf{Z}^T)^{-1}\mathbf{Z}^H$	$\lvert\mathbf{Z}\mathbf{Z}^*\rvert\mathbf{Z}^T(\mathbf{Z}^H\mathbf{Z}^T)^{-1}$
$\lvert\mathbf{Z}^*\mathbf{Z}\rvert$	$\lvert\mathbf{Z}^*\mathbf{Z}\rvert\,\mathrm{tr}((\mathbf{Z}^*\mathbf{Z})^{-1}\mathbf{Z}^*\mathrm{d}\mathbf{Z}$ $+\mathbf{Z}(\mathbf{Z}^*\mathbf{Z})^{-1}\mathrm{d}\mathbf{Z}^*)$	$\lvert\mathbf{Z}^*\mathbf{Z}\rvert\mathbf{Z}^H(\mathbf{Z}^T\mathbf{Z}^H)^{-1}$	$\lvert\mathbf{Z}^*\mathbf{Z}\rvert(\mathbf{Z}^T\mathbf{Z}^H)^{-1}\mathbf{Z}^T$
$\lvert\mathbf{Z}\mathbf{Z}^H\rvert$	$\lvert\mathbf{Z}\mathbf{Z}^H\rvert\,\mathrm{tr}(\mathbf{Z}^H(\mathbf{Z}\mathbf{Z}^H)^{-1}\mathrm{d}\mathbf{Z}$ $+\mathbf{Z}^T(\mathbf{Z}^*\mathbf{Z}^T)^{-1}\mathrm{d}\mathbf{Z}^*)$	$\lvert\mathbf{Z}\mathbf{Z}^H\rvert(\mathbf{Z}^*\mathbf{Z}^T)^{-1}\mathbf{Z}^*$	$\lvert\mathbf{Z}\mathbf{Z}^H\rvert(\mathbf{Z}\mathbf{Z}^H)^{-1}\mathbf{Z}$
$\lvert\mathbf{Z}^H\mathbf{Z}\rvert$	$\lvert\mathbf{Z}^H\mathbf{Z}\rvert\,\mathrm{tr}((\mathbf{Z}^H\mathbf{Z})^{-1}\mathbf{Z}^H\mathrm{d}\mathbf{Z}$ $+(\mathbf{Z}^T\mathbf{Z}^*)^{-1}\mathbf{Z}^T\mathrm{d}\mathbf{Z}^*)$	$\lvert\mathbf{Z}^H\mathbf{Z}\rvert\mathbf{Z}^*(\mathbf{Z}^T\mathbf{Z}^*)^{-1}$	$\lvert\mathbf{Z}^H\mathbf{Z}\rvert\mathbf{Z}(\mathbf{Z}^H\mathbf{Z})^{-1}$
$\lvert\mathbf{Z}^k\rvert$	$k\lvert\mathbf{Z}\rvert^k\,\mathrm{tr}(\mathbf{Z}^{-1}\mathrm{d}\mathbf{Z})$	$k\lvert\mathbf{Z}\rvert^k\mathbf{Z}^{-T}$	$\mathbf{O}$

where $\mathrm{d}\mathbf{f}(\mathbf{z},\mathbf{z}^*) = [\mathrm{d}f_1(\mathbf{z},\mathbf{z}^*),\ldots,\mathrm{d}f_n(\mathbf{z},\mathbf{z}^*)]^T$, while

$$\mathrm{D}_{\mathbf{z}}\mathbf{f}(\mathbf{z},\mathbf{z}^*) = \frac{\partial\,\mathbf{f}(\mathbf{z},\mathbf{z}^*)}{\partial\,\mathbf{z}^T} = \begin{bmatrix} \frac{\partial f_1(\mathbf{z},\mathbf{z}^*)}{\partial z_1} & \cdots & \frac{\partial f_1(\mathbf{z},\mathbf{z}^*)}{\partial z_m} \\ \vdots & \ddots & \vdots \\ \frac{\partial f_n(\mathbf{z},\mathbf{z}^*)}{\partial z_1} & \cdots & \frac{\partial f_n(\mathbf{z},\mathbf{z}^*)}{\partial z_m} \end{bmatrix}, \tag{3.4.59}$$

$$\mathrm{D}_{\mathbf{z}^*}\mathbf{f}(\mathbf{z},\mathbf{z}^*) = \frac{\partial\,\mathbf{f}(\mathbf{z},\mathbf{z}^*)}{\partial\,\mathbf{z}^H} = \begin{bmatrix} \frac{\partial f_1(\mathbf{z},\mathbf{z}^*)}{\partial z_1^*} & \cdots & \frac{\partial f_1(\mathbf{z},\mathbf{z}^*)}{\partial z_m^*} \\ \vdots & \ddots & \vdots \\ \frac{\partial f_n(\mathbf{z},\mathbf{z}^*)}{\partial z_1^*} & \cdots & \frac{\partial f_n(\mathbf{z},\mathbf{z}^*)}{\partial z_m^*} \end{bmatrix} \tag{3.4.60}$$

are respectively the complex Jacobian matrix and the complex conjugate Jacobian matrix of the vector function $\mathbf{f}(\mathbf{z},\mathbf{z}^*)$.

The above result is easily extended to the case of complex matrix functions as follows. For a $p\times q$ matrix function $\mathbf{F}(\mathbf{Z},\mathbf{Z}^*)$ with $m\times n$ complex matrix variable $\mathbf{Z}$, if $\mathbf{F}(\mathbf{Z},\mathbf{Z}^*) = [\mathbf{f}_1(\mathbf{Z},\mathbf{Z}^*),\ldots,\mathbf{f}_q(\mathbf{Z},\mathbf{Z}^*)]$, then $\mathrm{d}\mathbf{F}(\mathbf{Z},\mathbf{Z}^*) = [\mathrm{d}\mathbf{f}_1(\mathbf{Z},\mathbf{Z}^*),\ldots,\mathrm{d}\mathbf{f}_q(\mathbf{Z},\mathbf{Z}^*)]$, and (3.4.58) holds for the vector functions $\mathbf{f}_i(\mathbf{Z},\mathbf{Z}^*), i=1,\ldots,q$. This implies that

$$\begin{bmatrix} \mathrm{d}\mathbf{f}_1(\mathbf{Z},\mathbf{Z}^*) \\ \vdots \\ \mathrm{d}\mathbf{f}_q(\mathbf{Z},\mathbf{Z}^*) \end{bmatrix} = \begin{bmatrix} \mathrm{D}_{\mathrm{vec}\,\mathbf{Z}}\mathbf{f}_1(\mathbf{Z},\mathbf{Z}^*) \\ \vdots \\ \mathrm{D}_{\mathrm{vec}\,\mathbf{Z}}\mathbf{f}_q(\mathbf{Z},\mathbf{Z}^*) \end{bmatrix} \mathrm{d}\,\mathrm{vec}\,\mathbf{Z} + \begin{bmatrix} \mathrm{D}_{\mathrm{vec}\,\mathbf{Z}^*}\mathbf{f}_1(\mathbf{Z},\mathbf{Z}^*) \\ \vdots \\ \mathrm{D}_{\mathrm{vec}\,\mathbf{Z}^*}\mathbf{f}_q(\mathbf{Z},\mathbf{Z}^*) \end{bmatrix} \mathrm{d}\,\mathrm{vec}\,\mathbf{Z}^*, \tag{3.4.61}$$

where

$$\mathrm{D}_{\mathrm{vec}\,\mathbf{Z}}\mathbf{f}_i(\mathbf{Z},\mathbf{Z}^*) = \frac{\partial\,\mathbf{f}_i(\mathbf{Z},\mathbf{Z}^*)}{\partial(\mathrm{vec}\,\mathbf{Z})^T} \in \mathbb{C}^{p\times mn},$$

$$\mathrm{D}_{\mathrm{vec}\,\mathbf{Z}^*}\mathbf{f}_i(\mathbf{Z},\mathbf{Z}^*) = \frac{\partial\,\mathbf{f}_i(\mathbf{Z},\mathbf{Z}^*)}{\partial(\mathrm{vec}\,\mathbf{Z}^*)^T} \in \mathbb{C}^{p\times mn}.$$

Equation (3.4.61) can be simply rewritten as

$$\mathrm{d}\,\mathrm{vec}\,\mathbf{F}(\mathbf{Z},\mathbf{Z}^*) = \mathbf{A}\,\mathrm{d}\,\mathrm{vec}\,\mathbf{Z} + \mathbf{B}\,\mathrm{d}\,\mathrm{vec}\,\mathbf{Z}^* \in \mathbb{C}^{pq}, \tag{3.4.62}$$

where

$$\begin{aligned}
\mathrm{d}\,\mathrm{vec}\,\mathbf{F}(\mathbf{Z},\mathbf{Z}^*)) &= [\mathrm{d}f_{11}(\mathbf{Z},\mathbf{Z}^*),\ldots,\mathrm{d}f_{p1}(\mathbf{Z},\mathbf{Z}^*),\ldots,\mathrm{d}f_{1q}(\mathbf{Z},\mathbf{Z}^*),\ldots,\mathrm{d}f_{pq}(\mathbf{Z},\mathbf{Z}^*)]^T,\\
\mathrm{d}\,\mathrm{vec}\,\mathbf{Z} &= [\mathrm{d}Z_{11},\ldots,\mathrm{d}Z_{m1},\ldots,\mathrm{d}Z_{1n},\ldots,\mathrm{d}Z_{mn}]^T,\\
\mathrm{d}\,\mathrm{vec}\,\mathbf{Z}^* &= [\mathrm{d}Z_{11}^*,\ldots,\mathrm{d}Z_{m1}^*,\ldots,\mathrm{d}Z_{1n}^*,\ldots,\mathrm{d}Z_{mn}^*]^T,
\end{aligned}$$

while

$$\begin{aligned}
\mathbf{A} &= \begin{bmatrix}
\frac{\partial f_{11}(\mathbf{Z},\mathbf{Z}^*)}{\partial Z_{11}} & \cdots & \frac{\partial f_{11}(\mathbf{Z},\mathbf{Z}^*)}{\partial Z_{m1}} & \cdots & \frac{\partial f_{11}(\mathbf{Z},\mathbf{Z}^*)}{\partial Z_{1n}} & \cdots & \frac{\partial f_{11}(\mathbf{Z},\mathbf{Z}^*)}{\partial Z_{mn}}\\
\vdots & \vdots & \vdots & \vdots & \vdots & \vdots & \vdots\\
\frac{\partial f_{p1}(\mathbf{Z},\mathbf{Z}^*)}{\partial Z_{11}} & \cdots & \frac{\partial f_{p1}(\mathbf{Z},\mathbf{Z}^*)}{\partial Z_{m1}} & \cdots & \frac{\partial f_{p1}(\mathbf{Z},\mathbf{Z}^*)}{\partial Z_{1n}} & \cdots & \frac{\partial f_{p1}(\mathbf{Z},\mathbf{Z}^*)}{\partial Z_{mn}}\\
\vdots & \vdots & \vdots & \vdots & \vdots & \vdots & \vdots\\
\frac{\partial f_{1q}(\mathbf{Z},\mathbf{Z}^*)}{\partial Z_{11}} & \cdots & \frac{\partial f_{1q}(\mathbf{Z},\mathbf{Z}^*)}{\partial Z_{m1}} & \cdots & \frac{\partial f_{1q}(\mathbf{Z},\mathbf{Z}^*)}{\partial Z_{1n}} & \cdots & \frac{\partial f_{1q}(\mathbf{Z},\mathbf{Z}^*)}{\partial Z_{mn}}\\
\vdots & \vdots & \vdots & \vdots & \vdots & \vdots & \vdots\\
\frac{\partial f_{pq}(\mathbf{Z},\mathbf{Z}^*)}{\partial Z_{11}} & \cdots & \frac{\partial f_{pq}(\mathbf{Z},\mathbf{Z}^*)}{\partial Z_{m1}} & \cdots & \frac{\partial f_{pq}(\mathbf{Z},\mathbf{Z}^*)}{\partial Z_{1n}} & \cdots & \frac{\partial f_{pq}(\mathbf{Z},\mathbf{Z}^*)}{\partial Z_{mn}}
\end{bmatrix}\\
&= \frac{\partial\,\mathrm{vec}\,\mathbf{F}(\mathbf{Z},\mathbf{Z}^*)}{\partial(\mathrm{vec}\,\mathbf{Z})^T} = \mathrm{D}_{\mathrm{vec}\,\mathbf{Z}}\mathbf{F}(\mathbf{Z},\mathbf{Z}^*),
\end{aligned} \tag{3.4.63}$$

$$\begin{aligned}
\mathbf{B} &= \begin{bmatrix}
\frac{\partial f_{11}(\mathbf{Z},\mathbf{Z}^*)}{\partial Z_{11}^*} & \cdots & \frac{\partial f_{11}(\mathbf{Z},\mathbf{Z}^*)}{\partial Z_{m1}^*} & \cdots & \frac{\partial f_{11}(\mathbf{Z},\mathbf{Z}^*)}{\partial Z_{1n}^*} & \cdots & \frac{\partial f_{11}(\mathbf{Z},\mathbf{Z}^*)}{\partial Z_{mn}^*}\\
\vdots & \vdots & \vdots & \vdots & \vdots & \vdots & \vdots\\
\frac{\partial f_{p1}(\mathbf{Z},\mathbf{Z}^*)}{\partial Z_{11}^*} & \cdots & \frac{\partial f_{p1}(\mathbf{Z},\mathbf{Z}^*)}{\partial Z_{m1}^*} & \cdots & \frac{\partial f_{p1}(\mathbf{Z},\mathbf{Z}^*)}{\partial Z_{1n}^*} & \cdots & \frac{\partial f_{p1}(\mathbf{Z},\mathbf{Z}^*)}{\partial Z_{mn}^*}\\
\vdots & \vdots & \vdots & \vdots & \vdots & \vdots & \vdots\\
\frac{\partial f_{1q}(\mathbf{Z},\mathbf{Z}^*)}{\partial Z_{11}^*} & \cdots & \frac{\partial f_{1q}(\mathbf{Z},\mathbf{Z}^*)}{\partial Z_{m1}^*} & \cdots & \frac{\partial f_{1q}(\mathbf{Z},\mathbf{Z}^*)}{\partial Z_{1n}^*} & \cdots & \frac{\partial f_{1q}(\mathbf{Z},\mathbf{Z}^*)}{\partial Z_{mn}^*}\\
\vdots & \vdots & \vdots & \vdots & \vdots & \vdots & \vdots\\
\frac{\partial f_{pq}(\mathbf{Z},\mathbf{Z}^*)}{\partial Z_{11}^*} & \cdots & \frac{\partial f_{pq}(\mathbf{Z},\mathbf{Z}^*)}{\partial Z_{m1}^*} & \cdots & \frac{\partial f_{pq}(\mathbf{Z},\mathbf{Z}^*)}{\partial Z_{1n}^*} & \cdots & \frac{\partial f_{pq}(\mathbf{Z},\mathbf{Z}^*)}{\partial Z_{mn}^*}
\end{bmatrix}\\
&= \frac{\partial\,\mathrm{vec}\,\mathbf{F}(\mathbf{Z},\mathbf{Z}^*)}{\partial(\mathrm{vec}\,\mathbf{Z}^*)^T} = \mathrm{D}_{\mathrm{vec}\,\mathbf{Z}^*}\mathbf{F}(\mathbf{Z},\mathbf{Z}^*).
\end{aligned} \tag{3.4.64}$$

Obviously the matrices $\mathbf{A}$ and $\mathbf{B}$ are the complex Jacobian matrix and the complex conjugate Jacobian matrix, respectively.

The complex gradient matrix and the complex conjugate gradient matrix of the matrix function $\mathbf{F}(\mathbf{Z},\mathbf{Z}^*)$ are respectively defined as

$$\nabla_{\mathrm{vec}\,\mathbf{Z}}\mathbf{F}(\mathbf{Z},\mathbf{Z}^*) = \frac{\partial(\mathrm{vec}\,\mathbf{F}(\mathbf{Z},\mathbf{Z}^*))^T}{\partial\,\mathrm{vec}\,\mathbf{Z}} = (\mathrm{D}_{\mathrm{vec}\,\mathbf{Z}}\mathbf{F}(\mathbf{Z},\mathbf{Z}^*))^T, \tag{3.4.65}$$

$$\nabla_{\mathrm{vec}\,\mathbf{Z}^*}\mathbf{F}(\mathbf{Z},\mathbf{Z}^*) = \frac{\partial(\mathrm{vec}\,\mathbf{F}(\mathbf{Z},\mathbf{Z}^*))^T}{\partial\,\mathrm{vec}\,\mathbf{Z}^*} = (\mathrm{D}_{\mathrm{vec}\,\mathbf{Z}^*}\mathbf{F}(\mathbf{Z},\mathbf{Z}^*))^T. \tag{3.4.66}$$

In particular, for a real scalar function $f(\mathbf{Z},\mathbf{Z}^*)$, Equation (3.4.62) reduces to Equation (3.4.34).

Summarizing the above discussions, from Equation (3.4.62) we get the following proposition.

Proposition 3.5 *For a complex matrix function* $\mathbf{F}(\mathbf{Z},\mathbf{Z}^*) \in \mathbb{C}^{p\times q}$ *with* $\mathbf{Z},\mathbf{Z}^* \in \mathbb{C}^{m\times n}$*, its complex Jacobian matrix and the conjugate Jacobian matrix can be identified as follows:*

$$\mathrm{d}\,\mathrm{vec}\,\mathbf{F}(\mathbf{Z},\mathbf{Z}^*) = \mathbf{A}\,\mathrm{d}\,\mathrm{vec}\,\mathbf{Z} + \mathbf{B}\,\mathrm{d}\,\mathrm{vec}\,\mathbf{Z}^*$$
$$\Leftrightarrow \begin{cases} \mathrm{D}_{\mathrm{vec}\,\mathbf{Z}}\mathbf{F}(\mathbf{Z},\mathbf{Z}^*) = \mathbf{A}, \\ \mathrm{D}_{\mathrm{vec}\,\mathbf{Z}^*}\mathbf{F}(\mathbf{Z},\mathbf{Z}^*) = \mathbf{B}, \end{cases} \tag{3.4.67}$$

and the complex gradient matrix and the conjugate gradient matrix can be identified by

$$\mathrm{d}\,\mathrm{vec}\,\mathbf{F}(\mathbf{Z},\mathbf{Z}^*) = \mathbf{A}\,\mathrm{d}\,\mathrm{vec}\,\mathbf{Z} + \mathbf{B}\,\mathrm{d}\,\mathrm{vec}\,\mathbf{Z}^*$$
$$\Leftrightarrow \begin{cases} \nabla_{\mathrm{vec}\,\mathbf{Z}}\mathbf{F}(\mathbf{Z},\mathbf{Z}^*) = \mathbf{A}^T, \\ \nabla_{\mathrm{vec}\,\mathbf{Z}^*}\mathbf{F}(\mathbf{Z},\mathbf{Z}^*) = \mathbf{B}^T. \end{cases} \tag{3.4.68}$$

If $\mathrm{d}(\mathbf{F}(\mathbf{Z},\mathbf{Z}^*)) = \mathbf{A}(\mathrm{d}\mathbf{Z})\mathbf{B} + \mathbf{C}(\mathrm{d}\mathbf{Z}^*)\mathbf{D}$, then the vectorization result is given by

$$\mathrm{d}\,\mathrm{vec}\,\mathbf{F}(\mathbf{Z},\mathbf{Z}^*) = (\mathbf{B}^T \otimes \mathbf{A})\,\mathrm{d}\,\mathrm{vec}\,\mathbf{Z} + (\mathbf{D}^T \otimes \mathbf{C})\,\mathrm{d}\,\mathrm{vec}\,\mathbf{Z}^*.$$

By Proposition 3.5 we have the following identification formula:

$$\mathrm{d}\mathbf{F}(\mathbf{Z},\mathbf{Z}^*) = \mathbf{A}(\mathrm{d}\mathbf{Z})\mathbf{B} + \mathbf{C}(\mathrm{d}\mathbf{Z}^*)\mathbf{D}$$
$$\Leftrightarrow \begin{cases} \mathrm{D}_{\mathrm{vec}\,\mathbf{Z}}\mathbf{F}(\mathbf{Z},\mathbf{Z}^*) = \mathbf{B}^T \otimes \mathbf{A}, \\ \mathrm{D}_{\mathrm{vec}\,\mathbf{Z}^*}\mathbf{F}(\mathbf{Z},\mathbf{Z}^*) = \mathbf{D}^T \otimes \mathbf{C}. \end{cases} \tag{3.4.69}$$

Similarly, if $\mathrm{d}\mathbf{F}(\mathbf{Z},\mathbf{Z}^*) = \mathbf{A}(\mathrm{d}\mathbf{Z})^T\mathbf{B} + \mathbf{C}(\mathrm{d}\mathbf{Z}^*)^T\mathbf{D}$ then we have the result

$$\begin{aligned} \mathrm{d}\,\mathrm{vec}\,\mathbf{F}(\mathbf{Z},\mathbf{Z}^*) &= (\mathbf{B}^T \otimes \mathbf{A})\,\mathrm{d}\,\mathrm{vec}\,\mathbf{Z}^T + (\mathbf{D}^T \otimes \mathbf{C})\,\mathrm{d}\,\mathrm{vec}\,\mathbf{Z}^H \\ &= (\mathbf{B}^T \otimes \mathbf{A})\mathbf{K}_{mn}\mathrm{d}\,\mathrm{vec}\,\mathbf{Z} + (\mathbf{D}^T \otimes \mathbf{C})\mathbf{K}_{mn}\mathrm{d}\,\mathrm{vec}\,\mathbf{Z}^*, \end{aligned}$$

where we have used the vectorization property $\mathrm{vec}\mathbf{X}_{m\times n}^T = \mathbf{K}_{mn}\mathrm{vec}\mathbf{X}$. By Proposition 3.5, the following identification formula is obtained:

$$\begin{aligned}&\mathrm{d}\mathbf{F}(\mathbf{Z},\mathbf{Z}^*)=\mathbf{A}(\mathrm{d}\mathbf{Z})^T\mathbf{B}+\mathbf{C}(\mathrm{d}\mathbf{Z}^*)^T\mathbf{D}\\ \Leftrightarrow\ &\begin{cases}\mathrm{D}_{\mathrm{vec}\,\mathbf{Z}}\mathbf{F}(\mathbf{Z},\mathbf{Z}^*)=(\mathbf{B}^T\otimes\mathbf{A})\mathbf{K}_{mn},\\ \mathrm{D}_{\mathrm{vec}\,\mathbf{Z}^*}\mathbf{F}(\mathbf{Z},\mathbf{Z}^*)=(\mathbf{D}^T\otimes\mathbf{C})\mathbf{K}_{mn}.\end{cases}\end{aligned}\tag{3.4.70}$$

The above equation shows that, as in the vector case, the key to identifying the gradient matrix and conjugate gradient matrix of a matrix function $\mathbf{F}(\mathbf{Z},\mathbf{Z}^*)$ is to write its matrix differential into the canonical form $\mathrm{d}\mathbf{F}(\mathbf{Z},\mathbf{Z}^*)=\mathbf{A}(\mathrm{d}\mathbf{Z})^T\mathbf{B}+\mathbf{C}(\mathrm{d}\mathbf{Z}^*)^T\mathbf{D}$.

Table 3.11 lists the corresponding relationships between the first-order complex matrix differential and the complex Jacobian matrix, where $\mathbf{z}\in\mathbb{C}^m,\mathbf{Z}\in\mathbb{C}^{m\times n},\mathbf{F}\in\mathbb{C}^{p\times q}$.

Table 3.11 Complex matrix differential and complex Jacobian matrix

Function	First-order matrix differential	Jacobian matrix
$f(z,z^*)$	$\mathrm{d}f(z,z^*)=a\mathrm{d}z+b\mathrm{d}z^*$	$\frac{\partial f}{\partial z}=a,\ \frac{\partial f}{\partial z^*}=b$
$f(\mathbf{z},\mathbf{z}^*)$	$\mathrm{d}f(\mathbf{z},\mathbf{z}^*)=\mathbf{a}^T\mathrm{d}\mathbf{z}+\mathbf{b}^T\mathrm{d}\mathbf{z}^*$	$\frac{\partial f}{\partial \mathbf{z}^T}=\mathbf{a}^T,\ \frac{\partial f}{\partial \mathbf{z}^H}=\mathbf{b}^T$
$f(\mathbf{Z},\mathbf{Z}^*)$	$\mathrm{d}f(\mathbf{Z},\mathbf{Z}^*)=\mathrm{tr}(\mathbf{A}\mathrm{d}\mathbf{Z}+\mathbf{B}\mathrm{d}\mathbf{Z}^*)$	$\frac{\partial f}{\partial \mathbf{Z}^T}=\mathbf{A},\ \frac{\partial f}{\partial \mathbf{Z}^H}=\mathbf{B}$
$\mathbf{F}(\mathbf{Z},\mathbf{Z}^*)$	$\mathrm{d}(\mathrm{vec}\,\mathbf{F})=\mathbf{A}\mathrm{d}(\mathrm{vec}\,\mathbf{Z})+\mathbf{B}\mathrm{d}(\mathrm{vec}\,\mathbf{Z}^*)$	$\frac{\partial\,\mathrm{vec}\,\mathbf{F}}{\partial(\mathrm{vec}\,\mathbf{Z})^T}=\mathbf{A},\ \frac{\partial\,\mathrm{vec}\,\mathbf{F}}{\partial(\mathrm{vec}\,\mathbf{Z}^*)^T}=\mathbf{B}$
	$\mathrm{d}\mathbf{F}=\mathbf{A}(\mathrm{d}\mathbf{Z})\mathbf{B}+\mathbf{C}(\mathrm{d}\mathbf{Z}^*)\mathbf{D},$	$\begin{cases}\frac{\partial\,\mathrm{vec}\,\mathbf{F}}{\partial(\mathrm{vec}\,\mathbf{Z})^T}=\mathbf{B}^T\otimes\mathbf{A}\\ \frac{\partial\,\mathrm{vec}\,\mathbf{F}}{\partial(\mathrm{vec}\,\mathbf{Z}^*)^T}=\mathbf{D}^T\otimes\mathbf{C}\end{cases}$
	$\mathrm{d}\mathbf{F}=\mathbf{A}(\mathrm{d}\mathbf{Z})^T\mathbf{B}+\mathbf{C}(\mathrm{d}\mathbf{Z}^*)^T\mathbf{D}$	$\begin{cases}\frac{\partial\mathrm{vec}\,\mathbf{F}}{\partial(\mathrm{vec}\,\mathbf{Z})^T}=(\mathbf{B}^T\otimes\mathbf{A})\mathbf{K}_{mn}\\ \frac{\partial\,\mathrm{vec}\mathbf{F}}{\partial\,(\mathrm{vec}\,\mathbf{Z}^*)^T}=(\mathbf{D}^T\otimes\mathbf{C})\mathbf{K}_{mn}\end{cases}$

3.5 Complex Hessian Matrices and Identification

The previous section analyzed the first-order complex derivative matrices (the complex Jacobian matrix and complex gradient matrix) and their identification. In this section we discuss the second-order complex derivative matrices of the function $f(\mathbf{Z},\mathbf{Z}^*)$ and their identification. These second-order complex matrices are a full Hessian matrix and four part Hessian matrices.

3.5.1 Complex Hessian Matrices

The differential of the real function $f = f(\mathbf{Z}, \mathbf{Z}^*)$ can be equivalently written as

$$\mathrm{d}f = (\mathrm{D}_{\mathbf{Z}} f)\mathrm{d}\,\mathrm{vec}\,\mathbf{Z} + (\mathrm{D}_{\mathbf{Z}^*} f)\mathrm{d}\,\mathrm{vec}\,\mathbf{Z}^*, \tag{3.5.1}$$

where

$$\mathrm{D}_{\mathbf{Z}} f = \frac{\partial f}{\partial(\mathrm{vec}\,\mathbf{Z})^T}, \quad \mathrm{D}_{\mathbf{Z}^*} f = \frac{\partial f}{\partial(\mathrm{vec}\,\mathbf{Z}^*)^T}. \tag{3.5.2}$$

Noting that the differentials of both $\mathrm{D}_{\mathbf{Z}} f$ and $\mathrm{D}_{\mathbf{Z}^*} f$ are row vectors, we have

$$\begin{aligned}
\mathrm{d}(\mathrm{D}_{\mathbf{Z}} f) &= \left(\frac{\partial \mathrm{D}_{\mathbf{Z}} f}{\partial\,\mathrm{vec}\,\mathbf{Z}}\mathrm{d}\,\mathrm{vec}\,\mathbf{Z} + \frac{\partial \mathrm{D}_{\mathbf{Z}} f}{\partial\,\mathrm{vec}\,\mathbf{Z}^*}\mathrm{d}\,\mathrm{vec}\,\mathbf{Z}^*\right)^T \\
&= (\mathrm{d}\,\mathrm{vec}\,\mathbf{Z})^T \frac{\partial^2 f}{\partial\,\mathrm{vec}\,\mathbf{Z}\,\partial(\mathrm{vec}\,\mathbf{Z})^T} + (\mathrm{d}\,\mathrm{vec}\,\mathbf{Z}^*)^T \frac{\partial^2 f}{\partial\,\mathrm{vec}\,\mathbf{Z}^*\partial(\mathrm{vec}\,\mathbf{Z})^T} \\
\mathrm{d}(\mathrm{D}_{\mathbf{Z}^*} f) &= \left(\frac{\partial \mathrm{D}_{\mathbf{Z}^*} f}{\partial\,\mathrm{vec}\,\mathbf{Z}}\mathrm{d}\,\mathrm{vec}\,\mathbf{Z} + \frac{\partial \mathrm{D}_{\mathbf{Z}^*} f}{\partial(\mathrm{vec}\,\mathbf{Z})^T}\mathrm{d}\,\mathrm{vec}\,\mathbf{Z}^*\right)^T \\
&= (\mathrm{d}\,\mathrm{vec}\,\mathbf{Z})^T \frac{\partial^2 f}{\partial\,\mathrm{vec}\,\mathbf{Z}\,\partial(\mathrm{vec}\,\mathbf{Z}^*)^T} + (\mathrm{d}(\mathrm{vec}\,\mathbf{Z}^*)^T) \frac{\partial^2 f}{\partial\,\mathrm{vec}\,\mathbf{Z}^*\partial(\mathrm{vec}\,\mathbf{Z}^*)^T}.
\end{aligned}$$

Since $\mathrm{d}\,\mathrm{vec}\,\mathbf{Z}$ is not a function of $\mathrm{vec}\,\mathbf{Z}$, and $\mathrm{d}\,\mathrm{vec}\,\mathbf{Z}^*$ is not a function of $\mathrm{vec}\,\mathbf{Z}^*$, it follows that

$$\mathrm{d}^2(\mathrm{vec}\,\mathbf{Z}) = \mathrm{d}(\mathrm{d}\,\mathrm{vec}\,\mathbf{Z}) = 0, \quad \mathrm{d}^2(\mathrm{vec}\,\mathbf{Z}^*) = \mathrm{d}(\mathrm{d}\,\mathrm{vec}\mathbf{Z}^*) = 0. \tag{3.5.3}$$

Hence, the second-order differential of $f = f(\mathbf{Z}, \mathbf{Z}^*)$ is given by

$$\begin{aligned}
\mathrm{d}^2 f &= \mathrm{d}(\mathrm{D}_{\mathbf{Z}} f)\mathrm{d}(\mathrm{vec}\,\mathbf{Z}) + \mathrm{d}(\mathrm{D}_{\mathbf{Z}^*} f)\mathrm{d}\,\mathrm{vec}\,\mathbf{Z}^* \\
&= (\mathrm{d}\,\mathrm{vec}\,\mathbf{Z})^T \frac{\partial^2 f}{\partial\,\mathrm{vec}\,\mathbf{Z}\,\partial(\mathrm{vec}\,\mathbf{Z})^T}\mathrm{d}\,\mathrm{vec}\,\mathbf{Z} \\
&\quad + (\mathrm{d}\,\mathrm{vec}\,\mathbf{Z}^*)^T \frac{\partial^2 f}{\partial\,\mathrm{vec}\,\mathbf{Z}^*\,\partial(\mathrm{vec}\,\mathbf{Z})^T}\mathrm{d}\,\mathrm{vec}\,\mathbf{Z} \\
&\quad + (\mathrm{d}\,\mathrm{vec}\,\mathbf{Z})^T \frac{\partial^2 f}{\partial\,\mathrm{vec}\,\mathbf{Z}\,\partial(\mathrm{vec}\,\mathbf{Z}^*)^T}\mathrm{d}\,\mathrm{vec}\,\mathbf{Z}^* \\
&\quad + (\mathrm{d}\,\mathrm{vec}\,\mathbf{Z}^*)^T \frac{\partial^2 f}{\partial\,\mathrm{vec}\,\mathbf{Z}^*\,\partial(\mathrm{vec}\,\mathbf{Z}^*)^T}\mathrm{d}\,\mathrm{vec}\,\mathbf{Z}^*,
\end{aligned}$$

which can be written as

$$\begin{aligned}
\mathrm{d}^2 f &= \left[(\mathrm{d}\,\mathrm{vec}\,\mathbf{Z}^*)^T, (\mathrm{d}\,\mathrm{vec}\,\mathbf{Z})^T\right] \begin{bmatrix} \mathbf{H}_{\mathbf{Z}^*,\mathbf{Z}} & \mathbf{H}_{\mathbf{Z}^*,\mathbf{Z}^*} \\ \mathbf{H}_{\mathbf{Z},\mathbf{Z}} & \mathbf{H}_{\mathbf{Z},\mathbf{Z}^*} \end{bmatrix} \begin{bmatrix} \mathrm{d}\,\mathrm{vec}\,\mathbf{Z} \\ \mathrm{d}\,\mathrm{vec}\,\mathbf{Z}^* \end{bmatrix} \\
&= \begin{bmatrix} \mathrm{d}\,\mathrm{vec}\,\mathbf{Z} \\ \mathrm{d}\,\mathrm{vec}\,\mathbf{Z}^* \end{bmatrix}^H \mathbf{H} \begin{bmatrix} \mathrm{d}\,\mathrm{vec}\,\mathbf{Z} \\ \mathrm{d}\,\mathrm{vec}\,\mathbf{Z}^* \end{bmatrix},
\end{aligned} \tag{3.5.4}$$

where

$$\mathbf{H} = \begin{bmatrix} \mathbf{H}_{\mathbf{Z}^*,\mathbf{Z}} & \mathbf{H}_{\mathbf{Z}^*,\mathbf{Z}^*} \\ \mathbf{H}_{\mathbf{Z},\mathbf{Z}} & \mathbf{H}_{\mathbf{Z},\mathbf{Z}^*} \end{bmatrix} \tag{3.5.5}$$

is known as the *full complex Hessian matrix* of the function $f(\mathbf{Z}, \mathbf{Z}^*)$, and

$$\left.\begin{aligned} \mathbf{H}_{\mathbf{Z}^*,\mathbf{Z}} &= \frac{\partial^2 f(\mathbf{Z}, \mathbf{Z}^*)}{\partial \operatorname{vec} \mathbf{Z}^* \, \partial (\operatorname{vec} \mathbf{Z})^T}, \\ \mathbf{H}_{\mathbf{Z}^*,\mathbf{Z}^*} &= \frac{\partial^2 f(\mathbf{Z}, \mathbf{Z}^*)}{\partial \operatorname{vec} \mathbf{Z}^* \, \partial (\operatorname{vec} \mathbf{Z}^*)^T}, \\ \mathbf{H}_{\mathbf{Z},\mathbf{Z}} &= \frac{\partial^2 f(\mathbf{Z}, \mathbf{Z}^*)}{\partial \operatorname{vec} \mathbf{Z} \partial (\operatorname{vec} \mathbf{Z})^T}, \\ \mathbf{H}_{\mathbf{Z},\mathbf{Z}^*} &= \frac{\partial^2 f(\mathbf{Z}, \mathbf{Z}^*)}{\partial \operatorname{vec} \mathbf{Z} \partial (\operatorname{vec} \mathbf{Z}^*)^T}, \end{aligned}\right\} \tag{3.5.6}$$

are respectively the *part complex Hessian matrices* of the function $f(\mathbf{Z}, \mathbf{Z}^*)$; of these, $\mathbf{H}_{\mathbf{Z}^*,\mathbf{Z}}$ is the *main complex Hessian matrix* of the function $f(\mathbf{Z}, \mathbf{Z}^*)$.

By the above definition formulas, it is easy to show that the Hessian matrices of a scalar function $f(\mathbf{Z}, \mathbf{Z}^*)$ have the following properties.

1. The part complex Hessian matrices $\mathbf{H}_{\mathbf{Z}^*,\mathbf{Z}}$ and $\mathbf{H}_{\mathbf{Z},\mathbf{Z}^*}$ are Hermitian, and are complex conjugates:

$$\mathbf{H}_{\mathbf{Z}^*,\mathbf{Z}} = \mathbf{H}^H_{\mathbf{Z}^*,\mathbf{Z}}, \quad \mathbf{H}_{\mathbf{Z},\mathbf{Z}^*} = \mathbf{H}^H_{\mathbf{Z},\mathbf{Z}^*}, \quad \mathbf{H}_{\mathbf{Z}^*,\mathbf{Z}} = \mathbf{H}^*_{\mathbf{Z},\mathbf{Z}^*}. \tag{3.5.7}$$

2. The other two part complex Hessian matrices are symmetric and complex conjugate each other:

$$\mathbf{H}_{\mathbf{Z},\mathbf{Z}} = \mathbf{H}^T_{\mathbf{Z},\mathbf{Z}}, \quad \mathbf{H}_{\mathbf{Z}^*,\mathbf{Z}^*} = \mathbf{H}^T_{\mathbf{Z}^*,\mathbf{Z}^*} \quad \mathbf{H}_{\mathbf{Z},\mathbf{Z}} = \mathbf{H}^*_{\mathbf{Z}^*,\mathbf{Z}^*}. \tag{3.5.8}$$

3. The complex full Hessian matrix is Hermitian, i.e., $\mathbf{H} = \mathbf{H}^H$.

From Equation (3.5.4) it can be seen that, because the second-order differential of the real function $f(\mathbf{Z}, \mathbf{Z}^*)$, $\mathrm{d}^2 f$, is a quadratic function, the positive definiteness of the full Hessian matrix is determined by $\mathrm{d}^2 f$:

(1) The full Hessian matrix is positive definite if $\mathrm{d}^2 f > 0$ holds for all $\operatorname{vec} \mathbf{Z}$.
(2) The full Hessian matrix is positive semi-definite if $\mathrm{d}^2 f \geq 0$ holds for all $\operatorname{vec} \mathbf{Z}$.
(3) The full Hessian matrix is negative definite if $\mathrm{d}^2 f < 0$ holds for all $\operatorname{vec} \mathbf{Z}$.
(4) The full Hessian matrix is negative semi-definite if $\mathrm{d}^2 f \leq 0$ holds for all $\operatorname{vec} \mathbf{Z}$.

For a real function $f = f(\mathbf{z}, \mathbf{z}^*)$, its second-order differential is given by

$$\mathrm{d}^2 f = \begin{bmatrix} \mathrm{d}\mathbf{z} \\ \mathrm{d}\mathbf{z}^* \end{bmatrix}^H \mathbf{H} \begin{bmatrix} \mathrm{d}\mathbf{z} \\ \mathrm{d}\mathbf{z}^* \end{bmatrix}, \tag{3.5.9}$$

in which $\mathbf{H}$ is the full complex Hessian matrix of the function $f(\mathbf{z}, \mathbf{z}^*)$, and is defined as

$$\mathbf{H} = \begin{bmatrix} \mathbf{H}_{\mathbf{z}^*,\mathbf{z}} & \mathbf{H}_{\mathbf{z}^*,\mathbf{z}^*} \\ \mathbf{H}_{\mathbf{z},\mathbf{z}} & \mathbf{H}_{\mathbf{z},\mathbf{z}^*} \end{bmatrix}. \tag{3.5.10}$$

The four part complex Hessian matrices are given by

$$\left.\begin{aligned} \mathbf{H}_{\mathbf{z}^*,\mathbf{z}} &= \frac{\partial^2 f(\mathbf{z}, \mathbf{z}^*)}{\partial \mathbf{z}^* \partial \mathbf{z}^T}, \\ \mathbf{H}_{\mathbf{z}^*,\mathbf{z}^*} &= \frac{\partial^2 f(\mathbf{z}, \mathbf{z}^*)}{\partial \mathbf{z}^* \partial \mathbf{z}^H}, \\ \mathbf{H}_{\mathbf{z},\mathbf{z}} &= \frac{\partial^2 f(\mathbf{z}, \mathbf{z}^*)}{\partial \mathbf{z} \partial \mathbf{z}^T}, \\ \mathbf{H}_{\mathbf{z},\mathbf{z}^*} &= \frac{\partial^2 f(\mathbf{z}, \mathbf{z}^*)}{\partial \mathbf{z} \partial \mathbf{z}^H}. \end{aligned}\right\} \tag{3.5.11}$$

Clearly, the full complex Hessian matrix is Hermitian, i.e., $\mathbf{H} = \mathbf{H}^H$; and the part complex Hessian matrices have the following relationships:

$$\mathbf{H}_{\mathbf{z}^*,\mathbf{z}} = \mathbf{H}_{\mathbf{z}^*,\mathbf{z}}^H, \quad \mathbf{H}_{\mathbf{z},\mathbf{z}^*} = \mathbf{H}_{\mathbf{z},\mathbf{z}^*}^H, \quad \mathbf{H}_{\mathbf{z}^*,\mathbf{z}} = \mathbf{H}_{\mathbf{z},\mathbf{z}^*}^*, \tag{3.5.12}$$

$$\mathbf{H}_{\mathbf{z},\mathbf{z}} = \mathbf{H}_{\mathbf{z},\mathbf{z}}^T, \quad \mathbf{H}_{\mathbf{z}^*,\mathbf{z}^*} = \mathbf{H}_{\mathbf{z}^*,\mathbf{z}^*}^T \quad \mathbf{H}_{\mathbf{z},\mathbf{z}} = \mathbf{H}_{\mathbf{z}^*,\mathbf{z}^*}^*. \tag{3.5.13}$$

The above analysis show that, in complex Hessian matrix identification, we just need to identify two of the part Hessian matrices $\mathbf{H}_{\mathbf{Z},\mathbf{Z}}$ and $\mathbf{H}_{\mathbf{Z},\mathbf{Z}^*}$, since $\mathbf{H}_{\mathbf{Z}^*,\mathbf{Z}^*} = \mathbf{H}_{\mathbf{Z},\mathbf{Z}}^*$ and $\mathbf{H}_{\mathbf{Z}^*,\mathbf{Z}} = \mathbf{H}_{\mathbf{Z},\mathbf{Z}^*}^*$.

3.5.2 Complex Hessian Matrix Identification

By Proposition 3.5 it is known that the complex differential of a real scalar function $f(\mathbf{Z}, \mathbf{Z}^*)$ can be written in the canonical form $\mathrm{d}f(\mathbf{Z}, \mathbf{Z}^*) = \mathrm{tr}(\mathbf{A}\,\mathrm{d}\mathbf{Z} + \mathbf{A}^*\mathrm{d}\mathbf{Z}^*)$, where $\mathbf{A} = \mathbf{A}(\mathbf{Z}, \mathbf{Z}^*)$ is usually a complex matrix function of the matrix variables $\mathbf{Z}$ and $\mathbf{Z}^*$. Hence, the complex matrix differential of $\mathbf{A}$ can be expressed as

$$\mathrm{d}\mathbf{A} = \mathbf{C}(\mathrm{d}\mathbf{Z})\mathbf{D} + \mathbf{E}(\mathrm{d}\mathbf{Z}^*)\mathbf{F} \quad \text{or} \quad \mathrm{d}\mathbf{A} = \mathbf{D}(\mathrm{d}\mathbf{Z})^T\mathbf{C} + \mathbf{F}(\mathrm{d}\mathbf{Z}^*)^T\mathbf{E}. \tag{3.5.14}$$

Substituting (3.5.14) into the second-order complex differential

$$\mathrm{d}^2 f(\mathbf{Z}, \mathbf{Z}^*) = \mathrm{d}(\mathrm{d}f(\mathbf{Z}, \mathbf{Z}^*)) = \mathrm{tr}(\mathrm{d}\mathbf{A}\,\mathrm{d}\mathbf{Z} + \mathrm{d}\mathbf{A}^*\mathrm{d}\mathbf{Z}^*), \tag{3.5.15}$$

we get

$$\begin{aligned} \mathrm{d}^2 f(\mathbf{Z}, \mathbf{Z}^*) = &\,\mathrm{tr}(\mathbf{C}(\mathrm{d}\mathbf{Z})\mathbf{D}\,\mathrm{d}\mathbf{Z}) + \mathrm{tr}(\mathbf{E}(\mathrm{d}\mathbf{Z}^*)\mathbf{F}\mathrm{d}\mathbf{Z}) \\ &+ \mathrm{tr}(\mathbf{C}^*(\mathrm{d}\mathbf{Z}^*)\mathbf{D}^*\mathrm{d}\mathbf{Z}^*) + \mathrm{tr}(\mathbf{E}^*(\mathrm{d}\mathbf{Z})\mathbf{F}^*\mathrm{d}\mathbf{Z}^*), \end{aligned} \tag{3.5.16}$$

or

$$\begin{aligned} \mathrm{d}^2 f(\mathbf{Z}, \mathbf{Z}^*) = &\,\mathrm{tr}(\mathbf{D}(\mathrm{d}\mathbf{Z})^T\mathbf{C}\mathrm{d}\mathbf{Z}) + \mathrm{tr}(\mathbf{F}(\mathrm{d}\mathbf{Z}^*)^T\mathbf{E}\mathrm{d}\mathbf{Z}) \\ &+ \mathrm{tr}(\mathbf{D}^*(\mathrm{d}\mathbf{Z}^*)^T\mathbf{C}^*\mathrm{d}\mathbf{Z}^*) + \mathrm{tr}(\mathbf{F}^*(\mathrm{d}\mathbf{Z})^T\mathbf{E}^*\mathrm{d}\mathbf{Z}^*). \end{aligned} \tag{3.5.17}$$

Case 1 Hessian matrix identification based on (3.5.16)

By the trace property $\mathrm{tr}(\mathbf{XYUV}) = (\mathrm{vec}(\mathbf{V}^T))^T(\mathbf{U}^T \otimes \mathbf{X})\mathrm{vec}\,\mathbf{Y}$, the vectorization property $\mathrm{vec}\,\mathbf{Z}^T = \mathbf{K}_{mn}\mathrm{vec}\,\mathbf{Z}$ and the commutation matrix property $\mathbf{K}_{mn}^T = \mathbf{K}_{nm}$, it is easy to see that

$$\begin{aligned}\mathrm{tr}(\mathbf{C}(\mathrm{d}\mathbf{Z})\mathbf{D}\mathrm{d}\mathbf{Z}) &= (\mathbf{K}_{mn}\mathrm{vec}\mathrm{d}\mathbf{Z})^T(\mathbf{D}^T \otimes \mathbf{C})\mathrm{vec}\,\mathrm{d}\mathbf{Z}\\ &= (\mathrm{d}\,\mathrm{vec}\,\mathbf{Z})^T\mathbf{K}_{nm}(\mathbf{D}^T \otimes \mathbf{C})\mathrm{d}\,\mathrm{vec}\,\mathbf{Z}.\end{aligned}$$

Similarly, we have

$$\begin{aligned}\mathrm{tr}(\mathbf{E}(\mathrm{d}\mathbf{Z}^*)\mathbf{F}\,\mathrm{d}\mathbf{Z}) &= (\mathrm{d}\,\mathrm{vec}\,\mathbf{Z})^T\mathbf{K}_{nm}(\mathbf{F}^T \otimes \mathbf{E})\mathrm{d}\,\mathrm{vec}\,\mathbf{Z}^*,\\ \mathrm{tr}(\mathbf{E}^*(\mathrm{d}\mathbf{Z})\mathbf{F}^*\mathrm{d}\mathbf{Z}^*) &= (\mathrm{d}\,\mathrm{vec}\,\mathbf{Z}^*)^T\mathbf{K}_{nm}(\mathbf{F}^H \otimes \mathbf{E}^*)\mathrm{d}\,\mathrm{vec}\,\mathbf{Z},\\ \mathrm{tr}(\mathbf{C}^*(\mathrm{d}\mathbf{Z}^*)\mathbf{D}^*\mathrm{d}\mathbf{Z}^*) &= (\mathrm{d}\,\mathrm{vec}\,\mathbf{Z}^*)^T\mathbf{K}_{nm}(\mathbf{D}^H \otimes \mathbf{C}^*)\mathrm{d}\,\mathrm{vec}\,\mathbf{Z}^*.\end{aligned}$$

Then, on the one hand, Equation (3.5.16) can be written as follows:

$$\begin{aligned}\mathrm{d}^2 f(\mathbf{Z},\mathbf{Z}^*) = &(\mathrm{d}\,\mathrm{vec}\,\mathbf{Z})^T\mathbf{K}_{nm}(\mathbf{D}^T \otimes \mathbf{C})\mathrm{d}\,\mathrm{vec}\,\mathbf{Z}\\ &+ (\mathrm{d}\,\mathrm{vec}\,\mathbf{Z})^T\mathbf{K}_{nm}(\mathbf{F}^T \otimes \mathbf{E})\mathrm{d}\,\mathrm{vec}\,\mathbf{Z}^*\\ &+ (\mathrm{d}\,\mathrm{vec}\,\mathbf{Z}^*)^T\mathbf{K}_{nm}(\mathbf{F}^H \otimes \mathbf{E}^*)\mathrm{d}\,\mathrm{vec}\,\mathbf{Z}\\ &+ (\mathrm{d}\,\mathrm{vec}\,\mathbf{Z}^*)^T\mathbf{K}_{nm}(\mathbf{D}^H \otimes \mathbf{C}^*)\mathrm{d}\,\mathrm{vec}\,\mathbf{Z}.\end{aligned} \tag{3.5.18}$$

On the other hand, Equation (3.5.4) can be rewritten as follows:

$$\begin{aligned}\mathrm{d}^2 f = &(\mathrm{d}\,\mathrm{vec}\,\mathbf{Z})^T\mathbf{H}_{\mathbf{Z},\mathbf{Z}}\mathrm{d}\,\mathrm{vec}\,\mathbf{Z} + (\mathrm{d}\,\mathrm{vec}\,\mathbf{Z})^T\mathbf{H}_{\mathbf{Z},\mathbf{Z}^*}\mathrm{d}\,\mathrm{vec}\,\mathbf{Z}^*\\ &+ (\mathrm{d}\,\mathrm{vec}\,\mathbf{Z}^*)^T\mathbf{H}_{\mathbf{Z}^*,\mathbf{Z}}\mathrm{d}\,\mathrm{vec}\,\mathbf{Z}) + (\mathrm{d}\,\mathrm{vec}\,\mathbf{Z}^*)^T\mathbf{H}_{\mathbf{Z}^*,\mathbf{Z}^*}\mathrm{d}\,\mathrm{vec}\,\mathbf{Z}^*.\end{aligned} \tag{3.5.19}$$

By comparing (3.5.19) with (3.5.18), it immediately follows that

$$\mathbf{H}_{\mathbf{Z},\mathbf{Z}} = \mathbf{H}_{\mathbf{Z}^*,\mathbf{Z}^*}^* = \frac{1}{2}\left(\mathbf{K}_{nm}(\mathbf{D}^T \otimes \mathbf{C}) + (\mathbf{K}_{nm}(\mathbf{D}^T \otimes \mathbf{C}))^H\right), \tag{3.5.20}$$

$$\mathbf{H}_{\mathbf{Z},\mathbf{Z}^*} = \mathbf{H}_{\mathbf{Z}^*,\mathbf{Z}}^* = \frac{1}{2}\left(\mathbf{K}_{nm}(\mathbf{F}^T \otimes \mathbf{E}) + (\mathbf{K}_{nm}(\mathbf{F}^T \otimes \mathbf{E}))^H\right). \tag{3.5.21}$$

Case 2 Hessian matrix identification based on (3.5.17)

By the trace property $\mathrm{tr}(\mathbf{XYUV}) = (\mathrm{vec}\,\mathbf{V}^T)^T(\mathbf{U}^T \otimes \mathbf{X})\mathrm{vec}(\mathbf{Y})$, it is easy to see that the second-order differential formula (3.5.17) can be rewritten as

$$\begin{aligned}\mathrm{d}^2 f(\mathbf{Z},\mathbf{Z}^*) = &(\mathrm{d}\,\mathrm{vec}\,\mathbf{Z})^T(\mathbf{D}^T \otimes \mathbf{C})\mathrm{d}\,\mathrm{vec}\,\mathbf{Z} + (\mathrm{d}\,\mathrm{vec}\,\mathbf{Z}^*)^T(\mathbf{F}^T \otimes \mathbf{E})\mathrm{d}\,\mathrm{vec}\,\mathbf{Z}\\ &+ (\mathrm{d}\,\mathrm{vec}\,\mathbf{Z})^T(\mathbf{F}^H \otimes \mathbf{E}^*)\mathrm{d}\,\mathrm{vec}\,\mathbf{Z}^*\\ &+ (\mathrm{d}\,\mathrm{vec}\,\mathbf{Z}^*)^T(\mathbf{D}^H \otimes \mathbf{C}^*)\mathrm{d}\,\mathrm{vec}\,\mathbf{Z}.\end{aligned} \tag{3.5.22}$$

Comparing (3.5.22) with (3.5.19), we have

$$\mathbf{H}_{\mathbf{Z},\mathbf{Z}} = \mathbf{H}_{\mathbf{Z}^*,\mathbf{Z}^*}^* = \frac{1}{2}\left(\mathbf{D}^T \otimes \mathbf{C} + (\mathbf{D}^T \otimes \mathbf{C})^H\right), \tag{3.5.23}$$

$$\mathbf{H}_{\mathbf{Z},\mathbf{Z}^*} = \mathbf{H}_{\mathbf{Z}^*,\mathbf{Z}}^* = \frac{1}{2}\left((\mathbf{F}^T \otimes \mathbf{E}) + (\mathbf{F}^T \otimes \mathbf{E})^H\right). \tag{3.5.24}$$

The results above can be summarized in the following theorem on complex Hessian matrix identification.

Theorem 3.4 *Let $f(\mathbf{Z},\mathbf{Z}^*)$ be a differentiable real function of $m \times n$ matrix variables $\mathbf{Z},\mathbf{Z}^*$. If the second-order differential $\mathrm{d}^2 f(\mathbf{Z},\mathbf{Z}^*)$ has the canonical form (3.5.16) or (3.5.17) then the complex Hessian matrix can be identified by*

$$\mathrm{tr}(\mathbf{C}(\mathrm{d}\mathbf{Z})\mathbf{D}\mathrm{d}\mathbf{Z}) \quad \Leftrightarrow \quad \mathbf{H}_{\mathbf{Z},\mathbf{Z}} = \frac{1}{2}\left(\mathbf{K}_{nm}(\mathbf{D}^T \otimes \mathbf{C}) + (\mathbf{K}_{nm}(\mathbf{D}^T \otimes \mathbf{C}))^T\right)$$

$$\mathrm{tr}(\mathbf{E}(\mathrm{d}\mathbf{Z})\mathbf{F}\mathrm{d}\mathbf{Z}^*) \quad \Leftrightarrow \quad \mathbf{H}_{\mathbf{Z},\mathbf{Z}^*} = \frac{1}{2}\left(\mathbf{K}_{nm}(\mathbf{F}^T \otimes \mathbf{E}) + (\mathbf{K}_{nm}(\mathbf{F}^T \otimes \mathbf{E}))^H\right)$$

or

$$\mathrm{tr}\big(\mathbf{C}(\mathrm{d}\mathbf{Z})\mathbf{D}(\mathrm{d}\mathbf{Z})^T\big) \quad \Leftrightarrow \quad \mathbf{H}_{\mathbf{Z},\mathbf{Z}} = \frac{1}{2}\left(\mathbf{D}^T \otimes \mathbf{C} + (\mathbf{D}^T \otimes \mathbf{C})^T\right)$$

$$\mathrm{tr}\big(\mathbf{E}(\mathrm{d}\mathbf{Z})\mathbf{F}(\mathrm{d}\mathbf{Z}^*)^T\big) \quad \Leftrightarrow \quad \mathbf{H}_{\mathbf{Z},\mathbf{Z}^*} = \frac{1}{2}\left((\mathbf{F}^T \otimes \mathbf{E}) + (\mathbf{F}^T \otimes \mathbf{E})^H\right)$$

together with $\mathbf{H}_{\mathbf{Z}^*,\mathbf{Z}^*} = \mathbf{H}^*_{\mathbf{Z},\mathbf{Z}}$ and $\mathbf{H}_{\mathbf{Z}^*,\mathbf{Z}} = \mathbf{H}^*_{\mathbf{Z},\mathbf{Z}^*}$.

The complex conjugate gradient matrix and the full complex Hessian matrix play an important role in the optimization of an objective function with a complex matrix as variable.

Exercises

3.1 Show that the matrix differential of the Kronecker product is given by $\mathrm{d}(\mathbf{X} \otimes \mathbf{Y}) = (\mathrm{d}\mathbf{X})\mathbf{Y} + \mathbf{X} \otimes \mathrm{d}\mathbf{Y}$.

3.2 Prove that $\mathrm{d}(\mathbf{X}*\mathbf{Y}) = (\mathrm{d}\mathbf{X})\mathbf{Y}+\mathbf{X}*\mathrm{d}\mathbf{Y}$, where $\mathbf{X}*\mathbf{Y}$ represents the Hadamard product of $\mathbf{X}$ and $\mathbf{Y}$.

3.3 Show that $\mathrm{d}(\mathbf{UVW}) = (\mathrm{d}\mathbf{U})\mathbf{VW} + \mathbf{U}(\mathrm{d}\mathbf{V})\mathbf{W} + \mathbf{UV}(\mathrm{d}\mathbf{W})$. More generally, show that $\mathrm{d}(\mathbf{U}_1\mathbf{U}_2\cdots\mathbf{U}_k) = (\mathrm{d}\mathbf{U}_1)\mathbf{U}_2\cdots\mathbf{U}_k + \cdots + \mathbf{U}_1\cdots\mathbf{U}_{k-1}(\mathrm{d}\mathbf{U}_k)$.

3.4 Compute $\mathrm{d}(\mathrm{tr}(\mathbf{A}^k))$, where k is an integer.

3.5 Let the eigenvalues of the matrix $\mathbf{X} \in \mathbb{R}^{n\times n}$ be $\lambda_1,\ldots,\lambda_n$. Use $|\mathbf{X}| = \lambda_1\cdots\lambda_n$ to show that $\mathrm{d}|\mathbf{X}| = |\mathbf{X}|\mathrm{tr}(\mathbf{X}^{-1}\mathrm{d}\mathbf{X})$. (*Hint:* Rewrite the determinant as $|\mathbf{X}| = \exp(\log\lambda_1 + \cdots + \log\lambda_n)$.)

3.6 Find the Jacobian matrices and the gradient matrices of the determinant logarithm functions $\log|\mathbf{X}^T\mathbf{AX}|, \log|\mathbf{XAX}^T|$ and $\log|\mathbf{XAX}|$.

3.7 Show the result $\mathrm{d}\,\mathrm{tr}(\mathbf{X}^T\mathbf{X}) = 2\mathrm{tr}(\mathbf{X}^T\mathrm{d}\mathbf{X})$.

3.8 Show that the matrix differential of a Kronecker product $\mathrm{d}(\mathbf{Y} \otimes \mathbf{Z})$ is $\mathrm{d}\mathbf{Y} \otimes \mathbf{Z} + \mathbf{Y} \otimes \mathrm{d}\mathbf{Z}$.

3.9 Given the trace function $\mathrm{tr}\big((\mathbf{X}^T\mathbf{CX})^{-1}\mathbf{A}\big)$, find its differential and gradient matrix.

3.10 Show that

$$\frac{\mathrm{d}\mathbf{A}^{-1}_{il}}{\mathrm{d}\mathbf{A}_{jk}} = -\sum_k\sum_j \mathbf{A}^{-1}_{ij}\mathbf{A}^{-1}_{kl}.$$

(*Hint:* Write $\mathbf{A}^{-1}\mathbf{A} = \mathbf{I}$ as $\sum_j \mathbf{A}_{ij}^{-1}\mathbf{A}_{jk} = \delta_{ik}$.)

3.11 Given the determinant logarithms $\log|\mathbf{X}^T\mathbf{A}\mathbf{X}|, \log|\mathbf{X}\mathbf{A}\mathbf{X}^T|$ and $\log|\mathbf{X}\mathbf{A}\mathbf{X}|$, find their Hessian matrices.

3.12 Identify the Jacobian matrices of the matrix functions $\mathbf{AXB}$ and $\mathbf{AX}^{-1}\mathbf{B}$.

3.13 Find the gradient matrices of the functions $\mathrm{tr}(\mathbf{AX}^{-1}\mathbf{B})$ and $\mathrm{tr}(\mathbf{AX}^T\mathbf{BXC})$.

3.14 Given the determinant functions $|\mathbf{X}^T\mathbf{AX}|, |\mathbf{XAX}^T|, |\mathbf{XAX}|$ and $|\mathbf{X}^T\mathbf{AX}^T|$, determine their gradient matrices.

3.15 Given a matrix $\mathbf{X} \in \mathbb{C}^{n\times n}$ with eigenvalues $\lambda_1, \ldots, \lambda_n$, use $|\mathbf{X}| = \lambda_1 \cdots \lambda_n$ to prove that $\mathrm{d}|\mathbf{X}| = |\mathbf{X}|\mathrm{tr}(\mathbf{X}^{-1}\mathrm{d}\mathbf{X})$.

3.16 Compute $\mathrm{d}|\mathbf{A}^{-k}|$ and $\mathrm{d}\,\mathrm{tr}(\mathbf{A}^{-k})$, where k is an integer.

3.17 Identify the Jacobian matrices and the gradient matrices of the determinant logarithms $\log|\mathbf{X}^T\mathbf{AX}|, \log|\mathbf{XAX}^T|$ and $\log|\mathbf{XAX}|$, respectively.

3.18 Show that if $\mathbf{F}$ is a matrix function and second differentiable then

$$\mathrm{d}^2 \log|\mathbf{F}| = -\mathrm{tr}(\mathbf{F}^{-1}\mathrm{d}\mathbf{F})^2 + \mathrm{tr}(\mathbf{F}^{-1})\mathrm{d}^2\mathbf{F}.$$

3.19 Use $\mathrm{d}(\mathbf{X}^{-1}\mathbf{X})$ to find the differential of the inverse matrix, $\mathrm{d}\mathbf{X}^{-1}$.

3.20 Show that, for a nonsingular matrix $\mathbf{X} \in \mathbb{R}^{n\times n}$, its inverse matrix $\mathbf{X}^{-1}$ is infinitely differentiable and its rth-order differential is given by

$$\mathrm{d}^{(r)}(\mathbf{X}^{-1}) = (-1)^r r!(\mathbf{X}^{-1}\mathrm{d}\mathbf{X})^r\mathbf{X}^{-1}, \quad r = 1, 2, \ldots$$

3.21 Let $\mathbf{X} \in \mathbb{R}^{m\times n}$, and let its Moore–Penrose inverse matrix be $\mathbf{X}^\dagger$. Show that

$$\mathrm{d}(\mathbf{X}^\dagger\mathbf{X}) = \mathbf{X}^\dagger(\mathrm{d}\mathbf{X})(\mathbf{I}_n - \mathbf{X}^\dagger\mathbf{X}) + \left(\mathbf{X}^\dagger(\mathrm{d}\mathbf{X})(\mathbf{I}_n - \mathbf{X}^\dagger\mathbf{X})\right)^T,$$
$$\mathrm{d}(\mathbf{X}\mathbf{X}^\dagger) = (\mathbf{I}_m - \mathbf{X}\mathbf{X}^\dagger)(\mathrm{d}\mathbf{X})\mathbf{X}^\dagger + \left((\mathbf{I}_m - \mathbf{X}\mathbf{X}^\dagger)(\mathrm{d}\mathbf{X})\mathbf{X}^\dagger\right)^T.$$

(*Hint:* The matrices $\mathbf{X}^\dagger\mathbf{X}$ and $\mathbf{X}\mathbf{X}^\dagger$ are idempotent and symmetric.)

3.22 Find the Hessian matrices of the real scalar functions $f(\mathbf{x}) = \mathbf{a}^T\mathbf{x}$ and $f(\mathbf{x}) = \mathbf{x}^T\mathbf{A}\mathbf{x}$.

3.23 Given a real scalar function $f(\mathbf{X}) = \mathrm{tr}(\mathbf{AXBX}^T)$, show that its Hessian matrix is given by $\mathbf{H}[f(\mathbf{X})] = \mathbf{B}^T \otimes \mathbf{A} + \mathbf{B} \otimes \mathbf{A}^T$.

3.24 Compute the complex matrix differential $\mathrm{d}\,\mathrm{tr}(\mathbf{XAX}^H\mathbf{B})$.

3.25 Given a complex-valued function $f(\mathbf{X}, \mathbf{X}^*) = |\mathbf{XAX}^H\mathbf{B}|$, find its gradient matrix $\nabla_{\mathbf{X}} f(\mathbf{X}, \mathbf{X}^*) = \dfrac{\partial f(\mathbf{X}, \mathbf{X}^*)}{\partial \mathbf{X}}$ and conjugate gradient matrix $\nabla_{\mathbf{X}^*} f(\mathbf{X}, \mathbf{X}^*) = \dfrac{\partial f(\mathbf{X}, \mathbf{X}^*)}{\partial \mathbf{X}^*}$.

3.26 Given a function $f(\mathbf{x}) = (|\mathbf{a}^H\mathbf{x}|^2 - 1)^2$, find its conjugate gradient vector $\dfrac{\partial f(\mathbf{x})}{\partial \mathbf{x}^*}$.

3.27 Let $f(\mathbf{X}) = \log|\det(\mathbf{X})|$, find its conjugate gradient vector $\dfrac{\partial f(\mathbf{X})}{\partial \mathbf{X}^*}$.

3.28 Let $f(\mathbf{Z}, \mathbf{Z}^*) = |\mathbf{ZZ}^H|$, find its gradient matrix and its four part Hessian matrices.

PART II
MATRIX ANALYSIS

4

Gradient Analysis and Optimization

Optimization theory mainly studies the extreme values (the maximal or minimal values) of an objective function, which is usually some real-valued function of a real vector or matrix variable. In many engineering applications, however, the variable of an objective function is a complex vector or matrix. Optimization theory mainly considers (1) the existence conditions for an extremum value (gradient analysis); (2) the design of optimization algorithms and convergence analysis.

4.1 Real Gradient Analysis

Consider an unconstrained minimization problem for $f(\mathbf{x}) : \mathbb{R}^n \to \mathbb{R}$ described by

$$\min_{\mathbf{x} \in S} f(\mathbf{x}), \tag{4.1.1}$$

where $S \in \mathbb{R}^n$ is a subset of the n-dimensional vector space $\mathbb{R}^n$; the vector variable $\mathbf{x} \in S$ is called the *optimization vector* and is chosen so as to fulfil Equation (4.1.1). The function $f : \mathbb{R}^n \to \mathbb{R}$ is known as the *objective function*; it represents the cost or price paid by selecting the optimization vector $\mathbf{x}$, and thus is also called the *cost function*. Conversely, the negative cost function $-f(\mathbf{x})$ can be understood as the *value function* or *utility function* of $\mathbf{x}$. Hence, solving the optimization problem (4.1.1) corresponds to minimizing the cost function or maximizing the value function. That is to say, the minimization problem $\min_{\mathbf{x} \in S} f(\mathbf{x})$ and the maximization problem $\max_{\mathbf{x} \in S}\{-f(\mathbf{x})\}$ of the negative objective function are equivalent.

The above optimization problem has no constraint conditions and thus is called an *unconstrained optimization problem*.

Most nonlinear programming methods for solving unconstrained optimization problems are based on the idea of relaxation and approximation [344].

- *Relaxation* The sequence $\{a_k\}_{k=0}^{\infty}$ is known as a *relaxation sequence*, if $a_{k+1} \leq a_k, \forall\, k \geq 0$. Hence, in the process of solving iteratively the optimization problem (4.1.1) it is necessary to generate a relaxation sequence of the cost function $f(\mathbf{x}_{k+1}) \leq f(\mathbf{x}_k), k = 0, 1, \ldots$

- *Approximation* Approximating an objective function means using a simpler objective function instead of the original objective function.

Therefore, by relaxation and approximation, we can achieve the following.

(1) If the objective function $f(\mathbf{x})$ is lower bounded in the definition domain $S \in \mathbb{R}^n$ then the sequence $\{f(\mathbf{x}_k)\}_{k=0}^{\infty}$ must converge.
(2) In any case, we can improve upon the initial value of the objective function $f(\mathbf{x})$.
(3) The minimization of a nonlinear objective function $f(\mathbf{x})$ can be implemented by numerical methods to a sufficiently high approximation accuracy.

4.1.1 Stationary Points and Extreme Points

The stationary points and extreme points of an objective function play a key role in optimization. The analysis of stationary points depends on the gradient (i.e., the first-order gradient) vector of the objective function, and the analysis of the extreme points depends on the Hessian matrix (i.e., the second-order gradient) of the objective function. Hence the gradient analysis of an objective function is divided into first-order gradient analysis (stationary-point analysis) and second-order gradient analysis (extreme-point analysis).

In optimization, a *stationary point* or *critical point* of a differentiable function of one variable is a point in the domain of the function where its derivative is zero. Informally, a stationary point is a point where the function stops increasing or decreasing (hence the name); it could be a saddle point, however, is usually expected to be a global minimum or maximum point of the objective function $f(\mathbf{x})$.

Definition 4.1 A point $\mathbf{x}^\star$ in the vector subspace $S \in \mathbb{R}^n$ is known as a *global minimum point* of the function $f(\mathbf{x})$ if

$$f(\mathbf{x}^\star) \leq f(\mathbf{x}), \quad \forall\, \mathbf{x} \in S,\ \mathbf{x} \neq \mathbf{x}^\star. \tag{4.1.2}$$

A global minimum point is also called an absolute minimum point. The value of the function at this point, $f(\mathbf{x}^\star)$, is called the *global minimum* or absolute minimum of the function $f(\mathbf{x})$ in the vector subspace S.

Let $\mathcal{D}$ denote the definition domain of a function $f(\mathbf{x})$, where $\mathcal{D}$ may be the whole vector space $\mathbb{R}^n$ or some subset of it. If

$$f(\mathbf{x}^\star) < f(\mathbf{x}), \quad \forall\, \mathbf{x} \in \mathcal{D}, \tag{4.1.3}$$

then $\mathbf{x}^\star$ is said to be a *strict global minimum point* or a strict absolute minimum point of the function $f(\mathbf{x})$.

Needless to say, the ideal goal of minimization is to find the global minimum of a given objective function. However, this desired object is often difficult to achieve. The reasons are as follows.

(1) It is usually difficult to know the global or complete information about a function $f(\mathbf{x})$ in the definition domain $\mathcal{D}$.
(2) It is usually impractical to design an algorithm for identifying a global extreme point, because it is almost impossible to compare the value $f(\mathbf{x}^\star)$ with all values of the function $f(\mathbf{x})$ in the definition domain $\mathcal{D}$.

In contrast, it is much easier to obtain local information about an objective function $f(\mathbf{x})$ in the vicinity of some point $\mathbf{c}$. At the same time, it is much simpler to design an algorithm for comparing the function value at some point $\mathbf{c}$ with other function values at the points near $\mathbf{c}$. Hence, most minimization algorithms can find only a local minimum point $\mathbf{c}$; the value of the objective function at $\mathbf{c}$ is the minimum of its values in the neighborhood of $\mathbf{c}$.

Definition 4.2 Given a point $\mathbf{c} \in \mathcal{D}$ and a positive number r, the set of all points $\mathbf{x}$ satisfying $\|\mathbf{x} - \mathbf{c}\|_2 < r$ is said to be an *open neighborhood* with radius r of the point $\mathbf{c}$, denoted

$$B_o(\mathbf{c}; r) = \{\mathbf{x} | \, \mathbf{x} \in \mathcal{D}, \ \|\mathbf{x} - \mathbf{c}\|_2 < r\}. \tag{4.1.4}$$

If

$$B_c(\mathbf{c}; r) = \{\mathbf{x} | \, \mathbf{x} \in \mathcal{D}, \ \|\mathbf{x} - \mathbf{c}\|_2 \leq r\}, \tag{4.1.5}$$

then we say that $B_c(\mathbf{c}; r)$ is a *closed neighborhood* of $\mathbf{c}$.

Definition 4.3 Consider a scalar $r > 0$, and let $\mathbf{x} = \mathbf{c} + \Delta\mathbf{x}$ be a point in the definition domain $\mathcal{D}$. If

$$f(\mathbf{c}) \leq f(\mathbf{c} + \Delta\mathbf{x}), \quad \forall\, 0 < \|\Delta\mathbf{x}\|_2 \leq r, \tag{4.1.6}$$

then the point $\mathbf{c}$ and the function value $f(\mathbf{c})$ are called a *local minimum point* and a *local minimum* (value) of the function $f(\mathbf{x})$, respectively. If

$$f(\mathbf{c}) < f(\mathbf{c} + \Delta\mathbf{x}), \quad \forall\, 0 < \|\Delta\mathbf{x}\|_2 \leq r, \tag{4.1.7}$$

then the point $\mathbf{c}$ and the function value $f(\mathbf{c})$ are known as a *strict local minimum point* and a *strict local minimum* of the function $f(\mathbf{x})$, respectively. If

$$f(\mathbf{c}) \leq f(\mathbf{x}), \quad \forall\, \mathbf{x} \in \mathcal{D}, \tag{4.1.8}$$

then the point $\mathbf{c}$ and the function value $f(\mathbf{c})$ are said to be a *global minimum point* and a *global minimum* of the function $f(\mathbf{x})$ in the definition domain $\mathcal{D}$, respectively. If

$$f(\mathbf{c}) < f(\mathbf{x}), \quad \forall\, \mathbf{x} \in \mathcal{D}, \, \mathbf{x} \neq \mathbf{c}, \tag{4.1.9}$$

then the point $\mathbf{c}$ and the function value $f(\mathbf{c})$ are referred to as a *strict global minimum point* and a *strict global minimum* of the function $f(\mathbf{x})$ in the definition domain $\mathcal{D}$, respectively.

Definition 4.4 Consider a scalar $r > 0$, and let $\mathbf{x} = \mathbf{c} + \Delta\mathbf{x}$ be a point in the definition domain $\mathcal{D}$. If

$$f(\mathbf{c}) \geq f(\mathbf{c} + \Delta\mathbf{x}), \quad \forall 0 < \|\Delta\mathbf{x}\|_2 \leq r, \tag{4.1.10}$$

then the point $\mathbf{c}$ and the function value $f(\mathbf{c})$ are called a *local maximum point* and a *local maximum* (value) of the function $f(\mathbf{x})$, respectively. If

$$f(\mathbf{c}) > f(\mathbf{c} + \Delta\mathbf{x}), \quad \forall 0 < \|\Delta\mathbf{x}\|_2 \leq r, \tag{4.1.11}$$

then the point $\mathbf{c}$ and the function value $f(\mathbf{c})$ are known as a *strict local maximum point* and a *strict local maximum* of the function $f(\mathbf{x})$, respectively. If

$$f(\mathbf{c}) \geq f(\mathbf{x}), \quad \forall \mathbf{x} \in \mathcal{D}, \tag{4.1.12}$$

then the point $\mathbf{c}$ and the function value $f(\mathbf{c})$ are said to be a *global maximum point* and a *global maximum* of the function $f(\mathbf{x})$ in the definition domain $\mathcal{D}$, respectively. If

$$f(\mathbf{c}) > f(\mathbf{x}), \quad \forall \mathbf{x} \in \mathcal{D}, \mathbf{x} \neq \mathbf{c}, \tag{4.1.13}$$

then the point $\mathbf{c}$ and the function value $f(\mathbf{c})$ are referred to as a *strict global maximum point* and a *strict global maximum* of the function $f(\mathbf{x})$ in the definition domain $\mathcal{D}$, respectively.

It should be emphasized that the minimum points and the maximum points are collectively called the *extreme points* of the function $f(\mathbf{x})$, and the minimal value and the maximal value are known as the *extrema* or extreme values of $f(\mathbf{x})$.

A local minimum (maximum) point and a strict local minimum (maximum) point are sometimes called a weak local minimum (maximum) point and a strong local minimum (maximum) point, respectively.

In particular, if some point $\mathbf{x}_0$ is a unique local extreme point of a function $f(\mathbf{x})$ in the neighborhood $B(\mathbf{c}; r)$ then it is said to be an *isolated local extreme point.*

In order to facilitate the understanding of the stationary points and other extreme points, Figure 4.1 shows the curve of a one-variable function $f(x)$ with definition domain $\mathcal{D} = (0, 6)$.

The points $x = 1$ and $x = 4$ are a local minimum point and a local maximum point, respectively; $x = 2$ and $x = 3$ are a strict local maximum point and a strict local minimum point, respectively; all these points are extrema, while $x = 5$ is only a saddle point.

4.1.2 Real Gradient Analysis of $f(\mathbf{x})$

In practical applications, it is still very troublesome to compare directly the value of an objective function $f(\mathbf{x})$ at some point with all possible values in its neighborhood. Fortunately, the Taylor series expansion provides a simple method for coping with this difficulty.

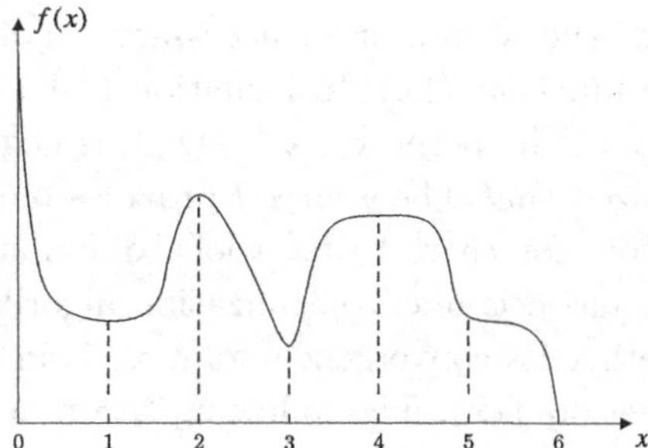

Figure 4.1 Extreme points of a one-variable function defined over the interval $(0, 6)$.

The second-order Taylor series approximation of $f(\mathbf{x})$ at the point $\mathbf{c}$ is given by

$$f(\mathbf{c}+\Delta\mathbf{x}) = f(\mathbf{c}) + (\nabla f(\mathbf{c}))^T \Delta\mathbf{x} + \frac{1}{2}(\Delta\mathbf{x})^T \mathbf{H}[f(\mathbf{c})]\Delta\mathbf{x}, \tag{4.1.14}$$

where

$$\nabla f(\mathbf{c}) = \frac{\partial f(\mathbf{c})}{\partial \mathbf{c}} = \left.\frac{\partial f(\mathbf{x})}{\partial \mathbf{x}}\right|_{\mathbf{x}=\mathbf{c}}, \tag{4.1.15}$$

$$\mathbf{H}[f(\mathbf{c})] = \frac{\partial^2 f(\mathbf{c})}{\partial \mathbf{c}\,\partial \mathbf{c}^T} = \left.\frac{\partial^2 f(\mathbf{x})}{\partial \mathbf{x}\,\partial \mathbf{x}^T}\right|_{\mathbf{x}=\mathbf{c}} \tag{4.1.16}$$

are respectively the gradient vector and the Hessian matrix of $f(\mathbf{x})$ at the point $\mathbf{c}$.

From (4.1.14) we get the following two necessary conditions of a local minimizer of $f(\mathbf{x})$.

Theorem 4.1 (First-order necessary conditions) [353] *If $\mathbf{x}_*$ is a local extreme point of a function $f(\mathbf{x})$ and $f(\mathbf{x})$ is continuously differentiable in the neighborhood $B(\mathbf{x}_*; r)$ of the point $\mathbf{x}_*$, then*

$$\nabla_{\mathbf{x}_*} f(\mathbf{x}_*) = \left.\frac{\partial f(\mathbf{x})}{\partial \mathbf{x}}\right|_{\mathbf{x}=\mathbf{x}_*} = \mathbf{0}. \tag{4.1.17}$$

Theorem 4.2 (Second-order necessary conditions) [311], [353] *If $\mathbf{x}_*$ is a local minimum point of $f(\mathbf{x})$ and the second-order gradient $\nabla^2_{\mathbf{x}} f(\mathbf{x})$ is continuous in an open neighborhood $B_o(\mathbf{x}_*; r)$ of $\mathbf{x}_*$, then*

$$\nabla_{\mathbf{x}_*} f(\mathbf{x}_*) = \left.\frac{\partial f(\mathbf{x})}{\partial \mathbf{x}}\right|_{\mathbf{x}=\mathbf{x}_*} = \mathbf{0} \quad \textit{and} \quad \nabla^2_{\mathbf{x}_*} f(\mathbf{x}_*) = \left.\frac{\partial^2 f(\mathbf{x})}{\partial \mathbf{x}\,\partial \mathbf{x}^T}\right|_{\mathbf{x}=\mathbf{x}_*} \succeq 0, \tag{4.1.18}$$

i.e., the gradient vector of $f(\mathbf{x})$ at the point $\mathbf{x}_*$ is a zero vector and the Hessian matrix $\nabla^2_{\mathbf{x}} f(\mathbf{x})$ at the point $\mathbf{x}_*$ is a positive semi-definite matrix.

If Equation (4.1.18) in Theorem 4.2 is replaced by

$$\nabla_{\mathbf{x}_*} f(\mathbf{x}_*) = \left.\frac{\partial f(\mathbf{x})}{\partial \mathbf{x}}\right|_{\mathbf{x}=\mathbf{x}_*} = \mathbf{0} \quad \text{and} \quad \nabla^2_{\mathbf{x}_*} f(\mathbf{x}_*) = \left.\frac{\partial^2 f(\mathbf{x})}{\partial \mathbf{x}\,\partial \mathbf{x}^T}\right|_{\mathbf{x}=\mathbf{x}_*} \preceq 0 \tag{4.1.19}$$

then Theorem 4.2 yields the second-order necessary condition for $\mathbf{x}_*$ to be a local maximum point of the function $f(\mathbf{x})$. In Equation (4.1.19), $\nabla^2_{\mathbf{x}_*} f(\mathbf{x}_*) \preceq 0$ means that the Hessian matrix at the point $\mathbf{x}_*$, $\nabla^2_{\mathbf{x}_*} f(\mathbf{x}_*)$, is negative semi-definite.

It should be emphasized that Theorem 4.2 provides only a necessary condition, not a sufficient condition, for there to be a local minimum point of the function $f(\mathbf{x})$. However, for an unconstrained optimization algorithm, we usually hope to determine directly whether its convergence point $\mathbf{x}_*$ is in fact an extreme point of the given objective function $f(\mathbf{x})$. The following theorem gives the answer to this question.

Theorem 4.3 (Second-order sufficient conditions) [311], [353] *Let $\nabla^2_{\mathbf{x}} f(\mathbf{x})$ be continuous in an open neighborhood of $\mathbf{x}_*$. If the conditions*

$$\nabla_{\mathbf{x}_*} f(\mathbf{x}_*) = \left.\frac{\partial f(\mathbf{x})}{\partial \mathbf{x}}\right|_{\mathbf{x}=\mathbf{x}_*} = \mathbf{0} \quad \textit{and} \quad \nabla^2_{\mathbf{x}_*} f(\mathbf{x}_*) = \left.\frac{\partial^2 f(\mathbf{x})}{\partial \mathbf{x} \partial \mathbf{x}^T}\right|_{\mathbf{x}=\mathbf{x}_*} \succ 0 \tag{4.1.20}$$

are satisfied then $\mathbf{x}_*$ is a strict local minimum point of the objection function $f(\mathbf{x})$. Here $\nabla^2_{\mathbf{x}} f(\mathbf{x}_*) \succ 0$ means that the Hessian matrix $\nabla^2_{\mathbf{x}_*} f(\mathbf{x}_*)$ at the point $\mathbf{x}_*$ is a positive definite matrix.

Note 1 Second-order necessary conditions for a local maximum point: If $\mathbf{x}_*$ is a local maximum point of $f(\mathbf{x})$, and the second-order gradient $\nabla^2_{\mathbf{x}} f(\mathbf{x})$ is continuous in an open neighborhood $B_o(\mathbf{x}_*; r)$ of $\mathbf{x}_*$, then

$$\nabla_{\mathbf{x}_*} f(\mathbf{x}_*) = \left.\frac{\partial f(\mathbf{x})}{\partial \mathbf{x}}\right|_{\mathbf{x}=\mathbf{x}_*} = \mathbf{0} \quad \text{and} \quad \nabla^2_{\mathbf{x}_*} f(\mathbf{x}_*) = \left.\frac{\partial^2 f(\mathbf{x})}{\partial \mathbf{x} \partial \mathbf{x}^T}\right|_{\mathbf{x}=\mathbf{x}_*} \preceq 0. \tag{4.1.21}$$

Note 2 Second-order sufficient conditions for a strict local maximum point: suppose that $\nabla^2_{\mathbf{x}} f(\mathbf{x})$ is continuous in an open neighborhood of $\mathbf{x}_*$; if the conditions

$$\nabla_{\mathbf{x}_*} f(\mathbf{x}_*) = \left.\frac{\partial f(\mathbf{x})}{\partial \mathbf{x}}\right|_{\mathbf{x}=\mathbf{x}_*} = \mathbf{0} \quad \text{and} \quad \nabla^2_{\mathbf{x}_*} f(\mathbf{x}_*) = \left.\frac{\partial^2 f(\mathbf{x})}{\partial \mathbf{x} \partial \mathbf{x}^T}\right|_{\mathbf{x}=\mathbf{x}_*} \prec 0 \tag{4.1.22}$$

are satisfied then $\mathbf{x}_*$ is a strict local maximum point of the objection function $f(\mathbf{x})$.

If $\nabla f(\mathbf{x}_*) = \mathbf{0}$ but $\nabla^2_{\mathbf{x}_*} f(\mathbf{x}_*)$ is an indefinite matrix, then $\mathbf{x}_*$ is a saddle point, not an extreme point, of $f(\mathbf{x})$ in the neighborhood $B(\mathbf{x}_*; r)$.

4.1.3 Real Gradient Analysis of $f(\mathbf{X})$

Now consider a real function $f(\mathbf{X}) : \mathbb{R}^{m\times n} \to \mathbb{R}$ with matrix variable $\mathbf{X} \in \mathbb{R}^{m\times n}$. First, perform the vectorization of the matrix variable to get an $mn \times 1$ vector $\text{vec}\,\mathbf{X}$.

Let $\mathcal{D}$ be the subset of the matrix space $\mathbb{R}^{m\times n}$ which is the definition domain of the $m \times n$ matrix variable $\mathbf{X}$, i.e., $\mathbf{X} \in \mathcal{D}$.

Definition 4.5 A *neighborhood* $B(\mathbf{X}_*;r)$ with center point $\text{vec}\,\mathbf{X}_*$ and the radius r is defined as

$$B(\mathbf{X}_*;r) = \{\mathbf{X}|\mathbf{X} \in \mathbb{R}^{m\times n},\ \|\text{vec}\,\mathbf{X} - \text{vec}\,\mathbf{X}_*\|_2 < r\}. \tag{4.1.23}$$

From Equation (4.1.16), the second-order Taylor series approximation formula of the function $f(\mathbf{X})$ at the point $\mathbf{X}_*$ is given by

$$\begin{aligned} f(\mathbf{X}_* + \Delta\mathbf{X}) &= f(\mathbf{X}_*) + \left(\frac{\partial f(\mathbf{X}_*)}{\partial \text{vec}\,\mathbf{X}_*}\right)^T \text{vec}(\Delta\mathbf{X}) \\ &\quad + \frac{1}{2}(\text{vec}(\Delta\mathbf{X}))^T \frac{\partial^2 f(\mathbf{X}_*)}{\partial \text{vec}\,\mathbf{X}_* \partial (\text{vec}\,\mathbf{X}_*)^T} \text{vec}(\Delta\mathbf{X}) \\ &= f(\mathbf{X}_*) + (\nabla_{\text{vec}\,\mathbf{X}_*} f(\mathbf{X}_*))^T \text{vec}(\Delta\mathbf{X}) \\ &\quad + \frac{1}{2}(\text{vec}(\Delta\mathbf{X}))^T \mathbf{H}[f(\mathbf{X}_*)]\text{vec}(\Delta\mathbf{X}), \end{aligned} \tag{4.1.24}$$

where

$$\begin{aligned} \nabla_{\text{vec}\,\mathbf{X}_*} f(\mathbf{X}_*) &= \left.\frac{\partial f(\mathbf{X})}{\partial \text{vec}\,\mathbf{X}}\right|_{\mathbf{X}=\mathbf{X}_*} \in \mathbb{R}^{mn} \\ \mathbf{H}[f(\mathbf{X}_*)] &= \left.\frac{\partial^2 f(\mathbf{X})}{\partial \text{vec}\,\mathbf{X} \partial (\text{vec}\,\mathbf{X})^T}\right|_{\mathbf{X}=\mathbf{X}_*} \in \mathbb{R}^{mn\times mn} \end{aligned}$$

are the gradient vector and the Hessian matrix of $f(\mathbf{X})$ at the point $\mathbf{X}_*$, respectively.

Comparing Equation (4.1.24) with Equation (4.1.14), we have the following conditions similar to Theorems 4.1–4.3.

(1) *First-order necessary condition for a minimizer* If $\mathbf{X}_*$ is a local extreme point of a function $f(\mathbf{X})$, and $f(\mathbf{X})$ is continuously differentiable in the neighborhood $B(\mathbf{X}_*;r)$ of the point $\mathbf{X}_*$, then

$$\nabla_{\text{vec}\,\mathbf{X}_*} f(\mathbf{X}_*) = \left.\frac{\partial f(\mathbf{X})}{\partial \text{vec}\,\mathbf{X}}\right|_{\mathbf{X}=\mathbf{X}_*} = \mathbf{0}. \tag{4.1.25}$$

(2) *Second-order necessary conditions for a local minimum point* If $\mathbf{X}_*$ is a local minimum point of $f(\mathbf{X})$, and the Hessian matrix $\mathbf{H}[f(\mathbf{X})]$ is continuous in an open neighborhood $B_o(\mathbf{X}_*;r)$ then

$$\nabla_{\text{vec}\,\mathbf{X}_*} f(\mathbf{X}_*) = \mathbf{0} \quad \text{and} \quad \left.\frac{\partial^2 f(\mathbf{X})}{\partial \text{vec}\,\mathbf{X} \partial (\text{vec}\,\mathbf{X})^T}\right|_{\mathbf{X}=\mathbf{X}_*} \succeq 0. \tag{4.1.26}$$

(3) *Second-order sufficient conditions for a strict local minimum point* It is assumed that $\mathbf{H}[f(\mathbf{x})]$ is continuous in an open neighborhood of $\mathbf{X}_*$. If the conditions

$$\nabla_{\text{vec}\,\mathbf{X}_*} f(\mathbf{X}_*) = \mathbf{0} \quad \text{and} \quad \left.\frac{\partial^2 f(\mathbf{X})}{\partial \text{vec}\,\mathbf{X} \partial (\text{vec}\,\mathbf{X})^T}\right|_{\mathbf{X}=\mathbf{X}_*} \succ 0 \tag{4.1.27}$$

are satisfied, then $\mathbf{X}_*$ is a strict local minimum point of $f(\mathbf{X})$.

(4) *Second-order necessary conditions for a local maximum point* If $\mathbf{X}_*$ is a local maximum point of $f(\mathbf{X})$, and the Hessian matrix $\mathbf{H}[f(\mathbf{X})]$ is continuous in an open neighborhood $B_o(\mathbf{X}_*; r)$, then

$$\nabla_{\mathrm{vec}\mathbf{X}_*} f(\mathbf{X}_*) = \mathbf{0} \quad \text{and} \quad \left.\frac{\partial^2 f(\mathbf{X})}{\partial \mathrm{vec}\mathbf{X}\partial(\mathrm{vec}\mathbf{X})^T}\right|_{\mathbf{X}=\mathbf{X}_*} \preceq 0. \tag{4.1.28}$$

(5) *Second-order sufficient conditions for a strict local maximum point* It is assumed that $\mathbf{H}[f(\mathbf{X})]$ is continuous in an open neighborhood of $\mathbf{X}_*$. If the conditions

$$\nabla_{\mathrm{vec}\mathbf{X}_*} f(\mathbf{X}_*) = \mathbf{0} \quad \text{and} \quad \left.\frac{\partial^2 f(\mathbf{X})}{\partial \mathrm{vec}\mathbf{X}\partial(\mathrm{vec}\mathbf{X})^T}\right|_{\mathbf{X}=\mathbf{X}_*} \prec 0 \tag{4.1.29}$$

are satisfied then $\mathbf{X}_*$ is a strict local maximum point of $f(\mathbf{X})$.

The point $\mathbf{X}_*$ is only a saddle point of $f(\mathbf{X})$, if

$$\nabla_{\mathrm{vec}\mathbf{X}_*} f(\mathbf{X}_*) = \mathbf{0} \quad \text{and} \quad \left.\frac{\partial^2 f(\mathbf{X})}{\partial \mathrm{vec}\mathbf{X}\partial(\mathrm{vec}\mathbf{X})^T}\right|_{\mathbf{X}=\mathbf{X}_*} \quad \text{is indefinite.}$$

4.2 Gradient Analysis of Complex Variable Function

In many engineering applications (such as wireless communications, radar and sonar), the signals encountered in practice are real-valued functions of complex vector variables. In this section we discuss the extreme point of a complex variable function and gradient analysis in unconstrained optimization.

4.2.1 Extreme Point of Complex Variable Function

Consider the unconstrained optimization of a real-valued function $f(\mathbf{z}, \mathbf{z}^*) : \mathbb{C}^n \times \mathbb{C}^n \to \mathbb{R}$. From the first-order differential

$$\begin{aligned} \mathrm{d}f(\mathbf{z}, \mathbf{z}^*) &= \frac{\partial f(\mathbf{z}, \mathbf{z}^*)}{\partial \mathbf{z}^T}\mathrm{d}\mathbf{z} + \frac{\partial f(\mathbf{z}, \mathbf{z}^*)}{\partial \mathbf{z}^H}\mathrm{d}\mathbf{z}^* \\ &= \left[\frac{\partial f(\mathbf{z}, \mathbf{z}^*)}{\partial \mathbf{z}^T}, \frac{\partial f(\mathbf{z}, \mathbf{z}^*)}{\partial \mathbf{z}^H}\right]\begin{bmatrix}\mathrm{d}\mathbf{z}\\ \mathrm{d}\mathbf{z}^*\end{bmatrix} \end{aligned} \tag{4.2.1}$$

and the second-order differential

$$\mathrm{d}^2 f(\mathbf{z}, \mathbf{z}^*) = [\mathrm{d}\mathbf{z}^H, \mathrm{d}\mathbf{z}^T]\begin{bmatrix} \dfrac{\partial^2 f(\mathbf{z}, \mathbf{z}^*)}{\partial \mathbf{z}^*\partial \mathbf{z}^T} & \dfrac{\partial^2 f(\mathbf{z}, \mathbf{z}^*)}{\partial \mathbf{z}^*\partial \mathbf{z}^H} \\ \dfrac{\partial^2 f(\mathbf{z}, \mathbf{z}^*)}{\partial \mathbf{z}\partial \mathbf{z}^T} & \dfrac{\partial^2 f(\mathbf{z}, \mathbf{z}^*)}{\partial \mathbf{z}\partial \mathbf{z}^H} \end{bmatrix}\begin{bmatrix}\mathrm{d}\mathbf{z}\\ \mathrm{d}\mathbf{z}^*\end{bmatrix}, \tag{4.2.2}$$

it can be easily seen that the Taylor series approximation of the function $f(\mathbf{z},\mathbf{z}^*)$ at the point $\mathbf{c}$ is given by

$$\begin{aligned} f(\mathbf{z},\mathbf{z}^*) &\approx f(\mathbf{c},\mathbf{c}^*) + \left[\frac{\partial f(\mathbf{c},\mathbf{c}^*)}{\partial \mathbf{c}}, \frac{\partial f(\mathbf{c},\mathbf{c}^*)}{\partial \mathbf{c}^*}\right]\begin{bmatrix}\Delta\mathbf{c}\\ \Delta\mathbf{c}^*\end{bmatrix} \\ &\quad + \frac{1}{2}[\Delta\mathbf{c}^H, \Delta\mathbf{c}^T]\begin{bmatrix}\dfrac{\partial^2 f(\mathbf{c},\mathbf{c}^*)}{\partial \mathbf{c}^*\partial \mathbf{c}^T} & \dfrac{\partial^2 f(\mathbf{c},\mathbf{c}^*)}{\partial \mathbf{c}^*\partial \mathbf{c}^H}\\ \dfrac{\partial^2 f(\mathbf{c},\mathbf{c}^*)}{\partial \mathbf{c}\partial \mathbf{c}^T} & \dfrac{\partial^2 f(\mathbf{c},\mathbf{c}^*)}{\partial \mathbf{c}\partial \mathbf{c}^H}\end{bmatrix}\begin{bmatrix}\Delta\mathbf{c}\\ \Delta\mathbf{c}^*\end{bmatrix} \\ &= f(\mathbf{c},\mathbf{c}^*) + (\nabla f(\mathbf{c},\mathbf{c}^*))^T\Delta\tilde{\mathbf{c}} + \frac{1}{2}(\Delta\tilde{\mathbf{c}})^H\mathbf{H}(f(\mathbf{c},\mathbf{c}^*))\Delta\tilde{\mathbf{c}}, \end{aligned} \tag{4.2.3}$$

where $\Delta\tilde{\mathbf{c}} = \begin{bmatrix}\Delta\mathbf{c}\\ \Delta\mathbf{c}^*\end{bmatrix}$ with $\Delta\mathbf{c} = \mathbf{z} - \mathbf{c}$ and $\Delta\mathbf{c}^* = \mathbf{z}^* - \mathbf{c}^*$; while the gradient vector is

$$\nabla f(\mathbf{c},\mathbf{c}^*) = \begin{bmatrix}\dfrac{\partial f(\mathbf{z},\mathbf{z}^*)}{\partial \mathbf{z}}\\ \dfrac{\partial f(\mathbf{z},\mathbf{z}^*)}{\partial \mathbf{z}^*}\end{bmatrix}_{\mathbf{x}=\mathbf{c}} \in \mathbb{C}^{2n}, \tag{4.2.4}$$

and the Hessian matrix is

$$\mathbf{H}[f(\mathbf{c},\mathbf{c}^*)] = \begin{bmatrix}\dfrac{\partial^2 f(\mathbf{z},\mathbf{z}^*)}{\partial \mathbf{z}^*\partial \mathbf{z}^T} & \dfrac{\partial^2 f(\mathbf{z},\mathbf{z}^*)}{\partial \mathbf{z}^*\partial \mathbf{z}^H}\\ \dfrac{\partial^2 f(\mathbf{z},\mathbf{z}^*)}{\partial \mathbf{z}\partial \mathbf{z}^T} & \dfrac{\partial^2 f(\mathbf{z},\mathbf{z}^*)}{\partial \mathbf{z}\partial \mathbf{z}^H}\end{bmatrix}_{\mathbf{x}=\mathbf{c}} \in \mathbb{C}^{2n\times 2n}. \tag{4.2.5}$$

Consider the neighborhood of the point $\mathbf{c}$

$$B(\mathbf{c};r) = \{\mathbf{x}|0 < \|\Delta\mathbf{c}\|_2 = \|\mathbf{x}-\mathbf{c}\|_2 \le r\},$$

where $\|\Delta\mathbf{c}\|_2$ is small enough so that the second-order term in Equation (4.2.3) can be ignored. This yields the first-order approximation to the function, as follows:

$$f(\mathbf{z},\mathbf{z}^*) \approx f(\mathbf{c},\mathbf{c}^*) + (\nabla f(\mathbf{c},\mathbf{c}^*))^T\Delta\tilde{\mathbf{c}} \quad (\text{for } \|\Delta\mathbf{c}\|_2 \text{ small enough}). \tag{4.2.6}$$

Clearly, in order that $f(\mathbf{c},\mathbf{c}^*)$ takes a minimal value or a maximal value, it is required that $\nabla f(\mathbf{c},\mathbf{c}^*)) = \mathbf{0}_{2n\times 1}$. This condition is equivalent to $\partial f(\mathbf{c},\mathbf{c}^*)/\partial\mathbf{c} = \mathbf{0}_{n\times 1}$ and $\partial f(\mathbf{c},\mathbf{c}^*)/\partial\mathbf{c}^* = \mathbf{0}_{n\times 1}$, or simply the condition that the conjugate derivative vector of the function $f(\mathbf{z},\mathbf{z}^*)$ at the point $\mathbf{c}$ is a zero vector, namely

$$\frac{\partial f(\mathbf{c},\mathbf{c}^*)}{\partial \mathbf{c}^*} = \left.\frac{\partial f(\mathbf{z},\mathbf{z}^*)}{\partial \mathbf{z}^*}\right|_{\mathbf{z}=\mathbf{c}} = \mathbf{0}_{n\times 1}. \tag{4.2.7}$$

The Taylor series approximation to the real-valued function $f(\mathbf{z},\mathbf{z}^*)$ at its stationary point $\mathbf{c}$ is given to second-order by

$$f(\mathbf{z},\mathbf{z}^*) \approx f(\mathbf{c},\mathbf{c}^*) + \frac{1}{2}(\Delta\tilde{\mathbf{c}})^H\mathbf{H}[f(\mathbf{c},\mathbf{c}^*)]\Delta\tilde{\mathbf{c}}. \tag{4.2.8}$$

From Equations (4.2.7) and (4.2.8) we have the following conditions on extreme points, similar to Theorems 4.1–4.3.

(1) *First-order necessary condition for a local extreme point* If $\mathbf{z}_*$ is a local extreme point of a function $f(\mathbf{z}, \mathbf{z}^*)$, and $f(\mathbf{z}, \mathbf{z}^*)$ is continuously differentiable in the neighborhood $B(\mathbf{c}; r)$ of the point $\mathbf{c}$, then

$$\nabla_{\mathbf{c}} f(\mathbf{c}, \mathbf{c}^*) = \left.\frac{\partial f(\mathbf{z}, \mathbf{z}^*)}{\partial \mathbf{z}}\right|_{\mathbf{z}=\mathbf{c}} = \mathbf{0}. \tag{4.2.9}$$

(2) *Second-order necessary conditions for a local minimum point* If $\mathbf{c}$ is a local minimum point of $f(\mathbf{z}, \mathbf{z}^*)$, and the Hessian matrix $\mathbf{H}[f(\mathbf{z}, \mathbf{z}^*)]$ is continuous in an open neighborhood $B_o(\mathbf{c}; r)$ of the point $\mathbf{c}$, then

$$\frac{\partial f(\mathbf{c}, \mathbf{c}^*)}{\partial \mathbf{c}^*} = \mathbf{0} \quad \text{and} \quad \mathbf{H}[f(\mathbf{c}, \mathbf{c}^*)] = \begin{bmatrix} \frac{\partial^2 f(\mathbf{z},\mathbf{z}^*)}{\partial \mathbf{z}^* \partial \mathbf{z}^T} & \frac{\partial^2 f(\mathbf{z},\mathbf{z}^*)}{\partial \mathbf{z}^* \partial \mathbf{z}^H} \\ \frac{\partial^2 f(\mathbf{z},\mathbf{z}^*)}{\partial \mathbf{z} \partial \mathbf{z}^T} & \frac{\partial^2 f(\mathbf{z},\mathbf{z}^*)}{\partial \mathbf{z} \partial \mathbf{z}^H} \end{bmatrix}_{\mathbf{z}=\mathbf{c}} \succeq 0. \tag{4.2.10}$$

(3) *Second-order sufficient conditions for a strict local minimum point* Suppose that $\mathbf{H}[f(\mathbf{z}, \mathbf{z}^*)]$ is continuous in an open neighborhood of $\mathbf{c}$. If the conditions

$$\frac{\partial f(\mathbf{c}, \mathbf{c}^*)}{\partial \mathbf{c}^*} = \mathbf{0} \quad \text{and} \quad \mathbf{H}[f(\mathbf{c}, \mathbf{c}^*)] \succ 0 \tag{4.2.11}$$

are satisfied then $\mathbf{c}$ is a strict local minimum point of $f(\mathbf{z}, \mathbf{z}^*)$.

(4) *Second-order necessary conditions for a local maximum point* If $\mathbf{c}$ is a local maximum point of $f(\mathbf{z}, \mathbf{z}^*)$, and the Hessian matrix $\mathbf{H}[f(\mathbf{z}, \mathbf{z}^*)]$ is continuous in an open neighborhood $B_o(\mathbf{c}; r)$ of the point $\mathbf{c}$, then

$$\frac{\partial f(\mathbf{c}, \mathbf{c}^*)}{\partial \mathbf{c}^*} = \mathbf{0} \quad \text{and} \quad \mathbf{H}[f(\mathbf{c}, \mathbf{c}^*)] \preceq 0. \tag{4.2.12}$$

(5) *Second-order sufficient conditions for a strict local maximum point* Suppose that $\mathbf{H}[f(\mathbf{z}, \mathbf{z}^*)]$ is continuous in an open neighborhood of $\mathbf{c}$. If the conditions

$$\frac{\partial f(\mathbf{c}, \mathbf{c}^*)}{\partial \mathbf{c}^*} = \mathbf{0} \quad \text{and} \quad \mathbf{H}[f(\mathbf{c}, \mathbf{c}^*)] \prec 0. \tag{4.2.13}$$

are satisfied then $\mathbf{c}$ is a strict local maximum point of $f(\mathbf{z}, \mathbf{z}^*)$.

If $\frac{\partial f(\mathbf{c}, \mathbf{c}^*)}{\partial \mathbf{c}^*} = \mathbf{0}$, but the Hessian matrix $\mathbf{H}[f(\mathbf{c}, \mathbf{c}^*)]$ is indefinite, then $\mathbf{c}$ is only a saddle point of $f(\mathbf{z}, \mathbf{z}^*)$, not an extremum.

For a real-valued function $f(\mathbf{Z}, \mathbf{Z}^*) : \mathbb{C}^{m\times n} \times \mathbb{C}^{m\times n} \to \mathbb{R}$ with complex matrix variable $\mathbf{Z}$, it is necessary to perform the vectorization of the matrix variables $\mathbf{Z}$ and $\mathbf{Z}^*$ respectively, to get $\mathrm{vec}\,\mathbf{Z}$ and $\mathrm{vec}\,\mathbf{Z}^*$.

From the first-order differential

$$\begin{aligned} \mathrm{d}f(\mathbf{Z}, \mathbf{Z}^*) &= \left(\frac{\partial f(\mathbf{Z}, \mathbf{Z}^*)}{\partial \mathrm{vec}\,\mathbf{Z}}\right)^T \mathrm{vec}(\mathrm{d}\mathbf{Z}) + \left(\frac{\partial f(\mathbf{Z}, \mathbf{Z}^*)}{\partial \mathrm{vec}\,\mathbf{Z}^*}\right)^T \mathrm{vec}(\mathrm{d}\mathbf{Z}^*) \\ &= \left[\frac{\partial f(\mathbf{Z}, \mathbf{Z}^*)}{\partial (\mathrm{vec}\,\mathbf{Z})^T}, \frac{\partial f(\mathbf{Z}, \mathbf{Z}^*)}{\partial (\mathrm{vec}\,\mathbf{Z}^*)^T)}\right] \begin{bmatrix} \mathrm{vec}(\mathrm{d}\mathbf{Z}) \\ \mathrm{vec}(\mathrm{d}\mathbf{Z})^* \end{bmatrix} \end{aligned} \tag{4.2.14}$$

and the second-order differential

$$\mathrm{d}^2 f(\mathbf{Z},\mathbf{Z}^*) = \begin{bmatrix} \mathrm{vec}(\mathrm{d}\mathbf{Z}) \\ \mathrm{vec}(\mathrm{d}\mathbf{Z}^*) \end{bmatrix}^H \begin{bmatrix} \frac{\partial^2 f(\mathbf{Z},\mathbf{Z}^*)}{\partial(\mathrm{vec}\,\mathbf{Z}^*)\partial(\mathrm{vec}\,\mathbf{Z})^T} & \frac{\partial^2 f(\mathbf{Z},\mathbf{Z}^*)}{\partial(\mathrm{vec}\,\mathbf{Z}^*)\partial(\mathrm{vec}\,\mathbf{Z}^*)^T} \\ \frac{\partial^2 f(\mathbf{Z},\mathbf{Z}^*)}{\partial(\mathrm{vec}\,\mathbf{Z})\partial(\mathrm{vec}\,\mathbf{Z})^T} & \frac{\partial^2 f(\mathbf{Z},\mathbf{Z}^*)}{\partial(\mathrm{vec}\,\mathbf{Z})\partial(\mathrm{vec}\,\mathbf{Z}^*)^T)} \end{bmatrix} \begin{bmatrix} \mathrm{vec}(\mathrm{d}\mathbf{Z}) \\ \mathrm{vec}(\mathrm{d}\mathbf{Z}^*) \end{bmatrix} \tag{4.2.15}$$

it follows that the second-order series approximation of $f(\mathbf{Z},\mathbf{Z}^*)$ at the point $\mathbf{C}$ is given by

$$\begin{aligned} f(\mathbf{Z},\mathbf{Z}^*) = f(\mathbf{C},\mathbf{C}^*) &+ (\nabla f(\mathbf{C},\mathbf{C}^*))^T \mathrm{vec}(\Delta\tilde{\mathbf{C}}) \\ &+ \frac{1}{2}(\mathrm{vec}(\Delta\tilde{\mathbf{C}}))^H \mathbf{H}[f(\mathbf{C},\mathbf{C}^*)]\mathrm{vec}(\Delta\tilde{\mathbf{C}}), \end{aligned} \tag{4.2.16}$$

where

$$\Delta\tilde{\mathbf{C}} = \begin{bmatrix} \Delta\mathbf{C} \\ \Delta\mathbf{C}^* \end{bmatrix} = \begin{bmatrix} \mathbf{Z}-\mathbf{C} \\ \mathbf{Z}^*-\mathbf{C}^* \end{bmatrix} \in \mathbb{C}^{2m\times n} \tag{4.2.17}$$

$$\nabla f(\mathbf{C},\mathbf{C}^*) = \begin{bmatrix} \frac{\partial f(\mathbf{Z},\mathbf{Z}^*)}{\partial(\mathrm{vec}\,\mathbf{Z})} \\ \frac{\partial f(\mathbf{Z},\mathbf{Z}^*)}{\partial(\mathrm{vec}\,\mathbf{Z}^*)} \end{bmatrix}_{\mathbf{Z}=\mathbf{C}} \in \mathbb{C}^{2mn\times 1}, \tag{4.2.18}$$

and

$$\begin{aligned} \mathbf{H}[f(\mathbf{C},\mathbf{C}^*)] &= \mathbf{H}[f(\mathbf{Z},\mathbf{Z}^*)]|_{\mathbf{Z}=\mathbf{C}} \\ &= \begin{bmatrix} \frac{\partial^2 f(\mathbf{Z},\mathbf{Z}^*)}{\partial(\mathrm{vec}\,\mathbf{Z}^*)\partial(\mathrm{vec}\,\mathbf{Z})^T} & \frac{\partial^2 f(\mathbf{Z},\mathbf{Z}^*)}{\partial(\mathrm{vec}\,\mathbf{Z}^*)\partial(\mathrm{vec}\,\mathbf{Z}^*)^T} \\ \frac{\partial^2 f(\mathbf{Z},\mathbf{Z}^*)}{\partial(\mathrm{vec}\,\mathbf{Z})\partial(\mathrm{vec}\,\mathbf{Z})^T} & \frac{\partial^2 f(\mathbf{Z},\mathbf{Z}^*)}{\partial(\mathrm{vec}\,\mathbf{Z})\partial(\mathrm{vec}\,\mathbf{Z}^*)^T} \end{bmatrix}_{\mathbf{Z}=\mathbf{C}} \\ &= \begin{bmatrix} \mathbf{H}_{\mathbf{Z}^*\mathbf{Z}} & \mathbf{H}_{\mathbf{Z}^*\mathbf{Z}^*} \\ \mathbf{H}_{\mathbf{Z}\mathbf{Z}} & \mathbf{H}_{\mathbf{Z}\mathbf{Z}^*} \end{bmatrix}_{\mathbf{Z}=\mathbf{C}} \in \mathbb{C}^{2mn\times 2mn}. \end{aligned} \tag{4.2.19}$$

Notice that

$$\begin{aligned} \nabla f(\mathbf{C},\mathbf{C}^*) = \mathbf{0}_{2mn\times 1} &\Leftrightarrow \frac{\partial f(\mathbf{Z},\mathbf{Z}^*)}{\partial \mathrm{vec}\,\mathbf{Z}} = \mathbf{0}_{mn\times 1}, \\ \frac{\partial f(\mathbf{Z},\mathbf{Z}^*)}{\partial \mathrm{vec}\,\mathbf{Z}^*} = \mathbf{0}_{mn\times 1} &\Leftrightarrow \frac{\partial f(\mathbf{Z},\mathbf{Z}^*)}{\partial \mathbf{Z}^*} = \mathbf{O}_{m\times n}. \end{aligned}$$

On comparing Equation (4.2.16) with Equation (4.2.3), we have the following conditions for the extreme points of $f(\mathbf{Z},\mathbf{Z}^*)$.

(1) *First-order necessary condition for a local extreme point* If $\mathbf{C}$ is a local extreme point of a function $f(\mathbf{Z},\mathbf{Z}^*)$, and $f(\mathbf{Z},\mathbf{Z}^*)$ is continuously differentiable in the neighborhood $B(\mathbf{C};r)$ of the point $\mathbf{C}$, then

$$\nabla f(\mathbf{C},\mathbf{C}^*) = \mathbf{0} \quad \text{or} \quad \left.\frac{\partial f(\mathbf{Z},\mathbf{Z}^*)}{\partial \mathbf{Z}^*}\right|_{\mathbf{Z}=\mathbf{C}} = \mathbf{O}. \tag{4.2.20}$$

(2) *Second-order necessary conditions for a local minimum point* If $\mathbf{C}$ is a local minimum point of $f(\mathbf{Z},\mathbf{Z}^*)$, and the Hessian matrix $\mathbf{H}[f(\mathbf{Z},\mathbf{Z}^*)]$ is continuous in an open neighborhood $B_o(\mathbf{C};r)$ of the point $\mathbf{C}$, then

$$\left.\frac{\partial f(\mathbf{Z},\mathbf{Z}^*)}{\partial \mathbf{Z}^*}\right|_{\mathbf{Z}=\mathbf{C}} = \mathbf{O} \quad \text{and} \quad \mathbf{H}[f(\mathbf{C},\mathbf{C}^*)] \succeq 0. \tag{4.2.21}$$

(3) *Second-order sufficient conditions for a strict local minimum point* Suppose that $\mathbf{H}[f(\mathbf{Z},\mathbf{Z}^*)]$ is continuous in an open neighborhood of $\mathbf{C}$. If the conditions

$$\left.\frac{\partial f(\mathbf{Z},\mathbf{Z}^*)}{\partial \mathbf{Z}^*}\right|_{\mathbf{Z}=\mathbf{C}} = \mathbf{O} \quad \text{and} \quad \mathbf{H}[f(\mathbf{C},\mathbf{C}^*)] \succ 0 \tag{4.2.22}$$

are satisfied, then $\mathbf{C}$ is a strict local minimum point of $f(\mathbf{Z},\mathbf{Z}^*)$.

(4) *Second-order necessary conditions for a local maximum point* If $\mathbf{C}$ is a local maximum point of $f(\mathbf{Z},\mathbf{Z}^*)$, and the Hessian matrix $\mathbf{H}[f(\mathbf{Z},\mathbf{Z}^*)]$ is continuous in an open neighborhood $B_o(\mathbf{C};r)$ of the point $\mathbf{C}$, then

$$\left.\frac{\partial f(\mathbf{Z},\mathbf{Z}^*)}{\partial \mathbf{Z}^*}\right|_{\mathbf{Z}=\mathbf{C}} = \mathbf{O} \quad \text{and} \quad \mathbf{H}[f(\mathbf{C},\mathbf{C}^*)] \preceq 0. \tag{4.2.23}$$

(5) *Second-order sufficient conditions for a strict local maximum point* Suppose that $\mathbf{H}[f(\mathbf{Z},\mathbf{Z}^*)]$ is continuous in an open neighborhood of $\mathbf{C}$, if the conditions

$$\left.\frac{\partial f(\mathbf{Z},\mathbf{Z}^*)}{\partial \mathbf{Z}^*}\right|_{\mathbf{Z}=\mathbf{C}} = \mathbf{O} \quad \text{and} \quad \mathbf{H}[f(\mathbf{C},\mathbf{C}^*)] \prec 0 \tag{4.2.24}$$

are satisfied, then $\mathbf{C}$ is a strict local maximum point of $f(\mathbf{Z},\mathbf{Z}^*)$.

If $\left.\frac{\partial f(\mathbf{Z},\mathbf{Z}^*)}{\partial \mathbf{Z}^*}\right|_{\mathbf{Z}=\mathbf{C}} = \mathbf{O}$, but the Hessian matrix $\mathbf{H}[f(\mathbf{C},\mathbf{C}^*)]$ is indefinite, then $\mathbf{C}$ is only a saddle point of $f(\mathbf{Z},\mathbf{Z}^*)$.

Table 4.1 summarizes the extreme-point conditions for three complex variable functions.

The Hessian matrices in Table 4.1 are defined as follows:

$$\mathbf{H}[f(c,c^*)] = \begin{bmatrix} \frac{\partial^2 f(z,z^*)}{\partial z^* \partial z} & \frac{\partial^2 f(z,z^*)}{\partial z^* \partial z^*} \\ \frac{\partial^2 f(z,z^*)}{\partial z \partial z} & \frac{\partial^2 f(z,z^*)}{\partial z \partial z^*} \end{bmatrix}_{z=c} \in \mathbb{C}^{2\times 2},$$

$$\mathbf{H}[f(\mathbf{c},\mathbf{c}^*)] = \begin{bmatrix} \frac{\partial^2 f(\mathbf{z},\mathbf{z}^*)}{\partial \mathbf{z}^* \partial \mathbf{z}^T} & \frac{\partial^2 f(\mathbf{z},\mathbf{z}^*)}{\partial \mathbf{z}^* \partial \mathbf{z}^H} \\ \frac{\partial^2 f(\mathbf{z},\mathbf{z}^*)}{\partial \mathbf{z} \partial \mathbf{z}^T} & \frac{\partial^2 f(\mathbf{z},\mathbf{z}^*)}{\partial \mathbf{z} \partial \mathbf{z}^H} \end{bmatrix}_{\mathbf{z}=\mathbf{c}} \in \mathbb{C}^{2n\times 2n},$$

$$\mathbf{H}[f(\mathbf{C},\mathbf{C}^*)] = \begin{bmatrix} \frac{\partial^2 f(\mathbf{Z},\mathbf{Z}^*)}{\partial(\text{vec}\,\mathbf{Z}^*)\partial(\text{vec}\,\mathbf{Z})^T} & \frac{\partial^2 f(\mathbf{Z},\mathbf{Z}^*)}{\partial(\text{vec}\,\mathbf{Z}^*)\partial(\text{vec}\,\mathbf{Z}^*)^T} \\ \frac{\partial^2 f(\mathbf{Z},\mathbf{Z}^*)}{\partial(\text{vec}\,\mathbf{Z})\partial(\text{vec}\,\mathbf{Z})^T} & \frac{\partial^2 f(\mathbf{Z},\mathbf{Z}^*)}{\partial(\text{vec}\,\mathbf{Z})\partial(\text{vec}\,\mathbf{Z}^*)^T} \end{bmatrix}_{\mathbf{Z}=\mathbf{C}} \in \mathbb{C}^{2mn\times 2mn}.$$

Table 4.1 Extreme point conditions for the complex variable functions

	$f(z,z^*):\mathbb{C}\to\mathbb{R}$	$f(\mathbf{z},\mathbf{z}^*):\mathbb{C}^n\to\mathbb{R}$	$f(\mathbf{Z},\mathbf{Z}^*):\mathbb{C}^{m\times n}\to\mathbb{R}$
Stationary point	$\left.\frac{\partial f(z,z^*)}{\partial z^*}\right\vert_{z=c}=0$	$\left.\frac{\partial f(\mathbf{z},\mathbf{z}^*)}{\partial \mathbf{z}^*}\right\vert_{\mathbf{z}=\mathbf{c}}=\mathbf{0}$	$\left.\frac{\partial f(\mathbf{Z},\mathbf{Z}^*)}{\partial \mathbf{Z}^*}\right\vert_{\mathbf{Z}=\mathbf{C}}=\mathbf{O}$
Local minimum	$\mathbf{H}[f(c,c^*)]\succeq 0$	$\mathbf{H}[f(\mathbf{c},\mathbf{c}^*)]\succeq 0$	$\mathbf{H}[f(\mathbf{C},\mathbf{C}^*)]\succeq 0$
Strict local minimum	$\mathbf{H}[f(c,c^*)]\succ 0$	$\mathbf{H}[f(\mathbf{c},\mathbf{c}^*)]\succ 0$	$\mathbf{H}[f(\mathbf{C},\mathbf{C}^*)]\succ 0$
Local maximum	$\mathbf{H}[f(c,c^*)]\preceq 0$	$\mathbf{H}[f(\mathbf{c},\mathbf{c}^*)]\preceq 0$	$\mathbf{H}[f(\mathbf{C},\mathbf{C}^*)]\preceq 0$
Strict local maximum	$\mathbf{H}[f(c,c^*)]\prec 0$	$\mathbf{H}[f(\mathbf{c},\mathbf{c}^*)]\prec 0$	$\mathbf{H}[f(\mathbf{C},\mathbf{C}^*)]\prec 0$
Saddle point	$\mathbf{H}[f(c,c^*)]$ indef.	$\mathbf{H}[f(\mathbf{c},\mathbf{c}^*)]$ indef.	$\mathbf{H}[f(\mathbf{C},\mathbf{C}^*)]$ indef.

4.2.2 Complex Gradient Analysis

Given a real-valued objective function $f(\mathbf{w},\mathbf{w}^*)$ or $f(\mathbf{W},\mathbf{W}^*)$, the gradient analysis of its unconstrained minimization problem can be summarized as follows.

(1) The conjugate gradient matrix determines a closed solution of the minimization problem.
(2) The sufficient conditions of local minimum points are determined by the conjugate gradient vector and the Hessian matrix of the objective function.
(3) The negative direction of the conjugate gradient vector determines the steepest-descent iterative algorithm for solving the minimization problem.
(4) The Hessian matrix gives the Newton algorithm for solving the minimization problem.

In the following, we discuss the above gradient analysis.

1. Closed Solution of the Unconstrained Minimization Problem

By letting the conjugate gradient vector (or matrix) be equal to a zero vector (or matrix), we can find a closed solution of the given unconstrained minimization problem.

EXAMPLE 4.1 For an over-determined matrix equation $\mathbf{Az}=\mathbf{b}$, define the error sum of squares,

$$\begin{aligned} J(\mathbf{z}) &= \|\mathbf{Az}-\mathbf{b}\|_2^2 = (\mathbf{Az}-\mathbf{b})^H(\mathbf{Az}-\mathbf{b}) \\ &= \mathbf{z}^H\mathbf{A}^H\mathbf{Az} - \mathbf{z}^H\mathbf{A}^H\mathbf{b} - \mathbf{b}^H\mathbf{Az} + \mathbf{b}^H\mathbf{b}, \end{aligned}$$

as the objective function. Let its conjugate gradient vector $\nabla_{\mathbf{z}^*}J(\mathbf{z}) = \mathbf{A}^H\mathbf{Az} -$

$\mathbf{A}^H\mathbf{b}$ be equal to a zero vector. Clearly, if $\mathbf{A}^H\mathbf{A}$ is nonsingular then

$$\mathbf{z} = (\mathbf{A}^H\mathbf{A})^{-1}\mathbf{A}^H\mathbf{b}. \tag{4.2.25}$$

This is the LS solution of the over-determined matrix equation $\mathbf{Az} = \mathbf{b}$.

EXAMPLE 4.2 For an over-determined matrix equation $\mathbf{Az} = \mathbf{b}$, define the log-likelihood function

$$l(\hat{\mathbf{z}}) = C - \frac{1}{\sigma^2}\mathbf{e}^H\mathbf{e} = C - \frac{1}{\sigma^2}(\mathbf{b} - \mathbf{A}\hat{\mathbf{z}})^H(\mathbf{b} - \mathbf{A}\hat{\mathbf{z}}), \tag{4.2.26}$$

where C is a real constant. Then, the conjugate gradient of the log-likelihood function is given by

$$\nabla_{\hat{\mathbf{z}}^*} l(\hat{\mathbf{z}}) = \frac{\partial l(\hat{\mathbf{z}})}{\partial \mathbf{z}^*} = \frac{1}{\sigma^2}\mathbf{A}^H\mathbf{b} - \frac{1}{\sigma^2}\mathbf{A}^H\mathbf{A}\hat{\mathbf{z}}.$$

By setting $\nabla_{\hat{\mathbf{z}}^*} l(\hat{\mathbf{z}}) = \mathbf{0}$, we obtain $\mathbf{A}^H\mathbf{A}\mathbf{z}_{\text{opt}} = \mathbf{A}^H\mathbf{b}$, where $\mathbf{z}_{\text{opt}}$ is the solution for maximizing the log-likelihood function $l(\hat{\mathbf{z}})$. Hence, if $\mathbf{A}^H\mathbf{A}$ is nonsingular then

$$\mathbf{z}_{\text{opt}} = (\mathbf{A}^H\mathbf{A})^{-1}\mathbf{A}^H\mathbf{b}. \tag{4.2.27}$$

This is the ML solution of the matrix equation $\mathbf{Az} = \mathbf{b}$. Clearly, the ML solution and the LS solution are the same for the matrix equation $\mathbf{Az} = \mathbf{b}$.

2. Sufficient Condition for Local Minimum Point

The sufficient conditions for a strict local minimum point are given by Equation (4.2.11). In particular, for a convex function $f(\mathbf{z}, \mathbf{z}^*)$, any local minimum point $\mathbf{z}_0$ is a global minimum point of the function. If the convex function $f(\mathbf{z}, \mathbf{z}^*)$ is differentiable then the stationary point $\mathbf{z}_0$ such that $\left.\frac{\partial f(\mathbf{z}, \mathbf{z}^*)}{\partial \mathbf{z}^*}\right|_{\mathbf{z}=\mathbf{z}_0} = \mathbf{0}$ is a global minimum point of $f(\mathbf{z}, \mathbf{z}^*)$.

3. Steepest Descent Direction of Real Objective Function

There are two choices in determining a stationary point $\mathbf{C}$ of an objective function $f(\mathbf{Z}, \mathbf{Z}^*)$ with a complex matrix as variable:

$$\left.\frac{\partial f(\mathbf{Z}, \mathbf{Z}^*)}{\partial \mathbf{Z}}\right|_{\mathbf{Z}=\mathbf{C}} = \mathbf{O}_{m\times n} \quad \text{or} \quad \left.\frac{\partial f(\mathbf{Z}, \mathbf{Z}^*)}{\partial \mathbf{Z}^*}\right|_{\mathbf{Z}=\mathbf{C}} = \mathbf{O}_{m\times n}. \tag{4.2.28}$$

The question is which gradient to select when designing a learning algorithm for an optimization problem. To answer this question, it is necessary to introduce the definition of the curvature direction.

DEFINITION 4.6 [158] If $\mathbf{H}$ is the Hessian operator acting on a nonlinear function $f(\mathbf{x})$, a vector $\mathbf{p}$ is said to be:

(1) the *direction of positive curvature* of the function f, if $\mathbf{p}^H\mathbf{H}\mathbf{p} > 0$;

(2) the *direction of zero curvature* of the function f, if $\mathbf{p}^H\mathbf{H}\mathbf{p} = 0$;

(3) the *direction of negative curvature* of the function f, if $\mathbf{p}^H\mathbf{H}\mathbf{p} < 0$.

The scalar $\mathbf{p}^H\mathbf{H}\mathbf{p}$ is called the *curvature* of the function f along the direction $\mathbf{p}$.

The *curvature direction* is the direction of the maximum rate of change of the objective function.

Theorem 4.4 [57] *Let $f(\mathbf{z})$ be a real-valued function of the complex vector $\mathbf{z}$. By regarding $\mathbf{z}$ and $\mathbf{z}^*$ as independent variables, the curvature direction of the real objective function $f(\mathbf{z}, \mathbf{z}^*)$ is given by the conjugate gradient vector $\nabla_{\mathbf{z}^*} f(\mathbf{z}, \mathbf{z}^*)$.*

Theorem 4.4 shows that each component of $\nabla_{\mathbf{z}^*} f(\mathbf{z}, \mathbf{z}^*)$, or $\nabla_{\text{vec}\,\mathbf{Z}^*} f(\mathbf{Z}, \mathbf{Z}^*)$, gives the rate of change of the objective function $f(\mathbf{z}, \mathbf{z}^*)$, or $f(\mathbf{Z}, \mathbf{Z}^*)$, in this direction:

- the conjugate gradient vector $\nabla_{\mathbf{z}^*} f(\mathbf{z}, \mathbf{z}^*)$, or $\nabla_{\text{vec}\,\mathbf{Z}^*} f(\mathbf{Z}, \mathbf{Z}^*)$, gives the fastest growing direction of the objective function;
- the negative gradient vector $-\nabla_{\mathbf{z}^*} f(\mathbf{z}, \mathbf{z}^*)$, or $-\nabla_{\text{vec}\,\mathbf{Z}^*} f(\mathbf{Z}, \mathbf{Z}^*)$, provides the steepest decreasing direction of the objective function.

Hence, along the negative conjugate gradient direction, the objective function reaches its minimum point at the fastest rate. This optimization algorithm is called the *gradient descent algorithm* or *steepest descent algorithm*.

As a geometric interpretation of Theorem 4.4, Figure 4.2 shows the gradient and the conjugate gradient of the function $f(z) = |z|^2$ at the points c_1 and c_2, where $\nabla_{c_i^*} f = \left.\dfrac{\partial f(z)}{\partial z^*}\right|_{z=c_i}, i = 1, 2.$

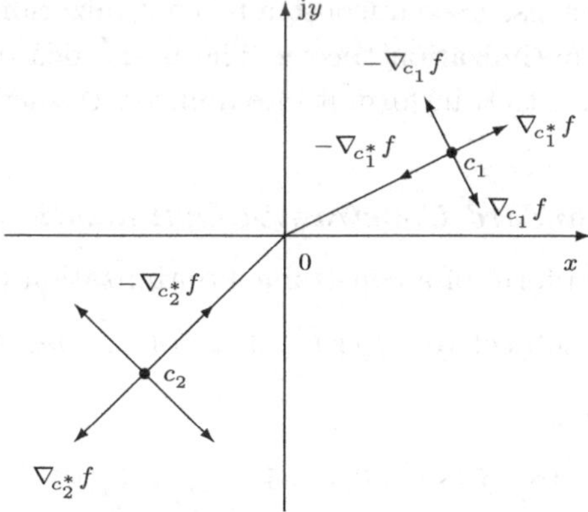

Figure 4.2 The gradient and the conjugate gradient of $f(z) = |z|^2$.

Clearly, only the negative direction of the conjugate gradient $-\nabla_{\mathbf{z}^*} f(\mathbf{z})$ points to the global minimum $z^\star = 0$ of the objective function. Hence, in minimization problems, the negative direction of the conjugate gradient is used as the updating direction:

$$\mathbf{z}_k = \mathbf{z}_{k-1} - \mu \nabla_{\mathbf{z}^*} f(\mathbf{z}), \quad \mu > 0. \tag{4.2.29}$$

That is to say, the correction amount $-\mu \nabla_{\mathbf{z}^*} f(\mathbf{z})$ of alternate solutions in the iterative process is proportional to the negative conjugate gradient of the objective function.

Equation (4.2.29) is called the *learning algorithm* for the alternate solution of the optimization problem. Because the direction of the negative conjugate gradient vector always points in the decreasing direction of an objection function, this type of learning algorithm is called a steepest descent algorithm.

The constant μ in a steepest descent algorithm is referred to as the *learning step* and determines the rate at which the alternate solution converges to the optimization solution.

4. Newton Algorithm for Solving Minimization Problems

The conjugate gradient vector contains only first-order information about an objective function. If, further, we adopt the second-order derivative information provided by the Hessian matrix of the objective function, then this promises an optimization algorithm with better performance. An optimization algorithm based on the Hessian matrix is the well-known Newton method.

The Newton method is a simple and efficient constrained optimization algorithm, and is widely applied.

4.3 Convex Sets and Convex Function Identification

In the above section we discussed unconstrained optimization problems. This section presents constrained optimization theory. The basic idea of solving a constrained optimization problem is to transform it into a unconstrained optimization problem.

4.3.1 Standard Constrained Optimization Problems

Consider the standard form of a constrained optimization problem

$$\min_{\mathbf{x}} f_0(\mathbf{x}) \quad \text{subject to} \quad f_i(\mathbf{x}) \le 0,\ i = 1, \ldots, m,\ \mathbf{A}\mathbf{x} = \mathbf{b} \tag{4.3.1}$$

which can be written as

$$\min_{\mathbf{x}} f_0(\mathbf{x}) \quad \text{subject to} \quad f_i(\mathbf{x}) \le 0, i = 1, \ldots, m,\ h_i(\mathbf{x}) = 0, i = 1, \ldots, q. \tag{4.3.2}$$

The variable $\mathbf{x}$ in constrained optimization problems is called the *optimization variable* or *decision variable* and the function $f_0(\mathbf{x})$ is the *objective function* or cost function, while

$$f_i(\mathbf{x}) \le 0,\ \mathbf{x} \in \mathcal{I}, \quad \text{and} \quad h_i(\mathbf{x}) = 0,\ \mathbf{x} \in \mathcal{E}, \tag{4.3.3}$$

are known as the *inequality constraint* and the *equality constraint*, respectively; $\mathcal{I}$ and $\mathcal{E}$ are respectively the definition domains of the inequality constraint function and the equality constraint function, i.e.,

$$\mathcal{I} = \bigcap_{i=1}^{m} \text{dom}\, f_i \quad \text{and} \quad \mathcal{E} = \bigcap_{i=1}^{q} \text{dom}\, h_i. \tag{4.3.4}$$

The inequality constraint and the equality constraint are collectively called *explicit constraints.* An optimization problem without explicit constraints (i.e., $m = q = 0$) reduces to an unconstrained optimization problem.

The inequality constraint $f_i(\mathbf{x}) \leq 0, i = 1, \ldots, m$ and the equality constraint $h_i(\mathbf{x}) = 0, i = 1, \ldots, q$, denote $m + q$ strict requirements or stipulations restricting the possible choices of $\mathbf{x}$. An objective function $f_0(\mathbf{x})$ represents the costs paid by selecting $\mathbf{x}$. Conversely, the negative objective function $-f_0(\mathbf{x})$ can be understood as the values or benefits got by selecting $\mathbf{x}$. Hence, solving the constrained optimization problem (4.3.2) amounts to choosing $\mathbf{x}$ given $m + q$ strict requirements such that the cost is minimized or the value is maximized.

The *optimal solution* of the constrained optimization problem (4.3.2) is denoted $p^\star$, and is defined as the *infimum* of the objective function $f_0(\mathbf{x})$:

$$p^\star = \inf\{f_0(\mathbf{x}) | f_i(\mathbf{x}) \leq 0,\ i = 1, \ldots, m,\ h_i(\mathbf{x}) = 0,\ i = 1, \ldots, q\}. \tag{4.3.5}$$

If $p^\star = \infty$, then the constrained optimization problem (4.3.2) is said to be *infeasible*, i.e., no point $\mathbf{x}$ can meet the $m + q$ constraint conditions. If $p^\star = -\infty$ then the constrained optimization problem (4.3.2) is *lower unbounded.* The following are the key steps in solving the constrained optimization (4.3.2).

(1) Search for the feasible points of a given constrained optimization problem.
(2) Search for the point such that the constrained optimization problem reaches the optimal value.
(3) Avoid turning the original constrained optimization into a lower unbounded problem.

The point $\mathbf{x}$ meeting all the inequality constraints and equality constraints is called a *feasible point.* The set of all feasible points is called the *feasible domain* or *feasible set*, denoted $\mathcal{F}$, and is defined as

$$\mathcal{F} \stackrel{\text{def}}{=} \mathcal{I} \cap \mathcal{E} = \{\mathbf{x} | f_i(\mathbf{x}) \leq 0,\ i = 1, \ldots, m,\ h_i(\mathbf{x}) = 0,\ i = 1, \ldots, q\}. \tag{4.3.6}$$

The points not in the feasible set are referred to as *infeasible points.*

The intersection set of the definition domain $\text{dom} f_0$ and the feasible domain $\mathcal{F}$ of an objective function,

$$\mathcal{D} = \text{dom} f_0 \cap \bigcap_{i=1}^{m} \text{dom} f_i \cap \bigcap_{i=1}^{p} \text{dom} h_i = \text{dom} f_0 \cap \mathcal{F}, \tag{4.3.7}$$

is called the *definition domain* of a constrained optimization problem.

A feasible point $\mathbf{x}$ is optimal if $f_0(\mathbf{x}) = p^\star$. It is usually difficult to solve a constrained optimization problem. When the number of decision variables in $\mathbf{x}$ is large, it is particularly difficult. This is due to the following reasons [211].

(1) A constrained optimization problem may be occupied by a local optimal solution in its definition domain.
(2) It is very difficult to find a feasible point.
(3) The stopping criteria in general unconstrained optimization algorithms usually fail in a constrained optimization problem.
(4) The convergence rate of a constrained optimization algorithm is usually poor.
(5) The size of numerical problem can make a constrained minimization algorithm either completely stop or wander, so that these algorithm cannot converge in the normal way.

The above difficulties in solving constrained optimization problems can be overcome by using the convex optimization technique. In essence, *convex optimization* is the minimization (or maximization) of a convex (or concave) function under the constraint of a convex set. Convex optimization is a fusion of optimization, convex analysis and numerical computation.

4.3.2 Convex Sets and Convex Functions

First, we introduce two basic concepts in convex optimization.

Definition 4.7 A set $S \in \mathbb{R}^n$ is *convex*, if the line connecting any two points $\mathbf{x}, \mathbf{y} \in S$ is also in the set S, namely

$$\mathbf{x}, \mathbf{y} \in S,\ \theta \in [0,1] \quad \Rightarrow \quad \theta\mathbf{x} + (1-\theta)\mathbf{y} \in S. \tag{4.3.8}$$

Figure 4.3 gives a schematic diagram of a convex set and a *nonconvex set.*

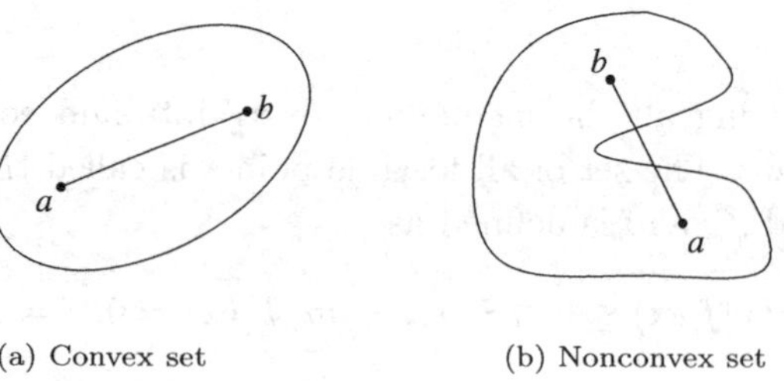

Figure 4.3 Convex set and nonconvex set.

Many familiar sets are convex, e.g., the *unit ball* $S = \{\mathbf{x} \,|\, \|\mathbf{x}\|_2 \leq 1\}$. However, the unit sphere $S = \{\mathbf{x} \,|\, \|\mathbf{x}\|_2 = 1\}$ is not a convex set because the line connecting two points on the sphere is obviously not itself on the sphere.

A convex set has the following important properties [344, Theorem 2.2.4]. Letting $S_1 \subseteq \mathbb{R}^n$ and $S_2 \subseteq \mathbb{R}^m$ be two convex sets, and $\mathcal{A}(\mathbf{x}) : \mathbb{R}^n \to \mathbb{R}^m$ be the linear operator such that $\mathcal{A}(\mathbf{x}) = \mathbf{A}\mathbf{x} + \mathbf{b}$, then:

1. The *intersection* $S_1 \cap S_2 = \{\mathbf{x} \in \mathbb{R}^n | \mathbf{x} \in S_1, \mathbf{x} \in S_2\}$ $(m = n)$ is a convex set.
2. The *sum of sets* $S_1 + S_2 = \{\mathbf{z} = \mathbf{x} + \mathbf{y} | \mathbf{x} \in S_1, \mathbf{y} \in S_2\}$ $(m = n)$ is a convex set.
3. The *direct sum* $S_1 \oplus S_2 = \{(\mathbf{x}, \mathbf{y}) \in \mathbb{R}^{n+m} | \mathbf{x} \in S_1, \mathbf{y} \in S_2\}$ is a convex set.
4. The *conic hull* $\mathcal{K}(S_1) = \{\mathbf{z} \in \mathbb{R}^n | \mathbf{z} = \beta\mathbf{x}, \mathbf{x} \in S_1, \beta \geq 0\}$ is a convex set.
5. The *affine image* $\mathcal{A}(S_1) = \{\mathbf{y} \in \mathbb{R}^m | \mathbf{y} = \mathcal{A}(\mathbf{x}), \mathbf{x} \in S_1\}$ is a convex set.
6. The *inverse affine image* $\mathcal{A}^{-1}(S_2) = \{\mathbf{x} \in \mathbb{R}^n | \mathbf{x} = \mathcal{A}^{-1}(\mathbf{y}), \mathbf{y} \in S_2\}$ is a convex set.
7. The following *convex hull* is a convex set:

$$\operatorname{conv}(S_1, S_2) = \{\mathbf{z} \in \mathbb{R}^n | \mathbf{z} = \alpha\mathbf{x} + (1-\alpha)\mathbf{y}, \mathbf{x} \in S_1, \mathbf{y} \in S_2; \alpha \in [0, 1]\}.$$

The most important property of a convex set is an extension of Property 1: the intersection of any number (or even an uncountable number) of convex sets is still a convex set. For instance, the intersection of the unit ball and nonnegative orthant $\mathbb{R}^n_+$, $S = \{\mathbf{x} | \, \|\mathbf{x}\|_2 \leq 1, x_i \geq 0\}$, retains the convexity. However, the union of two convex sets is not usually convex. For example, $S_1 = \{\mathbf{x} | \, \|\mathbf{x}\|_2 \leq 1\}$ and $S_2 = \{\mathbf{x} | \, \|\mathbf{x} - 3 \times \mathbf{1}\|_2 \leq 1\}$ (where $\mathbf{1}$ is a vector with all entries equal to 1) are convex sets, but their union $S_1 \cup S_2$ is not convex, because the segment connecting any two points in the two balls is not in $S_1 \cup S_2$.

Given a vector $\mathbf{x} \in \mathbb{R}^n$ and a constant $\rho > 0$, then

$$B_o(\mathbf{x}, \rho) = \{\mathbf{y} \in \mathbb{R}^n | \, \|\mathbf{y} - \mathbf{x}\|_2 < \rho\}, \tag{4.3.9}$$

$$B_c(\mathbf{x}, \rho) = \{\mathbf{y} \in \mathbb{R}^n | \, \|\mathbf{y} - \mathbf{x}\|_2 \leq \rho\}, \tag{4.3.10}$$

are called the *open ball* and the *closed ball* with center $\mathbf{x}$ and radius ρ.

A convex set $S \subseteq \mathbb{R}^n$ is known as a *convex cone* if all rays starting at the origin and all segments connecting any two points of these rays are still within the convex set, i.e.,

$$\mathbf{x}, \mathbf{y} \in S, \ \lambda, \mu \geq 0 \quad \Rightarrow \quad \lambda\mathbf{x} + \mu\mathbf{y} \in S. \tag{4.3.11}$$

The *nonnegative orthant* $\mathbb{R}^n_+ = \{\mathbf{x} \in \mathbb{R}^n | \mathbf{x} \succeq 0\}$ is a convex cone. The set of positive semi-definite matrices $\mathbf{X} \succeq 0$, $S^n_+ = \{\mathbf{X} \in \mathbb{R}^{n \times n} | \mathbf{X} \succeq 0\}$, is a convex cone as well, since the positive combination of any number of positive semi-definite matrices is still positive semi-definite. Hence, S^n_+ is called the *positive semi-definite cone*.

Definition 4.8 The vector function $\mathbf{f}(\mathbf{x}) : \mathbb{R}^n \to \mathbb{R}^m$ is known as an *affine function* if it has the form

$$\mathbf{f}(\mathbf{x}) = \mathbf{A}\mathbf{x} + \mathbf{b}. \tag{4.3.12}$$

Similarly, the matrix function $\mathbf{F}(\mathbf{x}) : \mathbb{R}^n \to \mathbb{R}^{p\times q}$ is called an affine function, if it has the form

$$\mathbf{F}(\mathbf{x}) = \mathbf{A}_0 + x_1\mathbf{A}_1 + \cdots + x_n\mathbf{A}_n, \tag{4.3.13}$$

where $\mathbf{A}_i \in \mathbb{R}^{p\times q}$. An affine function sometimes is roughly referred to as a linear function (see below).

Definition 4.9 [211] Given $n \times 1$ vectors $\mathbf{x}_i \in \mathbb{R}^n$ and real numbers $\theta_i \in \mathbb{R}$, then $\mathbf{y} = \theta_1\mathbf{x}_1 + \cdots + \theta_k\mathbf{x}_k$ is called:

(a) a *linear combination* for any real numbers θ_i;

(b) an *affine combination* if $\sum_i \theta_i = 1$;

(c) a *convex combination* if $\sum_i \theta_i = 1$ and all $\theta_i \geq 0$;

(d) a *conic combination* if $\theta_i \geq 0, i = 1, \ldots, k$.

An intersection of sets has the following properties [211].

1. The intersection of subspaces is also a subspace.
2. The intersection of affine functions is also an affine function.
3. The intersection of convex sets is also a convex set.
4. The intersection of convex cones is also a convex cone.

Definition 4.10 [344] Given a convex set $S \in \mathbb{R}^n$ and a function $f : S \to \mathbb{R}$, then we have the following.

- The function $f : \mathbb{R}^n \to \mathbb{R}$ is *convex* if and only if $S = \mathrm{dom} f$ is a convex set, and for all vectors $\mathbf{x}, \mathbf{y} \in S$ and each scalar $\alpha \in (0, 1)$, the function satisfies the *Jensen inequality*

$$f(\alpha\mathbf{x} + (1-\alpha)\mathbf{y}) \leq \alpha f(\mathbf{x}) + (1-\alpha)f(\mathbf{y}). \tag{4.3.14}$$

- The function $f(\mathbf{x})$ is known as a *strictly convex function* if and only if $S = \mathrm{dom} f$ is a convex set and, for all vectors $\mathbf{x}, \mathbf{y} \in S$ and each scalar $\alpha \in (0, 1)$, the function satisfies the inequality

$$f(\alpha\mathbf{x} + (1-\alpha)\mathbf{y}) < \alpha f(\mathbf{x}) + (1-\alpha)f(\mathbf{y}). \tag{4.3.15}$$

In optimization, it is often required that an objective function is *strongly convex*. A strongly convex function has the following three equivalent definitions:

(1) The function $f(\mathbf{x})$ is strongly convex if [344]

$$f(\alpha\mathbf{x} + (1-\alpha)\mathbf{y}) \leq \alpha f(\mathbf{x}) + (1-\alpha)f(\mathbf{y}) - \frac{\mu}{2}\alpha(1-\alpha)\|\mathbf{x}-\mathbf{y}\|_2^2 \tag{4.3.16}$$

holds for all vectors $\mathbf{x}, \mathbf{y} \in S$ and scalars $\alpha \in [0, 1]$ and $\mu > 0$.

(2) The function $f(\mathbf{x})$ is strongly convex if [43]

$$(\nabla f(\mathbf{x}) - \nabla f(\mathbf{y}))^T(\mathbf{x}-\mathbf{y}) \geq \mu\|\mathbf{x}-\mathbf{y}\|_2^2 \tag{4.3.17}$$

holds for all vectors $\mathbf{x}, \mathbf{y} \in S$ and some scalar $\mu > 0$.

(3) The function $f(\mathbf{x})$ is strongly convex if [344]

$$f(\mathbf{y}) \geq f(\mathbf{x}) + [\nabla f(\mathbf{x})]^T(\mathbf{y}-\mathbf{x}) + \frac{\mu}{2}\|\mathbf{y}-\mathbf{x}\|_2^2 \tag{4.3.18}$$

for some scalar $\mu > 0$.

In the above three definitions, the constant $\mu\,(> 0)$ is called the *convexity parameter* of the strongly convex function $f(\mathbf{x})$.

The following is the relationship between convex functions, strictly convex functions and strongly convex functions:

$$\text{strongly convex function} \Rightarrow \text{strictly convex function} \Rightarrow \text{convex function.} \tag{4.3.19}$$

Definition 4.11 [460] A function $f(\mathbf{x})$ is *quasi-convex* if, for all vectors $\mathbf{x}, \mathbf{y} \in S$ and a scalar $\alpha \in [0, 1]$, the inequality

$$f(\alpha\mathbf{x} + (1-\alpha)\mathbf{y}) \leq \max\{f(\mathbf{x}), f(\mathbf{y})\} \tag{4.3.20}$$

holds. A function $f(\mathbf{x})$ is *strongly quasi-convex* if, for all vectors $\mathbf{x}, \mathbf{y} \in S$, $\mathbf{x} \neq \mathbf{y}$ and the scalar $\alpha \in (0, 1)$, the strict inequality

$$f(\alpha\mathbf{x} + (1-\alpha)\mathbf{y}) < \max\{f(\mathbf{x}), f(\mathbf{y})\} \tag{4.3.21}$$

holds. A function $f(\mathbf{x})$ is referred to as *strictly quasi-convex* if the strict inequality (4.3.21) holds for all vectors $\mathbf{x}, \mathbf{y} \in S$, $f(\mathbf{x}) \neq f(\mathbf{y})$ and a scalar $\alpha \in (0, 1)$.

4.3.3 Convex Function Identification

Consider an objective function defined in the convex set S, $f(\mathbf{x}) : S \to \mathbb{R}$, a natural question to ask is how to determine whether the function is convex. Convex-function identification methods are divided into first-order gradient and second-order gradient identification methods.

Definition 4.12 [432] Given a convex set $S \in \mathbb{R}^n$, then the mapping function $\mathbf{F}(\mathbf{x}) : S \to \mathbb{R}^n$ is called:

(a) *a monotone function* in the convex set S if

$$\langle \mathbf{F}(\mathbf{x}) - \mathbf{F}(\mathbf{y}), \mathbf{x}-\mathbf{y}\rangle \geq 0, \quad \forall \mathbf{x}, \mathbf{y} \in S; \tag{4.3.22}$$

(b) *a strictly monotone function* in the convex set S if

$$\langle \mathbf{F}(\mathbf{x}) - \mathbf{F}(\mathbf{y}), \mathbf{x}-\mathbf{y}\rangle > 0, \quad \forall \mathbf{x}, \mathbf{y} \in S \text{ and } \mathbf{x} \neq \mathbf{y}; \tag{4.3.23}$$

(c) *a strongly monotone function* in the convex set S if

$$\langle \mathbf{F}(\mathbf{x}) - \mathbf{F}(\mathbf{y}), \mathbf{x}-\mathbf{y}\rangle \geq \mu\|\mathbf{x}-\mathbf{y}\|_2^2, \quad \forall \mathbf{x}, \mathbf{y} \in S. \tag{4.3.24}$$

If the gradient vector of a function $f(\mathbf{x})$ is taken as the mapping function, i.e., $\mathbf{F}(\mathbf{x}) = \nabla_{\mathbf{x}} f(\mathbf{x})$, then we have the following first-order necessary and sufficient conditions for identifying the convex function.

1. First-Order Necessary and Sufficient Conditions

Theorem 4.5 [432] *Let $f : S \to \mathbb{R}$ be a first differentiable function defined in the convex set S in the n-dimensional vector space $\mathbb{R}^n$, then for all vectors $\mathbf{x}, \mathbf{y} \in S$, we have*

$$\begin{aligned} f(\mathbf{x}) \text{ convex} &\Leftrightarrow \langle \nabla_{\mathbf{x}} f(\mathbf{x}) - \nabla_{\mathbf{x}} f(\mathbf{y}), \mathbf{x} - \mathbf{y} \rangle \geq 0, \\ f(\mathbf{x}) \text{ strictly convex} &\Leftrightarrow \langle \nabla_{\mathbf{x}} f(\mathbf{x}) - \nabla_{\mathbf{x}} f(\mathbf{y}), \mathbf{x} - \mathbf{y} \rangle > 0, \, \mathbf{x} \neq \mathbf{y}, \\ f(\mathbf{x}) \text{ strongly convex} &\Leftrightarrow \langle \nabla_{\mathbf{x}} f(\mathbf{x}) - \nabla_{\mathbf{x}} f(\mathbf{y}), \mathbf{x} - \mathbf{y} \rangle \geq \mu \|\mathbf{x} - \mathbf{y}\|_2^2. \end{aligned}$$

Theorem 4.6 [53] *If the function $f : S \to \mathbb{R}$ is differentiable in a convex definition domain then f is convex if and only if*

$$f(\mathbf{y}) \geq f(\mathbf{x}) + \langle \nabla_{\mathbf{x}} f(\mathbf{x}), \mathbf{y} - \mathbf{x} \rangle. \tag{4.3.25}$$

2. Second-Order Necessary and Sufficient Conditions

Theorem 4.7 [311] *Let $f : S \to \mathbb{R}$ be a function defined in the convex set $S \in \mathbb{R}^n$, and second differentiable, then $f(\mathbf{x})$ is a convex function if and only if the Hessian matrix is positive semi-definite, namely*

$$\mathbf{H}_{\mathbf{x}}[f(\mathbf{x})] = \frac{\partial^2 f(\mathbf{x})}{\partial \mathbf{x} \partial \mathbf{x}^T} \succeq 0, \quad \forall \mathbf{x} \in S. \tag{4.3.26}$$

Remark Let $f : S \to \mathbb{R}$ be a function defined in the convex set $S \in \mathbb{R}^n$, and second differentiable; then $f(\mathbf{x})$ is strictly convex if and only if the Hessian matrix is positive definite, namely

$$\mathbf{H}_{\mathbf{x}}[f(\mathbf{x})] = \frac{\partial^2 f(\mathbf{x})}{\partial \mathbf{x} \partial \mathbf{x}^T} \succ 0, \quad \forall \mathbf{x} \in S, \tag{4.3.27}$$

whereas the sufficient condition for a strict minimum point requires that the Hessian matrix is positive definite at one point $\mathbf{c}$, Equation (4.3.27) requires that the Hessian matrix is positive definite at all points in the convex set S.

The following basic properties are useful for identifying the convexity of a given function [211].

1. The function $f : \mathbb{R}^n \to \mathbb{R}$ is convex, if and only if $\tilde{f}(t) = f(\mathbf{x}_0 + t\mathbf{h})$ is convex for $t \in \mathbb{R}$ and all vectors $\mathbf{x}_0, \mathbf{h} \in \mathbb{R}^n$.
2. If $\alpha_1, \alpha_2 \geq 0$, and $f_1(\mathbf{x})$ and $f_2(\mathbf{x})$ are convex functions, then $f(\mathbf{x}) = \alpha_1 f_1(\mathbf{x}) + \alpha_2 f_2(\mathbf{x})$ is also convex.
3. If $p(y) \geq 0$ and $q(\mathbf{x}, y)$ is a convex function at the point $\mathbf{x} \in S$ then $\int p(y) q(\mathbf{x}, y) \mathrm{d}\, y$ is convex at all points $\mathbf{x} \in S$.

4. If $f(\mathbf{x})$ is a convex function then its affine transformation $f(\mathbf{Ax}+\mathbf{b})$ is also convex.

It should be pointed out that, apart from the ℓ_0-norm, the vector norms for all other $\ell_p\,(p\geq 1)$,

$$\|\mathbf{x}\|_p=\left(\sum_{i=1}^{n}|\,x_i|^p\right)^{1/p},\ p\geq 1;\quad \|\mathbf{x}\|_\infty=\max_i|\,x_i| \tag{4.3.28}$$

are convex functions.

4.4 Gradient Methods for Smooth Convex Optimization

Convex optimization is divided into first-order algorithms and second-order algorithms. Focusing upon smoothing functions, this section presents first-order algorithms for smooth convex optimization: the gradient method, the gradient-projection method, the conjugate gradient method and the Nesterov optimal gradient method.

4.4.1 Gradient Method

Consider the *smooth convex optimization problem* of an objective function $f:Q\to\mathbb{R}$, described by

$$\min_{\mathbf{x}\in Q} f(\mathbf{x}), \tag{4.4.1}$$

where $\mathbf{x}\in Q\subset\mathbb{R}^n$, and $f(\mathbf{x})$ is smooth and convex.

As shown in Figure 4.4, the vector $\mathbf{x}$ can be regarded as the input of a black box, the outputs of which are the function $f(\mathbf{x})$ and the gradient function $\nabla f(\mathbf{x})$.

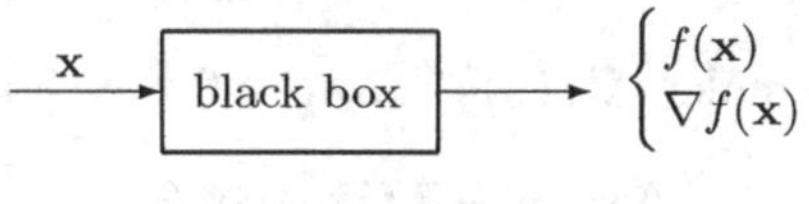

Figure 4.4 The first-order black-box method.

Let $\mathbf{x}_{\text{opt}}$ denote the optimal solution of $\min f(\mathbf{x})$. *First-order black-box optimization* uses only $f(\mathbf{x})$ and its gradient vector $\nabla f(\mathbf{x})$ for finding the vector $\mathbf{y}\in Q$ such that

$$f(\mathbf{y})-f(\mathbf{x}_{\text{opt}})\leq\epsilon,$$

where ϵ is a given accuracy error. The solution $\mathbf{y}$ satisfying the above condition is known as the *ϵ-suboptimal solution* of the objective function $f(\mathbf{x})$.

The first-order black-box optimization method includes two basic tasks:

(1) the design of a first-order iterative optimization algorithm;

(2) the convergence-rate or complexity analysis of the optimization algorithm.

First, we introduce optimization-algorithm design.

The descent method is a simplest first-order optimization method whose basic idea for solving the unconstrained minimization problem of a convex function $\min f(\mathbf{x})$ is as follows: when $Q = \mathbb{R}^n$, use the optimization sequence

$$\mathbf{x}_{k+1} = \mathbf{x}_k + \mu_k \Delta\mathbf{x}_k, \quad k = 1, 2, \ldots \tag{4.4.2}$$

to search for the optimal point $\mathbf{x}_{\text{opt}}$. In the above equation, $k = 1, 2, \ldots$ represents the iterative number and $\mu_k \geq 0$ is the step size or step length of the kth iteration; the concatenated symbol of Δ and $\mathbf{x}$, $\Delta\mathbf{x}$, denotes a vector in $\mathbb{R}^n$ known as the *step direction* or *search direction* of the objective function $f(\mathbf{x})$, while $\Delta\mathbf{x}_k = \mathbf{x}_{k+1} - \mathbf{x}_k$ represents the search direction at the kth iteration.

Because the minimization-algorithm design requires the objective function to decrease during the iteration process, i.e.,

$$f(\mathbf{x}_{k+1}) < f(\mathbf{x}_k), \quad \forall \mathbf{x}_k \in \operatorname{dom} f, \tag{4.4.3}$$

this method is called the *descent method.*

From the first-order approximation expression to the objective function at the point $\mathbf{x}_k$

$$f(\mathbf{x}_{k+1}) \approx f(\mathbf{x}_k) + (\nabla f(\mathbf{x}_k))^T \Delta\mathbf{x}_k, \tag{4.4.4}$$

it follows that if

$$(\nabla f(\mathbf{x}_k))^T \Delta\mathbf{x}_k < 0 \tag{4.4.5}$$

then $f(\mathbf{x}_{k+1}) < f(\mathbf{x}_k)$. Hence the search direction $\Delta\mathbf{x}_k$ such that $(\nabla f(\mathbf{x}_k))^T \Delta\mathbf{x}_k < 0$ is called the *descent step* or *descent direction* of the objective function $f(\mathbf{x})$ at the kth iteration.

Obviously, in order to make $(\nabla f(\mathbf{x}_k))^T \Delta\mathbf{x}_k < 0$, we should take

$$\Delta\mathbf{x}_k = -\nabla f(\mathbf{x}_k) \cos\theta, \tag{4.4.6}$$

where $0 \leq \theta < \pi/2$ is the acute angle between the descent direction and the negative gradient direction $-\nabla f(\mathbf{x}_k)$.

The case $\theta = 0$ implies $\Delta\mathbf{x}_k = -\nabla f(\mathbf{x}_k)$. That is to say, the negative gradient direction of the objective function f at the point $\mathbf{x}$ is directly taken as the search direction. In this case, the length of the descent step $\|\Delta\mathbf{x}_k\|_2 = \|\nabla f(\mathbf{x}_k)\|_2$ takes the maximum value, and thus the descent direction $\Delta\mathbf{x}_k$ has the maximal descent step or rate and the corresponding descent algorithm,

$$\mathbf{x}_{k+1} = \mathbf{x}_k - \mu_k \nabla f(\mathbf{x}_k), \quad k = 1, 2, \ldots, \tag{4.4.7}$$

is called the *steepest descent method.*

The steepest descent method can be explained using the quadratic approximation to the function. The second-order Taylor series expansion is given by

$$f(\mathbf{y}) \approx f(\mathbf{x}) + (\nabla f(\mathbf{x}))^T(\mathbf{y}-\mathbf{x}) + \frac{1}{2}(\mathbf{y}-\mathbf{x})^T \nabla^2 f(\mathbf{x})(\mathbf{y}-\mathbf{x}). \tag{4.4.8}$$

If $t^{-1}\mathbf{I}$ is used to replace the Hessian matrix $\nabla^2 f(\mathbf{x})$ then

$$f(\mathbf{y}) \approx f(\mathbf{x}) + (\nabla f(\mathbf{x}))^T(\mathbf{y}-\mathbf{x}) + \frac{1}{2t}\|\mathbf{y}-\mathbf{x}\|_2^2. \tag{4.4.9}$$

This is the quadratic approximation to the function $f(\mathbf{x})$ at the point $\mathbf{y}$.

It is easily found that the gradient vector $\nabla f(\mathbf{y}) = \frac{\partial f(\mathbf{y})}{\partial \mathbf{y}} = \nabla f(\mathbf{x}) + t^{-1}(\mathbf{y}-\mathbf{x})$. Letting $\nabla f(\mathbf{y}) = 0$, the solution is given by $\mathbf{y} = \mathbf{x} - t\nabla f(\mathbf{x})$. Then letting $\mathbf{y} = \mathbf{x}_{k+1}$ and $\mathbf{x} = \mathbf{x}_k$, we immediately get the updating formula of the steepest descent algorithm, $\mathbf{x}_{k+1} = \mathbf{x}_k - t\nabla f(\mathbf{x}_k)$.

The *steepest descent direction* $\Delta\mathbf{x} = -\nabla f(\mathbf{x})$ uses only first-order gradient information about the objective function $f(\mathbf{x})$. If second-order information, i.e., the Hessian matrix $\nabla^2 f(\mathbf{x}_k)$ of the objective function is used then we may find a better search direction. In this case, the optimal descent direction $\Delta\mathbf{x}$ is the solution minimizing the second-order Taylor approximation function of $f(\mathbf{x})$:

$$\min_{\Delta\mathbf{x}} \left\{ f(\mathbf{x}+\Delta\mathbf{x}) = f(\mathbf{x}) + (\nabla f(\mathbf{x}))^T \Delta\mathbf{x} + \frac{1}{2}(\Delta\mathbf{x})^T \nabla^2 f(\mathbf{x}) \Delta\mathbf{x} \right\}. \tag{4.4.10}$$

At the optimal point, the gradient vector of the objective function with respect to the parameter vector $\Delta\mathbf{x}$ has to equal zero, i.e.,

$$\begin{aligned} \frac{\partial f(\mathbf{x}+\Delta\mathbf{x})}{\partial \Delta\mathbf{x}} &= \nabla f(\mathbf{x}) + \nabla^2 f(\mathbf{x})\Delta\mathbf{x} = \mathbf{0} \\ \Leftrightarrow \quad \Delta\mathbf{x}_{\text{nt}} &= -\left(\nabla^2 f(\mathbf{x})\right)^{-1} \nabla f(\mathbf{x}), \end{aligned} \tag{4.4.11}$$

where $\Delta\mathbf{x}_{\text{nt}}$ is called the *Newton step* or the *Newton descent direction* and the corresponding optimization method is the *Newton* or *Newton–Raphson method.*

Algorithm 4.1 summarizes the gradient descent algorithm and its variants.

Algorithm 4.1 Gradient descent algorithm and its variants

---

given: An initial point $\mathbf{x}_1 \in \operatorname{dom} f$ and the allowed error $\epsilon > 0$. Put $k = 1$.

repeat

1. Compute the gradient $\nabla f(\mathbf{x}_k)$ (and the Hessian matrix $\nabla^2 f(\mathbf{x}_k)$).
2. Choose the descent direction

$$\Delta\mathbf{x}_k = \begin{cases} -\nabla f(\mathbf{x}_k) & \text{(steepest descent method),} \\ -\left(\nabla^2 f(\mathbf{x}_k)\right)^{-1} \nabla f(\mathbf{x}_k) & \text{(Newton method).} \end{cases}$$

3. Choose a step $\mu_k > 0$, and update $\mathbf{x}_{k+1} = \mathbf{x}_k + \mu_k \Delta\mathbf{x}_k$.
4. **exit if** $|f(\mathbf{x}_{k+1}) - f(\mathbf{x}_k)| \le \epsilon$.

return $k \leftarrow k+1$.

output: $\mathbf{x} \leftarrow \mathbf{x}_k$.

According to different choices of the step μ_k, gradient descent algorithms have the following common variants [344].

(1) Before the gradient algorithm is run, a step sequence $\{\mu_k\}_{k=0}^{\infty}$ is chosen, e.g.,

$$\mu_k = \mu \text{ (fixed step)} \quad \text{or} \quad \mu_k = h/\sqrt{k+1}.$$

(2) *Full relaxation* Take $\mu_k = \arg\min_{\mu \geq 0} f(\mathbf{x}_k - \mu \nabla f(\mathbf{x}_k))$.

(3) *Goldstein–Armijo rule* Find $\mathbf{x}_{k+1} = \mathbf{x}_k - \mu \nabla f(\mathbf{x}_k)$ such that

$$\alpha \langle \nabla f(\mathbf{x}_k), \mathbf{x}_k - \mathbf{x}_{k+1} \rangle \leq f(\mathbf{x}_k) - f(\mathbf{x}_{k+1}),$$
$$\beta \langle \nabla f(\mathbf{x}_k), \mathbf{x}_k - \mathbf{x}_{k+1} \rangle \geq f(\mathbf{x}_k) - f(\mathbf{x}_{k+1}),$$

where $0 < \alpha < \beta < 1$ are two fixed parameters.

Note that the variable vector $\mathbf{x}$ is unconstrained in the gradient algorithms, i.e., $\mathbf{x} \in \mathbb{R}^n$. When choosing a constrained variable vector $\mathbf{x} \in C$, where $C \subset \mathbb{R}^n$, then the update formula in the gradient algorithm should be replaced by a formula containing the projected gradient:

$$\mathbf{x}_{k+1} = \mathcal{P}_C(\mathbf{x}_k - \mu_k \nabla f(\mathbf{x}_k)). \tag{4.4.12}$$

This algorithm is called the *projected gradient method* or the *gradient-projection method.* In (4.4.12), $\mathcal{P}_C(\mathbf{y})$ is known as the *projection operator* and is defined as

$$\mathcal{P}_C(\mathbf{y}) = \arg\min_{\mathbf{x} \in C} \frac{1}{2} \|\mathbf{x} - \mathbf{y}\|_2^2. \tag{4.4.13}$$

The projection operator can be equivalently expressed as

$$\mathcal{P}_C(\mathbf{y}) = \mathbf{P}_C \mathbf{y}, \tag{4.4.14}$$

where $\mathbf{P}_C$ is the projection matrix onto the subspace C. If C is the column space of the matrix $\mathbf{A}$, then

$$\mathbf{P}_A = \mathbf{A}(\mathbf{A}^T\mathbf{A})^{-1}\mathbf{A}^T. \tag{4.4.15}$$

We will discuss the projection matrix in detail in Chapter 9 (on projection analysis).

In particular, if $C = \mathbb{R}^n$, i.e., the variable vector $\mathbf{x}$ is unconstrained, then the projection operator is equal to the identity matrix, namely $\mathcal{P}_C = \mathbf{I}$, and thus

$$\mathcal{P}_{\mathbb{R}^n}(\mathbf{y}) = \mathbf{P}\mathbf{y} = \mathbf{y}, \quad \forall \mathbf{y} \in \mathbb{R}^n.$$

In this case, the projected gradient algorithm simplifies to the actual gradient algorithm.

The following are projections of the vector $\mathbf{x}$ onto several typical sets [488].

1. The projection onto the affine set $C = \{\mathbf{x} | \mathbf{A}\mathbf{x} = \mathbf{b}\}$ with $\mathbf{A} \in \mathbb{R}^{p \times n}$, $\text{rank}(\mathbf{A}) = p$ is

$$\mathcal{P}_C(\mathbf{x}) = \mathbf{x} + \mathbf{A}^T(\mathbf{A}\mathbf{A}^T)^{-1}(\mathbf{b} - \mathbf{A}\mathbf{x}). \tag{4.4.16}$$

If $p \ll n$ or $\mathbf{A}\mathbf{A}^T = \mathbf{I}$, then the projection $\mathcal{P}_C(\mathbf{x})$ gives a low cost result.

2. The projection onto the *hyperplane* $C = \{\mathbf{x}|\mathbf{a}^T\mathbf{x} = b\}$ (where $\mathbf{a} \neq \mathbf{0}$) is

$$\mathcal{P}_C(\mathbf{x}) = \mathbf{x} + \frac{b - \mathbf{a}^T\mathbf{x}}{\|\mathbf{a}\|_2^2}\mathbf{a}. \tag{4.4.17}$$

3. The projection onto the *nonnegative orthant* $C = \mathbb{R}_+^n$ is

$$\mathcal{P}_C(\mathbf{x}) = (\mathbf{x})^+ \quad \Leftrightarrow \quad [(\mathbf{x})^+]_i = \max\{x_i, 0\}. \tag{4.4.18}$$

4. The projection onto the *half-space* $C = \{\mathbf{x}|\mathbf{a}^T\mathbf{x} \leq b\}$ (where $\mathbf{a} \neq \mathbf{0}$):

$$\mathcal{P}_C(\mathbf{x}) = \begin{cases} \mathbf{x} + \dfrac{b - \mathbf{a}^T\mathbf{x}}{\|\mathbf{a}\|_2^2}\mathbf{a}, & \text{if } \mathbf{a}^T\mathbf{x} > b, \\ \mathbf{x}, & \text{if } \mathbf{a}^T\mathbf{x} \leq b. \end{cases} \tag{4.4.19}$$

5. The projection onto the *rectangular set* $C = [\mathbf{a}, \mathbf{b}]$ (where $a_i \leq x_i \leq b_i$) is

$$\mathcal{P}_C(\mathbf{x}) = \begin{cases} a_i, & \text{if } x_i \leq a_i, \\ x_i, & \text{if } a_i \leq x_i \leq b_i, \\ b_i, & \text{if } x_i \geq b_i. \end{cases} \tag{4.4.20}$$

6. The projection onto the *second-order cones* $C = \{(\mathbf{x}, t)| \|\mathbf{x}\|_2 \leq t, \mathbf{x} \in \mathbb{R}^n\}$ is

$$\mathcal{P}_C(\mathbf{x}) = \begin{cases} (\mathbf{x}, t), & \text{if } \|\mathbf{x}\|_2 \leq t, \\ \dfrac{t + \|\mathbf{x}\|_2}{2\|\mathbf{x}\|_2}\begin{bmatrix}\mathbf{x} \\ t\end{bmatrix}, & \text{if } -t < \|\mathbf{x}\|_2 < t, \\ (0, 0), & \text{if } \|\mathbf{x}\|_2 \leq -t,\ \mathbf{x} \neq \mathbf{0}. \end{cases} \tag{4.4.21}$$

7. The projection onto the *Euclidean ball* $C = \{\mathbf{x}| \|\mathbf{x}\|_2 \leq 1\}$ is

$$\mathcal{P}_C(\mathbf{x}) = \begin{cases} \dfrac{1}{\|\mathbf{x}\|_2}\mathbf{x}, & \text{if } \|\mathbf{x}\|_2 > 1, \\ \mathbf{x}, & \text{if } \|\mathbf{x}\|_2 \leq 1. \end{cases} \tag{4.4.22}$$

8. The projection onto the *ℓ_1-norm ball* $C = \{\mathbf{x}| \|\mathbf{x}\|_1 \leq 1\}$ is

$$\mathcal{P}_C(\mathbf{x})_i = \begin{cases} x_i - \lambda, & \text{if } x_i > \lambda, \\ 0, & \text{if } -\lambda \leq x_i \leq \lambda, \\ x_i + \lambda, & \text{if } x_i < -\lambda, \end{cases} \tag{4.4.23}$$

where $\lambda = 0$ if $\|\mathbf{x}\|_1 \leq 1$; otherwise, λ is the solution of the equation

$$\sum_{i=1}^{n} \max\{|\,x_i| - \lambda, 0\} = 1.$$

9. The projection onto the *positive semi-definite cones* $C = \mathbb{S}_+^n$ is

$$\mathcal{P}_C(\mathbf{X}) = \sum_{i=1}^{n} \max\{0, \lambda_i\}\mathbf{q}_i\mathbf{q}_i^T, \tag{4.4.24}$$

where $\mathbf{X} = \sum_{i=1}^{n} \lambda_i\mathbf{q}_i\mathbf{q}_i^T$ is the eigenvalue decomposition of $\mathbf{X}$.

4.4.2 Conjugate Gradient Method

Consider the iteration solution of the matrix equation $\mathbf{Ax} = \mathbf{b}$, where $\mathbf{A} \in \mathbb{R}^{n\times n}$ is a nonsingular matrix. This matrix equation can be equivalently written as

$$\mathbf{x} = (\mathbf{I} - \mathbf{A})\mathbf{x} + \mathbf{b}. \tag{4.4.25}$$

This inspires the following iteration algorithm:

$$\mathbf{x}_{k+1} = (\mathbf{I} - \mathbf{A})\mathbf{x}_k + \mathbf{b}. \tag{4.4.26}$$

This algorithm is known as the *Richardson iteration* and can be rewritten in the more general form

$$\mathbf{x}_{k+1} = \mathbf{M}\mathbf{x}_k + \mathbf{c}, \tag{4.4.27}$$

where $\mathbf{M}$ is an $n \times n$ matrix, called the *iteration matrix*.

An iteration with the form of Equation (4.4.27) is known as a *stationary iterative method* and is not as effective as a *nonstationary iterative method.*

Nonstationary iteration methods comprise a class of iteration methods in which $\mathbf{x}_{k+1}$ is related to the former iteration solutions $\mathbf{x}_k, \mathbf{x}_{k-1}, \ldots, \mathbf{x}_0$. The most typical nonstationary iteration method is the *Krylov subspace method*, described by

$$\mathbf{x}_{k+1} = \mathbf{x}_0 + \mathcal{K}_k, \tag{4.4.28}$$

where

$$\mathcal{K}_k = \text{Span}(\mathbf{r}_0, \mathbf{A}\mathbf{r}_0, \ldots, \mathbf{A}^{k-1}\mathbf{r}_0) \tag{4.4.29}$$

is called the kth iteration Krylov subspace; $\mathbf{x}_0$ is the initial iteration value and $\mathbf{r}_0$ denotes the initial residual vector.

Krylov subspace methods have various forms, of which the three most common are the conjugate gradient method, the biconjugate gradient method and the preconditioned conjugate gradient method.

1. Conjugate Gradient Algorithm

The conjugate gradient method uses $\mathbf{r}_0 = \mathbf{A}\mathbf{x}_0 - \mathbf{b}$ as the initial residual vector.

The applicable object of the conjugate gradient method is limited to the symmetric positive definite equation, $\mathbf{Ax} = \mathbf{b}$, where $\mathbf{A}$ is an $n \times n$ symmetric positive definite matrix.

The nonzero vector combination $\{\mathbf{p}_0, \mathbf{p}_1, \ldots, \mathbf{p}_k\}$ is called $\mathbf{A}$-orthogonal or $\mathbf{A}$-conjugate if

$$\mathbf{p}_i^T\mathbf{A}\mathbf{p}_j = 0, \quad \forall\, i \neq j. \tag{4.4.30}$$

This property is known as $\mathbf{A}$-*orthogonality* or $\mathbf{A}$-*conjugacy*. Obviously, if $\mathbf{A} = \mathbf{I}$ then the $\mathbf{A}$-conjugacy reduces to general orthogonality.

All algorithms adopting the conjugate vector as the update direction are known as *conjugate direction algorithms*. If the conjugate vectors $\mathbf{p}_0, \mathbf{p}_1, \ldots, \mathbf{p}_{n-1}$ are not predetermined, but are updated by the gradient descent method in the updating process, then we say that the minimization algorithm for the objective function $f(\mathbf{x})$ is a *conjugate gradient algorithm.* Algorithm 4.2 gives a conjugate gradient algorithm.

Algorithm 4.2 Conjugate gradient algorithm [179], [243]

---

input: $\mathbf{A} = \mathbf{A}^T \in \mathbb{R}^{n\times n}$, $\mathbf{b} \in \mathbb{R}^n$, the largest iteration step number $k_{\max}$ and the allowed error ϵ.

initialization: Choose $\mathbf{x}_0 \in \mathbb{R}^n$, and let $\mathbf{r} = \mathbf{A}\mathbf{x}_0 - \mathbf{b}$ and $\rho_0 = \|\mathbf{r}\|_2^2$.

repeat

1. If $k = 1$ then $\mathbf{p} = \mathbf{r}$. Otherwise, let $\beta = \rho_{k-1}/\rho_{k-2}$ and $\mathbf{p} = \mathbf{r} + \beta\mathbf{p}$.
2. $\mathbf{w} = \mathbf{A}\mathbf{p}$.
3. $\alpha = \rho_{k-1}/(\mathbf{p}^T\mathbf{w})$.
4. $\mathbf{x} = \mathbf{x} + \alpha\mathbf{p}$.
5. $\mathbf{r} = \mathbf{r} - \alpha\mathbf{w}$.
6. $\rho_k = \|\mathbf{r}\|_2^2$.
7. **exit if** $\sqrt{\rho_k} < \epsilon\|\mathbf{b}\|_2$ or $k = k_{\max}$.

return $k \leftarrow k + 1$.

output: $\mathbf{x} \leftarrow \mathbf{x}_k$.

From Algorithm 4.2 it can be seen that in the iteration process of the conjugate gradient algorithm, the solution of the matrix equation $\mathbf{A}\mathbf{x} = \mathbf{b}$ is given by

$$\mathbf{x}_k = \sum_{i=1}^{k} \alpha_i \mathbf{p}_i = \sum_{i=1}^{k} \frac{\langle \mathbf{r}_{i-1}, \mathbf{r}_{i-1}\rangle}{\langle \mathbf{p}_i, \mathbf{A}\mathbf{p}_i\rangle}\mathbf{p}_i, \tag{4.4.31}$$

that is, $\mathbf{x}_k$ belongs to the kth Krylov subspace

$$\mathbf{x}_k \in \text{Span}\{\mathbf{p}_1, \mathbf{p}_2, \ldots, \mathbf{p}_k\} = \text{Span}\{\mathbf{r}_0, \mathbf{A}\mathbf{r}_0, \ldots, \mathbf{A}^{k-1}\mathbf{r}_0\}.$$

The fixed-iteration method requires the updating of the iteration matrix $\mathbf{M}$, but no matrix needs to be updated in Algorithm 4.2. Hence, the Krylov subspace method is also called the matrix-free method [243].

2. Biconjugate Gradient Algorithm

If the matrix $\mathbf{A}$ is not a real symmetric matrix, then we can use the *biconjugate gradient method* of Fletcher [156] for solving the matrix equation $\mathbf{A}\mathbf{x} = \mathbf{b}$. As the name implies, there are two search directions, $\mathbf{p}$ and $\bar{\mathbf{p}}$, that are $\mathbf{A}$-conjugate in this method:

$$\begin{aligned} \bar{\mathbf{p}}_i^T \mathbf{A}\mathbf{p}_j &= \mathbf{p}_i^T \mathbf{A}\bar{\mathbf{p}}_j = 0, \quad i \neq j, \\ \bar{\mathbf{r}}_i^T \mathbf{r}_j &= \mathbf{r}_i^T \bar{\mathbf{r}}_j^T = 0, \quad i \neq j, \\ \bar{\mathbf{r}}_i^T \mathbf{p}_j &= \mathbf{r}_i^T \bar{\mathbf{p}}_j^T = 0, \quad j < i. \end{aligned} \tag{4.4.32}$$

Algorithm 4.3 shows a biconjugate gradient algorithm.

Algorithm 4.3 Biconjugate gradient algorithm [156], [230]

initialization: $\mathbf{p}_1 = \mathbf{r}_1, \bar{\mathbf{p}}_1 = \bar{\mathbf{r}}_1$.

repeat

1. $\alpha_k = \bar{\mathbf{r}}_k^T \mathbf{r}_k / (\bar{\mathbf{p}}_k^T \mathbf{A} \mathbf{p}_k)$.
2. $\mathbf{r}_{k+1} = \mathbf{r}_k - \alpha_k \mathbf{A} \mathbf{p}_k$.
3. $\bar{\mathbf{r}}_{k+1} = \bar{\mathbf{r}}_k - \alpha_k \mathbf{A}^T \bar{\mathbf{p}}_k$.
4. $\beta_k = \bar{\mathbf{r}}_{k+1}^T \mathbf{r}_{k+1} / (\bar{\mathbf{r}}_k^T \mathbf{r}_k)$.
5. $\mathbf{p}_{k+1} = \mathbf{r}_{k+1} + \beta_k \mathbf{p}_k$.
6. $\bar{\mathbf{p}}_{k+1} = \bar{\mathbf{r}}_{k+1} + \beta_k \bar{\mathbf{p}}_k$.
7. **exit if** $k = k_{\max}$.

return $k \leftarrow k + 1$.

output: $\mathbf{x} \leftarrow \mathbf{x}_k + \alpha_k \mathbf{p}_k$, $\bar{\mathbf{x}}_{k+1} \leftarrow \bar{\mathbf{x}}_k + \alpha_k \bar{\mathbf{p}}_k$.

3. Preconditioned Conjugate Gradient Algorithm

Consider the symmetric indefinite saddle point problem

$$\begin{bmatrix} \mathbf{A} & \mathbf{B}^T \\ \mathbf{B} & \mathbf{O} \end{bmatrix} \begin{bmatrix} \mathbf{x} \\ \mathbf{q} \end{bmatrix} = \begin{bmatrix} \mathbf{f} \\ \mathbf{g} \end{bmatrix},$$

where $\mathbf{A}$ is an $n \times n$ real symmetric positive definite matrix, $\mathbf{B}$ is an $m \times n$ real matrix with full row rank m $(\leq n)$, and $\mathbf{O}$ is an $m \times m$ zero matrix. Bramble and Pasciak [56] developed a preconditioned conjugate gradient iteration method for solving the above problems.

The basic idea of the *preconditioned conjugate gradient iteration* is as follows: through a clever choice of the scalar product form, the preconditioned saddle matrix becomes a symmetric positive definite matrix.

To simplify the discussion, it is assumed that the matrix equation with large condition number $\mathbf{Ax} = \mathbf{b}$ needs to be converted into a new symmetric positive definite equation. To this end, let $\mathbf{M}$ be a symmetric positive definite matrix that can approximate the matrix $\mathbf{A}$ and for which it is easier to find the inverse matrix. Hence, the original matrix equation $\mathbf{Ax} = \mathbf{b}$ is converted into $\mathbf{M}^{-1}\mathbf{Ax} = \mathbf{M}^{-1}\mathbf{b}$ such that two matrix equations have the same solution. However, in the new matrix equation $\mathbf{M}^{-1}\mathbf{Ax} = \mathbf{M}^{-1}\mathbf{b}$ there exists a hidden danger: $\mathbf{M}^{-1}\mathbf{A}$ is generally not either symmetric or positive definite even if both $\mathbf{M}$ and $\mathbf{A}$ are symmetric positive definite. Therefore, it is unreliable to use the matrix $\mathbf{M}^{-1}$ directly as the preprocessor of the matrix equation $\mathbf{Ax} = \mathbf{b}$.

Let $\mathbf{S}$ be the square root of a symmetric matrix $\mathbf{M}$, i.e., $\mathbf{M} = \mathbf{SS}^T$, where $\mathbf{S}$ is symmetric positive definite. Now, use $\mathbf{S}^{-1}$ instead of $\mathbf{M}^{-1}$ as the *preprocessor* to convert the original matrix equation $\mathbf{Ax} = \mathbf{b}$ to $\mathbf{S}^{-1}\mathbf{Ax} = \mathbf{S}^{-1}\mathbf{b}$. If $\mathbf{x} = \mathbf{S}^{-T}\hat{\mathbf{x}}$, then the preconditioned matrix equation is given by

$$\mathbf{S}^{-1}\mathbf{AS}^{-T}\hat{\mathbf{x}} = \mathbf{S}^{-1}\mathbf{b}. \tag{4.4.33}$$

Compared with the matrix $\mathbf{M}^{-1}\mathbf{A}$, which is not symmetric positive definite, $\mathbf{S}^{-1}\mathbf{AS}^{-T}$ must be symmetric positive definite if $\mathbf{A}$ is symmetric positive definite. The symmetry of $\mathbf{S}^{-1}\mathbf{AS}^{-T}$ is easily seen, its positive definiteness can be verified as follows: by checking the quadratic function it easily follows that $\mathbf{y}^T(\mathbf{S}^{-1}\mathbf{AS}^{-T})\mathbf{y} = \mathbf{z}^T\mathbf{Az}$, where $\mathbf{z} = \mathbf{S}^{-T}\mathbf{y}$. Because $\mathbf{A}$ is symmetric positive definite, we have $\mathbf{z}^T\mathbf{Az} > 0, \forall\, \mathbf{z} \neq \mathbf{0}$, and thus $\mathbf{y}^T(\mathbf{S}^{-1}\mathbf{AS}^{-T})\mathbf{y} > 0, \forall\, \mathbf{y} \neq \mathbf{0}$. That is to say, $\mathbf{S}^{-1}\mathbf{AS}^{-T}$ must be positive definite.

At this point, the conjugate gradient method can be applied to solve the matrix equation (4.4.33) in order to get $\hat{\mathbf{x}}$, and then we can recover $\mathbf{x}$ via $\mathbf{x} = \mathbf{S}^{-T}\hat{\mathbf{x}}$.

Algorithm 4.4 shows a preconditioned conjugate gradient (PCG) algorithm with a preprocessor, developed in [440].

Algorithm 4.4 PCG algorithm with preprocessor

---

input: $\mathbf{A}$, $\mathbf{b}$, preprocessor $\mathbf{S}^{-1}$, maximal number of iterations $k_{\max}$, and allowed error $\epsilon < 1$.

initialization: $k = 0$, $\mathbf{r}_0 = \mathbf{Ax} - \mathbf{b}$, $\mathbf{d}_0 = \mathbf{S}^{-1}\mathbf{r}_0$, $\delta_{\text{new}} = \mathbf{r}_0^T\mathbf{d}_0$, $\delta_0 = \delta_{\text{new}}$.

repeat

1. $\mathbf{q}_{k+1} = \mathbf{Ad}_k$.
2. $\alpha = \delta_{\text{new}}/(\mathbf{d}_k^T\mathbf{q}_{k+1})$.
3. $\mathbf{x}_{k+1} = \mathbf{x}_k + \alpha\mathbf{d}_k$.
4. If k can be divided exactly by 50, then $\mathbf{r}_{k+1} = \mathbf{b} - \mathbf{Ax}_k$. Otherwise, update $\mathbf{r}_{k+1} = \mathbf{r}_k - \alpha\mathbf{q}_{k+1}$.
5. $\mathbf{s}_{k+1} = \mathbf{S}^{-1}\mathbf{r}_{k+1}$.
6. $\delta_{\text{old}} = \delta_{\text{new}}$.
7. $\delta_{\text{new}} = \mathbf{r}_{k+1}^T\mathbf{s}_{k+1}$.
8. **exit if** $k = k_{\max}$ or $\delta_{\text{new}} < \epsilon^2\delta_0$.
9. $\beta = \delta_{\text{new}}/\delta_{\text{old}}$.
10. $\mathbf{d}_{k+1} = \mathbf{s}_{k+1} + \beta\mathbf{d}_k$.

return $k \leftarrow k + 1$.

output: $\mathbf{x} \leftarrow \mathbf{x}_k$.

The use of a preprocessor can be avoided, because there are correspondences between the variables of the matrix equations $\mathbf{Ax} = \mathbf{b}$ and $\mathbf{S}^{-1}\mathbf{AS}^{-T}\hat{\mathbf{x}} = \mathbf{S}^{-1}\mathbf{b}$, as follows [243]:

$$\mathbf{x}_k = \mathbf{S}^{-1}\hat{\mathbf{x}}_k, \quad \mathbf{r}_k = \mathbf{S}\hat{\mathbf{r}}_k, \quad \mathbf{p}_k = \mathbf{S}^{-1}\hat{\mathbf{p}}_k, \quad \mathbf{z}_k = \mathbf{S}^{-1}\hat{\mathbf{r}}_k.$$

On the basis of these correspondences, it easy to develop a PCG algorithm without preprocessor [243], as shown in Algorithm 4.5.

Algorithm 4.5 PCG algorithm without preprocessor

input: $\mathbf{A} = \mathbf{A}^T \in \mathbb{R}^{n \times n}$, $\mathbf{b} \in \mathbb{R}^n$, maximal iteration number $k_{\max}$, and allowed error ϵ.
initialization: $\mathbf{x}_0 \in \mathbb{R}^n, \mathbf{r}_0 = \mathbf{A}\mathbf{x}_0 - \mathbf{b}, \rho_0 = \|\mathbf{r}\|_2^2$ and $\mathbf{M} = \mathbf{S}\mathbf{S}^T$. Put $k = 1$.
repeat
1. $\mathbf{z}_k = \mathbf{M}\mathbf{r}_{k-1}$.
2. $\tau_{k-1} = \mathbf{z}_k^T \mathbf{r}_{k-1}$.
3. If $k = 1$, then $\beta = 0$, $\mathbf{p}_1 = \mathbf{z}_1$. Otherwise, $\beta = \tau_{k-1}/\tau_{k-2}$, $\mathbf{p}_k \leftarrow \mathbf{z}_k + \beta \mathbf{p}_k$.
4. $\mathbf{w}_k = \mathbf{A}\mathbf{p}_k$.
5. $\alpha = \tau_{k-1}/(\mathbf{p}_k^T \mathbf{w}_k)$.
6. $\mathbf{x}_{k+1} = \mathbf{x}_k + \alpha \mathbf{p}_k$.
7. $\mathbf{r}_k = \mathbf{r}_{k-1} - \alpha \mathbf{w}_k$.
8. $\rho_k = \mathbf{r}_k^T \mathbf{r}_k$.
9. **exit if** $\sqrt{\rho_k} < \epsilon \|\mathbf{b}\|_2$ or $k = k_{\max}$.

return $k \leftarrow k + 1$.
output: $\mathbf{x} \leftarrow \mathbf{x}_k$.

A wonderful introduction to the conjugate gradient method is presented in [440].

Regarding the complex matrix equation $\mathbf{A}\mathbf{x} = \mathbf{b}$, where $\mathbf{A} \in \mathbb{C}^{n \times n}, \mathbf{x} \in \mathbb{C}^n, \mathbf{b} \in \mathbb{C}^n$, we can write it as the following real matrix equation:

$$\begin{bmatrix} \mathbf{A}_\mathrm{R} & -\mathbf{A}_\mathrm{I} \\ \mathbf{A}_\mathrm{I} & \mathbf{A}_\mathrm{R} \end{bmatrix} \begin{bmatrix} \mathbf{x}_\mathrm{R} \\ \mathbf{x}_\mathrm{I} \end{bmatrix} = \begin{bmatrix} \mathbf{b}_\mathrm{R} \\ \mathbf{b}_\mathrm{I} \end{bmatrix}. \tag{4.4.34}$$

If $\mathbf{A} = \mathbf{A}_\mathrm{R} + \mathrm{j}\mathbf{A}_\mathrm{I}$ is a Hermitian positive definite matrix, then Equation (4.4.34) is a symmetric positive definite matrix equation. Hence, $(\mathbf{x}_\mathrm{R}, \mathbf{x}_\mathrm{I})$ can be solved by adopting the conjugate gradient algorithm or the preconditioned conjugate gradient algorithm.

The preconditioned conjugate gradient method is a widely available technique for solving partial differential equations, and it has important applications in optimal control [208]. As a matter of fact, it is usually necessary to apply the preconditioned conjugate gradient algorithm to solve the KKT equation or the Newton equation in optimization problems in order to improve the numerical stability of solving the optimization search direction procedure.

In addition to the three conjugate gradient algorithms described above, there is a projected conjugate gradient algorithm [179]. This algorithm requires the use of a seed space $\mathcal{K}_k(\mathbf{A}, \mathbf{r}_0)$ to solve the matrix equation $\mathbf{A}\mathbf{x}_q = \mathbf{b}_q, q = 1, 2, \ldots$

The main advantage of gradient methods is that the computation of each iteration is very simple; however, their convergence is slow. Hence, an important question is how to accelerate a gradient method so as to get an optimal gradient method. To this end, it is necessary first to analyze the convergence rate of the gradient methods.

4.4.3 Convergence Rates

By the *convergence rate* is meant the number of iterations needed for an optimization algorithm to make the estimated error of the objective function achieve the required accuracy, or given an iteration number K, the accuracy that an optimization algorithm reaches. The inverse of the convergence rate is known as the *complexity* of an optimization algorithm.

Let $\mathbf{x}^\star$ be a local or global minimum point. The estimation error of an optimization algorithm is defined as the difference between the value of an objective function at iteration point $\mathbf{x}_k$ and its value at the global minimum point, i.e.,

$$\delta_k = f(\mathbf{x}_k) - f(\mathbf{x}^\star).$$

We are naturally interested in the convergence problems of an optimization algorithm:

(1) Given an iteration number K, what is the designed accuracy $\lim_{1\leq k\leq K} \delta_k$?

(2) Given an allowed accuracy ϵ, how many iterations does the algorithm need to achieve the designed accuracy $\min_k \delta_k \leq \epsilon$?

When analyzing the convergence problems of optimization algorithms, we often focus on the speed of the updating sequence $\{\mathbf{x}_k\}$ at which the objective function argument converges to its ideal minimum point $\mathbf{x}^\star$. In numerical analysis, the speed at which a sequence reaches its limit is called the convergence rate.

1. *Q*-Convergence Rates

Assume that a sequence $\{\mathbf{x}_k\}$ converges to $\mathbf{x}^\star$. If there are a real number $\alpha \geq 1$ and a positive constant μ independent of the iterations k such that

$$\mu = \lim_{k\to\infty} \frac{\|\mathbf{x}_{k+1} - \mathbf{x}^\star\|_2}{\|\mathbf{x}_k - \mathbf{x}^\star\|_2^\alpha} \tag{4.4.35}$$

then we say that $\{\mathbf{x}_k\}$ has an *α-order Q-convergence rate*. The Q-convergence rate means the quotient convergence rate. It has the following typical values [360]:

1. When $\alpha = 1$, the Q-convergence rate is called the limit-convergence rate of $\{\mathbf{x}_k\}$:

$$\mu = \lim_{k\to\infty} \frac{\|\mathbf{x}_{k+1} - \mathbf{x}^\star\|_2}{\|\mathbf{x}_k - \mathbf{x}^\star\|_2}. \tag{4.4.36}$$

   According to the value of μ, the limit-convergence rate of the sequence $\{\mathbf{x}_k\}$ can be divided into three types:

   (1) *sublinear convergence rate*, $\alpha = 1$, $\mu = 1$;

   (2) *linear convergence rate*, $\alpha = 1$, $\mu \in (0,1)$;

   (3) *superlinear convergence rate*, $\alpha = 1$, $\mu = 0$ or $1 < \alpha < 2$, $\mu = 0$.

2. When $\alpha = 2$, we say that the sequence $\{\mathbf{x}_k\}$ has a *quadratic convergence rate*.

3. When $\alpha = 3$, we say that the sequence $\{\mathbf{x}_k\}$ has a *cubic convergence rate.*

If $\{\mathbf{x}_k\}$ is sublinearly convergent, and

$$\lim_{k\to\infty} \frac{\|\mathbf{x}_{k+2} - \mathbf{x}_{k+1}\|_2}{\|\mathbf{x}_{k+1} - \mathbf{x}_k\|_2} = 1,$$

then the sequence $\{\mathbf{x}_k\}$ is said to have a *logarithmic convergence rate.*

Sublinear convergence comprises a class of slow convergence rates; the linear convergence comprises a class of fast convergence; and the superlinear convergence and quadratic convergence comprise classes of respectively very fast and extremely fast convergence. When designing an optimization algorithm, we often require it at least to be linearly convergent, preferably to be quadratically convergent. The ultra-fast cubic convergence rate is in general difficult to achieve.

2. Local Convergence Rates

The Q-convergence rate is a limited convergence rate. When we evaluate an optimization algorithm, a practical question is: how many iterations does it need to achieve the desired accuracy? The answer depends on the local convergence rate of the output objective function sequence of a given optimization algorithm.

The *local convergence rate* of the sequence $\{\mathbf{x}_k\}$ is denoted r_k and is defined by

$$r_k = \left\| \frac{\mathbf{x}_{k+1} - \mathbf{x}^\star}{\mathbf{x}_k - \mathbf{x}^\star} \right\|_2 . \tag{4.4.37}$$

The complexity of an optimization algorithm is defined as the inverse of the local convergence rate of the updating variable.

The following is a classification of local convergence rates [344].

(1) *Sublinear rate* This convergence rate is given by a power function of the iteration number k and is usually denoted as

$$f(\mathbf{x}_k) - f(\mathbf{x}^\star) \le \epsilon = O\left(\frac{1}{\sqrt{k}}\right). \tag{4.4.38}$$

(2) *Linear rate* This convergence rate is expressed as an exponential function of the iteration number k, and is usually defined as

$$f(\mathbf{x}_k) - f(\mathbf{x}^\star) \le \epsilon = O\left(\frac{1}{k}\right). \tag{4.4.39}$$

(3) *Quadratic rate* This convergence rate is a bi-exponential function of the iteration number k and is usually measured by

$$f(\mathbf{x}_k) - f(\mathbf{x}^\star) \le \epsilon = O\left(\frac{1}{k^2}\right). \tag{4.4.40}$$

For example, in order to achieve the approximation accuracy $f(\mathbf{x}_k) - f(\mathbf{x}^\star) \leq \epsilon = 0.0001$, optimization algorithms with sublinear, linear and quadratic rates need to run about $10^8, 10^4$ and 100 iterations, respectively.

THEOREM 4.8 [344] *Let $\epsilon = f(\mathbf{x}_k) - f(\mathbf{x}^\star)$ be the estimation error of the objective function given by the updating sequence of the gradient method $\mathbf{x}_{k+1} = \mathbf{x}_k - \alpha\nabla f(\mathbf{x}_k)$. For a convex function $f(\mathbf{x})$, the upper bound of the estimation error $\epsilon = f(\mathbf{x}_k) - f(\mathbf{x}^\star)$ is given by*

$$f(\mathbf{x}_k) - f(\mathbf{x}^\star) \leq \frac{2L\|\mathbf{x}_0 - \mathbf{x}^\star\|_2^2}{k+4}. \tag{4.4.41}$$

This theorem shows that the local convergence rate of the gradient method $\mathbf{x}_{k+1} = \mathbf{x}_k - \alpha\nabla f(\mathbf{x}_k)$ is the linear rate $O(1/k)$. Although this is a fast convergence rate, it is not by far optimal, as will be seen in the next section.

4.5 Nesterov Optimal Gradient Method

Let $Q \subset \mathbb{R}^n$ be a convex set in the vector space $\mathbb{R}^n$. Consider the unconstrained optimization problem $\min_{\mathbf{x}\in Q} f(\mathbf{x})$.

4.5.1 Lipschitz Continuous Function

DEFINITION 4.13 [344] We say that an objective function $f(\mathbf{x})$ is *Lipschitz continuous* in the definition domain Q if

$$|f(\mathbf{x}) - f(\mathbf{y})| \leq L\|\mathbf{x} - \mathbf{y}\|_2, \quad \forall\mathbf{x},\mathbf{y} \in Q \tag{4.5.1}$$

holds for some *Lipschitz constant* $L > 0$. Similarly, we say that the gradient vector $\nabla f(\mathbf{x})$ of a differentiable function $f(\mathbf{x})$ is Lipschitz continuous in the definition domain Q if

$$\|\nabla f(\mathbf{x}) - \nabla f(\mathbf{y})\|_2 \leq L\|\mathbf{x} - \mathbf{y}\|_2, \quad \forall\mathbf{x},\mathbf{y} \in Q \tag{4.5.2}$$

holds for some Lipschitz constant $L > 0$.

Hereafter a *Lipschitz continuous function* with the Lipschitz constant L will be denoted as an *L-Lipschitz continuous function.*

The function $f(\mathbf{x})$ is said to be continuous at the point $\mathbf{x}_0$, if $\lim_{\mathbf{x}\to\mathbf{x}_0} f(\mathbf{x}) = f(\mathbf{x}_0)$. When we say that $f(\mathbf{x})$ is a continuous function, this means the $f(\mathbf{x})$ is continuous at every point $\mathbf{x} \in Q$.

A Lipschitz continuous function $f(\mathbf{x})$ must be a continuous function, but a continuous function is not necessarily a Lipschitz continuous function.

Example 4.3 The function $f(x) = 1/\sqrt{x}$ is not a Lipschitz continuous function in the open interval $(0,1)$. If it is assumed to be a Lipschitz continuous function, then it must satisfy $|f(x_1) - f(x_2)| = \left|\frac{1}{\sqrt{x_1}} - \frac{1}{\sqrt{x_2}}\right| \leq L\left|\frac{1}{x_1} - \frac{1}{x_2}\right|$, namely $\left|\frac{1}{\sqrt{x_1}} + \frac{1}{\sqrt{x_2}}\right| \leq L$. However, when the point $x \to 0$, and $x_1 = 1/n^2$, $x_2 = 9/n^2$, we have $L \geq n/4 \to \infty$. Hence, the continuous function $f(x) = 1/\sqrt{x}$ is not a Lipschitz continuous function in the open interval $(0,1)$.

An everywhere differentiable function is called a *smooth function.* A smooth function must be continuous, but a continuous function is not necessarily smooth. A typical example occurs when the continuous function has "sharp" point at which it is nondifferentiable. Hence, a Lipschitz continuous function is not necessarily differentiable, but a function $f(\mathbf{x})$ with Lipschitz continuous gradient in the definition domain Q must be smooth in Q, because the definition of a Lipschitz continuous gradient stipulates that $f(\mathbf{x})$ is differentiable in the definition domain.

In convex optimization, the notation $C_L^{k,p}(Q)$ (where $Q \subseteq \mathbb{R}^n$) denotes the Lipschitz continuous function class with the following properties [344]:

- The function $f \in C_L^{k,p}(Q)$ is k times continuously differentiable in Q.
- The pth-order derivative of the function $f \in C_L^{k,p}(Q)$ is L-Lipschitz continuous:

$$\|f^{(p)}(\mathbf{x}) - f^{(p)}(\mathbf{y})\| \leq L\|\mathbf{x} - \mathbf{y}\|_2, \quad \forall \mathbf{x}, \mathbf{y} \in Q.$$

If $k \neq 0$, then $f \in C_L^{k,p}(Q)$ is said to be a differentiable function. Obviously, it always has $p \leq k$. If $q > k$ then $C_L^{q,p}(Q) \subseteq C_L^{k,p}(Q)$. For example, $C_L^{2,1}(Q) \subseteq C_L^{1,1}(Q)$.

The following are three common function classes $C_L^{k,p}(Q)$:

(1) $f(\mathbf{x}) \in C_L^{0,0}(Q)$ is L-Lipschitz continuous but nondifferentiable in Q;

(2) $f(\mathbf{x}) \in C_L^{1,0}(Q)$ is L-Lipschitz continuously differentiable in Q, but its gradient is not;

(3) $f(\mathbf{x}) \in C_L^{1,1}(Q)$ is L-Lipschitz continuously differentiable in Q, and its gradient $\nabla f(\mathbf{x})$ is L-Lipschitz continuous in Q.

The basic property of the $C_L^{k,p}(Q)$ function class is that if $f_1 \in C_{L_1}^{k,p}(Q)$, $f_2 \in C_{L_2}^{k,p}(Q)$ and $\alpha, \beta \in \mathbb{R}$ then

$$\alpha f_1 + \beta f_2 \in C_{L_3}^{k,p}(Q),$$

where $L_3 = |\alpha|\, L_1 + |\beta|\, L_2$.

Among all Lipschitz continuous functions, $C_L^{1,1}(Q)$, with the Lipschitz continuous gradient, is the most important function class, and is widely applied in convex optimization.

On the $C_L^{1,1}(Q)$ function class, one has the following two lemmas [344].

Lemma 4.1 *The function $f(\mathbf{x})$ belongs to $C_L^{2,1}(\mathbb{R}^n)$ if and only if*

$$\|f''(\mathbf{x})\|_F \leq L, \quad \forall \mathbf{x} \in \mathbb{R}^n. \tag{4.5.3}$$

Lemma 4.2 *If $f(\mathbf{x}) \in \mathcal{C}_L^{1,1}(Q)$ then*

$$|f(\mathbf{y}) - f(\mathbf{x}) - \langle \nabla f(\mathbf{x}), \mathbf{y} - \mathbf{x} \rangle| \leq \frac{L}{2} \|\mathbf{y} - \mathbf{x}\|_2^2, \quad \forall \mathbf{x}, \mathbf{y} \in Q. \tag{4.5.4}$$

Since the function $\mathcal{C}_L^{2,1}$ must also be a $\mathcal{C}_L^{1,1}$ function, Lemma 4.1 directly provides a simple way to determine whether a function belongs to the $\mathcal{C}_L^{1,1}$ function class.

Below we give some examples of the applications of Lemma 4.1 to determine the $\mathcal{C}_L^{1,1}(Q)$ function class.

Example 4.4 The linear function $f(\mathbf{x}) = \langle \mathbf{a}, \mathbf{x} \rangle + b$ belongs to the $\mathcal{C}_0^{1,1}$ function class; namely, the gradient of any linear function is not Lipschitz continuous because

$$f'(\mathbf{x}) = \mathbf{a}, \quad f''(\mathbf{x}) = \mathbf{O} \quad \Rightarrow \quad \|f''(\mathbf{x})\|_2 = 0.$$

Example 4.5 The quadratic function $f(\mathbf{x}) = \frac{1}{2}\mathbf{x}^T \mathbf{A} \mathbf{x} + \mathbf{a}^T \mathbf{x} + \mathbf{b}$ belongs to the $\mathcal{C}_{\|\mathbf{A}\|_F}^{1,1}(\mathbb{R}^n)$ function class, because

$$\nabla f(\mathbf{x}) = \mathbf{A}\mathbf{x} + \mathbf{a}, \quad \nabla^2 f(\mathbf{x}) = \mathbf{A} \quad \Rightarrow \quad \|f''(\mathbf{x})\|_2. = \|\mathbf{A}\|_F.$$

Example 4.6 The logarithmic function $f(x) = \ln(1 + \mathrm{e}^x)$ belongs to the $\mathcal{C}_{1/2}^{1,1}(\mathbb{R})$ function class, because

$$f'(x) = \frac{\mathrm{e}^x}{1 + \mathrm{e}^x}, \quad f''(x) = \frac{\mathrm{e}^x}{(1 + \mathrm{e}^x)^2} \quad \Rightarrow \quad |f''(x)| = \frac{1}{2} \left| 1 - \frac{1 + \mathrm{e}^{2x}}{(1 + \mathrm{e}^x)^2} \right| \leq \frac{1}{2},$$

where, in order to find $f''(x)$, we let $y = f'(x)$, yielding $y' = f''(x) = y(\ln y)'$.

Example 4.7 The function $f(x) = \sqrt{1 + x^2}$ belongs to the $\mathcal{C}_1^{1,1}(\mathbb{R})$ function class, because

$$f'(x) = \frac{x}{\sqrt{1 + x^2}}, \quad f''(x) = \frac{1}{(1 + x^2)^{3/2}} \quad \Rightarrow \quad |f''(x)| \leq 1.$$

Lemma 4.2 is a key inequality for analyzing the convergence rate of a $\mathcal{C}_L^{1,1}$ function $f(\mathbf{x})$ in gradient algorithms. By Lemma 4.2, it follows [536] that

$$\begin{aligned} f(\mathbf{x}_{k+1}) &\leq f(\mathbf{x}_k) + \langle \nabla f(\mathbf{x}_k), \mathbf{x}_{k+1} - \mathbf{x}_k \rangle + \frac{L}{2} \|\mathbf{x}_{k+1} - \mathbf{x}_k\|_2^2 \\ &\leq f(\mathbf{x}_k) + \langle \nabla f(\mathbf{x}_k), \mathbf{x} - \mathbf{x}_k \rangle + \frac{L}{2} \|\mathbf{x} - \mathbf{x}_k\|_2^2 - \frac{L}{2} \|\mathbf{x} - \mathbf{x}_{k+1}\|_2^2 \\ &\leq f(\mathbf{x}) + \frac{L}{2} \|\mathbf{x} - \mathbf{x}_k\|_2^2 - \frac{L}{2} \|\mathbf{x} - \mathbf{x}_{k+1}\|_2^2. \end{aligned}$$

Put $\mathbf{x} = \mathbf{x}^\star$ and $\delta_k = f(\mathbf{x}_k) - f(\mathbf{x}^\star)$; then

$$\begin{aligned} 0 \leq \frac{L}{2} \|\mathbf{x}^\star - \mathbf{x}_{k+1}\|_2^2 &\leq -\delta_{k+1} + \frac{L}{2} \|\mathbf{x}^\star - \mathbf{x}_k\|_2^2 \\ &\leq \cdots \leq -\sum_{i=1}^{k+1} \delta_i + \frac{L}{2} \|\mathbf{x}^\star - \mathbf{x}_0\|_2^2. \end{aligned}$$

From the estimation errors of the projected gradient algorithm $\delta_1 \geq \delta_2 \geq \cdots \geq \delta_{k+1}$, it follows that $-(\delta_1+\cdots+\delta_{k+1}) \leq -(k+1)\delta_{k+1}$, and thus the above inequality can be simplified to

$$0 \leq \frac{L}{2}\|\mathbf{x}^\star - \mathbf{x}_{k+1}\|_2^2 \leq -(k+1)\delta_k + \frac{L}{2}\|\mathbf{x}^\star - \mathbf{x}_0\|_2^2,$$

from which the upper bound of the convergence rate of the projected gradient algorithm is given by [536]

$$\delta_k = f(\mathbf{x}_k) - f(\mathbf{x}^\star) \leq \frac{L\|\mathbf{x}^\star - \mathbf{x}_0\|_2^2}{2(k+1)}. \tag{4.5.5}$$

This shows that, as for the (basic) gradient method, the local convergence rate of the projected gradient method is $O(1/k)$.

4.5.2 Nesterov Optimal Gradient Algorithms

Theorem 4.9 [343] *Let $f(\mathbf{x})$ be a convex function with L-Lipschitz gradient, if the updating sequence $\{\mathbf{x}_k\}$ meets the condition*

$$\mathbf{x}_k \in \mathbf{x}_0 + \mathrm{Span}\{\mathbf{x}_0, \ldots, \mathbf{x}_{k-1}\}$$

then the lower bound of the estimation error $\epsilon = f(\mathbf{x}_k) - f(\mathbf{x}^\star)$ achieved by any first-order optimization method is given by

$$f(\mathbf{x}_k) - f(\mathbf{x}^\star) \geq \frac{3L\|\mathbf{x}_0 - \mathbf{x}^\star\|_2^2}{32(k+1)^2}, \tag{4.5.6}$$

where $\mathrm{Span}\{\mathbf{u}_0, \ldots, \mathbf{u}_{k-1}\}$ *denotes the linear subspace spanned by $\mathbf{u}_0, \ldots, \mathbf{u}_{k-1}$, $\mathbf{x}_0$ is the initial value of the gradient method and $f(\mathbf{x}^\star)$ denotes the minimal value of the function f.*

Theorem 4.9 shows that the optimal convergence rate of any first-order optimization method is the quadratic rate $O(1/k^2)$.

Since the convergence rate of gradient methods is the linear rate $O(1/k)$, and the convergence rate of the optimal first-order optimization methods is the quadratic rate $O(1/k^2)$, the gradient methods are far from optimal.

The *heavy ball method* (HBM) can efficiently improve the convergence rate of the gradient methods.

The HBM is a two-step method: let $\mathbf{y}_0$ and $\mathbf{x}_0$ be two initial vectors and let α_k and β_k be two positive valued sequences; then the first-order method for solving an unconstrained minimization problem $\min_{\mathbf{x}\in\mathbb{R}^n} f(\mathbf{x})$ can use two-step updates [396]:

$$\left.\begin{aligned} \mathbf{y}_k &= \beta_k \mathbf{y}_{k-1} - \nabla f(\mathbf{x}_k), \\ \mathbf{x}_{k+1} &= \mathbf{x}_k + \alpha_k \mathbf{y}_k. \end{aligned}\right\} \tag{4.5.7}$$

In particular, letting $\mathbf{y}_0 = \mathbf{0}$, the above two-step updates can be rewritten as the one-step update

$$\mathbf{x}_{k+1} = \mathbf{x}_k - \alpha_k \nabla f(\mathbf{x}_k) + \beta_k(\mathbf{x}_k - \mathbf{x}_{k-1}), \tag{4.5.8}$$

where $\mathbf{x}_k - \mathbf{x}_{k-1}$ is called the *momentum*. As can be seen in (4.5.7), the HBM treats the iterations as a point mass with momentum $\mathbf{x}_k - \mathbf{x}_{k-1}$, which thus tends to continue moving in the direction $\mathbf{x}_k - \mathbf{x}_{k-1}$.

Let $\mathbf{y}_k = \mathbf{x}_k + \beta_k(\mathbf{x}_k - \mathbf{x}_{k-1})$, and use $\nabla f(\mathbf{y}_k)$ instead of $\nabla f(\mathbf{x}_k)$. Then the update equation (4.5.7) becomes

$$\left.\begin{aligned} \mathbf{y}_k &= \mathbf{x}_k + \beta_k(\mathbf{x}_k - \mathbf{x}_{k-1}), \\ \mathbf{x}_{k+1} &= \mathbf{y}_k - \alpha_k \nabla f(\mathbf{y}_k). \end{aligned}\right\}$$

This is the basic form of the optimal gradient method proposed by Nesterov in 1983 [343], usually called the Nesterov (first) optimal gradient algorithm and shown in Algorithm 4.6.

Algorithm 4.6 Nesterov (first) optimal gradient algorithm [343]

---

initialization: Choose $\mathbf{x}_{-1} = \mathbf{0}$, $\mathbf{x}_0 \in \mathbb{R}^n$ and $\alpha_0 \in (0,1)$. Set $k = 0$, $q = \mu/L$.

repeat

1. Compute $\alpha_{k+1} \in (0,1)$ from the equation $\alpha_{k+1}^2 = (1-\alpha_{k+1})\alpha_k^2 + q\alpha_{k+1}$.
2. Set $\beta_k = \dfrac{\alpha_k(1-\alpha_k)}{\alpha_k^2 + \alpha_{k+1}}$.
3. Compute $\mathbf{y}_k = \mathbf{x}_k + \beta_k(\mathbf{x}_k - \mathbf{x}_{k-1})$.
4. Compute $\mathbf{x}_{k+1} = \mathbf{y}_k - \alpha_{k+1}\nabla f(\mathbf{y}_k)$.
5. **exit if** $\mathbf{x}_k$ converges.

return $k \leftarrow k+1$.

output: $\mathbf{x} \leftarrow \mathbf{x}_k$.

It is easily seen that the Nesterov optimal gradient algorithm is intuitively like the heavy ball formula (4.5.7) but it is not identical; i.e., it uses $\nabla f(\mathbf{y})$ instead of $\nabla f(\mathbf{x})$.

In the Nesterov optimal gradient algorithm, the estimation sequence $\{\mathbf{x}_k\}$ is an *approximation solution sequence*, and $\{\mathbf{y}_k\}$ is a searching point sequence.

Theorem 4.10 [343] *Let f be a convex function with L-Lipschitz gradient. The Nesterov optimal gradient algorithm achieves*

$$f(\mathbf{x}_k) - f(\mathbf{x}^*) \le \frac{CL\|\mathbf{x}_k - \mathbf{x}^*\|_2^2}{(k+1)^2}. \tag{4.5.9}$$

Clearly, the convergence rate of the Nesterov optimal gradient method is the optimal rate for the first-order gradient methods up to constants. Hence, the Nesterov gradient method is indeed an optimal first-order minimization method.

It has been shown [180] that the use of a strong convexity constant μ^* is very effective in reducing the number of iterations of the Nesterov algorithm. A Nesterov algorithm with *adaptive convexity parameter*, using a decreasing sequence $\mu_k \to \mu^*$, was proposed in [180]. It is shown in Algorithm 4.7.

Algorithm 4.7 Nesterov algorithm with adaptive convexity parameter

---

input: Lipschitz constant L, and the convexity parameter guess $\mu^* \leq L$.

given: $\mathbf{x}_0, \mathbf{v}_0 = \mathbf{x}_0, \gamma_0 > 0, \beta > 1, \mu_0 \in [\mu^*, \gamma_0)$. Set $k = 0$, $\theta \in [0,1], \mu_+ = \mu_0$.

repeat

1. $\mathbf{d}_k = \mathbf{v}_k - \mathbf{x}_k$.
2. $\mathbf{y}_k = \mathbf{x}_k + \theta_k \mathbf{d}_k$.
3. **exit if** $\nabla f(\mathbf{y}_k) = \mathbf{0}$.
4. Steepest descent step: $\mathbf{x}_{k+1} = \mathbf{y}_k - \nu \nabla f(\mathbf{y}_k)$. If L is known, then $\nu \geq 1/L$.
5. If $(\gamma_k - \mu^*) < \beta(\mu_+ - \mu^*)$, then choose $\mu_+ \in [\mu^*, \gamma/\beta]$.
6. Compute $\tilde{\mu} = \dfrac{\|\nabla f(\mathbf{y}_k)\|^2}{2[f(\mathbf{y}_k) - f(\mathbf{x}_{k+1})]}$.
7. If $\mu_+ \geq \tilde{\mu}$, then $\mu_+ = \max\{\mu^*, \tilde{\mu}/10\}$.
8. Compute α_k as the largest root of $A\alpha^2 + B\alpha + C = 0$.
9. $\gamma_{k+1} = (1 - \alpha_k)\gamma_k + \alpha_k \mu_+$.
10. $\mathbf{v}_{k+1} = \dfrac{1}{\gamma_{k+1}}((1 - \alpha_k)\gamma_k \mathbf{v}_k + \alpha_k(\mu_+ \mathbf{y}_k - \nabla f(\mathbf{y}_k)))$.
11. **return** $k \leftarrow k + 1$.

output: $\mathbf{y}_k$ as an optimal solution.

The coefficients A, B, C in step 8 are given by

$$G = \gamma_k \left(\frac{\mu}{2} \|\mathbf{v}_k - \mathbf{y}_k\|^2 + (\nabla f(\mathbf{y}_k))^T (\mathbf{v}_k - \mathbf{y}_k) \right),$$

$$A = G + \frac{1}{2}\|\nabla f(\mathbf{y}_k)\|^2 + (\mu - \gamma_k)(f(\mathbf{x}_k) - f(\mathbf{y}_k)),$$

$$B = (\mu - \gamma_k)\left(f(\mathbf{x}_{k+1}) - f(\mathbf{x}_k)\right) - \gamma_k \left(f(\mathbf{y}_k) - f(\mathbf{x}_k)\right) - G,$$

$$C = \gamma_k \left(f(\mathbf{x}_{k+1}) - f(\mathbf{x}_k)\right), \quad \text{with } \mu = \mu_+$$

It is suggested that $\gamma_0 = L$ if L is known, $\mu_0 = \max\{\mu^*, \gamma_0/100\}$ and $\beta = 1.02$.

The Nesterov (first) optimal gradient algorithm has two limitations:

(1) $\mathbf{y}_k$ may be out of the definition domain Q, and thus the objective function $f(\mathbf{x})$ is required to be well-defined at every point in Q;

(2) this algorithm is available only for the minimization of the Euclidean norm $\|\mathbf{x}\|_2$.

To cope with these limitations, Nesterov developed two other optimal gradient algorithms [346], [345]. Denote by $d(\mathbf{x})$ a *prox-function* of the domain Q. It is assumed that $d(\mathbf{x})$ is continuous and strongly convex on Q. Depending on whether $\|\mathbf{x}\|_1$ or $\|\mathbf{x}\|_2$ is being minimized, there are different choices for the prox-function [345]:

$$\text{using} \quad \|\mathbf{x}\|_1 = \sum_{i=1}^n |x_i|, \quad d(\mathbf{x}) = \ln n + \sum_{i=1}^n x_i \ln |x_i|, \tag{4.5.10}$$

$$\text{using} \quad \|\mathbf{x}\|_2 = \left(\sum_{i=1}^n x_i^2\right)^{1/2}, \quad d(\mathbf{x}) = \frac{1}{2}\sum_{i=1}^n \left(x_i - \frac{1}{n}\right)^2. \tag{4.5.11}$$

Define the ith element of the $n \times 1$ mapping vector $V_Q(\mathbf{u},\mathbf{g})$ as [345]

$$V_Q^{(i)}(\mathbf{u},\mathbf{g}) = u_i e^{-g_i}\left(\sum_{j=1}^n u_j e^{-g_i}\right)^{-1}, \quad i = 1,\ldots,n. \tag{4.5.12}$$

Algorithm 4.8 shows the *Nesterov third optimal gradient algorithm.*

Algorithm 4.8 Nesterov (third) optimal gradient algorithm [345]

---

given: Choose $\mathbf{x}_0 \in \mathbb{R}^n$ and $\alpha \in (0,1)$. Set $k = 0$.

initialization: $\mathbf{y}_0 = \text{argmin}_{\mathbf{x}} \left\{\frac{L}{\alpha} d(\mathbf{x}) + \frac{1}{2}\left(f(\mathbf{x}_0) + \langle f'(\mathbf{x}_0), \mathbf{x} - \mathbf{x}_0\rangle\right) : \mathbf{x} \in Q\right\}$.

repeat

1. Find $\mathbf{z}_k = \text{argmin}_{\mathbf{x}} \left\{\frac{L}{\alpha} d(\mathbf{x}) + \sum_{i=0}^{k} \frac{i+1}{2}\left(f(\mathbf{x}_i) + \langle \nabla f(\mathbf{x}_i), \mathbf{x} - \mathbf{x}_i\rangle\right) : \mathbf{x} \in Q\right\}$.
2. Set $\tau_k = \frac{2}{k+3}$ and $\mathbf{x}_{k+1} = \tau_k \mathbf{z}_k + (1-\tau_k)\mathbf{y}_k$.
3. Use (4.5.12) to compute $\hat{\mathbf{x}}_{k+1}(i) = V_Q^{(i)}\left(\mathbf{z}_k, \frac{\alpha}{L}\tau_k \nabla f(\mathbf{x}_{k+1})\right), i = 1,\ldots,n$.
4. Set $\mathbf{y}_{k+1} = \tau_k \hat{\mathbf{x}}_{k+1} + (1-\tau_k)\mathbf{y}_k$.
5. **exit if** $\mathbf{x}_k$ converges.

return $k \leftarrow k+1$.

output: $\mathbf{x} \leftarrow \mathbf{x}_k$.

4.6 Nonsmooth Convex Optimization

The gradient methods require that for the objective function $f(\mathbf{x})$, the gradient $\nabla f(\mathbf{x})$ exists at the point $\mathbf{x}$; and the Nesterov optimal gradient method requires, further, that the objective function has L-Lipschitz continuous gradient. Hence, the gradient methods in general and the Nesterov optimal gradient method in particular are available only for smooth convex optimization. As an important extension of gradient methods, this section focuses upon the proximal gradient method which is available for nonsmooth convex optimization.

The following are two typical examples of nonsmooth convex minimization.

(1) *Basis pursuit de-noising* seeks a sparse near-solution to an under-determined matrix equation [95]

$$\min_{\mathbf{x}\in\mathbb{R}^n}\left\{\|\mathbf{x}\|_1+\frac{\lambda}{2}\|\mathbf{A}\mathbf{x}-\mathbf{b}\|_2^2\right\}. \tag{4.6.1}$$

This problem is essentially equivalent to the Lasso (sparse least squares) problem [462]

$$\min_{\mathbf{x}\in\mathbb{R}^n}\frac{\lambda}{2}\|\mathbf{A}\mathbf{x}-\mathbf{b}\|_2^2 \quad \text{subject to} \quad \|\mathbf{x}\|_1\leq K. \tag{4.6.2}$$

(2) *Robust principal component analysis* approximates an input data matrix $\mathbf{D}$ to a sum of a low-rank matrix $\mathbf{L}$ and a sparse matrix $\mathbf{S}$:

$$\min\left\{\|\mathbf{L}\|_*+\lambda\|\mathbf{S}\|_1+\frac{\gamma}{2}\|\mathbf{L}+\mathbf{S}-\mathbf{D}\|_F^2\right\}. \tag{4.6.3}$$

The key challenge of the above minimization problems is the nonsmoothness induced by the nondifferentiable ℓ_1 norm $\|\cdot\|_1$ and *nuclear norm* $\|\cdot\|_*$.

Consider the following *combinatorial optimization problem*

$$\min_{\mathbf{x}\in E}\left\{F(\mathbf{x})=f(\mathbf{x})+h(\mathbf{x})\right\}, \tag{4.6.4}$$

where $E\subset\mathbb{R}^n$ is a finite-dimensional real vector space and

$h:E\to\mathbb{R}$ is a convex function, but is nondifferentiable or nonsmooth in E.

$f:\mathbb{R}^n\to\mathbb{R}$ is a continuous smooth convex function, and its gradient is L-Lipschitz continuous:

$$\|\nabla f(\mathbf{x})-\nabla f(\mathbf{y})\|_2\leq L\,\|\mathbf{x}-\mathbf{y}\|_2, \quad \forall\,\mathbf{x},\mathbf{y}\in\mathbb{R}^n.$$

To address the challenge of nonsmoothness, there are two problems to be solved:

- how to cope with the nonsmoothness;
- how to design a Nesterov-like optimal gradient method for nonsmooth convex optimization.

4.6.1 Subgradient and Subdifferential

Because the gradient vector of a nonsmooth function $h(\mathbf{x})$ does not exist everywhere, neither a gradient method nor the Nesterov optimal gradient method is available. A natural question to ask is whether a nonsmooth function has some class of "generalized gradient" similar to a gradient vector.

For a twice continuous differentiable function $f(\mathbf{x})$, its second-order approximation is given by

$$f(\mathbf{x}+\Delta\mathbf{x})\approx f(\mathbf{x})+(\nabla f(\mathbf{x}))^T\Delta\mathbf{x}+(\Delta\mathbf{x})^T\mathbf{H}\Delta\mathbf{x}.$$

If the Hessian matrix $\mathbf{H}$ is positive semi-definite or positive definite then we have the inequality

$$f(\mathbf{x}+\Delta\mathbf{x}) \geq f(\mathbf{x}) + (\nabla f(\mathbf{x}))^T \Delta\mathbf{x},$$

or

$$f(\mathbf{y}) \geq f(\mathbf{x}) + (\nabla f(\mathbf{x}))^T (\mathbf{y}-\mathbf{x}), \quad \forall \mathbf{x},\mathbf{y} \in \operatorname{dom} f(\mathbf{x}). \tag{4.6.5}$$

Although a nonsmooth function $h(\mathbf{x})$ does not have a gradient vector $\nabla h(\mathbf{x})$, it is possible to find another vector $\mathbf{g}$ instead of the gradient vector $\nabla f(\mathbf{x})$ such that the inequality (4.6.5) still holds.

Definition 4.14 A vector $\mathbf{g} \in \mathbb{R}^n$ is said to be a *subgradient vector* of the function $h: \mathbb{R}^n \to \mathbb{R}$ at the point $\mathbf{x} \in \mathbb{R}^n$ if

$$h(\mathbf{y}) \geq h(\mathbf{x}) + \mathbf{g}^T (\mathbf{y}-\mathbf{x}), \quad \forall \mathbf{x},\mathbf{y} \in \operatorname{dom} h. \tag{4.6.6}$$

The set of all subgradient vectors of the function h at the point $\mathbf{x}$ is known as the *subdifferential* of the function h at the point $\mathbf{x}$, denoted $\partial h(\mathbf{x})$, and is defined as

$$\partial h(\mathbf{x}) \stackrel{\text{def}}{=} \left\{\mathbf{g} \middle| h(\mathbf{y}) \geq h(\mathbf{x}) + \mathbf{g}^T (\mathbf{y}-\mathbf{x}),\ \forall\, \mathbf{y} \in \operatorname{dom} h\right\}. \tag{4.6.7}$$

When $h(\mathbf{x})$ is differentiable, we have $\partial h(\mathbf{x}) = \{\nabla h(\mathbf{x})\}$, as the gradient vector of a smooth function is unique, so we can view the gradient operator ∇h of a convex and differentiable function as a point-to-point mapping, i.e., ∇h maps each point $\mathbf{x} \in \operatorname{dom} h$ to the point $\nabla h(\mathbf{x})$. In contrast, the subdifferential operator ∂h, defined in Equation (4.6.7), of a closed proper convex function h can be viewed as a point-to-set mapping, i.e., ∂h maps each point $\mathbf{x} \in \operatorname{dom} h$ to the set $\partial h(\mathbf{x})$.[1] Any point $\mathbf{g} \in \partial h(\mathbf{x})$ is called a subgradient of h at $\mathbf{x}$. Generally speaking, a function $h(\mathbf{x})$ may have one or several subgradient vectors at some point $\mathbf{x}$.

The function $h(\mathbf{x})$ is said to be *subdifferentiable* at the point $\mathbf{x}$ if it at least has one subgradient vector. More generally, the function $h(\mathbf{x})$ is said to be subdifferentiable in the definition domain $\operatorname{dom} h$, if it is subdifferentiable at all points $\mathbf{x} \in \operatorname{dom} h$.

The basic properties of the subdifferential are as follows [344], [54].

1. *Convexity* $\partial h(\mathbf{x})$ is always a closed convex set, even if $h(\mathbf{x})$ is not convex.
2. *Nonempty and boundedness* If $\mathbf{x} \in \operatorname{int}(\operatorname{dom} h)$ then the subdifferential $\partial h(\mathbf{x})$ is nonempty and bounded.
3. *Nonnegative factor* If $\alpha > 0$, then $\partial(\alpha h(\mathbf{x})) = \alpha \partial h(\mathbf{x})$.
4. *Subdifferential* If h is convex and differentiable at the point $\mathbf{x}$ then the subdifferential is a singleton $\partial h(\mathbf{x}) = \{\nabla h(\mathbf{x})\}$, namely its gradient is its unique subgradient. Conversely, if h is a convex function and $\partial h(\mathbf{x}) = \{\mathbf{g}\}$ then h is differentiable at the point $\mathbf{x}$ and $\mathbf{g} = \nabla h(\mathbf{x})$.

[1] A proper convex function $f(x)$ takes values in the extended real number line such that $f(x) < +\infty$ for at least one x and $f(x) > -\infty$ for every x.

5. *Minimum point of nondifferentiable function* The point $\mathbf{x}^\star$ is a minimum point of the convex function h if and only if h is subdifferentiable at $\mathbf{x}^\star$ and

$$\mathbf{0} \in \partial h(\mathbf{x}^\star). \tag{4.6.8}$$

This condition is known as the first-order optimality condition of the nonsmooth convex function $h(\mathbf{x})$. If h is differentiable, then the first-order optimality condition $\mathbf{0} \in \partial h(\mathbf{x})$ simplifies to $\nabla h(\mathbf{x}) = \mathbf{0}$.

6. *Subdifferential of a sum of functions* If $h_1, \ldots, h_m$ are convex functions then the subdifferential of $h(\mathbf{x}) = h_1(\mathbf{x}) + \cdots + h_m(\mathbf{x})$ is given by

$$\partial h(\mathbf{x}) = \partial h_1(\mathbf{x}) + \cdots + \partial h_m(\mathbf{x}).$$

7. *Subdifferential of affine transform* If $\phi(\mathbf{x}) = h(\mathbf{Ax}+\mathbf{b})$ then the subdifferential $\partial\phi(\mathbf{x}) = \mathbf{A}^T \partial h(\mathbf{Ax}+\mathbf{b})$.

8. *Subdifferential of pointwise maximal function* Let h be a pointwise maximal function of the convex functions $h_1, \ldots, h_m$, i.e., $h(\mathbf{x}) = \max\limits_{i=1,\ldots,m} h_i(\mathbf{x})$; then

$$\partial h(\mathbf{x}) = \mathrm{conv}\left(\cup \{\partial h_i(\mathbf{x}) | h_i(\mathbf{x}) = h(\mathbf{x})\}\right).$$

That is to say, the subdifferential of a pointwise maximal function $h(\mathbf{x})$ is the convex hull of the union set of subdifferentials of the "active function" $h_i(\mathbf{x})$ at the point $\mathbf{x}$.

The subgradient vector $\mathbf{g}$ of the function $h(\mathbf{x})$ is denoted as $\mathbf{g} = \tilde{\nabla} h(\mathbf{x})$.

Example 4.8 The function $h(x) = |x|$ is nonsmooth at $x = 0$ as its gradient $\nabla|x|$ does not exist there. To find the subgradient vectors of $h(x) = |x|$, rewrite the function as $h(s, x) = |x| = sx$ from which it follows that the gradient of the function $f(s, x)$ is given by $\partial f(s, x)/\partial x = s$; $s = -1$ if $x < 0$ and $s = +1$ if $x > 0$. Moreover, if $x = 0$ then the subgradient vector should satisfy Equation (4.6.6), i.e., $|y| \geq gy$ or $g \in [-1, +1]$. Hence, the subdifferential of the function $|x|$ is given by

$$\partial|x| = \begin{cases} \{-1\}, & x < 0, \\ \{+1\}, & x > 0, \\ [-1, 1], & x = 0. \end{cases} \tag{4.6.9}$$

Example 4.9 Consider the function $h(\mathbf{x}) = \sum_{i=1}^n |\mathbf{a}_i^T \mathbf{x} - b_i|$. Write

$$\begin{aligned} I_-(\mathbf{x}) &= \{i | \langle \mathbf{a}_i, \mathbf{x}\rangle - b_i < 0\}, \\ I_+(\mathbf{x}) &= \{i | \langle \mathbf{a}_i, \mathbf{x}\rangle - b_i > 0\}, \\ I_0(\mathbf{x}) &= \{i | \langle \mathbf{a}_i, \mathbf{x}\rangle - b_i = 0\}. \end{aligned}$$

Let $h(\mathbf{x}) = \sum_{i=1}^n |\mathbf{a}_i^T\mathbf{x} - b_i| = \sum_{i=1}^n s_i(\mathbf{a}_i^T\mathbf{x} - b_i)$; then

$$s_i = \begin{cases} -1, & \mathbf{x} \in I_-(\mathbf{x}), \\ 1, & \mathbf{x} \in I_+(\mathbf{x}), \\ [-1,1], & \mathbf{x} \in I_0(\mathbf{x}). \end{cases}$$

Hence the subdifferential of $h(\mathbf{x}) = \sum_{i=1}^n |\mathbf{a}_i^T\mathbf{x} - b_i|$ is given by

$$\partial h(\mathbf{x}) = \sum_{i=1}^n s_i\mathbf{a}_i = \sum_{i\in I_+(\mathbf{x})} \mathbf{a}_i - \sum_{i\in I_-(\mathbf{x})} \mathbf{a}_i + \sum_{i\in I_0(\mathbf{x})} \alpha_i\mathbf{a}_i, \tag{4.6.10}$$

where $\alpha_i = [-1,1]$.

Example 4.10 The ℓ_1-norm $h(\mathbf{x}) = \|\mathbf{x}\|_1 = |x_1| + \cdots + |x_m|$ is a special example of the function $h(\mathbf{x}) = \sum_{i=1}^n |\mathbf{a}_i^T\mathbf{x} - b_i|$ with $\mathbf{a}_i = \mathbf{e}_i$ and $b_i = 0$, where $\mathbf{e}_i$ is the basis vector with nonzero entry $e_i = 1$. Hence, by Equation (4.6.10), we have

$$\partial h(\mathbf{x}) = \sum_{i\in I_+(\mathbf{x})} \mathbf{e}_i - \sum_{i\in I_-(\mathbf{x})} \mathbf{e}_i + \sum_{i\in I_0(\mathbf{x})} \alpha_i\mathbf{e}_i, \quad \mathbf{x} \neq \mathbf{0},$$

where $I_-(\mathbf{x}) = \{i|x_i < 0\}$, $I_+(\mathbf{x}) = \{i|x_i > 0\}$ and $I_0(\mathbf{x}) = \{i|x_i = 0\}$. For $\mathbf{x} = \mathbf{0}$, by Definition 4.14 the subgradient should satisfy $\|\mathbf{y}\|_1 \geq \|\mathbf{0}\|_1 + \mathbf{g}^T(\mathbf{y} - \mathbf{0})$ with $h(\mathbf{0}) = 0$, i.e., $\|\mathbf{y}\|_1 \geq \mathbf{g}^T\mathbf{y}$, yielding $g_i = \max_{1\leq i\leq n} |y_i| \leq 1$, $\mathbf{y} \neq \mathbf{0}$. Therefore, the subdifferential of the ℓ_1-norm $h(\mathbf{x}) = \|\mathbf{x}\|_1$ is given by

$$\partial\|\mathbf{x}\|_1 = \begin{cases} \sum_{x_i>0} \mathbf{e}_i - \sum_{x_i<0} \mathbf{e}_i + \sum_{x_i=0} \alpha_i\mathbf{e}_i, & \mathbf{x} \neq \mathbf{0}, \\ \{\mathbf{y} \in \mathbb{R}^n |\max_{1\leq i\leq n} |y_i| \leq 1\}, & \mathbf{x} = \mathbf{0}, \mathbf{y} \neq \mathbf{0}. \end{cases} \tag{4.6.11}$$

Example 4.11 The Euclidean norm $h(\mathbf{x}) = \|\mathbf{x}\|_2$ is differentiable away from $\mathbf{x} = \mathbf{0}$, and thus

$$\partial\|\mathbf{x}\|_2 = \{\nabla\|\mathbf{x}\|_2\} = \{\nabla(\mathbf{x}^T\mathbf{x})^{1/2}\} = \{\mathbf{x}/\|\mathbf{x}\|_2\}, \quad \forall\, \mathbf{x} \neq \mathbf{0}.$$

At the point $\mathbf{x} = \mathbf{0}$, by Equation (4.6.6) we have

$$\mathbf{g} \in \partial h(\mathbf{0}) \quad \Leftrightarrow \quad \|\mathbf{y}\|_2 \geq \mathbf{0} + \mathbf{g}^T(\mathbf{y} - \mathbf{0}) \quad \Leftrightarrow \quad \frac{\langle \mathbf{g}, \mathbf{y}\rangle}{\|\mathbf{y}\|_2} \leq 1, \quad \forall \mathbf{y} \neq \mathbf{0}.$$

Hence, the subdifferential of the function $\|\mathbf{x}\|_2$ is given by

$$\partial\|\mathbf{x}\|_2 = \begin{cases} \{\mathbf{x}/\|\mathbf{x}\|_2\}, & \mathbf{x} \neq \mathbf{0}, \\ \{\mathbf{y} \in \mathbb{R}^n |\, \|\mathbf{y}\|_2 \leq 1\}, & \mathbf{x} = \mathbf{0}, \mathbf{y} \neq \mathbf{0}. \end{cases} \tag{4.6.12}$$

Let $\mathbf{X} \in \mathbb{R}^{m\times n}$ be any matrix with singular value decomposition $\mathbf{X} = \mathbf{U\Sigma V}^T$. Then the subdifferential of the nuclear norm of the matrix $\mathbf{X}$ (i.e., the sum of all singular values) is given by [79], [292], [503]

$$\partial\|\mathbf{X}\|_* = \big\{\mathbf{UV}^T + \mathbf{W}|\mathbf{W} \in \mathbb{R}^{m\times n}, \mathbf{U}^T\mathbf{W} = \mathbf{O}, \mathbf{WV} = \mathbf{O}, \|\mathbf{W}\|_{\text{spec}} \leq 1\big\}. \tag{4.6.13}$$

4.6.2 Proximal Operator

Let $C_i = \text{dom}\, f_i(\mathbf{x}), i = 1, \ldots, P$ be a closed convex set of the m-dimensional Euclidean space $\mathbb{R}^m$ and $C = \bigcap_{i=1}^P C_i$ be the intersection of these closed convex sets. Consider the combinatorial optimization problem

$$\min_{\mathbf{x} \in C} \sum_{i=1}^{P} f_i(\mathbf{x}), \tag{4.6.14}$$

where the closed convex sets $C_i, i = 1, \ldots, P$ express the constraints imposed on the *combinatorial optimization solution* $\mathbf{x}$.

The types of intersection C can be divided into the following three cases [73]:

(1) The intersection C is nonempty and "small" (all members of C are quite similar).
(2) The intersection C is nonempty and "large" (the differences between the members of C are large).
(3) The intersection C is empty, which means that the imposed constraints of intersecting sets are mutually contradictory.

It is difficult to solve the combinatorial optimization problem (4.6.14) directly. But, if

$$f_1(\mathbf{x}) = \|\mathbf{x} - \mathbf{x}_0\|, \quad f_i(\mathbf{x}) = I_{C_i}(\mathbf{x}) = \begin{cases} 0, & \mathbf{x} \in C_i, \\ +\infty, & \mathbf{x} \notin C_i, \end{cases}$$

then the original combinatorial optimization problem can be divided into separate problems

$$\min_{\mathbf{x} \in \bigcap_{i=2}^P C_i} \|\mathbf{x} - \mathbf{x}_0\|. \tag{4.6.15}$$

Differently from the combinatorial optimization problem (4.6.14), the separated optimization problems (4.6.15) can be solved using the projection method. In particular, when the C_i are convex sets, the projection of a convex objective function onto the intersection of these convex sets is closely related to the proximal operator of the objective function.

Definition 4.15 The *proximal operator* of a convex function $h(\mathbf{x})$ is defined as [331]

$$\mathbf{prox}_h(\mathbf{u}) = \arg\min_{\mathbf{x}} \left\{ h(\mathbf{x}) + \frac{1}{2}\|\mathbf{x} - \mathbf{u}\|_2^2 \right\} \tag{4.6.16}$$

or

$$\mathbf{prox}_{\mu h}(\mathbf{u}) = \arg\min_{\mathbf{x}} \left\{ h(\mathbf{x}) + \frac{1}{2\mu}\|\mathbf{x} - \mathbf{u}\|_2^2 \right\} \tag{4.6.17}$$

with scale parameter $\mu > 0$.

In particular, for a convex function $h(\mathbf{X})$, its proximal operator is defined as

$$\mathbf{prox}_{\mu h}(\mathbf{U}) = \arg\min_{\mathbf{X}} \left\{ h(\mathbf{X}) + \frac{1}{2\mu}\|\mathbf{X} - \mathbf{U}\|_F^2 \right\}. \tag{4.6.18}$$

The proximal operator has the following important properties [488]:

1. *Existence and uniqueness* The proximal operator $\mathbf{prox}_h(\mathbf{u})$ always exists, and is unique for all $\mathbf{x}$.
2. *Subgradient characterization* There is the following correspondence between the proximal mapping $\mathbf{prox}_h(\mathbf{u})$ and the subgradient $\partial h(\mathbf{x})$:

$$\mathbf{x} = \mathbf{prox}_h(\mathbf{u}) \quad \Leftrightarrow \quad \mathbf{x} - \mathbf{u} \in \partial h(\mathbf{x}). \tag{4.6.19}$$

3. *Nonexpansive mapping* The proximal operator $\mathbf{prox}_h(\mathbf{u})$ is a nonexpansive mapping with constant 1: if $\mathbf{x} = \mathbf{prox}_h(\mathbf{u})$ and $\hat{\mathbf{x}} = \mathbf{prox}_h(\hat{\mathbf{u}})$ then

$$(\mathbf{x} - \hat{\mathbf{x}})^T(\mathbf{u} - \hat{\mathbf{u}}) \geq \|\mathbf{x} - \hat{\mathbf{x}}\|_2^2.$$

4. *Proximal operator of a separable sum function* If $h : \mathbb{R}^{n_1} \times \mathbb{R}^{n_2} \to \mathbb{R}$ is a separable sum function, i.e., $h(\mathbf{x}_1, \mathbf{x}_2) = h_1(\mathbf{x}_1) + h_2(\mathbf{x}_2)$, then

$$\mathbf{prox}_h(\mathbf{x}_1, \mathbf{x}_2) = (\mathbf{prox}_{h_1}(\mathbf{x}_1), \mathbf{prox}_{h_2}(\mathbf{x}_2)).$$

5. *Scaling and translation of argument* If $h(\mathbf{x}) = f(\alpha\mathbf{x} + \mathbf{b})$, where $\alpha \neq 0$, then

$$\mathbf{prox}_h(\mathbf{x}) = \frac{1}{\alpha}\left(\mathbf{prox}_{\alpha^2 f}(\alpha\mathbf{x} + \mathbf{b}) - \mathbf{b}\right).$$

6. *Proximal operator of the conjugate function* If $h^*(\mathbf{x})$ is the conjugate function of the function $h(\mathbf{x})$ then, for all $\mu > 0$, the proximal operator of the conjugate function is given by

$$\mathbf{prox}_{\mu h^*}(\mathbf{x}) = \mathbf{x} - \mu\mathbf{prox}_{h/\mu}(\mathbf{x}/\mu).$$

   If $\mu = 1$ then the above equation simplifies to

$$\mathbf{x} = \mathbf{prox}_h(\mathbf{x}) + \mathbf{prox}_{h^*}(\mathbf{x}). \tag{4.6.20}$$

   This decomposition is called the *Moreau decomposition.*

An operator closely related to the proximal operator is the soft thresholding operator of a real variable.

Definition 4.16 The action of the *soft thresholding operator* on a real variable $x \in \mathbb{R}$ is denoted $\mathcal{S}_\tau[x]$ or $\text{soft}(x, \tau)$ and is defined as

$$\text{soft}(x, \tau) = \mathcal{S}_\tau[x] = \begin{cases} x - \tau, & x > \tau, \\ 0, & |x| \leq \tau, \\ x + \tau, & x < -\tau. \end{cases} \tag{4.6.21}$$

Here $\tau > 0$ is called the *soft threshold value* of the real variable x. The action of the soft thresholding operator can be equivalently written as

$$\begin{aligned}\text{soft}(x,\tau) &= (x-\tau)_+ - (-x-\tau)_+ = \max\{x-\tau,0\} - \max\{-x-\tau,0\}\\ &= (x-\tau)_+ + (x+\tau)_- = \max\{x-\tau,0\} + \min\{x+\tau,0\}.\end{aligned}$$

This operator is also known as the *shrinkage operator*, because it can reduce the variable x, the elements of the vector $\mathbf{x}$ and the matrix $\mathbf{X}$ to zero, thereby shrinking the range of elements. Hence, the action of the soft thresholding operator is sometimes written as [29], [52]

$$\text{soft}(x,\tau) = (|x|-\tau)_+\text{sign}(x) = (1-\tau/|x|)_+ x. \tag{4.6.22}$$

The soft thresholding operation on a real vector $\mathbf{x} \in \mathbb{R}^n$, denoted $\text{soft}(\mathbf{x},\tau)$, is defined as a vector with entries

$$\begin{aligned}\text{soft}(\mathbf{x},\tau)_i &= \max\{x_i-\tau,0\} + \min\{x_i+\tau,0\}\\ &= \begin{cases} x_i-\tau, & x_i > \tau,\\ 0, & |x_i| \le \tau,\\ x_i+\tau, & x_i < -\tau.\end{cases}\end{aligned} \tag{4.6.23}$$

The soft thresholding operation on a real matrix $\mathbf{X} \in \mathbb{R}^{m\times n}$, denoted $\text{soft}(\mathbf{X})$, is defined as an $m \times n$ real matrix with entries

$$\begin{aligned}\text{soft}(\mathbf{X},\tau)_{ij} &= \max\{x_{ij}-\tau,0\} + \min\{x_{ij}+\tau,0\}\\ &= \begin{cases} x_{ij}-\tau, & x_{ij} > \tau,\\ 0, & |x_{ij}| \le \tau,\\ x_{ij}+\tau, & x_{ij} < -\tau.\end{cases}\end{aligned} \tag{4.6.24}$$

Given a function $h(\mathbf{x})$, our goal is to find an explicit expression for

$$\mathbf{prox}_{\mu h}(\mathbf{u}) = \arg\min_{\mathbf{x}} \left\{ h(\mathbf{x}) + \frac{1}{2\mu}\|\mathbf{x}-\mathbf{u}\|_2^2 \right\}.$$

Theorem 4.11 [344, p. 129] *Let $\mathbf{x}^\star$ denote an optimal solution of the minimization problem $\min_{\mathbf{x}\in\text{dom}} \phi(\mathbf{x})$. If the function $\phi(\mathbf{x})$ is subdifferentiable then $\mathbf{x}^\star = \arg\min_{\mathbf{x}\in\text{dom}\phi} \phi(\mathbf{x})$ or $\phi(\mathbf{x}^\star) = \min_{\mathbf{x}\in\text{dom}\phi} \phi(\mathbf{x})$ if and only if $\mathbf{0} \in \partial\phi(\mathbf{x}^\star)$.*

By the above theorem, the first-order optimality condition of the function

$$\phi(\mathbf{x}) = h(\mathbf{x}) + \frac{1}{2\mu}\|\mathbf{x}-\mathbf{u}\|_2^2 \tag{4.6.25}$$

is given by

$$\mathbf{0} \in \partial h(\mathbf{x}^\star) + \frac{1}{\mu}(\mathbf{x}^\star - \mathbf{u}), \tag{4.6.26}$$

because $\partial \frac{1}{2\mu}\|\mathbf{x}-\mathbf{u}\|_2^2 = \frac{1}{\mu}(\mathbf{x}-\mathbf{u})$. Hence, if and only if $\mathbf{0} \in \partial h(\mathbf{x}^\star) + \mu^{-1}(\mathbf{x}^\star - \mathbf{u})$, we have

$$\mathbf{x}^\star = \mathbf{prox}_{\mu h}(\mathbf{u}) = \arg\min_{\mathbf{x}} \left\{ h(\mathbf{x}) + \frac{1}{2\mu}\|\mathbf{x}-\mathbf{u}\|_2^2 \right\}. \tag{4.6.27}$$

From Equation (4.6.26) we have

$$\mathbf{0} \in \mu\partial h(\mathbf{x}^\star) + (\mathbf{x}^\star - \mathbf{u}) \quad \Leftrightarrow \quad \mathbf{u} \in (I + \mu\partial h)\mathbf{x}^\star \quad \Leftrightarrow \quad (I+\mu\partial h)^{-1}\mathbf{u} \in \mathbf{x}^\star.$$

Since $\mathbf{x}^\star$ is only one point, $(I+\mu\partial h)^{-1}\mathbf{u} \in \mathbf{x}^\star$ should read as $(I+\mu\partial h)^{-1}\mathbf{u} = \mathbf{x}^\star$ and thus

$$\begin{aligned} \mathbf{0} \in \mu\partial h(\mathbf{x}^\star) + (\mathbf{x}^\star - \mathbf{u}) \quad &\Leftrightarrow \quad \mathbf{x}^\star = (I+\mu\partial h)^{-1}\mathbf{u} \\ &\Leftrightarrow \quad \mathbf{prox}_{\mu h}(\mathbf{u}) = (I+\mu\partial h)^{-1}\mathbf{u}. \end{aligned}$$

This shows that the proximal operator $\mathbf{prox}_{\mu h}$ and the subdifferential operator ∂h are related as follows:

$$\mathbf{prox}_{\mu h} = (I+\mu\partial h)^{-1}. \tag{4.6.28}$$

The (point-to-point) mapping $(I+\mu\partial h)^{-1}$ is known as the *resolvent* of the subdifferential operator ∂h with parameter $\mu > 0$, so the proximal operator $\mathbf{prox}_{\mu h}$ is the resolvent of the subdifferential operator ∂h.

Notice that the subdifferential $\partial h(\mathbf{x})$ is a point-to-set mapping, for which neither direction is unique, whereas the proximal operation $\mathbf{prox}_{\mu h}(\mathbf{u})$ is a point-to-point mapping: $\mathbf{prox}_{\mu h}(\mathbf{u})$ maps any point $\mathbf{u}$ to a unique point $\mathbf{x}$.

Example 4.12 Consider a linear function $h(\mathbf{x}) = \mathbf{a}^T\mathbf{x} + b$, where $\mathbf{a} \in \mathbb{R}^n$, $b \in \mathbb{R}$. By the first-order optimality condition

$$\frac{\partial}{\partial \mathbf{x}}\left(\mathbf{a}^T\mathbf{x} + b + \frac{1}{2\mu}\|\mathbf{x}-\mathbf{u}\|_2^2\right) = \mathbf{a} + \frac{1}{\mu}(\mathbf{x}^\star - \mathbf{u}) = \mathbf{0},$$

we get $\mathbf{prox}_{\mu h}(\mathbf{u}) = \mathbf{x}^\star = \mathbf{u} - \mu\mathbf{a}$.

Example 4.13 For the quadratic function $h(\mathbf{x}) = \frac{1}{2}\mathbf{x}^T\mathbf{A}\mathbf{x} - \mathbf{b}^T\mathbf{x} + \mathbf{c}$, where $\mathbf{A} \in \mathbb{R}^{n\times n}$ is positive definite. The first-order optimality condition of $\mathbf{prox}_{\mu h}(\mathbf{u})$ is

$$\frac{\partial}{\partial \mathbf{x}}\left(\frac{1}{2}\mathbf{x}^T\mathbf{A}\mathbf{x} - \mathbf{b}^T\mathbf{x} + \mathbf{c} + \frac{1}{2\mu}\|\mathbf{x}-\mathbf{u}\|_2^2\right) = \mathbf{A}\mathbf{x} - \mathbf{b} + \frac{1}{\mu}(\mathbf{x}-\mathbf{u}) = \mathbf{0},$$

with $\mathbf{x} = \mathbf{x}^\star = \mathbf{prox}_{\mu h}(\mathbf{u})$. Hence, we have

$$\mathbf{prox}_{\mu h}(\mathbf{u}) = (\mathbf{A} + \mu^{-1}\mathbf{I})^{-1}(\mu^{-1}\mathbf{u} + \mathbf{b}) = \mathbf{u} + (\mathbf{A} + \mu^{-1}\mathbf{I})^{-1}(\mathbf{b} - \mathbf{A}\mathbf{u}).$$

Example 4.14 If h is the *indicator function*

$$h(\mathbf{x}) = I_C(\mathbf{x}) = \begin{cases} 0, & \mathbf{x} \in C, \\ +\infty, & \mathbf{x} \notin C, \end{cases}$$

where C is a closed nonempty convex set, then the proximal operator of $h(\mathbf{x})$ reduces to Euclidean projection onto C, i.e., $\mathcal{P}_C(\mathbf{u}) = \arg\min_{\mathbf{x}\in C}\|\mathbf{x}-\mathbf{u}\|_2$.

Table 4.2 lists the proximal operators of several typical functions [109], [373].

Table 4.2 Proximal operators of several typical functions

Functions	Proximal operators
$h(\mathbf{x}) = \phi(\mathbf{x}-\mathbf{z})$	$\mathbf{prox}_{\mu h}(\mathbf{u}) = \mathbf{z} + \mathbf{prox}_{\mu\phi}(\mathbf{u}-\mathbf{z})$
$h(\mathbf{x}) = \phi(\mathbf{x}/\rho)$	$\mathbf{prox}_{h}(\mathbf{u}) = \rho\,\mathbf{prox}_{\phi/\rho^2}(\mathbf{u}/\rho)$
$h(\mathbf{x}) = \phi(-\mathbf{x})$	$\mathbf{prox}_{\mu h}(\mathbf{u}) = -\mathbf{prox}_{\mu\phi}(-\mathbf{u})$
$h(\mathbf{x}) = \phi^*(\mathbf{x})$	$\mathbf{prox}_{\mu h}(\mathbf{u}) = \mathbf{u} - \mathbf{prox}_{\mu\phi}(\mathbf{u})$
$h(\mathbf{x}) = I_C(\mathbf{x})$	$\mathbf{prox}_{\mu h}(\mathbf{u}) = \mathcal{P}_C(\mathbf{u}) = \arg\min_{\mathbf{x}\in C}\|\mathbf{x}-\mathbf{u}\|_2$
$h(\mathbf{x}) = \sup_{\mathbf{y}\in C}\mathbf{y}^T\mathbf{x}$	$\mathbf{prox}_{\mu h}(\mathbf{u}) = \mathbf{u} - \mu\mathcal{P}_C(\mathbf{u}/\mu)$
$h(\mathbf{x}) = \|\mathbf{x}\|_1$	$(\mathbf{prox}_{\mu h}(\mathbf{u}))_i = \begin{cases} u_i-\mu, & u_i>\mu,\\ 0, & \|u_i\|\leq\mu,\\ u_i+\mu, & u_i<-\mu.\end{cases}$
$h(\mathbf{x}) = \|\mathbf{x}\|_2$	$\mathbf{prox}_{\mu h}(\mathbf{u}) = \begin{cases}(1-\mu/\|\mathbf{u}\|_2)\mathbf{u}, & \|\mathbf{u}\|_2\geq\mu,\\ 0, & \|\mathbf{u}\|_2<\mu.\end{cases}$
$h(\mathbf{x}) = \mathbf{a}^T\mathbf{x}+\mathbf{b}$	$\mathbf{prox}_{\mu h}(\mathbf{u}) = \mathbf{u} - \mu\,\mathbf{a}$
$h(\mathbf{x}) = \frac{1}{2}\mathbf{x}^T\mathbf{A}\mathbf{x}+\mathbf{b}^T\mathbf{x}$	$\mathbf{prox}_{\mu h}(\mathbf{x}) = \mathbf{u} + (\mathbf{A}+\mu^{-1}\mathbf{I})^{-1}(\mathbf{b}-\mathbf{A}\mathbf{u})$
$h(\mathbf{x}) = -\sum_{i=1}^n \log x_i$	$(\mathbf{prox}_{\mu h}(\mathbf{u}))_i = \frac{1}{2}\left(u_i+\sqrt{u_i^2+4\mu}\right)$

In the following, we present the proximal gradient method for solving nonsmooth convex optimization problems.

4.6.3 Proximal Gradient Method

Consider a typical form of *nonsmooth convex optimization*,

$$\min_{\mathbf{x}} J(\mathbf{x}) = f(\mathbf{x}) + h(\mathbf{x}), \tag{4.6.29}$$

where $f(\mathbf{x})$ is a convex, smooth (i.e., differentiable) and L-Lipschitz function, and $h(\mathbf{x})$ is a convex but nonsmooth function (such as $\|\mathbf{x}\|_1$, $\|\mathbf{X}\|_*$ and so on).

1. Quadratic Approximation

Consider the quadratic approximation of an L-Lipschitz smooth function $f(\mathbf{x})$

around the point $\mathbf{x}_k$:

$$\begin{aligned} f(\mathbf{x}) &= f(\mathbf{x}_k) + (\mathbf{x}-\mathbf{x}_k)^T \nabla f(\mathbf{x}_k) + \frac{1}{2}(\mathbf{x}-\mathbf{x}_k)^T \nabla^2 f(\mathbf{x}_k)(\mathbf{x}-\mathbf{x}_k) \\ &\approx f(\mathbf{x}_k) + (\mathbf{x}-\mathbf{x}_k)^T \nabla f(\mathbf{x}_k) + \frac{L}{2}\|\mathbf{x}-\mathbf{x}_k\|_2^2, \end{aligned}$$

where $\nabla^2 f(\mathbf{x}_k)$ is approximated by a diagonal matrix $L\mathbf{I}$.

Minimize $J(\mathbf{x}) = f(\mathbf{x}) + h(\mathbf{x})$ via iteration to yield

$$\begin{aligned} \mathbf{x}_{k+1} &= \arg\min_{\mathbf{x}} \{f(\mathbf{x}) + h(\mathbf{x})\} \\ &\approx \arg\min_{\mathbf{x}} \left\{ f(\mathbf{x}_k) + (\mathbf{x}-\mathbf{x}_k)^T \nabla f(\mathbf{x}_k) + \frac{L}{2}\|\mathbf{x}-\mathbf{x}_k\|_2^2 + h(\mathbf{x}) \right\} \\ &= \arg\min_{\mathbf{x}} \left\{ h(\mathbf{x}) + \frac{L}{2}\left\| \mathbf{x} - \left(\mathbf{x}_k - \frac{1}{L}\nabla f(\mathbf{x}_k) \right) \right\|_2^2 \right\} \\ &= \mathbf{prox}_{L^{-1}h}\left(\mathbf{x}_k - \frac{1}{L}\nabla f(\mathbf{x}_k) \right). \end{aligned}$$

In practical applications, the Lipschitz constant L of $f(\mathbf{x})$ is usually unknown. Hence, a question of how to choose L in order to accelerate the convergence of $\mathbf{x}_k$ arises. To this end, let $\mu = 1/L$ and consider the fixed point iteration

$$\mathbf{x}_{k+1} = \mathbf{prox}_{\mu h}\left(\mathbf{x}_k - \mu \nabla f(\mathbf{x}_k)\right). \tag{4.6.30}$$

This iteration is called the *proximal gradient method* for solving the nonsmooth convex optimization problem, where the step μ is chosen to equal some constant or is determined by linear searching.

2. *Forward–Backward Splitting*

To derive a proximal gradient algorithm for nonsmooth convex optimization, let $A = \partial h$ and $B = \nabla f$ denote the subdifferential operator and the gradient operator, respectively. Then, the first-order optimality condition for the objective function $h(\mathbf{x}) + f(\mathbf{x})$, $\mathbf{0} \in (\partial h + \nabla f)\mathbf{x}^\star$, can be written in operator form as $\mathbf{0} \in (A+B)\mathbf{x}^\star$, and thus we have

$$\begin{aligned} \mathbf{0} \in (A+B)\mathbf{x}^\star \quad &\Leftrightarrow \quad (I-\mu B)\mathbf{x}^\star \in (I+\mu A)\mathbf{x}^\star \\ &\Leftrightarrow \quad (I+\mu A)^{-1}(I-\mu B)\mathbf{x}^\star = \mathbf{x}^\star. \end{aligned} \tag{4.6.31}$$

Here $(I-\mu B)$ is a forward operator and $(I+\mu A)^{-1}$ is a backward operator.

The backward operator $(I+\mu\partial h)^{-1}$ is sometimes known as the resolvent of ∂h with parameter μ.

By $\mathbf{prox}_{\mu h} = (I+\mu A)^{-1}$ and $(I-\mu B)\mathbf{x}_k = \mathbf{x}_k - \mu\nabla f(\mathbf{x}_k)$, Equation (4.6.31) can be written as the fixed-point iteration

$$\mathbf{x}_{k+1} = (I+\mu A)^{-1}(I-\mu B)\mathbf{x}_k = \mathbf{prox}_{\mu h}\left(\mathbf{x}_k - \mu\nabla f(\mathbf{x}_k)\right). \tag{4.6.32}$$

If we let

$$\mathbf{y}_k = (I - \mu B)\mathbf{x}_k = \mathbf{x}_k - \mu \nabla f(\mathbf{x}_k) \tag{4.6.33}$$

then Equation (4.6.32) reduces to

$$\mathbf{x}_{k+1} = (I + \mu A)^{-1}\mathbf{y}_k = \mathbf{prox}_{\mu h}(\mathbf{y}_k). \tag{4.6.34}$$

In other words, the proximal gradient algorithm can be split two iterations:

(1) *forward iteration* $\mathbf{y}_k = (I - \mu B)\mathbf{x}_k = \mathbf{x}_k - \mu \nabla f(\mathbf{x}_k)$;
(2) *backward iteration* $\mathbf{x}_{k+1} = (I + \mu A)^{-1}\mathbf{y}_k = \mathbf{prox}_{\mu h}(\mathbf{y}_k)$.

The forward iteration is an explicit iteration which is easily computed, whereas the backward iteration is an implicit iteration.

Example 4.15 For the ℓ_1-norm function $h(\mathbf{x}) = \|\mathbf{x}\|_1$, the backward iteration is

$$\begin{aligned}\mathbf{x}_{k+1} &= \mathbf{prox}_{\mu\|\mathbf{x}\|_1}(\mathbf{y}_k) = \mathrm{soft}_{\|\cdot\|_1}(\mathbf{y}_k, \mu)\\ &= [\mathrm{soft}_{\|\cdot\|_1}(y_k(1), \mu), \ldots, \mathrm{soft}_{\|\cdot\|_1}(y_k(n), \mu)]^T\end{aligned} \tag{4.6.35}$$

can be obtained by the soft thresholding operation [163]:

$$x_{k+1}(i) = (\mathrm{soft}_{\|\cdot\|_1}(\mathbf{y}_k, \mu))_i = \mathrm{sign}(y_k(i)) \max\{|y_k(i)| - \mu, 0\}, \tag{4.6.36}$$

for $i = 1, \ldots, n$. Here $x_{k+1}(i)$ and $y_k(i)$ are the ith entries of the vectors $\mathbf{x}_{k+1}$ and $\mathbf{y}_k$, respectively.

Example 4.16 For the ℓ_2-norm function $h(\mathbf{x}) = \|\mathbf{x}\|_2$, the backward iteration is

$$\begin{aligned}\mathbf{x}_{k+1} &= \mathbf{prox}_{\mu\|\mathbf{x}\|_2}(\mathbf{y}_k) = \mathrm{soft}_{\|\cdot\|_2}(\mathbf{y}_k, \mu)\\ &= [\mathrm{soft}_{\|\cdot\|_2}(y_k(1), \mu), \ldots, \mathrm{soft}_{\|\cdot\|_2}(y_k(n), \mu)]^T,\end{aligned} \tag{4.6.37}$$

where

$$\mathrm{soft}_{\|\cdot\|_2}(y_k(i), \mu) = \max\{|y_k(i)| - \mu, 0\}\frac{y_k(i)}{\|\mathbf{y}_k\|_2}, \quad i = 1, \ldots, n. \tag{4.6.38}$$

Example 4.17 For the nuclear norm of matrix $\|\mathbf{X}\|_* = \sum_{i=1}^{\min\{m,n\}} \sigma_i(\mathbf{X})$, the corresponding proximal gradient method becomes

$$\mathbf{X}_k = \mathbf{prox}_{\mu\|\cdot\|_*}\left(\mathbf{X}_{k-1} - \mu \nabla f(\mathbf{X}_{k-1})\right). \tag{4.6.39}$$

If $\mathbf{W} = \mathbf{U}\mathbf{\Sigma}\mathbf{V}^T$ is the SVD of the matrix $\mathbf{W} = \mathbf{X}_{k-1} - \mu \nabla f(\mathbf{X}_{k-1})$ then

$$\mathbf{prox}_{\mu\|\cdot\|_*}(\mathbf{W}) = \mathbf{U}\mathcal{D}_\mu(\mathbf{\Sigma})\mathbf{V}^T, \tag{4.6.40}$$

where $\mathcal{D}_\mu(\mathbf{\Sigma})$ is called the *singular value thresholding* operation, defined as

$$[\mathcal{D}_\mu(\mathbf{\Sigma})]_i = \begin{cases} \sigma_i(\mathbf{X}) - \mu, & \text{if } \sigma_i(\mathbf{X}) > \mu,\\ 0, & \text{otherwise.}\end{cases} \tag{4.6.41}$$

Example 4.18 If the nonsmooth function $h(\mathbf{x}) = I_C(\mathbf{x})$ is an indicator function, then the proximal gradient iteration $\mathbf{x}_{k+1} = \mathbf{prox}_{\mu h}(\mathbf{x}_k - \mu\nabla f(\mathbf{x}_k))$ reduces to the gradient projection iteration

$$\mathbf{x}_{k+1} = \mathcal{P}_C\left(\mathbf{x}_k - \mu\nabla f(\mathbf{x}_k)\right). \tag{4.6.42}$$

A comparison between other gradient methods and the proximal gradient method is summarized below.

- The update formulas are as follows:

  general gradient method, $\mathbf{x}_{k+1} = \mathbf{x}_k - \mu\nabla f(\mathbf{x}_k)$;
  Newton method, $\mathbf{x}_{k+1} = \mathbf{x}_k - \mu\mathbf{H}^{-1}(\mathbf{x}_k)\nabla f(\mathbf{x}_k)$;
  proximal gradient method, $\mathbf{x}_{k+1} = \mathbf{prox}_{\mu h}\left(\mathbf{x}_k - \mu\nabla f(\mathbf{x}_k)\right)$.

  General gradient methods and the Newton method use a low-level (explicit) update, whereas the proximal gradient method uses a high-level (implicit) operation $\mathbf{prox}_{\mu h}$.
- General gradient methods and the Newton method are available only for smooth and unconstrained optimization problems, while the proximal gradient method is available for smooth or nonsmooth and/or constrained or unconstrained optimization problems.
- The Newton method allows modest-sized problems to be addressed; the general gradient methods are applicable for large-size problems, and, sometimes, distributed implementations, and the proximal gradient method is available for all large-size problems and distributed implementations.

In particular, the proximal gradient method includes the general gradient method, the projected gradient method and the iterative soft thresholding method as special cases.

(1) *Gradient method* If $h(\mathbf{x}) = 0$, as $\mathbf{prox}_h(\mathbf{x}) = \mathbf{x}$ the proximal gradient algorithm (4.6.30) reduces to the general gradient algorithm $\mathbf{x}_k = \mathbf{x}_{k-1} - \mu_k\nabla f(\mathbf{x}_{k-1})$. That is to say, the gradient algorithm is a special case of the proximal gradient algorithm when the nonsmooth convex function $h(\mathbf{x}) = 0$.

(2) *Projected gradient method* For $h(\mathbf{x}) = I_C(\mathbf{x})$, an indictor function, the minimization problem (4.6.4) becomes the unconstrained minimization $\min_{\mathbf{x}\in C} f(\mathbf{x})$. Because $\mathbf{prox}_h(\mathbf{x}) = \mathcal{P}_C(\mathbf{x})$, the proximal gradient algorithm reduces to

$$\mathbf{x}_k = \mathcal{P}_C\left(\mathbf{x}_{k-1} - \mu_k\nabla f(\mathbf{x}_{k-1})\right) \tag{4.6.43}$$

$$= \arg\min_{\mathbf{u}\in C} \left\|\mathbf{u} - \mathbf{x}_{k-1} + \mu_k\nabla f(\mathbf{x}_{k-1})\right\|_2^2. \tag{4.6.44}$$

This is just the projected gradient method.

(3) *Iterative soft thresholding method* When $h(\mathbf{x}) = \|\mathbf{x}\|_1$, the minimization problem (4.6.4) becomes the unconstrained minimization problem $\min(f(\mathbf{x}) + \|\mathbf{x}\|_1)$. In this case, the proximal gradient algorithm becomes

$$\mathbf{x}_k = \mathbf{prox}_{\mu_k h}\left(\mathbf{x}_{k-1} - \mu_k \nabla f(\mathbf{x}_{k-1})\right). \tag{4.6.45}$$

This is called the *iterative soft thresholding method*, for which

$$\mathbf{prox}_{\mu h}(\mathbf{u})_i = \begin{cases} u_i - \mu, & u_i > \mu, \\ 0, & -\mu \le u_i \le \mu, \\ u_i + \mu, & u_i < -\mu. \end{cases}$$

From the viewpoint of convergence, the proximal gradient method is suboptimal. A Nesterov-like proximal gradient algorithm was developed by Beck and Teboulle [29], who called it the *fast iterative soft thresholding algorithm* (FISTA), as shown in Algorithm 4.9.

Algorithm 4.9 FISTA algorithm with fixed step [29]

---

input: The Lipschitz constant L of $\nabla f(\mathbf{x})$.

Initialize: $\mathbf{y}_1 = \mathbf{x}_0 \in \mathbb{R}^n$, $t_1 = 1$.

repeat

1. Compute $\mathbf{x}_k = \mathbf{prox}_{L^{-1}h}\left(\mathbf{y}_k - \frac{1}{L}\nabla f(\mathbf{y}_k)\right)$.
2. Compute $t_{k+1} = \dfrac{1 + \sqrt{1 + 4t_k^2}}{2}$.
3. Compute $\mathbf{y}_{k+1} = \mathbf{x}_k + \left(\dfrac{t_k - 1}{t_{k+1}}\right)(\mathbf{x}_k - \mathbf{x}_{k-1})$.
4. **exit if** $\mathbf{x}_k$ is converged.

return $k \leftarrow k + 1$.

output: $\mathbf{x} \leftarrow \mathbf{x}_k$.

Theorem 4.12 [29] *Let $\{\mathbf{x}_k\}$ and $\{\mathbf{y}_k\}$ be two sequences generated by the FISTA algorithm; then, for any iteration $k \ge 1$,*

$$F(\mathbf{x}_k) - F(\mathbf{x}^\star) \le \frac{2L\|\mathbf{x}_k - \mathbf{x}^\star\|_2^2}{(k+1)^2}, \quad \forall \mathbf{x}^\star \in X_\star,$$

where $\mathbf{x}^\star$ and $X_\star$ represent respectively the optimal solution and the optimal solution set of $\min(F(\mathbf{x}) = f(\mathbf{x}) + h(\mathbf{x}))$.

Theorem 4.12 shows that to achieve the ϵ-optimal solution $F(\bar{\mathbf{x}}) - F(\mathbf{x}^\star) \le \epsilon$, the FISTA algorithm needs at most $\lceil C/\sqrt{\epsilon} - 1 \rceil$ iterations, where $C = \sqrt{2L\|\mathbf{x}_0 - \mathbf{x}^\star\|_2^2}$. Since it has the same fast convergence rate as the optimal first-order algorithm, the FISTA is indeed an optimal algorithm.

4.7 Constrained Convex Optimization

Consider the *constrained minimization problem*

$$\min_{\mathbf{x}} f_0(\mathbf{x}) \quad \text{subject to} \quad f_i(\mathbf{x}) \leq 0, 1 \leq i \leq m, \; h_i(\mathbf{x}) = 0, 1 \leq i \leq q. \tag{4.7.1}$$

If the objective function $f_0(\mathbf{x})$ and the inequality constraint functions $f_i(\mathbf{x}), i = 1, \ldots, m$ are convex, and the equality constraint functions $h_i(\mathbf{x})$ have the affine form $\mathbf{h}(\mathbf{x}) = \mathbf{A}\mathbf{x} - \mathbf{b}$, then Equation (4.7.1) is called a *constrained convex optimization problem.*

The basic idea in solving a constrained optimization problem is to transform it into an unconstrained optimization problem. The transformation methods are of three types:

(1) the Lagrange multiplier method;
(2) the penalty function method;
(3) the augmented Lagrange multiplier method.

4.7.1 Lagrange Multiplier Method

Consider a simpler form of the constrained optimization problem (4.7.1)

$$\min f(\mathbf{x}) \quad \text{subject to} \quad \mathbf{A}\mathbf{x} = \mathbf{b}, \tag{4.7.2}$$

where $\mathbf{x} \in \mathbb{R}^n, \mathbf{A} \in \mathbb{R}^{m \times n}$ and the objective function $f : \mathbb{R}^n \to \mathbb{R}$ is convex.

The *Lagrange multiplier method* transforms the constrained optimization problem (4.7.2) into an unconstrained optimization problem with objective function

$$\mathcal{L}(\mathbf{x}, \boldsymbol{\lambda}) = f(\mathbf{x}) + \boldsymbol{\lambda}^T(\mathbf{A}\mathbf{x} - \mathbf{b}), \tag{4.7.3}$$

where $\boldsymbol{\lambda} \geq \mathbf{0}$ is known as the *dual variable* or the *Lagrange multiplier vector*, all of whose entries must be larger than or equal to zero.

The dual objective function of the original optimization problem (4.7.2) is given by

$$g(\boldsymbol{\lambda}) = \inf_{\mathbf{x}} \mathcal{L}(\mathbf{x}, \boldsymbol{\lambda}) = -f^*(-\mathbf{A}^T\boldsymbol{\lambda}) - \mathbf{b}^T\boldsymbol{\lambda}, \tag{4.7.4}$$

in which f^* is the convex conjugate of f.

By the Lagrange multiplier method, the original equality constrained minimization problem (4.7.2) becomes the dual maximization problem

$$\max_{\boldsymbol{\lambda} \in \mathbb{R}^m} \left\{ g(\boldsymbol{\lambda}) = -f^*(-\mathbf{A}^T\boldsymbol{\lambda}) - \mathbf{b}^T\boldsymbol{\lambda} \right\}. \tag{4.7.5}$$

The *dual ascent method* uses gradient ascent to solve the maximization problem (4.7.5). The dual ascent method consists of two steps:

$$\mathbf{x}_{k+1} = \arg\min_{\mathbf{x}} \mathcal{L}(\mathbf{x}, \boldsymbol{\lambda}_k), \tag{4.7.6}$$

$$\boldsymbol{\lambda}_{k+1} = \boldsymbol{\lambda}_k + \mu_k(\mathbf{A}\mathbf{x}_{k+1} - \mathbf{b}), \tag{4.7.7}$$

where (4.7.6) is the minimization step of the original variable $\mathbf{x}$ whereas (4.7.7) is the updating step of the dual variable $\boldsymbol{\lambda}$, with step size μ_k.

Because the dual variable $\boldsymbol{\lambda} \geq \mathbf{0}$ can be viewed as a "price" vector, the updating of the dual variable is also called the price-ascent or price-adjustment step. The object of price-ascent is to maximize the revenue function $\boldsymbol{\lambda}_k$.

The dual ascent method contains two aspects.

(1) The update of the dual variable $\boldsymbol{\lambda}$ adopts the gradient ascent method.
(2) The choice of step size μ_k should ensure the ascent of the dual objective function, namely $g(\boldsymbol{\lambda}_{k+1}) > g(\boldsymbol{\lambda}_k)$.

4.7.2 Penalty Function Method

The *penalty function method* is a widely used constrained optimization method, and its basic idea is this: by using the penalty function and/or the barrier function, a constrained optimization becomes an unconstrained optimization of the composite function consisting of the original objective function and the constraint conditions.

For the standard constrained optimization problem (4.7.1), the *feasible set* is defined as the set of points satisfying all the inequality and equality constraints, namely

$$\mathcal{F} = \{\mathbf{x} | f_i(\mathbf{x}) \leq 0, i = 1, \ldots, m,\ h_i(\mathbf{x}) = 0, i = 1, \ldots, q\}. \tag{4.7.8}$$

The set of points satisfying only the strict inequality constraints $f_i(\mathbf{x}) < 0$ and the equality constraints $h_i(\mathbf{x}) = 0$ is denoted by

$$\text{relint}(\mathcal{F}) = \{\mathbf{x} | f_i(\mathbf{x}) < 0, i = 1, \ldots, m,\ h_i(\mathbf{x}) = 0, i = 1, \ldots, q\} \tag{4.7.9}$$

and is called the *relative feasible interior set* or the *relative strictly feasible set.* The points in the feasible interior set are called *relative interior points.*

The penalty function method transforms the original constrained optimization problem into the unconstrained optimization problem

$$\min_{\mathbf{x} \in \mathcal{S}} \{L_\rho(\mathbf{x}) = f_0(\mathbf{x}) + \rho p(\mathbf{x})\}, \tag{4.7.10}$$

where the coefficient ρ is a penalty parameter that reflects the intensity of "punishment" via the weighting of the *penalty function* $p(\mathbf{x})$; the greater ρ is, the greater the value of the penalty term. The transformed optimization problem (4.7.10) is often called the *auxiliary optimization problem* of the original constrained optimization problem.

The main property of the penalty function is as follows: if $p_1(\mathbf{x})$ is the penalty for the closed set $\mathcal{F}_1$, and $p_2(\mathbf{x})$ is the penalty for the closed set $\mathcal{F}_2$, then $p_1(\mathbf{x}) + p_2(\mathbf{x})$ is the penalty for the intersection $\mathcal{F}_1 \cap \mathcal{F}_2$.

The following are two common penalty functions.

(1) *Exterior penalty function*

$$p(\mathbf{x}) = \rho_1 \sum_{i=1}^{m} \big(\max\{0, f_i(\mathbf{x})\}\big)^r + \rho_2 \sum_{i=1}^{q} |h_i(\mathbf{x})|^2, \tag{4.7.11}$$

where r is usually 1 or 2.

(2) *Interior penalty function*

$$p(\mathbf{x}) = \rho_1 \sum_{i=1}^{m} -\frac{1}{f_i(\mathbf{x})} \log(-f_i(\mathbf{x})) + \rho_2 \sum_{i=1}^{q} |h_i(\mathbf{x})|^2. \tag{4.7.12}$$

In the exterior penalty function, if $f_i(\mathbf{x}) \leq 0, \forall i = 1,\ldots,m$ and $h_i(\mathbf{x}) = 0, \forall i = 1,\ldots,q$, then $p(\mathbf{x}) = 0$, i.e., the penalty function has no effect on the interior points in the feasible set $\text{int}(\mathcal{F})$. On the contrary, if, for some iteration point $\mathbf{x}_k$, the inequality constraint $f_i(\mathbf{x}_k) > 0, i \in \{1,\ldots,m\}$, and/or $h(\mathbf{x}_k) \neq 0, i \in \{1,\ldots,q\}$, then the penalty term $p(\mathbf{x}_k) \neq 0$, that is, any point outside the feasible set $\mathcal{F}$ is "punished". Hence, the penalty function defined in (4.7.11) is known as an exterior penalty function.

The role of (4.7.12) is equivalent to establishing a fence on the feasible set boundary $\text{bnd}(\mathcal{F})$ to block any point in the feasible interior set $\text{int}(\mathcal{F})$ from crossing the boundary of the feasible set $\text{bnd}(\mathcal{F})$. Because the points in the relative feasible interior set $\text{relint}(\mathcal{F})$ are slightly punished, the penalty function given in (4.7.12) is called the interior penalty function; it is also known as the *barrier function*.

The following are three typical barrier functions for a closed set $\mathcal{F}$ [344]:

(1) *the power-function barrier function* $\phi(\mathbf{x}) = \sum_{i=1}^{m} -\frac{1}{(f_i(\mathbf{x}))^p}$, $p \geq 1$;

(2) *the logarithmic barrier function* $\phi(\mathbf{x}) = -\frac{1}{f_i(\mathbf{x})} \sum_{i=1}^{m} \log(-f_i(\mathbf{x}))$;

(3) *the exponential barrier function* $\phi(\mathbf{x}) = \sum_{i=1}^{m} \exp\left(-\frac{1}{f_i(\mathbf{x})}\right)$.

In other words, the logarithmic barrier function in (4.7.12) can be replaced by the power-function barrier function or the exponential barrier function.

When $p = 1$, the power-function barrier function $\phi(\mathbf{x}) = \sum_{i=1}^{m} -\frac{1}{f_i(\mathbf{x})}$ is called the inverse barrier function and was presented by Carroll in 1961 [86]; whereas

$$\phi(\mathbf{x}) = \mu \sum_{i=1}^{m} \frac{1}{\log(-f_i(\mathbf{x}))} \tag{4.7.13}$$

is known as the classical *Fiacco–McCormick logarithmic barrier function* [152], where μ is the *barrier parameter*.

A summary of the features of the external penalty function follows.

(1) The exterior penalty function method is usually known as the penalty function method, and the interior penalty function method is customarily called the *barrier method.*

(2) The exterior penalty function method punishes all points outside of the feasible set, and its solution satisfies all the inequality constraints $f_i(\mathbf{x}) \leq 0, i = 1,\ldots,m$, and all the equality constraints $h_i(\mathbf{x}) = 0, i = 1,\ldots,q$, and is thus an exact solution to the constrained optimization problem. In other words, the exterior penalty function method is an optimal design scheme. In contrast, the interior penalty function method or barrier method blocks all points on the boundary of the feasible set, and the solution that is found satisfies only the strict inequality $f_i(\mathbf{x}) < 0, i = 1,\ldots,m$, and the equality constraints $h_i(\mathbf{x}) = 0, i = 1,\ldots,q$, and hence is only approximate. That is to say, the interior penalty function method is a suboptimal design scheme.

(3) The exterior penalty function method can be started using an unfeasible point, and its convergence is slow, whereas the interior penalty function method requires the initial point to be a feasible interior point, so its selection is difficult; but it has a good convergence and approximation performance.

In evolutionary computations the exterior penalty function method is normally used, while the interior function method is NP-hard owing to the feasible initial point search.

Engineering designers, especially process controllers, prefer to use the interior penalty function method, because this method allows the designers to observe the changes in the objective function value corresponding to the design points in the feasible set in the optimization process. However, this facility cannot be provided by any exterior penalty function method.

In the strict sense of the penalty function classification, all the above kinds of penalty function belong to the "death penalty": the infeasible solution points $\mathbf{x} \in S \setminus F$ (the difference set of the search space S and the feasible set F) are completely ruled out by the penalty function $p(\mathbf{x}) = +\infty$. If the feasible search space is convex, or it is the rational part of the whole search space, then this death penalty works very well [325]. However, for genetic algorithms and evolutionary computations, the boundary between the feasible set and the infeasible set is unknown and thus it is difficult to determine the precise position of the feasible set. In these cases, other penalty functions should be used [529]: static, dynamic, annealing, adaptive or co-evolutionary penalties.

4.7.3 Augmented Lagrange Multiplier Method

According to [278], [41], the main deficiencies of the Lagrange multiplier method are as follows.

(1) Only when a constrained optimization problem has a locally convex structure, is the dual unconstrained optimization problem well-defined, so that the updating of the Lagrange multiplier $\boldsymbol{\lambda}_{k+1} = \boldsymbol{\lambda}_k + \alpha_k \mathbf{h}(\mathbf{x}_k)$ can be implemented.

(2) The convergence of the Lagrangian objective function is more time-consuming, as the updating of the Lagrange multiplier is an ascent iteration and its convergence is only moderately fast.

According to [41], the deficiency in the penalty function method is that its convergence is slow, and a large penalty parameter easily results in an ill-conditioned unconstrained optimization problem and thus causes numerical instability of the optimization algorithm.

A simple and efficient way to overcome the disadvantages of the Lagrange multiplier method and the penalty function method is to combine them into an *augmented Lagrange multiplier method.*

Consider the *augmented Lagrangian function*

$$\begin{aligned}\mathcal{L}(\mathbf{x}, \boldsymbol{\lambda}, \boldsymbol{\nu}) &= f_0(\mathbf{x}) + \sum_{i=1}^{m} \lambda_i f_i(\mathbf{x}) + \sum_{i=1}^{q} \nu_i h_i(\mathbf{x}) \\ &= f_0(\mathbf{x}) + \boldsymbol{\lambda}^T \mathbf{f}(\mathbf{x}) + \boldsymbol{\nu}^T \mathbf{h}(\mathbf{x}),\end{aligned} \tag{4.7.14}$$

where $\boldsymbol{\lambda} = [\lambda_1, \ldots, \lambda_m]^T$ and $\boldsymbol{\nu} = [\nu_1, \ldots, \nu_q]^T$ are the Lagrange multiplier vector and the *penalty parameter vector*, respectively, whereas $\mathbf{f}(\mathbf{x}) = [f_1(\mathbf{x}), \ldots, f_m(\mathbf{x})]^T$ and $\mathbf{h}(\mathbf{x}) = [h_1(\mathbf{x}), \ldots, h_q(\mathbf{x})]^T$ are the inequality constraint vector and the equality constraint vector, respectively.

If on the one hand $\boldsymbol{\nu} = \mathbf{0}$ then the augmented Lagrangian function simplifies to the Lagrangian function:

$$\mathcal{L}(\mathbf{x}, \boldsymbol{\lambda}, \mathbf{0}) = \mathcal{L}(\mathbf{x}, \boldsymbol{\lambda}) = f_0(\mathbf{x}) + \sum_{i=1}^{m} \lambda_i f_i(\mathbf{x}).$$

This implies that the augmented Lagrange multiplier method reduces to the Lagrange multiplier method for an inequality constrained-optimization problem. On the other hand, if $\boldsymbol{\lambda} = \mathbf{0}$ and $\boldsymbol{\nu} = \rho \mathbf{h}(\mathbf{x})$ with $\rho > 0$ then the augmented Lagrangian function reduces to the penalty function:

$$\mathcal{L}(\mathbf{x}, \mathbf{0}, \boldsymbol{\nu}) = \mathcal{L}(\mathbf{x}, \boldsymbol{\nu}) = f_0(\mathbf{x}) + \rho \sum_{i=1}^{q} |h_i(\mathbf{x})|^2.$$

This implies that the augmented Lagrange multiplier method includes the penalty function method for a equality constrained optimization problem as a special case.

The above two facts show that the augmented Lagrange multiplier method combines the Lagrange multiplier method and the penalty function method.

4.7.4 Lagrangian Dual Method

As in the Lagrange multiplier method, the Lagrange multiplier vector $\boldsymbol{\lambda}$ of the augmented Lagrange multiplier method has the nonnegative constraint $\boldsymbol{\lambda} \geq \mathbf{0}$. Under this constraint, let us consider how to solve the unconstrained minimization problem

$$\min \mathcal{L}(\mathbf{x}, \boldsymbol{\lambda}, \boldsymbol{\nu}) = \min_{\boldsymbol{\lambda} \geq \mathbf{0}, \boldsymbol{\nu}} \left\{ f_0(\mathbf{x}) + \sum_{i=1}^{m} \lambda_i f_i(\mathbf{x}) + \sum_{i=1}^{q} \nu_i h_i(\mathbf{x}) \right\}. \tag{4.7.15}$$

The vector $\mathbf{x}$ is known as the optimization variable or decision variable or primal variable, while $\boldsymbol{\lambda}$ is the dual variable.

The original constrained optimization problem (4.7.1) is simply called the *original problem*, and the unconstrained optimization problem (4.7.15) is simply known as the *dual problem*.

Owing to the nonnegativity of the Lagrange multiplier vector $\boldsymbol{\lambda}$, the augmented Lagrangian function $\mathcal{L}(\mathbf{x}, \boldsymbol{\lambda}, \boldsymbol{\nu})$ may tend to negative infinity when some λ_i equals a very large positive number. Therefore, we first need to maximize the original augmented Lagrangian function, to get

$$J_1(\mathbf{x}) = \max_{\boldsymbol{\lambda} \geq \mathbf{0}, \boldsymbol{\nu}} \left\{ f_0(\mathbf{x}) + \sum_{i=1}^{m} \lambda_i f_i(\mathbf{x}) + \sum_{i=1}^{q} \nu_i h_i(\mathbf{x}) \right\}. \tag{4.7.16}$$

The problem with using the unconstrained maximization (4.7.16) is that it is impossible to avoid *violation of the constraint* $f_i(\mathbf{x}) > 0$. This may result in $J_1(\mathbf{x})$ equalling positive infinity; namely,

$$J_1(\mathbf{x}) = \begin{cases} f_0(\mathbf{x}), & \text{if } \mathbf{x} \text{ meets all original constraints,} \\ (f_0(\mathbf{x}), +\infty), & \text{otherwise.} \end{cases} \tag{4.7.17}$$

From the above equation it follows that in order to minimize $f_0(\mathbf{x})$ subject to all inequality and equality constraints, we should minimize $J_1(\mathbf{x})$ to get the *primal cost function*

$$J_{\mathrm{P}}(\mathbf{x}) = \min_{\mathbf{x}} J_1(\mathbf{x}) = \min_{\mathbf{x}} \max_{\boldsymbol{\lambda} \geq \mathbf{0}, \boldsymbol{\nu}} \mathcal{L}(\mathbf{x}, \boldsymbol{\lambda}, \boldsymbol{\nu}). \tag{4.7.18}$$

This is a *minimax problem*, whose solution is the *supremum* of the Lagrangian function $\mathcal{L}(\mathbf{x}, \boldsymbol{\lambda}, \boldsymbol{\nu})$, namely

$$J_{\mathrm{P}}(\mathbf{x}) = \sup \left(f_0(\mathbf{x}) + \sum_{i=1}^{m} \lambda_i f_i(\mathbf{x}) + \sum_{i=1}^{q} \nu_i h_i(\mathbf{x}) \right). \tag{4.7.19}$$

From (4.7.19) and (4.7.17) it follows that the optimal value of the original constrained minimization problem is given by

$$p^\star = J_{\mathrm{P}}(\mathbf{x}^\star) = \min_{\mathbf{x}} f_0(\mathbf{x}) = f_0(\mathbf{x}^\star), \tag{4.7.20}$$

which is simply known as the *optimal primal value*.

However, the minimization of a nonconvex objective function cannot be converted into the minimization of another convex function. Hence, if $f_0(\mathbf{x})$ is a convex function then, even if we designed an optimization algorithm that can find a local extremum $\tilde{\mathbf{x}}$ of the original cost function, there is no guarantee that $\tilde{\mathbf{x}}$ is a global extremum point.

Fortunately, the minimization of a convex function $f(\mathbf{x})$ and the maximization of the concave function $-f(\mathbf{x})$ are equivalent. On the basis of this dual relation, it is easy to obtain a dual method for solving the optimization problem of a nonconvex objective function: convert the minimization of a nonconvex objective function into the maximization of a concave objective function.

For this purpose, construct another objective function from the Lagrangian function $\mathcal{L}(\mathbf{x}, \boldsymbol{\lambda}, \boldsymbol{\nu})$:

$$\begin{aligned} J_2(\boldsymbol{\lambda}, \boldsymbol{\nu}) &= \min_{\mathbf{x}} \mathcal{L}(\mathbf{x}, \boldsymbol{\lambda}, \boldsymbol{\nu}) \\ &= \min_{\mathbf{x}} \left\{ f_0(\mathbf{x}) + \sum_{i=1}^{m} \lambda_i f_i(\mathbf{x}) + \sum_{i=1}^{q} \nu_i h_i(\mathbf{x}) \right\}. \end{aligned} \tag{4.7.21}$$

From the above equation it is known that

$$\min_{\mathbf{x}} \mathcal{L}(\mathbf{x}, \boldsymbol{\lambda}, \boldsymbol{\nu}) = \begin{cases} \min_{\mathbf{x}} f_0(\mathbf{x}), & \text{if } \mathbf{x} \text{ meets all the original constraints,} \\ (-\infty, \min_{\mathbf{x}} f_0(\mathbf{x})), & \text{otherwise.} \end{cases} \tag{4.7.22}$$

Its maximization function

$$J_{\mathrm{D}}(\boldsymbol{\lambda}, \boldsymbol{\nu}) = \max_{\boldsymbol{\lambda} \geq \mathbf{0},\, \boldsymbol{\nu}} J_2(\boldsymbol{\lambda}, \boldsymbol{\nu}) = \max_{\boldsymbol{\lambda} \geq \mathbf{0},\, \boldsymbol{\nu}} \min_{\mathbf{x}} \mathcal{L}(\mathbf{x}, \boldsymbol{\lambda}, \boldsymbol{\nu}) \tag{4.7.23}$$

is called the *dual objective function* for the original problem. This is a *maximin problem* for the Lagrangian function $\mathcal{L}(\mathbf{x}, \boldsymbol{\lambda}, \boldsymbol{\nu})$.

Since the *maximin* of the Lagrangian function $\mathcal{L}(\mathbf{x}, \boldsymbol{\lambda}, \boldsymbol{\nu})$ is its *infimum*, we have

$$J_{\mathrm{D}}(\boldsymbol{\lambda}, \boldsymbol{\nu}) = \inf \left(f_0(\mathbf{x}) + \sum_{i=1}^{m} \lambda_i f_i(\mathbf{x}) + \sum_{i=1}^{q} \nu_i h_i(\mathbf{x}) \right). \tag{4.7.24}$$

The dual objective function defined by Equation (4.7.24) has the following characteristics.

(1) The function $J_{\mathrm{D}}(\boldsymbol{\lambda}, \boldsymbol{\nu})$ is an infimum of the augmented Lagrangian function $\mathcal{L}(\mathbf{x}, \boldsymbol{\lambda}, \boldsymbol{\nu})$.
(2) The function $J_{\mathrm{D}}(\boldsymbol{\lambda}, \boldsymbol{\nu})$ is a maximizing objective function, and thus it is a value or utility function rather than a cost function.
(3) The function $J_{\mathrm{D}}(\boldsymbol{\lambda}, \boldsymbol{\nu})$ is lower unbounded: its lower bound is $-\infty$. Hence, $J_{\mathrm{D}}(\boldsymbol{\lambda}, \boldsymbol{\nu})$ is a concave function of the variable $\mathbf{x}$ even if $f_0(\mathbf{x})$ is not a convex function.

THEOREM 4.13 [353, p. 16] *Any local minimum point $\mathbf{x}^\star$ of an unconstrained convex optimization function $f(\mathbf{x})$ is a global minimum point. If the convex function $f(\mathbf{x})$ is differentiable then the stationary point $\mathbf{x}^\star$ such that $\partial f(\mathbf{x})/\partial \mathbf{x} = \mathbf{0}$ is a global minimum point of $f(\mathbf{x})$.*

Theorem 4.13 shows that any extreme point of the concave function is a global extreme point. Therefore, the algorithm design of the standard constrained minimization problem (4.7.1) becomes the design of an unconstrained maximization algorithm for the dual objective function $J_{\rm D}(\boldsymbol{\lambda}, \boldsymbol{\nu})$. Such a method is called the *Lagrangian dual method.*

4.7.5 Karush–Kuhn–Tucker Conditions

The optimal value of the dual objective function is simply called the *optimal dual value*, denoted

$$d^\star = J_{\rm D}(\boldsymbol{\lambda}^\star, \boldsymbol{\nu}^\star). \tag{4.7.25}$$

From (4.7.22) and (4.7.23) it is immediately known that

$$d^\star \leq \min_{\mathbf{x}} f_0(\mathbf{x}) = p^\star. \tag{4.7.26}$$

The difference between the optimal primal value $p^\star$ and the optimal dual value $d^\star$, denoted $p^\star - d^\star$, is known as the *duality gap* between the original minimization problem and the dual maximization problem.

Equation (4.7.26) gives the relationship between the maximin and the minimax of the augmented Lagrangian function $\mathcal{L}(\mathbf{x}, \boldsymbol{\lambda}, \boldsymbol{\nu})$. In fact, for any nonnegative real-valued function $f(\mathbf{x}, \mathbf{y})$, there is the following inequality relation between the maximin and minimax:

$$\max_{\mathbf{x}} \min_{\mathbf{y}} f(\mathbf{x}, \mathbf{y}) \leq \min_{\mathbf{y}} \max_{\mathbf{x}} f(\mathbf{x}, \mathbf{y}). \tag{4.7.27}$$

If $d^\star \leq p^\star$ then the Lagrangian dual method is said to have *weak duality*, while when $d^\star = p^\star$, we say that the Lagrangian dual method satisfies *strong duality.*

Given an allowed dual gap ϵ, the points $\mathbf{x}$ and $(\boldsymbol{\lambda}, \boldsymbol{\nu})$ satisfying

$$\mathbf{p}^\star - f_0(\mathbf{x}) \leq \epsilon \tag{4.7.28}$$

are respectively called the *ϵ-suboptimal original point* and the *ϵ-suboptimal dual points* of the dual-concave-function maximization problem.

Let $\mathbf{x}^\star$ and $(\boldsymbol{\lambda}^\star, \boldsymbol{\nu}^\star)$ represent any original optimal point and dual optimal points with zero dual gap $\epsilon = 0$. Since $\mathbf{x}^\star$ among all original feasible points $\mathbf{x}$ minimizes the augmented Lagrangian objective function $\mathcal{L}(\mathbf{x}, \boldsymbol{\lambda}^\star, \boldsymbol{\nu}^\star)$, the gradient vector of $\mathcal{L}(\mathbf{x}, \boldsymbol{\lambda}^\star, \boldsymbol{\nu}^\star)$ at the point $\mathbf{x}^\star$ must be equal to the zero vector, namely

$$\nabla f_0(\mathbf{x}^\star) + \sum_{i=1}^{m} \lambda_i^\star \nabla f_i(\mathbf{x}^\star) + \sum_{i=1}^{q} \nu_i^\star \nabla h_i(\mathbf{x}^\star) = \mathbf{0}.$$

Therefore, the *Karush–Kuhn–Tucker (KKT) conditions* (i.e., the first-order necessary conditions) of the Lagrangian dual unconstrained optimization problem are given by [353]

$$\left.\begin{aligned} f_i(\mathbf{x}^\star) &\le 0, \quad i = 1,\ldots,m \quad \text{(original inequality constraints)},\\ h_i(\mathbf{x}^\star) &= 0, \quad i = 1,\ldots,q \quad \text{(original equality constraints)},\\ \lambda_i^\star &\ge 0, \quad i = 1,\ldots,m \quad \text{(nonnegativity)},\\ \lambda_i^\star f_i(\mathbf{x}^\star) &= 0, \quad i = 1,\ldots,m \quad \text{(complementary slackness)},\\ \nabla f_0(\mathbf{x}^\star) + \sum_{i=1}^m \lambda_i^\star \nabla f_i(\mathbf{x}^\star) &+ \sum_{i=1}^q \nu_i^\star \nabla h_i(\mathbf{x}^\star) = \mathbf{0} \quad \text{(stationary point)}. \end{aligned}\right\} \tag{4.7.29}$$

A point $\mathbf{x}$ satisfying the KKT conditions is called a KKT point.

Remark 1 The first KKT condition and the second KKT condition are the original inequality and equality constraint conditions, respectively.

Remark 2 The third KKT condition is the nonnegative condition of the Lagrange multiplier λ_i, which is a key constraint of the Lagrangian dual method.

Remark 3 The fourth KKT condition (complementary slackness) is also called dual complementary and is another key constraint of the Lagrangian dual method. This condition implies that, for a violated constraint $f_i(\mathbf{x}) > 0$, the corresponding Lagrange multiplier λ_i must equal zero, and thus we can avoid completely any violated constraint. Thus the role of this condition is to establish a barrier $f_i(\mathbf{x}) = 0, i = 1,\ldots,m$ on the boundary of inequality constraints, that prevents the occurrence of constraint violation $f_i(\mathbf{x}) > 0$.

Remark 4 The fifth KKT condition is the condition for there to be a stationary point at $\min_{\mathbf{x}} \mathcal{L}(\mathbf{x}, \boldsymbol{\lambda}, \boldsymbol{\nu})$.

Remark 5 If the inequality constraint $f_i(\mathbf{x}) \le 0, i = 1,\ldots,m$, in the constrained optimization (4.7.1) becomes $c_i(\mathbf{x}) \ge 0, i = 1,\ldots,m$, then the Lagrangian function should be modified to

$$\mathcal{L}(\mathbf{x}, \boldsymbol{\lambda}, \boldsymbol{\nu}) = f_0(\mathbf{x}) - \sum_{i=1}^m \lambda_i c_i(\mathbf{x}) + \sum_{i=1}^q \nu_i h_i(\mathbf{x}),$$

and all inequality constrained functions $f_i(\mathbf{x})$ in the KKT condition formula (4.7.29) should be replaced by $-c_i(\mathbf{x})$.

In the following, we discuss the necessary modifications of the KKT conditions under some assumptions.

Definition 4.17 For inequality constraints $f_i(\mathbf{x}) \le 0, i = 1,\ldots,m$, if $f_i(\bar{\mathbf{x}}) = 0$ at the point $\bar{\mathbf{x}}$ then the ith constraint is said to be an *active constraint* at the point $\bar{\mathbf{x}}$. If $f_i(\bar{\mathbf{x}}) < 0$ then the ith constraint is called an *inactive constraint* at the point $\bar{\mathbf{x}}$. If $f_i(\bar{\mathbf{x}}) > 0$ then the ith constraint is known as a *violated constraint* at the point $\bar{\mathbf{x}}$. The index set of all active constraints at the point $\bar{\mathbf{x}}$ is denoted $\mathcal{A}(\bar{\mathbf{x}}) = \{i | f_i(\bar{\mathbf{x}}) = 0\}$ and is referred to as the *active set* of the point $\bar{\mathbf{x}}$.

Let m inequality constraints $f_i(\mathbf{x}), i = 1, \ldots, m$, have k active constraints $f_{\mathcal{A}_1}(\mathbf{x}^\star), \ldots, f_{\mathcal{A}_k}(\mathbf{x}^\star)$ and $m - k$ inactive constraints at some KKT point $\mathbf{x}^\star$.

In order to satisfy the complementarity in the KKT conditions $\lambda_i f_i(\mathbf{x}^\star) = 0$, the Lagrange multipliers $\lambda_i^\star$ corresponding to the inactive constraints $f_i(\mathbf{x}^\star) < 0$ must be equal to zero. This implies that the last KKT condition in (4.7.29) becomes

$$\nabla f_0(\mathbf{x}^\star) + \sum_{i \in \mathcal{A}} \lambda_i^\star \nabla f_i(\mathbf{x}^\star) + \sum_{i=1}^{q} \nu_i^\star \nabla h_i(\mathbf{x}^\star) = \mathbf{0}$$

or

$$\begin{bmatrix} \dfrac{\partial f_0(\mathbf{x}^\star)}{\partial x_1^\star} \\ \vdots \\ \dfrac{\partial f_0(\mathbf{x}^\star)}{\partial x_n^\star} \end{bmatrix} + \begin{bmatrix} \dfrac{\partial h_1(\mathbf{x}^\star)}{\partial x_1^\star} & \cdots & \dfrac{\partial h_q(\mathbf{x}^\star)}{\partial x_1^\star} \\ \vdots & \ddots & \vdots \\ \dfrac{\partial h_1(\mathbf{x}^\star)}{\partial x_n^\star} & \cdots & \dfrac{\partial h_q(\mathbf{x}^\star)}{\partial x_n^\star} \end{bmatrix} \begin{bmatrix} \nu_1^\star \\ \vdots \\ \nu_q^\star \end{bmatrix}$$
$$= - \begin{bmatrix} \dfrac{\partial f_{\mathcal{A}1}(\mathbf{x}^\star)}{\partial x_1^\star} & \cdots & \dfrac{\partial f_{\mathcal{A}k}(\mathbf{x}^\star)}{\partial x_1^\star} \\ \vdots & \ddots & \vdots \\ \dfrac{\partial f_{\mathcal{A}1}(\mathbf{x}^\star)}{\partial x_n^\star} & \cdots & \dfrac{\partial f_{\mathcal{A}k}(\mathbf{x}^\star)}{\partial x_n^\star} \end{bmatrix} \begin{bmatrix} \lambda_{\mathcal{A}1}^\star \\ \vdots \\ \lambda_{\mathcal{A}k}^\star \end{bmatrix}$$

namely

$$\nabla f_0(\mathbf{x}^\star) + (\mathbf{J}_h(\mathbf{x}^\star))^T \boldsymbol{\nu}^\star = -(\mathbf{J}_{\mathcal{A}}(\mathbf{x}^\star))^T \boldsymbol{\lambda}_{\mathcal{A}}^\star, \tag{4.7.30}$$

where $\mathbf{J}_h(\mathbf{x}^\star)$ is the Jacobian matrix of the equality constraint function $h_i(\mathbf{x}) = 0, i = 1, \ldots, q$ at the point $\mathbf{x}^\star$ and

$$\mathbf{J}_{\mathcal{A}}(\mathbf{x}^\star) = \begin{bmatrix} \dfrac{\partial f_{\mathcal{A}1}(\mathbf{x}^\star)}{\partial x_1^\star} & \cdots & \dfrac{\partial f_{\mathcal{A}1}(\mathbf{x}^\star)}{\partial x_n^\star} \\ \vdots & \ddots & \vdots \\ \dfrac{\partial f_{\mathcal{A}k}(\mathbf{x}^\star)}{\partial x_1^\star} & \cdots & \dfrac{\partial f_{\mathcal{A}k}(\mathbf{x}^\star)}{\partial x_n^\star} \end{bmatrix} \in \mathbb{R}^{k \times n}, \tag{4.7.31}$$

$$\boldsymbol{\lambda}_{\mathcal{A}}^\star = [\lambda_{\mathcal{A}1}^\star, \ldots, \lambda_{\mathcal{A}k}^\star] \in \mathbb{R}^k, \tag{4.7.32}$$

are the Jacobian matrix and Lagrange multiplier vector respectively of the active constraint function.

Equation (4.7.30) shows that if the Jacobian matrix $\mathbf{J}_{\mathcal{A}}(\bar{\mathbf{x}})$ of the active constraint function at the feasible point $\bar{\mathbf{x}}$ is of full row rank, then the actively constrained Lagrange multiplier vector can be uniquely determined:

$$\boldsymbol{\lambda}_{\mathcal{A}}^\star = -(\mathbf{J}_{\mathcal{A}}(\bar{\mathbf{x}})\mathbf{J}_{\mathcal{A}}(\bar{\mathbf{x}})^T)^{-1}\mathbf{J}_{\mathcal{A}}(\bar{\mathbf{x}}) \left(\nabla f_0(\bar{\mathbf{x}}) + (\mathbf{J}_h(\bar{\mathbf{x}}))^T \boldsymbol{\nu}^\star\right). \tag{4.7.33}$$

In optimization-algorithm design we always want strong duality to be set up. A simple method for determining whether strong duality holds is Slater's theorem.

Define the relative interior of the feasible domain of the original inequality constraint function $\mathcal{F}$ as follows:

$$\mathrm{relint}(\mathcal{F}) = \{\mathbf{x}|f_i(\mathbf{x}) < 0, i = 1, \ldots, m,\ h_i(\mathbf{x}) = 0, i = 1, \ldots, p\}. \tag{4.7.34}$$

A point in the relative interior of the feasible domain $\bar{\mathbf{x}} \in \mathrm{relint}(\mathcal{F})$ is called a relative interior point.

In an optimization process, the constraint restriction that iterative points should be in the interior of the feasible domain is known as the *Slater condition.* Slater's theorem says that if the Slater condition is satisfied, and the original inequality optimization problem (4.7.1) is a convex optimization problem, then the optimal value $d^\star$ of the dual unconstrained optimization problem (4.7.23) is equal to the the optimal value $p^\star$ of the original optimization problem, i.e., strong duality holds.

The following summarizes the relationships between the original constrained optimization problem and the Lagrangian dual unconstrained convex optimization problem.

(1) Only when the inequality constraint functions $f_i(\mathbf{x}), i = 1, \ldots, m$, are convex, and the equality constraint functions $h_i(\mathbf{x})$, $i = 1, \ldots, q$, are affine, can an original constrained optimization problem be converted into a dual unconstrained maximization problem by the Lagrangian dual method.
(2) The maximization of a concave function is equivalent to the minimization of the corresponding convex function.
(3) If the original constrained optimization is a convex problem then the points of the Lagrangian objective function $\tilde{\mathbf{x}}$ and $(\tilde{\boldsymbol{\lambda}}, \tilde{\boldsymbol{\nu}})$ satisfying the KKT conditions are the original and dual optimal points, respectively. In other words, the optimal solution $\mathbf{d}^\star$ of the Lagrangian dual unconstrained optimization is the optimal solution $\mathbf{p}^\star$ of the original constrained convex optimization problem.
(4) In general, the optimal solution of the Lagrangian dual unconstrained optimization problem is not the optimal solution of the original constrained optimization problem but only a ϵ-suboptimal solution, where $\epsilon = f_0(\mathbf{x}^\star) - J_{\mathrm{D}}(\boldsymbol{\lambda}^\star, \nu^\star)$.

Consider the constrained optimization problem with equality and inequality constraints

$$\min_{\mathbf{x}}\ f(\mathbf{x}) \quad \text{subject to} \quad \mathbf{Ax} \leq \mathbf{h}, \mathbf{Bx} = \mathbf{b}. \tag{4.7.35}$$

Let the nonnegative vector $\mathbf{s} \geq \mathbf{0}$ be a *slack variable vector* such that $\mathbf{Ax} + \mathbf{s} = \mathbf{h}$. Hence, the inequality constraint $\mathbf{Ax} \leq \mathbf{h}$ becomes the equality constraint $\mathbf{Ax} + \mathbf{s} - \mathbf{h} = \mathbf{0}$. Taking the penalty function $\phi(\mathbf{g}(\mathbf{x})) = \frac{1}{2}\|\mathbf{g}(\mathbf{x})\|_2^2$, the augmented Lagrangian objective function is given by

$$\begin{aligned}\mathcal{L}_\rho(\mathbf{x}, \mathbf{s}, \boldsymbol{\lambda}, \boldsymbol{\nu}) &= f(\mathbf{x}) + \boldsymbol{\lambda}^T(\mathbf{Ax} + \mathbf{s} - \mathbf{h}) + \boldsymbol{\nu}^T(\mathbf{Bx} - \mathbf{b}) \\ &\quad + \frac{\rho}{2}\left(\|\mathbf{Bx} - \mathbf{b}\|_2^2 + \|\mathbf{Ax} + \mathbf{s} - \mathbf{h}\|_2^2\right),\end{aligned} \tag{4.7.36}$$

where the two Lagrange multiplier vectors $\boldsymbol{\lambda} \geq \mathbf{0}$ and $\boldsymbol{\nu} \geq \mathbf{0}$ are nonnegative vectors, and the penalty parameter $\rho > 0$.

The dual gradient ascent method for solving the original optimization problem (4.7.35) is given by

$$\mathbf{x}_{k+1} = \arg\min_{\mathbf{x}} \mathcal{L}_\rho(\mathbf{x}, \mathbf{s}_k, \boldsymbol{\lambda}_k, \boldsymbol{\nu}_k), \tag{4.7.37}$$

$$\mathbf{s}_{k+1} = \arg\min_{\mathbf{s}\geq\mathbf{0}} \mathcal{L}_\rho(\mathbf{x}_{k+1}, \mathbf{s}, \boldsymbol{\lambda}_k, \boldsymbol{\nu}_k), \tag{4.7.38}$$

$$\boldsymbol{\lambda}_{k+1} = \boldsymbol{\lambda}_k + \rho_k(\mathbf{A}\mathbf{x}_{k+1} + \mathbf{s}_{k+1} - \mathbf{h}), \tag{4.7.39}$$

$$\boldsymbol{\nu}_{k+1} = \boldsymbol{\nu}_k + \rho_k(\mathbf{B}\mathbf{x}_{k+1} - \mathbf{b}), \tag{4.7.40}$$

where the gradient vectors are

$$\frac{\partial \mathcal{L}_\rho(\mathbf{x}_{k+1}, \mathbf{s}_{k+1}, \boldsymbol{\lambda}_k, \boldsymbol{\nu}_k)}{\partial \boldsymbol{\lambda}_k} = \mathbf{A}\mathbf{x}_{k+1} + \mathbf{s}_{k+1} - \mathbf{h},$$

$$\frac{\partial \mathcal{L}_\rho(\mathbf{x}_{k+1}, \mathbf{s}_{k+1}, \boldsymbol{\lambda}_k, \boldsymbol{\nu}_k)}{\partial \boldsymbol{\nu}_k} = \mathbf{B}\mathbf{x}_{k+1} - \mathbf{b}.$$

Equations (4.7.37) and (4.7.38) are respectively the updates of the original variable $\mathbf{x}$ and the intermediate variable $\mathbf{s}$, whereas Equations (4.7.39) and (4.7.40) are the dual gradient ascent iterations of the Lagrange multiplier vectors $\boldsymbol{\lambda}$ and $\boldsymbol{\nu}$, taking into account the inequality constraint $\mathbf{A}\mathbf{x} \geq \mathbf{h}$ and the equality constraint $\mathbf{B}\mathbf{x} = \mathbf{b}$, respectively.

4.7.6 Alternating Direction Method of Multipliers

In applied statistics and machine learning we often encounter large-scale equality constrained optimization problems, where the dimension of $\mathbf{x} \in \mathbb{R}^n$ is very large. If the vector $\mathbf{x}$ may be decomposed into several subvectors, i.e., $\mathbf{x} = (\mathbf{x}_1, \ldots, \mathbf{x}_r)$, and the objective function may also be decomposed into

$$f(\mathbf{x}) = \sum_{i=1}^{r} f_i(\mathbf{x}_i),$$

where $\mathbf{x}_i \in \mathbb{R}^{n_i}$ and $\sum_{i=1}^{r} n_i = n$, then a large-scale optimization problem can be transformed into a few *distributed optimization problems*.

The *alternating direction multiplier method* (ADMM) is a simple and effective method for solving distributed optimization problems.

The ADMM method decomposes an optimization problem into smaller subproblems; then, their local solutions are restored or reconstructed into a large-scale optimization solution of the original problem.

The ADMM was proposed by Gabay and Mercier [167] and Glowinski and Marrocco [176] independently in the mid-1970s.

Corresponding to the composition of the objective function $f(\mathbf{x})$, the equality constraint matrix is blocked as follows:

$$\mathbf{A} = [\mathbf{A}_1, \ldots, \mathbf{A}_r], \qquad \mathbf{A}\mathbf{x} = \sum_{i=1}^{r} \mathbf{A}_i \mathbf{x}_i.$$

Hence the augmented Lagrangian objective function can be written as [52]

$$\begin{aligned}\mathcal{L}_\rho(\mathbf{x}, \boldsymbol{\lambda}) &= \sum_{i=1}^{r} \mathcal{L}_i(\mathbf{x}_i, \boldsymbol{\lambda}) \\ &= \sum_{i=1}^{r} \left(f_i(\mathbf{x}_i) + \boldsymbol{\lambda}^T \mathbf{A}_i \mathbf{x}_i\right) - \boldsymbol{\lambda}^T \mathbf{b} + \frac{\rho}{2}\left\| \sum_{i=1}^{r} (\mathbf{A}_i \mathbf{x}_i) - \mathbf{b} \right\|_2^2.\end{aligned}$$

Applying the dual ascent method to the augmented Lagrangian objective function yields a decentralized algorithm for parallel computing [52]:

$$\mathbf{x}_i^{k+1} = \arg\min_{\mathbf{x}_i \in \mathbb{R}^{n_i}} \mathcal{L}_i(\mathbf{x}_i, \boldsymbol{\lambda}_k), \quad i = 1, \ldots, r, \tag{4.7.41}$$

$$\boldsymbol{\lambda}_{k+1} = \boldsymbol{\lambda}_k + \rho_k \left(\sum_{i=1}^{r} \mathbf{A}_i \mathbf{x}_i^{k+1} - \mathbf{b} \right). \tag{4.7.42}$$

Here the updates $\mathbf{x}_i$ ($i = 1, \ldots, r$) can be run independently in parallel. Because $\mathbf{x}_i, i = 1, \ldots, r$ are updated in an alternating or sequential manner, this augmented multiplier method is known as the "alternating direction" method of multipliers.

In practical applications, the simplest decomposition is the objective function decomposition of $r = 2$:

$$\min \{f(\mathbf{x}) + h(\mathbf{z})\} \quad \text{subject to} \quad \mathbf{A}\mathbf{x} + \mathbf{B}\mathbf{z} = \mathbf{b}, \tag{4.7.43}$$

where $\mathbf{x} \in \mathbb{R}^n, \mathbf{z} \in \mathbb{R}^m, \mathbf{A} \in \mathbb{R}^{p \times n}, \mathbf{B} \in \mathbb{R}^{p \times m}, \mathbf{b} \in \mathbb{R}^p$.

The augmented Lagrangian cost function of the optimization problem (4.7.43) is given by

$$\begin{aligned}\mathcal{L}_\rho(\mathbf{x}, \mathbf{z}, \boldsymbol{\lambda}) &= f(\mathbf{x}) + h(\mathbf{z}) + \boldsymbol{\lambda}^T(\mathbf{A}\mathbf{x} + \mathbf{B}\mathbf{z} - \mathbf{b}) \\ &\quad + \frac{\rho}{2}\|\mathbf{A}\mathbf{x} + \mathbf{B}\mathbf{z} - \mathbf{b}\|_2^2.\end{aligned} \tag{4.7.44}$$

It is easily seen that the optimality conditions are divided into the original feasibility condition

$$\mathbf{A}\mathbf{x} + \mathbf{B}\mathbf{z} - \mathbf{b} = \mathbf{0} \tag{4.7.45}$$

and two dual feasibility conditions,

$$\mathbf{0} \in \partial f(\mathbf{x}) + \mathbf{A}^T\boldsymbol{\lambda} + \rho\mathbf{A}^T(\mathbf{A}\mathbf{x} + \mathbf{B}\mathbf{z} - \mathbf{b}) = \partial f(\mathbf{x}) + \mathbf{A}^T\boldsymbol{\lambda}, \tag{4.7.46}$$

$$\mathbf{0} \in \partial h(\mathbf{z}) + \mathbf{B}^T\boldsymbol{\lambda} + \rho\mathbf{B}^T(\mathbf{A}\mathbf{x} + \mathbf{B}\mathbf{z} - \mathbf{b}) = \partial h(\mathbf{z}) + \mathbf{B}^T\boldsymbol{\lambda}, \tag{4.7.47}$$

where $\partial f(\mathbf{x})$ and $\partial h(\mathbf{z})$ are the subdifferentials of the subobjective functions $f(\mathbf{x})$ and $h(\mathbf{z})$, respectively.

The updates of the ADMM for the optimization problem $\min \mathcal{L}_\rho(\mathbf{x}, \mathbf{z}, \boldsymbol{\lambda})$ are as follows:

$$\mathbf{x}_{k+1} = \arg\min_{\mathbf{x}\in\mathbb{R}^n} \mathcal{L}_\rho(\mathbf{x}, \mathbf{z}_k, \boldsymbol{\lambda}_k), \tag{4.7.48}$$

$$\mathbf{z}_{k+1} = \arg\min_{\mathbf{z}\in\mathbb{R}^m} \mathcal{L}_\rho(\mathbf{x}_{k+1}, \mathbf{z}, \boldsymbol{\lambda}_k), \tag{4.7.49}$$

$$\boldsymbol{\lambda}_{k+1} = \boldsymbol{\lambda}_k + \rho_k(\mathbf{A}\mathbf{x}_{k+1} + \mathbf{B}\mathbf{z}_{k+1} - \mathbf{b}). \tag{4.7.50}$$

The original feasibility cannot be strictly satisfied; its error

$$\mathbf{r}_k = \mathbf{A}\mathbf{x}_k + \mathbf{B}\mathbf{z}_k - \mathbf{b} \tag{4.7.51}$$

is known as the *original residual* (vector) in the kth iteration. Hence the update of the Lagrange multiplier vector can be simply written as

$$\boldsymbol{\lambda}_{k+1} = \boldsymbol{\lambda}_k + \rho_k \mathbf{r}_{k+1}. \tag{4.7.52}$$

Likewise, neither can dual feasibility be strictly satisfied. Because $\mathbf{x}_{k+1}$ is the minimization variable of $\mathcal{L}_\rho(\mathbf{x}, \mathbf{z}_k, \boldsymbol{\lambda}_k)$, we have

$$\begin{aligned}\mathbf{0} &\in \partial f(\mathbf{x}_{k+1}) + \mathbf{A}^T\boldsymbol{\lambda}_k + \rho\mathbf{A}^T(\mathbf{A}\mathbf{x}_{k+1} + \mathbf{B}\mathbf{z}_k - \mathbf{b}) \\ &= \partial f(\mathbf{x}_{k+1}) + \mathbf{A}^T(\boldsymbol{\lambda}_k + \rho\mathbf{r}_{k+1} + \rho\mathbf{B}(\mathbf{z}_k - \mathbf{z}_{k+1})) \\ &= \partial f(\mathbf{x}_{k+1}) + \mathbf{A}^T\boldsymbol{\lambda}_{k+1} + \rho\mathbf{A}^T\mathbf{B}(\mathbf{z}_k - \mathbf{z}_{k+1}).\end{aligned}$$

Comparing this result with the dual feasibility formula (4.7.46), it is easy to see that

$$\mathbf{s}_{k+1} = \rho\mathbf{A}^T\mathbf{B}(\mathbf{z}_k - \mathbf{z}_{k+1}) \tag{4.7.53}$$

is the error vector of the dual feasibility, and hence is known as the *dual residual* vector in the $(k+1)$th iteration.

The stopping criterion of the ADMM is as follows: the original residual and the dual residual in the $(k+1)$th iteration should be very small, namely [52]

$$\|\mathbf{r}_{k+1}\|_2 \le \epsilon_{\text{pri}}, \qquad \|\mathbf{s}_{k+1}\|_2 \le \epsilon_{\text{dual}}, \tag{4.7.54}$$

where ϵ_{pri} and ϵ_{dual} are respectively the allowed disturbances of the primal feasibility and the dual feasibility.

If we let $\boldsymbol{\nu} = (1/\rho)\boldsymbol{\lambda}$ be the Lagrange multiplier vector scaled by the ratio $1/\rho$, called the *scaled dual vector*, then Equations (4.7.48)–(4.7.50) become [52]

$$\mathbf{x}_{k+1} = \arg\min_{\mathbf{x}\in\mathbb{R}^n} \left\{ f(\mathbf{x}) + (\rho/2)\|\mathbf{A}\mathbf{x} + \mathbf{B}\mathbf{z}_k - \mathbf{b} + \boldsymbol{\nu}_k\|_2^2 \right\}, \tag{4.7.55}$$

$$\mathbf{z}_{k+1} = \arg\min_{\mathbf{z}\in\mathbb{R}^m} \left\{ h(\mathbf{z}) + (\rho/2)\|\mathbf{A}\mathbf{x}_{k+1} + \mathbf{B}\mathbf{z} - \mathbf{b} + \boldsymbol{\nu}_k\|_2^2 \right\}, \tag{4.7.56}$$

$$\boldsymbol{\nu}_{k+1} = \boldsymbol{\nu}_k + \mathbf{A}\mathbf{x}_{k+1} + \mathbf{B}\mathbf{z}_{k+1} - \mathbf{b} = \boldsymbol{\nu}_k + \mathbf{r}_{k+1}. \tag{4.7.57}$$

The scaled dual vector has an interesting interpretation [52]: from the residual at the kth iteration, $\mathbf{r}_k = \mathbf{A}\mathbf{x}_k + \mathbf{B}\mathbf{z}_k - \mathbf{b}$, it is easily seen that

$$\boldsymbol{\nu}_k = \boldsymbol{\nu}_0 + \sum_{i=1}^{k} \mathbf{r}^i. \tag{4.7.58}$$

That is, the scaled dual vector is the running sum of the original residuals of all k iterations.

Equations (4.7.55)–(4.7.57) are referred to as the scaled ADMM, while Equations (4.7.48)–(4.7.50) are the ADMM with no scaling.

4.8 Newton Methods

The first-order optimization algorithms use only the zero-order information $f(\mathbf{x})$ and the first-order information $\nabla f(\mathbf{x})$ about an objective function. It is known [305] that if the objective function is twice differentiable then the Newton method based on the Hessian matrix is of quadratic or more rapid convergence.

4.8.1 Newton Method for Unconstrained Optimization

For an unconstrained optimization $\min_{\mathbf{x}\in\mathbb{R}^n} f(\mathbf{x})$, if the Hessian matrix $\mathbf{H} = \nabla^2 f(\mathbf{x})$ is positive definite then, from the Newton equation $\nabla^2 f(\mathbf{x})\Delta\mathbf{x} = -\nabla f(\mathbf{x})$, we get the Newton step $\Delta\mathbf{x} = -(\nabla^2 f(\mathbf{x}))^{-1}\nabla f(\mathbf{x})$, which results in the gradient descent algorithm

$$\mathbf{x}_{k+1} = \mathbf{x}_k - \mu_k(\nabla^2 f(\mathbf{x}_k))^{-1}\nabla f(\mathbf{x}_k). \tag{4.8.1}$$

This is the well-known Newton method.

The Newton method may encounter the following two thorny issues in applications.

(1) The Hessian matrix $\mathbf{H} = \nabla^2 f(\mathbf{x})$ is difficult to find.

(2) Even if $\mathbf{H} = \nabla^2 f(\mathbf{x})$ can be found, its inverse $\mathbf{H}^{-1} = (\nabla^2 f(\mathbf{x}))^{-1}$ may be numerically unstable.

There are the following three methods for resolving the above two thorny issues.

1. Truncated Newton Method

Instead of using the inverse of the Hessian matrix directly, an iteration method for solving the Newton matrix equation $\nabla^2 f(\mathbf{x})\Delta\mathbf{x}_{\mathrm{nt}} = -\nabla f(\mathbf{x})$ is used to find an approximate solution for the Newton step $\Delta\mathbf{x}_{\mathrm{nt}}$.

The iteration method for solving an Newton matrix equation approximately is known as the *truncated Newton method* [122], where the conjugate gradient algorithm and the preconditioned conjugate gradient algorithm are two popular algorithms for such a case.

The truncated Newton method is especially useful for large-scale unconstrained and constrained optimization problems and interior-point methods.

2. Modified Newton Method

When the Hessian matrix is not positive definite, the Newton matrix equation can be modified to [158]

$$(\nabla^2 f(\mathbf{x}) + \mathbf{E})\Delta\mathbf{x}_{\text{nt}} = -\nabla f(\mathbf{x}), \tag{4.8.2}$$

where $\mathbf{E}$ is a positive semi-definite matrix that is usually taken as a diagonal matrix such that $\nabla^2 f(\mathbf{x}) + \mathbf{E}$ is symmetric positive definite. Such a method is called the *modified Newton method*. The typical modified Newton method takes $\mathbf{E} = \delta\mathbf{I}$, where $\delta > 0$ is small.

3. Quasi-Newton Method

Using a symmetric positive definite matrix $\mathbf{B}_k$ to approximate the inverse of the Hessian matrix $\mathbf{H}_k^{-1}$ in the Newton method, we have the following recursion:

$$\mathbf{x}_{k+1} = \mathbf{x}_k - \mathbf{B}_k \nabla f(\mathbf{x}_k). \tag{4.8.3}$$

The Newton method using a symmetric matrix to approximate the Hessian matrix is called the *quasi-Newton method*. Usually a symmetric matrix $\Delta\mathbf{H}_k = \mathbf{H}_{k+1} - \mathbf{H}_k$ is used to approximate the Hessian matrix, and we denote $\boldsymbol{\rho}_k = f'(\mathbf{x}_{k+1}) - f'(\mathbf{x}_k)$ and $\boldsymbol{\delta}_k = \mathbf{x}_{k+1} - \mathbf{x}_k$.

According to different approximations for the Hessian matrix, the quasi-Newton procedure employs the following three common methods [344].

(1) *Rank-1 updating method*

$$\Delta\mathbf{H}_k = \frac{(\boldsymbol{\delta}_k - \mathbf{H}_k\boldsymbol{\rho}_k)(\boldsymbol{\delta}_k - \mathbf{H}_k\boldsymbol{\rho}_k)^T}{\langle \boldsymbol{\delta}_k - \mathbf{H}_k\boldsymbol{\rho}_k, \boldsymbol{\rho}_k\rangle}.$$

(2) *DFP (Davidon–Fletcher–Powell) method*

$$\Delta\mathbf{H}_k = \frac{\boldsymbol{\delta}_k\boldsymbol{\delta}_k^T}{\langle \boldsymbol{\rho}_k, \boldsymbol{\delta}_k\rangle} - \frac{\mathbf{H}_k\boldsymbol{\rho}_k\boldsymbol{\rho}_k^T\mathbf{H}_k}{\langle \mathbf{H}_k\boldsymbol{\rho}_k, \boldsymbol{\rho}_k\rangle}.$$

(3) *BFGS (Broyden–Fletcher–Goldfarb–Shanno) method*

$$\Delta\mathbf{H}_k = \frac{\mathbf{H}_k\boldsymbol{\rho}_k\boldsymbol{\delta}_k^T + \boldsymbol{\delta}_k\boldsymbol{\rho}_k^T\mathbf{H}_k}{\langle \mathbf{H}_k\boldsymbol{\rho}_k, \boldsymbol{\rho}_k\rangle} - \beta_k \frac{\mathbf{H}_k\boldsymbol{\rho}_k\boldsymbol{\rho}_k^T\mathbf{H}_k}{\langle \mathbf{H}_k\boldsymbol{\rho}_k, \boldsymbol{\rho}_k\rangle},$$

where $\beta_k = 1 + \langle \boldsymbol{\rho}_k, \boldsymbol{\delta}_k\rangle / \langle \mathbf{H}_k\boldsymbol{\rho}_k, \boldsymbol{\rho}_k\rangle$.

An important property of the DFP and BFGS methods is that they preserve the positive definiteness of the matrices.

In the iterative process of the various Newton methods, it is usually required that a search is made of the optimal point along the direction of the straight line $\{\mathbf{x}+\mu\Delta\mathbf{x} \,|\, \mu \geq 0\}$. This step is called *linear search.* However, this choice of procedure can only minimize the objective function approximately, i.e., make it sufficiently small. Such a search for approximate minimization is called an *inexact line search.* In this search the step size μ_k on the search line $\{\mathbf{x} + \mu\Delta\mathbf{x} \,|\, \mu \geq 0\}$ must decrease the objective function $f(\mathbf{x}_k)$ sufficiently to ensure that

$$f(\mathbf{x}_k + \mu\Delta\mathbf{x}_k) < f(\mathbf{x}_k) + \alpha\mu(\nabla f(\mathbf{x}_k))^T\Delta\mathbf{x}_k, \quad \alpha \in (0,1). \tag{4.8.4}$$

The inequality condition (4.8.4) is sometimes called the *Armijo condition* [42], [157], [305], [353]. In general, for a larger step size μ, the Armijo condition is usually not met. Hence, Newton algorithms can start the search from $\mu = 1$; if the Armijo condition is not met then the step size μ needs to be reduced to $\beta\mu$, with the correction factor $\beta \in (0,1)$. If for step size $\beta\mu$ the Armijo condition is still not met, then the step size needs to be reduced again until the Aemijo condition is met. Such a search method is known as *backtracking line search.*

Algorithm 4.10 shows the Newton algorithm via backtracking line search.

Algorithm 4.10 Newton algorithm via backtracking line search

---

initialization: $\mathbf{x}_1 \in \operatorname{dom} f(\mathbf{x})$, and parameters $\alpha \in (0, 0.5), \beta \in (0,1)$. Put $k = 1$.

repeat

1. Compute $\mathbf{b}_k = \nabla f(\mathbf{x}_k)$ and $\mathbf{H}_k = \nabla^2 f(\mathbf{x}_k)$.
2. Solve the Newton equation $\mathbf{H}_k\Delta\mathbf{x}_k = -\mathbf{b}_k$.
3. Update $\mathbf{x}_{k+1} = \mathbf{x}_k + \mu\Delta\mathbf{x}_k$.
4. **exit if** $|f(\mathbf{x}_{k+1}) - f(\mathbf{x}_k)| < \epsilon$.

return $k \leftarrow k + 1$.

output: $\mathbf{x} \leftarrow \mathbf{x}_k$.

The backtracking line search method can ensure that the objective function satisfies $f(\mathbf{x}_{k+1}) < f(\mathbf{x}_k)$ and the step size μ may not need to be too small.

If the Newton equation is solved by other methods in step 2, then Algorithm 4.10 becomes the truncated Newton method, the modified Newton method or the quasi-Newton method, respectively.

Now consider the minimization $\min f(\mathbf{z})$, where $\mathbf{z} \in \mathbb{C}^n$ and $f : \mathbb{C}^n \to \mathbb{R}$.

By the second-order Taylor series approximation of a binary complex function,

$$\begin{aligned} f(x + h_1, y + h_2) &= f(x,y) + h_1\frac{\partial f(x,y)}{\partial x} + h_2\frac{\partial f(x,y)}{\partial y} \\ &\quad + \frac{1}{2!}\left(h_1^2\frac{\partial^2 f(x,y)}{\partial x\partial x} + 2h_1h_2\frac{\partial^2 f(x,y)}{\partial x\partial y} + h_2^2\frac{\partial^2 f(x,y)}{\partial y\partial y}\right), \end{aligned}$$

it is easily seen that the second-order Taylor series approximation of the holomorphic function $f(\mathbf{z},\mathbf{z}^*)$ is given by

$$\begin{aligned}&f(\mathbf{z}+\Delta\mathbf{z},\mathbf{z}^*+\Delta\mathbf{z}^*)\\&= f(\mathbf{z},\mathbf{z}^*)+\left[(\nabla_{\mathbf{z}}f(\mathbf{z},\mathbf{z}^*))^T,\ (\nabla_{\mathbf{z}^*}f(\mathbf{z},\mathbf{z}^*))^T\right]\begin{bmatrix}\Delta\mathbf{z}\\ \Delta\mathbf{z}^*\end{bmatrix}\\&\quad+\frac{1}{2}\left[(\Delta\mathbf{z})^H,\ (\Delta\mathbf{z})^T\right]\begin{bmatrix}\dfrac{\partial^2 f(\mathbf{z},\mathbf{z}^*)}{\partial\mathbf{z}^*\partial\mathbf{z}^T} & \dfrac{\partial^2 f(\mathbf{z},\mathbf{z}^*)}{\partial\mathbf{z}^*\partial\mathbf{z}^H}\\ \dfrac{\partial^2 f(\mathbf{z},\mathbf{z}^*)}{\partial\mathbf{z}\partial\mathbf{z}^T} & \dfrac{\partial^2 f(\mathbf{z},\mathbf{z}^*)}{\partial\mathbf{z}\partial\mathbf{z}^H}\end{bmatrix}\begin{bmatrix}\Delta\mathbf{z}\\ \Delta\mathbf{z}^*\end{bmatrix}.\end{aligned}\tag{4.8.5}$$

From the first-order optimality condition $\dfrac{\partial f(\mathbf{z}+\Delta\mathbf{z},\mathbf{z}^*+\Delta\mathbf{z}^*)}{\partial\begin{bmatrix}\Delta\mathbf{z}\\ \Delta\mathbf{z}^*\end{bmatrix}}=\begin{bmatrix}\mathbf{0}\\ \mathbf{0}\end{bmatrix}$, we obtain immediately the equation for the *complex Newton step*:

$$\begin{bmatrix}\mathbf{H}_{\mathbf{z}_k^*,\mathbf{z}_k} & \mathbf{H}_{\mathbf{z}_k^*,\mathbf{z}_k^*}\\ \mathbf{H}_{\mathbf{z}_k,\mathbf{z}_k} & \mathbf{H}_{\mathbf{z}_k,\mathbf{z}_k^*}\end{bmatrix}\begin{bmatrix}\Delta\mathbf{z}_{\mathrm{nt},k}\\ \Delta\mathbf{z}_{\mathrm{nt},k}^*\end{bmatrix}=-\begin{bmatrix}\nabla_{\mathbf{z}_k}f(\mathbf{z}_k,\mathbf{z}_k^*)\\ \nabla_{\mathbf{z}_k^*}f(\mathbf{z}_k,\mathbf{z}_k^*)\end{bmatrix},\tag{4.8.6}$$

where

$$\begin{aligned}\mathbf{H}_{\mathbf{z}_k^*,\mathbf{z}_k}&=\frac{\partial^2 f(\mathbf{z}_k,\mathbf{z}_k^*)}{\partial\mathbf{z}_k^*\partial\mathbf{z}_k^T}, & \mathbf{H}_{\mathbf{z}_k^*,\mathbf{z}_k^*}(k)&=\frac{\partial^2 f(\mathbf{z}_k,\mathbf{z}_k^*)}{\partial\mathbf{z}_k^*\partial\mathbf{z}_k^H},\\ \mathbf{H}_{\mathbf{z}_k,\mathbf{z}_k}&=\frac{\partial^2 f(\mathbf{z}_k,\mathbf{z}_k^*)}{\partial\mathbf{z}_k\partial\mathbf{z}_k^T}, & \mathbf{H}_{\mathbf{z}_k,\mathbf{z}_k^*}&=\frac{\partial^2 f(\mathbf{z}_k,\mathbf{z}_k^*)}{\partial\mathbf{z}_k\partial\mathbf{z}_k^H},\end{aligned}\tag{4.8.7}$$

are the part-Hessian matrices of the holomorphic function $f(\mathbf{z}_k,\mathbf{z}_k^*)$, respectively. Hence, the updating formula of the *complex Newton method* is described by

$$\begin{bmatrix}\mathbf{z}_{k+1}\\ \mathbf{z}_{k+1}^*\end{bmatrix}=\begin{bmatrix}\mathbf{z}_k\\ \mathbf{z}_k^*\end{bmatrix}+\mu\begin{bmatrix}\Delta\mathbf{z}_{\mathrm{nt},k}\\ \Delta\mathbf{z}_{\mathrm{nt},k}^*\end{bmatrix}.\tag{4.8.8}$$

4.8.2 Newton Method for Constrained Optimization

First, consider the equality constrained optimization problem with a real variable:

$$\min_{\mathbf{x}}\ f(\mathbf{x})\quad\text{subject to}\quad \mathbf{Ax}=\mathbf{b},\tag{4.8.9}$$

where $f:\mathbb{R}^n\to\mathbb{R}$ is a convex function and is twice differentiable, while $\mathbf{A}\in\mathbb{R}^{p\times n}$ and $\text{rank}(\mathbf{A})=p$ with $p<n$.

Let $\Delta\mathbf{x}_{\mathrm{nt}}$ denote the *Newton search direction*. Then the second-order Taylor approximation of the objective function $f(\mathbf{x})$ is given by

$$f(\mathbf{x}+\Delta\mathbf{x}_{\mathrm{nt}})=f(\mathbf{x})+(\nabla f(\mathbf{x}))^T\Delta\mathbf{x}_{\mathrm{nt}}+\frac{1}{2}(\Delta\mathbf{x}_{\mathrm{nt}})^T\nabla^2 f(\mathbf{x})\Delta\mathbf{x}_{\mathrm{nt}},$$

subject to $\mathbf{A}(\mathbf{x}+\Delta\mathbf{x}_{\mathrm{nt}})=\mathbf{b}$ and $\mathbf{A}\Delta\mathbf{x}_{\mathrm{nt}}=\mathbf{0}$. In other words, the Newton search direction can be determined by the equality constrained optimization problem:

$$\min_{\Delta\mathbf{x}_{\mathrm{nt}}}\left\{f(\mathbf{x})+(\nabla f(\mathbf{x}))^T\Delta\mathbf{x}_{\mathrm{nt}}+\frac{1}{2}(\Delta\mathbf{x}_{\mathrm{nt}})^T\nabla^2 f(\mathbf{x})\Delta\mathbf{x}_{\mathrm{nt}}\right\}\ \text{subject to}\ \mathbf{A}\Delta\mathbf{x}_{\mathrm{nt}}=\mathbf{0}.$$

Letting $\boldsymbol{\lambda}$ be the Lagrange multiplier vector multiplying the equality constraint $\mathbf{A}\Delta\mathbf{x}_{\rm nt}=\mathbf{0}$, we have the Lagrangian objective function

$$\mathcal{L}(\Delta\mathbf{x}_{\rm nt},\boldsymbol{\lambda})=f(\mathbf{x})+(\nabla f(\mathbf{x}))^T\Delta\mathbf{x}_{\rm nt}+\frac{1}{2}(\Delta\mathbf{x}_{\rm nt})^T\nabla^2 f(\mathbf{x})\Delta\mathbf{x}_{\rm nt}+\boldsymbol{\lambda}^T\mathbf{A}\Delta\mathbf{x}_{\rm nt}. \tag{4.8.10}$$

By the first-order optimal condition $\dfrac{\partial\mathcal{L}(\Delta\mathbf{x}_{\rm nt},\boldsymbol{\lambda})}{\partial\Delta\mathbf{x}_{\rm nt}}=\mathbf{0}$ and the constraint condition $\mathbf{A}\Delta\mathbf{x}_{\rm nt}=\mathbf{0}$, it is easily obtained that

$$\nabla f(\mathbf{x})+\nabla^2 f(\mathbf{x})\Delta\mathbf{x}_{\rm nt}+\mathbf{A}^T\boldsymbol{\lambda}=\mathbf{0}\quad\text{and}\quad\mathbf{A}\Delta\mathbf{x}_{\rm nt}=\mathbf{0}$$

which can be merged into

$$\begin{bmatrix}\nabla^2 f(\mathbf{x}) & \mathbf{A}^T\\ \mathbf{A} & \mathbf{O}\end{bmatrix}\begin{bmatrix}\Delta\mathbf{x}_{\rm nt}\\ \boldsymbol{\lambda}\end{bmatrix}=\begin{bmatrix}-\nabla f(\mathbf{x})\\ \mathbf{O}\end{bmatrix}. \tag{4.8.11}$$

Let $\mathbf{x}^\star$ be the optimal solution to the equality constrained optimization problem (4.8.9), and let the corresponding optimal value of objective function $f(\mathbf{x})$ be

$$\mathbf{p}^\star=\inf\{f(\mathbf{x})|\mathbf{A}\mathbf{x}=\mathbf{b}\}=f(\mathbf{x}^\star). \tag{4.8.12}$$

As a stopping criterion for the Newton algorithm, one can use [53]

$$\lambda^2(\mathbf{x})=(\Delta\mathbf{x}_{\rm nt})^T\nabla^2 f(\mathbf{x})\Delta\mathbf{x}_{\rm nt}. \tag{4.8.13}$$

Algorithm 4.11 takes a feasible point as the initial point, and thus is called the *feasible-start Newton algorithm.* This algorithm has the following features:

(1) It is a descent method, because backtracking line search can ensure objective function descent at each iteration, and that the equality constraint is satisfied at each iteration point $\mathbf{x}_k$.
(2) It requires a feasible point as an initial point.

However, in many cases it is not easy to find a feasible point as an initial point. In these cases, it is necessary to consider an infeasible start Newton method.

If $\mathbf{x}_k$ is an infeasible point, consider the equality constrained optimization problem

$$\min_{\Delta\mathbf{x}_k}\left\{f(\mathbf{x}_k+\Delta\mathbf{x}_k)=f(\mathbf{x}_k)+(\nabla f(\mathbf{x}_k))^T\Delta\mathbf{x}_k+\frac{1}{2}(\Delta\mathbf{x}_k)^T\nabla^2 f(\mathbf{x}_k)\Delta\mathbf{x}_k\right\}$$
$$\text{subject to}\quad\mathbf{A}(\mathbf{x}_k+\Delta\mathbf{x}_k)=\mathbf{b}.$$

Let $\boldsymbol{\lambda}_{k+1}(=\boldsymbol{\lambda}_k+\Delta\boldsymbol{\lambda}_k)$ be the Lagrange multiplier vector corresponding to the equality constraint $\mathbf{A}(\mathbf{x}_k+\Delta\mathbf{x}_k)=\mathbf{b}$. Then the Lagrangian objective function is

$$\begin{aligned}\mathcal{L}(\Delta\mathbf{x}_k,\boldsymbol{\lambda}_{k+1})&=f(\mathbf{x}_k)+(\nabla f(\mathbf{x}_k))^T\Delta\mathbf{x}_k+\frac{1}{2}(\Delta\mathbf{x}_k)^T\nabla^2 f(\mathbf{x}_k)\Delta\mathbf{x}_k\\&\quad+\boldsymbol{\lambda}_{k+1}^T(\mathbf{A}(\mathbf{x}_k+\Delta\mathbf{x}_k)-\mathbf{b}).\end{aligned}$$

Algorithm 4.11 Feasible-start Newton algorithm [53]

given: A feasible initial point $\mathbf{x}_1 \in \text{dom}\, f$ with $\mathbf{A}\mathbf{x}_1 = \mathbf{b}$, tolerance $\epsilon > 0$, parameters $\alpha \in (0, 0.5)$, $\beta \in (0, 1)$. Put $k = 1$.

repeat

1. Compute the gradient vector $\nabla f(\mathbf{x}_k)$ and the Hessian matrix $\nabla^2 f(\mathbf{x}_k)$.
2. Use the preconditioned conjugate gradient algorithm to solve
$$\begin{bmatrix} \nabla^2 f(\mathbf{x}_k) & \mathbf{A}^T \\ \mathbf{A} & \mathbf{O} \end{bmatrix} \begin{bmatrix} \Delta \mathbf{x}_{\text{nt},k} \\ \boldsymbol{\lambda}_{\text{nt}} \end{bmatrix} = \begin{bmatrix} -\nabla f(\mathbf{x}_k) \\ \mathbf{O} \end{bmatrix}.$$
3. Compute $\lambda^2(\mathbf{x}_k) = (\Delta \mathbf{x}_{\text{nt},k})^T \nabla^2 f(\mathbf{x}_k) \Delta \mathbf{x}_{\text{nt},k}$.
4. **exit if** $\lambda^2(\mathbf{x}_k) < \epsilon$.
5. Let $\mu = 1$.
6. **while** not converged **do**
7. Update $\mu \leftarrow \beta\mu$.
8. **break if** $f(\mathbf{x}_k + \mu \Delta \mathbf{x}_{\text{nt},k}) < f(\mathbf{x}_k) + \alpha\mu(\nabla f(\mathbf{x}_k))^T \Delta \mathbf{x}_{\text{nt},k}$.
9. **end while**
10. Update $\mathbf{x}_{k+1} \leftarrow \mathbf{x}_k + \mu \Delta \mathbf{x}_{\text{nt},k}$.

return $k \leftarrow k + 1$.

output: $\mathbf{x} \leftarrow \mathbf{x}_k$.

From the optimal condition $\dfrac{\partial \mathcal{L}(\Delta \mathbf{x}_k, \boldsymbol{\lambda}_{k+1})}{\partial \Delta \mathbf{x}_k} = \mathbf{0}$ and $\dfrac{\partial \mathcal{L}(\Delta \mathbf{x}_k, \boldsymbol{\lambda}_{k+1})}{\partial \boldsymbol{\lambda}_{k+1}} = \mathbf{0}$, it is easily known that

$$\nabla f(\mathbf{x}_k) + \nabla^2 f(\mathbf{x}_k)\Delta \mathbf{x}_k + \mathbf{A}^T \boldsymbol{\lambda}_{k+1} = \mathbf{0},$$
$$\mathbf{A}\Delta \mathbf{x}_k = -(\mathbf{A}\mathbf{x}_k - \mathbf{b}),$$

which can be merged into

$$\begin{bmatrix} \nabla^2 f(\mathbf{x}_k) & \mathbf{A}^T \\ \mathbf{A} & \mathbf{O} \end{bmatrix} \begin{bmatrix} \Delta \mathbf{x}_k \\ \boldsymbol{\lambda}_{k+1} \end{bmatrix} = -\begin{bmatrix} \nabla f(\mathbf{x}_k) \\ \mathbf{A}\mathbf{x}_k - \mathbf{b} \end{bmatrix}. \tag{4.8.14}$$

Substituting $\boldsymbol{\lambda}_{k+1} = \boldsymbol{\lambda}_k + \Delta \boldsymbol{\lambda}_k$ into the above equation, we have

$$\begin{bmatrix} \nabla^2 f(\mathbf{x}_k) & \mathbf{A}^T \\ \mathbf{A} & \mathbf{O} \end{bmatrix} \begin{bmatrix} \Delta \mathbf{x}_k \\ \Delta \boldsymbol{\lambda}_k \end{bmatrix} = -\begin{bmatrix} \nabla f(\mathbf{x}_k) + \mathbf{A}^T \boldsymbol{\lambda}_k \\ \mathbf{A}\mathbf{x}_k - \mathbf{b} \end{bmatrix}. \tag{4.8.15}$$

Algorithm 4.12 shows an infeasible-start Newton algorithm based on (4.8.15).

Algorithm 4.12 Infeasible-start Newton algorithm [53]

given: An initial point $\mathbf{x}_1 \in \mathbb{R}^n$, any initial Lagrange multiplier $\boldsymbol{\lambda}_1 \in \mathbb{R}^p$, an allowed tolerance $\epsilon > 0$, $\alpha \in (0, 0.5)$, and $\beta \in (0, 1)$. Put $k = 1$.

repeat

1. Compute the gradient vector $\nabla f(\mathbf{x}_k)$ and the Hessian matrix $\nabla^2 f(\mathbf{x}_k)$.
2. Compute $\mathbf{r}(\mathbf{x}_k, \boldsymbol{\lambda}_k)$ via Equation (4.8.16).
3. **exit if** $\mathbf{A}\mathbf{x}_k = \mathbf{b}$ and $\|\mathbf{r}(\mathbf{x}_k, \boldsymbol{\lambda}_k)\|_2 < \epsilon$.
4. Adopt the preconditioned conjugate gradient algorithm to solve the KKT equation (4.8.15), yielding the Newton step $(\Delta\mathbf{x}_k, \Delta\boldsymbol{\lambda}_k)$.
5. Let $\mu = 1$.
6. **while** not converged **do**
7. Update $\mu \leftarrow \beta\mu$.
8. **break if** $f(\mathbf{x}_k + \mu\Delta\mathbf{x}_{\text{nt},k}) < f(\mathbf{x}_k) + \alpha\mu(\nabla f(\mathbf{x}_k))^T \Delta\mathbf{x}_{\text{nt},k}$.
9. **end while**
10. Newton update

$$\begin{bmatrix} \mathbf{x}_{k+1} \\ \boldsymbol{\lambda}_{k+1} \end{bmatrix} = \begin{bmatrix} \mathbf{x}_k \\ \boldsymbol{\lambda}_k \end{bmatrix} + \mu \begin{bmatrix} \Delta\mathbf{x}_k \\ \Delta\boldsymbol{\lambda}_k \end{bmatrix}.$$

return $k \leftarrow k + 1$.

output: $\mathbf{x} \leftarrow \mathbf{x}_k$, $\boldsymbol{\lambda} \leftarrow \boldsymbol{\lambda}_k$.

Equation (4.8.15) reveals one of the stopping criterions of the Newton algorithms. Define the residual vector

$$\mathbf{r}(\mathbf{x}_k, \boldsymbol{\lambda}_k) = \begin{bmatrix} \mathbf{r}_{\text{dual}}(\mathbf{x}_k, \boldsymbol{\lambda}_k) \\ \mathbf{r}_{\text{pri}}(\mathbf{x}_k, \boldsymbol{\lambda}_k) \end{bmatrix} = \begin{bmatrix} \nabla f(\mathbf{x}_k) + \mathbf{A}^T \boldsymbol{\lambda}_k \\ \mathbf{A}\mathbf{x}_k - \mathbf{b} \end{bmatrix}, \tag{4.8.16}$$

where $\mathbf{r}_{\text{dual}}(\mathbf{x}_k, \boldsymbol{\lambda}_k) = \nabla f(\mathbf{x}_k) + \mathbf{A}^T \boldsymbol{\lambda}_k$ and $\mathbf{r}_{\text{pri}}(\mathbf{x}_k, \boldsymbol{\lambda}_k) = \mathbf{A}\mathbf{x}_k - \mathbf{b}$ represent the dual and original residual vectors, respectively.

Obviously, one stopping criterion of the infeasible-start Newton algorithm can be summarized as follows:

$$\begin{bmatrix} \Delta\mathbf{x}_k \\ \Delta\boldsymbol{\lambda}_k \end{bmatrix} \approx \begin{bmatrix} \mathbf{0} \\ \mathbf{0} \end{bmatrix} \quad \Leftrightarrow \quad \begin{bmatrix} \mathbf{r}_{\text{dual}}(\mathbf{x}_k, \boldsymbol{\lambda}_k) \\ \mathbf{r}_{\text{pri}}(\mathbf{x}_k, \boldsymbol{\lambda}_k) \end{bmatrix} \approx \begin{bmatrix} \mathbf{0} \\ \mathbf{0} \end{bmatrix} \quad \Leftrightarrow \quad \|\mathbf{r}(\mathbf{x}_k, \boldsymbol{\lambda}_k)\|_2 < \epsilon \tag{4.8.17}$$

holds for a very small disturbance error $\epsilon > 0$.

Another stopping criterion of the infeasible-start Newton algorithm is that the equality constraint must be met. This stopping criterion can ensure that the convergence point of the Newton algorithm is feasible, although its initial point and most iteration points are allowed to be infeasible.

Now consider the equality constrained minimization problem with complex variable:

$$\min_{\mathbf{z}} f(\mathbf{z}) \quad \text{subject to} \quad \mathbf{A}\mathbf{z} = \mathbf{b}, \tag{4.8.18}$$

where $\mathbf{z} \in \mathbb{C}^n, f : \mathbb{C}^n \to \mathbb{R}$ is the convex function and twice differentiable and $\mathbf{A} \in \mathbb{C}^{p\times n}$ with $\text{rank}(\mathbf{A}) = p$ and $p < n$.

Let $(\Delta\mathbf{z}_k, \Delta\mathbf{z}_k^*)$ denote the complex search direction at the kth iteration. Then the second-order Taylor expansion of $f(\mathbf{z}_k, \mathbf{z}_k^*)$ is given by

$$\begin{aligned} & f(\mathbf{z}_k + \Delta\mathbf{z}_k, \mathbf{z}_k^* + \Delta\mathbf{z}_k^*) \\ &= f(\mathbf{z}_k, \mathbf{z}_k^*) + (\nabla_{\mathbf{z}_k} f(\mathbf{z}_k, \mathbf{z}_k^*))^T \Delta\mathbf{z}_k + (\nabla_{\mathbf{z}_k^*} f(\mathbf{z}_k, \mathbf{z}_k^*))^T \Delta\mathbf{z}_k^* \\ &\quad + \frac{1}{2}\left[(\Delta\mathbf{z}_k)^H,\ (\Delta\mathbf{z}_k)^T\right] \begin{bmatrix} \dfrac{\partial^2 f(\mathbf{z}_k, \mathbf{z}_k^*)}{\partial \mathbf{z}_k^* \partial \mathbf{z}_k^T} & \dfrac{\partial^2 f(\mathbf{z}_k, \mathbf{z}_k^*)}{\partial \mathbf{z}_k^* \partial \mathbf{z}_k^H} \\ \dfrac{\partial^2 f(\mathbf{z}_k, \mathbf{z}_k^*)}{\partial \mathbf{z}_k \partial \mathbf{z}_k^T} & \dfrac{\partial^2 f(\mathbf{z}_k, \mathbf{z}_k^*)}{\partial \mathbf{z}_k \partial \mathbf{z}_k^H} \end{bmatrix} \begin{bmatrix} \Delta\mathbf{z}_k \\ \Delta\mathbf{z}_k^* \end{bmatrix} \end{aligned}$$

subject to $\mathbf{A}(\mathbf{z}_k + \Delta\mathbf{z}_k) = \mathbf{b}$ or $\mathbf{A}\Delta\mathbf{z}_k = \mathbf{0}$ if $\mathbf{z}_k$ is a feasible point. In other words, the complex Newton search direction can be determined by the equality constrained optimization problem:

$$\min_{\Delta\mathbf{z}_k, \Delta\mathbf{z}_k^*} f(\mathbf{z}_k + \Delta\mathbf{z}_k, \mathbf{z}_k^* + \Delta\mathbf{z}_k^*) \quad \text{subject to} \quad \mathbf{A}\Delta\mathbf{z}_k = \mathbf{0}. \tag{4.8.19}$$

Let $\boldsymbol{\lambda}_{k+1} = \boldsymbol{\lambda}_k + \Delta\boldsymbol{\lambda}_k \in \mathbb{R}^p$ be a Lagrange multiplier vector. Then the equality constrained minimization problem (4.8.19) can be transformed into the unconstrained minimization problem

$$\min_{\Delta\mathbf{z}_k, \Delta\mathbf{z}_k^*; \Delta\boldsymbol{\lambda}_k} \left\{ f(\mathbf{z}_k + \Delta\mathbf{z}_k, \mathbf{z}_k^* + \Delta\mathbf{z}_k^*) + (\boldsymbol{\lambda}_k + \Delta\boldsymbol{\lambda}_k)^T \mathbf{A}\Delta\mathbf{z}_k \right\}.$$

It is easy to obtain the complex Newton equation

$$\begin{bmatrix} \dfrac{\partial^2 f(\mathbf{z}_k, \mathbf{z}_k^*)}{\partial \mathbf{z}_k^* \partial \mathbf{z}_k^T} & \dfrac{\partial^2 f(\mathbf{z}_k, \mathbf{z}_k^*)}{\partial \mathbf{z}_k^* \partial \mathbf{z}_k^H} & \mathbf{A}^T \\ \dfrac{\partial^2 f(\mathbf{z}_k, \mathbf{z}_k^*)}{\partial \mathbf{z}_k \partial \mathbf{z}_k^T} & \dfrac{\partial^2 f(\mathbf{z}_k, \mathbf{z}_k^*)}{\partial \mathbf{z}_k \partial \mathbf{z}_k^H} & \mathbf{O} \\ \mathbf{A} & \mathbf{O} & \mathbf{O} \end{bmatrix} \begin{bmatrix} \Delta\mathbf{z}_k \\ \Delta\mathbf{z}_k^* \\ \Delta\boldsymbol{\lambda}_k \end{bmatrix} = -\begin{bmatrix} \nabla_{\mathbf{z}_k} f(\mathbf{z}_k, \mathbf{z}_k^*) + \mathbf{A}^T\boldsymbol{\lambda}_k \\ \nabla_{\mathbf{z}_k^*} f(\mathbf{z}_k, \mathbf{z}_k^*) \\ \mathbf{0} \end{bmatrix}. \tag{4.8.20}$$

Define the residual vector

$$\mathbf{r}(\mathbf{z}_k, \mathbf{z}_k^*, \boldsymbol{\lambda}_k) = \begin{bmatrix} \nabla_{\mathbf{z}_k} f(\mathbf{z}_k, \mathbf{z}_k^*) + \mathbf{A}^T\boldsymbol{\lambda}_k \\ \nabla_{\mathbf{z}_k^*} f(\mathbf{z}_k, \mathbf{z}_k^*) \end{bmatrix}. \tag{4.8.21}$$

The feasible-start complex Newton algorithm is shown in Algorithm 4.13.

This algorithm is easily extended into the infeasible start complex Newton algorithm. For this end, the equality constrained minimization problem corresponding to Equation (4.8.19) becomes

$$\min_{\Delta\mathbf{z}_k, \Delta\mathbf{z}_k^*} f(\mathbf{z}_k + \Delta\mathbf{z}_k, \mathbf{z}_k^* + \Delta\mathbf{z}_k^*) \quad \text{subject to} \quad \mathbf{A}(\mathbf{z}_k + \Delta\mathbf{z}_k) = \mathbf{b}. \tag{4.8.22}$$

Algorithm 4.13 Feasible-start complex Newton algorithm

given: A feasible initial point $\mathbf{z}_1 \in \operatorname{dom} f$ with $\mathbf{A}\mathbf{z}_1 = \mathbf{b}$, tolerance $\epsilon > 0$, parameters $\alpha \in (0, 0.5)$, $\beta \in (0, 1)$. Put $k = 1$.

repeat

1. Compute the residual vector $\mathbf{r}(\mathbf{z}_k, \mathbf{z}_k^*, \boldsymbol{\lambda}_k)$ via Equation (4.8.21).
2. **exit if** $\|\mathbf{r}(\mathbf{z}_k, \mathbf{z}_k^*, \boldsymbol{\lambda}_k)\|_2 < \epsilon$.
3. Compute $\nabla_{\mathbf{z}} f(\mathbf{z}_k, \mathbf{z}_k^*)$, $\nabla_{\mathbf{z}^*} f(\mathbf{z}_k, \mathbf{z}_k^*)$ and the full Hessian matrix

$$\mathbf{H}_k = \begin{bmatrix} \dfrac{\partial^2 f(\mathbf{z}_k, \mathbf{z}_k^*)}{\partial \mathbf{z}_k^* \partial \mathbf{z}_k^T} & \dfrac{\partial^2 f(\mathbf{z}_k, \mathbf{z}_k^*)}{\partial \mathbf{z}_k^* \partial \mathbf{z}_k^H} \\ \dfrac{\partial^2 f(\mathbf{z}_k, \mathbf{z}_k^*)}{\partial \mathbf{z}_k \partial \mathbf{z}_k^T} & \dfrac{\partial^2 f(\mathbf{z}_k, \mathbf{z}_k^*)}{\partial \mathbf{z}_k \partial \mathbf{z}_k^H} \end{bmatrix}.$$

4. Use the preconditioned conjugate gradient algorithm to solve the Newton equation (4.8.20), yielding the Newton step $(\Delta\mathbf{z}_{\text{nt},k}, \Delta\mathbf{z}_{\text{nt},k}^*, \Delta\boldsymbol{\lambda}_{\text{nt},k})$.
5. Let $\mu = 1$.
6. **while** not converged **do**
7. Update $\mu \leftarrow \beta\mu$.
8. **break if**

$$f(\mathbf{z}_k + \mu\Delta\mathbf{z}_{\text{nt},k}, \mathbf{z}_k^* + \mu\Delta\mathbf{z}_{\text{nt},k}^*) < f(\mathbf{z}_k, \mathbf{z}_k^*) + \alpha\mu \begin{bmatrix} \nabla_{\mathbf{z}} f(\mathbf{z}, \mathbf{z}^*) \\ \nabla_{\mathbf{z}^*} f(\mathbf{z}, \mathbf{z}^*) \end{bmatrix}^T \begin{bmatrix} \Delta\mathbf{z}_{\text{nt},k} \\ \Delta\mathbf{z}_{\text{nt},k}^* \end{bmatrix}.$$

9. **end while**
10. Newton update

$$\begin{bmatrix} \mathbf{z}_{k+1} \\ \mathbf{z}_{k+1}^* \end{bmatrix} = \begin{bmatrix} \mathbf{z}_k \\ \mathbf{z}_k^* \end{bmatrix} + \mu \begin{bmatrix} \Delta\mathbf{z}_{\text{nt},k} \\ \Delta\mathbf{z}_{\text{nt},k}^* \end{bmatrix}, \quad \boldsymbol{\lambda}_{k+1} = \boldsymbol{\lambda}_k + \Delta\boldsymbol{\lambda}_k.$$

return $k \leftarrow k + 1$.

output: $\mathbf{x} \leftarrow \mathbf{x}_k$.

In this case, the Newton equation (4.8.20) is modified to

$$\begin{bmatrix} \dfrac{\partial^2 f(\mathbf{z}_k, \mathbf{z}_k^*)}{\partial \mathbf{z}_k^* \partial \mathbf{z}_k^T} & \dfrac{\partial^2 f(\mathbf{z}_k, \mathbf{z}_k^*)}{\partial \mathbf{z}_k^* \partial \mathbf{z}_k^H} & \mathbf{A}^T \\ \dfrac{\partial^2 f(\mathbf{z}_k, \mathbf{z}_k^*)}{\partial \mathbf{z}_k \partial \mathbf{z}_k^T} & \dfrac{\partial^2 f(\mathbf{z}_k, \mathbf{z}_k^*)}{\partial \mathbf{z}_k \partial \mathbf{z}_k^H} & \mathbf{O} \\ \mathbf{A} & \mathbf{O} & \mathbf{O} \end{bmatrix} \begin{bmatrix} \Delta\mathbf{z}_k \\ \Delta\mathbf{z}_k^* \\ \Delta\boldsymbol{\lambda}_k \end{bmatrix} = - \begin{bmatrix} \nabla_{\mathbf{z}_k} f(\mathbf{z}_k, \mathbf{z}_k^*) + \mathbf{A}^T\boldsymbol{\lambda}_k \\ \nabla_{\mathbf{z}_k^*} f(\mathbf{z}_k, \mathbf{z}_k^*) \\ \mathbf{A}\mathbf{z}_k - \mathbf{b} \end{bmatrix}. \tag{4.8.23}$$

The corresponding residual vector is

$$\mathbf{r}(\mathbf{z}_k, \mathbf{z}_k^*, \boldsymbol{\lambda}_k) = \begin{bmatrix} \nabla_{\mathbf{z}_k} f(\mathbf{z}_k, \mathbf{z}_k^*) + \mathbf{A}^T\boldsymbol{\lambda}_k \\ \nabla_{\mathbf{z}_k^*} f(\mathbf{z}_k, \mathbf{z}_k^*) \\ \mathbf{A}\mathbf{z}_k - \mathbf{b} \end{bmatrix}. \tag{4.8.24}$$

Replacing the residual computation formula (4.8.21) in step 1 of Algorithm 4.13

with (4.8.24), and replacing the Newton equation (4.8.20) in step 4 with the Newton equation (4.8.23), Algorithm 4.13 becomes an infeasible-start complex Newton algorithm. From (4.8.24) it is known that when the residual vector approximates the zero vector closely, $\mathbf{A}\mathbf{z}_k$ closely approximates $\mathbf{b}$, and thus the convergence point of the new algorithm must be a feasible point.

4.9 Original–Dual Interior-Point Method

It has been recognized [158] that the term "interior-point method" was presented by Fiacco and McCormick as early as 1968 in their pioneering work [152, p. 41]. However, suffering from the lack of a low-complexity optimization algorithm, the interior-point method did not gain greater development in the next decade. However, when Karmarkar [239] in 1984 proposed a linear programming algorithm having polynomial complexity, the continuous optimization field underwent tremendous changes. This has been described as the "interior-point revolution" [158] and has led to a fundamental change in the thinking on continuous optimization problems. The unrelated field of optimization considered previously for many years actually provides a unified theoretical framework.

4.9.1 Original–Dual Problems

Consider the standard form of nonlinear optimization problem

$$\min f(\mathbf{x}) \quad \text{subject to} \quad \mathbf{d}(\mathbf{x},\mathbf{z}) = \mathbf{0},\ \mathbf{z} \geq \mathbf{0}, \tag{4.9.1}$$

where $\mathbf{x} \in \mathbb{R}^n$, $f : \mathbb{R}^n \to \mathbb{R}$, $\mathbf{d} : \mathbb{R}^n \times \mathbb{R}^m \to \mathbb{R}^m$ and $\mathbf{z} \in \mathbb{R}^m_+$; and $f(\mathbf{x})$ and $d_i(\mathbf{x},\mathbf{z}), i = 1,\ldots,m$, are convex functions and are twice continuously differentiable.

The nonlinear optimization model (4.9.1) includes the following common optimization models.

1. If $\mathbf{d}(\mathbf{x},\mathbf{z}) = \mathbf{h}(\mathbf{x}) - \mathbf{z}$, where $\mathbf{z} \geq \mathbf{0}$, then (4.9.1) gives the nonlinear optimization model in [490]:

$$\min\ f(\mathbf{x}) \quad \text{subject to} \quad h_i(\mathbf{x}) \geq 0, \quad i = 1,\ldots,m. \tag{4.9.2}$$

2. If $\mathbf{d}(\mathbf{x},\mathbf{z}) = \mathbf{h}(\mathbf{x}) + \mathbf{z}$, $z_i = 0$, $i = 1,\ldots,p$, $h_{p+i} = g_i$, $z_{p+i} \geq 0$, $i = 1,\ldots,m-p$, then (4.9.1) is consistent with the nonlinear optimization model in [72]:

$$\min f(\mathbf{x}) \quad \text{subject to} \quad h_i(\mathbf{x}) = 0, i = 1,\ldots,p,\ g_i(\mathbf{x}) \leq 0 \tag{4.9.3}$$

for $i = 1,\ldots,m-p$.

3. If $\mathbf{d}(\mathbf{x},\mathbf{z}) = \mathbf{c}(\mathbf{x}) - \mathbf{z}$ and $z_i = 0, i \in \mathcal{E}$, $z_i \geq 0$, $i \in \mathcal{I}$, then (4.9.1) gives the nonlinear optimization model in [158]:

$$\min\ f(\mathbf{x}) \quad \text{subject to} \quad c_i(\mathbf{x}) = 0,\ i \in \mathcal{E},\ c_i(\mathbf{x}) \geq 0,\ i \in \mathcal{I}, \tag{4.9.4}$$

for $i = 1,\ldots,m$.

4. If $\mathbf{d}(\mathbf{x},\mathbf{z}) = \mathbf{c}(\mathbf{x}), \mathbf{z} = \mathbf{x} \geq \mathbf{0}$, then Equation (4.9.1) is consistent with the nonlinear optimization model in [497]:

$$\min f(\mathbf{x}) \quad \text{subject to} \quad \mathbf{c}(\mathbf{x}) = \mathbf{0}, \mathbf{x} \geq \mathbf{0}. \tag{4.9.5}$$

5. If $f(\mathbf{x}) = \mathbf{c}^T\mathbf{x}, \mathbf{z} = \mathbf{x} \geq \mathbf{0}, m = n$ and $\mathbf{d}(\mathbf{x},\mathbf{z}) = \mathbf{A}\mathbf{x} - \mathbf{b}$ then (4.9.1) gives the linear programming model in [416]:

$$\min \mathbf{c}^T\mathbf{x} \quad \text{subject to} \quad \mathbf{A}\mathbf{x} = \mathbf{b},\ \mathbf{x} \geq \mathbf{0}. \tag{4.9.6}$$

For the original inequality constrained nonlinear optimization problem,

$$(P) \qquad \min f(\mathbf{x}) \quad \text{subject to} \quad \mathbf{h}(\mathbf{x}) \geq \mathbf{0} \quad (\text{where } \mathbf{h}: \mathbb{R}^n \to \mathbb{R}^m), \tag{4.9.7}$$

its dual problem can be represented as [490]

$$\begin{aligned} (D) \quad & \max \left\{ \mathcal{L}(\mathbf{x},\mathbf{y}) = f(\mathbf{x}) - \mathbf{y}^T\mathbf{h}(\mathbf{x}) + \left((\nabla\mathbf{h}(\mathbf{x}))^T\mathbf{y} - \nabla f(\mathbf{x}) \right]^T \mathbf{x} \right\} \\ & \text{subject to} \quad (\nabla\mathbf{h}(\mathbf{x}))^T\mathbf{y} = \nabla f(\mathbf{x}),\ \mathbf{y} \geq \mathbf{0}, \end{aligned} \tag{4.9.8}$$

where $\mathbf{x} \in \mathbb{R}^n$ is an original variable and $\mathbf{y} \in \mathbb{R}^m$ is a dual variable.

4.9.2 First-Order Original–Dual Interior-Point Method

In order to better understand the interior-point method, we take first the *linear programming (LP) problem*

$$\min \left\{ \mathbf{c}^T\mathbf{x} \middle| \mathbf{A}\mathbf{x} = \mathbf{b},\ \mathbf{x} \geq \mathbf{0} \right\} \tag{4.9.9}$$

as the object of discussion.

From Equation (4.9.8) it is known that the dual of the above linear programming problem is given by

$$\max \left\{ \mathbf{b}^T\mathbf{y} \middle| \mathbf{A}^T\mathbf{y} + \mathbf{z} = \mathbf{c},\ \mathbf{z} \geq \mathbf{0} \right\}, \tag{4.9.10}$$

where $\mathbf{x} \in \mathbb{R}_+^n, \mathbf{y} \in \mathbb{R}^m, \mathbf{z} \in \mathbb{R}_+^n$ are the original variable, the dual variable and the slack variable, respectively; here $\mathbf{A} \in \mathbb{R}^{m\times n}$ and $\mathbf{b} \in \mathbb{R}^m, \mathbf{c} \in \mathbb{R}^n$. Without loss of generality, we assume that the matrix $\mathbf{A}$ has full row rank, i.e., $\text{rank}(\mathbf{A}) = m$.

By the first-order optimal condition, it is easy to obtain the KKT equations for the original–dual problem as follows:

$$\mathbf{A}\mathbf{x} = \mathbf{b},\ \mathbf{x} \geq \mathbf{0}, \quad \mathbf{A}^T\mathbf{y} + \mathbf{z} = \mathbf{c},\ \mathbf{z} \geq \mathbf{0}, \quad x_i z_i = 0,\ i = 1,\dots,n.$$

The first two conditions are the first-order optimal condition and the feasibility condition for the original–dual problem, respectively, and the third condition is the complementarity condition. Since the first two conditions contain nonnegative requirements, these KKT equations can be solved only by iteration methods.

The complementarity condition $x_i z_i = 0$ can be replaced by a *centering condition* $x_i z_i = \mu, i = 1,\dots,n$, where $\mu > 0$. The relationship $x_i z_i = \mu, \forall\, i = 1,\dots,m$, is also called the *perturbed complementarity condition* or *complementary slackness.* Obviously, if $\mu \to 0$ then $x_i z_i \to 0$.

Replacing the complementarity condition with perturbed complementarity, the KKT equations can be equivalently written as

$$\begin{aligned} \mathbf{Ax} &= \mathbf{b}, \quad \mathbf{x} \geq \mathbf{0}, \\ \mathbf{A}^T\mathbf{y} + \mathbf{z} &= \mathbf{c}, \quad \mathbf{z} \geq \mathbf{0}, \\ \mathbf{x}^T\mathbf{z} &= n\mu. \end{aligned} \tag{4.9.11}$$

Definition 4.18 [417] If the original problem has a feasible solution $\mathbf{x} > \mathbf{0}$ and the dual problem has also a solution $(\mathbf{y}, \mathbf{z})$ with $\mathbf{z} > \mathbf{0}$ then the original–dual optimization problem is said to meet the *interior-point condition* (IPC).

If the interior-point condition is satisfied then the solution of Equation (4.9.11) is denoted $(\mathbf{x}(\mu), \mathbf{y}(\mu), \mathbf{z}(\mu))$ and is called the *μ-center* of the original problem (P) and the dual problem (D). The set of all μ-centers is known as the *center path* of the original–dual problem.

Substituting $\mathbf{x}_k = \mathbf{x}_{k-1} + \Delta\mathbf{x}_k, \mathbf{y}_k = \mathbf{y}_{k-1} + \Delta\mathbf{y}_k$ and $\mathbf{z}_k = \mathbf{z}_{k-1} + \Delta\mathbf{z}_k$ into Equation (4.9.11), we have [416]

$$\begin{aligned} \mathbf{A}\Delta\mathbf{x}_k &= \mathbf{b} - \mathbf{A}\mathbf{x}_{k-1}, \\ \mathbf{A}^T\Delta\mathbf{y}_k + \Delta\mathbf{z}_k &= \mathbf{c} - \mathbf{A}^T\mathbf{y}_{k-1} - \mathbf{z}_{k-1}, \\ \mathbf{z}_{k-1}^T\Delta\mathbf{x}_k + \mathbf{x}_k^T\Delta\mathbf{z}_{k-1} &= n\mu - \mathbf{x}_{k-1}^T\mathbf{z}_{k-1}, \end{aligned}$$

equivalently written as

$$\begin{bmatrix} \mathbf{A} & \mathbf{O} & \mathbf{O} \\ \mathbf{O} & \mathbf{A}^T & \mathbf{I} \\ \mathbf{z}_{k-1}^T & \mathbf{0}^T & \mathbf{x}_{k-1}^T \end{bmatrix} \begin{bmatrix} \Delta\mathbf{x}_k \\ \Delta\mathbf{y}_k \\ \Delta\mathbf{z}_k \end{bmatrix} = \begin{bmatrix} \mathbf{b} - \mathbf{A}\mathbf{x}_{k-1} \\ \mathbf{c} - \mathbf{A}^T\mathbf{y}_{k-1} - \mathbf{z}_{k-1} \\ n\mu - \mathbf{x}_{k-1}^T\mathbf{z}_{k-1} \end{bmatrix}. \tag{4.9.12}$$

The interior-point method based on the first-order optimal condition is called the *first-order interior-point method*, whose key step is to solve the KKT equation (4.9.12) to obtain the Newton step $(\Delta\mathbf{x}_k, \Delta\mathbf{y}_k, \Delta\mathbf{z}_k)$. This Newton step can be implemented by finding the inverse matrix directly, but an approach with better numerical performance is to use an iteration method for solving the KKT equation (4.9.12). Notice that since the leftmost matrix is not real symmetric, the conjugate gradient method and the preconditioned conjugate gradient method are unable to be used.

Algorithm 4.14 gives a feasible-start original–dual interior-point algorithm.

The above algorithm requires $(\mathbf{x}_0, \mathbf{y}_0, \mathbf{z}_0)$ to be a feasible point. This feasible point can be determined by using Algorithm 4.15.

Algorithm 4.14 Feasible-start original–dual interior-point algorithm [416]

input: Accuracy parameter $\epsilon > 0$, updating parameter $\theta, 0 < \theta < 1$, feasible point $(\mathbf{x}_0, \mathbf{y}_0, \mathbf{z}_0)$.
initialization: Take $\mu_0 = \mathbf{x}_0^T \mathbf{z}_0 / n$, and set $k = 1$.
repeat
1. Solve the KKT equation (4.9.12) to get the solution $(\Delta \mathbf{x}_k, \Delta \mathbf{y}_k, \Delta \mathbf{z}_k)$.
2. Update $\mu_k = (1 - \theta)\mu_{k-1}$.
3. Update the original variable, dual variable and Lagrange multiplier:
$$\mathbf{x}_k = \mathbf{x}_{k-1} + \mu_k \Delta \mathbf{x}_k, \quad \mathbf{y}_k = \mathbf{y}_{k-1} + \mu_k \Delta \mathbf{y}_k, \quad \mathbf{z}_k = \mathbf{z}_{k-1} + \mu_k \Delta \mathbf{z}_k.$$

exit if $\mathbf{x}_k^T \mathbf{z}_k < \epsilon$.
return $k \leftarrow k + 1$.
output: $(\mathbf{x}, \mathbf{y}, \mathbf{z}) \leftarrow (\mathbf{x}_k, \mathbf{y}_k, \mathbf{z}_k)$

Algorithm 4.15 Feasible-point algorithm [416]

input: Accuracy parameter $\epsilon > 0$, penalty update parameter $\theta\,(0 < \theta < 1)$ and threshold parameter $\tau > 0$.
initialization: $\mathbf{x}_0 > \mathbf{0}, \mathbf{z}_0 > \mathbf{0}, \mathbf{y}_0, \mathbf{x}_0^T \mathbf{z}_0 = n\mu_0$. Put $k = 1$.
repeat
1. Compute the residuals $\mathbf{r}_b^{k-1} = \mathbf{b} - \mathbf{A}^T \mathbf{x}_{k-1}$, $\mathbf{r}_c^{k-1} = \mathbf{c} - \mathbf{A}^T \mathbf{y}_{k-1} - \mathbf{z}_{k-1}$.
2. μ-update $\nu_{k-1} = \mu_{k-1}/\mu_0$.
3. Solve the KKT equation
$$\begin{bmatrix} \mathbf{A} & \mathbf{O} & \mathbf{O} \\ \mathbf{O} & \mathbf{A}^T & \mathbf{I} \\ \mathbf{z}_{k-1}^T & \mathbf{0}^T & \mathbf{x}_{k-1}^T \end{bmatrix} \begin{bmatrix} \Delta^f \mathbf{x}_k \\ \Delta^f \mathbf{y}_k \\ \Delta^f \mathbf{z}_k \end{bmatrix} = \begin{bmatrix} \theta \nu_{k-1} \mathbf{r}_b^0 \\ \theta \nu_{k-1} \mathbf{r}_c^0 \\ n\mu - \mathbf{x}_{k-1}^T \mathbf{z}_{k-1} \end{bmatrix}.$$
4. Update $(\mathbf{x}_k, \mathbf{y}_k, \mathbf{z}_k) = (\mathbf{x}_{k-1}, \mathbf{y}_{k-1}, \mathbf{z}_{k-1}) + (\Delta^f \mathbf{x}_k, \Delta^f \mathbf{y}_k, \Delta^f \mathbf{z}_k)$.
5. **exit if** $\max\{\mathbf{x}_k^T \mathbf{z}_k,\ \mathbf{b} - \mathbf{A}^T \mathbf{x}_k,\ \mathbf{c} - \mathbf{A}^T \mathbf{y}_k - \mathbf{z}_k\} \le \epsilon$.

return $k \leftarrow k + 1$.
output: $(\mathbf{x}, \mathbf{y}, \mathbf{z}) \leftarrow (\mathbf{x}_k, \mathbf{y}_k, \mathbf{z}_k)$.

4.9.3 Second-Order Original–Dual Interior-Point Method

In the first-order original–dual interior-point method there exists the following shortcomings.

- The matrix in the KKT equation is not symmetric, so it is difficult to apply the conjugate gradient method or the preconditioned conjugate gradient method for solving the KKT equation.
- The KKT equation uses only the first-order optimal condition; it does not use the second-order information provided by the Hessian matrix.
- It is difficult to ensure that iteration points are interior points.
- It is inconvenient to extend this method to the general nonlinear optimization problem.

In order to overcome the above shortcomings, the *original–dual interior-point method* for nonlinear optimization consists of the following three basic elements.

(1) *Barrier function* Limit the variable $\mathbf{x}$ to interior-points.
(2) *Newton method* Solve the KKT equation efficiently.
(3) *Backtracking line search* Determine a suitable step size.

This interior-point method, combining the barrier function with the Newton method, is known as the *second-order original–dual interior-point method.*

Define a slack variable $\mathbf{z} \in \mathbb{R}^m$ which is a nonnegative vector $\mathbf{z} \geq \mathbf{0}$ such that $\mathbf{h}(\mathbf{x}) - \mathbf{z} = \mathbf{0}$. Then the inequality constrained original problem (P) can be expressed as the equality constrained optimization problem

$$\min f(\mathbf{x}) \quad \text{subject to} \quad \mathbf{h}(\mathbf{x}) = \mathbf{z}, \; \mathbf{z} \geq \mathbf{0}. \tag{4.9.13}$$

To remove the inequality $\mathbf{z} \geq \mathbf{0}$, we introduce the Fiacco–McCormick logarithmic barrier

$$b_\mu(\mathbf{x}, \mathbf{z}) = f(\mathbf{x}) - \mu \sum_{i=1}^{m} \log z_i, \tag{4.9.14}$$

where $\mu > 0$ is the barrier parameter. For very small μ, the barrier function $b_\mu(\mathbf{x}, \mathbf{z})$ and the original objective function $f(\mathbf{x})$ have the same role at points for which the constraints are not close to zero.

The equality constrained optimization problem (4.9.13) can be expressed as the unconstrained optimization problem

$$\min_{\mathbf{x}, \mathbf{z}, \boldsymbol{\lambda}} \left\{ \mathcal{L}_\mu(\mathbf{x}, \mathbf{z}, \boldsymbol{\lambda}) = f(\mathbf{x}) - \mu \sum_{i=1}^{m} \log z_i - \boldsymbol{\lambda}^T(\mathbf{h}(\mathbf{x}) - \mathbf{z}) \right\}, \tag{4.9.15}$$

where $\mathcal{L}_\mu(\mathbf{x}, \mathbf{z}, \boldsymbol{\lambda})$ is the Lagrangian objective function and $\boldsymbol{\lambda} \in \mathbb{R}^m$ is the Lagrange multiplier or the dual variable.

Denote

$$\mathbf{Z} = \mathbf{Diag}(z_1, \ldots, z_m), \quad \boldsymbol{\Lambda} = \mathbf{Diag}(\lambda_1, \ldots, \lambda_m). \tag{4.9.16}$$

Let $\nabla_x \mathcal{L}_\mu = \dfrac{\partial \mathcal{L}_\mu}{\partial \mathbf{x}^T} = \mathbf{0}$, $\nabla_z \mathcal{L}_\mu = \dfrac{\partial \mathcal{L}_\mu}{\partial \mathbf{z}^T} = \mathbf{0}$ and $\nabla_\lambda \mathcal{L}_\mu = \dfrac{\partial \mathcal{L}_\mu}{\partial \boldsymbol{\lambda}^T} = \mathbf{0}$. Then, the first-order optimal condition is given by

$$\begin{aligned} \nabla f(\mathbf{x}) - (\nabla h(\mathbf{x}))^T \boldsymbol{\lambda} &= \mathbf{0}, \\ -\mu \mathbf{1} + \mathbf{Z}\boldsymbol{\Lambda}\mathbf{1} &= \mathbf{0}, \\ \mathbf{h}(\mathbf{x}) - \mathbf{z} &= \mathbf{0}. \end{aligned} \tag{4.9.17}$$

Here $\mathbf{1}$ is an m-dimensional vector with all entries equal to 1.

In order to derive the Newton equation for the unconstrained optimization problem (4.9.15), consider the Lagrangian objective function

$$\begin{aligned}\mathcal{L}_\mu(\mathbf{x}+\Delta\mathbf{x},\mathbf{z}+\Delta\mathbf{z},\boldsymbol{\lambda}+\Delta\boldsymbol{\lambda}) &= f(\mathbf{x}+\Delta\mathbf{x}) - \mu\sum_{i=1}^m \log(z_i+\Delta z_i)\\ &\quad - (\boldsymbol{\lambda}+\Delta\boldsymbol{\lambda})^T(\mathbf{h}(\mathbf{x}+\Delta\mathbf{x}) - \mathbf{z} - \Delta\mathbf{z}),\end{aligned}$$

where

$$\begin{aligned}f(\mathbf{x}+\Delta\mathbf{x}) &= f(\mathbf{x}) + (\nabla f(\mathbf{x}))^T\Delta\mathbf{x} + \frac{1}{2}(\Delta\mathbf{x})^T\nabla^2 f(\mathbf{x})\Delta\mathbf{x},\\ h_i(\mathbf{x}+\Delta\mathbf{x}) &= h_i(\mathbf{x}) + (\nabla h_i(\mathbf{x}))^T\Delta\mathbf{x} + \frac{1}{2}(\Delta\mathbf{x})^T\nabla^2 h_i(\mathbf{x})\Delta\mathbf{x},\end{aligned}$$

for $i = 1,\ldots,m$.

Letting $\nabla_x\mathcal{L}_\mu = \dfrac{\partial\mathcal{L}_\mu}{\partial(\Delta\mathbf{x})^T} = \mathbf{0}$, $\nabla_z\mathcal{L}_\mu = \dfrac{\partial\mathcal{L}_\mu}{\partial(\Delta\mathbf{z})^T} = \mathbf{0}$ and $\nabla_\lambda\mathcal{L}_\mu = \dfrac{\partial\mathcal{L}_\mu}{\partial(\Delta\boldsymbol{\lambda})^T} = \mathbf{0}$, the Newton equation is given by [490]

$$\begin{bmatrix}\mathbf{H}(\mathbf{x},\boldsymbol{\lambda}) & \mathbf{O} & -(\mathbf{A}(\mathbf{x}))^T\\ \mathbf{O} & \boldsymbol{\Lambda} & \mathbf{Z}\\ \mathbf{A}(\mathbf{x}) & -\mathbf{I} & \mathbf{O}\end{bmatrix}\begin{bmatrix}\Delta\mathbf{x}\\ \Delta\mathbf{z}\\ \Delta\boldsymbol{\lambda}\end{bmatrix} = \begin{bmatrix}-\nabla f(\mathbf{x}) + (\nabla h(\mathbf{x}))^T\boldsymbol{\lambda}\\ \mu\mathbf{1} - \mathbf{Z}\boldsymbol{\Lambda}\mathbf{1}\\ \mathbf{z} - \mathbf{h}(\mathbf{x})\end{bmatrix}, \tag{4.9.18}$$

where

$$\mathbf{H}(\mathbf{x},\boldsymbol{\lambda}) = \nabla^2 f(\mathbf{x}) - \sum_{i=1}^m \lambda_i\nabla^2 h_i(\mathbf{x}),\quad \mathbf{A}(\mathbf{x}) = \nabla\mathbf{h}(\mathbf{x}). \tag{4.9.19}$$

In (4.9.18), $\Delta\mathbf{x}$ is called the *optimality direction* and $\Delta\mathbf{z}$ the *centrality direction*, while $\Delta\boldsymbol{\lambda}$ is known as the *feasibility direction* [490]. The triple $(\Delta\mathbf{x},\Delta\mathbf{z},\Delta\boldsymbol{\lambda})$ gives the Newton step, i.e. the update search direction of the interior-point method.

From (4.9.18) it is easy to get [490]

$$\begin{bmatrix}-\mathbf{H}(\mathbf{x},\boldsymbol{\lambda}) & \mathbf{O} & (\mathbf{A}(\mathbf{x}))^T\\ \mathbf{O} & -\mathbf{Z}^{-1}\boldsymbol{\Lambda} & -\mathbf{I}\\ \mathbf{A}(\mathbf{x}) & -\mathbf{I} & \mathbf{O}\end{bmatrix}\begin{bmatrix}\Delta\mathbf{x}\\ \Delta\mathbf{z}\\ \Delta\boldsymbol{\lambda}\end{bmatrix} = \begin{bmatrix}\boldsymbol{\alpha}\\ -\boldsymbol{\beta}\\ \boldsymbol{\gamma}\end{bmatrix}, \tag{4.9.20}$$

where

$$\boldsymbol{\alpha} = \nabla f(\mathbf{x}) - (\nabla\mathbf{h}(\mathbf{x}))^T\boldsymbol{\lambda}, \tag{4.9.21}$$

$$\boldsymbol{\beta} = \mu\mathbf{Z}^{-1}\mathbf{1} - \boldsymbol{\lambda}, \tag{4.9.22}$$

$$\boldsymbol{\gamma} = \mathbf{z} - \mathbf{h}(\mathbf{x}). \tag{4.9.23}$$

The above three variables have the following meanings.

(1) The vector $\boldsymbol{\gamma}$ measures the original infeasibility: $\boldsymbol{\gamma} = \mathbf{0}$ means that $\mathbf{x}$ is a feasible point satisfying the equality constraint $\mathbf{h}(\mathbf{x}) - \mathbf{z} = \mathbf{0}$; otherwise, $\mathbf{x}$ is an infeasible point.

(2) The vector $\boldsymbol{\alpha}$ measures the dual infeasibility: $\boldsymbol{\alpha} = \mathbf{0}$ means that the dual variable $\boldsymbol{\lambda}$ satisfies the first-order optimal condition, and thus is feasible; otherwise, $\boldsymbol{\lambda}$ violates the first-order optimal condition and hence is infeasible.

(3) The vector $\boldsymbol{\beta}$ measures the complementary slackness: $\boldsymbol{\beta} = \mathbf{0}$, i.e., $\mu\mathbf{Z}^{-1}\mathbf{1} = \boldsymbol{\lambda}$, means that the complementary slackness $z_i\lambda_i = \mu, \forall\, i = 1, \ldots, m$ and complementarity is completely satisfied when $\mu = 0$. The smaller is the deviation of μ from zero, the smaller is the complementary slackness; the larger μ is, the larger is the complementary slackness.

Equation (4.9.20) can be decomposed into the following two parts:

$$\Delta\mathbf{z} = \boldsymbol{\Lambda}^{-1}\mathbf{Z}(\boldsymbol{\beta} - \Delta\boldsymbol{\lambda}) \tag{4.9.24}$$

and

$$\begin{bmatrix} -\mathbf{H}(\mathbf{x}, \boldsymbol{\lambda}) & (\mathbf{A}(\mathbf{x}))^T \\ \mathbf{A}(\mathbf{x}) & \mathbf{Z}\boldsymbol{\Lambda}^{-1} \end{bmatrix} \begin{bmatrix} \Delta\mathbf{x} \\ \Delta\boldsymbol{\lambda} \end{bmatrix} = \begin{bmatrix} \boldsymbol{\alpha} \\ \boldsymbol{\gamma} + \mathbf{Z}\boldsymbol{\Lambda}^{-1}\boldsymbol{\beta} \end{bmatrix}. \tag{4.9.25}$$

Importantly, the original Newton equation (4.9.20) becomes a sub-Newton equation (4.9.25) with smaller dimension.

The Newton equation (4.9.25) is easy to solve iteratively via the preconditioned conjugate gradient algorithm. Once the Newton step $(\Delta\mathbf{x}, \Delta\mathbf{z}, \Delta\boldsymbol{\lambda})$ is found from Equation (4.9.24) and the solution of Equation (4.9.25), we can perform the following updating formulae

$$\begin{aligned} \mathbf{x}_{k+1} &= \mathbf{x}_k + \eta\Delta\mathbf{x}_k, \\ \mathbf{z}_{k+1} &= \mathbf{z}_k + \eta\Delta\mathbf{z}_k, \\ \boldsymbol{\lambda}_{k+1} &= \boldsymbol{\lambda}_k + \eta\Delta\boldsymbol{\lambda}_k. \end{aligned} \tag{4.9.26}$$

Here η is the common step size of the updates.

1. *Modification of Convex Optimization Problems*

The key of the interior-point method is to ensure that each iteration point $\mathbf{x}_k$ is an interior point such that $\mathbf{h}(\mathbf{x}_k) > \mathbf{0}$. However, for nonquadratic convex optimization, to simply select step size is insufficient to ensure the positiveness of the nonnegative variables. For this reason, it is necessary to modify the convex optimization problem appropriately.

A simple modification of the convex optimization problem is achieved by introducing the *merit function*. Unlike the general cost function minimization, the minimization of the merit function has two purposes: as is necessary, to promote the appearance of iterative points close to the local minimum point of the objective function, but also to ensure that the iteration points are feasible points. Hence the main procedure using the merit function is twofold:

(1) transform the constrained optimization problem into an unconstrained optimization problem;

(2) reduce the merit function in order to move the iterative point closer to optimal solution of the original constrained optimization problem.

Consider the *classical Fiacco–McCormick merit function* [152]

$$\psi_{\rho,\mu}(\mathbf{x},\mathbf{z}) = f(\mathbf{x}) - \mu \sum_{i=1}^{m} \log z_i + \frac{\rho}{2}\|\mathbf{h}(\mathbf{x}) - \mathbf{z}\|_2^2. \tag{4.9.27}$$

Define the *dual normal matrix*

$$\mathbf{N}(\mathbf{x},\boldsymbol{\lambda},\mathbf{z}) = \mathbf{H}(\mathbf{x},\boldsymbol{\lambda}) + (\mathbf{A}(\mathbf{x}))^T \mathbf{Z}^{-1}\boldsymbol{\Lambda}\mathbf{A}(\mathbf{x}), \tag{4.9.28}$$

let $b(\mathbf{x},\boldsymbol{\lambda}) = f(\mathbf{x}) - \sum_{i=1}^{m} \lambda_i \log z_i$ represent the barrier function and let the dual normal matrix $\mathbf{N}(\mathbf{x},\boldsymbol{\lambda},\mathbf{z})$ be positive definite. Then it can be shown [490] that the search direction $(\Delta\mathbf{x},\Delta\boldsymbol{\lambda})$ determined by Equation (4.9.20) has the following properties.

- If $\gamma = \mathbf{0}$ then

$$\begin{bmatrix} \nabla_x b(\mathbf{x},\boldsymbol{\lambda}) \\ \nabla_\lambda b(\mathbf{x},\boldsymbol{\lambda}) \end{bmatrix}^T \begin{bmatrix} \Delta\mathbf{x} \\ \Delta\boldsymbol{\lambda} \end{bmatrix} \leq 0.$$

- There is a $\rho_{\min} \geq 0$ such that for each $\rho > \rho_{\min}$, the following inequality holds:

$$\begin{bmatrix} \nabla_x \psi_{\rho,\mu}(\mathbf{x},\mathbf{z}) \\ \nabla_z \psi_{\rho,\mu}(\mathbf{x},\mathbf{z}) \end{bmatrix}^T \begin{bmatrix} \Delta\mathbf{x} \\ \Delta\boldsymbol{\lambda} \end{bmatrix} \leq 0.$$

In the above two cases, the equality holds if and only if $(\mathbf{x},\mathbf{z})$ satisfies (4.9.17) for some $\boldsymbol{\lambda}$.

2. Modification of Nonconvex Optimization Problems

For a nonconvex optimization problem, the Hessian matrix $\mathbf{H}(\mathbf{x},\boldsymbol{\lambda})$ may not be positive semi-definite, hence the dual normal matrix $\mathbf{N}(\mathbf{x},\boldsymbol{\lambda},\mathbf{z})$ may not be positive definite. In this case, the Hessian matrix needs to be added a very small perturbation. That is to say, using $\tilde{\mathbf{H}}(\mathbf{x},\boldsymbol{\lambda}) = \mathbf{H}(\mathbf{x},\boldsymbol{\lambda}) + \delta\mathbf{I}$ to replace the Hessian matrix $\mathbf{H}(\mathbf{x},\boldsymbol{\lambda})$ in the dual normal matrix, we obtain

$$\tilde{\mathbf{N}}(\mathbf{x},\boldsymbol{\lambda},\mathbf{z}) = \mathbf{H}(\mathbf{x},\boldsymbol{\lambda}) + \delta\mathbf{I} + (\mathbf{A}(\mathbf{x}))^T \mathbf{Z}^{-1}\boldsymbol{\Lambda}\mathbf{A}(\mathbf{x}).$$

Although interior-point methods usually take only a few iterations to converge, they are difficult to handle for a higher-dimensional matrix because the complexity of computing the step direction is $O(m^6)$, where m is the dimension of the matrix. As a result, on a typical personal computer (PC), generic interior-point solvers cannot handle matrices with dimensions larger than $m = 10^2$. However, image and video processing often involve matrices of dimension $m = 10^4$ to 10^5 and web search and bioinformatics can easily involve matrices of dimension $m = 10^6$ and beyond. In these real applications, generic interior point solvers are too limited, and only first-order gradient methods are practical [298].

Exercises

4.1 Let $\mathbf{y}$ be a real-valued measurement data vector given by $\mathbf{y} = \alpha\mathbf{x}+\mathbf{v}$, where α is a real scalar, $\mathbf{x}$ represents a real-valued deterministic process, the additive noise vector $\mathbf{v}$ has zero-mean, and covariance matrix $\mathbf{R}_v = E\{\mathbf{v}\mathbf{v}^T\}$. Find an optimal filter vector $\mathbf{w}$ such that $\hat{\alpha} = \mathbf{w}^T\mathbf{y}$ is an unbiased estimator with minimum variance.

4.2 Let $f(t)$ be a given function. Consider the minimization of the quadratic function:

$$\min Q(x) = \min \int_0^1 (f(t) - x_0 - x_1 t - \cdots - x_n t^n)^2 \,dt.$$

Determine whether the matrix associated with the set of linear equations is ill-conditioned.

4.3 Given a function $f(\mathbf{X}) = \log|\det(\mathbf{X})|$, where $\mathbf{X} \in \mathbb{C}^{n\times n}$ and $|\det(\mathbf{X})|$ is the absolute value of the determinant $\det(\mathbf{X})$. Find the conjugate gradient matrix $\mathrm{d}f/\mathrm{d}\mathbf{X}^*$.

4.4 Consider the equation $\mathbf{y} = \mathbf{A}\theta + \mathbf{e}$, where $\mathbf{e}$ is an error vector. Define the weighted sum of squared errors $E_w \overset{\text{def}}{=} \mathbf{e}^H\mathbf{W}\mathbf{e}$, where $\mathbf{W}$ is a Hermitian positive definite matrix which weights the errors.

(a) Find the solution for the parameter vector θ such that E_w is minimized. This solution is called the weighted least squares estimate of θ.

(b) Use the LDL^H decomposition $\mathbf{W} = \mathbf{LDL}^H$ to prove that the weighted least squares criterion is equivalent to the prewhitening of the error or data vector.

4.5 Let a cost function be $f(\mathbf{w}) = \mathbf{w}^H\mathbf{R}_e\mathbf{w}$ and let the constraint on the filter be $\mathrm{Re}(\mathbf{w}^H\mathbf{x}) = b$, where b is a constant. Find the optimal filter vector $\mathbf{w}$.

4.6 Explain whether the following constrained optimization problems have solutions:

(a) $\min\,(x_1 + x_2)$ subject to $x_1^2 + x_2^2 = 2, 0 \le x_1 \le 1, 0 \le x_2 \le 1$;

(b) $\min\,(x_1 + x_2)$ subject to $x_1^2 + x_2^2 \le 1, x_1 + x_2 = 4$;

(c) $\min\,(x_1 x_2)$ subject to $x_1 + x_2 = 3$.

4.7 Consider the constrained optimization problem $\min\{(x-1)(y+1)\}$ subject to $x - y = 0$. Use the Lagrange multiplier method to show that the minimum point is $(1,1)$ and the Lagrange multiplier $\lambda = 1$. For a Lagrangian function $\psi(x,y) = (x-1)(y+1) - 1(x-y)$, show that $\psi(x,y)$ has a saddle point at $(0,0)$, i.e., the point $(0,0)$ cannot minimize $\psi(x,y)$.

4.8 Solve the constrained optimization problem $\min\{J(x,y,z) = x^2 + y^2 + z^2\}$ subject to $3x + 4y - z = 25$.

4.9 The steepest descent direction of an objective function $J(\mathbf{M})$ is given by $-\frac{\mathrm{d}J(\mathbf{M})}{\mathrm{d}\mathbf{M}}$. Given a positive definite matrix $\mathbf{W}$, determine which of the gradient matrices $-\mathbf{W}\frac{\mathrm{d}J(\mathbf{M})}{\mathrm{d}\mathbf{M}}$ and $-\frac{\mathrm{d}J(\mathbf{M})}{\mathrm{d}\mathbf{M}}\mathbf{W}$ is the steepest descent direction of $J(\mathbf{M})$, and write down the corresponding gradient descent algorithm.

4.10 Let an observation data vector be produced by the linear regression model $\mathbf{y} = \mathbf{X}\boldsymbol{\beta} + \boldsymbol{\epsilon}$, $E\{\boldsymbol{\epsilon}\} = \mathbf{0}$, $E\{\boldsymbol{\epsilon}\boldsymbol{\epsilon}^T\} = \sigma^2\mathbf{I}$. We want to design a filter matrix $\mathbf{A}$ whose output vector $\mathbf{e} = \mathbf{A}\mathbf{y}$ satisfies $E\{\mathbf{e}-\boldsymbol{\epsilon}\} = \mathbf{0}$, and minimizes $E\{(\mathbf{e}-\boldsymbol{\epsilon})^T(\mathbf{e}-\boldsymbol{\epsilon})\}$. Show that this optimization problem is equivalent to

$$\min\left\{\mathrm{tr}(\mathbf{A}^T\mathbf{A}) - 2\mathrm{tr}(\mathbf{A})\right\}$$

subject to $\mathbf{A}\mathbf{X} = \mathbf{O}$, where $\mathbf{O}$ is a zero matrix.

4.11 Show that the solution matrix of the optimization problem

$$\min\left\{\mathrm{tr}(\mathbf{A}^T\mathbf{A}) - 2\mathrm{tr}(\mathbf{A})\right\}$$

subject to $\mathbf{A}\mathbf{X} = \mathbf{O}$ (zero matrix) is given by $\hat{\mathbf{A}} = \mathbf{I} - \mathbf{X}\mathbf{X}^\dagger$.

4.12 Show that the unconstrained minimization problem

$$\min\left\{(\mathbf{y} - \mathbf{X}\boldsymbol{\beta})^T(\mathbf{V} + \mathbf{X}\mathbf{X}^T)^\dagger(\mathbf{y} - \mathbf{X}\boldsymbol{\beta})\right\}$$

and the constrained minimization problem

$$\min\left\{(\mathbf{y} - \mathbf{X}\boldsymbol{\beta})\mathbf{V}^\dagger(\mathbf{y} - \mathbf{X}\boldsymbol{\beta})\right\} \quad \text{subject to} \quad (\mathbf{I} - \mathbf{V}\mathbf{V}^\dagger)\mathbf{X}\boldsymbol{\beta} = (\mathbf{I} - \mathbf{V}\mathbf{V}^\dagger)\mathbf{y}$$

have the same solution vector $\boldsymbol{\beta}$.

4.13 Solve the constrained optimization problem

$$\min_{\mathbf{A}}\left\{\mathrm{tr}(\mathbf{A}^T\mathbf{A}) - 2\mathrm{tr}(\mathbf{A})\right\} \quad \text{subject to} \quad \mathbf{A}\mathbf{x} = \mathbf{0}.$$

4.14 Show that the constrained optimization problem

$$\min_{\mathbf{x}} \frac{1}{2}\mathbf{x}^T\mathbf{x} \quad \text{subject to} \quad \mathbf{C}\mathbf{x} = \mathbf{b}$$

has the unique solution $\mathbf{x}^* = \mathbf{C}^\dagger\mathbf{b}$.

4.15 Consider matrices $\mathbf{Y} \in \mathbb{R}^{n\times m}$, $\mathbf{Z} \in \mathbb{R}^{n\times(n-m)}$ and let their columns constitute a linearly independent set. If a solution vector subject to $\mathbf{A}\mathbf{x} = \mathbf{b}$ is expressed as $\mathbf{x} = \mathbf{Y}\mathbf{x}_Y + \mathbf{Z}\mathbf{x}_Z$, where $\mathbf{x}_Y$ and $\mathbf{x}_Z$ are some $m \times 1$ and $(n-m) \times 1$ vectors, respectively, show that $\mathbf{x} = \mathbf{Y}(\mathbf{A}\mathbf{Y})^{-1}\mathbf{b} + \mathbf{Z}\mathbf{x}_Z$.

4.16 If a constrained optimization problem is $\min\left\{\mathrm{tr}(\mathbf{A}\mathbf{V}\mathbf{A}^T)\right\}$ subject to $\mathbf{A}\mathbf{X} = \mathbf{W}$, show that $\mathbf{A} = \mathbf{W}(\mathbf{X}^T\mathbf{V}_0^\dagger\mathbf{X})^\dagger\mathbf{X}^T\mathbf{V}_0^\dagger + \mathbf{Q}(\mathbf{I} - \mathbf{V}_0\mathbf{V}_0^\dagger)$, where $\mathbf{V}_0 = \mathbf{V} + \mathbf{X}\mathbf{X}^T$ and $\mathbf{Q}$ is any matrix.

4.17 Assume for a matrix $\mathbf{X}$ that $\mathrm{rank}(\mathbf{X}) \leq r$. Show that, for all semi-orthogonal matrices $\mathbf{A}$ such that $\mathbf{A}^T\mathbf{A} = \mathbf{I}_r$ and all matrices $\mathbf{Z} \in \mathbb{R}^{n\times r}$, the following result is true:

$$\min_{\mathbf{X}} \mathrm{tr}\left\{(\mathbf{X} - \mathbf{Z}\mathbf{A}^T)(\mathbf{X} - \mathbf{Z}\mathbf{A}^T)^T\right\} = 0.$$

4.18 For $p > 1$ and $a_i \geq 0$, $i = 1, \ldots, n$, show that for each set of nonnegative real numbers $\{x_1, \ldots, x_n\}$ such that $\sum_{i=1}^n x_i^q = 1$, where $q = p/(p-1)$, the relationship

$$\sum_{i=1}^n a_i x_i \leq \left(\sum_{i=1}^n a_i^p \right)^{1/p}$$

is satisfied, and equality holds if and only if $a_1 = \cdots = a_n = 0$ or $x_i^q = a_i^p \left(\sum_{k=1}^p a_k^p\right)^{1/p}$, $i = 1, \ldots, n$. This relationship is called the representation theorem of $(\sum_{i=1}^p a_i^p)^{1/p}$.

4.19 Show that the solution matrix of the constrained optimization problem

$$\min \left\{ \mathrm{tr}(\mathbf{A}^T \mathbf{A}) - 2\mathrm{tr}(\mathbf{A}) \right\} \quad \text{subject to} \quad \mathbf{AX} = \mathbf{O} \quad \text{(the zero matrix)}$$

is given by $\hat{\mathbf{A}} = \mathbf{I} - \mathbf{XX}^\dagger$. (*Hint:* The matrix constraint condition $\mathbf{AX} = \mathbf{O}$ is equivalent to the scalar constraint condition $\mathrm{tr}(\mathbf{LAX}) = 0$, $\forall\, \mathbf{L}$.)

5

Singular Value Analysis

Beltrami (1835–1899) and Jordan (1838–1921) are recognized as the founders of singular value decomposition (SVD): Beltrami in 1873 published the first paper on SVD [34]. One year later, Jordan published his own independent derivation of SVD [236]. Now, the SVD and its generalizations are one of the most useful and efficient tools for numerical linear algebra, and are widely applied in statistical analysis, and physics and in applied sciences, such as signal and image processing, system theory, control, communication, computer vision and so on.

This chapter presents first the concept of the stability of numerical computations and the condition number of matrices in order to introduce the necessity of the SVD of matrices; then we discuss SVD and generalized SVD together with their numerical computation and applications.

5.1 Numerical Stability and Condition Number

In many applications such as information science and engineering, it is often necessary to consider an important problem: in the actual observation data there exist some uncertainties or errors, and, furthermore, numerical calculation of the data is always accompanied by error. What is the impact of these errors? Is a particular algorithm numerically stable for data processing? In order to answer these questions, the following two concepts are extremely important:

(1) the numerical stability of various kinds of algorithm;
(2) the condition number or perturbation analysis of the problem of interest.

Let f be some application problem, let $d^\star \in D$ be the data without noise or disturbance, where D denotes a data group, and let $f(d^\star) \in F$ represent the solution of f, where F is a solution set. Given observed data $d \in D$, we want to evaluate $f(d)$. Owing to background noise and/or observation error, $f(d)$ is usually different from $f(d^\star)$. If $f(d)$ is "close" to $f(d^\star)$ then the problem f is "well-conditioned". On the contrary, if $f(d)$ is obviously different from $f(d^\star)$ even when d is very close to $d^\star$ then we say that the problem f is "ill-conditioned". If there is no further information about the problem f, the term "approximation" cannot describe the situation accurately.

In perturbation theory, a method or algorithm for solving a problem f is said to be *numerically stable* if its sensitivity to a perturbation is not larger than the sensitivity inherent in the original problem. A stable algorithm can guarantee that the solution $f(d)$ with a slight perturbation Δd of d approximates closely the solution without the perturbation. More precisely, f is stable if the approximate solution $f(d)$ is close to the solution $f(d^\star)$ without perturbation for all $d \in D$ close to $d^\star$.

A mathematical description of numerical stability is discussed below.

First, consider the well-determined linear equation $\mathbf{Ax} = \mathbf{b}$, where the $n \times n$ coefficient matrix $\mathbf{A}$ has known entries and the $n \times 1$ data vector $\mathbf{b}$ is also known, while the $n \times 1$ vector $\mathbf{x}$ is an unknown parameter vector to be solved. Naturally, we will be interested in the stability of this solution: if the coefficient matrix $\mathbf{A}$ and/or the data vector $\mathbf{b}$ are perturbed, how will the solution vector $\mathbf{x}$ be changed? Does it maintain a certain stability? By studying the influence of perturbations of the coefficient matrix $\mathbf{A}$ and/or the data vector $\mathbf{b}$, we can obtain a numerical value, called the condition number, describing an important characteristic of the coefficient matrix $\mathbf{A}$.

For the convenience of analysis, we first assume that there is only a perturbation $\delta\mathbf{b}$ of the data vector $\mathbf{b}$, while the matrix $\mathbf{A}$ is stable. In this case, the exact solution vector $\mathbf{x}$ will be perturbed to $\mathbf{x} + \delta\mathbf{x}$, namely

$$\mathbf{A}(\mathbf{x} + \delta\mathbf{x}) = \mathbf{b} + \delta\mathbf{b}. \tag{5.1.1}$$

This implies that

$$\delta\mathbf{x} = \mathbf{A}^{-1}\delta\mathbf{b}, \tag{5.1.2}$$

since $\mathbf{Ax} = \mathbf{b}$. By the property of the Frobenius norm, Equation (5.1.2) gives

$$\|\delta\mathbf{x}\|_2 \leq \|\mathbf{A}^{-1}\|_F \|\delta\mathbf{b}\|_2. \tag{5.1.3}$$

Similarly, from the linear equation $\mathbf{Ax} = \mathbf{b}$ we get

$$\|\mathbf{b}\|_2 \leq \|\mathbf{A}\|_F \|\mathbf{x}\|_2. \tag{5.1.4}$$

From Equations (5.1.3) and (5.1.4) it is immediately clear that

$$\frac{\|\delta\mathbf{x}\|_2}{\|\mathbf{x}\|_2} \leq \left(\|\mathbf{A}\|_F \|\mathbf{A}^{-1}\|_F\right) \frac{\|\delta\mathbf{b}\|_2}{\|\mathbf{b}\|_2}. \tag{5.1.5}$$

Next, consider the influence of simultaneous perturbations $\delta\mathbf{x}$ and $\delta\mathbf{A}$. In this case, the linear equation becomes

$$(\mathbf{A} + \delta\mathbf{A})(\mathbf{x} + \delta\mathbf{x}) = \mathbf{b}.$$

From the above equation it can be derived that

$$\begin{aligned}\delta\mathbf{x} &= \left((\mathbf{A}+\delta\mathbf{A})^{-1}-\mathbf{A}^{-1}\right)\mathbf{b}\\ &= \{\mathbf{A}^{-1}\left(\mathbf{A}-(\mathbf{A}+\delta\mathbf{A})\right)(\mathbf{A}+\delta\mathbf{A})^{-1}\}\mathbf{b}\\ &= -\mathbf{A}^{-1}\delta\mathbf{A}(\mathbf{A}+\delta\mathbf{A})^{-1}\mathbf{b}\\ &= -\mathbf{A}^{-1}\delta\mathbf{A}(\mathbf{x}+\delta\mathbf{x})\end{aligned} \tag{5.1.6}$$

from which we have

$$\|\delta\mathbf{x}\|_2 \leq \|\mathbf{A}^{-1}\|_F\|\delta\mathbf{A}\|_F\|\mathbf{x}+\delta\mathbf{x}\|_2,$$

namely

$$\frac{\|\delta\mathbf{x}\|_2}{\|\mathbf{x}+\delta\mathbf{x}\|_2} \leq \left(\|\mathbf{A}\|_F\|\mathbf{A}^{-1}\|_F\right)\frac{\|\delta\mathbf{A}\|_F}{\|\mathbf{A}\|_F}. \tag{5.1.7}$$

Equations (5.1.5) and (5.1.7) show that the relative error of the solution vector $\mathbf{x}$ is proportional to the numerical value

$$\text{cond}(\mathbf{A}) = \|\mathbf{A}\|_F\|\mathbf{A}^{-1}\|_F, \tag{5.1.8}$$

where $\text{cond}(\mathbf{A})$ is called the *condition number* of the matrix $\mathbf{A}$ and is sometimes denoted $\kappa(\mathbf{A})$.

If a small perturbation of a coefficient matrix $\mathbf{A}$ causes only a small perturbation of the solution vector $\mathbf{x}$, then the matrix $\mathbf{A}$ is said to be a *well-conditioned.* If, however, a small perturbation of $\mathbf{A}$ causes a large perturbation of $\mathbf{x}$ then the matrix $\mathbf{A}$ is *ill-conditioned.* The condition number characterizes the influence of errors in the entries of $\mathbf{A}$ on the solution vector; hence it is an important index measuring the numerical stability of a set of linear equations.

Further, consider an over-determined linear equation $\mathbf{Ax} = \mathbf{b}$, where $\mathbf{A}$ is an $m \times n$ matrix with $m > n$. An over-determined equation has a unique linear least squares solution given by

$$\mathbf{A}^H\mathbf{A}\mathbf{x} = \mathbf{A}^H\mathbf{b}, \tag{5.1.9}$$

i.e., $\mathbf{x} = (\mathbf{A}^H\mathbf{A})^{-1}\mathbf{A}^H\mathbf{b}$. It is easy to show (see Subsection 5.2.2) that

$$\text{cond}\left(\mathbf{A}^H\mathbf{A}\right) = \left(\text{cond}(\mathbf{A})\right)^2. \tag{5.1.10}$$

From Equations (5.1.5) and (5.1.7) it can be seen that the effects of the data-vector error $\delta\mathbf{b}$ and the coefficient-matrix error $\delta\mathbf{A}$ on the solution-vector error for the over-determined equation (5.1.9) are each proportional to the square of the condition number of $\mathbf{A}$.

Example 5.1 Consider the matrix

$$\mathbf{A} = \begin{bmatrix} 1 & 1\\ \delta & 0\\ 0 & \delta \end{bmatrix},$$

where δ is small. The condition number of $\mathbf{A}$ is of the same order of magnitude as δ^{-1}. Since

$$\mathbf{B} = \mathbf{A}^H \mathbf{A} = \begin{bmatrix} 1+\delta^2 & 1 \\ 1 & 1+\delta^2 \end{bmatrix},$$

the condition number of $\mathbf{A}^H \mathbf{A}$ is of order δ^{-2}.

In addition, if we use the *QR decomposition* $\mathbf{A} = \mathbf{QR}$ to solve an over-determined equation $\mathbf{Ax} = \mathbf{b}$ then

$$\text{cond}(\mathbf{Q}) = 1, \quad \text{cond}(\mathbf{A}) = \text{cond}(\mathbf{Q}^H \mathbf{A}) = \text{cond}(\mathbf{R}), \tag{5.1.11}$$

since $\mathbf{Q}^H \mathbf{Q} = \mathbf{I}$. In this case the effects of the errors of $\mathbf{b}$ and $\mathbf{A}$ will be proportional to the condition number of $\mathbf{A}$, as shown in Equations (5.1.5) and (5.1.7), respectively.

The above fact tells us that, when solving an over-determined matrix equation, the QR decomposition method has better numerical stability (i.e., a smaller condition number) than the least squares method.

If the condition number of $\mathbf{A}$ is large then the matrix equation $\mathbf{Ax} = \mathbf{b}$ is ill-conditioned. In this case, for a data vector $\mathbf{b}$ close to the ideal vector $\mathbf{b}^\star$ without perturbation, the solution associated with $\mathbf{b}$ will be far from the ideal solution associated with $\mathbf{b}^\star$.

A more effective approach than QR decomposition for solving this class of ill-conditioned linear equations is the total least squares method (which will be presented in Chapter 6), whose basis is the singular value decomposition of matrices. This is the topic of the next section.

5.2 Singular Value Decomposition (SVD)

Singular value decomposition is one of the most basic and most important tools in modern numerical analysis (especially in numerical computations). This section presents the definition and geometric interpretation of SVD together with the properties of singular values.

5.2.1 Singular Value Decomposition

The SVD was proposed first by Beltrami in 1873 for a real square matrix $\mathbf{A}$ [34]. Consider the bilinear function

$$f(\mathbf{x}, \mathbf{y}) = \mathbf{x}^T \mathbf{A} \mathbf{y}, \quad \mathbf{A} \in \mathbb{R}^{n \times n};$$

by introducing the linear transformation $\mathbf{x} = \mathbf{U}\boldsymbol{\xi}$ and $\mathbf{y} = \mathbf{V}\boldsymbol{\eta}$ Beltrami transformed the bilinear function into $f(\mathbf{x}, \mathbf{y}) = \boldsymbol{\xi}^T \mathbf{S} \boldsymbol{\eta}$, where

$$\mathbf{S} = \mathbf{U}^T \mathbf{A} \mathbf{V}. \tag{5.2.1}$$

Beltrami observed that if both $\mathbf{U}$ and $\mathbf{V}$ are constrained to be orthogonal matrices then in their choice of entries there exist $n^2 - n$ degrees of freedom, respectively. Beltrami proposed using these degrees of freedom to make all off-diagonal entries of $\mathbf{S}$ zero, so that the matrix $\mathbf{S} = \mathbf{\Sigma} = \mathbf{Diag}(\sigma_1, \sigma_2, \ldots, \sigma_n)$ becomes diagonal. Then premultiplying (5.2.1) by $\mathbf{U}$, postmultiplying it by $\mathbf{V}^T$ and making use of the orthonormality of $\mathbf{U}$ and $\mathbf{V}$, we immediately get

$$\mathbf{A} = \mathbf{U}\mathbf{\Sigma}\mathbf{V}^T. \tag{5.2.2}$$

This is the *singular value decomposition* (SVD) of a real square matrix $\mathbf{A}$ obtained by Beltrami in 1873 [34]. In 1874, Jordan independently derived the SVD of real square matrices [236]. The history of the invention of SVD can be found in MacDuffee's book [309, p. 78] or in the paper of Stewart [454], which reviews the early development of SVD in detail.

Later, Autonne in 1902 [19] extended the SVD to complex square matrices, while Eckart and Young [141] in 1939 extended it further to general complex rectangular matrices; therefore, the SVD theorem for a complex rectangular matrix is usually called the Autonee–Eckart–Young theorem, stated below.

Theorem 5.1 (SVD) *Let $\mathbf{A} \in \mathbb{R}^{m\times n}$ (or $\mathbb{C}^{m\times n}$); then there exist orthogonal (or unitary) matrices $\mathbf{U} \in \mathbb{R}^{m\times m}$ (or $\mathbf{U} \in \mathbb{C}^{m\times m}$) and $\mathbf{V} \in \mathbb{R}^{n\times n}$ (or $\mathbf{V} \in \mathbb{C}^{n\times n}$) such that*

$$\mathbf{A} = \mathbf{U}\mathbf{\Sigma}\mathbf{V}^T \ (\text{or } \mathbf{U}\mathbf{\Sigma}\mathbf{V}^H), \quad \mathbf{\Sigma} = \begin{bmatrix} \mathbf{\Sigma}_1 & \mathbf{O} \\ \mathbf{O} & \mathbf{O} \end{bmatrix}, \tag{5.2.3}$$

where $\mathbf{\Sigma}_1 = \mathbf{Diag}(\sigma_1, \ldots, \sigma_r)$ with diagonal entries

$$\sigma_1 \geq \cdots \geq \sigma_r > 0, \quad r = \text{rank}(\mathbf{A}). \tag{5.2.4}$$

The above theorem was first shown by Eckart and Young [141] in 1939, but the proof by Klema and Laub [250] is simpler.

The numerical values $\sigma_1, \ldots, \sigma_r$ together with $\sigma_{r+1} = \cdots = \sigma_n = 0$ are called the *singular values* of the matrix $\mathbf{A}$.

Example 5.2 A singular value σ_i of an $m \times n$ matrix $\mathbf{A}$ is called a *single singular value* of $\mathbf{A}$ if $\sigma_i \neq \sigma_j$, $\forall j \neq i$.

Here are several explanations and remarks on singular values and SVD.

1. The SVD of a matrix $\mathbf{A}$ can be rewritten as the vector form

$$\mathbf{A} = \sum_{i=1}^{r} \sigma_i \mathbf{u}_i \mathbf{v}_i^H. \tag{5.2.5}$$

This expression is sometimes said to be the *dyadic decomposition* of $\mathbf{A}$ [179].

2. Suppose that the $n \times n$ matrix $\mathbf{V}$ is unitary. Postmultiply (5.2.3) by $\mathbf{V}$ to get $\mathbf{AV} = \mathbf{U\Sigma}$, whose column vectors are given by

$$\mathbf{A}\mathbf{v}_i = \begin{cases} \sigma_i \mathbf{u}_i, & i = 1, 2, \ldots, r, \\ 0, & i = r+1, r+2, \ldots, n. \end{cases} \tag{5.2.6}$$

Therefore the column vectors $\mathbf{v}_i$ of $\mathbf{V}$ are known as the *right singular vectors* of the matrix $\mathbf{A}$, and $\mathbf{V}$ is called the *right singular-vector matrix* of $\mathbf{A}$.

3. Suppose that the $m \times m$ matrix $\mathbf{U}$ is unitary. Premultiply (5.2.3) by $\mathbf{U}^H$ to yield $\mathbf{U}^H\mathbf{A} = \mathbf{\Sigma V}$, whose column vectors are given by

$$\mathbf{u}_i^H \mathbf{A} = \begin{cases} \sigma_i \mathbf{v}_i^T, & i = 1, 2, \ldots, r, \\ 0, & i = r+1, r+2, \ldots, n. \end{cases} \tag{5.2.7}$$

The column vectors $\mathbf{u}_i$ are called the *left singular vectors* of the matrix $\mathbf{A}$, and $\mathbf{U}$ is the *left singular-vector matrix* of $\mathbf{A}$.

4. When the matrix rank $r = \text{rank}(\mathbf{A}) < \min\{m, n\}$, because $\sigma_{r+1} = \cdots = \sigma_h = 0$ with $h = \min\{m, n\}$, the SVD formula (5.2.3) can be simplified to

$$\mathbf{A} = \mathbf{U}_r \mathbf{\Sigma}_r \mathbf{V}_r^H, \tag{5.2.8}$$

where $\mathbf{U}_r = [\mathbf{u}_1, \ldots, \mathbf{u}_r]$, $\mathbf{V}_r = [\mathbf{v}_1, \ldots, \mathbf{v}_r]$ and $\mathbf{\Sigma}_r = \mathbf{Diag}(\sigma_1, \ldots, \sigma_r)$. Equation (5.2.8) is called the *truncated singular value decomposition* or the *thin singular value decomposition* of the matrix $\mathbf{A}$. In contrast, Equation (5.2.3) is known as the *full singular value decomposition.*

5. Premultiplying (5.2.6) by $\mathbf{u}_i^H$, and noting that $\mathbf{u}_i^H \mathbf{u}_i = 1$, it is easy to obtain

$$\mathbf{u}_i^H \mathbf{A} \mathbf{v}_i = \sigma_i, \quad i = 1, 2, \ldots, \min\{m, n\}, \tag{5.2.9}$$

which can be written in matrix form as

$$\mathbf{U}^H \mathbf{A} \mathbf{V} = \mathbf{\Sigma} = \begin{bmatrix} \mathbf{\Sigma}_1 & \mathbf{O} \\ \mathbf{O} & \mathbf{O} \end{bmatrix}, \quad \mathbf{\Sigma}_1 = \mathbf{Diag}(\sigma_1, \ldots, \sigma_r). \tag{5.2.10}$$

Equations (5.2.3) and (5.2.10) are two definitive forms of SVD.

6. From (5.2.3) it follows that $\mathbf{A}\mathbf{A}^H = \mathbf{U}\mathbf{\Sigma}^2\mathbf{U}^H$. This shows that the singular value σ_i of an $m \times n$ matrix $\mathbf{A}$ is the positive square root of the corresponding nonnegative eigenvalue of the matrix product $\mathbf{A}\mathbf{A}^H$.

7. If the matrix $\mathbf{A}_{m \times n}$ has rank r then
   - the leftmost r columns of the $m \times m$ unitary matrix $\mathbf{U}$ constitute an orthonormal basis of the column space of the matrix $\mathbf{A}$, i.e., $\text{Col}(\mathbf{A}) = \text{Span}\{\mathbf{u}_1, \ldots, \mathbf{u}_r\}$;
   - the leftmost r columns of the $n \times n$ unitary matrix $\mathbf{V}$ constitute an orthonormal basis of the row space of $\mathbf{A}$ or the column space of $\mathbf{A}^H$, i.e., $\text{Row}(\mathbf{A}) = \text{Span}\{\mathbf{v}_1, \ldots, \mathbf{v}_r\}$;
   - the rightmost $n - r$ columns of $\mathbf{V}$ constitute an orthonormal basis of the null space of the matrix $\mathbf{A}$, i.e., $\text{Null}(\mathbf{A}) = \text{Span}\{\mathbf{v}_{r+1}, \ldots, \mathbf{v}_n\}$;

- the rightmost $m-r$ columns of $\mathbf{U}$ constitute an orthonormal basis of the null space of the matrix $\mathbf{A}^H$, i.e., $\text{Null}(\mathbf{A}^H)=\text{Span}\{\mathbf{u}_{r+1},\ldots,\mathbf{u}_m\}$.

5.2.2 Properties of Singular Values

Theorem 5.2 (Eckart–Young theorem) [141] *If the singular values of* $\mathbf{A} \in \mathbb{C}^{m\times n}$ *are given by*

$$\sigma_1 \geq \sigma_2 \geq \cdots \geq \sigma_r \geq 0, \quad r = \text{rank}(\mathbf{A}),$$

then

$$\sigma_k = \min_{\mathbf{E}\in\mathbb{C}^{m\times n}} \left(\|\mathbf{E}\|_{\text{spec}} | \text{rank}(\mathbf{A}+\mathbf{E}) \leq k-1\right), \quad k=1,\ldots,r, \tag{5.2.11}$$

and there is an error matrix $\mathbf{E}_k$ *with* $\|\mathbf{E}_k\|_{\text{spec}} = \sigma_k$ *such that*

$$\text{rank}(\mathbf{A}+\mathbf{E}_k) = k-1, \quad k=1,\ldots,r. \tag{5.2.12}$$

The Eckart–Young theorem shows that the singular value σ_k is equal to the minimum spectral norm of the error matrix $\mathbf{E}_k$ such that the rank of $\mathbf{A}+\mathbf{E}_k$ is $k-1$.

An important application of the Eckart–Young theorem is to the best rank-k approximation of the matrix $\mathbf{A}$, where $k < r = \text{rank}(\mathbf{A})$.

Define

$$\mathbf{A}_k = \sum_{i=1}^{k} \sigma_i \mathbf{u}_i \mathbf{v}_i^H, \quad k<r; \tag{5.2.13}$$

then $\mathbf{A}_k$ is the solution of the optimization problem

$$\mathbf{A}_k = \mathop{\arg\min}_{\text{rank}(\mathbf{X})=k} \|\mathbf{A}-\mathbf{X}\|_F^2, \quad k<r, \tag{5.2.14}$$

and the squared approximation error is given by

$$\|\mathbf{A}-\mathbf{A}_k\|_F^2 = \sigma_{k+1}^2 + \cdots + \sigma_r^2. \tag{5.2.15}$$

Theorem 5.3 [215], [226] *Let* $\mathbf{A}$ *be an* $m\times n$ *matrix with singular values* $\sigma_1 \geq \cdots \geq \sigma_r$*, where* $r = \min\{m,n\}$*. If the* $p\times q$ *matrix* $\mathbf{B}$ *is a submatrix of* $\mathbf{A}$ *with singular values* $\gamma_1 \geq \cdots \geq \gamma_{\min\{p,q\}}$ *then*

$$\sigma_i \geq \gamma_i, \quad i=1,\ldots,\min\{p,q\} \tag{5.2.16}$$

and

$$\gamma_i \geq \sigma_{i+(m-p)+(n-q)}, \quad i \leq \min\{p+q-m, p+q-n\}. \tag{5.2.17}$$

This is the *interlacing theorem for singular values.*

The singular values of a matrix are closely related to its norm, determinant and condition number.

1. Relationship between Singular Values and Norms

The spectral norm of a matrix $\mathbf{A}$ is equal to its maximum singular value, namely

$$\|\mathbf{A}\|_{\text{spec}} = \sigma_1. \tag{5.2.18}$$

Since the Frobenius norm $\|\mathbf{A}\|_F$ of a matrix $\mathbf{A}$ is of unitary invariance, i.e., $\|\mathbf{U}^H\mathbf{A}\mathbf{V}\|_F = \|\mathbf{A}\|_F$, we have

$$\|\mathbf{A}\|_F = \left(\sum_{i=1}^{m}\sum_{j=1}^{n}|a_{ij}|^2\right)^{1/2} = \|\mathbf{U}^H\mathbf{A}\mathbf{V}\|_F = \|\mathbf{\Sigma}\|_F = \sqrt{\sigma_1^2 + \cdots + \sigma_r^2}. \tag{5.2.19}$$

That is to say, the Frobenius norm of any matrix is equal to the positive square root of the sum of all nonzero squared singular values of the matrix.

2. Relationship between Singular Values and Determinant

Let $\mathbf{A}$ be an $n \times n$ matrix. Since the absolute value of the determinant of any unitary matrix is equal to 1, from Theorem 5.1 it follows that

$$|\det(\mathbf{A})| = |\det(\mathbf{U}\mathbf{\Sigma}\mathbf{V}^H)| = |\det\mathbf{\Sigma}| = \sigma_1 \cdots \sigma_n. \tag{5.2.20}$$

If all the σ_i are nonzero then $|\det(\mathbf{A})| \neq 0$, which shows that $\mathbf{A}$ is nonsingular. If there is at least one singular value $\sigma_i = 0\,(i > r)$ then $\det(\mathbf{A}) = 0$, namely $\mathbf{A}$ is singular. This is the reason why the $\sigma_i, i = 1, \ldots, \min\{m,n\}$ are known as the singular values.

3. Relationship between Singular Values and Condition Number

For a given $m \times n$ matrix $\mathbf{A}$, its condition number can be defined by its singular values as

$$\text{cond}(\mathbf{A}) = \sigma_1/\sigma_p, \quad p = \min\{m,n\}. \tag{5.2.21}$$

From the definition formula (5.2.21) it can be seen that the condition number is a positive number larger than or equal to 1 since $\sigma_1 \geq \sigma_p$. Obviously, because there is at least one singular value $\sigma_p = 0$, the condition number of a singular matrix is infinitely large. If the condition number is not infinite but is large then the matrix $\mathbf{A}$ is said to be nearly singular. This implies that when the condition number is large, the linear dependence between the row (or column) vectors is strong. On the other hand, the condition number of an orthogonal or unitary matrix is equal to 1 by Equation (5.1.8). In this sense, the orthogonal or unitary matrix is of "ideal condition".

For an over-determined linear equation $\mathbf{A}\mathbf{x} = \mathbf{b}$, the SVD of $\mathbf{A}^H\mathbf{A}$ is

$$\mathbf{A}^H\mathbf{A} = \mathbf{V}\mathbf{\Sigma}^2\mathbf{V}^H, \tag{5.2.22}$$

i.e., the maximum and minimum singular values of the matrix $\mathbf{A}^H\mathbf{A}$ are respectively the squares of the maximum and minimum singular values of $\mathbf{A}$, and thus

$$\text{cond}(\mathbf{A}^H\mathbf{A}) = \frac{\sigma_1^2}{\sigma_n^2} = (\text{cond}(\mathbf{A}))^2. \tag{5.2.23}$$

In other words, the condition number of the matrix $\mathbf{A}^H\mathbf{A}$ is the square of the condition number of $\mathbf{A}$. This implies that, for a given data matrix $\mathbf{X}$, its SVD has better numerical stability than the EVD of the covariance matrix $\mathbf{X}\mathbf{X}^H$.

The following are equality relationships for singular values [307].

1. $\mathbf{A}_{m\times n}$ and its Hermitian matrix $\mathbf{A}^H$ have the same nonzero singular values.
2. The nonzero singular values of a matrix $\mathbf{A}_{m\times n}$ are the positive square root of the nonzero eigenvalues of $\mathbf{A}\mathbf{A}^H$ or $\mathbf{A}^H\mathbf{A}$.
3. There is a single singular value $\sigma > 0$ of a matrix $\mathbf{A}_{m\times n}$ if and only if σ^2 is a single eigenvalue of $\mathbf{A}\mathbf{A}^H$ or $\mathbf{A}^H\mathbf{A}$.
4. If $p = \min\{m, n\}$, and $\sigma_1, \ldots, \sigma_p$ are the singular values of an $m \times n$ matrix $\mathbf{A}$, then $\text{tr}\left(\mathbf{A}^H\mathbf{A}\right) = \sum_{i=1}^{p} \sigma_i^2$.
5. The absolute value of the determinant of an $n \times n$ matrix is equal to the product of its singular values, namely $|\det(\mathbf{A})| = \sigma_1 \cdots \sigma_n$.
6. The spectral norm of a matrix $\mathbf{A}$ is equal to its maximum singular value, namely $\|\mathbf{A}\|_{\text{spec}} = \sigma_{\max}$.
7. If $m \geq n$ then, for a matrix $\mathbf{A}_{m\times n}$, one has

$$\sigma_{\min}(\mathbf{A}) = \min\left((\mathbf{x}^H\mathbf{A}^H\mathbf{A}\mathbf{x})^{1/2} \middle| \, \mathbf{x}^H\mathbf{x} = 1, \ \mathbf{x} \in \mathbb{C}^n\right),$$
$$\sigma_{\max}(\mathbf{A}) = \max\left((\mathbf{x}^H\mathbf{A}^H\mathbf{A}\mathbf{x})^{1/2} \middle| \, \mathbf{x}^H\mathbf{x} = 1, \ \mathbf{x} \in \mathbb{C}^n\right).$$

8. If an $m \times m$ matrix $\mathbf{A}$ is nonsingular, then

$$\frac{1}{\sigma_{\min}(\mathbf{A})} = \max\left(\left(\frac{\mathbf{x}^H(\mathbf{A}^{-1})^H\mathbf{A}^{-1}\mathbf{x}}{\mathbf{x}^H\mathbf{x}}\right)^{1/2} \middle| \, \mathbf{x} \neq \mathbf{0}, \ \mathbf{x} \in \mathbb{C}^n\right).$$

9. If $\sigma_1, \ldots, \sigma_p$ are the nonzero singular values of $\mathbf{A} \in \mathbb{C}^{m\times n}$, with $p = \min\{m, n\}$, then the matrix $\begin{bmatrix} \mathbf{O} & \mathbf{A} \\ \mathbf{A}^H & \mathbf{O} \end{bmatrix}$ has $2p$ nonzero singular values $\sigma_1, \ldots, \sigma_p, -\sigma_1, \ldots, -\sigma_p$ and $|m - n|$ zero singular values.
10. If $\mathbf{A} = \mathbf{U}\begin{bmatrix} \mathbf{\Sigma}_1 & \mathbf{O} \\ \mathbf{O} & \mathbf{O} \end{bmatrix}\mathbf{V}^H$ is the SVD of the $m \times n$ matrix $\mathbf{A}$ then the Moore–Penrose inverse matrix of $\mathbf{A}$ is given by

$$\mathbf{A}^\dagger = \mathbf{V}\begin{bmatrix} \mathbf{\Sigma}_1^{-1} & \mathbf{O} \\ \mathbf{O} & \mathbf{O} \end{bmatrix}\mathbf{U}^H.$$

11. When $\mathbf{P}$ and $\mathbf{Q}$ are respectively an $m \times m$ and an $n \times n$ unitary matrix, the SVD of $\mathbf{P}\mathbf{A}\mathbf{Q}^H$ is given by

$$\mathbf{P}\mathbf{A}\mathbf{Q}^H = \tilde{\mathbf{U}}\mathbf{\Sigma}\tilde{\mathbf{V}}^H \tag{5.2.24}$$

with $\tilde{\mathbf{U}} = \mathbf{P}\mathbf{U}$ and $\tilde{\mathbf{V}} = \mathbf{Q}\mathbf{V}$. That is to say, the two matrices $\mathbf{P}\mathbf{A}\mathbf{Q}^H$ and $\mathbf{A}$ have the same singular values, i.e., the singular values have the unitary invariance, but their singular vectors are different.

The following summarizes the inequality relations of singular values [214], [215], [275], [93], [307].

(1) If $\mathbf{A}$ and $\mathbf{B}$ are $m \times n$ matrices then, for $p = \min\{m, n\}$, one has

$$\sigma_{i+j-1}(\mathbf{A}+\mathbf{B}) \leq \sigma_i(\mathbf{A}) + \sigma_j(\mathbf{B}), \quad 1 \leq i, j \leq p, \; i+j \leq p+1.$$

In particular, when $j = 1$, $\sigma_i(\mathbf{A}+\mathbf{B}) \leq \sigma_i(\mathbf{A}) + \sigma_1(\mathbf{B})$ holds for $i = 1, \ldots, p$.

(2) Given $\mathbf{A}_{m\times n}$ and $\mathbf{B}_{m\times n}$, one has $\sigma_{\max}(\mathbf{A}+\mathbf{B}) \leq \sigma_{\max}(\mathbf{A}) + \sigma_{\max}(\mathbf{B})$.

(3) If $\mathbf{A}$ and $\mathbf{B}$ are the $m \times n$ matrices then

$$\sum_{j=1}^{p} [\sigma_j(\mathbf{A}+\mathbf{B}) - \sigma_j(\mathbf{A})]^2 \leq \|\mathbf{B}\|_F^2, \quad p = \min\{m, n\}.$$

(4) If the singular values of $\mathbf{A}$ satisfies $\sigma_1(\mathbf{A}) \geq \cdots \geq \sigma_m(\mathbf{A})$ then

$$\sum_{j=1}^{k} [\sigma_{m-k+j}(\mathbf{A})]^2 \leq \sum_{j=1}^{k} \mathbf{a}_j^H \mathbf{a}_j \leq \sum_{j=1}^{k} [\sigma_j(\mathbf{A})]^2, \quad k = 1, 2, \ldots, m.$$

(5) If $p = \min\{m, n\}$, and the singular values of $\mathbf{A}_{m\times n}$ and $\mathbf{B}_{m\times n}$ are arranged as $\sigma_1(\mathbf{A}) \geq \cdots \geq \sigma_p(\mathbf{A})$, $\sigma_1(\mathbf{B}) \geq \cdots \geq \sigma_p(\mathbf{B})$ and $\sigma_1(\mathbf{A}+\mathbf{B}) \geq \cdots \geq \sigma_p(\mathbf{A}+\mathbf{B})$, then

$$\sigma_{i+j-1}(\mathbf{A}\mathbf{B}^H) \leq \sigma_i(\mathbf{A})\sigma_j(\mathbf{B}), \quad 1 \leq i, j \leq p, \; i+j \leq p+1.$$

(6) If $\mathbf{B}$ is an $m \times (n-1)$ matrix obtained by deleting any column of an $m \times n$ matrix $\mathbf{A}$, and their singular values are arranged in descending order, then

$$\sigma_1(\mathbf{A}) \geq \sigma_1(\mathbf{B}) \geq \sigma_2(\mathbf{A}) \geq \sigma_2(\mathbf{B}) \geq \cdots \geq \sigma_h(\mathbf{A}) \geq \sigma_h(\mathbf{B}) \geq 0,$$

where $h = \min\{m, n-1\}$.

(7) The maximum singular value of a matrix $\mathbf{A}_{m\times n}$ satisfies the inequality

$$\sigma_{\max}(\mathbf{A}) \geq \left(\frac{1}{n}\mathrm{tr}(\mathbf{A}^H\mathbf{A})\right)^{1/2}.$$

5.2.3 Rank-Deficient Least Squares Solutions

In applications of singular value analysis, it is usually necessary to use a low-rank matrix to approximate a noisy or disturbed matrix. The following theorem gives an evaluation of the quality of approximation.

THEOREM 5.4 *Let $\mathbf{A} = \sum_{i=1}^{p} \sigma_i \mathbf{u}_i \mathbf{v}_i^T$ be the SVD of $\mathbf{A} \in \mathbb{R}^{m\times n}$, where $p =$ rank$(\mathbf{A})$. If $k < p$ and $\mathbf{A}_k = \sum_{i=1}^{k} \sigma_i \mathbf{u}_i \mathbf{v}_i^T$ is the rank-k approximation of $\mathbf{A}$, then the approximation quality can be respectively measured by the spectral norm or the Frobenius norm,*

$$\min_{\text{rank}(\mathbf{B})=k} \|\mathbf{A} - \mathbf{B}\|_{\text{spec}} = \|\mathbf{A} - \mathbf{A}_k\|_{\text{spec}} = \sigma_{k+1}, \tag{5.2.25}$$

or

$$\min_{\text{rank}(\mathbf{B})=k} \|\mathbf{A} - \mathbf{B}\|_F = \|\mathbf{A} - \mathbf{A}_k\|_F = \left(\sum_{i=k+1}^{q} \sigma_i^2\right)^{1/2} \tag{5.2.26}$$

with $q = \min\{m, n\}$.

Proof See, e.g., [140], [328], [226]. □

In signal processing and system theory, the most common system of linear equations $\mathbf{Ax} = \mathbf{b}$ is over-determined and rank-deficient; namely, the row number m of the matrix $\mathbf{A} \in \mathbb{C}^{m\times n}$ is larger than its column number n and $r = \text{rank}(\mathbf{A}) < n$. Let the SVD of $\mathbf{A}$ be given by $\mathbf{A} = \mathbf{U\Sigma V}^H$, where $\mathbf{\Sigma} = \mathbf{Diag}(\sigma_1, \ldots, \sigma_r, 0, \ldots, 0)$. Consider

$$\mathbf{G} = \mathbf{V\Sigma}^\dagger \mathbf{U}^H, \tag{5.2.27}$$

where $\mathbf{\Sigma}^\dagger = \mathbf{Diag}(1/\sigma_1, \ldots, 1/\sigma_r, 0, \ldots, 0)$. By the property of singular values it is known that $\mathbf{G}$ is the Moore–Penrose inverse matrix of $\mathbf{A}$. Hence

$$\hat{\mathbf{x}} = \mathbf{Gb} = \mathbf{V\Sigma}^\dagger \mathbf{U}^H \mathbf{b} \tag{5.2.28}$$

which can be represented as

$$\mathbf{x}_{\text{LS}} = \sum_{i=1}^{r} (\mathbf{u}_i^H \mathbf{b}/\sigma_i) \mathbf{v}_i. \tag{5.2.29}$$

The corresponding minimum residual is given by

$$\rho_{\text{LS}} = \|\mathbf{Ax}_{\text{LS}} - \mathbf{b}\|_2 = \left\|[\mathbf{u}_{r+1}, \ldots, \mathbf{u}_m]^H \mathbf{b}\right\|_2. \tag{5.2.30}$$

The approach to solving the least squares problem via SVD is simply known as the SVD method. Although, in theory, when $i > r$ the singular values $\sigma_i = 0$, the computed singular values $\hat{\sigma}_i$, $i > r$, are not usually equal to zero and sometimes even have quite a large perturbation. In these cases, an estimate of the matrix rank r is required. In signal processing and system theory, the rank estimate $\hat{r}$ is usually called the *effective rank*.

Effective-rank determination is carried out by one of the following two common methods.

1. Normalized Singular Value Method

Compute the *normalized singular values*

$$\bar{\sigma}_i = \frac{\hat{\sigma}_i}{\hat{\sigma}_1}, \tag{5.2.31}$$

and select the largest integer i satisfying the criterion $\bar{\sigma}_i \geq \epsilon$ as an estimate of the effective rank $\hat{r}$. Obviously, this criterion is equivalent to choosing the maximum integer i satisfying

$$\hat{\sigma}_i \geq \epsilon\hat{\sigma}_1 \tag{5.2.32}$$

as $\hat{r}$; here ϵ is a very small positive number, e.g., $\epsilon = 0.1$ or $\epsilon = 0.05$.

2. Norm Ratio Method

Let an $m \times n$ matrix $\mathbf{A}_k$ be the rank-k approximation to the original $m \times n$ matrix $\mathbf{A}$. Define the *Frobenius norm ratio* as

$$\nu(k) = \frac{\|\mathbf{A}_k\|_F}{\|\mathbf{A}\|_F} = \frac{\sqrt{\sigma_1^2 + \cdots + \sigma_k^2}}{\sqrt{\sigma_1^2 + \cdots + \sigma_h^2}}, \qquad h = \min\{m, n\}, \tag{5.2.33}$$

and choose the minimum integer k satisfying

$$\nu(k) \geq \alpha \tag{5.2.34}$$

as the effective rank estimate $\hat{r}$, where α is at the threshold of 1, e.g., $\alpha = 0.997$ or 0.998.

After the effective rank $\hat{r}$ has been determined via the above two criteria,

$$\hat{\mathbf{x}}_{\text{LS}} = \sum_{i=1}^{\hat{r}} (\hat{\mathbf{u}}_i^H \mathbf{b}/\hat{\sigma}_i)\hat{\mathbf{v}}_i \tag{5.2.35}$$

can be regarded as a reasonable approximation to the LS solution $\mathbf{x}_{\text{LS}}$. Clearly, this solution is the LS solution of the linear equation $\mathbf{A}_{\hat{r}}\mathbf{x} = \mathbf{b}$, where

$$\mathbf{A}_{\hat{r}} = \sum_{i=1}^{\hat{r}} \sigma_i \mathbf{u}_i \mathbf{v}_i^H. \tag{5.2.36}$$

In LS problems, using $\mathbf{A}_{\hat{r}}$ instead of $\mathbf{A}$ corresponds to filtering out the smaller singular values. This filtering is effective for a noisy matrix $\mathbf{A}$. It is easily seen that the LS solution $\hat{\mathbf{x}}_{\text{LS}}$ given by Equation (5.2.35) still contains n parameters. However, by the rank-deficiency of the matrix $\mathbf{A}$ we know that the unknown parameter vector $\mathbf{x}$ contains only r independent parameters; the other parameters are linearly dependent on r independent parameters. In many engineering applications, we naturally want to find r independent parameters other than the n parameters containing redundancy components. In other words, our objective is to estimate only the principal parameters and to eliminate minor components. This problem can be solved via the low-rank total least squares (TLS) method, which will be presented in Chapter 6.

5.3 Product Singular Value Decomposition (PSVD)

In the previous section we discussed the SVD of a single matrix. This section presents the SVD of the product of two matrices.

5.3.1 PSVD Problem

By the *product singular value decomposition* (PSVD) is meant the SVD of a matrix product $\mathbf{B}^T\mathbf{C}$.

Consider the matrix product

$$\mathbf{A} = \mathbf{B}^T\mathbf{C}, \quad \mathbf{B} \in \mathbb{C}^{p\times m}, \quad \mathbf{C} \in \mathbb{C}^{p\times n}, \quad \text{rank}(\mathbf{B}) = \text{rank}(\mathbf{C}) = p. \tag{5.3.1}$$

In principle, the PSVD is equivalent to the direct SVD of the matrix product. However, if we first compute the product of two matrices, and then calculate the PSVD then the smaller singular values are often considerably disturbed. To illustrate this, look at the following example [135].

Example 5.3 Let

$$\mathbf{B}^T = \begin{bmatrix} 1 & \xi \\ -1 & \xi \end{bmatrix}, \quad \mathbf{C} = \frac{1}{\sqrt{2}}\begin{bmatrix} 1 & 1 \\ -1 & 1 \end{bmatrix},$$

$$\mathbf{B}^T\mathbf{C} = \frac{1}{\sqrt{2}}\begin{bmatrix} 1-\xi & 1+\xi \\ -1-\xi & -1+\xi \end{bmatrix}. \tag{5.3.2}$$

Clearly, $\mathbf{C}$ is an orthogonal matrix and the two columns $[1,-1]^T$ and $[\xi,\xi]^T$ of $\mathbf{B}^T$ are orthogonal to each other. The real singular values of the matrix product $\mathbf{B}^T\mathbf{C}$ are $\sigma_1 = \sqrt{2}$ and $\sigma_2 = \sqrt{2}|\xi|$. However, if $|\xi|$ is smaller than the cutoff error ϵ, then the floating point calculation result of Equation (5.3.2) is $\mathbf{B}^T\mathbf{C} = \frac{1}{\sqrt{2}}\begin{bmatrix} 1 & 1 \\ -1 & -1 \end{bmatrix}$ whose singular values are $\hat{\sigma}_1 = \sqrt{2}$ and $\hat{\sigma}_2 = 0$. In this case, $\hat{\sigma}_1 = \sigma_1$ and $\hat{\sigma}_2 \approx \sigma_2$. If $|\xi| > 1/\epsilon$, then the floating point calculation result of the matrix product is $\mathbf{B}^T\mathbf{C} = \frac{1}{\sqrt{2}}\begin{bmatrix} -\xi & \xi \\ -\xi & \xi \end{bmatrix}$, whose computed singular values are $\hat{\sigma}_1 = \sqrt{2}\,|\xi|$ and $\hat{\sigma}_2 = 0$. In this case $\hat{\sigma}_1$ is clearly larger than σ_1, and $\hat{\sigma}_2$ is obviously different from σ_2.

Laub *et al.* [283] pointed out that when a linear system is close to uncontrollable and unobservable an accurate calculation for the small singular values is very important because, if a nonzero small singular value is calculated as zero, it may lead to a wrong conclusion: a minimum-phase system could be determined as a nonminimum phase system.

The above examples show that to use the direct SVD of the matrix product $\mathbf{B}^T\mathbf{C}$ is not numerically desirable. Hence, it is necessary to consider how to compute the SVD of $\mathbf{A} = \mathbf{B}^T\mathbf{C}$ such that the PSVD has an accuracy close to that of $\mathbf{B}$ and $\mathbf{C}$. This is the so-called *product singular value decomposition problem.*

Such a PSVD was first proposed by Fernando and Hammarling in 1988 [151], and is described in the following theorem.

THEOREM 5.5 (PSVD) [151] *Let $\mathbf{B}^T \in \mathbb{C}^{m\times p}$, $\mathbf{C} \in \mathbb{C}^{p\times n}$. Then there are two unitary matrices $\mathbf{U} \in \mathbb{C}^{m\times m}, \mathbf{V} \in \mathbb{C}^{n\times n}$ and a nonsingular matrix $\mathbf{Q} \in \mathbb{C}^{p\times p}$ such that*

$$\mathbf{U}\mathbf{B}^H\mathbf{Q} = \begin{bmatrix} \mathbf{I} & & \\ & \mathbf{O}_B & \\ & & \mathbf{\Sigma}_B \end{bmatrix}, \tag{5.3.3}$$

$$\mathbf{Q}^{-1}\mathbf{C}\mathbf{V}^H = \begin{bmatrix} \mathbf{O}_C & & \\ & \mathbf{I} & \\ & & \mathbf{\Sigma}_C \end{bmatrix}, \tag{5.3.4}$$

where $\mathbf{\Sigma}_B = \mathbf{Diag}(s_1, \ldots, s_r), 1 > s_1 \geq \cdots \geq s_r > 0$, $\mathbf{\Sigma}_C = \mathbf{Diag}(t_1, \ldots, t_r)$, $1 > t_1 \geq \cdots \geq t_r > 0$, and $s_i^2 + t_i^2 = 1$, $i = 1, \ldots, r$.

By Theorem 5.5, we can deduce that $\mathbf{U}\mathbf{B}^H\mathbf{C}\mathbf{V}^H = \mathbf{Diag}(\mathbf{O}_C, \mathbf{O}_B, \mathbf{\Sigma}_B\mathbf{\Sigma}_C)$. Hence the singular values of the matrix product $\mathbf{B}^H\mathbf{C}$ consist of both zero and nonzero singular values given by $s_i t_i$, $i = 1, \ldots, r$.

5.3.2 Accurate Calculation of PSVD

The basic idea of the accurate calculation algorithm for PSVD in [135] is as follows: after any matrix $\mathbf{A}$ is multiplied by an orthogonal matrix, the singular values of $\mathbf{A}$ remain unchanged. Hence, for matrices $\mathbf{B}$ and $\mathbf{C}$, if we let

$$\mathbf{B}' = \mathbf{T}\mathbf{B}\mathbf{U}, \qquad \mathbf{C}' = (\mathbf{T}^T)^{-1}\mathbf{C}\mathbf{V}, \tag{5.3.5}$$

where $\mathbf{T}$ is nonsingular and $\mathbf{U}$ and $\mathbf{V}$ are orthogonal matrices, then

$$\mathbf{B}'^T\mathbf{C}' = \mathbf{U}^T\mathbf{B}^T\mathbf{T}^T(\mathbf{T}^T)^{-1}\mathbf{C}\mathbf{V} = \mathbf{U}^T(\mathbf{B}^T\mathbf{C})\mathbf{V}$$

and $\mathbf{B}^T\mathbf{C}$ have the exactly same singular values (including zero singular values).

Given $\mathbf{B} \in \mathbb{R}^{p\times m}$ and $\mathbf{C} \in \mathbb{R}^{p\times n}$, $p \leq \min\{m, n\}$, denote the row vectors of $\mathbf{B}$ as $\mathbf{b}_i^T$, $i = 1, 2, \ldots, p$.

Drmac's PSVD algorithm is shown in Algorithm 5.1.

After performing Algorithm 5.1, we get $\mathbf{V}, \mathbf{Q}_F$ and $\begin{bmatrix} \mathbf{\Sigma} \oplus \mathbf{O} \\ \mathbf{O} \end{bmatrix}$, so the SVD of $\mathbf{B}^T\mathbf{C}$ is given by $\mathbf{U}(\mathbf{B}^T\mathbf{C})\mathbf{V}^T = \mathbf{\Sigma}$, where

$$\mathbf{U} = \begin{bmatrix} \mathbf{V}^T & \\ & \mathbf{I} \end{bmatrix} \mathbf{Q}_F^T, \quad \mathbf{\Sigma} = \begin{bmatrix} \mathbf{\Sigma} \oplus \mathbf{O} \\ \mathbf{O} \end{bmatrix}. \tag{5.3.6}$$

The above algorithm for computing the PSVD of two matrices has been extended to the accurate calculation of the PSVD of three matrices [137]; see Algorithm 5.2.

Algorithm 5.1 PSVD($\mathbf{B}, \mathbf{C}$) [135]

input: $\mathbf{B} \in \mathbb{R}^{p\times m}, \mathbf{C} \in \mathbb{R}^{p\times n}$, $p \leq \min\{m, n\}$. Denote $\mathbf{B}_\tau = [\mathbf{b}_1^\tau, \ldots, \mathbf{b}_p^\tau]$.

1. Compute $\mathbf{B}_\tau = \mathbf{Diag}(\|\mathbf{b}_1^\tau\|_2, \ldots, \|\mathbf{b}_p^\tau\|_2)$. Let $\mathbf{B}_1 = \mathbf{B}_\tau^\dagger \mathbf{B}, \mathbf{C}_1 = \mathbf{B}_\tau \mathbf{C}$.
2. Calculate the QR decomposition of $\mathbf{C}_1^T$
$$\mathbf{C}_1^T \mathbf{\Pi} = \mathbf{Q} \begin{bmatrix} \mathbf{R} \\ \mathbf{O} \end{bmatrix},$$
where $\mathbf{R} \in \mathbb{R}^{r\times p}$, $\text{rank}(\mathbf{R}) = r$ and $\mathbf{Q}$ is a $(n - r) \times p$ orthogonal matrix.
3. Compute $\mathbf{F} = \mathbf{B}_1^T \mathbf{\Pi} \mathbf{R}^T$.
4. Compute the QR decomposition of $\mathbf{F}$
$$\mathbf{F}\mathbf{\Pi}_F = \mathbf{Q}_F \begin{bmatrix} \mathbf{R}_F \\ \mathbf{O} \end{bmatrix}.$$
5. Compute the SVD $\mathbf{\Sigma} = \mathbf{V}^T \mathbf{R}_F \mathbf{W}$.

output: The SVD of $\mathbf{B}^T\mathbf{C}$ is
$$\begin{bmatrix} \mathbf{\Sigma} \oplus \mathbf{O} \\ \mathbf{O} \end{bmatrix} = \begin{bmatrix} \mathbf{V}^T & \\ & \mathbf{I} \end{bmatrix} \mathbf{Q}_F^T \left(\mathbf{B}^T\mathbf{C}\right) \left(\mathbf{Q}(\mathbf{W} \oplus \mathbf{I}_{n-p})\right),$$
where $\mathbf{A} \oplus \mathbf{D}$ represents the direct sum of $\mathbf{A}$ and $\mathbf{D}$.

Algorithm 5.2 PSVD of $\mathbf{B}^T\mathbf{S}\mathbf{C}$ [137]

input: $\mathbf{B} \in \mathbb{R}^{p\times m}$, $\mathbf{S} \in \mathbb{R}^{p\times q}$, $\mathbf{C} \in \mathbb{R}^{q\times n}$, $p \leq m$, $q \leq n$.

1. Compute
$\mathbf{B}_\tau = \mathbf{Diag}(\|\mathbf{b}_1^\tau\|_2, \ldots, \|\mathbf{b}_p^\tau\|_2)$ and $\mathbf{C}_\tau = \mathbf{Diag}(\|\mathbf{c}_1^\tau\|_2, \ldots, \|\mathbf{c}_q^\tau\|_2)$.
where $\mathbf{b}_i^\tau$ and $\mathbf{c}_j^\tau$ are the ith row vectors of $\mathbf{B}$ and $\mathbf{C}$, respectively.
2. Set $\mathbf{B}_1 = \mathbf{B}_\tau^\dagger \mathbf{B}, \mathbf{C}_1 = \mathbf{C}_\tau^\dagger \mathbf{C}, \mathbf{S}_1 = \mathbf{B}_\tau \mathbf{S} \mathbf{C}_\tau$.
3. Compute LU decomposition $\mathbf{\Pi}_1 \mathbf{S}_1 \mathbf{\Pi}_2 = \mathbf{L}\mathbf{U}$, where $\mathbf{L} \in \mathbb{R}^{p\times p}, \mathbf{U} \in \mathbb{R}^{p\times q}$.
4. Compute $\mathbf{M} = \mathbf{L}^T \mathbf{\Pi}_1 \mathbf{B}_1$ and $\mathbf{N} = \mathbf{U}\mathbf{\Pi}_2^T \mathbf{C}_1$.
5. Compute the SVD of $\mathbf{M}^T\mathbf{N}$ via Algorithm 5.1 to get $\mathbf{Q}, \mathbf{Q}_F, \mathbf{V}$ and $\mathbf{W}$.

output: The SVD of $\mathbf{B}^T\mathbf{S}\mathbf{C}$ is
$$\begin{bmatrix} \mathbf{\Sigma} \oplus \mathbf{O} \\ \mathbf{O} \end{bmatrix} = \begin{bmatrix} \mathbf{V}^T & \\ & \mathbf{I} \end{bmatrix} \mathbf{Q}_F^T \left(\mathbf{B}^T\mathbf{S}\mathbf{C}\right) \left(\mathbf{Q}(\mathbf{W} \oplus \mathbf{I}_{n-p})\right).$$

5.4 Applications of Singular Value Decomposition

The SVD has been widely applied for solving many engineering problems. This section presents two typical application examples of SVD, in static system modeling and in image compression.

5.4.1 Static Systems

Taking an electronic device as an example, we consider the SVD of a static system of voltages v_1 and v_2 and currents i_1 and i_2:

$$\underbrace{\begin{bmatrix} 1 & -1 & 0 & 0 \\ 0 & 0 & 1 & 1 \end{bmatrix}}_{\mathbf{F}} \begin{bmatrix} v_1 \\ v_2 \\ i_1 \\ i_2 \end{bmatrix} = \begin{bmatrix} 0 \\ 0 \end{bmatrix}. \tag{5.4.1}$$

In this model of a static system, the entries of the matrix $\mathbf{F}$ limits the allowable values of v_1, v_2, i_1, i_2.

If the measurement devices for voltages and currents have the same accuracy (for example 1%) then we can easily detect whether any set of measurements is a solution of Equation (5.4.1) within the desired range of accuracy. Let us assume that another method gives a static system model different from Equation (5.4.1):

$$\begin{bmatrix} 1 & -1 & 10^6 & 10^6 \\ 0 & 0 & 1 & 1 \end{bmatrix} \begin{bmatrix} v_1 \\ v_2 \\ i_1 \\ i_2 \end{bmatrix} = \begin{bmatrix} 0 \\ 0 \end{bmatrix}. \tag{5.4.2}$$

Obviously, only when the current measurements are very accurate, will a set of v_1, v_2, i_1, i_2 satisfy Equation (5.4.2) with suitable accuracy; for the general case where current measurements have perhaps 1% measurement errors, Equation (5.4.2) is quite different from the static system model (5.4.1). In this case, the voltage relationships given by Equations (5.4.1) and (5.4.2) are $v_1 - v_2 = 0$ and $v_1 - v_2 + 10^4 = 0$, respectively. However, from the algebraic viewpoint, Equations (5.4.1) and (5.4.2) are completely equivalent. Therefore, we hope to have some means to compare several algebraically equivalent model representations in order to determine which we want, i.e., one that can be used for the common static system model available in general rather than special circumstances. The basic mathematical tool for solving this problem is the SVD.

More generally, consider a matrix equation for a static system including n resistors [100]:

$$\mathbf{F} \begin{bmatrix} \mathbf{v} \\ \mathbf{i} \end{bmatrix} = 0. \tag{5.4.3}$$

Here $\mathbf{F}$ is an $m \times n$ matrix. In order to simplify the representation, some constant compensation terms are removed. Such a representation is very versatile, and can come from physical devices (e.g., linearized physical equations) and network equations. The action of the matrix $\mathbf{F}$ on the accurate and nonaccurate data parts can be analyzed using SVD. Let the SVD of $\mathbf{F}$ be

$$\mathbf{F} = \mathbf{U}^T \boldsymbol{\Sigma} \mathbf{V}. \tag{5.4.4}$$

Using a truncated SVD, the nonprecision parts of $\mathbf{F}$ can be removed, and thus the SVD of the matrix $\mathbf{F}$ will provide a design equation that is algebraically equivalent to but numerically more reliable than the original static system model. Since $\mathbf{U}$ is an orthogonal matrix, from Equations (5.4.3) and (5.4.4) we have

$$\boldsymbol{\Sigma}\mathbf{V}\begin{bmatrix}\mathbf{v}\\ \mathbf{i}\end{bmatrix} = \mathbf{0}. \tag{5.4.5}$$

If the diagonal matrix $\boldsymbol{\Sigma}$ is blocked,

$$\boldsymbol{\Sigma} = \begin{bmatrix}\boldsymbol{\Sigma}_1 & \mathbf{O}\\ \mathbf{O} & \mathbf{O}\end{bmatrix},$$

then making a corresponding blocking of the orthogonal matrix $\mathbf{V}$,

$$\mathbf{V} = \begin{bmatrix}\mathbf{A} & \mathbf{B}\\ \mathbf{C} & \mathbf{D}\end{bmatrix},$$

where $[\mathbf{A},\ \mathbf{B}]$ includes the top r rows of $\mathbf{V}$, Equation (5.4.5) can be rewritten as

$$\begin{bmatrix}\boldsymbol{\Sigma}_1 & \mathbf{O}\\ \mathbf{O} & \mathbf{O}\end{bmatrix}\begin{bmatrix}\mathbf{A} & \mathbf{B}\\ \mathbf{C} & \mathbf{D}\end{bmatrix}\begin{bmatrix}\mathbf{v}\\ \mathbf{i}\end{bmatrix} = \mathbf{0}.$$

Hence we obtain the new system model

$$[\mathbf{A},\ \mathbf{B}]\begin{bmatrix}\mathbf{v}\\ \mathbf{i}\end{bmatrix} = \mathbf{0}, \tag{5.4.6}$$

which is algebraically equivalent to, but numerically much more reliable than, the original system model.

5.4.2 Image Compression

The rapid development of science and technology and the popularity of network applications have produced a very large amount of digital information that needs to be stored, processed and transmitted. Image information, as an important multimedia resource, involves a large amount of data. For example, a 1024×768 24-bit BMP image uses about 2.25 MB. Large amounts of image information bring much compressive stress to the capacity of memorizers, the channel bandwidth of communications trunks and the processing speed of computers. Clearly, to solve this problem it is not realistic simply to increase the memory capacity and improve the channel bandwidth and processing speed of a computer. Hence, image compression is very necessary.

Image compression is feasible because the image data are highly correlated. There is a large correlation between adjacent pixels in most images, and thus an image has a lot of spatial redundancy. Also, there is a large correlation between the front and back frames of the image sequence (i.e., there is time redundancy). Furthermore, if the same code length is used to represent symbols with different probabilities of occurrence then a lot of bit numbers will be wasted, that is, the image representations will have symbol redundancy.

Spatial redundancy, time redundancy and symbol redundancy constitute the three main factors in image compression. It is also important that a certain amount of distortion is allowed in image compression.

Because an image has a matrix structure, the SVD can be applied to image compression. An important function of SVD is to reduce greatly the dimension of a matrix or image. In image and video processing the frames of an image are arranged into the columns of an image matrix $\mathbf{A} \in \mathbb{R}^{m\times n}$. If the rank of $\mathbf{A}$ is $r = \text{rank}(\mathbf{A})$, and its truncated SVD is given by

$$\mathbf{A} = \sum_{i=1}^{r} \sigma_i \mathbf{u}_i \mathbf{v}_i^T,$$

then the original image matrix $\mathbf{A}$ can be represented by an $m \times r$ orthogonal matrix $\mathbf{U}_r = [\mathbf{u}_i, \ldots, \mathbf{u}_r]$, an $n \times r$ orthogonal matrix $\mathbf{V}_r = [\mathbf{v}_1, \ldots, \mathbf{v}_r]^T$ and a diagonal vector composed of the singular values of $\mathbf{A}$, $\boldsymbol{\sigma} = [\sigma_1, \ldots, \sigma_r]^T$. In other words, the orthogonal matrices $\mathbf{U}_r$ and $\mathbf{V}_r$ contain mr and nr data entries, respectively. So, the $m \times n$ original image matrix $\mathbf{A}$ (total data mn) is compressed into $\mathbf{U}_r$, $\mathbf{V}_r$ and the diagonal vector $\boldsymbol{\sigma}$ (total data $(m+n+1)r$). Then, the image compression ratio is given by

$$\rho = \frac{mn}{r(m+n+1)}. \tag{5.4.7}$$

Hence, the basic idea of image compression based on SVD is to find the SVD of the image matrix and to use the k singular values and the corresponding left- and right-singular vectors to reconstruct the original image.

If $k \geq r$ then the corresponding image compression is known as the *lossless compression*. Conversely, if $k < r$ then the image compression is said to be *lossy compression*.

In the general case, the number of selected singular values should meet the condition

$$k(m+n+1) \ll mn. \tag{5.4.8}$$

Thus, during the process of transferring images, we need only transfer $k(m+n+1)$ data entries on selected singular values and the corresponding left- and right-singular vectors rather than mn data entries. At the receiving end, after receiving $k(m+n+1)$ data entries, we can reconstruct the original image matrix through

$$\mathbf{A}_k = \sum_{i=1}^{k} \sigma_i \mathbf{u}_i \mathbf{v}_i^T. \tag{5.4.9}$$

The error between the reconstructed image $\mathbf{A}_k$ and the original image $\mathbf{A}$ is measured by

$$\|\mathbf{A} - \mathbf{A}_k\|_F^2 = \sigma_{k+1}^2 + \sigma_{k+2}^2 + \cdots + \sigma_r^2. \tag{5.4.10}$$

The contribution of some singular value σ_i to the image can be defined as

$$\epsilon_i = \frac{\sigma_i^2}{\sigma_1^2 + \cdots + \sigma_r^2}. \tag{5.4.11}$$

For a given image, the larger a singular value is, the larger is its contribution to the image information. Conversely, the smaller a singular value is, the smaller is its contribution to the image information. For example, if $\sum_{i=1}^k \epsilon_i$ is close to 1 then the main information about the image is contained in $\mathbf{A}_k$.

On the basis of satisfying the visual requirements, if a suitable number of larger singular values $k\,(< r)$ is selected then the original image $\mathbf{A}$ can be recovered by using these $\mathbf{A}_k$. The smaller k is, the smaller is the data amount needed to approximate $\mathbf{A}$, and thus the larger is the compression ratio; as $k \to r$, the more similar to $\mathbf{A}$ is the compressed image matrix $\mathbf{A}_k$. In some applications, if the compression ratio is specified in advance then the number k of singular values to be used for image compression can be found, and thus the contribution of image compression can be computed from $\sum_{i=1}^k \epsilon_i$.

5.5 Generalized Singular Value Decomposition (GSVD)

Previous sections presented the SVD of a single matrix and the SVD of the product of two matrices $\mathbf{B}^T\mathbf{C}$. In this section we discuss the SVD of the matrix *pencil* or pair $(\mathbf{A}, \mathbf{B})$. This decomposition is called generalized SVD.

5.5.1 Definition and Properties

The *generalized singular value decomposition* (GSVD) method was proposed by Van Loan in 1976 [483].

Theorem 5.6 [483] *If* $\mathbf{A} \in \mathbb{C}^{m\times n}, m \geq n$, *and* $\mathbf{B} \in \mathbb{C}^{p\times n}$, *there exist two unitary matrices* $\mathbf{U} \in \mathbb{C}^{m\times m}$ *and* $\mathbf{V} \in \mathbb{C}^{p\times p}$ *together with a nonsingular matrix* $\mathbf{Q} \in \mathbb{C}^{n\times n}$ *such that*

$$\mathbf{UAQ} = [\underset{k}{\mathbf{\Sigma}_A},\ \underset{n-k}{\mathbf{O}}], \qquad \mathbf{\Sigma}_A = \begin{bmatrix} \mathbf{I}_r & & \\ & \mathbf{S}_A & \\ & & \mathbf{O}_A \end{bmatrix} \in \mathbb{R}^{m\times k}, \tag{5.5.1}$$

$$\mathbf{VBQ} = [\underset{k}{\mathbf{\Sigma}_B},\ \underset{n-k}{\mathbf{O}}], \qquad \mathbf{\Sigma}_B = \begin{bmatrix} \mathbf{O}_B & & \\ & \mathbf{S}_B & \\ & & \mathbf{I}_{k-r-s} \end{bmatrix} \in \mathbb{R}^{p\times k}, \tag{5.5.2}$$

where

$$\mathbf{S}_A = \mathbf{Diag}(\alpha_{r+1}, \ldots, \alpha_{r+s}),$$
$$\mathbf{S}_B = \mathbf{Diag}(\beta_{r+1}, \ldots, \beta_{r+s}),$$
$$1 > \alpha_{r+1} \geq \cdots \geq \alpha_{r+s} > 0,$$
$$0 < \beta_{r+1} \leq \cdots \leq \beta_{r+s} < 1,$$
$$\alpha_i^2 + \beta_i^2 = 1, \quad i = r+1, \ldots, r+s,$$

with integers

$$k = \text{rank}\begin{bmatrix}\mathbf{A}\\ \mathbf{B}\end{bmatrix}, \quad r = \text{rank}\begin{bmatrix}\mathbf{A}\\ \mathbf{B}\end{bmatrix} - \text{rank}(\mathbf{B}), \tag{5.5.3}$$

$$s = \text{rank}(\mathbf{A}) + \text{rank}(\mathbf{B}) - \text{rank}\begin{bmatrix}\mathbf{A}\\ \mathbf{B}\end{bmatrix}. \tag{5.5.4}$$

There is a variety of methods to prove this theorem, these can be found in Van Loan [483], Paige and Saunders [370], Golub and Van Loan [179] and Zha [540].

Using the method of [483], the diagonal entries of the diagonal matrices $\mathbf{\Sigma}_A$ in Equation (5.5.1) and $\mathbf{\Sigma}_B$ in Equation (5.5.2) constitute singular value pairs (α_i, β_i). Since $\mathbf{O}_A$ is a $(m-r-s) \times (k-r-s)$ zero matrix and $\mathbf{O}_B$ is a $(p-k+r) \times r$ zero matrix, the first k singular value pairs are respectively given by

$$\alpha_i = 1, \quad \beta_i = 0, \quad i = 1, \ldots, r,$$
$$\alpha_i, \beta_i \quad (\text{the entries of } \mathbf{S}_A \text{ and } \mathbf{S}_B), \quad i = r+1, \ldots, r+s,$$
$$\alpha_i = 0, \quad \beta_i = 1, \quad i = r+s+1, \ldots, k.$$

Hence the first k *generalized singular values* are given by

$$\gamma_i = \begin{cases} \alpha_i/\beta_i = 1/0 = \infty, & i = 1, \ldots, r, \\ \alpha_i/\beta_i, & i = r+1, \ldots, r+s, \\ \alpha_i/\beta_i = 0/1 = 0, & i = r+s+1, \ldots, k. \end{cases}$$

These k singular value pairs (α_i, β_i) are called the *nontrivial generalized singular value pairs* of the matrix pencil $(\mathbf{A}, \mathbf{B})$; $\alpha_i/\beta_i\, i = 1, 2, \ldots, k$, are known as the nontrivial generalized singular values of $(\mathbf{A}, \mathbf{B})$ and include the infinite, finite and zero values. In the contrary, the other $n-k$ pairs of generalized singular values, corresponding to the zero vectors in Equations (5.5.1) and (5.5.2), are called the *trivial generalized singular value pairs* of the matrix pencil $(\mathbf{A}, \mathbf{B})$.

The number of columns of the matrix $\mathbf{A}$ is restricted to be not greater than its number of rows in Theorem 5.6. Paige and Saunders [370] generalized Theorem 5.6 to propose the GSVD of any matrix pencil $(\mathbf{A}, \mathbf{B})$ with the same column number for $\mathbf{A}$ and $\mathbf{B}$.

THEOREM 5.7 (GSVD2) [370] *Let $\mathbf{A} \in \mathbb{C}^{m\times n}$ and $\mathbf{B} \in \mathbb{C}^{p\times n}$. Then for the block matrix*

$$\mathbf{K} = \begin{bmatrix}\mathbf{A}\\ \mathbf{B}\end{bmatrix}, \quad t = \mathrm{rank}(\mathbf{K}),$$

there are unitary matrices

$$\mathbf{U} \in \mathbb{C}^{m\times m}, \quad \mathbf{V} \in \mathbb{C}^{p\times p}, \quad \mathbf{W} \in \mathbb{C}^{t\times t}, \quad \mathbf{Q} \in \mathbb{C}^{n\times n}$$

such that

$$\mathbf{U}^H\mathbf{A}\mathbf{Q} = \mathbf{\Sigma}_A[\underbrace{\mathbf{W}^H\mathbf{R}}_{t}, \underbrace{\mathbf{O}}_{n-t}],$$

$$\mathbf{V}^H\mathbf{B}\mathbf{Q} = \mathbf{\Sigma}_B[\underbrace{\mathbf{W}^H\mathbf{R}}_{t}, \underbrace{\mathbf{O}}_{n-t}],$$

where

$$\underset{m\times t}{\mathbf{\Sigma}_A} = \begin{bmatrix}\mathbf{I}_A & & \\ & \mathbf{D}_A & \\ & & \mathbf{O}_A\end{bmatrix}, \quad \underset{p\times t}{\mathbf{\Sigma}_B} = \begin{bmatrix}\mathbf{I}_B & & \\ & \mathbf{D}_B & \\ & & \mathbf{O}_B\end{bmatrix} \tag{5.5.5}$$

and $\mathbf{R} \in \mathbb{C}^{t\times t}$ is nonsingular; its singular values are equal to the nonzero singular values of the matrix $\mathbf{K}$. The matrix $\mathbf{I}_A$ is an $r \times r$ identity matrix and $\mathbf{I}_B$ is a $(t-r-s)\times(t-r-s)$ identity matrix, where the values of r and s are dependent on the given data, and $\mathbf{O}_A$ and $\mathbf{O}_B$ are respectively the $(m-r-s)\times(t-r-s)$ and $(p-t+r)\times r$ zero matrices that may have not any row or column (i.e., $m-r-s=0$, $t-r-s=0$ or $p-t+r=0$ may occur) whereas

$$\mathbf{D}_A = \mathbf{Diag}(\alpha_{r+1}, \alpha_{r+2}, \ldots, \alpha_{r+s}), \ \mathbf{D}_B = \mathbf{Diag}(\beta_{r+1}, \beta_{r+2}, \ldots, \beta_{r+s})$$

satisfy

$$1 > \alpha_{r+1} \geq \alpha_{r+2} \geq \cdots \geq \alpha_{r+s} > 0, \quad 0 < \beta_{r+1} \leq \beta_{r+2} \leq \cdots \leq \beta_{r+s} < 1$$

and

$$\alpha_i^2 + \beta_i^2 = 1, \quad i = r+1, r+2, \ldots, r+s.$$

Proof See [218]. □

The following are a few remarks on the GSVD.

Remark 1 From the above definition it can be deduced that

$$\begin{aligned}\mathbf{A}\mathbf{B}^{-1} &= \mathbf{U}\mathbf{\Sigma}_A[\mathbf{W}^H\mathbf{R}, \mathbf{O}]\mathbf{Q}^H \cdot \mathbf{Q}[\mathbf{W}^H\mathbf{R}, \mathbf{O}]^{-1}\mathbf{\Sigma}_B^{-1}\mathbf{V}^H \\ &= \mathbf{U}\mathbf{\Sigma}_A\mathbf{\Sigma}_B^{-1}\mathbf{V}^H.\end{aligned}$$

This shows the equivalence between the GSVD of the matrix pencil $(\mathbf{A}, \mathbf{B})$ and the SVD of the matrix product $\mathbf{A}\mathbf{B}^{-1}$, because the generalized left and right singular-vector matrices $\mathbf{U}$ and $\mathbf{V}$ of $(\mathbf{A}, \mathbf{B})$ are equal to the left and right singular-vector matrices $\mathbf{U}$ and $\mathbf{V}$ of $\mathbf{A}\mathbf{B}^{-1}$, respectively.

Remark 2 From the equivalence between the GSVD of $(\mathbf{A}, \mathbf{B})$ and the SVD of $\mathbf{AB}^{-1}$ it is obvious that if $\mathbf{B} = \mathbf{I}$ is an identity matrix, then the GSVD simplifies to the general SVD. This result can be directly obtained from the definition of the GSVD, because all singular values of the identity matrix are equal to 1, and thus the generalized singular values of the matrix pencil $(\mathbf{A}, \mathbf{I})$ are equivalent to the singular values of $\mathbf{A}$.

Remark 3 Because $\mathbf{AB}^{-1}$ has the form of a quotient, and the generalized singular values are the quotients of the singular values of the matrices $\mathbf{A}$ and $\mathbf{B}$, the GSVD is sometimes called the *quotient singular value decomposition* (QSVD).

5.5.2 Algorithms for GSVD

If either $\mathbf{A}$ or $\mathbf{B}$ is an ill-conditioned matrix, then the calculation of $\mathbf{AB}^{-1}$ will lead usually to large numerical error, so the SVD of $\mathbf{AB}^{-1}$ is not recommended for computing the GSVD of the matrix pencil $(\mathbf{A}, \mathbf{B})$. A natural question to ask is whether we can get the GSVD of $(\mathbf{A}, \mathbf{B})$ directly without computing $\mathbf{AB}^{-1}$. It is entirely possible, because we have the following theorem.

THEOREM 5.8 [179] *If $\mathbf{A} \in \mathbb{C}^{m_1 \times n} (m_1 \geq n)$ and $\mathbf{B} \in \mathbb{C}^{m_2 \times n} (m_2 \geq n)$ then there exists a nonsingular matrix $\mathbf{X} \in \mathbb{C}^{n \times n}$ such that*

$$\mathbf{X}^H (\mathbf{A}^H \mathbf{A}) \mathbf{X} = \mathbf{D}_A = \mathbf{Diag}(\alpha_1, \alpha_2, \ldots, \alpha_n), \quad \alpha_k \geq 0,$$
$$\mathbf{X}^H (\mathbf{B}^H \mathbf{B}) \mathbf{X} = \mathbf{D}_B = \mathbf{Diag}(\beta_1, \beta_2, \ldots, \beta_n), \quad \beta_k \geq 0.$$

The quantities $\sigma_k = \sqrt{\alpha_k / \beta_k}$ are called the generalized singular values of the matrix pencil $(\mathbf{A}, \mathbf{B})$, and the columns $\mathbf{x}_k$ of $\mathbf{X}$ are known as the generalized singular vectors associated with the generalized singular values σ_k.

The above theorem provides various algorithms for computing the GSVD of the matrix pencil $(\mathbf{A}, \mathbf{B})$. In particular, we are interested in seeking the *generalized singular-vector matrix* $\mathbf{X}$ with $\mathbf{D}_B = \mathbf{I}$, because in this case the generalized singular values σ_k are given directly by $\sqrt{\alpha_k}$.

Algorithms 5.3 and 5.4 give two GSVD algorithms.

The main difference between Algorithms 5.3 and 5.4 is that the former requires computation of the matrix product $\mathbf{A}^H \mathbf{A}$ and $\mathbf{B}^H \mathbf{B}$, and the latter avoids this computation. Because computation of the matrix product may make the condition number become large, Algorithm 5.4 has a better numerical performance than Algorithm 5.3.

In 1998, Drmac [136] developed a tangent algorithm for computing the GSVD, as shown in Algorithm 5.5.

This algorithm is divided into two phases: in the first phase the matrix pencil $(\mathbf{A}, \mathbf{B})$ is simplified into a matrix $\mathbf{F}$; in the second phase the SVD of $\mathbf{F}$ is computed. The theoretical basis of the tangent algorithm is as follows: the GSVD is invariant

Algorithm 5.3 GSVD algorithm 1 [484]

input: $\mathbf{A} \in \mathbb{C}^{m_1 \times n}$, $\mathbf{B} \in \mathbb{C}^{m_2 \times n}$.

1. Compute $\mathbf{S}_1 = \mathbf{A}^H \mathbf{A}$ and $\mathbf{S}_2 = \mathbf{B}^H \mathbf{B}$.
2. Calculate the eigenvalue decomposition $\mathbf{U}_2^H \mathbf{S}_2 \mathbf{U}_2 = \mathbf{D} = \mathbf{Diag}(\gamma_1, \ldots, \gamma_n)$.
3. Compute $\mathbf{Y} = \mathbf{U}_2 \mathbf{D}^{-1/2}$ and $\mathbf{C} = \mathbf{Y}^H \mathbf{S}_1 \mathbf{Y}$.
4. Compute the EVD $\mathbf{Q}^H \mathbf{C} \mathbf{Q} = \mathbf{Diag}(\alpha_1, \ldots, \alpha_n)$, where $\mathbf{Q}^H \mathbf{Q} = \mathbf{I}$.

output: Generalized singular-vector matrix $\mathbf{X} = \mathbf{YQ}$ and generalized singular values $\sqrt{\alpha_k}$, $k = 1, \ldots, n$.

Algorithm 5.4 GSVD algorithm 2 [484]

input: $\mathbf{A} \in \mathbb{C}^{m_1 \times n}$, $\mathbf{B} \in \mathbb{C}^{m_2 \times n}$.

1. Compute the SVD $\mathbf{U}_2^H \mathbf{B} \mathbf{V}_2 = \mathbf{D} = \mathbf{Diag}(\gamma_1, \ldots, \gamma_n)$.
2. Calculate $\mathbf{Y} = \mathbf{V}_2 \mathbf{D}^{-1} \mathbf{V}_2 = \mathbf{Diag}(1/\gamma_1, \ldots, 1/\gamma_n)$.
3. Compute $\mathbf{C} = \mathbf{AY}$.
4. Compute the SVD $\mathbf{U}_1^H \mathbf{C} \mathbf{V}_1 = \mathbf{D}_A = \mathbf{Diag}(\alpha_1, \ldots, \alpha_n)$.

output: Generalized singular-vector matrix $\mathbf{X} = \mathbf{Y} \mathbf{V}_1$, generalized singular values α_k, $k = 1, \ldots, n$.

Algorithm 5.5 Tangent algorithm for GSVD [136]

input: $\mathbf{A} = [\mathbf{a}_1, \ldots, \mathbf{a}_n] \in \mathbb{R}^{m \times n}, \mathbf{B} \in \mathbb{R}^{p \times n}$, $m \geq n$, $\text{rank}(\mathbf{B}) = n$.

1. Compute
$$\mathbf{\Delta}_A = \mathbf{Diag}(\|\mathbf{a}_1\|_2, \ldots, \|\mathbf{a}_n\|_2),$$
$$\mathbf{A}_c = \mathbf{A}\mathbf{\Delta}_A^{-1}, \quad \mathbf{B}_1 = \mathbf{B}\mathbf{\Delta}_A^{-1}.$$
2. Compute the Householder QR decomposition $\begin{bmatrix} \mathbf{R} \\ \mathbf{O} \end{bmatrix} = \mathbf{Q}^T \mathbf{B}_1 \mathbf{\Pi}$.
3. Solve the matrix equation $\mathbf{FR} = \mathbf{A}_c \mathbf{\Pi}$ to get $\mathbf{F} = \mathbf{A}_c \mathbf{\Pi} \mathbf{R}^{-1}$.
4. Compute the SVD $\begin{bmatrix} \mathbf{\Sigma} \\ \mathbf{O} \end{bmatrix} = \mathbf{V}^T \mathbf{F} \mathbf{U}$.
5. Compute $\mathbf{X} = \mathbf{\Delta}_A^{-1} \mathbf{\Pi} \mathbf{R}^{-1} \mathbf{U}$ and $\mathbf{W} = \mathbf{Q} \begin{bmatrix} \mathbf{U} & \mathbf{O} \\ \mathbf{O} & \mathbf{I}_{p-n} \end{bmatrix}$.

output: The GSVD of $(\mathbf{A}, \mathbf{B})$ reads as
$$\mathbf{V}^T \mathbf{A} \mathbf{X} = \begin{bmatrix} \mathbf{\Sigma} \\ \mathbf{O} \end{bmatrix}, \quad \mathbf{W}^T \mathbf{B} \mathbf{X} = \begin{bmatrix} \mathbf{I} \\ \mathbf{O} \end{bmatrix}.$$
That is, the generalized singular values of $(\mathbf{A}, \mathbf{B})$ are given by the diagonal entries of $\mathbf{\Sigma}$, and the generalized left and right singular-vector matrices of $(\mathbf{A}, \mathbf{B})$ are given by $\mathbf{V}$ and $\mathbf{W}$, respectively.

under a equivalence transformation, namely

$$(\mathbf{A}, \mathbf{B}) \rightarrow (\mathbf{A}', \mathbf{B}') = (\mathbf{U}^T \mathbf{A} \mathbf{S}, \mathbf{V}^T \mathbf{B} \mathbf{S}), \tag{5.5.6}$$

where $\mathbf{U}, \mathbf{V}$ are any orthogonal matrices and $\mathbf{S}$ is any nonsingular matrix. Hence, by the definition, two matrix pencils $(\mathbf{A}, \mathbf{B})$ and $(\mathbf{A}', \mathbf{B}')$ have the same GSVD.

5.5.3 Two Application Examples of GSVD

The following are two application examples of GSVD.

1. Multi-Microphone Speech Enhancement

A noisy speech signal sampled by multi-microphone at a discrete time k can be described by the observation model

$$\mathbf{y}(k) = \mathbf{x}(k) + \mathbf{v}(k),$$

where $\mathbf{x}(k)$ and $\mathbf{v}(k)$ are the speech signal vector and the additive noise vector, respectively.

Letting $\mathbf{R}_{yy} = E\{\mathbf{y}(k)\mathbf{y}^T(k)\}$ and $\mathbf{R}_{vv} = E\{\mathbf{v}(k)\mathbf{v}^T(k)\}$ represent the autocorrelation matrices of the observation data vector $\mathbf{y}$ and the additive noise vector $\mathbf{v}$, respectively, one can make their joint diagonalization

$$\left.\begin{aligned} \mathbf{R}_{yy} &= \mathbf{Q}\mathbf{Diag}(\sigma_1^2, \ldots, \sigma_m^2)\mathbf{Q}^T, \\ \mathbf{R}_{vv} &= \mathbf{Q}\mathbf{Diag}(\eta_1^2, \ldots, \eta_m^2)\mathbf{Q}^T. \end{aligned}\right\} \tag{5.5.7}$$

Doclo and Moonen [127] showed in 2002 that, in order to achieve *multi-microphone speech enhancement*, the $M \times M$ optimal filter matrix $\mathbf{W}(k)$ with minimum mean-square error (MMSE) is given by

$$\begin{aligned} \mathbf{W}(k) &= \mathbf{R}_{yy}^{-1}(k)\mathbf{R}_{xx}(k) = \mathbf{R}_{yy}^{-1}(k)(\mathbf{R}_{yy}(k) - \mathbf{R}_{vv}(k)) \\ &= \mathbf{Q}^{-T}\mathbf{Diag}\left(1 - \frac{\eta_1^2}{\sigma_1^2}, 1 - \frac{\eta_2^2}{\sigma_2^2}, \ldots, 1 - \frac{\eta_M^2}{\sigma_M^2}\right)\mathbf{Q}. \end{aligned} \tag{5.5.8}$$

Let $\mathbf{Y}(k)$ be $p \times M$ speech-data matrices containing p speech-data vectors recorded during speech-and-noise periods, and let $\mathbf{V}(k')$ be the $q \times M$ additive noise matrix containing q noise data vectors recorded during noise-only periods:

$$\mathbf{Y}(k) = \begin{bmatrix} \mathbf{y}^T(k-p+1) \\ \vdots \\ \mathbf{y}^T(k-1) \\ \mathbf{y}^T(k) \end{bmatrix}, \qquad \mathbf{V}(k') = \begin{bmatrix} \mathbf{v}^T(k'-q+1) \\ \vdots \\ \mathbf{v}^T(k'-1) \\ \mathbf{v}^T(k') \end{bmatrix}, \tag{5.5.9}$$

where p and q are typically larger than M.

The GSVD of the matrix pencil $(\mathbf{Y}(k), \mathbf{V}(k'))$ is defined as

$$\mathbf{Y}(k) = \mathbf{U}_Y\mathbf{\Sigma}_Y\mathbf{Q}^T, \qquad \mathbf{V}(k') = \mathbf{U}_V\mathbf{\Sigma}_V\mathbf{Q}^T, \tag{5.5.10}$$

where $\mathbf{\Sigma}_Y = \mathbf{Diag}(\sigma_1, \ldots, \sigma_M)$, $\mathbf{\Sigma}_V = \mathbf{Diag}(\eta_1, \ldots, \eta_M)$, $\mathbf{U}_Y$ and $\mathbf{U}_V$ are orthogonal matrices, $\mathbf{Q}$ is an invertible but not necessarily orthogonal matrix containing the generalized singular vectors and σ_i/η_i are the generalized singular values. Substituting these results into Equation (5.5.8), the estimate of the optimal filter matrix is given by [127]

$$\hat{\mathbf{W}} = \mathbf{Q}^{-T}\mathbf{Diag}\left(1 - \frac{p}{q}\frac{\eta_1^2}{\sigma_1^2}, \ldots, \frac{p}{q}\frac{\eta_M^2}{\sigma_M^2}\right)\mathbf{Q}^T. \tag{5.5.11}$$

2. Dimension Reduction of Clustered Text Data

In signal processing, pattern recognition, image processing, information retrieval and so on, dimension reduction is imperative for efficiently manipulating the massive quantity of data.

Suppose that a given $m \times n$ *term-document matrix* $\mathbf{A} \in \mathbb{R}^{m\times n}$ is already partitioned into k clusters:

$$\mathbf{A} = [\mathbf{A}_1, \ldots, \mathbf{A}_k], \quad \text{where} \quad \mathbf{A}_i \in \mathbb{R}^{m\times n_i}, \quad \sum_{i=1}^{k} n_i = n.$$

Our goal is to find a lower-dimensional representation of the term-document matrix $\mathbf{A}$ such that the cluster structure existing in m-dimensional space is preserved in the lower-dimensional representation.

The centroid $\mathbf{c}_i$ of each cluster matrix $\mathbf{A}_i$ is defined as the average of the columns in $\mathbf{A}_i$, i.e.,

$$\mathbf{c}_i = \frac{1}{n_i}\mathbf{A}_i\mathbf{1}_i, \quad \text{where } \mathbf{1}_i = [1, \ldots, 1]^T \in \mathbb{R}^{n_i\times 1},$$

and the global centroid is given by

$$\mathbf{c} = \frac{1}{n}\mathbf{A}\mathbf{1}, \quad \text{where} \quad \mathbf{1} = [1, \ldots, 1]^T \in \mathbb{R}^{n\times 1}.$$

Denote the ith cluster as $\mathbf{A}_i = [\mathbf{a}_{i1}, \ldots, \mathbf{a}_{in_i}]$, and let $N_i = \{i1, \ldots, in_i\}$ denote the set of column indices that belong to the cluster i. Then the *within-class scatter matrix* $\mathbf{S}_w$ is defined as [218]

$$\mathbf{S}_w = \sum_{i=1}^{k}\sum_{j\in N_i}(\mathbf{a}_j - \mathbf{c}_i)(\mathbf{a}_j - \mathbf{c}_i)^T$$

and the *between-class scatter matrix* $\mathbf{S}_b$ is defined as

$$\mathbf{S}_b = \sum_{i=1}^{k} n_i(\mathbf{c}_i - \mathbf{c})(\mathbf{c}_i - \mathbf{c})^T.$$

Defining the matrices

$$\begin{aligned}\mathbf{H}_w &= [\mathbf{A}_1 - \mathbf{c}_1\mathbf{1}_1^T, \ldots, \mathbf{A}_k - \mathbf{c}_k\mathbf{1}_k^T] \in \mathbb{R}^{m\times n},\\ \mathbf{H}_b &= [\sqrt{n_1}(\mathbf{c}_1 - \mathbf{c}), \ldots, \sqrt{n_k}(\mathbf{c} - \mathbf{c}_k)] \in \mathbb{R}^{m\times k},\end{aligned} \tag{5.5.12}$$

the corresponding scatter matrices can be represented as

$$\mathbf{S}_w = \mathbf{H}_w \mathbf{H}_w^T \in \mathbb{R}^{m\times m}, \qquad \mathbf{S}_b = \mathbf{H}_b \mathbf{H}_b^T \in \mathbb{R}^{m\times m}.$$

It is assumed that the lower-dimensional representation of the term-document matrix $\mathbf{A} \in \mathbb{R}^{m\times n}$ consists of l vectors $\mathbf{u}_i \in \mathbb{R}^{m\times 1}$ with $\|\mathbf{u}_i\|_2 = 1$, where $i = 1, \ldots, l$ and $l < \min\{m, n, k\}$.

Consider the generalized Rayleigh quotient

$$\lambda_i = \frac{\|\mathbf{H}_b^T \mathbf{u}_i\|_2^2}{\|\mathbf{H}_w^T \mathbf{u}_i\|_2^2} = \frac{\mathbf{u}_i^T \mathbf{S}_b \mathbf{u}_i}{\mathbf{u}_i^T \mathbf{S}_w \mathbf{u}_i}; \tag{5.5.13}$$

its numerator and denominator denote the between-class scatter and the within-class scatter of the vector $\mathbf{u}_i$, respectively. As the measure of cluster quality, the between-class scatter should be as large as possible, while the within-class scatter should be as small as possible. Hence, for each $\mathbf{u}_i$, the generalized Rayleigh quotient, namely

$$\mathbf{u}_i^T \mathbf{S}_b \mathbf{u}_i = \lambda_i \mathbf{u}_i^T \mathbf{S}_w \mathbf{u}_i \quad \text{or} \quad \mathbf{u}_i^T \mathbf{H}_b \mathbf{H}_b^T \mathbf{u}_i = \lambda_i \mathbf{u}_i^T \mathbf{H}_w \mathbf{H}_w^T \mathbf{u}_i \tag{5.5.14}$$

should be maximized. This is the generalized eigenvalue decomposition of the cluster matrix pencil $(\mathbf{S}_b, \mathbf{S}_w)$.

Equation (5.5.14) can be equivalently written as

$$\mathbf{H}_b^T \mathbf{u}_i = \sigma_i \mathbf{H}_w^T \mathbf{u}_i, \quad \sigma_i = \sqrt{\lambda_i}. \tag{5.5.15}$$

This is just the GSVD of the matrix pencil $(\mathbf{H}_b^T, \mathbf{H}_w^T)$.

From the above analysis, the *dimension reduction method* for a given term-document matrix $\mathbf{A} \in \mathbb{R}^{m\times n}$ can be summarized as follows.

(1) Compute the matrices $\mathbf{H}_b$ and $\mathbf{H}_w$ using Equation (5.5.12).
(2) Perform the GSVD of the matrix pencil $(\mathbf{H}_b, \mathbf{H}_w)$ to get the nonzero generalized singular values $\sigma_i, i = 1, \ldots, l$ and the corresponding left-generalized singular vectors $\mathbf{u}_1, \ldots, \mathbf{u}_l$.
(3) The lower-dimensional representation of $\mathbf{A}$ is given by $[\mathbf{u}_1, \ldots, \mathbf{u}_l]^T$.

The GSVD has also been applied to the comparative analysis of genome-scale expression data sets of two different organisms [13], to discriminant analysis [219] etc.

5.6 Low-Rank–Sparse Matrix Decomposition

In the fields of applied science and engineering (such as image, voice and video processing, bioinformatics, network search and e-commerce), a data set is often high-dimensional, and its dimension can be even up to a million. It is important to discover and utilize the lower-dimensional structures in a high-dimensional data set.

5.6.1 Matrix Decomposition Problems

In many areas of engineering and applied sciences (such as machine learning, control, system engineering, signal processing, pattern recognition and computer vision), a data (image) matrix $\mathbf{A} \in \mathbb{R}^{m\times n}$ usually has the following special constructions.

(1) *High dimension* m, n are usually very large.
(2) *Low rank* $r = \text{rank}(\mathbf{A}) \ll \min\{m, n\}$.
(3) *Sparsity* Some observations are grossly corrupted, so that $\mathbf{A}$ becomes $\mathbf{A} + \mathbf{E}$; some entries of the error matrix $\mathbf{E}$ may be very large but most are equal to zero.
(4) *Incomplete data* Some observations may be missing too.

The following are several typical application areas with the above special constructions [90], [77].

1. *Graphical modeling* In many applications, because a small number of characteristic factors can explain most of the statistics of the observation data, a high-dimensional covariance matrix is commonly approximated by a low-rank matrix. In the graphical model [284] the inverse of the covariance matrix (also called the information matrix) is assumed to be sparse with respect to some graphic. Thus, in the statistical model setting, the data matrix is often decomposed into the sum of a low-rank matrix and a sparse matrix, to describe the roles of unobserved hidden variables and of the graphical model, respectively.
2. *Composite system identification* In system identification a composite system is often represented by the sum of a low-rank Hankel matrix and a sparse Hankel matrix, where the sparse Hankel matrix corresponds to a linear time-invariant system with sparse impulse response, and the low-rank Hankel matrix corresponds to the minimum realization system with a small order of model.
3. *Face recognition* Since images of a convex, Lambertian surface under different types of lighting span a lower-dimensional subspace [27], the lower-dimensional models are the most effective for image data. In particular, human face images can be well-approximated by a lower-dimensional subspace. Therefore, to retrieve this subspace correctly is crucial in many applications such as face recognition and alignment. However, actual face images often suffer from shadows and highlighting effects (specularities, or saturations in brightness), and hence it is necessary to remove such defects in face images by using low-rank matrix decomposition.
4. *Video surveillance* Given a sequence of surveillance video frames, it is often necessary to identify activities standing out from the background. If the video frames are stacked as columns of a data matrix $\mathbf{D}$ and a low-rank and sparse

decomposition $\mathbf{D} = \mathbf{A} + \mathbf{E}$ is made then the low-rank matrix $\mathbf{A}$ naturally corresponds to the stationary background and the sparse matrix $\mathbf{E}$ captures the moving objects in the foreground.

5. *Latent semantic indexing* Web search engines often need to index and retrieve the contents of a huge set of documents. A popular scheme is latent semantic indexing (LSI) [124]; its basic idea is to encode the relevance of a term or word in a document (for example, by its frequency) and to use these relevance indicators as elements of a document-versus-term matrix $\mathbf{D}$. If $\mathbf{D}$ can be decomposed into a sum of a low-rank matrix $\mathbf{A}$ and a sparse matrix $\mathbf{E}$ then $\mathbf{A}$ could capture the common words used in all the documents while $\mathbf{E}$ captures the few key words that best distinguish each document from others.

The above examples show that for many applications it is not enough to decompose a data matrix into the sum of a low-rank matrix and an error (or perturbation) matrix. A better approach is to decompose a data matrix $\mathbf{D}$ into the sum of a low-rank matrix $\mathbf{A}$ and a sparse matrix $\mathbf{E}$, i.e., $\mathbf{D} = \mathbf{A} + \mathbf{E}$, in order to recover its low-rank component matrix. Such a decomposition is called a *low-rank–sparse matrix decomposition.*

5.6.2 Singular Value Thresholding

Singular value thresholding is a key operation in matrix recovery and matrix completion. Before discussing in detail how to solve matrix recovery (and matrix completion) problems, it is necessary to introduce singular value thresholding.

Consider the truncated SVD of a low-rank matrix $\mathbf{W} \in \mathbb{R}^{m\times n}$:

$$\mathbf{W} = \mathbf{U}\boldsymbol{\Sigma}\mathbf{V}^T, \quad \boldsymbol{\Sigma} = \mathbf{Diag}(\sigma_1, \ldots, \sigma_r), \tag{5.6.1}$$

where $r = \text{rank}(\mathbf{W}) \ll \min\{m, n\}$, $\mathbf{U} \in \mathbb{R}^{m\times r}, \mathbf{V} \in \mathbb{R}^{n\times r}$.

Let the threshold value $\tau \geq 0$, then

$$\mathcal{D}_\tau(\mathbf{W}) = \mathbf{U}\mathcal{D}_\tau(\boldsymbol{\Sigma})\mathbf{V}^T \tag{5.6.2}$$

is called the *singular value thresholding* (SVT) of the matrix $\mathbf{W}$, where

$$\mathcal{D}_\tau(\boldsymbol{\Sigma}) = \text{soft}(\boldsymbol{\Sigma}, \tau) = \mathbf{Diag}\left((\sigma_1 - \tau)_+, \ldots, (\sigma_r - \tau)_+\right) \tag{5.6.3}$$

is known as the *soft thresholding* and

$$(\sigma_i - \tau)_+ = \begin{cases} \sigma_i - \tau, & \text{if } \sigma_i > \tau, \\ 0, & \text{otherwise}, \end{cases}$$

is the *soft thresholding operation.*

The relationship between the SVT and the SVD is as follows.

- If the soft threshold value $\tau = 0$ then the SVT reduces to the truncated SVD (5.6.1).

- All singular values are soft thresholded by the soft threshold value $\tau > 0$, which just changes the magnitudes of singular values; it does not change the left and right singular-vector matrices $\mathbf{U}$ and $\mathbf{V}$.

The proper choice of soft threshold value τ can effectively set some singular values to zero. In this sense the SVT transform is also called a *singular value shrinkage operator*. It is noted that if the soft threshold value τ is larger than most of the singular values, then the rank of the SVT operator $\mathcal{D}(\mathbf{W})$ will be much smaller than the rank of the original matrix $\mathbf{W}$.

The following are two key points for applying the SVT:

(1) How should an optimization problem be written in a standard form suitable for the SVT?
(2) How should the soft thresholding value τ be chosen?

The answers to the above two questions depend on the following proposition and theorem, respectively.

Proposition 5.1 *Let* $\langle \mathbf{X}, \mathbf{Y} \rangle = \text{tr}(\mathbf{X}^T\mathbf{Y})$ *denote the inner product of matrices* $\mathbf{X} \in \mathbb{R}^{m\times n}$ *and* $\mathbf{Y} \in \mathbb{R}^{m\times n}$. *Then we have*

$$\|\mathbf{X} \pm \mathbf{Y}\|_F^2 = \|\mathbf{X}\|_F^2 \pm 2\langle \mathbf{X}, \mathbf{Y} \rangle + \|\mathbf{Y}\|_F^2. \tag{5.6.4}$$

Proof By the relationship between the Frobenius norm and the trace of a matrix, $\|\mathbf{A}\|_F = \sqrt{\text{tr}(\mathbf{A}^H\mathbf{A})}$, we have

$$\begin{aligned}\|\mathbf{X} \pm \mathbf{Y}\|_F^2 &= \text{tr}[(\mathbf{X} \pm \mathbf{Y})^T(\mathbf{X} \pm \mathbf{Y})] \\ &= \text{tr}(\mathbf{X}^T\mathbf{X} \pm \mathbf{X}^T\mathbf{Y} \pm \mathbf{Y}^T\mathbf{X} + \mathbf{Y}^T\mathbf{Y}) \\ &= \|\mathbf{X}\|_F^2 \pm 2\text{tr}(\mathbf{X}^T\mathbf{Y}) + \|\mathbf{Y}\|_F^2 \\ &= \|\mathbf{X}\|_F^2 \pm 2\langle \mathbf{X}, \mathbf{Y} \rangle + \|\mathbf{Y}\|_F^2,\end{aligned}$$

where we have used the trace property $\text{tr}(\mathbf{Y}^T\mathbf{X}) = \text{tr}(\mathbf{X}^T\mathbf{Y})$. □

Theorem 5.9 *For each soft threshold value* $\mu > 0$ *and the matrix* $\mathbf{W} \in \mathbb{R}^{m\times n}$, *the SVT operator obeys* [75]

$$\mathbf{U}\text{soft}(\mathbf{\Sigma}, \mu)\mathbf{V}^T = \arg\min_{\mathbf{X}} \left\{ \mu\|\mathbf{X}\|_* + \frac{1}{2}\|\mathbf{X} - \mathbf{W}\|_F^2 \right\} = \mathbf{prox}_{\mu\|\cdot\|_*}(\mathbf{W}), \tag{5.6.5}$$

and [75]

$$\text{soft}(\mathbf{W}, \mu) = \arg\min_{\mathbf{X}} \left\{ \mu\|\mathbf{X}\|_1 + \frac{1}{2}\|\mathbf{X} - \mathbf{W}\|_F^2 \right\} = \mathbf{prox}_{\mu\|\cdot\|_1}(\mathbf{W}), \tag{5.6.6}$$

where $\mathbf{U\Sigma V}^T$ *is the SVD of* $\mathbf{W}$, *and the soft thresholding (shrinkage) operator*

$$[\text{soft}(\mathbf{W}, \mu)]_{ij} = \begin{cases} w_{ij} - \mu, & w_{ij} > \mu, \\ w_{ij} + \mu, & w_{ij} < -\mu, \\ 0, & \text{otherwise}, \end{cases}$$

with $w_{ij} \in \mathbb{R}$ *the* (i,j)*th entry of* $\mathbf{W} \in \mathbb{R}^{m\times n}$.

Theorem 5.9 tells us that when solving a matrix recovery or matrix completion problem, the underlying question is how to transform the corresponding optimization problem into the normalized form shown in Equations (5.6.5) and/or (5.6.6).

5.6.3 Robust Principal Component Analysis

Given an observation or data matrix $\mathbf{D} = \mathbf{A}+\mathbf{E}$, where $\mathbf{A}$ and $\mathbf{E}$ are unknown, but $\mathbf{A}$ is known to be low rank and $\mathbf{E}$ is known to be sparse, the aim is to recover $\mathbf{A}$. Because the low-rank matrix $\mathbf{A}$ can be regarded as the principal component of the data matrix $\mathbf{D}$, and the sparse matrix $\mathbf{E}$ may have a few gross errors or outlying observations, the SVD based principal component analysis (PCA) usually breaks down. Hence, this problem is a *robust principal component analysis* problem [298], [514].

So-called robust PCA is able to correctly recover underlying low-rank structure in the data matrix, even in the presence of gross errors or outlying observations. This problem is also called the *principal component pursuit* (PCP) problem [90] since it tracks down the principal components of the data matrix $\mathbf{D}$ by minimizing a combination of the nuclear norm $\|\mathbf{A}\|_*$ and the weighted ℓ_1-norm $\mu\|\mathbf{E}\|_1$:

$$\min_{\mathbf{A},\mathbf{E}} \{\|\mathbf{A}\|_* + \mu\|\mathbf{E}\|_1\} \quad \text{subject to} \quad \mathbf{D} = \mathbf{A}+\mathbf{E}. \tag{5.6.7}$$

By using the above minimization, an initially unknown low-rank matrix $\mathbf{A}$ and unknown sparse matrix $\mathbf{E}$ can be recovered from the data matrix $\mathbf{D} \in \mathbb{R}^{m\times n}$. Here, $\|\mathbf{A}\|_* = \sum_i^{\min\{m,n\}} \sigma_i(\mathbf{A})$ represents the nuclear norm of $\mathbf{A}$, i.e., the sum of its all singular values, which reflects the cost of the low-rank matrix $\mathbf{A}$ and $\|\mathbf{E}\|_1 = \sum_{i=1}^m \sum_{j=1}^n |E_{ij}|$ is the sum of the absolute values of all entries of the additive error matrix $\mathbf{E}$, where some of its entries may be arbitrarily large, whereas the role of the constant $\mu > 0$ is to balance the contradictory requirements of low rank and sparsity.

The unconstrained minimization of the robust PCA or the PCP problem is expressed as

$$\min_{\mathbf{A},\mathbf{E}} \left\{\|\mathbf{A}\|_* + \mu\|\mathbf{E}\|_1 + \frac{1}{2}(\mathbf{A}+\mathbf{E}-\mathbf{D})\right\}. \tag{5.6.8}$$

It can be shown [77] that, under rather weak assumptions, the PCP estimate can exactly recover the low-rank matrix $\mathbf{A}$ from the data matrix $\mathbf{D} = \mathbf{A}+\mathbf{E}$ with gross but sparse error matrix $\mathbf{E}$.

Thus, to solve the robust PCA or the PCP problem, consider a more general family of optimization problems of the form

$$F(\mathbf{X}) = f(\mathbf{X}) + \mu h(\mathbf{X}), \tag{5.6.9}$$

where $f(\mathbf{X})$ is a convex, smooth (i.e., differentiable) and L-Lipschitz function and $h(\mathbf{X})$ is a convex but nonsmooth function (such as $\|\mathbf{X}\|_1, \|\mathbf{X}\|_*$ and so on).

Instead of directly minimizing the composite function $F(\mathbf{X})$, we minimize its separable quadratic approximation $Q(\mathbf{X},\mathbf{Y})$, formed at specially chosen points $\mathbf{Y}$:

$$\begin{aligned} Q(\mathbf{X},\mathbf{Y}) &= f(\mathbf{Y}) + \langle \nabla f(\mathbf{Y}), \mathbf{X}-\mathbf{Y}\rangle + \frac{1}{2}(\mathbf{X}-\mathbf{Y})^T \nabla^2 f(\mathbf{Y})(\mathbf{X}-\mathbf{Y}) + \mu h(\mathbf{X}) \\ &= f(\mathbf{Y}) + \langle \nabla f(\mathbf{Y}), \mathbf{X}-\mathbf{Y}\rangle + \frac{L}{2}\|\mathbf{X}-\mathbf{Y}\|_F^2 + \mu h(\mathbf{X}), \end{aligned} \tag{5.6.10}$$

where $\nabla^2 f(\mathbf{Y})$ is approximated by $L\mathbf{I}$.

When minimizing $Q(\mathbf{X},\mathbf{Y})$ with respect to $\mathbf{X}$, the function term $f(\mathbf{Y})$ may be regarded as a constant term that is negligible. Hence, we have

$$\begin{aligned} \mathbf{X}_{k+1} &= \arg\min_{\mathbf{X}} \left\{ \mu h(\mathbf{X}) + \frac{L}{2}\left\| \mathbf{X} - \mathbf{Y}_k + \frac{1}{L}\nabla f(\mathbf{Y}_k)\right\|_F^2 \right\} \\ &= \mathbf{prox}_{\mu L^{-1} h}\left(\mathbf{Y}_k - \frac{1}{L}\nabla f(\mathbf{Y}_k)\right). \end{aligned} \tag{5.6.11}$$

If we let $f(\mathbf{Y}) = \frac{1}{2}\|\mathbf{A}+\mathbf{E}-\mathbf{D}\|_F^2$ and $h(\mathbf{X}) = \|\mathbf{A}\|_* + \lambda\|\mathbf{E}\|_1$ then the Lipschitz constant $L = 2$ and $\nabla f(\mathbf{Y}) = \mathbf{A}+\mathbf{E}-\mathbf{D}$. Hence

$$Q(\mathbf{X},\mathbf{Y}) = \mu h(\mathbf{X}) + f(\mathbf{Y}) = (\mu\|\mathbf{A}\|_* + \mu\lambda\|\mathbf{E}\|_1) + \frac{1}{2}\|\mathbf{A}+\mathbf{E}-\mathbf{D}\|_F^2$$

reduces to the quadratic approximation of the robust PCA problem.

By Theorem 5.9, one has

$$\mathbf{A}_{k+1} = \mathbf{prox}_{\mu/2\|\cdot\|_*}\left(\mathbf{Y}_k^A - \frac{1}{2}(\mathbf{A}_k + \mathbf{E}_k - \mathbf{D})\right) = \mathbf{U}\,\mathrm{soft}(\mathbf{\Sigma},\lambda)\mathbf{V}^T,$$

$$\mathbf{E}_{k+1} = \mathbf{prox}_{\mu\lambda/2\|\cdot\|_1}\left(\mathbf{Y}_k^E - \frac{1}{2}(\mathbf{A}_k + \mathbf{E}_k - \mathbf{D})\right) = \mathrm{soft}\left(\mathbf{W}_k^E, \frac{\mu\lambda}{2}\right),$$

where $\mathbf{U\Sigma V}^T$ is the SVD of $\mathbf{W}_k^A$ and

$$\mathbf{W}_k^A = \mathbf{Y}_k^A - \frac{1}{2}(\mathbf{A}_k + \mathbf{E}_k - \mathbf{D}), \tag{5.6.12}$$

$$\mathbf{W}_k^E = \mathbf{Y}_k^E - \frac{1}{2}(\mathbf{A}_k + \mathbf{E}_k - \mathbf{D}). \tag{5.6.13}$$

Algorithm 5.6 gives a robust PCA algorithm.

The robust PCA has the following convergence.

Theorem 5.10 [298] *Let $F(\mathbf{X}) = F(\mathbf{A},\mathbf{E}) = \mu\|\mathbf{A}\|_* + \mu\lambda\|\mathbf{E}\|_1 + \frac{1}{2}\|\mathbf{A}+\mathbf{E}-\mathbf{D}\|_F^2$. Then, for all $k > k_0 = C_1/\log(1/n)$ with $C_1 = \log(\mu_0/\mu)$, one has*

$$F(\mathbf{X}) - F(\mathbf{X}^\star) \leq \frac{4\|\mathbf{X}_{k_0} - \mathbf{X}^\star\|_F^2}{(k-k_0+1)^2}, \tag{5.6.14}$$

where $\mathbf{X}^\star$ is a solution to the robust PCA problem $\min_{\mathbf{X}} F(\mathbf{X})$.

Algorithm 5.6 Robust PCA via accelerated proximal gradient [298], [514]

input: Data matrix $\mathbf{D} \in \mathbb{R}^{m\times n}$, λ, allowed tolerance ϵ.
initialization: $\mathbf{A}_0, \mathbf{A}_{-1} \leftarrow \mathbf{O}; \mathbf{E}_0, \mathbf{E}_{-1} \leftarrow \mathbf{O}; t_0, t_{-1} \leftarrow 1; \bar{\mu} \leftarrow \delta\mu_0$.
repeat

1. $\mathbf{Y}_k^A \leftarrow \mathbf{A}_k + \frac{t_{k-1}-1}{t_k}(\mathbf{A}_k - \mathbf{A}_{k-1})$.
2. $\mathbf{Y}_k^E \leftarrow \mathbf{E}_k + \frac{t_{k-1}-1}{t_k}(\mathbf{E}_k - \mathbf{E}_{k-1})$.
3. $\mathbf{W}_k^A \leftarrow \mathbf{Y}_k^A - \frac{1}{2}(\mathbf{A}_k + \mathbf{E}_k - \mathbf{D})$.
4. $(\mathbf{U}, \mathbf{\Sigma}, \mathbf{V}) = \text{svd}(\mathbf{W}_k^A)$.
5. $r = \max\left\{j : \sigma_j > \frac{\mu_k}{2}\right\}$.
6. $\mathbf{A}_{k+1} = \sum_{i=1}^r \left(\sigma_i - \frac{\mu_k}{2}\mathbf{u}_i\mathbf{v}_i^T\right)$.
7. $\mathbf{W}_k^E \leftarrow \mathbf{Y}_k^E - \frac{1}{2}(\mathbf{A}_k + \mathbf{E}_k - \mathbf{D})$.
8. $\mathbf{E}_{k+1} = \text{soft}\left(\mathbf{W}_k^E, \frac{\lambda\mu_k}{2}\right)$.
9. $t_{k+1} \leftarrow \dfrac{1+\sqrt{4t_k^2+1}}{2}$.
10. $\mu_{k+1} \leftarrow \max(\eta\mu_k, \bar{\mu})$.
11. $\mathbf{S}_{k+1}^A = 2(\mathbf{Y}_k^A - \mathbf{A}_{k+1}) + (\mathbf{A}_{k+1} + \mathbf{E}_{k+1} - \mathbf{Y}_k^A - \mathbf{Y}_k^E)$.
12. $\mathbf{S}_{k+1}^E = 2(\mathbf{Y}_k^E - \mathbf{E}_{k+1}) + (\mathbf{A}_{k+1} + \mathbf{E}_{k+1} - \mathbf{Y}_k^A - \mathbf{Y}_k^E)$.
13. **exit if** $\|\mathbf{S}_{k+1}\|_F^2 = \|\mathbf{S}_{k+1}^A\|_F^2 + \|\mathbf{S}_{k+1}^E\|_F^2 \leq \epsilon$.

return $k \leftarrow k+1$.
output: $\mathbf{A} \leftarrow \mathbf{A}_k$, $\mathbf{E} \leftarrow \mathbf{E}_k$.

5.7 Matrix Completion

Finding a low-rank matrix to fit or approximate a higher-dimensional data matrix is a fundamental task in data analysis. In a slew of applications, two popular empirical approaches have been used to represent the target rank-k matrix.

(1) *Low-rank decomposition* A noisy data matrix $\mathbf{X} \in \mathbb{R}^{m\times n}$ is decomposed as the sum of two matrices $\mathbf{A} + \mathbf{E}$, where $\mathbf{A} \in \mathbb{R}^{m\times n}$ is a low-rank matrix with the same rank k as the target matrix to be estimated and $\mathbf{E} \in \mathbb{R}^{m\times n}$ is a fitting or approximating error matrix.
(2) *Low-rank bilinear modeling:* The data matrix $\mathbf{X} \in \mathbb{R}^{m\times n}$ is modeled in a bilinear form $\mathbf{X} = \mathbf{U}\mathbf{V}^T$, where $\mathbf{U} \in \mathbb{R}^{m\times k}$ and $\mathbf{V} \in \mathbb{R}^{n\times k}$ are full-column rank matrices.

When most entries of a data matrix are missing, its low-rank decomposition problem is called *matrix completion*, as presented by Candès and Recht in 2009 [79]. The low-rank bilinear modeling of a matrix is known as *matrix sensing* and was developed by Recht *et al.* [412] and Lee and Bresler [288].

Specifically, the matrix completion problem and the matrix sensing problem can be described as follows.

- *Matrix completion* involves completing a low-rank matrix by taking into account, i.e., observing, only a few of its elements.
- *Matrix sensing* refers to the problem of recovering a low-rank matrix $\mathbf{B} \in \mathbb{R}^{m \times n}$, given d linear measurements $b_i = \text{tr}(\mathbf{A}_i^T \mathbf{B})$ and measurement matrices $\mathbf{A}_i, i = 1, \ldots, d$.

In fact, matrix completion is a special case of the matrix sensing problem where each observed entry in the matrix completion problem represents a single-element measurement matrix $\mathbf{A}_i = A_i$.

This section focuses upon matrix completion.

5.7.1 Matrix Completion Problems

Matrix completion problems stem from the widespread practical applications.

1. *Dynamic scene reconstruction* It is a classical and hot issue in computer vision and graphics to make a dynamic scene reconstruction from multiview video sequences. Three-dimensional (3-D) motion estimation from multiview video sequences is of vital importance in achieving high-quality dynamic scene reconstruction and has great application potential in the virtual reality and future movie industry. Because 3-D motion vectors can be treated as an incomplete matrix according to their local distributions and frequently there are outliers in the known motion information, 3-D motion estimation is naturally a matrix completion problem.
2. *Radar imaging* For high-resolution radar imaging, the resulting high sampling rate poses difficulties for raw data transmission and storage, which results in the undersampling of stepped-frequency-radar data. Radar change imaging aims to recover the changed portion of an imaged region using the data collected before and after a change. Hence, radar change imaging with undersampled data can be cast as a matrix completion problem [45]. Tomographic synthetic aperture radar (TomoSAR) is a new SAR imaging modality, but the massive baselines involved make TomoSAR imaging an expensive and difficult task. However, the use of limited baselines with fixed-elevation aperture length to improve the recovery quality is worth researching. By exploiting multibaseline two-dimensional (2-D) focused SAR image data, the missing-data compensation method, based on matrix completion, can solve the 3-D focused TomoSAR imaging problem [46].
3. *MIMO radar* Multiple-input multiple-output (MIMO) radar has received significant attention owing to its improved resolution and target estimation performance as compared with conventional radars. For a sufficiently large number of transmission and reception antennas and a small number of targets, the data

matrix is low rank, and it can be recovered from a small number of its entries via matrix completion. This implies that, at each reception antenna, matched filtering does not need to be performed with all the transmit waveforms, but rather with just a small number of randomly selected waveforms from the waveform dictionary [237].

4. *Ranking and collaborative filtering* Predicting the user's preferences is increasingly important in e-commerce and advertisement. Companies now routinely collect user rankings for various products (such as movies, books or games). So-called ranking and collaborative filtering employs incomplete user rankings on some products for predicting any given user's preferences on another product. The best known ranking and collaborative filtering is perhaps the Netflix Prize for movie ranking. In this case the data matrix is incomplete, as the data collection process often lacks control or sometimes a small portion of the available rankings could be noisy and even have been subject to tampering. Therefore, it necessary to infer a low-rank matrix $\mathbf{L}$ from an incomplete and corrupted matrix [79], [78].
5. *Global positioning* Owing to power constraints, sensors in global positioning networks may only be able to construct reliable distance estimates from their immediate neighbors, and hence only a partially observed distance matrix can be formed. However, the problem is to infer all the pairwise distances from just a few observed data so that the locations of the sensors can be reliably estimated. This reduces to a matrix completion problem, where the unknown matrix is of rank 2 (if the sensors are located in the plane) or 3 (if they are located in space) [79], [78].
6. *Remote sensing* In array signal processing, it is necessary to estimate the direction of arrival (DOA) of incident signals in a coherent radio-frequency environment. The DOA is frequently estimated by using the covariance matrix for the observed data. However, in remote sensing applications, one may not be able to estimate or transmit all entries of the covariance matrix owing to power constraints. To infer a full covariance matrix from just a few observed partial correlations, one needs to solve a matrix completion problem in which the unknown signal covariance matrix has low rank since it is equal to the number of incident waves, which is usually much smaller than the number of sensors [79], [78].
7. *Multi-label image classification* In multi-label image classification, multiple labels are assigned to a given image; thus is useful in web-based image analysis. For example, keywords can be automatically assigned to an uploaded web image so that annotated images may be searched directly using text-based image retrieval systems. An ideal multi-label algorithm should be able to handle missing features, or cases where parts of the training data labels are unknown, and it should be robust against outliers and background noise. To this end, Luo *et*

al. [306] have developed recently a multiview matrix completion for handling multiview features in semi-supervised multi-label image classification.

8. *Computer vision and graphics* Many problems can be formulated as missing-value estimation problems, e.g., image in-painting, video decoding, video in-painting, scan completion and appearance acquisition completion [300].

The differences between matrix completion and matrix recovery (or low-rank–sparse matrix decomposition) are given below.

(1) Only a small number of matrix elements is known in matrix completion, while the whole data matrix is known in matrix recovery.
(2) Matrix completion reconstructs a higher-dimensional matrix, while matrix recovery extracts the lower-dimensional features of a higher-dimensional matrix.
(3) The capability of matrix completion is limited by noncoherence and the sampling rate, while the capability of the matrix recovery is limited by real matrix rank and data sampling method.

5.7.2 Matrix Completion Model and Incoherence

Let $\mathbf{D} \in \mathbb{R}^{m \times n}$ be a higher-dimensional incomplete data matrix, as discussed above: only a small amount of entries are known or observed and the index set of these known or sample entries is Ω, namely only the matrix entries $D_{ij}, (i,j) \in \Omega$ are known or observed.

The support of the index set is called the *base number*, denoted $|\Omega|$. The base number is the number of sample elements, and the ratio of the number of samples to the dimension number of the data matrix, $p = |\Omega|/(mn)$, is called the *sample density* of the data matrix. In some typical applications (e.g., the Netflix recommendation system), the sample density is only 1% or less.

As mentioned earlier, the mathematical problem of recovering a low-rank matrix from an incomplete data matrix is called *matrix completion*, and can be represented as

$$\hat{\mathbf{X}} = \arg\min_{\mathbf{X}} \ \text{rank}(\mathbf{X}) \tag{5.7.1}$$

subject to

$$\mathcal{P}_\Omega(\mathbf{X}) = \mathcal{P}_\Omega(\mathbf{D}) \quad \text{or} \quad \mathbf{X}_{ij} = \mathbf{D}_{ij}, \quad \text{for } (i,j) \in \Omega, \tag{5.7.2}$$

where $\Omega \subset \{1,\ldots,m\} \times \{1,\ldots,n\}$ is a subset of the complete set of entries and $\mathcal{P}_\Omega : \mathbb{R}^{m \times n} \to \mathbb{R}^{m \times n}$ is the projection onto the index set Ω:

$$[\mathcal{P}_\Omega(\mathbf{D})]_{ij} = \begin{cases} D_{ij}, & (i,j) \in \Omega, \\ 0, & \text{otherwise}. \end{cases} \tag{5.7.3}$$

Unfortunately, the rank minimization problem (5.7.1) is NP-hard owing to the nonconvexity and discontinuous nature of the rank function. It has been shown [412] that the nuclear norm of a matrix is the tightest convex lower bound of its rank function. Hence, a popular alternative is the convex relaxation of the rank function:

$$\hat{\mathbf{X}} = \arg\min_{\mathbf{X}} \|\mathbf{X}\|_* \quad \text{subject to} \quad \mathbf{X}_{ij} = \mathbf{D}_{ij}, \quad (i,j) \in \Omega, \tag{5.7.4}$$

where $\|\mathbf{X}\|_*$ is the nuclear norm of the matrix $\mathbf{X}$, defined as the sum of its singular values.

It was observed in [79] that if the entries of $\mathbf{D}$ are equal to zero in most rows or columns then it is impossible to complete $\mathbf{D}$ unless all of its entries are observed. An interesting question is how to guarantee the unique completion of an incomplete data matrix. The answer to this question depends on an additional property known as the incoherence of the data matrix $\mathbf{D}$.

Suppose that the SVD of an $m \times n$ matrix $\mathbf{M}$ with rank r is $\mathbf{M} = \mathbf{U}\boldsymbol{\Sigma}\mathbf{V}^T$. The matrix $\mathbf{M}$ is said to satisfy the *standard incoherence condition* with parameter μ_0 if [98]

$$\begin{aligned} &\max_{1\leq i\leq m} \left\{ \|\mathbf{U}^T\mathbf{e}_i\|_2 \leq \sqrt{\frac{\mu_0 r}{m}} \right\}, \\ &\max_{1\leq j\leq n} \left\{ \|\mathbf{V}^T\mathbf{e}_j\|_2 \leq \sqrt{\frac{\mu_0 r}{n}} \right\}, \end{aligned} \tag{5.7.5}$$

where $\mathbf{e}_i \in \mathbb{R}^{m\times 1}$ and $\mathbf{e}_j \in \mathbb{R}^{n\times 1}$ are the standard basis vectors, with only nonzero ith and jth entries, respectively.

The following theorem shows that only the standard incoherence condition is required for the unique recovery of an incomplete data matrix.

THEOREM 5.11 [98] *Suppose that a data matrix* $\mathbf{D}$ *satisfies the standard incoherence condition (5.7.5) with parameter* μ_0. *There exist universal constants* $c_0, c_1, c_2 > 0$ *such that if*

$$p \geq c_0 \frac{\mu_0 r \log^2(m+n)}{\min\{m,n\}}, \tag{5.7.6}$$

then the matrix $\mathbf{D}$ *is the unique optimal solution to (5.7.4) with probability at least* $1 - c_1(m+n)^{-c_2}$.

5.7.3 Singular Value Thresholding Algorithm

Consider the matrix completion problem

$$\min \left\{ \tau\|\mathbf{X}\|_* + \frac{1}{2}\|\mathbf{X}\|_F^2 \right\} \quad \text{subject to} \quad \mathcal{P}_\Omega(\mathbf{X}) = \mathcal{P}_\Omega(\mathbf{D}), \tag{5.7.7}$$

with Lagrangian objective function

$$L(\mathbf{X},\mathbf{Y}) = \tau\|\mathbf{X}\|_* + \frac{1}{2}\|\mathbf{X}\|_F^2 + \langle \mathbf{Y}, \mathcal{P}_\Omega(\mathbf{D}) - \mathcal{P}_\Omega(\mathbf{X})\rangle, \tag{5.7.8}$$

where the inner product $\langle \mathbf{A},\mathbf{B}\rangle = \text{tr}(\mathbf{A}^T\mathbf{B}) = \langle \mathbf{B},\mathbf{A}\rangle$.

Notice that $\langle \mathbf{Y}, \mathcal{P}_\Omega(\mathbf{D})\rangle$ and $\|\mathcal{P}_\Omega(\mathbf{Y})\|_F^2 = \langle \mathcal{P}_\Omega(\mathbf{Y}), \mathcal{P}_\Omega(\mathbf{Y})\rangle$ are constants with respect to $\mathbf{X}$. We delete the former term and add the latter term to yield

$$\begin{aligned}\min_{\mathbf{X}} L(\mathbf{X},\mathbf{Y}) &= \min_{\mathbf{X}} \left\{\tau\|\mathbf{X}\|_* + \frac{1}{2}\|\mathbf{X}\|_F^2 + \langle \mathbf{Y}, \mathcal{P}_\Omega(\mathbf{D}-\mathbf{X})\rangle\right\} \\ &= \min_{\mathbf{X}} \left\{\tau\|\mathbf{X}\|_* + \frac{1}{2}\|\mathbf{X}\|_F^2 - \langle \mathbf{X}, \mathcal{P}_\Omega(\mathbf{Y})\rangle + \frac{1}{2}\|\mathcal{P}_\Omega(\mathbf{Y})\|_F^2\right\}. \end{aligned} \tag{5.7.9}$$

Here we have $\langle \mathbf{Y}, \mathcal{P}_\Omega\mathbf{X}\rangle = \text{tr}(\mathbf{Y}^T\mathcal{P}_\Omega\mathbf{X}) = \text{tr}(\mathbf{X}^T\mathcal{P}_\Omega^T\mathbf{Y}) = \text{tr}(\mathbf{X}^T\mathcal{P}_\Omega\mathbf{Y}) = \langle \mathbf{X}, \mathcal{P}_\Omega\mathbf{Y}\rangle$, because $\mathcal{P}_\Omega^T = \mathcal{P}_\Omega$.

By Theorem 5.9, from Equation (5.7.9) we have

$$\begin{aligned}\mathbf{X}_{k+1} &= \arg\min_{\mathbf{X}} \left\{\tau\|\mathbf{X}\|_* + \frac{1}{2}\|\mathbf{X}\|_F^2 - \langle \mathbf{X}, \mathcal{P}_\Omega\mathbf{Y}_k\rangle + \frac{1}{2}\|\mathcal{P}_\Omega\mathbf{Y}_k\|_F^2\right\} \\ &= \arg\min_{\mathbf{X}} \left\{\|\mathbf{X}\|_* + \frac{1}{2}\|\mathbf{X} - \mathcal{P}_\Omega\mathbf{Y}_k\|_F^2\right\}. \end{aligned} \tag{5.7.10}$$

Because $\mathbf{Y}_k = \mathcal{P}_\Omega(\mathbf{Y}_k), \forall\, k \geq 0$, Theorem 5.9 gives

$$\mathbf{X}_{k+1} = \mathcal{D}_\tau(\mathcal{P}_\Omega\mathbf{Y}_k) = \mathbf{prox}_{\tau\|\cdot\|_*}(\mathbf{Y}_k) = \mathbf{U}\text{soft}(\mathbf{\Sigma},\tau)\mathbf{V}^T, \tag{5.7.11}$$

where $\mathbf{U\Sigma V}^T$ is the SVD of $\mathbf{Y}_k$.

Moreover, from Equation (5.7.8), the gradient of the Lagrangian function at the point $\mathbf{Y}$ is given by

$$\frac{\partial L(\mathbf{X},\mathbf{Y})}{\partial \mathbf{Y}^T} = \mathcal{P}_\Omega(\mathbf{X}-\mathbf{D}).$$

Hence the gradient descent update of $\mathbf{Y}_k$ is

$$\mathbf{Y}_{k+1} = \mathbf{Y}_k + \mu\mathcal{P}_\Omega(\mathbf{D} - \mathbf{X}_{k+1}), \tag{5.7.12}$$

where μ is the step size.

Equations (5.7.11) and (5.7.12) give the iteration sequence [75]

$$\left.\begin{aligned}\mathbf{X}_{k+1} &= \mathbf{U}\text{soft}(\mathbf{\Sigma},\tau)(\mathbf{V}^T), \\ \mathbf{Y}_{k+1} &= \mathbf{Y}_k + \mu\mathcal{P}_\Omega(\mathbf{D} - \mathbf{X}_{k+1}).\end{aligned}\right\} \tag{5.7.13}$$

The above iteration sequence has the following two features [75]:

(1) *Sparsity* For each $k \geq 0$, the entries of $\mathbf{Y}_k$ outside Ω are equal to zero, and thus $\mathbf{Y}_k$ is a sparse matrix. This fact can be used to evaluate the shrink function rapidly.

(2) *Low-rank property* The matrix $\mathbf{X}_k$ has low rank, and hence the iteration algorithm requires only a small amount of storage since we need to keep only the principal factors in the memory.

Algorithm 5.7 gives a singular value thresholding algorithm for matrix completion.

Algorithm 5.7 Singular value thresholding for matrix completion [75]

---

given: $\mathbf{D}_{ij}, (i,j) \in \Omega$, step size δ, tolerance ϵ, parameter τ, increment ℓ, maximum iteration count $k_{\max}$.

initialization: Define k_0 as the integer such that $\frac{\tau}{\delta \|\mathcal{P}_\Omega(\mathbf{D})\|_2} \in (k_0 - 1, k_0]$.

Set $\mathbf{Y}^0 = k_0 \delta \mathcal{P}_\Omega(\mathbf{D})$. Put r_0 and $k = 1$.

repeat

1. Compute the SVD $(\mathbf{U}, \mathbf{\Sigma}, \mathbf{V}) = \text{svd}(\mathbf{Y}^{k-1})$.
2. Set $r_k = \max\{j | \sigma_j > \tau\}$.
3. Update $\mathbf{X}^k = \sum_{j=1}^{r_k} (\sigma_j - \tau) \mathbf{u}_j \mathbf{v}_j^T$.
4. **exit if** $\|\mathcal{P}_\Omega(\mathbf{X}^k - \mathbf{D})\|_F / \|\mathcal{P}_\Omega \mathbf{D}\|_F \leq \epsilon$ or $k = k_{\max}$.
5. Compute $Y_{ij}^k = \begin{cases} Y_{ij}^{k-1} + \delta(D_{ij} - X_{ij}^k), & \text{if } (i,j) \in \Omega, \\ 0, & \text{if } (i,j) \notin \Omega. \end{cases}$

return $k \leftarrow k + 1$.

output: $\mathbf{X}_{\text{opt}} = \mathbf{X}_k$.

5.7.4 Fast and Accurate Matrix Completion

As compared with the nuclear norm, a better convex relaxation of the rank function is truncated nuclear norm regularization, proposed recently in [221].

Definition 5.1 [221] Given a matrix $\mathbf{X} \in \mathbb{R}^{m \times n}$ with $r = \text{rank}(\mathbf{X})$, its *truncated nuclear norm* $\|\mathbf{X}\|_{*,r}$ is defined as the sum of $\min\{m,n\} - r$ minimum singular values, i.e., $\|\mathbf{X}\|_{*,r} = \sum_{i=r+1}^{\min\{m,n\}} \sigma_i$.

Using the truncated nuclear norm instead of the nuclear norm, the standard matrix completion (5.7.4) becomes the *truncated nuclear norm minimization*

$$\min_{\mathbf{X}} \|\mathbf{X}\|_{*,r} \quad \text{subject to} \quad \mathcal{P}_\Omega(\mathbf{X}) = \mathcal{P}_\Omega(\mathbf{D}). \tag{5.7.14}$$

Unlike the convex nuclear norm, the truncated nuclear norm $\|\mathbf{X}\|_{*,r}$ is nonconvex, and thus (5.7.14) is difficult to solve. Fortunately, the following theorem can be used to overcome this difficulty.

Theorem 5.12 [221] *Given any matrix* $\mathbf{X} \in \mathbb{R}^{m\times n}$ *and* $r \leq \min\{m,n\}$, *for two matrices* $\mathbf{A} \in \mathbb{R}^{r\times m}, \mathbf{B} \in \mathbb{R}^{r\times n}$, *if* $\mathbf{A}\mathbf{A}^T = \mathbf{I}_{r\times r}$ *and* $\mathbf{B}\mathbf{B}^T = \mathbf{I}_{r\times r}$ *then*

$$\mathrm{tr}(\mathbf{A}\mathbf{X}\mathbf{B}^T) \leq \sum_{i=1}^{r} \sigma_i(\mathbf{X}). \tag{5.7.15}$$

A simple choice is $\mathbf{A} = [\mathbf{u}_1, \ldots, \mathbf{u}_r]^T$, $\mathbf{B} = [\mathbf{v}_1, \ldots, \mathbf{v}_r]^T$, and $\mathbf{X} = \sum_{i=1}^{r} \sigma_i(\mathbf{X})\mathbf{u}_i\mathbf{v}_i^T$ is the truncated SVD of $\mathbf{X}$. Under this choice, one has

$$\max_{\mathbf{A}\mathbf{A}^T=\mathbf{I},\, \mathbf{B}\mathbf{B}^T=\mathbf{I}} \left\{ \mathrm{tr}(\mathbf{A}\mathbf{X}\mathbf{B}^T) = \sum_{i=1}^{r} \sigma_i(\mathbf{X}) \right\}. \tag{5.7.16}$$

Hence the truncated nuclear norm minimization (5.7.14) can be relaxed as

$$\min_{\mathbf{X}} \left\{ \|\mathbf{X}\|_* - \max_{\mathbf{A}\mathbf{A}^T=\mathbf{I},\mathbf{B}\mathbf{B}^T=\mathbf{I}} \mathrm{tr}(\mathbf{A}\mathbf{X}\mathbf{B}^T) \right\} \quad \text{subject to} \quad \mathcal{P}_\Omega(\mathbf{X}) = \mathcal{P}_\Omega(\mathbf{D}). \tag{5.7.17}$$

If $\mathbf{A}_l$ and $\mathbf{B}_l$ are found then the above optimization problem becomes

$$\min_{\mathbf{X}} \left\{ \|\mathbf{X}\|_* - \mathrm{tr}(\mathbf{A}_l\mathbf{X}\mathbf{B}_l^T) \right\} \quad \text{subject to} \quad \mathcal{P}_\Omega(\mathbf{X}) = \mathcal{P}_\Omega(\mathbf{D}), \tag{5.7.18}$$

and its Lagrangian objective function is

$$L(\mathbf{X}) = \|\mathbf{X}\|_* - \mathrm{tr}(\mathbf{A}_l\mathbf{X}\mathbf{B}_l^T) + \frac{\lambda}{2}\|\mathcal{P}_\Omega\mathbf{X} - \mathcal{P}_\Omega\mathbf{D}\|_F^2, \tag{5.7.19}$$

where λ is a Lagrange multiplier.

Let

$$f(\mathbf{X}) = \frac{\lambda}{2}\|\mathcal{P}_\Omega\mathbf{X} - \mathcal{P}_\Omega\mathbf{D}\|_F^2 - \mathrm{tr}(\mathbf{A}_l\mathbf{X}\mathbf{B}_l^T) \quad \text{and} \quad h(\mathbf{X}) = \|\mathbf{X}\|_*,$$

then $f(\mathbf{X})$ is a convex, differentiable and L-Lipschitz function, and $h(\mathbf{X})$ is a convex but nonsmooth function.

Consider the quadratic approximation of $f(\mathbf{X})$ at a given point $\mathbf{Y}$:

$$\hat{f}(\mathbf{X}) = f(\mathbf{Y}) + \langle \mathbf{X} - \mathbf{Y}, \nabla f(\mathbf{Y})\rangle + \frac{L}{2}\|\mathbf{X} - \mathbf{Y}\|_F^2.$$

Then, the Lagrangian objective function $L(\mathbf{X}) = h(\mathbf{X}) + \hat{f}(\mathbf{X})$ can be reformulated as

$$Q(\mathbf{X}, \mathbf{Y}) = \|\mathbf{X}\|_* + f(\mathbf{Y}) + \langle \mathbf{X} - \mathbf{Y}, \nabla f(\mathbf{Y})\rangle + \frac{L}{2}\|\mathbf{X} - \mathbf{Y}\|_F^2. \tag{5.7.20}$$

In minimizing $Q(\mathbf{X}, \mathbf{Y})$ with respect to $\mathbf{X}$, the second additive term $f(\mathbf{Y})$ is constant and can be ignored. On the other hand, another constant term, $\frac{1}{2L}\|\nabla f(\mathbf{Y})\|_F^2 =$

$\frac{L}{2}\left\|\frac{1}{L}\nabla f(\mathbf{Y})\right\|_F^2$, can be added in $Q(\mathbf{X},\mathbf{Y})$. Then, we have

$$\begin{aligned}
&\min_{\mathbf{X}} Q(\mathbf{X},\mathbf{Y}) \\
&= \min_{\mathbf{X}} \left\{ \|\mathbf{X}\|_* + \frac{L}{2}\|\mathbf{X}-\mathbf{Y}\|_F^2 + L\langle \mathbf{X}-\mathbf{Y}, \frac{1}{L}\nabla f(\mathbf{Y})\rangle + \frac{L}{2}\|\frac{1}{L}\nabla f(\mathbf{Y})\|_F^2 \right\} \\
&= \min_{\mathbf{X}} \left\{ \|\mathbf{X}\|_* + \frac{L}{2}\|\mathbf{X}-\mathbf{Y}+\frac{1}{L}\nabla f(\mathbf{Y})\|_F^2 \right\} \quad \text{(by Proposition 5.1)}
\end{aligned}$$

which yields

$$\begin{aligned}
\mathbf{X}_{k+1} &= \arg\min_{\mathbf{X}} \left\{ \|\mathbf{X}\|_* + \frac{L}{2}\|\mathbf{X}-\mathbf{Y}_k+\frac{1}{L}\nabla f(\mathbf{Y}_k)\|_F^2 \right\} \\
&= \mathbf{prox}_{L^{-1}\|\cdot\|_*}\left(\mathbf{Y}_k - L^{-1}\nabla f(\mathbf{Y}_k)\right) \\
&= \mathbf{U}\,\text{soft}(\mathbf{\Sigma}, t)\,\mathbf{V}^T \quad \text{(by Theorem 5.9)},
\end{aligned} \tag{5.7.21}$$

where $t = 1/L$ and $\mathbf{U\Sigma V}^T$ is the SVD of the matrix $\mathbf{Y}_k - t_k\nabla f(\mathbf{Y}_k)$.

From $f(\mathbf{Y}) = \frac{1}{2}\lambda\|\mathcal{P}_\Omega\mathbf{Y} - \mathcal{P}_\Omega\mathbf{D}\|_F^2 - \text{tr}(\mathbf{A}_l\mathbf{Y}\mathbf{B}_l^T)$ it is easy to see that $\nabla f(\mathbf{Y}) = \lambda(\mathcal{P}_\Omega\mathbf{Y} - \mathcal{P}_\Omega\mathbf{D}) - \mathbf{A}_l^T\mathbf{B}_l$, and thus

$$(\mathbf{U},\mathbf{\Sigma},\mathbf{V}) = \text{svd}\left(\mathbf{Y}_k + t_k(\mathbf{A}_l^T\mathbf{B}_l - \lambda(\mathcal{P}_\Omega\mathbf{Y}_k - \mathcal{P}_\Omega\mathbf{D}))\right). \tag{5.7.22}$$

Algorithm 5.8 give the *accelerated proximal gradient line* (APGL) *search algorithm* for truncated nuclear norm minimization that was presented in [221].

Algorithm 5.8 Truncated nuclear norm minimization via APGL search

---

input: $\mathbf{A}_l, \mathbf{B}_l$; $\mathbf{D}_\Omega = \mathcal{P}_\Omega\mathbf{D}$ and tolerance ϵ.

initialization: $t_1 = 1, \mathbf{X}_1 = \mathbf{D}_\Omega, \mathbf{Y}_1 = \mathbf{X}_1$.

repeat

1. Compute $\mathbf{W}_{k+1} = \mathbf{Y}_k + t_k\left(\mathbf{A}_l^T\mathbf{B}_l - \lambda(\mathcal{P}_\Omega\mathbf{Y}_k - \mathcal{P}_\Omega\mathbf{D})\right)$.
2. Perform the SVD $(\mathbf{U},\mathbf{\Sigma},\mathbf{V}) = \text{svd}(\mathbf{W}_{k+1})$.
3. Compute $\mathbf{X}_{k+1} = \sum_{i=1}^{\min\{m,n\}} \max\{\sigma_i - t_k, 0\}\mathbf{u}_i\mathbf{v}_i^T$.
4. **exit if** $\|\mathbf{X}_{k+1} - \mathbf{X}_k\|_F \leq \epsilon$.
5. Compute $t_{k+1} = \dfrac{1+\sqrt{1+4t_k^2}}{2}$.
6. Update $\mathbf{Y}_{k+1} = \mathbf{X}_{k+1} + \dfrac{t_k - 1}{t_{k+1}}(\mathbf{X}_{k+1} - \mathbf{X}_k)$.

return $k \leftarrow k+1$.

output: $\mathbf{X}_{\text{opt}} = \mathbf{X}_k$.

However, the truncated nuclear norm minimization problem (5.7.19) can also be written in the standard form of the alternating-direction method of multipliers

(ADMM):

$$\min_{\mathbf{X},\mathbf{W}} \left\{\|\mathbf{X}\|_* - \mathrm{tr}(\mathbf{A}_l\mathbf{W}\mathbf{B}_l^T)\right\} \quad \text{subject to } \mathbf{X}=\mathbf{W},\ \mathcal{P}_\Omega\mathbf{W}=\mathcal{P}_\Omega\mathbf{D}. \tag{5.7.23}$$

The corresponding augmented Lagrangian objective function is given by

$$L(\mathbf{X},\mathbf{Y},\mathbf{W},\beta)=\|\mathbf{X}\|_* - \mathrm{tr}(\mathbf{A}_l\mathbf{W}\mathbf{B}_l^T)+\frac{\beta}{2}\|\mathbf{X}-\mathbf{W}\|_F^2+\langle\mathbf{Y},\mathbf{X}-\mathbf{W}\rangle,$$

where $\beta>0$ is the penalty parameter. Ignoring the constant term $\mathrm{tr}(\mathbf{A}_l\mathbf{W}\mathbf{B}_l^T)$ and adding the constant term $\beta^{-1}\|\mathbf{Y}\|_F^2$, we obtain

$$\begin{aligned}\mathbf{X}_{k+1}&=\arg\min_{\mathbf{X}}\left\{\|\mathbf{X}\|_*+\frac{\beta}{2}\|\mathbf{X}-\mathbf{W}_k\|_F^2+\langle\mathbf{Y}_k,\mathbf{X}-\mathbf{W}_k\rangle+\frac{1}{\beta}\|\mathbf{Y}_k\|_F^2\right\}\\&=\arg\min_{\mathbf{X}}\left\{\|\mathbf{X}\|_*+\frac{\beta}{2}\|\mathbf{X}-(\mathbf{W}_k-\frac{1}{\beta}\mathbf{Y}_k)\|_F^2\right\}\\&=\mathbf{prox}_{\beta^{-1}\|\cdot\|_*}(\mathbf{W}_k-\beta^{-1}\mathbf{Y}_k)=\mathbf{U}\mathrm{soft}(\boldsymbol{\Sigma},\beta^{-1})\mathbf{V}^T,\end{aligned} \tag{5.7.24}$$

where $\mathbf{U}\boldsymbol{\Sigma}\mathbf{V}^T$ is the SVD of the matrix $\mathbf{W}_k-\beta^{-1}\mathbf{Y}_k$.

Similarly, ignoring the constant term $\|\mathbf{X}_{k+1}\|_*$ and adding two constant terms $\langle\mathbf{X}_{k+1},\mathbf{A}_l^T\mathbf{B}_l\rangle$ and $(2\beta)^{-1}\|\mathbf{Y}_k-\mathbf{A}_l^T\mathbf{B}_l\|_F^2$, we have

$$\begin{aligned}\mathbf{W}_{k+1}&=\arg\min_{\mathcal{P}_\Omega\mathbf{W}=\mathcal{P}_\Omega\mathbf{D}} L(\mathbf{X}_{k+1},\mathbf{W},\mathbf{Y}_k,\beta)\\&=\arg\min_{\mathcal{P}_\Omega\mathbf{W}=\mathcal{P}_\Omega\mathbf{D}}\left\{-\mathrm{tr}(\mathbf{A}_l\mathbf{W}\mathbf{B}_l^T)+\frac{\beta}{2}\|\mathbf{X}_{k+1}-\mathbf{W}\|_F^2+\langle\mathbf{X}_{k+1},\mathbf{A}_l^T\mathbf{B}_l\rangle\right.\\&\qquad\left.+\langle\mathbf{Y}_k,\mathbf{X}_{k+1}-\mathbf{W}\rangle+\frac{1}{2\beta}\|\mathbf{Y}_k-\mathbf{A}_l^T\mathbf{B}_l\|_F^2\right\}\\&=\arg\min_{\mathcal{P}_\Omega\mathbf{W}=\mathcal{P}_\Omega\mathbf{D}}\left\{\frac{\beta}{2}\left\|\mathbf{W}-\left(\mathbf{X}_{k+1}-\frac{1}{\beta}(\mathbf{Y}_k+\mathbf{A}_l^T\mathbf{B}_l)\right)\right\|_F^2\right\}.\end{aligned} \tag{5.7.25}$$

Its closed solution is given by

$$\mathbf{W}_{k+1}=\mathbf{X}_{k+1}-\frac{1}{\beta}(\mathbf{Y}_k+\mathbf{A}_l^T\mathbf{B}_l), \tag{5.7.26}$$

$$\mathbf{W}_{k+1}=\mathcal{P}_\Omega(\mathbf{W}_{k+1})+\mathcal{P}_\Omega(\mathbf{D}). \tag{5.7.27}$$

Algorithm 5.9 lists the ADMM algorithm for truncated nuclear norm minimization.

Unlike traditional nuclear norm heuristics, which take into account all the singular values, the truncated nuclear norm regularization approach in [221] aims to minimize the sum of the smallest $\min\{m,n\}-r$ singular values, where r is the matrix rank. This helps to give a better approximation to the matrix rank when the original matrix has a low-rank structure.

Algorithm 5.9 Truncated nuclear norm minimization via ADMM [221]

input: $\mathbf{A}_l, \mathbf{B}_l; \mathbf{D}_{ij}, (i,j) \in \Omega$, tolerance ϵ and singular value threshold τ.
initialization: $\mathbf{X}_1 = \mathbf{D}_\Omega, \mathbf{W}_1 = \mathbf{X}_1, \mathbf{Y}_1 = \mathbf{X}_1$ and $\beta = 1$.
repeat
1. Compute the SVD $(\mathbf{U}, \boldsymbol{\Sigma}, \mathbf{V}) = \text{svd}(\mathbf{W}_k - \beta^{-1}\mathbf{Y}_k)$.
2. Compute $\mathbf{X}_{k+1} = \sum_{i=1}^{\min\{m,n\}} \max\{\sigma_i - \beta^{-1}, 0\}\mathbf{u}_i\mathbf{v}_i^T$.
3. **exit if** $\|\mathbf{X}_{k+1} - \mathbf{X}_k\|_F \le \epsilon$.
4. Update $\mathbf{W}_{k+1} = \mathbf{X}_{k+1} - \beta^{-1}(\mathbf{Y}_k + \mathbf{A}_l^T\mathbf{B}_l)$.
5. Update $\mathbf{W}_{k+1} = \mathcal{P}_\Omega(\mathbf{W}_{k+1}) + \mathcal{P}_\Omega(\mathbf{D})$.
6. Compute $\mathbf{Y}_{k+1} = \mathbf{Y}_k + \beta(\mathbf{X}_{k+1} - \mathbf{W}_{k+1})$.

return $k \leftarrow k+1$.
output: $\mathbf{X}_{\text{opt}} = \mathbf{X}_k$.

Exercises

5.1 Given the matrix

$$\mathbf{A} = \begin{bmatrix} 1 & 1 \\ 1 & 1 \\ 0 & 0 \end{bmatrix},$$

find its SVD.

5.2 Use the SVD of the matrix

$$\mathbf{A} = \begin{bmatrix} 1 & 0 \\ 0 & 1 \\ 1 & 1 \end{bmatrix}$$

to find its Moore–Penrose inverse $\mathbf{A}^\dagger$, and verify that it satisfies the four Moore–Penrose conditions.

5.3 Prove that if $\mathbf{A}$ is a normal matrix then each of its singular values is the modulus of the corresponding eigenvalue of $\mathbf{A}^T\mathbf{A}$.

5.4 Compute the SVD of the following two matrices:

$$\mathbf{A} = \begin{bmatrix} 1 & -1 \\ 3 & -3 \\ -3 & 3 \end{bmatrix} \quad \text{and} \quad \mathbf{A} = \begin{bmatrix} 3 & 4 & 5 \\ 2 & 1 & 7 \end{bmatrix}.$$

5.5 Given the matrix

$$\mathbf{A} = \begin{bmatrix} -149 & -50 & -154 \\ 537 & 180 & 546 \\ -27 & 9 & -25 \end{bmatrix},$$

find the singular values of $\mathbf{A}$ and the left and right singular vectors associated with the smallest singular value.

5.6 Given a complex matrix $\mathbf{A} = \mathbf{B} + \mathrm{j}\mathbf{C}$, where $\mathbf{B}, \mathbf{C} \in \mathbb{R}^{m \times n}$, use the SVD of the block matrix $\begin{bmatrix} \mathbf{B} & -\mathbf{C} \\ \mathbf{C} & \mathbf{B} \end{bmatrix}$ to represent the SVD of $\mathbf{A}$.

5.7 Let $\mathbf{A} = \mathbf{x}\mathbf{p}^H + \mathbf{y}\mathbf{q}^H$, where $\mathbf{x} \perp \mathbf{y}$ and $\mathbf{p} \perp \mathbf{q}$. Find the Frobenius norm $\|\mathbf{A}\|_F$. (*Hint:* Compute $\mathbf{A}^H\mathbf{A}$, and find the singular values of $\mathbf{A}$.)

5.8 Let $\mathbf{A} = \mathbf{U}\mathbf{\Sigma}\mathbf{V}^H$ be the SVD of $\mathbf{A}$. What is the relationship between the singular values of $\mathbf{A}^H$ and $\mathbf{A}$?

5.9 Given an $n \times n$ (where $n \geq 3$) matrix

$$\mathbf{A} = \begin{bmatrix} 1 & a & \cdots & a \\ a & 1 & \cdots & a \\ \vdots & \vdots & \ddots & \vdots \\ a & a & \cdots & 1 \end{bmatrix},$$

use the SVD to find the value of a such that $\mathrm{rank}(\mathbf{A}) = n - 1$.

5.10 Show that if $\mathbf{A}$ is a square matrix then $|\det(\mathbf{A})|$ is equal to the product of the singular values of $\mathbf{A}$.

5.11 Suppose that $\mathbf{A}$ is a invertible matrix. Find the SVD of $\mathbf{A}^{-1}$.

5.12 Show that if $\mathbf{A}$ is an $n \times n$ positive definite matrix then the singular values of $\mathbf{A}$ and its eigenvalues are the same.

5.13 Let $\mathbf{A}$ be an $m \times n$ matrix and $\mathbf{P}$ be an $m \times m$ positive definite matrix. Prove that the singular values of $\mathbf{PA}$ and $\mathbf{A}$ are the same. What is the relationship between the left and right singular vectors?

5.14 Let $\mathbf{A}$ be an $m \times n$ matrix and $\lambda_1, \ldots, \lambda_n$ be the eigenvalues of the matrix $\mathbf{A}^T\mathbf{A}$ with the corresponding eigenvectors $\mathbf{u}_1, \ldots, \mathbf{u}_n$. Prove that the singular values σ_i of $\mathbf{A}$ are equal to the norms $\|\mathbf{A}\mathbf{u}_i\|$, namely $\sigma_i = \|\mathbf{A}\mathbf{u}_i\|, i = 1, \ldots, n$.

5.15 Let $\lambda_1, \ldots, \lambda_n$ and $\mathbf{u}_1, \ldots, \mathbf{u}_n$ be the eigenvalues and eigenvectors of the matrix $\mathbf{A}^T\mathbf{A}$, respectively. Suppose that $\mathbf{A}$ has r nonzero singular values. Show that $\{\mathbf{A}\mathbf{u}_1, \ldots, \mathbf{A}\mathbf{u}_r\}$ is an orthogonal basis set of the column space $\mathrm{Col}(\mathbf{A})$ and that $\mathrm{rank}(\mathbf{A}) = r$.

5.16 Let $\mathbf{B}, \mathbf{C} \in \mathbb{R}^{m \times n}$. Find the relationship between the singular values and the singular vectors of the complex matrix $\mathbf{A} = \mathbf{B} + \mathrm{j}\,\mathbf{C}$ and those of the real block matrix $\begin{bmatrix} \mathbf{B} & -\mathbf{C} \\ \mathbf{C} & \mathbf{B} \end{bmatrix}$.

5.17 Use the singular vectors of the matrix $\mathbf{A} \in \mathbb{R}^{m \times n}$ $(m \geq n)$ to represent the eigenvectors of $\begin{bmatrix} \mathbf{O} & \mathbf{A}^T \\ \mathbf{A} & \mathbf{O} \end{bmatrix}$.

5.18 Prove that the distance from the point $\begin{pmatrix} \mathbf{a}_i \\ b_i \end{pmatrix} \in \mathbb{R}^{n+1}$ to the straight line $\mathbf{x}^T\mathbf{a} = b$ is equal to

$$d_i = \sqrt{\frac{|\mathbf{a}_i^T\mathbf{x} - b_i|^2}{\mathbf{x}^T\mathbf{x} + 1}}.$$

5.19 Adopt the MATLAB function [U, S, V] = svd(**X**) to solve the matrix equation $\mathbf{Ax} = \mathbf{b}$, where

$$\mathbf{A} = \begin{bmatrix} 1 & 1 & 1 \\ 3 & 1 & 3 \\ 1 & 0 & 1 \\ 2 & 2 & 1 \end{bmatrix}, \quad \mathbf{b} = \begin{bmatrix} 1 \\ 4 \\ 3 \\ 2 \end{bmatrix}.$$

5.20 Suppose that the observation data of a computer simulation are generated by

$$x(n) = \sqrt{20}\sin(2\pi\, 0.2n) + \sqrt{2}\sin(2\pi\, 0.215n) + w(n),$$

where $w(n)$ is a Gaussian white noise with mean 0 and covariance 1 and $n = 1, 2, \ldots, 128$. Perform 50 independent runs. Determine the effective rank of the autocorrelation matrix

$$\mathbf{R} = \begin{bmatrix} r(0) & r(-1) & \cdots & r(-2p) \\ r(1) & r(0) & \cdots & r(-2p+1) \\ \vdots & \vdots & \vdots & \vdots \\ r(M) & r(M-1) & \cdots & r(M-2p) \end{bmatrix},$$

where $M = 50$, $p = 10$, and $r(k) = \frac{1}{128}\sum_{i=1}^{128-k} x(i)x(i+k)$ represents sample autocorrelation functions of the observation data (unknown observation data are taken as zero).

5.21 [179] Apply the SVD to show that if $\mathbf{A} \in \mathbb{R}^{m\times n}$ $(m \geq n)$ then there are $\mathbf{Q} \in \mathbb{R}^{m\times n}$ and $\mathbf{P} \in \mathbb{R}^{n\times n}$ such that $\mathbf{A} = \mathbf{QP}$, where $\mathbf{Q}^T\mathbf{Q} = \mathbf{I}_n$, and $\mathbf{P}$ is symmetric and nonnegative definite. This decomposition is sometimes called the polar decomposition, because it is similar to the complex decomposition $z = |z|\mathrm{e}^{\mathrm{j}\arg(z)}$.

6

Solving Matrix Equations

In a wide variety of science and engineering disciplines, we often encounter three main types of matrix equation:

(1) *Over-determined matrix equations* (ODMEs) $\mathbf{A}_{m\times n}\mathbf{x}_{n\times 1} = \mathbf{b}_{m\times 1}$ with $m > n$; the data matrix $\mathbf{A}$ and the data vector $\mathbf{b}$ are known, but one or both may have errors or interference.

(2) *Blind matrix equations* (BMEs) Only the data matrix $\mathbf{X}$ is known, while the coefficient matrix $\mathbf{A}$ and the parameter matrix $\mathbf{S}$ are unknowns in the equation $\mathbf{A}_{m\times n}\mathbf{S}_{n\times p} = \mathbf{X}_{m\times p}$.

(3) *Under-determined sparse matrix equations* (UDSMEs) $\mathbf{Ax} = \mathbf{b}$ with $m < n$; the data vector $\mathbf{b}$ and the data matrix $\mathbf{A}$ are known, but the data vector $\mathbf{b}$ is a sparse vector.

In this chapter we discuss the methods for solving these three types of matrix equation:

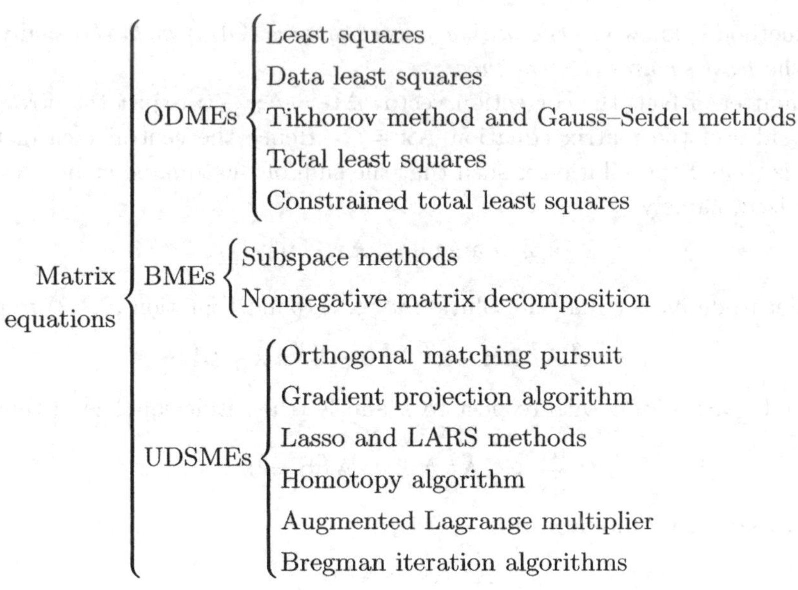

6.1 Least Squares Method

Linear parameter estimation is widely used in science and technology problems, and the least squares method is the most common.

6.1.1 Ordinary Least Squares Methods

Consider an over-determined matrix equation $\mathbf{Ax} = \mathbf{b}$, where $\mathbf{b}$ is an $m \times 1$ data vector, $\mathbf{A}$ is an $m \times n$ data matrix, and $m > n$.

Suppose that in the data vector there is additive observation error or noise, i.e., $\mathbf{b} = \mathbf{b}_0 + \mathbf{e}$, where $\mathbf{b}_0$ and $\mathbf{e}$ are the errorless data vector and the additive error vector, respectively.

In order to counterbalance the influence of the error $\mathbf{e}$ on the matrix equation solution, we introduce a correction vector $\Delta\mathbf{b}$ and use it to "perturb" the data vector $\mathbf{b}$. Our goal is to make the correction term $\Delta\mathbf{b}$ that compensates the uncertainty existing in the data vector $\mathbf{b}$ as small as possible so that $\mathbf{b} + \Delta\mathbf{b} \to \mathbf{b}_0$ and hence the following transformation is achieved:

$$\mathbf{Ax} = \mathbf{b} + \Delta\mathbf{b} \quad \Rightarrow \quad \mathbf{Ax} = \mathbf{b}_0. \tag{6.1.1}$$

That is to say, if we select the correction vector $\Delta\mathbf{b} = \mathbf{Ax} - \mathbf{b}$ directly, and make $\Delta\mathbf{b}$ as small as possible, we can achieve the solution of errorless matrix equation $\mathbf{Ax} = \mathbf{b}_0$.

This idea for solving the matrix equation can be described by the optimization problem

$$\min_{\mathbf{x}} \left\{ \|\Delta\mathbf{b}\|^2 = \|\mathbf{Ax} - \mathbf{b}\|_2^2 = (\mathbf{Ax} - \mathbf{b})^T(\mathbf{Ax} - \mathbf{b}) \right\}. \tag{6.1.2}$$

Such a method is known as the *ordinary least squares* (OLS) *method*, usually called simply the *least squares* (LS) *method.*

As a matter of fact, the correction vector $\Delta\mathbf{b} = \mathbf{Ax} - \mathbf{b}$ is just the error vector of both sides of the matrix equation $\mathbf{Ax} = \mathbf{b}$. Hence the central idea of the LS method is to find the solution $\mathbf{x}$ such that the sum of the squared error $\|\mathbf{Ax} - \mathbf{b}\|_2^2$ is minimized, namely

$$\hat{\mathbf{x}}_{\text{LS}} = \arg\min_{\mathbf{x}} \ \|\mathbf{Ax} - \mathbf{b}\|_2^2. \tag{6.1.3}$$

In order to derive an analytic solution for $\mathbf{x}$, expand Equation (6.1.2) to get

$$\phi = \mathbf{x}^T\mathbf{A}^T\mathbf{Ax} - \mathbf{x}^T\mathbf{A}^T\mathbf{b} - \mathbf{b}^T\mathbf{Ax} + \mathbf{b}^T\mathbf{b}.$$

Find the derivative of ϕ with respect to $\mathbf{x}$ and let the result equal zero; thus

$$\frac{\mathrm{d}\phi}{\mathrm{d}\mathbf{x}} = 2\mathbf{A}^T\mathbf{Ax} - 2\mathbf{A}^T\mathbf{b} = 0.$$

That is to say, the solution $\mathbf{x}$ must satisfy

$$\mathbf{A}^T\mathbf{Ax} = \mathbf{A}^T\mathbf{b}. \tag{6.1.4}$$

The above equation is "identifiable" or "unidentifiable" depending on the rank of the $m \times n$ matrix $\mathbf{A}$.

- *Identifiable* When $\mathbf{Ax} = \mathbf{b}$ is an over-determined equation, it has a unique solution

$$\mathbf{x}_{\mathrm{LS}} = (\mathbf{A}^T\mathbf{A})^{-1}\mathbf{A}^T\mathbf{b} \quad \text{if} \quad \mathrm{rank}(\mathbf{A}) = n, \tag{6.1.5}$$

or

$$\mathbf{x}_{\mathrm{LS}} = (\mathbf{A}^T\mathbf{A})^{\dagger}\mathbf{A}^T\mathbf{b} \quad \text{if} \quad \mathrm{rank}(\mathbf{A}) < n, \tag{6.1.6}$$

where $\mathbf{B}^{\dagger}$ denotes the Moore–Penrose inverse matrix of $\mathbf{B}$. In parameter estimation theory the unknown parameter vector $\mathbf{x}$ is said to be uniquely identifiable if it is uniquely determined.

- *Unidentifiable* For an under-determined equation $\mathbf{Ax} = \mathbf{b}$, if $\mathrm{rank}(\mathbf{A}) = m < n$, then different solutions for $\mathbf{x}$ give the same value of $\mathbf{Ax}$. Clearly, although the data vector $\mathbf{b}$ can provide some information about $\mathbf{Ax}$, we cannot distinguish the different parameter vectors $\mathbf{x}$ corresponding to the same $\mathbf{Ax}$. Such a unknown vector $\mathbf{x}$ is said to be unidentifiable.

6.1.2 Properties of Least Squares Solutions

In parameter estimation, an estimate $\hat{\boldsymbol{\theta}}$ of the parameter vector $\boldsymbol{\theta}$ is called an *unbiased estimator*, if its mathematical expectation is equal to the true unknown parameter vector, i.e., $E\{\hat{\boldsymbol{\theta}}\} = \boldsymbol{\theta}$. Further, an unbiased estimator is called the *optimal unbiased estimator* if it has the minimum variance. Similarly, for an over-determined matrix equation $\mathbf{A}\boldsymbol{\theta} = \mathbf{b} + \mathbf{e}$ with noisy data vector $\mathbf{b}$, if the mathematical expectation of the LS solution $\hat{\boldsymbol{\theta}}_{\mathrm{LS}}$ is equal to the true parameter vector $\boldsymbol{\theta}$, i.e., $E\{\hat{\boldsymbol{\theta}}_{\mathrm{LS}}\} = \boldsymbol{\theta}$, and has the minimum variance, then $\hat{\boldsymbol{\theta}}_{\mathrm{LS}}$ is the optimal unbiased solution.

Theorem 6.1 (Gauss–Markov theorem) *Consider the set of linear equations*

$$\mathbf{Ax} = \mathbf{b} + \mathbf{e}, \tag{6.1.7}$$

where the $m \times n$ matrix $\mathbf{A}$ and the $n \times 1$ vector $\mathbf{x}$ are respectively a constant matrix and the parameter vector; $\mathbf{b}$ is the $m \times 1$ vector with random error vector $\mathbf{e} = [e_1, \ldots, e_m]^T$ whose mean vector and covariance matrix are respectively

$$E\{\mathbf{e}\} = \mathbf{0}, \quad \mathrm{Cov}(\mathbf{e}) = E\{\mathbf{e}\mathbf{e}^H\} = \sigma^2\mathbf{I}.$$

Then the $n \times 1$ parameter vector $\mathbf{x}$ is the optimal unbiased solution $\hat{\mathbf{x}}$ if and only if $\mathrm{rank}(\mathbf{A}) = n$. *In this case the optimal unbiased solution is given by the LS solution*

$$\hat{\mathbf{x}}_{\mathrm{LS}} = (\mathbf{A}^H\mathbf{A})^{-1}\mathbf{A}^H\mathbf{b}, \tag{6.1.8}$$

and its covariance satisfies

$$\mathrm{Var}(\hat{\mathbf{x}}_{\mathrm{LS}}) \leq \mathrm{Var}(\tilde{\mathbf{x}}), \tag{6.1.9}$$

where $\tilde{\mathbf{x}}$ *is any other solution of the matrix equation* $\mathbf{Ax} = \mathbf{b} + \mathbf{e}$.

Proof From the assumption condition $E\{\mathbf{e}\} = \mathbf{0}$, it follows that

$$E\{\mathbf{b}\} = E\{\mathbf{Ax}\} - E\{\mathbf{e}\} = \mathbf{Ax}. \tag{6.1.10}$$

Using $\mathrm{Cov}(\mathbf{e}) = E\{\mathbf{ee}^H\} = \sigma^2\mathbf{I}$, and noticing that $\mathbf{Ax}$ is statistically uncorrelated with the error vector $\mathbf{e}$, we have

$$\begin{aligned} E\{\mathbf{bb}^H\} &= E\{(\mathbf{Ax} - \mathbf{e})(\mathbf{Ax} - \mathbf{e})^H\} = E\{\mathbf{Axx}^H\mathbf{A}^H\} + E\{\mathbf{ee}^H\} \\ &= \mathbf{Axx}^H\mathbf{A}^H + \sigma^2\mathbf{I}. \end{aligned} \tag{6.1.11}$$

Since $\mathrm{rank}(\mathbf{A}) = n$ and the matrix product $\mathbf{A}^H\mathbf{A}$ is nonsingular, we get

$$\begin{aligned} E\{\hat{\mathbf{x}}_{\mathrm{LS}}\} &= E\{(\mathbf{A}^H\mathbf{A})^{-1}\mathbf{A}^H\mathbf{b}\} = (\mathbf{A}^H\mathbf{A})^{-1}\mathbf{A}^H E\{\mathbf{b}\} \\ &= (\mathbf{A}^H\mathbf{A})^{-1}\mathbf{A}^H\mathbf{Ax} = \mathbf{x}, \end{aligned}$$

i.e., the LS solution $\hat{\mathbf{x}}_{\mathrm{LS}} = (\mathbf{A}^H\mathbf{A})^{-1}\mathbf{A}^H\mathbf{b}$ is an unbiased solution to the matrix equation $\mathbf{Ax} = \mathbf{b} + \mathbf{e}$.

Next, we show that $\hat{\mathbf{x}}_{\mathrm{LS}}$ has the minimum variance. To this end we suppose that $\mathbf{x}$ has an alternative solution $\tilde{\mathbf{x}}$ that is given by

$$\tilde{\mathbf{x}} = \hat{\mathbf{x}}_{\mathrm{LS}} + \mathbf{Cb} + \mathbf{d},$$

where $\mathbf{C}$ and $\mathbf{d}$ are a constant matrix and a constant vector, respectively. The solution $\tilde{\mathbf{x}}$ is unbiased, i.e.,

$$E\{\tilde{\mathbf{x}}\} = E\{\hat{\mathbf{x}}_{\mathrm{LS}}\} + E\{\mathbf{Cb}\} + \mathbf{d} = \mathbf{x} + \mathbf{CAx} + \mathbf{d} = \mathbf{x} + \mathbf{CAx} + \mathbf{d}, \quad \forall\mathbf{x},$$

if and only if

$$\mathbf{CA} = \mathbf{O} \text{ (the zero matrix)}, \quad \mathbf{d} = \mathbf{0}. \tag{6.1.12}$$

Using these constraint conditions, it can be seen that $E\{\mathbf{Cb}\} = \mathbf{C}E\{\mathbf{b}\} = \mathbf{CA}\boldsymbol{\theta} = \mathbf{0}$. Then we have

$$\begin{aligned} \mathrm{Cov}(\tilde{\mathbf{x}}) &= \mathrm{Cov}(\hat{\mathbf{x}}_{\mathrm{LS}} + \mathbf{Cb}) = E\left\{\left((\hat{\mathbf{x}}_{\mathrm{LS}} - \mathbf{x}) + \mathbf{Cb}\right)\left((\hat{\mathbf{x}}_{\mathrm{LS}} - \mathbf{x}) + \mathbf{Cb}\right)^H\right\} \\ &= \mathrm{Cov}(\hat{\mathbf{x}}_{\mathrm{LS}}) + E\left\{(\hat{\mathbf{x}}_{\mathrm{LS}} - \mathbf{x})(\mathbf{Cb})^H\right\} + E\left\{\mathbf{Cb}(\hat{\mathbf{x}}_{\mathrm{LS}} - \mathbf{x})^H\right\} \\ &\quad + E\left\{\mathbf{Cbb}^H\mathbf{C}^H\right\}. \end{aligned} \tag{6.1.13}$$

However, from (6.1.10)–(6.1.12), we have

$$\begin{aligned} E\{(\hat{\mathbf{x}}_{\rm LS}-\mathbf{x})(\mathbf{Cb})^H\} &= E\{(\mathbf{A}^H\mathbf{A})^{-1}\mathbf{A}^H\mathbf{b}\mathbf{b}^H\mathbf{C}^H\} - E\{\mathbf{x}\mathbf{b}^H\mathbf{C}^H\}\\ &= (\mathbf{A}^H\mathbf{A})^{-1}\mathbf{A}^H E\{\mathbf{b}\mathbf{b}^H\}\mathbf{C}^H - \mathbf{x}E\{\mathbf{b}^H\}\mathbf{C}^H\\ &= (\mathbf{A}^H\mathbf{A})^{-1}\mathbf{A}^H(\mathbf{A}\mathbf{x}\mathbf{x}^H\mathbf{A}^H+\sigma^2\mathbf{I})\mathbf{C}^H - \mathbf{x}\mathbf{x}^H\mathbf{A}^H\mathbf{C}^H\\ &= \mathbf{O},\\ E\{\mathbf{Cb}(\hat{\mathbf{x}}_{\rm LS}-\mathbf{x})^H\} &= \left(E\{(\hat{\mathbf{x}}_{\rm LS}-\mathbf{x})(\mathbf{Cb})^H\}\right)^H = \mathbf{O},\\ E\{\mathbf{C}\mathbf{b}\mathbf{b}^H\mathbf{C}^H\} &= \mathbf{C}E\{\mathbf{b}\mathbf{b}^H\}\mathbf{C}^H = \mathbf{C}(\mathbf{A}\mathbf{x}\mathbf{x}^H\mathbf{A}^H+\sigma^2\mathbf{I})\mathbf{C}^H\\ &= \sigma^2\mathbf{C}\mathbf{C}^H. \end{aligned}$$

Hence, (6.1.13) can be simplified to

$$\mathrm{Cov}(\tilde{\mathbf{x}}) = \mathrm{Cov}(\hat{\mathbf{x}}_{\rm LS}) + \sigma^2\mathbf{C}\mathbf{C}^H. \tag{6.1.14}$$

By the property $\mathrm{tr}(\mathbf{A}+\mathbf{B}) = \mathrm{tr}(\mathbf{A}) + \mathrm{tr}(\mathbf{B})$ of the trace function, and noticing that for a random vector $\mathbf{x}$ with zero mean, $\mathrm{tr}(\mathrm{Cov}(\mathbf{x})) = \mathrm{Var}(\mathbf{x})$, we can rewrite (6.1.14) as

$$\mathrm{Var}(\tilde{\mathbf{x}}) = \mathrm{Var}(\hat{\mathbf{x}}_{\rm LS}) + \sigma^2\mathrm{tr}(\mathbf{C}\mathbf{C}^H) \geq \mathrm{Var}(\hat{\mathbf{x}}_{\rm LS}),$$

because $\mathrm{tr}(\mathbf{C}\mathbf{C}^H) \geq 0$. This shows that $\hat{\mathbf{x}}_{\rm LS}$ has the minimum variance, and hence is the optimal unbiased solution to the matrix equation $\mathbf{Ax} = \mathbf{b} + \mathbf{e}$. □

Notice that the condition in the Gauss–Markov theorem $\mathrm{Cov}(\mathbf{e}) = \sigma^2\mathbf{I}$ implies that all components of the additive noise vector $\mathbf{e}$ are mutually uncorrelated and have the same variance σ^2. Only in this case is the LS solution unbiased and optimal.

Now consider the relationship between the LS solution and the maximum likelihood (ML) solution under the conditions of the Gauss–Markov theorem.

If the additive noise vector $\mathbf{e} = [e_1,\ldots,e_m]^T$ is a complex Gaussian random vector with independently identical distribution (iid) then from (1.5.18) it is known that the pdf of $\mathbf{e}$ is given by

$$f(\mathbf{e}) = \frac{1}{\pi^m|\boldsymbol{\Gamma}_e|}\exp\left(-(\mathbf{e}-\boldsymbol{\mu}_e)^H\boldsymbol{\Gamma}_e^{-1}(\mathbf{e}-\boldsymbol{\mu}_e)\right), \tag{6.1.15}$$

where $|\boldsymbol{\Gamma}_e| = \sigma_1^2\cdots\sigma_m^2$.

Under the conditions of the Gauss–Markov theorem (i.e., all the iid Gaussian random variables of the error vector have zero mean and the same variance σ^2), the pdf of the additive error vector reduces to

$$f(\mathbf{e}) = \frac{1}{(\pi\sigma^2)^m}\exp\left(-\frac{1}{\sigma^2}\mathbf{e}^H\mathbf{e}\right) = \frac{1}{(\pi\sigma^2)^m}\exp\left(-\frac{1}{\sigma^2}\|\mathbf{e}\|_2^2\right), \tag{6.1.16}$$

whose likelihood function is

$$L(\mathbf{e}) = \log f(\mathbf{e}) = -\frac{1}{\pi^m\sigma^{2(m+1)}}\|\mathbf{e}\|_2^2 = -\frac{1}{\pi^m\sigma^{2(m+1)}}\|\mathbf{Ax}-\mathbf{b}\|_2^2. \tag{6.1.17}$$

Thus the ML solution of the matrix equation $\mathbf{Ax} = \mathbf{b}$ is given by

$$\begin{aligned}\hat{\mathbf{x}}_{\rm ML} &= \arg\max_{\mathbf{x}} \left\{ -\frac{1}{\pi^m \sigma^{2(m+1)}} \|\mathbf{Ax} - \mathbf{b}\|_2^2 \right\} \\ &= \arg\min_{\mathbf{x}} \left\{ \frac{1}{2} \|\mathbf{Ax} - \mathbf{b}\|_2^2 \right\} = \hat{\mathbf{x}}_{\rm LS}. \end{aligned} \tag{6.1.18}$$

That is to say, under the conditions of the Gauss–Markov theorem, the ML solution $\hat{\mathbf{x}}_{\rm ML}$ and the LS solution $\hat{\mathbf{x}}_{\rm LS}$ are equivalent for the matrix equation $\mathbf{Ax} = \mathbf{b}$.

It is easily seen that when the error vector $\mathbf{e}$ is a zero-mean Gaussian random vector but its entries have differing variances since the covariance matrix $\mathbf{\Gamma}_e \neq \sigma^2 \mathbf{I}$, the ML solution in this case is given by

$$\hat{\mathbf{x}}_{\rm ML} = \arg\max_{\mathbf{x}} \left\{ \frac{1}{\pi^m \sigma_1^2 \cdots \sigma_m^2} \exp\left(-\mathbf{e}^H \mathbf{\Gamma}_e^{-1} \mathbf{e}\right) \right\},$$

which is obviously not equal to the LS solution $\hat{\mathbf{x}}_{\rm LS}$; namely, the LS solution is no longer optimal in the case of differing variances.

6.1.3 Data Least Squares

Consider again an over-determined matrix equation $\mathbf{Ax} = \mathbf{b}$. Unlike in an ordinary LS problem, it is assumed that the data vector $\mathbf{b}$ does not contain the observed error or noise; on the contrary, the data matrix $\mathbf{A} = \mathbf{A}_0 + \mathbf{E}$ contains the observed error or noise, and every element of the error matrix $\mathbf{E}$ obeys an independent Gaussian distribution with the zero-mean and equal variance.

The perturbation matrix $\Delta\mathbf{A}$ is used to correct the erroneous-data matrix $\mathbf{A}$ in such a way that $\mathbf{A} + \Delta\mathbf{A} = \mathbf{A}_0 + \mathbf{E} + \Delta\mathbf{A} \to \mathbf{A}_0$. As with the ordinary LS method, by forcing $(\mathbf{A} + \Delta\mathbf{A})\mathbf{x} = \mathbf{b}$ to compensate the errors in the data matrix, one can obtain

$$(\mathbf{A} + \Delta\mathbf{A})\mathbf{x} = \mathbf{b} \;\Rightarrow\; \mathbf{A}_0\mathbf{x} = \mathbf{b}.$$

In this case, the optimal solution of $\mathbf{x}$ is given by

$$\hat{\mathbf{x}}_{\rm DLS} = \arg\min_{\mathbf{x}} \|\Delta\mathbf{A}\|_2^2 \quad \text{subject to} \quad \mathbf{b} \in \text{Range}(\mathbf{A} + \Delta\mathbf{A}). \tag{6.1.19}$$

This is the *data least squares* (DLS) *method.* In the above equation, the constraint condition $\mathbf{b} \in \text{Range}(\mathbf{A} + \Delta\mathbf{A})$ implies that, for every given exact data vector $\mathbf{b} \in \mathbb{C}^m$ and the erroneous data matrix $\mathbf{A} \in \mathbb{C}^{m\times n}$, we can always find a vector $\mathbf{x} \in \mathbb{C}^n$ such that $(\mathbf{A}+\Delta\mathbf{A})\mathbf{x} = \mathbf{b}$. Hence the two constraint conditions $\mathbf{b} \in \text{Range}(\mathbf{A}+\Delta\mathbf{A})$ and $(\mathbf{A} + \Delta\mathbf{A})\mathbf{x} = \mathbf{b}$ are equivalent.

By using the Lagrange multiplier method, the constrained data LS problem (6.1.19) becomes the unconstrained optimization problem

$$\min \left\{ L(\mathbf{x}) = \text{tr}(\Delta\mathbf{A}(\Delta\mathbf{A})^H) + \boldsymbol{\lambda}^H(\mathbf{Ax} + \Delta\mathbf{A}\mathbf{x} - \mathbf{b}) \right\}. \tag{6.1.20}$$

Let the conjugate gradient matrix $\partial L(\mathbf{x})/(\partial \Delta \mathbf{A}^H) = \mathbf{O}$ (the zero matrix). Then we have $\Delta \mathbf{A} = -\boldsymbol{\lambda}\mathbf{x}^H$. Substitute $\Delta \mathbf{A} = -\boldsymbol{\lambda}\mathbf{x}^H$ into the constraint condition $(\mathbf{A} + \Delta \mathbf{A})\mathbf{x} = \mathbf{b}$ to yield $\boldsymbol{\lambda} = (\mathbf{A}\mathbf{x} - \mathbf{b})/(\mathbf{x}^H\mathbf{x})$, and thus $\Delta \mathbf{A} = -(\mathbf{A}\mathbf{x} - \mathbf{b})\mathbf{x}^H/(\mathbf{x}^H\mathbf{x})$. Therefore, the objective function is given by

$$J(\mathbf{x}) = \|\Delta \mathbf{A}\|_2^2 = \mathrm{tr}(\Delta \mathbf{A}(\Delta \mathbf{A})^H) = \mathrm{tr}\left(\frac{(\mathbf{A}\mathbf{x} - \mathbf{b})\mathbf{x}^H}{\mathbf{x}^H\mathbf{x}} \frac{\mathbf{x}(\mathbf{A}\mathbf{x} - \mathbf{b})^H}{\mathbf{x}^H\mathbf{x}}\right).$$

Then, using the trace function property $\mathrm{tr}(\mathbf{BC}) = \mathrm{tr}(\mathbf{CB})$, we immediately have

$$J(\mathbf{x}) = \frac{(\mathbf{A}\mathbf{x} - \mathbf{b})^H(\mathbf{A}\mathbf{x} - \mathbf{b})}{\mathbf{x}^H\mathbf{x}}, \tag{6.1.21}$$

which yields

$$\hat{\mathbf{x}}_{\mathrm{DLS}} = \arg\min_{\mathbf{x}} \frac{(\mathbf{A}\mathbf{x} - \mathbf{b})^H(\mathbf{A}\mathbf{x} - \mathbf{b})}{\mathbf{x}^H\mathbf{x}}. \tag{6.1.22}$$

This is the *data least squares solution* of the over-determined equation $\mathbf{A}\mathbf{x} = \mathbf{b}$.

6.2 Tikhonov Regularization and Gauss–Seidel Method

Suppose that we are solving an over-determined matrix equation $\mathbf{A}_{m\times n}\mathbf{x}_{n\times 1} = \mathbf{b}_{m\times 1}$ (where $m > n$). The ordinary LS method and the data LS method are based on two basic assumptions:

(1) the data matrix $\mathbf{A}$ is nonsingular or of full column rank;
(2) the data vector $\mathbf{b}$ or the data matrix $\mathbf{A}$ includes additive noise or error.

In this section we discuss the regularization method for solving matrix equations with rank deficiency or noise.

6.2.1 Tikhonov Regularization

When $m = n$ and $\mathbf{A}$ is nonsingular, the solution of the matrix equation $\mathbf{A}\mathbf{x} = \mathbf{b}$ is given by $\hat{\mathbf{x}} = \mathbf{A}^{-1}\mathbf{b}$; and when $m > n$ and $\mathbf{A}_{m\times n}$ is of full column rank, the solution is $\hat{\mathbf{x}}_{\mathrm{LS}} = \mathbf{A}^{\dagger}\mathbf{b} = (\mathbf{A}^H\mathbf{A})^{-1}\mathbf{A}^H\mathbf{b}$.

A problem is that the data matrix $\mathbf{A}$ is often rank deficient in engineering applications. In these cases, the solution $\hat{\mathbf{x}} = \mathbf{A}^{-1}\mathbf{b}$ or $\hat{\mathbf{x}}_{\mathrm{LS}} = (\mathbf{A}^H\mathbf{A})^{-1}\mathbf{A}^H\mathbf{b}$ may diverge; even if a solution exists, it may be a bad approximation to the unknown vector $\mathbf{x}$. If, however, we were lucky and happened to find a reasonable approximation of $\mathbf{x}$, the error estimate $\|\mathbf{x} - \hat{\mathbf{x}}\| \leq \|\mathbf{A}^{-1}\|\,\|\mathbf{A}\hat{\mathbf{x}} - \mathbf{b}\|$ or $\|\mathbf{x} - \hat{\mathbf{x}}\| \leq \|\mathbf{A}^{\dagger}\|\,\|\mathbf{A}\hat{\mathbf{x}} - \mathbf{b}\|$ would be greatly disappointing [347]. By observation, it can easily be seen that the problem lies in the inversion of the covariance matrix $\mathbf{A}^H\mathbf{A}$ of the rank-deficient data matrix $\mathbf{A}$.

As an improvement of the LS cost function $\frac{1}{2}\|\mathbf{A}\mathbf{x} - \mathbf{b}\|_2^2$, Tikhonov [464] in 1963 proposed the *regularized least squares cost function*

$$J(\mathbf{x}) = \frac{1}{2}\left(\|\mathbf{A}\mathbf{x} - \mathbf{b}\|_2^2 + \lambda\|\mathbf{x}\|_2^2\right), \tag{6.2.1}$$

where $\lambda \geq 0$ is the *regularization parameter*.

The conjugate gradient of the cost function $J(\mathbf{x})$ with respect to the argument $\mathbf{x}$ is given by

$$\frac{\partial J(\mathbf{x})}{\partial \mathbf{x}^H} = \frac{\partial}{\partial \mathbf{x}^H}\left((\mathbf{Ax}-\mathbf{b})^H(\mathbf{Ax}-\mathbf{b}) + \lambda \mathbf{x}^H\mathbf{x}\right) = \mathbf{A}^H\mathbf{Ax} - \mathbf{A}^H\mathbf{b} + \lambda\mathbf{x}.$$

Let $\partial J(\mathbf{x})/\partial \mathbf{x}^H = \mathbf{0}$; then

$$\hat{\mathbf{x}}_{\text{Tik}} = (\mathbf{A}^H\mathbf{A} + \lambda\mathbf{I})^{-1}\mathbf{A}^H\mathbf{b}. \tag{6.2.2}$$

This method using $(\mathbf{A}^H\mathbf{A} + \lambda\mathbf{I})^{-1}$ instead of the direct inverse of the covariance matrix $(\mathbf{A}^H\mathbf{A})^{-1}$ is called the *Tikhonov regularization method* (or simply the *regularized method*). In signal processing and image processing, the regularization method is sometimes known as the *relaxation method.*

The Tikhonov-regularization based solution to the matrix equation $\mathbf{Ax} = \mathbf{b}$ is called the *Tikhonov regularization solution*, denoted $\mathbf{x}_{\text{Tik}}$.

The idea of the Tikhonov regularization method is that by adding a very small perturbation λ to each diagonal entry of the covariance matrix $\mathbf{A}^H\mathbf{A}$ of a rank deficient matrix $\mathbf{A}$, the inversion of the singular covariance matrix $\mathbf{A}^H\mathbf{A}$ becomes the inversion of a nonsingular matrix $\mathbf{A}^H\mathbf{A} + \lambda\mathbf{I}$, thereby greatly improving the numerical stability of the solution process for the rank-deficient matrix equation $\mathbf{Ax} = \mathbf{b}$.

Obviously, if the data matrix $\mathbf{A}$ is of full column rank but includes error or noise, we must adopt the opposite method to Tikhonov regularization and add a very small negative perturbation, $-\lambda$, to each diagonal entry of the covariance matrix $\mathbf{A}^H\mathbf{A}$. Such a method using a very small negative perturbation matrix $-\lambda\mathbf{I}$ is called the *anti-Tikhonov regularization method* or *anti-regularized method*, and the solution is given by

$$\hat{\mathbf{x}} = (\mathbf{A}^H\mathbf{A} - \lambda\mathbf{I})^{-1}\mathbf{A}^H\mathbf{b}. \tag{6.2.3}$$

The total least squares method is a typical anti-regularized method and will be discussed later.

As stated above, the regularization parameter λ should take a small value so that $(\mathbf{A}^H\mathbf{A}+\lambda\mathbf{I})^{-1}$ can better approximate $(\mathbf{A}^H\mathbf{A})^{-1}$ and yet can avoid the singularity of $\mathbf{A}^H\mathbf{A}$; hence the Tikhonov regularization method can obviously improve the numerical stability in solving the singular and ill-conditioned equations. This is so because the matrix $\mathbf{A}^H\mathbf{A}$ is positive semi-definite, so the eigenvalues of $\mathbf{A}^H\mathbf{A} + \lambda\mathbf{I}$ lie in the interval $[\lambda, \lambda + \|\mathbf{A}\|_F^2]$, which gives a condition number

$$\text{cond}(\mathbf{A}^H\mathbf{A} + \lambda\mathbf{I}) \leq (\lambda + \|\mathbf{A}\|_F^2)/\lambda. \tag{6.2.4}$$

Clearly, as compared with the condition number $\text{cond}(\mathbf{A}^H\mathbf{A}) \leq \infty$, the condition number of the Tikhonov regularization is a great improvement.

In order to improve further the solution of singular and ill-conditioned matrix equations, one can adopt *iterative Tikhonov regularization* [347].

Let the initial solution vector $\mathbf{x}_0 = \mathbf{0}$ and the initial residual vector $\mathbf{r}_0 = \mathbf{b}$; then the solution vector and the residual vector can be updated via the following iteration formulas:

$$\begin{aligned} \mathbf{x}_k &= \mathbf{x}_{k-1} + (\mathbf{A}^H\mathbf{A} + \lambda\mathbf{I})^{-1}\mathbf{A}^H\mathbf{r}_{k-1}, \\ \mathbf{r}_k &= \mathbf{b} - \mathbf{A}\mathbf{x}_k, \quad k = 1, 2, \ldots \end{aligned} \tag{6.2.5}$$

Let $\mathbf{A} = \mathbf{U}\mathbf{\Sigma}\mathbf{V}^H$ be the SVD of the matrix $\mathbf{A}$, then $\mathbf{A}^H\mathbf{A} = \mathbf{V}\mathbf{\Sigma}^2\mathbf{V}^H$ and thus the ordinary LS solution and the *Tikhonov regularization solution* are respectively given by

$$\hat{\mathbf{x}}_{\text{LS}} = (\mathbf{A}^H\mathbf{A})^{-1}\mathbf{A}^H\mathbf{b} = \mathbf{V}\mathbf{\Sigma}^{-1}\mathbf{U}^H\mathbf{b}, \tag{6.2.6}$$

$$\hat{\mathbf{x}}_{\text{Tik}} = (\mathbf{A}^H\mathbf{A} + \sigma_{\min}^2\mathbf{I})^{-1}\mathbf{A}^H\mathbf{b} = \mathbf{V}(\mathbf{\Sigma}^2 + \sigma_{\min}^2\mathbf{I})^{-1}\mathbf{\Sigma}\mathbf{U}^H\mathbf{b}, \tag{6.2.7}$$

where $\sigma_{\min}$ is the minimal nonzero singular value of $\mathbf{A}$. If the matrix $\mathbf{A}$ is singular or ill-conditioned, i.e., $\sigma_n = 0$ or $\sigma_n \approx 0$, then some diagonal entry of $\mathbf{\Sigma}^{-1}$ will be $1/\sigma_n = \infty$, which thereby makes the LS solution diverge. On the contrary, the SVD based Tikhonov regularization solution $\hat{\mathbf{x}}_{\text{Tik}}$ has good numerical stability because the diagonal entries of the matrix

$$(\mathbf{\Sigma}^2 + \delta^2\mathbf{I})^{-1}\mathbf{\Sigma} = \mathbf{Diag}\left(\frac{\sigma_1}{\sigma_1^2 + \sigma_{\min}^2}, \ldots, \frac{\sigma_n}{\sigma_n^2 + \sigma_{\min}^2}\right) \tag{6.2.8}$$

lie in the interval $[0, \sigma_1(\sigma_1^2 + \sigma_{\min}^2)^{-1}]$.

When the regularization parameter λ varies in the definition interval $[0, \infty)$, the family of solutions for a regularized LS problem is known as its *regularization path*.

A Tikhonov regularization solution has the following important properties [248].

1. *Linearity* The Tikhonov regularization LS solution $\hat{\mathbf{x}}_{\text{Tik}} = (\mathbf{A}^H\mathbf{A}+\lambda\mathbf{I})^{-1}\mathbf{A}^H\mathbf{b}$ is a linear function of the observed data vector $\mathbf{b}$.
2. *Limit characteristic when $\lambda \to 0$* When the regularization parameter $\lambda \to 0$, the Tikhonov regularization LS solution converges to the ordinary LS solution or the Moore–Penrose solution $\lim_{\lambda \to 0} \hat{\mathbf{x}}_{\text{Tik}} = \hat{\mathbf{x}}_{\text{LS}} = \mathbf{A}^\dagger\mathbf{b} = (\mathbf{A}^H\mathbf{A})^{-1}\mathbf{A}^H\mathbf{b}$. The solution point $\hat{\mathbf{x}}_{\text{Tik}}$ has the minimum ℓ_2-norm among all the feasible points meeting $\mathbf{A}^H(\mathbf{A}\mathbf{x} - \mathbf{b}) = \mathbf{0}$:

$$\hat{\mathbf{x}}_{\text{Tik}} = \underset{\mathbf{A}^T(\mathbf{b}-\mathbf{A}\mathbf{x})=\mathbf{0}}{\arg\min} \|\mathbf{x}\|_2. \tag{6.2.9}$$

3. *Limit characteristic when $\lambda \to \infty$* When $\lambda \to \infty$, the optimal solution of the Tikhonov regularization LS problem converges to the zero vector: $\lim_{\lambda\to\infty} \hat{\mathbf{x}}_{\text{Tik}} = \mathbf{0}$.
4. *Regularization path* When the regularization parameter λ varies in $[0, \infty)$, the optimal solution of the Tikhonov regularization LS problem is a smooth function of the regularization parameter, i.e., when λ decreases to zero, the optimal solution converges to the Moore–Penrose solution and when λ increases, the optimal solution converges to the zero vector.

Tikhonov regularization can effectively prevent the divergence of the LS solution $\hat{\mathbf{x}}_{\rm LS} = (\mathbf{A}^T\mathbf{A})^{-1}\mathbf{A}^T\mathbf{b}$ when $\mathbf{A}$ is rank deficient; thereby it obviously improves the convergence property of the LS algorithm and the alternating LS algorithm, and so is widely applied.

6.2.2 Regularized Gauss–Seidel Method

Let $X_i \subseteq \mathbb{R}^{n_i}$ be the feasible set of the $n_i \times 1$ vector $\mathbf{x}_i$. Consider the nonlinear minimization problem

$$\min_{\mathbf{x}\in X} \{f(\mathbf{x}) = f(\mathbf{x}_1, \ldots, \mathbf{x}_m)\}, \tag{6.2.10}$$

where $\mathbf{x} \in X = X_1 \times X_2 \times \cdots \times X_m \subseteq \mathbb{R}^n$ is the Cartesian product of a closed nonempty convex set $X_i \subseteq \mathbb{R}^{n_i}, i = 1, \ldots, m$, and $\sum_{i=1}^m n_i = n$.

Equation (6.2.10) is an unconstrained optimization problem with m coupled variable vectors. An efficient approach for solving this class of *coupled optimization problems* is the *block nonlinear Gauss–Seidel method*, called simply the *GS method* [42], [190].

In every iteration of the GS method, $m-1$ variable vectors are regarded as known, and the remaining variable vector is minimized. This idea constitutes the basic framework of the GS method for solving nonlinear unconstrained optimization problem (6.2.10).

(1) Initialize $m-1$ variable vectors $\mathbf{x}_i, i = 2, \ldots, m$, and let $k = 0$.

(2) Find the solution of the separated suboptimization problem

$$\mathbf{x}_i^{k+1} = \arg\min_{\mathbf{y}\in X_i} f\big(\mathbf{x}_1^{k+1}, \ldots, \mathbf{x}_{i-1}^{k+1}, \mathbf{y}, \mathbf{x}_{i+1}^k, \ldots, \mathbf{x}_m^k\big), \quad i = 1, \ldots, m. \tag{6.2.11}$$

At the $(k+1)$th iteration of updating the vector $\mathbf{x}_i$, the vectors $\mathbf{x}_1, \ldots, \mathbf{x}_{i-1}$ have been updated as $\mathbf{x}_1^{k+1}, \ldots, \mathbf{x}_{i-1}^{k+1}$, so these updated subvectors and the vectors $\mathbf{x}_{i+1}^k, \ldots, \mathbf{x}_m^k$ yet be updated are to be regarded as known.

(3) Test whether the m variable vectors are all convergent. If they are convergent then output the optimization results $(\mathbf{x}_1^{k+1}, \ldots, \mathbf{x}_m^{k+1})$; otherwise, let $k \leftarrow k+1$, return to Equation (6.2.11) and continue to iterate until the convergence criterion is met.

If the objective function $f(\mathbf{x})$ of the optimization (6.2.10) is an LS error function (for example $\|\mathbf{A}\mathbf{x} - \mathbf{b}\|_2^2$), then the GS method is customarily called the *alternating least squares* (ALS) *method.*

Example 6.1 Consider the full-rank decomposition of an $m \times n$ known data matrix $\mathbf{X} = \mathbf{AB}$, where the $m \times r$ matrix $\mathbf{A}$ is of full column rank and the $r \times n$ matrix

$\mathbf{B}$ is of full row rank. Let the cost function of the matrix full-rank decomposition be

$$f(\mathbf{A}, \mathbf{B}) = \frac{1}{2}\|\mathbf{X} - \mathbf{A}\mathbf{B}\|_F^2. \tag{6.2.12}$$

Then the ALS algorithm first initializes the matrix $\mathbf{A}$. At the $(k+1)$th iteration, from the fixed matrix $\mathbf{A}_k$ we update the LS solution of $\mathbf{B}$ as follows:

$$\mathbf{B}_{k+1} = (\mathbf{A}_k^T \mathbf{A}_k)^{-1} \mathbf{A}_k^T \mathbf{X}. \tag{6.2.13}$$

Next, from the transpose of the matrix decomposition $\mathbf{X}^T = \mathbf{B}^T \mathbf{A}^T$, we can immediately update the LS solution of $\mathbf{A}^T$:

$$\mathbf{A}_{k+1}^T = (\mathbf{B}_{k+1} \mathbf{B}_{k+1}^T)^{-1} \mathbf{B}_{k+1} \mathbf{X}^T. \tag{6.2.14}$$

The above two kinds of LS procedures are performed alternately. Once the ALS algorithm converges, the optimization results of the matrix decomposition can be obtained.

To analyze the convergence of the GS method, we first introduce the concepts of the limit point and the critical point.

Let S be a subset in topological space X; a point x in the space X is called a *limit point* of the subset S if every neighbourhood of x contains at least one point in S apart from x itself. In other words, the limit point x is a point in topological space X approximated by a point in S (other than x).

A point on a function curve at which the derivative is equal to zero or does not exist plays an important role in the optimization problem. A point x is known as a *critical point* of the function $f(x)$ if x lies in the definition domain of the function and the derivative $f'(x) = 0$ or does not exist. The geometric interpretation of a critical point is that the tangent of the point on the curve is horizontal or vertical, or does not exist.

Consider a real function $f(x) = x^4 - 4x^2$ whose derivative $f'(x) = 4x^3 - 8x$. From $f'(x) = 0$ we get three critical points of $f(x)$: $x = 0, -\sqrt{2}$ and $\sqrt{2}$.

The above concepts on the limit point and critical points of a scalar variable are easily extended to a vector variable.

DEFINITION 6.1 The vector $\mathbf{x} \in \mathbb{R}^n$ is said to be a limit point of the vector sequence $\{\mathbf{x}_k\}_{k=1}^{\infty}$ in the vector space $\mathbb{R}^n$ if there is a subsequence of $\{\mathbf{x}_k\}_{k=1}^{\infty}$ that converges to $\mathbf{x}$.

DEFINITION 6.2 Let $f : X \to \mathbb{R}$ (where $X \subset \mathbb{R}^n$) be a real function; the vector $\bar{\mathbf{x}} \in \mathbb{R}^n$ is known as a critical point of the function $f(\mathbf{x})$ if the following condition is satisfied:

$$\mathbf{g}^T(\bar{\mathbf{x}})(\mathbf{y} - \bar{\mathbf{x}}) \geq 0, \quad \forall\, \mathbf{y} \in X, \tag{6.2.15}$$

where $\mathbf{g}^T(\bar{\mathbf{x}})$ represents the transpose of the vector function $\mathbf{g}(\bar{\mathbf{x}})$ and either $\mathbf{g}(\bar{\mathbf{x}}) =$

$\nabla f(\bar{\mathbf{x}})$ is the gradient vector of a continuous differentiable function $f(\mathbf{x})$ at the point $\bar{\mathbf{x}}$ or $\mathbf{g}(\bar{\mathbf{x}}) \in \partial f(\bar{\mathbf{x}})$ is the subgradient vector of a nondifferentiable nonsmooth function $f(\mathbf{x})$ at the point $\bar{\mathbf{x}}$.

If $X = \mathbb{R}^n$ or $\bar{\mathbf{x}}$ is an interior point in X then the critical-point condition (6.2.15) reduces to the stationary-point condition $\mathbf{0} = \nabla f(\bar{\mathbf{x}})$ (for a continuous differentiable objective function) or $\mathbf{0} \in \partial f(\bar{\mathbf{x}})$ (for a nonsmooth objective function) for the unconstrained minimization problem $\min f(\mathbf{x})$.

Let $\mathbf{x}^k = (\mathbf{x}_1^k, \ldots, \mathbf{x}_m^k)$ denote the iteration results generated by the GS algorithm; it is expected that the iterative sequence $\{\mathbf{x}^k\}_{k=1}^{\infty}$ has limit points and that each limit point is a critical point of the objective function f. For the optimization problem (6.2.10) this convergence performance of the GS algorithm depends on the quasi-convexity of the objective function f.

Put $\alpha \in (0,1)$ and $\mathbf{y}_i \neq \mathbf{x}_i$. The objective function $f(\mathbf{x})$ in the ALS problem (6.2.10) is called the *quasi-convex* with respect to $\mathbf{x}_i \in X_i$ if

$$
\begin{aligned}
&f(\mathbf{x}_1, \ldots, \mathbf{x}_{i-1}, \alpha\mathbf{x}_i + (1-\alpha)\mathbf{y}_i, \mathbf{x}_{i+1}, \ldots, \mathbf{x}_m) \\
&\quad \leq \max\{f(\mathbf{x}), f(\mathbf{x}_1, \ldots, \mathbf{x}_{i-1}, \mathbf{y}_i, \mathbf{x}_{i+1}, \ldots, \mathbf{x}_m)\}.
\end{aligned}
$$

The function $f(\mathbf{x})$ is known as *strictly quasi-convex*, if

$$
\begin{aligned}
&f(\mathbf{x}_1, \ldots, \mathbf{x}_{i-1}, \alpha\mathbf{x}_i + (1-\alpha)\mathbf{y}_i, \mathbf{x}_{i+1}, \ldots, \mathbf{x}_m) \\
&\quad < \max\{f(\mathbf{x}), f(\mathbf{x}_1, \ldots, \mathbf{x}_{i-1}, \mathbf{y}_i, \mathbf{x}_{i+1}, \ldots, \mathbf{x}_m)\}.
\end{aligned}
$$

The following give the convergence performance of the GS algorithm under different assumption conditions.

THEOREM 6.2 [190] *Let the function f be strictly quasi-convex with respect to the vectors $\mathbf{x}_i$ ($i = 1, \ldots, m-2$) in X, and let the sequence generated by the GS method, $\{\mathbf{x}^k\}$, have limit points; then every limit point $\bar{\mathbf{x}}$ of $\{\mathbf{x}^k\}$ is a critical point of the optimization problem (6.2.10).*

THEOREM 6.3 [42] *For the objective function f in Equation (6.2.10), it is assumed that, for every i and $\mathbf{x} \in X$, the minimum point of the optimization algorithm*

$$
\min_{\mathbf{y} \in X_i} f(\mathbf{x}_1, \ldots, \mathbf{x}_{i-1}, \mathbf{y}, \mathbf{x}_{i+1}, \ldots, \mathbf{x}_m)
$$

is uniquely determined. If $\{\mathbf{x}^k\}$ is the iterative sequence generated by the GS algorithm then every limit point of $\mathbf{x}^k$ is a critical point.

However, in practical applications, the objective function in the optimization (6.2.10) often does not satisfy the conditions of the above two theorems. For example, by Theorem 6.2, in the case of a rank-deficient matrix $\mathbf{A}$, the quadratic objective function $\|\mathbf{A}\mathbf{x} - \mathbf{b}\|_2^2$ is not quasi-convex, so the convergence is not be guaranteed.

The sequence generated by the GS method contains limit points, but they may not be the critical points of the optimization problem (6.2.10). The fact that the GS algorithm may not converge was observed by Powell in 1973 [399], who called it the "circle phenomenon" of the GS method. Recently, a lot of simulation experiences have shown [338], [293] that even it converges, the iterative process of the ALS method easily falls into a "swamp": an unusually large number of iterations leads to a very slow convergence rate. In particular, as long as the column rank of one variable matrix in a set of m variable matrices is deficient, or, although the m variable matrices are of full-column rank, there is collinearity among the column vectors of some of the variable matrices, one easily observes this swamp phenomenon [338], [293].

A simple and effective way for avoiding the circle and swamp phenomena of the GS method is to make a Tikhonov regularization of the objective function in the optimization problem (6.2.10); namely, the separated suboptimization algorithm (6.2.11) is regularized as

$$\mathbf{x}_i^{k+1} = \arg\min_{\mathbf{y}\in X_i} \left\{ f(\mathbf{x}_1^{k+1},\ldots,\mathbf{x}_{i-1}^{k+1},\mathbf{y},\mathbf{x}_{i+1}^{k},\ldots,\mathbf{x}_m^{k}) + \frac{1}{2}\tau_i\|\mathbf{y}-\mathbf{x}_i^k\|_2^2 \right\}, \tag{6.2.16}$$

where $i = 1,\ldots,m$.

The above algorithm is called the *proximal point version* of the GS methods [18], [44], abbreviated as the PGS method.

The role of the regularization term $\|\mathbf{y} - \mathbf{x}_i^k\|_2^2$ is to force the updated vector $\mathbf{x}_i^{k+1} = \mathbf{y}$ to be close to $\mathbf{x}_i^k$; this avoids any violent shock in the iterative process and so prevents the divergence of the algorithm.

The GS or PGS method is said to be well defined, if each suboptimization problem has an optimal solution [190].

Theorem 6.4 [190] *If the PGS method is well-defined, and in the sequence $\{\mathbf{x}^k\}$ there exist limit points, then every limit point $\bar{\mathbf{x}}$ of $\{\mathbf{x}^k\}$ is a critical point of the optimization problem (6.2.10).*

This theorem shows that the convergence performance of the PGS method is better than that of the GS method.

Many simulation experiments show [293] that under the condition of achieving the same error, the iterative number of the GS method in a swamp iteration is unusually large whereas the PGS method tends to converge quickly. The PGS method is also called the *regularized Gauss–Seidel method* in some literature.

The ALS method and the regularized ALS method have important applications in nonnegative matrix decomposition and tensor analysis; these will be discussed in later chapters.

6.3 Total Least Squares (TLS) Methods

Although its original name is different, the total least squares (TLS) method has a long history. The earliest ideas about TLS can be traced back to the paper of Pearson in 1901 [381] who considered an approximate method for solving the matrix equation $\mathbf{Ax} = \mathbf{b}$ when in both $\mathbf{A}$ and $\mathbf{b}$ there exist errors. However, as late as 1980, Golub and Van Loan [178] for the first time gave an overall treatment from the point of view of numerical analysis and formally referred to this method as the total least squares. In mathematical statistics the method is called orthogonal regression or errors-in-variables regression [175]. In system identification the TLS method is called the characteristic vector method or the Koopmans–Levin method [486]. Now the TLS method is widely used in statistics, physics, economics, biology and medicine, signal processing, automatic control, system science and many other disciplines and fields.

6.3.1 TLS Problems

Let $\mathbf{A}_0$ and $\mathbf{b}_0$ represent an unobservable error-free data matrix and an error-free data vector, respectively. The actual observed data matrix and data vector are respectively given by

$$\mathbf{A} = \mathbf{A}_0 + \mathbf{E}, \quad \mathbf{b} = \mathbf{b}_0 + \mathbf{e}, \tag{6.3.1}$$

where $\mathbf{E}$ and $\mathbf{e}$ express the error data matrix and the error data vector, respectively.

The basic idea of the TLS is not only to use a correction vector $\Delta\mathbf{b}$ to perturb the data vector $\mathbf{b}$, but also to use a correction matrix $\Delta\mathbf{A}$ to perturb the data matrix $\mathbf{A}$, thereby making a joint compensation for errors or noise in both $\mathbf{A}$ and $\mathbf{b}$:

$$\begin{aligned}\mathbf{b} + \Delta\mathbf{b} &= \mathbf{b}_0 + \mathbf{e} + \Delta\mathbf{b} \to \mathbf{b}_0,\\ \mathbf{A} + \Delta\mathbf{A} &= \mathbf{A}_0 + \mathbf{E} + \Delta\mathbf{A} \to \mathbf{A}_0.\end{aligned}$$

The purpose is to suppress the influence of the observation error or noise on the matrix equation solution in order to transform the solution of a noisy matrix equation into the solution of an error-free matrix equation:

$$(\mathbf{A} + \Delta\mathbf{A})\mathbf{x} = \mathbf{b} + \Delta\mathbf{b} \quad \Rightarrow \quad \mathbf{A}_0\mathbf{x} = \mathbf{b}_0. \tag{6.3.2}$$

Naturally, we want the correction data matrix and the correction data vectors to be as small as possible. Hence the TLS problem can be expressed as the constrained optimization problem

$$\text{TLS:} \quad \min_{\Delta\mathbf{A},\Delta\mathbf{b},\mathbf{x}} \|[\Delta\mathbf{A}, \Delta\mathbf{b}]\|_F^2 \quad \text{subject to} \quad (\mathbf{A} + \Delta\mathbf{A})\mathbf{x} = \mathbf{b} + \Delta\mathbf{b} \tag{6.3.3}$$

or

$$\text{TLS:} \quad \min_{\mathbf{z}} \|\mathbf{D}\|_F^2 \quad \text{subject to} \quad \mathbf{Dz} = -\mathbf{Bz}, \tag{6.3.4}$$

where $\mathbf{D} = [\Delta\mathbf{A}, \Delta\mathbf{b}]$, $\mathbf{B} = [\mathbf{A}, \mathbf{b}]$ and $\mathbf{z} = \begin{bmatrix}\mathbf{x}\\-1\end{bmatrix}$ is an $(n+1) \times 1$ vector.

Under the assumption $\|\mathbf{z}\|_2 = 1$, from $\|\mathbf{D}\|_2 \leq \|\mathbf{D}\|_F$ and $\|\mathbf{D}\|_2 = \sup_{\|\mathbf{z}\|_2=1} \|\mathbf{Dz}\|_2$ we have $\min_{\mathbf{z}} \|\mathbf{D}\|_F^2 = \min_{\mathbf{z}} \|\mathbf{Dz}\|_2^2$ and thus we can rewrite (6.3.4) as

$$\text{TLS:} \quad \min_{\mathbf{z}} \|\mathbf{Dz}\|_2^2 \quad \text{subject to} \quad \|\mathbf{z}\| = 1 \tag{6.3.5}$$

or

$$\text{TLS:} \quad \min_{\mathbf{z}} \|\mathbf{Bz}\|_2^2 \quad \text{subject to} \quad \|\mathbf{z}\| = 1, \tag{6.3.6}$$

since $\mathbf{Dz} = -\mathbf{Bz}$.

6.3.2 TLS Solution

There are two possible cases in the TLS solution of over-determined equations.

Case 1 Single smallest singular value

The singular value σ_n of $\mathbf{B}$ is significantly larger than the singular value σ_{n+1}, i.e., there is only one smallest singular value.

Equation (6.3.6) shows that the TLS problem can be summarized as follows. Find a perturbation matrix $\mathbf{D} \in \mathbb{C}^{m\times(n+1)}$ with the minimum norm squared such that $\mathbf{B} + \mathbf{D}$ is of non full rank (if full rank then there is only the trivial solution $\mathbf{z} = \mathbf{0}$).

The TLS problem (6.3.6) is easily solved via the Lagrange multiplier method. To this end, define the objective function

$$J(\mathbf{z}) = \|\mathbf{Bz}\|_2^2 + \lambda(1 - \mathbf{z}^H\mathbf{z}), \tag{6.3.7}$$

where λ is the Lagrange multiplier. Noticing that $\|\mathbf{Bz}\|_2^2 = \mathbf{z}^H\mathbf{B}^H\mathbf{Bz}$, from $\frac{\partial J(\mathbf{z})}{\partial \mathbf{z}^*} = \mathbf{0}$ it follows that

$$\mathbf{B}^H\mathbf{Bz} = \lambda\mathbf{z}. \tag{6.3.8}$$

This shows that the Lagrange multiplier should be selected as the smallest eigenvalue $\lambda_{\min}$ of the matrix $\mathbf{B}^H\mathbf{B} = [\mathbf{A}, \mathbf{b}]^H[\mathbf{A}, \mathbf{b}]$ (i.e., the square of the smallest singular value of $\mathbf{B}$), while the TLS solution vector $\mathbf{z}$ is the eigenvector corresponding to $\lambda_{\min}$. In other words, the TLS solution vector $\begin{bmatrix}\mathbf{x}\\-1\end{bmatrix}$ is the solution of the Rayleigh quotient minimization problem

$$\min_{\mathbf{x}} \left\{ J(\mathbf{x}) = \frac{\begin{bmatrix}\mathbf{x}\\-1\end{bmatrix}^H [\mathbf{A}, \mathbf{b}]^H [\mathbf{A}, \mathbf{b}] \begin{bmatrix}\mathbf{x}\\-1\end{bmatrix}}{\begin{bmatrix}\mathbf{x}\\-1\end{bmatrix}^H \begin{bmatrix}\mathbf{x}\\-1\end{bmatrix}} = \frac{\|\mathbf{Ax} - \mathbf{b}\|_2^2}{\|\mathbf{x}\|_2^2 + 1} \right\}. \tag{6.3.9}$$

Let the SVD of the $m \times (n+1)$ augmented matrix $\mathbf{B}$ be $\mathbf{B} = \mathbf{U\Sigma V}^H$; let its singular values be arranged in the order $\sigma_1 \geq \cdots \geq \sigma_{n+1}$ and let their corresponding right singular vectors be $\mathbf{v}_1, \ldots, \mathbf{v}_{n+1}$. Then, according to the above analysis, the TLS solution is $\mathbf{z} = \mathbf{v}_{n+1}$, namely

$$\mathbf{x}_{\text{TLS}} = -\frac{1}{v(n+1,n+1)} \begin{bmatrix} v(1,n+1) \\ \vdots \\ v(n,n+1) \end{bmatrix}, \tag{6.3.10}$$

where $v(i, n+1)$ is the ith entry of $(n+1)$th column of $\mathbf{V}$.

Remark If the augmented data matrix is given by $\mathbf{B} = [-\mathbf{b}, \mathbf{A}]$ then the TLS solution is provided by

$$\mathbf{x}_{\text{TLS}} = \frac{1}{v(1,n+1)} \begin{bmatrix} v(2,n+1) \\ \vdots \\ v(n+1,n+1) \end{bmatrix}. \tag{6.3.11}$$

Case 2 Multiple smallest singular values

In this case, there are multiple smallest singular values of $\mathbf{B}$, i.e., the smallest singular values are repeated or are very close.

Let

$$\sigma_1 \geq \sigma_2 \geq \cdots \geq \sigma_p > \sigma_{p+1} \approx \cdots \approx \sigma_{n+1}, \tag{6.3.12}$$

and $\mathbf{v}_i$ be any column in the subspace $S = \text{Span}\{\mathbf{v}_{p+1}, \mathbf{v}_{p+2}, \ldots, \mathbf{v}_{n+1}\}$. Then above any right singular vector $\mathbf{v}_{p+i} = \begin{bmatrix} \mathbf{y}_{p+i} \\ \alpha_{p+i} \end{bmatrix}$ gives a TLS solution

$$\mathbf{x}_i = -\mathbf{y}_{p+i}/\alpha_{p+i}, \quad i = 1, \ldots, n+1-p.$$

Hence there are $n+1-p$ TLS solutions. The interesting solution is the TLS solution that is unique in some sense. The possible unique TLS solutions are of two kinds:

(1) the *minimum norm solution*, composed of n parameters;
(2) the *optimal LS approximate solution*, containing only p parameters.

1. Minimum Norm Solution

The minimum norm solution of the matrix equation $\mathbf{A}_{m \times n}\mathbf{x}_{n \times 1} = \mathbf{b}_{m \times 1}$ is a TLS solution with n parameters. The TLS algorithm for finding the minimum norm solution was proposed by Golub and Van Loan [178]; it is given as Algorithm 6.1.

Remark If the augmented data matrix is given by $\mathbf{B} = [-\mathbf{b}, \mathbf{A}]$, then the Householder transformation in step 3 becomes

$$\mathbf{V}_1\mathbf{Q} = \left[\begin{array}{c:ccc} \alpha & 0 & \cdots & 0 \\ \hdashline \mathbf{y} & & \times & \end{array}\right], \tag{6.3.13}$$

Algorithm 6.1 TLS algorithm for minimum norm solution

input: $\mathbf{A} \in \mathbb{C}^{m\times n}, \mathbf{b} \in \mathbb{C}^m, \alpha > 0$.

repeat

1. Compute $\mathbf{B} = [\mathbf{A}, \mathbf{b}] = \mathbf{U\Sigma V}^H$, and save $\mathbf{V}$ and all singular values.
2. Determine the number p of principal singular values.
3. Put $\mathbf{V}_1 = [\mathbf{v}_{p+1}, \ldots, \mathbf{v}_{n+1}]$, and compute the Householder transformation

$$\mathbf{V}_1\mathbf{Q} = \begin{bmatrix} \mathbf{y} & \vdots & \times \\ \hline \alpha & \vdots & 0 \cdots 0 \end{bmatrix},$$

where α is a scalar, and $\times$ denotes the irrelevant block.

4. **exit if** $\alpha \neq 0$.

return $p \leftarrow p - 1$.

output: $\mathbf{x}_{\text{TLS}} = -\mathbf{y}/\alpha$.

---

and the output is given by $\mathbf{x}_{\text{TLS}} = \mathbf{y}/\alpha$.

It should be noticed that, like the unknown parameter vector $\mathbf{x}$ of the original matrix equation $\mathbf{Ax} = \mathbf{b}$, the minimum norm solution $\mathbf{x}_{\text{TLS}}$ contains n parameters. From this fact it is seen that even if the effective rank of $\mathbf{B}$ is $p < n$, the minimum norm solution still assumes that n unknown parameters of the vector $\mathbf{x}$ are mutually independent. As a matter of fact, because both the augmented matrix $\mathbf{B} = [\mathbf{A}, \mathbf{b}]$ and the original data matrix $\mathbf{A}$ have the same rank, the rank of $\mathbf{A}$ is also p. This implies that only p columns of $\mathbf{A}$ are linearly independent and thus that the number of the principal parameters in the original matrix equation $\mathbf{Ax} = \mathbf{b}$ is p rather than n.

To sum up, the minimum norm solution of the TLS problem contains some redundant parameters that is linearly dependent on other parameters. In signal processing and system theory, the unique TLS solution with no redundant parameters is more interesting, since it is the optimal LS approximate solution.

2. Optimal Least Squares Approximate Solution

First let the $m \times (n + 1)$ matrix $\hat{\mathbf{B}}$ be an optimal proximation with rank p of the augmented matrix $\mathbf{B}$, i.e.,

$$\hat{\mathbf{B}} = \mathbf{U\Sigma}_p\mathbf{V}^H, \tag{6.3.14}$$

where $\mathbf{\Sigma}_p = \mathbf{Diag}(\sigma_1, \ldots, \sigma_p, 0, \ldots, 0)$.

Then let the $m \times (p+1)$ matrix $\hat{\mathbf{B}}_j^{(p)}$ be a submatrix of the $m \times (n+1)$ optimal approximate matrix $\hat{\mathbf{B}}$, defined as

$$\hat{\mathbf{B}}_j^{(p)}: \text{submatrix consisting of the } j\text{th to the } (j+p)\text{th columns of } \hat{\mathbf{B}}. \tag{6.3.15}$$

Clearly, there are $(n+1-p)$ submatrices $\hat{\mathbf{B}}_1^{(p)}, \hat{\mathbf{B}}_2^{(p)}, \ldots, \hat{\mathbf{B}}_{n+1-p}^{(p)}$.

As stated before, the fact that the efficient rank of $\mathbf{B}$ is equal to p means that p components are linearly independent in the parameter vector $\mathbf{x}$. Let the $(p+1)\times 1$ vector be $\mathbf{a} = \begin{bmatrix}\mathbf{x}^{(p)}\\ -1\end{bmatrix}$, where $\mathbf{x}^{(p)}$ is the column vector consisting of the p linearly independent unknown parameters of the vector $\mathbf{x}$. Then, the original TLS problem becomes that of solving the following $n+1-p$ TLS problems:

$$\hat{\mathbf{B}}_j^{(p)}\mathbf{a} = 0, \qquad j = 1, 2, \ldots, n+1-p \tag{6.3.16}$$

or equivalently that of solving the synthetic TLS problem

$$\begin{bmatrix} \hat{\mathbf{B}}(1:p+1) \\ \vdots \\ \hat{\mathbf{B}}(n+1-p:n+1) \end{bmatrix} \mathbf{a} = \mathbf{0}, \tag{6.3.17}$$

where $\hat{\mathbf{B}}(i:p+i) = \hat{\mathbf{B}}_i^{(p)}$ is defined in (6.3.15). It is not difficult to show that

$$\hat{\mathbf{B}}(i:p+i) = \sum_{k=1}^{p} \sigma_k \mathbf{u}_k (\mathbf{v}_k^i)^H, \tag{6.3.18}$$

where $\mathbf{v}_k^i$ is a *windowed* segment of the kth column vector of $\mathbf{V}$, defined as

$$\mathbf{v}_k^i = [v(i,k), v(i+1,k), \ldots, v(i+p,k)]^T. \tag{6.3.19}$$

Here $v(i,k)$ is the (i,k)th entry of $\mathbf{V}$.

According to the least squares principle, finding the LS solution of Equation (6.3.17) is equivalent to minimizing the measure (or cost) function

$$\begin{aligned} f(\mathbf{a}) &= [\hat{\mathbf{B}}(1:p+1)\mathbf{a}]^H \hat{\mathbf{B}}(1:p+1)\mathbf{a} + [\hat{\mathbf{B}}(2:p+2)\mathbf{a}]^H \hat{\mathbf{B}}(2:p+2)\mathbf{a} \\ &\quad + \cdots + [\hat{\mathbf{B}}(n+1-p:n+1)\mathbf{a}]^H \hat{\mathbf{B}}(n+1-p:n+1)\mathbf{a} \\ &= \mathbf{a}^H \left(\sum_{i=1}^{n+1-p} [\hat{\mathbf{B}}(i:p+i)]^H \hat{\mathbf{B}}(i:p+i) \right) \mathbf{a}. \end{aligned} \tag{6.3.20}$$

Define the $(p+1)\times(p+1)$ matrix

$$\mathbf{S}^{(p)} = \sum_{i=1}^{n+1-p} [\hat{\mathbf{B}}(i:p+i)]^H \hat{\mathbf{B}}(i:p+i), \tag{6.3.21}$$

then the measure function can be simply written as

$$f(\mathbf{a}) = \mathbf{a}^H \mathbf{S}^{(p)} \mathbf{a}. \tag{6.3.22}$$

The minimal variable $\mathbf{a}$ of the measure function $f(\mathbf{a})$ is given by $\partial f(\mathbf{a})/\partial \mathbf{a}^* = \mathbf{0}$ below:

$$\mathbf{S}^{(p)}\mathbf{a} = \alpha \mathbf{e}_1, \tag{6.3.23}$$

where $\mathbf{e}_1 = [1, 0, \ldots, 0]^T$ and the constant $\alpha > 0$ represents the error energy. From (6.3.21) and (6.3.18) we have

$$\mathbf{S}^{(p)} = \sum_{j=1}^{p} \sum_{i=1}^{n+1-p} \sigma_j^2 \mathbf{v}_j^i (\mathbf{v}_j^i)^H. \tag{6.3.24}$$

Solving the matrix equation (6.3.23) is simple. If we let $\mathbf{S}^{-(p)}$ be the inverse matrix $\mathbf{S}^{(p)}$, then the solution vector $\mathbf{a}$ depends only on the first column of the inverse matrix $\mathbf{S}^{-(p)}$. it is easily seen that the ith entry of $\mathbf{x}^{(p)} = [x_{\mathrm{TLS}}(1), \ldots, x_{\mathrm{TLS}}(p)]^T$ in the TLS solution vector $\mathbf{a} = \begin{bmatrix} \mathbf{x}^{(p)} \\ -1 \end{bmatrix}$ is given by

$$x_{\mathrm{TLS}}(i) = -\mathbf{S}^{-(p)}(i, 1)/\mathbf{S}^{-(p)}(p+1, 1), \quad i = 1, \ldots, p. \tag{6.3.25}$$

This solution is known as the *optimal least squares approximate solution*. Because the number of parameters in this solution and the effective rank are the same, it is also called a *low-rank TLS solution* [74].

Notice that if the augmented matrix $\mathbf{B} = [-\mathbf{b}, \mathbf{A}]$ then

$$x_{\mathrm{TLS}}(i) = \mathbf{S}^{-(p)}(i+1, 1)/\mathbf{S}^{-(p)}(1, 1), \quad i = 1, 2, \ldots, p. \tag{6.3.26}$$

In summary, the algorithm for the low-rank TLS solution is given in Algorithm 6.2. The basic idea of this algorithm was proposed by Cadzow in 1982 [74].

Algorithm 6.2 SVD-TLS algorithm

---

input: $\mathbf{A} \in \mathbb{C}^{m \times n}, \mathbf{b} \in \mathbb{C}^n$.

1. Compute the SVD $\mathbf{B} = [\mathbf{A}, \mathbf{b}] = \mathbf{U}\boldsymbol{\Sigma}\mathbf{V}^H$, and save $\mathbf{V}$.
2. Determine the effective rank p of $\mathbf{B}$.
3. Use (6.3.24) and (6.3.19) to calculate the $(p+1) \times (p+1)$ matrix $\mathbf{S}^{(p)}$.
4. Compute $\mathbf{S}^{-(p)}$ and $x_{\mathrm{TLS}}(i) = -\mathbf{S}^{-(p)}(i, 1)/\mathbf{S}^{-(p)}(p+1, 1)$, $i = 1, \ldots, p$.

output: $\mathbf{x}_{\mathrm{TLS}}$.

6.3.3 Performances of TLS Solution

The TLS has two interesting interpretations: one is its geometric interpretation [178] and the other is a closed solution interpretation [510].

1. Geometric Interpretation of TLS Solution

Let $\mathbf{a}_i^T$ be the ith row of the matrix $\mathbf{A}$ and b_i be the ith entry of the vector $\mathbf{b}$. Then the TLS solution $\mathbf{x}_{\mathrm{TLS}}$ is the minimal vector such that

$$\min_{\mathbf{x}} \left\{ \frac{\|\mathbf{A}\mathbf{x} - \mathbf{b}\|_2^2}{\|\mathbf{x}\|_2^2 + 1} = \sum_{i=1}^{n} \frac{|\mathbf{a}_i^T \mathbf{x} - b_i|^2}{\mathbf{x}^T\mathbf{x} + 1} \right\}, \tag{6.3.27}$$

where $|\mathbf{a}_i^T\mathbf{x}-b_i|/(\mathbf{x}^T\mathbf{x}+1)$ is the distance from the point $\begin{bmatrix}\mathbf{a}_i\\ b_i\end{bmatrix}\in\mathbb{R}^{n+1}$ to the nearest point in the subspace P_x, which is defined as

$$P_x=\left\{\begin{bmatrix}\mathbf{a}\\ b\end{bmatrix}:\mathbf{a}\in\mathbb{R}^{n\times 1},b\in\mathbb{R},b=\mathbf{x}^T\mathbf{a}\right\}.\tag{6.3.28}$$

Hence the TLS solution can be expressed using the subspace P_x [178]: sum of the squared distances from the TLS solution point $\begin{bmatrix}\mathbf{a}_i\\ b_i\end{bmatrix}$ to points in the subspace P_x is minimized. Figure 6.1 shows, for comparison, the TLS solution and the LS solution.

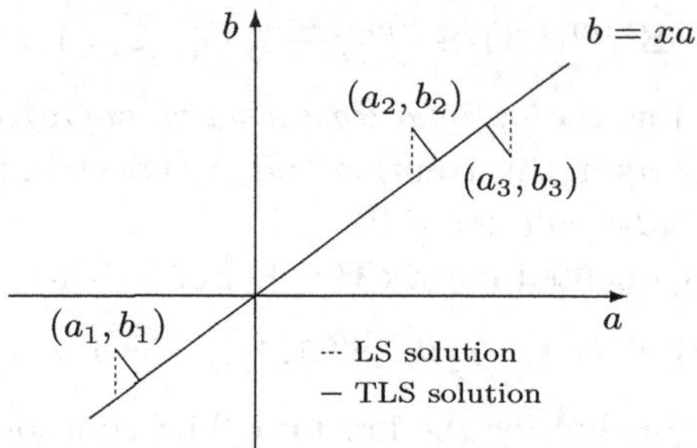

Figure 6.1 LS solution and TLS solution.

In Figure 6.1, each dotted line, which is a vertical distance parallel to the b-axis, is an LS solution; and each solid line, which starts at the point (a_i,b_i) and is the vertical distance to the straight line $b=xa$, is a TLS solution. From this geometric interpretation it can be concluded that the TLS solution is better than the LS solution, because the residual error of curve fitting given by the TLS solution is smaller.

2. Closed Solution of TLS Problems

If the singular values of the augmented matrix $\mathbf{B}$ are $\sigma_1\geq\cdots\geq\sigma_{n+1}$ then the TLS solution can be expressed as [510]

$$\mathbf{x}_{\text{TLS}}=(\mathbf{A}^H\mathbf{A}-\sigma_{n+1}^2\mathbf{I})^{-1}\mathbf{A}^H\mathbf{b}.\tag{6.3.29}$$

Compared with the Tikhonov regularization method, the TLS is a kind of anti-regularization method and can be interpreted as a least squares procedure with noise removal: it first removes the noise term $\sigma_{n+1}^2\mathbf{I}$ from the covariance matrix $\mathbf{A}^T\mathbf{A}$ and then finds the inverse matrix of $\mathbf{A}^T\mathbf{A}-\sigma_{n+1}^2\mathbf{I}$ to get the LS solution.

Letting the noisy data matrix be $\mathbf{A}=\mathbf{A}_0+\mathbf{E}$, its covariance matrix $\mathbf{A}^H\mathbf{A}=\mathbf{A}_0^H\mathbf{A}_0+\mathbf{E}^H\mathbf{A}_0+\mathbf{A}_0^H\mathbf{E}+\mathbf{E}^H\mathbf{E}$. Obviously, when the error matrix $\mathbf{E}$ has zero mean, the mathematical expectation of the covariance matrix is given by

$$E\{\mathbf{A}^H\mathbf{A}\}=E\{\mathbf{A}_0^H\mathbf{A}_0\}+E\{\mathbf{E}^H\mathbf{E}\}=\mathbf{A}_0^H\mathbf{A}_0+E\{\mathbf{E}^H\mathbf{E}\}.$$

If the column vectors of the error matrix are statistically independent and have the same variance, i.e., $E\{\mathbf{E}^T\mathbf{E}\} = \sigma^2\mathbf{I}$, then the smallest eigenvalue $\lambda_{n+1} = \sigma_{n+1}^2$ of the $(n+1)\times(n+1)$ covariance matrix $\mathbf{A}^H\mathbf{A}$ is the square of the singular value of the error matrix $\mathbf{E}$. Because the square of the singular value σ_{n+1}^2 happens to reflect the common variance σ^2 of each column vector of the error matrix, the covariance matrix $\mathbf{A}_0^H\mathbf{A}_0$ of the error-free data matrix can be retrieved from $\mathbf{A}^H\mathbf{A} - \sigma_{n+1}^2\mathbf{I}$, namely as $\mathbf{A}^T\mathbf{A} - \sigma_{n+1}^2\mathbf{I} = \mathbf{A}_0^H\mathbf{A}_0$. In other words, the TLS method can effectively restrain the influence of the unknown error matrix.

It should be pointed out that the main difference between the TLS method and the Tikhonov regularization method for solving the matrix equation $\mathbf{A}_{m\times n}\mathbf{x}_n = \mathbf{b}_m$ is that the TLS solution contains only $p = \text{rank}([\mathbf{A}, \mathbf{b}])$ principal parameters, and excludes the redundant parameters, whereas the Tikhonov regularization method can only provide all n parameters including the redundant parameters.

6.3.4 Generalized Total Least Squares

The ordinary LS, the data LS, Tikhonov regularization and the TLS method can be derived and explained by a unified theoretical framework.

Consider the minimization problem

$$\min_{\Delta\mathbf{A},\Delta\mathbf{b},\mathbf{x}} \left\{ \|[\Delta\mathbf{A}, \Delta\mathbf{b}]\|_F^2 + \lambda\|\mathbf{x}\|_2^2 \right\} \tag{6.3.30}$$

subject to

$$(\mathbf{A} + \alpha\Delta\mathbf{A})\mathbf{x} = \mathbf{b} + \beta\Delta\mathbf{b},$$

where α and β are the weighting coefficients of respectively the perturbation $\Delta\mathbf{A}$ of the data matrix $\mathbf{A}$ and the perturbation $\Delta\mathbf{b}$ of the data vector $\mathbf{b}$, and λ is the Tikhonov regularization parameter. The above minimization problem is called the *generalized total least squares* (GTLS) problem.

1. Comparison of Optimization Problems

(1) Ordinary least squares: $\alpha = 0, \beta = 1$ and $\lambda = 0$, which gives

$$\min_{\Delta\mathbf{A},\Delta\mathbf{b},\mathbf{x}} \|\Delta\mathbf{b}\|_2^2 \quad \text{subject to} \quad \mathbf{A}\mathbf{x} = \mathbf{b} + \Delta\mathbf{b}. \tag{6.3.31}$$

(2) Data least squares: $\alpha = 1, \beta = 0$ and $\lambda = 0$, which gives

$$\min_{\Delta\mathbf{A},\Delta\mathbf{b},\mathbf{x}} \|\Delta\mathbf{A}\|_F^2 \quad \text{subject to} \quad (\mathbf{A} + \Delta\mathbf{A})\mathbf{x} = \mathbf{b}. \tag{6.3.32}$$

(3) Tikhonov regularization: $\alpha = 0, \beta = 1$ and $\lambda > 0$, which gives

$$\min_{\Delta\mathbf{A},\Delta\mathbf{b},\mathbf{x}} \{\|\Delta\mathbf{b}\|_2^2 + \lambda\|\mathbf{x}\|_2^2\} \quad \text{subject to} \quad \mathbf{A}\mathbf{x} = \mathbf{b} + \Delta\mathbf{b}. \tag{6.3.33}$$

(4) Total least squares: $\alpha = \beta = 1$ and $\lambda = 0$, which gives

$$\min_{\Delta\mathbf{A},\Delta\mathbf{b},\mathbf{x}} \|\Delta\mathbf{A},\Delta\mathbf{b}\|_2^2 \quad \text{subject to} \quad (\mathbf{A}+\Delta\mathbf{A})\mathbf{x} = \mathbf{b}+\Delta\mathbf{b}. \tag{6.3.34}$$

2. Comparison of Solution Vectors

The constraint condition $(\mathbf{A}+\alpha\Delta\mathbf{A})\mathbf{x} = (\mathbf{b}+\beta\Delta\mathbf{b})$ can be represented as

$$\left([\alpha^{-1}\mathbf{A},\beta^{-1}\mathbf{b}] + [\Delta\mathbf{A},\Delta\mathbf{b}]\right)\begin{bmatrix}\alpha\mathbf{x}\\ -\beta\end{bmatrix} = 0.$$

Letting $\mathbf{D} = [\Delta\mathbf{A},\Delta\mathbf{b}]$ and $\mathbf{z} = \begin{bmatrix}\alpha\mathbf{x}\\ -\beta\end{bmatrix}$, the above equation becomes

$$\mathbf{D}\mathbf{z} = -\left[\alpha^{-1}\mathbf{A},\beta^{-1}\mathbf{b}\right]\mathbf{z}. \tag{6.3.35}$$

Under the assumption $\mathbf{z}^H\mathbf{z} = 1$, we have

$$\min\|\mathbf{D}\|_F^2 = \min\|\mathbf{D}\mathbf{z}\|_2^2 = \min\left\|[\alpha^{-1}\mathbf{A},\beta^{-1}\mathbf{b}]\mathbf{z}\right\|_2^2,$$

and thus the solution to the GTLS problem (6.3.30) can be rewritten as

$$\hat{\mathbf{x}}_{\text{GTLS}} = \arg\min_{\mathbf{z}}\left\{\frac{\left\|[\alpha^{-1}\mathbf{A},\beta^{-1}\mathbf{b}]\mathbf{z}\right\|_2^2}{\mathbf{z}^H\mathbf{z}} + \lambda\|\mathbf{x}\|_2^2\right\}. \tag{6.3.36}$$

Noting that $\mathbf{z}^H\mathbf{z} = \alpha^2\mathbf{x}^H\mathbf{x} + \beta^2$ and

$$[\alpha^{-1}\mathbf{A},\beta^{-1}\mathbf{b}]\mathbf{z} = [\alpha^{-1}\mathbf{A},\beta^{-1}\mathbf{b}]\begin{bmatrix}\alpha\mathbf{x}\\ -\beta\end{bmatrix} = \mathbf{A}\mathbf{x}-\mathbf{b},$$

the solution to the GTLS problem (6.3.36) is given by

$$\hat{\mathbf{x}}_{\text{GTLS}} = \arg\min_{\mathbf{x}}\left\{\frac{\|\mathbf{A}\mathbf{x}-\mathbf{b}\|_2^2}{\alpha^2\|\mathbf{x}\|_2^2+\beta^2} + \lambda\|\mathbf{x}\|_2^2\right\}. \tag{6.3.37}$$

When the weighting coefficients α, β and the Tikhonov regularization parameter λ take appropriate values, the GTLS solution (6.3.37) gives the following results:

$$\hat{\mathbf{x}}_{\text{LS}} = \arg\min_{\mathbf{x}} \|\mathbf{A}\mathbf{x}-\mathbf{b}\|_2^2 \qquad (\alpha=0,\beta=1,\lambda=0), \tag{6.3.38}$$

$$\hat{\mathbf{x}}_{\text{DLS}} = \arg\min_{\mathbf{x}} \frac{\|\mathbf{A}\mathbf{x}-\mathbf{b}\|_2^2}{\|\mathbf{x}\|_2^2} \qquad (\alpha=1,\beta=0,\lambda=0), \tag{6.3.39}$$

$$\hat{\mathbf{x}}_{\text{Tik}} = \arg\min_{\mathbf{x}} \left\{\|\mathbf{A}\mathbf{x}-\mathbf{b}\|_2^2 + \lambda\|\mathbf{x}\|_2^2\right\} \qquad (\alpha=0,\beta=1,\lambda>0), \tag{6.3.40}$$

$$\hat{\mathbf{x}}_{\text{TLS}} = \arg\min_{\mathbf{x}} \frac{\|\mathbf{A}\mathbf{x}-\mathbf{b}\|_2^2}{\|\mathbf{x}\|_2^2+1} \qquad (\alpha=1,\beta=1,\lambda=0). \tag{6.3.41}$$

3. Comparison of Perturbation Methods

(1) *Ordinary LS method* This uses the possible small correction term $\Delta\mathbf{b}$ to perturb the data vector $\mathbf{b}$ in such a way that $\mathbf{b}-\Delta\mathbf{b}\approx\mathbf{b}_0$, and thus compensates for the observed noise $\mathbf{e}$ in $\mathbf{b}$. The correction vector is selected as $\Delta\mathbf{b}=\mathbf{A}\mathbf{x}-\mathbf{b}$, and the analytical solution is $\hat{\mathbf{x}}_{\text{LS}}=(\mathbf{A}^H\mathbf{A})^{-1}\mathbf{A}^H\mathbf{b}$.
(2) *Data LS method* The correction term $\Delta\mathbf{A}=(\mathbf{A}\mathbf{x}-\mathbf{b})\mathbf{x}^H/(\mathbf{x}^H\mathbf{x})$ compensates the observed error matrix $\mathbf{E}$ in the data matrix $\mathbf{A}$. The data LS solution is

$$\hat{\mathbf{x}}_{\text{DLS}}=\arg\min_{\mathbf{x}}\frac{(\mathbf{A}\mathbf{x}-\mathbf{b})^H(\mathbf{A}\mathbf{x}-\mathbf{b})}{\mathbf{x}^H\mathbf{x}}.$$

(3) *Tikhonov regularization method* This adds the same perturbation term $\lambda>0$ to every diagonal entry of the matrix $\mathbf{A}^H\mathbf{A}$ to avoid the numerical instability of the LS solution $(\mathbf{A}^H\mathbf{A})^{-1}\mathbf{A}^H\mathbf{b}$. The analytical solution is $\hat{\mathbf{x}}_{\text{Tik}}=(\mathbf{A}^H\mathbf{A}+\lambda\mathbf{I})^{-1}\mathbf{A}^H\mathbf{b}$.
(4) *TLS method* By subtracting the perturbation matrix $\lambda\mathbf{I}$, the noise or perturbation in the covariance matrix of the original data matrix is restrained. There are three kinds of TLS solution: the minimum norm solution, the anti-regularization solution with n components $\hat{\mathbf{x}}_{\text{TLS}}=(\mathbf{A}^H\mathbf{A}-\lambda\mathbf{I})^{-1}\mathbf{A}^H\mathbf{b}$ and the SVD-TLS solution with only $p=\text{rank}([\mathbf{A},\mathbf{b}])$ principal parameters.

4. Comparison of Application Ranges

(1) The LS method is applicable for a data matrix $\mathbf{A}$ with full column rank and a data vector $\mathbf{b}$ containing iid Gaussian errors.
(2) The data LS method is applicable for a data vector $\mathbf{b}$ without error and a data matrix $\mathbf{A}$ that has full column rank and iid Gaussian error column vectors.
(3) The Tikhonov regularization method is applicable for a data matrix $\mathbf{A}$ with deficient column rank.
(4) The TLS method is applicable for a data matrix $\mathbf{A}$ with full column rank, where both $\mathbf{A}$ and the data vector $\mathbf{b}$ contain iid Gaussian error.

6.3.5 Total Least Squares Fitting

In the numerical analysis of science and engineering problems, it is usually necessary to fit a curve or a curved surface to a given set of points. Because these data points have generally been observed, they inevitably contain errors or are contaminated by noise. In such cases the TLS method is expected to provide better fitting results than the ordinary LS method.

Consider a data fitting process: given n data points $(x_1,y_1),\ldots,(x_n,y_n)$, we want to fit a straight line to these points. Assume the straight line equation is $ax+by-c=0$. If the straight line goes through the point (x_0,y_0) then $c=ax_0+by_0$.

Now consider fitting a straight line through the center $\bar{x}$, $\bar{y}$ of n known data points

$$\bar{x} = \frac{1}{n}\sum_{i=1}^{n} x_i, \quad \bar{y} = \frac{1}{n}\sum_{i=1}^{n} y_i. \tag{6.3.42}$$

Substituting $c = a\bar{x} + b\bar{y}$ into $ax + by - c = 0$, the straight line equation can be written as

$$a(x - \bar{x}) + b(y - \bar{y}) = 0 \tag{6.3.43}$$

or equivalently in the slope form

$$m(x - \bar{x}) + (y - \bar{y}) = 0. \tag{6.3.44}$$

The parameter vector $[a, b]^T$ is called the *normal vector* of the fitting straight line, and $-m = -a/b$ is its *slope*. Then, the straight line fitting problem becomes that of finding the normal vector $[a, b]^T$ or the *slope parameter* m.

Evidently, if we are substituting n known data points into a straight line equation then it cannot strictly be satisfied, and thus there are fitting errors. The procedure in LS fitting is to minimize the squared sum of the fitting errors; i.e., the cost function of LS fitting is

$$D_{\rm LS}^{(1)}(m, \bar{x}, \bar{y}) = \sum_{i=1}^{n}((x_i - \bar{x}) + m(y_i - \bar{y}))^2, \tag{6.3.45}$$

$$D_{\rm LS}^{(2)}(m, \bar{x}, \bar{y}) = \sum_{i=1}^{n}(m(x_i - \bar{x}) - (y_i - \bar{y}))^2. \tag{6.3.46}$$

Letting $\partial D_{\rm LS}^{(i)}(m, \bar{x}, \bar{y})/(\partial m) = 0$, $i = 1, 2$, we can find the slope m of the straight line. Then, substituting m into Equation (6.3.44), we can get the fitting equation of the straight line.

Unlike LS fitting, TLS fitting minimize the sum of the squared distances from the known data points to the linear equation $a(x - x_0) + b(y - y_0) = 0$.

The distance d of a point (p, q) from the straight line $ax+by-c = 0$ is determined by

$$d^2 = \frac{(ap + bq - c)^2}{a^2 + b^2} = \frac{(a(p - x_0) + b(q - y_0))^2}{a^2 + b^2}. \tag{6.3.47}$$

Then, the sum of the squared distances from the known n data points to the line $a(x - \bar{x}) + b(y - \bar{y}) = 0$ is given by

$$D(a, b, \bar{x}, \bar{y}) = \sum_{i=1}^{n} \frac{(a(x_i - \bar{x}) + b(y_i - \bar{y}))^2}{a^2 + b^2}. \tag{6.3.48}$$

Lemma 6.1 [350] *For a data point (x_0, y_0) on the line $a(x - \bar{x})Cb(y - \bar{y}) = 0$ we have the relationship*

$$D(a, b, \bar{x}, \bar{y}) \leq D(a, b, x_0, y_0), \tag{6.3.49}$$

and the equality holds if and only if $x_0 = \bar{x}$ *and* $y_0 = \bar{y}$.

Lemma 6.1 shows that the TLS best-fit line must pass through the center of n data points in order to minimize the deviation D.

Consider how to minimize the deviation D. To this end, we write D as the norm squared of the product of the 2×1 unit vector $\mathbf{t} = (a^2+b^2)^{-1/2}[a,b]^T$ and the $n \times 2$ matrix $\mathbf{M}$, i.e.,

$$D(a,b,\bar{x},\bar{y}) = \|\mathbf{Mt}\|_2^2 = \left\| \begin{bmatrix} x_1 - \bar{x} & y_1 - \bar{y} \\ \vdots & \vdots \\ x_n - \bar{x} & y_n - \bar{y} \end{bmatrix} \frac{1}{\sqrt{a^2+b^2}} \begin{bmatrix} a \\ b \end{bmatrix} \right\|_2^2, \tag{6.3.50}$$

where

$$\mathbf{M} = \begin{bmatrix} x_1 - \bar{x} & y_1 - \bar{y} \\ \vdots & \vdots \\ x_n - \bar{x} & y_n - \bar{y} \end{bmatrix}. \tag{6.3.51}$$

From Equation (6.3.50) one can directly obtain the following result.

Proposition 6.1 [350] *The distance squared sum* $D(a,b,\bar{x},\bar{y})$ *reaches a minimum for the unit normal vector* $\mathbf{t} = (a^2+b^2)^{-1/2}[a,b]^T$. *In this case, the mapping* $\mathbf{t} \mapsto \|\mathbf{Mt}\|_2$ *reaches the minimum value in the unit sphere* $\mathcal{S}^1 = \{\mathbf{t} \in \mathbb{R}^2 \,|\, \|\mathbf{t}\|_2 = 1\}$.

Proposition 6.1 shows that the distance squared sum $D(a,b,\bar{x},\bar{y})$ has a minimum. The following theorem provides a way for finding this minimal distance squared sum.

Theorem 6.5 [350] *If the* 2×1 *normal vector* $\mathbf{t}$ *is the eigenvector corresponding to the smallest eigenvalue of the* 2×2 *matrix* $\mathbf{M}^T\mathbf{M}$, *then the distance squared sum* $D(a,b,\bar{x},\bar{y})$ *takes the minimum value* σ_2^2.

Here we gives a proof of the above theorem that is simpler than the proof in [350]. Using the fact that $\mathbf{t}$ is a unit vector, we have $\|\mathbf{t}\|_2 = 1$, and thus the distance squared sum $D(a,b,\bar{x},\bar{y})$ can be written as

$$D(a,b,\bar{x},\bar{y}) = \frac{\mathbf{t}^T\mathbf{M}^T\mathbf{M}\mathbf{t}}{\mathbf{t}^T\mathbf{t}}. \tag{6.3.52}$$

This is the typical Rayleigh quotient form. Clearly, the condition that $D(a,b,\bar{x},\bar{y})$ takes a minimum value is that the normal vector $\mathbf{t}$ is the eigenvector corresponding to the smallest eigenvalue of the matrix $\mathbf{M}^T\mathbf{M}$.

Example 6.2 Given three data points $(2,1), (2,4), (5,1)$, compute their central point to get

$$\bar{x} = \tfrac{1}{3}(2+2+5) = 3, \quad \bar{y} = \tfrac{1}{3}(1+4+1) = 2.$$

By subtracting the mean values from the data points, we get the zero-mean data matrix

$$\mathbf{M} = \begin{bmatrix} 2-3 & 1-2 \\ 2-3 & 4-2 \\ 5-3 & 1-2 \end{bmatrix} = \begin{bmatrix} -1 & -1 \\ -1 & 2 \\ 2 & -1 \end{bmatrix},$$

and hence

$$\mathbf{M}^T\mathbf{M} = \begin{bmatrix} 6 & -3 \\ -3 & 6 \end{bmatrix}$$

whose EVD is given by

$$\mathbf{M}^T\mathbf{M} = \begin{bmatrix} \frac{1}{\sqrt{2}} & \frac{1}{\sqrt{2}} \\ -\frac{1}{\sqrt{2}} & \frac{1}{\sqrt{2}} \end{bmatrix} \begin{bmatrix} 9 & 0 \\ 0 & 3 \end{bmatrix} \begin{bmatrix} \frac{1}{\sqrt{2}} & -\frac{1}{\sqrt{2}} \\ \frac{1}{\sqrt{2}} & \frac{1}{\sqrt{2}} \end{bmatrix} = \begin{bmatrix} 6 & -3 \\ -3 & 6 \end{bmatrix}.$$

Therefore the normal vector $\mathbf{t} = [a,b]^T = [1/\sqrt{2}, 1/\sqrt{2}]^T$. The TLS best-fit equation is as follows:

$$a(x-\bar{x}) + b(y-\bar{y}) = 0 \quad \Rightarrow \quad \frac{1}{\sqrt{2}}(x-3) + \frac{1}{\sqrt{2}}(y-2) = 0,$$

i.e., $y = -x + 5$. In this case, the distance squared sum is

$$D_{\text{TLS}}(a,b,\bar{x},\bar{y}) = \|\mathbf{Mt}\|_2^2 = \left\| \begin{bmatrix} -1 & -1 \\ -1 & 2 \\ 2 & -1 \end{bmatrix} \begin{bmatrix} \frac{1}{\sqrt{2}} \\ \frac{1}{\sqrt{2}} \end{bmatrix} \right\|_2^2 = 3.$$

In contrast with the TLS fitting, the cost function of the LS fitting takes the form

$$\begin{aligned} D_{\text{LS}}^{(1)}(m,\bar{x},\bar{y}) &= \frac{1}{m^2+1} \sum_{i=1}^{3} [m(x_i-\bar{x}) + (y_i-\bar{y})]^2 \\ &= \frac{1}{m^2+1} [(-m-1)^2 + (-m+2)^2 + (2m-1)^2]. \end{aligned}$$

From $\partial D_{\text{LS}}^{(1)}(m,\bar{x},\bar{y})/(\partial m) = 6m - 3 = 0$ we get $m = 1/2$, i.e., its slope is $-1/2$. In this case, the LS best-fit equation is $\frac{1}{2}(x-3) + (y-2) = 0$, i.e., $x + 2y - 7 = 0$, and the corresponding distance squared sum $D_{\text{LS}}^{(1)}(m,\bar{x},\bar{y}) = 3.6$.

Similarly, if instead the LS fitting uses the cost function

$$\begin{aligned} D_{\text{LS}}^{(2)}(m,\bar{x},\bar{y}) &= \frac{1}{m^2+1} \sum_{i=1}^{3} (m(y_i-\bar{y}) + (x_i-\bar{x}))^2 \\ &= \frac{1}{m^2+1} \left((-m-1)^2 + (2m-1)^2 + (-m+2)^2\right), \end{aligned}$$

then the minimal slope such that $D_{\text{LS}}^{(2)}(m,\bar{x},\bar{y})$ is minimized is $m = \frac{1}{2}$, i.e., the best-fit equation is $2x - y - 4 = 0$, and the corresponding distance squared sum is $D_{\rm LS}^{(2)}(m, \bar{x}, \bar{y}) = 3.6$.

Figure 6.2 plots the results of the TLS fitting method and the two LS fitting methods.

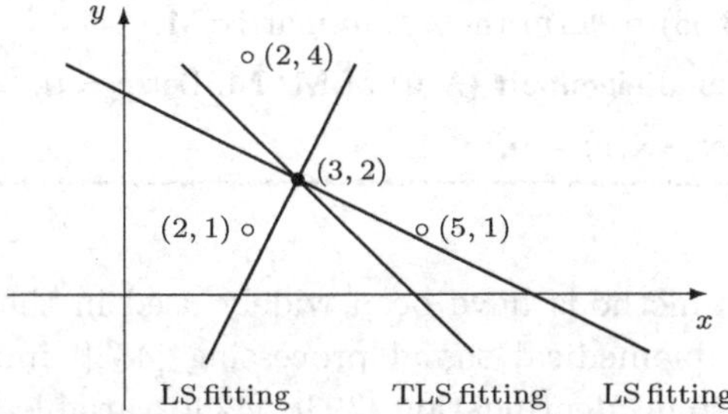

Figure 6.2 The LS fitting lines and the TLS fitting line.

For this example, $D_{\rm LS}^{(1)}(m, \bar{x}, \bar{y}) = D_{\rm LS}^{(2)}(m, \bar{x}, \bar{y}) > D_{\rm TLS}(a, b, \bar{x}, \bar{y})$, i.e., the two LS fittings have the same fitting deviations, which are larger than the fitting deviation of the TLS. It can be seen that the TLS fitting is thus more accurate than the LS fitting.

Theorem 6.5 is easily extended to higher-dimensional cases. Let n data vectors $\mathbf{x}_i = [x_{1i}, \ldots, x_{mi}]^T, i = 1, \ldots, n$ represent m-dimensional data points, and

$$\bar{\mathbf{x}} = \frac{1}{n}\sum_{i=1}^{n} \mathbf{x}_i = [\bar{x}_1, \ldots, \bar{x}_m]^T \tag{6.3.53}$$

be the mean (i.e., central) vector, where $\bar{x}_j = \sum_{i=1}^{n} x_{ji}$. Now use an m-dimensional normal vector $\mathbf{r} = [r_1, \ldots, r_m]^T$ to fit a hyperplane $\mathbf{x}$ to the known data vector, i.e., $\mathbf{x}$ satisfies the normal equation

$$\langle \mathbf{x} - \bar{\mathbf{x}}, \mathbf{r} \rangle = 0. \tag{6.3.54}$$

From the $n \times m$ matrix

$$\mathbf{M} = \begin{bmatrix} \mathbf{x}_1 - \bar{\mathbf{x}} \\ \vdots \\ \mathbf{x}_n - \bar{\mathbf{x}} \end{bmatrix} = \begin{bmatrix} x_{11} - \bar{x}_1 & x_{12} - \bar{x}_2 & \cdots & x_{1m} - \bar{x}_m \\ \vdots & \vdots & \vdots & \vdots \\ x_{n1} - \bar{x}_1 & x_{n2} - \bar{x}_2 & \cdots & x_{nm} - \bar{x}_m \end{bmatrix}, \tag{6.3.55}$$

one can fit the m-dimensional hyperplane, as shown in Algorithm 6.3.

It should be noted that if the smallest eigenvalues of the matrix $\mathbf{M}^T\mathbf{M}$ (or the smallest singular values of $\mathbf{M}$) have multiplicity, then correspondingly there are multiple eigenvectors, leading to the result that the hyperplane fitting problem has multiple solutions. The occurrence of this kind of situation shows that a linear data-fitting model may not be appropriate, and thus one should try other, nonlinear, fitting models.

Algorithm 6.3 TLS algorithm for fitting m-dimensional syperplane [350]

input: n data vectors $\mathbf{x}_1, \ldots, \mathbf{x}_n$.

1. Compute the mean vector $\bar{\mathbf{x}} = \frac{1}{n}\sum_{i=1}^{n} \mathbf{x}_i$.
2. Use Equation (6.3.55) to form the $n \times m$ matrix $\mathbf{M}$.
3. Compute the minimal eigenpair $(\lambda, \mathbf{u})$ of $\mathbf{M}^T\mathbf{M}$. Let $\mathbf{r} = \mathbf{u}$.

output: Hyperplane $\langle \mathbf{x} - \bar{\mathbf{x}}, \mathbf{r} \rangle = 0$.

Total least squares methods have been widely used in the following areas: signal processing [241], biomedical signal processing [482], image processing [348], frequency domain system identification [393], [429], variable-error modeling [481], [457], subspace identification of linear systems [485], radar systems [150], astronomy [58], communications [365], fault detection [223] and so on.

6.3.6 Total Maximum Likelihood Method

In the previous sections, we presented the LS method, the DLS method, Tikhonov regularization and the TLS method for solving the matrix equation $\mathbf{Ax} = \mathbf{b} + \mathbf{w}$, where the data matrix $\mathbf{A}$ is assumed to be deterministic and known. But, in some important applications, $\mathbf{A}$ is random and unobservable.

Consider the matrix equation

$$\mathbf{Ax} = \mathbf{b} + \mathbf{w}, \tag{6.3.56}$$

where $\mathbf{A} \in \mathbb{R}^{m\times n}$ is the data matrix, $\mathbf{x} \in \mathbb{R}^n$ is an unknown deterministic vector and $\mathbf{w} \in \mathbb{R}^m$ is an unknown perturbation or noise vector subject to the Gaussian distribution $N(0, \sigma_w^2)$.

The data matrix $\mathbf{A} = [a_{ij}]$ is a *random matrix* with random Gaussian variables whose distribution is assumed to be

$$a_{ij} = N(\bar{a}_{ij}, \sigma_A^2), \quad [\bar{a}_{ij}] = \bar{\mathbf{A}}. \tag{6.3.57}$$

Recently, under the assumption that the measurement matrix consists of random Gaussian variables, an alternative approach to solving the matrix equation $\mathbf{Ax} = \mathbf{b} + \mathbf{w}$ has been proposed, [508], [509]. This technique is referred to as the *total maximum likelihood* (TML) *method* in [30].

The TML solution of the *random matrix equation* $\mathbf{Ax} = \mathbf{b} + \mathbf{w}$ is determined by [508]

$$\hat{\mathbf{x}}_{\text{TML}} = \arg\min_{\mathbf{x}} \log p(\mathbf{b}; \mathbf{x}), \tag{6.3.58}$$

where $\mathbf{b} \sim N\left(\bar{\mathbf{A}}\mathbf{x}, \sigma_A^2(\|\mathbf{x}\|^2 + \sigma_w^2)\mathbf{I}\right)$. Hence, the solution can be rewritten as

$$\hat{\mathbf{x}}_{\text{TML}} = \arg\min_{\mathbf{x}} \left\{ \frac{\|\mathbf{b} - \bar{\mathbf{A}}\mathbf{x}\|_2^2}{\sigma_A^2\|\mathbf{x}\|_2^2 + \sigma_w^2} + m\log\left(\sigma_A^2\|\mathbf{x}\|_2^2 + \sigma_w^2\right) \right\}. \tag{6.3.59}$$

Theorem 6.6 *For any $t \geq 0$, let*

$$f(t) = \min_{\mathbf{x}:\|\mathbf{x}\|_2^2 = t} \|\mathbf{A}\mathbf{x} - \mathbf{b}\|_2^2, \tag{6.3.60}$$

and denote the optimal argument by $\mathbf{x}(t)$. Then, the maximum likelihood estimator of $\mathbf{x}$ in the random matrix equation (6.3.56) is $\mathbf{x}(t^*)$, where t^* is the solution to the following unimodal optimization problem:

$$\min_{t\geq 0} \left\{ \frac{f(t)}{\sigma_A^2 t + \sigma_w^2} + m\log(\sigma_A^2 t + \sigma_w^2) \right\}. \tag{6.3.61}$$

Proof See [508]. □

Theorem 6.6 allows for an efficient solution of the ML problem. This is so because [508]:

(1) there are standard methods for evaluating $f(t)$ in (6.3.60) for any $t \geq 0$;
(2) the line search in (6.3.61) is unimodal in $t \geq 0$, and thus any simple one-dimensional search algorithm can efficiently find its global minima.

For the solution of the optimization problem (6.3.60), one has the following lemma:

Lemma 6.2 [159], [508] *The solution of optimization problem in (6.3.60),*

$$\min_{\mathbf{x}:\|\mathbf{x}\|_2^2 = t} \|\mathbf{A}\mathbf{x} - \mathbf{b}\|_2^2, \tag{6.3.62}$$

has the solution

$$\mathbf{x}(t) = \left(\bar{\mathbf{A}}^T\bar{\mathbf{A}} + \alpha^*\mathbf{I}\right)^\dagger \bar{\mathbf{A}}\mathbf{b}, \tag{6.3.63}$$

where $\alpha^* \geq -\lambda_{\min}(\bar{\mathbf{A}}^T\bar{\mathbf{A}})$ is the unique root of the equation

$$\|\mathbf{x}(t)\|_2^2 = \left\|\left(\bar{\mathbf{A}}^T\bar{\mathbf{A}} + \alpha^*\mathbf{I}\right)^\dagger \bar{\mathbf{A}}\mathbf{b}\right\|_2^2 = t. \tag{6.3.64}$$

Remark Using the eigenvalue decomposition of $\bar{\mathbf{A}}^T\bar{\mathbf{A}}$, one can calculate $\|(\bar{\mathbf{A}}^T\bar{\mathbf{A}} + \alpha^*\mathbf{I})^\dagger\bar{\mathbf{A}}\mathbf{b}\|_2^2$ for different values of α. The monotonicity of this squared norm enables us to find the α that satisfies (6.3.64) using a simple line search.

The above discussion is summarized as Algorithm 6.4.

The TML method has been extended to the case in which the measurement matrix is structured, so that the perturbations are not arbitrary but rather follow a fixed pattern. For this case, a structured TML (STML) strategy was proposed in [30].

Algorithm 6.4 Total maximum likelihood algorithm [508]

input: $\mathbf{b}, \bar{\mathbf{A}}, t_0, \Delta t$.

initialization: $k = 1$.

repeat

1. Compute $t_k = t_0 + \Delta t$.
2. Use one-dimensional line search to solve the equation
$$\left\| \left(\bar{\mathbf{A}}^T\bar{\mathbf{A}} + \alpha\mathbf{I}\right)^\dagger \bar{\mathbf{A}}\mathbf{b} \right\|_2^2 = t_k$$
in order to get the optimal solution α^*.
3. $\mathbf{x}_k = (\bar{\mathbf{A}}^T\bar{\mathbf{A}} + \alpha^*\mathbf{I})^\dagger \bar{\mathbf{A}}\mathbf{b}$.
4. $f(t_k) = \|\mathbf{b} - \bar{\mathbf{A}}\mathbf{x}(t_k)\|_2^2$.
5. $J(t_k) = \dfrac{f(t_k)}{\sigma_A^2 t_k + \sigma_w^2} + m\log(\sigma_A^2 t_k + \sigma_w^2)$.
6. **exit if** $J(t_k) > J(t_{k-1})$.

return $k \leftarrow k + 1$.

output: $\mathbf{x}_{\text{TML}} \leftarrow \mathbf{x}_{k-1}$.

It is interesting and enlightening to compare the TML with the GTLS from the viewpoint of optimization:

(1) The optimization solutions of the GTLS problem and of the TML problem are respectively given by

$$\text{GTLS}: \quad \hat{\mathbf{x}}_{\text{GTLS}} = \arg\min_{\mathbf{x}} \left\{ \frac{\|\mathbf{b} - \mathbf{A}\mathbf{x}\|_2^2}{\alpha^2\|\mathbf{x}\|_2^2 + \beta^2} + \lambda\|\mathbf{x}\|_2^2 \right\}, \tag{6.3.65}$$

$$\text{TML}: \quad \hat{\mathbf{x}}_{\text{TML}} = \arg\min_{\mathbf{x}} \left\{ \frac{\|\mathbf{b} - \bar{\mathbf{A}}\mathbf{x}\|_2^2}{\sigma_A^2\|\mathbf{x}\|_2^2 + \sigma_w^2} + m\log\left(\sigma_A^2\|\mathbf{x}\|_2^2 + \sigma_w^2\right) \right\}. \tag{6.3.66}$$

They consist of a cost function and a penalty function. Taking $\alpha = \sigma_A$ and $\beta = \sigma_w$ and replacing the deterministic data matrix $\mathbf{A}$ in the GTLS by the deterministic term $\bar{\mathbf{A}}$ of the random data matrix $\mathbf{A}$, the cost function of the GTLS problem becomes the cost function of the TML problem.

(2) The penalty function in the GTLS problem is the regularization term $\lambda\|\mathbf{x}\|_2^2$, whereas the penalty function in the TML problem is the barrier function term $m\log(\sigma_A^2\|\mathbf{x}\|_2^2 + \sigma_w^2)$.

6.4 Constrained Total Least Squares

The data LS method and the TLS method for solving the matrix equation $\mathbf{A}\mathbf{x} = \mathbf{b}$ consider the case where the data matrix includes observed errors or noise, but the two methods assume that the errors are iid random varibles and have the same variances. However, in some important applications, the noise coefficients of the data matrix $\mathbf{A}$ may be statistically correlated or, although statistically independent, they have different variances. In this section we discuss the solution of over-determined matrix equations when the column vectors of the noise matrix are statistically correlated.

6.4.1 Constrained Total Least Squares Method

The matrix equation $\mathbf{A}_{m\times n}\mathbf{x}_n = \mathbf{b}_m$ can be rewritten as

$$[\mathbf{A},\, \mathbf{b}]\begin{bmatrix}\mathbf{x}\\ -1\end{bmatrix} = \mathbf{0} \quad \text{or} \quad \mathbf{C}\begin{bmatrix}\mathbf{x}\\ -1\end{bmatrix} = \mathbf{0}, \tag{6.4.1}$$

where $\mathbf{C} = [\mathbf{A},\, \mathbf{b}] \in \mathbb{C}^{m\times(n+1)}$ is the augmented data matrix.

Consider the noise matrix $\mathbf{D} = [\mathbf{E}, \mathbf{e}]$ of $[\mathbf{A},\, \mathbf{b}]$. If the column vectors of the noise matrix are statistically correlated then the column vectors of the correction matrix should also be statistically correlated in order to use an augmented correction matrix $\Delta\mathbf{C} = [\Delta\mathbf{A}, \Delta\mathbf{b}]$ to suppress the effects of the noise matrix $\mathbf{D} = [\mathbf{E}, \mathbf{e}]$.

A simple way to make the column vectors of the correction matrix $\Delta\mathbf{C}$ statistically correlated is to let every column vector be linearly dependent on the same vector (e.g., $\mathbf{u}$):

$$\Delta\mathbf{C} = [\mathbf{G}_1\mathbf{u}, \ldots, \mathbf{G}_{n+1}\mathbf{u}] \ \in \mathbb{R}^{m\times(n+1)}, \tag{6.4.2}$$

where $\mathbf{G}_i \in \mathbb{R}^{m\times m}, i = 1, \ldots, n{+}1$, are known matrices, while $\mathbf{u}$ is to be determined.

The *constrained TLS problem* can be stated as follows [1]: determine a solution vector $\mathbf{x}$ and a minimum-norm perturbation vector $\mathbf{u}$ such that

$$\big(\mathbf{C} + [\mathbf{G}_1\mathbf{u}, \ldots, \mathbf{G}_{n+1}\mathbf{u}]\big)\begin{bmatrix}\mathbf{x}\\ -1\end{bmatrix} = \mathbf{0}, \tag{6.4.3}$$

which can be equivalently expressed as the constrained optimization problem

$$\min_{\mathbf{u},\mathbf{x}} \mathbf{u}^T\mathbf{W}\mathbf{u} \quad \text{subject to} \quad \big(\mathbf{C} + [\mathbf{G}_1\mathbf{u}, \ldots, \mathbf{G}_{n+1}\mathbf{u}]\big)\begin{bmatrix}\mathbf{x}\\ -1\end{bmatrix} = \mathbf{0}, \tag{6.4.4}$$

where $\mathbf{W}$ is a weighting matrix and is usually diagonal or the identity matrix.

The correction matrix $\Delta\mathbf{A}$ is constrained as $\Delta\mathbf{A} = [\mathbf{G}_1\mathbf{u}, \ldots, \mathbf{G}_n\mathbf{u}]$, while the correction vector $\Delta\mathbf{b}$ is constrained as $\Delta\mathbf{b} = \mathbf{G}_{n+1}\mathbf{u}$. In constrained TLS problems, the linear correlation structure between column vectors of the augmented correction matrix $[\Delta\mathbf{A}, \Delta\mathbf{b}]$ is kept by selecting appropriate matrices $\mathbf{G}_i$ $(i = 1, \ldots, n+1)$. The key to the constrained TLS method is how to choose appropriate matrices $\mathbf{G}_1, \ldots, \mathbf{G}_{n+1}$, depending on the application.

Theorem 6.7 [1] *Let*

$$\mathbf{W}_x = \sum_{i=1}^{n} x_i\mathbf{G}_i - \mathbf{G}_{n+1}. \tag{6.4.5}$$

Then the constrained TLS solution is given by

$$\min_{\mathbf{x}} \left\{ F(\mathbf{x}) = \begin{bmatrix} \mathbf{x} \\ -1 \end{bmatrix}^H \mathbf{C}^H (\mathbf{W}_x \mathbf{W}_x^H)^{\dagger} \mathbf{C} \begin{bmatrix} \mathbf{x} \\ -1 \end{bmatrix} \right\}, \tag{6.4.6}$$

where $\mathbf{W}_x^{\dagger}$ *is the Moore–Penrose inverse matrix of* $\mathbf{W}_x$.

A complex form of the Newton method was proposed in [1] for calculating the constrained TLS solution.

The matrix $F(\mathbf{x})$ is regarded as a complex analytic function with $2n$ complex variables $x_1, \ldots, x_n, x_1^*, \ldots, x_n^*$. Then the Newton recursive formulas are given by

$$\mathbf{x} = \mathbf{x}_0 + (\mathbf{A}^* \mathbf{B}^{-1} \mathbf{A} - \mathbf{B}^*)^{-1} (\mathbf{a}^* - \mathbf{A}^* \mathbf{B}^{-1} \mathbf{a}), \tag{6.4.7}$$

where

$$\left. \begin{aligned} \mathbf{a} &= \frac{\partial F}{\partial \mathbf{x}} = \left[\frac{\partial F}{\partial x_1}, \frac{\partial F}{\partial x_2}, \ldots, \frac{\partial F}{\partial x_n} \right]^T = \text{complex gradient of } F, \\ \mathbf{A} &= \frac{\partial^2 F}{\partial \mathbf{x} \partial \mathbf{x}^T} = \text{nonconjugate complex Hessian matrix of } F, \\ \mathbf{B} &= \frac{\partial^2 F}{\partial \mathbf{x}^* \partial \mathbf{x}^T} = \text{conjugate complex Hessian matrix of } F. \end{aligned} \right\} \tag{6.4.8}$$

The (k,l)th entries of the two $n \times n$ part Hessian matrices are defined as

$$\left[\frac{\partial^2 F}{\partial \mathbf{x} \partial \mathbf{x}^T} \right]_{k,l} = \frac{\partial^2 F}{\partial x_k \partial x_l} = \frac{1}{4} \left(\frac{\partial F}{\partial x_{k\mathrm{R}}} - \mathrm{j} \frac{\partial F}{\partial x_{k\mathrm{I}}} \right) \left(\frac{\partial F}{\partial x_{l\mathrm{R}}} - \mathrm{j} \frac{\partial F}{\partial x_{l\mathrm{I}}} \right), \tag{6.4.9}$$

$$\left[\frac{\partial^2 F}{\partial \mathbf{x}^* \partial \mathbf{x}^T} \right]_{k,l} = \frac{\partial^2 F}{\partial x_k^* \partial x_l} = \frac{1}{4} \left(\frac{\partial F}{\partial x_{k\mathrm{R}}} + \mathrm{j} \frac{\partial F}{\partial x_{k\mathrm{I}}} \right) \left(\frac{\partial F}{\partial x_{l\mathrm{R}}} - \mathrm{j} \frac{\partial F}{\partial x_{l\mathrm{I}}} \right). \tag{6.4.10}$$

Here $x_{k\mathrm{R}}$ and $x_{k\mathrm{I}}$ are respectively the real part and the imaginary part of x_k.

Put

$$\mathbf{u} = (\mathbf{W}_x \mathbf{W}_x^H)^{-1} \mathbf{C} \begin{bmatrix} \mathbf{x} \\ -1 \end{bmatrix}, \tag{6.4.11}$$

$$\tilde{\mathbf{B}} = \mathbf{C} \mathbf{I}_{n+1,n} - \left[\mathbf{G}_1 \mathbf{W}_x^H \mathbf{u}, \ldots, \mathbf{G}_n \mathbf{W}_x^H \mathbf{u} \right], \tag{6.4.12}$$

$$\tilde{\mathbf{G}} = \left[\mathbf{G}_1^H \mathbf{u}, \ldots, \mathbf{G}_n^H \mathbf{u} \right], \tag{6.4.13}$$

where $\mathbf{I}_{n+1,n}$ is a $(n+1) \times n$ diagonal matrix with diagonal entries equal to 1. Hence, $\mathbf{a}$, $\mathbf{A}$ and $\mathbf{B}$ can be calculated as follows:

$$\mathbf{a} = \left(\mathbf{u}^H \tilde{\mathbf{B}} \right)^T, \tag{6.4.14}$$

$$\mathbf{A} = -\tilde{\mathbf{G}}^H \mathbf{W}_x^H \left(\mathbf{W}_x \mathbf{W}_x^H \right)^{-1} \tilde{\mathbf{B}} - \left(\tilde{\mathbf{G}}^H \mathbf{W}_x^H (\mathbf{W}_x \mathbf{W}_x^H)^{-1} \tilde{\mathbf{B}} \right)^T, \tag{6.4.15}$$

$$\mathbf{B} = \left(\tilde{\mathbf{B}}^H (\mathbf{W}_x \mathbf{W}_x^H)^{-1} \tilde{\mathbf{B}} \right)^T + \tilde{\mathbf{G}}^H \left(\mathbf{W}_x^H (\mathbf{W}_x \mathbf{W}_x^H)^{-1} \mathbf{W}_x - \mathbf{I} \right) \tilde{\mathbf{G}}. \tag{6.4.16}$$

It is shown in [1] that the constrained TLS estimate is equivalent to the constrained maximum likelihood estimate.

6.4.2 Harmonic Superresolution

Abatzoglou *et al.* [1] presented applications of the constrained TLS method in harmonic superresolution. Assume that L narrowband wavefront signals (harmonics) irradiate a uniform linear array of N elements. The array output signals satisfy the forward linear prediction equation [271]

$$\mathbf{C}_k \begin{bmatrix} \mathbf{x} \\ -1 \end{bmatrix} = \mathbf{0}, \qquad k = 1, 2, \ldots, M, \tag{6.4.17}$$

where M is the number of snapshots and

$$\mathbf{C}_k = \left[\begin{array}{cccc} y_k(1) & y_k(2) & \cdots & y_k(L+1) \\ y_k(2) & y_k(3) & \cdots & y_k(L+2) \\ \vdots & \vdots & \vdots & \vdots \\ y_k(N-L) & y_k(N-L+1) & \cdots & y_k(N) \\ \hdashline y_k^*(L+1) & y_k^*(L) & \cdots & y_k^*(1) \\ \vdots & \vdots & \vdots & \vdots \\ y_k^*(N) & y_k^*(N-1) & \cdots & y_k^*(N-L) \end{array}\right]. \tag{6.4.18}$$

Here $y_k(i)$ is the output observation of the kth array element at the time i. The matrix $\mathbf{C}_k$ is referred to as the data matrix at the kth snapshot. Combining all the data matrices into one data matrix $\mathbf{C}$, we get

$$\mathbf{C} = \begin{bmatrix} \mathbf{C}_1 \\ \vdots \\ \mathbf{C}_M \end{bmatrix}. \tag{6.4.19}$$

Then, the *harmonic superresolution problem* can be summarized as follows: use the constrained TLS method to solve the matrix equation

$$\mathbf{C} \begin{bmatrix} \mathbf{x} \\ -1 \end{bmatrix} = \mathbf{0},$$

while the estimation result of $\mathbf{W}_x$ is given by

$$\hat{\mathbf{W}}_x = \begin{bmatrix} \mathbf{W}_1 & 0 \\ 0 & \mathbf{W}_2 \end{bmatrix},$$

where

$$\mathbf{W}_1 = \begin{bmatrix} x_1 & x_2 & \cdots & x_L & -1 & & 0 \\ & \ddots & \ddots & & \ddots & \ddots & \\ 0 & & x_1 & x_2 & \cdots & x_L & -1 \end{bmatrix}$$

and

$$\mathbf{W}_2 = \begin{bmatrix} -1 & x_L & \cdots & x_2 & x_1 & & 0 \\ & \ddots & \ddots & & \ddots & \ddots & \\ 0 & & -1 & x_L & \cdots & x_2 & x_1 \end{bmatrix}.$$

Moreover,

$$\begin{aligned} F(\mathbf{x}) &= \begin{bmatrix} \mathbf{x} \\ -1 \end{bmatrix}^H \mathbf{C}^H (\mathbf{W}_x \mathbf{W}_x^H)^{-1} \mathbf{C} \begin{bmatrix} \mathbf{x} \\ -1 \end{bmatrix} \\ &= \begin{bmatrix} \mathbf{x} \\ -1 \end{bmatrix}^H \sum_{m=1}^{M} \mathbf{C}_m^H (\hat{\mathbf{W}}_x \hat{\mathbf{W}}_x^H)^{-1} \mathbf{C}_m \begin{bmatrix} \mathbf{x} \\ -1 \end{bmatrix}. \end{aligned} \tag{6.4.20}$$

In order to estimate the angles of arrival ϕ_i of the L space signals at the linear uniform array, the constrained TLS method consists of the following three steps [1].

(1) Use the Newton method to minimize $F(\mathbf{x})$ in (6.4.20), yielding $\mathbf{x}$.
(2) Find the roots $z_i, i = 1, \ldots, L$, of the linear-prediction-coefficient polynomial

$$\sum_{k=1}^{L} x_k z^{k-1} - z^L = 0. \tag{6.4.21}$$

(3) Estimate the angles of arrival $\phi_i = \arg(z_i)$, $i = 1, \ldots, L$.

6.4.3 Image Restoration

It is important to be able to recover lost information from degraded image data. The aim of image restoration is to find an optimal solution for the original image in the case where there is known record data and some prior knowledge.

Let the $N \times 1$ *point-spread function* (PSF) be expressed by

$$\mathbf{h} = \bar{\mathbf{h}} + \Delta\mathbf{h}, \tag{6.4.22}$$

where $\bar{\mathbf{h}}$ and $\Delta\mathbf{h} \in \mathbb{R}^N$ are respectively the known part and (unknown) error part of the PSF. The error components $\Delta h(i), i = 0, 1, \ldots, N-1$, of $\Delta\mathbf{h} = [\Delta h(0), \Delta h(1), \ldots, \Delta h(N-1)]^T$ are independent identically distributed (iid) noises with zero mean and the same variance σ_h.

The observed degraded image is expressed by $\mathbf{g}$, and the imaging equation can be expressed as

$$\mathbf{g} = \mathbf{H}\mathbf{f} + \Delta\mathbf{g}, \tag{6.4.23}$$

where $\mathbf{f}$ and $\Delta\mathbf{g} \in \mathbb{R}^N$ are respectively the original image and the additive noise of the observed image. The additive noise vector $\Delta\mathbf{g} = [\Delta g(0), \Delta g(1), \ldots, \Delta g(N-1)]^T$ is an iid random vector and is statistically independent of the PSF error component

$\Delta\mathbf{h}$. The matrix $\mathbf{H} \in \mathbb{R}^{N\times N}$ denotes the PSF, which consists of the known part $\bar{\mathbf{H}}$ and the noise part:

$$\mathbf{H} = \bar{\mathbf{H}} + \Delta\mathbf{H}. \tag{6.4.24}$$

The TLS solution of Equation (6.4.23) is

$$\mathbf{f} = \arg \min_{[\hat{\mathbf{H}},\hat{\mathbf{g}}]\in \mathbb{R}^{N\times(N+1)}} \left\| [\mathbf{H},\mathbf{g}] - [\hat{\mathbf{H}},\hat{\mathbf{g}}] \right\|_F^2, \tag{6.4.25}$$

where the constraint condition of $\hat{\mathbf{g}}$ is

$$\hat{\mathbf{g}} \in \text{Range}(\hat{\mathbf{H}}). \tag{6.4.26}$$

By defining an unknown regularized noise vector

$$\mathbf{u} = \left[\frac{\Delta h(0)}{\sigma_h},\ldots,\frac{\Delta h(N-1)}{\sigma_h},\frac{\Delta g(0)}{\sigma_g},\ldots,\frac{\Delta g(N-1)}{\sigma_g}\right]^T, \tag{6.4.27}$$

Mesarovic *et al.* [323] proposed the constrained TLS-based image restoration algorithm

$$\mathbf{f} = \arg\min_{\mathbf{f}} \|\mathbf{u}\|_2^2 \quad \text{subject to} \quad \bar{\mathbf{H}}\mathbf{f} - \mathbf{g} + \mathbf{L}\mathbf{u} = \mathbf{0}, \tag{6.4.28}$$

where $\mathbf{L}$ is an $N \times 2N$ matrix defined as

$$\mathbf{L} = \begin{bmatrix} \sigma_h f(0) & \sigma_h f(N-1) & \cdots & \sigma_h f(1) & \vdots & \sigma_g & 0 & \cdots & 0 \\ \sigma_h f(1) & \sigma_h f(0) & \cdots & \sigma_h f(2) & \vdots & 0 & \sigma_g & \cdots & 0 \\ \vdots & \vdots & \ddots & \vdots & \vdots & \vdots & \vdots & \ddots & \vdots \\ \sigma_h f(N-1) & \sigma_h f(N-2) & \cdots & \sigma_h f(0) & \vdots & 0 & 0 & \cdots & \sigma_g \end{bmatrix}. \tag{6.4.29}$$

In the case where there is a given data vector $\mathbf{g}$ and a part of the PSF matrix $\bar{\mathbf{H}}$ is known, solving for the original image $\mathbf{f}$ in Equation (6.4.23) is a typical inverse problem: the solution of the image restoration problem corresponds mathematically to the existence and uniqueness of the inverse transformation of Equation (6.4.23). If the inverse transformation does not exist then the image restoration is called a *singular inverse problem.* Moreover, even if the inverse transformation exists, its solution may be not unique. For a practical physical problem, such a nonunique solution is not acceptable. In this case, the image restoration is said to be an *ill-conditioned inverse problem.* This implies that a very small perturbation in the observed data vector $\mathbf{g}$ may lead to a large perturbation in the image restoration [16], [465].

An effective way to overcome the ill-conditioned problem in image restoration is to use the regularization method [465], [123], which yields the regularized constrained TLS algorithm [123], [168].

The basic idea of the regularized constrained TLS image restoration algorithm is to introduce a regularizing operator $\mathbf{Q}$ and a regularizing parameter $\lambda > 0$, and to replace the objective function by the sum of two complementary functions; i.e., (6.4.28) becomes

$$\mathbf{f} = \arg\min_{\mathbf{f}} \left\{ \|\mathbf{u}\|_2^2 + \lambda \|\mathbf{Qf}\|_2^2 \right\} \tag{6.4.30}$$

subject to

$$\bar{\mathbf{H}}\mathbf{f} - \mathbf{g} + \mathbf{Lu} = \mathbf{0}.$$

The choice of regularizing parameter λ needs to take into account both the fidelity to the observed data and the smoothness of the solution. In order to improve further the performance of the regularized constrained TLS image restoration algorithm, Chen *et al.* [97] proposed a method for adaptively choosing the regularizing parameter λ, this method is called the adaptively regularized constrained TLS image restoration. The solution of this algorithm is given by

$$\mathbf{f} = \arg\min_{\mathbf{f}} \left\{ \|\mathbf{u}\|_2^2 + \lambda(\mathbf{f}) \|\mathbf{Qf}\|_2^2 \right\} \tag{6.4.31}$$

subject to

$$\bar{\mathbf{H}}\mathbf{f} - \mathbf{g} + \mathbf{Lu} = \mathbf{0}.$$

6.5 Subspace Method for Solving Blind Matrix Equations

Consider a blind matrix equation

$$\mathbf{X} = \mathbf{AS}, \tag{6.5.1}$$

where $\mathbf{X} \in \mathbb{C}^{N\times M}$ is a complex matrix whose entries are the observed data and the two complex matrices $\mathbf{A} \in \mathbb{C}^{N\times d}$ and $\mathbf{S} \in \mathbb{C}^{d\times M}$ are unknown.

The question is: when only $\mathbf{X}$ is known, the solution of the unknown matrix $\mathbf{S}$ of blind matrix equation can be found? The answer is yes, but it is necessary to assume two conditions, that the matrix $\mathbf{A}$ is of full column rank and the matrix $\mathbf{S}$ is of full row rank. These assumptions are often satisfied in engineering problems. For example, in array signal processing, full column rank of the matrix $\mathbf{A}$ means that the directions of arrival (DOA) of all source signals are mutually independent, while full row rank of the matrix $\mathbf{S}$ requires that each source signal is independently transmitted.

Assume that N is the data length, d is the number of sources, M is the number of sensors, usually $M \geq d$ and $N > M$. Define the truncated SVD of the data matrix $\mathbf{X}$ as

$$\mathbf{X} = \hat{\mathbf{U}}\hat{\mathbf{\Sigma}}\hat{\mathbf{V}}^H, \tag{6.5.2}$$

where $\hat{\mathbf{\Sigma}}$ is a $d \times d$ diagonal matrix consisting of the d principal singular values of $\mathbf{X}$.

Since $\text{Col}(\mathbf{A}) = \text{Col}(\hat{\mathbf{U}})$, i.e., the matrices $\mathbf{A}$ and $\hat{\mathbf{U}}$ span the same signal subspace, we have

$$\hat{\mathbf{U}} = \mathbf{AT}, \tag{6.5.3}$$

where $\mathbf{T}$ is a $d \times d$ nonsingular matrix.

Let $\mathbf{W}$ be a $d \times N$ complex matrix that represents a neural network or filter. Premultiply Equation (6.5.1) by $\mathbf{W}$:

$$\mathbf{WX} = \mathbf{WAS}. \tag{6.5.4}$$

Adjusting the matrix $\mathbf{W}$ so that $\mathbf{WA} = \mathbf{I}_d$, the solution of the above equation is given by

$$\mathbf{S} = \mathbf{WX}. \tag{6.5.5}$$

In order to find $\mathbf{W}$, we compute

$$\mathbf{W}\hat{\mathbf{U}} = \mathbf{WAT} = \mathbf{T}$$

and thus obtain

$$\mathbf{W} = \mathbf{T}\hat{\mathbf{U}}^H. \tag{6.5.6}$$

Summarizing the above discussion, we have the following method for solving the blind matrix (6.5.1):

$$\left.\begin{array}{ll}\text{Data model} & \mathbf{X} = \mathbf{AS}, \\ \text{Truncated SVD} & \mathbf{X} = \hat{\mathbf{U}}\hat{\boldsymbol{\Sigma}}\hat{\mathbf{V}}, \\ \text{Solving} & \hat{\mathbf{U}} = \mathbf{AT} \text{ for } \mathbf{T}, \\ \text{Solution of matrix equation} & \mathbf{S} = \left(\mathbf{T}\hat{\mathbf{U}}^T\right)\mathbf{X}.\end{array}\right\} \tag{6.5.7}$$

This method is called the subspace method because it is based on the signal subspace. Hence, the key problem in solving a blind matrix equation is how to find the nonsingular matrix $\mathbf{T}$ from $\hat{\mathbf{U}} = \mathbf{AT}$ in the case where both $\mathbf{A}$ and $\mathbf{T}$ are unknown.

Here we consider an example of solving the blind matrix equation (6.5.1) in wireless communications.

Without taking into consideration the multipath transmission in wireless communication, the matrix in Equation (6.5.1) is given by [493]

$$\mathbf{X} = \mathbf{A}_\theta \mathbf{B}, \tag{6.5.8}$$

with

$$\mathbf{A}_\theta = [\mathbf{a}(\theta_1), \ldots, \mathbf{a}(\theta_d)] = \begin{bmatrix} 1 & 1 & \cdots & 1 \\ \theta_1 & \theta_2 & \cdots & \theta_d \\ \vdots & \vdots & \vdots & \vdots \\ \theta_1^{N-1} & \theta_2^{N-1} & \cdots & \theta_d^{N-1} \end{bmatrix},$$

$$\mathbf{B} = \mathbf{Diag}(\beta_1, \beta_2, \ldots, \beta_d),$$

in which θ_i and β_i are the unknown direction of arrival and the unknown attenuation coefficient of the ith user, respectively.

Define the diagonal matrix

$$\mathbf{\Theta} = \mathbf{Diag}(\theta_1, \ldots, \theta_d) \tag{6.5.9}$$

and the $(M-1) \times M$ selection matrices

$$\mathbf{J}_1 = [\mathbf{I}_{M-1}, \mathbf{0}], \quad \mathbf{J}_2 = [\mathbf{0}, \mathbf{I}_{M-1}]$$

which select respectively the upper $M-1$ rows and the lower $M-1$ rows of the matrix $\mathbf{A}_\theta$. It is easy to see that

$$(\mathbf{J}_1\mathbf{A}_\theta)\mathbf{\Theta} = \mathbf{J}_2\mathbf{A}_\theta. \tag{6.5.10}$$

Hence

$$\hat{\mathbf{U}} = \mathbf{XT} = \mathbf{A}_\theta\mathbf{BT}. \tag{6.5.11}$$

In order to find the nonsingular matrix $\mathbf{T}$, if we premultiply (6.5.11) by the selection matrices $\mathbf{J}_1$ and $\mathbf{J}_2$, respectively, and let $\mathbf{A}'_\theta = \mathbf{J}_1\mathbf{A}_\theta$ then, using (6.5.10), we get

$$\hat{\mathbf{U}}_1 = \mathbf{J}_1\hat{\mathbf{U}} = (\mathbf{J}_1\mathbf{A}_\theta)\mathbf{BT} = \mathbf{A}'_\theta\mathbf{BT}, \tag{6.5.12}$$

$$\hat{\mathbf{U}}_2 = \mathbf{J}_2\hat{\mathbf{U}} = (\mathbf{J}_2\mathbf{A}_\theta)\mathbf{BT} = \mathbf{A}'_\theta\mathbf{\Theta}\mathbf{BT}. \tag{6.5.13}$$

Since $\mathbf{B}$ and $\mathbf{\Theta}$ are diagonal matrices, we have $\mathbf{\Theta B} = \mathbf{B\Theta}$ and thus

$$\hat{\mathbf{U}}_2 = \mathbf{A}'_\theta\mathbf{B\Theta T} = \mathbf{A}'_\theta\mathbf{BTT}^{-1}\mathbf{\Theta T} = \hat{\mathbf{U}}_1\mathbf{T}^{-1}\mathbf{\Theta T},$$

which can be written as

$$\hat{\mathbf{U}}_1^\dagger\hat{\mathbf{U}}_2 = \mathbf{T}^{-1}\mathbf{\Theta T}, \tag{6.5.14}$$

where $\hat{\mathbf{U}}_1^\dagger = (\hat{\mathbf{U}}_1^H\hat{\mathbf{U}}_1)^{-1}\hat{\mathbf{U}}_1^H$ is the generalized inverse of the matrix $\hat{\mathbf{U}}_1$.

Because $\mathbf{\Theta}$ is a diagonal matrix, it is easily seen that Equation (6.5.14) is a typical similarity transformation. Hence, through this similarity transformation of the matrix $\hat{\mathbf{U}}_1^\dagger\hat{\mathbf{U}}_2$, we can obtain the nonsingular matrix $\mathbf{T}$.

The above discussion is summarized in Algorithm 6.5 for solving the blind matrix equation $\mathbf{X} = \mathbf{A}_\theta\mathbf{B}$.

Although we describe only the case of single-path transmission, the subspace method for solving the matrix equation $\mathbf{AS} = \mathbf{X}$ is also applicable to the case of multipath transmission. The difference is in the form of the matrix $\mathbf{A}$, so that the method of finding the nonsingular matrix $\mathbf{T}$ is also different. The interested reader may refer to [493].

Algorithm 6.5 Solving blind matrix equation $\mathbf{X} = \mathbf{A}_\theta \mathbf{B}$

---

input: $\mathbf{X} \in \mathbb{C}^{N \times M}$.

1. Compute the truncated SVD $\mathbf{X} = \hat{\mathbf{U}}\hat{\boldsymbol{\Sigma}}\hat{\mathbf{V}}^H$.
2. Calculate $\hat{\mathbf{U}}_1 = \hat{\mathbf{U}}\mathbf{J}_1$ and $\hat{\mathbf{U}}_2 = \hat{\mathbf{U}}\mathbf{J}_2$.
3. Make the similarity transformation $\mathbf{T}^{-1}\hat{\mathbf{U}}_1^\dagger \hat{\mathbf{U}}_2 \mathbf{T}$ to get $\mathbf{T}$.

output: $\mathbf{B} = (\hat{\mathbf{U}}^H \mathbf{T})\mathbf{X}$.

6.6 Nonnegative Matrix Factorization: Optimization Theory

A matrix with nonnegative real entries is called a *nonnegative matrix.* Consider the *nonnegative blind matrix equation* $\mathbf{X} = \mathbf{AS}$, where a known matrix $\mathbf{X}$ and two unknown matrices $\mathbf{A}$ and $\mathbf{S}$ are nonnegative matrices. This blind matrix equation is widespread in engineering application problems.

6.6.1 Nonnegative Matrices

Definition 6.3 An $n \times n$ matrix $\mathbf{A}$ or an $n \times 1$ vector $\mathbf{a}$ is said to be

(1) *positive (or elementwise positive)*, denoted $\mathbf{A} > \mathbf{O}$ or $\mathbf{a} > \mathbf{0}$, if all its entries are positive;
(2) *nonnegative (or elementwise nonnegative)*, denoted $\mathbf{A} \geq \mathbf{O}$ or $\mathbf{a} \geq \mathbf{0}$, if its all entries are nonnegative.

We use the notation $\mathbf{A} > \mathbf{B}$ ($\mathbf{A} \geq \mathbf{B}$) to mean that $\mathbf{A} - \mathbf{B}$ is a positive (nonnegative) matrix, i.e., $A_{ij} > B_{ij}$ ($A_{ij} \geq B_{ij}$) for all i, j.

Definition 6.4 We say that a matrix $\mathbf{A} \in \mathbb{R}^{n \times n}$ is *reducible* if there exists a permutation matrix $\mathbf{P}$ such that

$$\mathbf{B} = \mathbf{PAP}^T = \begin{bmatrix} \mathbf{A}_{11} & \mathbf{A}_{12} \\ \mathbf{O} & \mathbf{A}_{21} \end{bmatrix}, \tag{6.6.1}$$

where $\mathbf{A}_{11} \in \mathbb{R}^{r \times r}, \mathbf{A}_{22} \in \mathbb{R}^{n-r,n-r}, \mathbf{A}_{12} \in \mathbb{R}^{r \times (n-r)}$ and $\mathbf{O}$ is an $(n-r) \times r$ zero matrix.

Definition 6.5 A matrix $\mathbf{A} \in \mathbb{R}^{n \times n}$ is said to be irreducible, if it is not reducible.

Definition 6.6 An $n \times n$ nonnegative matrix $\mathbf{A}$ is said to be *regular or primitive*, if there is a $k \geq 1$ such that all the entries of $\mathbf{A}^k$ are positive.

Example 6.3 Any positive matrix $\mathbf{A}$ is regular since, for $k = 1$, $\mathbf{A} > \mathbf{O}$.

EXAMPLE 6.4 The following matrices are not regular:

$$\mathbf{A} = \begin{bmatrix} 1 & 1 \\ 0 & 1 \end{bmatrix}, \quad \mathbf{A} = \begin{bmatrix} 1 & 0 \\ 1 & 1 \end{bmatrix} \quad \text{and} \quad \mathbf{A} = \begin{bmatrix} 0 & 1 \\ 1 & 0 \end{bmatrix}$$

because, for any $k \geq 1$, $\mathbf{A}^k \geq \mathbf{O}$. But, the matrix

$$\mathbf{A} = \begin{bmatrix} 1 & 1 \\ 1 & 0 \end{bmatrix}$$

is regular, since

$$\mathbf{A} = \begin{bmatrix} 1 & 1 \\ 0 & 1 \end{bmatrix}^2 = \begin{bmatrix} 2 & 1 \\ 1 & 1 \end{bmatrix} > \mathbf{O}.$$

EXAMPLE 6.5 The following matrix is regular:

$$\mathbf{A} = \begin{bmatrix} 1 & 0 & 1 \\ 1 & 0 & 0 \\ 0 & 1 & 0 \end{bmatrix}$$

because, for $k = 4$, we have

$$\mathbf{A}^4 = \begin{bmatrix} 2 & 1 & 1 \\ 2 & 1 & 1 \\ 1 & 1 & 1 \end{bmatrix} > \mathbf{O}.$$

There is the following relationship between an irreducible matrix $\mathbf{A}$ and a regular matrix.

LEMMA 6.3 *If $\mathbf{A}$ is a nonnegative and irreducible $n \times n$ matrix then $(\mathbf{I}+\mathbf{A})^{n-1} > \mathbf{O}$, i.e., $\mathbf{I} + \mathbf{A}$ is regular or primitive.*

Proof [355] It suffices to prove that $(\mathbf{I} + \mathbf{A})^{n-1}\mathbf{x} > \mathbf{O}$ for any $\mathbf{x} \geq \mathbf{0}, \mathbf{x} \neq \mathbf{0}$. Define the sequence $\mathbf{x}_{k+1} = (\mathbf{I} + \mathbf{A})\mathbf{x}_k \geq \mathbf{0}, k = 0, 1, \ldots, n-2, \mathbf{x}_0 = \mathbf{x}$. Since $\mathbf{x}_{k+1} = \mathbf{x}_k + \mathbf{A}\mathbf{x}_k$, $\mathbf{x}_{k+1}$ has no more zero entries than $\mathbf{x}_k$. In order to prove by contradiction that $\mathbf{x}_{k+1}$ has fewer zero entries than $\mathbf{x}_k$, suppose that $\mathbf{x}_{k+1}$ and $\mathbf{x}_k$ have exactly the same number of zero entries. Then, there exists a permutation matrix $\mathbf{P}$ such that

$$\mathbf{P}\mathbf{x}_{k+1} = \begin{bmatrix} \mathbf{y} \\ \mathbf{0} \end{bmatrix}, \quad \mathbf{P}\mathbf{x}_k = \begin{bmatrix} \mathbf{z} \\ \mathbf{0} \end{bmatrix}, \quad \mathbf{y}, \mathbf{z} \in \mathbb{R}^m, \mathbf{y}, \mathbf{z} > \mathbf{0},\ 1 \leq m < n.$$

Then

$$\begin{aligned} \mathbf{P}\mathbf{x}_{k+1} = \begin{bmatrix} \mathbf{y} \\ \mathbf{0} \end{bmatrix} &= \mathbf{P}(\mathbf{x}_k + \mathbf{A}\mathbf{x}_k) = \mathbf{P}\mathbf{x}_k + \mathbf{P}\mathbf{A}\mathbf{P}^T\mathbf{P}\mathbf{x}_k \\ &= \begin{bmatrix} \mathbf{z} \\ \mathbf{0} \end{bmatrix} + \begin{bmatrix} \mathbf{A}_{11} & \mathbf{A}_{12} \\ \mathbf{A}_{21} & \mathbf{A}_{22} \end{bmatrix} \begin{bmatrix} \mathbf{z} \\ \mathbf{0} \end{bmatrix}. \end{aligned}$$

This implies that $\mathbf{A}_{21} = \mathbf{O}$, which contradicts the assumption condition that $\mathbf{A}$ is irreducible. Thus, $\mathbf{x}_0 = \mathbf{x}$ has at most $n-1$ zero entries, $\mathbf{x}_k$ has at most $n-k-1$ zero entries, and hence $\mathbf{x}_{n-1} = (\mathbf{I}+\mathbf{A})^{n-1}\mathbf{x}_0$ has no zero entry, i.e., $\mathbf{x}_{n-1}$ is a positive vector, which completes the proof. □

Theorem 6.8 (Perron–Frobenius theorem for regular matrices) *Suppose that $\mathbf{A} \in \mathbb{R}^{n\times n}$ is a nonnegative and regular matrix, i.e., $\mathbf{A}^k > 0$ for some $k \geq 1$; then*

- *there is a Perron–Frobenius (PF) eigenvalue λ_{PF} of $\mathbf{A}$ that is real and positive, together with positive left and right eigenvectors;*
- *any other eigenvalue λ satisfies $|\lambda| < \lambda_{\mathrm{PF}}$;*
- *the eigenvalue λ_{PF} is simple, i.e., has multiplicity 1, and corresponds to a 1×1 Jordan block.*

The Perron–Frobenius theorem was proved by Oskar Perron in 1907 [385] for positive matrices and was extended by Georg Frobenius in 1912 [165] to nonnegative and irreducible matrices.

The Perron–Frobenius theorem has important applications to probability theory (Markov chains), the theory of dynamical systems, economics, population dynamics (the Leslie population-age-distribution model), power control and so on. Interested readers can refer to the books [39], [324]. In particular, the Perron–Frobenius theorem has been extended to tensors [39].

6.6.2 Nonnegativity and Sparsity Constraints

In many engineering applications it is necessary to impose two constraints on the data: a nonnegativity constraint and a sparsity constraint.

1. Nonnegativity Constraint

As the name suggests, the *nonnegativity constraint* constrains the data to be nonnegative. A lot of actual data are nonnegative, and so constitute the nonnegative data matrices. Such nonnegative matrices exist widely in daily life. The following are four important actual examples of nonnegative matrices [277].

1. In document collections, documents are stored as vectors. Each element of a document vector is a count (possibly weighted) of the number of times a corresponding term appears in that document. Stacking document vectors one after the other creates a nonnegative term-by-document matrix that represents the entire document collection numerically.
2. In image collections, each image is represented by a vector, and each element of the vector corresponds to a pixel. The intensity and color of a pixel is given by a nonnegative number, thereby creating a nonnegative pixel-by-image matrix.
3. For item sets or recommendation systems, the information for a purchase history of customers or ratings on a subset of items is stored in a nonnegative sparse matrix.

4. In gene expression analysis, gene-by-experiment matrices are formed from observations of the gene sequences produced under various experimental conditions.

In addition, in pattern recognition and signal processing, for some particular pattern or target signal a linear combination of all the feature vectors (an "all-combination") may not be appropriate. On the contrary, a part-combination of some feature vectors is more suitable. For example, in face recognition, the combination of the specific parts of the eye, nose, mouth and so on is often more effective.

In an all-combination, the positive and negative combination coefficients emphasize respectively the positive and negative effects of some features, while a zero combination coefficient implies a characteristics that does not play a role. In a part-combination, there are only two kinds of characteristics: those play or do not play a role. Therefore, in order to emphasize the role of some main features, it is natural to add nonnegative constraints to the elements in the coefficient vector.

2. Sparsity Constraint

By a *sparsity constraint* is meant an assumption that the data are not dense but sparse, i.e., most data take a zero value, only a few take nonzero values. A matrix for which most entries are zero and only a few entries are nonzero is called a sparse matrix, while a sparse matrix whose entries take only nonnegative nonzero values is known as a *nonnegative sparse matrix.*

For example, in the commodity recommendation system, a matrix that is composed of the customer's purchases or scores is a nonnegative sparse matrix. In economics, a lot of variables and data (such as volume and price) are not only sparse but also nonnegative.

Sparsity constraints can increase the effectiveness of investment portfolios, while nonnegative constraints can improve investment efficiency and also reduce the investment risks [436], [538].

Although many natural signals and images are not themselves sparse, after a certain transformation they are sparse in the transform domain. For example, the discrete cosine transform (DCT) of a face and a medical image is a set of typical sparse data. The short-time Fourier transform (STFT) of a speech signal is also sparse in the time domain.

6.6.3 Nonnegative Matrix Factorization Model

The basic problem of linear data analysis is as follows. By an appropriate transformation or factorization, a higher-dimensional original data vector can be represented as a linear combination of a set of low-dimensional vectors. Because the nature or character of the original data vector is extracted, it can be used for pattern recognition. So these low-dimensional vectors are often called "*pattern vectors*" or "*basis vectors*" or "*feature vectors*".

In the process of data analysis, modeling and processing, the two basic requirements of a pattern vector must be considered.

(1) *Interpretability* The components of a pattern vector should have definite physical or physiological meaning.
(2) *Statistical fidelity* When the data are consistent and do not have too much error or noise, the components of a pattern vector should be able to explain the variance of the data (in the main energy distribution).

Vector quantization (VQ) and *principal component analysis* (PCA) are two widely used unsupervised learning algorithms; they adopt fundamentally different ways to encoding of data.

1. Vector Quantization Method

The vector quantization (VQ) method uses a stored prototype vector as the *code vector*. Let $\mathbf{c}_n$ be a k-dimensional code vector and let there be N such stored code vectors, i.e.,

$$\mathbf{c}_n = [c_{n,1}, \ldots, c_{n,k}]^T, \quad n = 1, \ldots, N.$$

The collection of N code vectors $\{\mathbf{c}_1, \ldots, \mathbf{c}_N\}$ comprises a *codebook*.

The subset consisting of the data vectors closest to the stored pattern vectors or the code vectors $\mathbf{c}_n$ is called the *encoding region* of the code vectors $\mathbf{c}_n$, denoted S_n and defined as

$$S_n = \left\{\mathbf{x} \,\middle|\, \|\mathbf{x} - \mathbf{c}_n\|^2 \leq \|\mathbf{x} - \mathbf{c}_{n'}\|^2, \forall\, n' = 1, \ldots, N\right\}. \tag{6.6.2}$$

The formulation of the *vector quantization problem* is as follows: given M $k \times 1$ data vectors $\mathbf{x}_i = [x_{i,1}, \ldots, x_{i,k}]^T, i = 1, \ldots, M$, determine the encoding region of these vectors, i.e., their corresponding code vectors.

Let $\mathbf{X} = [\mathbf{x}_1, \ldots, \mathbf{x}_M] \in \mathbb{R}^{k\times M}$ be a data matrix and $\mathbf{C} = [\mathbf{c}_1, \ldots, \mathbf{c}_N] \in \mathbb{R}^{k\times N}$ denote a *codebook matrix*. Then the VQ of the data matrix can be described by the model

$$\mathbf{X} = \mathbf{CS}, \tag{6.6.3}$$

where $\mathbf{S} = [\mathbf{s}_1, \ldots, \mathbf{s}_M] \in \mathbb{R}^{N\times M}$ is a *quantization coefficient matrix*, its columns are called *quantization coefficient vectors*.

From the viewpoint of optimization, the VQ optimization criterion is "winner-take-all". According to this criterion, the input data is clustered into mutually exclusive patterns [172]. From the viewpoint of encoding, the VQ is a "grandmother cell coding": all data are explained or clustered simply by a basis vector [506].

Specifically, each quantization coefficient vector is an $N \times 1$ basis vector with only a single entry 1 and the rest zero entries. Therefore, if the (i, j)th entry of the codebook matrix is equal to 1 then the data vector $\mathbf{x}_j$ is judged to be closest to the code vector $\mathbf{c}_i$, that is, the data vector corresponds to only one code vector. The

VQ method can capture the nonlinear structure of the input data but its capturing ability is weak, because the data vector and the code vector are in one-to-one correspondence in this method. If the dimension of the data matrix is large, a large number of code vectors are needed to represent the input data.

2. Principal Component Analysis Method

The linear data model is a widely used data model. Principal component analysis (PCA), linear discriminant analysis (LDA), independent component analysis (ICA) and other multivariate data analysis methods adopt this linear data model.

As we saw, the VQ method uses code vectors. Similarly, the base matrix $\mathbf{A}$ in the PCA method consists of a set of orthogonal basis vectors $\mathbf{a}_i$ corresponding to the principal components. These basis vectors are called the pattern or feature vectors. For a data vector $\mathbf{x}$, PCA uses the principle of shared constraints to make the optimization and uses a linear combination of the pattern vectors $\mathbf{x} = \mathbf{As}$ to represent the input data. From the encoding point of view, PCA provides a *distributed encoding*, as compared with the grandmother-cell-encoding-based VQ method, thus the PCA method requires only a small group of basic vectors to represent higher-dimensional data.

The disadvantages of the PCA method are as follows.

1. It cannot capture any nonlinear structure in the input data.
2. The basis vectors can be statistically interpreted as the directions of maximum difference, but many directions do not have a clear visual interpretation. The reason is that the entries of the basis matrix $\mathbf{A}$ and the quantization coefficient vector $\mathbf{s}$ can take zero, positive or negative signs. Since the basis vectors are used in the linear combination, and this combination involves a complex cancellation of positive and negative numbers, many individual basis vectors lose their intuitive physical meaning due to this cancellation and do not have an explanatory role for nonnegative data (such as the pixel values of a color image). On the one hand, the entries of a nonnegative pattern vector should all be nonnegative values, but on the other hand mutually orthogonal eigenvectors must contain negative entries: if all the entries of the eigenvector $\mathbf{u}_1$ corresponding to the maximum eigenvalue are nonnegative, then any other eigenvector orthogonal to $\mathbf{u}_1$ must contain at least one negative entry, otherwise the orthogonality condition of two vectors $\langle \mathbf{u}_1, \mathbf{u}_j \rangle = 0, j \neq 1$, cannot be met. This fact indicates that the mutually orthogonal eigenvectors cannot be used as pattern vectors or basis vectors in nonnegative data analysis.

In the PCA, LDA and ICA methods, the coefficient vector elements usually take positive or negative values; few take a zero value. That is to say, in these methods, all the basis vectors are involved in the fitting or regression of the data vectors.

3. Nonnegative Matrix Factorization Method

Unlike in the VQ, PCA, LDA and ICA methods, the basis vectors and the elements of the coefficient vector are treated as nonnegative constraints in nonnegative matrix factorization (NMF). It can be seen that the number of basis vectors involved in the fitting or regression of a data vector is certainly less in the NMF. From this perspective, NMF basis vectors have the role of extracting the principal basis vectors.

Another prominent advantage of NMF is that the nonnegative constraint on combination factors facilitates the generation of sparse coding, so that many encoded values are zero. In biology, the human brain encodes information in this sparse coding way [153]. Therefore, as an alternative method of linear data analysis, we should use nonnegative instead of orthogonal constraints on the basis vectors.

The NMF method gives a multilinear nonnegative approximation to the data.

Let $\mathbf{x}(j) = [x_1(j), \ldots, x_I(j)]^T \in \mathbb{R}_+^{I\times 1}$ and $\mathbf{s}(j) = [s_1(j), \ldots, s_K(j)]^T \in \mathbb{R}_+^{K\times 1}$ denote respectively the nonnegative data vector and K-dimensional nonnegative coefficient vector measured by I sensors at the discrete time j, where $\mathbb{R}_+$ denotes the nonnegative quadrant.

The mathematical model for nonnegative vectors is described by

$$\begin{bmatrix} x_1(j) \\ \vdots \\ x_I(j) \end{bmatrix} = \begin{bmatrix} a_{11} & \cdots & a_{1K} \\ \vdots & \ddots & \vdots \\ a_{I1} & \cdots & a_{IK} \end{bmatrix} \begin{bmatrix} s_1(j) \\ \vdots \\ s_K(j) \end{bmatrix} \quad \text{or} \quad \mathbf{x}(j) = \mathbf{A}\mathbf{s}(j), \tag{6.6.4}$$

in which $\mathbf{A} = [\mathbf{a}_1, \ldots, \mathbf{a}_K] \in \mathbb{R}^{I\times K}$ is the basis matrix and $\mathbf{a}_k, k = 1, \ldots, K$, are the basis vectors.

Since the measurement vectors $\mathbf{x}(j)$ at different times are expressed in the same set of basis vectors, so these I-dimensional basis vectors can be imagined as the building blocks of data representation, while the element $s_k(j)$ of the K-dimensional coefficient vector represents the strength of the kth basis vector (building block) $\mathbf{a}_k$ in the data vector $\mathbf{x}(j)$, reflecting the contribution of $\mathbf{a}_k$ in the fitting or regression of $\mathbf{x}$. Therefore, the elements $s_k(j)$ of the coefficient vectors are often referred to as *fitting coefficients*, regression coefficients or combination coefficients:

(1) $s_k(j) > 0$ represents a positive contribution of the basis vector $\mathbf{a}_k$ to the additive combination;
(2) $s_k(j) = 0$ represents a null contribution of the corresponding basis vector, i.e., it is not involved in the fitting or regression;
(3) $s_k(j) < 0$ implies a negative contribution of the basis vector $\mathbf{a}_k$.

If the nonnegative data vectors measured at the discrete times $j = 1, \ldots, J$ are arranged as a nonnegative observation matrix, then

$$[\mathbf{x}(1), \ldots, \mathbf{x}(J)] = \mathbf{A}[\mathbf{s}(1), \ldots, \mathbf{s}(J)] \quad \Rightarrow \quad \mathbf{X} = \mathbf{AS}. \tag{6.6.5}$$

The matrix $\mathbf{S}$ is called the coefficient matrix. It is essentially an encoding of the basis matrix.

The problem of solving the blind nonnegative matrix equation $\mathbf{X} = \mathbf{AS}$ can be described as: given a nonnegative matrix $\mathbf{X} \in \mathbb{R}_+^{I \times J}$ (its entries $x_{ij} \geq 0$) with a low rank $r < \min\{I, J\}$, find the basis matrix $\mathbf{A} \in \mathbb{R}_+^{I \times r}$ and the coefficient matrix $\mathbf{S} \in \mathbb{R}_+^{r \times J}$ for $\mathbf{X}$ such that

$$\mathbf{X} = \mathbf{AS} + \mathbf{N} \tag{6.6.6}$$

or

$$\mathbf{X}_{ij} = [\mathbf{AS}]_{ij} + \mathbf{N}_{ij} = \sum_{k=1}^{r} a_{ik} s_{kj} + n_{ij}, \tag{6.6.7}$$

where $\mathbf{N} \in \mathbb{R}^{I \times J}$ is the approximate error matrix. The coefficient matrix $\mathbf{S}$ is also called the *encoding variable matrix*, its entries are unknown hidden nonnegative components.

The problem of decomposing a nonnegative data matrix into the product of a nonnegative basis matrix and a nonnegative coefficient matrix is called *nonnegative matrix factorization* (NMF) and was proposed by Lee and Seung in 1999 in *Nature* [286]. The NMF is essentially a multilinear nonnegative data representation. If the data matrix $\mathbf{X}$ is a positive matrix then the basis matrix $\mathbf{A}$ and the coefficient matrix $\mathbf{S}$ are required to be positive matrices. Such a matrix factorization is called *positive matrix factorization* (PMF) and was proposed by Paatero and Tapper in 1994 [367].

Equation (6.6.6) shows that when the data-matrix rank $r = \text{rank}(\mathbf{X}) < \min\{I, J\}$, the *nonnegative matrix approximation* $\mathbf{AS}$ can be regarded as a compression and de-noising form of the data matrix $\mathbf{X}$.

Table 6.1 shows the connections and differences between VQ, PCA and NMF.

Table 6.1 Comparison of VQ, PCA and NMF

	VQ	PCA	NMF
Constraints	winner-take-all	all share	minority share
Form	$\begin{cases}\text{codebook matrix } \mathbf{C}\\ \text{coeff. matrix } \mathbf{S}\end{cases}$	$\begin{cases}\text{basis matrix } \mathbf{A}\\ \text{coeff. matrix } \mathbf{S}\end{cases}$	$\begin{cases}\text{basis matrix } \mathbf{A}\\ \text{coeff. matrix } \mathbf{S}\end{cases}$
Model	clustering $\mathbf{X} = \mathbf{CS}$	$\begin{cases}\text{combination}\\ \mathbf{X} = \mathbf{AS}\end{cases}$	NMF $\mathbf{X} = \mathbf{AS}$
Analysis ability	nonlinear	linear	multilinear
Encoding way	grandmother cell	distributive	distributive + nonnegative
Machine learning	single mode	distributed	parts combination

In the following, we address the features of NMF.

1. *Distributed nonnegative encoding* Nonnegative matrix factorization does not allow negative elements in the factorization matrices **A** and **S**. Unlike the single constraint of the VQ, the nonnegative constraint allows the use of a combination of basis images, or eigenfaces, to represent a face image. Unlike the PCA, the NMF allows only additive combinations; because the nonzero elements of **A** and **S** are all positive, the occurrence of any subtraction between basis images can be avoided. In terms of optimization criteria, the VQ adopts the "winner-take-all" constraint and the PCA is based on the "all-sharing" constraint, while the NMF adopts a "group-sharing" constraint together with a nonnegative constraint. From the encoding point of view, the NMF is a distributed nonnegative encoding, which often leads to sparse encoding.
2. *Parts combination* The NMF gives the intuitive impression that it is not a combination of all features; rather, just some of the features, (simply called the parts), are combined into a (target) whole. From the perspective of machine learning, the NMF is a kind of machine learning method based on the combination of parts which has the ability to extract the main features.
3. *Multilinear data analysis capability* The PCA uses a linear combination of all the basis vectors to represent the data and can extract only the linear structure of the data. In contrast, the NMF uses combinations of different numbers of and different labels of the basis vectors (parts) to represent the data and can extract its multilinear structure; thereby it has certain nonlinear data analysis capabilities.

6.6.4 Divergences and Deformed Logarithm

The NMF is essentially an optimization problem and often uses a divergence as its cost function.

The distance $D(p\|g)$ between probability densities p and q is called the *divergence* if it meets only the nonnegativity and positive definite condition $D(p\|g) \geq 0$ (the equality holds if and only if $p = g$).

According to the prior knowledge of the statistical noise distribution, common divergences in NMF are the Kullback–Leibler divergence and the alpha–beta (AB) divergence and so on.

1. Squared Euclidean Distance

Consider the NMF approximation $\mathbf{X} \approx \mathbf{AS}$. The distance between the nonnegative measurement matrix $\mathbf{X}$ and the NMF $\mathbf{AS}$ is denoted $D(\mathbf{X}\|\mathbf{AS})$. When the approximation error follows a normal distribution, the NMF generally uses the squared

Euclidean distance of the error matrix as the cost function:

$$D_E(\mathbf{X}\|\mathbf{AS}) = \|\mathbf{X} - \mathbf{AS}\|_2^2 = \frac{1}{2}\sum_{i=1}^{I}\sum_{j=1}^{J}(x_{ij} - [\mathbf{AS}]_{ij})^2. \tag{6.6.8}$$

In many applications, in addition to the nonnegative constraint $a_{ik} \geq 0$, $s_{kj} \geq 0, \forall\, i,j,k$, it is usually required that the expectation solution of $\mathbf{S}$ or of $\mathbf{A}$ has some characteristics associated with the cost functions $J_S(\mathbf{S})$ or $J_A(\mathbf{A})$ respectively, which yields the cost function

$$D_E(\mathbf{X}\|\mathbf{AS}) = \frac{1}{2}\|\mathbf{X} - \mathbf{AS}\|_2^2 + \alpha_A J_A(\mathbf{A}) + \alpha_S J_S(\mathbf{S}) \tag{6.6.9}$$

called the *squared Euclidean distance* between parameters α_S and α_A.

2. AB Divergence

Let $\mathbf{P}, \mathbf{G} \in \mathbb{R}^{I\times J}$ be a nonnegative measurement matrix and its approximation matrix; then their alpha–beta divergence is simply called the AB divergence, and is defined as (see e.g., [14], [101], [102])

$$D_{AB}^{(\alpha,\beta)}(\mathbf{P}\|\mathbf{G})$$

$$= \begin{cases} -\dfrac{1}{\alpha\beta}\displaystyle\sum_{i=1}^{I}\sum_{j=1}^{J}\left(p_{ij}^{\alpha}g_{ij}^{\beta} - \tfrac{\alpha}{\alpha+\beta}p_{ij}^{\alpha+\beta} - \tfrac{\beta}{\alpha+\beta}g_{ij}^{\alpha+\beta}\right), & \alpha,\beta,\alpha+\beta \neq 0,\\[2mm] \dfrac{1}{\alpha^2}\displaystyle\sum_{i=1}^{I}\sum_{j=1}^{J}\left(p_{ij}^{\alpha}\ln\frac{p_{ij}^{\alpha}}{g_{ij}^{\alpha}} - p_{ij}^{\alpha} + g_{ij}^{\alpha}\right), & \alpha\neq 0, \beta = 0,\\[2mm] \dfrac{1}{\alpha^2}\displaystyle\sum_{i=1}^{I}\sum_{j=1}^{J}\left(\ln\frac{g_{ij}^{\alpha}}{p_{ij}^{\alpha}} + \left(\frac{g_{ij}^{\alpha}}{p_{ij}^{\alpha}}\right)^{-1} - 1\right), & \alpha = -\beta \neq 0,\\[2mm] \dfrac{1}{\beta^2}\displaystyle\sum_{i=1}^{I}\sum_{j=1}^{J}\left(g_{ij}^{\beta}\ln\frac{g_{ij}^{\beta}}{p_{ij}^{\beta}} - g_{ij}^{\beta} + p_{ij}^{\beta}\right), & \alpha = 0, \beta\neq 0,\\[2mm] \dfrac{1}{2}\displaystyle\sum_{i=1}^{I}\sum_{j=1}^{J}\left(\ln p_{ij} - \ln g_{ij}\right)^2, & \alpha = 0, \beta = 0. \end{cases}$$

The AB divergence is a constant function with α and β as parameters, and contains most divergences as special cases.

(1) *Alpha divergence* When $\alpha + \beta = 1$, the AB divergence reduces to the alpha divergence (α *divergence*):

$$\begin{aligned} D_\alpha(\mathbf{P}\|\mathbf{G}) &= D_{AB}^{(\alpha,1-\alpha)}(\mathbf{P}\|\mathbf{G})\\ &= \frac{1}{\alpha(\alpha-1)}\sum_{i=1}^{I}\sum_{j=1}^{J}\left(p_{ij}^{\alpha}g_{ij}^{1-\alpha} - \alpha p_{ij} + (\alpha-1)g_{ij}\right), \end{aligned} \tag{6.6.10}$$

where $\alpha \neq 0$ and $\alpha \neq 1$. The following are some common α divergences [102]:

- When $\alpha = 2$, the α divergence reduces to the Pearson χ^2 distance.
- When $\alpha = 0.5$, the α divergence reduces to the Hellinger distance.

- When $\alpha = -1$, the α divergence reduces to the Neyman χ^2 distance.
- As $\alpha \to 0$, the limit of the α divergence is the KL divergence of $\mathbf{G}$ from $\mathbf{P}$, i.e.,

$$\lim_{\alpha \to 0} D_\alpha(\mathbf{P}\|\mathbf{G}) = D_{\text{KL}}(\mathbf{G}\|\mathbf{P}).$$

- As $\alpha \to 1$, the limit of the α is the KL divergence of $\mathbf{P}$ from $\mathbf{G}$, i.e.,

$$\lim_{\alpha \to 1} D_\alpha(\mathbf{P}\|\mathbf{G}) = D_{\text{KL}}(\mathbf{P}\|\mathbf{G}).$$

(2) *Beta divergence* When $\alpha = 1$, the AB divergence reduces to the beta divergence (*β divergence*):

$$\begin{aligned} D_\beta(\mathbf{P}\|\mathbf{G}) &= D_{AB}^{(1,\beta)}(\mathbf{P}\|\mathbf{G}) \\ &= -\frac{1}{\beta}\sum_{i=1}^{I}\sum_{j=1}^{J}\left(p_{ij}g_{ij}^{\beta} - \frac{1}{1+\beta}p_{ij}^{1+\beta} - \frac{\beta}{1+\beta}g_{ij}^{1+\beta}\right), \end{aligned} \tag{6.6.11}$$

where $\beta \neq 0$. In particular, if $\beta = 1$ then

$$D_{\beta=1}(\mathbf{P}\|\mathbf{G}) = D_{AB}^{(1,1)}(\mathbf{P}\|\mathbf{G}) = \frac{1}{2}\sum_{i=1}^{I}\sum_{j=1}^{J}(p_{ij} - g_{ij})^2 \tag{6.6.12}$$

reduces to the squared Euclidean distance.

(3) *Kullback–Leibler* (KL) *divergence* When $\alpha = 1$ and $\beta = 0$, the AB divergence reduces to the standard KL divergence, i.e.,

$$D_{AB}^{(1,0)}(\mathbf{P}\|\mathbf{G}) = D_{\beta=0}(\mathbf{P}\|\mathbf{G}) = D_{\text{KL}}(\mathbf{P}\|\mathbf{G}).$$

(4) *Itakura–Saito* (IS) *divergence* When $\alpha = 1$ and $\beta = -1$, the AB divergence gives the standard Itakura–Saito divergence

$$D_{\text{IS}}(\mathbf{P}\|\mathbf{G}) = D_{AB}^{(1,-1)}(\mathbf{P}\|\mathbf{G}) = \sum_{i=1}^{I}\sum_{j=1}^{J}\left(\ln\frac{g_{ij}}{p_{ij}} + \frac{p_{ij}}{g_{ij}} - 1\right). \tag{6.6.13}$$

3. Kullback–Leibler Divergence

Let $\phi : \mathcal{D} \to \mathbb{R}$ be a continuous differentiable convex function defined in the closed convex set $\mathcal{D} \subseteq \mathbb{R}_+^K$. The *Bregman distance* between two vectors $\mathbf{x}, \mathbf{g} \in \mathcal{D}$ associated with the function ϕ is denoted $B_\phi(\mathbf{x}\|\mathbf{g})$, and defined as

$$B_\phi(\mathbf{x}\|\mathbf{g}) \stackrel{\text{def}}{=} \phi(\mathbf{x}) - \phi(\mathbf{g}) - \langle \nabla\phi(\mathbf{x}), \mathbf{x} - \mathbf{g}\rangle. \tag{6.6.14}$$

Here $\nabla\phi(\mathbf{x})$ is the gradient of the function ϕ at $\mathbf{x}$. In particular, if

$$\phi(\mathbf{x}) = \sum_{i=1}^{K} x_i \ln x_i \tag{6.6.15}$$

then the Bregman distance is called the *Kullback–Leibler* (KL) *divergence*, denoted

$D_{\text{KL}}(\mathbf{x}\|\mathbf{g})$. In probability and information theory, the KL divergence is also called the information divergence, the information gain, or the relative entropy.

For two probability distribution matrices of a stochastic process $\mathbf{P}, \mathbf{G} \in \mathbb{R}^{I\times J}$, if they are nonnegative then their KL divergence $D_{\text{KL}}(\mathbf{P}\|\mathbf{G})$ is defined as

$$D_{\text{KL}}(\mathbf{P}\|\mathbf{G}) = \sum_{i=1}^{I}\sum_{j=1}^{J}\left(p_{ij}\ln\frac{p_{ij}}{g_{ij}} - p_{ij} + g_{ij}\right). \tag{6.6.16}$$

Clearly, we have $D_{\text{KL}}(\mathbf{P}\|\mathbf{G}) \geq 0$ and $D_{\text{KL}}(\mathbf{P}\|\mathbf{G}) \neq D_{\text{KL}}(\mathbf{G}\|\mathbf{P})$, i.e., the KL divergence does not have symmetry.

4. Entropy, Tsallis Statistics and Deformed Logarithms

In mathematical statistics and information theory, for a set of probabilities $\{p_i\}$ such that $\sum_i p_i = 1$, the *Shannon entropy* is defined as

$$S = -\sum_i p_i \log_2 p_i, \tag{6.6.17}$$

while the *Boltzmann–Gibbs entropy* is defined as

$$S_{\text{BG}} = -k\sum_i p_i \ln p_i. \tag{6.6.18}$$

For two independent systems A and B, their joint probability density $p(A,B) = p(A)p(B)$, and their Shannon entropy $S(\cdot)$ and Boltzmann–Gibbs entropy $S_{BG}(\cdot)$ have the additivity property

$$S(A+B) = S(A) + S(B), \quad S_{\text{BG}}(A+B) = S_{\text{BG}}(A) + S_{\text{BG}}(B),$$

so the Shannon entropy and the Boltzmann–Gibbs entropy are each *extensive entropies*.

In physics, the *Tsallis entropy* is an extension of the standard Boltzmann–Gibbs entropy. The Tsallis entropy was proposed by Tsallis in 1988, and is also called the q-entropy [473]; it is defined as

$$S_q(p_i) = \frac{1 - \sum_i^q (p_i)^q}{q-1}. \tag{6.6.19}$$

The Boltzmann–Gibbs entropy is the limit of the Tsallis entropy when $q \to 1$, i.e.,

$$S_{\text{BG}} = \lim_{q\to 1} D_q(p_i).$$

The Tsallis entropy has only pseudo-additivity:

$$S_q(A+B) = S_q(A) + S_q(B) + (1-q)S_q(A)S_q(B).$$

Hence it is a *nonextensive* or *nonadditive entropy* [474].

The mathematical statistics defined by the Tsallis entropy is often called the *Tsallis mathematical statistics*. The major mathematical tools of Tsallis mathematical statistics are the q-logarithm and the q-exponential. In particular, the important q-Gaussian distribution is defined by the q-exponential.

For nonnegative real numbers q and x, the function

$$\ln_q x = \begin{cases} \frac{x^{(1-q)}-1}{1-q}, & q \neq 1 \\ \ln x, & q = 1 \end{cases} \tag{6.6.20}$$

is called the *Tsallis logarithm* of x [473], also called the *q-logarithm*. For all $x \geq 0$, the Tsallis logarithm is an analytic, increasing, concave function. The inverse function of the q-logarithm is called the *q-exponential*, defined as

$$\exp_q(x) = \begin{cases} (1+(1-q)x)^{1/(1-q)}, & 1+(1-q)x > 0, \\ 0, & q < 1, \\ +\infty, & q > 1, \\ \exp(x), & q = 1. \end{cases} \tag{6.6.21}$$

The relationship between the q-exponential and the Tsallis logarithm is given by

$$\exp_q(\ln_q x) = x, \tag{6.6.22}$$

$$\ln_q \exp_q(x) = x. \tag{6.6.23}$$

The probability density distribution $f(x)$ is known as a *q-Gaussian distribution*, if

$$f(x) = \frac{\sqrt{\beta}}{C_q} \exp_q(-\beta x^2), \tag{6.6.24}$$

where $\exp_q(x) = [1+(1-q)x]^{1/(1-q)}$ is the q-exponential; the normalized factor C_q is given by

$$C_q = \begin{cases} \frac{2\sqrt{\pi}\,\Gamma\left(\frac{1}{1-q}\right)}{(3-q)\sqrt{1-q}\,\Gamma\left(\frac{1}{1-q}\right)}, & -\infty < q < 1, \\ \sqrt{\pi}, & q = 1, \\ \frac{\sqrt{\pi}\,\Gamma\left(\frac{3-q}{2(q-1)}\right)}{\sqrt{q-1}\,\Gamma\left(\frac{1}{q-1}\right)}, & 1 < q < 3. \end{cases}$$

When $q \to 1$, the limit of the q-Gaussian distribution is the Gaussian distribution; the q-Gaussian distribution is widely applied in statistical mechanics, geology, anatomy, astronomy, economics, finance and machine learning.

Compared with the Gaussian distribution, a prominent feature of the q-Gaussian distribution with $1 < q < 3$ is its obvious tailing. Owing to this feature, the q-logarithm (i.e., the Tsallis logarithm) and the q-exponential are very well suitable for using the AB-divergence as the cost function of the NMF optimization problems.

In order to facilitate the application of the q-logarithm and the q-exponential in

NMF, define the *deformed logarithm*

$$\phi(x) = \ln_{1-\alpha} x = \begin{cases} (x^\alpha - 1)/\alpha, & \alpha \neq 0, \\ \ln x, & \alpha = 0. \end{cases} \tag{6.6.25}$$

The inverse transformation of the deformed logarithm,

$$\phi^{-1}(x) = \exp_{1-\alpha}(x) = \begin{cases} \exp(x), & \alpha = 0, \\ (1+\alpha x)^{1/\alpha}, & \alpha \neq 0,\ 1+\alpha x \geq 0, \\ 0, & \alpha \neq 0,\ 1+\alpha x < 0 \end{cases} \tag{6.6.26}$$

is known as the *deformed exponential.*

The deformed logarithm and the deformed exponential have important applications in the optimization algorithms of NMF.

6.7 Nonnegative Matrix Factorization: Optimization Algorithms

The NMF is a minimization problem with nonnegative constraints. By the classification of Berry *et al.* [40], the NMF has three kinds of basic algorithm:

(1) multiplication algorithms;

(2) gradient descent algorithms;

(3) alternating least squares (ALS) algorithms.

These algorithms belong to the category of first-order optimization algorithms, where the multiplication algorithms are essentially gradient descent algorithms as well, but with a suitable choice of step size, a multiplication algorithm can transform the subtraction update rule of the general gradient descent method into a multiplicative update.

Later, on the basis of the classification of Berry *et al.* [40], Cichocki *et al.* [104] added the quasi-Newton method (a second-order optimization algorithm) and the multilayer decomposition method.

In the following, the above five representative methods are introduced in turn.

6.7.1 Multiplication Algorithms

The gradient descent algorithm is a widely applied optimization algorithm; its basic idea is that a correction term is added during the updating of the variable to be optimized. The step size is a key parameter for this kind of algorithm, and determines the magnitude of the correction.

Consider the general updating rule of the gradient descent algorithm for the unconstrained minimization problem $\min f(\mathbf{X})$:

$$x_{ij} \leftarrow x_{ij} - \eta_{ij} \nabla f(x_{ij}), \quad i = 1, \ldots, I; j = 1, \ldots, J, \tag{6.7.1}$$

where x_{ij} is the entry of the variable matrix $\mathbf{X}$, and $\nabla f(x_{ij}) = [\partial f(\mathbf{X})/\partial \mathbf{X}]_{ij}$ is the gradient matrix of the cost function $f(\mathbf{X})$ at the point x_{ij}. If the step size η_{ij} is chosen in such a way that the addition term x_{ij} in the additive update rule is eliminated, then the original gradient descent algorithm with additive operations becomes the gradient descent algorithm with multiplication operations. This multiplication algorithm was proposed by Lee and Seung [287] for the NMF but is available for other many optimization problems as well.

1. Multiplication Algorithm for Squared Euclidean Distance Minimization

Consider an unconstrained problem $\min D_E(\mathbf{X}\|\mathbf{AS}) = \frac{1}{2}\|\mathbf{X} - \mathbf{AS}\|_2^2$ using the typical squared Euclidean distance as the cost function; its gradient descent algorithm is given by

$$a_{ik} \leftarrow a_{ik} - \mu_{ik}\frac{\partial D_E(\mathbf{X}\|\mathbf{AS})}{\partial a_{ik}}, \tag{6.7.2}$$

$$s_{kj} \leftarrow s_{kj} - \eta_{kj}\frac{\partial D_E(\mathbf{X}\|\mathbf{AS})}{\partial s_{kj}}, \tag{6.7.3}$$

where

$$\frac{\partial D_E(\mathbf{X}\|\mathbf{AS})}{\partial a_{ik}} = -\left[(\mathbf{X} - \mathbf{AS})\mathbf{S}^T\right]_{ik},$$

$$\frac{\partial D_E(\mathbf{X}\|\mathbf{AS})}{\partial s_{kj}} = -\left[\mathbf{A}^T(\mathbf{X} - \mathbf{AS})\right]_{kj}$$

are the gradients of the cost function with respect to the entry a_{ik} of the variable matrix $\mathbf{A}$ and the entry s_{kj} of the variable matrix $\mathbf{S}$, respectively.

If we choose

$$\mu_{ik} = \frac{a_{ik}}{[\mathbf{ASS}^T]_{ik}}, \qquad \eta_{kj} = \frac{s_{kj}}{[\mathbf{A}^T\mathbf{AS}]_{kj}}, \tag{6.7.4}$$

then the gradient descent algorithm becomes a multiplication algorithm:

$$a_{ik} \leftarrow a_{ik}\frac{[\mathbf{XS}^T]_{ik}}{[\mathbf{ASS}^T]_{ik}}, \qquad i = 1, \ldots, I,\ k = 1, \ldots, K, \tag{6.7.5}$$

$$s_{kj} \leftarrow s_{kj}\frac{[\mathbf{A}^T\mathbf{X}]_{kj}}{[\mathbf{A}^T\mathbf{AS}]_{kj}}, \qquad k = 1, \ldots, K,\ j = 1, \ldots, J. \tag{6.7.6}$$

Regarding on multiplication algorithms, there are four important remarks to be made.

Remark 1 The theoretical basis of the multiplication algorithm is the auxiliary function in the *expectation-maximization* (EM) *algorithm*. Hence, the multiplication algorithm is practically an *expectation maximization maximum likelihood* (EMML) *algorithm* [102].

Remark 2 The elementwise multiplication algorithm above is easily rewritten as a multiplication algorithm in matrix form:

$$\mathbf{A} \leftarrow \mathbf{A} * \left[(\mathbf{X}\mathbf{S}^T) \oslash (\mathbf{A}\mathbf{S}\mathbf{S}^T)\right], \tag{6.7.7}$$

$$\mathbf{S} \leftarrow \mathbf{S} * \left[(\mathbf{A}^T\mathbf{X}) \oslash (\mathbf{A}^T\mathbf{A}\mathbf{S})\right], \tag{6.7.8}$$

where $\mathbf{B} * \mathbf{C}$ represents the componentwise product (Hadamard product) of two matrices, while $\mathbf{B} \oslash \mathbf{C}$ denotes the componentwise division of two matrices, namely

$$[\mathbf{B} * \mathbf{C}]_{ik} = b_{ik}c_{ik}, \quad [\mathbf{B} \oslash \mathbf{C}]_{ik} = b_{ik}/c_{ik}. \tag{6.7.9}$$

Remark 3 In the gradient descent algorithm, a fixed step length or an adaptive step length is usually taken; this is independent of the index of the updated variable. In other words, the step size may change with the time but at a given update time. In the same update time, the different entries of the variable matrix are updated with the same step size. It contrast, the multiplication algorithm adopts different steps μ_{ik} for the different entries of the variable matrix. Therefore, this kind of step length is adaptive to the matrix entries. This is an important reason why the multiplication algorithm can improve on the performance of the gradient descent algorithm.

Remark 4 The divergence $D(\mathbf{X}\|\mathbf{A}\mathbf{S})$ is nonincreasing in the different multiplicative update rules [287], and this nonincreasing property is likely to have the result that the algorithm cannot converge to a stationary point [181].

For the regularized NMF cost function

$$J(\mathbf{A}, \mathbf{S}) = \frac{1}{2}\|\mathbf{X} - \mathbf{A}\mathbf{S}\|_2^2 + \alpha_A J_A(\mathbf{A}) + \alpha_S J_S(\mathbf{S}), \tag{6.7.10}$$

since the gradient vectors are respectively given by

$$\nabla_{a_{ik}} J(\mathbf{A}, \mathbf{S}) = \frac{\partial D_E(\mathbf{X}\|\mathbf{A}\mathbf{S})}{\partial a_{ik}} + \alpha_A \frac{\partial J_A(\mathbf{A})}{\partial a_{ik}},$$

$$\nabla_{s_{kj}} J(\mathbf{A}, \mathbf{S}) = \frac{\partial D_E(\mathbf{X}\|\mathbf{A}\mathbf{S})}{\partial s_{kj}} + \alpha_A \frac{\partial J_S(\mathbf{S})}{\partial s_{kj}},$$

the multiplication algorithm should be modified to [287]

$$a_{ik} \leftarrow a_{ik} \frac{\left[[\mathbf{X}\mathbf{S}^T]_{ik} - \alpha_A \nabla J_A(a_{ik})\right]_+}{[\mathbf{A}\mathbf{S}\mathbf{S}^T]_{ik} + \epsilon}, \tag{6.7.11}$$

$$s_{kj} \leftarrow s_{kj} \frac{\left[[\mathbf{A}^T\mathbf{X}]_{kj} - \alpha_S \nabla J_S(s_{kj})\right]_+}{[\mathbf{A}^T\mathbf{A}\mathbf{S}]_{kj} + \epsilon}, \tag{6.7.12}$$

where $\nabla J_A(a_{ik}) = \partial J_A(\mathbf{A})/\partial a_{ik}$, $\nabla J_S(s_{kj}) = \partial J_S(\mathbf{S})/\partial s_{kj}$, and $[u]_+ = \max\{u, \epsilon\}$. The parameter ϵ is usually a very small positive number that prevents the emergence of a zero denominator, as needed to ensure the convergence and numerical stability of the multiplication algorithms.

The elementwise multiplication algorithm above can also be rewritten in the matrix form

$$\mathbf{A} \leftarrow \mathbf{A} * \left((\mathbf{X}\mathbf{S}^T - \alpha_A \boldsymbol{\Psi}_A) \oslash (\mathbf{A}\mathbf{S}\mathbf{S}^T + \epsilon \mathbf{I}) \right), \tag{6.7.13}$$

$$\mathbf{S} \leftarrow \mathbf{S} * \left((\mathbf{A}^T\mathbf{X} - \alpha_S \boldsymbol{\Psi}_S) \oslash (\mathbf{A}^T\mathbf{A}\mathbf{S} + \epsilon \mathbf{I}) \right), \tag{6.7.14}$$

where $\boldsymbol{\Psi}_A = \partial J_A(\mathbf{A})/\partial \mathbf{A}$ and $\boldsymbol{\Psi}_S = \partial J_S(\mathbf{S})/\partial \mathbf{S}$ are two gradient matrices, while $\mathbf{I}$ is the identity matrix.

2. Multiplication Algorithms for KL Divergence Minimization

Consider the minimization of the KL divergence

$$D_{\text{KL}}(\mathbf{X}\|\mathbf{A}\mathbf{S}) = \sum_{i_1=1}^{I} \sum_{j_1=1}^{J} \left(x_{i_1 j_1} \ln \frac{x_{i_1 j_1}}{[\mathbf{A}\mathbf{S}]_{i_1 j_1}} - x_{i_1 j_1} + [\mathbf{A}\mathbf{S}]_{i_1 j_1} \right). \tag{6.7.15}$$

Since

$$\frac{\partial D_{\text{KL}}(\mathbf{X}\|\mathbf{A}\mathbf{S})}{\partial a_{ik}} = -\sum_{j=1}^{J} \left(\frac{s_{kj} x_{ij}}{[\mathbf{A}\mathbf{S}]_{ij}} + s_{kj} \right),$$

$$\frac{\partial D_{\text{KL}}(\mathbf{X}\|\mathbf{A}\mathbf{S})}{\partial s_{kj}} = -\sum_{i=1}^{I} \left(\frac{a_{ik} x_{ij}}{[\mathbf{A}\mathbf{S}]_{ij}} + a_{ik} \right),$$

the gradient descent algorithm is given by

$$a_{ik} \leftarrow a_{ik} - \mu_{ik} \times \left(-\sum_{j=1}^{J} \left(\frac{s_{kj} x_{ij}}{[\mathbf{A}\mathbf{S}]_{ij}} + s_{kj} \right) \right),$$

$$s_{kj} \leftarrow s_{kj} - \eta_{kj} \times \left(-\sum_{i=1}^{I} \left(\frac{a_{ik} x_{ij}}{[\mathbf{A}\mathbf{S}]_{ij}} + a_{ik} \right) \right).$$

Selecting

$$\mu_{ik} = \frac{1}{\sum_{j=1}^{J} s_{kj}}, \quad \eta_{kj} = \frac{1}{\sum_{i=1}^{I} a_{ik}},$$

the gradient descent algorithm can be rewritten as a multiplication algorithm [287]:

$$a_{ik} \leftarrow a_{ik} \frac{\sum_{j=1}^{J} s_{kj} x_{ik}/[\mathbf{A}\mathbf{S}]_{ik}}{\sum_{j=1}^{J} s_{kj}}, \tag{6.7.16}$$

$$s_{kj} \leftarrow s_{kj} \frac{\sum_{i=1}^{I} a_{ik} x_{ik}/[\mathbf{A}\mathbf{S}]_{ik}}{\sum_{i=1}^{I} a_{ik}}. \tag{6.7.17}$$

The corresponding matrix form is

$$\mathbf{A} \leftarrow \left(\mathbf{A} \oslash (\mathbf{1}_I \otimes (\mathbf{S}\mathbf{1}_K)^T)\right) * \left((\mathbf{X} \oslash (\mathbf{AS}))\,\mathbf{S}^T\right), \tag{6.7.18}$$

$$\mathbf{S} \leftarrow \left(\mathbf{S} \oslash ((\mathbf{A}^T\mathbf{1}_I) \otimes \mathbf{1}_K)\right) * \left(\mathbf{A}^T\,(\mathbf{X} \oslash (\mathbf{AS}))\right), \tag{6.7.19}$$

where $\mathbf{1}_I$ is an $I \times 1$ vector whose entries are all equal to 1.

3. Multiplication Algorithm for AB Divergence Minimization

For AB divergence (where $\alpha, \beta, \alpha+\beta \neq 0$) we have

$$D_{AB}^{(\alpha,\beta)}(\mathbf{X}\|\mathbf{AS}) = -\frac{1}{\alpha\beta}\sum_{i=1}^{I}\sum_{j=1}^{J}\left(x_{ij}^{\alpha}[\mathbf{AS}]_{ij}^{\beta} - \frac{\alpha}{\alpha+\beta}x_{ij}^{\alpha+\beta} - \frac{\beta}{\alpha+\beta}[\mathbf{AS}]_{ij}^{\alpha+\beta}\right); \tag{6.7.20}$$

its gradient is given by

$$\frac{\partial D_{AB}^{(\alpha,\beta)}(\mathbf{X}\|\mathbf{AS})}{\partial a_{ik}} = -\frac{1}{\alpha}\sum_{j=1}^{J}\left(s_{kj}x_{ij}^{\alpha}[\mathbf{AS}]_{ij}^{1-\beta} - [\mathbf{AS}]_{ij}^{\alpha+\beta-1}s_{kj}\right), \tag{6.7.21}$$

$$\frac{\partial D_{AB}^{(\alpha,\beta)}(\mathbf{X}\|\mathbf{AS})}{\partial s_{kj}} = -\frac{1}{\alpha}\sum_{i=1}^{I}\left(a_{ik}x_{ij}^{\alpha}[\mathbf{AS}]_{ij}^{1-\beta} - [\mathbf{AS}]_{ij}^{\alpha+\beta-1}a_{ik}\right). \tag{6.7.22}$$

Hence, letting the step sizes be

$$\mu_{ik} = \frac{\alpha a_{ik}}{\sum_{j=1}^{J} s_{kj}[\mathbf{AS}]_{ij}^{\alpha+\beta-1}},$$

$$\eta_{kj} = \frac{\alpha s_{kj}}{\sum_{i=1}^{I} a_{ik}[\mathbf{AS}]_{ij}^{\alpha+\beta-1}},$$

the multiplication algorithm for the AB divergence minimization is given by

$$a_{ik} \leftarrow a_{ik}\frac{\sum_{j=1}^{J} s_{kj}x_{ij}^{\alpha}[\mathbf{AS}]_{ij}^{\beta-1}}{\sum_{j=1}^{J} s_{kj}[\mathbf{AS}]_{ij}^{\alpha+\beta-1}},$$

$$s_{kj} \leftarrow s_{kj}\frac{\sum_{i=1}^{I} a_{ik}x_{ij}^{\alpha}[\mathbf{AS}]_{ij}^{\beta-1}}{\sum_{i=1}^{I} a_{ik}[\mathbf{AS}]_{ij}^{\alpha+\beta-1}}.$$

In order to speed up the convergence of the algorithm, one can use instead the following updating rules [101]

$$a_{ik} \leftarrow a_{ik}\left(\frac{\sum_{j=1}^{J} s_{kj}x_{ij}^{\alpha}[\mathbf{AS}]_{ij}^{\beta-1}}{\sum_{j=1}^{J} s_{kj}[\mathbf{AS}]_{ij}^{\alpha+\beta-1}}\right)^{1/\alpha}, \tag{6.7.23}$$

$$s_{kj} \leftarrow s_{kj}\left(\frac{\sum_{i=1}^{I} a_{ik}x_{ij}^{\alpha}[\mathbf{AS}]_{ij}^{\beta-1}}{\sum_{i=1}^{I} a_{ik}[\mathbf{AS}]_{ij}^{\alpha+\beta-1}}\right)^{1/\alpha}, \tag{6.7.24}$$

where the positive relaxation parameter $1/\alpha$ is used to improve the convergence of the algorithm.

When $\beta = 1-\alpha$, from the multiplication algorithm for the AB divergence minimization we get the multiplication algorithm for the α divergence minimization,

$$a_{ik} \leftarrow a_{ik} \left(\frac{\sum_{j=1}^{J} (x_{ij}/[\mathbf{AS}]_{ij})^{\alpha} s_{kj}}{\sum_{j=1}^{J} s_{kj}} \right)^{1/\alpha}, \tag{6.7.25}$$

$$s_{kj} \leftarrow s_{kj} \left(\frac{\sum_{i=1}^{I} a_{ik} (x_{ij}/[\mathbf{AS}]_{ij})^{\alpha}}{\sum_{i=1}^{I} a_{ik}} \right)^{1/\alpha}. \tag{6.7.26}$$

For the general cases of $\alpha = 0$ and/or $\beta = 0$, the gradient of the AB divergence is given by [101]

$$\frac{\partial D_{AB}^{(\alpha,\beta)}(\mathbf{X}\|\mathbf{AS})}{\partial a_{ik}} = -\sum_{j=1}^{J} [\mathbf{AS}]_{ij}^{\lambda-1} s_{kj} \ln_{1-\alpha} \frac{x_{ij}}{[\mathbf{AS}]_{ij}}, \tag{6.7.27}$$

$$\frac{\partial D_{AB}^{(\alpha,\beta)}(\mathbf{X}\|\mathbf{AS})}{\partial s_{kj}} = -\sum_{i=1}^{I} [\mathbf{AS}]_{ij}^{\lambda-1} a_{ik} \ln_{1-\alpha} \frac{x_{ij}}{[\mathbf{AS}]_{ij}}, \tag{6.7.28}$$

where $\lambda = \alpha + \beta$. Then, by the relationship between the deformed logarithm and the deformed exponential $\exp_{1-\alpha}(\ln_{1-\alpha} x) = x$, it is easy to see that the gradient algorithm for AB divergence minimization is given by

$$a_{ik} \leftarrow \exp_{1-\alpha} \left(\ln_{1-\alpha} a_{ik} - \mu_{ik} \frac{\partial D_{AB}^{(\alpha,\beta)}(\mathbf{X}\|\mathbf{AS})}{\partial \ln_{1-\alpha} a_{ik}} \right), \tag{6.7.29}$$

$$s_{kj} \leftarrow \exp_{1-\alpha} \left(\ln_{1-\alpha} s_{kj} - \eta_{kj} \frac{\partial D_{AB}^{(\alpha,\beta)}(\mathbf{X}\|\mathbf{AS})}{\partial \ln_{1-\alpha} s_{kj}} \right). \tag{6.7.30}$$

Choosing

$$\mu_{ik} = \frac{a_{ik}^{2\alpha-1}}{\sum_{j=1}^{J} s_{kj} [\mathbf{AS}]_{ij}^{\lambda-1}},$$

$$\eta_{kj} = \frac{s_{kj}^{2\alpha-1}}{\sum_{i=1}^{I} a_{ik} [\mathbf{AS}]_{ij}^{\lambda-1}},$$

then the multiplication algorithm for AB divergence-based NMF is as follows [101]:

$$a_{ik} \leftarrow a_{ik} \exp_{1-\alpha} \left(\sum_{j=1}^{J} \frac{s_{kj} [\mathbf{AS}]_{ij}^{\lambda-1}}{\sum_{j=1}^{J} s_{kj} [\mathbf{AS}]_{ij}^{\lambda-1}} \ln_{1-\alpha} \frac{x_{ij}}{[\mathbf{AS}]_{ij}} \right), \tag{6.7.31}$$

$$s_{kj} \leftarrow s_{kj} \exp_{1-\alpha} \Bigg(\sum_{i=1}^{I} \underbrace{\frac{a_{ik} [\mathbf{AS}]_{ij}^{\lambda-1}}{\sum_{i=1}^{I} a_{ik} [\mathbf{AS}]_{ij}^{\lambda-1}}}_{\text{weighting coefficient}} \underbrace{\ln_{1-\alpha} \frac{x_{ij}}{[\mathbf{AS}]_{ij}}}_{\alpha\text{-zooming}} \Bigg). \tag{6.7.32}$$

By α-zooming is meant that, owing to the zoom function of the parameter α, the relative errors of updating the elements of the basis matrix $\mathbf{A} = [a_{ik}]$ and the coefficient matrix $\mathbf{S} = [s_{kj}]$ are mainly controlled by the deformed logarithm $\ln_{1-\alpha}(x_{ij}/[\mathbf{AS}]_{ij})$, as follows:

(1) When $\alpha > 1$, the deformed logarithm $\ln_{1-\alpha}(x_{ij}/[\mathbf{AS}]_{ij})$ has the "zoom-out" function and emphasizes the role of the larger ratio $x_{ij}/[\mathbf{AS}]_{ij}$. Because the smaller ratio is reduced, it can be more easily neglected.
(2) When $\alpha < 1$, the "zoom-in" function of the deformed logarithm highlights the smaller ratio $x_{ij}/[\mathbf{AS}]_{ij}$.

Because the AB divergence contains several divergences as special cases, the AB multiplication algorithm for NMF is correspondingly available for a variety of NMF algorithms.

In the updating formulas of various multiplication algorithms for NMF, it is usually necessary to add a very small perturbation (such as $\epsilon = 10^{-9}$) in order to prevent the denominator becoming zero.

6.7.2 Nesterov Optimal Gradient Algorithm

The gradient descent NMF algorithm in fact consists of two gradient descent algorithms,

$$\mathbf{A}_{k+1} = \mathbf{A}_k - \mu_A \frac{\partial f(\mathbf{A}_k, \mathbf{S}_k)}{\partial \mathbf{A}_k},$$

$$\mathbf{S}_{k+1} = \mathbf{S}_k - \mu_S \frac{\partial f(\mathbf{A}_k, \mathbf{S}_k)}{\partial \mathbf{S}_k}.$$

In order to ensure the nonnegativity of $\mathbf{A}_k$ and $\mathbf{S}_k$, all elements of the updated matrices $\mathbf{A}_{k+1}$ and $\mathbf{S}_{k+1}$ need to be projected onto the nonnegative quadrant in each updating step, which constitutes the projected gradient algorithm for NMF [297]:

$$\mathbf{A}_{k+1} = \left[\mathbf{A}_k - \mu_A \frac{\partial f(\mathbf{A}_k, \mathbf{S}_k)}{\partial \mathbf{A}_k}\right]_+, \tag{6.7.33}$$

$$\mathbf{S}_{k+1} = \left[\mathbf{S}_k - \mu_S \frac{\partial f(\mathbf{A}_k, \mathbf{S}_k)}{\partial \mathbf{S}_k}\right]_+. \tag{6.7.34}$$

Consider the factorization of a low-rank nonnegative matrix $\mathbf{X} \in \mathbb{R}^{m \times n}$

$$\min \frac{1}{2}\|\mathbf{X} - \mathbf{AS}\|_F^2 \quad \text{subject to} \quad \mathbf{A} \in \mathbb{R}_+^{m \times r},\ \mathbf{S} \in \mathbb{R}_+^{r \times n}, \tag{6.7.35}$$

where $r = \text{rank}(\mathbf{X}) < \min\{m, n\}$.

Since Equation (6.7.35) is a nonconvex minimization problem, one can use alternating nonnegative least squares to represent the local solutions of the minimization problem (6.7.35):

$$\mathbf{S}_{k+1} = \arg\min_{\mathbf{S}\geq\mathbf{O}} \left\{ F(\mathbf{A}_k, \mathbf{S}) = \frac{1}{2}\|\mathbf{X} - \mathbf{A}_k\mathbf{S}\|_F^2 \right\}, \tag{6.7.36}$$

$$\mathbf{A}_{k+1}^T = \arg\min_{\mathbf{A}\geq\mathbf{O}} \left\{ F(\mathbf{S}_{k+1}^T, \mathbf{A}^T) = \frac{1}{2}\|\mathbf{X}^T - \mathbf{S}_{k+1}^T\mathbf{A}^T\|_F^2 \right\}. \tag{6.7.37}$$

Here k denotes the kth data block.

Recently, Guan *et al.* [191] showed that the NMF objective function satisfies the two conditions of the Nesterov optimal gradient method:

(1) the objective function $F(\mathbf{A}_t, \mathbf{S}) = \frac{1}{2}\|\mathbf{X} - \mathbf{A}_t\mathbf{S}\|_F^2$ is a convex function;
(2) the gradient of the objective function $F(\mathbf{A}_t, \mathbf{S})$ is Lipschitz continuous with Lipschitz constant $L = \|\mathbf{A}_t^T\mathbf{A}_t\|_F$.

On the basis of the above results, a Nesterov nonnegative matrix factorization (NeNMF) algorithm was proposed in [191], as shown in Algorithm 6.6, where $\mathrm{OGM}(\mathbf{A}_t, \mathbf{S})$ is given in Algorithm 6.7.

Algorithm 6.6 NeNMF algorithm [191]

---

input: Data matrix $\mathbf{X} \in \mathbb{R}_+^{m\times n}, 1 \leq r \leq \min\{m, n\}$.

initialization: $\mathbf{A}_1 \geq \mathbf{O}, \mathbf{S}_1 \geq \mathbf{O}, k = 1$.

repeat

1. Update $\mathbf{S}_{k+1} = \mathrm{OGM}(\mathbf{A}_k, \mathbf{S})$ and $\mathbf{A}_{k+1} = \mathrm{OGM}(\mathbf{S}_{k+1}^T, \mathbf{A}^T)$.
2. **exit if** $\nabla_S^P F(\mathbf{A}_k, \mathbf{S}_k) = \mathbf{O}$ and $\nabla_A^P F(\mathbf{A}_k, \mathbf{S}_k) = \mathbf{O}$, where

$$\left[\nabla_S^P F(\mathbf{A}_k, \mathbf{S}_k)\right]_{ij} = \begin{cases} [\nabla_S F(\mathbf{A}_k, \mathbf{S}_k)]_{ij}, & [\mathbf{S}_k]_{ij} > 0, \\ \min\left\{0, [\nabla_S F(\mathbf{A}_k, \mathbf{S}_k)]_{ij}\right\}, & [\mathbf{S}_k]_{ij} = 0. \end{cases}$$

$$\left[\nabla_A^P F(\mathbf{A}_k, \mathbf{S}_k)\right]_{ij} = \begin{cases} [\nabla_A F(\mathbf{A}_k, \mathbf{S}_k)]_{ij}, & [\mathbf{A}_k]_{ij} > 0, \\ \min\left\{0, [\nabla_A F(\mathbf{A}_k, \mathbf{S}_k)]_{ij}\right\}, & [\mathbf{A}_k]_{ij} = 0. \end{cases}$$

return $k \leftarrow k + 1$.

output: Basis matrix $\mathbf{A} \in \mathbb{R}_+^{m\times r}$, coefficient matrix $\mathbf{S} \in \mathbb{R}_+^{r\times n}$.

Remark If $\mathbf{A}_k \to \mathbf{S}_k^T$ and $\mathbf{S}_k \to \mathbf{A}_k^T$ in Algorithm 6.7, and the Lipschitz constant is replaced by $L = \|\mathbf{S}_k\mathbf{S}_k^T\|_F$, then Algorithm 6.7 will give the optimal gradient method $\mathrm{OGM}(\mathbf{S}_{k+1}^T, \mathbf{A}^T)$, yielding the output $\mathbf{A}_{k+1}^T = \mathbf{A}_t^T$.

It has been shown [191] that owing to the introduction of structural information, when minimizing one matrix with another matrix held fixed, the NeNMF algorithm can converge at the rate $O(1/k^2)$.

Algorithm 6.7 Optimal gradient method OGM($\mathbf{A}_k, \mathbf{S}$) [191]

input: $\mathbf{A}_k$ and $\mathbf{S}_k$.
initialization: $\mathbf{Y}_0 = \mathbf{S}_k \geq \mathbf{O}, \alpha_0 = 1, L = \|\mathbf{A}_k^T \mathbf{A}_k\|_F, t = 0$.
repeat

1. Update $\mathbf{S}_t = \mathcal{P}_+\big(\mathbf{Y}_t - L^{-1}\nabla_S F(\mathbf{A}_k, \mathbf{Y}_t)\big)$.
2. Update $\alpha_{t+1} = \dfrac{1}{2}\big(1 + \sqrt{4\alpha_t^2 + 1}\big)$.
3. update $\mathbf{Y}_{t+1} = \mathbf{S}_t + \dfrac{\alpha_t - 1}{\alpha_{t+1}}(\mathbf{S}_t - \mathbf{S}_{t-1})$.
4. **exit if** $\nabla_S^P F(\mathbf{A}_k, \mathbf{S}_t) = \mathbf{O}$.

return $t \leftarrow t + 1$.
output: $\mathbf{S}_{k+1} = \mathbf{S}_t$.

6.7.3 Alternating Nonnegative Least Squares

The alternating least squares method was used for NMF first by Paatero and Tapper [367]. The resulting method is called the *alternating nonnegative least squares* (ANLS) *method.*

The NMF optimization problem for $\mathbf{X}_{I\times J} = \mathbf{A}_{I\times K}\mathbf{S}_{K\times J}$, described by

$$\min_{\mathbf{A},\mathbf{S}} \frac{1}{2}\|\mathbf{X} - \mathbf{AS}\|_F^2 \quad \text{subject to} \quad \mathbf{A}, \mathbf{S} \geq \mathbf{O}, \tag{6.7.38}$$

can be decomposed into two ANLS subproblems [367]

$$\text{ANLS1:} \quad \min_{\mathbf{S}\geq\mathbf{O}} \left\{ f_1(\mathbf{S}) = \frac{1}{2}\|\mathbf{AS} - \mathbf{X}\|_F^2 \right\} \quad (\mathbf{A}\text{ fixed}), \tag{6.7.39}$$

$$\text{ANLS2:} \quad \min_{\mathbf{A}\geq\mathbf{O}} \left\{ f_2(\mathbf{A}^T) = \frac{1}{2}\|\mathbf{S}^T\mathbf{A}^T - \mathbf{X}^T\|_F^2 \right\} \quad (\mathbf{S}\text{ fixed}). \tag{6.7.40}$$

These two ANLS subproblems correspond to using the LS method to solve alternately the matrix equations $\mathbf{AS} = \mathbf{X}$ and $\mathbf{S}^T\mathbf{A}^T = \mathbf{X}^T$, whose LS solutions are respectively given by

$$\mathbf{S} = \mathcal{P}_+\big((\mathbf{A}^T\mathbf{A})^\dagger \mathbf{A}^T\mathbf{X}\big), \tag{6.7.41}$$

$$\mathbf{A}^T = \mathcal{P}_+\big((\mathbf{SS}^T)^\dagger \mathbf{SX}^T\big). \tag{6.7.42}$$

If $\mathbf{A}$ and/or $\mathbf{S}$ are singular in the iteration process, the algorithm cannot converge.

In order to overcome the shortcoming of the ALS algorithm due to numerical stability, Langville *et al.* [277] and Pauca *et al.* [378] independently proposed a *constrained nonnegative matrix factorization* (CNMF):

$$\text{CNMF:} \ \min_{\mathbf{A},\mathbf{S}} \left\{ \frac{1}{2}\left(\|\mathbf{X} - \mathbf{AS}\|_F^2 + \alpha\|\mathbf{A}\|_F^2 + \beta\|\mathbf{S}\|_F^2\right) \right\} \quad \text{subject to} \quad \mathbf{A}, \mathbf{S} \geq \mathbf{O}, \tag{6.7.43}$$

where $\alpha \geq 0$ and $\beta \geq 0$ are two regularization parameters that suppress $\|\mathbf{A}\|_F^2$ and $\|\mathbf{S}\|_F^2$, respectively.

This constrained NMF is a typical application of the regularization least squares method of Tikhonov [464].

The regularization NMF problem can be decomposed into two *alternating regularization nonnegative least squares* (ARNLS) problems

$$\text{ARNLS1:} \quad \min_{\mathbf{S}\in\mathbb{R}_+^{J\times K}} \left\{ J_1(\mathbf{S}) = \frac{1}{2}\|\mathbf{AS}-\mathbf{X}\|_F^2 + \frac{1}{2}\beta\|\mathbf{S}\|_F^2 \right\} \ (\mathbf{A}\text{ fixed}), \tag{6.7.44}$$

$$\text{ARNLS2:} \quad \min_{\mathbf{A}\in\mathbb{R}_+^{I\times J}} \left\{ J_2(\mathbf{A}^T) = \frac{1}{2}\|\mathbf{S}^T\mathbf{A}^T-\mathbf{X}^T\|_F^2 + \frac{1}{2}\alpha\|\mathbf{A}\|_F^2 \right\} \ (\mathbf{S}\text{ fixed}), \tag{6.7.45}$$

equivalently written as

$$\text{ARNLS1:} \quad \min_{\mathbf{S}\in\mathbb{R}_+^{J\times K}} \left\{ J_1(\mathbf{S}) = \frac{1}{2}\left\| \begin{bmatrix} \mathbf{A} \\ \sqrt{\beta}\mathbf{I}_J \end{bmatrix} \mathbf{S} - \begin{bmatrix} \mathbf{X} \\ \mathbf{O}_{J\times K} \end{bmatrix} \right\|_F^2 \right\}, \tag{6.7.46}$$

$$\text{ARNLS2:} \quad \min_{\mathbf{A}\in\mathbb{R}_+^{I\times J}} \left\{ J_2(\mathbf{A}^T) = \frac{1}{2}\left\| \begin{bmatrix} \mathbf{S}^T \\ \sqrt{\alpha}\mathbf{I}_J \end{bmatrix} \mathbf{A}^T - \begin{bmatrix} \mathbf{X}^T \\ \mathbf{O}_{J\times I} \end{bmatrix} \right\|_F^2 \right\}. \tag{6.7.47}$$

From the matrix differentials

$$\begin{aligned} \mathrm{d}J_1(\mathbf{S}) &= \frac{1}{2}\mathrm{d}\left(\mathrm{tr}((\mathbf{AS}-\mathbf{X})^T(\mathbf{AS}-\mathbf{X})) + \beta\,\mathrm{tr}(\mathbf{S}^T\mathbf{S})\right) \\ &= \mathrm{tr}\left((\mathbf{S}^T\mathbf{A}^T\mathbf{A} - \mathbf{X}^T\mathbf{A} + \beta\mathbf{S}^T)\mathrm{d}\mathbf{S}\right), \\ \mathrm{d}J_2(\mathbf{A}^T) &= \frac{1}{2}\mathrm{d}\left(\mathrm{tr}((\mathbf{AS}-\mathbf{X})(\mathbf{AS}-\mathbf{X})^T) + \alpha\,\mathrm{tr}(\mathbf{A}^T\mathbf{A})\right) \\ &= \mathrm{tr}\left((\mathbf{ASS}^T - \mathbf{XS}^T + \alpha\mathbf{A})\mathrm{d}\mathbf{A}^T\right), \end{aligned}$$

we get the gradient matrices

$$\frac{\partial J_1(\mathbf{S})}{\partial \mathbf{S}} = -\mathbf{A}^T\mathbf{X} + \mathbf{A}^T\mathbf{AS} + \beta\mathbf{S}, \tag{6.7.48}$$

$$\frac{\partial J_2(\mathbf{A}^T)}{\partial \mathbf{A}^T} = -\mathbf{SX}^T + \mathbf{SS}^T\mathbf{A}^T + \alpha\mathbf{A}^T. \tag{6.7.49}$$

From $\partial J_1(\mathbf{S})/\partial\mathbf{S} = \mathbf{O}$ and $\partial J_2(\mathbf{A}^T)/\partial\mathbf{A}^T = \mathbf{O}$, it follows that the solutions of the two regularization least squares problems are respectively given by

$$(\mathbf{A}^T\mathbf{A} + \beta\mathbf{I}_J)\mathbf{S} = \mathbf{A}^T\mathbf{X} \quad \text{or} \quad \mathbf{S} = (\mathbf{A}^T\mathbf{A} + \beta\mathbf{I}_J)^{-1}\mathbf{A}^T\mathbf{X}, \tag{6.7.50}$$

$$(\mathbf{SS}^T + \alpha\mathbf{I}_J)\mathbf{A}^T = \mathbf{SX}^T \quad \text{or} \quad \mathbf{A}^T = (\mathbf{SS}^T + \alpha\mathbf{I}_J)^{-1}\mathbf{SX}^T. \tag{6.7.51}$$

The LS method for solving the above two problems is called the *alternate constrained least squares method*, whose basic framework is as follows [277].

(1) Use the nonzero entries to initialize $\mathbf{A} \in \mathbb{R}^{I\times K}$.

(2) Find iteratively the regularization LS solutions (6.7.50) and (6.7.51), and force the matrices $\mathbf{S}$ and $\mathbf{A}$ to be nonnegative:

$$s_{kj} = [\mathbf{S}]_{kj} = \max\{0, s_{kj}\} \quad \text{and} \quad a_{ik} = [\mathbf{A}]_{ik} = \max\{0, a_{ik}\}. \tag{6.7.52}$$

(3) Now normalize each column of $\mathbf{A}$ and each row of $\mathbf{S}$ to the unit Frobenius norm. Then return to (2), and repeat the iterations until some convergence criteria are met.

A better way, however, is to use the multiplication algorithms to solve two alternating least squares problems. From the gradient formulas (6.7.48) and (6.7.49), it immediately follows that the alternating gradient algorithm is given by

$$s_{kj} \leftarrow s_{kj} + \eta_{kj}\left[\mathbf{A}^T\mathbf{X} - \mathbf{A}^T\mathbf{A}\mathbf{S} - \beta\mathbf{S}\right]_{kj},$$

$$a_{ik} \leftarrow a_{ik} + \mu_{ik}\left[\mathbf{X}\mathbf{S}^T - \mathbf{A}\mathbf{S}\mathbf{S}^T - \alpha\mathbf{A}\right]_{ik}.$$

Choosing

$$\eta_{kj} = \frac{s_{kj}}{\left[\mathbf{A}^T\mathbf{A}\mathbf{S} + \beta\mathbf{S}\right]_{kj}}, \qquad \mu_{ik} = \frac{a_{ik}}{\left[\mathbf{A}\mathbf{S}\mathbf{S}^T + \alpha\mathbf{A}\right]_{ik}},$$

the gradient algorithm becomes the multiplication algorithm

$$s_{kj} \leftarrow s_{kj}\frac{\left[\mathbf{A}^T\mathbf{X}\right]_{kj}}{\left[\mathbf{A}^T\mathbf{A}\mathbf{S} + \beta\mathbf{S}\right]_{kj}}, \tag{6.7.53}$$

$$a_{ik} \leftarrow a_{ik}\frac{\left[\mathbf{X}\mathbf{S}^T\right]_{ik}}{\left[\mathbf{A}\mathbf{S}\mathbf{S}^T + \alpha\mathbf{A}\right]_{ik}}. \tag{6.7.54}$$

As long as the matrices $\mathbf{A}$ and $\mathbf{S}$ are initialized using nonnegative values, the above-described iteration ensures the nonnegativity of the two matrices.

By [378], choosing step sizes

$$\eta_{kj} = \frac{s_{kj}}{\left[\mathbf{A}^T\mathbf{A}\mathbf{S}\right]_{kj}}, \qquad \mu_{ik} = \frac{a_{ik}}{\left[\mathbf{A}\mathbf{S}\mathbf{S}^T\right]_{ik}}$$

gives the multiplication algorithm

$$s_{kj} \leftarrow s_{kj}\frac{\left[\mathbf{A}^T\mathbf{X} - \beta\mathbf{S}\right]_{kj}}{\left[\mathbf{A}^T\mathbf{A}\mathbf{S}\right]_{kj} + \epsilon}, \tag{6.7.55}$$

$$a_{ik} \leftarrow a_{ik}\frac{\left[\mathbf{X}\mathbf{S}^T - \alpha\mathbf{A}\right]_{ik}}{\left[\mathbf{A}\mathbf{S}\mathbf{S}^T\right]_{ik} + \epsilon}. \tag{6.7.56}$$

Notice that the above algorithm cannot ensure the nonnegativity of the matrix entries, owing to the subtraction operations in the numerators.

6.7.4 Quasi-Newton Method

1. Basic Quasi-Newton Method [539]

Consider solving the over-determined matrix equation $\mathbf{S}_{K\times J}^T\mathbf{A}_{I\times K}^T = \mathbf{X}_{I\times K}^T$ (where $J \gg I$) for the unknown matrix $\mathbf{A}^T$, an effective way is the quasi-Newton method.

The Hessian matrix $\mathbf{H}_A = \nabla_A^2(D_E) = \mathbf{I}_{I\times I}\otimes\mathbf{SS}^T \in \mathbb{R}^{IK\times IK}$ of the cost function $D_E(\mathbf{X}\|\mathbf{AS}) = (1/2)\|\mathbf{X}-\mathbf{AS}\|_2^2$ is a block diagonal matrix whose block matrix on the diagonal is $\mathbf{SS}^T$. Then the quasi-Newton method takes the update

$$\mathbf{A} \leftarrow \left[\mathbf{A} - \nabla_A\big(D_E(\mathbf{X}\|\mathbf{AS})\big)\mathbf{H}_A^{-1}\right],$$

here $\nabla_A\left(D_E(\mathbf{X}\|\mathbf{AS})\right) = (\mathbf{AS}-\mathbf{X})\mathbf{S}^T$ is the gradient matrix of the cost function $D_E(\mathbf{X}\|\mathbf{AS})$. Hence, the quasi-Newton algorithm is given by

$$\mathbf{A} \leftarrow \left[\mathbf{A} - (\mathbf{AS}-\mathbf{X})\mathbf{S}^T\left(\mathbf{SS}^T\right)^{-1}\right].$$

In order to prevent the matrix $\mathbf{SS}^T$ from being singular or having a large condition number, one can use the relaxation method

$$\mathbf{A} \leftarrow \left[\mathbf{A} - (\mathbf{AS}-\mathbf{X})\mathbf{S}^T\left(\mathbf{SS}^T + \lambda\mathbf{I}_{K\times K}\right)^{-1}\right],$$

where λ is a Tikhonov regularization parameter.

2. Multilayer Decomposition Method [105], [106]

The basic idea of the multilayer decomposition method is as follows. The largest error is in the first decomposition result $\mathbf{X} \approx \mathbf{A}^{(1)}\mathbf{S}^{(1)}$; hence $\mathbf{S}^{(1)}$ is regarded as a new data matrix, and then the NMF of the second layer $\mathbf{S}^{(1)} \approx \mathbf{A}^{(2)}\mathbf{S}^{(2)}$ is constructed. The second layer decomposition result is still subject to error, and so it is necessary to construct the third layer NMF $\mathbf{S}^{(2)} \approx \mathbf{A}^{(3)}\mathbf{S}^{(3)}$, and to continue this process to form an L-layer NMF:

$$\begin{aligned} \mathbf{X} &\approx \mathbf{A}^{(1)}\mathbf{S}^{(1)} \in \mathbb{R}^{I\times J} \quad (\text{where } \mathbf{A}^{(1)} \in \mathbb{R}^{I\times K}),\\ \mathbf{S}^{(1)} &\approx \mathbf{A}^{(2)}\mathbf{S}^{(2)} \in \mathbb{R}^{K\times J} \quad (\text{where } \mathbf{A}^{(2)} \in \mathbb{R}^{K\times K}),\\ &\vdots\\ \mathbf{S}^{(L-1)} &\approx \mathbf{A}^{(L)}\mathbf{S}^{(L)} \in \mathbb{R}^{K\times J} \quad (\text{where } \mathbf{A}^{(L)} \in \mathbb{R}^{K\times K}). \end{aligned}$$

The decomposition of every layer is made by NMF, and any algorithm described above can be adopted. The final multilayer decomposition result is

$$\mathbf{X} \approx \mathbf{A}^{(1)}\mathbf{A}^{(2)}\cdots\mathbf{A}^{(L)}\mathbf{S}^{(L)},$$

from which the NMF is given by $\mathbf{A} = \mathbf{A}^{(1)}\mathbf{A}^{(2)}\cdots\mathbf{A}^{(L)}$ and $\mathbf{S} = \mathbf{S}^{(L)}$.

6.7.5 Sparse Nonnegative Matrix Factorization

When a sparse representation of the data is expected to obtained by NMF, it is necessary to consider NMF with a sparse constraint.

Given a vector $\mathbf{x} \in \mathbb{R}^n$, Hoyer [220] proposed using the ratio of the ℓ_1-norm and the ℓ_2-norm,

$$\text{sparseness}\,(\mathbf{x}) = \frac{\sqrt{n} - \|\mathbf{x}\|_1/\|\mathbf{x}\|_2}{\sqrt{n} - 1}, \tag{6.7.57}$$

as the measure of the *sparseness* of the vector. Clearly, if $\mathbf{x}$ has only one nonzero element then its sparseness is equal to 1; if and only if the absolute values of all the elements of $\mathbf{x}$ are equal, the sparseness of the vector is zero. The sparseness of any vector lies in the region $[0, 1]$.

An NMF with nonnegative constraint is defined as follows [220]: given a nonnegative data matrix $\mathbf{X} \in \mathbb{R}_+^{I \times J}$, find the nonnegative basis matrix $\mathbf{A} \in \mathbb{R}_+^{I \times K}$ and the nonnegative coefficient matrix $\mathbf{S} \in \mathbb{R}_+^{K \times J}$ such that

$$L(\mathbf{A}, \mathbf{S}) = \|\mathbf{X} - \mathbf{A}\mathbf{S}\|_F^2 \tag{6.7.58}$$

is minimized and $\mathbf{A}$ and $\mathbf{S}$ satisfy respectively the following sparse constraints:

$$\text{sparseness}\,(\mathbf{a}_k) = S_a, \quad \text{sparseness}\,(\mathbf{s}_k) = S_s, \quad k = 1, \ldots, K.$$

Here $\mathbf{a}_k$ and $\mathbf{s}_k$ are respectively the kth column of the nonnegative matrix $\mathbf{A}$ and the kth row of $\mathbf{S}$, K is the number of components, S_a and S_s are respectively the (expected) sparseness of the columns of $\mathbf{A}$ and the rows of $\mathbf{S}$.

The following is the basic framework of nonnegative matrix factorization with sparseness constraint [220]:

1. Use two random positive matrices to initialize $\mathbf{A}$ and $\mathbf{S}$.
2. If adding a sparseness constraint to the matrix $\mathbf{A}$ then:
   - let $\mathbf{A} \leftarrow \mathbf{A} - \mu_A(\mathbf{A}\mathbf{S} - \mathbf{X})\mathbf{S}^T$;
   - project every column vector of $\mathbf{A}$ onto a new nonnegative column vector with unchanged ℓ_2-norm but ℓ_1-norm equal to the expected sparseness.
3. If not adding a sparseness constraint to $\mathbf{A}$ then $\mathbf{A} \leftarrow \mathbf{A} * (\mathbf{X}\mathbf{S}^T) \oslash (\mathbf{A}\mathbf{S}\mathbf{S}^T)$.
4. If adding a sparseness constraint to the matrix $\mathbf{S}$ then:
   - let $\mathbf{S} \leftarrow \mathbf{S} - \mu_S \mathbf{A}^T(\mathbf{A}\mathbf{S} - \mathbf{X})$;
   - project every row vector of $\mathbf{S}$ onto a new nonnegative row vector with unchanged ℓ_2-norm but ℓ_1-norm equal to the expected sparseness.
5. If not adding a sparseness constraint to $\mathbf{S}$ then $\mathbf{S} \leftarrow \mathbf{S} * (\mathbf{A}^T\mathbf{X}) \oslash (\mathbf{A}^T\mathbf{A}\mathbf{S})$.

In the following, we consider a regularization NMF with sparseness constraint on the columns of $\mathbf{S}$:

$$\min_{\mathbf{A},\mathbf{S}} \frac{1}{2}\left\{\|\mathbf{AS}-\mathbf{X}\|_F^2+\alpha\|\mathbf{A}\|_F^2+\beta\sum_{j=1}^J\|\mathbf{S}_{:,j}\|_1^2\right\}$$
$$\text{subject to}\quad \mathbf{A},\mathbf{S}\geq\mathbf{O}, \tag{6.7.59}$$

where $\mathbf{S}_{:,j}$ is the jth column of the matrix $\mathbf{S}$.

The sparse NMF problem (6.7.59) can be decomposed into two alternating LS subproblems [247]

$$\min_{\mathbf{A}\in\mathbb{R}_+^{I\times K}}\left\{J_3(\mathbf{S})=\frac{1}{2}\left\|\begin{bmatrix}\mathbf{A}\\ \sqrt{\beta}\mathbf{1}_K^T\end{bmatrix}\mathbf{S}-\begin{bmatrix}\mathbf{X}\\ \mathbf{O}_{K\times J}\end{bmatrix}\right\|_F^2\right\}, \tag{6.7.60}$$

$$\min_{\mathbf{S}\in\mathbb{R}_+^{K\times J}}\left\{J_4(\mathbf{A}^T)=\frac{1}{2}\left\|\begin{bmatrix}\mathbf{S}^T\\ \sqrt{\alpha}\mathbf{I}_K\end{bmatrix}\mathbf{A}^T-\begin{bmatrix}\mathbf{X}^T\\ \mathbf{O}_{K\times I}\end{bmatrix}\right\|_F^2\right\}. \tag{6.7.61}$$

From the matrix differentials

$$\begin{aligned}\mathrm{d}J_3(\mathbf{S})&=\frac{1}{2}\mathrm{d}\Big(\mathrm{tr}((\mathbf{AS}-\mathbf{X})^T(\mathbf{AS}-\mathbf{X})+\beta\mathbf{S}^T\mathbf{1}_K\mathbf{1}_K^T\mathbf{S})\Big)\\ &=\mathrm{tr}\Big((\mathbf{S}^T\mathbf{A}^T\mathbf{A}-\mathbf{X}^T\mathbf{A}+\beta\mathbf{S}^T\mathbf{E}_K)\mathrm{d}\mathbf{S}\Big),\\ \mathrm{d}J_4(\mathbf{A}^T)&=\mathrm{d}J_2(\mathbf{A}^T)=\mathrm{tr}\left((\mathbf{ASS}^T-\mathbf{XS}^T+\alpha\mathbf{A})\mathrm{d}\mathbf{A}^T\right),\end{aligned}$$

it follows that the gradient matrices of the objective function of the sparse nonnegative LS problem are given by

$$\frac{\partial J_3(\mathbf{S})}{\partial\mathbf{S}}=-\mathbf{A}^T\mathbf{X}+\mathbf{A}^T\mathbf{AS}+\beta\mathbf{E}_J\mathbf{S}, \tag{6.7.62}$$

$$\frac{\partial J_4(\mathbf{A}^T)}{\partial\mathbf{A}^T}=-\mathbf{SX}^T+\mathbf{SS}^T\mathbf{A}^T+\alpha\mathbf{A}^T. \tag{6.7.63}$$

Here $\mathbf{E}_K=\mathbf{1}_K\mathbf{1}_K^T$ is a $K\times K$ matrix with all entries equal to 1.

Then, the alternating sparse nonnegative LS solutions are respectively given by

$$\begin{aligned}(\mathbf{A}^T\mathbf{A}+\beta\mathbf{E}_J)\mathbf{S}&=\mathbf{A}^T\mathbf{X},\\ (\mathbf{SS}^T+\alpha\mathbf{I}_J)\mathbf{A}^T&=\mathbf{SX}^T,\end{aligned}$$

namely

$$\mathbf{S}=(\mathbf{A}^T\mathbf{A}+\beta\mathbf{E}_J)^{-1}\mathbf{A}^T\mathbf{X}, \tag{6.7.64}$$

$$\mathbf{A}^T=(\mathbf{SS}^T+\alpha\mathbf{I}_J)^{-1}\mathbf{SX}^T. \tag{6.7.65}$$

However, the gradient algorithm for the sparse NMF is

$$\begin{aligned}[\mathbf{S}]_{kj}&\leftarrow[\mathbf{S}]_{kj}+\eta_{kj}[\mathbf{A}^T\mathbf{X}-\mathbf{A}^T\mathbf{AS}-\beta\mathbf{E}_K\mathbf{S}]_{kj},\\ [\mathbf{A}^T]_{ki}&\leftarrow[\mathbf{A}^T]_{ki}+\mu_{ki}[\mathbf{SX}^T-\mathbf{SS}^T\mathbf{A}^T-\alpha\mathbf{A}^T]_{ki}.\end{aligned}$$

Choosing the step sizes

$$\eta_{kj} = \frac{[\mathbf{S}]_{kj}}{[\mathbf{A}^T\mathbf{A}\mathbf{S} + \beta\mathbf{E}_J\mathbf{S}]_{kj}}, \quad \mu_{ki} = \frac{[\mathbf{A}^T]_{ki}}{[\mathbf{S}\mathbf{S}^T\mathbf{A}^T + \alpha\mathbf{A}^T]_{ki}},$$

the gradient algorithm becomes the multiplication algorithm

$$[\mathbf{S}]_{kj} \leftarrow [\mathbf{S}]_{kj} \frac{[\mathbf{A}^T\mathbf{X}]_{kj}}{[\mathbf{A}^T\mathbf{A}\mathbf{S} + \beta\mathbf{E}_J\mathbf{S}]_{kj}}, \tag{6.7.66}$$

$$[\mathbf{A}^T]_{ki} \leftarrow [\mathbf{A}^T]_{ki} \frac{[\mathbf{S}\mathbf{X}^T]_{ki}}{[\mathbf{S}\mathbf{S}^T\mathbf{A}^T + \alpha\mathbf{A}^T]_{ki}}. \tag{6.7.67}$$

This is the alternating LS multiplication algorithm for a sparse NMF [247].

The NMF is widely used, and here we present a few typical applications.

In image processing the data can be represented as an $m \times n$ nonnegative matrix $\mathbf{X}$ each of whose columns is an image described by m nonnegative pixel values. Then NMF gives two factor matrices $\mathbf{A}$ and $\mathbf{S}$ such that $\mathbf{X} \approx \mathbf{AS}$, where $\mathbf{A}$ is the basis matrix and $\mathbf{S}$ is the coding matrix. In pattern recognition applications such as face recognition, each column of the basis matrix $\mathbf{A}$ can be regarded as one key part of the whole face, such as nose, eye, ear etc., and each column of the coding matrix $\mathbf{S}$ gives a weight by which the corresponding image can be reconstructed as a linear combination of different parts. Then, NMF can discover the common basis hidden behind an observed image and hence perform face recognition. However, for an image such as a face or brain, etc., the standard version of NMF does not necessarily provide a correct part-of-whole representation. In such cases, it is necessary to add the sparsity constraint of NMF to the face or brain recognition in order to improve the part-of-whole representation of the image.

The main purpose of clustering text, image or biology data is to discover patterns from data automatically. Given an $m \times n$ nonnegative matrix $\mathbf{X} = [\mathbf{x}_1, \ldots, \mathbf{x}_n]$, suppose that there are r cluster patterns. Our problem is to determine to which cluster the $m \times 1$ data vector $\mathbf{x}_j$ belongs. Via NMF, we can find two factor matrices $\mathbf{A} \in \mathbb{R}_+^{m\times r}$ and $\mathbf{S} \in \mathbb{R}_+^{r\times n}$ such that $\mathbf{X} \approx \mathbf{AS}$, where r is the cluster number. The factor matrix $\mathbf{A}$ can be regarded as a cluster centroid matrix, while $\mathbf{S}$ is a cluster membership indicator matrix.

The NMF model $\mathbf{X} = \mathbf{AS}$ can be rewritten as

$$[\mathbf{x}_1, \ldots, \mathbf{x}_n] = [\mathbf{a}_1, \ldots, \mathbf{a}_r] \begin{bmatrix} S_{11} & \cdots & S_{1n} \\ \vdots & \ddots & \vdots \\ S_{r1} & \cdots & S_{rn} \end{bmatrix} \tag{6.7.68}$$

or

$$\mathbf{x}_j = \sum_{i=1}^{r} \mathbf{a}_i S_{ij}, \quad j = 1, \ldots, n. \tag{6.7.69}$$

This shows that if S_{kj} is the largest value of the jth column of $\mathbf{S}$ then the jth data vector $\mathbf{x}_j$ can be determined as belonging to the kth cluster pattern.

Nonnegative matrix factorization has been successfully applied in different fields such as metagenes and molecular pattern discovery in bioinformatics [70], text or document clustering [379], [434], the analysis of financial data [134] and so on.

The use of K-means provides one of the most famous and traditional methods in clustering analysis, and probabilistic latent semantic indexing (PLSI) is one of the state-of-the-art unsupervised learning models in data mining. The NMF method is closely related to K-means and the PLSI [544] as follows.

(1) A model using soft K-means can be rewritten as a symmetric NMF model, and hence K-means and NMF are equivalent. This equivalence justifies the ability of NMF for data clustering, but this does not mean that K-means and NMF will generate identical cluster results since they employ different algorithms.
(2) The PLSI and NMF methods optimize the same objective function (the KL divergence), but the PLSI has additional constraints. The algorithms for the two models can generate equivalent solutions, but they are different in essence.

6.8 Sparse Matrix Equation Solving: Optimization Theory

In sparse representation and compressed sensing (also known as compressive sensing, compressive sampling, or sparse sampling), it is necessary to solve an under-determined sparse matrix equation $\mathbf{Ax} = \mathbf{b}$, where $\mathbf{x}$ is a sparse vector with only a few nonzero entries. In this section we discuss the use of optimization theory for solving sparse matrix equations.

6.8.1 ℓ_1-Norm Minimization

A full row rank under-determined matrix equation $\mathbf{A}_{m\times n}\mathbf{x}_{n\times 1} = \mathbf{b}_{m\times 1}$ with $m < n$ has infinitely many solutions. Suppose that we seek the sparsest solution, with the fewest nonzero entries. Can it ever be unique? How do we find the sparsest solution? This section answers the first question; the second question will be discussed in the next section.

For any positive number $p > 0$, the ℓ_p-norm of the vector $\mathbf{x}$ is defined as

$$\|\mathbf{x}\|_p = \left(\sum_{i\in \text{support}(\mathbf{x})} |x_i|^p \right)^{1/p}. \tag{6.8.1}$$

Thus the ℓ_0-norm of an $n \times 1$ vector $\mathbf{x}$ can be defined as

$$\|\mathbf{x}\|_0 = \lim_{p\to 0} \|\mathbf{x}\|_p^p = \lim_{p\to 0} \sum_{i=1}^{n} |x_i|^p = \sum_{i=1}^{n} 1(x_i \neq 0) = \#\{i|x_i \neq 0\}. \tag{6.8.2}$$

Thus if $\|\mathbf{x}\|_0 \ll n$ then $\mathbf{x}$ is sparse.

As described in Chapter 1, the core problem of sparse representation is the ℓ_0-norm minimization

$$(P_0) \qquad \min_{\mathbf{x}} \|\mathbf{x}\|_0 \quad \text{subject to} \quad \mathbf{b} = \mathbf{A}\mathbf{x}, \tag{6.8.3}$$

where $\mathbf{A} \in \mathbb{R}^{m\times n}, \mathbf{x} \in \mathbb{R}^n, \mathbf{b} \in \mathbb{R}^m$.

As an observation signal is usually contaminated by noise, the equality constraint in the above optimization problem should be relaxed to ℓ_0-norm minimization with the inequality constraint $\epsilon \geq 0$ which allows a certain error perturbation:

$$\min_{\mathbf{x}} \|\mathbf{x}\|_0 \quad \text{subject to} \quad \|\mathbf{A}\mathbf{x} - \mathbf{b}\|_2 \leq \epsilon. \tag{6.8.4}$$

A key term coined and defined in [130] that is crucial for the study of uniqueness is the *spark* of the matrix $\mathbf{A}$.

Definition 6.7 [130] Given a matrix $\mathbf{A}$, its *sparsity* $\sigma = \text{Spark}(\mathbf{A})$ is defined as the smallest possible number such that there exists a subgroup of σ columns from $\mathbf{A}$ that are linearly dependent.

The spark gives a simple criterion for the uniqueness of a sparse solution of an under-determined system of linear equations $\mathbf{A}\mathbf{x} = \mathbf{b}$, as shown below.

Theorem 6.9 [182], [130], [69] *If a system of linear equations* $\mathbf{A}\mathbf{x} = \mathbf{b}$ *has a solution* $\mathbf{x}$ *obeying* $\|\mathbf{x}\|_0 < \text{spark}(\mathbf{A})/2$, *this solution is necessarily the sparsest solution.*

To solve the optimization problem (P_0) directly, one must screen out all nonzero elements in the coefficient vector $\mathbf{x}$. However, this method is intractable, or nondeterministic polynomial-time hard (NP-hard), since the searching space is too large [313], [118], [337].

The index set of nonzero elements of the vector $\mathbf{x} = [x_i, \ldots, x_n]^T$ is called its support, denoted support$(\mathbf{x}) = \{i | x_i \neq 0\}$. The length of the support (i.e., the number of nonzero elements) is measured by the ℓ_0-norm

$$\|\mathbf{x}\|_0 = |\text{support}(\mathbf{x})|. \tag{6.8.5}$$

A vector $\mathbf{x} \in \mathbb{R}^n$ is said to be *K-sparse* if $\|\mathbf{x}\|_0 \leq K$, where $K \in \{1, \ldots, n\}$.

The set of K-sparse vectors is denoted by

$$\Sigma_K = \left\{\mathbf{x} \in \mathbb{R}^{n\times 1} \middle| \|\mathbf{x}\|_0 \leq K\right\}. \tag{6.8.6}$$

If $\hat{\mathbf{x}} \in \Sigma_K$ then the vector $\hat{\mathbf{x}} \in \mathbb{R}^n$ is known as the *K-sparse approximation* or K-term approximation of $\mathbf{x} \in \mathbb{R}^n$.

A given set of L $m\times 1$ real input vectors $\{\mathbf{y}_1, \ldots, \mathbf{y}_L\}$ constitutes the data matrix $\mathbf{Y} = [\mathbf{y}_1, \ldots, \mathbf{y}_L] \in \mathbb{R}^{m\times L}$. The *sparse coding problem* involves determining n $m \times 1$ basis vectors $\mathbf{a}_1, \ldots, \mathbf{a}_n \in \mathbb{R}^m$, and, for each input vector $\mathbf{b}_l$ determining an $n \times 1$

sparse weighting vector, or coefficient vector, $\mathbf{s}_l \in \mathbb{R}^n$ such that the weighted linear combination of a few of these basis vectors can approximate the original input vector, namely

$$\mathbf{y}_l = \sum_{i=1}^{n} s_{l,i}\mathbf{a}_i = \mathbf{A}\mathbf{s}_l, \quad l = 1, \ldots, L, \tag{6.8.7}$$

where $s_{l,i}$ is the ith entry of the sparse weighting vector $\mathbf{s}_l$.

Sparse coding can be regarded as a form of neural coding: as the weight vector is sparse, so for each input vector, only a small number of neurons (basis vectors) are strongly excited. If the input vectors are different then the excited neurons are also different.

The vector $\hat{\mathbf{x}}$ is said to be the *optimal K-sparse approximation* of $\mathbf{x}$ in the ℓ_p-norm condition if the ℓ_p-norm of the approximation error vector $\mathbf{x} - \hat{\mathbf{x}}$ reaches an infimum, i.e.,

$$\|\mathbf{x} - \hat{\mathbf{x}}\|_p = \inf_{\mathbf{z} \in \Sigma_K} \|\mathbf{x} - \mathbf{z}\|_p.$$

Obviously, there is a close relationship between the ℓ_0-norm definition formula (6.8.5) and the ℓ_p-norm definition formula (6.8.1): as $p \to 0$, $\|\mathbf{x}\|_0 = \lim_{p \to 0} \|\mathbf{x}\|_p^p$. Because, if and only if $p \geq 1$, $\|\mathbf{x}\|_p$ is a convex function, the ℓ_1-norm is the objective function closest to ℓ_0-norm. Thus, from the viewpoint of optimization, the ℓ_1-norm is said to be the *convex relaxation* of the ℓ_0-norm. Therefore, the ℓ_0-norm minimization problem (P_0) can be transformed into the convex relaxed *ℓ_1-norm minimization* problem

$$(P_1) \qquad \min_{\mathbf{x}} \|\mathbf{x}\|_1 \quad \text{subject to} \quad \mathbf{b} = \mathbf{A}\mathbf{x}. \tag{6.8.8}$$

This is a convex optimization problem because the ℓ_1-norm $\|\mathbf{x}\|_1$, as the objective function, is itself convex and the equality constraint $\mathbf{b} = \mathbf{A}\mathbf{x}$ is affine.

In an actual situation of existing observation noise, the equality constrained optimization problem (P_1) can be relaxed to the inequality constrained optimization problem

$$(P_{10}) \qquad \min_{\mathbf{x}} \|\mathbf{x}\|_1 \quad \text{subject to} \quad \|\mathbf{b} - \mathbf{A}\mathbf{x}\|_2 \leq \epsilon. \tag{6.8.9}$$

The ℓ_1-norm minimization in Equation (6.8.9) is also called *basis pursuit* (BP). This is a *quadratically constrained linear programming* (QCLP) *problem.*

If $\mathbf{x}_1$ is the solution to (P_1) and $\mathbf{x}_0$ is the solution to (P_0), then [129]

$$\|\mathbf{x}_1\|_1 \leq \|\mathbf{x}_0\|_1, \tag{6.8.10}$$

because $\mathbf{x}_0$ is only a feasible solution to (P_1), while $\mathbf{x}_1$ is the optimal solution to (P_1). The direct relationship between $\mathbf{x}_0$ and $\mathbf{x}_1$ is given by $\mathbf{A}\mathbf{x}_1 = \mathbf{A}\mathbf{x}_0$.

Similarly to the inequality constrained ℓ_0-norm minimization expression (6.8.4), the inequality constrained ℓ_1-norm minimization expression (6.8.9) also has two variants:

(1) Since $\mathbf{x}$ is constrained as a K-sparse vector, the inequality constrained ℓ_1-norm minimization becomes the inequality constrained ℓ_2-norm minimization

$$(P_{11}) \qquad \min_{\mathbf{x}} \frac{1}{2}\|\mathbf{b}-\mathbf{A}\mathbf{x}\|_2^2 \quad \text{subject to} \quad \|\mathbf{x}\|_1 \le q. \tag{6.8.11}$$

This is a *quadratic programming* (QP) *problem.*

(2) Using the Lagrange multiplier method, the inequality constrained ℓ_1-norm minimization problem (P_{11}) becomes

$$(P_{12}) \qquad \min_{\lambda,\mathbf{x}} \left\{\frac{1}{2}\|\mathbf{b}-\mathbf{A}\mathbf{x}\|_2^2 + \lambda\|\mathbf{x}\|_1\right\}. \tag{6.8.12}$$

This optimization is called *basis pursuit de-noising* (BPDN) [96].

The optimization problems (P_{10}) and (P_{11}) are respectively called error constrained ℓ_1-norm minimization and ℓ_1-penalty minimization [471]. The ℓ_1-penalty minimization is also known as regularized ℓ_1 linear programming or ℓ_1-norm regularization least squares.

The Lagrange multiplier acts as a regularization parameter that controls the sparseness of the sparse solution; the greater λ is, the more sparse the solution $\mathbf{x}$ is. When the regularization parameter λ is large enough, the solution $\mathbf{x}$ is a zero vector. With the gradual decrease of λ, the sparsity of the solution vector $\mathbf{x}$ also gradually decreases. As λ tends to 0, the solution vector $\mathbf{x}$ becomes the vector such that $\|\mathbf{b}-\mathbf{A}\mathbf{x}\|_2^2$ is minimized. That is to say, $\lambda > 0$ can balance the twin objectives by minimizing the error squared sum cost function $\frac{1}{2}\|\mathbf{b}-\mathbf{A}\mathbf{x}\|_2^2$ and the ℓ_1-norm cost function $\|\mathbf{x}\|_1$:

$$J(\lambda,\mathbf{x}) = \frac{1}{2}\|\mathbf{b}-\mathbf{A}\mathbf{x}\|_2^2 + \lambda\|\mathbf{x}\|_1. \tag{6.8.13}$$

6.8.2 Lasso and Robust Linear Regression

Minimizing the squared error usually leads to sensitive solutions. Many regularization methods have been proposed to decrease this sensitivity. Among them, Tikhonov regularization [465] and Lasso [462], [143] are two widely known and cited algorithms, as pointed out in [515].

The problem of regression analysis is a fundamental problem within many fields, such as statistics, supervised machine learning, optimization and so on. In order to reduce the computational complexity of solving the optimization problem (P_1) directly, consider a *linear regression* problem: given an observed data vector $\mathbf{b} \in \mathbb{R}^m$ and an observed data matrix $\mathbf{A} \in \mathbb{R}^{m\times n}$, find a fitting coefficient vector $\mathbf{x} \in \mathbb{R}^n$ such that

$$\hat{b}_i = x_1 a_{i1} + x_2 a_{i2} + \cdots + x_n a_{in}, \quad i = 1,\ldots,m, \tag{6.8.14}$$

or

$$\hat{\mathbf{b}} = \sum_{i=1}^{n} x_i \mathbf{a}_i = \mathbf{A}\mathbf{x}, \tag{6.8.15}$$

where

$$\mathbf{A} = [\mathbf{a}_1, \ldots, \mathbf{a}_n] = \begin{bmatrix} a_{11} & \cdots & a_{1n} \\ \vdots & \vdots & \vdots \\ a_{m1} & \cdots & a_{mn} \end{bmatrix},$$
$$\mathbf{x} = [x_1, \ldots, x_n]^T,$$
$$\mathbf{b} = [b_1, \ldots, b_m]^T.$$

As a preprocessing of the linear regression, it is assumed that

$$\sum_{i=1}^{m} b_i = 0, \quad \sum_{i=1}^{m} a_{ij} = 0, \quad \sum_{i=1}^{m} a_{ij}^2 = 1, \quad j = 1, \ldots, n, \tag{6.8.16}$$

and that the column vectors of the input matrix $\mathbf{A}$ are linearly independent. The preprocessed input matrix $\mathbf{A}$ is called an *orthonormal input matrix*, its column vectors $\mathbf{a}_i$ are known as a *predictors* and the vector $\mathbf{x}$ is simply called the coefficient vector.

Tibshirani [462] in 1996 proposed the *least absolute shrinkage and selection operator* (Lasso) *algorithm* for solving the linear regression. The basic idea of the Lasso is as follows: under the constraint that the ℓ_1-norm of the prediction vector does not exceed an upper bound q, the squared sum of the prediction errors is minimized, namely

$$\text{Lasso:} \qquad \min_{\mathbf{x}} \|\mathbf{b} - \mathbf{A}\mathbf{x}\|_2^2 \quad \text{subject to} \quad \|\mathbf{x}\|_1 = \sum_{i=1}^{n} |x_i| \leq q. \tag{6.8.17}$$

Obviously, the Lasso model and the QP problem (6.8.11) have exactly the same form.

The bound q is a tuning parameter. When q is large enough, the constraint has no effect on $\mathbf{x}$ and the solution is just the usual multiple linear least squares regression of $\mathbf{x}$ on $a_{i1}, \ldots, a_{in}$ and $b_i, i = 1, \ldots, m$. However, for smaller values of q ($q \geq 0$), some of the coefficients x_j will take the value zero. Choosing a suitable q will lead to a sparse coefficient vector $\mathbf{x}$.

The regularized Lasso problem is given by

$$\text{Regularized Lasso:} \qquad \min_{\mathbf{x}} \left\{ \|\mathbf{b} - \mathbf{A}\mathbf{x}\|_2^2 + \lambda \|\mathbf{x}\|_1 \right\}. \tag{6.8.18}$$

The Lasso problem involves both the ℓ_1-norm constrained fitting for statistics and data mining.

The Lasso has the following two basic functions.

(1) *Contraction function* The Lasso shrinks the range of parameters to be estimated; only a small number of selected parameters are estimated at each step.
(2) *Selection function* The Lasso automatically selects a very small part of the variables for linear regression, yielding a spare solution.

Therefore, the Lasso is a shrinkage and selection method for linear regression. It has a close relation to the soft thresholding of wavelet coefficients, forward stagewise regression, boosting methods and so on.

The Lasso achieves a better prediction accuracy by shrinkage and variable selection.

In practical linear regression, the observed matrix is usually corrupted by some potentially malicious perturbation. In order to find the optimal solution in the worst case, *robust linear regression* solves the following minimax problem [515]

$$\min_{\mathbf{x}\in\mathbb{R}^n} \max_{\Delta\mathbf{A}\in\mathcal{U}} \|\mathbf{b}-(\mathbf{A}+\Delta\mathbf{A})\mathbf{x}\|_2, \tag{6.8.19}$$

where $\Delta\mathbf{A} = [\boldsymbol{\delta}_1,\ldots,\boldsymbol{\delta}_n]$ denotes the perturbation of the matrix $\mathbf{A}$ and

$$\mathcal{U} \stackrel{\text{def}}{=} \left\{[\boldsymbol{\delta}_1,\ldots,\boldsymbol{\delta}_n] \Big| \|\boldsymbol{\delta}_i\|_2 \leq c_i,\ i=1,\ldots,n\right\} \tag{6.8.20}$$

is called the *uncertainty set*, or the set of admissible perturbations of the matrix $\mathbf{A}$. This uncertainty set is said to be featurewise uncoupled, in contrast with coupled uncertainty sets, which require the perturbations of different features to satisfy some joint constraints.

Uncoupled ℓ_1 norm-bounded uncertainty sets lead to an easily solvable optimization problem, as shown in the following theorem.

Theorem 6.10 [515] *The robust regression problem (6.8.19) with uncertainty set $\mathcal{U}$ given by (6.8.20) is equivalent to the following ℓ_1-regularized regression problem:*

$$\min_{\mathbf{x}\in\mathbb{R}^n} \left\{ \|\mathbf{b}-\mathbf{A}\mathbf{x}\|_2 + \sum_{i=1}^n c_i|x_i| \right\}. \tag{6.8.21}$$

With the ℓ_1-norm changed to an arbitrary norm $\|\cdot\|_a$, Theorem 6.10 can be extended to the following result.

Theorem 6.11 [515] *Let*

$$\mathcal{U}_a \stackrel{\text{def}}{=} \left\{[\boldsymbol{\delta}_1,\ldots,\boldsymbol{\delta}_n] \Big| \|\boldsymbol{\delta}_i\|_a \leq c_i,\ i=1,\ldots,n\right\}. \tag{6.8.22}$$

Then the robust regression problem (6.8.19) with uncertainty set $\mathcal{U}_a$ is equivalent to the following arbitrary norm $\|\cdot\|_a$-regularized regression problem:

$$\min_{\mathbf{x}\in\mathbb{R}^n} \left\{ \max_{\Delta\mathbf{A}\in\mathcal{U}_a} \|\mathbf{b}-(\mathbf{A}+\Delta\mathbf{A})\mathbf{x}\|_a \right\}. \tag{6.8.23}$$

THEOREM 6.12 *The optimization problems*

$$\min_{\mathbf{x}\in\mathbb{R}^n} \left(\|\mathbf{b}-\mathbf{A}\mathbf{x}\|_2+\lambda_1\|\mathbf{x}\|_1\right) \quad \text{and} \quad \min_{\mathbf{x}\in\mathbb{R}^n} \left(\|\mathbf{b}-\mathbf{A}\mathbf{x}\|_2^2+\lambda_2\|\mathbf{x}\|_1\right) \tag{6.8.24}$$

are equivalent, up to changes in the tradeoff parameters λ_1 and λ_2.

Proof See Appendix A in [515]. □

Taking $c_i=\lambda$ and orthonormalizing the column vector $\mathbf{a}_i$ of $\mathbf{A}$ for all $i=1,\ldots,n$, the ℓ_1 regularized regression problem (6.8.21) becomes

$$\min_{\mathbf{x}\in\mathbb{R}^n} \left(\|\mathbf{b}-\mathbf{A}\mathbf{x}\|_2+\lambda\|\mathbf{x}\|_1\right). \tag{6.8.25}$$

By Theorem 6.12, this optimization problem is equivalent to the regularized Lasso problem (6.8.18). That is to say, a robust linear regression problem is equivalent to a regularized Lasso problem.

Regarding the solution to Lasso as the solution of a robust least squares problem has two important consequences [515].

(1) Robustness provides a connection of the regularizer to a physical property, namely, protection from noise.
(2) Perhaps most significantly, robustness is a strong property that can itself be used as an avenue for investigating different properties of the solution: robustness of the solution can explain why the solution is sparse; moreover, a robust solution is, by definition, the optimal solution under a worst-case perturbation.

The Lasso has many generalizations; here are a few examples. In machine learning, sparse kernel regression is referred to as *generalized Lasso* [418]. The multidimensional shrinkage-thresholding method is known as *group Lasso* [402]. The sparse multiview feature selection method via low-rank analysis is called *MRM-Lasso* (multiview rank minimization-based Lasso) [528]. *Distributed Lasso* solves distributed sparse linear regression problems [316].

6.8.3 Mutual Coherence and RIP Conditions

The ℓ_1-norm minimization problem (P_1) is a convex relaxation of the ℓ_0-norm minimization (P_0). Unlike the ℓ_0-norm minimization problem, the ℓ_1-norm minimization problem has the tractability. A natural question to ask is "what is the relationship between these two kinds of optimization problems?"

DEFINITION 6.8 (see [313], [131], [69]) The *mutual coherence* of a given matrix $\mathbf{A}$ is the largest absolute normalized inner product between different columns from $\mathbf{A}$. Denoting the kth column in $\mathbf{A}$ by $\mathbf{a}_k$, the mutual coherence is thus defined as

$$\mu(\mathbf{A}) = \max_{1\leq k,j\leq n, k\neq j} \frac{|\mathbf{a}_k^T\mathbf{a}_j|}{\|\mathbf{a}_k\|_2\,\|\mathbf{a}_j\|_2}. \tag{6.8.26}$$

THEOREM 6.13 [130], [69] *For any matrix $\mathbf{A} \in \mathbb{R}^{m\times n}$, one has*

$$\text{spark}(\mathbf{A}) \geq 1 + \frac{1}{\mu(\mathbf{A})}. \tag{6.8.27}$$

LEMMA 6.4 [130], [69] *For an under-determined equation $\mathbf{Ax} = \mathbf{b}$ ($\mathbf{A} \in \mathbb{R}^{m\times n}$ with full rank), if there exists a solution $\mathbf{x}$ such that*

$$\|\mathbf{x}\|_0 < \frac{1}{2}\left(1 + \frac{1}{\mu(\mathbf{A})}\right) \tag{6.8.28}$$

then $\mathbf{x}$ is not only the unique solution of (P_1) but also the unique solution of (P_0).

DEFINITION 6.9 A matrix $\mathbf{A}$ is said to satisfy a Kth-order *restricted isometry property* (RIP) condition, if

$$\|\mathbf{x}\|_0 \leq K \quad \Rightarrow \quad (1-\delta_K)\|\mathbf{x}\|_2^2 \leq \|\mathbf{A}_K\mathbf{x}\|_2^2 \leq (1+\delta_K)\|\mathbf{x}\|_2^2, \tag{6.8.29}$$

where $0 \leq \delta_K < 1$ is a constant related to the sparseness of $\mathbf{A}$, while $\mathbf{A}_K$ is a submatrix consisting of any K columns of the dictionary matrix $\mathbf{A}$.

The RIP condition was presented by Candès and Tao [81] in 2006 and was refined by Foucart and Lai [160] in 2009.

When the RIP condition is satisfied, the nonconvex ℓ_0-norm minimization (P_0) is equivalent to the convex ℓ_1-norm minimization (P_1).

Let $I = \{i|x_i \neq 0\} \subset \{1,\ldots,n\}$ denote the support of the nonzero elements of the sparse vector $\mathbf{x}$; then $|I|$ denotes the length of the support, i.e., the number of nonzero elements of the sparse vector $\mathbf{x}$.

The Kth-order RIP condition with parameter δ_K is denoted by RIP(K,δ_K), where δ_K is called the *restricted isometry constant* (RIC) and is defined as the infimum of all the parameters δ such that RIP(K,δ_K) holds, i.e.,

$$\delta_K = \inf\left\{\delta \middle| (1-\delta)\|\mathbf{z}\|_2^2 \leq \|\mathbf{A}_I\mathbf{z}\|_2^2 \leq (1+\delta)\|\mathbf{z}\|_2^2,\ \forall\,|I| \leq K,\ \forall\,\mathbf{z} \in \mathbb{R}^{|I|}\right\}. \tag{6.8.30}$$

By Definition 6.9 it is known that if the matrix $\mathbf{A}_K$ is orthogonal, then $\delta_K = 0$ since $\|\mathbf{A}_K\mathbf{x}\|_2 = \|\mathbf{x}\|_2$. Therefore, a nonzero value of the RIC δ_K of a matrix can be used to evaluate the nonorthogonality of the matrix.

It is shown in [76] that a more compact lower bound of the RIC is

$$\delta_K < 0.307. \tag{6.8.31}$$

Under this condition, if there is no noise then a K-sparse signal can be exactly recovered by ℓ_1-norm minimization and, in the case of noise, the K-sparse signals can be robustly estimated by ℓ_1-norm minimization.

The RIC meets the monotonicity condition [76]

$$\delta_K \leq \delta_{K_1} \quad \text{if } K \leq K_1. \tag{6.8.32}$$

6.8.4 Relation to Tikhonov Regularization

In Tikhonov regularization LS problems, when using the ℓ_1-norm of the unknown coefficient vector $\mathbf{x}$ in the regularization term instead of the ℓ_2-norm, we have the ℓ_1 regularization LS problem

$$\min_{\mathbf{x}} \left\{ \frac{1}{2}\|\mathbf{b} - \mathbf{A}\mathbf{x}\|_2^2 + \lambda\|\mathbf{x}\|_1 \right\}. \tag{6.8.33}$$

The ℓ_1 regularization LS problem always has a solution, but not necessarily a unique solution.

The following are the ℓ_1 regularization properties, which reflect the similarity and differences between ℓ_1-regularization and Tikhonov regularization [248].

1. *Nonlinearity* The solution vector $\mathbf{x}$ of a Tikhonov regularization problem is a linear function of the observed data vector $\mathbf{b}$, but the solution vector of the ℓ_1-regularization problem is not.
2. *Regularization path* When the regularization parameter λ varies in the interval $[0, \infty)$, the optimal solution of the Tikhonov regularization problem is a smooth function of the regularization parameter. However, the solution family of the ℓ_1-regularization problem has a piecewise linear solution path [143]: there are regularization parameters $\lambda_1, \ldots, \lambda_k$ (where $0 = \lambda_k < \cdots < \lambda_1 = \lambda_{\max}$) such that the solution vector of the ℓ_1-regularization problem is piecewise linear:
$$\mathbf{x}_{\ell_1} = \frac{\lambda_i - \lambda}{\lambda_i - \lambda_{i+1}}\mathbf{x}_{\ell_1}^{(i+1)} - \frac{\lambda - \lambda_{i+1}}{\lambda_i - \lambda_{i+1}}\mathbf{x}_{\ell_1}^{(i)}. \tag{6.8.34}$$
Here $\lambda_{i+1} \leq \lambda \leq \lambda_i$, $i = 1, \ldots, k-1$, and $\mathbf{x}_{\ell_1}^{(i)}$ is the solution vector of the ℓ_1-regularization problem when the regularization parameter is λ_i, while $\mathbf{x}_{\ell_1}$ is the optimal solution of the ℓ_1-regularization problem. Therefore, $\mathbf{x}_{\ell_1}^{(1)} = \mathbf{0}$ and $\mathbf{x}_{\ell_1} = \mathbf{0}$ when $\lambda \geq \lambda_1$.
3. *Limit characteristic of* $\lambda \to 0$ When $\lambda \to 0$, the limit point of the solution of the Tikhonov regularization problem has the smallest ℓ_2-norm $\|\mathbf{x}\|_2$ among all the feasible points such that $\mathbf{A}^H(\mathbf{b} - \mathbf{A}\mathbf{x}) = \mathbf{0}$. In contrast the limit of the solution of the ℓ_1-regularization problem when $\lambda \to 0$ has the smallest ℓ_1-norm $\|\mathbf{x}\|_1$ among all the feasible points such that $\mathbf{A}^H(\mathbf{b} - \mathbf{A}\mathbf{x}) = \mathbf{0}$.
4. *Limit characteristics of* $\lambda \geq \lambda_{\max}$ When $\lambda \to \infty$, the optimal solution of the Tikhonov regularization problem converges to a zero vector. However, only if
$$\lambda \geq \lambda_{\max} = \|\mathbf{A}^H\mathbf{b}\|_\infty, \tag{6.8.35}$$
does the solution of the ℓ_1-regularization problem converge to a zero vector. In the above equation, $\|\mathbf{A}^H\mathbf{b}\|_\infty = \max\{[\mathbf{A}^H\mathbf{b}]_i\}$ is the ℓ_∞-norm of the vector.

The most fundamental difference between the ℓ_1-regularization LS problem and the Tikhonov regularization LS problem is that the solution of the former is usually a sparse vector while the coefficients of the solution of the latter are generally nonzero.

6.8.5 Gradient Analysis of ℓ_1-Norm Minimization

The signum function of a real-valued variable $t \in \mathbb{R}$ is defined as

$$\text{sign}(t) = \begin{cases} +1, & t > 0, \\ 0, & t = 0, \\ -1, & t < 0. \end{cases} \tag{6.8.36}$$

The *signum multifunction* of $t \in \mathbb{R}$, denoted $\text{SGN}(t)$, also called the set-valued function, is defined as [194]

$$\text{SGN}(t) = \frac{\partial |t|}{\partial t} = \begin{cases} \{+1\}, & t > 0, \\ [-1, +1], & t = 0, \\ \{-1\}, & t < 0. \end{cases} \tag{6.8.37}$$

The signum multifunction is also the subdifferential of $|t|$.

For the ℓ_1-norm optimization problem

$$\min_{\mathbf{x}} J(\lambda, \mathbf{x}) = \min_{\mathbf{x}} \left\{ \frac{1}{2} \|\mathbf{b} - \mathbf{A}\mathbf{x}\|_2^2 + \lambda \|\mathbf{x}\|_1 \right\}, \tag{6.8.38}$$

the gradient vector of the objective function is given by

$$\begin{aligned} \nabla_{\mathbf{x}} J(\lambda, \mathbf{x}) &= \frac{\partial J(\lambda, \mathbf{x})}{\partial \mathbf{x}} = -\mathbf{A}^T(\mathbf{b} - \mathbf{A}\mathbf{x}) + \lambda \nabla_{\mathbf{x}} \|\mathbf{x}\|_1 \\ &= -\mathbf{c} + \lambda \nabla_{\mathbf{x}} \|\mathbf{x}\|_1. \end{aligned} \tag{6.8.39}$$

Here $\mathbf{c} = \mathbf{A}^T(\mathbf{b} - \mathbf{A}\mathbf{x})$ is the vector of residual correlation vector, and $\nabla_{\mathbf{x}} \|\mathbf{x}\|_1 = [\nabla_{x_1} \|\mathbf{x}\|_1, \ldots, \nabla_{x_n} \|\mathbf{x}\|_1]^T$ is the gradient vector of the ℓ_1-norm $\|\mathbf{x}\|_1$, with ith entry

$$\nabla_{x_i} \|\mathbf{x}\|_1 = \frac{\partial \|\mathbf{x}\|_1}{\partial x_i} = \begin{cases} \{+1\}, & x_i > 0, \\ \{-1\}, & x_i < 0, \\ [-1, +1], & x_i = 0, \end{cases} \tag{6.8.40}$$

for $i = 1, \ldots, n$.

From (6.8.39) it can be seen that the stationary point of the ℓ_1-norm minimization problem (P_{12}) is determined by the condition $\nabla_{\mathbf{x}} J(\lambda, \mathbf{x}) = -\mathbf{c} + \lambda \nabla_{\mathbf{x}} \|\mathbf{x}\|_1 = \mathbf{0}$, i.e.,

$$\mathbf{c} = \lambda \nabla_{\mathbf{x}} \|\mathbf{x}\|_1. \tag{6.8.41}$$

Letting $\mathbf{c} = [c(1), \ldots, c(n)]^T$ and substituting (6.8.40) into (6.8.41), the stationary-point condition can be rewritten as

$$c(i) = \begin{cases} \{+\lambda\}, & x_i > 0, \\ \{-\lambda\}, & x_i < 0, \\ [-\lambda, \lambda], & x_i = 0, \end{cases} \tag{6.8.42}$$

for $i = 1, \ldots, n$. Since ℓ_1-norm minimization is a convex optimization problem, the above stationary-point condition is in practice a sufficient and necessary condition for the optimal solution of the ℓ_1-norm minimization.

The stationary-point condition (6.8.42) can be represented using the vector of residual correlations as

$$\mathbf{c}(I) = \lambda\text{sign}(\mathbf{x}) \quad \text{and} \quad |\mathbf{c}(I^c)| \leq \lambda, \tag{6.8.43}$$

in which $I^c = \{1, \ldots, n\} \backslash I$ is the complementary set of the support I. This shows that the amplitude of the residual correlation is equal to λ and the sign is consistent with the sign of the corresponding element of the vector $\mathbf{x}$.

Equation (6.8.43) can be equivalently written as

$$|c(j)| = \lambda, \quad \forall\, j \in I \quad \text{and} \quad |c(j)| \leq \lambda \quad \forall\, j \in I^c. \tag{6.8.44}$$

That is to say, the absolute value of the residual correlation within the support is equal to λ, while that outside the support is less than or equal to λ, namely $\|\mathbf{c}\|_\infty = \max\{c(j)\} = \lambda$.

6.9 Sparse Matrix Equation Solving: Optimization Algorithms

The common basic idea behind ℓ_1-norm minimization algorithms is as follows: by recognizing the support region of sparse vectors, the solution of an under-determined sparse matrix equation is transformed into the solution of an over-determined (nonsparse) matrix equation.

6.9.1 Basis Pursuit Algorithms

The orthogonal matching pursuit method is a kind of greedy stepwise least squares method for fitting sparse models.

The general method for finding the whole optimal solution of an under-determined matrix equation $\mathbf{A}_{m\times n}\mathbf{x}_{n\times 1} = \mathbf{b}_{m\times 1}$ $(m \ll n)$ with sparseness s is first to find the LS solutions of all the over-determined matrix equations $\mathbf{A}_{m\times s}\tilde{\mathbf{x}}_{s\times 1} = \mathbf{b}$ (usually $m \gg s$) and then to determine the optimal solution among these solutions. Because there are C_m^s possible combinations of the over-determined equations, the overall solution process is time-consuming and tedious.

The basic idea of a *greedy algorithm* [470] is not to seek a global optimal solution but to try to find a local optimal solution in some sense as quickly as possible. The greedy method cannot get a global optimal solution for all problems, but it may provide a global optimal solution or its approximation in most cases.

Typical greedy algorithms use the following matching pursuit methods:

1. *Basic matching pursuit* (MP) *method* This was proposed by Mallat and Zhang [313] in 1993; its basic idea is not to minimize some cost function, but to iteratively construct a sparse solution $\mathbf{x}$ using a linear combination of a few column

vectors (called atoms) of a dictionary matrix $\mathbf{A}$ to make a sparse approximation to $\mathbf{x}$ to get $\mathbf{Ax} = \mathbf{b}$. In each iteration the column vector in the dictionary matrix $\mathbf{A}$ that is most similar to the present residual vector $\mathbf{r} = \mathbf{Ax} - \mathbf{b}$ is selected as a new column of the action set. If the residual is decreasing with iteration, the convergence of the MP algorithm can be guaranteed.

2. *Orthogonal matching pursuit* (OMP) Matching pursuit can construct a residual vector orthogonal to every column vector of the selected dictionary matrix at each iteration, but the residual vector generally is not orthogonal to the previously selected column vectors of the dictionary matrix. The OMP [377], [118], [173] ensures that the residual vector is orthogonal to all the previously selected columns vectors after each iteration, and thus guarantees the optimality of the iteration process, and reduces the number of iterations, making the performance more robust. The complexity of the OMP algorithm is $O(mn)$, and it can produce a coefficient vector with the sparseness $K \leq m/(2 \log n)$.
3. *Regularization orthogonal matching pursuit* (ROMP) On the basis of the OMP algorithm, the regularization procedure is added [340], [341] by first selecting multiple atoms from the relevant atoms as the candidate set and then selecting a fraction of atoms from the candidate set in accordance with the regularization principle. Finally, these atoms are incorporated into the final support set to achieve a fast and effective choice of atoms.
4. *Stagewise orthogonal matching pursuit* (StOMP) This is a reduced form of the OMP algorithm [133]. At the expense of approximation accuracy, it further improves the computation speed and has complexity $O(n)$, which is more suitable for solving large-scale sparse approximation problems.
5. *Compression sampling matching pursuit* (CoSaMP) This is an improvement of ROMP [342]. As compared with the OMP algorithm, the approximate accuracy of the CoSaMP algorithm is higher, its complexity $O(n \log^2 n)$ is lower and the sparseness of the sparse solution is $K \leq m/(2 \log(1 + n/K))$.

Algorithm 6.8 shows the orthogonal matching pursuit algorithm.

The following are three different stopping criteria [472].

(1) Stop running after a fixed number of iterations.
(2) Stop running when the residual energy (the sum of the squared amplitudes of the residual vector's components) $\|\mathbf{r}_k\|_2$ is less than a given value ϵ, i.e., $\|\mathbf{r}_k\|_2 \leq \epsilon$.
(3) Stop running when no column of $\mathbf{A}$ has any significant residual energy $\mathbf{r}_k$, i.e., $\|\mathbf{A}^H \mathbf{r}_k\|_\infty \leq \epsilon$.

A subspace pursuit algorithm for compressed-sensing signal reconstruction is shown in Algorithm 6.9.

At the kth iteration, the OMP, stagewise OMP and regular OMP methods combine the index of the new candidate column of the dictionary matrix $\mathbf{A}$ with the

Algorithm 6.8 Orthogonal matching pursuit algorithm [377], [118], [69]

task: Approximate the solution of (P_0): $\underset{\mathbf{x}}{\text{argmin}} \|\mathbf{x}\|_0$ subject to $\mathbf{Ax} = \mathbf{b}$.

input: $\mathbf{A} \in \mathbb{R}^{m\times n}$, $\mathbf{b} \in \mathbb{R}^m$ and the error threshold ϵ.

initialization: Initial solution $\mathbf{x}_0 = \mathbf{0}$, initial residual $\mathbf{r}_0 = \mathbf{A}\mathbf{x}_0 - \mathbf{b} = \mathbf{b}$, initial solution support $\Omega_0 = \text{support}(\mathbf{x}_0) = \emptyset$. Put $k = 1$.

repeat

1. Sweep: use $z_j = \mathbf{a}_j^T \mathbf{r}_{k-1}/\|\mathbf{a}\|_j^2$ to compute $\epsilon(j) = \min_{z_j} \|\mathbf{a}_j z_j - \mathbf{r}_{k-1}\|_2^2, \forall j$.
2. Update support: find a minimizer j_0 of $\epsilon(j)$ satisfying $\epsilon(j_0) \leq \epsilon(j), \forall j \notin \Omega_{k-1}$, and update $\Omega_k = \Omega_{k-1} \cup j_0$.
3. Update provisional solution: compute $\mathbf{x}_k = (\mathbf{A}_{\Omega_k}^H \mathbf{A}_{\Omega_k})^{-1} \mathbf{A}_{\Omega_k}^H \mathbf{b}$.
4. Update residual: calculate $\mathbf{r}_k = \mathbf{b} - \mathbf{A}_{\Omega_k} \mathbf{x}_k$.
5. **exit if** the stopping criterion is satisfied.

return $k \leftarrow k + 1$.

output: The sparse coefficient vector $\mathbf{x}(i) = \begin{cases} \mathbf{x}_k(i), & i \in \Omega_k, \\ 0, & i \notin \Omega_k. \end{cases}$

Algorithm 6.9 Subspace pursuit algorithm [115]

task: Approximate the solution of (P_0): $\underset{\mathbf{x}}{\text{argmin}} \|\mathbf{x}\|_0$ subject to $\mathbf{Ax} = \mathbf{b}$.

input: Sparsity K, $\mathbf{A} \in \mathbb{R}^{m\times n}$, $\mathbf{b} \in \mathbb{R}^m$.

initialization:

1. $\Omega_0 = \{K|$ indexes corresponding to the largest-magnitude entries in $\mathbf{A}^T\mathbf{b}\}$.
2. $\mathbf{r}_0 = \mathbf{b} - \mathbf{A}_{\Omega_0} \mathbf{A}_{\Omega_0}^\dagger \mathbf{b}$.

repeat

1. $\tilde{\Omega}_k = \Omega_{k-1} \bigcup \{K|$ indexes corresponding to the largest magnitude entries in the vector $\mathbf{A}_{\Omega_{k-1}}^T \mathbf{r}_{k-1}\}$.
2. Compute $\mathbf{x}_p = \mathbf{A}_{\tilde{\Omega}_k}^\dagger \mathbf{b}$.
3. $\Omega_k = \{K|$ indexes corresponding to the largest magnitude entries in $\mathbf{x}_p\}$.
4. $\mathbf{r}_k = \mathbf{b} - \mathbf{A}_{\Omega_k} \mathbf{A}_{\Omega_k}^\dagger \mathbf{b}$.
5. **exit if** $\|\mathbf{r}_k\|_2 \leq \|\mathbf{r}_{k-1}\|_2$.
6. Let $\Omega_k = \Omega_{k-1}$.

return $k \leftarrow k + 1$.

output: $\mathbf{x} = \mathbf{x}_p$.

index set at the $(k-1)$th iteration Ω_{k-1}. Once a candidate is selected, it will remain in the list of selected columns until the algorithm ends. Unlike these methods, for a K-sparse signal, Algorithm 6.9 keeps the index set of K candidate columns unchanged and allows the candidate column to be continuously updated during the iteration process.

6.9.2 First-Order Augmented Lagrangian Algorithm

In practice, solving the LP problem (P_1) is hard, because the matrix $\mathbf{A}$ is large and dense and an LP problem is often ill-conditioned. In typical compressed sensing (CS) applications, the dimension of $\mathbf{A}$ is large ($n \approx 10^6$), hence general interior-point methods are not practical for solving (P_1) in CS applications owing to the need to factorize an $m \times n$ matrix $\mathbf{A}$.

However, the LP problem (P_1) can be efficiently solved in the Lagrangian form (see e.g., [154], [194], [20])

$$\min_{\mathbf{x}\in\mathbb{R}^n} \left\{ \lambda\|\mathbf{x}\|_1 + \frac{1}{2}\|\mathbf{Ax}-\mathbf{b}\|_2^2 \right\} \tag{6.9.1}$$

with penalty parameter $\lambda \searrow 0$.

An augmented Lagrange function for the above LP problem is given by [20]

$$J(\mathbf{x}) = \lambda\|\mathbf{x}\|_1 - \lambda\boldsymbol{\theta}^T(\mathbf{Ax}-\mathbf{b}) + \frac{1}{2}\|\mathbf{Ax}-\mathbf{b}\|_2^2, \tag{6.9.2}$$

where $\boldsymbol{\theta}$ is a Lagrange multiplier vector for the constraints $\mathbf{Ax} = \mathbf{b}$.

Algorithm 6.10 shows the *first-order augmented Lagrangian* (FAL) *algorithm*. This algorithm solves the LP problem by inexactly solving a sequence of optimization problems of the form

$$\min_{\mathbf{x}\in\mathbb{R}^n \mid \|\mathbf{x}\|_1\le\eta_k} \left\{ \lambda_k\|\mathbf{x}\|_1 - \lambda_k\boldsymbol{\theta}_k^T(\mathbf{Ax}-\mathbf{b}) + \frac{1}{2}\|\mathbf{Ax}-\mathbf{b}\|_2^2 \right\}, \tag{6.9.3}$$

for an appropriately chosen sequence of $\{\lambda_k, \boldsymbol{\theta}_k, \eta_k\}, k = 0, 1, \ldots$

The APG stopping criterion APGstop is given by step 8 in Algorithm 6.10, and the FAL stopping criterion FALstop is given by

$$\text{FALstop} = \left\{ \left\|\mathbf{u}^{(l)} - \mathbf{u}^{(l-1)}\right\|_\infty \le \gamma \right\} \tag{6.9.4}$$

for noiseless measurements or

$$\text{FALstop} = \frac{\left\|\mathbf{u}^{(l)} - \mathbf{u}^{(l-1)}\right\|_2}{\left\|\mathbf{u}^{(l-1)}\right\|_2} \le \gamma \tag{6.9.5}$$

for noisy measurements, where γ is the threshold value.

Algorithm FAL produces $\mathbf{x}_{\text{sol}} = \mathbf{u}^{(l)}$ when FALstop is true.

The APG function used in step 9 in Algorithm 6.10 is given by Algorithm 6.11.

6.9.3 Barzilai–Borwein Gradient Projection Algorithm

Consider the standard *bound-constrained quadratic program (BCQP) problem*

$$\min_{\mathbf{z}} \left\{ q(\mathbf{x}) = \frac{1}{2}\mathbf{x}^T\mathbf{Ax} - \mathbf{b}^T\mathbf{x} \right\} \quad \text{subject to } \mathbf{v} \le \mathbf{x} \le \mathbf{u}. \tag{6.9.6}$$

Algorithm 6.10 FAL $(\{\lambda_k, \epsilon_k, \tau_k\}_{k\in\mathbb{Z}_+}, \eta)$ [20]

task: Find $\mathbf{x} = \underset{\mathbf{x}}{\text{argmin}} \left\{\lambda_k \|\mathbf{x}\|_1 - \lambda_k \boldsymbol{\theta}_k^T(\mathbf{Ax} - \mathbf{b}) + \frac{1}{2}\|\mathbf{Ax} - \mathbf{b}\|_2^2\right\}$.

input: $\mathbf{x}_0, L = \sigma_{\max}(\mathbf{AA}^T)$.

initialization: $\boldsymbol{\theta}_1 = \mathbf{0}, k = 0$.

1. **while** (FALSTOP not true) **do**
2. $k \leftarrow k + 1$.
3. $p_k = \lambda_k \|\mathbf{x}\|_1, \quad f_k(\mathbf{x}) = \frac{1}{2}\|\mathbf{Ax} - \mathbf{b} - \lambda_k \boldsymbol{\theta}_k\|_2^2$.
4. $h_k(\mathbf{x}) = \frac{1}{2}\|\mathbf{x} - \mathbf{x}_{k-1}\|_2^2$.
5. $\eta_k \leftarrow \eta + (\lambda_k/2)\|\boldsymbol{\theta}_k\|_2^2$.
6. $F_k = \{\mathbf{x} \in \mathbb{R}^n | \; \|\mathbf{x}\|_1 \leq \eta_k\}$.
7. $\ell_{k,\max} \leftarrow \sigma_{\max}(\mathbf{A})(\eta_k + \|\mathbf{x}_{k-1}\|_2)\sqrt{2/\epsilon_k}$.
8. APGSTOP $= \{\ell \geq \ell_{k,\max}\}$ or $\{\exists \mathbf{g} \in \partial P_k(\mathbf{x})|_{v_\ell}$ with $\|\mathbf{g}\|_2 \leq \tau_k\}$.
9. $\mathbf{x}_k \leftarrow$ APG $(p_k, f_k, L, F_k, \mathbf{x}_{k-1}, h_k$, APGSTOP$)$.
10. $\boldsymbol{\theta}_{k+1} \leftarrow \boldsymbol{\theta}_k - (\mathbf{Ax}_k - \mathbf{b})/\lambda_k$.
11. **end while**
12. **output:** $\mathbf{x} \leftarrow \mathbf{x}_k$.

Algorithm 6.11 APG $(p_k, f_k, L, F_k, \mathbf{x}_{k-1}, h_k$, APGSTOP$)$ [20]

initialization: $\mathbf{u}_0 = \mathbf{x}_0, \mathbf{w}_0 \leftarrow \text{argmin}_{\mathbf{x}\in p} h(\mathbf{x}), \theta_0 \leftarrow 1, l \leftarrow 0$.

1. **while** (APGSTOP not true) **do**
2. $\mathbf{v}_l \leftarrow (1 - \theta_l)\mathbf{u}_l + \theta_l \mathbf{w}_l$.
3. $\mathbf{w}_{l+1} \leftarrow \text{argmin}\left\{\sum_{i=0}^{l} \theta_i^{-1} \left(p(\mathbf{z}) + [\nabla f(v_l)]^T \mathbf{z}\right) + (L/c)h(\mathbf{z}) \middle| \mathbf{z} \in F\right\}$.
4. $\hat{\mathbf{u}}_{l+1} \leftarrow (1 - \theta_l)\mathbf{u}_l + \theta_l \mathbf{w}_{l+1}$.
5. $H_l(\mathbf{x}) \leftarrow p(\mathbf{x}) + [\nabla f(\mathbf{v}_l)]^T \mathbf{x} + \frac{L}{2}\|\mathbf{x} - \mathbf{v}_l\|_2^2$.
6. $\mathbf{u}_{l+1} \leftarrow \text{argmin}\{H_l(\mathbf{x}) | \mathbf{x} \in F\}$.
7. $\theta_{l+1} \leftarrow \frac{1}{2}\left(\sqrt{\theta_l^4 - 4\theta_l^2} - \theta_l^2\right)$.
8. $l \leftarrow l + 1$.
9. **end while**
10. **return** $\mathbf{u}_l$ or $\mathbf{v}_l$ depending on APGSTOP.

Gradient projection (GP) or projected gradient methods provide an alternative way of solving large-scale BCQP problems.

Let Ω be the feasible set of (6.9.6) such that

$$\Omega = \{\mathbf{x} \in \mathbb{R}^n | \mathbf{v} \leq \mathbf{x} \leq \mathbf{u}\}, \tag{6.9.7}$$

and let $\mathcal{P}$ denote the projection operator onto Ω, that is

$$\mathcal{P}(\mathbf{x}) = \text{mid}(\mathbf{v}, \mathbf{x}, \mathbf{u}), \tag{6.9.8}$$

where $\text{mid}(\mathbf{v}, \mathbf{x}, \mathbf{u})$ is the vector whose ith component is the median of the set $\{v_i, x_i, u_i\}$. Suppose that a feasible point $\mathbf{x}_k$ is generated, the gradient projection method computes the next point as [114]

$$\mathbf{x}_{k+1} = \mathcal{P}(\mathbf{x}_k - \alpha_k \mathbf{g}_k), \tag{6.9.9}$$

where $\alpha_k > 0$ is some step size and $\mathbf{g}_k = \mathbf{A}\mathbf{x}_k - \mathbf{b}$.

Barzilai and Borwein [26] proposed a method for greatly improving the effectiveness of gradient projection methods by appropriate choices of the step size α_k. Two useful choices of the step size were given in [26]:

$$\alpha_k^{BB1} = \frac{\mathbf{s}_{k-1}^T \mathbf{s}_{k-1}}{\mathbf{s}_{k-1}^T \mathbf{y}_{k-1}}, \qquad \alpha_k^{BB2} = \frac{\mathbf{s}_{k-1}^T \mathbf{y}_{k-1}}{\mathbf{y}_{k-1}^T \mathbf{y}_{k-1}}, \tag{6.9.10}$$

where $\mathbf{s}_{k-1} = \mathbf{x}_k - \mathbf{x}_{k-1}$ and $\mathbf{y}_{k-1} = \mathbf{g}_k - \mathbf{g}_{k-1}$.

Gradient projection using the Barzilai–Borwein step size is called the *Barzilai–Borwein gradient projection* (BBGP) *method*. One of its features is that it is nonmonotonic, i.e., $q(\mathbf{x}_k)$ may increase on some iterations. Nevertheless, the BBGP method is simple and easy to implement and avoids any matrix factorization.

Interestingly, the convex unconstrained optimization problem

$$\min_{\mathbf{x}} \left\{ \frac{1}{2} \|\mathbf{b} - \mathbf{A}\mathbf{x}\|_2^2 + \tau \|\mathbf{x}\|_1 \right\}, \quad \mathbf{x} \in \mathbb{R}^n, \mathbf{A} \in \mathbb{R}^{m \times n}, \mathbf{b} \in \mathbb{R}^m \tag{6.9.11}$$

can be transformed into a bound-constrained quadratic program (BCQP) problem.

To this end, define $u_i = (x_i)_+$ and $v_i = (-x_i)_+$ for all $i = 1, \ldots, n$, where $(x)_+ = \max\{x, 0\}$ denotes the *positive-part operator*. Then, the coefficient vector $\mathbf{x}$ can be split into a difference between nonnegative vectors:

$$\mathbf{x} = \mathbf{u} - \mathbf{v}, \quad \mathbf{u} \geq \mathbf{0}, \quad \mathbf{v} \geq \mathbf{0}. \tag{6.9.12}$$

The above equation can also be written as $\mathbf{v} \leq \mathbf{x} \leq \mathbf{u}$, where, for vectors $\mathbf{a}, \mathbf{b}$, $\mathbf{a} \leq \mathbf{b}$ means $a_i \leq b_i, \forall\, i$.

Obviously, $\|\mathbf{x}\|_1 = \mathbf{1}_n^T \mathbf{u} + \mathbf{1}_n^T \mathbf{v}$, where $\mathbf{1}_n = [1, \ldots, 1]^T$ is an $n \times 1$ summing vector consisting of n ones. By setting $\mathbf{x} = \mathbf{u} - \mathbf{v}$, the convex unconstrained optimization problem (6.9.11) can be rewritten as the following BCQP problem [154]:

$$\min_{\mathbf{z}} \left\{ \mathbf{c}^T \mathbf{z} + \frac{1}{2} \mathbf{z}^T \mathbf{B} \mathbf{z} \equiv F(\mathbf{z}) \right\} \quad \text{subject to} \quad \mathbf{z} \geq \mathbf{0}. \tag{6.9.13}$$

Here

$$\mathbf{z} = \begin{bmatrix} \mathbf{u} \\ \mathbf{v} \end{bmatrix}, \quad \mathbf{c} = \tau \mathbf{1}_{2n} + \begin{bmatrix} -\mathbf{A}^T\mathbf{b} \\ \mathbf{A}^T\mathbf{b} \end{bmatrix}, \quad \mathbf{B} = \begin{bmatrix} \mathbf{A}^T\mathbf{A} & -\mathbf{A}^T\mathbf{A} \\ -\mathbf{A}^T\mathbf{A} & \mathbf{A}^T\mathbf{A} \end{bmatrix}. \tag{6.9.14}$$

Algorithm 6.12 shows the *GPSR-BB algorithm* (Barzilai–Borwein gradient projection for sparse reconstruction), which was proposed in [154].

Algorithm 6.12 GPSR-BB algorithm [154]

---

task: Solve $\min\limits_{\mathbf{x}} \left\{ \frac{1}{2}\|\mathbf{b} - \mathbf{A}\mathbf{x}\|_2^2 + \tau\|\mathbf{x}\|_1 \right\}$ or $\min\limits_{\mathbf{z} \geq \mathbf{0}} \left\{ \mathbf{c}^T\mathbf{z} + \frac{1}{2}\mathbf{z}^T\mathbf{B}\mathbf{z} \equiv F(\mathbf{z}) \right\}$.

input: $\mathbf{B}$ and $F(\mathbf{z})$.

initialization: Given $\mathbf{z}_0$, choose $\alpha_{\min}, \alpha_{\max}, \alpha_0 \in [\alpha_{min}, \alpha_{\max}]$, and set $k = 0$.

repeat

1. Compute $\boldsymbol{\delta}_k = (\mathbf{z}_k - \alpha_k \nabla F(\mathbf{z}_k))_+ - \mathbf{z}_k$.
2. Compute $\gamma_k = \boldsymbol{\delta}_k^T \mathbf{B} \boldsymbol{\delta}_k$.
3. Line search: If $\gamma_k = 0$, let $\lambda_k = 1$, otherwise
   $\lambda_k = \text{mid}\left\{0, \boldsymbol{\delta}_k^T \nabla F(\mathbf{z}_k)/\gamma_k, 1\right\}$.
4. Update $\mathbf{z}_{k+1} = \mathbf{z}_k + \lambda_k \boldsymbol{\delta}_k$.
5. **exit if**
   $\|\mathbf{z}_{k+1} - (\mathbf{z}_{k+1} - \bar{\alpha}\nabla F(\mathbf{z}_{k+1}))_+\|_2 \leq \text{tol}$,
   where tol is a small parameter and $\bar{\alpha}$ is a positive constant.
6. if $\gamma_k = 0$, let $\alpha_{k+1} = \alpha_{\max}$, otherwise
   $\alpha_{k+1} = \text{mid}\left\{\alpha_{\min}, \|\boldsymbol{\delta}_k\|_2^2/\gamma_k, \alpha_{\max}\right\}$.
7. **return** $k \leftarrow k + 1$.

output: $\mathbf{z} = \mathbf{z}_{k+1}$.

6.9.4 ADMM Algorithms for Lasso Problems

The alternating direction method of multipliers (ADMM) form of the Lasso problem $\min\left\{\frac{1}{2}\|\mathbf{A}\mathbf{x} - \mathbf{b}\|_2^2 + \lambda\|\mathbf{x}\|_1\right\}$ is given by

$$\min\left\{\frac{1}{2}\|\mathbf{A}\mathbf{x} - \mathbf{b}\|_2^2 + \lambda\|\mathbf{z}\|_1\right\} \quad \text{subject to} \quad \mathbf{x} - \mathbf{z} = \mathbf{0}. \tag{6.9.15}$$

The corresponding ADMM algorithm is [52]:

$$\mathbf{x}^{k+1} = (\mathbf{A}^T\mathbf{A} + \rho\mathbf{I})^{-1}(\mathbf{A}^T\mathbf{b} + \rho\mathbf{z}^k - \mathbf{y}^k), \tag{6.9.16}$$

$$\mathbf{z}^{k+1} = S_{\lambda/\rho}(\mathbf{x}^{k+1} + \mathbf{y}^k/\rho), \tag{6.9.17}$$

$$\mathbf{y}^{k+1} = \mathbf{y}^k + \rho(\mathbf{x}^{k+1} - \mathbf{z}^{k+1}), \tag{6.9.18}$$

where the soft thresholding operator S is defined as

$$S_{\lambda/\rho}(x) = \begin{cases} x - \lambda/\rho, & x > \lambda/\rho, \\ 0, & |x| \leq \lambda/\rho, \\ x + \lambda/\rho, & x < \lambda/\rho. \end{cases} \tag{6.9.19}$$

The Lasso problem can be generalized to

$$\min \left\{ \frac{1}{2}\|\mathbf{Ax} - \mathbf{b}\|_2^2 + \lambda \|\mathbf{Fx}\|_1 \right\}, \tag{6.9.20}$$

where $\mathbf{F}$ is an arbitrary linear transformation. The above problem is called the *generalized Lasso problem.*

The ADMM form of the generalized Lasso problem can be written as

$$\min \left\{ \frac{1}{2}\|\mathbf{Ax} - \mathbf{b}\|_2^2 + \lambda \|\mathbf{z}\|_1 \right\} \quad \text{subject to} \quad \mathbf{Fx} - \mathbf{z} = \mathbf{0}. \tag{6.9.21}$$

The corresponding ADMM algorithm can be described as [52]

$$\mathbf{x}^{k+1} = (\mathbf{A}^T\mathbf{A} + \rho\mathbf{F}^T\mathbf{F})^{-1}(\mathbf{A}^T\mathbf{b} + \rho\mathbf{F}^T\mathbf{z}^k - \mathbf{y}^k), \tag{6.9.22}$$

$$\mathbf{z}^{k+1} = S_{\lambda/\rho}(\mathbf{Fx}^{k+1} + \mathbf{y}^k/\rho), \tag{6.9.23}$$

$$\mathbf{y}^{k+1} = \mathbf{y}^k + \rho(\mathbf{Fx}^{k+1} - \mathbf{z}^{k+1}). \tag{6.9.24}$$

Consider the *group Lasso problem* [537]

$$\min \left\{ \frac{1}{2}\|\mathbf{Ax} - \mathbf{b}\|_2^2 + \lambda \sum_{i=1}^{n} \|\mathbf{x}_i\|_2 \right\} \tag{6.9.25}$$

with feature groups $\mathbf{Ax} = \sum_{i=1}^N \mathbf{A}_i\mathbf{x}_i$, where $\mathbf{A}_i \in \mathbb{R}^{m\times n_i}, \mathbf{x}_i \in \mathbb{R}^{n_i}$, $i = 1, \ldots, N$ and $\sum_{i=1}^N n_i = n$. Another difference from the generalized Lasso is the Euclidean-norm (not squared) regularization of $\mathbf{x}$. The group Lasso is also called the *sum-of-norms regularization* [356].

Put $\overline{\mathbf{Ax}}^k = N^{-1}\sum_{i=1}^N \mathbf{A}_i\mathbf{x}_i^k$. The ADMM algorithm for the group Lasso is given by [52]

$$\mathbf{x}_i^{k+1} = \arg\min_{\mathbf{x}_i} \left\{ \frac{\rho}{2}\|\mathbf{A}_i\mathbf{x}_i - \mathbf{A}_i\mathbf{x}_i^k - \bar{\mathbf{z}}^k + \overline{\mathbf{Ax}}^k + \mathbf{y}^k\|_2^2 + \lambda\|\mathbf{x}_i\|_2 \right\}, \tag{6.9.26}$$

$$\bar{\mathbf{z}}^{k+1} = \frac{1}{N+\rho}(\mathbf{b} + \rho\overline{\mathbf{Ax}}^{k+1} + \rho\mathbf{y}^k), \tag{6.9.27}$$

$$\mathbf{y}^{k+1} = \mathbf{y}^k + \overline{\mathbf{Ax}}^{k+1} - \bar{\mathbf{z}}^{k+1}). \tag{6.9.28}$$

6.9.5 LARS Algorithms for Lasso Problems

An efficient approach for solving Lasso problems is the *least angle regressive* (LARS) *algorithm* [143]. Algorithm 6.13 shows the LARS algorithm with Lasso modification.

Algorithm 6.13 LARS algorithm with Lasso modification [143]

given: The data vector $\mathbf{b} \in \mathbb{R}^m$ and the input matrix $\mathbf{A} \in \mathbb{R}^{m\times n}$.

initialization: $\Omega_0 = \emptyset, \hat{\mathbf{b}} = \mathbf{0}$ and $\mathbf{A}_{\Omega_0} = \mathbf{A}$. Put $k = 1$.

repeat

1. Compute the correlation vector $\hat{\mathbf{c}}_k = \mathbf{A}_{\Omega_{k-1}}^T(\mathbf{b} - \hat{\mathbf{b}}_{k-1})$.
2. Update the active set $\Omega_k = \Omega_{k-1} \bigcup \{j^{(k)} \mid |\hat{c}_k(j)| = C\}$ with $C = \max\{|\hat{c}_k(1)|, \ldots, |\hat{c}_k(n)|\}$.
3. Update the input matrix $\mathbf{A}_{\Omega_k} = [s_j \mathbf{a}_j, j \in \Omega_k]$, where $s_j = \text{sign}(\hat{c}_k(j))$.
4. Find the direction of the current minimum angle
$$\mathbf{G}_{\Omega_k} = \mathbf{A}_{\Omega_k}^T \mathbf{A}_{\Omega_k} \in \mathbb{R}^{k\times k},$$
$$\alpha_{\Omega_k} = (\mathbf{1}_k^T \mathbf{G}_{\Omega_k} \mathbf{1}_k)^{-1/2},$$
$$\mathbf{w}_{\Omega_k} = \alpha_{\Omega_k} \mathbf{G}_{\Omega_k}^{-1} \mathbf{1}_k \in \mathbb{R}^k,$$
$$\boldsymbol{\mu}_k = \mathbf{A}_{\Omega_k} \mathbf{w}_{\Omega_k} \in \mathbb{R}^k.$$
5. Compute $\mathbf{b} = \mathbf{A}_{\Omega_k}^T \boldsymbol{\mu}_k = [b_1, \ldots, b_m]^T$ and estimate the coefficient vector
$$\hat{\mathbf{x}}_k = (\mathbf{A}_{\Omega_k}^T \mathbf{A}_{\Omega_k})^{-1} \mathbf{A}_{\Omega_k}^T = \mathbf{G}_{\Omega_k}^{-1} \mathbf{A}_{\Omega_k}^T.$$
6. Compute $\hat{\gamma} = \min\limits_{j\in\Omega_k^c} \left\{ \frac{C - \hat{c}_k(j)}{\alpha_{\Omega_k} - b_j}, \frac{C + \hat{c}_k(j)}{\alpha_{\Omega_k} + b_j} \right\}^+, \tilde{\gamma} = \min\limits_{j\in\Omega_k} \left\{ -\frac{x_j}{w_j} \right\}^+$, where w_j is the jth of entry $\mathbf{w}_{\Omega_k} = [w_1, \ldots, w_n]^T$ and $\min\{\cdot\}^+$ denotes the positive minimum term. If there is not positive term then $\min\{\cdot\}^+ = \infty$.
7. If $\tilde{\gamma} < \hat{\gamma}$ then the fitted vector and the active set are modified as follows:
$$\hat{\mathbf{b}}_k = \hat{\mathbf{b}}_{k-1} + \tilde{\gamma}\boldsymbol{\mu}_k,\ \Omega_k = \Omega_k - \{\tilde{j}\},$$
where the removed index $\tilde{j}$ is the index $j \in \Omega_k$ such that $\tilde{\gamma}$ is a minimum.
Conversely, if $\hat{\gamma} < \tilde{\gamma}$ then $\hat{\mathbf{b}}_k$ and Ω_k are modified as follows:
$$\hat{\mathbf{b}}_k = \hat{\mathbf{b}}_{k-1} + \hat{\gamma}\boldsymbol{\mu}_k,\ \Omega_k = \Omega_k \cup \{\hat{j}\},$$
where the added index $\hat{j}$ is the index $j \in \Omega_k$ such that $\hat{\gamma}$ is a minimum.
8. **exit** If some stopping criterion is satisfied.
9. **return** $k \leftarrow k + 1$.

output: The coefficient vector $\mathbf{x} = \hat{\mathbf{x}}_k$.

The LARS algorithm is a stepwise regression method, and its basic idea is this: while ensuring that the current residual is equal to the selected correlation coefficient, select as the solution path the projection of the current residual onto the space constituted by the selected variables. Then, continue to search on this solution path, absorbing the new added variable, and adjusting the solution path.

Suppose that Ω_k is the active set of variables at the beginning of the kth iteration step, and let $\mathbf{x}_{\Omega_k}$ be the coefficient vector for these variables at this step.

Letting the initial active set $\Omega_0 = \emptyset$, the initial value of the residual vector is given by $\mathbf{r} = \mathbf{b}$. At the first iteration find the column (i.e., the predictor) of the input matrix $\mathbf{A}$ that is correlated with the residual vector $\mathbf{r} = \mathbf{b}$; denote it $\mathbf{a}_i^{(1)}$, and enter its index in the active set Ω_1. Thus, one obtains the first regression-variable set $\mathbf{A}_{\Omega_1}$.

The basic step of the LARS algorithm is as follows: suppose that after $k-1$ LARS steps the regression variable set is $\mathbf{A}_{\Omega_{k-1}}$. Then, one has the vector expression

$$\mathbf{w}_{\Omega_{k-1}} = \left(\mathbf{1}_{\Omega_{k-1}}^T(\mathbf{A}_{\Omega_{k-1}}^T\mathbf{A}_{\Omega_{k-1}})^{-1}\mathbf{1}_{\Omega_{k-1}}\right)^{-1/2}(\mathbf{A}_{\Omega_{k-1}}^T\mathbf{A}_{\Omega_{k-1}})^{-1}\mathbf{1}_{\Omega_{k-1}}. \tag{6.9.29}$$

Hence $\mathbf{A}_{\Omega_{k-1}}\mathbf{w}_{\Omega_{k-1}}$ is the solution path of the LARS algorithm in the current regression variable set Ω_{k-1}, and $\mathbf{w}_{\Omega_{k-1}}$ is the path for which $\mathbf{x}$ continues to search.

The core idea of the LARS algorithm with Lasso modification, as shown in Algorithm 6.13, is "one at a time": at the every iteration step, one must add or remove a variable. Specifically, from the existing regression variable set and the current residuals, a solution path is determined. The maximum possible step forward on this path is denoted by $\hat{\gamma}$ and the maximum step needed to find a new variable is denoted by $\tilde{\gamma}$. If $\tilde{\gamma} < \hat{\gamma}$ then the new variable $x_j(\gamma)$ corresponding to the LARS estimate is not a Lasso estimate, and hence this variable should be deleted from the regression variable set. Conversely, if $\tilde{\gamma} > \hat{\gamma}$ then the new estimate $x_j(\gamma)$ corresponding to the LARS estimate should become a new Lasso estimate, and hence this variable should be added to the regression set. After deleting or adding a variable, one should stop searching on the original solution path. Then, one recomputes the correlation coefficient of the current residual and current new variable set to determine a new solution path, and continues the "one at a time" form of the LARS iteration steps. This process is repeated until all Lasso estimates are obtained by using the LARS algorithm.

6.9.6 Covariance Graphical Lasso Method

The estimation of a covariance matrix via a sample of vectors drawn from a multivariate Gaussian distribution is among the most fundamental problems in statistics [48]. Suppose n samples of p normal random vectors are given by $\mathbf{x}_1, \ldots, \mathbf{x}_n \sim N_p(\mathbf{0}, \mathbf{\Sigma})$. The log likelihood is [48]

$$L(\mathbf{\Sigma}) = -\frac{np}{2} - \frac{n}{2}\log\det\mathbf{\Sigma} - \frac{n}{2}\mathrm{tr}(\mathbf{\Sigma}^{-1}\mathbf{S}),$$

where $\mathbf{S} = n^{-1}\sum_{i=1}^n \mathbf{x}_i\mathbf{x}_i^T$ is the sample covariance matrix. This approximates the true covariance matrix $\mathbf{\Sigma} = [\sigma_{ij}]$ with $\sigma_{ij} = E\{\mathbf{x}_i^T\mathbf{x}_j\}$.

Bien and Tibshirani [48] in 2011 proposed a covariance graphical Lasso method using a Lasso penalty on the elements of the covariance matrix.

The basic version of the covariance graphical Lasso problem minimizes the objective function [48], [498]:

$$\min_{\mathbf{\Sigma}} g(\mathbf{\Sigma}) = \min_{\mathbf{\Sigma}} \left\{ \rho\|\mathbf{\Sigma}\|_1 + \log(\det \mathbf{\Sigma}) + \text{tr}(\mathbf{S}\mathbf{\Sigma}^{-1}) \right\} \tag{6.9.30}$$

over the space of positive definite matrices $\mathbf{M}^+$ with shrinkage parameter $\rho \geq 0$. Then, covariance graphical Lasso estimation is used to solve the optimization problem (6.9.30).

The covariance graphical Lasso method is particularly useful owing to the following two facts [498]:

(1) by using the ℓ_1-norm term, the covariance graphical Lasso is able to set some off-diagonal elements of $\mathbf{\Sigma}$ exactly equal to zero at the minimum point of (6.9.30);
(2) the zeros in $\mathbf{\Sigma}$ encode the marginal independence structures among the components of a multivariate normal random vector with covariance matrix $\mathbf{\Sigma}$.

However, the objective function is not convex, which makes the optimization challenging. To address this challenge, Bien and Tibshirani [48] proposed a majorize–minimize approach to minimize (6.9.30) approximately, and Wang [498] in 2014 proposed a coordinate descent algorithm for covariance graphical Lasso. In comparison with the majorize–minimize algorithm of Bien and Tibshirani, Wang's coordinate descent algorithm is simpler to implement, substantially faster to run and numerically more stable as shown in experimental results.

The idea of coordinate descent is simple: to update one column and one row of $\mathbf{\Sigma}$ at a time while holding all the rest of the elements in $\mathbf{\Sigma}$ fixed. For example, to update the last column and row, $\mathbf{\Sigma}$ and $\mathbf{S}$ are partitioned as follows:

$$\mathbf{\Sigma} = \begin{bmatrix} \mathbf{\Sigma}_{11} & \boldsymbol{\sigma}_{12} \\ \boldsymbol{\sigma}_{12}^T & \sigma_{22} \end{bmatrix}, \quad \mathbf{S} = \begin{bmatrix} \mathbf{S}_{11} & \mathbf{s}_{12} \\ \mathbf{s}_{12}^T & s_{22} \end{bmatrix}, \tag{6.9.31}$$

where $\mathbf{\Sigma}_{11}$ and $\mathbf{S}_{11}$ are respectively the covariance matrix and the sample covariance matrix of the first $p-1$ variables, and $\boldsymbol{\sigma}_{12}$ and $\mathbf{s}_{12}$ are respectively the covariance and the sample covariance between the first $p-1$ variables and the last variable while σ_{22} and s_{22} respectively are the variance and the sample variance of the last variable.

Letting $\boldsymbol{\beta} = \boldsymbol{\sigma}_{12}$ and $\gamma = \sigma_{22} - \boldsymbol{\sigma}_{12}^T\mathbf{\Sigma}_{11}^{-1}\boldsymbol{\sigma}_{12}$, we have

$$\mathbf{\Sigma}^{-1} = \begin{bmatrix} \mathbf{\Sigma}_{11}^{-1} + \mathbf{\Sigma}_{11}^{-1}\boldsymbol{\beta}\boldsymbol{\beta}^T\mathbf{\Sigma}_{11}^{-1}\gamma^{-1} & -\mathbf{\Sigma}_{11}^{-1}\boldsymbol{\beta}^T\gamma^{-1} \\ -\boldsymbol{\beta}^T\mathbf{\Sigma}_{11}^{-1}\gamma^{-1} & \gamma^{-1} \end{bmatrix}. \tag{6.9.32}$$

Therefore, the three terms in (6.9.30) can be represented as follows [498]:

$$\begin{aligned} \beta\|\mathbf{\Sigma}\|_1 &= 2\rho\|\boldsymbol{\beta}\|_1 + \rho(\boldsymbol{\beta}^T\mathbf{\Sigma}_{11}^{-1}\boldsymbol{\beta} + \gamma) + c_1, \\ \log(\det \mathbf{\Sigma}) &= \log\gamma + c_2, \\ \text{tr}(\mathbf{S}\mathbf{\Sigma}^{-1}) &= \boldsymbol{\beta}^T\mathbf{\Sigma}_{11}^{-1}\mathbf{S}\mathbf{\Sigma}_{11}^{-1}\boldsymbol{\beta}\gamma^{-1} - 2\mathbf{s}_{12}^T\mathbf{\Sigma}_{11}^{-1}\boldsymbol{\beta}\gamma^{-1} + s_{22}\gamma^{-1} + c_3. \end{aligned}$$

Dropping the constants c_1, c_2, c_3, we get the objection function in (6.9.30):

$$f(\boldsymbol{\beta},\gamma) = 2\rho\|\boldsymbol{\beta}\|_1 + \rho(\boldsymbol{\beta}^T\boldsymbol{\Sigma}_{11}^{-1}\boldsymbol{\beta}+\gamma) + \boldsymbol{\beta}^T\boldsymbol{\Sigma}_{11}^{-1}\mathbf{S}\boldsymbol{\Sigma}_{11}^{-1}\boldsymbol{\beta}\gamma^{-1} \\ - 2\mathbf{s}_{12}^T\boldsymbol{\Sigma}_{11}^{-1}\boldsymbol{\beta}\gamma^{-1} + s_{22}\gamma^{-1} + \log\gamma. \tag{6.9.33}$$

This objective function of the covariance graphical Lasso method can be written as two separated objective functions

$$f_1(\gamma) = \log\gamma + a\gamma^{-1} + \rho\gamma, \tag{6.9.34}$$

$$f_2(\boldsymbol{\beta}) = 2\rho\|\boldsymbol{\beta}\|_1 + \boldsymbol{\beta}^T\mathbf{V}\boldsymbol{\beta} - 2\mathbf{u}^T\boldsymbol{\beta}, \tag{6.9.35}$$

where $a = \boldsymbol{\beta}^T\boldsymbol{\Sigma}_{11}^{-1}\mathbf{S}\boldsymbol{\Sigma}_{11}^{-1}\boldsymbol{\beta} - 2\mathbf{s}_{12}^T\boldsymbol{\Sigma}_{11}^{-1}\boldsymbol{\beta} + s_{22}$, $\mathbf{V} = \gamma^{-1}\boldsymbol{\Sigma}_{11}^{-1}\mathbf{S}\boldsymbol{\Sigma}_{11}^{-1} + \rho\boldsymbol{\Sigma}_{11}^{-1}$ and $\mathbf{u} = \gamma^{-1}\boldsymbol{\Sigma}_{11}^{-1}\mathbf{s}_{12}$.

From (6.9.34) it can be seen that

$$\gamma = \arg\min_{\gamma} f_1(\gamma) = \begin{cases} a, & \rho = 0, \\ \left(-1+\sqrt{1+4a\rho}\right)/(2\rho), & \rho \neq 0. \end{cases} \tag{6.9.36}$$

Equation (6.9.35) shows that $f_2(\boldsymbol{\beta})$ is still a Lasso problem and can be efficiently solved by coordinate descent algorithms. For $j \in \{1,\ldots,p-1\}$, the minimum point of (6.9.35) along the coordinate direction in which β_j varies is given by

$$\hat{\beta}_j = \text{soft}\left(\mathbf{u}_j - \sum_{k=1,k\neq j}^{p-1} v_{kj}\hat{\beta}_k, \rho\right) \Big/ v_{jj}, \quad j = 1,\ldots,p-1 \tag{6.9.37}$$

where $\text{soft}(x,\rho) = \text{sign}(x)(|x|-\rho)_+$ is the soft thresholding operator.

A coordinate descent algorithm for covariance graphical Lasso is proposed in [498] based on the above results, as shown in Algorithm 6.14.

6.9.7 Homotopy Algorithm

In topology, the concept of homotopy describes a "continuous change" between two objects. The homotopy algorithm is a kind of searching algorithm that starts from a simple solution and uses iterative calculation to find the desired complex solution. Therefore, the key of a homotopy algorithm is to determine its initial simple solution appropriately.

Consider the relationship between the ℓ_1-norm minimization problem (P_1) and the unconstrained ℓ_2-minimization problem (P_{12}). Suppose that there is a corresponding unique solution $\mathbf{x}_\lambda$ for every minimization problem (P_{12}) with $\lambda \in [0,\infty)$. Then the set $\{\mathbf{x}_\lambda | \lambda \in [0,\infty)\}$ determines a solution path and has $\mathbf{x}_\lambda = \mathbf{0}$ for a sufficiently large λ; and when $\lambda \to 0$, the solution $\tilde{\mathbf{x}}_\lambda$ of (P_{12}) converges to the solution of the ℓ_1-norm minimization problem (P_1). Hence, $\mathbf{x}_\lambda = \mathbf{0}$ is the initial solution of the homotopy algorithm for solving the minimization problem (P_1).

The homotopy algorithm for solving the unconstrained ℓ_2-norm minimization

Algorithm 6.14 Coordinate descent algorithm [498]

task: Find $\boldsymbol{\Sigma} = \underset{\boldsymbol{\Sigma}}{\text{argmin}}\{\rho\|\boldsymbol{\Sigma}\|_1 + \log(\det \boldsymbol{\Sigma}) + \text{tr}(\mathbf{S}\boldsymbol{\Sigma}^{-1})\}$.

input: $\mathbf{S} = \mathbf{Y}^T\mathbf{Y}, \rho$.

initialization: Given $\boldsymbol{\Sigma}_0 = \mathbf{S}$ and partition $\mathbf{S} = \begin{bmatrix} \mathbf{S}_{11} & \mathbf{s}_{12} \\ \mathbf{s}_{12}^T & s_{22} \end{bmatrix}$. Put $k = 0$.

repeat

1. $\boldsymbol{\Sigma}_{k+1} = \boldsymbol{\Sigma}_k$, and Partition $\boldsymbol{\Sigma}_{k+1} = \begin{bmatrix} \boldsymbol{\Sigma}_{11} & \boldsymbol{\sigma}_{12} \\ \boldsymbol{\sigma}_{12}^T & \sigma_{22} \end{bmatrix}$.
2. Put $\boldsymbol{\beta} = \boldsymbol{\sigma}_{12}$.
3. $a = \boldsymbol{\beta}^T\boldsymbol{\Sigma}_{11}^{-1}\mathbf{S}\boldsymbol{\Sigma}_{11}^{-1}\boldsymbol{\beta} - 2\mathbf{s}_{12}^T\boldsymbol{\Sigma}_{11}^{-1}\boldsymbol{\beta} + s_{22}$.
4. $\gamma = \begin{cases} a, & \rho = 0, \\ (-1 + \sqrt{1 + 4a\rho})/(2\rho), & \rho \neq 0. \end{cases}$
5. $\mathbf{V} = \gamma^{-1}\boldsymbol{\Sigma}_{11}^{-1}\mathbf{S}\boldsymbol{\Sigma}_{11}^{-1} + \rho\boldsymbol{\Sigma}_{11}^{-1}$.
6. $\mathbf{u} = \gamma^{-1}\boldsymbol{\Sigma}_{11}^{-1}\mathbf{s}_{12}$.
7. Compute $\hat{\beta}_j = \text{soft}\left(\mathbf{u}_j - \sum_{k \neq j} v_{kj}\hat{\beta}_k, \rho\right) \Big/ v_{jj}, j = 1, \ldots, p$.
8. Calculate $\boldsymbol{\Sigma}_k = \begin{bmatrix} \boldsymbol{\Sigma}_{11}^{-1} + \boldsymbol{\Sigma}_{11}^{-1}\boldsymbol{\beta}\boldsymbol{\beta}^T\boldsymbol{\Sigma}_{11}^{-1}\gamma^{-1} & -\boldsymbol{\Sigma}_{11}^{-1}\boldsymbol{\beta}\gamma^{-1} \\ -\boldsymbol{\beta}^T\boldsymbol{\Sigma}_{11}^{-1}\gamma^{-1} & \gamma^{-1} \end{bmatrix}^{-1}$.
9. **exit if** $\|\boldsymbol{\Sigma}_{k+1} - \boldsymbol{\Sigma}_k\|_2 \leq \epsilon$.
10. **return** $k \leftarrow k + 1$.

output: $\boldsymbol{\Sigma}_k$.

---

problem (P_{12}) starts from the initial value $\mathbf{x}_0 = \mathbf{0}$, and runs in an iteration form to compute the solutions $\mathbf{x}_k$ at the kth step for $k = 1, 2, \ldots$ In the whole calculation, the following active set is kept unchanged:

$$I = \big\{j \big| |c_k(j)| = \|\mathbf{c}_k\|_\infty = \lambda\big\}. \tag{6.9.38}$$

Algorithm 6.15 shows the homotopy algorithm for solving the ℓ_1-norm minimization problem. With the reduction of λ, the objective function of (P_{11}) will go through a homotopy process from the ℓ_1-norm constraint to the ℓ_2-norm objective function. It has been shown [143], [363], [132] that the homotopy algorithm is an efficient solver for the ℓ_1-norm minimization problem (P_1).

6.9.8 Bregman Iteration Algorithms

The sparse optimization models for solving the matrix equation $\mathbf{Au} = \mathbf{b}$ discussed above can be summarized as follows:

Algorithm 6.15 Homotopy algorithm [132]

input: Observed vector $\mathbf{b} \in \mathbb{R}^m$, input matrix $\mathbf{A}$ and parameter λ.

initialization: $\mathbf{x}_0 = \mathbf{0}$, $\mathbf{c}_0 = \mathbf{A}^T\mathbf{b}$.

repeat

1. Use (6.9.38) to form the active set I, and put $\mathbf{c}_k(I) = [c_k(i), i \in I]$ and $\mathbf{A}_I = [\mathbf{a}_i, i \in I]$.
2. Compute the residual correlation vector $\mathbf{c}_k(I) = \mathbf{A}_I^T(\mathbf{b} - \mathbf{A}_I\mathbf{x}_k)$.
3. Solve the matrix equation $\mathbf{A}_I^T\mathbf{A}_I\mathbf{d}_k(I) = \text{sign}(\mathbf{c}_k(I))$ for $\mathbf{d}_k(I)$.
4. Compute
$$\gamma_k^+ = \min_{i\in I^c}{}^+ \left\{ \frac{\lambda - c_k(i)}{1 - \boldsymbol{\phi}_i^T\mathbf{v}_k}, \frac{\lambda + c_k(i)}{1 + \boldsymbol{\phi}_i^T\mathbf{v}_k} \right\},$$
$$\gamma_k^- = \min_{i\in I} \left\{ -\frac{x_k(i)}{d_k(i)} \right\}.$$
5. Determine the breakpoint $\gamma_k = \min\{\gamma_k^+, \gamma_k^-\}$.
6. Update the solution vector $\mathbf{x}_k = \mathbf{x}_{k-1} + \gamma_k\mathbf{d}_k$.
7. **exit if** $\|\mathbf{x}_k\|_\infty = 0$.
8. **return** $k \leftarrow k + 1$.

output: $\mathbf{x} = \mathbf{x}_k$.

(1) *ℓ_0 norm minimization model* $\min_{\mathbf{u}} \|\mathbf{u}\|_0$ subject to $\mathbf{Au} = \mathbf{b}$;

(2) *basis-pursuit* (BP)/*compressed sensing model* $\min_{\mathbf{u}} \|\mathbf{u}\|_1$ subject to $\mathbf{Au} = \mathbf{b}$;

(3) *basis-pursuit de-noising model* $\min_{\mathbf{u}} \|\mathbf{u}\|_1$ subject to $\|\mathbf{Au} - \mathbf{b}\|_2 < \epsilon$;

(4) *Lasso model* $\min_{\mathbf{u}} \|\mathbf{Au} - \mathbf{b}\|_2$ subject to $\|\mathbf{u}\|_1 \le s$.

Among these models, the BP model is a relaxed form of the ℓ_0-norm minimization model, and the Lasso model is a linear prediction representation equivalent to the basis pursuit de-noising model.

Now, consider the general form of the above four optimization models:

$$\mathbf{u}_{k+1} = \arg\min_{\mathbf{u}} \{J(\mathbf{u}) + \lambda H(\mathbf{u})\}, \tag{6.9.39}$$

where $J : X \to \mathbb{R}$ and $H : X \to \mathbb{R}$ are nonnegative convex functions (X is a closed convex set), but J is a nonsmooth, while H is differentiable.

Figuratively speaking, a function is of the *bounded-variation* type if the oscillation of its graphics (i.e. its swing or variation) is manageable or "tamed" in a certain range.

The *bounded-variation norm* of a vector $\mathbf{u}$, denoted $\|\mathbf{u}\|_{\text{BV}}$, is defined as [8]

$$\|\mathbf{u}\|_{\text{BV}} = \|\mathbf{u}\|_1 + J_0(\mathbf{u}) = \|\mathbf{u}\|_1 + \int_\Omega |\nabla\mathbf{u}| dx, \tag{6.9.40}$$

where $J_0(\mathbf{u})$ denotes the total variation of $\mathbf{u}$.

Let $J(\mathbf{u}) = \|\mathbf{u}\|_{\mathrm{BV}}$ and $H(\mathbf{u}) = \frac{1}{2}\|\mathbf{u} - \mathbf{f}\|_2^2$. Then the optimization model (6.9.39) becomes a total variation or *Rudin–Osher–Fatemi* (ROF) *de-noising model* [420]:

$$\mathbf{u}_{k+1} = \arg\min_{\mathbf{u}} \left\{ \|\mathbf{u}\|_{\mathrm{BV}} + \frac{\lambda}{2}\|\mathbf{u} - \mathbf{f}\|_2^2 \right\}. \tag{6.9.41}$$

A well-known iteration method for solving the optimization problem (6.9.39) is *Bregman iteration*, based on the Bregman distance [59].

Definition 6.10 Let $J(\mathbf{u})$ be a convex function, let the vectors $\mathbf{u}, \mathbf{v} \in X$ and let $\mathbf{g} \in \partial J(\mathbf{v})$ be the subgradient vector of the function J at the point $\mathbf{v}$. The *Bregman distance* between the points $\mathbf{u}$ and $\mathbf{v}$, denoted $D_J^g(\mathbf{u}, \mathbf{v})$, is defined as

$$D_J^g(\mathbf{u}, \mathbf{v}) = J(\mathbf{u}) - J(\mathbf{v}) - \langle \mathbf{g}, \mathbf{u} - \mathbf{v} \rangle. \tag{6.9.42}$$

The Bregman distance is not a distance in traditional sense, because $D_J^g(\mathbf{u}, \mathbf{v}) \neq D_J^g(\mathbf{v}, \mathbf{u})$. However, it has the following properties that make it an efficient tool for solving the ℓ_1-norm regularization problem:

(1) For all $\mathbf{u}, \mathbf{v} \in X$ and $\mathbf{g} \in \partial J(\mathbf{v})$, the Bregman distance is nonnegative, i.e., $D_J^g(\mathbf{u}, \mathbf{v}) \geq 0$.

(2) The Bregman distance for the same point is zero, i.e., $D_J^g(\mathbf{v}, \mathbf{v}) = 0$.

(3) The Bregman distance can measure the closeness of two points $\mathbf{u}$ and $\mathbf{v}$ since $D_J^g(\mathbf{u}, \mathbf{v}) \geq D_J^g(\mathbf{w}, \mathbf{v})$ for any point $\mathbf{w}$ on the line connecting $\mathbf{u}$ and $\mathbf{v}$.

Consider the first-order Taylor series approximation of the nonsmooth function $J(\mathbf{u})$ at the kth iteration point $\mathbf{u}_k$, denoted $J(\mathbf{u}) = J(\mathbf{u}_k) + \langle \mathbf{g}_k, \mathbf{u} - \mathbf{u}_k \rangle$. The approximate error is measured by the Bregman distance

$$D_J^{g_k}(\mathbf{u}, \mathbf{u}_k) = J(\mathbf{u}) - J(\mathbf{u}_k) - \langle \mathbf{g}_k, \mathbf{u} - \mathbf{u}_k \rangle. \tag{6.9.43}$$

Early in 1965, Bregman proposed that the unconstrained optimization problem (6.9.39) can be modified as [59]

$$\begin{aligned} \mathbf{u}_{k+1} &= \arg\min_{\mathbf{u}} \left\{ D_J^{g_k}(\mathbf{u}) + \lambda H(\mathbf{u}) \right\} \\ &= \arg\min_{\mathbf{u}} \left\{ J(\mathbf{u}) - \langle \mathbf{g}_k, \mathbf{u} - \mathbf{u}_k \rangle + \lambda H(\mathbf{u}) \right\}. \end{aligned} \tag{6.9.44}$$

This is the well-known Bregman iteration. In what follows we give the Bregman iteration algorithm and its generalization.

The objective function of the Bregman iterative optimization problem is denoted by $L(\mathbf{u}) = J(\mathbf{u}) - \langle \mathbf{g}_k, \mathbf{u} - \mathbf{u}_k \rangle + \lambda H(\mathbf{u})$. By the stationary-point condition $\mathbf{0} \in \partial L(\mathbf{u})$ it follows that $\mathbf{0} \in \partial J(\mathbf{u}) - \mathbf{g}_k + \lambda \nabla H(\mathbf{u})$. Hence, at the $(k+1)$th iterative point $\mathbf{u}_{k+1}$ one has

$$\mathbf{g}_{k+1} = \mathbf{g}_k - \lambda \nabla H(\mathbf{u}_{k+1}), \quad \mathbf{g}_{k+1} \in \partial J(\mathbf{u}_{k+1}). \tag{6.9.45}$$

Equations (6.9.44) and (6.9.45) constitute the *Bregman iterative algorithm*, which was proposed by Osher *et al.* in 2005 [364] for image processing.

THEOREM 6.14 [364] *Suppose that J and H are convex functions and that H is differentiable. If a solution to Equation (6.9.44) exists then the following convergence results are true.*

- *The function H is monotonically decreasing in the whole iterative process, namely $H(\mathbf{u}_{k+1}) \leq H(\mathbf{u}_k)$.*
- *The function H will converge to the optimal solution $H(\mathbf{u}^\star)$, since $H(\mathbf{u}_k) \leq H(\mathbf{u}^\star) + J(\mathbf{u}^\star)/k$.*

The Bregman iterative algorithm has two versions [532]:

Version 1 $k = 0$, $\mathbf{u}_0 = \mathbf{0}$, $\mathbf{g}_0 = \mathbf{0}$.
Iterations:

$$\mathbf{u}_{k+1} = \arg\min_{\mathbf{u}} \left\{ D_J^{g_k}(\mathbf{u}, \mathbf{u}_k) + \frac{1}{2}\|\mathbf{Au} - \mathbf{b}\|_2^2 \right\},$$
$$\mathbf{g}_{k+1} = \mathbf{g}_k - \mathbf{A}^T(\mathbf{Au}_{k+1} - \mathbf{b}).$$

If $\mathbf{u}_k$ does not converge then let $k \leftarrow k+1$ and return to the iteration.

Version 2 $k = 0$, $\mathbf{b}_0 = \mathbf{0}$, $\mathbf{u}_0 = \mathbf{0}$.
Iterations:

$$\mathbf{b}_{k+1} = \mathbf{b} + (\mathbf{b}_k - \mathbf{Au}_k).$$
$$\mathbf{u}_{k+1} = \arg\min_{\mathbf{u}} \left\{ J(\mathbf{u}) + \frac{1}{2}\|\mathbf{Au} - \mathbf{b}_{k+1}\|_2^2 \right\}.$$

If $\mathbf{u}_k$ does not converge then let $k \leftarrow k+1$ and return to the iteration.

It has been shown [532] that the above two versions are equivalent.

The Bregman iterative algorithm provides an efficient tool for optimization, but, at every step, one must minimize the objective function $D_J^{g_k}(\mathbf{u}, \mathbf{u}_k) + H(\mathbf{u})$. In order to improve the computational efficiency of the Bregman iterative algorithm, Yin *et al.* [532] proposed a linearized Bregman iterative algorithm.

The basic idea of the linearized Bregman iteration is as follows: for the Bregman iteration, use a first-order Taylor series expansion to linearize the nonlinear function $H(\mathbf{u})$ to $H(\mathbf{u}) = H(\mathbf{u}_k) + \langle \nabla H(\mathbf{u}_k), \mathbf{u} - \mathbf{u}_k \rangle$ at the point $\mathbf{u}_k$. Then, the optimization problem (6.9.39) with $\lambda = 1$ becomes

$$\mathbf{u}_{k+1} = \arg\min_{\mathbf{u}} \left\{ D_J^{g_k}(\mathbf{u}, \mathbf{u}_k) + H(\mathbf{u}_k) + \langle \nabla H(\mathbf{u}_k), \mathbf{u} - \mathbf{u}_k \rangle \right\}.$$

Note that the first-order Taylor series expansion is exact only for the neighborhood of $\mathbf{u}$ at the point $\mathbf{u}_k$, and the additive constant term $H(\mathbf{u}_k)$ can be made negative in the optimization with respect to $\mathbf{u}$, so a more exact expression of the above optimization problem is

$$\mathbf{u}_{k+1} = \arg\min_{\mathbf{u}} \left\{ D_J^{g_k}(\mathbf{u}, \mathbf{u}_k) + \langle \nabla H(\mathbf{u}_k), \mathbf{u} - \mathbf{u}_k \rangle + \frac{1}{2\delta} \|\mathbf{u} - \mathbf{u}_k\|_2^2 \right\}. \tag{6.9.46}$$

Importantly, the above equation can be equivalently written as

$$\mathbf{u}_{k+1} = \arg\min_{\mathbf{u}} \left\{ D_J^{g_k}(\mathbf{u}, \mathbf{u}_k) + \frac{1}{2\delta} \|\mathbf{u} - (\mathbf{u}_k - \delta \nabla H(\mathbf{u}_k))\|_2^2 \right\}, \tag{6.9.47}$$

because (6.9.46) and (6.9.47) differ by only a constant term that is independent of $\mathbf{u}$.

In particular, if $H(\mathbf{u}) = \frac{1}{2}\|\mathbf{A}\mathbf{u} - \mathbf{b}\|_2^2$ then from $\nabla H(\mathbf{u}) = \mathbf{A}^T(\mathbf{A}\mathbf{u} - \mathbf{b})$ one can write (6.9.47) as

$$\mathbf{u}_{k+1} = \arg\min_{\mathbf{u}} \left\{ D_J^{g_k}(\mathbf{u}, \mathbf{u}_k) + \frac{1}{2\delta} \left\|\mathbf{u} - \left(\mathbf{u}_k - \delta \mathbf{A}^T(\mathbf{A}\mathbf{u}_k - \mathbf{b})\right)\right\|_2^2 \right\}. \tag{6.9.48}$$

Consider the objective function in (6.9.48) as

$$L(\mathbf{u}) = J(\mathbf{u}) - J(\mathbf{u}_k) - \langle \mathbf{g}_k, \mathbf{u} - \mathbf{u}_k \rangle + \frac{1}{2\delta} \left\|\mathbf{u} - \left(\mathbf{u}_k - \delta \mathbf{A}^T(\mathbf{A}\mathbf{u}_k - \mathbf{b})\right)\right\|_2^2.$$

By the subdifferential stationary-point condition $\mathbf{0} \in \partial L(\mathbf{u})$ it is known that

$$\mathbf{0} \in \partial J(\mathbf{u}) - \mathbf{g}_k + \frac{1}{\delta}\left(\mathbf{u} - \left(\mathbf{u}_k - \delta \mathbf{A}^T(\mathbf{A}\mathbf{u}_k - \mathbf{b})\right)\right).$$

Denoting $\mathbf{g}_{k+1} \in \partial J(\mathbf{u}_{k+1})$, from the above equation one has [532]

$$\begin{aligned} \mathbf{g}_{k+1} &= \mathbf{g}_k - \mathbf{A}^T(\mathbf{A}\mathbf{u}_k - \mathbf{b}) - \frac{\mathbf{u}_{k+1} - \mathbf{u}_k}{\delta} = \cdots \\ &= \sum_{i=1}^{k} \mathbf{A}^T(\mathbf{b} - \mathbf{A}\mathbf{u}_i) - \frac{\mathbf{u}_{k+1}}{\delta}. \end{aligned} \tag{6.9.49}$$

Letting

$$\mathbf{v}_k = \sum_{i=1}^{k} \mathbf{A}^T(\mathbf{b} - \mathbf{A}\mathbf{u}_i), \tag{6.9.50}$$

one can obtain two important iterative formulas.

First, from (6.9.49) and (6.9.50) we get the update formula of the variable $\mathbf{u}$ at the kth iteration:

$$\mathbf{u}_{k+1} = \delta(\mathbf{v}_k - \mathbf{g}_{k+1}). \tag{6.9.51}$$

Then, Equation (6.9.50) directly yields an iterative formula for the intermediate variable $\mathbf{v}_k$:

$$\mathbf{v}_{k+1} = \mathbf{v}_k + \mathbf{A}^T(\mathbf{b} - \mathbf{A}\mathbf{u}_{k+1}). \tag{6.9.52}$$

Equations (6.9.51) and (6.9.52) constitute the linearized Bregman iterative algorithm [532] for solving the optimization problem

$$\min_{\mathbf{u}} \left\{ J(\mathbf{u}) + \frac{1}{2}\|\mathbf{A}\mathbf{u} - \mathbf{b}\|_2^2 \right\}. \tag{6.9.53}$$

In particular, for the optimization problem

$$\min_{\mathbf{u}} \left\{ \mu\|\mathbf{u}\|_1 + \frac{1}{2}\|\mathbf{A}\mathbf{u} - \mathbf{b}\|_2^2 \right\}, \tag{6.9.54}$$

we have

$$\partial(\|\mathbf{u}\|_1)_i = \begin{cases} \{+1\}, & \text{if}\, u_i > 0, \\ [-1,+1], & \text{if}\, u_i = 0, \\ \{-1\}, & \text{if}\, u_i < 0. \end{cases} \tag{6.9.55}$$

Hence Equation (6.9.51) can be written in the component form as

$$u_{k+1}(i) = \delta(v_k(i) - g_{k+1}(i)) = \delta\,\text{shrink}(v_k(i), \mu), \quad i = 1, \ldots, n, \tag{6.9.56}$$

where

$$\text{shrink}(y, \alpha) = \text{sign}(y)\max\{|y| - \alpha, 0\} = \begin{cases} y - \alpha, & y \in (\alpha, \infty), \\ 0, & y \in [-\alpha, \alpha], \\ y + \alpha, & y \in (-\infty, -\alpha) \end{cases}$$

is the shrink operator.

The above results can be summarized as the *linearized Bregman iterative algorithm* for solving the basis-pursuit de-noising or total variation de-noising problem [532], see Algorithm 6.16.

Consider the optimization problem

$$\mathbf{u}_{k+1} = \arg\min_{\mathbf{u}} \{\|\mathbf{A}\mathbf{u}\|_1 + H(\mathbf{u})\}. \tag{6.9.57}$$

Introducing the intermediate variable $\mathbf{z} = \mathbf{A}\mathbf{u}$, the unconstrained optimization problem (6.9.39) can be written as the constrained optimization

$$(\mathbf{u}_{k+1}, \mathbf{z}_{k+1}) = \arg\min_{\mathbf{u},\mathbf{z}} \{\|\mathbf{z}\|_1 + H(\mathbf{u})\} \quad \text{subject to} \quad \mathbf{z} = \mathbf{A}\mathbf{u}. \tag{6.9.58}$$

By adding an ℓ_2-norm penalty term, this constrained optimization problem becomes the unconstrained optimization problem

$$(\mathbf{u}_{k+1}, \mathbf{z}_{k+1}) = \arg\min_{\mathbf{u},\mathbf{z}} \left\{ \|\mathbf{z}\|_1 + H(\mathbf{u}) + \frac{\lambda}{2}\|\mathbf{z} - \mathbf{A}\mathbf{u}\|_2^2 \right\}. \tag{6.9.59}$$

Algorithm 6.16 Linearized Bregman iterative algorithm [532]

task: Solve basis-pursuit/total variation de-noising problem $\min_{\mathbf{u}} \{\mu\|\mathbf{u}\|_1 + \frac{1}{2}\|\mathbf{Au}-\mathbf{b}\|_2^2\}$.

input: Input matrix $\mathbf{A} \in \mathbb{R}^{m\times n}$ and observed vector $\mathbf{b} \in \mathbb{R}^m$.

initialization: $k = 0$, $\mathbf{u}_0 = \mathbf{0}$, $\mathbf{v}_0 = \mathbf{0}$.

repeat

1. **for** $i = 1, \ldots, n$ **do**
2. $u_{k+1}(i) = \delta\,\text{shrink}(v_k(i), \mu)$.
3. **end for**
4. $\mathbf{v}_{k+1} = \mathbf{v}_k + \mathbf{A}^T(\mathbf{b} - \mathbf{Au}_{k+1})$.
5. **exit if** $\|\mathbf{u}_{k+1} - \mathbf{u}_k\|_2 \leq \epsilon$.

return $k \leftarrow k + 1$.

output: $\mathbf{u}_k$.

The *split Bregman iterative algorithm* is given by [177]

$$\mathbf{u}_{k+1} = \arg\min_{\mathbf{u}} \left\{ H(\mathbf{u}) + \frac{\lambda}{2}\|\mathbf{z}_k - \mathbf{Au} - \mathbf{b}_k\|_2^2 \right\}, \tag{6.9.60}$$

$$\mathbf{z}_{k+1} = \arg\min_{\mathbf{z}} \left\{ \|\mathbf{z}\|_1 + \frac{\lambda}{2}\|\mathbf{z} - \mathbf{Au}_{k+1} - \mathbf{b}_k\|_2^2 \right\}, \tag{6.9.61}$$

$$\mathbf{b}_{k+1} = \mathbf{b}_k + [\mathbf{Au}_{k+1} - \mathbf{z}_{k+1}]. \tag{6.9.62}$$

The three iterations of the split Bregman iteration algorithm have the following characteristics.

(1) The first iteration is a differentiable optimization problem that can be solved by the Gauss–Seidel method.
(2) The second iteration can be solved efficiently using the shrink operator.
(3) The third iteration is a explicit calculation.

Exercises

6.1 Consider the matrix equation $\mathbf{Ax} + \boldsymbol{\epsilon} = \mathbf{x}$, where $\boldsymbol{\epsilon}$ is an additive color noise vector satisfying the conditions $E\{\boldsymbol{\epsilon}\} = \mathbf{0}$ and $E\{\boldsymbol{\epsilon}\boldsymbol{\epsilon}^T\} = \mathbf{R}$. Let $\mathbf{R}$ be known, and use the weighting error function $Q(\mathbf{x}) = \boldsymbol{\epsilon}^T\mathbf{W}\boldsymbol{\epsilon}$ as the cost function for finding the optimal estimate $\hat{\mathbf{x}}_{\text{WLS}}$. Such a method is called the weighting least squares method. Show that

$$\hat{\mathbf{x}}_{\text{WLS}} = (\mathbf{A}^T\mathbf{WA})^{-1}\mathbf{A}^T\mathbf{Wx},$$

where the optimal choice of the weighting matrix $\mathbf{W}$ is $\mathbf{W}_{\text{opt}} = \mathbf{R}^{-1}$.

6.2 Given an over-determined linear equation $\mathbf{Z}_t^T \mathbf{X}_t = \mathbf{Z}_t^T \mathbf{Y}_t \mathbf{x}$ with $\mathbf{Z}_t \in \mathbb{R}^{(t+1)\times K}$ is known as the instrumental variable matrix and $t+1 > K$.

(a) Letting the estimate of the parameter vector $\mathbf{x}$ at time t be $\hat{\mathbf{x}}$, find an expression for it. This is called the instrumental variable method.

(b) Setting

$$\mathbf{Y}_{t+1} = \begin{bmatrix} \mathbf{Y}_t \\ \mathbf{b}_{t+1} \end{bmatrix}, \quad \mathbf{Z}_{t+1} = \begin{bmatrix} \mathbf{Z}_t \\ \mathbf{z}_{t+1} \end{bmatrix}, \quad \mathbf{X}_{t+1} = \begin{bmatrix} \mathbf{X}_t \\ \mathbf{x}_{t+1} \end{bmatrix},$$

find a recursive computation formula for $\mathbf{x}_{t+1}$.

6.3 Consider the matrix equation $\mathbf{y} = \mathbf{A}\boldsymbol{\theta} + \mathbf{e}$, where $\mathbf{e}$ is an error vector. Define the weighting-error squared sum $E_w = \mathbf{e}^H \mathbf{W} \mathbf{e}$, where the weighting matrix $\mathbf{W}$ is Hermitian positive definite. Find a solution such that E_w is minimized. Such a solution is called the *weighted least squares* (WLS) *solution.*

6.4 Let $\lambda > 0$, and let $\mathbf{A}\mathbf{x} = \mathbf{b}$ is an over-determined matrix equation. Show that the anti-Tikhonov regularized optimization problem

$$\min \left\{ \frac{1}{2} \|\mathbf{A}\mathbf{x} - \mathbf{b}\|_2^2 - \frac{1}{2}\lambda \|\mathbf{x}\|_2^2 \right\}$$

has optimal solution $\mathbf{x} = (\mathbf{A}^H \mathbf{A} - \lambda \mathbf{I})^{-1} \mathbf{A}^H \mathbf{b}$.

6.5 [179] The TLS solution of the linear equation $\mathbf{A}\mathbf{x} = \mathbf{b}$ can be also expressed as

$$\min_{\mathbf{b}+\mathbf{e} \in \text{Range}(\mathbf{A}+\mathbf{E})} \|\mathbf{D}[\mathbf{E}, \mathbf{e}]\mathbf{T}\|_F, \quad \mathbf{E} \in \mathbb{R}^{m\times n}, \quad \mathbf{e} \in \mathbb{R}^m,$$

where $\mathbf{D} = \mathbf{Diag}(d_1, \ldots, d_m)$ and $\mathbf{T} = \mathbf{Diag}(t_1, \ldots, t_{n+1})$ are nonsingular. Show the following results.

(a) If $\text{rank}(\mathbf{A}) < n$ then the above TLS problem has one solution if and only if $\mathbf{b} \in \text{Range}(\mathbf{A})$.

(b) If $\text{rank}(\mathbf{A}) = n$, $\mathbf{A}^T \mathbf{D}^2 \mathbf{b} = \mathbf{0}$, $|t_{n+1}| \|\mathbf{D}\mathbf{b}\|_2 \geq \sigma_n(\mathbf{D}\mathbf{A}\mathbf{T}_1)$, where $\mathbf{T}_1 = \mathbf{Diag}(t_1, \ldots, t_n)$ then the TLS problem has no solution. Here $\sigma_n(\mathbf{C})$ denotes the nth singular value of the matrix $\mathbf{C}$.

6.6 Consider the TLS in the previous problem. Show that if $\mathbf{C} = \mathbf{D}[\mathbf{A}, \mathbf{b}]\mathbf{T} = [\mathbf{A}_1, \mathbf{d}]$ and $\sigma_n(\mathbf{C}) > \sigma_{n+1}(\mathbf{C})$ then the TLS solution satisfies $(\mathbf{A}_1^T \mathbf{A}_1 - \sigma_{n+1}^2(\mathbf{C})\mathbf{I})\mathbf{x} = \mathbf{A}_1^T \mathbf{d}$.

6.7 Given the data points $(1,3)$, $(3,1)$, $(5,7)$, $(4,6)$, $(7,4)$, find respectively TLS and LS linear fittings and analyze their squared sum of distances.

6.8 Prove that the solution of the optimization problem

$$\min_{\mathbf{A}} \left\{ \text{tr}(\mathbf{A}^T \mathbf{A}) - 2\text{tr}(\mathbf{A}) \right\} \quad \text{subject to} \quad \mathbf{A}\mathbf{X} = \mathbf{O}$$

is given by $\hat{\mathbf{A}} = \mathbf{I} - \mathbf{X}\mathbf{X}^\dagger$, where $\mathbf{O}$ is a null matrix.

6.9 Consider the harmonic recovery problem in additive white noise

$$x(n) = \sum_{i=1}^{p} A_i \sin(2\pi f_i n + \phi_i) + e(n),$$

where A_i, f_i and ϕ_i are the amplitude, frequency and phase of the ith harmonic, and $e(n)$ is the additive white noise. The above harmonic process obeys the modified Yule–Walker (MYW) equation

$$R_x(k) + \sum_{i=1}^{2p} a_i R_x(k-i) = 0, \quad \forall\, k,$$

and the harmonic frequency can be recovered from

$$f_i = \frac{1}{2\pi} \arctan\left(\frac{\operatorname{Im} z_i}{\operatorname{Re} z_i}\right), \quad i = 1, 2, \ldots, p,$$

where z_i is a root in the conjugate root pair (z_i, z_i^*) of the characteristic polynomial $A(z) = 1 + \sum_{i=1}^{2p} a_i z^{-i}$. If

$$x(n) = \sqrt{20}\sin(2\pi 0.2n) + \sqrt{2}\sin(2\pi 0.213n) + e(n),$$

where $e(n)$ is a Gaussian white noise with mean 0 and variance 1, take $n = 1, 2, \ldots, 128$ and use the LS method and SVD-TLS algorithm to estimate the AR parameters a_i of the ARMA model and the harmonic frequencies f_1 and f_2. Assume that there are 40 MYW equations, and that $p = 2$ and $p = 3$ respectively in the LS method, while the number of unknown parameters is 14 in the TLS algorithm. When using the TLS algorithm, the number of harmonics is determined by the number of leading singular values and then the roots of the characteristic polynomial are computed. Compare the computer simulation results of the LS method and the SVD-TLS algorithm.

6.10 The TLS problem for solving the matrix equation $\mathbf{Ax} = \mathbf{b}$ can be equivalently represented as

$$\min_{\mathbf{b}+\Delta\mathbf{b} \in \text{Range}(\mathbf{A}+\Delta\mathbf{A})} \|\mathbf{C}[\Delta\mathbf{A}, \Delta\mathbf{b}]\mathbf{T}\|_F^2, \quad \Delta\mathbf{A} \in \mathbb{R}^{m\times n}, \quad \Delta\mathbf{b} \in \mathbb{R}^m,$$

where $\mathbf{C} = \mathbf{Diag}(c_1, \ldots, d_m)$ and $\mathbf{T} = \mathbf{Diag}(t_1, \ldots, t_{n+1})$. Show the following results.

(a) If $\text{rank}(\mathbf{A}) < n$ then the above TLS problem has a solution if and only if $\mathbf{b} \in \text{Range}(\mathbf{A})$.

(b) If $\text{rank}(\mathbf{A}) = n$, $\mathbf{A}^T\mathbf{Cb} = \mathbf{0}$ and $|t_{n+1}|\,\|\mathbf{Cb}\|_2 \geq \sigma_n(\mathbf{CAT}_1)$ with $\mathbf{T}_1 = \mathbf{Diag}(t_1, \ldots, t_n)$ then the TLS problem has no solutions, $\sigma_n(\mathbf{CAT}_1)$ is the nth singular value of the matrix $\mathbf{CAT}_1$.

6.11 Show the nonnegativity of the KL divergence

$$D_{\rm KL}(\mathbf{P}\|\mathbf{Q}) = \sum_{i=1}^{I}\sum_{j=1}^{K}\left(p_{ij}\log\frac{p_{ij}}{q_{ij}} - p_{ij} + q_{ij}\right)$$

and show that the KL divergence is equal to zero if and only if $\mathbf{P} = \mathbf{Q}$.

6.12 Letting $D_E(\mathbf{X}\|\mathbf{AS}) = \frac{1}{2}\|\mathbf{X} - \mathbf{AS}\|_2^2$, show that

$$\nabla D_E(\mathbf{X}\|\mathbf{AS}) = \frac{\partial D_E(\mathbf{X}\|\mathbf{AS})}{\partial \mathbf{A}} = -(\mathbf{X} - \mathbf{AS})\mathbf{S}^T,$$

$$\nabla D_E(\mathbf{X}\|\mathbf{AS}) = \frac{\partial D_E(\mathbf{X}\|\mathbf{AS})}{\partial \mathbf{S}} = -\mathbf{A}^T(\mathbf{X} - \mathbf{AS}).$$

7
Eigenanalysis

Focusing on the *eigenanalysis* of matrices, i.e. analyzing matrices by means of eigenvalues, in this chapter we first discuss the eigenvalue decomposition (EVD) of matrices and then present various generalizations of EVD: generalized eigenvalue decomposition, the Rayleigh quotient, the generalized Rayleigh quotient, quadratic eigenvalue problems and the joint diagonalization of multiple matrices. In order to facilitate readers' understanding, this chapter will present some typical applications.

7.1 Eigenvalue Problem and Characteristic Equation

The eigenvalue problem not only is a very interesting theoretical problem but also has a wide range of applications.

7.1.1 Eigenvalue Problem

If $\mathcal{L}[\mathbf{w}] = \mathbf{w}$ holds for any nonzero vector $\mathbf{w}$, then $\mathcal{L}$ is called an *identity transformation.* More generally, if, when a linear operator acts on a vector, the output is a multiple of this vector, then the linear operator is said to have an *input reproducing* characteristic. This input reproducing has two cases.

(1) For any nonzero input vector, the output of a linear operator is always completely the same as the input vector (e.g., it is the identity operator).
(2) For some special input vector only, the output vector of a linear operator is the same as the input vector up to a constant factor.

Definition 7.1 Suppose that a nonzero vector $\mathbf{u}$ is the input of the linear operator $\mathcal{L}$. If the output vector is the same as the input vector $\mathbf{u}$ up to a constant factor λ, i.e.,

$$\mathcal{L}[\mathbf{u}] = \lambda\mathbf{u}, \quad \mathbf{u} \neq \mathbf{0}, \tag{7.1.1}$$

then the vector $\mathbf{u}$ is known as an *eigenvector* of the linear operator $\mathcal{L}$ and the scalar λ is the corresponding *eigenvalue* of $\mathcal{L}$.

The most commonly used linear operator in engineering applications is undoubtedly a linear time-invariant system. From the above definition, it is known that if each eigenvector $\mathbf{u}$ is regarded as an input of the linear time-invariant system, then the eigenvalue λ associated with eigenvector $\mathbf{u}$ is equivalent to the gain of the linear system $\mathcal{L}$ with $\mathbf{u}$ as the input. Only when the input of the system is an eigenvector $\mathbf{u}$, its output is the same as the input up to a constant factor. Thus an eigenvector can be viewed as describing the characteristics of the system, and hence is also called a characteristic vector. This is a physical explanation of eigenvectors in relation to linear systems.

If a linear transformation $\mathbf{w} = \mathcal{L}[\mathbf{x}]$ can be represented as $\mathbf{w} = \mathbf{A}\mathbf{x}$, then $\mathbf{A}$ is called a standard matrix of the linear transformation. Clearly, if $\mathbf{A}$ is a standard matrix of a linear transformation, then its eigenvalue problem (7.1.1) can be written as

$$\mathbf{A}\mathbf{u} = \lambda\mathbf{u}, \quad \mathbf{u} \neq \mathbf{0}. \tag{7.1.2}$$

The scalar λ is known as an eigenvalue of the matrix $\mathbf{A}$, and the vector $\mathbf{u}$ as an eigenvector associated with the eigenvalue λ. Equation (7.1.2) is sometimes called the *eigenvalue–eigenvector equation.*

Example 7.1 Consider a linear time-invariant system $h(k)$ with transfer function $H(\mathrm{e}^{\mathrm{j}\omega}) = \sum_{k=-\infty}^{\infty} h(k)\mathrm{e}^{-\mathrm{j}\omega k}$. When inputting a complex exponential or harmonic signal $\mathrm{e}^{\mathrm{j}\omega n}$, the system output is given by

$$\mathcal{L}[\mathrm{e}^{\mathrm{j}\omega n}] = \sum_{k=-\infty}^{\infty} h(n-k)\mathrm{e}^{\mathrm{j}\omega k} = \sum_{k=-\infty}^{\infty} h(k)\mathrm{e}^{\mathrm{j}\omega(n-k)} = H(\mathrm{e}^{\mathrm{j}\omega})\mathrm{e}^{\mathrm{j}\omega n}.$$

Letting $\mathbf{u}(\omega) = [1, \mathrm{e}^{\mathrm{j}\omega}, \ldots, \mathrm{e}^{\mathrm{j}\omega(N-1)}]^T$ be an input vector, the system output is given by

$$\mathcal{L}\begin{bmatrix} 1 \\ \mathrm{e}^{\mathrm{j}\omega} \\ \vdots \\ \mathrm{e}^{\mathrm{j}\omega(N-1)} \end{bmatrix} = H(\mathrm{e}^{\mathrm{j}\omega})\begin{bmatrix} 1 \\ \mathrm{e}^{\mathrm{j}\omega} \\ \vdots \\ \mathrm{e}^{\mathrm{j}\omega(N-1)} \end{bmatrix} \quad \Rightarrow \quad \mathcal{L}[\mathbf{u}(\omega)] = H(\mathrm{e}^{\mathrm{j}\omega})\mathbf{u}(\omega).$$

This shows that the harmonic signal vector $\mathbf{u}(\omega) = [1, \mathrm{e}^{\mathrm{j}\omega}, \ldots, \mathrm{e}^{\mathrm{j}\omega(N-1)}]^T$ is an eigenvector of the linear time-invariant system, and the system transfer function $H(\mathrm{e}^{\mathrm{j}\omega})$ is the eigenvalue associated with $\mathbf{u}(\omega)$.

From Equation (7.1.2) it is easily seen that if $\mathbf{A} \in \mathbb{C}^{n\times n}$ is a Hermitian matrix then its eigenvalue λ must be a real number, and

$$\mathbf{A} = \mathbf{U}\boldsymbol{\Sigma}\mathbf{U}^H, \tag{7.1.3}$$

where $\mathbf{U} = [\mathbf{u}_1, \ldots, \mathbf{u}_n]^T$ is a unitary matrix, and $\boldsymbol{\Sigma} = \mathbf{Diag}(\lambda_1, \ldots, \lambda_n)$. Equation (7.1.3) is called the *eigenvalue decomposition* (EVD) of $\mathbf{A}$.

Since an eigenvalue λ and its associated eigenvector $\mathbf{u}$ often appear in pairs, the two-tuple $(\lambda, \mathbf{u})$ is called an *eigenpair* of the matrix $\mathbf{A}$. Although an eigenvalue can take a zero value, an eigenvector cannot be a zero vector.

Equation (7.1.2) means that the linear transformation $\mathbf{Au}$ does not "change the direction" of the input vector $\mathbf{u}$. Hence the linear transformation $\mathbf{Au}$ is a mapping "keeping the direction unchanged". In order to determine the vector $\mathbf{u}$, we can rewrite (7.1.2) as

$$(\mathbf{A} - \lambda\mathbf{I})\mathbf{u} = \mathbf{0}. \tag{7.1.4}$$

If the above equation is assumed to hold for certain nonzero vectors $\mathbf{u}$, then the only condition for a nonzero solution to Equation (7.1.4) to exist is that, for those vectors $\mathbf{u}$, the determinant of the matrix $\mathbf{A} - \lambda\mathbf{I}$ is equal to zero, namely

$$\det(\mathbf{A} - \lambda\mathbf{I}) = 0. \tag{7.1.5}$$

It should be pointed out that some eigenvalues λ may take the same value. The repetition number of an eigenvalue is said to be the *eigenvalue multiplicity*. For example, all n eigenvalues of an $n \times n$ identity matrix $\mathbf{I}$ are equal to 1, so its multiplicity is n.

It is easily concluded that if the eigenvalue problem (7.1.4) has a nonzero solution $\mathbf{u} \neq \mathbf{0}$, then the scalar λ must make the $n \times n$ matrix $\mathbf{A} - \lambda\mathbf{I}$ singular. Hence the eigenvalue problem solving consists of the following two steps:

(1) find all scalars λ (eigenvalues) such that the matrix $\mathbf{A} - \lambda\mathbf{I}$ is singular;
(2) given an eigenvalue λ such that the matrix $\mathbf{A} - \lambda\mathbf{I}$ is singular, find all nonzero vectors $\mathbf{u}$ satisfying $(\mathbf{A} - \lambda\mathbf{I})\mathbf{u} = \mathbf{0}$; these are the eigenvector(s) corresponding to the eigenvalue λ.

7.1.2 Characteristic Polynomial

As discussed above, the matrix $(\mathbf{A} - \lambda\mathbf{I})$ is singular if and only if its determinant $\det(\mathbf{A} - \lambda\mathbf{I}) = 0$, i.e.,

$$(\mathbf{A} - \lambda\mathbf{I}) \text{ singular} \quad \Leftrightarrow \quad \det(\mathbf{A} - \lambda\mathbf{I}) = 0, \tag{7.1.6}$$

where the matrix $\mathbf{A} - \lambda\mathbf{I}$ is called the *characteristic matrix* of $\mathbf{A}$.

The determinant

$$\begin{aligned} p(x) = \det(\mathbf{A} - x\mathbf{I}) &= \begin{vmatrix} a_{11} - x & a_{12} & \cdots & a_{1n} \\ a_{21} & a_{22} - x & \cdots & a_{2n} \\ \vdots & \vdots & \ddots & \vdots \\ a_{n1} & a_{n2} & \cdots & a_{nn} - x \end{vmatrix} \\ &= p_n x^n + p_{n-1}x^{n-1} + \cdots + p_1 x + p_0 \end{aligned} \tag{7.1.7}$$

is known as the *characteristic polynomial* of $\mathbf{A}$, and

$$p(x) = \det(\mathbf{A} - x\mathbf{I}) = 0 \tag{7.1.8}$$

is said to be the *characteristic equation* of $\mathbf{A}$.

The roots of the characteristic equation $\det(\mathbf{A} - x\mathbf{I}) = 0$ are known as the eigenvalues, characteristic values, latent values, the characteristic roots or latent roots.

Obviously, computing the n eigenvalues λ_i of an $n \times n$ matrix $\mathbf{A}$ and finding the n roots of the nth-order characteristic polynomial $p(x) = \det(\mathbf{A} - x\mathbf{I}) = 0$ are two equivalent problems. An $n \times n$ matrix $\mathbf{A}$ generates an nth-order characteristic polynomial. Likewise, each nth-order polynomial can also be written as the characteristic polynomial of an $n \times n$ matrix.

Theorem 7.1 [32] *Any polynomial*

$$p(\lambda) = \lambda^n + a_1\lambda^{n-1} + \cdots + a_{n-1}\lambda + a_n$$

can be written as the characteristic polynomial of the $n \times n$ matrix

$$\mathbf{A} = \begin{bmatrix} -a_1 & -a_2 & \cdots & -a_{n-1} & -a_n \\ -1 & 0 & \cdots & 0 & 0 \\ 0 & -1 & \cdots & 0 & 0 \\ \vdots & \vdots & \ddots & \vdots & \vdots \\ 0 & 0 & \cdots & -1 & 0 \end{bmatrix},$$

namely $p(\lambda) = \det(\lambda\mathbf{I} - \mathbf{A})$.

7.2 Eigenvalues and Eigenvectors

In this section we discuss the computation and properties of the eigenvalues and eigenvectors of an $n \times n$ matrix $\mathbf{A}$.

7.2.1 Eigenvalues

Even if an $n \times n$ matrix $\mathbf{A}$ is real, the n roots of its characteristic equation may be complex, and the root multiplicity can be arbitrary or even equal to n. These roots are collectively referred to as the eigenvalues of the matrix $\mathbf{A}$.

Regarding eigenvalues, it is necessary to introduce the following terminology [422, p. 15].

(1) An eigenvalue λ of a matrix $\mathbf{A}$ is said to have *algebraic multiplicity* μ, if λ occurs μ times as a root of the characteristic polynomial $\det(\mathbf{A} - z\mathbf{I}) = 0$.

(2) If the algebraic multiplicity of an eigenvalue λ is 1, then it is called a *single eigenvalue*. A nonsingle eigenvalue is said to be a *multiple eigenvalue*.

(3) An eigenvalue λ of $\mathbf{A}$ is said to have *geometric multiplicity* γ if the number of linearly independent eigenvectors corresponding to the eigenvalue is γ. In other words, the geometric multiplicity γ is the dimension of the eigenspace $\text{Null}(\mathbf{A} - \lambda\mathbf{I})$.

(4) A matrix $\mathbf{A}$ is known as a *derogatory matrix* if there is at least one of its eigenvalues with geometric multiplicity greater than 1.

(5) An eigenvalue is referred to as *semi-single*, if its algebraic multiplicity is equal to its geometric multiplicity. A semi-single eigenvalue is also called a *defective eigenvalue.*

It is well known that any nth-order polynomial $p(x)$ can be written in the factorized form

$$p(x) = a(x - x_1)(x - x_2) \cdots (x - x_n). \tag{7.2.1}$$

The n roots of the characteristic polynomial $p(x)$, denoted $x_1, x_2, \ldots, x_n$, are not necessarily different from each other, and also are not necessarily real.

In general the eigenvalues of a matrix $\mathbf{A}$ are different from each other, but a characteristic polynomial has multiple roots then we say that the matrix $\mathbf{A}$ has *degenerate eigenvalues.*

It has already been noted that even if $\mathbf{A}$ is a real matrix, its eigenvalues may be complex. Taking the Givens rotation matrix

$$\mathbf{A} = \begin{bmatrix} \cos\theta & -\sin\theta \\ \sin\theta & \cos\theta \end{bmatrix}$$

as an example, its characteristic equation is

$$\det(\mathbf{A} - \lambda\mathbf{I}) = \begin{vmatrix} \cos\theta - \lambda & -\sin\theta \\ \sin\theta & \cos\theta - \lambda \end{vmatrix} = (\cos\theta - \lambda)^2 + \sin^2\theta = 0.$$

However, if θ is not an integer multiple of π then $\sin^2\theta > 0$. In this case, the characteristic equation cannot give a real value for λ, i.e., the two eigenvalues of the rotation matrix are complex, and the two corresponding eigenvectors are complex vectors.

The eigenvalues of an $n \times n$ matrix (not necessarily Hermitian) $\mathbf{A}$ have the following properties [214].

1. An $n \times n$ matrix $\mathbf{A}$ has a total of n eigenvalues, where multiple eigenvalues count according to their multiplicity.
2. If a nonsymmetric real matrix $\mathbf{A}$ has complex eigenvalues and/or complex eigenvectors, then they must appear in the form of a complex conjugate pair.
3. If $\mathbf{A}$ is a real symmetric matrix or a Hermitian matrix then its all eigenvalues are real numbers.
4. The eigenvalues of a diagonal matrix and a triangular matrix satisfy the following:

- if $\mathbf{A} = \mathbf{Diag}(a_{11}, \ldots, a_{nn})$ then its eigenvalues are given by $a_{11}, \ldots, a_{nn}$;
- if $\mathbf{A}$ is a triangular matrix then all its diagonal entries are eigenvalues.

5. Given an $n \times n$ matrix $\mathbf{A}$:
   - if λ is an eigenvalue of $\mathbf{A}$ then λ is also an eigenvalue of $\mathbf{A}^T$;
   - if λ is an eigenvalue of $\mathbf{A}$ then λ^* is an eigenvalue of $\mathbf{A}^H$;
   - if λ is an eigenvalue of $\mathbf{A}$ then $\lambda + \sigma^2$ is an eigenvalue of $\mathbf{A} + \sigma^2\mathbf{I}$;
   - if λ is an eigenvalue of $\mathbf{A}$ then $1/\lambda$ is an eigenvalue of its inverse matrix $\mathbf{A}^{-1}$.
6. All eigenvalues of an idempotent matrix $\mathbf{A}^2 = \mathbf{A}$ are 0 or 1.
7. If $\mathbf{A}$ is a real orthogonal matrix, then all its eigenvalues are located on the unit circle.
8. The relationship between eigenvalues and the matrix singularity is as follows:
   - if $\mathbf{A}$ is singular then it has at least one zero eigenvalue;
   - if $\mathbf{A}$ is nonsingular then its all eigenvalues are nonzero.
9. The relationship between eigenvalues and trace: the sum of all eigenvalues of $\mathbf{A}$ is equal to its trace, namely $\sum_{i=1}^{n} \lambda_i = \text{tr}(\mathbf{A})$.
10. A Hermitian matrix $\mathbf{A}$ is positive definite (semi-definite) if and only if its all eigenvalues are positive (nonnegative).
11. The relationship between eigenvalues and determinant: the product of all eigenvalues of $\mathbf{A}$ is equal to its determinant, namely $\prod_{i=1}^{n} \lambda_i = \det(\mathbf{A}) = |\mathbf{A}|$.
12. If the eigenvalues of $\mathbf{A}$ are different from each other then there must be a similar matrix $\mathbf{S}$ (see Subsection 7.3.1) such that $\mathbf{S}^{-1}\mathbf{AS} = \mathbf{D}$ (diagonal matrix); the diagonal entries of $\mathbf{D}$ are the eigenvalues of $\mathbf{A}$.
13. The relationship between eigenvalues and rank is as follows:
    - if an $n \times n$ matrix $\mathbf{A}$ has r nonzero eigenvalues then $\text{rank}(\mathbf{A}) \geq r$;
    - if 0 is a single eigenvalue of an $n \times n$ matrix $\mathbf{A}$ then $\text{rank}(\mathbf{A}) = n - 1$;
    - if $\text{rank}(\mathbf{A} - \lambda\mathbf{I}) \leq n - 1$ then λ is an eigenvalue of the matrix $\mathbf{A}$.
14. The geometric multiplicity of any eigenvalue λ of an $n \times n$ matrix $\mathbf{A}$ cannot be greater than the algebraic multiplicity of λ.
15. The Cayley–Hamilton theorem: if $\lambda_1, \lambda_2, \ldots, \lambda_n$ are the eigenvalues of an $n \times n$ matrix $\mathbf{A}$ then
$$\prod_{i=1}^{n} (\mathbf{A} - \lambda_i \mathbf{I}) = 0.$$
16. On the eigenvalues of similar matrices:
    - if λ is an eigenvalue of an $n \times n$ matrix $\mathbf{A}$ and another $n \times n$ matrix $\mathbf{B}$ is nonsingular then λ is also an eigenvalue of the matrix $\mathbf{B}^{-1}\mathbf{AB}$;
    - if λ is an eigenvalue of an $n \times n$ matrix $\mathbf{A}$ and another $n \times n$ matrix $\mathbf{B}$ is unitary then λ is also an eigenvalue of the matrix $\mathbf{B}^H\mathbf{AB}$;
    - if λ is an eigenvalue of an $n \times n$ matrix $\mathbf{A}$ and another $n \times n$ matrix $\mathbf{B}$ is orthogonal then λ is also an eigenvalue of $\mathbf{B}^T\mathbf{AB}$.

17. The eigenvalues λ_i of the correlation matrix $\mathbf{R} = E\{\mathbf{x}(t)\mathbf{x}^H(t)\}$ of a random vector $\mathbf{x}(t) = [x_1(t), \ldots, x_n(t)]^T$ are bounded by the maximum power $P_{\max} = \max_i E\{|x_i(t)|^2\}$ and the minimum power $P_{\min} = \min_i E\{|x_i(t)|^2\}$ of the signal components, namely
$$P_{\min} \leq \lambda_i \leq P_{\max}. \tag{7.2.2}$$
18. The eigenvalue spread of the correlation matrix $\mathbf{R}$ of a random vector $\mathbf{x}(t)$ is
$$\mathcal{X}(\mathbf{R}) = \frac{\lambda_{\max}}{\lambda_{\min}}. \tag{7.2.3}$$
19. The matrix products $\mathbf{A}_{m\times n}\mathbf{B}_{n\times m}$ and $\mathbf{B}_{n\times m}\mathbf{A}_{m\times n}$ have the same nonzero eigenvalues.
20. If an eigenvalue of a matrix $\mathbf{A}$ is λ then the corresponding eigenvalue of the matrix polynomial $f(\mathbf{A}) = \mathbf{A}^n + c_1\mathbf{A}^{n-1} + \cdots + c_{n-1}\mathbf{A} + c_n\mathbf{I}$ is given by
$$f(\lambda) = \lambda^n + c_1\lambda^{n-1} + \cdots + c_{n-1}\lambda + c_n. \tag{7.2.4}$$
21. If λ is an eigenvalue of a matrix $\mathbf{A}$ then e^λ is an eigenvalue of the matrix exponential function $e^{\mathbf{A}}$.

7.2.2 Eigenvectors

If a matrix $\mathbf{A}_{n\times n}$ is a complex matrix, and λ is one of its eigenvalues then the vector $\mathbf{v}$ satisfying
$$(\mathbf{A} - \lambda\mathbf{I})\mathbf{v} = \mathbf{0} \quad \text{or} \quad \mathbf{A}\mathbf{v} = \lambda\mathbf{v} \tag{7.2.5}$$
is called the *right eigenvector* of the matrix $\mathbf{A}$ associated with the eigenvalue λ, while the vector $\mathbf{u}$ satisfying
$$\mathbf{u}^H(\mathbf{A} - \lambda\mathbf{I}) = \mathbf{0}^T \quad \text{or} \quad \mathbf{u}^H\mathbf{A} = \lambda\mathbf{u}^H \tag{7.2.6}$$
is known as the *left eigenvector* of $\mathbf{A}$ associated with the eigenvalue λ.

If a matrix $\mathbf{A}$ is Hermitian then its all eigenvalues are real, and hence from Equation (7.2.5) it is immediately known that $((\mathbf{A} - \lambda\mathbf{I})\mathbf{v})^T = \mathbf{v}^T(\mathbf{A} - \lambda\mathbf{I}) = \mathbf{0}^T$, yielding $\mathbf{v} = \mathbf{u}$; namely, the right and left eigenvectors of any Hermitian matrix are the same.

It is useful to compare the similarities and differences of the SVD and the EVD of a matrix.

(1) The SVD is available for any $m \times n$ (where $m \geq n$ or $m < n$) matrix, while the EVD is available only for square matrices.

(2) For an $n \times n$ non-Hermitian matrix $\mathbf{A}$, its kth singular value is defined as the spectral norm of the error matrix $\mathbf{E}_k$ making the rank of the original matrix $\mathbf{A}$ decreased by 1:
$$\sigma_k = \min_{\mathbf{E}\in\mathbb{C}^{m\times n}} \left\{\|\mathbf{E}\|_{\text{spec}} : \text{rank}(\mathbf{A} + \mathbf{E}) \leq k - 1\right\}, \quad k = 1, \ldots, \min\{m, n\}, \tag{7.2.7}$$

while its eigenvalues are defined as the roots of the characteristic polynomial $\det(\mathbf{A} - \lambda\mathbf{I}) = 0$. There is no inherent relationship between the singular values and the eigenvalues of the same square matrix, but each nonzero singular value of an $m \times n$ matrix $\mathbf{A}$ is the positive square root of some nonzero eigenvalue of the $n \times n$ Hermitian matrix $\mathbf{A}^H\mathbf{A}$ or the $m \times m$ Hermitian matrix $\mathbf{A}\mathbf{A}^H$.

(3) The left singular vector $\mathbf{u}_i$ and the right singular vector $\mathbf{v}_i$ of an $m \times n$ matrix $\mathbf{A}$ associated with the singular value σ_i are defined as the two vectors satisfying $\mathbf{u}_i^H\mathbf{A}\mathbf{v}_i = \sigma_i$, while the left and right eigenvectors are defined by $\mathbf{u}^H\mathbf{A} = \lambda_i\mathbf{u}^H$ and $\mathbf{A}\mathbf{v}_i = \lambda_i\mathbf{v}_i$, respectively. Hence, for the same $n \times n$ non-Hermitian matrix $\mathbf{A}$, there is no inherent relationship between its (left and right) singular vectors and its (left and right) eigenvectors. However, the left singular vector $\mathbf{u}_i$ and right singular vector $\mathbf{v}_i$ of a matrix $\mathbf{A} \in \mathbb{C}^{m\times n}$ are respectively the eigenvectors of the $m \times m$ Hermitian matrix $\mathbf{A}\mathbf{A}^H$ and of the $n \times n$ matrix $\mathbf{A}^H\mathbf{A}$.

From Equation (7.1.2) it is easily seen that after multiplying an eigenvector $\mathbf{u}$ of a matrix $\mathbf{A}$ by any nonzero scalar μ, then $\mu\mathbf{u}$ is still an eigenvector of $\mathbf{A}$. For convenience, it is generally assumed that eigenvectors have unit norm, i.e., $\|\mathbf{u}\|_2 = 1$.

Using eigenvectors we can introduce the condition number of any single eigenvalue.

Definition 7.2 [422, p. 93] The condition number of a single eigenvalue λ of any matrix $\mathbf{A}$ is defined as

$$\text{cond}(\lambda) = \frac{1}{\cos\theta(\mathbf{u},\mathbf{v})}, \tag{7.2.8}$$

where $\theta(\mathbf{u},\mathbf{v})$ represents the acute angle between the left and right eigenvectors associated with the eigenvalue λ.

Definition 7.3 The set of all eigenvalues $\lambda \in \mathbb{C}$ of a matrix $\mathbf{A} \in \mathbb{C}^{n\times n}$ is called the *spectrum* of the matrix $\mathbf{A}$, denoted $\lambda(\mathbf{A})$. The *spectral radius* of a matrix $\mathbf{A}$, denoted $\rho(\mathbf{A})$, is a nonnegative real number and is defined as

$$\rho(\mathbf{A}) = \max|\lambda| : \lambda \in \lambda(\mathbf{A}). \tag{7.2.9}$$

Definition 7.4 The *inertia* of a symmetric matrix $\mathbf{A} \in \mathbb{R}^{n\times n}$, denoted $\text{In}(\mathbf{A})$, is defined as the triplet

$$\text{In}(\mathbf{A}) = (i_+(\mathbf{A}), i_-(\mathbf{A}), i_0(\mathbf{A})),$$

where $i_+(\mathbf{A}), i_-(\mathbf{A})$ and $i_0(\mathbf{A})$ are respectively the numbers of the positive, negative and zero eigenvalues of $\mathbf{A}$ (each multiple eigenvalue is counted according to its multiplicity). Moreover, the quality $i_+(\mathbf{A}) - i_-(\mathbf{A})$ is known as the *signature* of $\mathbf{A}$.

The following summarize the properties of the eigenpair $(\lambda, \mathbf{u})$ [214].

1. If $(\lambda, \mathbf{u})$ is an eigenpair of a matrix $\mathbf{A}$ then $(c\lambda, \mathbf{u})$ is an eigenpair of the matrix $c\mathbf{A}$, where c is a nonzero constant.

2. If $(\lambda, \mathbf{u})$ is an eigenpair of a matrix $\mathbf{A}$ then $(\lambda, c\mathbf{u})$ is also an eigenpair of the matrix $\mathbf{A}$, where c is a nonzero constant.
3. If $(\lambda_i, \mathbf{u}_i)$ and $(\lambda_j, \mathbf{u}_j)$ are two eigenpairs of a matrix $\mathbf{A}$, and $\lambda_i \neq \lambda_j$, then the eigenvector $\mathbf{u}_i$ is linearly independent of $\mathbf{u}_j$.
4. The eigenvectors of a Hermitian matrix associated with different eigenvalues are orthogonal to each other, namely $\mathbf{u}_i^H \mathbf{u}_j = 0$ for $\lambda_i \neq \lambda_j$.
5. If $(\lambda, \mathbf{u})$ is an eigenpair of a matrix $\mathbf{A}$ then $(\lambda^k, \mathbf{u})$ is the eigenpair of the matrix $\mathbf{A}^k$.
6. If $(\lambda, \mathbf{u})$ is an eigenpair of a matrix $\mathbf{A}$ then $(\mathrm{e}^\lambda, \mathbf{u})$ is an eigenpair of the matrix exponential function $\mathrm{e}^{\mathbf{A}}$.
7. If $\lambda(\mathbf{A})$ and $\lambda(\mathbf{B})$ are respectively the eigenvalues of the matrices $\mathbf{A}$ and $\mathbf{B}$, and $\mathbf{u}(\mathbf{A})$ and $\mathbf{u}(\mathbf{B})$ are respectively the eigenvectors associated with $\lambda(\mathbf{A})$ and $\lambda(\mathbf{B})$, then $\lambda(\mathbf{A})\lambda(\mathbf{B})$ is the eigenpair of the matrix Kronecker product $\mathbf{A} \otimes \mathbf{B}$, and $\mathbf{u}(\mathbf{A}) \otimes \mathbf{u}(\mathbf{B})$ is the eigenvector associated with the eigenvalue $\lambda(\mathbf{A})\lambda(\mathbf{B})$.

The SVD of an $m \times n$ matrix $\mathbf{A}$ can be transformed to the EVD of the corresponding matrix. There are two main methods to achieve this transformation.

Method 1

The nonzero singular value σ_i of a matrix $\mathbf{A}_{m\times n}$ is the positive square root of the nonzero eigenvalue λ_i of the $m \times m$ matrix $\mathbf{A}\mathbf{A}^T$ or the $n \times n$ matrix $\mathbf{A}^T\mathbf{A}$, and the left singular vector $\mathbf{u}_i$ and the right singular vector $\mathbf{v}_i$ of $\mathbf{A}$ associated with σ_i are the eigenvectors of $\mathbf{A}\mathbf{A}^T$ and $\mathbf{A}^T\mathbf{A}$ associated with the nonzero eigenvalue λ_i.

Method 2

The SVD of a matrix $\mathbf{A}_{m\times n}$ is transformed to the EVD of an $(m+n) \times (m+n)$ augmented matrix

$$\begin{bmatrix} \mathbf{O} & \mathbf{A} \\ \mathbf{A}^T & \mathbf{O} \end{bmatrix}. \tag{7.2.10}$$

Theorem 7.2 (Jordan–Wielandt theorem) [455, Theorem I.4.2] *If $\sigma_1 \geq \sigma_2 \geq \cdots \geq \sigma_{p-1} \geq \sigma_p$ are the singular values of $\mathbf{A}_{m\times n}$ (where $p = \min\{m, n\}$) then the augmented matrix in (7.2.10) has the eigenvalues*

$$-\sigma_1, \ldots, -\sigma_p, \underbrace{0, \ldots, 0}_{|m-n|}, \sigma_p, \ldots, \sigma_1,$$

and the eigenvectors associated with $\pm\sigma_j$ are given by

$$\begin{bmatrix} \mathbf{u}_j \\ \pm\mathbf{v}_j \end{bmatrix}, \quad j = 1, 2, \ldots, p.$$

If $m \neq n$ then we have

$$\begin{bmatrix} \mathbf{u}_j \\ \mathbf{0} \end{bmatrix}, \; n+1 \leq j \leq m, \; m > n \quad \text{or} \quad \begin{bmatrix} \mathbf{0} \\ \mathbf{v}_j \end{bmatrix}, \; m+1 \leq j \leq n, \; m < n.$$

On the eigenvalues of the matrix sum $\mathbf{A}+\mathbf{B}$, one has the following result.

Theorem 7.3 (Weyl theorem) [275] *Let $\mathbf{A},\mathbf{B}\in\mathbb{C}^{n\times n}$ be Hermitian matrices, and let their eigenvalues be arranged in ascending order:*

$$\begin{aligned}\lambda_1(\mathbf{A})&\leq\lambda_2(\mathbf{A})\leq\cdots\leq\lambda_n(\mathbf{A}),\\ \lambda_1(\mathbf{B})&\leq\lambda_2(\mathbf{B})\leq\cdots\leq\lambda_n(\mathbf{B}),\\ \lambda_1(\mathbf{A}+\mathbf{B})&\leq\lambda_2(\mathbf{A}+\mathbf{B})\leq\cdots\leq\lambda_n(\mathbf{A}+\mathbf{B}).\end{aligned}$$

Then

$$\lambda_i(\mathbf{A}+\mathbf{B})\geq\begin{cases}\lambda_i(\mathbf{A})+\lambda_1(\mathbf{B}),\\ \lambda_{i-1}(\mathbf{A})+\lambda_2(\mathbf{B}),\\ \quad\vdots\\ \lambda_1(\mathbf{A})+\lambda_i(\mathbf{B}),\end{cases}\tag{7.2.11}$$

$$\lambda_i(\mathbf{A}+\mathbf{B})\leq\begin{cases}\lambda_i(\mathbf{A})+\lambda_n(\mathbf{B}),\\ \lambda_{i+1}(\mathbf{A})+\lambda_{n-1}(\mathbf{B}),\\ \quad\vdots\\ \lambda_n(\mathbf{A})+\lambda_i(\mathbf{B}),\end{cases}\tag{7.2.12}$$

where $i=1,2,\ldots,n$.

In particular, when $\mathbf{A}$ is a real symmetric matrix and $\mathbf{B}=\alpha\mathbf{z}\mathbf{z}^T$ (see below), then one has the following *interlacing eigenvalue theorem* [179, Theorem 8.1.8].

Theorem 7.4 *Let $\mathbf{A}\in\mathbb{R}^{n\times n}$ be a symmetric matrix with eigenvalues $\lambda_1,\ldots,\lambda_n$ satisfying $\lambda_1\geq\lambda_2\geq\cdots\geq\lambda_n$, and let $\mathbf{z}\in\mathbb{R}^n$ be a vector with norm $\|\mathbf{z}\|=1$. Suppose that α is a real number, and the eigenvalues of the matrix $\mathbf{A}+\alpha\mathbf{z}\mathbf{z}^T$ are arranged as $\xi_1\geq\xi_2\geq\cdots\geq\xi_n$. Then one has*

$$\xi_1\geq\lambda_1\geq\xi_2\geq\lambda_2\geq\cdots\geq\xi_n\geq\lambda_n\quad(\text{for }\alpha>0)\tag{7.2.13}$$

or

$$\lambda_1\geq\xi_1\geq\lambda_2\geq\xi_2\geq\cdots\geq\lambda_n\geq\xi_n\quad(\text{for }\alpha<0),\tag{7.2.14}$$

and, whether $\alpha>0$ or $\alpha<0$, the following result is true:

$$\sum_{i=1}^n(\xi_i-\lambda_i)=\alpha.\tag{7.2.15}$$

7.3 Similarity Reduction

The normalized representation of a matrix is called the *canonical form* (or normal or standard form) of the matrix. In most fields, a canonical form specifies a unique representation while a normal form simply specifies the form without the requirement of uniqueness. The simplest canonical form of a matrix is its diagonalized representation. However, many matrices are not diagonalizable. The problem is that in many applications it is necessary to reduce a given matrix to as simple a form as possible. In these applications the similarity reduction of a matrix is a very natural choice.

The standard form of the similarity reduction is the Jordan canonical form. Therefore, the core problem of matrix similarity reduction is how to obtain the Jordan canonical form.

There are two different ways to find the Jordan canonical form of a matrix.

(1) For a constant matrix, use a direct similarity reduction of the matrix itself.
(2) First make a balanced reduction of the polynomial matrix corresponding to the original constant matrix, and then transform the Smith normal form of the balanced reduction into the Jordan canonical form of the similarity reduction of the original matrix.

In this section we discuss the implementation of the first way, and in the next section we describe similarity reduction based on the second way.

7.3.1 Similarity Transformation of Matrices

The theoretical basis and mathematical tools of matrix similarity reduction are provided by the similarity transformation of matrices.

Let $\mathbf{P} \in \mathbb{C}^{m \times m}$ be a nonsingular matrix, and use it to make the linear transformation of a matrix $\mathbf{A} \in \mathbb{C}^{m \times m}$:

$$\mathbf{B} = \mathbf{P}^{-1}\mathbf{A}\mathbf{P}. \tag{7.3.1}$$

Suppose that an eigenvalue of the linear transformation $\mathbf{B}$ is λ and that a corresponding eigenvector is $\mathbf{y}$, i.e.,

$$\mathbf{B}\mathbf{y} = \lambda\mathbf{y}. \tag{7.3.2}$$

Substitute (7.3.1) into (7.3.2) to get $\mathbf{P}^{-1}\mathbf{A}\mathbf{P}\mathbf{y} = \lambda\mathbf{y}$ or $\mathbf{A}(\mathbf{P}\mathbf{y}) = \lambda(\mathbf{P}\mathbf{y})$. Letting $\mathbf{x} = \mathbf{P}\mathbf{y}$ or $\mathbf{y} = \mathbf{P}^{-1}\mathbf{x}$, we have

$$\mathbf{A}\mathbf{x} = \lambda\mathbf{x}. \tag{7.3.3}$$

Comparing (7.3.2) with (7.3.3), it can be seen that the two matrices $\mathbf{A}$ and $\mathbf{B} = \mathbf{P}^{-1}\mathbf{A}\mathbf{P}$ have the same eigenvalues and the eigenvector $\mathbf{y}$ is a linear transformation of the eigenvector $\mathbf{x}$ of the matrix $\mathbf{A}$, i.e., $\mathbf{y} = \mathbf{P}^{-1}\mathbf{x}$. Because the eigenvalues of the two matrices $\mathbf{A}$ and $\mathbf{B} = \mathbf{P}^{-1}\mathbf{A}\mathbf{P}$ are the same and their eigenvectors have a linear transformation relationship, the matrices $\mathbf{A}$ and $\mathbf{B}$ are said to be similar.

Definition 7.5 If there is a nonsingular matrix $\mathbf{P} \in \mathbb{C}^{m \times m}$ such that $\mathbf{B} = \mathbf{P}^{-1}\mathbf{A}\mathbf{P}$ then the matrix $\mathbf{B} \in \mathbb{C}^{m \times m}$ is said to be *similar* to the matrix $\mathbf{A} \in \mathbb{C}^{m \times m}$ and $\mathbf{P}$ is known as the *similarity transformation matrix*.

"$\mathbf{B}$ is similar to $\mathbf{A}$" is denoted as $\mathbf{B} \sim \mathbf{A}$.

Similar matrices have the following basic properties.

(1) *Reflexivity* $\mathbf{A} \sim \mathbf{A}$, i.e., any matrix is similar to itself.
(2) *Symmetry* If $\mathbf{A}$ is similar to $\mathbf{B}$, then $\mathbf{B}$ is also similar to $\mathbf{A}$.
(3) *Transitivity* If $\mathbf{A}$ is similar to $\mathbf{B}$, and $\mathbf{B}$ is similar to $\mathbf{C}$, then $\mathbf{A}$ is similar to $\mathbf{C}$, i.e., $\mathbf{A} \sim \mathbf{C}$.

In addition, similar matrices have also the following important properties.

1. Similar matrices $\mathbf{B} \sim \mathbf{A}$ have the same determinant, i.e., $|\mathbf{B}| = |\mathbf{A}|$.
2. If the matrix $\mathbf{P}^{-1}\mathbf{A}\mathbf{P} = \mathbf{T}$ (an upper triangle matrix) then the diagonal entries of $\mathbf{T}$ give the eigenvalues λ_i of $\mathbf{A}$.
3. Two similar matrices have exactly the same eigenvalues.
4. For the similar matrix $\mathbf{B} = \mathbf{P}^{-1}\mathbf{A}\mathbf{P}$ we have $\mathbf{B}^2 = \mathbf{P}^{-1}\mathbf{A}\mathbf{P}\mathbf{P}^{-1}\mathbf{A}\mathbf{P} = \mathbf{P}^{-1}\mathbf{A}^2\mathbf{P}$, and thus $\mathbf{B}^k = \mathbf{P}^{-1}\mathbf{A}^k\mathbf{P}$. This is to say, if $\mathbf{B} \sim \mathbf{A}$ then $\mathbf{B}^k \sim \mathbf{A}^k$. This property is called the *power property* of similar matrices.
5. If both $\mathbf{B} = \mathbf{P}^{-1}\mathbf{A}\mathbf{P}$ and $\mathbf{A}$ are invertible then $\mathbf{B}^{-1} = \mathbf{P}^{-1}\mathbf{A}^{-1}\mathbf{P}$, i.e., when two matrices are similar, their inverse matrices are similar as well.

If $\mathbf{P}$ is a unitary matrix then $\mathbf{B} = \mathbf{P}^{-1}\mathbf{A}\mathbf{P}$ is called the *unitary similarity transformation* of $\mathbf{A}$.

The following example shows how to perform the diagonalization of an $m \times m$ matrix $\mathbf{A}$.

Example 7.2 Find the similarity transformation of the 3×3 real matrix

$$\mathbf{A} = \begin{bmatrix} 1 & 1 & 1 \\ 0 & 3 & 3 \\ -2 & 1 & 1 \end{bmatrix}.$$

Solution Direct computation gives the characteristic polynomial

$$|\lambda\mathbf{I} - \mathbf{A}| = \begin{vmatrix} \lambda - 1 & -1 & -1 \\ 0 & \lambda - 3 & -3 \\ 2 & -1 & \lambda - 1 \end{vmatrix} = \lambda(\lambda - 2)(\lambda - 3).$$

On solving the characteristic equation $|\lambda\mathbf{I} - \mathbf{A}| = 0$, the three eigenvalues of $\mathbf{A}$ are given by $\lambda = 0, 2, 3$.

(a) For the eigenvalue $\lambda = 0$, we have $(0\mathbf{I} - \mathbf{A})\mathbf{x} = \mathbf{0}$, i.e.,

$$\begin{aligned} x_1 + x_2 + x_3 &= 0, \\ 3x_2 + 3x_3 &= 0, \\ 2x_1 - x_2 - x_3 &= 0, \end{aligned}$$

with solution $x_1 = 0$ and $x_2 = -x_3$, where x_3 is arbitrary. Hence, the eigenvector associated with the eigenvalue $\lambda = 0$ is

$$\mathbf{x} = \begin{bmatrix} 0 \\ -a \\ a \end{bmatrix} = a \begin{bmatrix} 0 \\ -1 \\ 1 \end{bmatrix}, \quad a \neq 0.$$

Taking $a = 1$ yields the eigenvector $\mathbf{x}_1 = [0, \ -1, \ 1]^T$.

(b) For $\lambda = 2$, the characteristic polynomial is given by $(2\mathbf{I} - \mathbf{A})\mathbf{x} = \mathbf{0}$, i.e.,

$$\begin{aligned} x_1 - x_2 - x_3 &= 0, \\ x_2 + 3x_3 &= 0, \\ 2x_1 - x_2 + x_3 &= 0, \end{aligned}$$

whose solution is $x_1 = -2x_3$, $x_2 = -3x_3$, where x_3 is arbitrary. Thus we get the eigenvector

$$\mathbf{x} = \begin{bmatrix} -2a \\ -3a \\ a \end{bmatrix} = \begin{bmatrix} -2 \\ -3 \\ 1 \end{bmatrix}, \quad a = 1.$$

(c) Similarly, the eigenvector associated with the eigenvalue $\lambda = 3$ is $\mathbf{x}_3 = [1, \ 2, \ 0]^T$. The three eigenvectors constitute the similarity transformation matrix, which is given, along with its inverse, by

$$\mathbf{P} = \begin{bmatrix} 0 & -2 & 1 \\ -1 & -3 & 2 \\ 1 & 1 & 0 \end{bmatrix}, \quad \mathbf{P}^{-1} = \begin{bmatrix} 1 & -0.5 & 0.5 \\ -1 & 0.5 & 0.5 \\ -1 & 1.0 & 1.0 \end{bmatrix}.$$

Therefore, the diagonalized matrix corresponding to $\mathbf{A}$ is given by

$$\begin{aligned} \mathbf{P}^{-1}\mathbf{A}\mathbf{P} &= \begin{bmatrix} 1 & -0.5 & 0.5 \\ -1 & 0.5 & 0.5 \\ -1 & 1.0 & 1.0 \end{bmatrix} \begin{bmatrix} 1 & 1 & 1 \\ 0 & 3 & 3 \\ -2 & 1 & 1 \end{bmatrix} \begin{bmatrix} 0 & -2 & 1 \\ -1 & -3 & 2 \\ 1 & 1 & 0 \end{bmatrix} \\ &= \begin{bmatrix} 1 & -0.5 & 0.5 \\ -1 & 0.5 & 0.5 \\ -1 & 1.0 & 1.0 \end{bmatrix} \begin{bmatrix} 0 & -4 & 3 \\ 0 & -6 & 6 \\ 0 & 2 & 0 \end{bmatrix} \\ &= \begin{bmatrix} 0 & 0 & 0 \\ 0 & 2 & 0 \\ 0 & 0 & 3 \end{bmatrix}, \end{aligned}$$

which is the diagonal matrix consisting of the three different eigenvalues of the matrix $\mathbf{A}$.

Definition 7.6 An $m \times m$ real matrix $\mathbf{A}$ is said to be a *diagonalization matrix* if it is similar to a diagonal matrix.

The following theorem gives the necessary and sufficient condition for the diagonalization of the matrix $\mathbf{A} \in \mathbb{C}^{m \times m}$.

Theorem 7.5 *An $m \times m$ matrix $\mathbf{A}$ is diagonalizable if and only if $\mathbf{A}$ has m linearly independent eigenvectors.*

The following theorem gives the necessary and sufficient condition for all eigenvectors to be linearly independent, and thus it is the necessary and sufficient condition for matrix diagonalization.

Theorem 7.6 (Diagonalization theorem) [433, p. 307] *Given a matrix $\mathbf{A} \in \mathbb{C}^{m \times m}$ whose eigenvalues λ_k have the algebraic multiplicity d_k, $k = 1, \ldots, p$, where $\sum_{k=1}^{p} d_k = m$. The matrix $\mathbf{A}$ has m linearly independent eigenvectors if and only if* $\text{rank}(\mathbf{A} - \lambda_k \mathbf{I}) = m - d_k$, $k = 1, \ldots, p$. *In this case, the matrix $\mathbf{U}$ in $\mathbf{AU} = \mathbf{U\Sigma}$ is nonsingular, and $\mathbf{A}$ is diagonalized as $\mathbf{U}^{-1}\mathbf{AU} = \mathbf{\Sigma}$.*

7.3.2 Similarity Reduction of Matrices

A nondiagonalizable matrix $\mathbf{A} \in \mathbb{C}^{m \times m}$ with multiple eigenvalues can be reduced to the Jordan canonical form via a similarity transformation.

Definition 7.7 Let a matrix $\mathbf{A}$ with $r = \text{rank}(\mathbf{A})$ have d different nonzero eigenvalues, and let the nonzero eigenvalue λ_i have multiplicity m_i, i.e., $m_1 + \cdots + m_d = r$. If the similarity transformation has the form

$$\mathbf{P}^{-1}\mathbf{AP} = \mathbf{J} = \mathbf{Diag}(\mathbf{J}_1, \ldots, \mathbf{J}_d, \underbrace{\mathbf{0}, \ldots, \mathbf{0}}_{m-r}), \tag{7.3.4}$$

then $\mathbf{J}$ is called the *Jordan canonical form* of the matrix $\mathbf{A}$ and $\mathbf{J}_i, i = 1, \ldots, d$, are known as *Jordan blocks*.

A Jordan block corresponding to an eigenvalue λ_i with multiplicity 1 is of first order and is a 1×1 Jordan block matrix $\mathbf{J}_{1\times 1} = \lambda_i$. A kth-order Jordan block is defined as

$$\mathbf{J}_{k\times k} = \begin{bmatrix} \lambda & 1 & & 0 \\ & \lambda & \ddots & \\ & & \ddots & 1 \\ 0 & & & \lambda \end{bmatrix} \in \mathbb{C}^{k\times k}, \tag{7.3.5}$$

in which the k entries on the main diagonal line are λ, and the $k - 1$ entries on the subdiagonal on the right of main diagonal are equal to 1, while all other entries are equal to zero. For example, second-order and third-order Jordan blocks are as follows:

$$\mathbf{J}_{2\times 2} = \begin{bmatrix} \lambda & 1 \\ 0 & \lambda \end{bmatrix}, \quad \mathbf{J}_{3\times 3} = \begin{bmatrix} \lambda & 1 & 0 \\ 0 & \lambda & 1 \\ 0 & 0 & \lambda \end{bmatrix}.$$

The Jordan canonical form $\mathbf{J}$ has the following properties.

1. $\mathbf{J}$ is an upper bidiagonal matrix.
2. $\mathbf{J}$ is a diagonal matrix in the special case of n Jordan blocks of size $n_k = 1$.
3. $\mathbf{J}$ is unique up to permutations of the blocks (so it is called the Jordan canonical form).
4. $\mathbf{J}$ can have multiple blocks with the same eigenvalue.

A multiple eigenvalue λ with algebraic multiplicity m_i may have one or more Jordan blocks, depending on the geometric multiplicity of λ. For example, if a 3×3 matrix $\mathbf{A}$ has an eigenvalue λ_0 with algebraic multiplicity 3, then the Jordan canonical form of the matrix $\mathbf{A}$ may have three forms:

$$\mathbf{J}_1 = \begin{bmatrix} \mathbf{J}_{1\times 1} & 0 & 0 \\ 0 & \mathbf{J}_{1\times 1} & 0 \\ 0 & 0 & \mathbf{J}_{1\times 1} \end{bmatrix} = \begin{bmatrix} \lambda_0 & 0 & 0 \\ 0 & \lambda_0 & 0 \\ 0 & 0 & \lambda_0 \end{bmatrix} \quad \text{(geometric multiplicity 3)},$$

$$\mathbf{J}_2 = \begin{bmatrix} \mathbf{J}_{1\times 1} & 0 \\ 0 & \mathbf{J}_{2\times 2} \end{bmatrix} = \begin{bmatrix} \lambda_0 & 0 & 0 \\ 0 & \lambda_0 & 1 \\ 0 & 0 & \lambda_0 \end{bmatrix} \quad \text{(geometric multiplicity 2)},$$

$$\mathbf{J}_3 = \mathbf{J}_{3\times 3} = \begin{bmatrix} \lambda_0 & 1 & 0 \\ 0 & \lambda_0 & 1 \\ 0 & 0 & \lambda_0 \end{bmatrix} \quad \text{(geometric multiplicity 1)},$$

because the meaning of the geometric multiplicity α of the eigenvalue λ is that the number of linearly independent eigenvectors corresponding to λ is α.

Jordan canonical forms with different permutation orders of the Jordan blocks are regarded as the same Jordan canonical form, as stated above. For example, the second-order Jordan canonical form $\mathbf{J}_2$ can also be arranged as

$$\mathbf{J}_2 = \begin{bmatrix} \mathbf{J}_{2\times 2} & 0 \\ 0 & \mathbf{J}_{1\times 1} \end{bmatrix} = \begin{bmatrix} \lambda_0 & 1 & 0 \\ 0 & \lambda_0 & 0 \\ 0 & 0 & \lambda_0 \end{bmatrix}.$$

In practical applications, a nondiagonalizable $m \times m$ matrix usually has multiple eigenvalues. Because the number of Jordan blocks associated with an eigenvalue λ_i is equal to its geometric multiplicity α_i, a natural question is how to determine α_i for a given eigenvalue λ_i?

The method for determining the geometric multiplicity (i.e. the number of Jordan blocks) of a given $m \times m$ matrix $\mathbf{A}$ is as follows.

(1) The number of Jordan blocks with order greater than 1 (i.e., ≥ 2) is determined by $\alpha_1 = \text{rank}(\lambda_i \mathbf{I} - \mathbf{A}) - \text{rank}(\lambda_i \mathbf{I} - \mathbf{A})^2$.

(2) The number of Jordan blocks with order greater than 2 (i.e., ≥ 3) is given by $\alpha_2 = \text{rank}(\lambda_i \mathbf{I} - \mathbf{A})^2 - \text{rank}(\lambda_i \mathbf{I} - \mathbf{A})^3$.

(3) More generally, the number of Jordan blocks with the order greater than $k-1$ (i.e., $\geq k$) is determined by $\alpha_{k-1} = \text{rank}(\lambda_i \mathbf{I} - \mathbf{A})^{k-1} - \text{rank}(\lambda_i \mathbf{I} - \mathbf{A})^k$.

(4) The sum of the orders of the Jordan blocks corresponding to an eigenvalue λ_i is equal to its algebraic multiplicity.

Example 7.3 Find the Jordan canonical form of a 3×3 matrix

$$\mathbf{A} = \begin{bmatrix} 1 & 0 & 4 \\ 2 & -1 & 4 \\ -1 & 0 & -3 \end{bmatrix}. \tag{7.3.6}$$

Solution From the characteristic determinant

$$|\lambda \mathbf{I} - \mathbf{A}| = \begin{vmatrix} \lambda - 1 & 0 & -4 \\ -2 & \lambda + 1 & -4 \\ 1 & 0 & \lambda + 3 \end{vmatrix} = (\lambda + 1)^3 = 0,$$

we find that get the eigenvalue of the matrix $\mathbf{A}$ is $\lambda = -1$ with (algebraic) multiplicity 3.

For $\lambda = -1$, since

$$\lambda \mathbf{I} - \mathbf{A} = \begin{bmatrix} -2 & 0 & -4 \\ -2 & 0 & -4 \\ 1 & 0 & 2 \end{bmatrix}, \quad (\lambda \mathbf{I} - \mathbf{A})^2 = \begin{bmatrix} 0 & 0 & 0 \\ 0 & 0 & 0 \\ 0 & 0 & 0 \end{bmatrix},$$

the number of Jordan blocks with order ≥ 2 is given by $\text{rank}(\lambda \mathbf{I} - \mathbf{A}) - \text{rank}(\lambda \mathbf{I} - \mathbf{A})^2 = 1 - 0 = 1$, i.e., there is one Jordan block with order ≥ 2. Moreover, since $\text{rank}(\lambda \mathbf{I} - \mathbf{A})^2 - \text{rank}(\lambda \mathbf{I} - \mathbf{A})^3 = 0 - 0 = 0$, there is no third-order Jordan block. Therefore, the eigenvalue -1 with multiplicity 3 corresponds to two Jordan blocks, of which one Jordan block has order 2, and another has order 1. In other words, the Jordan canonical form of the given matrix $\mathbf{A}$ is

$$\mathbf{J} = \begin{bmatrix} \mathbf{J}_{1\times 1} & 0 \\ 0 & \mathbf{J}_{2\times 2} \end{bmatrix} = \begin{bmatrix} -1 & 0 & 0 \\ 0 & -1 & 1 \\ 0 & 0 & -1 \end{bmatrix}.$$

The Jordan canonical form $\mathbf{P}^{-1}\mathbf{A}\mathbf{P} = \mathbf{J}$ can be written as $\mathbf{A}\mathbf{P} = \mathbf{P}\mathbf{J}$. When $\mathbf{A}$ has d different eigenvalues λ_i with multiplicity m_i, $i = 1, \ldots, d$, and $m_1 + \cdots + m_d = r$ with $r = \text{rank}(\mathbf{A})$, the Jordan canonical form is $\mathbf{J} = \mathbf{Diag}(\mathbf{J}_1, \ldots, \mathbf{J}_d, \mathbf{0}, \ldots, \mathbf{0})$, so $\mathbf{AP} = \mathbf{PJ}$ simplifies to

$$\mathbf{A}[\mathbf{P}_1, \ldots, \mathbf{P}_d] = [\mathbf{P}_1, \ldots, \mathbf{P}_d] \begin{bmatrix} \mathbf{J}_1 & \cdots & \mathbf{O} \\ \vdots & \ddots & \vdots \\ \mathbf{O} & \cdots & \mathbf{J}_d \end{bmatrix}$$
$$= [\mathbf{P}_1\mathbf{J}_1, \ldots, \mathbf{P}_d\mathbf{J}_d], \tag{7.3.7}$$

where the Jordan blocks are given by

$$\mathbf{J}_i = \begin{bmatrix} \lambda_i & 1 & \cdots & 0 \\ 0 & \ddots & \ddots & \vdots \\ \vdots & \ddots & \lambda_i & 1 \\ 0 & \cdots & 0 & \lambda_i \end{bmatrix} \in \mathbb{R}^{m_i \times m_i}, \quad i = 1, \ldots, d.$$

Letting $\mathbf{P}_i = [\mathbf{p}_{i,1}, \ldots, \mathbf{p}_{i,m_i}]$, Equation (7.3.7) gives

$$\mathbf{A}\mathbf{p}_{i,j} = \mathbf{p}_{i,j-1} + \lambda_i \mathbf{p}_{i,j}, \quad i = 1, \ldots, d,\ j = 1, \ldots, m_i, \tag{7.3.8}$$

where $\mathbf{p}_{i,0} = \mathbf{0}$.

Equation (7.3.8) provides a method of computation for the similarity transformation matrix $\mathbf{P}$.

Algorithm 7.1 shows how to find the Jordan canonical form of a matrix $\mathbf{A} \in \mathbb{C}^{m\times m}$ via similarity reduction.

Algorithm 7.1 Similarity reduction of matrix $\mathbf{A}$

---

input: $\mathbf{A} \in \mathbb{R}^{m\times m}$.

1. Solve $|\lambda\mathbf{I} - \mathbf{A}| = 0$ for λ_i with multiplicity m_i, where $i = 1, \ldots, d$, and $m_1 + \cdots + m_d = m$.
2. Solve Equation (7.3.8) for $\mathbf{P}_i = [\mathbf{p}_{i,1}, \ldots, \mathbf{p}_{i,m_i}]$, $i = 1, \ldots, d$.
3. Set $\mathbf{P} = [\mathbf{P}_1, \ldots, \mathbf{P}_d]$, and find its inverse matrix $\mathbf{P}^{-1}$.
4. Compute $\mathbf{P}^{-1}\mathbf{AP} = \mathbf{J}$.

output: Similarity transformation matrix $\mathbf{P}$ and Jordan canonical form $\mathbf{J}$.

Example 7.4 Use similarity reduction to find the Jordan canonical form of the matrix

$$\mathbf{A} = \begin{bmatrix} 1 & 0 & 4 \\ 2 & -1 & 4 \\ -1 & 0 & -3 \end{bmatrix}.$$

Solution From $|\lambda\mathbf{I} - \mathbf{A}| = 0$ we get the single eigenvalue $\lambda = -1$ with algebraic multiplicity 3.

Let $j = 1$, and solve Equation (7.3.8):

$$\begin{bmatrix} 1 & 0 & 4 \\ 2 & -1 & 4 \\ -1 & 0 & -3 \end{bmatrix} \begin{bmatrix} p_{11} \\ p_{21} \\ p_{31} \end{bmatrix} = - \begin{bmatrix} p_{11} \\ p_{21} \\ p_{31} \end{bmatrix}$$

from which we have $p_{11} = -p_{31}$; p_{21} can take an arbitrary value. Take $p_{11} = p_{31} = 0$ and $p_{21} = 1$.

Solving Equation (7.3.8) for $j = 2$,

$$\begin{bmatrix} 1 & 0 & 4 \\ 2 & -1 & 4 \\ -1 & 0 & -3 \end{bmatrix} \begin{bmatrix} p_{12} \\ p_{22} \\ p_{32} \end{bmatrix} = \begin{bmatrix} 0 \\ 1 \\ 0 \end{bmatrix} - \begin{bmatrix} p_{12} \\ p_{22} \\ p_{32} \end{bmatrix},$$

we get $p_{12} = -2p_{32}$; p_{22} can take an arbitrary value. Take $p_{12} = 2$, $p_{22} = 2$ and $p_{32} = -1$.

Similarly, from Equation (7.3.8) with $j = 3$,

$$\begin{bmatrix} 1 & 0 & 4 \\ 2 & -1 & 4 \\ -1 & 0 & -3 \end{bmatrix} \begin{bmatrix} p_{13} \\ p_{23} \\ p_{33} \end{bmatrix} = \begin{bmatrix} 2 \\ 2 \\ -1 \end{bmatrix} - \begin{bmatrix} p_{13} \\ p_{23} \\ p_{33} \end{bmatrix},$$

we obtain $p_{13} + 2p_{33} = 1$; p_{32} can take an arbitrary value. Take $p_{13} = 1$, $p_{23} = 0$ and $p_{33} = 0$. Hence, the similarity transformation matrix and its inverse are respectively as follows:

$$\mathbf{P} = \begin{bmatrix} 0 & 2 & 1 \\ 1 & 2 & 0 \\ 0 & -1 & 0 \end{bmatrix}, \quad \mathbf{P}^{-1} = \begin{bmatrix} 0 & 1 & 2 \\ 0 & 0 & -1 \\ 1 & 0 & 2 \end{bmatrix}. \tag{7.3.9}$$

Therefore the similarity reduction of $\mathbf{A}$ is given by

$$\mathbf{J} = \mathbf{P}^{-1}\mathbf{A}\mathbf{P} = \begin{bmatrix} -1 & 0 & 0 \\ 0 & -1 & 1 \\ 0 & 0 & -1 \end{bmatrix}. \tag{7.3.10}$$

7.3.3 Similarity Reduction of Matrix Polynomials

Consider a polynomial

$$f(x) = a_n x^n + a_{n-1} x^{n-1} + \cdots + a_1 x + a_0. \tag{7.3.11}$$

When $a_n \neq 0$, n is called the order of the polynomial $f(x)$. An nth-order polynomial is known as a *monic polynomial* if the coefficient of x^n is equal to 1.

Given a matrix $\mathbf{A} \in \mathbb{C}^{m\times m}$ and a polynomial function $f(x)$, we say that

$$f(\mathbf{A}) = a_n \mathbf{A}^n + a_{n-1} \mathbf{A}^{n-1} + \cdots + a_1 \mathbf{A} + a_0 \mathbf{I} \tag{7.3.12}$$

is an *nth-order matrix polynomial* of $\mathbf{A}$.

Suppose that the characteristic polynomial $p(\mathbf{A}) = \lambda\mathbf{I} - \mathbf{A}$ has d different roots which give the eigenvalues $\lambda_1, \ldots, \lambda_d$ of the matrix $\mathbf{A}$ and that the multiplicity of the eigenvalue λ_i is m_i, i.e., $m_1 + \cdots + m_d = m$.

If the Jordan canonical form of the matrix $\mathbf{A}$ is $\mathbf{J}$, i.e.,

$$\mathbf{A} = \mathbf{PJP}^{-1} = \mathbf{P}\mathbf{Diag}(\mathbf{J}_1(\lambda_1), \ldots, \mathbf{J}_d(\lambda_d))\mathbf{P}^{-1}, \tag{7.3.13}$$

then

$$\begin{aligned} f(\mathbf{A}) &= a_n\mathbf{A}^n + a_{n-1}\mathbf{A}^{n-1} + \cdots + a_1\mathbf{A} + a_0\mathbf{I} \\ &= a_n(\mathbf{PJP}^{-1})^n + a_{n-1}(\mathbf{PJP}^{-1})^{n-1} + \cdots + a_1(\mathbf{PJP}^{-1}) + a_0\mathbf{I} \\ &= \mathbf{P}(a_n\mathbf{J}^n + a_{n-1}\mathbf{J}^{n-1} + \cdots + a_1\mathbf{J} + a_0\mathbf{I})\mathbf{P}^{-1} \\ &= \mathbf{P}f(\mathbf{J})\mathbf{P}^{-1} \end{aligned} \tag{7.3.14}$$

is known as the similarity reduction of the matrix polynomial $f(\mathbf{A})$, where

$$\begin{aligned} f(\mathbf{J}) &= a_n\mathbf{J}^n + a_{n-1}\mathbf{J}^{n-1} + \cdots + a_1\mathbf{J} + a_0\mathbf{I} \\ &= a_n\begin{bmatrix} \mathbf{J}_1 & & \mathbf{O} \\ & \ddots & \\ \mathbf{O} & & \mathbf{J}_p \end{bmatrix}^n + \cdots + a_1\begin{bmatrix} \mathbf{J}_1 & & \mathbf{O} \\ & \ddots & \\ \mathbf{O} & & \mathbf{J}_d \end{bmatrix} + a_0\begin{bmatrix} \mathbf{I}_1 & & \mathbf{O} \\ & \ddots & \\ \mathbf{O} & & \mathbf{I}_d \end{bmatrix} \\ &= \mathbf{Diag}(a_n\mathbf{J}_1^n + \cdots + a_1\mathbf{J}_1 + a_0\mathbf{I}_1, \ldots, a_n\mathbf{J}_d^n + \cdots + a_1\mathbf{J}_d + a_0\mathbf{I}_d) \\ &= \mathbf{Diag}(f(\mathbf{J}_1), \ldots, f(\mathbf{J}_d)). \end{aligned} \tag{7.3.15}$$

Substitute Equation (7.3.15) into Equation (7.3.14) to yield

$$f(\mathbf{A}) = \mathbf{P}\mathbf{Diag}(f(\mathbf{J}_1), \ldots, f(\mathbf{J}_d))\mathbf{P}^{-1}, \tag{7.3.16}$$

where $f(\mathbf{J}_i) \in \mathbb{C}^{m_i \times m_i}$ is the *Jordan representation* of the matrix function $f(\mathbf{A})$ associated with the eigenvalue λ_i and is defined as

$$f(\mathbf{J}_i) = \begin{bmatrix} f(\lambda_i) & f'(\lambda_i) & \frac{1}{2!}f''(\lambda) & \cdots & \frac{1}{(m_i-1)!}f^{(m_i-1)}(\lambda_i) \\ & f(\lambda_i) & f'(\lambda_i) & \cdots & \frac{1}{(m_i-2)!}f^{(m_i-2)}(\lambda_i) \\ & & \ddots & \ddots & \vdots \\ & & & f(\lambda_i) & f'(\lambda_i) \\ 0 & & & & f(\lambda_i) \end{bmatrix}, \tag{7.3.17}$$

where $f^{(k)}(x) = \mathrm{d}^k f(x)/\mathrm{d}x^k$ is the kth-order derivative of the function $f(x)$.

Equation (7.3.16) is derived under the assumption that $f(\mathbf{A})$ is a the matrix polynomial. However, using $(\mathbf{PAP}^{-1})^k = \mathbf{PJ}^k\mathbf{P}^{-1}$, it is easy to see that Equation (7.3.16) applies to other common matrix functions $f(\mathbf{A})$ as well.

(1) *Powers of a matrix*

$$\mathbf{A}^K \stackrel{\text{def}}{=} \mathbf{PJ}^K\mathbf{P}^{-1} = \mathbf{P}f(\mathbf{J})\mathbf{P}^{-1}.$$

In this case, $f(x) = x^K$.

(2) *Matrix logarithm*

$$\ln(\mathbf{I}+\mathbf{A}) \stackrel{\text{def}}{=} \sum_{n=1}^{\infty} \frac{(-1)^{n-1}}{n} \mathbf{A}^n = \mathbf{P}\left(\sum_{n=1}^{\infty} \frac{(-1)^{n-1}}{n} \mathbf{A}^n\right)\mathbf{P}^{-1}$$
$$= \mathbf{P}f(\mathbf{J})\mathbf{P}^{-1}.$$

In this case, $f(x) = \ln(1+x)$.

(3) *Sine and cosine functions*

$$\sin \mathbf{A} \stackrel{\text{def}}{=} \sum_{n=0}^{\infty} \frac{(-1)^n}{(2n+1)!} \mathbf{A}^{2n+1} = \mathbf{P}\left(\sum_{n=0}^{\infty} \frac{(-1)^n}{(2n+1)!} \mathbf{J}^{2n+1}\right)\mathbf{P}^{-1}$$
$$= \mathbf{P}f(\mathbf{J})\mathbf{P}^{-1},$$
$$\cos \mathbf{A} \stackrel{\text{def}}{=} \sum_{n=0}^{\infty} \frac{(-1)^n}{(2n)!} \mathbf{A}^{2n} = \mathbf{P}\left(\sum_{n=0}^{\infty} \frac{(-1)^n}{(2n)!} \mathbf{J}^{2n}\right)\mathbf{P}^{-1} = \mathbf{P}f(\mathbf{J})\mathbf{P}^{-1},$$

where $f_1(x) = \sin x$ and $f_2(x) = \cos x$, respectively.

(4) *Matrix exponentials*

$$\mathrm{e}^{\mathbf{A}} \stackrel{\text{def}}{=} \sum_{n=0}^{\infty} \frac{1}{n!} \mathbf{A}^n = \mathbf{P}\left(\sum_{n=0}^{\infty} \frac{1}{n!} \mathbf{J}^n\right)\mathbf{P}^{-1} = \mathbf{P}f(\mathbf{J})\mathbf{P}^{-1},$$
$$\mathrm{e}^{-\mathbf{A}} \stackrel{\text{def}}{=} \sum_{n=0}^{\infty} \frac{1}{n!}(-1)^n \mathbf{A}^n = \mathbf{P}\left(\sum_{n=0}^{\infty} \frac{1}{n!}(-1)^n \mathbf{J}^n\right)\mathbf{P}^{-1} = \mathbf{P}f(\mathbf{J})\mathbf{P}^{-1}.$$

Here the scalar functions corresponding to the matrix exponentials $f_1(\mathbf{A}) = \mathrm{e}^{\mathbf{A}}$ and $f_2(\mathbf{A}) = \mathrm{e}^{-\mathbf{A}}$ are respectively $f_1(x) = \mathrm{e}^x$ and $f_2(x) = \mathrm{e}^{-x}$.

(5) *Matrix exponential functions*

$$\mathrm{e}^{\mathbf{A}t} \stackrel{\text{def}}{=} \sum_{n=0}^{\infty} \frac{1}{n!} \mathbf{A}^n t^n = \mathbf{P}\left(\sum_{n=0}^{\infty} \frac{1}{n!} \mathbf{J}^n t^n\right)\mathbf{P}^{-1} = \mathbf{P}f(\mathbf{J})\mathbf{P}^{-1},$$
$$\mathrm{e}^{-\mathbf{A}t} \stackrel{\text{def}}{=} \sum_{n=0}^{\infty} \frac{1}{n!}(-1)^n \mathbf{A}^n t^n = \mathbf{P}\left(\sum_{n=0}^{\infty} \frac{1}{n!}(-1)^n \mathbf{J}^n t^n\right)\mathbf{P}^{-1} = \mathbf{P}f(\mathbf{J})\mathbf{P}^{-1},$$

where the scalar functions corresponding to the matrix exponential functions $f_1(\mathbf{A}) = \mathrm{e}^{\mathbf{A}t}$ and $f_2(\mathbf{A}) = \mathrm{e}^{-\mathbf{A}t}$ are respectively $f_1(x) = \mathrm{e}^{xt}$ and $f_2(x) = \mathrm{e}^{-xt}$.

The following example shows how to compute matrix functions using (7.3.16) and (7.3.17).

Example 7.5 Given the matrix

$$\mathbf{A} = \begin{bmatrix} 1 & 0 & 4 \\ 2 & -1 & 4 \\ -1 & 0 & -3 \end{bmatrix},$$

compute (a) the matrix polynomial $f(\mathbf{A}) = \mathbf{A}^4 - 3\mathbf{A}^3 + \mathbf{A} - \mathbf{I}$, (b) the matrix power $f(\mathbf{A}) = \mathbf{A}^{1000}$, (c) the matrix exponential $f(\mathbf{A}) = \mathrm{e}^{\mathbf{A}}$, (d) the matrix exponential function $f(\mathbf{A}) = \mathrm{e}^{\mathbf{A}t}$ and (e) the matrix trigonometric function $\sin \mathbf{A}$.

Solution In Example 7.4, we found the similarity transformation matrix $\mathbf{P}$, its inverse matrix $\mathbf{P}^{-1}$ and the Jordan canonical form of the above matrix, as follows:

$$\mathbf{P} = \begin{bmatrix} 0 & 2 & 1 \\ 1 & 2 & 0 \\ 0 & -1 & 0 \end{bmatrix}, \quad \mathbf{P}^{-1} = \begin{bmatrix} 0 & 1 & 2 \\ 0 & 0 & -1 \\ 1 & 0 & 2 \end{bmatrix}, \quad \mathbf{J} = \mathbf{P}^{-1}\mathbf{A}\mathbf{P} = \begin{bmatrix} -1 & 0 & 0 \\ 0 & -1 & 1 \\ 0 & 0 & -1 \end{bmatrix}.$$

Thus we have the matrix polynomial

$$\begin{aligned} f(\mathbf{A}) &= \mathbf{P} f(\mathbf{J}) \mathbf{P}^{-1} \\ &= \begin{bmatrix} 0 & 2 & 1 \\ 1 & 2 & 0 \\ 0 & -1 & 0 \end{bmatrix} \begin{bmatrix} f(-1) & 0 & 0 \\ 0 & f(-1) & f'(-1) \\ 0 & 0 & f(-1) \end{bmatrix} \begin{bmatrix} 0 & 1 & 2 \\ 0 & 0 & -1 \\ 1 & 0 & 2 \end{bmatrix} \\ &= \begin{bmatrix} f(-1) + 2f'(-1) & 0 & 4f'(-1) \\ 2f'(-1) & f(-1) & 4f'(-1) \\ -f'(-1) & 0 & f(-1) - 2f'(-1) \end{bmatrix}. \end{aligned} \tag{7.3.18}$$

(a) To compute the matrix polynomial $f(\mathbf{A}) = \mathbf{A}^4 - 3\mathbf{A}^3 + \mathbf{A} - \mathbf{I}$, consider the corresponding polynomial function $f(x) = x^4 - 3x^3 + x - 1$ with first-order derivative $f'(x) = 4x^3 - 9x^2 + 1$. Hence, for the triple eigenvalue $\lambda = -1$, $f(-1) = 5$ and the first-order derivative $f'(-1) = -12$. Substituting $f(-1) = 5$ and $f'(-1) = -12$ into Equation (7.3.18), we have

$$f(\mathbf{A}) = \mathbf{A}^4 - 3\mathbf{A}^3 + \mathbf{A} - \mathbf{I} = \begin{bmatrix} -19 & 0 & -48 \\ -24 & 5 & -48 \\ 12 & 0 & 29 \end{bmatrix}.$$

(b) To compute the matrix power $f(\mathbf{A}) = \mathbf{A}^{1000}$, consider the corresponding polynomial function $f(x) = x^{1000}$, whose first-order derivative $f'(x) = 1000x^{999}$. Substitute $f(-1) = 1$ and $f'(-1) = -1000$ into Equation (7.3.18) to yield directly

$$\mathbf{A}^{1000} = \begin{bmatrix} -1999 & 0 & -4000 \\ -2000 & 1 & -4000 \\ 1000 & 0 & 2001 \end{bmatrix}.$$

(c) To compute the matrix exponential $f(\mathbf{A}) = \mathrm{e}^{\mathbf{A}}$, consider the polynomial functions $f(x) = \mathrm{e}^x$ and $f'(x) = \mathrm{e}^x$; we have $f(-1) = \mathrm{e}^{-1}$ and $f'(-1) = \mathrm{e}^{-1}$. From these values and Equation (7.3.18), we immediately get

$$\mathrm{e}^{\mathbf{A}} = \begin{bmatrix} 3\mathrm{e}^{-1} & 0 & 4\mathrm{e}^{-1} \\ 2\mathrm{e}^{-1} & \mathrm{e}^{-1} & 4\mathrm{e}^{-1} \\ -\mathrm{e}^{-1} & 0 & -\mathrm{e}^{-1} \end{bmatrix}.$$

(d) To compute the matrix exponential function $f(\mathbf{A}) = \mathrm{e}^{\mathbf{A}t}$, consider the polynomial functions $f(x) = \mathrm{e}^{xt}$ and $f'(x) = t\mathrm{e}^{xt}$; it is well known that $f(-1) = \mathrm{e}^{-t}$, $f'(-1) = t\mathrm{e}^{-t}$. Substituting these two values into (7.3.18), we get

$$\mathrm{e}^{\mathbf{A}t} = \begin{bmatrix} \mathrm{e}^{-t} + 2t\mathrm{e}^{-t} & 0 & 4t\mathrm{e}^{-t} \\ 2t\mathrm{e}^{-t} & \mathrm{e}^{-t} & 4t\mathrm{e}^{-t} \\ -t\mathrm{e}^{-t} & 0 & \mathrm{e}^{-t} - 2t\mathrm{e}^{-t} \end{bmatrix}.$$

(e) To compute the matrix trigonometric function $\sin \mathbf{A}$, consider the polynomial function $f(x) = \sin x$, whose first-order derivative is $f'(x) = \cos x$. Substituting $f(-1) = \sin(-1)$ and $f'(-1) = \cos(-1)$ into Equation (7.3.18), we immediately get

$$\sin \mathbf{A} = \begin{bmatrix} \sin(-1) + 2\cos(-1) & 0 & 4\cos(-1) \\ 2\cos(-1) & \sin(-1) & 4\cos(-1) \\ -\cos(-1) & 0 & \sin(-1) - 2\cos(-1) \end{bmatrix}.$$

7.4 Polynomial Matrices and Balanced Reduction

The matrix polynomial and its similarity reduction presented in the above section can be used to calculate matrix powers and the matrix functions, but the determination of the Jordan canonical forms requires three key steps.

(1) Solve the characteristic equation $|\lambda \mathbf{I} - \mathbf{A}| = 0$ to get the eigenvalues $\lambda_1, \ldots, \lambda_m$ of $\mathbf{A} \in \mathbb{C}^{m\times m}$.
(2) Solve $\mathbf{A}\mathbf{p}_{i,j} = \mathbf{p}_{i,j-1} + \lambda_i \mathbf{p}_{i,j}, i = 1, \ldots, d, j = 1, \ldots, m_i$ to determine the similarity transformation sub-matrices $\mathbf{P}_i = [\mathbf{p}_{i,1}, \ldots, \mathbf{p}_{i,m_i}], i = 1, \ldots, d$.
(3) Find the inverse matrix $\mathbf{P}^{-1} = [\mathbf{P}_1, \ldots, \mathbf{P}_d]^{-1}$.

This section presents an alternative way to determine the Jordan canonical form of $\mathbf{A} \in \mathbb{C}^{m\times m}$. The basic starting point of this idea is to transform the similarity reduction of the constant matrix $\mathbf{A}$ into a balanced reduction of the characteristic polynomial matrix $\lambda\mathbf{I} - \mathbf{A}$ and then to transform its Smith normal form into the Jordan canonical form of $\mathbf{A}$.

7.4.1 Smith Normal Forms

Definition 7.8 An $m \times n$ matrix $\mathbf{A}(x) = [a_{ij}(x)]$, with polynomials of the argument x as entries, is called a *polynomial matrix.*

Denote $\mathbb{R}[x]^{m\times n}$ as the collection of $m\times n$ real polynomial matrices and $\mathbb{C}[x]^{m\times n}$ as the collection of $m \times n$ complex polynomial matrices. It is easily to verify that $\mathbb{C}[x]^{m\times n}$ and $\mathbb{R}[x]^{m\times n}$ are respectively linear spaces in the complex field $\mathbb{C}$ and the real field $\mathbb{R}$.

The polynomial matrix and the matrix polynomial are two different concepts: the matrix polynomial $f(\mathbf{A}) = a_n\mathbf{A}^n + a_{n-1}\mathbf{A}^{n-1} + \cdots + a_1\mathbf{A} + a_0\mathbf{I}$ is a polynomial with the matrix $\mathbf{A}$ as its argument, while the polynomial matrix $\mathbf{A}(x)$ is a matrix with polynomials of x as entries $a_{ij}(x)$.

Definition 7.9 Let $\mathbf{A}(\lambda) \in \mathbb{C}[\lambda]^{m\times n}$; then r is called the *rank of the polynomial matrix* $\mathbf{A}(x)$, denoted $r = \text{rank}[\mathbf{A}(x)]$, if any $k \geq (r+1)$th-order minor is equal to zero and there is at least one kth-order minor that is the nonzero polynomial in $\mathbb{C}[x]$.

In particular, if $m = \text{rank}[\mathbf{A}(x)]$, then an mth-order polynomial matrix $\mathbf{A}(x) \in \mathbb{C}[x]^{m\times m}$ is said to be a *full-rank polynomial matrix* or a *nonsingular polynomial matrix.*

Definition 7.10 Let $\mathbf{A}(x) \in \mathbb{C}[x]^{m\times m}$. If there is another polynomial matrix $\mathbf{B}(x) \in \mathbb{C}[x]^{m\times m}$ such that $\mathbf{A}(x)\mathbf{B}(x) = \mathbf{B}(x)\mathbf{A}(x) = \mathbf{I}_{m\times m}$, then the polynomial matrix $\mathbf{A}(x)$ is said to be invertible, and $\mathbf{B}(x)$ is called the *inverse of the polynomial matrix* $\mathbf{A}(x)$, denoted $\mathbf{B}(x) = \mathbf{A}^{-1}(x)$.

For a constant matrix, its nonsingularity and invertibility are two equivalent concepts. However, the nonsingularity and invertibility of a polynomial matrix are two different concepts: nonsingularity is weaker than invertibility, i.e., a nonsingular polynomial matrix is not necessarily invertible while an invertible polynomial matrix must be nonsingular.

A constant matrix $\mathbf{A} \in \mathbb{C}^{m\times m}$ can be reduced to the Jordan canonical form $\mathbf{J}$ via a similarity transformation. A natural question to ask is: what is the standard form of polynomial matrix reduction? To answer this question, we first discuss how to reduce a polynomial matrix.

Definition 7.11 Let $\mathbf{A} \in \mathbb{C}[x]^{m\times n}, \mathbf{P} \in \mathbb{R}^{m\times m}$ and $d(x) \in \mathbb{R}[x]$. *Elementary row matrices* are of the following three types.

The *Type-I elementary row matrix*

$$\mathbf{P}_{ij} = [\mathbf{e}_1, \ldots, \mathbf{e}_{i-1}, \mathbf{e}_j, \mathbf{e}_{i+1}, \ldots, \mathbf{e}_{j-1}, \mathbf{e}_i, \mathbf{e}_{j+1}, \ldots, \mathbf{e}_m]$$

is such that $\mathbf{P}_{ij}\mathbf{A}$ interchanges rows i and j of $\mathbf{A}$.

The *Type-II elementary row matrix*

$$\mathbf{P}_i(\alpha) = [\mathbf{e}_1, \ldots, \mathbf{e}_{i-1}, \alpha\mathbf{e}_i, \mathbf{e}_{i+1}, \ldots, \mathbf{e}_m]$$

is such that $\mathbf{P}_i(\alpha)\mathbf{A}$ multiplies row i of $\mathbf{A}$ by $\alpha \neq 0$.

The *Type-III elementary row matrix*

$$\mathbf{P}_{ij}[d(x)] = [\mathbf{e}_1, \ldots, \mathbf{e}_i, \ldots, \mathbf{e}_{j-1}, \mathbf{e}_j + d(x)\mathbf{e}_i, \ldots, \mathbf{e}_m]$$

is such that $\mathbf{P}_{ij}[d(x)]\mathbf{A}$ adds $d(x)$ times row i to row j.

Similarly, letting $\mathbf{Q} \in \mathbb{R}^{n\times n}$, we can define three types of elementary column matrices, $\mathbf{Q}_{ij}$, $\mathbf{Q}_i(\alpha)$ and $\mathbf{Q}_{ij}[d(x)]$, respectively, as follows.

Definition 7.12 Two $m \times n$ polynomial matrices $\mathbf{A}(x)$ and $\mathbf{B}(x)$ are said to be *balanced*, denoted $\mathbf{A}(x) \cong \mathbf{B}(x)$, if there is a sequence of elementary row matrices $\mathbf{P}_1(x), \ldots, \mathbf{P}_s(x)$ and a sequence of elementary column matrices $\mathbf{Q}_1(x), \ldots, \mathbf{Q}_t(x)$ such that

$$\mathbf{B}(x) = \mathbf{P}_s(x) \cdots \mathbf{P}_1(x)\mathbf{A}(x)\mathbf{Q}_1(x) \cdots \mathbf{Q}_t(x) = \mathbf{P}(x)\mathbf{A}(x)\mathbf{Q}(x), \tag{7.4.1}$$

where $\mathbf{P}(x)$ is the product of the elementary row matrices and $\mathbf{Q}(x)$ is the product of the elementary column matrices.

Balanced matrices have the following basic properties.

(1) *Reflexivity* Any polynomial matrix $\mathbf{A}(x)$ is balanced with respect to itself: $\mathbf{A}(x) \cong \mathbf{A}(x)$.
(2) *Symmetry* $\mathbf{B}(x) \cong \mathbf{A}(x) \Leftrightarrow \mathbf{A}(x) \cong \mathbf{B}(x)$.
(3) *Transitivity* If $\mathbf{C}(x) \cong \mathbf{B}(x)$ and $\mathbf{B}(x) \cong \mathbf{A}(x)$ then $\mathbf{C}(x) \cong \mathbf{A}(x)$.

The balance of polynomial matrices is also known as their equivalence. A reduction which transforms a polynomial matrix into another, simpler, balanced polynomial matrix is known as *balanced reduction.*

It is self-evident that, given a polynomial matrix $\mathbf{A}(x) \in \mathbb{C}[x]^{m\times n}$, it is desirable that it is balanced to the simplest possible form. This simplest polynomial matrix is the well-known *Smith normal form.*

Theorem 7.7 *If the rank of a polynomial matrix $\mathbf{A}(x) \in \mathbb{C}[x]^{m\times n}$ is equal to r, then $\mathbf{A}(x)$ can be balanced to the Smith normal form $\mathbf{S}(x)$ as follows:*

$$\mathbf{A}(x) \cong \mathbf{S}(x) = \mathbf{Diag}[\sigma_1(x), \sigma_2(x), \ldots, \sigma_r(x), 0, \ldots, 0], \tag{7.4.2}$$

where $\sigma_i(x)$ divides $\sigma_{i+1}(x)$ for $1 \leq i \leq r-1$.

Example 7.6 Given a polynomial matrix

$$\mathbf{A}(x) = \begin{bmatrix} x+1 & 2 & -6 \\ 1 & x & -3 \\ 1 & 1 & x-4 \end{bmatrix},$$

perform the elementary transformations of $\mathbf{A}(x)$ to reduce it.

Solution

$$\mathbf{A}(x) = \begin{bmatrix} x+1 & 2 & -6 \\ 1 & x & -3 \\ 1 & 1 & x-4 \end{bmatrix} \to \begin{bmatrix} x-1 & 0 & -2x+2 \\ 1 & x & -3 \\ 1 & 1 & x-4 \end{bmatrix} \to \begin{bmatrix} x-1 & 0 & 0 \\ 1 & x & -1 \\ 1 & 1 & x-2 \end{bmatrix}$$

$$\to \begin{bmatrix} x-1 & 0 & 0 \\ 0 & x-1 & -x+1 \\ 1 & 1 & x-2 \end{bmatrix} \to \begin{bmatrix} x-1 & 0 & 0 \\ 0 & x-1 & 0 \\ 1 & 1 & x-1 \end{bmatrix}$$

$$\to \begin{bmatrix} x-1 & 0 & 0 \\ 0 & x-1 & 0 \\ x-1 & x-1 & (x-1)^2 \end{bmatrix} \to \begin{bmatrix} x-1 & 0 & 0 \\ 0 & x-1 & 0 \\ 0 & 0 & (x-1)^2 \end{bmatrix} = \mathbf{B}(x).$$

The matrix $\mathbf{B}(x)$ is not in Smith normal form, because $|\mathbf{B}(x)| = (x-1)^4$ is a fourth-order polynomial of x while $|\mathbf{A}(x)|$ is clearly a third-order polynomial of x. Since b_{11}, b_{22}, b_{33} contain the first-order factor $x-1$, the first row, say, of $\mathbf{B}(x)$ can be divided by this factor. After this elementary row transformation, the matrix $\mathbf{B}(x)$ is further reduced to

$$\mathbf{A}(x) \cong \begin{bmatrix} 1 & 0 & 0 \\ 0 & x-1 & 0 \\ 0 & 0 & (x-1)^2 \end{bmatrix}.$$

This is the Smith normal form of the polynomial matrix $x\mathbf{I} - \mathbf{A}$.

Corollary 7.1 *The Smith normal form of a polynomial matrix $\mathbf{A}(x) \in \mathbb{C}[x]^{m\times n}$ is unique. That is, every polynomial matrix $\mathbf{A}(x)$ can be balanced to precisely one matrix in Smith normal form.*

The major shortcoming of the elementary transformation method for reducing a polynomial matrix $\mathbf{A}(x)$ to the Smith normal form is that a sequence of elementary transformations requires a certain skill that is not easy to program, and is sometimes more trouble, than just using the matrix $\mathbf{A}(x)$ as it stands.

7.4.2 Invariant Factor Method

To overcome the shortcoming of the elementary transformation method, it is necessary to introduce an alternative balanced reduction method.

Extract any k rows and any k columns of an $m \times m$ matrix $\mathbf{A}$ to obtain a square matrix whose determinant is the kth minor of $\mathbf{A}$.

Definition 7.13 The *determinant rank* of $\mathbf{A}$ is defined to be the largest integer r for which there exists a nonzero $r \times r$ minor of $\mathbf{A}$.

Definition 7.14 Given a polynomial matrix $\mathbf{A}(x) \in \mathbb{C}[x]^{m\times n}$, if its rank is $\text{rank}[\mathbf{A}(x)] = r$ then, for a natural number $k \le r$, there is at least one nonzero kth minor of $\mathbf{A}(x)$. The greatest common divisor of all nonzero kth minors, denoted $d_k(x)$, is known as the *kth determinant divisor* of the polynomial matrix $\mathbf{A}(x)$.

Theorem 7.8 *The kth determinant divisors $d_k(x) \neq 0$ for $1 \le k \le r$ with $r = \text{rank}(\mathbf{A})$. Also, $d_{k-1}(x)$ divides $d_k(x)$, denoted $d_{k-1}|d_k$, for $1 \le k \le r$.*

Proof (from [319]) Let $r = \text{rank}(\mathbf{A})$. Then there exists an $r \times r$ nonzero minor of $\mathbf{A}$ and hence $d_r(x) \neq 0$. Then, because each $r \times r$ minor is a linear combination of $(r-1)\times(r-1)$ minors of $\mathbf{A}$, it follows that some $(r-1)\times(r-1)$ minor of $\mathbf{A}$ is also nonzero and thus $d_{r-1}(x) \neq 0$. Also, $d_{r-1}(x)$ divides each minor of size $r-1$ and consequently divides each minor of size r, and hence $d_{r-1}(x)$ divides $d_r(x)$, the greatest common divisor of all minors of size r. This argument can be repeated with r replaced by $r-1$ and so on. □

The kth determinant divisors $d_k(x)$, $k = 1, \ldots, r$ in Theorem 7.8 are the polynomials with leading coefficient 1 (i.e., monic polynomials).

Definition 7.15 If $\text{rank}(\mathbf{A}(x)) = r$ then the *invariant factors* of $\mathbf{A}(x)$ are defined as $\sigma_i(x) = d_i(x)/d_{i-1}(x), 1 \le i \le r$, where $d_0(x) = 1$.

The numbers of the various minors of a given $m \times m$ polynomial matrix are as follows:

(1) $C_m^1 \times C_m^1 = m^2$ first minors;
(2) $C_m^2 \times C_m^2$ second minors;
(3) $C_m^k \times C_m^k = \left[\frac{m(m-1)\cdots(m-k+1)}{2\times 3\times\cdots\times k}\right]^2$ kth minors, where $k = 1, \ldots, m-1$;
(4) $C_m^m = 1$ mth minor.

For example, for the polynomial matrix

$$\mathbf{A}(x) = \begin{bmatrix} x & 0 & 1 \\ x^2+1 & x & 0 \\ x-1 & -x & x+1 \end{bmatrix},$$

its various minors are as follows:

The number of first minors is 9, in which the number of nonzero minors is 7.
The number of second minors is $C_3^2 \times C_3^2 = 9$ and all are nonzero minors.
The number of third minors is just 1, and $|\mathbf{A}(x)| = 0$.

Example 7.7 Given a polynomial matrix

$$\mathbf{A}(x) = \begin{bmatrix} 0 & x(x-1) & 0 \\ x & 0 & x+1 \\ 0 & 0 & -x+1 \end{bmatrix},$$

find its Smith normal form.

Solution The nonzero first minors are given by

$$|x(x-1)| = x^2 - x, \quad |x| = x, \quad |x+1| = x+1, \quad |-x+1| = -x+1.$$

So, the first determinant divisor is $d_1(x) = 1$.

The nonzero second minors,

$$\begin{vmatrix} 0 & x(x-1) \\ x & 0 \end{vmatrix} = -x^2(x-1), \quad \begin{vmatrix} x & x+1 \\ 0 & -x+2 \end{vmatrix} = -x(x-2),$$
$$\begin{vmatrix} x(x-1) & 0 \\ 0 & x+1 \end{vmatrix} = x(x-1)(x+1),$$

have greatest common divisor x, so the second determinant divisor is $d_2(x) = x$.

The third determinant,

$$\begin{vmatrix} 0 & x(x-1) & 0 \\ x & 0 & x+1 \\ 0 & 0 & -x+1 \end{vmatrix} = x^2(x-1)(x-2),$$

yields the third determinant divisor $d_3(x) = x^2(x-1)(x-2)$ directly.

Hence, the invariant factors are as follows:

$$\sigma_1(x) = \frac{d_1(x)}{d_0(x)} = 1, \quad \sigma_2(x) = \frac{d_2(x)}{d_1(x)} = x,$$
$$\sigma_3(x) = \frac{d_3(x)}{d_2(x)} = x(x-1)(x-2).$$

Thus the Smith normal form is

$$\mathbf{S}(x) = \begin{bmatrix} 1 & 0 & 0 \\ 0 & x & 0 \\ 0 & 0 & x(x-1)(x-2) \end{bmatrix}.$$

The most common polynomial matrix is $\lambda\mathbf{I} - \mathbf{A}$, where λ is an eigenvalue of the $m \times m$ matrix $\mathbf{A}$. The polynomial matrix $\lambda\mathbf{I} - \mathbf{A}$, in automatic control, signal processing, system engineering and so on, is often called the *λ-matrix* of $\mathbf{A}$, and is denoted $\mathbf{A}(\lambda) = \lambda\mathbf{I} - \mathbf{A}$.

Example 7.8 Find the Smith normal form of the λ-matrix of the 3×3 matrix

$$\mathbf{A}(\lambda) = \lambda\mathbf{I} - \mathbf{A} = \begin{bmatrix} \lambda+1 & 2 & -6 \\ 1 & \lambda & -3 \\ 1 & 1 & \lambda-4 \end{bmatrix}.$$

Solution Every entry consists of a first minor, and their common divisor is 1; hence the first determinant divisor $d_1(\lambda) = 1$.

The second minors are respectively given by

$$\begin{vmatrix} \lambda+1 & 2 \\ 1 & \lambda \end{vmatrix} = (\lambda-1)(\lambda+2), \quad \begin{vmatrix} \lambda+1 & -6 \\ 1 & -3 \end{vmatrix} = -3(\lambda-1),$$

$$\begin{vmatrix} \lambda+1 & 2 \\ 1 & 1 \end{vmatrix} = \lambda-1, \quad \begin{vmatrix} \lambda+1 & -6 \\ 1 & \lambda-4 \end{vmatrix} = (\lambda-1)(\lambda-2),$$

$$\begin{vmatrix} 2 & -6 \\ 1 & \lambda-4 \end{vmatrix} = 2(\lambda-1), \quad \begin{vmatrix} 2 & -6 \\ \lambda & -3 \end{vmatrix} = 6(\lambda-1), \quad \begin{vmatrix} 1 & \lambda \\ 1 & 1 \end{vmatrix} = -(\lambda-1),$$

$$\begin{vmatrix} 1 & -3 \\ 1 & \lambda-4 \end{vmatrix} = \lambda-1, \quad \begin{vmatrix} \lambda & -3 \\ 1 & \lambda-4 \end{vmatrix} = (\lambda-1)(\lambda-3).$$

The greatest common divisor is $\lambda - 1$, so the second determinant divisor $d_2(\lambda) = \lambda - 1$.

The only third minor has

$$\begin{vmatrix} \lambda+1 & 2 & -6 \\ 1 & \lambda & -3 \\ 1 & 1 & \lambda-4 \end{vmatrix} = (\lambda-1)^3,$$

which gives the third determinant divisor $d_3(\lambda) = (\lambda-1)^3$.

From the above results, it is known that the invariant factors are given by

$$\sigma_1(\lambda) = \frac{d_1(\lambda)}{d_0(\lambda)} = 1, \quad \sigma_2(\lambda) = \frac{d_2(\lambda)}{d_1(\lambda)} = \lambda - 1, \quad \sigma_3(\lambda) = \frac{d_3(\lambda)}{d_2(\lambda)} = (\lambda-1)^2,$$

which gives the Smith normal form

$$\mathbf{S}(\lambda) = \begin{bmatrix} 1 & 0 & 0 \\ 0 & \lambda-1 & 0 \\ 0 & 0 & (\lambda-1)^2 \end{bmatrix}.$$

Example 7.9 Find the invariant factors and the Smith normal form of the λ-matrix

$$\mathbf{A}(\lambda) = \begin{bmatrix} \lambda-1 & 0 & -4 \\ -2 & \lambda+1 & -4 \\ 1 & 0 & \lambda+3 \end{bmatrix}.$$

Solution Its determinant divisors are as follows.

(1) The greatest common divisor of the nonzero first minors is 1, so the first determinant divisor $d_1(\lambda) = 1$.

(2) The greatest common divisor of the nonzero second minors is $\lambda+1$, which yields directly $d_2(\lambda) = \lambda + 1$.

(3) The determinant

$$\begin{vmatrix} \lambda-1 & 0 & -4 \\ -2 & \lambda+1 & -4 \\ 1 & 0 & \lambda+3 \end{vmatrix} = (\lambda+1)^3$$

gives directly the third determinant divisor $d_3(\lambda) = (\lambda+1)^3$.

Hence the invariant factors are

$$\sigma_1(\lambda) = \frac{d_1}{d_0} = 1, \quad \sigma_2(\lambda) = \frac{d_2}{d_1} = \lambda + 1, \quad \sigma_3(\lambda) = \frac{d_3}{d_2} = (\lambda+1)^2.$$

From these invariant factors we have the Smith normal form

$$\mathbf{S}(\lambda) = \begin{bmatrix} 1 & 0 & 0 \\ 0 & \lambda+1 & 0 \\ 0 & 0 & (\lambda+1)^2 \end{bmatrix}.$$

7.4.3 Conversion of Jordan Form and Smith Form

We have presented the similarity reduction of matrices and the balanced reduction of polynomial matrices, respectively. An interesting issue is the relationship between the Jordan canonical form of similarity reduction and the Smith normal form of balanced reduction.

Via the linear transformation

$$\mathbf{A}(\lambda) = \lambda\mathbf{I} - \mathbf{A} = \begin{bmatrix} \lambda - a_{11} & -a_{12} & \cdots & -a_{1m} \\ -a_{21} & \lambda - a_{22} & \cdots & -a_{2m} \\ \vdots & \vdots & \ddots & \vdots \\ -a_{m1} & -a_{m2} & \cdots & \lambda - a_{mm} \end{bmatrix}, \tag{7.4.3}$$

a constant matrix $\mathbf{A}$ easily becomes a polynomial matrix $\mathbf{A}(\lambda)$.

Theorem 7.9 *Let $\mathbf{A}$ and $\mathbf{B}$ be two $m \times m$ (constant) matrices. Then*

$$\begin{aligned} \mathbf{A} \sim \mathbf{B} \quad &\Leftrightarrow \quad (x\mathbf{I}_m - \mathbf{A}) \cong (x\mathbf{I}_m - \mathbf{B}) \\ &\Leftrightarrow \quad x\mathbf{I}_m - \mathbf{A} \text{ and } x\mathbf{I}_m - \mathbf{B} \text{ have the same Smith normal form.} \end{aligned}$$

Proof [319] For the forward implication: if $\mathbf{P}^{-1}\mathbf{A}\mathbf{P} = \mathbf{B}$, where $\mathbf{P} \in \mathbb{R}^{m\times m}$ then

$$\mathbf{P}^{-1}(x\mathbf{I}_m - \mathbf{A})\mathbf{P} = x\mathbf{I}_m - \mathbf{P}^{-1}\mathbf{A}\mathbf{P} = x\mathbf{I}_m - \mathbf{B}.$$

For the backward implication: if $x\mathbf{I}_m - \mathbf{A}$ and $x\mathbf{I}_m - \mathbf{B}$ are balanced then they have the same invariant factors and so have the same nontrivial invariant factors. That is, $\mathbf{A}$ and $\mathbf{B}$ have the same invariant factors, and hence are similar. □

Theorem 7.9 establishes the relationship between the Jordan canonical form $\mathbf{J}$ of the matrix $\mathbf{A}$ and the Smith normal form of the λ-matrix $\lambda\mathbf{I} - \mathbf{A}$.

It may be noticed that if the polynomial matrix $\mathbf{A}(x)$ is not in the form of $x\mathbf{I} - \mathbf{A}$ then Theorem 7.9 no longer holds: $\mathbf{A} \sim \mathbf{B}$ does not mean $\mathbf{A}(x) \cong \mathbf{B}(x)$ if $\mathbf{A}(x) \neq x\mathbf{I} - \mathbf{A}$.

Let the Jordan canonical form of a matrix $\mathbf{A}$ be $\mathbf{J}$, i.e., $\mathbf{A} \sim \mathbf{J}$. Hence, by Theorem 7.9, we have

$$\lambda\mathbf{I} - \mathbf{A} \cong \lambda\mathbf{I} - \mathbf{J}. \tag{7.4.4}$$

If the Smith normal form of the λ-matrix $\lambda\mathbf{I} - \mathbf{A}$ is $\mathbf{S}(\lambda)$, i.e.,

$$\lambda\mathbf{I} - \mathbf{A} \cong \mathbf{S}(\lambda), \tag{7.4.5}$$

then from Equations (7.4.4) and (7.4.5) and the transitivity of balanced matrices, it can immediately be seen that

$$\mathbf{S}(\lambda) \cong \lambda\mathbf{I} - \mathbf{J}. \tag{7.4.6}$$

This is the relationship between the Jordan canonical form $\mathbf{J}$ of the matrix $\mathbf{A}$ and the Smith normal form $\mathbf{S}(\lambda)$ of the λ-matrix $\lambda\mathbf{I} - \mathbf{A}$.

Let $\mathbf{J} = \mathbf{Diag}(\mathbf{J}_1, \ldots, \mathbf{J}_d, \mathbf{0}, \ldots, \mathbf{0})$ denote the Jordan canonical form of a constant matrix $\mathbf{A} \in \mathbb{C}^{m\times m}$.

From Equation (7.4.6) it is known that the Smith normal form of the λ-matrix $\mathbf{A}(\lambda) = \lambda\mathbf{I} - \mathbf{J}$ should also have d Smith blocks:

$$\mathbf{S}(\lambda) = \mathbf{Diag}[\mathbf{S}_1(\lambda), \ldots, \mathbf{S}_d(\lambda), \mathbf{0} \ldots, \mathbf{0}], \tag{7.4.7}$$

where the Smith block $\mathbf{S}_i(\lambda) = \mathbf{Diag}(\sigma_{i,1}, \ldots, \sigma_{i,m_i})$ and $m_1 + \cdots + m_d = r$. Thus the Smith block $\mathbf{S}_i$ and the Jordan blocks $\mathbf{J}_i$ have the same dimensions, $m_i \times m_i$.

Therefore, the conversion relationship between the Jordan canonical form and the Smith normal form becomes that between the Jordan block and the Smith block corresponding to the same eigenvalue λ_i.

7.4.4 Finding Smith Blocks from Jordan Blocks

Our problem is: given the Jordan blocks $\mathbf{J}_i, i = 1, \ldots, d$ of a constant matrix $\mathbf{A}$, find the Smith blocks $\mathbf{S}_i(\lambda), i = 1, \ldots, d$ of the λ-matrix $\lambda\mathbf{I} - \mathbf{A}$.

According to the respective eigenvalue multiplicities, there are different corresponding relationships between Jordan blocks and Smith blocks.

(1) The first-order Jordan block corresponding to the single eigenvalue λ_i is given by $\mathbf{J}_i = \lambda_i$. From Equation (7.4.6) it is known that $\mathbf{J}_i = [\lambda] \Rightarrow \mathbf{S}_i(\lambda) = [\lambda - \lambda_i]$.

(2) The second-order Jordan block $\mathbf{J}_i$ corresponding to the eigenvalue λ_i with algebraic multiplicity $m_i = 2$ is given by

$$\mathbf{J}_{i2}^2 = \begin{bmatrix} \lambda_i & 0 \\ 0 & \lambda_i \end{bmatrix} \quad \Rightarrow \quad \mathbf{S}_{i2}^2(\lambda) \cong \lambda\mathbf{I}_2 - \mathbf{J}_{i2}^2 = \begin{bmatrix} \lambda - \lambda_i & 0 \\ 0 & \lambda - \lambda_i \end{bmatrix},$$

$$\mathbf{J}_{i2}^1 = \begin{bmatrix} \lambda_i & 1 \\ 0 & \lambda_i \end{bmatrix} \quad \Rightarrow \quad \mathbf{S}_{i2}^1(\lambda) \cong \lambda\mathbf{I}_2 - \mathbf{J}_{i1}^1 = \begin{bmatrix} \lambda - \lambda_i & -1 \\ 0 & \lambda - \lambda_i \end{bmatrix}$$

$$\cong \begin{bmatrix} 1 & 0 \\ 0 & (\lambda - \lambda_i)^2 \end{bmatrix},$$

because the determinant divisors of $\lambda\mathbf{I}_2 - \mathbf{J}_{i1}$ are $d_1 = 1$ and $d_2 = (\lambda - \lambda_i)^2$, and hence $\sigma_1(\lambda) = d_1/d_0 = 1$ and $\sigma_2(\lambda) = d_2/d_1 = (\lambda - \lambda_i)^2$.

(3) The third-order Jordan block $\mathbf{J}_i$ is given by

$$\mathbf{J}_{i3}^3 = \begin{bmatrix} \lambda_i & 0 & 0 \\ 0 & \lambda_i & 0 \\ 0 & 0 & \lambda_i \end{bmatrix} \quad \Rightarrow \quad \mathbf{S}_{i3}^3(\lambda) \cong \lambda\mathbf{I}_3 - \mathbf{J}_{31} \cong \begin{bmatrix} \lambda - \lambda_i & 0 & 0 \\ 0 & \lambda - \lambda_i & 0 \\ 0 & 0 & \lambda - \lambda_i \end{bmatrix},$$

$$\mathbf{J}_{i3}^2 = \begin{bmatrix} \lambda_i & 0 & 0 \\ 0 & \lambda_i & 1 \\ 0 & 0 & \lambda_i \end{bmatrix} \quad \Rightarrow \quad \mathbf{S}_{i3}^2(\lambda) \cong \lambda\mathbf{I}_3 - \mathbf{J}_{32} \cong \begin{bmatrix} 1 & 0 & 0 \\ 0 & \lambda - \lambda_i & 0 \\ 0 & 0 & (\lambda - \lambda_i)^2 \end{bmatrix},$$

and

$$\mathbf{J}_{i3}^1 = \begin{bmatrix} \lambda_i & 1 & 0 \\ 0 & \lambda_i & 1 \\ 0 & 0 & \lambda_i \end{bmatrix} \quad \Rightarrow \quad \mathbf{S}_{i3}^1(\lambda) \cong \lambda \mathbf{I}_3 - \mathbf{J}_{33} \cong \begin{bmatrix} 1 & 0 & 0 \\ 0 & 1 & 0 \\ 0 & 0 & (\lambda - \lambda_i)^3 \end{bmatrix},$$

because the invariant factors of $\lambda \mathbf{I}_3 - \mathbf{J}_{i3}^2$ are

$$\sigma_1 = d_1/d_0 = 1, \quad \sigma_2 = d_2/d_1 = \lambda - \lambda_i, \quad \sigma_3 = d_3/d_2 = (\lambda - \lambda_i)^2,$$

while the invariant factors of $\lambda \mathbf{I}_3 - \mathbf{J}_{i3}^1$ are

$$\sigma_1 = 1, \quad \sigma_2 = 1, \quad \sigma_3 = (\lambda - \lambda_i)^3.$$

Summarizing the above analysis, we have the corresponding relationship between the Jordan canonical form and the Smith canonical form:

$$\mathbf{J}_1 = [\lambda_i] \quad \Leftrightarrow \quad \mathbf{S}_1(\lambda) = [\lambda - \lambda_i]; \tag{7.4.8}$$

$$\mathbf{J}_{21} = \begin{bmatrix} \lambda_i & 0 \\ 0 & \lambda_i \end{bmatrix} \quad \Leftrightarrow \quad \mathbf{S}_{21}(\lambda) = \begin{bmatrix} \lambda - \lambda_i & 0 \\ 0 & \lambda - \lambda_i \end{bmatrix}, \tag{7.4.9}$$

$$\mathbf{J}_{22} = \begin{bmatrix} \lambda_i & 1 \\ 0 & \lambda_i \end{bmatrix} \quad \Leftrightarrow \quad \mathbf{S}_{22}(\lambda) = \begin{bmatrix} 1 & 0 \\ 0 & (\lambda - \lambda_i)^2 \end{bmatrix}; \tag{7.4.10}$$

and

$$\mathbf{J}_{31} = \begin{bmatrix} \lambda_i & 0 & 0 \\ 0 & \lambda_i & 0 \\ 0 & 0 & \lambda_i \end{bmatrix} \quad \Leftrightarrow \quad \mathbf{S}_{31}(\lambda) = \begin{bmatrix} \lambda - \lambda_i & 0 & 0 \\ 0 & \lambda - \lambda_i & 0 \\ 0 & 0 & \lambda - \lambda_i \end{bmatrix}, \tag{7.4.11}$$

$$\mathbf{J}_{32} = \begin{bmatrix} \lambda_i & 0 & 0 \\ 0 & \lambda_i & 1 \\ 0 & 0 & \lambda_i \end{bmatrix} \quad \Leftrightarrow \quad \mathbf{S}_{32}(\lambda) = \begin{bmatrix} 1 & 0 & 0 \\ 0 & \lambda - \lambda_i & 0 \\ 0 & 0 & (\lambda - \lambda_i)^2 \end{bmatrix}, \tag{7.4.12}$$

$$\mathbf{J}_{33} = \begin{bmatrix} \lambda_i & 1 & 0 \\ 0 & \lambda_i & 1 \\ 0 & 0 & \lambda_i \end{bmatrix} \quad \Leftrightarrow \quad \mathbf{S}_{33}(\lambda) = \begin{bmatrix} 1 & 0 & 0 \\ 0 & 1 & 0 \\ 0 & 0 & (\lambda - \lambda_i)^3 \end{bmatrix}. \tag{7.4.13}$$

7.4.5 Finding Jordan Blocks from Smith Blocks

Conversely, given the Smith normal form $\mathbf{S}(\lambda)$ of a λ-matrix $\lambda \mathbf{I} - \mathbf{A}$, how do we find the Jordan canonical form $\mathbf{J}$ of the matrix $\mathbf{A}$?

Equations (7.4.8)–(7.4.13) can also be used to find the Jordan blocks from the Smith blocks. The question is how to separate the Smith blocks from the Smith normal form.

Using block-divided method of the Smith normal form, we exclude the zero-order invariant factor 1 and then separate the Smith blocks from the Smith normal form according the following two cases.

Case 1 Single eigenvalue

If some ith invariant factor dependents only on one eigenvalue λ_k, i.e., $\sigma_i(\lambda) = (\lambda - \lambda_k)^{n_i}$, where $n_i \geq 1$, then

$$\mathbf{S}_i(\lambda) = \begin{bmatrix} 1 & & & 0 \\ & \ddots & & \\ & & 1 & \\ 0 & & & (\lambda - \lambda_j)^{n_i} \end{bmatrix} \Rightarrow \mathbf{J}_i = \begin{bmatrix} \lambda_j & 1 & & 0 \\ & \ddots & \ddots & \\ & & \lambda_j & 1 \\ 0 & & & \lambda_j \end{bmatrix} \in \mathbb{C}^{n_i \times n_i}. \tag{7.4.14}$$

For example,

$$\mathbf{S}(\lambda) = \begin{bmatrix} 1 & & 0 \\ & (\lambda - \lambda_0) & \\ 0 & & (\lambda - \lambda_0)^2 \end{bmatrix} \Leftrightarrow \begin{cases} \mathbf{S}_1(\lambda) = \lambda - \lambda_0, \\ \mathbf{S}_2(\lambda) = \begin{bmatrix} 1 & 0 \\ 0 & (\lambda - \lambda_0)^2 \end{bmatrix}, \end{cases}$$

$$\Leftrightarrow \begin{cases} \mathbf{J}_1 = \lambda_0 \\ \mathbf{J}_2 = \begin{bmatrix} \lambda_0 & 1 \\ 0 & \lambda_0 \end{bmatrix}, \end{cases} \Leftrightarrow \mathbf{J}(\lambda) = \begin{bmatrix} \lambda_0 & 0 & 0 \\ 0 & \lambda_0 & 1 \\ 0 & 0 & \lambda_0 \end{bmatrix}.$$

Case 2 Multiple eigenvalues

If the kth invariant factor is a product of factors depending on s eigenvalues,

$$\sigma_k(\lambda) = (\lambda - \lambda_{j1})^{n_{k,1}} (\lambda - \lambda_{j2})^{n_{k,2}} \cdots (\lambda - \lambda_{js})^{n_{k,s}}, \quad i = 1, \ldots, s,$$

then the invariant factor $\sigma_k(\lambda)$ can be decomposed into s different eigenvalue factors $\sigma_{k1}(\lambda) = (\lambda - \lambda_{j1})^{n_{k,1}}, \ldots, \sigma_{ks}(\lambda) = (\lambda - \lambda_{js})^{n_{k,s}}$. Then, using $\sigma_{ki}(\lambda) = (\lambda - \lambda_{ji})^{n_{k,i}}$, the Jordan block can be obtained from the Smith block.

Example 7.10 The third invariant factor $\sigma_3(\lambda) = (\lambda - \lambda_1)(\lambda - \lambda_2)^2$ is decomposed into two single-eigenvalue factors $\sigma_{31}(\lambda) = \lambda - \lambda_1$ and $\sigma_{32}(\lambda) = (\lambda - \lambda_2)^2$; hence

$$\mathbf{S}(\lambda) = \begin{bmatrix} 1 & & 0 \\ & 1 & \\ 0 & & (\lambda - \lambda_1)(\lambda - \lambda_2)^2 \end{bmatrix} \Leftrightarrow \begin{cases} \mathbf{S}_1(\lambda) = \lambda - \lambda_1, \\ \mathbf{S}_2(\lambda) = \begin{bmatrix} 1 & 0 \\ 0 & (\lambda - \lambda_2)^2 \end{bmatrix}, \end{cases}$$

$$\Leftrightarrow \begin{cases} \mathbf{J}_1 = \lambda_1, \\ \mathbf{J}_2 = \begin{bmatrix} \lambda_2 & 1 \\ 0 & \lambda_2 \end{bmatrix}, \end{cases} \Leftrightarrow \mathbf{J}(\lambda) = \begin{bmatrix} \lambda_1 & 0 & 0 \\ 0 & \lambda_2 & 1 \\ 0 & 0 & \lambda_2 \end{bmatrix}.$$

Example 7.11 The nonzero-order invariant factor of the 5×5 Smith normal form is $(\lambda - \lambda_1)^2(\lambda - \lambda_2)^3$, which is decomposed into two single-eigenvalue factors

$\sigma_{51}(\lambda) = (\lambda - \lambda_1)^2$ and $\sigma_{52}(\lambda) = (\lambda - \lambda_2)^3$. Hence we have

$$\mathbf{S}(\lambda) = \mathbf{Diag}[1, 1, 1, 1, (\lambda - \lambda_1)^2(\lambda - \lambda_2)^3]$$

$$\Leftrightarrow \begin{cases} \mathbf{S}_1(\lambda) = \begin{bmatrix} 1 & 0 \\ 0 & (\lambda - \lambda_1)^2 \end{bmatrix}, \\ \mathbf{S}_2(\lambda) = \begin{bmatrix} 1 & 0 & 0 \\ 0 & 1 & 0 \\ 0 & 0 & (\lambda - \lambda_2)^3 \end{bmatrix}, \end{cases}$$

$$\Leftrightarrow \begin{cases} \mathbf{J}_1 = \begin{bmatrix} \lambda_1 & 1 \\ 0 & \lambda_1 \end{bmatrix}, \\ \mathbf{J}_2 = \begin{bmatrix} \lambda_2 & 1 & 0 \\ 0 & \lambda_2 & 1 \\ 0 & 0 & \lambda_2 \end{bmatrix}, \end{cases} \Leftrightarrow \mathbf{J}(\lambda) = \begin{bmatrix} \lambda_1 & 1 & 0 & 0 & 0 \\ 0 & \lambda_1 & 0 & 0 & 0 \\ 0 & 0 & \lambda_2 & 1 & 0 \\ 0 & 0 & 0 & \lambda_2 & 1 \\ 0 & 0 & 0 & 0 & \lambda_2 \end{bmatrix}.$$

In addition to determinant divisors and invariant factors, there is another divisor of a polynomial matrix.

Definition 7.16 Decompose an invariant factor into the product of a few nonzero-order factors, i.e., write $\sigma_k(\lambda) = (\lambda - \lambda_{k1})^{c_{k1}} \cdots (\lambda - \lambda_{kl_k})^{c_{kl_k}}$, where $c_{ij} > 0$, $j = 1, \ldots, l_k$; then every factor $(\lambda - \lambda_{kl_k})^{c_{kl_k}}$ is called an *elementary divisor* of $\mathbf{A}(\lambda)$.

The elementary divisors are arranged in no particular order. All elementary divisors of a polynomial matrix $\mathbf{A}(\lambda)$ constitute the elementary divisor set of $\mathbf{A}(\lambda)$.

Example 7.12 Given a polynomial matrix $\mathbf{A}(\lambda) \in \mathbb{C}[\lambda]^{5\times 5}$ with rank 4 and Smith normal form

$$\mathbf{S}(\lambda) = \mathbf{Diag}(1, \lambda, \lambda^2(\lambda - 1)(\lambda + 1), \lambda^2(\lambda - 1)(\lambda + 1)^3, 0),$$

find the elementary divisors of $\mathbf{A}(\lambda)$.

Solution From the second invariant factor $\sigma_2(\lambda) = \lambda$ it is known that λ is an elementary divisor of $\mathbf{A}(\lambda)$.

From the third invariant factor $\sigma_3(\lambda) = \lambda^2(\lambda - 1)(\lambda + 1)$ it follows that other three elementary divisors of $\mathbf{A}(\lambda)$ are $\lambda^2, \lambda - 1$ and $\lambda + 1$.

From the fourth invariant factor $\sigma_4(\lambda) = \lambda^2(\lambda - 1)(\lambda + 1)^3$ we have the corresponding three elementary divisors of $\mathbf{A}(\lambda)$, given by $\lambda^2, \lambda - 1$ and $(\lambda + 1)^3$.

Hence there are seven elementary divisors of the polynomial matrix $\mathbf{A}(\lambda)$: $\lambda, \lambda^2, \lambda^2, \lambda - 1, \lambda - 1, \lambda + 1, (\lambda + 1)^3$.

Given all the elementary divisors of a polynomial matrix $\mathbf{A}(\lambda)$, its invariant factors can be reconstructed by the following method.

(1) Use all the elementary divisors to form a rectangular array: the same order divisors are arranged as a row, in order from low to high according to their

powers. If the number of the same-order divisors is less than $r = \text{rank}(\mathbf{A}(\lambda))$, then make up the row with the zero-order divisor 1.

(2) The invariant factors of $\mathbf{A}(\lambda)$ are obtained by respectively multiplying along the columns.

EXAMPLE 7.13 Given a polynomial matrix $\mathbf{A}(\lambda)$ with rank $r = 4$, and seven elementary divisors, $\lambda, \lambda^2, \lambda^2, \lambda - 1, \lambda - 1, \lambda + 1, (\lambda + 1)^3$, find the invariant divisors of $\mathbf{A}(\lambda)$.

Solution Use the given elementary divisors to form the rectangular array

$$\begin{array}{cccc} 1 & \lambda & \lambda^2 & \lambda^2 \\ 1 & 1 & \lambda - 1 & \lambda - 1 \\ 1 & 1 & \lambda + 1 & (\lambda + 1)^3 \end{array}$$

Multiply respectively down the columns to obtain the invariant factors of $\mathbf{A}(\lambda)$:

$$\sigma_1(\lambda) = 1, \quad \sigma_2(\lambda) = \lambda, \quad \sigma_3(\lambda) = \lambda^2(\lambda - 1)(\lambda + 1),$$
$$\sigma_4(\lambda) = \lambda^2(\lambda - 1)(\lambda + 1)^3.$$

Once the Smith normal form of the λ-matrix $\lambda\mathbf{I} - \mathbf{A}$ is transformed to the Jordan canonical form of the matrix polynomial, the matrix function can be computed. However, although this method overcomes the disadvantage of the elementary transformations, it still needs to use the similarity transformation matrix $\mathbf{P}$ and its inverse matrix $\mathbf{P}^{-1}$. Can we avoid both elementary transformations and also the similarity transformation matrix together with its inverse matrix to calculate a matrix function? This is the topic of the next section.

7.5 Cayley–Hamilton Theorem with Applications

The eigenvalues of an $n \times n$ matrix $\mathbf{A}$ are determined by roots of its characteristic polynomial $\det(\mathbf{A} - \lambda\mathbf{I})$. Moreover, the characteristic polynomial is also closely related to the calculation of the matrix inverse, matrix power and matrix exponential function, so a more in-depth discussion and analysis on characteristic polynomial is needed.

7.5.1 Cayley–Hamilton Theorem

The Cayley–Hamilton theorem is an important result of regarding the characteristic polynomial of a general matrix. Using this theorem many problems can be easily solved, such as finding the inverse of matrices and the computation of several matrix functions. In order to introduce the Cayley–Hamilton theorem, we first present a few important concepts on polynomials.

As stated earlier, when $p_n \neq 0$, n is called the order of the polynomial $p(x) =$

$p_n x^n + p_{n-1}x^{n-1} + \cdots + p_1 x + p_0$. An nth-order polynomial is said to be *monic* if the coefficient of x^n is equal to 1.

If

$$p(\mathbf{A}) = p_n\mathbf{A}^n + p_{n-1}\mathbf{A}^{n-1} + \cdots + p_1\mathbf{A} + p_0\mathbf{I} = \mathbf{O},$$

then $p(x) = p_n x^n + p_{n-1}x^{n-1} + \cdots + p_1 x + p_0$ is the polynomial "annihilating" the matrix $\mathbf{A}$, and it is simply called the *annihilating polynomial.*

For an $n \times n$ matrix $\mathbf{A}$, let m be the smallest integer such that the powers $\mathbf{I}, \mathbf{A}, \ldots, \mathbf{A}^m$ are linearly dependent. Hence, we have

$$p_m\mathbf{A}^m + p_{m-1}\mathbf{A}^{m-1} + \cdots + p_1\mathbf{A} + p_0\mathbf{I} = \mathbf{O}_{n\times n}. \tag{7.5.1}$$

The polynomial $p(x) = p_m x^m + p_{m-1}x^{m-1} + \cdots + p_1 x + p_0$ is said to be the *minimal polynomial* of the matrix $\mathbf{A}$.

As atated in the following theorem, the characteristic polynomial $p(x) = \det(\mathbf{A} - x\mathbf{I})$ is the annihilating polynomial of the $n \times n$ matrix $\mathbf{A}$.

THEOREM 7.10 (Cayley–Hamilton theorem) *Any square matrix $\mathbf{A}_{n\times n}$ satisfies its own characteristic equation: if the characteristic equation is $p(x) = |\mathbf{A} - x\mathbf{I}| = p_n x^n + p_{n-1}x^{n-1} + \cdots + p_1 x + p_0 = 0$ then*

$$p_n\mathbf{A}^n + p_{n-1}\mathbf{A}^{n-1} + \cdots + p_1\mathbf{A} + p_0\mathbf{I} = \mathbf{O}, \tag{7.5.2}$$

where $\mathbf{I}$ and $\mathbf{O}$ are the $n \times n$ identity matrix and null matrix, respectively.

Proof Let $p(\mathbf{A}) = p_n\mathbf{A}^n + p_{n-1}\mathbf{A}^{n-1} + \cdots + p_1\mathbf{A} + p_0\mathbf{I}$. If $(\lambda, \mathbf{u})$ is an eigenpair of the matrix $\mathbf{A}$, i.e., $\mathbf{A}\mathbf{u} = \lambda\mathbf{u}$, then

$$\mathbf{A}^2\mathbf{u} = \lambda\mathbf{A}\mathbf{u} = \lambda^2\mathbf{u}, \quad \mathbf{A}^3\mathbf{u} = \lambda\mathbf{A}^2\mathbf{u} = \lambda^3\mathbf{u}, \quad \cdots, \quad \mathbf{A}^n\mathbf{u} = \lambda^n\mathbf{u}.$$

Therefore we have

$$\begin{aligned} p(\mathbf{A})\mathbf{u} &= (p_n\mathbf{A}^n + p_{n-1}\mathbf{A}^{n-1} + \cdots + p_1\mathbf{A} + p_0\mathbf{I})\mathbf{u} \\ &= p_n\mathbf{A}^n\mathbf{u} + p_{n-1}\mathbf{A}^{n-1}\mathbf{u} + \cdots + p_1\mathbf{A}\mathbf{u} + p_0\mathbf{u} \\ &= p_n\lambda^n\mathbf{u} + p_{n-1}\lambda^{n-1}\mathbf{u} + \cdots + p_1\lambda\mathbf{u} + p_0\mathbf{u} \\ &= (p_n\lambda^n + p_{n-1}\lambda^{n-1} + \cdots + p_1\lambda + p_0)\mathbf{u}. \end{aligned} \tag{7.5.3}$$

Because the eigenvalue λ is a root of the characteristic equation $p(x) = 0$, we get

$$p(\lambda) = |\mathbf{A} - \lambda\mathbf{I}| = p_n\lambda^n + p_{n-1}\lambda^{n-1} + \cdots + p_1\lambda + p_0 = 0. \tag{7.5.4}$$

Substitute this result into Equation (7.5.3) to yield

$$p(\mathbf{A})\mathbf{u} = 0 \cdot \mathbf{u} = \mathbf{0}, \quad \text{for any eigenvector } \mathbf{u} \neq \mathbf{0}. \tag{7.5.5}$$

Hence, $p(\mathbf{A}) = \mathbf{O}$. □

The Cayley–Hamilton theorem has many interesting and important applications. For example, by using this theorem, it can be shown directly that two similar matrices have the same eigenvalues.

Consider the characteristic polynomials of two similar matrices. Let $\mathbf{B} = \mathbf{S}^{-1}\mathbf{AS}$ be a matrix similar to $\mathbf{A}$, and let its characteristic polynomial $p(x) = \det(\mathbf{A} - x\mathbf{I}) = p_n x^n + p_{n-1}x^{n-1} + \cdots + p_1 x + p_0$. By the Cayley–Hamilton theorem, we know that $p(\mathbf{A}) = p_n\mathbf{A}^n + p_{n-1}\mathbf{A}^{n-1} + \cdots + p_1\mathbf{A} + p_0\mathbf{I} = \mathbf{O}$.

For the similar matrix $\mathbf{B}$, because

$$\mathbf{B}^k = (\mathbf{S}^{-1}\mathbf{AS})(\mathbf{S}^{-1}\mathbf{AS})\cdots(\mathbf{S}^{-1}\mathbf{AS}) = \mathbf{S}^{-1}\mathbf{A}^k\mathbf{S},$$

we have

$$\begin{aligned} p(\mathbf{B}) &= p_n\mathbf{B}^n + p_{n-1}\mathbf{B}^{n-1} + \cdots + p_1\mathbf{B} + p_0\mathbf{I} \\ &= p_n\mathbf{S}^{-1}\mathbf{A}^n\mathbf{S} + p_{n-1}\mathbf{S}^{-1}\mathbf{A}^{n-1}\mathbf{S} + \cdots + p_1\mathbf{S}^{-1}\mathbf{AS} + p_0\mathbf{I} \\ &= \mathbf{S}^{-1}\left(p_n\mathbf{A}^n + p_{n-1}\mathbf{A}^{n-1} + \cdots + p_1\mathbf{A} + p_0\mathbf{I}\right)\mathbf{S} \\ &= \mathbf{S}^{-1}p(\mathbf{A})\mathbf{S} \\ &= \mathbf{O} \end{aligned}$$

since $p(\mathbf{A}) = \mathbf{O}$ by the Cayley–Hamilton theorem. In other words, two similar matrices $\mathbf{A} \sim \mathbf{B}$ have the same characteristic polynomials and hence have the same eigenvalues.

In the following, we discuss some important applications of the Cayley–Hamilton theorem.

7.5.2 Computation of Inverse Matrices

If a matrix $\mathbf{A}_{n\times n}$ is nonsingular, then premultiply (or postmultiply) both sides of Equation (7.5.2) by $\mathbf{A}^{-1}$ to yield directly

$$p_n\mathbf{A}^{n-1} + p_{n-1}\mathbf{A}^{n-2} + \cdots + p_2\mathbf{A} + p_1\mathbf{I} + p_0\mathbf{A}^{-1} = \mathbf{O},$$

from which we get a computation formula for the inverse matrix:

$$\mathbf{A}^{-1} = -\frac{1}{p_0}(p_n\mathbf{A}^{n-1} + p_{n-1}\mathbf{A}^{n-2} + \cdots + p_2\mathbf{A} + p_1\mathbf{I}). \tag{7.5.6}$$

Example 7.14 Use the Cayley–Hamilton theorem to compute the inverse of the matrix

$$\mathbf{A} = \begin{bmatrix} 2 & 0 & 1 \\ 0 & 2 & 0 \\ 0 & 0 & 3 \end{bmatrix}.$$

Solution The second power of $\mathbf{A}$ is given by

$$\mathbf{A}^2 = \begin{bmatrix} 2 & 0 & 1 \\ 0 & 2 & 0 \\ 0 & 0 & 3 \end{bmatrix} \begin{bmatrix} 2 & 0 & 1 \\ 0 & 2 & 0 \\ 0 & 0 & 3 \end{bmatrix} = \begin{bmatrix} 4 & 0 & 5 \\ 0 & 4 & 0 \\ 0 & 0 & 9 \end{bmatrix}$$

and the characteristic polynomial is

$$\begin{aligned} \det(\mathbf{A} - x\mathbf{I}) &= \begin{vmatrix} 2-x & 0 & 1 \\ 0 & 2-x & 0 \\ 0 & 0 & 3-x \end{vmatrix} = (2-x)^2(3-x) \\ &= -x^3 + 7x^2 - 16x + 12, \end{aligned}$$

i.e., $p_0 = 12$, $p_1 = -16$, $p_2 = 7$, $p_3 = -1$. Substitute these values together with the matrix $\mathbf{A}$ and its second power $\mathbf{A}^2$ into the matrix inverse formula (7.5.6):

$$\begin{aligned} \mathbf{A}^{-1} &= -\frac{1}{12}\left(-\mathbf{A}^2 + 7\mathbf{A} - 16\mathbf{I}\right) \\ &= -\frac{1}{12}\left(- \begin{bmatrix} 4 & 0 & 5 \\ 0 & 4 & 0 \\ 0 & 0 & 9 \end{bmatrix} + 7 \begin{bmatrix} 2 & 0 & 1 \\ 0 & 2 & 0 \\ 0 & 0 & 3 \end{bmatrix} - 16 \begin{bmatrix} 1 & 0 & 0 \\ 0 & 1 & 0 \\ 0 & 0 & 1 \end{bmatrix} \right) \\ &= \frac{1}{6}\begin{bmatrix} 3 & 0 & -1 \\ 0 & 3 & 0 \\ 0 & 0 & 2 \end{bmatrix}. \end{aligned}$$

The Cayley–Hamilton theorem also can be used to compute the generalized inverse matrix of any complex matrix.

Theorem 7.11 [119] *Let* $\mathbf{A}$ *be an* $m \times n$ *complex matrix, and*

$$f(\lambda) = (-1)^m(a_0\lambda^m + a_1\lambda^{m-1} + \cdots + a_{m-1}\lambda + a_m), \quad a_0 = 1, \tag{7.5.7}$$

be the characteristic polynomial $\det(\mathbf{AA}^H - \lambda\mathbf{I})$ *of the matrix product* $\mathbf{AA}^H$. *If* k *is the largest integer such that* $a_k \neq 0$ *then the generalized inverse matrix of* $\mathbf{A}$ *is determined by*

$$\mathbf{A}^{\dagger} = -a_k^{-1}\mathbf{A}^H\left((\mathbf{AA}^H)^{k-1} + a_1(\mathbf{AA}^H)^{k-2} + \cdots + a_{k-2}(\mathbf{AA}^H) + a_{k-1}\mathbf{I}\right). \tag{7.5.8}$$

When $k = 0$ *is the largest integer such that* $a_k \neq 0$, *the generalized inverse matrix* $\mathbf{A}^{\dagger} = \mathbf{O}$.

The above theorem is sometimes called the *Decell theorem.* Notice that the largest integer k in the Decell theorem is the rank of the matrix $\mathbf{A}$, i.e., $k = \text{rank}(\mathbf{A})$. From this theorem, Decell presented the following method for computing the Moore–Penrose inverse matrix [119].

(1) Form the matrix sequence $\mathbf{A}_0, \mathbf{A}_1, \ldots, \mathbf{A}_k$, where the scalars q_i are defined in the second column and where the matrices $\mathbf{B}_i$ are recursively computed from the first two columns:

$$\begin{aligned}
&\mathbf{A}_0 = \mathbf{O}, && -1 = q_0, && \mathbf{B}_0 = \mathbf{I}\\
&\mathbf{A}_1 = \mathbf{A}\mathbf{A}^H, && \mathrm{tr}(\mathbf{A}_1) = q_1, && \mathbf{B}_1 = \mathbf{A}_1 - q_1\mathbf{I}\\
&\mathbf{A}_2 = \mathbf{A}\mathbf{A}^H\mathbf{B}_1, && \frac{\mathrm{tr}(\mathbf{A}_2)}{2} = q_2, && \mathbf{B}_2 = \mathbf{A}_2 - q_2\mathbf{I}\\
&\vdots && \vdots && \vdots\\
&\mathbf{A}_{k-1} = \mathbf{A}\mathbf{A}^H\mathbf{B}_{k-2}, && \frac{\mathrm{tr}(\mathbf{A}_{k-1})}{k-1} = q_{k-1}, && \mathbf{B}_{k-1} = \mathbf{A}_{k-1} - q_{k-1}\mathbf{I}\\
&\mathbf{A}_k = \mathbf{A}\mathbf{A}^H\mathbf{B}_{k-1}, && \frac{\mathrm{tr}(\mathbf{A}_k)}{k} = q_k, && \mathbf{B}_k = \mathbf{A}_k - q_k\mathbf{I}.
\end{aligned}$$

Faddeev [149, pp. 260–265] showed that the coefficients constructed in this way satisfy $q_i = -a_i$, $i = 1, \ldots, k$.

(2) Compute the Moore–Penrose inverse matrix

$$\begin{aligned}
\mathbf{A}^\dagger &= -a_k^{-1}\mathbf{A}^H[(\mathbf{A}\mathbf{A}^H)^{k-1} + a_1(\mathbf{A}\mathbf{A}^H)^{k-2} + \cdots + a_{k-2}(\mathbf{A}\mathbf{A}^H) + a_{k-1}\mathbf{I}]\\
&= -a_k^{-1}\mathbf{A}^H\mathbf{B}_{k-1}.
\end{aligned} \tag{7.5.9}$$

7.5.3 Computation of Matrix Powers

Given a matrix $\mathbf{A}_{n\times n}$ and an integer k, we say that $\mathbf{A}^k$ is the kth power of the matrix $\mathbf{A}$. If k is large, calculation of the matrix power $\mathbf{A}^k$ is a very tedious process. Fortunately, the Cayley–Hamilton theorem provides a simplest solution to this computation problem.

Consider the polynomial division $f(x)/g(x)$, where $g(x) \neq 0$. According to Euclidean division, it is known that there exist two polynomials $q(x)$ and $r(x)$ such that

$$f(x) = g(x)q(x) + r(x),$$

where $q(x)$ and $r(x)$ are respectively called the *quotient* and the *remainder*, and either the order of the remainder $r(x)$ is less than the order of the quotient $g(x)$ or $r(x) = 0$.

Let the characteristic polynomial of the matrix $\mathbf{A}$ be given by

$$p(x) = p_n x^n + p_{n-1}x^{n-1} + \cdots + p_1 x + p_0.$$

Then, for any x, its Kth power is given by

$$x^K = p(x)q(x) + r(x). \tag{7.5.10}$$

When x is a root of the characteristic equation $p(x) = 0$, the above equation becomes

$$x^K = r(x) = r_0 + r_1 x + \cdots + r_{n-1} x^{n-1}, \tag{7.5.11}$$

because the order of $r(x)$ is less than the order n of $p(x)$.

Replacing the scalar x by the matrix $\mathbf{A}$, Equation (7.5.10) becomes

$$\mathbf{A}^K = p(\mathbf{A})q(\mathbf{A}) + r(\mathbf{A}). \tag{7.5.12}$$

According to the Cayley–Hamilton theorem, if $p(x)$ is the characteristic polynomial of the matrix $\mathbf{A}$ then $p(\mathbf{A}) = \mathbf{O}$. Hence Equation (7.5.12) reduces to

$$\mathbf{A}^K = r(\mathbf{A}) = r_0 \mathbf{I} + r_1 \mathbf{A} + \cdots + r_{n-1} \mathbf{A}^{n-1}. \tag{7.5.13}$$

Equation (7.5.13) provides Algorithm 7.2 for computing $\mathbf{A}^K$.

Algorithm 7.2 Calculation of matrix powers [229, p. 172]

---

input: $\mathbf{A} \in \mathbb{R}^{n \times n}$ and the integer K.

1. Compute n roots (eigenvalues) $\lambda_1, \ldots, \lambda_n$ of the characteristic polynomial
   $p(x) = \det(\mathbf{A} - x\mathbf{I}) = p_n x^n + p_{n-1} x^{n-1} + \cdots + p_1 x + p_0$.
2. Solve the system of linear equations
   $r_0 + \lambda_1 r_1 + \cdots + \lambda_1^{n-1} r_{n-1} = \lambda_1^K,$
   $r_0 + \lambda_2 r_1 + \cdots + \lambda_2^{n-1} r_{n-1} = \lambda_2^K,$
   $\vdots$
   $r_0 + \lambda_n r_1 + \cdots + \lambda_n^{n-1} r_{n-1} = \lambda_n^K,$
   to obtain the solutions $r_0, r_1, \ldots, r_{n-1}$.
3. Calculate the matrix power $\mathbf{A}^K = r_0 \mathbf{I} + r_1 \mathbf{A} + \cdots + r_{n-1} \mathbf{A}^{n-1}$.

output: $\mathbf{A}^K$.

Example 7.15 Given the matrix $\mathbf{A} = \begin{bmatrix} 1 & 1 \\ 2 & 1 \end{bmatrix}$, calculate the matrix power $\mathbf{A}^{731}$.

Solution Form the characteristic polynomial

$$p(x) = \det(\mathbf{A} - x\mathbf{I}) = \begin{vmatrix} 1-x & 1 \\ 2 & 1-x \end{vmatrix} = x^2 - 2x - 1.$$

Put $p(x) = 0$ to get the eigenvalues of $\mathbf{A}$: $\lambda_1 = 1 + \sqrt{2}$ and $\lambda_2 = 1 - \sqrt{2}$.

Solve the system of linear equations

$$r_0 + (1 + \sqrt{2}) r_1 = (1 + \sqrt{2})^{731},$$
$$r_0 + (1 - \sqrt{2}) r_1 = (1 - \sqrt{2})^{731},$$

to get the solutions

$$r_0 = \left(\frac{1}{2} - \frac{\sqrt{2}}{4}\right)(1+\sqrt{2})^{731} + \left(\frac{1}{2} + \frac{\sqrt{2}}{4}\right)(1-\sqrt{2})^{731},$$

$$r_1 = \frac{\sqrt{2}}{4}[(1+\sqrt{2})^{731} - (1-\sqrt{2})^{731}].$$

Finally, $\mathbf{A}^{731} = r_0\mathbf{I}_2 + r_1\mathbf{A}$.

7.5.4 Calculation of Matrix Exponential Functions

Mimicking the scalar exponential function

$$e^{at} = 1 + at + \frac{1}{2!}a^2t^2 + \cdots + \frac{1}{k!}a^kt^k + \cdots,$$

for a given matrix $\mathbf{A}$, one can define the matrix exponential function as

$$e^{\mathbf{A}t} = \mathbf{I} + \mathbf{A}t + \frac{1}{2!}\mathbf{A}^2t^2 + \cdots + \frac{1}{k!}\mathbf{A}^kt^k + \cdots. \tag{7.5.14}$$

The matrix exponential function can be used to represent the solutions of certain first-order differential equations. In engineering applications we often meet the first-order linear differential equation

$$\dot{\mathbf{x}}(t) = \mathbf{A}\mathbf{x}(t), \quad \mathbf{x}(0) = \mathbf{x}_0,$$

where $\mathbf{A}$ is a constant matrix. The solution of the above differential equation can be written as $\mathbf{x}(t) = e^{\mathbf{A}t}\mathbf{x}_0$. Hence, solving the linear first-order differential equation is equivalent to computing the matrix exponential function $e^{\mathbf{A}t}$.

Using the Cayley–Hamilton theorem, it can be shown that the solution of an nth-order linear matrix differential equation is uniquely determined.

Theorem 7.12 [289] *Let $\mathbf{A}$ be an $n \times n$ constant matrix; its characteristic polynomial is*

$$p(\lambda) = \det(\lambda\mathbf{I} - \mathbf{A}) = \lambda^n + c_{n-1}\lambda^{n-1} + \cdots + c_1\lambda + c_0. \tag{7.5.15}$$

If the nth-order matrix differential equation

$$\mathbf{\Phi}^{(n)}(t) + c_{n-1}\mathbf{\Phi}^{(n-1)}(t) + \cdots + c_1\mathbf{\Phi}'(t) + c_0\mathbf{\Phi}(t) = \mathbf{O} \tag{7.5.16}$$

satisfies the initial conditions

$$\mathbf{\Phi}(0) = \mathbf{I},\ \mathbf{\Phi}'(0) = \mathbf{A},\ \mathbf{\Phi}''(0) = \mathbf{A}^2,\ \ \ldots\ \ ,\ \mathbf{\Phi}^{(n-1)} = \mathbf{A}^{n-1} \tag{7.5.17}$$

then $\mathbf{\Phi}(t) = e^{\mathbf{A}t}$ is the unique solution of the nth-order matrix differential equation (7.5.16).

Theorem 7.13 [289] *Let* $\mathbf{A}$ *be an* $n \times n$ *constant matrix and its characteristic polynomial be*

$$p(\lambda) = \det(\lambda \mathbf{I} - \mathbf{A}) = \lambda^n + c_{n-1}\lambda^{n-1} + \cdots + c_1\lambda + c_0,$$

then the solution of the matrix differential equation (7.5.16) satisfying the initial condition (7.5.17) is given by

$$\mathbf{\Phi}(t) = \mathrm{e}^{\mathbf{A}t} = x_1(t)\mathbf{I} + x_2(t)\mathbf{A} + x_3(t)\mathbf{A}^2 + \cdots + x_n(t)\mathbf{A}^{n-1}, \tag{7.5.18}$$

where $x_k(t)$ *is the solution of the nth-order scalar differential equation*

$$x^{(n)}(t) + c_{n-1}x^{(n-1)}(t) + \cdots + c_1 x'(t) + c_0 x(t) = 0 \tag{7.5.19}$$

satisfying the conditions

$$\left.\begin{matrix} x_1(0) = 1 \\ x_1'(0) = 0 \\ \vdots \\ x_1^{(n-1)}(0) = 0 \end{matrix}\right\}, \quad \left.\begin{matrix} x_2(0) = 0 \\ x_2'(0) = 1 \\ \vdots \\ x_2^{(n-1)}(0) = 0 \end{matrix}\right\}, \quad \cdots, \quad \left.\begin{matrix} x_n(0) = 0 \\ x_n'(0) = 0 \\ \vdots \\ x_n^{(n-1)}(0) = 1 \end{matrix}\right\}.$$

Equation (7.5.18) gives an efficient method for computing the matrix exponential function $\mathrm{e}^{\mathbf{A}t}$.

Example 7.16 Given a matrix

$$\mathbf{A} = \begin{bmatrix} 1 & 0 & 1 \\ 0 & 1 & 0 \\ 0 & 0 & 2 \end{bmatrix},$$

use Theorem 7.13 to compute $\mathrm{e}^{\mathbf{A}t}$.

Solution From the characteristic equation

$$\det(\lambda \mathbf{I} - \mathbf{A}) = \begin{vmatrix} \lambda - 1 & 0 & -1 \\ 0 & \lambda - 1 & 0 \\ 0 & 0 & \lambda - 2 \end{vmatrix} = (\lambda - 1)^2(\lambda - 2) = 0,$$

we get the eigenvalues of $\mathbf{A}$: $\lambda_1 = 1, \lambda_2 = 1, \lambda_3 = 2$, where the multiplicity of the eigenvalue 1 is 2.

From Theorem 7.13 it is known that the matrix exponential function is given by

$$\mathrm{e}^{\mathbf{A}t} = x_1(t)\mathbf{I} + x_2(t)\mathbf{A} + x_3(t)\mathbf{A}^2,$$

where $x_1(t), x_2(t), x_3(t)$ are the solutions of the third-order scalar differential equation

$$x'''(t) + c_2 x''(t) + c_1 x'(t) + c_0 x(t) = 0$$

subject to the initial conditions

$$\left.\begin{array}{r}x_1(0)=1\\x_1'(0)=0\\x_1''(0)=0\end{array}\right\},\quad \left.\begin{array}{r}x_2(0)=0\\x_2'(0)=1\\x_2''(0)=0\end{array}\right\},\quad \left.\begin{array}{r}x_3(0)=0\\x_3'(0)=0\\x_3''(0)=1\end{array}\right\}.$$

From the eigenvalues $\lambda_1 = 1, \lambda_2 = 1, \lambda_3 = 2$ it is known that the general solution of the scalar differential equation $x'''(t) + c_2x''(t) + c_1x'(t) + c_0x(t) = 0$ is given by

$$x(t) = a_1te^t + a_2e^t + a_3e^{2t}.$$

From the general solution formula it is easy to proved as follows.

(a) Substitute the initial condition $x_1(0) = 1, x_1'(0) = 0, x_1''(0) = 0$ into the general solution formula to give

$$\begin{cases} a_2 + a_3 = 1 \\ a_1 + a_2 + 2a_3 = 0 \\ 2a_1 + a_2 + 4a_3 = 0 \end{cases} \Rightarrow \begin{cases} a_1 = -2 \\ a_2 = 0 \\ a_3 = 1 \end{cases}$$

i.e., the special solution satisfying those initial conditions is $x_1(t) = -2te^t + e^{2t}$.

(b) From the initial conditions $x_2(0) = 0, x_2'(0) = 1, x_2''(0) = 0$ we obtain

$$\begin{cases} a_2 + a_3 = 0 \\ a_1 + a_2 + 2a_3 = 1 \\ 2a_1 + a_2 + 4a_3 = 0 \end{cases} \Rightarrow \begin{cases} a_1 = 1 \\ a_2 = 2 \\ a_3 = -1 \end{cases}$$

and hence we have that the special solution satisfying those initial conditions is $x_2(t) = te^t + 2e^t - e^{2t}$.

(c) From the initial conditions $x_3(0) = 0, x_3'(0) = 0, x_3''(0) = 1$ we get

$$\begin{cases} a_2 + a_3 = 0 \\ a_1 + a_2 + 2a_3 = 0 \\ 2a_1 + a_2 + 4a_3 = 1 \end{cases} \Rightarrow \begin{cases} a_1 = -1 \\ a_2 = -1 \\ a_3 = 1. \end{cases}$$

In other words, the special solution satisfying those initial conditions is $x_3(t) = -te^t - e^t + e^{2t}$.

Moreover, from

$$\mathbf{A}^2 = \begin{bmatrix} 1 & 0 & 1 \\ 0 & 1 & 0 \\ 0 & 0 & 2 \end{bmatrix}\begin{bmatrix} 1 & 0 & 1 \\ 0 & 1 & 0 \\ 0 & 0 & 2 \end{bmatrix} = \begin{bmatrix} 1 & 0 & 3 \\ 0 & 1 & 0 \\ 0 & 0 & 4 \end{bmatrix}$$

and Theorem 7.13 we have

$$\begin{aligned}
e^{\mathbf{A}t} &= x_1(t)\mathbf{I} + x_2(t)\mathbf{A} + x_3(t)\mathbf{A}^2 \\
&= (-2te^t + e^{2t}) \begin{bmatrix} 1 & 0 & 0 \\ 0 & 1 & 0 \\ 0 & 0 & 1 \end{bmatrix} + (te^t + 2e^t - e^{2t}) \begin{bmatrix} 1 & 0 & 1 \\ 0 & 1 & 0 \\ 0 & 0 & 2 \end{bmatrix} \\
&\quad + (-te^t - e^t + e^{2t}) \begin{bmatrix} 1 & 0 & 3 \\ 0 & 1 & 0 \\ 0 & 0 & 4 \end{bmatrix},
\end{aligned}$$

which gives

$$e^{\mathbf{A}t} = \begin{bmatrix} -2te^t + e^t + e^{2t} & 0 & -2te^t - e^t + 2e^{2t} \\ 0 & -2te^t + e^t + e^{2t} & 0 \\ 0 & 0 & -4te^t + 3e^{2t} \end{bmatrix}.$$

Summarizing the above discussions, we may conclude that the Cayley–Hamliton theorem provides a very effective tool for matrix inversion, computation of matrix powers, solving linear differential equations (or equivalently computing the matrix exponential function) and so on.

7.6 Application Examples of Eigenvalue Decomposition

Eigenvalue decomposition has wide applications. In this section we present three typical applications of EVD: Pisarenko harmonic decomposition, the Karhunen–Loeve transformation and principal component analysis.

7.6.1 Pisarenko Harmonic Decomposition

Harmonic processes are often encountered in many engineering applications, and often it is necessary to determine the frequencies and powers of these harmonics. This is so-called harmonic retrieval. Here we present the harmonic retrieval method of Pisarenko [394].

Consider a harmonic process consisting of p real sinusoids:

$$x(n) = \sum_{i=1}^{p} A_i \sin(2\pi f_i n + \theta_i). \tag{7.6.1}$$

When each phase θ_i is a constant, the above harmonic process is a deterministic process that is nonstationary. In order to guarantee the stationarity of the harmonic process, it is generally assumed that the phases θ_i are random numbers distributed uniformly in $[-\pi, \pi]$. In this case, the harmonic process becomes stochastic.

A harmonic process can be described by a difference equation. First consider a single sinusoid. For simplicity, let the harmonic signal be $x(n) = \sin(2\pi fn + \theta)$. Recall the trigonometric identity

$$\sin(2\pi fn + \theta) + \sin(2\pi f(n-2) + \theta) = 2\cos(2\pi f)\sin(2\pi f(n-1) + \theta).$$

Substituting $x(n) = \sin(2\pi fn+\theta)$ into the above identity, we have the second-order difference equation

$$x(n) - 2\cos(2\pi f)x(n-1) + x(n-2) = 0.$$

Make the z transformation $X(z) = \sum_{n=-\infty}^{\infty} x(n)z^{-n}$ to get

$$(1 - 2\cos(2\pi f)z^{-1} + z^{-2})X(z) = 0.$$

Hence the characteristic equation is

$$1 - 2\cos(2\pi f)z^{-1} + z^{-2} = 0$$

which has a pair of complex conjugate roots, i.e.,

$$z = \cos(2\pi f) \pm \mathrm{j}\,\sin(2\pi f) = \mathrm{e}^{\pm \mathrm{j}\,2\pi f}.$$

Note that the modulus of the conjugate roots is equal to 1, i.e., $|z_1| = |z_2| = 1$. From the roots we can determine the frequencies of the sinusoids:

$$f_i = \frac{1}{2\pi}\arctan\left(\frac{\mathrm{Im}\, z_i}{\mathrm{Re}\, z_i}\right). \tag{7.6.2}$$

If p real sinusoidal signals have no repeated frequencies then the p frequencies are determined by the roots of the characteristic polynomial

$$\prod_{i=1}^{p}(z - z_i)(z - z_i^*) = \sum_{i=0}^{2p} a_i z^{2p-i} = 0$$

or

$$1 + a_1 z^{-1} + \cdots + a_{2p-1}z^{-(2p-1)} + a_{2p}z^{-2p} = 0. \tag{7.6.3}$$

It is easily seen that all the moduli of these roots are equal to 1. Because these roots appear in the form of conjugate pairs, in the coefficients of the characteristic polynomial (7.6.3) there exist the symmetries

$$a_i = a_{2p-i}, \quad i = 0, 1, \ldots, p, \tag{7.6.4}$$

and the difference equation corresponding to (7.6.3) is

$$x(n) + \sum_{i=1}^{2p} a_i x(n-i) = 0, \tag{7.6.5}$$

where n is a time variable.

A sinusoidal signal is generally observed to have additive white noise. Letting the additive white noise be $e(n)$, then the observed signal is given by

$$y(n) = x(n) + e(n) = \sum_{i=1}^{p} A_i \sin(2\pi f_i n + \theta_i) + e(n), \tag{7.6.6}$$

where $e(n) \sim N(0, \sigma_e^2)$ is a Gaussian white noise that is statistically independent of the sinusoidal signal $x(n)$. Substituting $x(n) = y(n) - e(n)$ into (7.6.5), we immediately have a difference equation satisfied by the sinusoidal signal with white noise as follows:

$$y(n) + \sum_{i=1}^{2p} a_i y(n-i) = e(n) + \sum_{i=1}^{2p} a_i e(n-i). \tag{7.6.7}$$

This is a special case of an autoregressive moving-average (ARMA) process.

To derive the *normal equation* obeyed by the autoregressive parameters of this special ARMA process, define the vector

$$\begin{aligned} \mathbf{y}(n) &= [y(n), y(n-1), \ldots, y(n-2p)]^T, \\ \mathbf{w} &= [1, a_1, \ldots, a_{2p}]^T, \\ \mathbf{e}(n) &= [e(n), e(n-1), \ldots, e(n-2p)]^T. \end{aligned} \tag{7.6.8}$$

Then, (7.6.7) can be written as

$$\mathbf{y}^T(n)\mathbf{w} = \mathbf{e}^T(n)\mathbf{w}. \tag{7.6.9}$$

Premultiplying (7.6.9) by the vector $\mathbf{y}(n)$ and taking the expectation yields

$$E\{\mathbf{y}(n)\mathbf{y}^T(n)\}\mathbf{w} = E\{\mathbf{y}(n)\mathbf{e}^T(n)\}\mathbf{w}. \tag{7.6.10}$$

Letting $R_y(k) = E\{y(n+k)y(n)\}$ denote the autocorrelation function of the observation data $y(n)$, we have

$$E\{\mathbf{y}(n)\mathbf{y}^T(n)\} = \begin{bmatrix} R_y(0) & R_y(-1) & \cdots & R_y(-2p) \\ R_y(1) & R_y(0) & \cdots & R_y(-2p+1) \\ \vdots & \vdots & \ddots & \vdots \\ R_y(2p) & R_y(2p-1) & \cdots & R_y(0) \end{bmatrix} \stackrel{\text{def}}{=} \mathbf{R},$$

$$E\{\mathbf{y}(n)\mathbf{e}^T(n)\} = E\{[\mathbf{x}(n) + \mathbf{e}(n)]\mathbf{e}^T(n)\} = E\{\mathbf{e}(n)\mathbf{e}^T(n)\} = \sigma_e^2\mathbf{I}.$$

Substituting the above two relations into (7.6.10), we get the important *normal equation*

$$\mathbf{R}\mathbf{w} = \sigma_e^2\mathbf{w}. \tag{7.6.11}$$

Here σ_e^2 is the smallest eigenvalue of the autocorrelation matrix $\mathbf{R} = E\{\mathbf{y}(n)\mathbf{y}^T(n)\}$ of the observation process $\{y(n)\}$, and the coefficient vector $\mathbf{w}$ is the eigenvector associated with this smallest eigenvalue. This is the basis of the Pisarenko harmonic decomposition method.

Note that the smallest eigenvalue σ_e^2 of the autocorrelation matrix $\mathbf{R}$ is just the variance of the white noise $e(n)$, and the other eigenvalues correspond to the powers of the harmonic signals. When the signal-to-noise ratio is large, the eigenvalue σ_e^2 is clearly smaller than other eigenvalues. Hence, the Pisarenko harmonic decomposition informs us that the harmonic retrieval problem becomes the EVD of the sample correlation matrix $\mathbf{R}$: the coefficient vector $\mathbf{w}$ of the characteristic polynomial of the harmonic process is the eigenvector corresponding to the smallest eigenvalue σ_e^2 of $\mathbf{R}$.

7.6.2 Discrete Karhunen–Loeve Transformation

In many engineering applications, it is necessary to represent observed samples of a random signal as a set of coefficients, and it is to be hoped that this new signal representation has some desirable properties. For example, for encoding it is desired that the signal can be represented by a small number of coefficients and that these coefficients represent the power of the original signal. For optimal filtering, it is desirable that the converted samples are statistically uncorrelated, so that we can reduce the complexity of the filter or enhance the signal to noise ratio. The common way to achieve these goals is to expand the signal as a linear combination of orthogonal basis functions.

If a set of orthogonal basis functions is selected according to the autocorrelation matrix of the signal observation samples, it is possible to obtain the signal representation with the minimum mean square error (MMSE) among all possible orthogonal basis functions. In the MMSE sense, such a signal representation is the optimal signal representation. This signal transformation, proposed by Karhunen and Loeve for a continuous random signal, is called the *Kauhunen–Loeve* (KL) *transformation.* Later, Hotelling extended this transformation to a discrete random signal, so it is also called the *Hotelling transformation.* However, in most of the literature, this is still called the *discrete Karhunen–Loeve transformation.*

Let $\mathbf{x} = [x_1, \ldots, x_M]^T$ be a zero-mean random vector whose autocorrelation matrix is $\mathbf{R}_x = E\{\mathbf{x}\mathbf{x}^H\}$. Consider the linear transformation

$$\mathbf{w} = \mathbf{Q}^H \mathbf{x}, \tag{7.6.12}$$

where $\mathbf{Q}$ is a unitary matrix, i.e., $\mathbf{Q}^{-1} = \mathbf{Q}^H$. Then, the original random signal vector $\mathbf{x}$ can be expressed as the linear combination of the coefficient of the vector $\mathbf{w}$, namely

$$\mathbf{x} = \mathbf{Q}\mathbf{w} = \sum_{i=1}^{M} w_i \mathbf{q}_i, \quad \mathbf{q}_i^H \mathbf{q}_j = 0, \quad i \neq j, \tag{7.6.13}$$

where $\mathbf{q}_i$ are the column vectors of $\mathbf{Q}$.

In order to reduce the number of transform coefficients w_i, it is assumed that only the first m coefficients $w_1, \ldots, w_m$, $m = 1, \ldots, M$ are used to approximate the random signal vector $\mathbf{x}$, i.e.,

$$\hat{\mathbf{x}} = \sum_{i=1}^{m} w_i \mathbf{q}_i, \quad 1 \leq m \leq M. \tag{7.6.14}$$

Then, the error of the mth-order approximation of a random-signal vector is given by

$$\mathbf{e}_m = \mathbf{x} - \hat{\mathbf{x}} = \sum_{i=1}^{M} w_i \mathbf{q}_i - \sum_{i=1}^{m} w_i \mathbf{q}_i = \sum_{i=m+1}^{M} w_i \mathbf{q}_i. \tag{7.6.15}$$

This yields the mean squared error

$$E_m = E\{\mathbf{e}_m^H \mathbf{e}_m\} = \sum_{i=m+1}^{M} \mathbf{q}_i^H E\{|w_i|^2\} \mathbf{q}_i = \sum_{i=m+1}^{M} E\{|w_i|^2\} \mathbf{q}_i^H \mathbf{q}_i. \tag{7.6.16}$$

From $w_i = \mathbf{q}_i^H \mathbf{x}$ it is easily seen that $E\{|w_i|^2\} = \mathbf{q}_i^H \mathbf{R}_x \mathbf{q}_i$. If we have the constraint $\mathbf{q}_i^H \mathbf{q}_i = 1$ then the mean squared error expressed by (7.6.16) can be rewritten as

$$E_m = \sum_{i=m+1}^{M} E\{|w_i|^2\} = \sum_{i=m+1}^{M} \mathbf{q}_i^H \mathbf{R}_x \mathbf{q}_i \tag{7.6.17}$$

subject to

$$\mathbf{q}_i^H \mathbf{q}_i = 1, \quad i = m+1, m+2, \ldots, M.$$

In order to minimize the mean squared error, use the Lagrange multiplier method to form the cost function

$$J = \sum_{i=m+1}^{M} \mathbf{q}_i^H \mathbf{R}_x \mathbf{q}_i + \sum_{i=m+1}^{M} \lambda_i (1 - \mathbf{q}_i^H \mathbf{q}_i).$$

Let $\dfrac{\partial J}{\partial \mathbf{q}_i^*} = 0$, $i = m+1, m+2, \ldots, M$, i.e.,

$$\frac{\partial}{\partial \mathbf{q}_i^*} \left(\sum_{i=m+1}^{M} \mathbf{q}_i^H \mathbf{R}_x \mathbf{q}_i + \sum_{i=m+1}^{M} \lambda_i (1 - \mathbf{q}_i^H \mathbf{q}_i) \right) = \mathbf{R}_x \mathbf{q}_i - \lambda_i \mathbf{q}_i = 0$$

for $i = m+1, m+2, \ldots, M$; then we have

$$\mathbf{R}_x \mathbf{q}_i = \lambda_i \mathbf{q}_i, \quad i = m+1, m+2, \ldots, M. \tag{7.6.18}$$

This transformation is known as the Karhunen–Loeve (KL) transformation.

The above discussion shows that when using (7.6.14) to approximate a random signal vector $\mathbf{x}$, in order to minimize the mean squared error of approximation, we should select Lagrange multipliers λ_i and the orthogonal basis vector of the cost function $\mathbf{g}_i$ to be respectively the last $M - m$ eigenvalues and the corresponding eigenvectors of $\mathbf{R}_x$. In other words, the orthogonal basis for the random-signal vector in (7.6.14) should be the first m eigenvectors of $\mathbf{R}_x$.

Let the EVD of the $M \times M$ autocorrelation $\mathbf{R}_x$ be

$$\mathbf{R}_x = \sum_{i=1}^{M} \lambda_i \mathbf{u}_i \mathbf{u}_i^H, \tag{7.6.19}$$

then the selected orthogonal basis vectors in (7.6.14) are $\mathbf{q}_i = \mathbf{u}_i$, $i = 1, \ldots, m$.

If the autocorrelation matrix $\mathbf{R}_x$ has only K large eigenvalues, and the other $M - K$ eigenvalues are negligible, then the approximation order m in (7.6.14) should equal K, and hence the Kth-order *discrete Kauhunen–Loeve expansion* is obtained:

$$\hat{\mathbf{x}} = \sum_{i=1}^{K} w_i \mathbf{u}_i, \tag{7.6.20}$$

where $w_i, i = 1, \ldots, K$, is the ith element of the $K \times 1$ vector

$$\mathbf{w} = \mathbf{U}_1^H \mathbf{x}. \tag{7.6.21}$$

Here $\mathbf{U}_1 = [\mathbf{u}_1, \ldots, \mathbf{u}_K]$ consists of the eigenvectors corresponding to the K leading eigenvalues of the autocorrelation matrix. In this case, the mean squared error of the Kth-order discrete *Karhunen–Loeve expansion* is

$$E_K = \sum_{i=K+1}^{M} \mathbf{u}_i^H \mathbf{R}_x \mathbf{u}_i = \sum_{i=K+1}^{M} \mathbf{u}_i^H \left(\sum_{j=1}^{M} \lambda_j \mathbf{u}_j \mathbf{u}_j^H \right) \mathbf{u}_i = \sum_{i=K+1}^{M} \lambda_i. \tag{7.6.22}$$

Because λ_i, $i = K + 1, \ldots, M$, are the minor eigenvalues of the autocorrelation matrix $\mathbf{R}_x$, the mean squared error E_K is very small.

Suppose that the original data to be transmitted are $x_1, \ldots, x_M$, the direct transmitting of these data arriving at the transmitter may bring two problems: they may easily be received by others; and the data length M can be very large in many cases. For example, an image needs to be converted into data by rows, and then all these data are synthesized to a long segment. By using the discrete Karhunen–Loeve expansion, we can avoid the above two difficulties of direct data emission. Let M samples of the image or voice signals to be compressed and transmitted be $x_c(0), x_c(1), \ldots, x_c(M - 1)$, where M is large. If we perform an EVD of the autocorrelation matrix of the given data and determine the number K of principal eigenvalues, then we obtain K linear transformation coefficients $w_1, \ldots, w_K$ and K orthogonal eigenvectors $\mathbf{u}_1, \ldots, \mathbf{u}_K$. Thus we need only to transmit K coefficients $w_1, \ldots, w_K$ at the transmitter. If the K eigenvectors are stored at the receiver then we can utilize

$$\hat{\mathbf{x}} = \sum_{i=1}^{K} w_i \mathbf{u}_i \tag{7.6.23}$$

to reconstruct the M pieces of transmitted data $x_c(i)$, $i = 0, 1, \ldots, M-1$.

The process of transforming the M pieces of signal data $x_c(0), x_c(1), \ldots, x_c(M-1)$ into K coefficients $w_1, \ldots, w_K$ is called *signal encoding* or *data compression*, while the process of reconstructing M data signals from K coefficients is known as *signal decoding*. Figure 7.1 shows this signal encoding and decoding schematically.

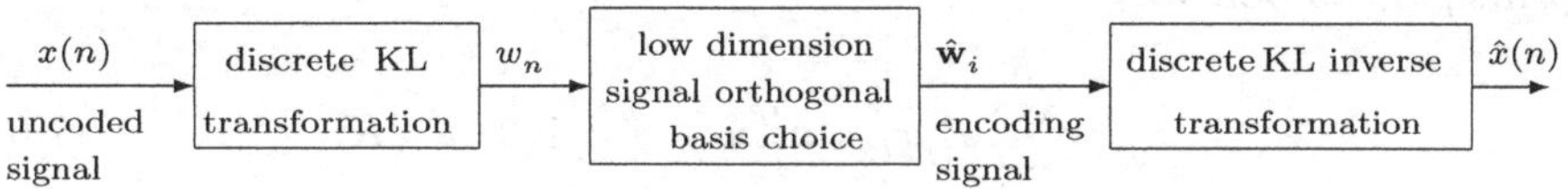

Figure 7.1 Discrete KL transformation based signal encoding.

The ratio M/K is called the *compression ratio*. If K is much smaller than M then the compression ratio is large. Obviously, when the discrete KL transformation is used to encode the original data, not only is the length of the transmitted data greatly compressed but also, even if the K encoding coefficients are received by others, it is difficult for them to accurately reconstruct the original data without knowing the K principal eigenvectors.

7.6.3 Principal Component Analysis

Consider an index set of P statistically correlated characters $\{x_1, \ldots, x_P\}$. Due to correlation, there is redundant information among the P characters. We hope to find a new set of K mutually orthogonal characters $\{\tilde{x}_{p_1}, \ldots, \tilde{x}_{p_K}\}$ where $\tilde{x}_{p_i}, i = 1, \ldots, K$, is the ith principal component in which there is no redundant information. Such a process is known as *feature extraction*. Via feature extraction, a higher-dimensional vector space with redundant information becomes a lower-dimensional vector space without redundant information. Such an operation is called *dimensionality reduction*. A typical dimensionality reduction method is principal component analysis.

Through an orthogonal transformation, P statistically correlated original characters are transformed to K mutually orthogonal new characters. Because the K new characters $\{\tilde{x}_{p_1}, \ldots, \tilde{x}_{p_K}\}$ have the larger powers $|\tilde{x}_{p_1}|^2, \ldots, |\tilde{x}_{p_K}|^2$, they can be regarded as the principal components of the P original character indexes $x_1, \ldots, x_P$. Hence, data or signal analysis based on K principal components is known as the *principal component analysis* (PCA).

Thus the main objective of PCA is to use K ($< P$) principal components to express P statistically correlated variables. In order to fully reflect the useful information carried by the P variables, each principal component should be a linear combination of the original P variables.

Definition 7.17 Let $\mathbf{R}_x$ be the autocorrelation matrix of a $P \times 1$ data vector

$\mathbf{x}$. If the $P \times P$ matrix $\mathbf{R}_x$ has K principal eigenvalues then their corresponding K eigenvectors are called the *principal components* of the data vector $\mathbf{x}$.

The main steps and ideas of the PCA are as follows.

(1) *Dimensionality reduction* The P original variables are synthesized into K principal components:

$$\tilde{x}_j = \sum_{i=1}^{P} a_{ij}^* x_i = \mathbf{a}_j^H \mathbf{x}, \quad j = 1, 2, \ldots, K, \tag{7.6.24}$$

where $\mathbf{a}_j = [a_{1j}, \ldots, a_{Pj}]^T$ and $\mathbf{x} = [x_1, \ldots, x_P]^T$.

(2) *Orthogonalization* In order to make K principal components orthonormal, i.e.,

$$\langle \tilde{x}_i, \tilde{x}_j \rangle = \mathbf{x}^H \mathbf{x} \mathbf{a}_i^H \mathbf{a}_j = \delta_{ij},$$

the coefficient vectors $\mathbf{a}_i$ must satisfy the orthonormal condition $\mathbf{a}_i^H \mathbf{a}_j = \delta_{ij}$, as all components of $\mathbf{x}$ are statistically correlated, i.e., $\mathbf{x}^H \mathbf{x} \neq 0$.

(3) *Power maximization* Selecting $\mathbf{a}_i = \mathbf{u}_i$, $i = 1, \ldots, K$, where $\mathbf{u}_i$ are the K eigenvectors corresponding to the K leading eigenvalues $\lambda_1 \geq \cdots \geq \lambda_K$ of the autocorrelation matrix $\mathbf{R}_x = E\{\mathbf{x}\mathbf{x}^H\}$, it is easy to compute the energy of each nonredundant component:

$$E_{\tilde{x}_i} = E\{|\tilde{x}_i|^2\} = E\{\mathbf{a}_i^H \mathbf{x} (\mathbf{a}_i^H \mathbf{x})^*\} = \mathbf{u}_i^H E\{\mathbf{x}\mathbf{x}^H \mathbf{u}_i\} = \mathbf{u}_i^H \mathbf{R}_x \mathbf{u}_i = \lambda_i.$$

Because the eigenvalues are arranged in nondescending order $E_{\tilde{x}_1} \geq E_{\tilde{x}_2} \geq \cdots \geq E_{\tilde{x}_k}$, $\tilde{x}_1$ is often referred to as the first principal component of $\mathbf{x}$, $\tilde{x}_2$ as the second principal component of $\mathbf{x}$ and so on.

For the $P \times P$ autocorrelation matrix

$$\mathbf{R}_x = E\{\mathbf{x}\mathbf{x}^H\} = \begin{bmatrix} E\{|x_1|^2\} & E\{x_1 x_2^*\} & \cdots & E\{x_1 x_P^*\} \\ E\{x_2 x_1^*\} & E\{|x_2|^2\} & \cdots & E\{x_2 x_P^*\} \\ \vdots & \vdots & \ddots & \vdots \\ E\{x_P x_1^*\} & E\{x_P x_2^*\} & \cdots & E\{|x_P|^2\} \end{bmatrix}, \tag{7.6.25}$$

by the definition and property of the matrix trace, it can be seen that

$$\mathrm{tr}(\mathbf{R}_x) = E\{|x_1|^2\} + \cdots + E\{|x_P|^2\} = \lambda_1 + \cdots + \lambda_P. \tag{7.6.26}$$

If the autocorrelation matrix $\mathbf{R}_x$ has only K leading eigenvalues then

$$E\{|x_1|^2\} + \cdots + E\{|x_P|^2\} \approx \lambda_1 + \cdots + \lambda_K. \tag{7.6.27}$$

Summarizing the above discussion, we may conclude that the basic idea of PCA is to transform the original P statistically correlated random data into K mutually orthogonal principal components by using dimensionality reduction, orthogonalization and energy maximization. The sum of the energies of these principal components is approximately equal to the sum of the energies of the original P random data.

Definition 7.18 [520] Let $\mathbf{R}_x$ be the autocorrelation matrix of a $P \times 1$ data vector $\mathbf{x}$ with K principal eigenvalues and $P - K$ minor eigenvalues (i.e., small eigenvalues). The $P - K$ eigenvectors corresponding to these minor eigenvalues are called the *minor components* of the data vector $\mathbf{x}$.

Data or signal analysis based on the $P - K$ minor components is known as *minor component analysis* (MCA).

The PCA can provide the contour and main information about a signal or an image. In contrast, MCA can provide the details of a signal or the texture of an image. It has been widely used in many fields such as frequency estimation [317], [318], blind beamforming [189], moving-target display [251], clutter cancellation [24], etc. In signal or image analysis and pattern recognition, when the PCA cannot distinguish two object signals or images, the MCA is used to compare their details.

7.7 Generalized Eigenvalue Decomposition (GEVD)

The EVD of an $n \times n$ single matrix can be extended to the EVD of a matrix pair or pencil consisting of two matrices. The EVD of a matrix pencil is called *generalized eigenvalue decomposition* (GEVD). As a matter of fact, the EVD is a special example of the GEVD.

7.7.1 Generalized Eigenvalue Decomposition

The basis of the EVD is the eigensystem expressed by the linear transformation $\mathcal{L}[\mathbf{u}] = \lambda\mathbf{u}$: taking the linear transformation $\mathcal{L}[\mathbf{u}] = \mathbf{A}\mathbf{u}$, one gets the EVD $\mathbf{A}\mathbf{u} = \lambda\mathbf{u}$.

Now consider the generalization of the eigensystem composed of two linear systems $\mathcal{L}_a$ and $\mathcal{L}_b$ both of whose inputs are the vector $\mathbf{u}$, but the output $\mathcal{L}_a[\mathbf{u}]$ of the first system $\mathcal{L}_a$ is some constant times λ the output $\mathcal{L}_b[\mathbf{u}]$ of the second system $\mathcal{L}_b$. Hence the eigensystem is generalized as [231]

$$\mathcal{L}_a[\mathbf{u}] = \lambda\mathcal{L}_b[\mathbf{u}] \quad (\mathbf{u} \neq \mathbf{0}); \tag{7.7.1}$$

this is called a *generalized eigensystem*, denoted $(\mathcal{L}_a, \mathcal{L}_b)$. The constant λ and the nonzero vector $\mathbf{u}$ are known as the *generalized eigenvalue* and the *generalized eigenvector*, respectively.

In particular, if two linear transformations are respectively $\mathcal{L}_a[\mathbf{u}] = \mathbf{A}\mathbf{u}$ and $\mathcal{L}_b[\mathbf{u}] = \mathbf{B}\mathbf{u}$ then the generalized eigensystem becomes

$$\mathbf{A}\mathbf{u} = \lambda\mathbf{B}\mathbf{u}. \tag{7.7.2}$$

The two $n \times n$ matrices $\mathbf{A}$ and $\mathbf{B}$ compose a *matrix pencil* (or *matrix pair*), denoted $(\mathbf{A}, \mathbf{B})$. The constant λ and the nonzero vector $\mathbf{u}$ are respectively referred to as the generalized eigenvalue and the generalized eigenvector of the matrix pencil $(\mathbf{A}, \mathbf{B})$.

A generalized eigenvalue and its corresponding generalized eigenvector are collectively called a *generalized eigenpair*, denoted $(\lambda, \mathbf{u})$. Equation (7.7.2) shows that the ordinary eigenvalue problem is a special example of the generalized eigenvalue problem when the matrix pencil is $(\mathbf{A}, \mathbf{I})$.

Although the generalized eigenvalue and the generalized eigenvector always appear in pairs, the generalized eigenvalue can be found independently. To find the generalized eigenvalue independently, we rewrite the generalized characteristic equation (7.7.2) as

$$(\mathbf{A} - \lambda\mathbf{B})\mathbf{u} = \mathbf{0}. \tag{7.7.3}$$

In order to ensure the existence of a nonzero solution $\mathbf{u}$, the matrix $\mathbf{A} - \lambda\mathbf{B}$ cannot be nonsingular, i.e., the determinant must be equal to zero:

$$(\mathbf{A} - \lambda\mathbf{B}) \text{ singular} \quad \Leftrightarrow \quad \det(\mathbf{A} - \lambda\mathbf{B}) = 0. \tag{7.7.4}$$

The polynomial $\det(\mathbf{A} - \lambda\mathbf{B})$ is called the *generalized characteristic polynomial* of the matrix pencil $(\mathbf{A}, \mathbf{B})$, and $\det(\mathbf{A} - \lambda\mathbf{B}) = 0$ is known as the *generalized characteristic equation* of the matrix pencil $(\mathbf{A}, \mathbf{B})$. Hence, the matrix pencil $(\mathbf{A}, \mathbf{B})$ is often expressed as $\mathbf{A} - \lambda\mathbf{B}$.

The generalized eigenvalues of the matrix pencil $(\mathbf{A}, \mathbf{B})$ are all solutions z of the generalized characteristic equation $\det(\mathbf{A} - z\mathbf{B}) = 0$, including the generalized eigenvalues equal to zero. If the matrix $\mathbf{B} = \mathbf{I}$ then the generalized characteristic polynomial reduces to the characteristic polynomial $\det(\mathbf{A} - \lambda\mathbf{I}) = 0$ of the matrix $\mathbf{A}$. In other words, the generalized characteristic polynomial is a generalization of the characteristic polynomial, whereas the characteristic polynomial is an special example of the generalized characteristic polynomial of the matrix pencil $(\mathbf{A}, \mathbf{B})$ when $\mathbf{B} = \mathbf{I}$.

The generalized eigenvalues $\lambda(\mathbf{A}, \mathbf{B})$ are defined as

$$\lambda(\mathbf{A}, \mathbf{B}) = \big\{ z \in \mathbb{C} \big| \det(\mathbf{A} - z\mathbf{B}) = 0 \big\}. \tag{7.7.5}$$

Theorem 7.14 [224] *The pair $\lambda \in \mathbb{C}$ and $\mathbf{u} \in \mathbb{C}^n$ are respectively the generalized eigenvalue and the generalized eigenvector of the $n \times n$ matrix pencil $(\mathbf{A}, \mathbf{B})$ if and only if $|\mathbf{A} - \lambda\mathbf{B}| = 0$ and $\mathbf{u} \in \mathrm{Null}(\mathbf{A} - \lambda\mathbf{B})$ with $\mathbf{u} \neq \mathbf{0}$.*

The following are the properties of the GEVD $\mathbf{A}\mathbf{u} = \lambda\mathbf{B}\mathbf{u}$ [230, pp. 176–177].

1. If $\mathbf{A}$ and $\mathbf{B}$ are exchanged, then every generalized eigenvalue will become its reciprocal, but the generalized eigenvector is unchanged, namely

$$\mathbf{A}\mathbf{u} = \lambda\mathbf{B}\mathbf{u} \quad \Rightarrow \quad \mathbf{B}\mathbf{u} = \frac{1}{\lambda}\mathbf{A}\mathbf{u}.$$

2. If $\mathbf{B}$ is nonsingular then the GEVD becomes the standard EVD of $\mathbf{B}^{-1}\mathbf{A}$:

$$\mathbf{A}\mathbf{u} = \lambda\mathbf{B}\mathbf{u} \quad \Rightarrow \quad (\mathbf{B}^{-1}\mathbf{A})\mathbf{u} = \lambda\mathbf{u}.$$

3. If $\mathbf{A}$ and $\mathbf{B}$ are real positive definite matrices then the generalized eigenvalues must be positive.
4. If $\mathbf{A}$ is singular then $\lambda = 0$ must be a generalized eigenvalue.
5. If both $\mathbf{A}$ and $\mathbf{B}$ are positive definite Hermitian matrices then the generalized eigenvalues must be real, and

$$\mathbf{u}_i^H\mathbf{A}\mathbf{u}_j = \mathbf{u}_i^H\mathbf{B}\mathbf{u}_j = 0, \quad i \neq j.$$

Strictly speaking, the generalized eigenvector $\mathbf{u}$ described above is called the *right generalized eigenvector* of the matrix pencil $(\mathbf{A}, \mathbf{B})$. The *left generalized eigenvector* corresponding to the generalized eigenvalue λ is defined as the column vector $\mathbf{v}$ such that

$$\mathbf{v}^H\mathbf{A} = \lambda\mathbf{v}^H\mathbf{B}. \tag{7.7.6}$$

Let both $\mathbf{X}$ and $\mathbf{Y}$ be nonsingular matrices; then from (7.7.2) and (7.7.6) one has

$$\mathbf{X}\mathbf{A}\mathbf{u} = \lambda\mathbf{X}\mathbf{B}\mathbf{u}, \quad \mathbf{v}^H\mathbf{A}\mathbf{Y} = \lambda\mathbf{v}^H\mathbf{B}\mathbf{Y}. \tag{7.7.7}$$

This shows that premultiplying the matrix pencil $(\mathbf{A}, \mathbf{B})$ by a nonsingular matrix $\mathbf{X}$ does not change the right generalized eigenvector $\mathbf{u}$ of the matrix pencil $(\mathbf{A}, \mathbf{B})$, while postmultiplying $(\mathbf{A}, \mathbf{B})$ by any nonsingular matrix $\mathbf{Y}$ does not change the left generalized eigenvector $\mathbf{v}$.

Algorithm 7.3 uses a contraction mapping to compute the generalized eigenpairs $(\lambda, \mathbf{u})$ of a given $n \times n$ real symmetric matrix pencil $(\mathbf{A}, \mathbf{B})$.

Algorithm 7.3 Lanczos algorithm for GEVD [422, p. 298]

---

given: $n \times n$ real symmetric matrices $\mathbf{A}, \mathbf{B}$.
initialization: Choose $\mathbf{u}_1$ such that $\|\mathbf{u}_1\|_2 = 1$, and set $\alpha_1 = 0$, $\mathbf{z}_0 = \mathbf{u}_0 = \mathbf{0}$, $\mathbf{z}_1 = \mathbf{B}\mathbf{u}_1$ and $i = 1$.

repeat

1. Compute

$\mathbf{u} = \mathbf{A}\mathbf{u}_i - \alpha_i\mathbf{z}_{i-1}$.
$\beta_i = \langle \mathbf{u}, \mathbf{u}_i \rangle$.
$\mathbf{u} = \mathbf{u} - \beta_i\mathbf{z}_i$.
$\mathbf{w} = \mathbf{B}^{-1}\mathbf{u}$.
$\alpha_{i+1} = \sqrt{\langle \mathbf{w}, \mathbf{u} \rangle}$.
$\mathbf{u}_{i+1} = \mathbf{w}/\alpha_{i+1}$.
$\mathbf{z}_{i+1} = \mathbf{u}/\alpha_{i+1}$.
$\lambda_i = \beta_{i+1}/\alpha_{i+1}$.

2. **exit if** $i = n$.

return $i \leftarrow i + 1$.
output: $(\lambda_i, \mathbf{u}_i), i = 1, \ldots, n$.

Algorithm 7.4 is a tangent algorithm for computing the GEVD of a symmetric positive definite matrix pencil.

Algorithm 7.4 Tangent algorithm for computing the GEVD [136]

---

given: $n \times n$ real symmetric matrices $\mathbf{A}, \mathbf{B}$.

repeat

1. Compute $\boldsymbol{\Delta}_A = \mathbf{Diag}(A_{11}, \ldots, A_{nn})^{-1/2}, \mathbf{A}_s = \boldsymbol{\Delta}_A \mathbf{A} \boldsymbol{\Delta}_A$, $\mathbf{B}_1 = \boldsymbol{\Delta}_A \mathbf{B} \boldsymbol{\Delta}_A$.
2. Calculate the Cholesky decomposition $\mathbf{R}_A^T \mathbf{R}_A = \mathbf{A}_s$, $\mathbf{R}_B^T \mathbf{R}_B = \boldsymbol{\Pi}^T \mathbf{B}_1 \boldsymbol{\Pi}$.
3. Solve the matrix equation $\mathbf{F}\mathbf{R}_B = \mathbf{A}\boldsymbol{\Pi}$ for $\mathbf{F} = \mathbf{A}\boldsymbol{\Pi}\mathbf{R}_B^{-1}$.
4. Compute the SVD $\boldsymbol{\Sigma} = \mathbf{V}\mathbf{F}\mathbf{U}^T$ of $\mathbf{F}$.
5. Compute $\mathbf{X} = \boldsymbol{\Delta}_A \boldsymbol{\Pi} \mathbf{R}_B^{-1} \mathbf{U}$.
6. **exit if** $\mathbf{A}\mathbf{X} = \mathbf{B}\mathbf{X}\boldsymbol{\Sigma}^2$.

repeat

output: $\mathbf{X}, \boldsymbol{\Sigma}$.

When the matrix $\mathbf{B}$ is singular, Algorithms 7.3 and 7.4 will be unstable. The GEVD algorithm for the matrix pencil $(\mathbf{A}, \mathbf{B})$ with singular matrix $\mathbf{B}$ was proposed by Nour-Omid *et al.* [354], this is shown in Algorithm 7.5. The main idea of this algorithm is to introduce a shift factor σ such that $\mathbf{A} - \sigma\mathbf{B}$ is nonsingular.

Algorithm 7.5 GEVD algorithm for singular matrix $\mathbf{B}$ [354], [422]

---

initialization: Choose a basis $\mathbf{w}$, $\mathbf{z}_1 = \mathbf{B}\mathbf{w}$, $\alpha_1 = \sqrt{\langle \mathbf{w}, \mathbf{z}_1 \rangle}$. Set $\mathbf{u}_0 = \mathbf{0}$, $i = 1$.

repeat

1. Compute

$$\begin{aligned}
&\mathbf{u}_i = \mathbf{w}/\alpha_i.\\
&\mathbf{z}_i = (\mathbf{A} - \sigma\mathbf{B})^{-1}\mathbf{w}.\\
&\mathbf{w} = \mathbf{w} - \alpha_i \mathbf{u}_{i-1}.\\
&\beta_i = \langle \mathbf{w}, \mathbf{z}_i \rangle.\\
&\mathbf{z}_{i+1} = \mathbf{B}\mathbf{w}.\\
&\alpha_{i+1} = \sqrt{\langle \mathbf{z}_{i+1}, \mathbf{w} \rangle}.\\
&\lambda_i = \beta_i / \alpha_{i+1}.
\end{aligned}$$

2. **exit if** $i = n$.

return $i \leftarrow i + 1$.

output: $(\lambda_i, \mathbf{u}_i), i = 1, \ldots, n$.

Definition 7.19 Two matrix pencils with the same generalized eigenvalues are known as *equivalent matrix pencils*.

From the definition of the generalized eigenvalue $\det(\mathbf{A} - \lambda\mathbf{B}) = 0$ and the properties of a determinant it is easy to see that

$$\det(\mathbf{XAY} - \lambda\mathbf{XBY}) = 0 \quad \Leftrightarrow \quad \det(\mathbf{A} - \lambda\mathbf{B}) = 0.$$

Therefore, premultiplying and/or postmultiplying one or more nonsingular matrices, the generalized eigenvalues of the matrix pencil are unchanged. This result can be described as the following proposition.

Proposition 7.1 *If* $\mathbf{X}$ *and* $\mathbf{Y}$ *are two nonsingular matrices then the matrix pencils* $(\mathbf{XAY}, \mathbf{XBY})$ *and* $(\mathbf{A}, \mathbf{B})$ *are equivalent.*

The GEVD can be equivalently written as $\alpha\mathbf{Au} = \beta\mathbf{Bu}$ as well. In this case, the generalized eigenvalue is defined as $\lambda = \beta/\alpha$.

7.7.2 Total Least Squares Method for GEVD

In applications of the GEVD we are only interested in the nonzero generalized eigenvalues, because their number reflects the number of signal components and the nonzero generalized eigenvalues themselves often imply useful information about the signal parameters. However, the number of signal components is usually unknown and needs to be estimated. Generally, the dimension of the matrix pencil $(\mathbf{A}, \mathbf{B})$ is larger than the number of signal components. Moreover, the two matrices $\mathbf{A}$ and $\mathbf{B}$ are usually the autocorrelation and cross-correlation matrices consisting of the observed data vectors, respectively. Because the observation data are short and noisy, the sample autocorrelation and cross-correlation matrices have the large estimation errors. Owing to the large dimension of the matrix pencil and the large estimation errors of the sample autocorrelation and cross-correlation matrices, the estimation of the nonzero generalized eigenvalues of the matrix pencil is a least squares problem.

As shown by Roy and Kailath [419], use of a least squares estimator can lead to some potential numerical difficulties in solving the generalized eigenvalue problem. To overcome this difficulty, a higher-dimensional ill-conditioned LS problem needs to be transformed into a lower-dimensional well-conditioned LS problem.

Without changing its nonzero generalized eigenvalues, a higher-dimensional matrix pencil can be transformed into a lower-dimensional matrix pencil by using a truncated SVD.

Consider the GEVD of the matrix pencil $(\mathbf{A}, \mathbf{B})$. Let the SVD of $\mathbf{A}$ be given by

$$\mathbf{A} = \mathbf{U\Sigma V}^H = [\mathbf{U}_1, \mathbf{U}_2] \begin{bmatrix} \mathbf{\Sigma}_1 & \mathbf{O} \\ \mathbf{O} & \mathbf{\Sigma}_2 \end{bmatrix} \begin{bmatrix} \mathbf{V}_1^H \\ \mathbf{V}_2^H \end{bmatrix}, \tag{7.7.8}$$

where $\mathbf{\Sigma}_1$ is composed of p principal singular values. Under the condition that the generalized eigenvalues remain unchanged, we can premultiply the matrix pencil $\mathbf{A} - \gamma\mathbf{B}$ by $\mathbf{U}_1^H$ and postmultiply it by $\mathbf{V}_1$ to get [543]:

$$\mathbf{U}_1^H(\mathbf{A} - \gamma\mathbf{B})\mathbf{V}_1 = \mathbf{\Sigma}_1 - \gamma\mathbf{U}_1^H\mathbf{BV}_1. \tag{7.7.9}$$

Hence the original $n \times n$ matrix pencil $(\mathbf{A}, \mathbf{B})$ becomes a $p \times p$ matrix pencil $(\mathbf{\Sigma}_1, \mathbf{U}_1^H \mathbf{B} \mathbf{V}_1)$. The GEVD of the lower-dimensional matrix pencil $(\mathbf{\Sigma}_1, \mathbf{U}_1^H \mathbf{B} \mathbf{V}_1)$ is called the total least squares (TLS) method for the GEVD of the higher-dimensional matrix pencil $(\mathbf{A}, \mathbf{B})$, which was developed in [543].

7.7.3 Application of GEVD: ESPRIT

Consider a linear combination $x(n)$ at time n of p harmonic signals subject to white noise:

$$x(n) = \sum_{i=1}^{p} s_i \mathrm{e}^{\mathrm{j}n\omega_i} + w(n), \tag{7.7.10}$$

where s_i and $\omega_i \in (-\pi, \pi)$ are the complex amplitude and frequency of the ith harmonic signal. Suppose that $w(n)$ is a complex Gaussian white noise with zero mean and variance σ^2:

$$E\{w(k)w^*(l)\} = \sigma^2 \delta(k-l), \quad E\{w(k)w(l)\} = 0, \quad \forall\, k, l.$$

Our problem is to estimate the number p and frequencies $\omega_1, \ldots, \omega_p$ of the harmonic signals using only the observation data $x(1), \ldots, x(N)$.

Define $y(n) \stackrel{\text{def}}{=} x(n+1)$. Choose $m > p$, and introduce the $m \times 1$ vectors

$$\begin{aligned}
\mathbf{x}(n) &\stackrel{\text{def}}{=} [x(n), x(n+1), \ldots, x(n+m-1)]^T, \\
\mathbf{w}(n) &\stackrel{\text{def}}{=} [w(n), w(n+1), \ldots, w(n+m-1)]^T, \\
\mathbf{y}(n) &\stackrel{\text{def}}{=} [y(n), y(n+1), \ldots, y(n+m-1)]^T \\
&= [x(n+1), x(n+2), \ldots, x(n+m)]^T, \\
\mathbf{a}(\omega_i) &\stackrel{\text{def}}{=} [1, \mathrm{e}^{\mathrm{j}\omega_i}, \ldots, \mathrm{e}^{\mathrm{j}(m-1)\omega_i}]^T.
\end{aligned}$$

Then Equation (7.7.10) can be written as in vector form

$$\mathbf{x}(n) = \mathbf{A}\mathbf{s}(n) + \mathbf{w}(n). \tag{7.7.11}$$

Moreover,

$$\mathbf{y}(n) = \mathbf{A}\mathbf{\Phi}\mathbf{s}(n) + \mathbf{w}(n+1), \tag{7.7.12}$$

where

$$\begin{aligned}
\mathbf{A} &\stackrel{\text{def}}{=} [\mathbf{a}(\omega_1), \mathbf{a}(\omega_2), \ldots, \mathbf{a}(\omega_p)], \\
\mathbf{s}(n) &\stackrel{\text{def}}{=} [s_1 \mathrm{e}^{\mathrm{j}\omega_1 n}, s_2 \mathrm{e}^{\mathrm{j}\omega_2 n}, \ldots, s_p \mathrm{e}^{\mathrm{j}\omega_p n}]^T, \\
\mathbf{\Phi} &\stackrel{\text{def}}{=} \mathbf{Diag}(\mathrm{e}^{\mathrm{j}\omega_1}, \mathrm{e}^{\mathrm{j}\omega_2}, \ldots, \mathrm{e}^{\mathrm{j}\omega_p}).
\end{aligned}$$

Because $\mathbf{y}(n) = \mathbf{x}(n+1)$, the vector $\mathbf{y}(n)$ can be regarded as a shifted version of $\mathbf{x}(n)$. Hence $\mathbf{\Phi}$ is called the *rotation operator* since the shift is a simple rotation.

The autocorrelation matrix of the observation vector $\mathbf{x}(n)$ is given by

$$\mathbf{R}_{xx} = E\{\mathbf{x}(n)\mathbf{x}^H(n)\} = \mathbf{APA}^H + \sigma^2\mathbf{I}, \tag{7.7.13}$$

where $\mathbf{P} = E\{\mathbf{s}(n)\mathbf{s}^H(n)\}$ is the autocorrelation matrix of the signal vector $\mathbf{s}(n)$.

The cross-correlation matrix of $\mathbf{x}(n)$ and $\mathbf{y}(n)$ is defined as

$$\mathbf{R}_{xy} = E\{\mathbf{x}(n)\mathbf{y}^H(n)\} = \mathbf{AP\Phi}^H\mathbf{A}^H + \sigma^2\mathbf{Z}, \tag{7.7.14}$$

where $\sigma^2\mathbf{Z} = E\{\mathbf{w}(n)\mathbf{w}^H(n+1)\}$. It is easy to verify that $\mathbf{Z}$ is an $m \times m$ special matrix given by

$$\mathbf{Z} = \begin{bmatrix} 0 & & & 0 \\ 1 & 0 & & \\ & \ddots & \ddots & \\ 0 & & 1 & 0 \end{bmatrix}. \tag{7.7.15}$$

That is, all entries on the subdiagonal below the main diagonal are equal to 1 while all other entries are equal to 0.

The problem is to estimate the number p, the frequencies ω_i and the harmonic powers $|s_i|^2$ $(i = 1, \ldots, p)$ of the harmonic signals by using the autocorrelation matrix $\mathbf{R}_{xx}$ and the cross-correlation matrix $\mathbf{R}_{xy}$.

Importantly, the data vector $\mathbf{x}$ and its shifted vector $\mathbf{y}$ have the same autocorrelation matrices, i.e., $\mathbf{R}_{xx} \stackrel{\text{def}}{=} E\{\mathbf{x}(n)\mathbf{x}^H(n)\} = E\{\mathbf{x}(n+1)\mathbf{x}^H(n+1)\} \stackrel{\text{def}}{=} \mathbf{R}_{yy}$.

Making the EVD of $\mathbf{R}_{xx}$, we get its minimum eigenvalue $\lambda_{\min} = \sigma^2$. Construct a new matrix pair

$$\mathbf{C}_{xx} = \mathbf{R}_{xx} - \lambda_{\min}\mathbf{I} = \mathbf{R}_{xx} - \sigma^2\mathbf{I} = \mathbf{APA}^H, \tag{7.7.16}$$

$$\mathbf{C}_{xy} = \mathbf{R}_{xy} - \lambda_{\min}\mathbf{Z} = \mathbf{R}_{xy} - \sigma^2\mathbf{Z} = \mathbf{AP\Phi}^H\mathbf{A}^H. \tag{7.7.17}$$

Now use $\mathbf{C}_{xx}$ and $\mathbf{C}_{xy}$ to constitute a matrix pencil $(\mathbf{C}_{xx}, \mathbf{C}_{xy})$. We have

$$\mathbf{C}_{xx} - \gamma\mathbf{C}_{xy} = \mathbf{AP}(\mathbf{I} - \gamma\mathbf{\Phi}^H)\mathbf{A}^H. \tag{7.7.18}$$

Owing to the full column rank of $\mathbf{A}$ and the nonsingularity of $\mathbf{P}$, the above equation can be written as

$$\text{rank}(\mathbf{C}_{xx} - \gamma\mathbf{C}_{xy}) = \text{rank}(\mathbf{I} - \gamma\mathbf{\Phi}^H). \tag{7.7.19}$$

When $\gamma \neq \mathrm{e}^{\mathrm{j}\omega_i}, i = 1, \ldots, p$, the matrix $\mathbf{I} - \gamma\mathbf{\Phi}$ is nonsingular, whereas it is singular or rank-deficit when $\gamma = \mathrm{e}^{\mathrm{j}\omega_i}$, because $\gamma\mathrm{e}^{-\mathrm{j}\omega_i} = 1$. This shows that $\mathrm{e}^{\mathrm{j}\omega_i}, i = 1, \ldots, p$, are the generalized eigenvalues of the matrix pencil $(\mathbf{C}_{xx}, \mathbf{C}_{xy})$. This result is given in the following theorem.

Theorem 7.15 [419] *Define $\mathbf{\Gamma}$ as the generalized eigenvalue matrix of a matrix pencil $(\mathbf{C}_{xx}, \mathbf{C}_{xy})$, where $\mathbf{C}_{xx} = \mathbf{R}_{xx} - \lambda_{\min}\mathbf{I}, \mathbf{C}_{xy} = \mathbf{R}_{xy} - \lambda_{\min}\mathbf{Z}$ and $\lambda_{\min}$ is the minimum eigenvalue of the autocorrelation matrix $\mathbf{R}_{xx}$. If the matrix $\mathbf{P}$ is* *nonsingular then the matrix* $\mathbf{\Gamma}$ *and the rotation operator matrix* $\mathbf{\Phi}$ *have the following relationship:*

$$\mathbf{\Gamma} = \begin{bmatrix} \mathbf{\Phi} & \mathbf{O} \\ \mathbf{O} & \mathbf{O} \end{bmatrix}, \tag{7.7.20}$$

i.e., the nonzero entries of $\mathbf{\Gamma}$ *are an arrangement of the entries of the rotation operator* $\mathbf{\Phi}$.

Roy and Kailath [419] proposed a method for *estimating signal parameters via rotational invariance techniques* (ESPRIT). See Algorithm 7.6.

Algorithm 7.6 Basic ESPRIT algorithm 1 [419]

---

input: $x(1), \ldots, x(N)$.

1. Estimate $R_{xx}(i,j) = \frac{1}{N}\sum_{n=1}^{N} x(n)x^*(n-i+j)$, $R_{xy}(i,j) = \frac{1}{N}\sum_{n=1}^{N} x(n)x^*(n+1-i+j)$ for $i,j = 0,1,\ldots,m-1$.
2. Form $\mathbf{R}_{xx} = [R_{xx}(i,j)]_{i,j=0}^{m-1,m-1}$, $\mathbf{R}_{xy} = [R_{xy}(i,j)]_{i,j=0}^{m-1,m-1}$.
3. Make the EVD of $\mathbf{R}_{xx}$ to estimate the minimum eigenvalue as σ^2.
4. Compute $\mathbf{C}_{xx} = \mathbf{R}_{xx} - \sigma^2\mathbf{I}$ and $\mathbf{C}_{xy} = \mathbf{R}_{xy} - \sigma^2\mathbf{Z}$.
5. Make the GEVD of $(\mathbf{C}_{xx}, \mathbf{C}_{xy})$ to find p generalized eigenvalues $e^{j\omega_i}$.

output: p and $\omega_1, \ldots, \omega_p$.

This basic ESPRIT method is sometimes called the LS-ESPRIT algorithm. The TLS-ESPRIT algorithm, shown in Algorithm 7.7, can improve the numerical performance of the LS-ESPRIT algorithm.

Algorithm 7.7 TLS–ESPRIT algorithm [543]

---

input: $x(1), \ldots, x(N)$.

1. Estimate $R_{xx}(i,j) = \frac{1}{N}\sum_{n=1}^{N} x(n)x^*(n-i+j)$, $R_{xy}(i,j) = \frac{1}{N}\sum_{n=1}^{N} x(n)x^*(n+1-i+j)$ for $i,j = 0,1,\ldots,m-1$.
2. Form $\mathbf{R}_{xx} = [R_{xx}(i,j)]_{i,j=0}^{m-1,m-1}$, $\mathbf{R}_{xy} = [R_{xy}(i,j)]_{i,j=0}^{m-1,m-1}$.
3. Make the EVD of $\mathbf{R}_{xx}$ to estimate the minimum eigenvalue σ^2.
4. Compute $\mathbf{C}_{xx} = \mathbf{R}_{xx} - \sigma^2\mathbf{I}$ and $\mathbf{C}_{xy} = \mathbf{R}_{xy} - \sigma^2\mathbf{Z}$.
5. Make the SVD of $\mathbf{C}_{xx}$, and determine p and the matrices $\mathbf{\Sigma}_1, \mathbf{U}_1, \mathbf{V}_1$.
6. Make the GEVD of $(\mathbf{\Sigma}_1, \mathbf{U}_1^H\mathbf{C}_{xy}\mathbf{V}_1)$ to estimate the frequencies $\omega_1, \ldots, \omega_p$.

output: p and $\omega_1, \ldots, \omega_p$.

It has been shown that the LS-ESPRIT and the TLS-ESPRIT algorithms give the same asymptotic estimation accuracy for large samples, but the estimation accuracy of the TLS-ESPRIT is better than that of the LS-ESPRIT for small samples.

7.7.4 Similarity Transformation in GEVD

Consider an isometric line array that consists of m elements, as shown in Figure 7.2.

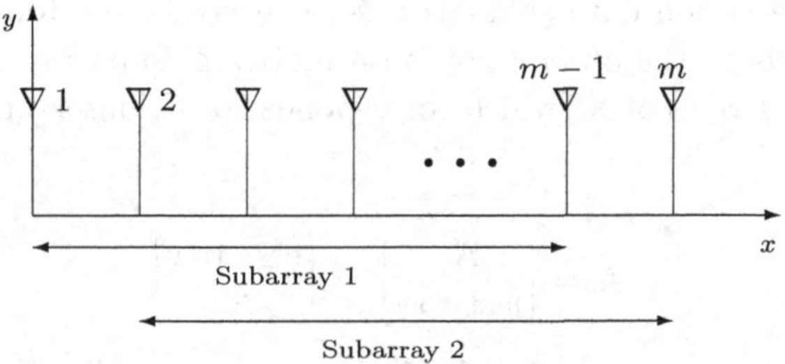

Figure 7.2 An array is divided into two subarrays.

This isometric line array is divided into two subarrays, where subarray 1 consists of the first to the $(m-1)$th elements, and subarray 2 consists of the second to the mth elements.

Let the $m \times N$ matrix $\mathbf{X} = [\mathbf{x}(1), \ldots, \mathbf{x}(N)]$ denote the observation data matrix of the original array, where $\mathbf{x}(n) = [x_1(n), \ldots, x_m(n)]^T$ is the observation data vector, consisting of m elements, at time n and N is the data length, i.e., $n = 1, \ldots, N$.

Letting

$$\mathbf{S} = [\mathbf{s}(1), \ldots, \mathbf{s}(N)] \tag{7.7.21}$$

denote the signal matrix, where $\mathbf{s}(n) = [s_1(n), \ldots, s_p(n)]^T$ are the signal vectors, then for N snapshots, Equation (7.7.10) can be represented in matrix form as

$$\mathbf{X} = [\mathbf{x}(1), \ldots, \mathbf{x}(N)] = \mathbf{AS}, \tag{7.7.22}$$

where $\mathbf{A}$ is an $m \times p$ array response matrix.

Let $\mathbf{J}_1$ and $\mathbf{J}_2$ be two $(m-1) \times m$ selection matrices such that

$$\mathbf{J}_1 = [\mathbf{I}_{m-1}, \mathbf{0}_{m-1}], \tag{7.7.23}$$

$$\mathbf{J}_2 = [\mathbf{0}_{m-1}, \mathbf{I}_{m-1}], \tag{7.7.24}$$

where $\mathbf{I}_{m-1}$ is a $(m-1) \times (m-1)$ identity matrix, and $\mathbf{0}_{m-1}$ is a $(m-1) \times 1$ zero vector.

Premultiply the observation matrix $\mathbf{X}$ respectively by the selection matrices $\mathbf{J}_1$

and $\mathbf{J}_2$:

$$\mathbf{X}_1 = \mathbf{J}_1\mathbf{X} = [\mathbf{x}_1(1), \ldots, \mathbf{x}_1(N)], \tag{7.7.25}$$

$$\mathbf{X}_2 = \mathbf{J}_2\mathbf{X} = [\mathbf{x}_2(1), \ldots, \mathbf{x}_2(N)], \tag{7.7.26}$$

where

$$\mathbf{x}_1(n) = [x_1(n), x_2(n), \ldots, x_{m-1}(n)]^T, \quad n = 1, \ldots, N, \tag{7.7.27}$$

$$\mathbf{x}_2(n) = [x_2(n), x_3(n), \ldots, x_m(n)]^T, \quad n = 1, \ldots, N. \tag{7.7.28}$$

That is, the observation data submatrix $\mathbf{X}_1$ consists of the top $m-1$ rows of $\mathbf{X}$, which corresponds to the observation data matrix of subarray 1; and $\mathbf{X}_2$ consists of the lower $m-1$ rows of $\mathbf{X}$, which corresponds to the observation data matrix of subarray 2.

Put

$$\mathbf{A} = \begin{bmatrix} \mathbf{A}_1 \\ \text{last row} \end{bmatrix} = \begin{bmatrix} \text{first row} \\ \mathbf{A}_2 \end{bmatrix}; \tag{7.7.29}$$

then, from the array response matrix $\mathbf{A}$ of the isometric line array it can be seen that there is the following relationship between the submatrices $\mathbf{A}_1$ and $\mathbf{A}_2$:

$$\mathbf{A}_2 = \mathbf{A}_1\mathbf{\Phi}, \tag{7.7.30}$$

where $\mathbf{\Phi}$ is a rotation operator matrix defined earlier. It can be shown that

$$\mathbf{X}_1 = \mathbf{A}_1\mathbf{S}, \tag{7.7.31}$$

$$\mathbf{X}_2 = \mathbf{A}_2\mathbf{S} = \mathbf{A}_1\mathbf{\Phi}\mathbf{S}. \tag{7.7.32}$$

Because $\mathbf{\Phi}$ is a unitary matrix, $\mathbf{X}_1$ and $\mathbf{X}_2$ have the same signal subspace and the same noise subspace; namely subarray 1 and subarray 2 have the same observation space (signal subspace + noise subspace). This is the physical explanation of the shift invariance of an isometric line array.

From Equation (7.7.13) we get the following EVD:

$$\mathbf{R}_{xx} = \mathbf{A}\mathbf{P}\mathbf{A}^H + \sigma^2\mathbf{I} = [\mathbf{U}_s \;\; \mathbf{U}_n] \begin{bmatrix} \mathbf{\Sigma}_s & \mathbf{O} \\ \mathbf{O} & \sigma^2\mathbf{I} \end{bmatrix} \begin{bmatrix} \mathbf{U}_s^H \\ \mathbf{U}_n^H \end{bmatrix}$$

$$= \begin{bmatrix} \mathbf{U}_s\mathbf{\Sigma}_s & \sigma^2\mathbf{U}_n \end{bmatrix} \begin{bmatrix} \mathbf{U}_s^H \\ \mathbf{U}_n^H \end{bmatrix} = \mathbf{U}_s\mathbf{\Sigma}_s\mathbf{U}_s^H + \sigma^2\mathbf{U}_n\mathbf{U}_n^H. \tag{7.7.33}$$

Since $[\mathbf{U}_s, \mathbf{U}_n][\mathbf{U}_s, \mathbf{U}_n]^H = \mathbf{I}$, i.e., $\mathbf{I} - \mathbf{U}_n\mathbf{U}_n^H = \mathbf{U}_s\mathbf{U}_s^H$, Equation (7.7.33) yields

$$\mathbf{A}\mathbf{P}\mathbf{A}^H + \sigma^2\mathbf{U}_s\mathbf{U}_s^H = \mathbf{U}_s\mathbf{\Sigma}_s\mathbf{U}_s^H. \tag{7.7.34}$$

Postmultiplying both sides of the above equation by $\mathbf{U}_s$, and noticing that $\mathbf{U}_s^H\mathbf{U}_s = \mathbf{I}$, we have

$$\mathbf{U}_s = \mathbf{A}\mathbf{T}, \tag{7.7.35}$$

where $\mathbf{T}$ is a nonsingular matrix given by

$$\mathbf{T} = \mathbf{A}^H \mathbf{U}_s (\boldsymbol{\Sigma}_s - \sigma^2 \mathbf{I})^{-1}. \tag{7.7.36}$$

Although $\mathbf{T}$ is unknown, it is simply a virtual parameter in the following analysis in which we use only its nonsingularity. Postmultiply (7.7.29) by $\mathbf{T}$ to yield

$$\mathbf{AT} = \begin{bmatrix} \mathbf{A}_1 \mathbf{T} \\ \text{last row} \end{bmatrix} = \begin{bmatrix} \text{first row} \\ \mathbf{A}_2 \mathbf{T} \end{bmatrix}. \tag{7.7.37}$$

Adopting the same block form, we divide $\mathbf{U}_s$ into

$$\mathbf{U}_s = \begin{bmatrix} \mathbf{U}_1 \\ \text{last row} \end{bmatrix} = \begin{bmatrix} \text{first row} \\ \mathbf{U}_2 \end{bmatrix}. \tag{7.7.38}$$

Because $\mathbf{AT} = \mathbf{U}_s$, comparing (7.7.37) with (7.7.38), we immediately have

$$\mathbf{U}_1 = \mathbf{A}_1 \mathbf{T} \quad \text{and} \quad \mathbf{U}_2 = \mathbf{A}_2 \mathbf{T}. \tag{7.7.39}$$

Substituting (7.7.30) into (7.7.39), we get

$$\mathbf{U}_2 = \mathbf{A}_1 \boldsymbol{\Phi} \mathbf{T}. \tag{7.7.40}$$

By (7.7.39) and (7.7.40), we have also

$$\mathbf{U}_1 \mathbf{T}^{-1} \boldsymbol{\Phi} \mathbf{T} = \mathbf{A}_1 \mathbf{T} \mathbf{T}^{-1} \boldsymbol{\Phi} \mathbf{T} = \mathbf{A}_1 \boldsymbol{\Phi} \mathbf{T} = \mathbf{U}_2. \tag{7.7.41}$$

Define

$$\boldsymbol{\Psi} = \mathbf{T}^{-1} \boldsymbol{\Phi} \mathbf{T}; \tag{7.7.42}$$

then the matrix $\boldsymbol{\Psi}$ is similar to $\boldsymbol{\Phi}$ and thus they have the same eigenvalues, i.e., the eigenvalues of $\boldsymbol{\Psi}$ are also $\mathrm{e}^{\mathrm{j}\phi_m}$, $m = 1, 2, \ldots, M$.

Finally, substituting (7.7.42) into (7.7.41), we get an important relationship:

$$\mathbf{U}_2 = \mathbf{U}_1 \boldsymbol{\Psi}. \tag{7.7.43}$$

Equation (7.7.43) suggests an alternative to the basic ESPRIT algorithm, which is given in Algorithm 7.8.

Algorithm 7.8 Basic ESPRIT algorithm 2

---

input: $x(1), \ldots, x(N)$.

1. Estimate $R_{xx}(i,j) = \frac{1}{N} \sum_{n=1}^{N} x(n) x^*(n - i + j)$, $R_{xy}(i,j) = \frac{1}{N} \sum_{n=1}^{N} x(n) x^*(n + 1 - i + j)$ for $i, j = 0, 1, \ldots, m-1$.
2. Form $\mathbf{R}_{xx} = [R_{xx}(i,j)]_{i,j=0}^{m-1,m-1}$, $\mathbf{R}_{xy} = [R_{xy}(i,j)]_{i,j=0}^{m-1,m-1}$.
3. Compute the EVD $\mathbf{R}_{xx} = \mathbf{U} \boldsymbol{\Sigma} \mathbf{U}^H$.
4. Use $\mathbf{U}$ and the p principal eigenvalues of $\mathbf{R}_{xx}$ to form $\mathbf{U}_s$.
5. Extract the upper and lower $m-1$ rows of $\mathbf{U}_s$ to constitute $\mathbf{U}_1$ and $\mathbf{U}_2$.
6. Compute the EVD of $\boldsymbol{\Psi} = (\mathbf{U}_1^H \mathbf{U}_1)^{-1} \mathbf{U}_1^H \mathbf{U}_2$ to get the p eigenvalues $\mathrm{e}^{\mathrm{j}\omega_i}$.

output: p and $\omega_1, \ldots, \omega_p$.

7.8 Rayleigh Quotient

In physics and information technology, one often meets the maximum or minimum of the quotient of the quadratic function of a Hermitian matrix. This quotient has two forms: the Rayleigh quotient (also called the Rayleigh–Ritz ratio) of a Hermitian matrix and the generalized Rayleigh quotient (or generalized Rayleigh–Ritz ratio) of two Hermitian matrices.

7.8.1 Definition and Properties of Rayleigh Quotient

In his study of the small oscillations of a vibrational system, in order to find the appropriate generalized coordinates Rayleigh [411] in the 1930s proposed a special form of quotient, later called the Rayleigh quotient.

Definition 7.20 The *Rayleigh quotient* or *Rayleigh–Ritz ratio* $R(\mathbf{x})$ of a Hermitian matrix $\mathbf{A} \in \mathbb{C}^{n\times n}$ is a scalar and is defined as

$$R(\mathbf{x}) = R(\mathbf{x}, \mathbf{A}) = \frac{\mathbf{x}^H \mathbf{A}\mathbf{x}}{\mathbf{x}^H \mathbf{x}}, \tag{7.8.1}$$

where $\mathbf{x}$ is a vector to be selected such that the Rayleigh quotient is maximized or minimized.

The Rayleigh quotient has the following important properties [374], [375], [93], [205].

1. *Homogeneity* If α and β are scalars then

$$R(\alpha\mathbf{x}, \beta\mathbf{A}) = \beta R(\mathbf{x}, \mathbf{A}). \tag{7.8.2}$$

2. *Shift invariance* $R(\mathbf{x}, \mathbf{A} - \alpha\mathbf{I}) = R(\mathbf{x}, \mathbf{A}) - \alpha$.
3. *Orthogonality* $\mathbf{x} \perp (\mathbf{A} - R(\mathbf{x})\mathbf{I})\mathbf{x}$.
4. *Boundedness* When the vector $\mathbf{x}$ lies in the range of all nonzero vectors, the Rayleigh quotient $R(\mathbf{x})$ falls in a complex plane region (called the range of $\mathbf{A}$). This region is closed, bounded and convex. If $\mathbf{A}$ is a Hermitian matrix such that $\mathbf{A} = \mathbf{A}^H$ then this region is a closed interval $[\lambda_1, \lambda_n]$.
5. *Minimum residual* For all vectors $\mathbf{x} \neq \mathbf{0}$ and all scalars μ,

$$\|(\mathbf{A} - R(\mathbf{x})\mathbf{I})\mathbf{x}\| \leq \|(\mathbf{A} - \mu\mathbf{I})\mathbf{x}\|. \tag{7.8.3}$$

The boundedness of the Rayleigh quotient of a Hermitian matrix can be strictly described as in the following theorem.

Theorem 7.16 (Rayleigh–Ritz theorem) *Let $\mathbf{A} \in \mathbb{C}^{n\times n}$ be a Hermitian matrix whose eigenvalues are arranged in increasing order:*

$$\lambda_{\min} = \lambda_1 \leq \lambda_2 \leq \cdots \leq \lambda_{n-1} \leq \lambda_n = \lambda_{\max}; \tag{7.8.4}$$

then

$$\max_{\mathbf{x}\neq\mathbf{0}} \frac{\mathbf{x}^H\mathbf{A}\mathbf{x}}{\mathbf{x}^H\mathbf{x}} = \max_{\mathbf{x}^H\mathbf{x}=1} \frac{\mathbf{x}^H\mathbf{A}\mathbf{x}}{\mathbf{x}^H\mathbf{x}} = \lambda_{\max}, \quad \textit{if } \mathbf{A}\mathbf{x} = \lambda_{\max}\mathbf{x}, \tag{7.8.5}$$

and

$$\min_{\mathbf{x}\neq\mathbf{0}} \frac{\mathbf{x}^H\mathbf{A}\mathbf{x}}{\mathbf{x}^H\mathbf{x}} = \min_{\mathbf{x}^H\mathbf{x}=1} \frac{\mathbf{x}^H\mathbf{A}\mathbf{x}}{\mathbf{x}^H\mathbf{x}} = \lambda_{\min}, \quad \textit{if } \mathbf{A}\mathbf{x} = \lambda_{\min}\mathbf{x}. \tag{7.8.6}$$

More generally, the eigenvectors and eigenvalues of $\mathbf{A}$ *correspond respectively to the critical points and the critical values of the Rayleigh quotient* $R(\mathbf{x})$.

This theorem has various proofs, see e.g., [179], [205] or [107].

7.8.2 Rayleigh Quotient Iteration

Let $\mathbf{A} \in \mathbb{C}^{n\times n}$ be a diagonalizable matrix, and let its eigenvalues be λ_i and the corresponding eigenvectors be $\mathbf{u}_i$. For convenience, suppose that the matrix $\mathbf{A}$ is nonsingular, its first eigenvalue is larger than the others and $\lambda_1 > \lambda_2 \geq \cdots \geq \lambda_n$; then the eigenvalue λ_1 and its corresponding eigenvector $\mathbf{u}_1$ are called the principal eigenvalue and the principal eigenvector of $\mathbf{A}$, respectively.

Since $\lambda_i = R(\mathbf{x}_i)$, Rayleigh proposed an iteration method for computing the principal eigenvalues and the principal eigenvectors when $\mathbf{A}$ is a Hermitian matrix. This method was later called *Rayleigh quotient iteration*.

Algorithm 7.9 shows the standard *Rayleigh quotient iteration algorithm* for the Hermitian matrix $\mathbf{A}$.

Algorithm 7.9 Rayleigh quotient iteration algorithm

---

input: Hermitian matrix $\mathbf{A}$.

initialization: Choose $\mathbf{x}_1$ with $\|\mathbf{x}_1\|_2 = 1$.

repeat

1. Compute $R_k = R(\mathbf{x}_k) = \mathbf{x}_k^H\mathbf{A}\mathbf{x}_k$.
2. **exit if** $\mathbf{A} - R_k\mathbf{I}$ is singular.
3. Calculate $\mathbf{x}_{k+1} = \dfrac{(\mathbf{A}_k - R_k\mathbf{I})^{-1}\mathbf{x}_k}{\|(\mathbf{A}_k - R_k\mathbf{I})^{-1}\mathbf{x}_k\|}$.

return $k \leftarrow k+1$.

Find the nonzero solution of $(\mathbf{A} - R_k\mathbf{I})\mathbf{x}_{k+1} = \mathbf{0}$.

output: $\mathbf{x} = \mathbf{x}_{k+1}$.

The iterative result $\{R_k, \mathbf{x}_k\}$ is called the *Rayleigh sequence* generated by Rayleigh iteration. The Rayleigh sequence has the following properties [374].

(1) *Scale invariance* Two matrices $\alpha\mathbf{A}$ $(\alpha \neq 0)$ and $\mathbf{A}$ generate the same Rayleigh sequence.

(2) *Shift invariance* If $\mathbf{A}$ generates a Rayleigh sequence $\{R_k, \mathbf{x}_k\}$ then the Rayleigh sequence generated by the matrix $\mathbf{A} - \alpha\mathbf{I}$ is $\{R_k - \alpha, \mathbf{x}_k\}$.
(3) *Unitary similarity* Let $\mathbf{U}$ be a unitary matrix, then the Rayleigh sequence generated by $\mathbf{UAU}^H$ is $\{R_k, \mathbf{Ux}_k\}$.

The extended Rayleigh quotient iteration algorithm [374] is available for a general matrix $\mathbf{A}$, as shown in Algorithm 7.10. If $\mathbf{A} = \mathbf{A}^H$, then the above extension algorithm reduces to the standard Rayleigh quotient iteration algorithm.

Algorithm 7.10 Rayleigh quotient iteration for general matrix [374]

---

input: $\mathbf{A}$.
initialization: Choose $\mathbf{x}_0$ with $\|\mathbf{x}_0\|_2 = 1$.
repeat
1. Compute $R_k = R(\mathbf{x}_k) = \mathbf{x}_k^H \mathbf{A}\mathbf{x}_k$.
2. Solve $\mathbf{x}_{k+1}^H(\mathbf{A} - R_k\mathbf{I}) = \tau_k \mathbf{x}_k^H$ for $\mathbf{x}_{k+1}$, and let $\mathbf{x}_{k+1} = \dfrac{\mathbf{x}_{k+1}}{\|\mathbf{x}_{k+1}\|_2}$.
3. Calculate $R_{k+1} = R(\mathbf{x}_{k+1}) = \mathbf{x}_{k+1}^H \mathbf{A}\mathbf{x}_{k+1}$.
4. **exit if** $\mathbf{A} - R_k\mathbf{I}$ or $\mathbf{A} - R_{k+1}\mathbf{I}$ is singular.
5. Calculate $\mathbf{x}_{k+1} = \dfrac{(\mathbf{A}_k - R_k\mathbf{I})^{-1}\mathbf{x}_k}{\|(\mathbf{A}_k - R_k\mathbf{I})^{-1}\mathbf{x}_k\|_2}$.
6. Solve $(\mathbf{A} - R_{k+1}\mathbf{I})\mathbf{x}_{k+2} = \tau_{k+1}\mathbf{x}_{k+1}$ for $\mathbf{x}_{k+2}$, and let $\mathbf{x}_{k+2} = \dfrac{\mathbf{x}_{k+2}}{\|\mathbf{x}_{k+2}\|_2}$.

return $k \leftarrow k+2$.
Find the nonzero solution of $(\mathbf{A} - R_k\mathbf{I})\mathbf{x}_{k+1} = \mathbf{0}$ or $(\mathbf{A} - R_{k+1}\mathbf{I})\mathbf{x}_{k+1} = \mathbf{0}$.
output: $\mathbf{x} = \mathbf{x}_{k+1}$.

7.8.3 Algorithms for Rayleigh Quotient

Choose the negative direction of the gradient of the Rayleigh quotient as the gradient flow of the vector $\mathbf{x}$, i.e.,

$$\dot{\mathbf{x}} = -(\mathbf{A} - R(\mathbf{x})\mathbf{I})\mathbf{x}; \tag{7.8.7}$$

then the vector $\mathbf{x}$ can be iteratively computed using a gradient algorithm as follows [205]:

$$\mathbf{x}_{k+1} = \mathbf{x}_k + \mu\dot{\mathbf{x}}_k = \mathbf{x}_k - \mu(\mathbf{A} - R(\mathbf{x}_k)\mathbf{I})\mathbf{x}_k. \tag{7.8.8}$$

As described in the following theorem, the gradient algorithm for solving the Rayleigh quotient problem has a faster convergence rate than the standard Rayleigh quotient iteration algorithm.

Theorem 7.17 *Suppose that $\lambda_1 > \lambda_2$. For an initial value $\mathbf{x}_0$ such that $\|\mathbf{x}_0\| = 1$, the vector $\mathbf{x}_k$ computed iteratively by the gradient algorithm converges exponentially to the maximum eigenvector $\mathbf{u}_1$ or $-\mathbf{u}_1$ of the matrix $\mathbf{A}$ at a rate $\lambda_1 - \lambda_2$.*

Proof [205, p. 18]. □

We now consider a conjugate gradient algorithm for solving the Rayleigh quotient problem (7.8.1), where $\mathbf{A}$ is a real symmetric matrix.

Starting from some initial vector $\mathbf{x}_0$, the conjugate gradient algorithm uses the iteration formula

$$\mathbf{x}_{k+1} = \mathbf{x}_k + \alpha_k \mathbf{p}_k \tag{7.8.9}$$

to update and approximate the minimum (or maximum) eigenvector of the matrix $\mathbf{A}$. The real coefficient α_k is given by [437], [526]

$$\alpha_k = \pm \frac{1}{2D}\left(-B + \sqrt{B^2 - 4CD}\right), \tag{7.8.10}$$

where the positive sign is applied to update the minimum eigenvector, and the negative sign corresponds to the update of the maximum eigenvector.

The calculation formulas of the parameters D, B, C in Equation (7.8.10) are as follows:

$$D = P_b(k)P_c(k) - P_a(k)P_d(k), \tag{7.8.11}$$
$$B = P_b(k) - \lambda_k P_d(k), \tag{7.8.12}$$
$$C = P_a(k) - \lambda_k P_c(k), \tag{7.8.13}$$

where

$$P_a(k) = \mathbf{p}_k^T \mathbf{A}\mathbf{x}_k/(\mathbf{x}_k^T\mathbf{x}_k), \tag{7.8.14}$$
$$P_b(k) = \mathbf{p}_k^T \mathbf{A}\mathbf{p}_k/(\mathbf{x}_k^T\mathbf{x}_k), \tag{7.8.15}$$
$$P_c(k) = \mathbf{p}_k^T \mathbf{x}_k/(\mathbf{x}_k^T\mathbf{x}_k), \tag{7.8.16}$$
$$P_d(k) = \mathbf{p}_k^T \mathbf{p}_k/(\mathbf{x}_k^T\mathbf{x}_k), \tag{7.8.17}$$
$$\lambda_k = R(\mathbf{x}_k) = \mathbf{x}_k^T \mathbf{A}\mathbf{x}_k/(\mathbf{x}_k^T\mathbf{x}_k). \tag{7.8.18}$$

At the $(k+1)$th iteration, the search direction is selected in the following way:

$$\mathbf{p}_{k+1} = \mathbf{r}_{k+1} + b(k)\mathbf{p}_k, \tag{7.8.19}$$

where $b(-1) = 0$ and $\mathbf{r}_{k+1}$ is the residual error vector at the $(k+1)$th iteration and is given by

$$\mathbf{r}_{k+1} = -\frac{1}{2}\nabla_{\mathbf{x}^*} R(\mathbf{x}_{k+1}) = (\lambda_{k+1}\mathbf{x}_{k+1} - \mathbf{A}\mathbf{x}_{k+1})/(\mathbf{x}_{k+1}^T\mathbf{x}_{k+1}). \tag{7.8.20}$$

The selection of $b(k)$ in Equation (7.8.19) should make the search directions $\mathbf{p}_{k+1}$ and $\mathbf{p}_k$ conjugate or orthogonal with respect to the Hessian matrix of the Rayleigh quotient:

$$\mathbf{p}_{k+1}^T \mathbf{H}\mathbf{p}_k = 0. \tag{7.8.21}$$

Compute the Hessian matrix of the Rayleigh quotient directly to get [526]:

$$\mathbf{H}(\mathbf{x}) = \frac{2}{\mathbf{x}^T\mathbf{x}}\left(\mathbf{A} - \frac{\partial R(\mathbf{x})}{\partial \mathbf{x}}\mathbf{x}^T - \mathbf{x}\left(\frac{\partial R(\mathbf{x})}{\partial \mathbf{x}}\right)^T - R(\mathbf{x})\mathbf{I}\right), \tag{7.8.22}$$

where

$$\frac{\partial R(\mathbf{x})}{\partial \mathbf{x}} = -\frac{\mathbf{x}}{(\mathbf{x}^T\mathbf{x})^2}\mathbf{x}^T\mathbf{A}\mathbf{x} + \frac{2}{\mathbf{x}^T\mathbf{x}}\mathbf{A}\mathbf{x} = -\frac{R(\mathbf{x})}{\mathbf{x}^T\mathbf{x}}\mathbf{x} + \frac{2}{\mathbf{x}^T\mathbf{x}}\mathbf{A}\mathbf{x}. \tag{7.8.23}$$

When $\mathbf{x} = \mathbf{x}_{k+1}$, denote the Hessian matrix as $\mathbf{H}_{k+1} = \mathbf{H}(\mathbf{x}_{k+1})$. In this case, the parameter $b(k)$ is given by

$$b(k) = -\frac{\mathbf{r}_{k+1}^T\mathbf{H}_{k+1}\mathbf{p}_k}{\mathbf{p}_k\mathbf{H}_{k+1}\mathbf{p}_k} = -\frac{\mathbf{r}_{k+1}^T\mathbf{A}\mathbf{p}_k + (\mathbf{r}_{k+1}^T\mathbf{r}_{k+1})(\mathbf{x}_{k+1}^T\mathbf{p}_k)}{\mathbf{p}_k^T(\mathbf{A}\mathbf{p}_k - \lambda_{k+1}\mathbf{I})\mathbf{p}_k}. \tag{7.8.24}$$

Equations (7.8.9)–(7.8.19) and Equation (7.8.24) together form a conjugate gradient algorithm for solving the Rayleigh quotient problem (7.8.1). This algorithm was proposed by Yang *et al.* [526].

7.9 Generalized Rayleigh Quotient

The extended form of the Rayleigh quotient is the generalized Rayleigh quotient. This section presents the definition of and the solution method for the generalized Rayleigh quotient together with its typical applications in pattern recognition and wireless communications.

7.9.1 Definition and Properties

Definition 7.21 Let $\mathbf{A}$ and $\mathbf{B}$ be two $n \times n$ Hermitian matrices, and let $\mathbf{B}$ be positive definite. The *generalized Rayleigh quotient* or *generalized Rayleigh–Ritz ratio* $R(\mathbf{x})$ of the matrix pencil $(\mathbf{A}, \mathbf{B})$ is a scalar, and is defined as

$$R(\mathbf{x}) = \frac{\mathbf{x}^H\mathbf{A}\mathbf{x}}{\mathbf{x}^H\mathbf{B}\mathbf{x}}, \tag{7.9.1}$$

where $\mathbf{x}$ is a vector to be selected such that the generalized Rayleigh quotient is maximized or minimized.

In order to find the generalized Rayleigh quotient define a new vector $\tilde{\mathbf{x}} = \mathbf{B}^{1/2}\mathbf{x}$, where $\mathbf{B}^{1/2}$ denotes the square root of the positive definite matrix $\mathbf{B}$. Substitute $\mathbf{x} = \mathbf{B}^{-1/2}\tilde{\mathbf{x}}$ into the definition formula of the generalized Rayleigh quotient (7.9.1); then we have

$$R(\tilde{\mathbf{x}}) = \frac{\tilde{\mathbf{x}}^H\left(\mathbf{B}^{-1/2}\right)^H\mathbf{A}\left(\mathbf{B}^{-1/2}\right)\tilde{\mathbf{x}}}{\tilde{\mathbf{x}}^H\tilde{\mathbf{x}}}. \tag{7.9.2}$$

This shows that the generalized Rayleigh quotient of the matrix pencil $(\mathbf{A}, \mathbf{B})$ is equivalent to the Rayleigh quotient of the matrix product $\mathbf{C} = (\mathbf{B}^{-1/2})^H\mathbf{A}(\mathbf{B}^{-1/2})$. By the Rayleigh–Ritz theorem, it is known that, when the vector $\tilde{\mathbf{x}}$ is selected as the eigenvector corresponding to the minimum eigenvalue λ_{min} of $\mathbf{C}$, the generalized Rayleigh quotient takes a minimum value λ_{min}, while when the vector $\tilde{\mathbf{x}}$ is selected as the eigenvector corresponding to the maximum eigenvalue λ_{max} of $\mathbf{C}$, the generalized Rayleigh quotient takes the maximum value λ_{max}.

Consider the EVD of the matrix product $(\mathbf{B}^{-1/2})^H\mathbf{A}(\mathbf{B}^{-1/2})$:

$$(\mathbf{B}^{-1/2})^H\mathbf{A}(\mathbf{B}^{-1/2})\tilde{\mathbf{x}} = \lambda\tilde{\mathbf{x}}. \tag{7.9.3}$$

If $\mathbf{B} = \sum_{i=1}^n \beta_i \mathbf{v}_i\mathbf{v}_i^H$ is the EVD of the matrix $\mathbf{B}$ then

$$\mathbf{B}^{1/2} = \sum_{i=1}^n \sqrt{\beta_i}\mathbf{v}_i\mathbf{v}_i^H$$

and $\mathbf{B}^{1/2}\mathbf{B}^{1/2} = \mathbf{B}$. Because the matrix $\mathbf{B}^{1/2}$ and its inverse $\mathbf{B}^{-1/2}$ have the same eigenvector and reciprocal eigenvalues, we have

$$\mathbf{B}^{-1/2} = \sum_{i=1}^n \frac{1}{\sqrt{\beta_i}}\mathbf{v}_i\mathbf{v}_i^H. \tag{7.9.4}$$

This shows that $\mathbf{B}^{-1/2}$ is also a Hermitian matrix, so that $(\mathbf{B}^{-1/2})^H = \mathbf{B}^{-1/2}$.

Premultiplying Equation (7.9.3) by $\mathbf{B}^{-1/2}$ and substituting $(\mathbf{B}^{-1/2})^H = \mathbf{B}^{-1/2}$, it follows that

$$\mathbf{B}^{-1}\mathbf{A}\mathbf{B}^{-1/2}\tilde{\mathbf{x}} = \lambda\mathbf{B}^{-1/2}\tilde{\mathbf{x}} \quad \text{or} \quad \mathbf{B}^{-1}\mathbf{A}\mathbf{x} = \lambda\mathbf{x},$$

as $\mathbf{x} = \mathbf{B}^{-1/2}\tilde{\mathbf{x}}$. Hence, the EVD of the matrix product $(\mathbf{B}^{-1/2})^H\mathbf{A}(\mathbf{B}^{-1/2})$ is equivalent to the EVD of $\mathbf{B}^{-1}\mathbf{A}$. Because the EVD of $\mathbf{B}^{-1}\mathbf{A}$ is the GEVD of the matrix pencil $(\mathbf{A}, \mathbf{B})$, the above analysis can be summarized as follows: the conditions for the maximum and the minimum of the generalized Rayleigh quotient are:

$$R(\mathbf{x}) = \frac{\mathbf{x}^H\mathbf{A}\mathbf{x}}{\mathbf{x}^H\mathbf{B}\mathbf{x}} = \lambda_{\text{max}}, \quad \text{if we choose } \mathbf{A}\mathbf{x} = \lambda_{\text{max}}\mathbf{B}\mathbf{x}, \tag{7.9.5}$$

$$R(\mathbf{x}) = \frac{\mathbf{x}^H\mathbf{A}\mathbf{x}}{\mathbf{x}^H\mathbf{B}\mathbf{x}} = \lambda_{\text{min}}, \quad \text{if we choose } \mathbf{A}\mathbf{x} = \lambda_{\text{min}}\mathbf{B}\mathbf{x}. \tag{7.9.6}$$

That is to say, in order to maximize the generalized Rayleigh quotient, the vector $\mathbf{x}$ must be taken as the generalized eigenvector corresponding to the maximum generalized eigenvalue of the matrix pencil $(\mathbf{A}, \mathbf{B})$. In contrast, for the generalized Rayleigh quotient to be minimized, the vector $\mathbf{x}$ should be taken as the generalized eigenvector corresponding to the minimum generalized eigenvalue of the matrix pencil $(\mathbf{A}, \mathbf{B})$.

In the following, two applications of the generalized Rayleigh quotient are presented.

7.9.2 Effectiveness of Class Discrimination

Pattern recognition is widely applied in recognitions of human characteristics (such as a person's face, fingerprint, iris) and various radar targets (such as aircraft, ships). In these applications, the extraction of signal features is crucial. For example, when a target is considered as a linear system, the target parameter is a feature of the target signal.

The divergence is a measure of the distance or dissimilarity between two signals and is often used in feature discrimination and evaluation of the effectiveness of class discrimination.

Let Q denote the common dimension of the signal-feature vectors extracted by various methods. Assume that there are c classes of signals; the Fisher class separability measure, called simply the *Fisher measure*, between the ith and the jth classes is used to determine the ranking of feature vectors. Consider the class discrimination of c classes of signals. Let $\mathbf{v}_k^{(l)}$ denote the kth sample feature vector of the lth class of signals, where $l = 1, \ldots, c$ and $k = 1, \cdots, l_K$, with l_K the number of sample feature vectors in the lth signal class. Under the assumption that the prior probabilities of the random vectors $\mathbf{v}^{(l)} = \mathbf{v}_k^{(l)}$ are the same for $k = 1, \ldots, l_K$ (i.e., equal probabilities), the Fisher measure is defined as

$$m^{(i,j)} = \frac{\sum_{l=i,j}\left(\text{mean}_k(\mathbf{v}_k^{(l)}) - \text{mean}_l(\text{mean}(\mathbf{v}_k^{(l)}))\right)^2}{\sum_{l=i,j}\text{var}_k(\mathbf{v}_k^{(l)})}, \tag{7.9.7}$$

where:

$\text{mean}_k(\mathbf{v}_k^{(l)})$ is the mean (centroid) of all signal-feature vectors of the lth class;
$\text{var}(\mathbf{v}_k^{(l)})$ is the variance of all signal-feature vectors of the lth class;
$\text{mean}_l(\text{mean}_k(\mathbf{v}_k^{(l)}))$ is the total sample centroid of all classes.

As an extension of the Fisher measure, consider the projection of all $Q \times 1$ feature vectors onto the $(c-1)$th dimension of class discrimination space.

Put $N = N_1 + \cdots + N_c$, where N_i represents the number of feature vectors of the ith class signal extracted in the training phase. Suppose that

$$\mathbf{s}_{i,k} = [s_{i,k}(1), \ldots, s_{i,k}(Q)]^T$$

denotes the $Q \times 1$ feature vector obtained by the kth group of observation data of the ith signal in the training phase, while

$$\mathbf{m}_i = [m_i(1), \ldots, m_i(Q)]^T$$

is the sample mean of the ith signal-feature vector, where

$$m_i(q) = \frac{1}{N_i}\sum_{k=1}^{N_i} s_{i,k}(q), \quad i = 1, \ldots, c,\ q = 1, \ldots, Q.$$

Similarly, let

$$\mathbf{m} = [m(1), \ldots, m(Q)]^T$$

denote the ensemble mean of all the signal-feature vectors obtained from all the observation data, where

$$m(q) = \frac{1}{c}\sum_{i=1}^{c} m_i(q), \quad q = 1, \ldots, Q.$$

Using the above vectors, one can define a $Q \times Q$ within-class scatter matrix [138]

$$\mathbf{S}_w \stackrel{\text{def}}{=} \frac{1}{c}\sum_{i=1}^{c}\left(\frac{1}{N_i}\sum_{k=1}^{N_i}(\mathbf{s}_{i,k} - \mathbf{m}_i)(\mathbf{s}_{i,k} - \mathbf{m}_i)^T\right) \tag{7.9.8}$$

and a $Q \times Q$ between-class scatter matrix [138]

$$\mathbf{S}_b \stackrel{\text{def}}{=} \frac{1}{c}\sum_{i=1}^{c}(\mathbf{m}_i - \mathbf{m})(\mathbf{m}_i - \mathbf{m})^T. \tag{7.9.9}$$

For the optimal class discrimination matrix $\mathbf{U}$ to be determined, define a criterion function

$$J(\mathbf{U}) \stackrel{\text{def}}{=} \frac{\prod_{\text{diag}} \mathbf{U}^T\mathbf{S}_b\mathbf{U}}{\prod_{\text{diag}} \mathbf{U}^T\mathbf{S}_w\mathbf{U}}, \tag{7.9.10}$$

where $\prod_{\text{diag}} \mathbf{A}$ denotes the product of diagonal entries of the matrix $\mathbf{A}$. As it is a measure of class discrimination ability, J should be maximized. We refer to $\text{Span}(\mathbf{U})$ as the *class discrimination space*, if

$$\mathbf{U} = \arg\max_{\mathbf{U}\in\mathbb{R}^{Q\times Q}} J(\mathbf{U}) = \frac{\prod_{\text{diag}} \mathbf{U}^T\mathbf{S}_b\mathbf{U}}{\prod_{\text{diag}} \mathbf{U}^T\mathbf{S}_w\mathbf{U}}. \tag{7.9.11}$$

This optimization problem can be equivalently written as

$$[\mathbf{u}_1, \ldots, \mathbf{u}_Q] = \arg\max_{\mathbf{u}_i\in\mathbb{R}^Q}\left\{\frac{\prod_{i=1}^{Q}\mathbf{u}_i^T\mathbf{S}_b\mathbf{u}_i}{\prod_{i=1}^{Q}\mathbf{u}_i^T\mathbf{S}_w\mathbf{u}_i} = \prod_{i=1}^{Q}\frac{\mathbf{u}_i^T\mathbf{S}_b\mathbf{u}_i}{\mathbf{u}_i^T\mathbf{S}_w\mathbf{u}_i}\right\} \tag{7.9.12}$$

whose solution is given by

$$\mathbf{u}_i = \arg\max_{\mathbf{u}_i\in\mathbb{R}^Q}\frac{\mathbf{u}_i^T\mathbf{S}_b\mathbf{u}_i}{\mathbf{u}_i^T\mathbf{S}_w\mathbf{u}_i}, \quad i = 1, \ldots, Q. \tag{7.9.13}$$

This is just the maximization of the generalized Rayleigh quotient. The above equation has a clear physical meaning: the column vectors of the matrix $\mathbf{U}$ constituting the optimal class discrimination subspace which should maximize the between-class scatter and minimize the within-class scatter, namely which should maximize the generalized Rayleigh quotient.

For c classes of signals, the optimal class discrimination subspace is $(c-1)$-dimensional. Therefore Equation (7.9.13) is maximized only for $c-1$ generalized Rayleigh quotients. In other words, it is necessary to solve the generalized eigenvalue problem

$$\mathbf{S}_b \mathbf{u}_i = \lambda_i \mathbf{S}_w \mathbf{u}_i, \quad i = 1, 2, \ldots, c-1, \tag{7.9.14}$$

for $c-1$ generalized eigenvectors $\mathbf{u}_1, \ldots, \mathbf{u}_{c-1}$. These generalized eigenvectors constitute the $Q \times (c-1)$ matrix

$$\mathbf{U}_{c-1} = [\mathbf{u}_1, \ldots, \mathbf{u}_{c-1}] \tag{7.9.15}$$

whose columns span the optimal class discrimination subspace.

After obtaining the $Q \times (c-1)$ matrix $\mathbf{U}_{c-1}$, we can find its projection onto the optimal class discrimination subspace,

$$\mathbf{y}_{i,k} = \mathbf{U}_{c-1}^T \mathbf{s}_{i,k}, \quad i = 1, \ldots, c, \ k = 1, \ldots, N_i, \tag{7.9.16}$$

for every signal-feature vector obtained in the training phase.

When there are only three classes of signals ($c = 3$), the optimal class discrimination subspace is a plane, and the projection of every feature vector onto the optimal class discrimination subspace is a point. These projections directly reflect the discrimination ability of different feature vectors in signal classification.

7.9.3 Robust Beamforming

In wireless communications, if a base station uses an antenna array which is composed of multiple antennas (called array elements), it can separate multiple users in a common channel through spatial processing and thereby detect the desired user signals.

Consider an array consisting of M omnidirectional antennas, and suppose that K narrowband signals are located in the far field. The observation signal vector received by the M array elements at time n is given by

$$\mathbf{y}(n) = \mathbf{d}(n) + \mathbf{i}(n) + \mathbf{e}(n), \tag{7.9.17}$$

where $\mathbf{e}(n)$ is the vector whose entries are the additive white noises at the M array elements, while

$$\mathbf{d}(n) = \mathbf{a}(\theta_0(n)) s_0(n), \tag{7.9.18}$$

$$\mathbf{i}(n) = \sum_{k=1}^{K-1} \mathbf{a}(\theta_k(n)) s_k(n), \tag{7.9.19}$$

are respectively the desired signal vector received by the M array elements and the interference signal vector from the other $K-1$ users; $\theta_0(n)$ and $\theta_k(n)$ respectively denote the directions of arrival (DOAs) of the desired user and the kth interference user, and the vectors $\mathbf{a}(\theta_0(n))$ and $\mathbf{a}(\theta_k(n))$ are respectively the array response vectors of the desired user and the kth interference user.

Assume that these signal sources are statistically independent of each other, and that the additive white noises of different array elements are statistically independent of each other and have the same variance σ^2. Therefore the autocorrelation matrix of the observation signal is given by

$$\mathbf{R}_y = E\{\mathbf{y}(n)\mathbf{y}^H(n)\} = \mathbf{R}_d + \mathbf{R}_{i+e}, \tag{7.9.20}$$

where

$$\mathbf{R}_d = E\{\mathbf{d}(n)\mathbf{d}^H(n)\} = P_0\mathbf{a}(\theta_0)\mathbf{a}^H(\theta_0), \tag{7.9.21}$$

$$\begin{aligned}\mathbf{R}_{i+e} &= E\{(\mathbf{i}(n)+\mathbf{e}(n))(\mathbf{i}(n)+\mathbf{e}(n))^H\} \\ &= \sum_{k=1}^{K-1} P_k\mathbf{a}(\theta_k)\mathbf{a}^H(\theta_k) + \sigma^2\mathbf{I};\end{aligned} \tag{7.9.22}$$

the constants P_0 and P_k represent the powers of the desired user and of the kth interference user, respectively.

Let $\mathbf{w}(n)$ be the weighting vector of the beamformer at time n; then its output is

$$z(n) = \mathbf{w}^H(n)\mathbf{y}(n). \tag{7.9.23}$$

It is easy to find the signal-to-interference-plus-noise ratio (SINR) of the beamformer output:

$$\text{SINR}(\mathbf{w}) = \frac{E\{|\mathbf{w}^H\mathbf{d}(n)|^2\}}{E\{|\mathbf{w}^H(\mathbf{i}(n)+\mathbf{e}(n))|^2\}} = \frac{\mathbf{w}^H\mathbf{R}_d\mathbf{w}}{\mathbf{w}^H\mathbf{R}_{i+e}\mathbf{w}}. \tag{7.9.24}$$

In order to achieve the purpose of interference suppression, we need to maximize the generalized Rayleigh quotient $\text{SINR}(\mathbf{w})$. That is to say, the optimal beamformer for interference suppression should be selected as the generalized eigenvector corresponding to the maximum generalized eigenvalue of the matrix pencil $\{\mathbf{R}_d, \mathbf{R}_{i+e}\}$. However, in this method we need to compute respectively the autocorrelation matrix $\mathbf{R}_d$ of the desired user and autocorrelation matrix $\mathbf{R}_{i+e}$ of the interference plus noise. These autocorrelation matrices are hard to compute directly.

Substituting $\mathbf{R}_d = P_0\mathbf{a}(\theta_0)\mathbf{a}^H(\theta_0)$ into the SINR formula (7.9.24), the unconstrained optimization problem can be represented as

$$\max\ \text{SINR}(\mathbf{w}) = \max \frac{P_0|\mathbf{w}^H\mathbf{a}(\theta_0)|^2}{\mathbf{w}^H\mathbf{R}_{i+e}\mathbf{w}}. \tag{7.9.25}$$

Adding the constraint condition $\mathbf{w}^H\mathbf{a}(\theta_0) = 1$, the unconstrained optimization problem becomes the constrained optimization problem

$$\min \mathbf{w}^H \mathbf{R}_{i+e}\mathbf{w} \quad \text{subject to} \quad \mathbf{w}^H \mathbf{a}(\theta_0) = 1. \tag{7.9.26}$$

This optimization problem is still not easy to solve, because $\mathbf{R}_{i+e}$ cannot be calculated.

Note that

$$\mathbf{w}^H \mathbf{R}_y \mathbf{w} = \mathbf{w}^H \mathbf{R}_d \mathbf{w} + \mathbf{w}^H \mathbf{R}_{i+e}\mathbf{w} = P_0 + \mathbf{w}^H \mathbf{R}_{i+e}\mathbf{w}$$

and that the power of the desired signal P_0 is a constant that is independent of the beamformer, so the optimization problem in Equation (7.9.26) is also equivalent to

$$\min\ \mathbf{w}^H \mathbf{R}_y \mathbf{w} \quad \text{subject to} \quad \mathbf{w}^H \mathbf{a}(\theta_0) = 1. \tag{7.9.27}$$

In contrast with Equation (7.9.26), the autocorrelation matrix of the signal vector $\mathbf{R}_y$ is easily estimated. Using the Lagrange multiplier method, it is easy to find that the solution of the constrained optimization problem (7.9.27) is

$$\mathbf{w}_{\text{opt}}(n) = \frac{\mathbf{R}_y^{-1}\mathbf{a}(\theta_0(n))}{\mathbf{a}^H(\theta_0(n))\mathbf{R}_y^{-1}\mathbf{a}(\theta_0(n))}. \tag{7.9.28}$$

This is the robust beamformer for interference suppression proposed in [414].

7.10 Quadratic Eigenvalue Problems

We often encounter a common problem – the quadratic eigenvalue problem (QEP) – in linear stability studies in fluid mechanics [61], [204], the dynamic analysis of acoustic systems, the vibrational analysis of structural systems in structural mechanics, circuit simulation [466], the mathematical modeling of microelectronic mechanical systems (MEMSs) [108], biomedical signal processing, time series prediction, the linear-prediction coding of voices [116], multiple input–multiple output (MIMO) system analysis [466], finite element analysis of partial differential equations in industrial applications [260], [359], and in a number of applications of linear algebra problems.

The quadratic eigenvalue problem is closely linked to the eigenvalue problem (especially the generalized eigenvalue problem), but there is a clear difference.

7.10.1 Description of Quadratic Eigenvalue Problems

Consider a system with viscous damping and without any external forces; its motion equation is the differential equation [230], [422]

$$\mathbf{M}\ddot{\mathbf{x}} + \mathbf{C}\dot{\mathbf{x}} + \mathbf{K}\mathbf{x} = \mathbf{0}, \tag{7.10.1}$$

where $\mathbf{M}$, $\mathbf{C}$, $\mathbf{K}$ are respectively $n \times n$ mass, damping and stiffness matrices, while the vectors $\ddot{\mathbf{x}}$, $\dot{\mathbf{x}}$ and $\mathbf{x}$ are respectively the acceleration vector, the velocity vector and the displacement vector.

In vibrational analysis, the general solution of the homogeneous linear equation (7.10.1) is given by $\mathbf{x} = \mathrm{e}^{\lambda t}\mathbf{u}$, where $\mathbf{u}$ is usually a complex vector while λ is an eigenvalue that is generally a complex number. Substituting $\mathbf{x} = \mathrm{e}^{\lambda t}\mathbf{u}$ and its derivatives with respect to time into (7.10.1), one gets the *quadratic characteristic equation*

$$(\lambda^2\mathbf{M} + \lambda\mathbf{C} + \mathbf{K})\mathbf{u} = \mathbf{0}, \quad \text{or} \quad \mathbf{v}^H(\lambda^2\mathbf{M} + \lambda\mathbf{C} + \mathbf{K}) = \mathbf{0}^T$$

if instead we write $\mathbf{x} = \mathbf{v}^H\mathrm{e}^{\lambda t}$.

The determinant $|\lambda^2\mathbf{M} + \lambda\mathbf{C} + \mathbf{K}|$ is known as the *quadratic characteristic polynomial*, and $|\lambda^2\mathbf{M} + \lambda\mathbf{C} + \mathbf{K}| = 0$ is called the *quadratic characteristic equation.*

In general, the *quadratic eigenvalue problem* can be described as follows [274]: given $n \times n$ matrices $\mathbf{M}, \mathbf{C}, \mathbf{K}$, find scalars λ_i and nonzero vectors $\mathbf{u}_i, \mathbf{v}_i$ such that

$$(\lambda_i^2\mathbf{M} + \lambda_i\mathbf{C} + \mathbf{K})\mathbf{u}_i = \mathbf{0}, \tag{7.10.2}$$

$$\mathbf{v}_i^H(\lambda_i^2\mathbf{M} + \lambda_i\mathbf{C} + \mathbf{K}) = \mathbf{0}^T. \tag{7.10.3}$$

The scalars λ_i satisfying the above equation are called *quadratic eigenvalues*, and the nonzero vectors $\mathbf{u}_i$ and $\mathbf{v}_i$ are respectively known as the *right quadratic eigenvector* and the *left quadratic eigenvector* corresponding to the quadratic eigenvalue λ_i. Every quadratic eigenvalue λ_i and its corresponding quadratic eigenvectors $\mathbf{u}_i, \mathbf{v}_i$ constitute a *quadratic eigentriple* $(\lambda_i, \mathbf{u}_i, \mathbf{v}_i)$ of the quadratic eigenvalue problem.

A quadratic eigenvalue problem has $2n$ eigenvalues (finite or infinitely large), $2n$ right eigenvectors and $2n$ left eigenvectors. In engineering applications the eigenvalues are usually complex; the imaginary parts are the resonant frequencies and the real parts denote exponential damping. We hope to find all the eigenvalues in the frequency range of interest. In some applications the eigentriples usually number dozens to hundreds.

The quadratic eigenvalue problem is an important subclass of nonlinear eigenvalue problems. Let

$$\mathbf{Q}(\lambda) = \lambda^2\mathbf{M} + \lambda\mathbf{C} + \mathbf{K}. \tag{7.10.4}$$

This is a quadratic $n \times n$ matrix polynomial. In other words, the coefficients of the matrix $\mathbf{Q}(\lambda)$ constitute the quadratic polynomial of λ. As discussed in Subsection 7.4.2, the matrix $\mathbf{Q}(\lambda)$ is called the λ-matrix [273]. The quadratic eigenvalues are given by the solutions to the characteristic equation

$$|\mathbf{Q}(z)| = |z^2\mathbf{M} + z\mathbf{C} + \mathbf{K}| = 0. \tag{7.10.5}$$

Definition 7.22 [466] The matrix $\mathbf{Q}(\lambda)$ is called a *regular λ-matrix* if the characteristic polynomial $|\mathbf{Q}(z)|$ is not identically zero for all z. Conversely, if $|\mathbf{Q}(z)| \equiv 0$ for all z, then say that the matrix $\mathbf{Q}(z)$ is *nonregular λ-matrix.*

The λ-matrix $\mathbf{Q}(\lambda)$ is usually assumed to be regular. For a regular λ-matrix $\mathbf{Q}(\lambda)$, two different eigenvalues are likely to have the same eigenvector.

Table 7.1 summarizes the properties of the eigenvalues and eigenvectors of the quadratic eigenvalue problem (QEP).

Table 7.1 Properties of quadratic eigenvalues [466]

No.	Matrix	Eigenvalue	Eigenvectors
1	$\mathbf{M}$ nonsingular	$2n$, finite	
2	$\mathbf{M}$ singular	finite and infinite	
3	$\mathbf{M}, \mathbf{C}, \mathbf{K}$: real	real or (λ, λ^*) pairs	$(\lambda, \mathbf{u}), (\lambda^*, \mathbf{u}^*)$
4	$\mathbf{M}, \mathbf{C}, \mathbf{K}$: Hermitian	real or (λ, λ^*) pairs	$(\lambda, \mathbf{u}), (\lambda, \mathbf{u}^*)$
5	$\mathbf{M}$: Hermitian pdf; $\mathbf{C}, \mathbf{K}$: Hermitian psdf	$\mathrm{Re}(\lambda) \leq 0$	
6	$\mathbf{M}, \mathbf{C}$: symmetric pdf; $\mathbf{K}$: symmetric psdf; $\gamma(\mathbf{M}, \mathbf{C}, \mathbf{K}) > 0$	$\lambda \neq 0$, n principal and n minor eigenvalues	$2n$ eigenvectors are independent
7	$\mathbf{M}, \mathbf{K}$: Hermitian; $\mathbf{M}$: pdf, $\mathbf{C} = -\mathbf{C}^H$	imaginary eigenvalues or conjugate pair (λ, λ^*)	$(\lambda, \mathbf{u})$: right eigenpair; $(-\lambda^*, \mathbf{u})$: left eigenpair
8	$\mathbf{M}, \mathbf{K}$: real symmetric pdf, $\mathbf{C} = -\mathbf{C}^T$	imaginary eigenvalues	

In Table 7.1, pdf means positive definite, psdf means positive semi-definite and

$$\gamma(\mathbf{M}, \mathbf{C}, \mathbf{K}) = \min\left\{ \left(\mathbf{u}^H \mathbf{C} \mathbf{u}\right)^2 - 4\left(\mathbf{u}^H \mathbf{M} \mathbf{u}\right)\left(\mathbf{u}^H \mathbf{K} \mathbf{u}\right) \mid \|\mathbf{u}\|_2 = 1 \right\}.$$

According to the value of the quadratic function, the QEP can be further classified as follows [197]:

(1) *Elliptic QEP* The QEP (7.10.1) with $(\mathbf{u}^H \mathbf{C} \mathbf{u})^2 < 4(\mathbf{u}^H \mathbf{M} \mathbf{u})(\mathbf{u}^H \mathbf{K} \mathbf{u})$ is known as an elliptic quadratic eigenvalue problem.

(2) *Hyperbolic QEP* The QEP (7.10.1) is called a hyperbolic quadratic eigenvalue problem if the function $(\mathbf{u}^H \mathbf{C} \mathbf{u})^2 > 4(\mathbf{u}^H \mathbf{M} \mathbf{u})(\mathbf{u}^H \mathbf{K} \mathbf{u})$.

7.10.2 Solving Quadratic Eigenvalue Problems

There are two main methods in solving quadratic eigenvalue problems:

(1) *Decomposition method:* On the basis of the generalized Bezout theorem, a quadratic eigenvalue problem can be divided into two first-order eigenvalue problems.

(2) *Linearized method:* Via linearization, a nonlinear quadratic eigenvalue problem becomes a linear generalized eigenvalue problem.

1. Decomposition Method

Define the matrix

$$\mathbf{Q}(\mathbf{S}) = \mathbf{M}\mathbf{S}^2 + \mathbf{C}\mathbf{S} + \mathbf{K}, \quad \mathbf{S} \in \mathbb{C}^{n\times n}; \tag{7.10.6}$$

then the difference between $\mathbf{Q}(\mathbf{S})$ and $\mathbf{Q}(\lambda)$ is

$$\begin{aligned}\mathbf{Q}(\lambda) - \mathbf{Q}(\mathbf{S}) &= \mathbf{M}(\lambda^2\mathbf{I} - \mathbf{S}^2) + \mathbf{C}(\lambda\mathbf{I} - \mathbf{S}) \\ &= (\lambda\mathbf{M} + \mathbf{M}\mathbf{S} + \mathbf{C})(\lambda\mathbf{I} - \mathbf{S}).\end{aligned} \tag{7.10.7}$$

This result is known as the *generalized Bezout theorem* for a second-order matrix polynomial [171].

If for the second-order matrix equation

$$\mathbf{Q}(\mathbf{S}) = \mathbf{M}\mathbf{S}^2 + \mathbf{C}\mathbf{S} + \mathbf{K} = \mathbf{O} \tag{7.10.8}$$

there exists one solution $\mathbf{S} \in \mathbb{C}^{n\times n}$, then we say that this solution is the right solvent of the second-order matrix equation. Similarly, the solution of the equation $\mathbf{S}^2\mathbf{M} + \mathbf{S}\mathbf{C} + \mathbf{K} = \mathbf{O}$ is called the left solvent of the second-order matrix equation [466]. Clearly, if $\mathbf{S}$ is a solution of the second-order matrix equation $\mathbf{Q}(\mathbf{S}) = \mathbf{O}$ then the generalized Bezout theorem (7.10.7) simplifies to

$$\mathbf{Q}(\lambda) = (\lambda\mathbf{M} + \mathbf{M}\mathbf{S} + \mathbf{C})(\lambda\mathbf{I} - \mathbf{S}). \tag{7.10.9}$$

Equation (7.10.9) shows that if $\mathbf{S}$ is the solution of the second-order matrix equation (7.10.8) then the quadratic eigenvalue problem $|\mathbf{Q}(\lambda)| = |\lambda^2\mathbf{M}+\lambda\mathbf{C}+\mathbf{K}| = 0$ is equivalent to $|\mathbf{Q}(\lambda)| = |(\lambda\mathbf{M} + \mathbf{M}\mathbf{S} + \mathbf{C})(\lambda\mathbf{I} - \mathbf{S})| = 0$. From the determinant property $|\mathbf{A}\mathbf{B}| = |\mathbf{A}|\,|\mathbf{B}|$ it is known that the quadratic eigenvalue problem becomes two (linear) eigenvalue subproblems:

- n of the quadratic eigenvalues are solutions of the characteristic equation $|\lambda\mathbf{M} + \mathbf{M}\mathbf{S} + \mathbf{C}| = 0$;
- other n quadratic eigenvalues are solutions of the characteristic equation $|\lambda\mathbf{I} - \mathbf{S}| = 0$.

Because $|\lambda\mathbf{M} + \mathbf{M}\mathbf{S} + \mathbf{C}| = |\mathbf{M}\mathbf{S} + \mathbf{C} - \lambda(-\mathbf{M})|$, the eigenvalue problem $|\lambda\mathbf{M} + \mathbf{M}\mathbf{S} + \mathbf{C}| = 0$ is equivalent to the generalized eigenvalue problem $(\mathbf{M}\mathbf{S} + \mathbf{C})\mathbf{u} = \lambda(-\mathbf{M})\mathbf{u}$.

Summarizing the above discussion, it may be concluded that, by the generalized Bezout theorem, the quadratic eigenvalue problem $(\lambda^2\mathbf{M} + \lambda\mathbf{C} + \mathbf{K})\mathbf{u} = \mathbf{0}$ can be decomposed into two linear eigenvalue subproblems:

(1) a generalized eigenvalue problem for the matrix pencil $(\mathbf{M}\mathbf{S} + \mathbf{C}, -\mathbf{M})$, i.e., $(\mathbf{M}\mathbf{S} + \mathbf{C})\mathbf{u} = -\lambda\mathbf{M}\mathbf{u}$;

(2) the standard eigenvalue problem $\mathbf{Su} = \lambda\mathbf{u}$ for the matrix $\mathbf{S}$.

2. Linearized Method

A common way to solve nonlinear equations is to linearize them. This idea is also available for solving quadratic eigenvalue problems, because they themselves are an important subclass of nonlinear equations.

Let $\mathbf{z} = \begin{bmatrix} \lambda\mathbf{x} \\ \mathbf{x} \end{bmatrix}$; then the quadratic characteristic equation $(\lambda^2\mathbf{M} + \lambda\mathbf{C} + \mathbf{K})\mathbf{x} = \mathbf{0}$ can be written as the equivalent form $\mathbf{L}_c(\lambda)\mathbf{z} = \mathbf{0}$, where

$$\mathbf{L}_c(\lambda) = \lambda \begin{bmatrix} \mathbf{M} & \mathbf{O} \\ \mathbf{O} & \mathbf{I} \end{bmatrix} - \begin{bmatrix} -\mathbf{C} & -\mathbf{K} \\ \mathbf{I} & \mathbf{O} \end{bmatrix} \tag{7.10.10}$$

or

$$\mathbf{L}_c(\lambda) = \lambda \begin{bmatrix} \mathbf{M} & \mathbf{C} \\ \mathbf{O} & \mathbf{I} \end{bmatrix} - \begin{bmatrix} \mathbf{O} & -\mathbf{K} \\ \mathbf{I} & \mathbf{O} \end{bmatrix}. \tag{7.10.11}$$

Here $\mathbf{L}_c(\lambda)$ is called the *companion form* or *linearized* λ*-matrix* of $\mathbf{Q}(\lambda)$.

The companion forms $\lambda\mathbf{B} - \mathbf{A}$ are divided into the *first companion form*

$$L1: \qquad \mathbf{A} = \begin{bmatrix} -\mathbf{C} & -\mathbf{K} \\ \mathbf{I} & \mathbf{O} \end{bmatrix}, \quad \mathbf{B} = \begin{bmatrix} \mathbf{M} & \mathbf{O} \\ \mathbf{O} & \mathbf{I} \end{bmatrix} \tag{7.10.12}$$

and the *second companion form*

$$L2: \qquad \mathbf{A} = \begin{bmatrix} \mathbf{O} & -\mathbf{K} \\ \mathbf{I} & \mathbf{O} \end{bmatrix}, \quad \mathbf{B} = \begin{bmatrix} \mathbf{M} & \mathbf{C} \\ \mathbf{O} & \mathbf{I} \end{bmatrix}. \tag{7.10.13}$$

Via linearization, a quadratic eigenvalue problem $\mathbf{Q}(\lambda)\mathbf{x} = \mathbf{0}$ becomes a generalized eigenvalue problem $\mathbf{L}_c(\lambda)\mathbf{z} = \mathbf{0}$ or $\mathbf{Az} = \lambda\mathbf{Bz}$.

As stated in Section 7.7, in order to ensure that the GEVD of a matrix pencil $(\mathbf{A}, \mathbf{B})$ is uniquely determined, the matrix $\mathbf{B}$ has to be nonsingular. From (7.10.12) and (7.10.13) it can be seen that this is equivalent to the matrix $\mathbf{M}$ being nonsingular.

Algorithm 7.11 shows a linearized algorithm for the quadratic eigenvalue problem (QEP).

The QZ decomposition in step 2 of Algorithm 7.11 can be run by using the qz function in MATLIB directly. It may be noticed that when some entry of the upper triangular matrix $\mathbf{S}$ is equal to zero then the eigenvalue $\lambda = \infty$.

When $\mathbf{M}, \mathbf{C}$ and $\mathbf{K}$ are symmetric matrices, and $\mathbf{A} - \lambda\mathbf{B}$ is the symmetric linearization of $\mathbf{P}(\lambda) = \lambda^2\mathbf{M} + \lambda\mathbf{C} + \mathbf{K}$, Algorithm 7.11 will not ensure the symmetry of $\mathbf{A} - \lambda\mathbf{B}$ because the QZ decomposition in step 2 cannot ensure the symmetry of $(\mathbf{A}, \mathbf{B})$. In this case, one needs to use the following method [466].

(1) First compute the Cholesky decomposition $\mathbf{B} = \mathbf{LL}^T$, where $\mathbf{L}$ is a lower triangular matrix.

Algorithm 7.11 Linearized algorithm for QEP

input: λ-matrix $\mathbf{P}(\lambda) = \lambda^2\mathbf{M} + \lambda\mathbf{C} + \mathbf{K}$.

1. Form the matrix pencil $(\mathbf{A}, \mathbf{B})$ via linearization in (7.10.12) or (7.10.13).
2. Use QZ decomposition to compute the generalized Schur decomposition
$$\mathbf{T} = \mathbf{Q}^H\mathbf{A}\mathbf{Z}, \quad \mathbf{S} = \mathbf{Q}^H\mathbf{B}\mathbf{Z}.$$
3. Calculate the quadratic eigenvalues and eigenvectors:

for $k = 1 : 2n$ (i.e., $k = 1, \ldots, 2n$),

$\lambda_k = t_{kk}/s_{kk}$.

Solve $(\mathbf{T} - \lambda_k\mathbf{S})\boldsymbol{\phi} = \mathbf{0}$, and let $\boldsymbol{\xi} = \mathbf{Z}\boldsymbol{\phi}$.

$\boldsymbol{\xi}_1 = \boldsymbol{\xi}(1 : n)$; $\boldsymbol{\xi}_2 = \boldsymbol{\xi}(n + 1 : 2n)$.

$\mathbf{r}_1 = \mathbf{P}(\lambda_k)\boldsymbol{\xi}_1/\|\boldsymbol{\xi}_1\|$; $\mathbf{r}_2 = \mathbf{P}(\lambda_2)\boldsymbol{\xi}_2/\|\boldsymbol{\xi}_2\|$.

$$\mathbf{u}_k = \begin{cases} \boldsymbol{\xi}(1 : n), & \text{if } \|\mathbf{r}_1\| \le \|\mathbf{r}_2\|, \\ \boldsymbol{\xi}(n + 1 : 2n), & \text{otherwise.} \end{cases}$$

endfor

output: $(\lambda_k, \mathbf{u}_k)$, $k = 1, \ldots, 2n$.

(2) Substitute $\mathbf{B} = \mathbf{L}\mathbf{L}^T$ into the symmetric GEVD problem $\mathbf{A}\boldsymbol{\xi} = \lambda\mathbf{B}\boldsymbol{\xi}$ to get the standard eigenvalue problem $\mathbf{L}^{-1}\mathbf{A}\mathbf{L}^{-T}\boldsymbol{\phi} = \lambda\boldsymbol{\phi}$, where $\boldsymbol{\phi} = \mathbf{L}^T\boldsymbol{\xi}$.

(3) Use the symmetric QR decomposition to compute the eigenpair $(\lambda_i, \boldsymbol{\phi}_i)$ of the symmetric matrix $\mathbf{L}^{-1}\mathbf{A}\mathbf{L}^{-T}$. These eigenpairs are the quadratic eigenpairs of the original quadratic eigenvalue problem.

Meerbergen [321] presented a generalized Davidson algorithm, a modified Davidson algorithm and a second-order residual-error iteration method, etc., for the linearized quadratic eigenvalue problem.

Algorithm 7.11 is only applicable to the case of a nonsingular matrix $\mathbf{M}$. However, in some industrial applications the matrix $\mathbf{M}$ is singular. For example, in the finite element analysis of a damping structure, a so-called massless degree of freedom is often encountered, which corresponds to some zero column vectors of the mass matrix $\mathbf{M}$ [260].

When the matrix $\mathbf{M}$ is singular, and thus $\mathbf{B}$ is also singular, there are two ways to improve the original quadratic eigenvalue problem [260].

The first way uses spectral transformation, i.e., the introduction of an appropriate shift λ_0 to change the eigenvalue to μ as follows:

$$\mu = \lambda - \lambda_0. \tag{7.10.14}$$

Then, the Type I linearization becomes

$$\begin{bmatrix} -\mathbf{C} - \lambda_0\mathbf{M} & -\mathbf{K} \\ \mathbf{I} & -\lambda_0\mathbf{I} \end{bmatrix} \begin{bmatrix} \dot{\mathbf{u}} \\ \mathbf{u} \end{bmatrix} = \mu \begin{bmatrix} \mathbf{M} & \mathbf{O} \\ \mathbf{O} & \mathbf{I} \end{bmatrix} \begin{bmatrix} \dot{\mathbf{u}} \\ \mathbf{u} \end{bmatrix}. \tag{7.10.15}$$

The shift parameter λ_0 is chosen to ensure that the matrix $\begin{bmatrix} -\mathbf{C}-\lambda_0\mathbf{M} & -\mathbf{K} \\ \mathbf{I} & -\lambda_0\mathbf{I} \end{bmatrix}$ is nonsingular.

The alternative way is to set

$$\alpha = \frac{1}{\mu}. \tag{7.10.16}$$

Substitute $\mu = 1/\alpha$ into Equation (7.10.15) and rearrange the result, obtaining

$$\begin{bmatrix} -\mathbf{C}-\lambda_0\mathbf{M} & -\mathbf{K} \\ \mathbf{I} & -\lambda_0\mathbf{I} \end{bmatrix}^{-1} \begin{bmatrix} \mathbf{M} & \mathbf{O} \\ \mathbf{O} & \mathbf{I} \end{bmatrix} \begin{bmatrix} \dot{\mathbf{u}} \\ \mathbf{u} \end{bmatrix} = \alpha \begin{bmatrix} \dot{\mathbf{u}} \\ \mathbf{u} \end{bmatrix}. \tag{7.10.17}$$

The above equation is the standard eigenvalue decomposition

$$\mathbf{Ax} = \alpha\mathbf{x}, \tag{7.10.18}$$

where

$$\mathbf{A} = \begin{bmatrix} -\mathbf{C}-\lambda_0\mathbf{M} & -\mathbf{K} \\ \mathbf{I} & -\lambda_0\mathbf{I} \end{bmatrix}^{-1} \begin{bmatrix} \mathbf{M} & \mathbf{O} \\ \mathbf{O} & \mathbf{I} \end{bmatrix}, \quad \mathbf{x} = \begin{bmatrix} \dot{\mathbf{u}} \\ \mathbf{u} \end{bmatrix}. \tag{7.10.19}$$

Therefore the solution to the eigenvalue problem $\mathbf{Ax} = \alpha\mathbf{x}$ yields the eigenvalue α and its corresponding eigenvector. Then we use

$$\lambda = \mu + \lambda_0 = \frac{1}{\alpha} + \lambda_0 \tag{7.10.20}$$

to determine the eigenvalues of the quadratic eigenvalue problem.

It is necessary to point out that any higher-order eigenvalue problem

$$(\lambda^m \mathbf{A}_m + \lambda^{m-1}\mathbf{A}_{m-1} + \cdots + \lambda\mathbf{A}_1 + \mathbf{A}_0)\mathbf{u} = \mathbf{0} \tag{7.10.21}$$

can be linearized [321]:

$$\begin{bmatrix} -\mathbf{A}_0 & & & \\ & \mathbf{I} & & \\ & & \ddots & \\ & & & \mathbf{I} \end{bmatrix} \begin{bmatrix} \mathbf{u} \\ \lambda\mathbf{u} \\ \vdots \\ \lambda^{m-1}\mathbf{u} \end{bmatrix} = \lambda \begin{bmatrix} \mathbf{A}_1 & \mathbf{A}_2 & \cdots & \mathbf{A}_m \\ \mathbf{I} & \mathbf{O} & \cdots & \mathbf{O} \\ \vdots & \ddots & \ddots & \vdots \\ \mathbf{O} & \cdots & \mathbf{I} & \mathbf{O} \end{bmatrix} \begin{bmatrix} \mathbf{u} \\ \lambda\mathbf{u} \\ \vdots \\ \lambda^{m-1}\mathbf{u} \end{bmatrix}. \tag{7.10.22}$$

That is, an mth-order quadratic eigenvalue problem can be linearized to a standard generalized eigenvalue problem $\mathbf{Ax} = \lambda\mathbf{Bx}$.

7.10.3 Application Examples

In the following, three application examples of quadratic eigenvalues are presented.

1. Autoregressive Parameter Estimation

Consider the modeling of a real random process as an autoregressive (AR) process

$$x(n) + a(1)x(n-1) + \cdots + a(p)x(n-p) = e(n), \tag{7.10.23}$$

where $a(1),\ldots,a(p)$ are the AR parameters, p is the AR order and $e(n)$ is an unobservable excitation signal that is usually a white noise with variance σ_e^2.

Multiplying Equation (7.10.23) by $x(n-\tau), \tau \geq 1$, and taking the mathematical expectation, we get the linear normal equation

$$R_x(\tau) + a(1)R_x(\tau-1) + \cdots + a(p)R_x(\tau-p) = 0, \quad \tau = 1,2,\ldots, \tag{7.10.24}$$

where $R_x(\tau) = E\{x(n)x(n-\tau)\}$ is the autocorrelation of the AR process. This normal equation is called the Yule–Walker equation.

Taking $\tau = 1,\ldots,p$, then Equation (7.10.24) can be written as

$$\begin{bmatrix} R_x(0) & R_x(-1) & \cdots & R_x(-p+1) \\ R_x(1) & R_x(0) & \cdots & R_x(-p+2) \\ \vdots & \vdots & \ddots & \vdots \\ R_x(p-1) & R_x(p-2) & \cdots & R_x(0) \end{bmatrix} \begin{bmatrix} a(1) \\ a(2) \\ \vdots \\ a(p) \end{bmatrix} = - \begin{bmatrix} R_x(1) \\ R_x(2) \\ \vdots \\ R_x(p) \end{bmatrix}. \tag{7.10.25}$$

However, in many cases there is an observation noise $v(n)$, i.e., the observation signal is

$$y(n) = x(n) + v(n),$$

where $v(n)$ is a white noise with variance σ^2 and is statistically independent of $x(n)$. Under this assumption, there is the following relationship between the observation signal $y(n)$ and the AR random process $x(n)$:

$$R_x(\tau) = R_y(\tau) - \sigma^2\delta(\tau) = \begin{cases} R_y(0) - \sigma^2, & \tau = 0, \\ R_y(\tau), & \tau \neq 0. \end{cases}$$

After substituting this relation into Equation (7.10.25), the Yule–Walker equation becomes

$$\begin{bmatrix} R_y(0)-\sigma^2 & R_y(-1) & \cdots & R_y(-p+1) \\ R_y(1) & R_y(0)-\sigma^2 & \cdots & R_y(-p+2) \\ \vdots & \vdots & \ddots & \vdots \\ R_y(p-1) & R_y(p-2) & \cdots & R_y(0)-\sigma^2 \end{bmatrix} \begin{bmatrix} a(1) \\ a(2) \\ \vdots \\ a(p) \end{bmatrix} = - \begin{bmatrix} R_y(1) \\ R_y(2) \\ \vdots \\ R_y(p) \end{bmatrix}, \tag{7.10.26}$$

which is the Yule–Walker equation with noise compensation [466].

Let $\mathbf{a} = [-a(1),\ldots,-a(p)]^T, \mathbf{r}_1 = [R_y(1),\ldots,R_y(p)]^T$ and

$$\mathbf{R}_y = \begin{bmatrix} R_y(0) & R_y(-1) & \cdots & R_y(-p+1) \\ R_y(1) & R_y(0) & \cdots & R_y(-p+2) \\ \vdots & \vdots & \ddots & \vdots \\ R_y(p-1) & R_y(p-2) & \cdots & R_y(0) \end{bmatrix}.$$

Then Equation (7.10.26) can be rewritten as

$$(\mathbf{R}_y - \sigma^2\mathbf{I}_p)\mathbf{a} = \mathbf{r}_1. \tag{7.10.27}$$

Since $R_y(\tau) = R_x(\tau)$, $\tau \geq 1$, if we let $\tau = p+1, p+2, \ldots, p+q$ then from Equation (7.10.24) it follows that

$$\mathbf{g}_i^T \mathbf{a} = R_y(p+i), \quad i = 1, \ldots, q, \tag{7.10.28}$$

where $\mathbf{g}_i = [R_y(p+i-1), R_y(p+i-2), \ldots, R_y(i)]^T$. Now combine Equations (7.10.27) and (7.10.28) to yield the matrix equation

$$(\bar{\mathbf{R}}_y - \lambda \mathbf{D})\mathbf{v} = \mathbf{0}_{p+q}, \tag{7.10.29}$$

where $\bar{\mathbf{R}}_y$ and $\mathbf{D}$ are $(p+q)\times(p+1)$ matrices and $\mathbf{v}$ is a $(p+1)\times 1$ vector, defined respectively as

$$\bar{\mathbf{R}}_y = \begin{bmatrix} R_y(1) & R_y(0) & R_y(-1) & \cdots & R_y(-p+1) \\ R_y(2) & R_y(1) & R_y(0) & \cdots & R_y(-p+2) \\ \vdots & \vdots & \vdots & \ddots & \vdots \\ R_y(p) & R_y(p-1) & R_y(p-2) & \cdots & R_y(0) \\ R_y(p+1) & R_y(p) & R_y(2) & \cdots & R_y(1) \\ \vdots & \vdots & \vdots & \ddots & \vdots \\ R_y(p+q) & R_y(p+q-1) & R_y(p+q-2) & \cdots & R_y(q) \end{bmatrix},$$

$$\mathbf{D} = \begin{bmatrix} 0 & 1 & 0 & \cdots & 0 \\ 0 & 0 & 1 & \cdots & 0 \\ \vdots & \vdots & \vdots & \ddots & \vdots \\ 0 & 0 & 0 & \cdots & 1 \\ 0 & 0 & 0 & \cdots & 0 \\ \vdots & \vdots & \vdots & \ddots & \vdots \\ 0 & 0 & 0 & \cdots & 0 \end{bmatrix} \quad \text{and} \quad \mathbf{v} = \begin{bmatrix} 1 \\ a(1) \\ a(2) \\ \vdots \\ a(p) \end{bmatrix}.$$

Premultiplying Equation (7.10.29) by $(\bar{\mathbf{R}}_y - \lambda \mathbf{D})^T$, it follows that

$$(\lambda^2 \mathbf{M} + \lambda \mathbf{C} + \mathbf{K})\mathbf{v} = \mathbf{0}_{p+1}, \tag{7.10.30}$$

where

$$\mathbf{M} = \bar{\mathbf{R}}_y^T \bar{\mathbf{R}}_y, \quad \mathbf{C} = -(\bar{\mathbf{R}}_y^T \mathbf{D} + \mathbf{D}^T \bar{\mathbf{R}}_y), \quad \mathbf{K} = \mathbf{D}^T \mathbf{D}. \tag{7.10.31}$$

Because the matrices $\mathbf{M}$, $\mathbf{C}$ and $\mathbf{K}$ are symmetric, Equation (7.10.30) is a symmetric quadratic eigenvalue problem.

As before, let

$$\mathbf{A} = \begin{bmatrix} \mathbf{K} & \mathbf{O} \\ \mathbf{O} & \mathbf{I} \end{bmatrix}, \quad \mathbf{B} = \begin{bmatrix} -\mathbf{C} & -\mathbf{M} \\ \mathbf{I} & \mathbf{O} \end{bmatrix}. \tag{7.10.32}$$

Then the quadratic eigenvalue problem becomes a $2(p+1)$ generalized eigenvalue problem

$$(\mathbf{A} - \lambda \mathbf{B})\mathbf{u} = \mathbf{0}. \tag{7.10.33}$$

Because the eigenvalues are real or appear in conjugate pairs (λ, λ^*), the right eigenvectors corresponding to the conjugate pair (λ, λ^*) are the conjugate pair $(\mathbf{u}, \mathbf{u}^*)$. Among the $2(p+1)$ eigenpairs $(\lambda_i, \mathbf{u}_i)$, the eigenvalue λ and its corresponding eigenvector $\mathbf{u}$ satisfying both Equation (7.10.27) and Equation (7.10.28) are respectively estimates of the variance σ^2 of the observation noise and the AR parameter vector $[1, a(1), \ldots, a(p)]^T$.

The above method for estimating AR parameters from observation data noisy with the additive white noise was proposed by Davila in 1998 [116].

2. Constrained Least Squares

Consider the constrained least squares problem

$$\mathbf{x} = \arg\min \left\{\mathbf{x}^T\mathbf{A}\mathbf{x} - 2\mathbf{b}^T\mathbf{x}\right\} \quad \text{subject to} \;\; \mathbf{x}^T\mathbf{x} = c^2, \tag{7.10.34}$$

where $\mathbf{A} \in \mathbb{R}^{n\times n}$ is a symmetric matrix and c is a real nonzero constant that takes the value $c = 1$ in many applications.

This constrained optimization problem can be solved by the Lagrange multiplier method. To this end, let the cost function be

$$J(\mathbf{x}, \lambda) = \mathbf{x}^T\mathbf{A}\mathbf{x} - 2\mathbf{b}^T\mathbf{x} + \lambda(c^2 - \mathbf{x}^T\mathbf{x}). \tag{7.10.35}$$

From $\dfrac{\partial J(\mathbf{x},\lambda)}{\partial \mathbf{x}} = \mathbf{0}$ and $\dfrac{\partial J(\mathbf{x},\lambda)}{\partial \lambda} = 0$, we have

$$(\mathbf{A} - \lambda\mathbf{I})\mathbf{x} = \mathbf{b} \quad \text{and} \quad \mathbf{x}^T\mathbf{x} = c^2. \tag{7.10.36}$$

Let $\mathbf{x} = (\mathbf{A} - \lambda\mathbf{I})\mathbf{y}$; substituting this into the first formula of Equation (7.10.36), we get

$$(\mathbf{A} - \lambda\mathbf{I})^2\mathbf{y} = \mathbf{b} \quad \text{or} \quad (\lambda^2\mathbf{I} - 2\lambda\mathbf{A} + \mathbf{A}^2)\mathbf{y} - \mathbf{b} = \mathbf{0}. \tag{7.10.37}$$

Using the above assumption that $\mathbf{x} = (\mathbf{A} - \lambda\mathbf{I})\mathbf{y}$ and $\mathbf{A}$ are the symmetric matrices, it is easily seen that

$$\begin{aligned}\mathbf{x}^T\mathbf{x} &= \mathbf{y}^T(\mathbf{A} - \lambda\mathbf{I})^T(\mathbf{A} - \lambda\mathbf{I})\mathbf{y} = \mathbf{y}^T(\mathbf{A} - \lambda\mathbf{I})[(\mathbf{A} - \lambda\mathbf{I})\mathbf{y}]\\ &= \mathbf{y}^T(\mathbf{A} - \lambda\mathbf{I})\mathbf{x} = \mathbf{y}^T\mathbf{b}.\end{aligned}$$

Then, the second formula of Equation (7.10.36) can be equivalently written as

$$\mathbf{y}^T\mathbf{b} = c^2 \quad \text{or} \quad 1 = \mathbf{y}^T\mathbf{b}/c^2,$$

from which we have

$$\mathbf{b} = \mathbf{b}\mathbf{y}^T\mathbf{b}/c^2 = \mathbf{b}\mathbf{b}^T\mathbf{y}/c^2,$$

because $\mathbf{y}^T\mathbf{b} = \mathbf{b}^T\mathbf{y}$. Substitute the above equation into Equation (7.10.37) to yield

$$\left(\lambda^2\mathbf{I} - 2\lambda\mathbf{A} + \left(\mathbf{A}^2 - c^{-2}\mathbf{b}\mathbf{b}^T\right)\right)\mathbf{y} = \mathbf{0}. \tag{7.10.38}$$

This is just a symmetric quadratic eigenvalue problem.

To sum up the above discussion, the solution steps for constrained least squares problem (7.10.34) are as follows [169].

(1) Solve the quadratic eigenvalue problem (7.10.38) to get the eigenvalues λ_i (the eigenvectors may not be needed).
(2) Determine the minimum eigenvalue $\lambda_{\min}$.
(3) The solution of the constrained least squares problem (7.10.34) is given by $\mathbf{x} = (\mathbf{A} - \lambda_{\min}\mathbf{I})^{-1}\mathbf{b}$.

3. *Multiple-Input–Multiple-Output System*

The *multiple-input–multiple-output* (MIMO) system is a linear system often encountered in communication, radar, signal processing, automatic control and system engineering. Consider a linear controlled system with m input and n outputs:

$$\mathbf{M}\ddot{\mathbf{q}}(t) + \mathbf{C}\dot{\mathbf{q}}(t) + \mathbf{K}\mathbf{q}(t) = \mathbf{B}\mathbf{u}(t), \tag{7.10.39}$$

$$\mathbf{y}(t) = \mathbf{L}\mathbf{q}(t), \tag{7.10.40}$$

where $\mathbf{u}(t) \in \mathbb{C}^m, m \leq r$, is some input signal vector, $\mathbf{q}(t) \in \mathbb{C}^r$ is the state vector of the system, $\mathbf{y}(t) \in \mathbb{C}^n$ is the system output vector, $\mathbf{B} \in \mathbb{C}^{r\times m}$ is the input function matrix, $\mathbf{L} \in \mathbb{C}^{n\times r}$ is the output function matrix and $\mathbf{M}, \mathbf{C}, \mathbf{K}$ are $r \times r$ matrices.

Making a Laplace transformation of the MIMO system, and assuming a zero initial condition, we have

$$s^2\mathbf{M}\bar{\mathbf{q}}(s) + s\mathbf{C}\bar{\mathbf{q}}(s) + \mathbf{K}\bar{\mathbf{q}}(s) = \mathbf{B}\bar{\mathbf{u}}(s), \tag{7.10.41}$$

$$\bar{\mathbf{y}}(s) = \mathbf{L}\bar{\mathbf{q}}(s). \tag{7.10.42}$$

Hence the transfer function matrix of the MIMO system is given by

$$\mathbf{G}(s) = \frac{\bar{\mathbf{y}}(s)}{\bar{\mathbf{u}}(s)} = \mathbf{L}(s^2\mathbf{M} + s\mathbf{C} + \mathbf{K})^{-1}\mathbf{B}. \tag{7.10.43}$$

Put $\mathbf{Q}(s) = s^2\mathbf{M} + s\mathbf{C} + \mathbf{K}$; then, from Equations (7.10.12) and (7.10.13) one gets the inverse matrices in the cases of the first companion form $L1$ and the second companion form $L2$:

$$L1: \quad (s^2\mathbf{M} + s\mathbf{C} + \mathbf{K})^{-1} = \mathbf{U}(s\mathbf{I} - \boldsymbol{\Lambda})^{-1}\boldsymbol{\Lambda}\mathbf{V}^H = \sum_{i=1}^{2n} \frac{\lambda_i \mathbf{u}_i \mathbf{v}_i^H}{s - \lambda_i},$$

$$L2: \quad (s^2\mathbf{M} + s\mathbf{C} + \mathbf{K})^{-1} = \mathbf{U}(s\mathbf{I} - \boldsymbol{\Lambda})^{-1}\mathbf{V}^H = \sum_{i=1}^{2n} \frac{\mathbf{u}_i \mathbf{v}_i^H}{s - \lambda_i}.$$

That is to say, the eigenvalues of the quadratic characteristic polynomial $|s^2\mathbf{M} + s\mathbf{C} + \mathbf{K}|$ provide the poles of the controlled MIMO system transfer function. Hence, the quadratic eigenvalues provide the basis for studying the control performances and response performances of a controlled MIMO system [466].

Many applications of quadratic eigenvalue problems can be found in [466], which is a wonderful review of quadratic eigenvalue problems.

7.11 Joint Diagonalization

The EVD is the diagonalization of one Hermitian matrix, and the GEVD can be viewed as the joint diagonalization of two matrices. A natural question to ask is whether the more than two matrices can be simultaneous diagonalized or jointly diagonalized? This is the theme of this section.

7.11.1 Joint Diagonalization Problems

Consider the following model for an array that is receiving signals

$$\mathbf{x}(t) = \mathbf{A}\mathbf{s}(t) + \mathbf{v}(t), \quad t = 1, 2, \ldots, \tag{7.11.1}$$

in which $\mathbf{x}(t) = [x_1(t), \ldots, x_m(t)]^T$ is the observed signal vector, m is the number of sensors observing the spatial signals, $\mathbf{s}(t) = [s_1(t), \ldots, s_n(t)]^T$ is the source vector, $\mathbf{v}(t) = [v_1(t), \ldots, v_m(t)]^T$ is the additive noise vector on the sensor array, while $\mathbf{A} \in \mathbb{C}^{m \times n}$ $(m > n)$ is the matrix representing the mixing of the n sources, called the mixture matrix.

The so-called *blind source separation* (BSS) *problem* uses only the observed data vector $\mathbf{x}(t)$ to determine the unknown mixture matrix $\mathbf{A}$ and then to separate the source vector $\mathbf{s}(t) = \mathbf{W}\mathbf{x}(t)$ via the separation or demixing matrix $\mathbf{W} = \mathbf{A}^\dagger \in \mathbb{C}^{n \times m}$. Hence the key to BSS is the determination of the mixture matrix $\mathbf{A}$. To this end, the following assumptions are made.

(1) The additive noise is time-domain white and spatial-domain colored, i.e., its correlation matrix is

$$\begin{aligned}
\mathbf{R}_v(k) &= E\{\mathbf{v}(t)\mathbf{v}^H(t-k)\} \\
&= \delta(k)\mathbf{R}_v \\
&= \begin{cases} \mathbf{R}_v, & k = 0 \text{ (spatial-domain colored)}, \\ \mathbf{O}, & k \neq 0 \text{ (time-domain white)}. \end{cases}
\end{aligned}$$

Here "time-domain white" means that the additive noise on every sensor is white noise, while "spatial-domain colored" means that the additive white noise on different sensors may be correlated.

(2) The n sources are statistically independent, i.e., $E\left\{\mathbf{s}(t)\mathbf{s}^H(t-k)\right\} = \mathbf{D}_k$ (a diagonal matrix).

(3) The source vector $\mathbf{s}(t)$ is statistically independent of the noise vector $\mathbf{v}(t)$, i.e., $E\left\{\mathbf{s}(t)\mathbf{v}^H(t-k)\right\} = \mathbf{O}$ (zero matrix).

Under the above three assumptions, the autocorrelation matrix of the array output vector is given by

$$\begin{aligned}\mathbf{R}_x(k) &= E\{\mathbf{x}(t)\mathbf{x}^H(t-k)\} \\ &= E\{(\mathbf{A}\mathbf{s}(t)+\mathbf{v}(t))(\mathbf{A}\mathbf{s}(t-k)+\mathbf{v}(t-k))^H\} \\ &= \mathbf{A}E\{\mathbf{s}(t)\mathbf{s}^H(t-k)\}\mathbf{A}^H + E\{\mathbf{v}(t)\mathbf{v}^H(t-k)\} \\ &= \begin{cases}\mathbf{A}\mathbf{D}_0\mathbf{A}^H + \mathbf{R}_v, & k=0, \\ \mathbf{A}\mathbf{D}_k\mathbf{A}^H, & k\neq 0.\end{cases}\end{aligned} \tag{7.11.2}$$

Hence, for K autocorrelation matrices $\mathbf{R}_x(k)$, $k=1,\ldots,K$ with no noise (time lag $k\neq 0$), making a *joint diagonalization*

$$\mathbf{R}_x(k) = \mathbf{U}\mathbf{\Sigma}_k\mathbf{U}^H, \tag{7.11.3}$$

then

$$\mathbf{R}_x(k) = \mathbf{U}\mathbf{\Sigma}_k\mathbf{U}^H = \mathbf{A}\mathbf{D}_k\mathbf{A}^H, \quad k=1,\ldots,K. \tag{7.11.4}$$

The question is whether the joint diagonalization result $\mathbf{U}$ is the mixture matrix $\mathbf{A}$? The answer is no, because the $\mathbf{\Sigma}_k$ do not happen to be the same as $\mathbf{D}_k, k=1,\ldots,K$.

On the other hand, in the signal model (7.11.1) there exists an uncertainty

$$\mathbf{x}(t) = \sum_{i=1}^{n} \alpha_i \mathbf{a}_i \cdot \frac{s_i(t)}{\alpha_i} + \mathbf{v}(t) = \mathbf{A}\mathbf{G}\mathbf{G}^{-1}\mathbf{s}(t) + \mathbf{v}(t), \tag{7.11.5}$$

where $\mathbf{G}$ is an $n\times n$ generalized permutation matrix. In this case, the autocorrelation matrix of the array output vector is given by

$$\begin{aligned}\mathbf{R}_x(k) &= E\{\mathbf{x}(t)\mathbf{x}^H(t-k)\} \\ &= E\left\{\left(\mathbf{A}\mathbf{G}\mathbf{G}^{-1}\mathbf{s}(t)+\mathbf{v}(t)\right)\left(\mathbf{A}\mathbf{G}\mathbf{G}^{-1}\mathbf{s}(t-k)+\mathbf{v}(t-k)\right)^H\right\} \\ &= \mathbf{A}\mathbf{G}\mathbf{G}^{-1}E\{\mathbf{s}(t)\mathbf{s}^H(t-k)\}(\mathbf{A}\mathbf{G}\mathbf{G}^{-1})^H + E\{\mathbf{v}(t)\mathbf{v}^H(t-k)\} \\ &= \begin{cases}\mathbf{A}\mathbf{G}\mathbf{G}^{-1}\mathbf{D}_0\mathbf{G}^{-H}\mathbf{G}^H\mathbf{A}^H + \mathbf{R}_v, & k=0, \\ \mathbf{A}\mathbf{G}(\mathbf{G}^{-1}\mathbf{D}_k\mathbf{G}^{-H})\mathbf{G}^H\mathbf{A}^H, & k\neq 0.\end{cases}\end{aligned} \tag{7.11.6}$$

Thus, there is the following relationship between the joint diagonalization result and the model parameters:

$$\mathbf{R}_x(k) = \mathbf{U}\mathbf{\Sigma}_k\mathbf{U}^H = \mathbf{A}\mathbf{G}(\mathbf{G}^{-1}\mathbf{D}_k\mathbf{G}^{-H})\mathbf{G}^H\mathbf{A}^H \tag{7.11.7}$$

for $k=1,\ldots,K$. This shows that $\mathbf{U}=\mathbf{A}\mathbf{G}$, i.e., we can only identify $\mathbf{A}\mathbf{G}$ through the joint diagonalization result $\mathbf{U}$.

Definition 7.23 Two matrices $\mathbf{M}$ and $\mathbf{N}$ are said to be *essentially equal matrices*, denoted $\mathbf{M} \doteq \mathbf{N}$, if there exists a generalized permutation matrix $\mathbf{G}$ such that $\mathbf{M}=\mathbf{N}\mathbf{G}$.

When using $\mathbf{U} = \mathbf{AG}$ as the separating matrix, the separated output is given by

$$\hat{\mathbf{s}}(t) = \mathbf{U}^\dagger \mathbf{x}(t) = \mathbf{G}^{-1}\mathbf{A}^\dagger \mathbf{x}(t) = \mathbf{G}^{-1}\mathbf{s}(t) + \mathbf{G}^{-1}\mathbf{A}^\dagger \mathbf{v}(t). \tag{7.11.8}$$

Under an appropriate SNR the effect of the additive noise $\mathbf{v}$ can be ignored, so the separated output is given by

$$\hat{\mathbf{s}}(t) \approx \mathbf{G}^{-1}\mathbf{s}(t), \tag{7.11.9}$$

which is just a copy of the source vector up to factor uncertainty and ordering uncertainty. Such a separation result is allowed by the source separation. This is the reason why the separating matrix matrix $\mathbf{U}$ and the mixture matrix $\mathbf{A}$ are called essentially equal matrices.

The joint diagonalization of multiple matrices was first proposed by Flury in 1984 in the principal component analysis of K covariance matrices. Later, Cardoso and Souloumiac [85] in 1996 and Belochrani *et al.* [33] in 1997 respectively proposed the approximate joint diagonalization of multiple cumulant matrices or covariance matrices from the BSS angle. From then on, joint diagonalization has been widely studied and applied.

The mathematical problem of BSS is as follows: given K matrices $\mathbf{A}_1, \ldots, \mathbf{A}_K \in \mathbb{C}^{m\times m}$, the joint diagonalization problem seeks a nonsingular *diagonalizing matrix* $\mathbf{W} \in \mathbb{C}^{m\times m}$ and K associated diagonal matrices $\mathbf{\Lambda}_1, \ldots, \mathbf{\Lambda}_K \in \mathbb{C}^{m\times m}$ such that

$$\mathbf{A}_k = \mathbf{W}\mathbf{\Lambda}_k\mathbf{W}^H, \quad k = 1, \ldots, K, \tag{7.11.10}$$

where $\mathbf{W}$ is called sometimes the *joint diagonalizer*.

The above diagonalization is called *exact joint diagonalization*. However, a practical form of joint diagonalization is *approximate joint diagonalization*: this approach seeks a joint diagonalizer $\mathbf{W}$ and K associated diagonal matrices $\mathbf{\Lambda}_1, \ldots, \mathbf{\Lambda}_K$ to minimize the objective function [84], [85] according to

$$\min J_1(\mathbf{W}, \mathbf{\Lambda}_1, \ldots, \mathbf{\Lambda}_K) = \min \sum_{k=1}^{K} \alpha_k \left\| \mathbf{W}^H \mathbf{A}_k \mathbf{W} - \mathbf{\Lambda}_k \right\|_F^2 \tag{7.11.11}$$

or [504], [530]

$$\min J_2(\mathbf{W}, \mathbf{\Lambda}_1, \ldots, \mathbf{\Lambda}_K) = \min \sum_{k=1}^{K} \alpha_k \left\| \mathbf{A}_k - \mathbf{W}\mathbf{\Lambda}_k\mathbf{W}^H \right\|_F^2, \tag{7.11.12}$$

where $\alpha_1, \ldots, \alpha_K$ are positive weighting coefficients. To simplify the description, we assume that $\alpha_1 = \cdots = \alpha_K = 1$.

7.11.2 Orthogonal Approximate Joint Diagonalization

In many engineering applications only the joint diagonalizer $\mathbf{W}$ is used, while the diagonal matrices $\mathbf{\Lambda}_1, \ldots, \mathbf{\Lambda}_K$ are not needed. In this case, it is an interesting problem to transform the objective function of approximate joint diagonalization into an objective function containing only the joint diagonalizer $\mathbf{W}$. This optimization problem has the following solution methods.

1. Off-Function Minimization Method

In numerical analysis, the sum of the squared absolute values of all off-diagonal entries of an $m \times m$ matrix $\mathbf{B} = [B_{ij}]$ is defined as its *off function*, namely

$$\text{off}(\mathbf{B}) \stackrel{\text{def}}{=} \sum_{i=1,\, i\neq j}^{m} \sum_{j=1}^{n} |B_{ij}|^2. \tag{7.11.13}$$

The $m \times m$ matrix $\mathbf{M}$ constituting of all the off-diagonal entries,

$$[\mathbf{M}_{\text{off}}]_{ij} = \begin{cases} 0, & i = j, \\ M_{ij}, & i \neq j, \end{cases} \tag{7.11.14}$$

is the *off matrix* of the original matrix $\mathbf{B}$. Therefore the off function is the square of the Frobenius norm of the off matrix:

$$\text{off}(\mathbf{M}) = \|\mathbf{M}_{\text{off}}\|_F^2. \tag{7.11.15}$$

Using the off function, the *orthogonal approximate joint diagonalization* problem can be represented as [84], [85]

$$\min \left\{ J_{1a}(\mathbf{W}) = \sum_{k=1}^{K} \text{off}(\mathbf{W}^H \mathbf{A}_k \mathbf{W}) = \sum_{k=1}^{K} \sum_{i=1,\, i\neq j}^{n} \sum_{j=1}^{n} |(\mathbf{W}^H \mathbf{A}_k \mathbf{W})_{ij}|^2 \right\}. \tag{7.11.16}$$

Making a Givens rotation of the off-diagonal entries of $\mathbf{A}_1, \ldots, \mathbf{A}_K$, one can implement the orthogonal joint diagonalization of these matrices. The product of all the Givens rotation matrices gives the joint diagonalizer $\mathbf{W}$. This is the Jacobi algorithm for approximate joint diagonalization proposed by Cardoso and Souloumiac [84], [85].

2. Diagonal Function Maximization Method

A diagonal function of a square matrix may be a scalar, vector or matrix.

(1) *Diagonal function* $\text{diag}(\mathbf{B}) \in \mathbb{R}$ is the diagonal function of the $m \times m$ matrix $\mathbf{B}$, and is defined as

$$\text{diag}(\mathbf{B}) \stackrel{\text{def}}{=} \sum_{i=1}^{m} |B_{ii}|^2. \tag{7.11.17}$$

(2) *Diagonal vector function* The diagonal vector function of an $m \times m$ matrix $\mathbf{B}$, denoted $\mathbf{diag}(\mathbf{B}) \in \mathbb{C}^m$, is a column vector consisting of the diagonal entries of $\mathbf{B}$ arranged in order, i.e.,

$$\mathbf{diag}(\mathbf{B}) \stackrel{\text{def}}{=} [B_{11}, \ldots, B_{mm}]^T. \tag{7.11.18}$$

(3) *Diagonal matrix function* The diagonal matrix function of an $m \times m$ matrix $\mathbf{B}$, denoted $\mathbf{Diag}(\mathbf{B}) \in \mathbb{C}^{m\times m}$, is a diagonal matrix consisting of the diagonal entries of $\mathbf{B}$, i.e.,

$$\mathbf{Diag}(\mathbf{B}) \stackrel{\text{def}}{=} \begin{bmatrix} B_{11} & & 0 \\ & \ddots & \\ 0 & & B_{mm} \end{bmatrix}. \tag{7.11.19}$$

Minimizing $\text{off}(\mathbf{B})$ is equivalent to maximizing the diagonal function $\text{diag}(\mathbf{B})$, i.e.,

$$\min \text{off}(\mathbf{B}) = \max \text{diag}(\mathbf{B}). \tag{7.11.20}$$

So, Equation (7.11.16) can be rewritten as [504]

$$\max\left\{ J_{1b}(\mathbf{W}) = \sum_{k=1}^{K} \text{diag}(\mathbf{W}^H \mathbf{A}_k \mathbf{W}) = \sum_{k=1}^{K}\sum_{i=1}^{n} \left| [\mathbf{W}^H \mathbf{A}_k \mathbf{W}]_{ii} \right|^2 \right\}. \tag{7.11.21}$$

As a matter of fact, $\mathbf{W}^H \mathbf{A}_k \mathbf{W}$ should be a diagonal matrix, and for any $n \times n$ diagonal matrix $\mathbf{D}$, we have

$$\sum_{i=1}^{n} |D_{ii}|^2 = \text{tr}(\mathbf{D}^H \mathbf{D}). \tag{7.11.22}$$

Therefore the optimization problem (7.11.21) can be written as the trace form

$$\max\left\{ J_{1b}(\mathbf{W}) = \sum_{k=1}^{K} \text{tr}(\mathbf{W}^H \mathbf{A}_k^H \mathbf{W}\mathbf{W}^H \mathbf{A}_k \mathbf{W}) \right\}. \tag{7.11.23}$$

In particular, in the case of orthogonal approximate joint diagonalization, since $\mathbf{W}\mathbf{W}^H = \mathbf{W}^H\mathbf{W} = \mathbf{I}$ and $\mathbf{A}_k^H = \mathbf{A}_k, k = 1, \ldots, K$, the above equation reduces to

$$\max\left\{ J_{1b}(\mathbf{W}) = \sum_{k=1}^{K} \text{tr}(\mathbf{W}^H \mathbf{A}_k^2 \mathbf{W}) \right\}. \tag{7.11.24}$$

3. *Subspace Method*

In the case of orthogonal approximate joint diagonalization, the joint diagonalizer $\mathbf{W}$ is the unitary matrix whose column vectors have unit norm, i.e., $\|\mathbf{u}_k\|_2 = 1$.

Define $\mathbf{m}_k = \mathbf{diag}(\mathbf{\Lambda}_k)$ as the vector consisting of the diagonal entries of the diagonal matrix $\mathbf{\Lambda}$, and let

$$\hat{\mathbf{A}} = [\text{vec}(\mathbf{A}_1), \ldots, \text{vec}(\mathbf{A}_K)], \quad \mathbf{M} = [\mathbf{m}_1, \ldots, \mathbf{m}_K];$$

then the cost function of the joint diagonalization can be equivalently written as

$$\begin{aligned}\sum_{k=1}^{K} \|\mathbf{A}_k - \mathbf{W}\mathbf{\Lambda}_k\mathbf{W}^H\|_F^2 &= \sum_{k=1}^{K} \|\text{vec}\,\mathbf{A}_k - (\mathbf{W}^* \odot \mathbf{W})\mathbf{Diag}(\mathbf{\Lambda}_k)\|_F^2 \\ &= \|\hat{\mathbf{A}} - (\mathbf{W}^* \odot \mathbf{W})\mathbf{M}\|_F^2, \end{aligned} \tag{7.11.25}$$

where $\mathbf{R} \odot \mathbf{D}$ is the Khatri–Rao product of two matrices $\mathbf{R}$ and $\mathbf{D}$, which is the Kronecker product of the column vectors of $\mathbf{R}$ and $\mathbf{D}$:

$$\mathbf{R} \odot \mathbf{D} = [\mathbf{c}_1 \otimes \mathbf{d}_1, \mathbf{c}_2 \otimes \mathbf{d}_2, \ldots, \mathbf{c}_n \otimes \mathbf{d}_n]. \tag{7.11.26}$$

Hence the solution of the joint diagonalization (7.11.25) can be rewritten as

$$\{\mathbf{W}, \mathbf{M}\} = \arg \min_{\mathbf{W}, \mathbf{M}} \|\hat{\mathbf{A}} - \mathbf{B}\mathbf{M}\|_F^2, \quad \mathbf{B} = \mathbf{W}^* \odot \mathbf{W}. \tag{7.11.27}$$

This optimization problem is separable, since the least squares solution of $\mathbf{M}$ is

$$\mathbf{M} = (\mathbf{B}^H\mathbf{B})^{-1}\mathbf{B}^H\hat{\mathbf{A}}. \tag{7.11.28}$$

Substituting (7.11.28) into (7.11.27), we can cancel $\mathbf{M}$ to get

$$\mathbf{W} = \arg\min_{\mathbf{W}} \|\hat{\mathbf{A}} - \mathbf{B}(\mathbf{B}^H\mathbf{B})^{-1}\mathbf{B}^H\hat{\mathbf{A}}\|_F^2 = \arg \min_{\mathbf{W}} \|\mathbf{P}_B^{\perp}\hat{\mathbf{A}}\|_F^2, \tag{7.11.29}$$

where $\mathbf{P}_B^{\perp} = \mathbf{I} - \mathbf{B}(\mathbf{B}^H\mathbf{B})^{-1}\mathbf{B}^H$ is the orthogonal projection matrix. An algorithm for solving the optimization problem (7.11.29) can be found in [480].

The three methods described above belong in the orthogonal approximate joint diagonalization category, because the joint diagonalizer provided by these methods is a unitary matrix.

7.11.3 Nonorthogonal Approximate Joint Diagonalization

In orthogonal joint diagonalization, the solution cannot be trivial or ill-conditioned because of the unitarity of $\mathbf{W}$. The limitation of orthogonal joint diagonalization is that it is available only for a whitened observation data vector.

However, the prewhitening of data has the following two disadvantages.

(1) The whitening operation, as a preprocessing, can adversely affect the performance of the separated signals, since the statistical error involved in this preprocessing stage cannot be corrected in the "effective separation" stage [83].
(2) In practice the whitening can distort the weighting least squares criterion, leading to the exact diagonalization of one selected matrix at the possible cost of the poor diagonalization of others [530].

Nonorthogonal joint diagonalization is joint diagonalization without the constraint $\mathbf{W}^H\mathbf{W} = \mathbf{I}$, and has become a mainstream joint diagonalization method. The advantage of nonorthogonal joint diagonalization is that it avoids two disadvantages of data prewhitening, while the disadvantage is that it may give a trivial or a degenerate solution.

Typically, algorithms for nonorthogonal joint diagonalization are of the iterative type.

- The iterative algorithm in [387] uses an information-theoretic criterion.

- The subspace-fitting algorithm in [480] uses Newton-type iterations.
- The AC–DC algorithm in [530] minimizes the weighted least squared (WLS) error.
- The gradient search algorithm in [233], the Newton search algorithm in [234] and the penalty function approach in [501] minimize the sum of the WLS error and some penalty term.
- The quadratic optimization algorithm in [496] minimizes the WLS error under various constraints on the diagonalizing matrix.

The AC–DC algorithm separates the coupled optimization problem into two single optimization problems:

$$\begin{aligned} J_{\mathrm{WLS2}}(\mathbf{W}, \mathbf{\Lambda}_1, \ldots, \mathbf{\Lambda}_K) &= \sum_{k=1}^{K} w_k \|\mathbf{A}_k - \mathbf{W}\mathbf{\Lambda}_k\mathbf{W}^H\|_F^2 \\ &= \sum_{k=1}^{K} w_k \left\| \mathbf{A}_k - \sum_{n=1}^{m} \lambda_n^{[k]} \mathbf{w}_k \mathbf{w}_k^H \right\|_F^2 . \end{aligned}$$

The AC–DC algorithm consists of two phases.

(1) *Alternating columns* (AC) *phase* Fix all but one column of $\mathbf{W}$ and the matrices $\mathbf{\Lambda}_1, \ldots, \mathbf{\Lambda}_K$ and minimize the objective function $J_{\mathrm{WLS2}}(\mathbf{W})$ with respect to the remaining column vector of $\mathbf{W}$.
(2) *Diagonal centers* (DC) *phase* Fix $\mathbf{W}$ and minimize the cost function $J_{\mathrm{WLS2}}(\mathbf{W}, \mathbf{\Lambda}_1, \ldots, \mathbf{\Lambda}_K)$ with respect to the matrices $\mathbf{\Lambda}_1, \ldots, \mathbf{\Lambda}_K$.

The simplest way of avoiding a trivial solution is to add the constraint condition $\mathbf{Diag}(\mathbf{W}) = \mathbf{I}$. However, the biggest drawback of the nonorthogonal joint diagonalization is the possibility that its K diagonal matrices $\mathbf{\Lambda}_1, \ldots, \mathbf{\Lambda}_K$ and/or the joint diagonalizer $\mathbf{W}$ may be singular; this leads to *degenerate solutions* of the nonorthogonal joint diagonalization problem [295].

In order to avoid both trivial solutions and degenerate solutions of the nonorthogonal joint diagonalization problem, Li and Zhang [295] proposed another objective function:

$$\min \left\{ J(\mathbf{W}) = \sum_{k=1}^{K} \alpha_k \sum_{i=1}^{m} \sum_{j=1, j\neq i}^{m} \left| [\mathbf{W}^H \mathbf{A}_k \mathbf{W}]_{ij} \right|^2 - \beta \log \left| \det(\mathbf{W}) \right| \right\}, \tag{7.11.30}$$

where $\alpha_k (1 \leq k \leq K)$ are the positive weighting coefficients, and $\beta > 0$.

The above cost function can be divided into the sum of a squared off-diagonal error term,

$$J_1(\mathbf{W}) = \sum_{k=1}^{K} \alpha_k \sum_{i=1}^{m} \sum_{j=1, j\neq i}^{m} \left| [\mathbf{W}^H \mathbf{A}_k \mathbf{W}]_{ij} \right|^2, \tag{7.11.31}$$

and a negative log determinant term,

$$J_2(\mathbf{W}) = -\log\big|\det(\mathbf{W})\big|. \tag{7.11.32}$$

A significant advantage of the cost function (7.11.30) is that when $\mathbf{W}$ is equal a zero matrix or is singular, $J_2(\mathbf{W}) \to +\infty$. Hence, minimization of the cost function $J(\mathbf{W})$ can avoid both a trivial solution and a degenerate solution.

Proposition 7.2 [295] *The criterion (7.11.30) is lower unbounded if and only if there is a nonsingular matrix* $\mathbf{W}$ *such that the given matrices* $\mathbf{A}_1,\ldots,\mathbf{A}_K$ *are all exactly diagonalized.*

When optimizing at the mth iteration and collecting all the terms related to $\mathbf{w}_i$ in $J_1(\mathbf{W})$, the cost function in (7.11.30) can be rewritten as

$$J(\mathbf{W}) = \mathbf{w}_i^H\mathbf{Q}_i\mathbf{w}_i - \beta\log\big|\det(\mathbf{W})\big| + C_i, \tag{7.11.33}$$

where C_i is a quantity independent of $\mathbf{w}_i$ and the matrix $\mathbf{Q}_i$ is defined as

$$\mathbf{Q}_i = \sum_{k=1}^{K}\sum_{j=1,j\neq i}^{m} \alpha_k\big(\mathbf{A}_k\mathbf{w}_j\mathbf{w}_j^H\mathbf{A}_k^H + \mathbf{A}_k^H\mathbf{w}_j\mathbf{w}_j^H\mathbf{A}_k\big). \tag{7.11.34}$$

Here $\mathbf{Q}_i$ is always Hermitian and (semi-)positive definite.

Based on the cost function (7.11.30), a *fast approximate joint diagonalization* (FAJD) *algorithm* was developed in [295], as shown in Algorithm 7.12.

Proposition 7.3 [295] *When the matrix* $\mathbf{Q}_i$ *is positive definite and the constant* $\beta > 0$, *the global minimum point of the cost function* $J(\mathbf{w}_i) = \mathbf{w}_i^H\mathbf{Q}_i\mathbf{w}_i - \beta\log|\mathbf{c}_s^H\mathbf{w}_i|$ *is determined by*

$$\mathbf{w}_i^{(\text{opt})} = e^{\mathrm{j}\phi}\frac{\sqrt{\beta}\mathbf{Q}_i^{-1}\mathbf{c}_s}{\sqrt{2\mathbf{c}_s^H\mathbf{Q}_i^{-1}\mathbf{c}_s}}, \tag{7.11.35}$$

where ϕ *is an ambiguous rotation angle (with real-valued numbers if* ϕ *is* 0 *or* π *and with complex-valued numbers for* $0 \le \phi < 2\pi$*).*

Proposition 7.3 shows that the minimization of the cost function $J(\mathbf{W})$ is independent of the penalty parameter β, which implies that β can be any finite large positive number; the simplest choice is $\beta = 1$. This fact avoids the influence of the choice of penalty parameter on the performance of the penalty function method.

Joint diagonalization is widely applied in blind source separation (see, e.g. [10], [33], [332], [530], [295]), blind beamforming [84], time delay estimation [531], frequency estimation [326], array signal processing [505], MIMO blind equalization [112], blind MIMO system identification [94], and so on.

Algorithm 7.12 Fast approximate joint diagonalization algorithm [295]

input: Target matrices $\mathbf{A}_1, \ldots, \mathbf{A}_K \in \mathbb{C}^{m\times m}$ and weights $\alpha_1, \ldots, \alpha_K$ and β.
initialization: $\mathbf{W}$ (a unitary matrix or any invertible matrix). Set $i = 1$.
repeat

1. Compute $\mathbf{Q}_i = \sum_{k=1}^{K}\sum_{j=1,j\neq i}^{m} \alpha_k(\mathbf{A}_k\mathbf{w}_j\mathbf{w}_j^H\mathbf{A}_k^H + \mathbf{A}_k^H\mathbf{w}_j\mathbf{w}_j^H\mathbf{A}_k)$.
2. Form the exchange matrix
$$[\mathbf{E}_{s,t}]_{ij} = \begin{cases} 1, & \text{when } i = s, j = t \text{ or } i = t, j = s, \\ 1, & \text{when } i = j, i \neq s, t, \\ 0, & \text{otherwise.} \end{cases}$$
3. For some integer $s, 1 \le s \le m-1$ make the matrix partition
$$\mathbf{E}_{s,m}\mathbf{W}\mathbf{E}_{i,m} = \begin{bmatrix} \mathbf{B} & \mathbf{u} \\ \mathbf{b}^H & v \end{bmatrix}$$
such that the block matrix $\mathbf{B}$ is invertible.
4. Calculate $\mathbf{c}_s^H = [-\mathbf{b}^H\mathbf{B}^{-1}, 1]\mathbf{E}_{s,m}$.
5. Replace the ith column in $\mathbf{W}$ with $\mathbf{w}_i^{\text{opt}} = (\sqrt{\beta}\mathbf{Q}_i^{-1}\mathbf{c}_s)\Big/\sqrt{2\mathbf{c}_s^H\mathbf{Q}_i^{-1}\mathbf{c}_s}$.
6. **exit if** $i = m$.

return $i \leftarrow i + 1$.
output: Diagonalizing matrix $\mathbf{W}$.

Exercises

7.1 Show the following properties: if λ is an eigenvalue of the $n \times n$ matrix $\mathbf{A}$ then:

(a) λ^k is an eigenvalue of the matrix $\mathbf{A}^k$.
(b) If $\mathbf{A}$ is nonsingular, then $\mathbf{A}^{-1}$ has the eigenvalue $1/\lambda$.
(c) An eigenvalue of the matrix $\mathbf{A} + \sigma^2\mathbf{I}$ is $\lambda + \sigma^2$.

7.2 Show that when $\mathbf{A}$ is an idempotent matrix, the matrices $\mathbf{BA}$ and $\mathbf{ABA}$ have the same eigenvalues.

7.3 Assuming that all the entries of the $n \times n$ matrix $\mathbf{A}$ are equal to 1, find the n eigenvalues of $\mathbf{A}$.

7.4 Consider the matrix
$$\mathbf{A} = \begin{bmatrix} 0 & 1 & 0 & 0 \\ 1 & 0 & 0 & 0 \\ 0 & 0 & y & 1 \\ 0 & 0 & 1 & 2 \end{bmatrix}.$$

(a) An eigenvalue of $\mathbf{A}$ is known to be 3; find the value of y.
(b) Find the matrix $\mathbf{P}$ such that $(\mathbf{AP})^T\mathbf{AP}$ is diagonal.

7.5 For initial values $u(0) = 2, v(0) = 8$, use the EVD to solve the coupled differential equations

$$u'(t) = 3u(t) + v(t),$$
$$v'(t) = -2u(t) + v(t).$$

7.6 A 4×4 Hessenberg matrix has the form

$$\mathbf{H} = \begin{bmatrix} a_1 & b_1 & c_1 & d_1 \\ a_2 & b_2 & c_2 & d_2 \\ 0 & b_3 & c_3 & d_3 \\ 0 & 0 & c_4 & d_4 \end{bmatrix}.$$

Show the following two results.

(a) If a_2, b_3, c_4 are nonzero, and any eigenvalue λ of $\mathbf{H}$ is a real number, then the geometric multiplicity of λ must be equal to 1.

(b) If $\mathbf{H}$ is similar to the symmetric matrix $\mathbf{A}$, and the algebraic multiplicity of some eigenvalue λ of $\mathbf{A}$ is larger than 1, then of a_2, b_3, c_4 at least one equals zero.

7.7 Show the following results:

(a) If $\mathbf{A} = \mathbf{A}^2 = \mathbf{A}^H \in \mathbb{C}^{n\times n}$ and $\text{rank}(\mathbf{A}) = r < n$ then there is an $n \times n$ unitary matrix $\mathbf{V}$ such that $\mathbf{V}^H\mathbf{AV} = \mathbf{Diag}(\mathbf{I}_r, \mathbf{0})$.

(b) If $\mathbf{A} = \mathbf{A}^H$ and $\mathbf{A}^2 = \mathbf{I}_n$ then there is an unitary matrix $\mathbf{V}$ such that $\mathbf{V}^H\mathbf{AV} = \mathbf{Diag}(\mathbf{I}_r, \mathbf{I}_{n-r})$.

7.8 Let $\mathbf{H} = \mathbf{I} - 2\mathbf{uu}^H$ be a Householder transformation matrix.

(a) Show that the vector $\mathbf{u}$ with unit norm is an eigenvector of the Householder transformation matrix, and find its corresponding eigenvalue.

(b) If a vector $\mathbf{w}$ is orthogonal to $\mathbf{u}$, show that $\mathbf{w}$ is also an eigenvector of the matrix $\mathbf{H}$, and find its corresponding eigenvalue.

7.9 Assume $\mathbf{A}$ and $\mathbf{B}$ are similar matrices, where

$$\mathbf{A} = \begin{bmatrix} -2 & 0 & 0 \\ 2 & x & 2 \\ 3 & 1 & 1 \end{bmatrix}, \quad \mathbf{B} = \begin{bmatrix} -1 & 0 & 0 \\ 0 & 2 & 0 \\ 0 & 0 & y \end{bmatrix}.$$

(a) Find the values of x and y.

(b) Find the invertible matrix $\mathbf{P}$ such that $\mathbf{P}^T\mathbf{AP} = \mathbf{B}$.

7.10 A block matrix can be used to reduce the dimension of an eigenvalue problem. Let

$$\mathbf{A} = \begin{bmatrix} \mathbf{B} & \mathbf{X} \\ \mathbf{O} & \mathbf{R} \end{bmatrix} \quad (\mathbf{O} \text{ is the zero matrix}).$$

Show that $\det(\mathbf{A} - \lambda\mathbf{I}) = \det(\mathbf{B} - \lambda\mathbf{I})\det(\mathbf{R} - \lambda\mathbf{I})$.

7.11 Let $\mathbf{u}$ be an eigenvector associated with the eigenvalue λ of the matrix $\mathbf{A}$.

(a) Show that $\mathbf{u}$ is also an eigenvector of the matrix $\mathbf{A} - 3\mathbf{I}$.

(b) Show that $\mathbf{u}$ is also an eigenvector of the matrix $\mathbf{A}^2 - 4\mathbf{A} + 3\mathbf{I}$.

7.12 [229] Let

$$\mathbf{A} = \begin{bmatrix} -2 & 4 & 3 \\ 0 & 0 & 0 \\ -1 & 5 & 2 \end{bmatrix}, \quad f(x) = x^{593} - 2x^{15}.$$

Show that

$$\mathbf{A}^{593} - 2\mathbf{A}^{15} = -\mathbf{A} = \begin{bmatrix} 2 & -4 & -3 \\ 0 & 0 & 0 \\ 1 & -5 & -2 \end{bmatrix}.$$

7.13 Given the matrix

$$\mathbf{A} = \begin{bmatrix} 2 & 0 & 1 \\ 0 & 2 & 0 \\ 0 & 0 & 3 \end{bmatrix},$$

find $e^{\mathbf{A}t}$.

7.14 [229] Given the matrix

$$\mathbf{A} = \begin{bmatrix} 1 & 1 & 2 \\ -1 & 2 & 1 \\ 0 & 1 & 3 \end{bmatrix},$$

find the nonsingular matrix $\mathbf{S}$ such that $\mathbf{B} = \mathbf{S}^{-1}\mathbf{A}\mathbf{S}$ is diagonal.

7.15 Show that the matrix $\mathbf{A}_{n\times n}$ satisfying $\mathbf{A}^2 - \mathbf{A} = 2\mathbf{I}$ is diagonalizable.

7.16 Given that $\mathbf{u} = [1, 1, -1]^T$ is an eigenvector of the matrix

$$\mathbf{A} = \begin{bmatrix} 2 & -1 & 2 \\ 5 & a & 3 \\ -1 & b & -2 \end{bmatrix},$$

(a) find a, b and the eigenvalue corresponding to the eigenvector $\mathbf{u}$;

(b) investigate whether the matrix $\mathbf{A}$ is similar to a diagonal matrix and explain the reason for your answer.

7.17 Show that a nonzero eigenvalue $\lambda(\mathbf{A} \otimes \mathbf{B})$ of the Kronecker product $\mathbf{A} \otimes \mathbf{B}$ of the matrices $\mathbf{A}$ and $\mathbf{B}$ is equal to the product of an eigenvalue of $\mathbf{A}$ and an eigenvalue of $\mathbf{B}$, i.e., $\lambda(\mathbf{A} \otimes \mathbf{B}) = \lambda(\mathbf{A})\lambda(\mathbf{B})$.

7.18 Suppose that both $\mathbf{A}$ and $\mathbf{B}$ are Hermitian matrices, and λ_i and μ_i are respectively the eigenvalues of the matrices $\mathbf{A}$ and $\mathbf{B}$. Show that if $c_1\mathbf{A} + c_2\mathbf{B}$ has the eigenvalue $c_1\lambda_i + c_2\mu_i$, where c_1, c_2 are any scalars, then $\mathbf{AB} = \mathbf{BA}$.

7.19 Show that if $\mathbf{A}, \mathbf{B}, \mathbf{C}$ are positive definite then the root of $|\mathbf{A}\lambda^2 + \mathbf{B}\lambda + \mathbf{C}| = 0$ has a negative real part.

7.20 Let $\mathbf{P}, \mathbf{Q}, \mathbf{R}, \mathbf{X}$ be 2×2 matrices. Show that every eigenvalue of the solution $\mathbf{X}$ of the matrix equation $\mathbf{P}\mathbf{X}^2 + \mathbf{Q}\mathbf{X} + \mathbf{R} = \mathbf{O}$ (the zero matrix) is a root of $|\mathbf{P}\lambda^2 + \mathbf{Q}\lambda + \mathbf{C}| = 0$.

7.21 Put

$$\mathbf{A} = \begin{bmatrix} -1 & 0 \\ 0 & 1 \end{bmatrix}, \qquad \mathbf{B} = \begin{bmatrix} 0 & 1 \\ 1 & 0 \end{bmatrix}.$$

Defining the generalized eigenvalues of the matrix pencil $(\mathbf{A}, \mathbf{B})$ as the scalar quantities α and β satisfying $\det(\beta\mathbf{A} - \alpha\mathbf{B}) = 0$, find α and β.

7.22 Let the matrix $\mathbf{G}$ be the generalized inverse of a matrix $\mathbf{A}$, and let both $\mathbf{A}$ and $\mathbf{GA}$ be symmetric. Show that the reciprocal of a nonzero eigenvalue of $\mathbf{A}$ is an eigenvalue of the generalized inverse matrix $\mathbf{G}$.

7.23 Use the characteristic equation to show that if λ is a real eigenvalue of the matrix $\mathbf{A}$ then the absolute value of the eigenvalue of $\mathbf{A} + \mathbf{A}^{-1}$ is equal to or larger than 2.

7.24 Let $\mathbf{A}_{4\times 4}$ satisfy three conditions $|3\mathbf{I}_4 + \mathbf{A}| = 0, \mathbf{A}\mathbf{A}^T = 2\mathbf{I}_4$ and $|\mathbf{A}| < 0$. Find an eigenvalue of the adjoint matrix $\text{adj}(\mathbf{A}) = \det(\mathbf{A})\mathbf{A}^{-1}$ of the matrix $\mathbf{A}$.

7.25 Show that the maximum of the quadratic form $f = \mathbf{x}^T\mathbf{A}\mathbf{x}$ at $\|\mathbf{x}\| = 1$ is equal to the maximal eigenvalue of the symmetric matrix $\mathbf{A}$. (*Hint:* Write f in the standard quadratic form.)

7.26 Let the eigenvalues of $\mathbf{A}_{n\times n}$ be $\lambda_1, \lambda_2, \ldots, \lambda_n$ and the eigenvector corresponding to λ_i be $\mathbf{u}_i$. Find:

(a) the eigenvalues of $\mathbf{P}^{-1}\mathbf{A}\mathbf{P}$ and the corresponding eigenvectors;
(b) the eigenvalues of $(\mathbf{P}^{-1}\mathbf{A}\mathbf{P})^T$ and the corresponding eigenvectors.

7.27 Let $p(z) = a_0 + a_1 z + \cdots + a_n z^n$ be a polynomial. Show that each eigenvector of the matrix $\mathbf{A}$ must be an eigenvector of the matrix polynomial $p(\mathbf{A})$, but that the eigenvectors of $p(\mathbf{A})$ are not necessarily eigenvectors of $\mathbf{A}$.

7.28 Let $\mathbf{A}$ be an orthogonal matrix and λ be an eigenvalue of $\mathbf{A}$ not equal to ± 1, but with modulus 1. If $\mathbf{u} + \mathrm{j}\,\mathbf{v}$ is the eigenvector corresponding to λ, where $\mathbf{u}$ and $\mathbf{v}$ are real vectors, show that $\mathbf{u}$ and $\mathbf{v}$ are orthogonal.

7.29 Show that an $n\times n$ real symmetric matrix $\mathbf{A}$ can be written as $\mathbf{A} = \sum_{i=1}^n \lambda_i \mathbf{Q}_i$, where λ_i is an eigenvalue of $\mathbf{A}$ while $\mathbf{Q}_i$ is a nonnegative definite matrix that not only meets the orthogonality condition

$$\mathbf{Q}_i\mathbf{Q}_j = \mathbf{O}, \quad i \neq j,$$

but also is an idempotent matrix, i.e., $\mathbf{Q}_i^2 = \mathbf{Q}_i$. This representation of the matrix $\mathbf{A}$ is known as its spectral factorization [32, p. 64].

7.30 Given

$$\mathbf{A} = \begin{bmatrix} 3 & -1 & 0 \\ 0 & 5 & -2 \\ 0 & 0 & 9 \end{bmatrix},$$

find a nonsingular matrix $\mathbf{S}$ such that the similarity transformation $\mathbf{S}^{-1}\mathbf{AS} = \mathbf{B}$ is diagonal, and find $\mathbf{B}$.

7.31 The matrix

$$\mathbf{A} = \begin{bmatrix} 1 & -1 & 1 \\ x & 4 & y \\ -3 & -3 & 5 \end{bmatrix}$$

has three linearly independent eigenvectors and eigenvalues $\lambda_1 = \lambda_2 = 2$. Find the invertible matrix $\mathbf{P}$ such that $\mathbf{P}^{-1}\mathbf{AP}$ is a diagonal matrix.

7.32 Let the vectors $\boldsymbol{\alpha} = [\alpha_1, \ldots, \alpha_n]^T$ and $\boldsymbol{\beta} = [\beta_1, \ldots, \beta_n]^T$ be orthogonal nonzero vectors. Put $\mathbf{A} = \boldsymbol{\alpha}\boldsymbol{\beta}^T$ and find:

(a) $\mathbf{A}^2$;

(b) the eigenvalues and eigenvectors of $\mathbf{A}$.

7.33 Given the matrix

$$\mathbf{A} = \begin{bmatrix} 1 & 0 & 1 \\ 0 & 2 & 0 \\ 1 & 0 & 1 \end{bmatrix}$$

and $\mathbf{B} = (k\mathbf{I} + \mathbf{A})^2$, where k is real number,

(a) find the diagonalized matrix $\mathbf{\Lambda}$ corresponding to the matrix $\mathbf{B}$;

(b) find for which values of k the matrix is positive definite.

7.34 Given a matrix

$$\mathbf{A} = \begin{bmatrix} -1 & 1 & 0 \\ -4 & 3 & 0 \\ 1 & 0 & 2 \end{bmatrix},$$

find its Jordan canonical form and its similarity transformation matrix.

7.35 Use the Jordan canonical form to solve the linear differential equations

$$\frac{dx_1}{dt} = -x_1 + x_2,$$
$$\frac{dx_2}{dt} = -4x_1 + 3x_2,$$
$$\frac{dx_3}{dt} = x_1 + 2x_3.$$

7.36 Given a matrix

$$\mathbf{A} = \begin{bmatrix} 2 & -1 & -2 \\ -1 & 2 & 2 \\ 0 & 0 & 1 \end{bmatrix},$$

find

$$g(\mathbf{A}) = \mathbf{A}^8 - 9\mathbf{A}^6 + \mathbf{A}^4 - 3\mathbf{A}^3 + 4\mathbf{A}^2 - \mathbf{I}.$$

7.37 Let

$$|\mathbf{A}| = \begin{vmatrix} a & -1 & c \\ 5 & b & 3 \\ 1-c & 0 & -a \end{vmatrix} = -1,$$

and its adjoint matrix $\text{adj}(\mathbf{A}) = |\mathbf{A}|\mathbf{A}^{-1}$ have an eigenvalue λ and corresponding eigenvector $\mathbf{u} = [-1, -1, 1]^T$. Find the values of a, b, c and λ.

7.38 Find an orthogonal basis of the matrix

$$\mathbf{A} = \begin{bmatrix} 1 & -1 & -1 & -1 \\ -1 & 1 & -1 & -1 \\ -1 & -1 & 1 & -1 \\ -1 & -1 & -1 & 1 \end{bmatrix}.$$

7.39 For a solution of the generalized eigenvalue problem $\mathbf{Ax} = \lambda\mathbf{Bx}$, the matrix $\mathbf{B}$ must be nonsingular. Now assume that $\mathbf{B}$ is singular and its generalized inverse matrix is $\mathbf{B}^\dagger$.

(a) Let $(\lambda, \mathbf{x})$ be an eigenpair of the matrix $\mathbf{B}^\dagger\mathbf{A}$. Show that this eigenpair is a generalized eigenpair of the matrix pencil $(\mathbf{A}, \mathbf{B})$ if $\mathbf{Ax}$ is the eigenvector corresponding to the eigenvalue 1 of the matrix $\mathbf{BB}^\dagger$.

(b) Let $\lambda, \mathbf{x}$ satisfy $\mathbf{Ax} = \lambda\mathbf{Bx}$. Show that if $\mathbf{x}$ is also the eigenvector corresponding to the eigenvalue 1 of the matrix $\mathbf{B}^\dagger\mathbf{B}$ then $(\lambda, \mathbf{x})$ is an eigenpair of the matrix $\mathbf{B}^\dagger\mathbf{A}$.

7.40 Let $\mathbf{u}_1$ and $\mathbf{v}_1$ be respectively the right and left eigenvectors corresponding to the generalized eigenvalue λ_1 of the matrix pencil $(\mathbf{A}, \mathbf{B})$, and $\langle \mathbf{Bu}_1, \mathbf{Bv}_1 \rangle = 1$. Show that two matrix pencils $(\mathbf{A}, \mathbf{B})$ and $(\mathbf{A}_1, \mathbf{B}_1)$, where

$$\mathbf{A}_1 = \mathbf{A} - \sigma_1 \mathbf{Bu}_1\mathbf{v}_1^H\mathbf{B}^H, \qquad \mathbf{B}_1 = \mathbf{B} - \sigma_2 \mathbf{Au}_1\mathbf{v}_1^H\mathbf{B}^H,$$

have the same left and right eigenvectors; it may be assumed that the shift factors σ_1 and σ_2 satisfy the condition $1 - \sigma_1\sigma_2 \neq 0$.

7.41 If both $\mathbf{A}$ and $\mathbf{B}$ are positive definite Hermitian matrices, show that any generalized eigenvalue must be real and that the generalized eigenvectors corresponding to different generalized eigenvalues are orthogonal with respect to the positive matrices $\mathbf{A}$ and $\mathbf{B}$, respectively, i.e.,

$$\mathbf{x}_i^H\mathbf{Ax}_j = \mathbf{x}_i^H\mathbf{Bx}_j = 0, \quad i \neq j.$$

7.42 If $\mathbf{M}, \mathbf{C}, \mathbf{K}$ are symmetric and positive definite, show that the roots of the quadratic eigenvalue problem

$$|\lambda^2\mathbf{M} + \lambda\mathbf{C} + \mathbf{K}| = 0$$

have a negative real part.

7.43 Assume that the eigenvalues of the $n \times n$ Hermitian matrix $\mathbf{A}$ are arranged in the order that $\lambda_1(\mathbf{A}) \geq \lambda_2(\mathbf{A}) \geq \cdots \geq \lambda_n(\mathbf{A})$. Use the Rayleigh quotient to show that

(a) $\lambda_1(\mathbf{A}+\mathbf{B}) \geq \lambda_1(\mathbf{A}) + \lambda_n(\mathbf{B})$,
(b) $\lambda_n(\mathbf{A}+\mathbf{B}) \geq \lambda_n(\mathbf{A}) + \lambda_n(\mathbf{B})$.

7.44 Use the Rayleigh quotient to show that for any $n\times n$ symmetric matrix $\mathbf{A}$ and any $n \times n$ positive semi-definite matrix $\mathbf{B}$, the eigenvalues obey the inequality

$$\lambda_k(\mathbf{A}+\mathbf{B}) \geq \lambda_k(\mathbf{A}), \quad k = 1, 2, \ldots, n,$$

and that if $\mathbf{B}$ is positive definite then

$$\lambda_k(\mathbf{A}+\mathbf{B}) > \lambda_k(\mathbf{A}), \quad k = 1, 2, \ldots, n.$$

7.45 Let $\mathbf{A}$ be an $n \times n$ matrix (not necessarily positive definite). Show that, for any $n \times 1$ vector $\mathbf{x}$, the following two inequalities hold:

$$(\mathbf{x}^T\mathbf{A}\mathbf{x})^2 \leq (\mathbf{x}^T\mathbf{A}\mathbf{A}^T\mathbf{x})(\mathbf{x}^T\mathbf{x})$$

and

$$\frac{1}{2}\left|\frac{\mathbf{X}^T(\mathbf{A}+\mathbf{A}^T)\mathbf{x}}{\mathbf{x}^T\mathbf{x}}\right| \leq \left(\frac{\mathbf{x}^T\mathbf{A}\mathbf{A}^T\mathbf{x}}{\mathbf{x}^T\mathbf{x}}\right)^{1/2}.$$

7.46 Consider the symmetric matrix sequence

$$\mathbf{A}_r = [a_{ij}], \quad i, j = 1, 2, \ldots, r,$$

where $r = 1, 2, \ldots, n$. Let $\lambda_i(\mathbf{A}_r)$, $i = 1, 2, \ldots, r$, be the ith eigenvalue of the matrix $\mathbf{A}_r$ and let $\lambda_1(\mathbf{A}_r) \geq \lambda_2(\mathbf{A}_r) \geq \cdots \geq \lambda_r(\mathbf{A}_r)$; then

$$\lambda_{k+1}(\mathbf{A}_{i+1}) \leq \lambda_k(\mathbf{A}_{i},) \leq \lambda_k(\mathbf{A}_{i+1}).$$

This result is called the Sturmian separating theorem [32, p. 117]. Use the Rayleigh quotient to show this theorem.

7.47 Show that Equation (7.10.1) is a hyperbolic quadratic eigenvalue problem if and only if $\mathbf{Q}(\lambda)$ is negative definite for some eigenvalues $\lambda \in R$.

7.48 For any two square matrices $\mathbf{A}, \mathbf{B}$ with the same dimension, show that

(a) $2(\mathbf{A}\mathbf{A}^T + \mathbf{B}\mathbf{B}^T) - (\mathbf{A}+\mathbf{B})(\mathbf{A}+\mathbf{B})^T$ is positive semi-definite.
(b) $\mathrm{tr}((\mathbf{A}+\mathbf{B})(\mathbf{A}+\mathbf{B})^T) \leq 2(\mathrm{tr}(\mathbf{A}\mathbf{A}^T) + \mathrm{tr}(\mathbf{B}\mathbf{B}^T))$.
(c) $\lambda((\mathbf{A}+\mathbf{B})(\mathbf{A}+\mathbf{B})^T) \leq 2(\lambda(\mathbf{A}\mathbf{A}^T) + \lambda(\mathbf{B}\mathbf{B}^T))$.

7.49 Let $\mathbf{A}$ and $\mathbf{B}$ be two positive semi-definite matrices with the same dimension; show that [525]

$$\sqrt{\mathrm{tr}(\mathbf{A}\mathbf{B})} \leq \frac{1}{2}(\mathrm{tr}(\mathbf{A}) + \mathrm{tr}(\mathbf{B}))$$

and that the equality holds if and only if $\mathbf{A} = \mathbf{B}$ and $\mathrm{rank}(\mathbf{A}) \leq 1$.

7.50 For the polynomial $p(x) = x^n + a_{n-1}x^{n-1} + \cdots + a_1 x + a_0$, the $n \times n$ matrix

$$\mathbf{C}_p = \begin{bmatrix} 0 & 1 & 0 & \cdots & 0 \\ 0 & 0 & 1 & \cdots & 0 \\ \vdots & \vdots & \vdots & \ddots & \vdots \\ 0 & 0 & 0 & \cdots & 1 \\ -a_0 & -a_1 & -a_2 & \cdots & -a_{n-1} \end{bmatrix}$$

is known as the companion matrix of the polynomial $p(x)$. Use mathematical induction to prove that for $n \geq 2$, one has

$$\det(\mathbf{C}_p - \lambda \mathbf{I}) = (-1)^n (a_0 + a_1 \lambda + \cdots + a_{n-1}\lambda^{n-1} + \lambda^n) = (-1)^n p(\lambda).$$

7.51 Let $p(x) = x^3 + a_2 x^2 + a_1 x + a_0$ and let λ be a zero point of the polynomial $p(x)$.

(a) Write down the companion matrix $\mathbf{C}_p$ of the polynomial $p(x)$.

(b) Explain why $\lambda^3 = -a_2\lambda^2 - a_1\lambda - a_0$, and prove that $(1, \lambda, \lambda^2)$ are the eigenvalues of the companion matrix $\mathbf{C}_p$ of the polynomial $p(x)$.

7.52 Show that, for any an $n \times n$ symmetric matrix $\mathbf{A}$ and any positive semi-definite matrix $\mathbf{B}$, the eigenvalue inequality

$$\lambda_k(\mathbf{A} + \mathbf{B}) \geq \lambda_k(\mathbf{A}), \quad k = 1, 2, \ldots, n$$

is true, and if $\mathbf{B}$ is positive definite then

$$\lambda_k(\mathbf{A} + \mathbf{B}) > \lambda_k(\mathbf{A}), \quad k = 1, 2, \ldots, n.$$

(*Hint:* Use the maximum–minimum principle of the Rayleigh quotient.)

8

Subspace Analysis and Tracking

In many considerations involving approximation, optimization, differential equations, communications, signal processing and system science, subspaces play a very important role. The framework of vector subspaces can help us answer some important questions: for example, how to get a good polynomial approximation for complex functions, how to get a good approximate solution of a differential equation, how to design a better signal processor. Questions like these are a core problem in many engineering applications. The vector subspace provides a class of effective methods for addressing these issues, which are known as subspace methods.

This chapter focuses on the theoretical analysis of subspaces, and presents some typical applications of subspace methods. In many engineering problems, subspaces are time-variant, there is a need for real-time processing of the received signal or for a real-time control system. In these cases we need to keep track of the subspaces. Therefore, this chapter will focus on the subspace tracking and updating.

8.1 General Theory of Subspaces

Before discussing subspaces, it is necessary to present the basic concepts of subspaces and the algebraic and geometric relationships between subspaces.

8.1.1 Bases of Subspaces

Let $V = \mathbb{C}^n$ be an n-dimensional complex vector space. Consider a subset of m n-dimensional vectors with $m < n$.

Definition 8.1 If $S = \{\mathbf{u}_1, \ldots, \mathbf{u}_m\}$ is a vector subset in the vector space V, then the set W of the linear combinations of $\mathbf{u}_1, \ldots, \mathbf{u}_m$ is called the *subspace* spanned by $\mathbf{u}_1, \ldots, \mathbf{u}_m$ and is defined as

$$W = \text{Span}\{\mathbf{u}_1, \ldots, \mathbf{u}_m\} = \big\{\mathbf{u}\big|\mathbf{u} = a_1\mathbf{u}_1 + \cdots + a_m\mathbf{u}_m\big\}. \tag{8.1.1}$$

Every vector spanning the subspace W is called a *subspace generator* of W and the set consisting of all the generators $\{\mathbf{u}_1, \ldots, \mathbf{u}_m\}$ is known as the *spanning set* of the subspace. A vector subspace containing only the zero vector is said to be a *trivial subspace.*

Theorem 8.1 (Spanning set theorem) [285, p. 234] *Let $S = \{\mathbf{u}_1, \ldots, \mathbf{u}_m\}$ be a subset in the vector space V, and $W = \text{Span}\{\mathbf{u}_1, \ldots, \mathbf{u}_p\}$ (where $p \leq m$) be a subspace spanned by p column vectors in S.*

- *If some vector (say $\mathbf{u}_k$) in S is a linear combination of other vectors, then after removing the vector $\mathbf{u}_k$ from S, the other vectors still span the subspace W.*
- *If $W \neq \{\mathbf{0}\}$, i.e., W is a nontrivial subspace, then there must be some subset in S consisting of p linearly independent vectors which span the subspace W.*

Let all redundant vectors be deleted from a vector set $S = \{\mathbf{u}_1, \ldots, \mathbf{u}_m\}$, and the remaining p linearly independent vectors be $\{\mathbf{u}_1, \ldots, \mathbf{u}_p\}$; then they still span the subspace W. In the sense of spanning the same subspace W, $\{\mathbf{u}_1, \ldots, \mathbf{u}_p\}$ and $\{\mathbf{u}_1, \ldots, \mathbf{u}_m\}$ are said to be equivalent spanning sets.

Definition 8.2 Let W be a vector subspace. A vector set $\{\mathbf{u}_1, \ldots, \mathbf{u}_p\}$ is called a set of basis vectors or a basis set or just a basis if the following two conditions are satisfied:

(a) the subspace $W = \text{Span}\{\mathbf{u}_1, \ldots, \mathbf{u}_p\}$;
(b) the vector set $B = \{\mathbf{u}_1, \ldots, \mathbf{u}_p\}$ is a set of linearly independent vectors.

On the basis vectors of a subspace, the following two observations are useful.

(1) When using the spanning set theorem to delete some vector from a set S, once S becomes a set of linearly independent vectors, we have to stop deleting vectors from S. If we delete a supposedly supernumerary vector that is not in fact a linear combination of the remaining vectors, then the smaller vector set will not span the original subspace W. Therefore, a basis set is the possible small spanning set. In other words, the number of basis vectors spanning the subspace W cannot be fewer.
(2) A basis set is also the largest possible set of linearly independent vectors. Let S be a basis set of the subspace W. If adding a vector (say $\mathbf{w}$) from the subspace W to S, then the new vector set cannot be linearly independent, because S spans the subspace W, and the vector $\mathbf{w}$ in W itself is a linear combination of all the basis vectors in S. Thus, the number of basis vectors spanning the subspace W cannot be greater.

It should be noted that, when referring to a vector space basis set, we do not imply this is unique, it should be emphasized that this is just one basis set. Although there is a variety of choices for the basis set of a vector subspace, all the basis sets necessarily have the same number of linearly independent vectors; a spanning set containing more vectors cannot be counted as a basis set. From this discussion, it is easy to introduce the concept of subspace dimension.

Definition 8.3 The number of vectors in any basis set of a subspace W is called the *subspace dimension* of W, denoted $\dim(W)$. If any basis set does not consist of a finite number of linearly independent vectors then W is said to be an *infinite-dimensional vector subspace.*

Since any zero vector is linearly dependent on other vectors, without loss of generality it is usually assumed that the spanning set of a subspace does not contain any zero vector.

For a given spanning set $\{\mathbf{u}_1, \ldots, \mathbf{u}_n\}$, it is easy to construct a basis set for the subspace $\mathrm{Span}\{\mathbf{u}_1, \ldots, \mathbf{u}_n\}$; see the discussion in Section 8.2.

An important reason for specifying a basis set for the subspace W is to provide coordinates for the subspace W. The following theorem shows the existence of such a coordinate system.

Theorem 8.2 [285] *Let W be an n-dimensional vector subspace, and $B = \{\mathbf{b}_1, \ldots, \mathbf{b}_n\}$ be a basis set for W. Then, for any vector $\mathbf{x} \in W$, there is a unique set of scalars $c_1, \ldots, c_n$ such that $\mathbf{x}$ can be expressed as*

$$\mathbf{x} = c_1\mathbf{b}_1 + \cdots + c_n\mathbf{b}_n. \tag{8.1.2}$$

This is called the *unique representation theorem* for subspace vectors. The coefficients $c_1, \ldots, c_n$ are unique, so that they form n coordinates that represent the subspace W and thus constitute a *coordinate system* for the subspace W.

8.1.2 Disjoint Subspaces and Orthogonal Complement

In subspace analysis, the relation between subspaces is described by the relation between the subspace elements (i.e., vectors). In the following, we discuss the algebraic relations between subspaces.

The intersection of the subspaces $S_1, \ldots, S_n$,

$$S = S_1 \cap \cdots \cap S_n, \tag{8.1.3}$$

is the set consisting of all common vectors of the subspaces $S_1, \ldots, S_n$. If the unique common vector of these subspaces is the zero vector, i.e., $S = S_1 \cap \cdots \cap S_n = \{\mathbf{0}\}$, then the subspaces $S_1, \ldots, S_n$ are said to be *disjoint*. The union of the disjoint subspaces $S = S_1 \cup \cdots \cup S_n$ is called the *direct sum of subspaces*, also denoted as

$$S = S_1 \oplus \cdots \oplus S_n. \tag{8.1.4}$$

In this case, every vector $\mathbf{x} \in S$ has the unique decomposition representation $\mathbf{x} = \mathbf{a}_1 + \cdots + \mathbf{a}_n$, where $\mathbf{a}_i \in S_i$.

If a vector is orthogonal to all the vectors in subspace S then we say that the vector is orthogonal to the subspace S. By extension, the subspaces $S_1, \ldots, S_n$ are called *orthogonal subspaces*, denoted $S_i \perp S_j$, $i \neq j$, if $\mathbf{a}_i \perp \mathbf{a}_j$ holds for all $\mathbf{a}_i \in S_i, \mathbf{a}_j \in S_j$ $(i \neq j)$.

In particular, the set of all vectors orthogonal to the subspace S comprises a vector subspace called the *orthogonal complement subspace* of S, denoted $S^\perp$. Specifically, let S be a vector space, then the vector space $S^\perp$ is the orthogonal complement of S if

$$S^\perp = \{\mathbf{x}|\mathbf{x}^T\mathbf{y} = 0, \ \forall \mathbf{y} \subset S\}. \tag{8.1.5}$$

The dimensions of both the subspace S and its orthogonal complement $S^\perp$ satisfy the relationship

$$\dim(S) + \dim(S^\perp) = \dim(V). \tag{8.1.6}$$

As the name suggests, the orthogonal complement subspace $S^\perp$ of the subspace S in the vector space V has the double attributes of orthogonality and complementarity.

(1) The subspace $S^\perp$ is orthogonal to S.
(2) The vector space V is the direct sum of the subspaces S and $S^\perp$, i.e., $V = S \oplus S^\perp$. In other words, the vector space V is the combination of the subspace S and its complement $S^\perp$.

In particular, every vector $\mathbf{u}$ in the vector space $\mathbb{R}^m$ can be uniquely decomposed into the sum of a vector $\mathbf{x}$ in the subspace S and a vector $\mathbf{y}$ in the orthogonal complement $S^\perp$, namely

$$\mathbf{u} = \mathbf{x} + \mathbf{y}, \qquad \mathbf{x} \perp \mathbf{y}. \tag{8.1.7}$$

This decomposition form is called the *orthogonal decomposition* of a vector and has the wide applications in signal processing, pattern recognition, automatic control, wireless communications, system science and so on.

Example 8.1 A function $u(t)$ is called *strictly square integrable*, denoted $u(t) \in L^2(R)$, if

$$\int_{-\infty}^{\infty} |u(t)|^2 \mathrm{d}t < \infty.$$

In wavelet analysis, one normally uses multiple resolutions to approximate a square integrable function or signal $u(t) \in L^2(R)$. This is known as the multiple resolution analysis of functions or signals. In multiple resolution analysis it is necessary to construct a subspace sequence or chain $\{V_j|j \in Z\}$ in the space $L^2(R)$ such that the subspace sequence has some desired properties. For example, this subspace sequence must have the inclusion

$$\cdots \subset V_{-2} \subset V_{-1} \subset V_0 \subset V_1 \subset V_2 \subset \cdots.$$

From this inclusion it is seen that V_j is a subspace of V_{j+1}. Therefore, there must exist the orthogonal complement $W_j = V_j^\perp$ such that

$$V_{j+1} = V_j \oplus W_j,$$

where V_j and W_j are respectively called the *scale* subspace and the *wavelet* subspace at resolution 2^{-j}. The multiple resolution analysis satisfying the above condition is known as orthogonal multiple resolution analysis.

The following are relationships among disjoint subspaces, orthogonal subspaces and orthogonal complement subspaces.

(1) Disjointness is a weaker condition than orthogonality, because that two subspaces are disjoint means only that they have no pair of nonzero common vectors; it does not imply any other relation of these subspaces. On the contrary, when subspaces S_1 and S_2 are mutually orthogonal, any two vectors $\mathbf{x} \in S_1$ and $\mathbf{y} \in S_2$ are orthogonal and therefore have no correlated components, i.e., S_1 and S_2 must be disjoint. Therefore, two disjoint subspaces are not necessarily orthogonal but two orthogonal subspaces must be disjoint.
(2) The concept of an orthogonal complement subspace is a stricter concept than that of an orthogonal subspace: the orthogonal complement $S^\perp$ of a subspace S in the vector space V must be orthogonal to S but a subspace orthogonal to S is generally not the orthogonal complement of S. There may be more subspaces $S_1, \ldots, S_p$ orthogonal to S in V, but there is only one orthogonal complement $S^\perp$; it is uniquely determined.

A subspace set $\{S_1, \ldots, S_m\}$ satisfying the relation $\{0\} \subset S_1 \subset \cdots \subset S_m$ is called a *subspace nest.*

Definition 8.4 [179] For a matrix $\mathbf{A}$, a subspace $S \subseteq \mathbb{C}^n$ is called $\mathbf{A}$-invariant if

$$\mathbf{x} \in S \quad \Rightarrow \quad \mathbf{A}\mathbf{x} \in S.$$

Example 8.2 Let the eigenvectors of the $n \times n$ matrix $\mathbf{A}$ be $\mathbf{u}_1, \ldots, \mathbf{u}_n$, and let $S = \text{Span}\{\mathbf{u}_1, \ldots, \mathbf{u}_n\}$; then because $\mathbf{A}\mathbf{u}_i = \lambda_i \mathbf{u}_i, i = 1, 2, \ldots, n$, we have

$$\mathbf{u}_i \in S \quad \Rightarrow \quad \mathbf{A}\mathbf{u}_i \in S, \quad i = 1, \ldots, n.$$

This shows that the subspace S spanned by the eigenvectors of $\mathbf{A}$ is an $\mathbf{A}$-invariant subspace.

For any eigenvalue λ of an $n \times n$ matrix $\mathbf{A}$, the subspace $\text{Null}(\mathbf{A} - \lambda\mathbf{I})$ is an $\mathbf{A}$-invariant subspace because

$$\mathbf{u} \in \text{Null}(\mathbf{A} - \lambda\mathbf{I}) \quad \Rightarrow \quad (\mathbf{A} - \lambda\mathbf{I})\mathbf{u} = \mathbf{0} \quad \Rightarrow \quad \mathbf{A}\mathbf{u} = \lambda\mathbf{u} \in \text{Null}(\mathbf{A} - \lambda\mathbf{I}).$$

The null space $\text{Null}(\mathbf{A} - \lambda\mathbf{I})$ is called the *eigenspace* of the matrix $\mathbf{A}$ corresponding to the eigenvalue λ.

Let $\mathbf{A} \in \mathbb{C}^{n\times n}$, $\mathbf{B} \in \mathbb{C}^{k\times k}$, $\mathbf{X} \in \mathbb{C}^{n\times k}$ and $\mathbf{X} = [\mathbf{x}_1, \ldots, \mathbf{x}_k]$; then the jth column of $\mathbf{AX} = \mathbf{XB}$ is

$$(\mathbf{AX})_j = \begin{bmatrix} b_{1j}x_{11} + b_{2j}x_{12} + \cdots + b_{kj}x_{1k} \\ b_{1j}x_{21} + b_{2j}x_{22} + \cdots + b_{kj}x_{2k} \\ \vdots \\ b_{1j}x_{n1} + b_{2j}x_{n2} + \cdots + b_{kj}x_{nk} \end{bmatrix} = b_{1j}\mathbf{x}_1 + b_{2j}\mathbf{x}_2 + \cdots + b_{kj}\mathbf{x}_k.$$

Hence, if $S = \text{Span}\{\mathbf{x}_1, \ldots, \mathbf{x}_k\}$ then

$$\mathbf{Ax}_j \in S, \quad j = 1, \ldots, k.$$

In other words, the subspace $S = \text{Span}\{\mathbf{x}_1, \ldots, \mathbf{x}_k\}$ is an $\mathbf{A}$-invariant subspace, if

$$\mathbf{AX} = \mathbf{XB}, \quad \mathbf{A} \in \mathbb{C}^{n\times n}, \ \mathbf{B} \in \mathbb{C}^{k\times k}, \ \mathbf{X} \in \mathbb{C}^{n\times k}. \tag{8.1.8}$$

In this case, if $\mathbf{X}$ has full column rank, and $(\lambda, \mathbf{u})$ is an eigenpair of the matrix $\mathbf{B}$, i.e., $\mathbf{Bu} = \lambda\mathbf{u}$, then premultiplying $\mathbf{Bu} = \lambda\mathbf{u}$ by the full column rank matrix $\mathbf{X}$ gives

$$\mathbf{Bu} = \lambda\mathbf{u} \quad \Rightarrow \quad \mathbf{XBu} = \lambda\mathbf{Xu} \quad \Rightarrow \quad \mathbf{A}(\mathbf{Xu}) = \lambda(\mathbf{Xu}). \tag{8.1.9}$$

Then, $\lambda(\mathbf{B}) \subseteq \lambda(\mathbf{A})$, and the equality holds if and only if $\mathbf{X}$ is a square nonsingular matrix. That is to say, if $\mathbf{X}$ is a nonsingular matrix, then $\mathbf{B} = \mathbf{X}^{-1}\mathbf{AX}$ is similar to $\mathbf{A}$, and $\lambda(\mathbf{B}) = \lambda(\mathbf{A})$. This shows again, from the viewpoint of invariant subspaces, that two similar matrices may have the same eigenvalues but their eigenvectors could still be different.

The concept of invariant subspaces plays an important role in using the subspaces to track and update iteratively the eigenvalues of a large-dimensional sparse matrix.

8.2 Column Space, Row Space and Null Space

Before presenting vector subspace analysis, it is necessary to discuss the basic spaces relating to matrices. These basic spaces are column space, row space and null space.

8.2.1 Definitions and Properties

As previously, for a matrix $\mathbf{A} \in \mathbb{C}^{m\times n}$, its m row vectors are denoted as

$$\mathbf{r}_1 = [a_{11}, \ldots, a_{1n}], \quad \ldots, \quad \mathbf{r}_m = [a_{m1}, \ldots, a_{mn}],$$

and its n column vectors are denoted as

$$\mathbf{a}_1 = \begin{bmatrix} a_{11} \\ \vdots \\ a_{m1} \end{bmatrix}, \quad \ldots, \quad \mathbf{a}_n = \begin{bmatrix} a_{1n} \\ \vdots \\ a_{mn} \end{bmatrix}.$$

Definition 8.5 If $\mathbf{A} = [\mathbf{a}_1, \ldots, \mathbf{a}_n] \in \mathbb{C}^{m\times n}$ then the set of all linear combinations of its column vectors comprises a subspace that is called the *column space* or *column span* of the matrix $\mathbf{A}$, denoted $\text{Col}(\mathbf{A})$ or $\text{Span}\{\mathbf{a}_1, \ldots, \mathbf{a}_n\}$, i.e.,

$$\text{Col}(\mathbf{A}) = \text{Span}\{\mathbf{a}_1, \ldots, \mathbf{a}_n\} \tag{8.2.1}$$

$$= \left\{\mathbf{y} \in \mathbb{C}^m \middle| \mathbf{y} = \sum_{j=1}^{n} \alpha_j \mathbf{a}_j,\ \alpha_j \in \mathbb{C}\right\}. \tag{8.2.2}$$

Similarly, the set of all linear combination of complex conjugate row vectors $\mathbf{r}_1^*, \ldots, \mathbf{r}_m^* \in \mathbb{C}^n$ of the matrix $\mathbf{A}$ is known as the *row space* or *row span* of the matrix $\mathbf{A}$, denoted $\text{Row}(\mathbf{A})$ or $\text{Span}\{\mathbf{r}_1^*, \ldots, \mathbf{r}_m^*\}$, namely

$$\text{Row}(\mathbf{A}) = \text{Span}\{\mathbf{r}_1^*, \ldots, \mathbf{r}_m^*\} \tag{8.2.3}$$

$$= \left\{\mathbf{y} \in \mathbb{C}^n \middle| \mathbf{y} = \sum_{i=1}^{m} \beta_i \mathbf{r}_i^*,\ \beta_i \in C\right\}. \tag{8.2.4}$$

The column space of the matrix $\mathbf{A}$ is sometimes denoted as $\text{Span}\{\mathbf{A}\}$, i.e.,

$$\text{Col}(\mathbf{A}) = \text{Span}(\mathbf{A}) = \text{Span}\{\mathbf{a}_1, \ldots, \mathbf{a}_n\}. \tag{8.2.5}$$

Similarly, the symbol $\text{Span}(\mathbf{A}^H)$ represents the column space of the complex conjugate transpose matrix $\mathbf{A}^H$. Because the column vectors of $\mathbf{A}^H$ are the conjugate row vectors of $\mathbf{A}$, one has

$$\text{Row}(\mathbf{A}) = \text{Col}(\mathbf{A}^H) = \text{Span}(\mathbf{A}^H) = \text{Span}\{\mathbf{r}_1^*, \ldots, \mathbf{r}_m^*\}. \tag{8.2.6}$$

That is, the row space of the complex matrix $\mathbf{A}$ and the column space of the complex conjugate transpose matrix $\mathbf{A}^H$ are equivalent.

The row space and the column space are two vector subspaces defined directly by the matrix $\mathbf{A}_{m\times n}$. In addition, there are other two subspaces defined directly by the matrix transform $\mathbf{Ax}$ rather than $\mathbf{A}$. These two subspaces are the range and the null space of the mapping or transform.

In Chapter 1, the range of the mapping T was defined as the set of all values of $T(\mathbf{x}) \neq \mathbf{0}$, while the kernel or null space is defined as the set of all nonzero vectors satisfying $T(\mathbf{x}) = \mathbf{0}$. Naturally, if the linear mapping $\mathbf{y} = T(\mathbf{x})$ is a matrix transform from the space $\mathbb{C}^n$ to the space $\mathbb{C}^m$, i.e., $\mathbf{y}_{m\times 1} = \mathbf{A}_{m\times n}\mathbf{x}_{n\times 1}$, then, for a given matrix $\mathbf{A}$, the range of the matrix transform $\mathbf{Ax}$ is defined as the set of all possible values of the vectors $\mathbf{y} = \mathbf{Ax}$ and the null space is defined as the set of the vectors $\mathbf{x}$ satisfying $\mathbf{Ax} = \mathbf{0}$. In some publications (especially in engineering), the range and the null space are defined as follows.

Definition 8.6 If $\mathbf{A}$ is an $m \times n$ complex matrix then the *range* of $\mathbf{A}$ is defined as

$$\text{Range}(\mathbf{A}) = \left\{\mathbf{y} \in \mathbb{C}^m \middle| \mathbf{Ax} = \mathbf{y},\ \mathbf{x} \in \mathbb{C}^n\right\}. \tag{8.2.7}$$

The *null space* of $\mathbf{A}$ is also called the kernel of $\mathbf{A}$ and is defined as the set of all solution vectors of the *homogeneous linear equation* $\mathbf{Ax} = \mathbf{0}$, i.e.,

$$\text{Null}(\mathbf{A}) = \text{Ker}(\mathbf{A}) = \{\mathbf{x} \in \mathbb{C}^n | \mathbf{Ax} = \mathbf{0}\}. \tag{8.2.8}$$

Similarly, the *null space* of the complex conjugate transpose matrix $\mathbf{A}^H$ is defined as

$$\text{Null}(\mathbf{A}^H) = \text{Ker}(\mathbf{A}^H) = \{\mathbf{x} \in \mathbb{C}^m | \mathbf{A}^H\mathbf{x} = \mathbf{0}\}. \tag{8.2.9}$$

The dimension of the null space is called the *nullity* of $\mathbf{A}$, namely

$$\text{nullity}(\mathbf{A}) = \dim[\text{Null}(\mathbf{A})]. \tag{8.2.10}$$

Let $\mathbf{A} = [\mathbf{a}_1, \ldots, \mathbf{a}_n]$ and $\mathbf{x} = [\alpha_1, \ldots, \alpha_n]^T$, then $\mathbf{Ax} = \sum_{j=1}^n \alpha_j \mathbf{a}_j$, so we have

$$\text{Range}(\mathbf{A}) = \left\{ \mathbf{y} \in \mathbb{C}^m \middle| \mathbf{y} = \sum_{j=1}^n \alpha_j \mathbf{a}_j,\ \alpha_j \in C \right\} = \text{Span}\{\mathbf{a}_1, \ldots, \mathbf{a}_n\}.$$

Therefore we have

$$\text{Range}(\mathbf{A}) = \text{Col}(\mathbf{A}) = \text{Span}\{\mathbf{a}_1, \ldots, \mathbf{a}_n\}, \tag{8.2.11}$$

$$\text{Range}(\mathbf{A}^H) = \text{Col}(\mathbf{A}^H) = \text{Span}\{\mathbf{r}_1^*, \ldots, \mathbf{r}_m^*\}. \tag{8.2.12}$$

Theorem 8.3 *If* $\mathbf{A}$ *is an* $m \times n$ *complex matrix then the orthogonal complement* $(\text{Row}(\mathbf{A}))^\perp$ *of the row space of* $\mathbf{A}$ *is the null space of* $\mathbf{A}$, *and the orthogonal complement* $(\text{Col}(\mathbf{A}))^\perp$ *of the column space of* $\mathbf{A}$ *is the null space of* $\mathbf{A}^H$, *i.e.,*

$$(\text{Row}(\mathbf{A}))^\perp = \text{Null}(\mathbf{A}), \qquad (\text{Col}\,\mathbf{A})^\perp = \text{Null}(\mathbf{A}^H). \tag{8.2.13}$$

Proof Let the row vectors of $\mathbf{A}$ be $\mathbf{r}_i, i = 1, \ldots, m$. By the matrix multiplication rule it is known that if $\mathbf{x}$ lies in the null space Null($\mathbf{A}$), i.e., $\mathbf{Ax} = \mathbf{0}$, then $\mathbf{r}_i\mathbf{x} = 0$. This result can be written in the standard orthogonal form of two column vectors: $(\mathbf{r}_i^H)^H\mathbf{x} = 0, i = 1, \ldots, m$. This implies that the vector $\mathbf{x}$ is orthogonal to the linear span of $\mathbf{r}_1^H, \ldots, \mathbf{r}_m^H$, i.e.,

$$\mathbf{x} \perp \text{Span}\{\mathbf{r}_1^H, \ldots, \mathbf{r}_m^H\} = \text{Col}(\mathbf{A}^H) = \text{Row}(\mathbf{A}).$$

Hence Null($\mathbf{A}$) and Row($\mathbf{A}$) are orthogonal. On the contrary, if $\mathbf{x}$ and Row($\mathbf{A}$) = Col($\mathbf{A}^H$) are orthogonal then $\mathbf{x}$ is also orthogonal to $\mathbf{r}_i^H, i = 1, \ldots, m$, i.e., $(\mathbf{r}_i^H)^H\mathbf{x} = 0$; this is equivalent to $\mathbf{r}_i\mathbf{x} = 0, i = 1, \ldots, m$, or $\mathbf{Ax} = \mathbf{0}$. This shows that the set of vectors $\mathbf{x}$ orthogonal to Row($\mathbf{A}$) is the null space of $\mathbf{A}$, and thus $(\text{Row}(\mathbf{A}))^\perp = \text{Null}(\mathbf{A})$.

In the general equality $\left(\text{Row}(\mathbf{B}_{n\times m})\right)^\perp = \text{Null}(\mathbf{B})$, let $\mathbf{B} = \mathbf{A}_{m\times n}^H$; then

$$\left(\text{Row}(\mathbf{A}^H)\right)^\perp = \text{Null}(\mathbf{A}^H) \Rightarrow \left(\text{Col}(\mathbf{A})\right)^\perp = \text{Null}(\mathbf{A}^H).$$

This completes the proof of the theorem. □

In summary, there are the following relationships between the vector subspaces of a matrix $\mathbf{A}$.

(1) The range and column space of a matrix are equal, i.e.,

$$\text{Range}(\mathbf{A}) = \text{Col}(\mathbf{A}) = \text{Span}(\mathbf{A}) = \text{Span}\{\mathbf{a}_1, \ldots, \mathbf{a}_n\}.$$

(2) The row space of $\mathbf{A}$ and the column space of $\mathbf{A}^H$ are the same, i.e.,

$$\text{Row}(\mathbf{A}) = \text{Col}(\mathbf{A}^H) = \text{Span}(\mathbf{A}^H) = \text{Range}(\mathbf{A}^H).$$

(3) The orthogonal complement of the row space of $\mathbf{A}$ is equal to its null space:

$$(\text{Row}(\mathbf{A}))^{\perp} = \text{Null}(\mathbf{A}).$$

(4) The orthogonal complements of the column space of $\mathbf{A}$ and the null space of $\mathbf{A}^H$ are the same, i.e.,

$$(\text{Col}(\mathbf{A}))^{\perp} = \text{Null}(\mathbf{A}^H).$$

Since the column space Col($\mathbf{A}$) of the matrix $\mathbf{A}$ is the collection of linear combinations of its column vectors, Col($\mathbf{A}$) is determined only by the linearly independent column vectors $\mathbf{a}_{i_1}, \ldots, \mathbf{a}_{i_k}$, while the other column vectors, which are linearly dependent on $\mathbf{a}_{i_1}, \ldots, \mathbf{a}_{i_k}$, are redundant for generating the column space.

The subset $\{\mathbf{a}_{i_1}, \ldots, \mathbf{a}_{i_k}\}$ is the maximal linearly independent subset of the column vector set $\{\mathbf{a}_1, \ldots, \mathbf{a}_n\}$ if $\mathbf{a}_{i_1}, \ldots, \mathbf{a}_{i_k}$ are linearly independent and these linearly independent column vectors are not contained in any other linearly independent subset of $\{\mathbf{a}_1, \ldots, \mathbf{a}_n\}$.

If $\{\mathbf{a}_{i_1}, \ldots, \mathbf{a}_{i_k}\}$ is the maximal linearly independent subset, then

$$\text{Span}\{\mathbf{a}_1, \ldots, \mathbf{a}_n\} = \text{Span}\{\mathbf{a}_{i_1}, \ldots, \mathbf{a}_{i_k}\}, \tag{8.2.14}$$

and the maximal linearly independent subset $\{\mathbf{a}_{i_1}, \ldots, \mathbf{a}_{i_k}\}$ is known as a *basis* of the column space Col($\mathbf{A}$). Clearly, for a given matrix $\mathbf{A}_{m\times n}$, its basis can have different combination forms, but all the bases must have the same number of basis vectors. This common number of vectors is called the *dimension* of the column space Col($\mathbf{A}$), denoted dim[Col($\mathbf{A}$)]. Since the rank of $\mathbf{A}$ is defined as the number of its linearly independent column vectors, the rank of $\mathbf{A}$ and the dimension of the column space Col($\mathbf{A}$) are the same quantity. In other words, the rank can also be defined as

$$\text{rank}(\mathbf{A}) = \dim[\text{Col}(\mathbf{A})] = \dim[\text{Range}(\mathbf{A})]. \tag{8.2.15}$$

For a given matrix $\mathbf{A} \in \mathbb{C}^{m\times n}$, the following is a comparison between column space and null space [285, p. 226].

1. The column space Col($\mathbf{A}$) is a subspace of $\mathbb{C}^n$, while the null space Null($\mathbf{A}$) is a subspace of $\mathbb{C}^m$.

2. Col(**A**) is explicitly defined and is spanned by all the columns of **A**, while Null(**A**) is implicitly defined, and has no relation with column vectors of **A**.
3. Each basis vector of Col(**A**) is a pivot column (see Definition 1.13) of **A**, while any basis vector **v** of Null(**A**) must meet the condition $\mathbf{Av} = \mathbf{0}$.
4. Each column of **A** is in Col(**A**), while there is no obvious relation of Null(**A**) to the entries of **A**.
5. A typical vector **v** in Col(**A**) has the property that $\mathbf{Ax} = \mathbf{v}$ has an exact solution **x**, i.e., $[\mathbf{A}, \mathbf{v}]$ and **A** have the same rank; while a typical vector $\mathbf{v} \in \text{Null}(\mathbf{A})$ satisfies $\mathbf{Av} = \mathbf{0}$.
6. $\text{Col}(\mathbf{A}) = \mathbb{R}^m$ if and only if the equation $\mathbf{Ax} = \mathbf{b}$ has a solution for all $\mathbf{b} \in \mathbb{R}^m$, while $\text{Null}(\mathbf{A}) = \{\mathbf{0}\}$ if and only if the equation $\mathbf{Ax} = \mathbf{0}$ has only a zero solution.

8.2.2 Subspace Basis Construction

As stated above, the column space and the row space of the matrix $\mathbf{A}_{m\times n}$ are spanned by n column vectors and m row vectors of **A**, respectively. But, if the matrix rank is $r = \text{rank}(\mathbf{A})$, then only r linearly independent column or row vectors can generate the column space Span(**A**) and the row space $\text{Span}(\mathbf{A}^H)$, respectively. Obviously, the use of basis vectors is a more economical and better way for subspace representation. So, how do we find the required basis vectors? In the following, we discuss the constructions of basis vectors for the row space Row(**A**), the column space Col(**A**) and the null spaces Null(**A**) and $\text{Null}(\mathbf{A}^H)$.

It is easy to show the following two results [285].

(1) Elementary row transformations do not change the row space Row(**A**) or the null space Null(**A**) of the matrix **A**.
(2) Elementary column transformations do not change the column space Col(**A**) of **A** or the null space $\text{Null}(\mathbf{A}^H)$ of $\mathbf{A}^H$.

The following theorem provides a method for using elementary row or column transformations to construct the required subspaces.

Theorem 8.4 [285] *Let the matrix* $\mathbf{A}_{m\times n}$ *become the echelon matrix* **B** *via elementary row transformations; then*

- *the nonzero rows of the echelon matrix* **B** *constitute a basis set for the row space of* **A***;*
- *the pivot columns constitute a basis set for the column space* Col(**A**) *of* **A***.*

The above discussion can be summarized in terms of the elementary transformation methods for constructing the basis vectors of the column space and the row space for a given matrix.

1. Elementary Row Transformation Method

Let a matrix $\mathbf{A}$ become the reduced echelon matrix $\mathbf{B}_r$ via the elementary row transformation, then

(1) The nonzero rows corresponding to the pivots of the reduced echelon matrix $\mathbf{B}_r$ constitute the basis vectors of the row space $\text{Row}(\mathbf{A})$.

(2) The pivot columns of the matrix $\mathbf{A}$ constitute the basis of the column space $\text{Col}(\mathbf{A})$.

(3) The nonpivot columns of the matrix $\mathbf{A}$ constitute the basis of the null space $\text{Null}(\mathbf{A}^H)$.

2. Elementary Column Transformation Method

Let a matrix $\mathbf{A}$ become the reduced echelon matrix in column form $\mathbf{B}_c$; then

(1) The nonzero columns corresponding to the pivots in the columns of the reduced echelon matrix $\mathbf{B}_c$ constitute the basis vectors of the column space $\text{Col}(\mathbf{A})$.

(2) The pivot rows of the matrix $\mathbf{A}$ constitute the basis of the row space $\text{Row}(\mathbf{A})$.

(3) The nonpivot rows of the matrix $\mathbf{A}$ constitute the basis of the null space $\text{Null}(\mathbf{A})$.

Example 8.3 Find the row space and the column space of a 3×3 matrix

$$\mathbf{A} = \begin{bmatrix} 1 & 2 & 1 \\ -1 & -1 & 1 \\ 1 & 4 & 5 \end{bmatrix}.$$

Solution *Method 1:* By the elementary column transformations $C_2 - 2C_1$ (multiplying the first column by -2, then adding it to the second column), $C_3 - C_1, C_1 + C_2$ and $C_3 - 2C_2$, we have

$$\mathbf{B}_c = \begin{bmatrix} 1 & 0 & 0 \\ 0 & 1 & 0 \\ 3 & 2 & 0 \end{bmatrix}.$$

This yields two linearly independent column vectors $\mathbf{c}_1 = [1, 0, 3]^T$ and $\mathbf{c}_2 = [0, 1, 2]^T$ which are two basis vectors of the column space $\text{Col}(\mathbf{A})$, i.e.,

$$\text{Col}(\mathbf{A}) = \text{Span}\left\{ \begin{bmatrix} 1 \\ 0 \\ 3 \end{bmatrix}, \begin{bmatrix} 0 \\ 1 \\ 2 \end{bmatrix} \right\}.$$

From the pivots in the columns of the reduced echelon matrix in column form, $\mathbf{B}_c$, the pivot rows of the matrix $\mathbf{A}$ are the first row and the second row, i.e., the row space $\text{Row}(\mathbf{A})$ can be written as

$$\text{Row}(\mathbf{A}) = \text{Span}\{[1, 2, 1], [-1, -1, 1]\}.$$

Method 2: Making the elementary row transformations: $R_2 + R_1$ (the first row is added to the second row), $R_3 - R_1$ and $R_3 - 2R_2$, we have

$$\mathbf{B}_r = \begin{bmatrix} 1 & 2 & 1 \\ 0 & 1 & 2 \\ 0 & 0 & 0 \end{bmatrix},$$

which yields two linearly independent row vectors, $\mathbf{r}_1 = [1, 2, 1]$ and $\mathbf{r}_2 = [0, 1, 2]$. They constitute the basis vectors of the row space $\text{Row}(\mathbf{A})$, namely

$$\text{Row}(\mathbf{A}) = \text{Span}\{[1, 2, 1], [0, 1, 2]\}.$$

The pivot columns of the matrix $\mathbf{A}$ are its first and second columns, and they constitute the basis vectors of the column space $\text{Col}(\mathbf{A})$, i.e.,

$$\text{Col}(\mathbf{A}) = \text{Span}\left\{ \begin{bmatrix} 1 \\ -1 \\ 1 \end{bmatrix}, \begin{bmatrix} 2 \\ -1 \\ 4 \end{bmatrix} \right\}.$$

As a matter of fact, the two solutions are equivalent, because making elementary column transformations of the basis vectors of the column space in method 2, we have

$$\begin{bmatrix} 1 & 2 \\ -1 & -1 \\ 1 & 4 \end{bmatrix} \xrightarrow{-C_1+C_2} \begin{bmatrix} 1 & 1 \\ -1 & 0 \\ 1 & 3 \end{bmatrix} \xrightarrow{C_2-C_1} \begin{bmatrix} 0 & 1 \\ 1 & 0 \\ 2 & 3 \end{bmatrix} \xrightarrow{C_1 \leftrightarrow C_2} \begin{bmatrix} 1 & 0 \\ 0 & 1 \\ 3 & 2 \end{bmatrix}.$$

These are the same as the basis vectors of the column space obtained from method 1. Similarly, it can be shown that the basis vectors of the row spaces obtained in method 1 and method 2 are equivalent as well.

8.2.3 SVD-Based Orthonormal Basis Construction

Because the basis sets obtained by elementary row transformations and by elementary column transformations are equivalent, we can choose any elementary transformation. Elementary row transformations are usually applied. However, if the number of matrix columns is significantly less than the number of rows, then elementary column transformations require fewer computations.

Elementary transformation methods give only the linearly independent basis vectors. However, in many applications, it is required to obtain the orthonormal basis vectors of the basic spaces (column space, row space and null space) of a given matrix. Of course, the Gram–Schmidt orthogonalization of the linearly independent basis vectors will give the orthonormal basis vectors, but a simpler method is based on the SVD of a matrix.

Let the matrix $\mathbf{A}_{m \times n}$ with the rank $\text{rank}(\mathbf{A}) = r$ have the SVD

$$\mathbf{A} = \mathbf{U}\boldsymbol{\Sigma}\mathbf{V}^H, \tag{8.2.16}$$

where

$$\mathbf{U} = [\mathbf{U}_r, \tilde{\mathbf{U}}_r], \quad \mathbf{V} = [\mathbf{V}_r, \tilde{\mathbf{V}}_r], \quad \boldsymbol{\Sigma} = \begin{bmatrix} \boldsymbol{\Sigma}_r & \mathbf{O}_{r\times(n-r)} \\ \mathbf{O}_{(m-r)\times(n-r)} & \mathbf{O}_{(n-r)\times(n-r)} \end{bmatrix}.$$

Here $\mathbf{U}_r$ and $\tilde{\mathbf{U}}_r$ are respectively $m \times r$ and $m \times (m-r)$ matrices, $\mathbf{V}_r$ and $\tilde{\mathbf{V}}_r$ are respectively $n \times r$ and $n \times (n-r)$ matrices and $\boldsymbol{\Sigma}_r = \mathbf{Diag}(\sigma_1, \ldots, \sigma_r)$.

Obviously, the SVD of $\mathbf{A}$ can be simplified as follows:

$$\mathbf{A} = \mathbf{U}_r \boldsymbol{\Sigma}_r \mathbf{V}_r^H = \sum_{i=1}^{r} \sigma_i \mathbf{u}_i \mathbf{v}_i^H, \tag{8.2.17}$$

$$\mathbf{A}^H = \mathbf{V}_r \boldsymbol{\Sigma}_r \mathbf{U}_r^H = \sum_{i=1}^{r} \sigma_i \mathbf{v}_i \mathbf{u}_i^H. \tag{8.2.18}$$

1. Orthonormal Basis Construction of Column Space

Substituting (8.2.17) into the definition formula of Range($\mathbf{A}$), we have

$$\begin{aligned}
\text{Range}(\mathbf{A}) &= \{\mathbf{y} \in \mathbb{C}^m \,|\, \mathbf{y} = \mathbf{A}\mathbf{x},\ \mathbf{x} \in \mathbb{C}^n\} \\
&= \left\{ \mathbf{y} \in \mathbb{C}^m \,\middle|\, \mathbf{y} = \sum_{i=1}^{r} \sigma_i \mathbf{u}_i \mathbf{v}_i^H \mathbf{x},\ \mathbf{x} \in \mathbb{C}^n \right\} \\
&= \left\{ \mathbf{y} \in \mathbb{C}^m \,\middle|\, \mathbf{y} = \sum_{i=1}^{r} \mathbf{u}_i (\sigma_i \mathbf{v}_i^H \mathbf{x}),\ \mathbf{x} \in \mathbb{C}^n \right\} \\
&= \left\{ \mathbf{y} \in \mathbb{C}^m \,\middle|\, \mathbf{y} = \sum_{i=1}^{r} \alpha_i \mathbf{u}_i,\ \alpha_i = \sigma_i \mathbf{v}_i^H \mathbf{x} \in C \right\} \\
&= \text{Span}\{\mathbf{u}_1, \ldots, \mathbf{u}_r\}.
\end{aligned}$$

Using the equivalent relation between the range and the column space, we have

$$\text{Col}(\mathbf{A}) = \text{Range}(\mathbf{A}) = \text{Span}\{\mathbf{u}_1, \ldots, \mathbf{u}_r\}.$$

This shows that r left singular vectors $\mathbf{u}_1, \ldots, \mathbf{u}_r$ constitute a basis set (a group of basis vectors) for the column space Col($\mathbf{A}$).

2. Orthonormal Basis Construction of Row Space

Computing the range of the complex conjugate transposed matrix $\mathbf{A}^H$, we get

$$\begin{aligned}
\text{Range}(\mathbf{A}^H) &= \{\mathbf{y} \in \mathbb{C}^n \,|\, \mathbf{y} = \mathbf{A}^H\mathbf{x},\ \mathbf{x} \in \mathbb{C}^m\} \\
&= \left\{ \mathbf{y} \in \mathbb{C}^n \,\middle|\, \mathbf{y} = \sum_{i=1}^{r} \sigma_i \mathbf{v}_i \mathbf{u}_i^H \mathbf{x},\ \mathbf{x} \in \mathbb{C}^m \right\} \\
&= \left\{ \mathbf{y} \in \mathbb{C}^n \,\middle|\, \mathbf{y} = \sum_{i=1}^{r} \alpha_i \mathbf{v}_i,\ \alpha_i = \sigma_i \mathbf{u}_i^H \mathbf{x} \in C \right\} \\
&= \text{Span}\{\mathbf{v}_1, \ldots, \mathbf{v}_r\},
\end{aligned}$$

and hence

$$\text{Row}(\mathbf{A}) = \text{Range}(\mathbf{A}^H) = \text{Span}\{\mathbf{v}_1, \ldots, \mathbf{v}_r\}.$$

That is, the right singular vectors $\mathbf{v}_1, \ldots, \mathbf{v}_r$ corresponding to the r nonzero singular values constitute a basis set (group of basis vectors) for the row space Row($\mathbf{A}$).

3. Orthonormal Basis Construction of Null Space

Since the rank of $\mathbf{A}$ is assumed to be r, the dimension of the null space Null($\mathbf{A}$) is equal to $n - r$. Hence we need to search for $n - r$ linearly independent orthonormal vectors as the orthonormal basis of the null space. To this end, consider the vectors satisfying $\mathbf{Ax} = \mathbf{0}$. From the property of singular vectors we get $\mathbf{v}_i^H \mathbf{v}_j = 0,\ \forall\, i = 1, \ldots, r,\ j = r + 1, \ldots, n$. From this it can be seen that

$$\mathbf{A}\mathbf{v}_j = \sum_{i=1}^{r} \sigma_i \mathbf{u}_i \mathbf{v}_i^H \mathbf{v}_j = \mathbf{0}, \quad \forall\, j = r + 1, \ldots, n.$$

Because the $n - r$ right singular vectors $\mathbf{v}_{r+1}, \ldots, \mathbf{v}_n$ corresponding to the zero singular values are linearly independent, and satisfy $\mathbf{Ax} = \mathbf{0}$, they constitute the basis of the null space Null($\mathbf{A}$), i.e.,

$$\text{Null}(\mathbf{A}) = \text{Span}\{\mathbf{v}_{r+1}, \ldots, \mathbf{v}_n\}.$$

Similarly, we have

$$\mathbf{A}^H \mathbf{u}_j = \sum_{i=1}^{r} \sigma_i \mathbf{v}_i \mathbf{u}_i^H \mathbf{u}_j = \mathbf{0}, \quad \forall j = r + 1, \ldots, m.$$

Owing to the linear independence of the $m - r$ right singular vectors $\mathbf{u}_{r+1}, \ldots, \mathbf{u}_m$ and the condition $\mathbf{A}^H \mathbf{x} = \mathbf{0}$, these vectors constitute the basis of the null space Null($\mathbf{A}^H$), namely

$$\text{Null}(\mathbf{A}^H) = \text{Span}\{\mathbf{u}_{r+1}, \ldots, \mathbf{u}_m\}.$$

Because the left singular-vector matrix $\mathbf{U}$ and the right singular-vector matrix $\mathbf{V}$ of the matrix $\mathbf{A} \in \mathbb{C}^{m \times n}$ are unitary, the basis set provided by the above SVD methods is an orthonormal basis of the column space, the row space or the null space of $\mathbf{A}$. Summarizing the above discussions, we have the following conclusions on a complex matrix $\mathbf{A} \in \mathbb{C}^{m \times n}$ with rank r.

(1) The r left singular vectors $\mathbf{u}_1, \ldots, \mathbf{u}_r$ corresponding to the nonzero singular values form an orthonormal basis for the column space Col($\mathbf{A}$), i.e.,

$$\text{Col}(\mathbf{A}) = \text{Span}\{\mathbf{u}_1, \ldots, \mathbf{u}_r\}. \tag{8.2.19}$$

(2) The $m - r$ left singular vectors $\mathbf{u}_{r+1}, \ldots, \mathbf{u}_m$ corresponding to the zero singular values form an orthonormal basis for the null space Null($\mathbf{A}^H$), namely

$$\text{Null}(\mathbf{A}^H) = (\text{Col}(\mathbf{A}))^{\perp} = \text{Span}\{\mathbf{u}_{r+1}, \ldots, \mathbf{u}_m\}. \tag{8.2.20}$$

(3) The r right singular vectors $\mathbf{v}_1, \ldots, \mathbf{v}_r$ corresponding to the nonzero singular values form an orthonormal basis for the row space Row($\mathbf{A}$):

$$\text{Row}(\mathbf{A}) = \text{Span}\{\mathbf{v}_1, \ldots, \mathbf{v}_r\}. \tag{8.2.21}$$

(4) The $n-r$ right singular vectors $\mathbf{v}_{r+1}, \ldots, \mathbf{v}_n$ corresponding to the zero singular values form an orthonormal basis for the null space Null($\mathbf{A}$), i.e.,

$$\text{Null}(\mathbf{A}) = \big(\text{Row}(\mathbf{A})\big)^{\perp} = \text{Span}\{\mathbf{v}_{r+1}, \ldots, \mathbf{v}_n\}. \tag{8.2.22}$$

By [502], the QR decomposition $\mathbf{A} = \mathbf{QR}$ is an alternative method for constructing an orthogonal basis for the column space of the matrix $\mathbf{A}$. Let the $n \times n$ matrix $\mathbf{A} = [\mathbf{a}_1, \ldots, \mathbf{a}_n]$ be nonsingular, and let $\mathbf{Q} = [\mathbf{q}_1, \ldots, \mathbf{q}_n]$. From $\mathbf{a}_1 = r_{11}\mathbf{q}_1$, $\mathbf{a}_2 = r_{12}\mathbf{q}_1 + r_{22}\mathbf{q}_2$ and so on, we have

$$\mathbf{a}_k = r_{1k}\mathbf{q}_1 + \cdots + r_{kk}\mathbf{q}_k, \quad k = 1, \ldots, n.$$

Then it follows that $\text{Span}\{\mathbf{a}_1\} = \text{Span}\{\mathbf{q}_1\}$, $\text{Span}\{\mathbf{a}_1, \mathbf{a}_2\} = \text{Span}\{\mathbf{q}_1, \mathbf{q}_2\}$ and

$$\text{Span}\{\mathbf{a}_1, \ldots, \mathbf{a}_k\} = \text{Span}\{\mathbf{q}_1, \ldots, \mathbf{q}_k\}, \quad k = 1, \ldots, n.$$

Finally, we get $\text{Col}(\mathbf{A}) = \text{Col}(\mathbf{Q})$. In other words, the column vectors of the unitary matrix $\mathbf{Q}$ are an orthonormal basis set for the column space of the matrix $\mathbf{A}$.

8.2.4 Basis Construction of Subspaces Intersection

We now consider how to construct an orthonormal basis for the intersection of the null spaces $\text{Null}(\mathbf{A}) \cap \text{Null}(\mathbf{B})$ for two given matrices $\mathbf{A} \in \mathbb{C}^{m\times n}$ and $\mathbf{B} \in \mathbb{C}^{p\times n}$.

Obviously, letting

$$\mathbf{C} = \begin{bmatrix} \mathbf{A} \\ \mathbf{B} \end{bmatrix} \in \mathbb{C}^{(m+p)\times n},$$

then

$$\mathbf{Cx} = \mathbf{0} \quad \Leftrightarrow \quad \mathbf{Ax} = \mathbf{0} \;\text{ and }\; \mathbf{Bx} = \mathbf{0}.$$

That is, the null space of $\mathbf{C}$ is equal to the intersection of the null space of $\mathbf{A}$ and the null space of $\mathbf{B}$, i.e.,

$$\text{Null}(\mathbf{C}) = \text{Null}(\mathbf{A}) \cap \text{Null}(\mathbf{B}).$$

This shows that if the rank of an $(m+p) \times n$ matrix $\mathbf{C}$ is $r = \text{rank}(\mathbf{C})$, then, among its right singular vectors $\mathbf{v}_1, \ldots, \mathbf{v}_n$, the right singular vectors $\mathbf{v}_{r+1}, \ldots, \mathbf{v}_n$ corresponding to the $n-r$ zero singular values constitute an orthonormal basis for the intersection of the null spaces $\text{Null}(\mathbf{A}) \cap \text{Null}(\mathbf{B})$. However, this is related to the SVD of the $(m+p) \times n$ matrix $\mathbf{C}$.

Theorem 8.5 [179, p. 583] *Let $\mathbf{A} \in \mathbb{R}^{m\times n}$ and $\{\mathbf{z}_1,\ldots,\mathbf{z}_t\}$ be an orthogonal basis set for the null space* Null($\mathbf{A}$)*. Denote $\mathbf{Z} = [\mathbf{z}_1,\ldots,\mathbf{z}_t]$, and define $\{\mathbf{w}_1,\ldots,\mathbf{w}_q\}$ as an orthogonal basis set for the null space* Null($\mathbf{BZ}$)*, where $\mathbf{B} \in \mathbb{R}^{p\times n}$. If $\mathbf{W} = [\mathbf{w}_1,\ldots,\mathbf{w}_q]$ then the column vectors of $\mathbf{ZW}$ constitute an orthogonal basis set for the intersection of the null space* Null($\mathbf{A}$) $\cap$ Null($\mathbf{B}$).

Given the matrices $\mathbf{A}_{m\times n}$, $\mathbf{B}_{p\times n}$, the above theorem provides the construction method for an orthogonal basis set for Null($\mathbf{A}$) $\cap$ Null($\mathbf{B}$).

(1) Compute the SVD $\mathbf{A} = \mathbf{U}_A\mathbf{\Sigma}_A\mathbf{V}_A^T$, determine the effective rank r of the matrix $\mathbf{A}$ and thus obtain orthogonal basis vectors $\mathbf{v}_{r+1},\ldots,\mathbf{v}_n$ for the null space Null($\mathbf{A}$), where $\mathbf{v}_i$ is the right singular vector of $\mathbf{A}$. Let $\mathbf{Z} = [\mathbf{v}_{r+1},\ldots,\mathbf{v}_n]$.
(2) Calculate the matrix product $\mathbf{C}_{p\times(n-r)} = \mathbf{BZ}$ and its SVD $\mathbf{C} = \mathbf{U}_C\mathbf{\Sigma}_C\mathbf{V}_{\mathbb{C}}^T$ and determine its effective rank q; this yields orthogonal basis vectors $\mathbf{w}_{q+1},\ldots,\mathbf{w}_{n-r}$ for the null space Null($\mathbf{BZ}$), where $\mathbf{w}_i$ is the right singular vector of $\mathbf{C} = \mathbf{BZ}$. Put $\mathbf{W} = [\mathbf{w}_{q+1},\ldots,\mathbf{w}_{n-r}]$.
(3) Compute the matrix product $\mathbf{ZW}$ whose column vectors give the orthogonal basis vectors of the intersection of null spaces Null($\mathbf{A}$) $\cap$ Null($\mathbf{B}$) (as $\mathbf{Z}$ and $\mathbf{W}$ are respectively the matrices consisting of the right singular vectors of $\mathbf{A}$ and of $\mathbf{BZ}$, $\mathbf{ZW}$ has the orthogonality property).

8.3 Subspace Methods

An observed data matrix $\mathbf{X}$ inevitably contains observation errors and/or noises as well as the true data matrix $\mathbf{A}$:

$$\mathbf{X} = \mathbf{A} + \mathbf{W} = [\mathbf{x}_1,\ldots,\mathbf{x}_n] \in \mathbb{C}^{m\times n}, \tag{8.3.1}$$

where $\mathbf{x}_i \in \mathbb{C}^{m\times 1}$ is an observation data vector and $\mathbf{W}$ represents the additive observation error matrix.

In engineering, the column space of an observed data matrix,

$$\text{Span}(\mathbf{X}) = \text{Span}\{\mathbf{x}_1,\ldots,\mathbf{x}_n\}, \tag{8.3.2}$$

is called the *observation data space*, while the column space of the observation error matrix,

$$\text{Span}(\mathbf{W}) = \text{Span}\{\mathbf{w}_1,\ldots,\mathbf{w}_n\}, \tag{8.3.3}$$

is known as the *noise subspace*.

8.3.1 Signal Subspace and Noise Subspace

Define the autocorrelation matrix

$$\mathbf{R}_X = E\{\mathbf{X}^H\mathbf{X}\} = E\{(\mathbf{A}+\mathbf{W})^H(\mathbf{A}+\mathbf{W})\}. \tag{8.3.4}$$

Assume that the error matrix $\mathbf{W} = [\mathbf{w}_1, \ldots, \mathbf{w}_n]$ and the true data matrix $\mathbf{A}$ are statistically uncorrelated; then

$$\mathbf{R}_X = E\{\mathbf{X}^H\mathbf{X}\} = E\{\mathbf{A}^H\mathbf{A}\} + E\{\mathbf{W}^H\mathbf{W}\}. \tag{8.3.5}$$

Put $E\{\mathbf{A}^H\mathbf{A}\} = \mathbf{R}$ and $E\{\mathbf{W}^H\mathbf{W}\} = \sigma_w^2\mathbf{I}$ (i.e., the observation noises are statistically mutually uncorrelated, and have the same variances σ_w^2). Then

$$\mathbf{R}_X = \mathbf{R} + \sigma_w^2\mathbf{I}.$$

Letting $\text{rank}(\mathbf{A}) = r$, the EVD of the autocorrelation matrix $\mathbf{R}_X$ is given by

$$\mathbf{R}_X = \mathbf{U}\boldsymbol{\Lambda}\mathbf{U}^H + \sigma_w^2\mathbf{I} = \mathbf{U}(\boldsymbol{\Lambda} + \sigma_w^2\mathbf{I})\mathbf{U}^H = \mathbf{U}\boldsymbol{\Pi}\mathbf{U}^H,$$

where

$$\boldsymbol{\Pi} = \boldsymbol{\Sigma} + \sigma_w^2\mathbf{I} = \mathbf{Diag}(\sigma_1^2 + \sigma_w^2, \ldots, \sigma_r^2 + \sigma_w^2, \sigma_w^2, \ldots, \sigma_w^2)$$

with $\boldsymbol{\Sigma} = \mathbf{Diag}(\sigma_1^2, \ldots, \sigma_r^2, 0, \ldots, 0)$; $\sigma_1^2 \geq \cdots \geq \sigma_r^2$ are the nonzero eigenvalues of the true autocorrelation matrix $E\{\mathbf{A}^H\mathbf{A}\}$.

Obviously, if the signal-to-noise ratio is large enough, i.e., σ_r^2 is significantly greater than σ_w^2, then the r largest eigenvalues of the noisy autocorrelation matrix $\mathbf{R}_X$,

$$\lambda_1 = \sigma_1^2 + \sigma_w^2, \quad \ldots \quad, \lambda_r = \sigma_r^2 + \sigma_w^2$$

are called the principal eigenvalues and other $n - r$ smaller eigenvalues,

$$\lambda_{r+1} = \sigma_w^2, \quad \ldots \quad, \lambda_n = \sigma_w^2,$$

are known as the minor eigenvalues.

Then the EVD of the autocorrelation matrix $\mathbf{R}_X$ can be written as

$$\mathbf{R}_X = [\mathbf{U}_s, \mathbf{U}_n] \begin{bmatrix} \boldsymbol{\Sigma}_s & \mathbf{O} \\ \mathbf{O} & \boldsymbol{\Sigma}_n \end{bmatrix} \begin{bmatrix} \mathbf{U}_s^H \\ \mathbf{U}_n^H \end{bmatrix} = \mathbf{S}\boldsymbol{\Sigma}_s\mathbf{S}^H + \mathbf{G}\boldsymbol{\Sigma}_n\mathbf{G}^H, \tag{8.3.6}$$

where

$$\begin{aligned}
\mathbf{S} &\stackrel{\text{def}}{=} [\mathbf{s}_1, \ldots, \mathbf{s}_r] = [\mathbf{u}_1, \ldots, \mathbf{u}_r], \\
\mathbf{G} &\stackrel{\text{def}}{=} [\mathbf{g}_1, \ldots, \mathbf{g}_{n-r}] = [\mathbf{u}_{r+1}, \ldots, \mathbf{u}_n], \\
\boldsymbol{\Sigma}_s &= \mathbf{Diag}(\sigma_1^2 + \sigma_w^2, \ldots, \sigma_r^2 + \sigma_w^2), \\
\boldsymbol{\Sigma}_n &= \mathbf{Diag}(\sigma_w^2, \ldots, \sigma_w^2).
\end{aligned}$$

Definition 8.7 Let $\mathbf{X} = \mathbf{U}\boldsymbol{\Sigma}\mathbf{V}^H$ be the SVD of the observed data matrix $\mathbf{X}$ with $\mathbf{U} = [\mathbf{S}, \mathbf{G}]$. If $\mathbf{S}$ is the eigenvector matrix corresponding to the r principal eigenvalues $\lambda_1, \ldots, \lambda_r$ of $\mathbf{X}$, then $\text{Span}(\mathbf{S}) = \text{Span}\{\mathbf{u}_1, \ldots, \mathbf{u}_r\}$ is called the *signal subspace* of the observation data space $\text{Span}(\mathbf{X})$ and the column space $\text{Span}(\mathbf{G}) = \text{Span}\{\mathbf{u}_{r+1}, \ldots, \mathbf{u}_n\}$ is known as the *noise subspace* of the observation data space.

Let us analyze the geometric significance of the signal and noise subspaces.

From the characteristics of the construction of subspaces and unitary matrices, we know that the signal and noise subspaces are mutually orthogonal, i.e.,

$$\text{Span}\{\mathbf{s}_1, \ldots, \mathbf{s}_r\} \perp \text{Span}\{\mathbf{g}_1, \ldots, \mathbf{g}_{n-r}\}. \tag{8.3.7}$$

Since $\mathbf{U}$ is a unitary matrix, we have

$$\mathbf{U}\mathbf{U}^H = [\mathbf{S}, \mathbf{G}]\begin{bmatrix}\mathbf{S}^H \\ \mathbf{G}^H\end{bmatrix} = \mathbf{S}\mathbf{S}^H + \mathbf{G}\mathbf{G}^H = \mathbf{I},$$

namely

$$\mathbf{G}\mathbf{G}^H = \mathbf{I} - \mathbf{S}\mathbf{S}^H. \tag{8.3.8}$$

Define the projection matrix onto the signal subspace as

$$\mathbf{P}_s \stackrel{\text{def}}{=} \mathbf{S}\langle\mathbf{S}, \mathbf{S}\rangle^{-1}\mathbf{S}^H = \mathbf{S}\mathbf{S}^H, \tag{8.3.9}$$

since $\langle\mathbf{S}, \mathbf{S}\rangle = \mathbf{S}^H\mathbf{S} = \mathbf{I}$. Then, $\mathbf{P}_s\mathbf{x} = \mathbf{S}\mathbf{S}^H\mathbf{x}$ can be viewed as the projection of the vector $\mathbf{x}$ onto the signal subspace, so $\mathbf{S}\mathbf{S}^H$ is often spoken of as the signal subspace.

Moreover, $(\mathbf{I} - \mathbf{P}_s)\mathbf{x}$ represents the projection of the vector $\mathbf{x}$ that is orthogonal to the signal subspace. From $\langle\mathbf{G}, \mathbf{G}\rangle = \mathbf{G}^H\mathbf{G} = \mathbf{I}$ we get the projection matrix onto the noise subspace $\mathbf{P}_n = \mathbf{G}\langle\mathbf{G}, \mathbf{G}\rangle^{-1}\mathbf{G}^H = \mathbf{G}\mathbf{G}^H$. Therefore we often say that

$$\mathbf{P}_n = \mathbf{G}\mathbf{G}^H = \mathbf{I} - \mathbf{S}\mathbf{S}^H = \mathbf{I} - \mathbf{P}_s \tag{8.3.10}$$

is the noise subspace.

Signal analysis methods using only the signal subspace $\mathbf{S}\mathbf{S}^H$ or the noise subspace $\mathbf{G}\mathbf{G}^H$ are called the *signal subspace method* and the *noise subspace method*, respectively. In pattern recognition, the signal subspace method is known as the principal component analysis (PCA) method while the noise subspace method is called the minor component analysis (MCA) method.

Subspace applications have the following characteristics [522].

(1) The signal subspace method and the noise subspace method need only a few singular vectors or eigenvectors. If the number of the large singular values (or eigenvalues) of $\mathbf{A}_{m\times n}$ is less than the number of its small singular values (or eigenvalues) then one should use the signal subspace with the smaller dimension; otherwise use the noise subspace.

(2) In many applications, it is not necessary to know singular values or eigenvalues; it is sufficient to know the rank of the matrix and the singular vectors or eigenvectors.

(3) In most cases, it is not necessary to know the singular vectors or eigenvectors exactly; it is sufficient to know the basis vectors spanning the signal subspace or noise subspace.

(4) Conversion between the signal subspace $\mathbf{SS}^H$ and the noise subspace $\mathbf{GG}^H$ can be made using $\mathbf{GG}^H = \mathbf{I} - \mathbf{SS}^H$.

In the following, we present two typical applications of subspace methods.

8.3.2 Multiple Signal Classification (MUSIC)

Let $\mathbf{x}(t)$ be a data vector observed at time snapshot t. In array signal processing and spatial spectrum estimation, the vector $\mathbf{x}(t) = [x_1(t), x_2(t), \ldots, x_n(t)]^T$ consists of the observation data of n array elements (antennas or sensors). In time-domain spectrum estimation, $\mathbf{x}(t) = [x(t), x(t-1), \ldots, x(t-n-1)]^T$ consists of n observation data samples.

Assume that the data vector $\mathbf{x}(t)$ is an observation data vector of an array with n elements illuminated by r narrowband signals that form a superposition of incoherent complex harmonics:

$$\mathbf{x}(t) = \sum_{i=1}^{r} s_i(t)\mathbf{a}(\omega_i) + \mathbf{v}(t) = \mathbf{A}\mathbf{s}(t) + \mathbf{v}(t), \tag{8.3.11}$$

where $\mathbf{A} = [\mathbf{a}(\omega_1), \ldots, \mathbf{a}(\omega_r)]$ is an $n \times r$ array response matrix and $\mathbf{a}(\omega_i) = [1, \mathrm{e}^{\mathrm{j}\omega_i}, \cdots, \mathrm{e}^{\mathrm{j}(n-1)\omega_i}]^T$ is a direction vector or frequency vector; $\mathbf{s}(t) = [s_1(t), \ldots, s_r(t)]^T$ is the random signal vector with zero mean vector and autocorrelation matrix $\mathbf{R}_s = E\{\mathbf{s}(t)\mathbf{s}^H(t)\}$; $\mathbf{v}(t) = [v_1(t), \ldots, v_n(t)]^T$ is the additive noise vector whose components are Gaussian white noises, each with the zero mean and the same variance σ^2. In harmonic retrieval, the parameters ω_i are the frequencies of complex harmonics; in array signal processing, ω_i is a spatial parameter equal to $2\pi(d/\lambda)\sin\theta_i$, where d is the distance between the adjacent array elements (assuming the array elements are uniformly arranged on a straight line), λ is the wavelength and θ_i denotes the incident direction of the ith narrowband signal arriving at the array element, referred to as the direction of arrival (DOA).

The question now is how to use N snapshots of the observation data vector $\mathbf{x}(t)$ $(t = 1, \ldots, N)$ to estimate the r signal parameters ω_i.

Suppose that the noise vector $\mathbf{v}(t)$ and the signal vector $\mathbf{s}(t)$ are statistically uncorrelated, and let the EVD of the autocorrelation matrix $\mathbf{R}_{xx} = E\{\mathbf{x}(t)\mathbf{x}^H(t)\}$ be given by

$$\begin{aligned}\mathbf{R}_{xx} &= \mathbf{A}\mathbf{P}_{ss}\mathbf{A}^H + \sigma^2\mathbf{I} = \mathbf{U}\boldsymbol{\Sigma}\mathbf{U}^H \\ &= [\mathbf{S}, \mathbf{G}]\begin{bmatrix}\boldsymbol{\Sigma} & \mathbf{O} \\ \mathbf{O} & \sigma^2\mathbf{I}_{n-r}\end{bmatrix}\begin{bmatrix}\mathbf{S}^H \\ \mathbf{G}^H\end{bmatrix},\end{aligned} \tag{8.3.12}$$

where $\mathbf{P}_{ss} = E\{\mathbf{s}(t)\mathbf{s}^H(t)\}$ and $\boldsymbol{\Sigma}$ contains r large or principal eigenvalues that are significantly larger than σ^2.

Consider on the one hand

$$\begin{aligned}\mathbf{R}_{xx}\mathbf{G} &= [\mathbf{S},\,\mathbf{G}]\begin{bmatrix}\boldsymbol{\Sigma} & \mathbf{O}\\ \mathbf{O} & \sigma^2\mathbf{I}_{n-r}\end{bmatrix}\begin{bmatrix}\mathbf{S}^H\\ \mathbf{G}^H\end{bmatrix}\mathbf{G}\\ &= [\mathbf{S},\,\mathbf{G}]\begin{bmatrix}\boldsymbol{\Sigma} & \mathbf{O}\\ \mathbf{O} & \sigma^2\mathbf{I}_{n-r}\end{bmatrix}\begin{bmatrix}\mathbf{O}\\ \mathbf{I}\end{bmatrix} = \sigma^2\mathbf{G}. \end{aligned}\tag{8.3.13}$$

On the other hand, from $\mathbf{R}_{xx} = \mathbf{A}\mathbf{P}_{ss}\mathbf{A}^H + \sigma^2\mathbf{I}$ we have $\mathbf{R}_{xx}\mathbf{G} = \mathbf{A}\mathbf{P}_{ss}\mathbf{A}^H\mathbf{G} + \sigma^2\mathbf{G}$. Using the result of (8.3.13), we immediately have $\mathbf{A}\mathbf{P}_{ss}\mathbf{A}^H\mathbf{G} = \mathbf{O}$, and hence

$$\mathbf{G}^H\mathbf{A}\mathbf{P}_{ss}\mathbf{A}^H\mathbf{G} = \mathbf{O}. \tag{8.3.14}$$

It is well known that, when $\mathbf{Q}$ is nonsingular, $\mathbf{t}^H\mathbf{Q}\mathbf{t} = 0$ if and only if $\mathbf{t} = \mathbf{0}$, so the necessary and sufficient condition for (8.3.14) to hold is

$$\mathbf{A}^H\mathbf{G} = \mathbf{O}, \tag{8.3.15}$$

as $\mathbf{P}_{ss} = E\{\mathbf{s}(t)\mathbf{s}^H(t)\}$ is nonsingular. Equation (8.3.15) can be rewritten as

$$\mathbf{a}^H(\omega)\mathbf{G} = \mathbf{0}^T, \quad \omega = \omega_1, \ldots, \omega_p. \tag{8.3.16}$$

Clearly, when $\omega \neq \omega_1, \ldots, \omega_p$, we have $\mathbf{a}^H(\omega)\mathbf{G} \neq \mathbf{0}^T$.

Rewriting (8.3.16) in a scalar form, one can define a function similar to the spectrum:

$$P(\omega) = \frac{1}{\|\mathbf{G}^H\mathbf{a}(\omega)\|_2^2} = \frac{1}{\mathbf{a}^H(\omega)\mathbf{G}\mathbf{G}^H\mathbf{a}(\omega)}. \tag{8.3.17}$$

The p peak values $P(\omega_1), \ldots, P(\omega_p)$ give the DOA of the p signals $\omega_i = 2\pi(d/\lambda)\sin\theta_i$, $i = 1, \ldots, p$.

The function $P(\omega)$ defined in (8.3.17) describes the distribution of the spatial parameters, so is called the spatial spectrum. The spatial-spectrum-based signal classification method is called the *multiple signal classification* (MUSIC) method, which was proposed independently by Schmidt [426] and Biemvenu and Kopp [47] in 1979. Later, in 1986 Schmidt [427] republished his paper.

Substituting (8.3.10) into (8.3.17) yields

$$P(\omega) = \frac{1}{\mathbf{a}^H(\omega)(\mathbf{I} - \mathbf{S}\mathbf{S}^H)\mathbf{a}(\omega)}. \tag{8.3.18}$$

Since $\mathbf{G}\mathbf{G}^H$ and $\mathbf{S}\mathbf{S}^H$ represent the signal subspace and the noise subspace, respectively, Equations (8.3.17) and (8.3.18) can be viewed as the noise subspace method and the signal subspace method, respectively.

In practical applications, ω is usually divided into hundreds of equal interval units,

$$\omega_i = 2\pi i\Delta f. \tag{8.3.19}$$

For example, for $\Delta f = 0.5/500 = 0.001$, $P(\omega)$ is computed for every ω_i to determine the ω values corresponding to all the spectrum peaks. Hence, the MUSIC algorithm needs to make a global search in the frequency axis.

8.3.3 Subspace Whitening

Let $\mathbf{a}$ be an $m \times 1$ random vector with zero mean and covariance matrix $\mathbf{C}_a = E\{\mathbf{a}\mathbf{a}^H\}$. If the $m \times m$ covariance matrix $\mathbf{C}_a$ is nonsingular, and is not equal to the identity matrix, then the random vector $\mathbf{a}$ is said to be a colored or nonwhite random vector.

Assume that the EVD of the covariance matrix is $\mathbf{C}_a = E\{\mathbf{a}\mathbf{a}^H\} = \mathbf{V}\mathbf{D}\mathbf{V}^H$ and set

$$\mathbf{W} = \mathbf{V}\mathbf{D}^{-1/2}\mathbf{V}^H = \mathbf{C}_a^{-1/2}; \tag{8.3.20}$$

then the covariance matrix of the linear transform

$$\mathbf{b} = \mathbf{W}\mathbf{a} = \mathbf{C}_a^{-1/2}\mathbf{a} \tag{8.3.21}$$

is equal to the identity matrix, i.e.,

$$\mathbf{C}_b = E\{\mathbf{b}\mathbf{b}^H\} = \mathbf{W}\mathbf{C}_a\mathbf{W}^H = \mathbf{C}_a^{-1/2}\mathbf{V}\mathbf{D}\mathbf{V}^H\left(\mathbf{C}_a^{-1/2}\right)^H = \mathbf{I}, \tag{8.3.22}$$

as $\mathbf{C}_a^{-1/2} = \mathbf{V}\mathbf{D}^{-1/2}\mathbf{V}^H$ is a Hermitian matrix. The above equation shows that, after the linear transformation $\mathbf{W}\mathbf{a}$, the original colored random vector becomes a white random vector. The linear transformation matrix $\mathbf{W} = \mathbf{C}_a^{-1/2}$ is called the *whitening matrix* of the random vector $\mathbf{a}$.

However, if the $m \times m$ covariance matrix $\mathbf{C}_a$ is singular or rank-deficient, for example $\text{rank}(\mathbf{C}_a) = n < m$, then there is no whitening matrix $\mathbf{W}$ such that $\mathbf{W}\mathbf{C}_a\mathbf{W}^H = \mathbf{I}$. In this case, one should consider whitening a random vector $\mathbf{a}$ in the range space $V = \text{Range}(\mathbf{C}_a) = \text{Col}(\mathbf{C}_a)$. This whitening is known as *subspace whitening* and was proposed by Eldar and Oppenheim in 2003 [145].

Let the EVD of the rank-deficient covariance matrix $\mathbf{C}_a$ be

$$\mathbf{C}_a = [\mathbf{V}_1, \mathbf{V}_2] \begin{bmatrix} \mathbf{D}_{n\times n} & \mathbf{O}_{n\times(m-n)} \\ \mathbf{O}_{(m-n)\times n} & \mathbf{O}_{(m-n)\times(m-n)} \end{bmatrix} \begin{bmatrix} \mathbf{V}_1^H \\ \mathbf{V}_2^H \end{bmatrix}, \tag{8.3.23}$$

and let

$$\mathbf{W} = \mathbf{V}_1\mathbf{D}^{-1/2}\mathbf{V}_1^H. \tag{8.3.24}$$

Then the covariance matrix of the linear transform $\mathbf{b} = \mathbf{W}\mathbf{a} = \mathbf{V}_1\mathbf{D}^{-1/2}\mathbf{V}_1^H\mathbf{a}$ is given by

$$\begin{aligned}
\mathbf{C}_b &= E\{\mathbf{b}\mathbf{b}^H\} = \mathbf{W}\mathbf{C}_a\mathbf{W}^H \\
&= \mathbf{V}_1\mathbf{D}^{-1/2}\mathbf{V}_1^H[\mathbf{V}_1, \mathbf{V}_2]\begin{bmatrix}\mathbf{D} & \mathbf{O}\\ \mathbf{O} & \mathbf{O}\end{bmatrix}\begin{bmatrix}\mathbf{V}_1^H\\ \mathbf{V}_2^H\end{bmatrix}\mathbf{V}_1\mathbf{D}^{-1/2}\mathbf{V}_1^H \\
&= [\mathbf{V}_1\mathbf{D}^{-1/2}, \mathbf{O}]\begin{bmatrix}\mathbf{D} & \mathbf{O}\\ \mathbf{O} & \mathbf{O}\end{bmatrix}\begin{bmatrix}\mathbf{D}^{-1/2}\mathbf{V}_1^H\\ \mathbf{O}\end{bmatrix} \\
&= \begin{bmatrix}\mathbf{I}_n & \mathbf{O}\\ \mathbf{O} & \mathbf{O}\end{bmatrix}.
\end{aligned}$$

That is, $\mathbf{b} = \mathbf{W}\mathbf{a}$ is a white random vector in the subspace $\text{Range}(\mathbf{C}_a)$. Therefore, $\mathbf{W} = \mathbf{V}_1\mathbf{D}^{-1/2}\mathbf{V}_1^H$ is called the *subspace whitening matrix.*

8.4 Grassmann Manifold and Stiefel Manifold

Consider the minimization of the objective function $J(\mathbf{W})$, where $\mathbf{W}$ is an $n \times r$ matrix. The common constraints on $\mathbf{W}$ are of two types, as follows.

(1) *Orthogonality constraint* $\mathbf{W}$ is required to satisfy an orthogonality condition, either $\mathbf{W}^H\mathbf{W} = \mathbf{I}_r$ $(n \geq r)$ or $\mathbf{W}\mathbf{W}^H = \mathbf{I}_n$ $(n < r)$. The matrix $\mathbf{W}$ satisfying such a condition is called *semi-orthogonal.*

(2) *Homogeneity constraint* It is required that $J(\mathbf{W}) = J(\mathbf{W}\mathbf{Q})$, where $\mathbf{Q}$ is an $r \times r$ orthogonal matrix.

There are two classes of optimization problems: in one class both the orthogonality constraint and the homogeneity constraint are adapted; the other uses only the orthogonality constraint.

8.4.1 Equivalent Subspaces

Definition 8.8 (Linear manifold) Let H be a subspace in the space V, and let $\mathcal{L}$ denote the set of linear combinations of finite elements in H, i.e.,

$$\mathcal{L} = \left\{\xi \,\middle|\, \xi = \sum_{i=1}^{n} a_i\eta_i,\ \eta_i \in H\right\}.$$

Then $\mathcal{L}$ is known as the *linear manifold* spanned by H.

Two matrices are said to be equivalent, if the subspaces spanned by their column vectors are the same. In other words, sets of equivalent matrices have the same column spaces, i.e., such a subspace is invariant with respect to any choice of basis. Therefore, this class of subspaces is known as *invariant subspaces.*

In many engineering applications, the projection matrix $\mathbf{P}_H$ onto a subspace H

and the orthogonal projection matrix $\mathbf{P}_H^{\perp} = \mathbf{I} - \mathbf{P}_H$ play a key role. In particular, invariant subspaces can be described using a projection matrix.

1. Tall-Skinny Matrix

When $n > r$, the $n \times r$ matrix $\mathbf{W}$ is called a *tall-skinny matrix*. Let $\mathbf{M}$ be an $r \times r$ nonsingular matrix, and $\mathbf{W}_1 = \mathbf{W}_2\mathbf{M}$ be a tall-skinny matrix. If $H = \text{Span}(\mathbf{W}_1) = \text{Col}(\mathbf{W}_1)$ and $S = \text{Col}(\mathbf{W}_2)$ then the projection matrix onto the subspace H is given by

$$\begin{aligned}\mathbf{P}_H &= \mathbf{W}_1(\mathbf{W}_1^H\mathbf{W}_1)^\dagger\mathbf{W}_1^H \\ &= \mathbf{W}_2\mathbf{M}(\mathbf{M}^H\mathbf{W}_2^H\mathbf{W}_2\mathbf{M})^\dagger\mathbf{M}^H\mathbf{W}_2^H \\ &= \mathbf{W}_2(\mathbf{W}_2^H\mathbf{W}_2)^\dagger\mathbf{W}_2^H \\ &= \mathbf{P}_S.\end{aligned}$$

Because the projection matrix $\mathbf{P}_H$ onto the subspace H is equal to the projection matrix $\mathbf{P}_S$, one has

$$\mathbf{P}_H\mathbf{x} = \mathbf{P}_S\mathbf{x}, \quad \forall \mathbf{x} \in \mathbb{C}^n. \tag{8.4.1}$$

The column subspaces H and S such that $\mathbf{P}_H\mathbf{x} = \mathbf{P}_S\mathbf{x}$ are called *equivalent subspaces.* In other words, if $\mathbf{M}$ is an $r\times r$ nonsingular matrix, then the two tall-skinny matrices $\mathbf{W}$ and $\mathbf{WM}$ span the same or equivalent subspaces.

2. Short Matrix

When $n < r$, the $n \times r$ matrix $\mathbf{W}$ is called a *short matrix.* Let $\mathbf{M}$ be an $n \times n$ nonsingular matrix, and $\mathbf{W}_1 = \mathbf{M}\mathbf{W}_2$ be a short matrix. If $H = \text{Row}(\mathbf{W}_1)$ and $S = \text{Row}(\mathbf{W}_2)$ then the projection matrix onto the subspace H is given by

$$\begin{aligned}\mathbf{P}_H &= \mathbf{W}_1^H(\mathbf{W}_1\mathbf{W}_1^H)^\dagger\mathbf{W}_1 \\ &= \mathbf{W}_2^H\mathbf{M}^H(\mathbf{M}\mathbf{W}_2\mathbf{W}_2^H\mathbf{M}^H)^\dagger\mathbf{M}\mathbf{W}_2 \\ &= \mathbf{W}_2^H(\mathbf{W}_2\mathbf{W}_2^H)^\dagger\mathbf{W}_2 \\ &= \mathbf{P}_S.\end{aligned}$$

Hence, for any row vector $\mathbf{x} \in \mathbb{C}^{1\times r}$, one has $\mathbf{x}\mathbf{P}_H = \mathbf{x}\mathbf{P}_S$. This shows that the row subspaces $H = \text{Row}(\mathbf{W}_1) = \text{Row}(\mathbf{M}\mathbf{W}_2)$ and $S = \text{Row}(\mathbf{W}_2)$ are equivalent subspaces.

8.4.2 Grassmann Manifold

In optimization problems, a full column (or row) rank matrix $\mathbf{M}$ is usually constrained to be semi-orthogonal.

1. Tall-Skinny Semi-Orthogonal Matrix

If a tall-skinny matrix $\mathbf{W}\in\mathbb{C}^{n\times r}$ satisfies the semi-orthogonality constraint $\mathbf{W}^H\mathbf{W}=$

$\mathbf{I}_r$, then $\mathbf{W}$ is known as a *tall-skinny semi-orthogonal matrix*. For any a tall-skinny semi-orthogonal matrix $\mathbf{W}$, its column space $H = \mathrm{Col}(\mathbf{W})$ can be equivalently represented by the projection matrix

$$\mathbf{P}_H = \mathbf{W}(\mathbf{W}^H\mathbf{W})^{-1}\mathbf{W}^H = \mathbf{W}\mathbf{W}^H. \tag{8.4.2}$$

For an unconstrained optimization problem $\min J(\mathbf{W})$ with tall-skinny matrix $\mathbf{W}$, its solution is a single matrix $\mathbf{W}$. However, for an optimization problem subject to the semi-orthogonality constraint $\mathbf{W}^H\mathbf{W} = \mathbf{I}_r$ and the homogeneity constraint $J(\mathbf{W}) = J(\mathbf{WQ})$, where $\mathbf{Q}$ is an $r \times r$ orthogonal matrix, i.e.,

$$\min J(\mathbf{W}) \quad \text{subject to} \quad \mathbf{W}^H\mathbf{W} = \mathbf{I}_r,\ J(\mathbf{W}) = J(\mathbf{WQ}), \tag{8.4.3}$$

its solution is any member of the set of tall-skinny semi-orthogonal matrices defined by

$$Gr(n,r) = \{\mathbf{W} \in \mathbb{C}^{n\times r} \,|\, \mathbf{W}^H\mathbf{W} = \mathbf{I}_r,\ \mathbf{W}_1\mathbf{W}_1^H = \mathbf{W}_2\mathbf{W}_2^H\}. \tag{8.4.4}$$

This set is known as the *Grassmann manifold* of the tall-skinny semi-orthogonal matrices.

The Grassmann manifold was proposed by Grassmann in 1848, but it only became widely known in the 1980s [2]. The original definition of the Grassmann manifold can be found in [184, Chapter 3, Section 1].

2. Short Semi-Orthogonal Matrix

Let $\mathbf{W} \in \mathbb{C}^{n\times r}$ with $n < r$ be a short matrix. If the short matrix $\mathbf{W}$ meets the semi-orthogonality constraint $\mathbf{W}\mathbf{W}^H = \mathbf{I}_n$ then $\mathbf{W}$ is called a *short semi-orthogonal matrix*.

For the unconstrained optimization problem $\min J(\mathbf{W})$ with short matrix $\mathbf{W}$, its solution is a single short matrix $\mathbf{W}$. However, for an optimization problem subject to the semi-orthogonality constraint $\mathbf{W}\mathbf{W}^H = \mathbf{I}_n$ and the homogeneity constraint $J(\mathbf{W}) = J(\mathbf{QW})$, where $\mathbf{Q}$ is an $n \times n$ orthogonal matrix, i.e.,

$$\min J(\mathbf{W}) \quad \text{subject to} \quad \mathbf{W}\mathbf{W}^H = \mathbf{I}_n,\ J(\mathbf{W}) = J(\mathbf{QW}), \tag{8.4.5}$$

its solution is any member of the set of short semi-orthogonal matrices given by

$$Gr(n,r) = \{\mathbf{W} \in \mathbb{C}^{n\times r} \,|\, \mathbf{W}\mathbf{W}^H = \mathbf{I}_n,\ \mathbf{W}_1^H\mathbf{W}_1 = \mathbf{W}_2^H\mathbf{W}_2\}, \tag{8.4.6}$$

which is the Grassmann manifold of short semi-orthogonal matrices.

Clearly, both the semi-orthogonality constraint and the homogeneous constraint greatly ease the solution of an optimization problem.

The Grassmann manifold has important applications in sciences and engineering. On the geometric characteristics of the Grassmann manifold, Edelman *et al.* [142] in 1998 gave a systematic interpretation.

8.4.3 Stiefel Manifold

Next we consider an optimization problem with only the semi-orthogonality constraint:

$$\mathbf{W} = \underset{\mathbf{W}\in\mathbb{C}^{n\times r}, n>r}{\text{argmin}} \; J(\mathbf{W}) \quad \text{subject to} \quad \mathbf{W}^H\mathbf{W} = \mathbf{I}_r, \tag{8.4.7}$$

or

$$\mathbf{W} = \underset{\mathbf{W}\in\mathbb{C}^{n\times r}, n<r}{\text{argmin}} \; J(\mathbf{W}) \quad \text{subject to} \quad \mathbf{W}\mathbf{W}^H = \mathbf{I}_n. \tag{8.4.8}$$

The solution of this optimization problem is a set of $n\times r$ semi-orthogonal matrices. The set of all $n \times r$ semi-orthogonal matrices is known as the *Stiefel manifold*, denoted $St(n, r)$, namely

$$St(n,r) = \left\{\mathbf{W} \in \mathbb{C}^{n\times r} \,\middle|\, \mathbf{W}^H\mathbf{W} = \mathbf{I}_r, n > r\right\} \tag{8.4.9}$$

or

$$St(n,r) = \left\{\mathbf{W} \in \mathbb{C}^{n\times r} \,\middle|\, \mathbf{W}\mathbf{W}^H = \mathbf{I}_n, n < r\right\}. \tag{8.4.10}$$

This manifold was proposed by Stiefel [456] in the 1930s in a topological study (Stiefel together with Hestenes [210] proposed the famous conjugate gradient algorithm in 1952).

There are the following relationships between the Grassmann manifold and the Stiefel manifold.

(1) The Stiefel manifold $St(n, r)$ is a set of $n\times r$ semi-orthogonal matrices $\mathbf{W}^H\mathbf{W} = \mathbf{I}_r (n > r)$ or $\mathbf{W}\mathbf{W}^H = \mathbf{I}_n (n < r)$. Any point on the Stiefel manifold $St(n, r)$ represents a matrix in the set of $n \times r$ semi-orthogonal matrices.
(2) The Grassmann manifold $Gr(n, r)$ consists of the matrices spanning the same column space $\mathbf{W}_1\mathbf{W}_1^H = \mathbf{W}_2\mathbf{W}_2^H (n > r)$ or the same row space $\mathbf{W}_1^H\mathbf{W}_1 = \mathbf{W}_2^H\mathbf{W}_2 (n < r)$ in the Stiefel manifold $St(n, r)$.

The set of all the $r\times r$ orthogonal matrices $\mathbf{Q}$ forms an *orthogonal group*, denoted O_r, and is defined as

$$O_r = \left\{\mathbf{Q}_r \in \mathbb{C}^{r\times r} \,\middle|\, \mathbf{Q}_r^H\mathbf{Q}_r = \mathbf{Q}_r\mathbf{Q}_r^H = \mathbf{I}_r\right\}. \tag{8.4.11}$$

The orthogonal group, the Grassmann manifold and the Stiefel manifold are three classes of subspace manifolds.

To discuss the relations among these three classes of subspace manifolds, let $\mathbf{W}$ be a point on the Stiefel manifold, i.e., $\mathbf{W} \in St(n, r)$ is an $n \times r$ semi-orthogonal matrix. Collecting the $n \times (n - r)$ matrices $\mathbf{W}_\perp \in St(n, n - r)$ satisfying the orthogonality conditions $\mathbf{W}_\perp^H\mathbf{W}_\perp = \mathbf{I}_{n-r}$ and $\mathbf{W}_\perp^H\mathbf{W} = \mathbf{O}_{(n-r)\times r}$, then $[\mathbf{W}, \mathbf{W}_\perp]$ constitutes an orthogonal group O_n. If we let $\mathbf{Q}$ be any $(n-r)\times(n-r)$ orthogonal matrix such that $\mathbf{W}_\perp\mathbf{Q} = \mathbf{O}$ then the set of matrices $\mathbf{Q}$ forms another orthogonal group, O_{n-r}. Notice that matrix product $[\mathbf{W}, \mathbf{W}_\perp]\begin{bmatrix}\mathbf{I}\\ \mathbf{Q}\end{bmatrix} = \mathbf{W}$. This shows that the semi-orthogonal matrix $\mathbf{W}_{n\times r}$ can be determined by the $n \times n$ orthogonal group O_n and the $(n-r)\times(n-r)$ orthogonal group O_{n-r}. From the matrix multiplication in the above equation it can be seen that a point on the Stiefel manifold $St(n,r)$ can be represented by the quotient of two orthogonal groups O_n/O_{n-r}:

$$St(n,r) = O_n/O_{n-r}. \tag{8.4.12}$$

Moreover, using the semi-orthogonal matrix $\mathbf{W}_{n\times r}$ to represent a point on the Stiefel manifold, the matrices $\mathbf{U}_s$ such that $\mathbf{W} = \mathbf{U}_s\mathbf{Q}$ (where $\mathbf{Q}$ is any $r \times r$ orthogonal matrix) or $\mathbf{U}_s = \mathbf{W}\mathbf{Q}^{-1}$ constitute a point (equivalent to a subspace class) on the Grassmann manifold $Gr(n,r)$. Therefore, if the inverse matrix operation is regarded as matrix division then the Grassmann manifold $Gr(n,r)$ can be represented as the quotient of the Stiefel manifold $St(n,r)$ and the orthogonal matrix $\mathbf{Q}$, i.e.,

$$Gr(n,r) = St(n,r)/O_r, \tag{8.4.13}$$

where O_r denotes an $r \times r$ orthogonal group. If we substitute (8.4.12) into (8.4.13) then we see that the Grassmann manifold can also be represented as the quotient of two orthogonal groups:

$$Gr(n,r) = O_n/(O_r \times O_{n-r}). \tag{8.4.14}$$

Table 8.1 summarizes the relationships among the Grassmann manifold, the Stiefel manifold and the orthogonal group.

Table 8.1 Subspace manifold representations [142]

Subspace manifold	Symbol	Matrix form	Quotient
Orthogonal group	O_n	$n \times n$ matrix	—
Stiefel manifold	$St(n,r)$	$n \times r$ matrix	O_n/O_{n-r}
Grassmann manifold	$Gr(n,r)$	—	$\begin{cases} St(n,r)/O_r \text{ or} \\ O_n/(O_r \times O_{n-r}) \end{cases}$

8.5 Projection Approximation Subspace Tracking (PAST)

The tracking and updating of eigenspaces are mainly applied in real-time signal processing, so they should be fast. The following factors need to be taken into account in the design of a fast subspace tracking algorithm.

(1) The subspace at time n can be updated from the subspace at time $n-1$.

(2) The change in the covariance matrix from time $n-1$ to time n should be of as low a rank change as possible (best are rank-1 or rank-2 changes).

(3) Only the low-dimensional subspaces need to be tracked.

Eigenspace tracking and updating methods fall into four classes, as follows.

1. *Orthogonal basis tracking* This method uses only the orthogonal basis without using the eigenvectors themselves, which can simplify the adaptive tracking problem of eigenspaces.
2. *Rank-1 updating* The covariance matrix of a nonstationary signal at time k can be regarded as the sum of the covariance matrix at time $k-1$ and another rank-1 matrix. Therefore, the tracking of the EVD of the covariance matrix is closely related to the rank-1 updating.
3. *Projection approximation* The determination of an eigenspace is solved by unconstrained optimization; this method is called projection approximation subspace tracking [522].
4. *Lanczos subspace tracking* The subspace tracking of a time-varying data matrix can be performed using the Lanczos-type iteration and random approximation [166]. Xu *et al.* [516], [518] proposed the tri-Lanczos and the bi-Lanczos subspace tracking algorithms: the tri-Lanczos algorithm tracks the EVD of a covariance matrix, and the bi-Lanczos algorithm tracks the SVD of a data matrix, and the numbers of principal eigenvalues and principal singular values can be detected and estimated.

This section focuses upon method 3, the projection approximation subspace tracking method.

8.5.1 Basic PAST Theory

The minimization of an objective function $J(\mathbf{W})$ with orthogonality constraint $\mathbf{W}_{n\times r}^H\mathbf{W}_{n\times r}=\mathbf{I}_r$ and homogeneity constraint $J(\mathbf{W})=J(\mathbf{WQ}_{r\times r})$ can be made equivalent to an unconstrained optimization problem.

Let $\mathbf{C}=E\{\mathbf{xx}^H\}$ denote the autocorrelation matrix of an $n\times 1$ random vector, and let the objective function be

$$\begin{aligned} J(\mathbf{W}) &= E\Big\{\|\mathbf{x}-\mathbf{WW}^H\mathbf{x}\|_2^2\Big\} \\ &= E\Big\{\big(\mathbf{x}-\mathbf{WW}^H\mathbf{x}\big)^H\big(\mathbf{x}-\mathbf{WW}^H\mathbf{x}\big)\Big\} \\ &= E\{\mathbf{x}^H\mathbf{x}\}-2E\{\mathbf{x}^H\mathbf{WW}^H\mathbf{x}\}+E\{\mathbf{x}^H\mathbf{WW}^H\mathbf{WW}^H\mathbf{x}\}. \end{aligned} \tag{8.5.1}$$

Because

$$E\{\mathbf{x}^H\mathbf{x}\} = \sum_{i=1}^{n} E\{|x_i|^2\} = \mathrm{tr}\left(E\{\mathbf{x}\mathbf{x}^H\}\right) = \mathrm{tr}(\mathbf{C}),$$

$$E\{\mathbf{x}^H\mathbf{W}\mathbf{W}^H\mathbf{x}\} = \mathrm{tr}\Big(E\{\mathbf{W}^H\mathbf{x}\mathbf{x}^H\mathbf{W}\}\Big) = \mathrm{tr}(\mathbf{W}^H\mathbf{C}\mathbf{W}),$$

$$\begin{aligned} E\{\mathbf{x}^H\mathbf{W}\mathbf{W}^H\mathbf{W}\mathbf{W}^H\mathbf{x}\} &= \mathrm{tr}\left(E\{\mathbf{W}^H\mathbf{x}\mathbf{x}^H\mathbf{W}\mathbf{W}^H\mathbf{W}\}\right) \\ &= \mathrm{tr}(\mathbf{W}^H\mathbf{C}\mathbf{W}\mathbf{W}^H\mathbf{W}), \end{aligned}$$

the objective function can be represented as the trace form

$$J(\mathbf{W}) = \mathrm{tr}(\mathbf{C}) - 2\,\mathrm{tr}(\mathbf{W}^H\mathbf{C}\mathbf{W}) + \mathrm{tr}(\mathbf{W}^H\mathbf{C}\mathbf{W}\mathbf{W}^H\mathbf{W}), \tag{8.5.2}$$

where $\mathbf{W}$ is an $n \times r$ matrix whose rank is assumed to be r.

Regarding the minimization problem $\min J(\mathbf{W})$, important questions are:

- Is there a global minimizer $\mathbf{W}$ of $J(\mathbf{W})$?
- What is the relationship between the minimizer $\mathbf{W}$ and the signal subspace of the autocorrelation matrix $\mathbf{C}$?
- Are there other local minima of $J(\mathbf{W})$?

The following two theorems provide answers to the above three questions.

Theorem 8.6 [522] *The matrix $\mathbf{W}$ is a stationary point of $J(\mathbf{W})$ if and only if $\mathbf{W} = \mathbf{U}_r\mathbf{Q}$, where $\mathbf{U}_r \in \mathbb{C}^{n\times r}$ consists of r different eigenvectors of the autocorrelation matrix $\mathbf{C}$, and $\mathbf{Q} \in \mathbb{C}^{r\times r}$ is any unitary matrix. At every stationary point, the value of the objective function $J(\mathbf{W})$ is equal to the sum of the eigenvalues whose corresponding eigenvectors are not in $\mathbf{U}_r$.*

Theorem 8.7 [522] *All its stationary points are saddle points of the objective function $J(\mathbf{W})$ unless $\mathbf{U}_r$ consists of r principal eigenvectors of the autocorrelation matrix $\mathbf{C}$. In this particular case $J(\mathbf{W})$ reaches a global minimum.*

Theorems 8.6 and 8.7 show the following facts.

1. In the unconstrained minimization of the objective function $J(\mathbf{W})$ in (8.5.1), it is not required that the columns of $\mathbf{W}$ are orthogonal, but the above two theorems show that minimization of the objective function $J(\mathbf{W})$ in (8.5.1) will automatically result in a semi-orthogonal matrix $\mathbf{W}$ such that $\mathbf{W}^H\mathbf{W} = \mathbf{I}$.
2. Theorem 8.7 shows that when the column space of $\mathbf{W}$ is equal to the signal subspace, i.e., $\mathrm{Col}(\mathbf{W}) = \mathrm{Span}(\mathbf{U}_r)$, the objective function $J(\mathbf{W})$ reaches a global minimum, and it has no other local minimum.
3. From the definition formula (8.5.1) it is easy to see that $J(\mathbf{W}) = J(\mathbf{W}\mathbf{Q})$ holds for all $r\times r$ unitary matrices $\mathbf{Q}$, i.e., the objective function automatically satisfies the homogeneity constraint.

4. Because the objective function defined by (8.5.1) automatically satisfies the homogeneity constraint, and its minimization automatically has the result that $\mathbf{W}$ satisfies the orthogonality constraint $\mathbf{W}^H\mathbf{W} = \mathbf{I}$, the solution $\mathbf{W}$ is not uniquely determined but is a point on the Grassmann manifold.
5. The projection matrix $\mathbf{P} = \mathbf{W}(\mathbf{W}^H\mathbf{W})^{-1}\mathbf{W}^H = \mathbf{W}\mathbf{W}^H = \mathbf{U}_r\mathbf{U}_r^H$ is uniquely determined. That is to say, the different solutions $\mathbf{W}$ span the same column space.
6. When $r = 1$, i.e., the objective function is a real function of a vector $\mathbf{w}$, the solution $\mathbf{w}$ of the minimization of $J(\mathbf{w})$ is the eigenvector corresponding to the largest eigenvalue of the autocorrelation matrix $\mathbf{C}$.

Therefore, the minimization of the objective function $J(\mathbf{W})$ with orthogonality constraint and homogeneity constraint becomes an SVD or EVD problem, as follows.

(1) Use the data vector $\mathbf{x}(k)$ to form the data matrix $\mathbf{X} = [\mathbf{x}(1), \ldots, \mathbf{x}(N)]$; then compute the SVD of $\mathbf{X}$ and determine its effective rank r, which yields r principal singular values and the corresponding left singular-vector matrix $\mathbf{U}_r$. The optimal solution of the minimization problem is given by $\mathbf{W} = \mathbf{U}_r$.
(2) Compute the EVD of the autocorrelation matrix $\mathbf{C} = \mathbf{X}\mathbf{X}^H$ to get the eigenvector matrix $\mathbf{U}_r$ corresponding to the r principal eigenvalues. This eigenvector matrix $\mathbf{U}_r$ is the optimal solution of the minimization problem.

However, in practical applications, the autocorrelation matrix $\mathbf{C}$ is usually time-varying, and thus its eigenvalues and eigenvectors are time-varying as well. From Equation (8.5.2) it can be seen that in the time-varying case the matrix differential of the objective function $J(\mathbf{W}(t))$ is given by

$$\begin{aligned}
&\mathrm{d}J(\mathbf{W}(t))\\
&= -2\mathrm{tr}\Big(\mathbf{W}^H(t)\mathbf{C}(t)\mathrm{d}\mathbf{W}(t) + (\mathbf{C}(t)\mathbf{W}(t))^T\mathrm{d}\mathbf{W}^*(t)\Big)\\
&\quad + \mathrm{tr}\Big((\mathbf{W}^H(t)\mathbf{W}(t)\mathbf{W}^H(t)\mathbf{C}(t) + \mathbf{W}(t)\mathbf{C}(t)\mathbf{W}(t)\mathbf{W}^H(t))\,\mathrm{d}\mathbf{W}(t)\\
&\quad + (\mathbf{C}(t)\mathbf{W}(t)\mathbf{W}^H(t)\mathbf{W}(t) + \mathbf{W}(t)\mathbf{W}^H(t)\mathbf{C}(t)\mathbf{W}(t))^T\mathrm{d}\mathbf{W}^*(t)\Big),
\end{aligned}$$

which yields the gradient matrix

$$\begin{aligned}
\nabla_W J(\mathbf{W}(t)) &= -2\mathbf{C}(t)\mathbf{W}(t) + \mathbf{C}(t)\mathbf{W}(t)\mathbf{W}^H(t)\mathbf{W}(t)\\
&\quad + \mathbf{W}(t)\mathbf{W}^H(t)\mathbf{C}(t)\mathbf{W}(t)\\
&= \mathbf{W}(t)\mathbf{W}^H(t)\mathbf{C}(t)\mathbf{W}(t) - \mathbf{C}(t)\mathbf{W}(t),
\end{aligned}$$

where the semi-orthogonality constraint $\mathbf{W}^H(t)\mathbf{W}(t) = \mathbf{I}$ has been used.

Substituting $\mathbf{C}(t) = \mathbf{x}(t)\mathbf{x}^H(t)$ into the above gradient matrix formula, one obtains a gradient descent algorithm for solving the minimization problem $\mathbf{W}(t) =$

$\mathbf{W}(t-1) - \mu\nabla_W J(\mathbf{W}(t))$, as follows:

$$\mathbf{y}(t) = \mathbf{W}^H(t)\mathbf{x}(t) \tag{8.5.3}$$
$$\mathbf{W}(t) = \mathbf{W}(t-1) + \mu(\mathbf{x}(t) - \mathbf{W}(t-1)\mathbf{y}(t))\mathbf{y}^H(t). \tag{8.5.4}$$

However, this gradient descent algorithm for updating $\mathbf{W}(t)$ has a low convergence rate, and its ability to track a time-varying subspace is poor. A better method is the following *recursive least squares* (RLS) *algorithm.*

Define an exponential weighting objective function

$$\begin{aligned} J_1(\mathbf{W}(t)) &= \sum_{i=1}^{t} \beta^{t-i} \left\|\mathbf{x}(i) - \mathbf{W}(t)\mathbf{W}^H(t)\mathbf{x}(i)\right\|_2^2 \\ &= \sum_{i=1}^{t} \beta^{t-i} \left\|\mathbf{x}(i) - \mathbf{W}(t)\mathbf{y}(i)\right\|_2^2, \end{aligned} \tag{8.5.5}$$

where $0 < \beta \leq 1$ is a *forgetting factor* and $\mathbf{y}(i) = \mathbf{W}^H(t)\mathbf{x}(i)$.

By the adaptive filtering theorem, the optimal solution of the minimization problem $\min J_1(\mathbf{W})$ is the Wiener filter, given by

$$\mathbf{W}(t) = \mathbf{C}_{xy}(t)\mathbf{C}_{yy}^{-1}(t), \tag{8.5.6}$$

where

$$\mathbf{C}_{xy}(t) = \sum_{i=1}^{t} \beta^{t-i}\mathbf{x}(i)\mathbf{y}^H(i) = \beta\mathbf{C}_{xy}(t-1) + \mathbf{x}(t)\mathbf{y}^H(t), \tag{8.5.7}$$

$$\mathbf{C}_{yy}(t) = \sum_{i=1}^{t} \beta^{t-i}\mathbf{y}(i)\mathbf{y}^H(i) = \beta\mathbf{C}_{yy}(t-1) + \mathbf{y}(t)\mathbf{y}^H(t). \tag{8.5.8}$$

8.5.2 PAST Algorithms

Substituting Equations (8.5.7) and (8.5.8) into Equation (8.5.6), and applying the matrix inverse lemma, one obtains the *projection approximation subspace tracking* (PAST) *algorithm* [522]; see Algorithm 8.1.

In Algorithm 8.1, Tri[**A**] denotes an operation computing only the upper (or lower) triangular part of **A** then copying this as the lower (or upper) triangular part of **A**.

The PAST algorithm extracts the signal subspace from the data vector and is a PCA method. In particular, if the first formula in the above algorithm is replaced by

$$\mathbf{y}(t) = g\left(\mathbf{W}^H(t-1)\mathbf{x}(t)\right), \tag{8.5.9}$$

where $g(\mathbf{z}(t)) = [g(z_1(t)), g(z_2(t)), \ldots, g(z_n(t))]^T$ is the nonlinear vector function, then the PAST algorithm becomes the blind source separation algorithm based on

Algorithm 8.1 PAST algorithm [522]

task: Solve $\min_{\mathbf{W}} E\{\|\mathbf{x} - \mathbf{W}\mathbf{W}^H\mathbf{x}\|_2^2\}$.

input: Observation vector $\mathbf{x}(t)$.

initialize: $\mathbf{P}(0)$ and $\mathbf{W}(0)$.

repeat

1. $\mathbf{y}(t) = \mathbf{W}^H(t-1)\mathbf{x}(t)$.
2. $\mathbf{h}(t) = \mathbf{P}(t-1)\mathbf{y}(t)$.
3. $\mathbf{g}(t) = \mathbf{h}(t)/(\beta + \mathbf{y}^H(t)\mathbf{h}(t))$.
4. $\mathbf{P}(t) = \beta^{-1}\text{Tri}[\mathbf{P}(t-1) - \mathbf{g}(t)\mathbf{h}^H(t)]$.
5. $\mathbf{e}(t) = \mathbf{x}(t) - \mathbf{W}(t-1)\mathbf{y}(t)$.
6. $\mathbf{W}(t) = \mathbf{W}(t-1) + \mathbf{e}(t)\mathbf{g}^H(t)$.
7. **exit if** $\|\mathbf{W}(t) - \mathbf{W}(t-1)\|_F \le \epsilon$.

return $t \leftarrow t+1$.

output: $\mathbf{W} = \mathbf{W}(t)$.

nonlinear PCA. The LMS algorithm and the RLS algorithm for nonlinear PCA were developed in [358] and [371], respectively.

Algorithm 8.2 shows the *PAST via deflation* (PASTd) *algorithm*.

Algorithm 8.2 PAST via deflation (PASTd) algorithm [522]

task: Solve $\min_{\mathbf{W}} E\{\|\mathbf{x} - \mathbf{W}\mathbf{W}^H\mathbf{x}\|_2^2\}$.

input: Observation vector $\mathbf{x}(t)$.

initialize: $\mathbf{d}_i(0)$ and $\mathbf{w}_i(0)$.

repeat

1. $\mathbf{x}_1(t) = \mathbf{x}(t)$.
2. **for** $i = 1, \ldots, r$ **do**
   - 2.1 $y_i(t) = \mathbf{w}_i^H(t-1)\mathbf{x}_i(t)$.
   - 2.2 $d_i(t) = \beta d_i(t-1) + |y_i(t)|^2$.
   - 2.3 $\mathbf{e}_i(t) = \mathbf{x}_i(t) - \mathbf{w}_i(t-1)y_i(t)$.
   - 2.4 $\mathbf{w}_i(t) = \mathbf{w}_i(t-1) + \mathbf{e}_i(t)\big(y_i^*(t)/d_i(t)\big)$.
   - 2.5 $\mathbf{x}_{i+1}(t) = \mathbf{x}_i(t) - \mathbf{w}_i(t)y_i(t)$.
3. **end for**
4. **exit if** $\sum_{i=1}^r \|\mathbf{w}_i(t) - \mathbf{w}_i(t-1)\|_2 \le \epsilon$.

return $t \leftarrow t+1$.

output: $\mathbf{W} = [\mathbf{w}_1(t), \ldots, \mathbf{w}_r(t)]$.

The PASTd algorithm can be generalized to an algorithm for tracking both the rank and the subspace; see [523].

8.6 Fast Subspace Decomposition

In the terminology of Krylov subspaces, the signal subspace tracking of a sample autocorrelation matrix $\hat{\mathbf{R}}$ becomes the tracking of the Rayleigh–Ritz (RR) vectors of $\hat{\mathbf{R}}$. The starting point of this method is that both the span of the principal eigenvectors of a sample autocorrelation matrix $\hat{\mathbf{R}}$ and the span of its RR vectors are asymptotically equivalent estimates of the signal subspace of $\mathbf{R}$. Because RR vectors can be found efficiently using the Lanczos algorithm, this method can implement the fast decomposition of signal subspaces.

8.6.1 Rayleigh–Ritz Approximation

Given an observed data vector $\mathbf{x}(t) = \mathbf{s}(t) + \mathbf{n}(t) \in \mathbb{C}^M$, where $\mathbf{s}(t)$ and $\mathbf{n}(t)$ are the signal and additive noise vectors, respectively; and the entries of $\mathbf{n}(t)$ have the zero mean and the same variance σ. Consider the autocorrelation matrix of the observed data vector,

$$\mathbf{R}_{xx} = E\{\mathbf{x}(t)\mathbf{x}^H(t)\} = E\{\mathbf{s}(t)\mathbf{s}^H(t)\} + E\{\mathbf{n}(t)\mathbf{n}^H(t)\} = \mathbf{R}_{ss} + \sigma\mathbf{I} \tag{8.6.1}$$

whose sample autocorrelation matrix is

$$\begin{aligned}\hat{\mathbf{R}}_{xx}(n) &= \frac{1}{n}\sum_{t=1}^{n}\mathbf{x}(t)\mathbf{x}^H(t)\\ &= \frac{n-1}{n}\hat{\mathbf{R}}_{xx}(n-1) + \frac{1}{n}\mathbf{x}(n)\mathbf{x}^H(n)\\ &= \hat{\mathbf{R}}_{ss}(n) + \sigma\mathbf{I}.\end{aligned} \tag{8.6.2}$$

Thus the sample autocorrelation matrix $\hat{\mathbf{R}}_{xx}(n)$ at the time n can be updated in time by

$$\hat{\mathbf{R}}_{xx}(n) = \frac{n-1}{n}\hat{\mathbf{R}}_{xx}(n-1) + \frac{1}{n}\mathbf{x}(n)\mathbf{x}^H(n), \tag{8.6.3}$$

and the sample autocorrelation matrix of the unknown signal vector $\hat{\mathbf{R}}_{ss}(n)$ is directly estimated by $\hat{\mathbf{R}}_{ss}(n) = \hat{\mathbf{R}}_{xx}(n) - \sigma\mathbf{I}$. The question now is how to track the signal subspace $\text{Span}\{\hat{\mathbf{R}}_{ss}(n)\}$ or equivalently how to track the principal eigenvectors of $\hat{\mathbf{R}}_{ss}(n)$ using the sample autocorrelation matrix $\hat{\mathbf{R}}_{xx}(n)$. The direct EVD of $\hat{\mathbf{R}}_{xx}(n)$ at any given time does not meet the requirement of real-time tracking of the principal eigenvectors or the signal subspace. The Krylov subspace and the Rayleigh–Ritz (RR) approximation provide a fast subspace decomposition.

For convenience, let $\mathbf{A} = \hat{\mathbf{R}}_{xx}(n)$ and $\mathbf{B} = \hat{\mathbf{R}}_{ss}(n) + \sigma\mathbf{I}$. Then (8.6.2) becomes

$$\mathbf{A} = \mathbf{B} + \sigma\mathbf{I} \quad \text{or} \quad \mathbf{B} = \mathbf{A} - \sigma\mathbf{I}. \tag{8.6.4}$$

Let the EVD of the Hermitian matrix $\mathbf{A} \in \mathbb{C}^{M\times M}$ be

$$\mathbf{A} = \sum_{k=1}^{M} \lambda_k \mathbf{u}_k \mathbf{u}_k^H, \tag{8.6.5}$$

where $(\lambda_k, \mathbf{u}_k)$ are the kth eigenvalue and eigenvector of $\mathbf{A}$ and it is assumed that $\lambda_1 > \cdots > \lambda_d > \lambda_{d+1} = \cdots = \lambda_M = \sigma$. That is to say, $\{\lambda_k - \sigma, \mathbf{u}_k\}_{k=1}^{d}$ are respectively the signal (i.e., principal) eigenvalues and the signal eigenvectors of $\mathbf{B}$, owing to $\mathbf{B} = \mathbf{A} - \sigma\mathbf{I}$.

Now consider the RR approximation problem for the signal eigenvalues and the signal eigenvectors. To this end, introduce the following two definitions.

Definition 8.9 For an m-dimensional subspace S^m, if

$$\mathbf{A}\mathbf{y}_i^{(m)} - \theta_i^{(m)}\mathbf{y}_i^{(m)} \perp S^m \tag{8.6.6}$$

then $\theta_i^{(m)}$ and $\mathbf{y}_i^{(m)}$ are respectively the *RR values* and the *RR vectors* of the $M\times M$ Hermitian matrix $\mathbf{A}$ with respect to the subspace S^m.

The RR values and the RR vectors are combined in *Ritz pairs* $(\theta_i^{(m)}, \mathbf{y}_i^{(m)})$.

Definition 8.10 The order-m *Krylov subspace* $\mathcal{K}^m(\mathbf{A}, \mathbf{f})$ generated by an $M\times M$ matrix $\mathbf{A}$ and a vector $\mathbf{f}$ of dimension M is the linear subspace spanned by the images of $\mathbf{f}$ under the first m powers of $\mathbf{A}$ (starting from $\mathbf{A}^0 = \mathbf{I}$), that is,

$$\mathcal{K}^m(\mathbf{A}, \mathbf{f}) = \text{Span}\{\mathbf{K}_m(\mathbf{A}, \mathbf{f})\} = \text{Span}\{\mathbf{f}, \mathbf{A}\mathbf{f}, \ldots, \mathbf{A}^{m-1}\mathbf{f}\}, \tag{8.6.7}$$

where $\mathbf{K}_m(\mathbf{A}, \mathbf{f}) = [\mathbf{f}, \mathbf{A}\mathbf{f}, \ldots, \mathbf{A}^{m-1}\mathbf{f}]$ is called the *Krylov matrix*.

An important property of the Krylov subspace is that [519]

$$\mathcal{K}^m(\mathbf{A}, \mathbf{f}) = \mathcal{K}^m(\mathbf{A} - \sigma\mathbf{I}, \mathbf{f}) \tag{8.6.8}$$

for any scalar α.

Choose the order-m Krylov subspace $\mathcal{K}^m(\mathbf{A}, \mathbf{f})$ as the subspace S^m in (8.6.6), and suppose that $\mathbf{f}$ is not orthogonal to any of the d principal eigenvectors or to at least one of the nonprincipal eigenvectors and that the d principal eigenvalues are distinct. Then, by [517], $\dim(\mathcal{K}^{d+1}(\mathbf{A}, \mathbf{f})) = d + 1$.

Lemma 8.1 [375] *Let $(\theta_i^{(m)}, \mathbf{y}_i^{(m)})\, i = 1, \ldots, m$, be the RR values and the RR vectors of a Hermitian matrix $\mathbf{A}$ with respect to the subspace S^m, and let $\mathbf{Q} = [\mathbf{q}_1, \ldots, \mathbf{q}_m]$ be an orthogonal basis of the same subspace. If $(\alpha_i, \mathbf{u}_i)$ is the ith eigenpair of the $m \times m$ matrix $\mathbf{Q}^H\mathbf{A}\mathbf{Q}$, where $i = 1, \ldots, m$, then*

$$\theta_i^{(m)} = \alpha_i \quad \textit{and} \quad \mathbf{y}_i^{(m)} = \mathbf{Q}\mathbf{u}_i. \tag{8.6.9}$$

If $d = \text{rank}(\mathbf{A})$ then $\text{Span}\{\mathbf{y}_1^{(m)}, \ldots, \mathbf{y}_d^{(m)}\} = \text{Span}\{\mathbf{u}_1, \ldots, \mathbf{u}_d\}$, i.e., for an m-dimensional subspace $S^m = \mathcal{K}^m(\mathbf{A}, \mathbf{f})$, the signal subspace can be approximated by the span of its RR vectors $\mathbf{y}_1^{(m)}, \ldots, \mathbf{y}_d^{(m)}$. This approximation is called the *Rayleigh–Ritz approximation*, simply called the RR approximation.

The RR pairs $(\theta_i, \mathbf{y}_i^{(m)})$ of an $M \times M$ higher-dimensional matrix $\mathbf{A}$ can be equivalently estimated via (8.6.9) by the eigenpairs $(\alpha_i, \mathbf{u}_i)$ of an $m \times m$ low-dimensional matrix $\mathbf{Q}^H \mathbf{A} \mathbf{Q}$. The remaining issue is how to construct the orthogonal basis matrix $\mathbf{Q} \in \mathbb{C}^{M \times m}$ of the Krylov subspace $\mathcal{K}^m(\mathbf{A}, \mathbf{f})$.

The properties of the Rayleigh–Ritz approximation are evaluated by the asymptotic properties of the RR values and the RR vectors: for $m > d$, their errors are given respectively by

$$\theta_k^{(m)} - \hat{\lambda}_k = O(N^{-m-d}), \quad k = 1, \ldots, d, \tag{8.6.10}$$

$$\mathbf{y}_k^{(m)} - \hat{\mathbf{u}}_k = O(N^{-(m-d)/2}), \quad k = 1, \ldots, d. \tag{8.6.11}$$

Here N is the data length. Therefore, once $m \geq d + 2$, then

$$\lim_{N \to \infty} \sqrt{N}(\mathbf{y}_k^{(m)} - \mathbf{u}_k) = \lim_{N \to \infty} \sqrt{N}(\hat{\mathbf{u}}_k - \mathbf{u}_k), \quad k = 1, \ldots, d, \tag{8.6.12}$$

i.e., both $\text{Span}\{\mathbf{y}_1^{(m)}, \ldots, \mathbf{y}_d^{(m)}\}$ and $\text{Span}\{\hat{\mathbf{u}}_1, \ldots, \hat{\mathbf{u}}_d\}$ are asymptotically equivalent estimates of the signal subspace $\text{Span}\{\mathbf{u}_1, \ldots, \mathbf{u}_d\}$, so the tracking of the signal subspace of a Hermitian matrix $\mathbf{A}$ becomes the tracking of the RR vectors $\mathbf{y}_1^{(m)}, \ldots, \mathbf{y}_d^{(m)}$ of $\mathbf{A}$.

8.6.2 Fast Subspace Decomposition Algorithm

By Lemma 8.1, the tracking of the RR vectors $\mathbf{y}_1^{(m)}, \ldots, \mathbf{y}_d^{(m)}$ of $\mathbf{A}$ is equivalent to the tracking of the principal eigenvectors $\mathbf{u}_1^{(m)}, \ldots, \mathbf{u}_d^{(m)}$ of $\mathbf{Q}_m \mathbf{A} \mathbf{Q}_m$.

The Lanczos method provides an efficient way of finding an orthonormal basis $\mathbf{Q}_m = [\mathbf{q}_1, \ldots, \mathbf{q}_m]$ for $\mathcal{K}^m(\mathbf{A}, \mathbf{f})$. The orthonormal basis constructed by the Lanczos method is called the *Lanczos basis*. The use of a Lanczos basis means that the RR pairs $(\theta_i^{(m)}, \mathbf{y}_i^{(m)}), i = 1, \ldots, d$ are closely related to the eigenpairs of the tridiagonalized matrix of $\mathbf{A}$.

Let $\mathbf{Q}_m = [\mathbf{q}_1, \ldots, \mathbf{q}_m]$ be a Lanczos basis. It is known from [375] that

$$\mathbf{Q}_m^H \hat{\mathbf{A}} \mathbf{Q}_m = \mathbf{T}_m = \begin{bmatrix} \alpha_1 & \beta_1 & & & 0 \\ \beta_1 & \alpha_2 & \beta_2 & & \\ & \ddots & \ddots & \ddots & \\ & & \ddots & \alpha_{m-1} & \beta_{m-1} \\ 0 & & & \beta_{m-1} & \alpha_m \end{bmatrix}. \tag{8.6.13}$$

The iteration based on (8.6.13) is called the *tri-Lanczos iteration* for tridiagonalization of the Hermitian matrix $\mathbf{A}$. Once the tridiagonalization $\mathbf{T}_m$ of the Hermitian matrix $\mathbf{A}$ has been achieved, one can obtain the eigenpairs $(\alpha_i^{(m)}, \mathbf{u}_i^{(m)}), i =$

$1, \ldots, m$, by the EVD of the $m \times m$ tridiagonal matrix $\mathbf{T}_m$, and thus find the RR vectors $\mathbf{y}_1^{(m)}, \ldots, \mathbf{y}_d^{(m)}$ of $\mathbf{A}$ by (8.6.9).

The most attractive property of a Lanczos algorithm is that the original larger problem of finding the expected eigenvalues and eigenvectors of an $M \times M$ (complex) sample covariance (Hermitian) matrix $\hat{\mathbf{A}}$ is transformed into the smaller problem of computing the eigenvalue decomposition of an $m \times m$ (real) tridiagonal matrix $\mathbf{T}_m$, because m is usually much smaller than M.

Lanczos algorithms are divided into two kinds: one is the tri-Lanczos iteration for the tridiagonalization of the Hermitian matrix, and another is the bi-Lanczos iteration for bidiagonalization of any matrix.

The tri-Lanczos iteration is available only for the tridiagonalization of a sample autocorrelation matrix $\hat{\mathbf{A}} = N^{-1} \sum_{n=1}^{N} \mathbf{x}(n)\mathbf{x}^H(n)$, and is not applicable for any nonsquare data matrix. For an $N \times M$ data matrix $\mathbf{X}_N = N^{-1/2}[\mathbf{x}(1), \ldots, \mathbf{x}(N)]^H$ such that $\hat{\mathbf{A}} = \mathbf{X}_N^H \mathbf{X}_N$, the bi-Lanczos iteration can be used directly for the data matrix $\mathbf{X}_N$, as shown in Algorithm 8.3. The bi-Lanczos iteration gives the left Lanczos basis $\mathbf{U}_j = [\mathbf{u}_1, \ldots, \mathbf{u}_j]$, the right Lanczos basis $\mathbf{V}_j = [\mathbf{v}_1, \ldots, \mathbf{v}_j]$ and the bidiagonal matrix

$$\mathbf{B}_j = \begin{bmatrix} \alpha_1^{(b)} & \beta_1^{(b)} & & 0 \\ & \alpha_2^{(b)} & \ddots & \\ & & \ddots & \beta_{j-1}^{(b)} \\ 0 & & & \alpha_j^{(b)} \end{bmatrix}, \quad j = 1, \ldots, m. \tag{8.6.14}$$

Algorithm 8.4 shows the bi-Lanczos-iteration-based fast subspace decomposition algorithm developed in [516].

Regarding the RR approximation, Xu and Kailath [518] showed that the d largest RR values found at the mth Lanczos iteration can be used to replace the signal eigenvalues. However, it is necessary first to estimate d. To this end, construct the testing statistic

$$\phi_{\hat{d}} = N(M - \hat{d}) \log \frac{\sqrt{(M-\hat{d})^{-1}\left(\|\hat{\mathbf{A}}\|_F^2 - \sum_{k=1}^{\hat{d}} \theta_k^{(m)^2}\right)}}{(M-\hat{d})^{-1}\left(\mathrm{tr}(\hat{\mathbf{A}}) - \sum_{k=1}^{\hat{d}} \theta_k^{(m)}\right)}, \tag{8.6.15}$$

where

$$\mathrm{tr}(\hat{\mathbf{A}}) = \sum_{k=1}^{M} \hat{\lambda}_k, \quad \|\hat{\mathbf{A}}\|_F^2 = \sum_{k=1}^{M} \hat{\lambda}_k^2. \tag{8.6.16}$$

For applications of the fast subspace decomposition in signal processing and wireless communications, see [516].

Algorithm 8.3 Bi-Lanczos iteration [179]

input: Data matrix $\mathbf{X}_N$.
initialize: $\mathbf{p}_0 = \mathbf{f}$ (unit-norm vector); $\beta_0 = 1; \mathbf{u}_0 = \mathbf{0}; j = 0$.
while $(\beta_j^{(b)} \neq 0)$ **do**
1. $\mathbf{v}_{j+1} = \mathbf{p}_j / \beta_j^{(b)}$.
2. $j = j + 1$.
3. $\mathbf{r}_j = \mathbf{X}_N \mathbf{v}_j - \beta_{j-1}^{(b)} \mathbf{u}_{j-1}$.
4. $\alpha_j^{(b)} = \|\mathbf{r}_j\|$.
5. $\mathbf{u}_j = \mathbf{r}_j / \alpha_j^{(b)}$.
6. $\mathbf{p}_j = \mathbf{X}_N^H \mathbf{u}_j - \alpha_j^{(b)} \mathbf{v}$.
7. $\beta_j^{(b)} = \|\mathbf{p}_j\|$.

end

Algorithm 8.4 Fast subspace decomposition algorithm

input: Data matrix $\mathbf{X}_N$.
initialize: Select $\mathbf{r}_0 = \mathbf{f}$ (unit-norm vector). Let $m = 1, \beta_0 = \|\mathbf{r}_0\| = 1$, $\hat{d} = 1$.
repeat
1. Carry out the $j = m$th bi-Lanczos iteration (Algorithm 8.3).
2. Use α and β estimated by Algorithm 8.3 to form an $m \times m$ bidiagonal matrix $\mathbf{B}_m$ and find its singular values $\theta_i^{(m)}, i = 1, \ldots, m$.
3. For $\hat{d} = 1, \ldots, m - 1$, calculate the testing statistics
$$\phi_{\hat{d}} = \sqrt{N} |\log(\hat{\sigma}_{\hat{d}} / \hat{\sigma}_{\hat{d}+1})|$$
where $\hat{\sigma}_i = \frac{1}{M - j} \left(\left\| \frac{1}{\sqrt{N}} \mathbf{X}_N \right\|_2^2 - \sum_{i=1}^{j} (\theta_i^{(m)})^2 \right)$.
4. If $\phi_{\hat{d}} < \gamma_{\hat{d}} \sqrt{\log N}$, then let $d = \hat{d}$, and go to step 5; otherwise let $m \leftarrow m + 1$, and return to step 1.
5. Compute the SVD of the $m \times m$ bidiagonal matrix $\mathbf{B}_m$ to get d principal right singular vectors $v_k^{(m)}$ associated with the Krylov subspace $\mathcal{K}^m(\hat{\mathbf{A}}, \mathbf{f})$.

output: Signal subspace estimate $\text{Span}\{\mathbf{v}_1^{(m)}, \ldots, \mathbf{v}_d^{(m)}\}$.

Exercises

8.1 Let V be the vector space of all the 2×2 matrices; show that the subspace
$$W = \left\{ \mathbf{A} \middle| \mathbf{A} = \begin{bmatrix} a & b \\ c & d \end{bmatrix}, \; ad = 0, \, bc = 0 \right\}$$

is not a subspace in V.

8.2 Let V be the vector space of all the 2×2 matrices, and

$$W = \left\{ \mathbf{A} \middle| \mathbf{A} = \begin{bmatrix} a & 0 \\ 0 & b \end{bmatrix}, \ a, b \text{ are any real numbers} \right\}$$

is a subspace in V. If

$$\mathbf{B}_1 = \begin{bmatrix} 0 & 2 \\ 1 & 0 \end{bmatrix}, \quad \mathbf{B}_2 = \begin{bmatrix} 0 & 3 \\ 0 & 0 \end{bmatrix}, \quad \mathbf{B}_3 = \begin{bmatrix} 0 & 3 \\ 2 & 0 \end{bmatrix},$$

(a) show that the matrix set $\{\mathbf{B}_1, \mathbf{B}_2, \mathbf{B}_3\}$ is linearly dependent, and represent $\mathbf{B}_3$ as a linear combination of $\mathbf{B}_1$ and $\mathbf{B}_2$;
(b) show that $\{\mathbf{B}_1, \mathbf{B}_2\}$ is a linearly independent matrix set.

8.3 Let $\mathbf{u}_1, \ldots, \mathbf{u}_p$ be vectors in a finite-dimensional vector space V, and $S = \{\mathbf{u}_1, \ldots, \mathbf{u}_p\}$ be a vector set. Determine whether the following results are true or false.

(a) The set of all the linear combinations of $\mathbf{u}_1, \ldots, \mathbf{u}_p$ is a vector space.
(b) If $\{\mathbf{u}_1, \ldots, \mathbf{u}_{p-1}\}$ is linearly independent, then S is also the linearly independent vector set.
(c) If the vector set S is linearly independent, then S is a basis set in V.
(d) If $V = \text{Span}\{\mathbf{u}_1, \ldots, \mathbf{u}_p\}$, then some subset of S is a basis set in V.
(e) If $\dim(V) = p$ and $V = \text{Span}\{\mathbf{u}_1, \ldots, \mathbf{u}_p\}$ then the vector set S cannot be linearly dependent.

8.4 Determine whether the following results are true.

(a) The row space of a matrix $\mathbf{A}$ is the same as the column space of $\mathbf{A}^T$.
(b) The dimensions of the row space and the column space of the matrix $\mathbf{A}$ are the same, even if $\mathbf{A}$ is not a square matrix.
(c) The sum of the dimensions of the row space and the column space of a matrix $\mathbf{A}$ is equal to the number of rows of $\mathbf{A}$.
(d) The row space of the matrix $\mathbf{A}^T$ and the column space of $\mathbf{A}$ are the same.

8.5 Let $\mathbf{A}$ be an $m \times n$ matrix. There are several different subspaces among the subspaces $\text{Row}(\mathbf{A})$, $\text{Col}(\mathbf{A})$, $\text{Null}(\mathbf{A})$, $\text{Row}(\mathbf{A}^T)$, $\text{Col}(\mathbf{A}^T)$ and $\text{Null}(\mathbf{A}^T)$. Which are located in $\mathbb{R}^m$ and $\mathbb{R}^n$, respectively?

8.6 Let $\mathbf{A} \in \mathbb{C}^{m \times n}$, $\text{Row}(\mathbf{A})$ be the row space of $\mathbf{A}$ and $\text{Col}(\mathbf{A})$ be the column space of $\mathbf{A}$. Show that:

(a) the orthogonal complement $(\text{Row}(\mathbf{A}))^\perp$ is the null space of $\mathbf{A}$, namely, $(\text{Row}(\mathbf{A}))^\perp = \text{Null}(\mathbf{A})$;
(b) the orthogonal complement $(\text{Col}(\mathbf{A}))^\perp$ is the null space of $\mathbf{A}^H$, i.e.,

$$(\text{Col}(\mathbf{A}))^\perp = \text{Null}(\mathbf{A}^H).$$

8.7 For each case, show that the vector set W is a vector subspace or give a counter example to show that it is not a vector subspace:

(a) $W = \left\{ \begin{bmatrix} a \\ b \\ c \\ d \end{bmatrix} \middle| \, 2a + b = c, \, a + b + c = d \right\}$;

(b) $W = \left\{ \begin{bmatrix} a - b \\ 3b \\ 3a - 2b \\ a \end{bmatrix} \middle| \, a, b \text{ are real numbers} \right\}$;

(c) $W = \left\{ \begin{bmatrix} 2a + 3b \\ c + a - 2b \\ 4c + a \\ 3c - a - b \end{bmatrix} \middle| \, a, b, c \text{ are real numbers} \right\}$.

8.8 Given

$$\mathbf{A} = \begin{bmatrix} 8 & 2 & 9 \\ -3 & -2 & -4 \\ 5 & 0 & 5 \end{bmatrix}, \quad \mathbf{w} = \begin{bmatrix} 2 \\ 1 \\ -2 \end{bmatrix},$$

determine whether $\mathbf{w}$ is in the column space $\text{Col}(\mathbf{A})$ or in the null space $\text{Null}(\mathbf{A})$.

8.9 In statistical theory a matrix is often required to be of full rank. If the matrix $\mathbf{A}$ is an $m \times n$ matrix, where $m > n$, explain why the condition for $\mathbf{A}$ to be full rank is that its columns are linearly independent.

8.10 Show that $\mathbf{v}$ is in $\text{Col}(\mathbf{A})$ of the matrix $\mathbf{A}$ if $\mathbf{A}\mathbf{v} = \lambda\mathbf{v}$ and $\lambda \neq 0$.

8.11 Letting the column vectors of $\mathbf{V}_1$ and $\mathbf{V}_2$ be respectively the orthogonal basis in the same subspace of $\mathbb{C}^n$, show that $\mathbf{V}_1\mathbf{V}_1^H\mathbf{x} = \mathbf{V}_2\mathbf{V}_2^H\mathbf{x}$, $\forall \mathbf{x}$.

8.12 Assume that V is a subspace and S is the generator or spanning set of V. Given that

$$S = \left\{ \begin{bmatrix} 1 \\ 0 \\ -1 \end{bmatrix}, \begin{bmatrix} -1 \\ -2 \\ -3 \end{bmatrix}, \begin{bmatrix} 1 \\ 1 \\ 2 \end{bmatrix}, \begin{bmatrix} 2 \\ -1 \\ 0 \end{bmatrix} \right\},$$

find the basis of V, and compute $\dim(V)$.

8.13 Let the matrix $\mathbf{A} \in \mathbb{C}^{m \times n}$. Show that

$$\min_{\text{rank}(\mathbf{X}) \leq k} \left(\|\mathbf{A} - \mathbf{X}\|_F = \left(\sum_{i=k+1}^{n} \sigma_i^2 \right)^{1/2} \right),$$

where $\mathbf{X} \in \mathbb{C}^{m \times n}, k < n$, and $\sigma_1 \geq \cdots \geq \sigma_n$ are the n singular values of $\mathbf{A}$.

8.14 The cosine of the angle of two vectors $\mathbf{x}$ and $\mathbf{y}$ is defined as

$$\cos\theta(\mathbf{x}, \mathbf{y}) = \frac{|\langle \mathbf{x}, \mathbf{y} \rangle|}{\|\mathbf{x}\|_2 \|\mathbf{y}\|_2}, \quad 0 \leq \theta(\mathbf{x}, \mathbf{y}) \leq \frac{\pi}{2}.$$

Suppose that $\mathbf{X} = [\mathbf{x}_i, \ldots, \mathbf{x}_k], \mathbf{Y} = [\mathbf{y}_1, \ldots, \mathbf{y}_k] \in \mathbb{C}^{n \times k} (n \geq k)$ and $\mathbf{X}^H\mathbf{X} =$

$\mathbf{Y}^H\mathbf{Y} = \mathbf{I}_k$. If the cosine function of the ith angle between the spaces Col($\mathbf{X}$) and Col($\mathbf{Y}$) is defined as $\cos\theta_i(\text{Col}(\mathbf{X}), \text{Col}(\mathbf{Y})) = \cos\theta(\mathbf{x}_i, \mathbf{y}_i)$, prove that $\cos\theta_i(\text{Col}(\mathbf{X}), \text{Col}(\mathbf{Y}))$ is the ith singular value of the matrix $\mathbf{X}^H\mathbf{Y}$ (where the singular values are in decreasing order).

8.15 Assume that $\mathbf{X}, \mathbf{Y} \in \mathbb{C}^{m\times n}$ $(m \geq n)$, and $\mathbf{X}^H\mathbf{X} = \mathbf{Y}^H\mathbf{Y} = \mathbf{I}_n$. Define the distance between the subspaces $S_1 = \text{Col}(\mathbf{X})$ and $S_2 = \text{Col}(\mathbf{Y})$ as $d(S_1, S_2) = \|\mathbf{P}_{S_1} - \mathbf{P}_{S_2}\|_F$.

(a) Find the projection matrices $\mathbf{P}_{S_1}$ and $\mathbf{P}_{S_2}$.

(b) Show that

$$d(S_1, S_2) = \left(1 - \sigma_{\min}(\mathbf{X}^H\mathbf{Y})\right)^{1/2},$$

where $\sigma_{\min}(\mathbf{X}^H\mathbf{Y})$ is the minimum singular value of the matrix $\mathbf{X}^H\mathbf{Y}$.

8.16 Let $\mathbf{A}$ and $\mathbf{B}$ be two $m \times n$ matrices, and $m \geq n$. Show that

$$\min_{\mathbf{Q}^T\mathbf{Q}=\mathbf{I}_n} \|\mathbf{A} - \mathbf{B}\mathbf{Q}\|_F^2 = \sum_{i=1}^{n} \left((\sigma_i(\mathbf{A}))^2 - 2\sigma_i(\mathbf{B}^T\mathbf{A}) + (\sigma_i(\mathbf{B}))^2\right),$$

where $\sigma_i(\mathbf{A})$ is the ith singular value of the matrix $\mathbf{A}$.

8.17 Assume that T is a one-to-one linear transformation and that $T(\mathbf{u}) = T(\mathbf{v})$ always means $\mathbf{u} = \mathbf{v}$. Prove that if the set $\{T(\mathbf{u}_1), T(\mathbf{u}_2), \ldots, T(\mathbf{u}_p)\}$ is linearly dependent then the vector set $\{\mathbf{u}_1, \mathbf{u}_2, \ldots, \mathbf{u}_p\}$ is linearly dependent. (Note that this result shows that a one-to-one linear transformation maps a linearly independent vector set into another linearly independent vector set.)

8.18 Given the matrix

$$\mathbf{A} = \begin{bmatrix} 2 & 5 & -8 & 0 & 17 \\ 1 & 3 & -5 & 1 & 5 \\ -3 & -11 & 19 & -7 & -1 \\ 1 & 7 & -13 & 5 & -3 \end{bmatrix},$$

find the bases of its column space, row space and null space.

8.19 Letting

$$\mathbf{A} = \begin{bmatrix} -2 & 1 & -1 & 6 & -8 \\ 1 & -2 & -4 & 3 & -2 \\ 7 & -8 & -10 & -3 & 10 \\ 4 & -5 & -7 & 0 & 4 \end{bmatrix}$$

and

$$\mathbf{B} = \begin{bmatrix} 1 & -2 & -4 & 3 & -2 \\ 0 & 3 & 9 & -12 & 12 \\ 0 & 0 & 0 & 0 & 0 \\ 0 & 0 & 0 & 0 & 0 \end{bmatrix}$$

be two row-equivalent matrices, find

(a) the rank of the matrix $\mathbf{A}$ and the dimension of the null space $\text{Null}(\mathbf{A})$,
(b) bases of the column space $\text{Col}(\mathbf{A})$ and the row space $\text{Row}(\mathbf{A})$,
(c) a basis of the null space $\text{Null}(\mathbf{A})$,
(d) the number of pivot columns in the row echelon of $\mathbf{A}^T$.

8.20 Let $\mathbf{A}$ be an $n \times n$ symmetric matrix. Show that
(a) $(\text{Col}(\mathbf{A}))^{\perp} = \text{Null}(\mathbf{A})$,
(b) every vector $\mathbf{x}$ in $\mathbb{R}^n$ can be written as $\mathbf{x} = \hat{\mathbf{x}} + \mathbf{z}$, where $\hat{\mathbf{x}} \in \text{Col}(\mathbf{A})$ and $\mathbf{z} \in \text{Null}(\mathbf{A})$.

9

Projection Analysis

In many engineering applications (such as wireless communication, radar, sonar, time series analysis, signal processing, and so on), the optimal solution of many problems consists of extracting a desired signal and restraining all other interferences, clutters or noises. Projection is one of the most important mathematical tools for solving this kind of problem There are two types of projection, orthogonal projection and oblique projection, as we will see. This chapter focuses upon the projection analysis and applications.

9.1 Projection and Orthogonal Projection

In the process of learning mechanics, we become familiar with the decomposition of the force of gravity of an object on a slope, shown in Figure 9.1.

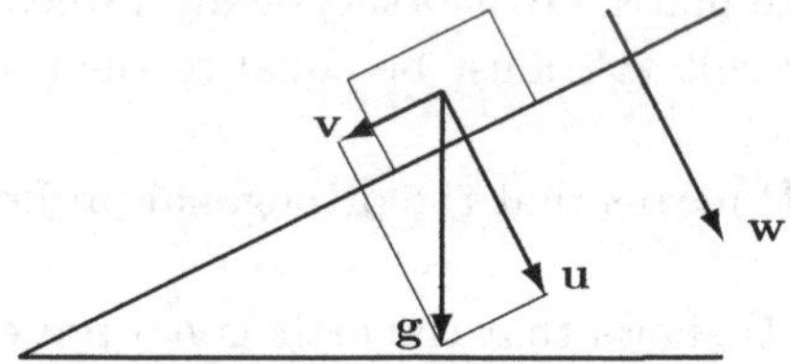

Figure 9.1 Decomposition of the force of gravity on an object.

In Figure 9.1, the force of gravity $\mathbf{g}$ on the object is vertically downward. It can be decomposed into two components: one is perpendicular to the slant and acts as the normal pressure force $\mathbf{u}$ on the object; the other is parallel to the slope and acts as sliding force $\mathbf{v}$ on the object. Thus $\mathbf{g} = \mathbf{u} + \mathbf{v}$. Since the pressure force $\mathbf{u}$ and the sliding force $\mathbf{v}$ are perpendicular to each other, the decomposition $\mathbf{g} = \mathbf{u} + \mathbf{v}$ is an example of *orthogonal decomposition.* We can describe this as follows. The *projection* of $\mathbf{g}$ onto the direction of $\mathbf{u}$ is obtained as the arrowhead of $\mathbf{g}$ moves along the direction of $\mathbf{v}$. In Figure 9.1 $\mathbf{u}$ and $\mathbf{v}$ are orthogonal but this need not be the case provided that $\mathbf{u}$ and $\mathbf{v}$ still satisfy $\mathbf{u}+\mathbf{v}=\mathbf{g}$.

9.1.1 Projection Theorem

More generally, we now consider projection, and in particular the orthogonal projection, in a vector space.

Let $\mathbf{x} \in \mathbb{R}^n$, and let S and H be any two subspaces in $\mathbb{R}^n$. Consider using a linear matrix transform $\mathbf{P}$ to map the vector $\mathbf{x}$ in $\mathbb{R}^n$ onto a vector $\mathbf{x}_1$ in the subspace S. Such a linear matrix transform is called a projector onto subspace S along the "direction" of subspace H, denoted $\mathbf{P}_{S|H}$, and $\mathbf{P}_{S|H}\mathbf{x}$ is said to be a projection of $\mathbf{x}$ onto S along H.

In particular, if the subspace $H = S^{\perp} = \mathbb{R}^n \setminus S$ is the orthogonal complement of S in $\mathbb{R}^n$ then $\mathbf{P}_{S|S^{\perp}}\mathbf{x}$ is called the *orthogonal projection* of $\mathbf{x}$ onto S and is simply denoted $\mathbf{P}_S\mathbf{x}$ or $\mathbf{P}\mathbf{x}$.

Definition 9.1 [179, p. 75] The matrix $\mathbf{P} \in \mathbb{C}^{n\times n}$ is known as an *orthogonal projector* onto a subspace S, if:

(1) $\text{Range}(\mathbf{P}) = S$;
(2) $\mathbf{P}^2 = \mathbf{P}$;
(3) $\mathbf{P}^H = \mathbf{P}$.

Regarding the above three conditions, we have the following interpretations.

- The condition $\text{Range}(\mathbf{P}) = S$ means that the column space of $\mathbf{P}$ must be equal to the subspace S. If the subspace S is the subspace spanned by the n column vectors of an $m\times n$ matrix $\mathbf{A}$, i.e., $S = \text{Span}(\mathbf{A})$, then $\text{Range}(\mathbf{P}) = \text{Span}(\mathbf{A}) = \text{Range}(\mathbf{A})$. This means that if the matrix $\mathbf{A}$ is orthogonally projected onto the subspace S then the projection result $\mathbf{PA}$ must be equal to the original matrix $\mathbf{A}$, namely $\mathbf{PA} = \mathbf{A}$.
- The condition $\mathbf{P}^2 = \mathbf{P}$ means that the orthogonal projector must be an idempotent operator.
- The condition $\mathbf{P}^H = \mathbf{P}$ shows that the orthogonal projector must have complex conjugate symmetry, i.e., Hermiticity.

For a given orthogonal projector $\mathbf{P}$ onto a subspace S, $\mathbf{P}^{\perp} = \mathbf{I} - \mathbf{P}$ is called the orthogonal projector onto the orthogonal complement $S^{\perp}$ of S, because

$$(\mathbf{I}-\mathbf{P})\mathbf{x} = \mathbf{x} - \mathbf{P}\mathbf{x} \in \mathbb{R}^n \setminus S = S^{\perp}.$$

Let $\mathbf{P}_1$ and $\mathbf{P}_2$ be two distinct orthogonal projectors onto the subspace S; then, for any vector $\mathbf{x} \in \mathbb{R}^n$, we have

$$\begin{aligned} \mathbf{P}_1\mathbf{x} \in S,\ (\mathbf{I}-\mathbf{P}_2)\mathbf{x} \in S^{\perp} \quad &\Rightarrow \quad (\mathbf{P}_1\mathbf{x})^H(\mathbf{I}-\mathbf{P}_2)\mathbf{x} = 0, \\ \mathbf{P}_2\mathbf{x} \in S,\ (\mathbf{I}-\mathbf{P}_1)\mathbf{x} \in S^{\perp} \quad &\Rightarrow \quad (\mathbf{P}_2\mathbf{x})^H(\mathbf{I}-\mathbf{P}_1)\mathbf{x} = 0. \end{aligned}$$

Hence

$$\begin{aligned}\|(\mathbf{P}_1-\mathbf{P}_2)\mathbf{x}\|_2^2 &= (\mathbf{P}_1\mathbf{x}-\mathbf{P}_2\mathbf{x})^H(\mathbf{P}_1\mathbf{x}-\mathbf{P}_2\mathbf{x})\\ &= (\mathbf{P}_1\mathbf{x})^H(\mathbf{I}-\mathbf{P}_2)\mathbf{x}+(\mathbf{P}_2\mathbf{x})^H(\mathbf{I}-\mathbf{P}_1)\mathbf{x}\\ &\equiv 0,\quad \forall\,\mathbf{x}.\end{aligned}$$

That is to say, $\mathbf{P}_1 = \mathbf{P}_2$, namely the orthogonal projector onto a subspace is uniquely determined.

It is well known that in elementary geometry, the shortest distance from a point to a straight line is the perpendicular distance. By extension, if $\mathbf{x} \in V$, while S is a subspace of the vector space V and $\mathbf{x}$ is not in the subspace S then the shortest distance problem from $\mathbf{x}$ to S involves finding the vector $\mathbf{y} \in S$ such that the length of the vector $\mathbf{x}-\mathbf{y}$ is shortest. In other words, given a vector $\mathbf{x} \in V$, we want to find a vector $\hat{\mathbf{x}} \in S$ such that

$$\|\mathbf{x}-\hat{\mathbf{x}}\|_2 \leq \|\mathbf{x}-\mathbf{y}\|_2,\quad \forall\,\mathbf{y}\in S. \tag{9.1.1}$$

Theorem 9.1 (Projection theorem) *Let V be an n-dimensional vector space and S be an n-dimensional subspace. If for the vector $\mathbf{x} \in V$, there is a vector $\hat{\mathbf{x}}$ in the subspace S such that $\mathbf{x}-\hat{\mathbf{x}}$ is orthogonal to every vector $\mathbf{y} \in S$, i.e.,*

$$\langle \mathbf{x}-\hat{\mathbf{x}}, \mathbf{y}\rangle = 0,\quad \forall\,\mathbf{y}\in S, \tag{9.1.2}$$

then the relation $\|\mathbf{x}-\hat{\mathbf{x}}\|_2 \leq \|\mathbf{x}-\mathbf{y}\|_2$ holds for all the vectors $\mathbf{y} \in S$, but the equality holds if and only if $\mathbf{y} = \hat{\mathbf{x}}$.

Proof Compute the square of the vector norm:

$$\begin{aligned}\|\mathbf{x}-\mathbf{y}\|_2^2 &= \|(\mathbf{x}-\hat{\mathbf{x}})+(\hat{\mathbf{x}}-\mathbf{y})\|_2^2\\ &= (\mathbf{x}-\hat{\mathbf{x}})^T(\mathbf{x}-\hat{\mathbf{x}})+2(\mathbf{x}-\hat{\mathbf{x}})^T(\hat{\mathbf{x}}-\mathbf{y})+(\hat{\mathbf{x}}-\mathbf{y})^T(\hat{\mathbf{x}}-\mathbf{y}).\end{aligned}$$

Since $(\mathbf{x}-\hat{\mathbf{x}})^T\mathbf{y} = \langle \mathbf{x}-\hat{\mathbf{x}},\mathbf{y}\rangle = 0$ holds for every vector $\mathbf{y} \in M$, this equation holds for the vector $\hat{\mathbf{x}} \in M$ as well, i.e., $(\mathbf{x}-\hat{\mathbf{x}})^T\hat{\mathbf{x}} = 0$. Then

$$(\mathbf{x}-\hat{\mathbf{x}})^T(\hat{\mathbf{x}}-\mathbf{y}) = (\mathbf{x}-\hat{\mathbf{x}})^T\hat{\mathbf{x}} - (\mathbf{x}-\hat{\mathbf{x}})^T\mathbf{y} = 0.$$

From the above two equation we immediately have $\|\mathbf{x}-\mathbf{y}\|_2^2 = \|\mathbf{x}-\hat{\mathbf{x}}\|_2^2+\|\hat{\mathbf{x}}-\mathbf{y}\|_2^2 \geqslant \|\mathbf{x}-\hat{\mathbf{x}}\|_2^2$, and the equality holds if and only if $\mathbf{y} = \hat{\mathbf{x}}$. □

The solution to Equation (9.1.2) in Theorem 9.1 is given by $\hat{\mathbf{x}} = \mathbf{P}_S\mathbf{x}$, i.e., $\hat{\mathbf{x}}$ is the orthogonal projection of the vector $\mathbf{x}$ onto the subspace S. In this case, $\mathbf{x}-\hat{\mathbf{x}} = \mathbf{x}-\mathrm{P}_S\mathbf{x} \in V\setminus S = S^{\perp}$, which implies that $\mathbf{x}-\hat{\mathbf{x}}$ is orthogonal to any vector $\mathbf{y}$ in S, i.e., Equation (9.1.2) holds.

The orthogonal projection $\mathbf{P}_S\mathbf{x}$ has the following properties [67]:

1. $\mathbf{P}_S(\alpha\mathbf{x}+\beta\mathbf{y}) = \alpha\mathbf{P}_S\mathbf{x}+\beta\mathbf{P}_S\mathbf{y}$, for $\mathbf{x},\mathbf{y} \in V$ and $\alpha,\beta \in \mathbb{C}$.
2. $\|\mathbf{x}\|^2 = \|\mathbf{P}_S\mathbf{x}\|^2 + \|(\mathbf{I}-\mathbf{P}_S)\mathbf{x}\|^2$.

3. Every $\mathbf{x} \in V$ has the following unique representation

$$\mathbf{x} = \mathbf{P}_S\mathbf{x} + (\mathbf{I} - \mathbf{P}_S)\mathbf{x}. \tag{9.1.3}$$

   That is, $\mathbf{x}$ can be uniquely decomposed into the sum of the component $\mathbf{P}_S\mathbf{x}$ on S and the component $(\mathbf{I} - \mathbf{P}_S)\mathbf{x}$ on $S^\perp$.
4. $\mathbf{P}_S\mathbf{x}_n \to \mathbf{P}_S\mathbf{x}$ if $\|\mathbf{x}_n - \mathbf{x}\| \to 0$.
5. $\mathbf{x} \in S$ if and only if $\mathbf{P}_S\mathbf{x} = \mathbf{x}$.
6. $\mathbf{x} \in S^\perp$ if and only if $\mathbf{P}_S\mathbf{x} = \mathbf{0}$.

9.1.2 Mean Square Estimation

A set $M \subseteq L_2$ is called an *orthogonal system*, if $\boldsymbol{\xi} \perp \boldsymbol{\eta}$ for every $\boldsymbol{\xi}, \boldsymbol{\eta} \in M\,(\boldsymbol{\xi} \neq \boldsymbol{\eta})$. In particular, M is an *orthonormal system* if $\|\boldsymbol{\xi}\| = 1$ for every $\boldsymbol{\xi} \in M$.

Consider n data vectors $\boldsymbol{\eta}_1, \ldots, \boldsymbol{\eta}_n$. Using the linear combination $\hat{\boldsymbol{\xi}} = \sum_{i=1}^n a_i\boldsymbol{\eta}_i$ to fit an unknown random variable $\boldsymbol{\xi}$, the fitted (or estimated) error variable is given by

$$\boldsymbol{\epsilon} = \boldsymbol{\xi} - \hat{\boldsymbol{\xi}} = \boldsymbol{\xi} - \sum_{i=1}^n a_i\boldsymbol{\eta}_i. \tag{9.1.4}$$

Many engineering problems can be described as follows: find n constants $a_1, \ldots, a_n$ such that the mean square of the fitted error variable

$$P = E\left\{\left|\boldsymbol{\xi} - \sum_{i=1}^n a_i\boldsymbol{\eta}_i\right|^2\right\} \tag{9.1.5}$$

is minimized. In parameter estimation theory [329], such an estimate is called the *best linear mean square estimate* of $\boldsymbol{\xi}$.

Theorem 9.2 [329] *If the data vectors $\boldsymbol{\eta}_1, \ldots, \boldsymbol{\eta}_n$ constitute an orthonormal system, then the best mean square estimate of the random variable $\boldsymbol{\xi}$ is determined by*

$$\hat{\boldsymbol{\xi}} = \sum_{i=1}^n \langle\boldsymbol{\xi}, \boldsymbol{\eta}_i\rangle\boldsymbol{\eta}_i. \tag{9.1.6}$$

Consider the decomposition

$$\boldsymbol{\xi} = \hat{\boldsymbol{\xi}} + (\boldsymbol{\xi} - \hat{\boldsymbol{\xi}}). \tag{9.1.7}$$

It can be shown that the above decomposition is orthogonal, i.e., $\hat{\boldsymbol{\xi}} \perp (\boldsymbol{\xi} - \hat{\boldsymbol{\xi}})$.

The proof is simple:

$$
\begin{aligned}
E\{\hat{\boldsymbol{\xi}}(\boldsymbol{\xi}-\hat{\boldsymbol{\xi}})\} &= E\left\{\left(\sum_{j=1}^{n}\langle\boldsymbol{\xi},\boldsymbol{\eta}_j\rangle\boldsymbol{\eta}_j\right)\left(\boldsymbol{\xi}-\sum_{i=1}^{n}\langle\boldsymbol{\xi},\boldsymbol{\eta}_i\rangle\boldsymbol{\eta}_i\right)\right\} \\
&= E\left\{\boldsymbol{\xi}\sum_{j=1}^{n}\langle\boldsymbol{\xi},\boldsymbol{\eta}_j\rangle\boldsymbol{\eta}_j\right\} - E\left\{\sum_{i=1}^{n}\langle\boldsymbol{\xi},\boldsymbol{\eta}_i\rangle\boldsymbol{\eta}_i\sum_{j=1}^{n}\langle\boldsymbol{\xi},\boldsymbol{\eta}_j\rangle\boldsymbol{\eta}_j\right\} \\
&= \sum_{j=1}^{n}\left(E\{\langle\boldsymbol{\xi},\boldsymbol{\eta}_j\rangle\}\right)^2 - \sum_{i=1}^{n}\sum_{j=1}^{n}E\{\langle\boldsymbol{\xi},\boldsymbol{\eta}_i\rangle\}E\{\langle\boldsymbol{\xi},\boldsymbol{\eta}_j\rangle\}E\{\langle\boldsymbol{\eta}_i,\boldsymbol{\eta}_j\rangle\} \\
&= \sum_{j=1}^{n}\left(E\{\langle\boldsymbol{\xi},\boldsymbol{\eta}_j\rangle\}\right)^2 - \sum_{i=1}^{n}\left(E\{\langle\boldsymbol{\xi},\boldsymbol{\eta}_i\rangle\}\right)^2 \\
&= 0.
\end{aligned}
$$

Figure 9.2 shows, for a simple case, the orthogonal decomposition (9.1.7); see also (9.1.4).

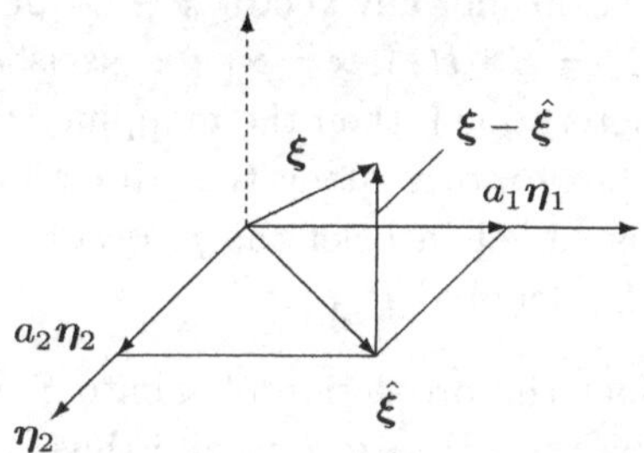

Figure 9.2 Orthogonal decomposition.

The projection $\text{Proj}\{\boldsymbol{\xi}\,|\,\boldsymbol{\eta}_1,\ldots,\boldsymbol{\eta}_n\}$ denotes the mean square estimate of the unknown parameter vector $\boldsymbol{\xi}$ for known data vectors $\boldsymbol{\eta}_1,\ldots,\boldsymbol{\eta}_n$. This estimate is sometimes written as $\hat{E}\{\boldsymbol{\xi}\,|\,\boldsymbol{\eta}_1,\ldots,\boldsymbol{\eta}_n\}$.

The above projection theorem reveals a method for finding the best linear mean square estimate, but it requires all the given data vectors $\boldsymbol{\eta}_1,\ldots,\boldsymbol{\eta}_n$ to be orthonormal. When the known data vectors are not mutually orthogonal, we need to use the prewhitening to render the original nonorthogonal data vectors as standard orthonormal white noise vectors with zero mean and unit variance. Then, we apply the projection theorem to the whitened data vectors to find a mean square estimate.

In some cases, we can easily find the orthogonal projection of the vector $\mathbf{x}$ onto the subspace M.

Theorem 9.3 [232] *Let H be an inner product space, and $\mathbf{x}$ be a vector in H. If M is an n-dimensional subspace in H, and $\{\mathbf{u}_1,\ldots,\mathbf{u}_n\}$ is a set of orthogonal* *basis vectors of the subspace* M *then* $\|\mathbf{x}-\hat{\mathbf{x}}\| \leq \|\mathbf{x}-\mathbf{y}\|$ *if and only if*

$$\hat{\mathbf{x}} = \frac{\langle \mathbf{x}, \mathbf{u}_1\rangle}{\langle \mathbf{u}_1, \mathbf{u}_1\rangle}\mathbf{u}_1 + \frac{\langle \mathbf{x}, \mathbf{u}_2\rangle}{\langle \mathbf{u}_2, \mathbf{u}_2\rangle}\mathbf{u}_2 + \cdots + \frac{\langle \mathbf{x}, \mathbf{u}_n\rangle}{\langle \mathbf{u}_n, \mathbf{u}_n\rangle}\mathbf{u}_n. \tag{9.1.8}$$

The significance of this theorem is that when M is a finite-dimensional subspace in the inner product space H, we can first find a group of orthogonal basis vectors $\{\mathbf{u}_1, \ldots, \mathbf{u}_n\}$ in the subspace M (for example, using Gram–Schmidt orthogonalization) and use (9.1.8) to compute the projection $\hat{\mathbf{x}}$ of the vector $\mathbf{x}$ onto the subspace M.

9.2 Projectors and Projection Matrices

The tool for projection is the projection matrix.

9.2.1 Projector and Orthogonal Projector

Definition 9.2 [410] Consider any vector $\mathbf{x} \in \mathbb{C}^n$ in the direct sum decomposition of the vector space $\mathbb{C}^n = S \oplus H$. If $\mathbf{x} = \mathbf{x}_1 + \mathbf{x}_2$ satisfies $\mathbf{x}_1 \in S$ and $\mathbf{x}_2 \in H$, and $\mathbf{x}_1$ and $\mathbf{x}_2$ are uniquely determined, then the mapping $\mathbf{Px} = \mathbf{x}_1$ is the projection of the vector $\mathbf{x}$ as its arrowhead moves along the "direction" of the subspace H onto the subspace S, and $\mathbf{P}$ is called in brief the *projector onto* subspace S *along* the "direction" of subspace H, denoted $\mathbf{P}_{S|H}$.

Let $\mathbf{y} = \mathbf{P}_{S|H}\mathbf{x}$ represent the projection of $\mathbf{x}$ onto S along H, then $\mathbf{y} \in S$. If we project $\mathbf{y}$ onto S along H, then $\mathbf{P}_{S|H}\mathbf{y} = \mathbf{y}$, and thus

$$\begin{aligned} \mathbf{P}_{S|H}\mathbf{y} = \mathbf{y} \quad &\Rightarrow \quad \mathbf{P}_{S|H}(\mathbf{P}_{S|H}\mathbf{x}) = \mathbf{P}_{S|H}\mathbf{P}_{S|H}\mathbf{x} = \mathbf{P}_{S|H}\mathbf{x} \\ &\Rightarrow \quad \mathbf{P}_{S|H}^2 = \mathbf{P}_{S|H}. \end{aligned}$$

Definition 9.3 A homogeneous linear operator $\mathbf{P}$ is called a *projection operator* if it is idempotent, i.e., $\mathbf{P}^2 = \mathbf{PP} = \mathbf{P}$.

The unique decomposition

$$\mathbf{x} = \mathbf{x}_1 + \mathbf{x}_2 = \mathbf{Px} + (\mathbf{I} - \mathbf{P})\mathbf{x} \tag{9.2.1}$$

maps any vector $\mathbf{x}$ in $\mathbb{C}^n$ onto the component $\mathbf{x}_1$ in the subspace S and the component $\mathbf{x}_2$ in the subspace H. The unique decomposition in (9.2.1) does not ensure that $\mathbf{x}_1$ and $\mathbf{x}_2$ are mutually orthogonal. However, in many practical applications it is required that the projections $\mathbf{x}_1$ and $\mathbf{x}_2$ of any vector $\mathbf{x}$ in complex vector space $\mathbb{C}^n$ onto two subspaces are mutually orthogonal. From this orthogonality requirement on $\mathbf{x}_1$ and $\mathbf{x}_2$,

$$\langle \mathbf{x}_1, \mathbf{x}_2\rangle = (\mathbf{Px})^H(\mathbf{I} - \mathbf{P})\mathbf{x} = 0 \quad \Rightarrow \quad \mathbf{x}^H\mathbf{P}^H(\mathbf{I} - \mathbf{P})\mathbf{x} = 0,\ \forall\, \mathbf{x} \neq \mathbf{0},$$

we immediately have

$$\mathbf{P}^H(\mathbf{I}-\mathbf{P})=\mathbf{O} \quad \Rightarrow \quad \mathbf{P}^H\mathbf{P}=\mathbf{P}^H=\mathbf{P}.$$

From this we can introduce the following definition.

DEFINITION 9.4 The projector $\mathbf{P}^{\perp}=\mathbf{I}-\mathbf{P}$ is known as the *orthogonal projector* of $\mathbf{P}$, if $\mathbf{P}$ is not only an idempotent matrix but is also Hermitian.

An orthogonal projector has the following properties [382].

1. If $\mathbf{I}-\mathbf{P}$ is the orthogonal projector of $\mathbf{P}$ then $\mathbf{P}$ is also the orthogonal projector of $\mathbf{I}-\mathbf{P}$.
2. If both $\mathbf{E}_1$ and $\mathbf{E}_2$ are orthogonal projectors, and $\mathbf{E}_1$ and $\mathbf{E}_2$ have no cross-term, i.e., $\mathbf{E}_1\mathbf{E}_2=\mathbf{E}_2\mathbf{E}_1=\mathbf{O}$, then $\mathbf{E}_1+\mathbf{E}_2$ is an orthogonal projector.
3. If both $\mathbf{E}_1$ and $\mathbf{E}_2$ are orthogonal projectors, and $\mathbf{E}_1\mathbf{E}_2=\mathbf{E}_2\mathbf{E}_1=\mathbf{E}_2$ then $\mathbf{E}_1-\mathbf{E}_2$ is an orthogonal projector.
4. If both $\mathbf{E}_1$ and $\mathbf{E}_2$ are orthogonal projectors and $\mathbf{E}_1\mathbf{E}_2=\mathbf{E}_2\mathbf{E}_1$ then $\mathbf{E}_1\mathbf{E}_2$ is an orthogonal projector.

In the following, we discuss orthogonal projectors from the viewpoint of signal processing.

As shown in Figure 9.3, a discrete time filter can be regarded as a projector, denoted $\mathbf{P}$. Let $\mathbf{x}(n)=[x(1),\ldots,x(n)]^T=\mathbf{s}(n)+\mathbf{v}(n)$ be the input vector of the filter at the discrete time n, where $\mathbf{s}(n)$ is the signal vector and $\mathbf{v}(n)$ is the additive white noise vector.

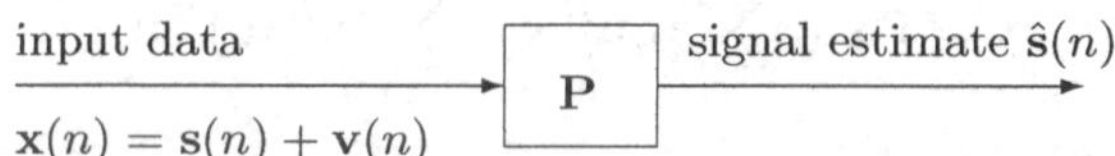

Figure 9.3 Projector representation of the filter.

The filter outputs a signal estimate $\hat{\mathbf{s}}(n)=\mathbf{P}\mathbf{x}(n)$. For simplicity we omit the time variable in the vectors $\mathbf{s}(n)$ and $\mathbf{x}(n)$ and denote them simply as $\mathbf{s}$ and $\mathbf{x}$.

The following are the basic requirements of the filter operator $\mathbf{P}$.

(1) In order to ensure that the signal through the filter does not incur "distortion", the projector $\mathbf{P}$ must be a linear operator.

(2) If the filter output $\hat{\mathbf{s}}$ is again passed through the filter, the signal estimate $\hat{\mathbf{s}}$ should not change. This implies that $\mathbf{PPx}=\mathbf{Px}=\hat{\mathbf{s}}$ must be satisfied. This condition is equivalent to $\mathbf{P}^2 \stackrel{\text{def}}{=} \mathbf{PP}=\mathbf{P}$, i.e., the projector $\mathbf{P}$ must be an idempotent operator.

(3) The signal estimate is $\hat{\mathbf{s}}=\mathbf{Px}$, so $\mathbf{x}-\mathbf{Px}$ denotes the estimation error of the filter operator $\mathbf{P}$. By the orthogonality-principle lemma, when a filter is optimal, its estimation error $\mathbf{x}-\mathbf{Px}$ should be orthogonal to the estimate $\mathbf{Px}$

of the expected response, i.e., $(\mathbf{x} - \mathbf{Px}) \perp \mathbf{Px}$. This implies that $\mathbf{P}^T = \mathbf{P}$, i.e., the projector must have Hermiticity.

Here we give a proof that $(\mathbf{x} - \mathbf{Px}) \perp \mathbf{Px} \Rightarrow \mathbf{P}^T = \mathbf{P}$. First, $(\mathbf{x} - \mathbf{Px}) \perp \mathbf{Px}$ can be equivalently written as

$$
\begin{aligned}
(\mathbf{Px})^T(\mathbf{x} - \mathbf{Px}) = 0 \quad &\Leftrightarrow \quad \mathbf{x}^T\mathbf{P}^T\mathbf{x} - \mathbf{x}^T\mathbf{P}^T\mathbf{Px} = 0 \\
&\Leftrightarrow \quad \mathbf{x}^T(\mathbf{P}^T - \mathbf{P}^T\mathbf{P})\mathbf{x} = 0, \ \forall \mathbf{x} \neq \mathbf{0},
\end{aligned}
$$

and thus $\mathbf{P}^T = \mathbf{P}^T\mathbf{P}$, whose transpose is $\mathbf{P} = \mathbf{P}^T\mathbf{P}$. Therefore, since $\mathbf{P}^T = \mathbf{P}^T\mathbf{P}$ and $\mathbf{P} = \mathbf{P}^T\mathbf{P}$, we have $\mathrm{P}^T = \mathbf{P}$.

9.2.2 Projection Matrices

The matrix form of a projector is called a *projection matrix*. Consider the construction of a projection matrix: it is an idempotent matrix that has only the eigenvalues 1 and 0. Let an $m \times m$ projection matrix $\mathbf{P}$ have r eigenvalues equal to 1, other $m - r$ eigenvalues equal to 0. Then, the projection matrix can be, in EVD form, written as

$$
\mathbf{P} = \sum_{i=1}^{m} \lambda_i \mathbf{u}_i \mathbf{u}_i^H = \sum_{i=1}^{r} \mathbf{u}_i \mathbf{u}_i^H. \tag{9.2.2}
$$

Consider the projection $\mathbf{y} = \mathbf{Px}$ of any $m \times 1$ vector $\mathbf{x}$:

$$
\mathbf{y} = \mathbf{Px} = \sum_{i=1}^{r} \mathbf{u}_i \mathbf{u}_i^H \mathbf{x} = \sum_{i=1}^{r} (\mathbf{x}^H \mathbf{u}_i)^H \mathbf{u}_i. \tag{9.2.3}
$$

Equation (9.2.3) reveals the two essential roles of a projection matrix:

(1) After the projection of a vector $\mathbf{x}$ by the projection matrix $\mathbf{P}$, the components of $\mathbf{x}$ correlated with the eigenvectors corresponding to the eigenvalue 1 of $\mathbf{P}$, $\mathbf{x}^H\mathbf{u}_i, i = 1, 2, \ldots, r$, are fully retained in the projection result $\mathbf{Px}$.

(2) The other components of $\mathbf{x}$, correlated with the eigenvectors corresponding to the eigenvalues 0 of $\mathbf{P}$, $\mathbf{x}^H\mathbf{u}_i, i = r+1, r+2, \ldots, m$, are fully cancelled, and do not appear in the projection result $\mathbf{Px}$.

Therefore, when, for example, a signal-to-noise ratio is high enough, the signal components of the observed data vector are fully retained by the projection matrix, and the noise components are fully cancelled by the projection matrix.

Let an $m \times n$ matrix $\mathbf{A}$ be a matrix with full column rank, i.e., $\mathrm{rank}(\mathbf{A}) = n$. Denote the column space of $\mathbf{A}$ by $A = \mathrm{Col}(\mathbf{A}) = \mathrm{Range}(\mathbf{A})$. An interesting question is, how do we construct a projection matrix $\mathbf{P_A}$ onto the column space A?

Because the rank of the matrix $\mathbf{A}$ is n, there are only n nonzero singular values

$\sigma_1, \ldots, \sigma_n$ in the SVD of $\mathbf{A}$:

$$\mathbf{A} = \mathbf{U}\boldsymbol{\Sigma}\mathbf{V}^H = [\mathbf{U}_1, \mathbf{U}_2] \begin{bmatrix} \boldsymbol{\Sigma}_1 \\ \mathbf{O}_{(m-n)\times n} \end{bmatrix} \mathbf{V}^H$$

$$= \mathbf{U}_1 \boldsymbol{\Sigma}_1 \mathbf{V}^H, \tag{9.2.4}$$

where the diagonal matrix $\boldsymbol{\Sigma}_1 = \mathbf{Diag}(\sigma_1, \ldots, \sigma_n)$, and

$$\mathbf{U}_1 = [\mathbf{u}_1, \ldots, \mathbf{u}_n], \quad \mathbf{U}_2 = [\mathbf{u}_{n+1}, \ldots, \mathbf{u}_m] \tag{9.2.5}$$

are respectively the singular-vector matrices corresponding to the n nonzero singular values and the $m - n$ zero singular values.

In Chapter 8, we obtained two important results on column space as follows.

(1) The n left singular vectors $\mathbf{u}_1, \ldots, \mathbf{u}_n$ corresponding to the nonzero singular values are the orthonormal basis vectors of the column space $\text{Col}(\mathbf{A})$, i.e.,

$$\text{Col}(\mathbf{A}) = \text{Span}\{\mathbf{u}_1, \ldots, \mathbf{u}_n\} = \text{Span}(\mathbf{U}_1). \tag{9.2.6}$$

(2) The $m-n$ left singular vectors $\mathbf{u}_{n+1}, \ldots, \mathbf{u}_m$ corresponding to the zero singular values are the orthonormal basis vectors of the null space $\text{Null}(\mathbf{A}^H)$, i.e.,

$$\text{Null}(\mathbf{A}^H) = (\text{Col}\,\mathbf{A})^\perp = \text{Span}\{\mathbf{u}_{n+1}, \ldots, \mathbf{u}_m\} = \text{Span}(\mathbf{U}_2). \tag{9.2.7}$$

Using the matrix $\mathbf{A}$ to make a linear transformation of the $n \times 1$ vector $\mathbf{x}$, we have the result

$$\mathbf{y} = \mathbf{A}\mathbf{x} = \sum_{i=1}^{n} \sigma_i \mathbf{u}_i \mathbf{v}_i^H \mathbf{x} = \sum_{i=1}^{n} \sigma_i \alpha_i \mathbf{u}_i, \tag{9.2.8}$$

where $\alpha_i = \mathbf{v}_i^H \mathbf{x}$ is the inner product of the eigenvector $\mathbf{v}$ and the data vector $\mathbf{x}$. Equation (9.2.8) shows that the linear transformation result $\mathbf{y} = \mathbf{A}\mathbf{x}$ is a linear combination of the left singular vectors $\mathbf{u}_1, \ldots, \mathbf{u}_n$ of $\mathbf{A}$.

The projection matrix $\mathbf{P_A}$ of the $m \times n$ linear transformation matrix $\mathbf{A}$ projects all the $m \times 1$ data vectors $\mathbf{x}$ onto the subspace defined by the matrix $\mathbf{A}$. The projection matrix $\mathbf{P_A}$ and the matrix $\mathbf{A}$ have the same eigenvectors: n of these vectors correspond to the nonzero eigenvalues of $\mathbf{P_A}$ or $\mathbf{A}$ and the other $m - n$ eigenvectors correspond to the zero eigenvalues. Because the projection matrix $\mathbf{P}$ has only the eigenvalues 1 and 0, n eigenvectors of $\mathbf{P}$ correspond to its eigenvalues 1 and the other $m - n$ eigenvectors correspond to its eigenvalues 0. In other words, the EVD of the projection matrix has the form

$$\mathbf{P_A} = [\mathbf{U}_1, \mathbf{U}_2] \begin{bmatrix} \mathbf{I}_n & \mathbf{O}_{n\times(m-n)} \\ \mathbf{O}_{(m-n)\times n} & \mathbf{O}_{(m-n)\times(m-n)} \end{bmatrix} \begin{bmatrix} \mathbf{U}_1^H \\ \mathbf{U}_2^H \end{bmatrix}$$

$$= \mathbf{U}_1 \mathbf{U}_1^H. \tag{9.2.9}$$

On the other hand, from (9.2.4) it follows that

$$\begin{aligned}\mathbf{A}\langle\mathbf{A},\mathbf{A}\rangle^{-1}\mathbf{A}^H &= \mathbf{A}(\mathbf{A}^H\mathbf{A})^{-1}\mathbf{A}^H\\ &= \mathbf{U}_1\boldsymbol{\Sigma}_1\mathbf{V}^H(\mathbf{V}\boldsymbol{\Sigma}_1\mathbf{U}_1^H\mathbf{U}_1\boldsymbol{\Sigma}_1\mathbf{V}^H)^{-1}\mathbf{V}\boldsymbol{\Sigma}_1\mathbf{U}_1^H\\ &= \mathbf{U}_1\boldsymbol{\Sigma}_1\mathbf{V}^H(\mathbf{V}\boldsymbol{\Sigma}_1^2\mathbf{V}^H)^{-1}\mathbf{V}\boldsymbol{\Sigma}_1\mathbf{U}_1^H\\ &= \mathbf{U}_1\boldsymbol{\Sigma}_1\mathbf{V}^H\mathbf{V}\boldsymbol{\Sigma}_1^{-2}\mathbf{V}^H\mathbf{V}\boldsymbol{\Sigma}_1\mathbf{U}_1^H\\ &= \mathbf{U}_1\mathbf{U}_1^H.\end{aligned}\tag{9.2.10}$$

Comparing (9.2.9) with (9.2.10), one immediately has a definition for the projection matrix $\mathbf{P}_\mathbf{A}$ of the linear transformation matrix $\mathbf{A}$:

$$\mathbf{P}_\mathbf{A} = \mathbf{A}\langle\mathbf{A},\mathbf{A}\rangle^{-1}\mathbf{A}^H.\tag{9.2.11}$$

The above derivation of the definition formula (9.2.11) of the projection matrix $\mathbf{P}_\mathbf{A}$ is the method commonly used in the mathematical literature. The key of the derivation is first to guess that the projection matrix has the form of $\mathbf{A}\langle\mathbf{A},\mathbf{A}\rangle^{-1}\mathbf{A}^H$, and then to verify this.

Here we present another derivation that is based on some basic facts about the Moore–Penrose inverse matrix and the projection matrix.

Postmultiply both sides of the mapping function $\mathbf{PU}=\mathbf{V}$ by $\mathbf{U}^\dagger\mathbf{U}$:

$$\mathbf{PUU}^\dagger\mathbf{U} = \mathbf{PU} = \mathbf{VU}^\dagger\mathbf{U},\ \forall\mathbf{U}.$$

Hence

$$\mathbf{PU}=\mathbf{V}\quad\Rightarrow\quad\mathbf{P}=\mathbf{VU}^\dagger = \mathbf{V}(\mathbf{U}^H\mathbf{U})^\dagger\mathbf{U}^H.\tag{9.2.12}$$

Let $A = \mathrm{Col}(\mathbf{A})$ be the column space of the matrix $\mathbf{A}\in\mathbb{C}^{m\times n}$, and let $A^\perp = (\mathrm{Col}(\mathbf{A}))^\perp = \mathrm{Null}(\mathbf{A}^H)$ be the orthogonal complement of the column space A. Hence, if $\mathbf{S}\in A^\perp$ then $\mathbf{S}$ and $\mathbf{A}$ are mutually orthogonal, i.e., $\mathbf{S}\perp\mathbf{A}$.

Therefore, we have two basic facts of the projection:

$$\mathbf{P}_\mathbf{A}\mathbf{A}=\mathbf{A},\quad \mathbf{P}_\mathbf{A}\mathbf{S}=\mathbf{O},\ \forall\mathbf{S}\in\mathrm{Range}(\mathbf{A})^\perp,\tag{9.2.13}$$

which can be rewritten as

$$\mathbf{P}_\mathbf{A}[\mathbf{A},\mathbf{S}] = [\mathbf{A},\mathbf{O}],\tag{9.2.14}$$

where $\mathbf{O}$ is the zero matrix. From (9.2.12) and (9.2.14) we immediately have

$$\begin{aligned}\mathbf{P}_\mathbf{A} &= [\mathbf{A},\mathbf{O}]([\mathbf{A},\mathbf{S}]^H[\mathbf{A},\mathbf{S}])^\dagger[\mathbf{A},\mathbf{S}]^H\\ &= [\mathbf{A},\mathbf{O}]\begin{bmatrix}\mathbf{A}^H\mathbf{A} & \mathbf{A}^H\mathbf{S}\\ \mathbf{S}^H\mathbf{A} & \mathbf{S}^H\mathbf{S}\end{bmatrix}^\dagger\begin{bmatrix}\mathbf{A}^H\\ \mathbf{S}^H\end{bmatrix}\\ &= [\mathbf{A},\mathbf{O}]\begin{bmatrix}\mathbf{A}^H\mathbf{A} & \mathbf{O}\\ \mathbf{O} & \mathbf{S}^H\mathbf{S}\end{bmatrix}^\dagger\begin{bmatrix}\mathbf{A}^H\\ \mathbf{S}^H\end{bmatrix}\\ &= \mathbf{A}(\mathbf{A}^H\mathbf{A})^\dagger\mathbf{A}^H.\end{aligned}\tag{9.2.15}$$

It is easy to be verify that the projection matrix $\mathbf{P}_\mathbf{A} = \mathbf{A}(\mathbf{A}^H\mathbf{A})^\dagger\mathbf{A}^H$ has idempotence, $\mathbf{P}_\mathbf{A}\mathbf{P}_\mathbf{A} = \mathbf{P}_\mathbf{A}$ and Hermiticity, $\mathbf{P}_\mathbf{A}^H = \mathbf{P}_\mathbf{A}$.

The projection matrix orthogonal to $\mathbf{P}_\mathbf{A}$ can be defined as follows:

$$\mathbf{P}_\mathbf{A}^\perp = \mathbf{I} - \mathbf{P}_\mathbf{A} = \mathbf{I} - \mathbf{A}\langle\mathbf{A},\mathbf{A}\rangle^{-1}\mathbf{A}^H. \tag{9.2.16}$$

9.2.3 Derivatives of Projection Matrix

Let a projection matrix be the function of some vector $\boldsymbol{\theta}$:

$$\mathbf{P}_\mathbf{A}(\boldsymbol{\theta}) = \mathbf{A}(\boldsymbol{\theta})\left(\mathbf{A}^H(\boldsymbol{\theta})\mathbf{A}(\boldsymbol{\theta})\right)^{-1}\mathbf{A}^H(\boldsymbol{\theta}) = \mathbf{A}(\boldsymbol{\theta})\mathbf{A}^\dagger(\boldsymbol{\theta})$$

where $\mathbf{A}^\dagger(\boldsymbol{\theta}) = \left(\mathbf{A}^H(\boldsymbol{\theta})\mathbf{A}(\boldsymbol{\theta})\right)^{-1}\mathbf{A}^H(\boldsymbol{\theta})$ is the pseudo-inverse matrix of $\mathbf{A}(\boldsymbol{\theta})$. For simplicity of representation, $\mathbf{P}_\mathbf{A}(\boldsymbol{\theta})$ will be denoted simply as $\mathbf{P}$.

Consider the first-order and the second-order derivatives of the projection matrix $\mathbf{P}$ with respect to the elements θ_i of the vector $\boldsymbol{\theta} = [\theta_1,\ldots,\theta_n]^T$.

Definite the first-order partial derivative of the projection matrix with respect to θ_i as

$$\mathbf{P}_i \overset{\text{def}}{=} \frac{\partial\mathbf{P}}{\partial\theta_i}. \tag{9.2.17}$$

Using the chain rule for the derivatives, one has

$$\mathbf{P}_i = \mathbf{A}_i\mathbf{A}^\dagger + \mathbf{A}\mathbf{A}_i^\dagger, \tag{9.2.18}$$

where

$$\mathbf{A}_i \overset{\text{def}}{=} \frac{\partial\mathbf{A}}{\partial\theta_i}, \quad \mathbf{A}_i^\dagger \overset{\text{def}}{=} \frac{\partial\mathbf{A}^\dagger}{\partial\theta_i} \tag{9.2.19}$$

are respectively the partial derivatives of the matrix $\mathbf{A}(\boldsymbol{\theta})$ and of its Moore–Penrose inverse matrix $\mathbf{A}^\dagger(\boldsymbol{\theta})$ with respect to θ_i.

The first-order partial derivative of the pseudo-inverse matrix $\mathbf{A}^\dagger$ is given by

$$\mathbf{A}_i^\dagger = (\mathbf{A}^H\mathbf{A})^{-1}\mathbf{A}_i^H\mathbf{P}^\perp - \mathbf{A}^\dagger\mathbf{A}_i\mathbf{A}^\dagger. \tag{9.2.20}$$

Combine (9.2.18) and (9.2.20):

$$\mathbf{P}_i = \mathbf{P}^\perp\mathbf{A}_i\mathbf{A}^\dagger + (\mathbf{P}^\perp\mathbf{A}_i\mathbf{A}^\dagger)^H. \tag{9.2.21}$$

The second-order partial derivative of the projection matrix is given by

$$\begin{aligned}\mathbf{P}_{i,j} =& \mathbf{P}_j^\perp\mathbf{A}_i\mathbf{A}^\dagger + \mathbf{P}^\perp\mathbf{A}_{i,j}\mathbf{A}^\dagger + \mathbf{P}^\perp\mathbf{A}_i\mathbf{A}_j^H \\ &+ (\mathbf{P}_j^\perp\mathbf{A}_i\mathbf{A}^\dagger + \mathbf{P}^\perp\mathbf{A}_{i,j}\mathbf{A}^\dagger + \mathbf{P}^\perp\mathbf{A}_i\mathbf{A}_j^H)^H.\end{aligned} \tag{9.2.22}$$

Noting that $\mathbf{P}_j^{\perp} = -\mathbf{P}_j$ and using (9.2.20), (9.2.22) can be expressed as [495]

$$\begin{aligned}\mathbf{P}_{i,j} = & -\mathbf{P}^{\perp}\mathbf{A}_j\mathbf{A}^{\dagger}\mathbf{A}_i\mathbf{A}^{\dagger} - (\mathbf{A}^{\dagger})^H\mathbf{A}_j^H\mathbf{P}^{\perp}\mathbf{A}_i\mathbf{A}^H + \mathbf{P}^{\perp}\mathbf{A}_{i,j}\mathbf{A}^{\dagger} \\ & + \mathbf{P}^{\perp}\mathbf{A}_i(\mathbf{A}^H\mathbf{A})^{-1}\mathbf{A}_j^H\mathbf{P}^{\perp} - \mathbf{P}^{\perp}\mathbf{A}_i\mathbf{A}^{\dagger}\mathbf{A}_j\mathbf{A}^{\dagger} \\ & + \Big(-\mathbf{P}^{\perp}\mathbf{A}_j\mathbf{A}^{\dagger}\mathbf{A}_i\mathbf{A}^{\dagger} - (\mathbf{A}^{\dagger})^H\mathbf{A}_j^H\mathbf{P}^{\perp}\mathbf{A}_i\mathbf{A}^H + \mathbf{P}^{\perp}\mathbf{A}_{i,j}\mathbf{A}^{\dagger} \\ & + \mathbf{P}^{\perp}\mathbf{A}_i(\mathbf{A}^H\mathbf{A})^{-1}\mathbf{A}_j^H\mathbf{P}^{\perp} - \mathbf{P}^{\perp}\mathbf{A}_i\mathbf{A}^{\dagger}\mathbf{A}_j\mathbf{A}^{\dagger}\Big)^H . \end{aligned} \tag{9.2.23}$$

The derivative formulas of the projection matrix are very useful in analyzing the statistical properties of some estimators relating to the projection matrix. Interested readers can refer to the literature [495].

9.3 Updating of Projection Matrices

The adaptive filter is a filter whose coefficients can be adapted over time. This kind of adaptive adjustment depends on the use of a simple time updating formula, as complex calculations cannot meet the demands of real-time signal processing. Therefore, in order to apply the projection matrix and the orthogonal projection matrix to the adaptive filter, it is necessary to deduce time-updating formulas for these two matrices.

9.3.1 Updating Formulas for Projection Matrices

Suppose that the present data space is $\{\mathbf{U}\}$, the projection matrix corresponding to $\mathbf{U}$ is $\mathbf{P}_U$ and the orthogonal projection matrix is $\mathbf{P}_U^{\perp}$. Here, $\mathbf{U}$ takes the form $\mathbf{X}_{1,m}(n)$ or $\mathbf{X}_{0,m-1}(n)$ etc. Now we assume that a new data vector $\mathbf{u}$ is added to the original basis set of $\{\mathbf{U}\}$. Generally, the new data vector $\mathbf{u}$ will provide some new information not included in the original basis set of $\{\mathbf{U}\}$. Because the data subspace is enlarged from $\{\mathbf{U}\}$ to $\{\mathbf{U},\mathbf{u}\}$, we should search for a "new" projection matrix $\mathbf{P}_{U,u}$, and its orthogonal projection matrix $\mathbf{P}_{U,u}^{\perp}$, corresponding to the new subspace.

From the perspective of adaptive updating, the simplest method for finding the updated projection matrix $\mathbf{P}_{U,u}$ from the known projection matrix $\mathbf{P}_U$ is to decompose $\mathbf{P}_{U,u}$ into two parts: one is the nonadaptive or known part, while the other is the adaptive updating part. There is a particularly useful way of decomposition, i.e., the nonadaptive part and the adaptive updating part are required to be mutually orthogonal. Specifically, the orthogonal decomposition of the projection matrix $\mathbf{P}_{U,u}$ is

$$\mathbf{P}_{U,u} = \mathbf{P}_U + \mathbf{P}_w, \tag{9.3.1}$$

where $\mathbf{P}_w = \mathbf{w}\langle\mathbf{w},\mathbf{w}\rangle^{-1}\mathbf{w}^H$ should be selected to satisfy the orthogonal condition

$$\langle\mathbf{P}_U,\mathbf{P}_w\rangle = (\mathrm{vec}\,\mathbf{P}_U)^H\mathrm{vec}\,\mathbf{P}_w = 0, \tag{9.3.2}$$

simply denoted as $\mathbf{P}_w \perp \mathbf{P}_U$.

Because the updating is implemented through the original projection matrix $\mathbf{P}_U$ and the new data vector $\mathbf{u}$, while $\mathbf{P}_U$ contain no component of the new data vector, the updating part $\mathbf{P}_w$ in the orthogonal decomposition should contain the new data vector $\mathbf{u}$.

Let $\mathbf{w} = \mathbf{Xu}$, i.e., $\mathbf{P}_w = \mathbf{Xu}\langle \mathbf{Xu}, \mathbf{Xu}\rangle^{-1}(\mathbf{Xu})^T$. Substituting $\mathbf{P}_w$ into the orthogonal condition formula (9.3.2), we obtain

$$\langle \mathbf{P}_U, \mathbf{P}_w \rangle = \mathbf{P}_U \mathbf{Xu}\langle \mathbf{Xu}, \mathbf{Xu}\rangle^{-1}(\mathbf{Xu})^T = \mathbf{0}.$$

As $\mathbf{w} = \mathbf{Xu} \neq \mathbf{0}$, the Moore–Penrose inverse matrix $\langle \mathbf{Xu}, \mathbf{Xu}\rangle^{-1}(\mathbf{Xu})^T$ cannot be a zero matrix, and thus the above equation implies that $\mathbf{P}_U\mathbf{Xu} = \mathbf{0}, \forall\, \mathbf{u}$, holds, namely $\mathbf{P}_U\mathbf{X} = \mathbf{O}$. That is to say, $\mathbf{X}$ should be the orthogonal projection matrix $\mathbf{P}_U^\perp$.

Summarizing the above discussion, we obtain the vector $\mathbf{w}$ as follows:

$$\mathbf{w} = \mathbf{P}_U^\perp \mathbf{u}. \tag{9.3.3}$$

That is, $\mathbf{w}$ is the orthogonal projection of the data vector $\mathbf{u}$ onto the column space of the data matrix $\mathbf{U}$.

By using the definition of the projection matrix and the symmetry of the orthogonal projection matrix $[\mathbf{P}_U^\perp]^T = \mathbf{P}_U^\perp$, we easily get

$$\mathbf{P}_w = \mathbf{w}\langle \mathbf{w}, \mathbf{w}\rangle^{-1}\mathbf{w}^T = \mathbf{P}_U^\perp \mathbf{u}\langle \mathbf{P}_U^\perp \mathbf{u}, \mathbf{P}_U^\perp \mathbf{u}\rangle^{-1}\mathbf{u}^T \mathbf{P}_U^\perp. \tag{9.3.4}$$

Substituting (9.3.4) into the orthogonal decomposition (9.3.1), we obtain an updating formula giving the "new" projection matrix:

$$\mathbf{P}_{U,u} = \mathbf{P}_U + \mathbf{P}_U^\perp \mathbf{u}\langle \mathbf{P}_U^\perp \mathbf{u}, \mathbf{P}_U^\perp \mathbf{u}\rangle^{-1}\mathbf{u}^T \mathbf{P}_U^\perp. \tag{9.3.5}$$

From $\mathbf{P}_{U,u}^\perp = \mathbf{I} - \mathbf{P}_{U,u}$, we immediately have the "new" orthogonal projection matrix

$$\mathbf{P}_{U,u}^\perp = \mathbf{P}_U^\perp - \mathbf{P}_U^\perp \mathbf{u}\langle \mathbf{P}_U^\perp \mathbf{u}, \mathbf{P}_U^\perp \mathbf{u}\rangle^{-1}\mathbf{u}^T \mathbf{P}_U^\perp. \tag{9.3.6}$$

By postmultiplying Equations (9.3.5) and (9.3.6) by a vector $\mathbf{y}$, we get the updating formulas

$$\mathbf{P}_{U,u}\mathbf{y} = \mathbf{P}_U\mathbf{y} + \mathbf{P}_U^\perp \mathbf{u}\langle \mathbf{P}_U^\perp \mathbf{u}, \mathbf{P}_U^\perp \mathbf{u}\rangle^{-1}\langle \mathbf{u}, \mathbf{P}_U^\perp \mathbf{y}\rangle, \tag{9.3.7}$$

$$\mathbf{P}_{U,u}^\perp\mathbf{y} = \mathbf{P}_U^\perp\mathbf{y} - \mathbf{P}_U^\perp \mathbf{u}\langle \mathbf{P}_U^\perp \mathbf{u}, \mathbf{P}_U^\perp \mathbf{u}\rangle^{-1}\langle \mathbf{u}, \mathbf{P}_U^\perp \mathbf{y}\rangle. \tag{9.3.8}$$

Premultiplying (9.3.7) and (9.3.8) respectively by a vector $\mathbf{z}^H$, we obtain

$$\langle \mathbf{z}, \mathbf{P}_{U,u}\mathbf{y}\rangle = \langle \mathbf{z}, \mathbf{P}_U\mathbf{y}\rangle + \langle \mathbf{z}, \mathbf{P}_U^\perp \mathbf{u}\rangle\langle \mathbf{P}_U^\perp \mathbf{u}, \mathbf{P}_U^\perp \mathbf{u}\rangle^{-1}\langle \mathbf{u}, \mathbf{P}_U^\perp \mathbf{y}\rangle, \tag{9.3.9}$$

$$\langle \mathbf{z}, \mathbf{P}_{U,u}^\perp\mathbf{y}\rangle = \langle \mathbf{z}, \mathbf{P}_U^\perp\mathbf{y}\rangle - \langle \mathbf{z}, \mathbf{P}_U^\perp \mathbf{u}\rangle\langle \mathbf{P}_U^\perp \mathbf{u}, \mathbf{P}_U^\perp \mathbf{u}\rangle^{-1}\langle \mathbf{u}, \mathbf{P}_U^\perp \mathbf{y}\rangle. \tag{9.3.10}$$

We conclude with the following remarks:

(1) Equations (9.3.5) and (9.3.6) are respectively the updating formulas for a projection matrix and its orthogonal projection matrix.
(2) Equations (9.3.7) and (9.3.8) are respectively the updating formulas for the projection and the orthogonal projection of a vector.
(3) Equations (9.3.9) and (9.3.10) are respectively the updating formulas for a scalar related to a projection matrix and its orthogonal projection matrix.

In the following, we present an application of the above updating formulas in prediction filter updating.

9.3.2 Prediction Filters

Assume that the data input and weighting tap coefficients are real numbers. For convenience, first introduce the *time-shifting operator* z^j:

$$z^{-j}\mathbf{x}(n) = [0, \ldots, 0, x(1), \ldots, x(n-j)]^T. \tag{9.3.11}$$

Here $z^{-j}\mathbf{x}(n)$ denotes a *time-shifting operation* rather than a multiplication. In addition, the starting point of time is assumed to be 1, i.e., $x(n) = 0$ for all $n \le 0$.

Consider the mth-order forward prediction filter

$$\hat{x}(k) = \sum_{i=1}^{m} w_i^{\mathrm{f}}(n) x(k-i), \quad k = 1, 2, \ldots, n, \tag{9.3.12}$$

where $w_i^{\mathrm{f}}(n), i = 1, \ldots, m$, denotes the weighting coefficient of the filter at time n. The above equation can be rewritten as a matrix equation:

$$\begin{bmatrix} 0 & 0 & \cdots & 0 \\ x(1) & 0 & \cdots & 0 \\ \vdots & \vdots & \vdots & \vdots \\ x(n-1) & x(n-2) & \cdots & x(n-m) \end{bmatrix} \begin{bmatrix} w_1^{\mathrm{f}}(n) \\ w_2^{\mathrm{f}}(n) \\ \vdots \\ w_m^{\mathrm{f}}(n) \end{bmatrix} = \begin{bmatrix} \hat{x}(1) \\ \hat{x}(2) \\ \vdots \\ \hat{x}(n) \end{bmatrix}. \tag{9.3.13}$$

Define a data matrix

$$\begin{aligned} \mathbf{X}_{1,m}(n) &\stackrel{\text{def}}{=} [z^{-1}\mathbf{x}(n), z^{-2}\mathbf{x}(n), \ldots, z^{-m}\mathbf{x}(n)] \\ &= \begin{bmatrix} 0 & 0 & \cdots & 0 \\ x(1) & 0 & \cdots & 0 \\ \vdots & \vdots & \vdots & \vdots \\ x(n-1) & x(n-2) & \cdots & x(n-m) \end{bmatrix}, \end{aligned}$$

and define respectively an mth-order forward prediction coefficient vector $\mathbf{w}_m^{\mathrm{f}}(n)$ and a forward predictor vector $\hat{\mathbf{x}}(n)$:

$$\mathbf{w}_m^{\mathrm{f}}(n) \stackrel{\text{def}}{=} [w_1^{\mathrm{f}}(n), w_2^{\mathrm{f}}(n), \ldots, w_m^{\mathrm{f}}(n)]^T, \tag{9.3.14}$$

$$\hat{\mathbf{x}}(n) \stackrel{\text{def}}{=} [\hat{x}(1), \hat{x}(2), \ldots, \hat{x}(n)]^T; \tag{9.3.15}$$

then (9.3.13) can be rewritten in the simpler form

$$\mathbf{X}_{1,m}(n)\mathbf{w}_m^{\mathrm{f}}(n) = \hat{\mathbf{x}}(n). \tag{9.3.16}$$

In order to find an LS estimate of the forward predictor vector, use $\mathbf{x}(n)$ instead of $\hat{\mathbf{x}}(n)$ in (9.3.16):

$$\mathbf{w}_m^{\mathrm{f}}(n) = \langle \mathbf{X}_{1,m}(n), \mathbf{X}_{1,m}(n)\rangle^{-1}\mathbf{X}_{1,m}^T(n)\mathbf{x}(n). \tag{9.3.17}$$

By substituting (9.3.17) into (9.3.16), we can express the forward predictor vector as

$$\hat{\mathbf{x}}(n) = \mathbf{P}_{1,m}(n)\mathbf{x}(n), \tag{9.3.18}$$

where $\mathbf{P}_{1,m}(n) = \mathbf{X}_{1,m}(n)\langle \mathbf{X}_{1,m}(n), \mathbf{X}_{1,m}(n)\rangle^{-1}\mathbf{X}_{1,m}^T(n)$ denotes the projection matrix of the data matrix $\mathbf{X}_{1,m}(n)$.

Defining the forward prediction error vector as

$$\mathbf{e}_m^{\mathrm{f}}(n) = [e_m^{\mathrm{f}}(1), \ldots, e_m^{\mathrm{f}}(n)]^T = \mathbf{x}(n) - \hat{\mathbf{x}}(n), \tag{9.3.19}$$

where $e_m^{\mathrm{f}}(k), k = 1, \ldots, n$, is the forward prediction error of the filter at time k, then from (9.3.18) and the definition of the orthogonal projection matrix, we have

$$\mathbf{e}_m^{\mathrm{f}}(n) = \mathbf{P}_{1,m}^{\perp}(n)\mathbf{x}(n). \tag{9.3.20}$$

The geometric interpretations of (9.3.18) and (9.3.20) are that the forward predictor vector $\hat{\mathbf{x}}(n)$ and the forward prediction error vector $\mathbf{e}_m^{\mathrm{f}}(n)$ are respectively the projection and the orthogonal projection of the data vector $\mathbf{x}(n)$ onto the subspace spanned by the data matrix $\mathbf{X}_{1,m}(n)$.

Now consider the backward prediction filter

$$\hat{x}(k-m) = \sum_{i=1}^{m} w_i^{\mathrm{b}}(n)x(k-m+i), \quad k = 1, \ldots, n; \tag{9.3.21}$$

here $w_i^{\mathrm{b}}(n), i = 1, \ldots, m$ are the weighting coefficients of an mth-order backward prediction filter at the time n. The above equation can be rewritten in the matrix–vector form

$$\begin{bmatrix} x(1) & 0 & \cdots & 0 \\ \vdots & \vdots & \vdots & \vdots \\ x(m) & x(m-1) & \cdots & 0 \\ x(m+1) & x(m) & \cdots & x(1) \\ \vdots & \vdots & \vdots & \vdots \\ x(n) & x(n-1) & \cdots & x(n-m+1) \end{bmatrix} \begin{bmatrix} w_m^{\mathrm{b}}(n) \\ w_{m-1}^{\mathrm{b}}(n) \\ \vdots \\ w_1^{\mathrm{b}}(n) \end{bmatrix} = \begin{bmatrix} 0 \\ \vdots \\ 0 \\ \hat{x}(1) \\ \vdots \\ \hat{x}(n-m) \end{bmatrix} \tag{9.3.22}$$

or

$$\mathbf{X}_{0,m-1}(n)\mathbf{w}_m^{\mathrm{b}}(n) = z^{-m}\hat{\mathbf{x}}(n), \tag{9.3.23}$$

where

$$\mathbf{X}_{0,m-1}(n) = [z^0\mathbf{x}(n), z^{-1}\mathbf{x}(n), \ldots, z^{-m+1}\mathbf{x}(n)]$$
$$= \begin{bmatrix} x(1) & 0 & \cdots & 0 \\ x(2) & x(1) & \cdots & 0 \\ \vdots & \vdots & \vdots & \vdots \\ x(n) & x(n-1) & \cdots & x(n-m+1) \end{bmatrix}, \tag{9.3.24}$$

$$\mathbf{w}_m^{\mathrm{b}}(n) = [w_m^{\mathrm{b}}(n), w_{m-1}^{\mathrm{b}}(n), \ldots, w_1^{\mathrm{b}}(n)]^T, \tag{9.3.25}$$

$$\hat{\mathbf{x}}(n-m) = z^{-m}\hat{\mathbf{x}}(n) = [0, \ldots, 0, \hat{x}(1), \ldots, \hat{x}(n-m)]^T. \tag{9.3.26}$$

Using the known data vector $\mathbf{x}(n)$ instead of the unknown predictor vector $\hat{\mathbf{x}}(n)$ in (9.3.23), we get an LS solution for the weighting vector of the backward prediction filter:

$$\mathbf{w}_m^{\mathrm{b}}(n) = \langle \mathbf{X}_{0,m-1}(n), \mathbf{X}_{0,m-1}(n)\rangle^{-1}\mathbf{X}_{0,m-1}^T(n)\mathbf{x}(n-m). \tag{9.3.27}$$

Substituting (9.3.27) into (9.3.23), the backward predictor vector can be expressed as follows:

$$\hat{\mathbf{x}}(n-m) = \mathbf{P}_{0,m-1}(n)\mathbf{x}(n-m) = \mathbf{P}_{0,m-1}(n)z^{-m}\mathbf{x}(n), \tag{9.3.28}$$

where

$$\mathbf{P}_{0,m-1}(n) = \mathbf{X}_{0,m-1}(n)\langle \mathbf{X}_{0,m-1}(n), \mathbf{X}_{0,m-1}(n)\rangle^{-1}\mathbf{X}_{0,m-1}^T(n)$$

is the prediction matrix of $\mathbf{X}_{0,m-1}(n)$.

Define the backward prediction error vector

$$\mathbf{e}_m^{\mathrm{b}}(n) = [e_m^{\mathrm{b}}(1), \ldots, e_m^{\mathrm{b}}(n)]^T = \mathbf{x}(n-m) - \hat{\mathbf{x}}(n-m) \tag{9.3.29}$$

in which $e_m^{\mathrm{b}}(k), k = 1, \ldots, n$, is the backward prediction error at time k. Then, by substituting (9.3.28) into (9.3.29), we can use the orthogonal prediction matrix to express the backward prediction error vector:

$$\mathbf{e}_m^{\mathrm{b}}(n) = \mathbf{P}_{0,m-1}^{\perp}(n)z^{-m}\mathbf{x}(n). \tag{9.3.30}$$

The geometric interpretation of (9.3.28) and (9.3.29) is as follows: the backward predictor vector $\hat{\mathbf{x}}(n-m)$ and the backward prediction error vector $\mathbf{e}_m^{\mathrm{b}}(n)$ are respectively the projection and the orthogonal projection of the shifted data vector $z^{-m}\mathbf{x}(n)$ onto the subspace spanned by the data matrix $\mathbf{X}_{0,m-1}(n)$.

Define an $n \times 1$ unit vector

$$\boldsymbol{\pi} = [0, \ldots, 0, 1]^T, \tag{9.3.31}$$

called an *extracting vector* since the last entry of $\mathbf{u} = [u_1, \ldots, u_n]^T$ is extracted by $\boldsymbol{\pi}^T\mathbf{u}$. Then, the current forward and backward prediction errors can be expressed using the inner products of the prediction error vectors and the extracting vector $\boldsymbol{\pi}$:

$$e^{\rm f}_m(n) = \langle \boldsymbol{\pi}, \mathbf{e}^{\rm f}_m(n)\rangle = \langle \boldsymbol{\pi}, \mathbf{P}^{\perp}_{1,m}(n)\mathbf{x}(n)\rangle, \tag{9.3.32}$$

$$e^{\rm b}_m(n) = \langle \boldsymbol{\pi}, \mathbf{e}^{\rm b}_m(n)\rangle = \langle \boldsymbol{\pi}, \mathbf{P}^{\perp}_{0,m-1}(n)z^{-m}\mathbf{x}(n)\rangle. \tag{9.3.33}$$

The forward and backward prediction residuals at time n are defined respectively as

$$\epsilon^{\rm f}_m(n) = \langle \mathbf{e}^{\rm f}_m(n), \mathbf{e}^{\rm f}_m(n)\rangle, \tag{9.3.34}$$

$$\epsilon^{\rm b}_m(n) = \langle \mathbf{e}^{\rm b}_m(n), \mathbf{e}^{\rm b}_m(n)\rangle. \tag{9.3.35}$$

Using the shifting property of z^{-1}, we obtain the forward and backward prediction residuals at the time $n-1$:

$$\epsilon^{\rm f}_m(n-1) = \langle z^{-1}\mathbf{e}^{\rm f}_m(n), z^{-1}\mathbf{e}^{\rm f}_m(n)\rangle, \tag{9.3.36}$$

$$\epsilon^{\rm b}_m(n-1) = \langle z^{-1}\mathbf{e}^{\rm b}_m(n), z^{-1}\mathbf{e}^{\rm b}_m(n)\rangle. \tag{9.3.37}$$

9.3.3 Updating of Lattice Adaptive Filter

A lattice adaptive filter is a least squares type of filter consisting of coupled forward and backward prediction filters.

The forward and backward prediction error equations of a lattice filter are given by

$$e^{\rm f}_{m+1}(n) = e^{\rm f}_m(n) - K^{\rm b}_{m+1}(n)e^{\rm b}_m(n-1), \tag{9.3.38}$$

$$e^{\rm b}_{m+1}(n) = -K^{\rm f}_{m+1}(n)e^{\rm f}_m(n) + e^{\rm b}_m(n-1), \tag{9.3.39}$$

where $K^f_{m+1}(n)$ and $K^b_{m+1}(n)$ are respectively the forward- and backward-reflecting coefficients of the $(m+1)$th-order lattice filter.

The lattice adaptive filter design is to derive the recursive formulas of the forward- and backward-reflecting coefficients. Owing to the coupling of the forward and backward predictions, the updating design of a lattice adaptive filter is complicated. Fortunately, the updating formulas for the projection matrix and the orthogonal projection matrix provide an effective method for designing a lattice adaptive filter [12].

Before deriving the recursive formulae for the lattice filter, we first prove an important relationship:

$$z^{-1}\mathbf{P}^{\perp}_{0,m-1}(n)z^{-m} = \mathbf{P}^{\perp}_{1,m}(n)z^{-m-1}. \tag{9.3.40}$$

Proof Using the shifting property of z^{-1}, it is easy to see that for any vector $\mathbf{y}(n)$,

$$\begin{aligned} z^{-1}\mathbf{P}^{\perp}_{0,m-1}(n)z^{-m}\mathbf{y}(n) &= \mathbf{P}^{\perp}_{0,m-1}(n-1)z^{-m}\mathbf{y}(n-1) \\ &= \mathbf{P}^{\perp}_{0,m-1}(n-1)z^{-m-1}\mathbf{y}(n). \end{aligned} \tag{9.3.41}$$

Noting that

$$\begin{aligned}\mathbf{X}_{0,m-1}(n-1) &= z^{-1}[z^0\mathbf{x}(n), z^{-1}\mathbf{x}(n), \ldots, z^{-m+1}\mathbf{x}(n)]\\ &= [z^{-1}\mathbf{x}(n), z^{-2}\mathbf{x}(n), \ldots, z^{-m}\mathbf{x}(n)]\\ &= \mathbf{X}_{1,m}(n),\end{aligned}$$

we have

$$\begin{aligned}&\mathbf{P}^{\perp}_{0,m-1}(n-1)\\ &= \mathbf{I} - \mathbf{X}_{0,m-1}(n-1)\langle \mathbf{X}_{0,m-1}(n-1), \mathbf{X}_{0,m-1}(n-1)\rangle^{-1}\mathbf{X}^T_{0,m-1}(n-1)\\ &= \mathbf{I} - \langle \mathbf{X}_{1,m}(n)\langle \mathbf{X}_{1,m}(n), \mathbf{X}_{1,m}(n)\rangle^{-1}\mathbf{X}^T_{1,m}(n)\\ &= \mathbf{P}^{\perp}_{1,m}(n).\end{aligned}$$

Then, Equation (9.3.41) can be written as

$$z^{-1}\mathbf{P}^{\perp}_{0,m-1}(n)z^{-m}\mathbf{y}(n) = \mathbf{P}^{\perp}_{1,m}(n)z^{-m-1}\mathbf{y}(n).$$

As the above equation holds for any nonzero vector $\mathbf{y}(n)$, Equation (9.3.40) is true. □

For the recursion of the forward prediction error $e^{\rm f}_m(n) = \langle \boldsymbol{\pi}, \mathbf{P}^{\perp}_{1,m}(n)\mathbf{x}(n)\rangle$ in (9.3.32), from (9.3.40) it follows that

$$\begin{aligned}\mathbf{P}^{\perp}_{1,m}(n)z^{-m-1}\mathbf{x}(n) &= z^{-1}\mathbf{P}^{\perp}_{0,m-1}z^{-m}\mathbf{x}(n)\\ &= z^{-1}\mathbf{e}^{\rm b}_m(n)\\ &= \mathbf{e}^{\rm b}_m(n-1).\end{aligned} \tag{9.3.42}$$

Letting $\mathbf{z} = \boldsymbol{\pi}, \mathbf{U} = \mathbf{X}_{1,m}(n), \mathbf{u} = z^{-m-1}\mathbf{x}(n)$ and $\mathbf{y} = \mathbf{x}(n)$ in (9.3.10), we have

$$\begin{aligned}\{\mathbf{U}, \mathbf{u}\} &= \{\mathbf{X}_{1,m}(n), z^{-m-1}\mathbf{x}(n)\}\\ &= \{\mathbf{X}_{1,m+1}(n)\},\end{aligned} \tag{9.3.43}$$

and hence

$$\left.\begin{aligned}\mathbf{P}^{\perp}_{U} &= \mathbf{P}^{\perp}_{1,m}(n),\\ \mathbf{P}^{\perp}_{U,u} &= \mathbf{P}^{\perp}_{1,m+1}(n),\\ \mathbf{P}^{\perp}_{U}\mathbf{u} &= \mathbf{P}^{\perp}_{1,m}(n)z^{-m-1}\mathbf{x}(n) = \mathbf{e}^{\rm b}_m(n-1),\\ \mathbf{P}^{\perp}_{U}\mathbf{y} &= \mathbf{P}^{\perp}_{1,m}(n)\mathbf{x}(n),\\ \mathbf{P}^{\perp}_{U,u}\mathbf{y} &= \mathbf{P}^{\perp}_{1,m+1}(n)\mathbf{x}(n).\end{aligned}\right\} \tag{9.3.44}$$

Substituting (9.3.44) into (9.3.10) and using (9.3.42), we get

$$\begin{aligned} e^{\rm f}_{m+1}(n) &= \langle \boldsymbol{\pi}, \mathbf{P}^{\perp}_{1,m+1}(n)\mathbf{x}(n)\rangle \\ &= \langle \boldsymbol{\pi}, \mathbf{P}^{\perp}_{1,m}(n)\mathbf{x}(n)\rangle - \frac{\langle \boldsymbol{\pi}, \mathbf{e}^{\rm b}_m(n-1)\rangle\langle \mathbf{P}^{\perp}_m(n)\mathbf{x}(n), z^{-m-1}\mathbf{x}(n)\rangle}{\langle \mathbf{e}^{\rm b}_m(n-1), \mathbf{e}^{\rm b}_m(n-1)\rangle} \\ &= e^{\rm f}_m(n) - \frac{\langle \boldsymbol{\pi}, \mathbf{e}^{\rm b}_m(n-1)\rangle\langle \mathbf{P}^{\perp}_m(n)\mathbf{x}(n), \mathbf{P}^{\perp}_m(n)z^{-m-1}\mathbf{x}(n)\rangle}{\langle \mathbf{e}^{\rm b}_m(n-1), \mathbf{e}^{\rm b}_m(n-1)\rangle} \\ &= e^{\rm f}_m(n) - \frac{\langle \boldsymbol{\pi}, \mathbf{e}^{\rm b}_m(n-1)\rangle\langle \mathbf{P}^{\perp}_m(n)\mathbf{x}(n), \mathbf{e}^{\rm b}_m(n-1)\rangle}{\langle \mathbf{e}^{\rm b}_m(n-1), \mathbf{e}^{\rm b}_m(n-1)\rangle} \\ &= e^{\rm f}_m(n) - \frac{\langle \mathbf{e}^{\rm f}_m(n), z^{-1}\mathbf{e}^{\rm b}_m(n)\rangle}{\epsilon^{\rm b}_m(n-1)} e^{\rm b}_m(n-1). \end{aligned} \tag{9.3.45}$$

By comparing (9.3.45) with (9.3.39), the backward reflecting coefficient is given by

$$K^{\rm b}_{m+1}(n) = \frac{\langle \mathbf{e}^{\rm f}_m(n), z^{-1}\mathbf{e}^{\rm b}_m(n)\rangle}{\epsilon^{\rm b}_m(n-1)} = \frac{\Delta_{m+1}(n)}{\epsilon^{\rm b}_m(n-1)}, \tag{9.3.46}$$

where

$$\Delta_{m+1}(n) = \langle \mathbf{e}^{\rm f}_m(n), z^{-1}\mathbf{e}^{\rm b}_m(n)\rangle \tag{9.3.47}$$

is called the partial correlation coefficient.

Mimicking the above derivation, the forward-reflecting coefficient is found as

$$K^{\rm f}_{m+1}(n) = \frac{\Delta_{m+1}(n)}{\epsilon^{\rm f}_m(n)}. \tag{9.3.48}$$

Summarizing the above discussions, we have the following conclusion: the design of a lattice adaptive filter amounts to deriving recursive formulas for the partial correlation coefficient $\Delta_{m+1}(n)$, the forward prediction residual $\epsilon^{\rm f}_m(n)$, the backward prediction residual $\epsilon^{\rm b}_m(n)$, the forward reflecting coefficient $K^{\rm f}_{m+1}(n)$ and the backward reflecting coefficient $K^{\rm b}_{m+1}(n)$.

In order to derive the recursive formula for the forward prediction residual $\epsilon^{\rm f}_m(n)$, substitute $\mathbf{e}^{\rm f}_m(n) = \mathbf{P}^{\perp}_{1,m}(n)\mathbf{x}(n)$ into (9.3.34):

$$\begin{aligned} \epsilon^{\rm f}_m(n) &= \langle \mathbf{P}^{\perp}_{1,m}(n)\mathbf{x}(n), \mathbf{P}^{\perp}_{1,m}(n)\mathbf{x}(n)\rangle \\ &= \mathbf{x}^T(n)[\mathbf{P}^{\perp}_{1,m}(n)]^T\mathbf{P}^{\perp}_{1,m}(n)\mathbf{x}(n) \\ &= \mathbf{x}^T(n)\mathbf{P}^{\perp}_{1,m}(n)\mathbf{x}(n) \\ &= \langle \mathbf{x}(n), \mathbf{P}^{\perp}_{1,m}(n)\mathbf{x}(n)\rangle. \end{aligned} \tag{9.3.49}$$

Similarly, from (9.3.37) it can be shown that

$$\epsilon^{\rm b}_m(n) = \langle z^{-m}\mathbf{x}(n), \mathbf{P}^{\perp}_{0,m-1}(n)\mathbf{x}(n-m)\rangle. \tag{9.3.50}$$

Letting $\mathbf{z} = \mathbf{y} = \mathbf{x}, \mathbf{U} = \mathbf{X}_{1,m}(n)$ and $\mathbf{u} = z^{-m-1}\mathbf{x}(n)$ in (9.3.10), from (9.3.49) it follows that

$$\epsilon_{m+1}^{\mathrm{f}}(n) = \epsilon_m^{\mathrm{f}}(n) - \frac{\Delta_{m+1}^2(n)}{\epsilon_m^{\mathrm{b}}(n-1)}. \tag{9.3.51}$$

Similarly, letting $\mathbf{z} = \mathbf{y} = z^{-m-1}\mathbf{x}, \mathbf{U} = \mathbf{X}_{1,m}(n)$ and $\mathbf{u} = \mathbf{x}(n)$ in (9.3.10), then from (9.3.50) we have

$$\epsilon_{m+1}^{\mathrm{b}}(n) = \epsilon_m^{\mathrm{f}}(n-1) - \frac{\Delta_{m+1}^2(n)}{\epsilon_m^{\mathrm{f}}(n)}. \tag{9.3.52}$$

By [12], there is the following relationship between the projection matrix $\mathbf{P}_{U,\pi}(n)$ of the subspace $\{\mathbf{U}, \boldsymbol{\pi}\}$ and the projection matrix $\mathbf{P}_U(n-1)$ of the subspace $\{\mathbf{U}\}$:

$$\mathbf{P}_{U,\pi}(n) = \begin{bmatrix} \mathbf{P}_U(n-1) & \mathbf{0} \\ \mathbf{O} & 1 \end{bmatrix}, \tag{9.3.53}$$

Likewise, we have

$$\mathbf{P}_{U,\pi}^{\perp}(n) = \begin{bmatrix} \mathbf{P}_U^{\perp}(n-1) & \mathbf{0} \\ \mathbf{O} & 0 \end{bmatrix}. \tag{9.3.54}$$

Letting $\mathbf{z} = \mathbf{x}(n), \mathbf{y} = z^{-m-1}\mathbf{x}(n), \mathbf{U} = \mathbf{X}_{1,m}(n)$ and $\mathbf{u} = \boldsymbol{\pi}$, and using (9.3.54), the partial correlation coefficient can be recursively computed from

$$\Delta_{m+1}(n) = \Delta_{m+1}(n-1) + \frac{\epsilon_m^{\mathrm{f}}(n)\epsilon_m^{\mathrm{b}}(n-1)}{\gamma_m(n-1)}, \tag{9.3.55}$$

where

$$\gamma_m(n-1) = \langle \boldsymbol{\pi}, \mathbf{P}_{1,m}^{\perp}\mathbf{x}(n)\boldsymbol{\pi} \rangle \tag{9.3.56}$$

is known as the angular parameter.

Letting $\mathbf{z} = \mathbf{y} = \boldsymbol{\pi}, \mathbf{U} + \mathbf{X}_{1,m}(n)$ and $\mathbf{u} = z^{-m-1}\mathbf{x}(n)$ in (9.3.10), (9.3.56) then yields a recursive formula for the angular parameter:

$$\gamma_{m+1}(n-1) = \gamma_m(n-1) - \frac{\left(e_m^{\mathrm{b}}(n-1)\right)^2}{\epsilon_m^{\mathrm{b}}(n-1)}. \tag{9.3.57}$$

9.4 Oblique Projector of Full Column Rank Matrix

As described in the previous sections, the orthogonal projection of a vector $\mathbf{z}$ onto a subspace H can be decomposed into the sum of its orthogonal projections onto the subspaces H_1 and H_2, where H_2 is required to be the orthogonal complement of H_1, and the orthogonal projector must be idempotent and Hermitian. If the subspace H_2 is not the orthogonal complement of H_1, then the idempotent operator is no longer Hermitian. An idempotent operator without Hermiticity is an *oblique projector*.

The oblique projector was first proposed by Murray [335] and Lorch [303] in the 1930s. Later, Afriat [11], Lyantse [308], Rao and Mitra [410], Halmos [195], Kato [240] and Takeuchi *et al.* [461] made further studies. In 1989, Kayalar and Weinert

[242] presented an application of the oblique projector to array signal processing and derived new formulas and an iterative algorithm for computing the oblique projector. In 1994, Behrens and Scharf [31] derived more practical computation formulas for the oblique projector onto the column space of a matrix. In 2000, Vandaele and Moonen [487] presented formulas for oblique projection formulas onto the row space of a matrix using the LQ decomposition of a matrix.

Owing to the wide applications of the oblique projector, in this section and the next we discuss the theory, methods and applications of oblique projectors of matrices with full column rank and full row rank, respectively.

9.4.1 Definition and Properties of Oblique Projectors

Let V be a Hilbert space and H be a closed subspace in V spanned by the observation vectors. Assume the data are divided into two subsets which span respectively the closed subspaces H_1 and H_2 such that $H = H_1 + H_2$, where the intersection of H_1 and H_2 satisfies $H_1 \cap H_2 = \{0\}$, i.e., the two subspaces H_1 and H_2 are nonoverlapping or disjoint. Note the fact that the two subspaces are disjoint shows only they have no common nonzero element, and does not implies their orthogonality.

Let $\mathbf{P}, \mathbf{P}_1, \mathbf{P}_2$ respectively be orthogonal projectors onto the subspaces H, H_1, H_2. Given a vector $\mathbf{z} \in V$, we hope compute the orthogonal projection of $\mathbf{z}$ onto the subspace H, denoted $\mathbf{Pz}$. Aronszajn [17] presented a method for computing the projection $\mathbf{Pz}$: first find the orthogonal projections of $\mathbf{z}$ onto the subspaces H_1 and H_2 to get $\mathbf{P}_1\mathbf{z}$ and $\mathbf{P}_2\mathbf{z}$, respectively. Then use the synthesizing formula

$$\mathbf{Pz} = (\mathbf{I} - \mathbf{P}_2)(\mathbf{I} - \mathbf{P}_1\mathbf{P}_2)^{-1}\mathbf{P}_1\mathbf{z} + (\mathbf{I} - \mathbf{P}_1)(\mathbf{I} - \mathbf{P}_2\mathbf{P}_1)^{-1}\mathbf{P}_2\mathbf{z} \tag{9.4.1}$$

to yield $\mathbf{Pz}$.

The above Aronszajn synthetic formula can be directly used to solve some statistical interpolation problems [9], [380], [423]. And, as pointed in [242], the dual filter smoothing formula is simply a special case of the Aronszajn synthetic formula.

It should be pointed out that when the subspace H_2 is the orthogonal complement of H_1, the orthogonal projectors are $\mathbf{P}_2 = \mathbf{P}_1^{\perp}$ and $\mathbf{P}_1 = \mathbf{P}_2^{\perp}$. In this case, the Aronszajn synthetic formula (9.4.1) reduces to

$$\mathbf{Pz} = \mathbf{P}_1\mathbf{z} + \mathbf{P}_1^{\perp}\mathbf{z} = \mathbf{P}_2\mathbf{z} + \mathbf{P}_2^{\perp}\mathbf{z}, \tag{9.4.2}$$

i.e., a typical orthogonal decomposition. Hence, the Aronszajn synthetic formula is a generalization of the orthogonal decomposition.

The Aronsajn synthetic method applies two orthogonal projectors $\mathbf{P}_1$ and $\mathbf{P}_2$ onto two nonorthogonal subspaces to compute the orthogonal projection $\mathbf{Pz}$. This method has the following two disadvantages:

(1) It needs a large amount of calculation.

(2) It cannot easily be generalized to cases where the subspace H is decomposed into three or more subspaces.

Two inherent disadvantages of the Aronszajn synthetic method can be avoided by the use of nonorthogonal projection. This nonorthogonal projection is known as oblique projection, to be discussed below.

In order to make the calculation of the orthogonal projection $\mathbf{Pz}$ as simple as possible, the easiest way is to let $\mathbf{Pz}$ be the direct sum of the projections of the vector $\mathbf{z}$ onto the subspaces H_1 and H_2, i.e.,

$$\mathbf{Pz} = \mathbf{E}_1\mathbf{z} + \mathbf{E}_2\mathbf{z}, \tag{9.4.3}$$

where $\mathbf{E}_1$ and $\mathbf{E}_2$ are projectors. Although the above equation still synthesizes the orthogonal projection $\mathbf{Pz}$, it is different from the Aronszajn synthetic formula (9.4.2) in that the two projection matrices in (9.4.3) are no longer orthogonal: $\mathbf{E}_1 \neq \mathbf{E}_2^{\perp}$, i.e., the two subspaces H_1 and H_2 are not mutually orthogonal. In this case, we say that

$$H = H_1 \oplus H_2 \tag{9.4.4}$$

is the *direct sum decomposition* of the subspace H. Note that the symbols $\mathbf{P}$ and $\mathbf{E}$ are used to represent respectively an orthogonal projector and an nonorthogonal projector. A nonorthogonal projector is known as an *oblique projector.*

It should be noted that both orthogonal projectors and oblique projectors must satisfy the idempotence of any projector. It is easy to verify that $(\mathbf{I} - \mathbf{P}_2)(\mathbf{I} - \mathbf{P}_1\mathbf{P}_2)^{-1}\mathbf{P}_1$ and $(\mathbf{I}-\mathbf{P}_1)(\mathbf{I}-\mathbf{P}_2\mathbf{P}_1)^{-1}\mathbf{P}_2$ in the Aronszajn synthetic formula (9.4.1) are not orthogonal or oblique projectors, because they are not idempotent operators.

The question is, how do we construct the two oblique projectors $\mathbf{E}_1$ and $\mathbf{E}_2$ in (9.4.3)? To this end, let us consider a full column rank matrix.

Let the $n \times m$ matrix $\mathbf{H}$ be a full column rank matrix whose range (space) is

$$\text{Range}(\mathbf{H}) = \{\mathbf{y} \in \mathbb{C}^m | \mathbf{y} = \mathbf{Hx},\ \mathbf{x} \in \mathbb{C}^n\}, \tag{9.4.5}$$

and whose null space is

$$\text{Null}(\mathbf{H}) = \{\mathbf{x} \in \mathbb{C}^n | \mathbf{Hx} = \mathbf{0}\}. \tag{9.4.6}$$

Now let the $n \times m$ full column rank matrix $\mathbf{H}$ and an $n \times k$ full row rank matrix be combined into an $n \times (m + k)$ matrix $[\mathbf{H}, \mathbf{S}]$, where $m + k < n$ such that the column rank of the matrix $[\mathbf{H}, \mathbf{S}]$ is less than the number of its rows and the column vectors of $\mathbf{H}$ and the column vectors of $\mathbf{S}$ are linearly independent. Owing to this linear independence, the two ranges Range($\mathbf{H}$) and Range($\mathbf{S}$) are disjoint.

According to the definition of the projection matrix, the orthogonal projector onto the range space Range$(\mathbf{H}, \mathbf{S})$ is given by

$$\mathbf{P}_{HS} = [\mathbf{H}, \mathbf{S}] \langle [\mathbf{H}, \mathbf{S}], [\mathbf{H}, \mathbf{S}] \rangle^{-1} [\mathbf{H}, \mathbf{S}]^H = [\mathbf{H}, \mathbf{S}] \begin{bmatrix} \mathbf{H}^H\mathbf{H} & \mathbf{H}^H\mathbf{S} \\ \mathbf{S}^H\mathbf{H} & \mathbf{S}^H\mathbf{S} \end{bmatrix}^{-1} \begin{bmatrix} \mathbf{H}^H \\ \mathbf{S}^H \end{bmatrix}. \tag{9.4.7}$$

Equation (9.4.7) shows that the orthogonal projector $\mathbf{P}_{HS}$ can be decomposed as follows [31]:

$$\mathbf{P}_{HS} = \mathbf{E}_{H|S} + \mathbf{E}_{S|H}, \tag{9.4.8}$$

where

$$\mathbf{E}_{H|S} = [\mathbf{H}, \mathbf{O}] \begin{bmatrix} \mathbf{H}^H\mathbf{H} & \mathbf{H}^H\mathbf{S} \\ \mathbf{S}^H\mathbf{H} & \mathbf{S}^H\mathbf{S} \end{bmatrix}^{-1} \begin{bmatrix} \mathbf{H}^H \\ \mathbf{S}^H \end{bmatrix}, \tag{9.4.9}$$

$$\mathbf{E}_{S|H} = [\mathbf{O}, \mathbf{S}] \begin{bmatrix} \mathbf{H}^H\mathbf{H} & \mathbf{H}^H\mathbf{S} \\ \mathbf{S}^H\mathbf{H} & \mathbf{S}^H\mathbf{S} \end{bmatrix}^{-1} \begin{bmatrix} \mathbf{H}^H \\ \mathbf{S}^H \end{bmatrix}, \tag{9.4.10}$$

with zero matrix $\mathbf{O}$.

From the the inversion lemma for a block matrix (1.7.11), it is easy to find that

$$\begin{bmatrix} \mathbf{H}^H\mathbf{H} & \mathbf{H}^H\mathbf{S} \\ \mathbf{S}^H\mathbf{H} & \mathbf{S}^H\mathbf{S} \end{bmatrix}^{-1} = \begin{bmatrix} (\mathbf{H}^H\mathbf{P}_S^\perp\mathbf{H})^{-1} & -(\mathbf{H}^H\mathbf{P}_S^\perp\mathbf{H})^{-1}\mathbf{H}^H\mathbf{S}(\mathbf{S}^H\mathbf{S})^{-1} \\ -(\mathbf{S}^H\mathbf{P}_H^\perp\mathbf{S})^{-1}\mathbf{S}^H\mathbf{H}(\mathbf{H}^H\mathbf{H})^{-1} & (\mathbf{S}^H\mathbf{P}_H^\perp\mathbf{S})^{-1} \end{bmatrix}, \tag{9.4.11}$$

where

$$\mathbf{P}_S^\perp = \mathbf{I} - \mathbf{P}_S = \mathbf{I} - \mathbf{S}(\mathbf{S}^H\mathbf{S})^{-1}\mathbf{S}^H,$$
$$\mathbf{P}_H^\perp = \mathbf{I} - \mathbf{P}_H = \mathbf{I} - \mathbf{H}(\mathbf{H}^H\mathbf{H})^{-1}\mathbf{H}^H.$$

By substituting (9.4.11) into (9.4.9), we have

$$\mathbf{E}_{H|S} = [\mathbf{H}, \mathbf{O}] \begin{bmatrix} (\mathbf{H}^H\mathbf{P}_S^\perp\mathbf{H})^{-1} & -(\mathbf{H}^H\mathbf{P}_S^\perp\mathbf{H})^{-1}\mathbf{H}^H\mathbf{S}(\mathbf{S}^H\mathbf{S})^{-1} \\ -(\mathbf{S}^H\mathbf{P}_H^\perp\mathbf{S})^{-1}\mathbf{S}^H\mathbf{H}(\mathbf{H}^H\mathbf{H})^{-1} & (\mathbf{S}^H\mathbf{P}_H^\perp\mathbf{S})^{-1} \end{bmatrix} \begin{bmatrix} \mathbf{H}^H \\ \mathbf{S}^H \end{bmatrix}$$

namely

$$\mathbf{E}_{H|S} = \mathbf{H}(\mathbf{H}^H\mathbf{P}_S^\perp\mathbf{H})^{-1}\mathbf{H}^H - \mathbf{H}(\mathbf{H}^H\mathbf{P}_S^\perp\mathbf{H})^{-1}\mathbf{H}^H\mathbf{P}_S$$

or

$$\mathbf{E}_{H|S} = \mathbf{H}(\mathbf{H}^H\mathbf{P}_S^\perp\mathbf{H})^{-1}\mathbf{H}^H\mathbf{P}_S^\perp. \tag{9.4.12}$$

Similarly, substitute (9.4.11) into (9.4.10) to yield

$$\mathbf{E}_{S|H} = \mathbf{S}(\mathbf{S}^H\mathbf{P}_H^\perp\mathbf{S})^{-1}\mathbf{S}^H\mathbf{P}_H^\perp. \tag{9.4.13}$$

Equations (9.4.12) and (9.4.13) were obtained by Behrens and Scharf in 1994 [31].

From (9.4.12) and (9.4.13) we see that the idempotent operators $\mathbf{E}_{H|S}$ and $\mathbf{E}_{S|H}$ are not Hermitian, so they are oblique projectors rather than orthogonal projectors.

Definition 9.5 An idempotent operator $\mathbf{E}$ without complex conjugate symmetry is called an *oblique projector.*

By definition, both the operator $\mathbf{E}_{H|S}$ defined in (9.4.12) and the operator $\mathbf{E}_{S|H}$ defined in (9.4.13) are oblique projectors.

The oblique projector $\mathbf{E}_{H|S}$ reads as "the projector onto the subspace Range($\mathbf{H}$) 'along' the subspace Range($\mathbf{S}$)".

Similarly, the oblique projector $\mathbf{E}_{S|H}$ reads as "the projector onto the subspace Range($\mathbf{S}$) 'along' the subspace Range($\mathbf{H}$)".

The geometric interpretation of an oblique projector is as follows: for an oblique projector $\mathbf{E}_{H|S}$, its projection direction is along to the subspace Range($\mathbf{S}$). Thus the subspace of the projector $\mathbf{E}_{H|S}$ and the subspace Range($\mathbf{S}$) cannot have any intersection. Because $\mathbf{E}_{H|S}\mathbf{H}$ is the projection of $\mathbf{H}$ onto the subspace Range($\mathbf{H}$) along the subspace Range($\mathbf{S}$), while $\mathbf{H}$ itself lies in the subspace Range($\mathbf{H}$), the result of the oblique projection is $\mathbf{E}_{H|S}\mathbf{H} = \mathbf{H}$.

The oblique projector is an extension of the orthogonal projector, while the orthogonal projector is a special case of the oblique projector. The following summarizes the important properties of oblique projectors [31].

1. Both $\mathbf{E}_{H|S}$ and $\mathbf{E}_{S|H}$ are idempotent operators, i.e.,
$$\mathbf{E}_{H|S}^2 = \mathbf{E}_{H|S}, \quad \mathbf{E}_{S|H}^2 = \mathbf{E}_{S|H}.$$
2. $\mathbf{E}_{H|S}[\mathbf{H},\mathbf{S}] = [\mathbf{H},\mathbf{O}]$ and $\mathbf{E}_{S|H}[\mathbf{H},\mathbf{S}] = [\mathbf{O},\mathbf{S}]$, or equivalently
$$\mathbf{E}_{H|S}\mathbf{H} = \mathbf{H}, \quad \mathbf{E}_{H|S}\mathbf{S} = \mathbf{O},$$
$$\mathbf{E}_{S|H}\mathbf{H} = \mathbf{O}, \quad \mathbf{E}_{S|H}\mathbf{S} = \mathbf{S}.$$
3. The cross-terms of the oblique projectors $\mathbf{E}_{H|S}$ and $\mathbf{E}_{S|H}$ are equal to zero, i.e.,
$$\mathbf{E}_{H|S}\mathbf{E}_{S|H} = \mathbf{O}, \quad \mathbf{E}_{S|H}\mathbf{E}_{H|S} = \mathbf{O}.$$
4. Oblique projection followed by orthogonal projection does not change the original oblique projection:
$$\mathbf{E}_{H|S} = \mathbf{P}_H\mathbf{E}_{H|S}, \quad \mathbf{E}_{S|H} = \mathbf{P}_S\mathbf{E}_{S|H}.$$
5. Let $\mathbf{B}^\dagger = (\mathbf{B}^H\mathbf{B})^{-1}\mathbf{B}^H$ denote the generalized inverse matrix of $\mathbf{B}$; then
$$\mathbf{H}^\dagger\mathbf{E}_{H|S} = (\mathbf{P}_S^\perp\mathbf{H})^\dagger, \quad \mathbf{S}^\dagger\mathbf{E}_{S|H} = (\mathbf{P}_H^\perp\mathbf{S})^\dagger.$$
6. There exists the following relationship between an oblique projection matrix and the corresponding orthogonal projection matrix:
$$\mathbf{P}_S^\perp\mathbf{E}_{H|S}\mathbf{P}_S^\perp = \mathbf{P}_S^\perp\mathbf{P}_{\mathbf{P}_S^\perp\mathbf{H}}\mathbf{P}_S^\perp = \mathbf{P}_{\mathbf{P}_S^\perp\mathbf{H}},$$

where $\mathbf{P}_{\mathbf{P}_S^\perp \mathbf{H}} = \mathbf{P}_S^\perp \mathbf{H} \langle \mathbf{P}_S^\perp \mathbf{H}, \mathbf{P}_S^\perp \mathbf{H} \rangle^{-1} (\mathbf{P}_S^\perp \mathbf{H})^H$.

7. If the subspaces Range(**H**) and Range(**S**) are orthogonal, i.e., Range(**H**) $\perp$ Range(**S**), then

$$\mathbf{E}_{H|S} = \mathbf{P}_H, \quad \mathbf{E}_{S|H} = \mathbf{P}_H^\perp.$$

On the above properties, we have the following four remarks:

Remark 1 Property 2 shows that:

(1) the range of $\mathbf{E}_{H|S}$ is Range(**H**) and the null space of $\mathbf{E}_{H|S}$ contains Range(**S**), i.e.,

$$\text{Range}(\mathbf{E}_{H|S}) = \text{Range}(\mathbf{H}), \quad \text{Range}(\mathbf{S}) \subset \text{Null}(\mathbf{E}_{H|S}); \tag{9.4.14}$$

(2) the range of $\mathbf{E}_{S|H}$ is Range(**S**) and the null space of $\mathbf{E}_{S|H}$ contains Range(**H**), i.e.,

$$\text{Range}(\mathbf{E}_{S|H}) = \text{Range}(\mathbf{S}), \quad \text{Range}(\mathbf{H}) \subset \text{Null}(\mathbf{E}_{S|H}). \tag{9.4.15}$$

In other words, the range of the oblique projector $\mathbf{E}_{H|S}$ is in the null space of $\mathbf{E}_{S|H}$, whereas the range of $\mathbf{E}_{S|H}$ is in the null space of $\mathbf{E}_{H|S}$.

Remark 2 Since the cross-terms of the oblique projectors $\mathbf{E}_{H|S}$ and $\mathbf{E}_{S|H}$ are zero, Range($\mathbf{E}_{H|S}$) and Range($\mathbf{E}_{S|H}$) are disjoint.

Remark 3 Property 4 shows that $\mathbf{E}_{H|S}$ is the oblique projection onto the subspace Range(**H**) along the subspace Range(**S**), i.e., Range($\mathbf{E}_{H|S}$) = Range(**H**). Therefore, if the oblique projection $\mathbf{E}_{H|S}$ is again projected onto the subspace Range(**H**) then the projection result will not change.

Remark 4 Properties 5 and 6 can be regarded as two relationships between the oblique projector and the orthogonal projector.

9.4.2 Geometric Interpretation of Oblique Projectors

From the above discussions, we can summarize three projectors below.

(1) The orthogonal projector $\mathbf{P}_{HS}$ is a projector onto the subspace Range(**H**, **S**) of the composite matrix [**H**, **S**].
(2) The oblique projector $\mathbf{E}_{H|S}$ is a projector onto the subspace Range(**H**) along another range Range(**S**).
(3) The oblique projector $\mathbf{E}_{S|H}$ is a projector onto the subspace Range(**S**) along the range Range(**H**).

As explained, the range of the oblique projector $\mathbf{E}_{H|S}$ is Range(**H**), and its null space contains Range(**S**). In order to describe the null space of the oblique projector

$\mathbf{E}_{H|S}$, define the matrix $\mathbf{A}$ as the matrix orthogonal to the composite matrix $[\mathbf{H},\mathbf{S}]$, i.e., $[\mathbf{H},\mathbf{S}]^H\mathbf{A}=\mathbf{O}$ and thus

$$\mathbf{H}^H\mathbf{A}=\mathbf{O},\quad \mathbf{S}^H\mathbf{A}=\mathbf{O}. \tag{9.4.16}$$

Premultiplying the second equation above by the full column rank matrix $\mathbf{S}(\mathbf{S}^H\mathbf{S})^{-1}$, $\mathbf{S}^H\mathbf{A}=\mathbf{O}$ can be equivalently written as

$$\mathbf{P}_S\mathbf{A}=\mathbf{O}\quad\Rightarrow\quad \mathbf{P}_S^\perp\mathbf{A}=\mathbf{A}. \tag{9.4.17}$$

Therefore, from (9.4.12) we have

$$\mathbf{E}_{H|S}\mathbf{A}=\mathbf{H}(\mathbf{H}^H\mathbf{P}_S^\perp\mathbf{H})^{-1}\mathbf{H}^H\mathbf{P}_S^\perp\mathbf{A}=\mathbf{H}(\mathbf{H}^H\mathbf{P}_S^\perp\mathbf{H})^{-1}\mathbf{H}^H\mathbf{A}=\mathbf{O}.$$

This shows that the subspace Range($\mathbf{A}$) spanned by the columns of $\mathbf{A}$ is in the null space of the oblique projector $\mathbf{E}_{H|S}$, i.e., Range($\mathbf{A}$) $\subset$ Null($\mathbf{E}_{H|S}$).

Equations (9.4.16) show that the subspace Range($\mathbf{A}$) is orthogonal to the two subspaces Range($\mathbf{H}$) and Range($\mathbf{S}$), respectively. Therefore, we can use three space directions Range($\mathbf{H}$), Range($\mathbf{S}$) and Range($\mathbf{A}$) as coordinate axes of the Euclidean space $\mathcal{C}^{m+k}$ spanned by the column vectors of the $n\times(m+k)$ matrices $[\mathbf{H},\mathbf{S}]$, as shown in Figure 9.4.

In Figure 9.4(b), the coordinate axis Range($\mathbf{A}$) is perpendicular to each of the other two coordinate axes, Range($\mathbf{H}$) and Range($\mathbf{S}$), but Range($\mathbf{H}$) and Range($\mathbf{S}$) are not mutually perpendicular. Nevertheless, the Euclidean space can be decomposed into the direct sum of the three directions:

$$\mathcal{C}^{m+k}=\text{Range}(\mathbf{H})\oplus\text{Range}(\mathbf{S})\oplus\text{Range}(\mathbf{A}). \tag{9.4.18}$$

The orthogonal projection in Figure 9.4(a) has been explained earlier. The oblique projection in Figure 9.4(b) is explained as follows: when the vector $\mathbf{y}$ is in the horizontal plane defined by the two coordinate axes Range($\mathbf{H}$) and Range($\mathbf{S}$), the oblique projection $\mathbf{E}_{H|S}\mathbf{y}$ of $\mathbf{y}$ onto Range($\mathbf{H}$) along Range($\mathbf{S}$) satisfies the following two conditions:

- the oblique projection $\mathbf{E}_{H|S}\mathbf{y}$ lies in the coordinate axis Range($\mathbf{H}$), i.e., $\mathbf{E}_{H|S}\mathbf{y}\in$ Range($\mathbf{H}$);
- the connection line from the arrowhead of the oblique projection vector $\mathbf{E}_{H|S}\mathbf{y}$ to the arrowhead of the vector $\mathbf{y}$ (when in the horizontal plane) is parallel to the coordinate axis Range($\mathbf{S}$), i.e., $\mathbf{E}_{H|S}\mathbf{y}\notin$ Range($\mathbf{S}$).

As shown in Figure 9.4(b), for the vector $\mathbf{y}$ in the Euclidean space, the construction of the oblique projection $\mathbf{E}_{H|S}\mathbf{y}$ can be divided into the following two steps.

(1) The vector $\mathbf{y}$ is first projected orthogonally onto the plane defined by the coordinate axes Range($\mathbf{H}$) and Range($\mathbf{S}$), i.e., the projection $\mathbf{P}_{HS}\mathbf{y}$ is the orthogonal projection of the vector $\mathbf{y}$ onto the range Range($\mathbf{H},\mathbf{S}$) spanned by the column vectors of the matrix $[\mathbf{H},\mathbf{S}]$, denoted $\mathbf{y}_{HS}=\mathbf{P}_{HS}\mathbf{y}$.

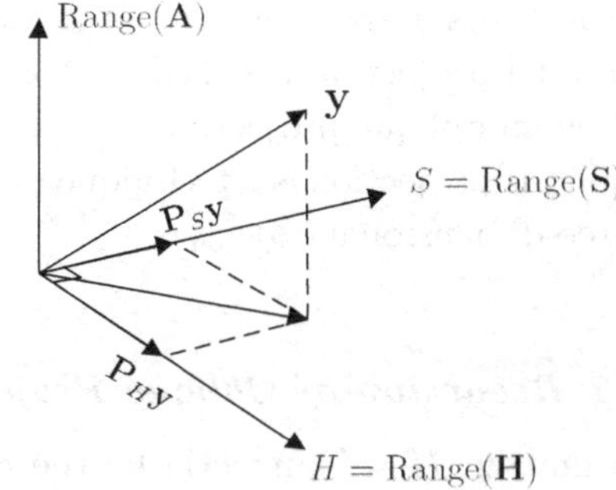

(a) Geometric interpretation of orthogonal projection

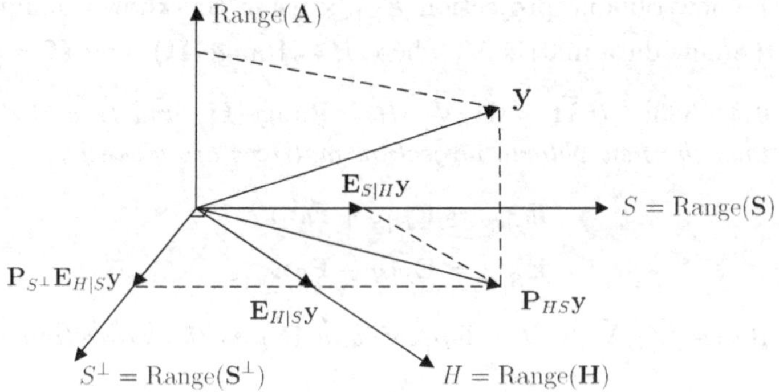

(b) Geometric interpretation of oblique projections

Figure 9.4 Orthogonal projection and oblique projections [383].

(2) The method illustrated in Figure 9.4(b) is used to find the oblique projection of the orthogonal projection of $\mathbf{y}_{HS}$ onto Range($\mathbf{H}$) along Range($\mathbf{S}$), denoted $\mathbf{E}_{H|S}\mathbf{y}_{HS}$.

The physical meaning of the oblique projection operator is obtained as follows: the sentence "the oblique projection $\mathbf{E}_{H|S}\mathbf{y}$ of a vector $\mathbf{y}$ is its projection onto Range($\mathbf{H}$) 'along' Range($\mathbf{S}$)" is translated into signal processing language as "the oblique projection $\mathbf{E}_{H|S}\mathbf{y}$ extracts the components of the vector $\mathbf{y}$ in the special direction Range($\mathbf{H}$) and fully cancels all the components of $\mathbf{y}$ in another direction, Range($\mathbf{S}$)".

From Figure 9.4 it can be deduced that the projection $\mathbf{P}_A\mathbf{y}$ is the orthogonal projection of the vector $\mathbf{y}$ onto the range $A = \text{Range}(\mathbf{A})$, where the projection $\mathbf{P}_A\mathbf{y}$ is not only orthogonal to $\mathbf{E}_{H|S}\mathbf{y}$ but also is orthogonal to $\mathbf{E}_{S|H}\mathbf{y}$, i.e., $\mathbf{P}_A\mathbf{y} \perp \mathbf{E}_{H|S}\mathbf{y}$ and $\mathbf{P}_A\mathbf{y} \perp \mathbf{E}_{S|H}\mathbf{y}$.

In mathematics, oblique projections are a type of parallel projection. The parallel projection of a three-dimensional point (x, y, z) onto the plane (x, y) is given by

$(x+az, y+bz, 0)$. The constants a and b determine uniquely a parallel projection. If $a=b=0$, then the parallel projection is called an "orthographic" or orthogonal projection. Otherwise, it is an oblique projection.

In image processing, oblique projection is a technique providing a two-dimensional picture or image of a three-dimensional object.

9.4.3 Recursion of Oblique Projectors

Let $\mathbf{H}$ be a known data matrix, $H = \text{Range}(\mathbf{H})$ be the range (i.e., column space) of $\mathbf{H}$ and $\mathbf{E}_{H|S}\mathbf{x}$ be the oblique projection of the vector $\mathbf{x}$ onto H along a disjoint subspace S. If a new data matrix $\mathbf{V}$ is added, the question is how to compute recursively the new oblique projection $E_{\tilde{H}|S}\mathbf{x}$ using the known oblique projection $\mathbf{E}_{H|S}\mathbf{x}$ and the new data matrix $\mathbf{V}$, where $\tilde{H} = \text{Range}(\tilde{\mathbf{H}})$, and $\tilde{\mathbf{H}} = [\mathbf{H}, \mathbf{V}]$.

Theorem 9.4 [383] *If $\tilde{\mathbf{H}} = [\mathbf{H}, \mathbf{V}]$, $\tilde{H} = \text{Range}(\tilde{\mathbf{H}})$ and $\tilde{H}$ and $S = \text{Range}(\mathbf{S})$ are disjoint then the new oblique projection matrices are given by*

$$\mathbf{E}_{\tilde{H}|S} = \mathbf{E}_{H|S} + \mathbf{E}_{\tilde{V}|S}, \tag{9.4.19}$$

$$\mathbf{E}_{S|\tilde{H}} = \mathbf{E}_{S|H} - \mathbf{P}_S\mathbf{E}_{\tilde{V}|S}, \tag{9.4.20}$$

where $\tilde{V} = \text{Range}(\tilde{\mathbf{V}}), \tilde{\mathbf{V}} = \mathbf{V} - \mathbf{E}_{H|S}\mathbf{V}$ and $\mathbf{P}_H$ is the projection matrix of the matrix $\mathbf{H}$.

The following three remarks on Theorem 9.4 help us understand the role of oblique projection matrices and the relationship between oblique projection recursion and orthogonal projection recursion.

Remark 1 If two subspaces $\tilde{H}$ and S are orthogonal then

$$\mathbf{E}_{\tilde{H}|S} = \mathbf{P}_{\tilde{H}}, \quad \mathbf{E}_{H|S} = \mathbf{P}_H,$$
$$\mathbf{E}_{S|\tilde{H}} = \mathbf{P}_{\tilde{H}}^{\perp}, \quad \mathbf{E}_{S|H} = \mathbf{P}_H^{\perp}.$$

In this case, $\tilde{\mathbf{V}} = (\mathbf{I} - \mathbf{P}_H)\mathbf{V} = \mathbf{P}_H^{\perp}\mathbf{V}$, and hence

$$\mathbf{E}_{\tilde{V}} = \mathbf{P}_{\tilde{V}}, \quad \mathbf{P}_S\mathbf{P}_{\tilde{V}} = \mathbf{P}_{\tilde{V}}.$$

Substitute the above relations into (9.4.19) and (9.4.20) to yield

$$\mathbf{P}_{\tilde{H}} = \mathbf{P}_H + \mathbf{P}_H^{\perp}\mathbf{V}\langle\mathbf{P}_H^{\perp}, \mathbf{P}_H^{\perp}\mathbf{V}\rangle^{-1}\mathbf{V}^T\mathbf{P}_H^{\perp},$$
$$\mathbf{P}_{\tilde{H}}^{\perp} = \mathbf{P}_H^{\perp} - \mathbf{P}_H^{\perp}\mathbf{V}\langle\mathbf{P}_H^{\perp}, \mathbf{P}_H^{\perp}\mathbf{V}\rangle^{-1}\mathbf{V}^T\mathbf{P}_H^{\perp}.$$

These are just the recursive formulas for the projection matrix and the orthogonal projection matrix, respectively. Therefore Theorem 9.4 is a generalization of the definitions of projection matrix and the orthogonal projection matrix to the case of nonorthogonal subspaces $\tilde{H}$ and S.

Remark 2 Theorem 9.4 can be explained from the angle of *innovation processes.* By the Kalman filtering theorem, in the case of an orthogonal projection, $\mathbf{P}_H\mathbf{V}$ represents the mean square estimate of the data matrix $\mathbf{V}$ and $\tilde{\mathbf{V}} = \mathbf{V} - \mathbf{P}_H\mathbf{V}$ can be regarded as the innovation matrix of the data matrix $\mathbf{V}$, while Range($\tilde{\mathbf{V}}$) denotes the innovation subspace in orthogonal projection. Similarly, $\mathbf{E}_{H|S}\mathbf{V}$ is the mean square estimate of the data matrix $\mathbf{V}$ in the subspace H along the disjoint subspace S, and $\tilde{\mathbf{V}} = \mathbf{V} - \mathbf{E}_{H|S}\mathbf{V}$ can be regarded as the innovation matrix of the data matrix $\mathbf{V}$ in H along S, while Range($\tilde{\mathbf{V}}$) denotes the innovation subspace in oblique projection.

Remark 3 From the viewpoint of subspaces, the vector space $\mathcal{C}^n$ equals $H \oplus S \oplus (H \oplus S)^{\perp}$, where H, S and $(H \oplus S)^{\perp}$ represent respectively the desired signal (the range), the structured noise (the interference) and the unstructured noise subspace. By Theorem 9.4 and $\mathbf{P}_{HS} = \mathbf{E}_{H|S} + \mathbf{E}_{S|H}$, it is easy to see that

$$\tilde{\mathbf{V}} = (\mathbf{I} - \mathbf{E}_{H|S})\mathbf{V} = \mathbf{E}_{S|H}\mathbf{V} + \mathbf{P}_{HS}^{\perp}\mathbf{V} = \mathbf{V}_S + \mathbf{V}_{(HS)^{\perp}}.$$

That is, the innovation matrix $\tilde{\mathbf{V}}$ of the data matrix $\mathbf{V}$ in H along S consists of two components: $\mathbf{V}_S = \mathbf{E}_{S|H}\mathbf{V}$ is the innovation matrix of the structured noise in the subspace S and $\mathbf{V}_{(H,S)^{\perp}}\mathbf{V}$ is the innovation matrix of the unstructured noise in the orthogonal complement subspace $(H \oplus S)^{\perp}$. If the structured noise is negligible compared with the unstructured noise then the innovation matrix $\tilde{\mathbf{V}} = \mathbf{V}_{(HS)^{\perp}}$. In this case, $\mathbf{P}_S^{\perp}\tilde{\mathbf{V}} \approx \mathbf{P}_S^{\perp}\mathbf{V}_{(HS)^{\perp}} = \mathbf{V}_{(HS)^{\perp}}$ and

$$\mathbf{E}_{\tilde{V}|S} = \tilde{\mathbf{V}}(\tilde{\mathbf{V}}^H\mathbf{P}_S^{\perp}\tilde{\mathbf{V}})^{-1}\tilde{\mathbf{V}}^H\mathbf{P}_S^{\perp} = \mathbf{V}_{(HS)^{\perp}}(\mathbf{V}_{(HS)^{\perp}}^H\mathbf{V}_{(HS)^{\perp}})^{-1}\mathbf{V}_{(HS)^{\perp}}^H$$

which is just the projection matrix of $\mathbf{V}_{(HS)^{\perp}}$.

9.5 Oblique Projector of Full Row Rank Matrices

In the previous sections, our focus was the action of orthogonal projectors and the oblique projectors upon full column rank matrices. If the matrix is of full row rank, then the corresponding orthogonal projector and oblique projector have different definitions and forms of expression.

9.5.1 Definition and Properties

If $\mathbf{D} \in \mathbb{C}^{m\times k}$ has the full row rank, i.e., rank($\mathbf{D}$) $= m$ $(m < k)$ then its projection $\mathbf{P}_D$ is a $k \times k$ matrix and is defined as

$$\mathbf{P}_D = \mathbf{D}^H(\mathbf{D}\mathbf{D}^H)^{-1}\mathbf{D}. \tag{9.5.1}$$

Let the matrices $\mathbf{B} \in \mathbb{C}^{m\times k}$ and $\mathbf{C} \in \mathbb{C}^{n\times k}$ be two full row rank matrices, i.e., rank($\mathbf{B}$) $= m$, rank($\mathbf{C}$) $= n$, and let their combined matrix $\begin{bmatrix}\mathbf{B}\\ \mathbf{C}\end{bmatrix} \in \mathbb{C}^{(m+n)\times k}$ be also of full row rank, i.e., $m + n$, where $m + n < k$. This implies that the row vectors of the matrix $\mathbf{B}$ and the row vectors of $\mathbf{C}$ are linearly independent. Hence the range subspace spanned by row vectors of the matrix $\mathbf{B}$, denoted $\mathcal{Z}_B$ and simply called the row subspace of $\mathbf{B}$, and the row subspace $\mathcal{Z}_C$ of the matrix $\mathbf{C}$ are disjoint.

Putting $\mathbf{D} = \begin{bmatrix}\mathbf{B}\\ \mathbf{C}\end{bmatrix}$ and substituting $\mathbf{D}$ into (9.5.1), we find the orthogonal projector

$$\mathbf{P}_D = [\mathbf{B}^H, \mathbf{C}^H] \begin{bmatrix} \mathbf{B}\mathbf{B}^H & \mathbf{B}\mathbf{C}^H \\ \mathbf{C}\mathbf{B}^H & \mathbf{C}\mathbf{C}^H \end{bmatrix}^{-1} \begin{bmatrix}\mathbf{B}\\ \mathbf{C}\end{bmatrix}, \tag{9.5.2}$$

which can be decomposed as follows:

$$\mathbf{P}_D = \mathbf{E}_{\mathcal{Z}_B|\mathcal{Z}_C} + \mathbf{E}_{\mathcal{Z}_C|\mathcal{Z}_B}, \tag{9.5.3}$$

where

$$\mathbf{E}_{\mathcal{Z}_B|\mathcal{Z}_C} = [\mathbf{B}^H, \mathbf{C}^H] \begin{bmatrix} \mathbf{B}\mathbf{B}^H & \mathbf{B}\mathbf{C}^H \\ \mathbf{C}\mathbf{B}^H & \mathbf{C}\mathbf{C}^H \end{bmatrix}^{-1} \begin{bmatrix}\mathbf{B}\\ \mathbf{O}\end{bmatrix}, \tag{9.5.4}$$

$$\mathbf{E}_{\mathcal{Z}_C|\mathcal{Z}_B} = [\mathbf{B}^H, \mathbf{C}^H] \begin{bmatrix} \mathbf{B}\mathbf{B}^H & \mathbf{B}\mathbf{C}^H \\ \mathbf{C}\mathbf{B}^H & \mathbf{C}\mathbf{C}^H \end{bmatrix}^{-1} \begin{bmatrix}\mathbf{O}\\ \mathbf{C}\end{bmatrix}. \tag{9.5.5}$$

Premultiplying the above two equations by the matrix $\begin{bmatrix}\mathbf{B}\\ \mathbf{C}\end{bmatrix}$, then

$$\begin{bmatrix}\mathbf{B}\\ \mathbf{C}\end{bmatrix} \mathbf{E}_{\mathcal{Z}_B|\mathcal{Z}_C} = \begin{bmatrix}\mathbf{B}\\ \mathbf{O}\end{bmatrix}, \tag{9.5.6}$$

$$\begin{bmatrix}\mathbf{B}\\ \mathbf{C}\end{bmatrix} \mathbf{E}_{\mathcal{Z}_C|\mathcal{Z}_B} = \begin{bmatrix}\mathbf{O}\\ \mathbf{C}\end{bmatrix}. \tag{9.5.7}$$

These results can be equivalently written as

$$\mathbf{B}\mathbf{E}_{\mathcal{Z}_B|\mathcal{Z}_C} = \mathbf{B}, \quad \mathbf{C}\mathbf{E}_{\mathcal{Z}_B|\mathcal{Z}_C} = \mathbf{O}, \tag{9.5.8}$$

$$\mathbf{B}\mathbf{E}_{\mathcal{Z}_C|\mathcal{Z}_B} = \mathbf{O}, \quad \mathbf{C}\mathbf{E}_{\mathcal{Z}_C|\mathcal{Z}_B} = \mathbf{C}. \tag{9.5.9}$$

This shows that the operators $\mathbf{E}_{\mathcal{Z}_B|\mathcal{Z}_C}$ and $\mathbf{E}_{\mathcal{Z}_C|\mathcal{Z}_B}$ have no cross-term, i.e.,

$$\mathbf{E}_{\mathcal{Z}_B|\mathcal{Z}_C}\mathbf{E}_{\mathcal{Z}_C|\mathcal{Z}_B} = \mathbf{O} \quad \text{and} \quad \mathbf{E}_{\mathcal{Z}_C|\mathcal{Z}_B}\mathbf{E}_{\mathcal{Z}_B|\mathcal{Z}_C} = \mathbf{O}. \tag{9.5.10}$$

By the idempotence of the orthogonal projector $\mathbf{P}_D$, we have

$$\begin{aligned} \mathbf{P}_D^2 &= (\mathbf{E}_{\mathcal{Z}_B|\mathcal{Z}_C} + \mathbf{E}_{\mathcal{Z}_C|\mathcal{Z}_B})(\mathbf{E}_{\mathcal{Z}_B|\mathcal{Z}_C} + \mathbf{E}_{\mathcal{Z}_C|\mathcal{Z}_B}) \\ &= \mathbf{E}^2_{\mathcal{Z}_B|\mathcal{Z}_C} + \mathbf{E}^2_{\mathcal{Z}_C|\mathcal{Z}_B} = \mathbf{P}_D \\ &= \mathbf{E}_{\mathcal{Z}_B|\mathcal{Z}_C} + \mathbf{E}_{\mathcal{Z}_C|\mathcal{Z}_B} \end{aligned}$$

which yields

$$\mathbf{E}^2_{\mathcal{Z}_B|\mathcal{Z}_C} = \mathbf{E}_{\mathcal{Z}_B|\mathcal{Z}_C}, \quad \mathbf{E}^2_{\mathcal{Z}_C|\mathcal{Z}_B} = \mathbf{E}_{\mathcal{Z}_C|\mathcal{Z}_B}. \tag{9.5.11}$$

As they are idempotence, $\mathbf{E}_{\mathcal{Z}_B|\mathcal{Z}_C}$, defined in (9.5.4) and $\mathbf{E}_{\mathcal{Z}_C|\mathcal{Z}_B}$, defined in (9.5.5), are regarded as oblique projectors of the full row rank matrices $\mathbf{B}$ and $\mathbf{C}$, respectively.

In particular, if the row subspace of $\mathbf{B}$ and the row subspace of $\mathbf{C}$ are orthogonal, i.e., $\mathbf{BC}^H = \mathbf{O}$ and $\mathbf{CB}^H = \mathbf{O}$, then the oblique projectors are reduced to orthogonal projectors [487]:

$$\mathbf{E}_{\mathcal{Z}_B|\mathcal{Z}_C} = \mathbf{B}^H(\mathbf{BB}^H)^{-1}\mathbf{B} = \mathbf{P}_B, \tag{9.5.12}$$

$$\mathbf{E}_{\mathcal{Z}_C|\mathcal{Z}_B} = \mathbf{C}^H(\mathbf{CC}^H)^{-1}\mathbf{C} = \mathbf{P}_C. \tag{9.5.13}$$

Proof Substituting the orthogonality condition $\mathbf{BC}^H = \mathbf{O}$ and $\mathbf{CB}^H = \mathbf{O}$ into (9.5.4), then

$$\begin{aligned}
\mathbf{E}_{\mathcal{Z}_B|\mathcal{Z}_C} &= [\mathbf{B}^H, \mathbf{C}^H]\begin{bmatrix}\mathbf{BB}^H & \mathbf{O}\\ \mathbf{O} & \mathbf{CC}^H\end{bmatrix}^{-1}\begin{bmatrix}\mathbf{B}\\ \mathbf{O}\end{bmatrix}\\
&= [\mathbf{B}^H, \mathbf{C}^H]\begin{bmatrix}(\mathbf{BB}^H)^{-1} & \mathbf{O}\\ \mathbf{O} & (\mathbf{CC}^H)^{-1}\end{bmatrix}\begin{bmatrix}\mathbf{B}\\ \mathbf{O}\end{bmatrix}\\
&= \mathbf{B}^H(\mathbf{BB}^H)^{-1}\mathbf{B}\\
&= \mathbf{P}_B.
\end{aligned}$$

Similarly, we can show that $\mathbf{E}_{\mathcal{Z}_C|\mathcal{Z}_B} = \mathbf{C}^H(\mathbf{CC}^H)^{-1}\mathbf{C} = \mathbf{P}_C$. □

9.5.2 Calculation of Oblique Projection

The QR decomposition of an $m \times n$ $(m > n)$ real matrix $\mathbf{A}$ is defined as

$$\mathbf{Q}^T\mathbf{A} = \begin{bmatrix}\mathbf{R}\\ \mathbf{O}\end{bmatrix}, \tag{9.5.14}$$

where $\mathbf{Q}$ is an $m \times m$ orthogonal matrix and $\mathbf{R}$ is an upper triangular matrix.

If $\mathbf{B}$ is a real matrix with the number of columns greater than the number of rows then, letting $\mathbf{B} = \mathbf{A}^T$ and taking the transpose of the QR decomposition formula (9.5.14), we obtain the LQ decomposition of $\mathbf{B}$:

$$\mathbf{BQ} = \mathbf{A}^T\mathbf{Q} = [\mathbf{L}, \mathbf{O}] \tag{9.5.15}$$

or

$$\mathbf{B} = [\mathbf{L}, \mathbf{O}]\mathbf{Q}^T, \tag{9.5.16}$$

where $\mathbf{L} = \mathbf{R}^T$ is a lower triangular matrix.

Putting $\mathbf{B} \in \mathbb{R}^{m\times k}, \mathbf{C} \in \mathbb{R}^{n\times k}$ and $\mathbf{A} \in \mathbb{R}^{p\times k}$, where $m + n + p < k$, the LQ decomposition of the matrix $\left[\mathbf{B}^T, \mathbf{C}^T, \mathbf{A}^T\right]^T$ is given by

$$\begin{bmatrix}\mathbf{B}\\ \mathbf{C}\\ \mathbf{A}\end{bmatrix}[\mathbf{Q}_1, \mathbf{Q}_2, \mathbf{Q}_3] = \begin{bmatrix}\mathbf{L}_{11} & & \\ \mathbf{L}_{21} & \mathbf{L}_{22} & \\ \mathbf{L}_{31} & \mathbf{L}_{32} & \mathbf{L}_{33}\end{bmatrix}, \tag{9.5.17}$$

where $\mathbf{Q}_1 \in \mathbb{R}^{k\times m}, \mathbf{Q}_2 \in \mathbb{R}^{k\times n}, \mathbf{Q}_3 \in \mathbb{R}^{k\times p}$ are orthogonal matrices, i.e., $\mathbf{Q}_i^T\mathbf{Q}_i = \mathbf{I}$ and $\mathbf{Q}_i^T\mathbf{Q}_j = \mathbf{O}, i \neq j$, and $\mathbf{L}_{ij}$ is a lower triangular matrix.

Equation (9.5.17) can be equivalently written as

$$\begin{bmatrix} \mathbf{B} \\ \mathbf{C} \\ \mathbf{A} \end{bmatrix} = \begin{bmatrix} \mathbf{L}_{11} & & \\ \mathbf{L}_{21} & \mathbf{L}_{22} & \\ \mathbf{L}_{31} & \mathbf{L}_{32} & \mathbf{L}_{33} \end{bmatrix} \begin{bmatrix} \mathbf{Q}_1^T \\ \mathbf{Q}_2^T \\ \mathbf{Q}_3^T \end{bmatrix}. \tag{9.5.18}$$

By the oblique projection definition (9.5.4), and using $\mathbf{Q}_i^T\mathbf{Q}_i = \mathbf{I}$ and $\mathbf{Q}_i^T\mathbf{Q}_j = \mathbf{O}, i \neq j$, then

$$\begin{aligned} \mathbf{E}_{\mathcal{Z}_B|\mathcal{Z}_C} &= [\mathbf{B}^T, \mathbf{C}^T] \begin{bmatrix} \mathbf{B}\mathbf{B}^T & \mathbf{B}\mathbf{C}^T \\ \mathbf{C}\mathbf{B}^T & \mathbf{C}\mathbf{C}^T \end{bmatrix}^{-1} \begin{bmatrix} \mathbf{B} \\ \mathbf{O} \end{bmatrix} \\ &= [\mathbf{Q}_1, \mathbf{Q}_2] \begin{bmatrix} \mathbf{L}_{11} & \\ \mathbf{L}_{21} & \mathbf{L}_{22} \end{bmatrix}^T \\ &\quad \times \left(\begin{bmatrix} \mathbf{L}_{11} & \\ \mathbf{L}_{21} & \mathbf{L}_{22} \end{bmatrix} \begin{bmatrix} \mathbf{Q}_1^T\mathbf{Q}_1 & \mathbf{Q}_1^T\mathbf{Q}_2 \\ \mathbf{Q}_2^T\mathbf{Q}_1 & \mathbf{Q}_2^T\mathbf{Q}_2 \end{bmatrix} \begin{bmatrix} \mathbf{L}_{11} & \\ \mathbf{L}_{21} & \mathbf{L}_{22} \end{bmatrix}^T \right)^{-1} \begin{bmatrix} \mathbf{B} \\ \mathbf{O} \end{bmatrix} \\ &= [\mathbf{Q}_1, \mathbf{Q}_2] \begin{bmatrix} \mathbf{L}_{11} & \\ \mathbf{L}_{21} & \mathbf{L}_{22} \end{bmatrix}^{-1} \begin{bmatrix} \mathbf{B} \\ \mathbf{O} \end{bmatrix}. \end{aligned} \tag{9.5.19}$$

Similarly, one has

$$\mathbf{E}_{\mathcal{Z}_C|\mathcal{Z}_B} = [\mathbf{Q}_1, \mathbf{Q}_2] \begin{bmatrix} \mathbf{L}_{11} & \\ \mathbf{L}_{21} & \mathbf{L}_{22} \end{bmatrix}^{-1} \begin{bmatrix} \mathbf{O} \\ \mathbf{C} \end{bmatrix}. \tag{9.5.20}$$

Note that the LQ decomposition of $\mathbf{A}$ is

$$\mathbf{A} = [\mathbf{L}_{31}, \mathbf{L}_{32}, \mathbf{L}_{33}] \begin{bmatrix} \mathbf{Q}_1^T \\ \mathbf{Q}_2^T \\ \mathbf{Q}_3^T \end{bmatrix}.$$

From (9.5.19) one can find the oblique projection of the $p \times k$ matrix $\mathbf{A}$ along the direction parallel to the row space of the $n \times k$ matrix $\mathbf{C}$ onto the row space of the $m \times k$ matrix $\mathbf{B}$ as follows:

$$\begin{aligned} \mathbf{A}\mathbf{E}_{\mathcal{Z}_B|\mathcal{Z}_C} &= [\mathbf{L}_{31}, \mathbf{L}_{32}, \mathbf{L}_{33}] \begin{bmatrix} \mathbf{Q}_1^T \\ \mathbf{Q}_2^T \\ \mathbf{Q}_3^T \end{bmatrix} [\mathbf{Q}_1, \mathbf{Q}_2] \begin{bmatrix} \mathbf{L}_{11} & \\ \mathbf{L}_{21} & \mathbf{L}_{22} \end{bmatrix}^{-1} \begin{bmatrix} \mathbf{B} \\ \mathbf{O} \end{bmatrix} \\ &= [\mathbf{L}_{31}, \mathbf{L}_{32}] \begin{bmatrix} \mathbf{L}_{11} & \\ \mathbf{L}_{21} & \mathbf{L}_{22} \end{bmatrix}^{-1} \begin{bmatrix} \mathbf{B} \\ \mathbf{O} \end{bmatrix}. \end{aligned}$$

If we let

$$[\mathbf{L}_{31}, \mathbf{L}_{32}] \begin{bmatrix} \mathbf{L}_{11} & \\ \mathbf{L}_{21} & \mathbf{L}_{22} \end{bmatrix}^{-1} = [\mathbf{L}_B, \mathbf{L}_C], \tag{9.5.21}$$

then [487]

$$\mathbf{AE}_{\mathcal{Z}_B|\mathcal{Z}_C} = [\mathbf{L}_B, \mathbf{L}_C]\begin{bmatrix}\mathbf{B}\\ \mathbf{O}\end{bmatrix} = \mathbf{L}_B\mathbf{B}. \tag{9.5.22}$$

Similarly, the oblique projection of the $p\times k$ matrix $\mathbf{A}$, along the direction parallel to the row space of the $m \times k$ matrix $\mathbf{B}$, onto the row space of the $n \times k$ matrix $\mathbf{C}$ is given by

$$\mathbf{AE}_{\mathcal{Z}_C|\mathcal{Z}_B} = [\mathbf{L}_B, \mathbf{L}_C]\begin{bmatrix}\mathbf{O}\\ \mathbf{C}\end{bmatrix} = \mathbf{L}_C\mathbf{C}. \tag{9.5.23}$$

From (9.5.21) one has

$$[\mathbf{L}_{31}, \mathbf{L}_{32}] = [\mathbf{L}_B, \mathbf{L}_C]\begin{bmatrix}\mathbf{L}_{11} & \\ \mathbf{L}_{21} & \mathbf{L}_{22}\end{bmatrix},$$

namely

$$\mathbf{L}_C = \mathbf{L}_{32}\mathbf{L}_{22}^{-1}, \tag{9.5.24}$$

$$\mathbf{L}_B = (\mathbf{L}_{31} - \mathbf{L}_C\mathbf{L}_{21})\mathbf{L}_{11}^{-1} = (\mathbf{L}_{31} - \mathbf{L}_{32}\mathbf{L}_{22}^{-1}\mathbf{L}_{21})\mathbf{L}_{11}^{-1}. \tag{9.5.25}$$

Substituting the above two equations into (9.5.22) and (9.5.23), we have

$$\begin{aligned}\mathbf{AE}_{\mathcal{Z}_B|\mathcal{Z}_C} &= (\mathbf{L}_{31} - \mathbf{L}_{32}\mathbf{L}_{22}^{-1}\mathbf{L}_{21})\mathbf{L}_{11}^{-1}\mathbf{L}_{11}\mathbf{Q}_1^T\\ &= (\mathbf{L}_{31} - \mathbf{L}_{32}\mathbf{L}_{22}^{-1}\mathbf{L}_{21})\mathbf{Q}_1^T,\end{aligned} \tag{9.5.26}$$

$$\begin{aligned}\mathbf{AE}_{\mathcal{Z}_C|\mathcal{Z}_B} &= \mathbf{L}_{32}\mathbf{L}_{22}^{-1}[\mathbf{L}_{21}, \mathbf{L}_{22}]\begin{bmatrix}\mathbf{Q}_1^T\\ \mathbf{Q}_2^T\end{bmatrix}\\ &= \mathbf{L}_{32}\mathbf{L}_{22}^{-1}\mathbf{L}_{21}\mathbf{Q}_1^T + \mathbf{L}_{32}\mathbf{Q}_2^T.\end{aligned} \tag{9.5.27}$$

9.5.3 Applications of Oblique Projectors

Oblique projectors have been used in generalized image restoration [534], solving large unsymmetric systems [421], fast system identification [424], multivariate analysis [461], partial correlation (PARCOR) estimation [249], impulse noise cancellation [512], error correction coding [315], burst error correction decoding [270], model reduction [227], system modeling [31], blind channel estimation [487], joint channel and symbol estimation [535] and so on.

In actual situations of system identification, parameter estimation, signal detection and so on, in addition to the signal of interest (referred to as the desired signal), there are often other interfering signals or additive colored noise. Moreover, measurement errors are always inevitable, and they are usually expressed as Gaussian white noise. Let $\boldsymbol{\theta}$ be the parameter vector of the signal $\mathbf{x}$ to be estimated, and let the linear system $\mathbf{H}$ provide this signal, $\mathbf{x} = \mathbf{H}\boldsymbol{\theta}$. Assume that other interference signal vectors and/or the additive colored noise vectors $\mathbf{i}$ are generated by another synthesizing linear system $\mathbf{S}$ with parameter vector $\boldsymbol{\phi}$, i.e., $\mathbf{i} = \mathbf{S}\boldsymbol{\phi}$. If the observation data vector is $\mathbf{y}$, and the additive white observation error vector is $\mathbf{e}$ then

$$\mathbf{y} = \mathbf{H}\boldsymbol{\theta} + \mathbf{S}\boldsymbol{\phi} + \mathbf{e}. \tag{9.5.28}$$

Figure 9.5 shows a block diagram of this observation model.

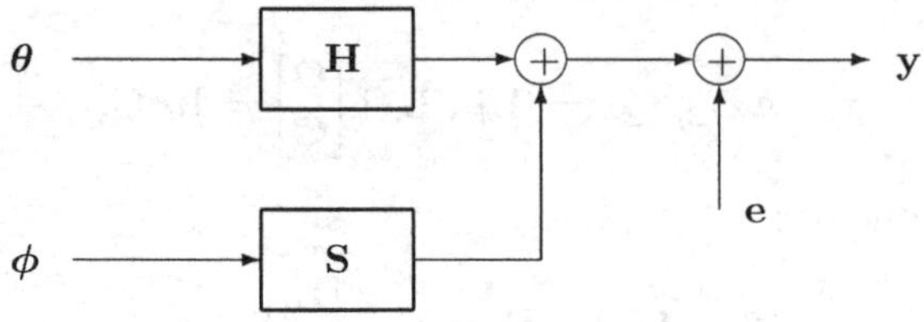

Figure 9.5 Observation model.

In general, the column vectors of the system matrix $\mathbf{H}$ generating the desired signal are not only linearly independent of each other but also are linearly independent of the column vectors of the linear system matrix $\mathbf{S}$ generating interference or colored noise. Therefore, the range Range($\mathbf{H}$) of $\mathbf{H}$ and the range Range($\mathbf{S}$) of $\mathbf{S}$ are disjoint, but in general they will be nonorthogonal.

Any undesired signal generated by a linear system is known as structured noise. Assuming that this structured noise is orthogonal to the additive Gaussian white noise $\mathbf{e}$, we have

$$\langle \mathbf{S}\boldsymbol{\phi}, \mathbf{e}\rangle = 0 \quad \Rightarrow \quad \boldsymbol{\phi}^H\mathbf{S}^H\mathbf{e} = 0,\ \forall\, \boldsymbol{\phi} \neq \mathbf{0} \quad \Rightarrow \quad \mathbf{S}^H\mathbf{e} = \mathbf{0}. \tag{9.5.29}$$

Similarly, letting the desired signal also be orthogonal to the additive Gaussian white noise, then

$$\mathbf{H}^H\mathbf{e} = \mathbf{0}. \tag{9.5.30}$$

Given two matrices $\mathbf{H}$ and $\mathbf{S}$, the objective of system modeling is to estimate the system parameter vector $\boldsymbol{\theta}$ relating to the desired signal. To this end, premultiplying (9.5.28) by the orthogonal projection matrix $\mathbf{P}_S^\perp$, we have

$$\mathbf{P}_S^\perp\mathbf{y} = \mathbf{P}_S^\perp\mathbf{H}\boldsymbol{\theta} + \mathbf{e}. \tag{9.5.31}$$

Here the relation $\mathbf{P}_S^\perp\mathbf{S} = \mathbf{O}$ and $\mathbf{P}_S^\perp\mathbf{e} = [\mathbf{I} - \mathbf{S}(\mathbf{S}^H\mathbf{S})^{-1}\mathbf{S}^H]\mathbf{e} = \mathbf{e}$ have been used.

Premultiplying both sides of (9.5.31) by the matrix $\mathbf{H}^H$, and using $\mathbf{H}^H\mathbf{e} = \mathbf{0}$, then

$$\boldsymbol{\theta} = (\mathbf{H}^H\mathbf{P}_S^\perp\mathbf{H})^{-1}\mathbf{H}^H\mathbf{P}_S^\perp\mathbf{y}. \tag{9.5.32}$$

Hence an estimate of the desired signal is given by

$$\hat{\mathbf{x}} = \mathbf{H}\boldsymbol{\theta} = \mathbf{H}(\mathbf{H}^H\mathbf{P}_S^\perp\mathbf{H})^{-1}\mathbf{H}^H\mathbf{P}_S^\perp\mathbf{y} = \mathbf{E}_{H|S}\mathbf{y}, \tag{9.5.33}$$

i.e., the estimate of the desired signal is the oblique projection of the observation data vector $\mathbf{y}$ onto the column space of $\mathbf{H}$ along the direction parallel to the column space of $\mathbf{S}$.

Exercises

9.1 Let $\mathbf{P}$ be a projector, and $\text{Range}(\mathbf{P}) = \text{Col}(\mathbf{U}), \text{Null}(\mathbf{P}) = \text{Col}(\mathbf{V})$. Use the matrices $\mathbf{U}$ and $\mathbf{V}$ to represent the projector $\mathbf{P}$.

9.2 Let both $\mathbf{E}_1$ and $\mathbf{E}_2$ be orthogonal projectors, and let $\mathbf{E}_1\mathbf{E}_2 = \mathbf{E}_2\mathbf{E}_1$. Show that $\mathbf{E}_1\mathbf{E}_2$ is an orthogonal projector.

9.3 Let $\mathbf{X}$ denote an observation data matrix. Use it to estimate a vector $\mathbf{y}$ as follows. Given two observation data vectors $\mathbf{x}_1 = [2,1,2,3]^T, \mathbf{x}_2 = [1,2,1,1]^T$ in $\mathbf{X}$, use them to estimate $\mathbf{y} = [1,2,3,2]^T$ and find the sum of the squared estimation errors. (*Hint:* Let the optimal filter vector be $\mathbf{w}_{\text{opt}}$; then the observation equation is $\mathbf{X}\mathbf{w}_{\text{opt}} = \mathbf{y}$.)

9.4 Assuming that $\mathbf{V}_1, \mathbf{V}_2$ are two sets of orthonormal basis vectors of the subspace W of a complex vector space $\mathbb{C}^n$, prove that $\mathbf{V}_1\mathbf{V}_1^H\mathbf{x} = \mathbf{V}_2\mathbf{V}_2^H\mathbf{x}$ holds for all vectors $\mathbf{x}$.

9.5 Show that if the projectors $\mathbf{P}_1$ and $\mathbf{P}_2$ are commute, i.e., $\mathbf{P}_1\mathbf{P}_2 = \mathbf{P}_2\mathbf{P}_1$, then their product $\mathbf{P} = \mathbf{P}_1\mathbf{P}_2$ is also a projector and find the range $\text{Range}(\mathbf{P})$ and the null space $\text{Null}(\mathbf{P})$ of $\mathbf{P}$, respectively.

9.6 Given the matrices $\mathbf{A} = \begin{bmatrix} 6 & 2 \\ -7 & 6 \end{bmatrix}$ and $\mathbf{A} = \begin{bmatrix} 0 & 1 & 1 \\ 1 & 1 & 0 \end{bmatrix}$, find their respective Moore–Penrose inverse matrices $\mathbf{A}^\dagger$, and explain why $\mathbf{A}\mathbf{A}^\dagger$ and $\mathbf{A}^\dagger\mathbf{A}$ are orthogonal projections respectively onto the column space and the row space of the matrix $\mathbf{A}$.

9.7 Assuming two basis vectors $\mathbf{u}_1 = [-1,2,-4,3,1]^T$ and $\mathbf{u}_2 = [5,6,2,-2,-1]^T$, generate the column space $\text{Range}(\mathbf{U}) = \text{Span}\{\mathbf{u}_1, \mathbf{u}_2\}$. Is the vector $\mathbf{v} = [-31,-18,-34,28,11]^T$ in the vector space $\text{Range}(\mathbf{U})$? Explain your answer.

9.8 Use the inverse matrix of inner product $\langle \mathbf{X}_{1,p}(n-1), \mathbf{X}_{1,p}(n-1)\rangle^{-1}$ to represent the inverse matrix of inner product $\langle \mathbf{X}_{1,p}(n), \mathbf{X}_{1,p}(n)\rangle^{-1}$.

9.9 For matrices $\mathbf{H}, \mathbf{S}$, define

$$\begin{aligned} \mathbf{E}_{H|S} &= \mathbf{H}(\mathbf{H}^H\mathbf{P}_S^\perp\mathbf{H})^{-1}\mathbf{H}^H\mathbf{P}_S^\perp \\ \mathbf{E}_{S|H} &= \mathbf{S}(\mathbf{S}^H\mathbf{P}_H^\perp\mathbf{S})^{-1}\mathbf{S}^H\mathbf{P}_H^\perp. \end{aligned}$$

Show that

(a) $\mathbf{E}_{H|S}$ and $\mathbf{E}_{S|H}$ are idempotent operators, i.e., $\mathbf{E}_{H|S}^2 = \mathbf{E}_{H|S}$ and $\mathbf{E}_{S|H}^2 = \mathbf{E}_{S|H}$;

(b) the oblique projection of $[\mathbf{H}, \mathbf{S}]$ along the direction of $\text{Range}(\mathbf{H})$ onto $\text{Range}(\mathbf{S})$ is given by $\mathbf{E}_{H|S}[\mathbf{H}, \mathbf{S}] = [\mathbf{H}, \mathbf{O}]$ and $\mathbf{E}_{S|H}[\mathbf{H}, \mathbf{S}] = [\mathbf{O}, \mathbf{S}]$, or equivalently by

$$\mathbf{E}_{H|S}\mathbf{H} = \mathbf{H} \quad \text{and} \quad \mathbf{E}_{H|S}\mathbf{S} = \mathbf{O},$$
$$\mathbf{E}_{S|H}\mathbf{H} = \mathbf{O} \quad \text{and} \quad \mathbf{E}_{S|H}\mathbf{S} = \mathbf{S}.$$

9.10 Prove the following results:

(a) $\mathbf{P}_H\mathbf{E}_{H|S} = \mathbf{E}_{H|S}$ and $\mathbf{P}_S\mathbf{E}_{S|H} = \mathbf{E}_{S|H}$;
(b) $\mathbf{E}_{H|S}\mathbf{P}_H = \mathbf{P}_H$ and $\mathbf{E}_{S|H}\mathbf{P}_S = \mathbf{P}_S$.

9.11 Let $H = \text{Range}(\mathbf{H})$ and $S = \text{Range}(\mathbf{S})$. Prove the following results:

$$\mathbf{P}_S^\perp\mathbf{E}_{H|S}\mathbf{P}_S^\perp = \mathbf{P}_S^\perp\mathbf{P}_{\mathbf{P}_S^\perp\mathbf{H}}\mathbf{P}_S^\perp = \mathbf{P}_{\mathbf{P}_S^\perp\mathbf{H}}.$$

9.12 If $H = \text{Range}(\mathbf{H}), S = \text{Range}(\mathbf{S})$, and $H \perp S$ show that $\mathbf{E}_{H|S} = \mathbf{P}_H$ and $\mathbf{E}_{S|H} = \mathbf{P}_S$.

9.13 Prove that $\mathbf{E}_{S|H} = \mathbf{P}_S - \mathbf{P}_S\mathbf{E}_{H|S} = \mathbf{P}_S(\mathbf{I} - \mathbf{E}_{H|S})$.

9.14 Prove that the products of the generalized inverse matrices $\mathbf{H}^\dagger$ and $\mathbf{S}^\dagger$ and their oblique projection satisfy $\mathbf{H}^\dagger\mathbf{E}_{H|S} = (\mathbf{P}_S^\perp\mathbf{H})^\dagger$ and $\mathbf{S}^\dagger\mathbf{E}_{S|H} = (\mathbf{P}_H^\perp\mathbf{S})^\dagger$, where $\mathbf{B}^\dagger = (\mathbf{B}^H\mathbf{B})^{-1}\mathbf{B}^H$ denotes the generalized inverse matrix of $\mathbf{B}$.

9.15 Prove that the cross-terms of the oblique projectors $\mathbf{E}_{H|S}$ and $\mathbf{E}_{S|H}$ is equal to zero, i.e., $\mathbf{E}_{H|S}\mathbf{E}_{S|H} = \mathbf{O}$ and $\mathbf{E}_{S|H}\mathbf{E}_{H|S} = \mathbf{O}$.

PART III
HIGHER-ORDER MATRIX ANALYSIS

10

Tensor Analysis

In many disciplines, for more and more problems three or more subscripts are needed to describe the data. Data with three or more subscripts are referred to as multi-channel data, whose representation is the tensor.

Data analysis based on vectors and matrices belongs to linear data analysis, while data analysis based on tensors is known as *tensor analysis*, which belongs to the category of multilinear data analysis. Linear data analysis is a single-factor analysis method, and multilinear data analysis is a multi-factor analysis method. Just as tensors are the generalization of matrices, multilinear data analysis is a natural extension of linear data analysis.

10.1 Tensors and their Presentation

A column or row vector, with data arranged all in the same direction, is known as a *one-way* data array, while a matrix is a *two-way* data array, with data arranged along both the horizontal and vertical directions. A tensor is a *multi-way* data array that is an extension of a matrix. Thus a three-way tensor has data arranged in three directions and an N-way tensor has N "directions" or modes.

The tensor in mathematics refers specifically to such a multi-way data array and should not be confused with the tensors in physics and engineering, which are often referred to as tensor fields [447].

10.1.1 Tensors

Here, tensors will be represented by mathcal symbols, such as $\mathcal{T}, \mathcal{A}, \mathcal{X}$ and so on. A tensor represented in an N-way array is called an Nth-order tensor, and is defined as a multilinear function on an N-dimensional Cartesian product vector space, denoted $\mathcal{T} \in \mathbb{K}^{I_1 \times I_2 \times \cdots \times I_N}$, where $\mathbb{K}$ denotes either the real field $\mathbb{R}$ or the complex field $\mathbb{C}$ and where I_n is the number of entries or the dimension in the nth "direction" or mode. For example, for a third-order tensor $\mathcal{T} \in \mathbb{K}^{4 \times 2 \times 5}$, the first, second and third modes have dimensions respectively equal to 4, 2 and 5. A scalar is a zero-order tensor, a vector is a first-order tensor and a matrix is a second-order tensor. An Nth-order tensor is a multilinear mapping

$$\mathcal{T}: \mathbb{K}^{I_1} \times \mathbb{K}^{I_2} \times \cdots \times \mathbb{K}^{I_N} \mapsto \mathbb{K}^{I_1 \times I_2 \times \cdots \times I_N}. \tag{10.1.1}$$

Because $\mathcal{T} \in \mathbb{K}^{I_1 \times \cdots \times I_N}$ can be regarded as an Nth-order matrix, tensor analysis is essentially *higher-order matrix analysis*.

With development of modern applications, multi-way higher-order data models have been successfully applied across many fields, which include, among others, chemometrics, chemistry, large data analysis and fusion, medicine and neuroscience, signal processing, text mining, clustering, internet traffic, EEG, computer vision, communication records, large-scale social networks etc. The following are typical applications of multi-way data models in various subjects [7].

1. *Social network analysis/Web mining* Multi-way data analysis is often used to extract relationships in social networks. The purpose of social network analysis is to study and discover the hidden structure in social networks, for example, to extract the patterns of communication between people or between people and organizations. In [6], chat room communication data is arranged into a three-way array "user × keyword × time sample", and the performance of this multi-way model in capturing the potential user-group structure is compared with the two-way model. Not only chat rooms, but also e-mail communication data can also be expressed using the three-way model "sender × receiver × time". In the analysis of network links, by combining the hyperlink with the anchor text information, the page graphic data are rearranged into a sparse three-way model "page × page × anchor text" [255], [256]. In Web personalization, click data are arranged into a three-way array "user × query word × page" [458].
2. *Computer vision* Tensor approximation has been shown to have important applications in computer vision, such as in image compression and face recognition. In computer vision and computer graphics, the data are usually multi-way. For example, a color image is a three-way array: its x and y coordinates are two modes and the color is the third mode. Fast image encoding technology regards an image as a vector or matrix. It has been shown that when an image is expressed as a tensor, a rank-1 approximation to the tensor obtained by iteration can be used to compress the image [499]. In addition, not only can this tensor structure of an image retain the two-dimensional features of the image but also it avoids the loss of image information [202]. Therefore, in image compression, a tensor representation is more effective than a matrix representation [435].
3. *Hyperspectral image* A hyperspectral image is usually a set of two-dimensional images, and thus it can be represented by three-way data: two-way spatial data and one-way spectral data [376], [91], [290]: the two-way data determine the location of the pixel and the one-way spectral data are related to the spectral band.

4. *Medicine and neuroscience* Multi-way electroencephalography (EEG) data are typically represented as an $m \times n$ matrix whose elements are the signal values from m time samples and n electrodes. However, in order to find the hidden power structure of a brain, we need to consider the frequency components of the EEG signals (such as the instantaneous signal powers at a particular frequency p). In this case, the EEG data can be arranged in a three-way data set. A three-way data array corresponds to channel or space (different electrode position), time of the data sample and frequency component. If adding theme and condition, then one obtains a five-way array or tensor representing "channel $\times$ time $\times$ frequency $\times$ theme $\times$ condition". Multi-way models were used to study the effects of new drugs on brain activity in 2001 [148]. In this study, the experimental data from the EEG, and different dosages of the drug for several patients, in some cases, were arranged in a six-way array corresponding to EEG, dosage, patient, condition etc. The results showed that through the multi-way model rather than a two-way model (such as used in principal component analysis), important information was successfully extracted from complex drug data.
5. *Multilinear image analysis of face recognition* [491], [492] Faces are modeled using a fifth-order tensor $\mathcal{D} \in \mathbb{R}^{28\times 5\times 3\times 3\times 7943}$, which represents the face images from 28 personal shots in five viewpoints, three illuminations and three expressions, while every image contain 7943 pixels.
6. *Epilepsy tensors* By identifying the spatial, spectrum and temporal features of epileptic seizure and false appearance, one can determine the focal point of epileptic seizure or exclude the false appearance of epilepsy. Therefore, it is necessary to use a three-way array consisting of time, frequency and space (electrode position) to model epilepsy signals [5].
7. *Chemistry* In the chemical, pharmaceutical and food sciences, commonly used fluorescence excitation–emission data contain several chemical compositions at different concentrations. The fluorescence spectra generate a three-way data set with the model "sample $\times$ excitation $\times$ emission". The main purpose of analyzing this type of data is to determine the chemical components and their relative concentration in every sample.

Figure 10.1 plots examples of a three-way array and a five-way tensor modeling a face.

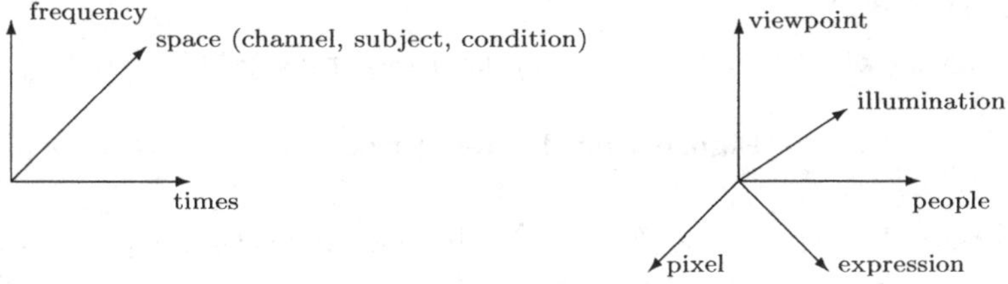

Figure 10.1 Three-way array (left) and a tensor modeling a face (right).

The main objective of this chapter is to investigate the relationship between matrices and tensors and to address the following questions: how to matricize (tensorize) data and how to apply the inherent tensor structure in data in high-dimensional problems.

10.1.2 Tensor Representation

In linear algebra (i.e., matrix algebra in a finite-dimensional vector space) a linear operator is defined in a finite-dimensional vector space. Because a matrix expresses a second-order tensor, linear algebra can also be regarded as the algebra of second-order tensors. In multiple linear algebra (i.e., higher-order tensor algebra) a multiple linear operator is also defined in a finite-dimensional vector space.

A matrix $\mathbf{A} \in \mathbb{K}^{m\times n}$ is represented by its entries and the symbol $[\cdot]$ as $\mathbf{A} = [a_{ij}]_{i,j=1}^{m,n}$. Similarly, an nth-order tensor $\mathcal{A} \in \mathbb{K}^{I_1\times I_2\times\cdots\times I_n}$ is, using a dual matrix symbol $[\![\cdot]\!]$, represented as $\mathcal{A} = [\![a_{i_1\cdots i_n}]\!]_{i_1,\ldots,i_n=1}^{I_1,\ldots,I_n}$, where $a_{i_1 i_2\cdots i_n}$ is the $(i_1,\ldots,i_n)$th entry of the tensor. An nth-order tensor is sometimes called an *n-dimensional hypermatrix* [111]. The set of all $(I_1\times I_2\times\cdots\times I_n)$-dimensional tensors is often denoted $\mathcal{T}(I_1,I_2,\ldots,I_n)$.

The most common tensor is a third-order tensor, $\mathcal{A} = [\![a_{ijk}]\!]_{i,j,k}^{I,J,K} \in \mathbb{K}^{I\times J\times K}$. A third-order tensor is sometimes called a three-dimensional matrix [476]. A square third-order tensor $\mathcal{X} \in \mathbb{K}^{I\times I\times I}$ is said to be *cubical.* In particular, a cubical tensor is *supersymmetric* [110], [257], if its entries have the following symmetry:

$$x_{ijk} = x_{ikj} = x_{jik} = x_{jki} = x_{kij} = x_{kji}, \quad \forall\, i,j,k = 1,\ldots,I.$$

Figure 10.2(a) depicts a third-order tensor $\mathcal{A} \in \mathbb{R}^{I\times J\times K}$.

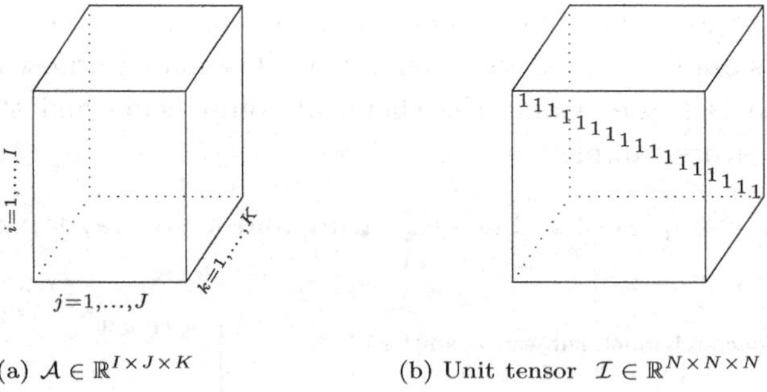

(a) $\mathcal{A} \in \mathbb{R}^{I\times J\times K}$ (b) Unit tensor $\mathcal{I} \in \mathbb{R}^{N\times N\times N}$

Figure 10.2 Third-order tensors.

For a tensor $\mathcal{A} = [\![a_{i_1\cdots i_m}]\!] \in \mathbb{R}^{N\times\cdots\times N}$, the line connecting $a_{11\cdots 1}$ to $a_{NN\cdots N}$ is called the *superdiagonal line* of the tensor $\mathcal{A}$. A tensor $\mathcal{A}$ is known as *unit tensor*, if all its entries on the superdiagonal line are equal to 1 and all other entries are equal to zero, i.e.,

$$a_{i_1\cdots i_m} = \delta_{i_1\cdots i_m} = \begin{cases} 1, & \text{if } i_1 = \cdots = i_m, \\ 0, & \text{otherwise.} \end{cases} \tag{10.1.2}$$

A unit third-order tensor is denoted $\mathcal{I} \in \mathbb{R}^{N\times N\times N}$, as shown in Figure 10.2(b).

In tensor analysis, it is convenient to view a third-order tensor as a set of vectors or matrices. Suppose that the ith entry of a vector $\mathbf{a}$ is denoted by a_i and that a matrix $\mathbf{A} = [a_{ij}] \in \mathbb{K}^{I\times J}$ has I row vectors $\mathbf{a}_{i:}, i = 1,\ldots,I$ and J column vectors $\mathbf{a}_{:j}, j = 1,\ldots,J$; the ith row vector $\mathbf{a}_{i:}$ is denoted by $[a_{i1},\ldots,a_{iJ}]$, and the jth column vector $\mathbf{a}_{:j}$ is denoted by $[a_{1j},\ldots,a_{Ij}]^T$. However, the concepts of row vectors and column vectors are no longer directly applicable for a higher-order tensor.

The three-way arrays of a third-order tensor are not called row vectors and column vectors, but are renamed *tensor fibers*. A fiber is a one-way array with one subscript that is variable and the other subscripts fixed. The fibers of a third-order tensor are vertical, horizontal and "depth" fibers. The vertical fibers of the third-order tensor $\mathcal{A} \in \mathbb{K}^{I\times J\times K}$ are also called *column fibers*, denoted $\mathbf{a}_{:jk}$; the horizontal fibers are also known as the *row fibers*, denoted $\mathbf{a}_{i:k}$; the depth fibers are also called the *tube fiber*, denoted $\mathbf{a}_{ij:}$.

Figure 10.3(a)–(c) show respectively the column fibers, row fibers and tube fibers of a third-order tensor.

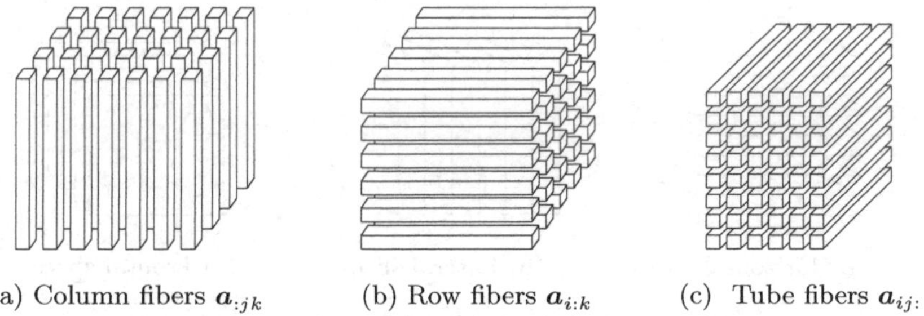

(a) Column fibers $\boldsymbol{a}_{:jk}$ (b) Row fibers $\boldsymbol{a}_{i:k}$ (c) Tube fibers $\boldsymbol{a}_{ij:}$

Figure 10.3 The fiber patterns of a third-order tensor.

Obviously, a third-order tensor $\mathcal{A} \in \mathbb{K}^{I\times J\times K}$ has respectively JK column fibers, KI row fibers and IJ tube fibers. An Nth-order tensor has N different kinds of fibers, called mode-n fibers or *mode-n vectors*, where $n = 1,\ldots,N$.

Definition 10.1 [280] Let $\mathcal{A} = [\![a_{i_1\cdots i_N}]\!] \in \mathbb{K}^{I_1\times\cdots\times I_N}$ be an Nth-order tensor; its mode-n vectors, for $n = 1,\ldots,N$, are I_n-dimensional vectors with i_n as the entry subscript that is variable and all the other subscripts $\{i_1,\ldots,i_N\} \setminus i_n$ fixed for a given mode-n vector, denoted as $\mathbf{a}_{i_1\cdots i_{n-1}:i_{n+1}\cdots i_N}$.

The difference between the order and the dimensions of a tensor $\mathcal{A} \in \mathbb{R}^{I_1 \times \cdots \times I_N}$ is that N is the order of the tensor, while I_n is the dimension of the mode-n vectors.

For a matrix, its column vectors are the mode-1 vectors and its row vectors are the mode-2 vectors. For example, for a 2×3 matrix, with two rows and three columns, the mode-1 vectors (the three column vectors) have dimension 2 and the mode-2 vectors (the two row vectors) have dimension 3. In a third-order tensor $\mathcal{A} \in \mathbb{K}^{I \times J \times K}$, the column fibers $\mathbf{a}_{:jk}$, row fibers $\mathbf{a}_{i:k}$ and tube fibers $\mathbf{a}_{ij:}$ are the mode-1, mode-2 and mode-3 vectors, respectively. There are JK mode-1 vectors, denoted $\mathbf{a}_{:jk}$, and every mode-1 vector has I entries, i.e., $\mathbf{a}_{:jk} = (a_{1jk}, \ldots, a_{Ijk})$; there are KI mode-2 vectors, denoted $\mathbf{a}_{i:k}$, and every mode-2 vector consists of J entries, i.e., $\mathbf{a}_{i:k} = (a_{i1k}, \ldots, a_{iJk})$; while there are IJ mode-3 vectors, denoted $\mathbf{a}_{ij:} = (a_{ij1}, \ldots, a_{ijK})$. The subspace spanned by the mode-n vectors is called a mode-n space. Generally, $\mathbf{a}_{:i_2,\ldots,i_N}, \mathbf{a}_{i_1:i_3,\ldots,i_N}, \mathbf{a}_{i_1,\ldots,i_{n-1}:i_{n+1},\ldots,i_N}$ are respectively the mode-1, mode-2 and mode-n vectors of the Nth-order ($N > 3$) tensor $\mathcal{A} \in \mathbb{K}^{I_1 \times I_2 \times \cdots \times I_N}$. Clearly, an Nth-order tensor has a total of $I_2 \cdots I_N$ mode-1 vectors and $I_1 \cdots I_{n-1} I_{n+1} \cdots I_N$ mode-n vectors.

Note that we do not write the mode-n vector representation as a column vector $[\cdots]^T$ or row vector $[\cdots]$, but deliberately use the symbol $(\cdots)$ because the same mode vector is sometimes a column vector and sometimes a row vector, depending on the slice matrices (see below) of the tensor.

A higher-order tensor can also be represented by matrix sets. For a third-order tensor, there exist three kinds of *slice matrices*: the *horizontal slice*, the *lateral slice* and the *frontal slice*, as shown in Figure 10.4.

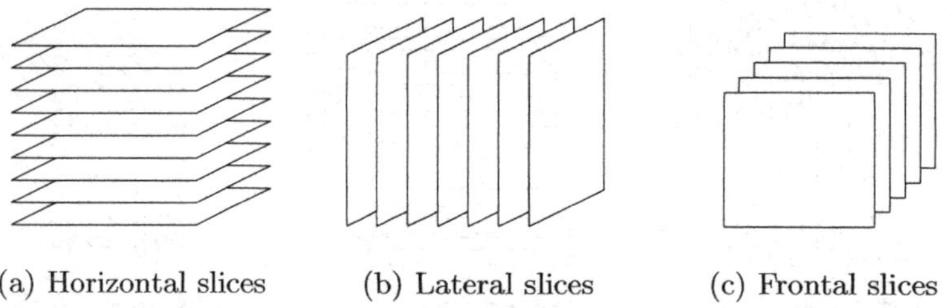

(a) Horizontal slices (b) Lateral slices (c) Frontal slices

Figure 10.4 Slices of a third-order tensor.

The horizontal slices, lateral slices and frontal slices of a third-order tensor are respectively denoted by the matrix notations $\mathbf{A}_{i::}$, $\mathbf{A}_{:j:}$ and $\mathbf{A}_{::k}$. Figure 10.5 shows representations of these three slice matrices.

In Figure 10.5, the arrow direction denotes the column direction for each slice matrix. In fact, the indexes i, j, k of a third-order tensor constitute three index sets $(i, j), (j, k), (k, i)$ in order. The first and second elements of each set represent respectively the row and column indexes of the corresponding slice matrix.

The relationships between the rows and columns of each slice matrix of a third-order tensor are as follows:

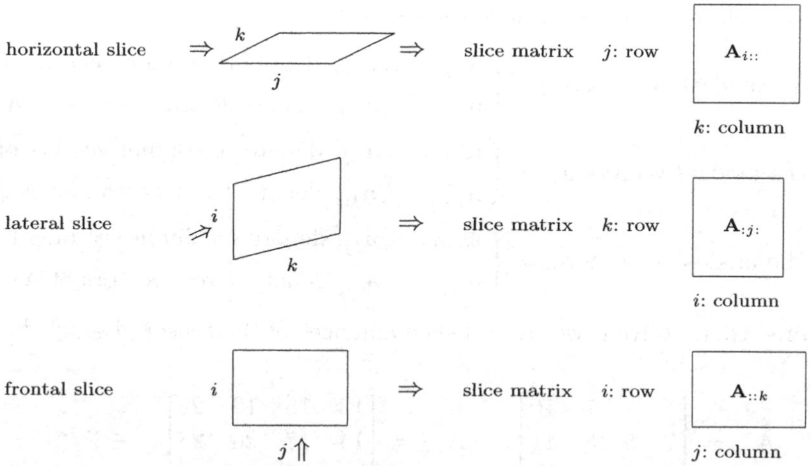

horizontal slice ⇒ k, j ⇒ slice matrix j: row $\mathbf{A}_{i::}$ k: column

lateral slice ⇒ i, k ⇒ slice matrix k: row $\mathbf{A}_{:j:}$ i: column

frontal slice i, j ⇑ ⇒ slice matrix i: row $\mathbf{A}_{::k}$ j: column

Figure 10.5 Representation of slice matrices.

For the horizontal slice matrix $\mathbf{A}_{i::}$, j is the row index, k is the column index;
For the lateral slice matrix $\mathbf{A}_{:j:}$, k is the row index, i is the column index;
For the frontal slice matrix $\mathbf{A}_{::k}$, i is the row index, j is the column index.

The following are mathematical representations of the three kinds of slice matrices of a third-order tensor.

(1) The third-order tensor $\mathcal{A} \in \mathbb{K}^{I\times J\times K}$ has I horizonal slice matrices

$$\mathbf{A}_{i::} \stackrel{\text{def}}{=} \begin{bmatrix} a_{i11} & \cdots & a_{i1K} \\ \vdots & \ddots & \vdots \\ a_{iJ1} & \cdots & a_{iJK} \end{bmatrix} = [\mathbf{a}_{i:1}, \ldots, \mathbf{a}_{i:K}] = \begin{bmatrix} \mathbf{a}_{i1:} \\ \vdots \\ \mathbf{a}_{iJ:} \end{bmatrix} \tag{10.1.3}$$

for $i = 1, \ldots, I$.

(2) The third-order tensor $\mathcal{A} \in \mathbb{K}^{I\times J\times K}$ has J lateral slice matrices

$$\mathbf{A}_{:j:} \stackrel{\text{def}}{=} \begin{bmatrix} a_{1j1} & \cdots & a_{Ij1} \\ \vdots & \ddots & \vdots \\ a_{1jK} & \cdots & a_{IjK} \end{bmatrix} = [\mathbf{a}_{1j:}, \ldots, \mathbf{a}_{Ij:}] = \begin{bmatrix} \mathbf{a}_{:j1} \\ \vdots \\ \mathbf{a}_{:jK} \end{bmatrix} \tag{10.1.4}$$

for $j = 1, \ldots, J$.

(3) The third-order tensor $\mathcal{A} \in \mathbb{K}^{I\times J\times K}$ has K frontal slice matrices

$$\mathbf{A}_{::k} \stackrel{\text{def}}{=} \begin{bmatrix} a_{11k} & \cdots & a_{1Jk} \\ \vdots & \ddots & \vdots \\ a_{I1k} & \cdots & a_{IJk} \end{bmatrix} = [\mathbf{a}_{:1k}, \ldots, \mathbf{a}_{:Jk}] = \begin{bmatrix} \mathbf{a}_{1:k} \\ \vdots \\ \mathbf{a}_{I:k} \end{bmatrix} \tag{10.1.5}$$

for $k = 1, \ldots, K$.

From the above analysis we have the following results:

$$\text{for the mode-1 vectors } \mathbf{a}_{:jk}: \begin{cases} \mathbf{a}_{:1k}, \ldots, \mathbf{a}_{:Jk} \text{ denote } J \text{ column vectors of } \mathbf{A}_{::k}, \\ \mathbf{a}_{:j1}, \ldots, \mathbf{a}_{:jK} \text{ denote } K \text{ row vectors of } \mathbf{A}_{:j:}; \end{cases}$$

$$\text{for the mode-2 vectors } \mathbf{a}_{i:k}: \begin{cases} \mathbf{a}_{i:1}, \ldots, \mathbf{a}_{i:K} \text{ denote } K \text{ column vectors of } \mathbf{A}_{i::}, \\ \mathbf{a}_{1:k}, \ldots, \mathbf{a}_{I:k} \text{ denote } I \text{ row vectors of } \mathbf{A}_{::k}; \end{cases}$$

$$\text{for the mode-3 vectors } \mathbf{a}_{ij:}: \begin{cases} \mathbf{a}_{1j:}, \ldots, \mathbf{a}_{Ij:} \text{ denote } I \text{ column vectors of } \mathbf{A}_{:j:}, \\ \mathbf{a}_{i1:}, \ldots, \mathbf{a}_{iJ:} \text{ denote } J \text{ row vectors of } \mathbf{A}_{i::}. \end{cases}$$

Example 10.1 Given two frontal slice matrices of the tensor $\mathcal{A} \in \mathbb{R}^{3\times 4\times 2}$ [254], i.e.,

$$\mathbf{A}_{::1} = \begin{bmatrix} 1 & 4 & 7 & 10 \\ 2 & 5 & 8 & 11 \\ 3 & 6 & 9 & 12 \end{bmatrix}, \quad \mathbf{A}_{::2} = \begin{bmatrix} 13 & 16 & 19 & 22 \\ 14 & 17 & 20 & 23 \\ 15 & 18 & 21 & 24 \end{bmatrix} \in \mathbb{R}^{3\times 4},$$

find all the mode-n vectors and the horizontal and lateral slice matrices of $\mathcal{A}$.

Solution The total number of mode-1 vectors (i.e., column fibers) is $JK = 4\times 2 = 8$; they are

$$\begin{bmatrix} 1 \\ 2 \\ 3 \end{bmatrix}, \begin{bmatrix} 4 \\ 5 \\ 6 \end{bmatrix}, \begin{bmatrix} 7 \\ 8 \\ 9 \end{bmatrix}, \begin{bmatrix} 10 \\ 11 \\ 12 \end{bmatrix}, \begin{bmatrix} 13 \\ 14 \\ 15 \end{bmatrix}, \begin{bmatrix} 16 \\ 17 \\ 18 \end{bmatrix}, \begin{bmatrix} 19 \\ 20 \\ 21 \end{bmatrix}, \begin{bmatrix} 22 \\ 23 \\ 24 \end{bmatrix},$$

and they span the mode-1 space of the tensor $\mathcal{A} \in \mathbb{R}^{3\times 4\times 2}$. The total number of mode-2 vectors is $KI = 2 \times 3 = 6$; they are

$$[1,4,7,10],\ [2,5,8,11],\ [3,6,9,12];\ [13,16,19,22],\ [14,17,20,23],\ [15,18,21,24].$$

The total number of mode-3 vectors is $IJ = 3 \times 4 = 12$; they are

$$[1,13],\ [4,16],\ [7,19],\ [10,22];\ [2,14],\ [5,17],\ [8,20],\ [11,23];\ [3,15],\ [6,18],\ [9,21],\ [12,24].$$

Moreover, the three horizontal slice matrices of the tensor are

$$\mathbf{A}_{1::} = \begin{bmatrix} 1 & 13 \\ 4 & 16 \\ 7 & 19 \\ 10 & 22 \end{bmatrix}, \quad \mathbf{A}_{2::} = \begin{bmatrix} 2 & 14 \\ 5 & 17 \\ 8 & 20 \\ 11 & 23 \end{bmatrix}, \quad \mathbf{A}_{3::} = \begin{bmatrix} 3 & 15 \\ 6 & 18 \\ 9 & 21 \\ 12 & 24 \end{bmatrix} \in \mathbb{R}^{4\times 2}.$$

The four lateral slice matrices are

$$\left.\begin{aligned} \mathbf{A}_{:1:} &= \begin{bmatrix} 1 & 2 & 3 \\ 13 & 14 & 15 \end{bmatrix}, \quad \mathbf{A}_{:2:} = \begin{bmatrix} 4 & 5 & 6 \\ 16 & 17 & 18 \end{bmatrix} \\ \mathbf{A}_{:3:} &= \begin{bmatrix} 7 & 8 & 9 \\ 19 & 20 & 21 \end{bmatrix}, \quad \mathbf{A}_{:4:} = \begin{bmatrix} 10 & 11 & 12 \\ 22 & 23 & 24 \end{bmatrix} \end{aligned}\right\} \in \mathbb{R}^{2\times 3}.$$

By extending the concept of the slices of a third-order tensor, we can divide a fourth-order tensor $\mathcal{A} \in \mathbb{K}^{I\times J\times K\times L}$ into L third-order tensor slices $\mathcal{A}_1, \ldots, \mathcal{A}_L \in \mathbb{K}^{I\times J\times K}$. By analogy, a fifth-order tensor $\mathcal{A}^{I\times J\times K\times L\times M}$ can be divided into M fourth-order tensor slices, and each fourth-order tensor slice can be further divided into L third-order tensor slices. In this sense, the third-order tensor is the basis of tensor analysis.

10.2 Vectorization and Matricization of Tensors

A third-order tensor $\mathcal{A} \in \mathbb{K}^{I\times J\times K}$ has I horizontal slice matrices $\mathbf{A}_{i::} \in \mathbb{K}^{J\times K}, i = 1, \ldots, I$, J lateral slice matrices $\mathbf{A}_{:j:} \in \mathbb{K}^{K\times I}, j = 1, \ldots, J$ and K frontal slice matrices $\mathbf{A}_{::k} \in \mathbb{K}^{I\times J}, k = 1, \ldots, K$. However, in tensor analysis and computation, it is frequently desirable to use a single matrix to represent a third-order tensor. In this case, we need an operation such that a third-order tensor (a three-way array) is reorganized or rearranged into a matrix (a two-way array). The transformation of a three-way or N-way array into a matrix form is called *tensor matricization* [245]. The matricization of a tensor $\mathcal{A}$ is denoted by $\mathrm{unvec}(\mathcal{A})$.

The matricization of a tensor is sometimes known as *tensor unfolding* or *tensor flattening* [65]. It should be noted that the unfolding or flattening here refers only to the unfolding of a cubic or three-dimensional array into a planar or two-dimensional array.

In addition to a unique matrix formulation, a unique vector formulation of a higher-order tensor is also of interest in many applications. *Tensor vectorization* is a transformation that rearranges the tensor as a unique vector. The vectorization of a tensor $\mathcal{A}$ is denoted by $\mathrm{vec}(\mathcal{A})$.

10.2.1 Vectorization and Horizontal Unfolding

As mentioned above, the *tensor vectorization* of an Nth-order tensor $\mathcal{A} = [\![\mathcal{A}_{i_1,\ldots,i_N}]\!] \in \mathbb{K}^{I_1\times\cdots\times I_N}$ is denoted by $\mathbf{a} = \mathrm{vec}(\mathcal{A})$, that is, a mapping $\mathbf{a} : \mathbb{K}^{I_1\times\cdots\times I_N} \mapsto \mathbb{K}^{I_1 I_2\cdots I_N\times 1}$.

For all the indices $i_n \in \{1, \ldots, I_n\}, n = 1, \ldots, N$, the entries of the vector $\mathbf{a}$ are given by

$$a_l = \mathcal{A}_{i_1,i_2,\ldots,i_N}, \quad l = i_1 + \sum_{n=2}^{N}\left((i_n - 1)\prod_{k=1}^{n-1} I_k\right). \tag{10.2.1}$$

For example, $a_1 = \mathcal{A}_{1,1,\ldots,1}, a_{I_1} = \mathcal{A}_{I_1,1,\ldots,1}, a_{I_2} = \mathcal{A}_{1,I_2,1,\ldots,1}, a_{I_1 I_2\cdots I_N} = \mathcal{A}_{I_1,I_2,\ldots,I_N}$.

Given an Nth-order tensor $\mathcal{A} \in \mathbb{K}^{I_1\times\cdots\times I_N}$, the matrix $\mathbf{A}_{(n)} = \mathbb{K}^{I_n\times(I_1\cdots I_{n-1}I_{n+1}\cdots I_N)}$ is known as the (mode-n) *horizontal unfolding matrix* of the tensor $\mathcal{A}$; and the matrix $\mathbf{A}^{(n)} = \mathbb{K}^{(I_1\cdots I_{n-1}I_{n+1}\cdots I_N)\times I_n}$ is called the (mode-n) *longitudinal unfolding matrix* of $\mathcal{A}$. Hence, the mappings $\mathcal{A} \mapsto \mathbf{A}_{(n)}$ and $\mathcal{A} \mapsto \mathbf{A}^{(n)}$ are said to be the *horizontal unfolding* and *longitudinal unfolding* of the tensor $\mathcal{A}$, respectively.

We discuss first the horizontal unfolding of a tensor here. In a *horizontal unfolding* of a third-order tensor, matrices from the same slice set are arranged side by side in a row in the same plane.

There are three horizontal unfolding methods.

1. Kiers Horizontal Unfolding Method

This is the tensor matricization method proposed by Kiers in 2000 [245]. In this method, a third-order tensor $\mathcal{A} \in \mathbb{K}^{I\times J\times K}$ is matricized as one of the following three unfolding matrices:

$$\left.\begin{aligned} a^{\text{Kiers}}_{i,(k-1)J+j} = a_{ijk} \quad &\Leftrightarrow \quad \mathbf{A}^{(I\times JK)}_{\text{Kiers}} = \mathbf{A}_{\text{Kiers}(1)} = [\mathbf{A}_{::1},\ldots,\mathbf{A}_{::K}],\\ a^{\text{Kiers}}_{j,(i-1)K+k} = a_{ijk} \quad &\Leftrightarrow \quad \mathbf{A}^{(J\times KI)}_{\text{Kiers}} = \mathbf{A}_{\text{Kiers}(2)} = [\mathbf{A}_{1::},\ldots,\mathbf{A}_{I::}],\\ a^{\text{Kiers}}_{k,(j-1)I+i} = a_{ijk} \quad &\Leftrightarrow \quad \mathbf{A}^{(K\times IJ)}_{\text{Kiers}} = \mathbf{A}_{\text{Kiers}(3)} = [\mathbf{A}_{:1:},\ldots,\mathbf{A}_{:J:}]. \end{aligned}\right\} \tag{10.2.2}$$

Figure 10.6 plots the Kiers horizontal unfolding of a third-order tensor $\mathcal{A} \in \mathbb{K}^{I\times J\times K}$. The three unfoldings are pictured on the right of the figure.

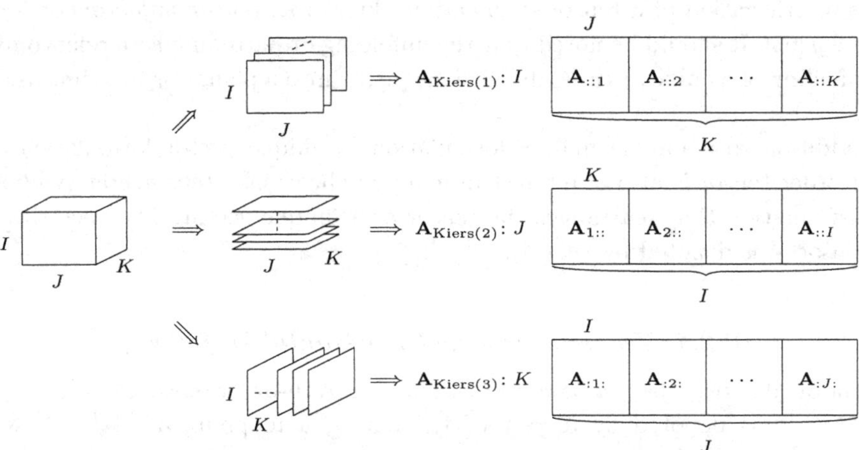

Figure 10.6 The horizontal unfolding of a tensor (Kiers method).

By extension, given an Nth-order tensor $\mathcal{A} \in \mathbb{K}^{I_1\times I_2\times\cdots\times I_N}$, the Kiers horizontal unfolding method maps its entry $a_{i_1 i_2\cdots i_N}$ to the (i_n, j)th entry of the matrix $\mathbf{A}_{\text{Kiers}(n)} \in \mathbb{K}^{I_n\times(I_1\cdots I_{n-1}I_{n+1}\cdots I_N)}$, i.e.,

$$\mathbf{A}_{\text{Kiers}(n)}(i_n, j) = a^{\text{Kiers}}_{i_n,j} = a_{i_1,i_2,\ldots,i_N}, \tag{10.2.3}$$

where $i_n = 1,\ldots,I_n$ and

$$j = \sum_{p=1}^{N-2}\left((i_{N+n-p}-1)\prod_{q=n+1}^{N+n-p-1} I_q\right) + i_{n+1}, \quad n = 1,\ldots,N \tag{10.2.4}$$

with $I_{N+m} = I_m$ and $i_{N+m} = i_m$ $(m > 0)$.

2. LMV Horizontal Unfolding Method

Lathauwer, Moor and Vanderwalle [281] in 2000 proposed the following horizontal unfolding of a third-order tensor (simply called the LMV method):

$$\left.\begin{array}{lll} a^{\text{LMV}}_{i,(j-1)K+k} = a_{ijk} & \Leftrightarrow & \mathbf{A}^{(I\times JK)}_{\text{LMV}} = \mathbf{A}_{\text{LMV}(1)} = [\mathbf{A}^T_{:1:}, \ldots, \mathbf{A}^T_{:J:}], \\ a^{\text{LMV}}_{j,(k-1)I+i} = a_{ijk} & \Leftrightarrow & \mathbf{A}^{(J\times KI)}_{\text{LMV}} = \mathbf{A}_{\text{LMV}(2)} = [\mathbf{A}^T_{::1}, \ldots, \mathbf{A}^T_{::K}], \\ a^{\text{LMV}}_{k,(i-1)J+j} = a_{ijk} & \Leftrightarrow & \mathbf{A}^{(K\times IJ)}_{\text{LMV}} = \mathbf{A}_{\text{LMV}(3)} = [\mathbf{A}^T_{1::}, \ldots, \mathbf{A}^T_{I::}]. \end{array}\right\} \quad (10.2.5)$$

Figure 10.7 shows the LMV horizontal unfolding of a third-order tensor.

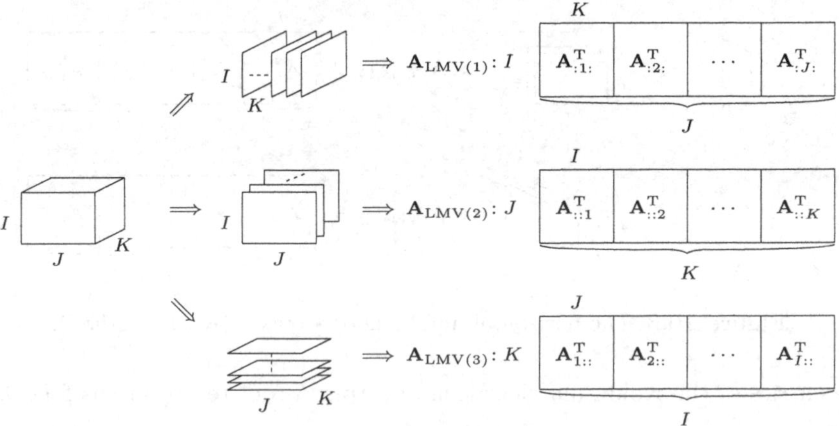

Figure 10.7 The horizontal unfolding of a tensor (LMV method).

This unfolding maps the entry $a_{i_1,\ldots,i_N}$ of an Nth-order tensor $\mathcal{A} \in \mathbb{K}^{I_1\times\cdots\times I_N}$ to the entry $a^{\text{LMV}}_{i_n,j}$ of the mode-n matrix $\mathbf{A}_{\text{LMV}(n)} \in \mathbb{K}^{I_n\times(I_1\cdots I_{n-1}I_{n+1}\cdots I_N)}$, where

$$\begin{aligned} j = &(i_{n+1}-1)I_{n+2}I_{n+3}\cdots I_N I_1 I_2\cdots I_{n-1} \\ &+ (i_{n+2}-1)I_{n+3}I_{n+4}\cdots I_N I_1 I_2 \cdots I_{n-1} + \cdots \\ &+ (i_N - 1)I_1 I_2\cdots I_{n-1} + (i_1-1)I_2 I_3\cdots I_{n-1} + (i_2-1)I_3I_4\cdots I_{n-1} \\ &+ \cdots + i_{n-1} \end{aligned}$$

or, written in a compact form,

$$j = i_{n-1} + \sum_{k=1}^{n-2}\left((i_k-1)\prod_{m=k+1}^{n-1} I_m\right) + \prod_{p=1}^{n-1} I_p\left(\sum_{k=n+1}^{N}\left((i_k-1)\prod_{q=k}^{N} I_{q+1}\right)\right), \quad (10.2.6)$$

where $I_q = 1$ if $q > N$.

3. Kolda Horizontal Unfolding Method

The Kolda horizontal unfolding [254] maps the entry $a_{i_1,\ldots,i_N}$ of an Nth-order tensor to the entry $a_{i_n,j}^{\text{Kolda}}$ of the mode-n matrix $\mathbf{A}_{\text{Kolda}(n)} \in \mathbb{K}^{I_n \times (I_1 \cdots I_{n-1} I_{n+1} \cdots I_N)}$, where

$$j = 1 + \sum_{k=1, k\neq n}^{N} \left((i_k - 1) \prod_{m=1, m\neq n}^{k-1} I_m \right). \tag{10.2.7}$$

Figure 10.8 shows the Kolda horizontal unfolding of a third-order tensor.

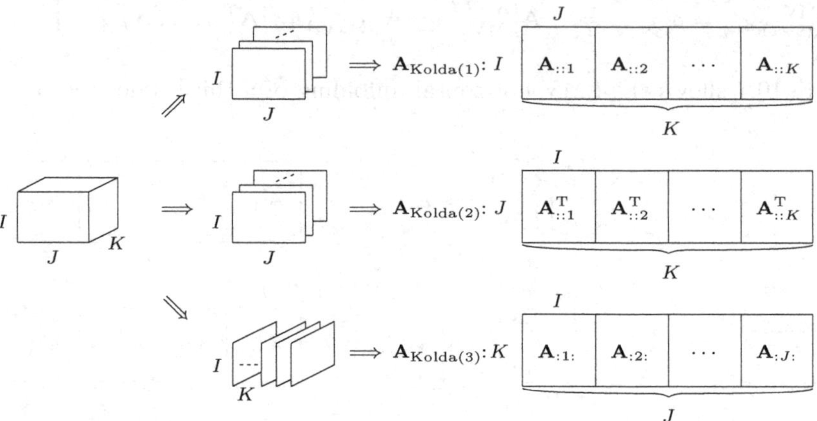

Figure 10.8 The horizontal unfolding of a tensor (Kolda method).

The entries of the Kolda matricization of a third-order tensor are as follows:

$$\left.\begin{aligned} a_{i,(k-1)J+j}^{\text{Kolda}} = a_{ijk} \quad &\Leftrightarrow \quad \mathbf{A}_{\text{Kolda}}^{(I\times JK)} = \mathbf{A}_{\text{Kolda}(1)} = [\mathbf{A}_{::1}, \ldots, \mathbf{A}_{::K}], \\ a_{j,(k-1)I+i}^{\text{Kolda}} = a_{ijk} \quad &\Leftrightarrow \quad \mathbf{A}_{\text{Kolda}}^{(J\times KI)} = \mathbf{A}_{\text{Kolda}(2)} = [\mathbf{A}_{::1}^T, \ldots, \mathbf{A}_{::K}^T], \\ a_{k,(j-1)I+j}^{\text{Kolda}} = a_{ijk} \quad &\Leftrightarrow \quad \mathbf{A}_{\text{Kolda}}^{(K\times IJ)} = \mathbf{A}_{\text{Kolda}(3)} = [\mathbf{A}_{:1:}, \ldots, \mathbf{A}_{:J:}]. \end{aligned}\right\} \tag{10.2.8}$$

Comparing the above three horizontal unfolding methods, we have their relationships.

(1) The mode-1 horizontal unfolding matrices of both the Kolda method and the Kiers method are the same, and their mode-3 horizontal unfolding matrices are also the same, namely

$$\mathbf{A}_{\text{Kolda}(1)} = \mathbf{A}_{\text{Kiers}(1)}, \qquad \mathbf{A}_{\text{Kolda}(3)} = \mathbf{A}_{\text{Kiers}(3)}. \tag{10.2.9}$$

(2) The mode-2 horizontal unfolding matrices of both the Kolda method and the LMV method are the same, i.e.,

$$\mathbf{A}_{\text{Kolda}(2)} = \mathbf{A}_{\text{LMV}(2)}. \tag{10.2.10}$$

Example 10.2 Consider a third-order tensor $\mathcal{A}$ with entries

$$A_{1,1,1} = 1, \quad A_{1,2,1} = 3, \quad A_{1,1,2} = 5, \quad A_{1,2,2} = 7,$$
$$A_{2,1,1} = 2, \quad A_{2,2,1} = 4, \quad A_{2,1,2} = 6, \quad A_{2,2,2} = 8,$$

by (10.2.1), we have the tensor vectorization $\mathbf{a} = [1,2,3,4,5,6,7,8]^T$. From (10.2.2) we get the Kiers horizontal unfolding matrices:

$$\mathbf{A}_{\text{Kiers}(1)} = [\mathbf{A}_{::1}, \mathbf{A}_{::2}] = \begin{bmatrix} A_{111} & A_{121} & | & A_{112} & A_{122} \\ A_{211} & A_{221} & | & A_{212} & A_{222} \end{bmatrix} = \begin{bmatrix} 1 & 3 & | & 5 & 7 \\ 2 & 4 & | & 6 & 8 \end{bmatrix},$$

$$\mathbf{A}_{\text{Kiers}(2)} = [\mathbf{A}_{:1:}, \mathbf{A}_{:2:}] = \begin{bmatrix} A_{111} & A_{112} & | & A_{121} & A_{122} \\ A_{211} & A_{212} & | & A_{221} & A_{222} \end{bmatrix} = \begin{bmatrix} 1 & 5 & | & 3 & 7 \\ 2 & 6 & | & 4 & 8 \end{bmatrix},$$

$$\mathbf{A}_{\text{Kiers}(3)} = [\mathbf{A}_{1::}, \mathbf{A}_{2::}] = \begin{bmatrix} A_{111} & A_{112} & | & A_{211} & A_{212} \\ A_{121} & A_{122} & | & A_{221} & A_{222} \end{bmatrix} = \begin{bmatrix} 1 & 5 & | & 2 & 6 \\ 3 & 7 & | & 4 & 8 \end{bmatrix}.$$

10.2.2 Longitudinal Unfolding of Tensors

If the same kind of slice matrices of a tensor are arranged in a longitudinal direction, then we get the *longitudinal unfolding* of a tensor. This is the transpose of the horizontal unfolding. Thus longitudinal unfolding has also three methods, which correspond to the three methods of horizontal unfolding. The mode-n longitudinal unfolding of an Nth-order tensor $\mathcal{A} \in \mathbb{K}^{I_1 \times I_2 \times \cdots \times I_N}$ is denoted by the notation $\mathbf{A}^{(n)} \in \mathbb{K}^{(I_1 \cdots I_{n-1} I_{n+1} \cdots I_N) \times I_n}$.

A third-order tensor $\mathcal{A} \in \mathbb{K}^{I \times J \times K}$ can be matricized as a $(JK) \times I$ matrix, a $(KI) \times J$ matrix or an $(IJ) \times K$ matrix, denoted

$$\mathbf{A}^{(1)} = \mathbf{A}^{(JK \times I)}, \quad \mathbf{A}^{(2)} = \mathbf{A}^{(KI \times J)}, \quad \mathbf{A}^{(3)} = \mathbf{A}^{(IJ \times K)}. \tag{10.2.11}$$

1. Kiers Longitudinal Unfolding Method

Given an Nth-order tensor $\mathcal{A} \in \mathbb{K}^{I_1 \times I_2 \times \cdots \times I_N}$, let $\mathbf{A}^{(I_1 \cdots I_{n-1} I_{n+1} \cdots I_N) \times I_n}$ be its *mode-n* longitudinal unfolding. Then, the entry of the jth row and the i_nth column of the longitudinal unfolding is defined as

$$a_{j,i_n}^{(I_1 \cdots I_{n-1} I_{n+1} \cdots I_N) \times I_n} = a_{i_1, i_2, \ldots, i_N}, \tag{10.2.12}$$

where j is given by (10.2.4).

More specifically, the Kiers method for the longitudinal unfolding of a third-order tensor gives

$$\left.\begin{aligned}
a^{(JK\times I)}_{(k-1)J+j,i} = a_{ijk} \quad &\Leftrightarrow \quad \mathbf{A}^{(JK\times I)} = \mathbf{A}^{(1)}_{\text{Kiers}} = \begin{bmatrix} \mathbf{A}^T_{::1} \\ \vdots \\ \mathbf{A}^T_{::K} \end{bmatrix}, \\
a^{(KI\times J)}_{(i-1)K+k,j} = a_{ijk} \quad &\Leftrightarrow \quad \mathbf{A}^{(KI\times J)} = \mathbf{A}^{(2)}_{\text{Kiers}} = \begin{bmatrix} \mathbf{A}^T_{1::} \\ \vdots \\ \mathbf{A}^T_{I::} \end{bmatrix}, \\
a^{(IJ\times K)}_{(j-1)I+i,k} = a_{ijk} \quad &\Leftrightarrow \quad \mathbf{A}^{(IJ\times K)} = \mathbf{A}^{(3)}_{\text{Kiers}} = \begin{bmatrix} \mathbf{A}^T_{:1:} \\ \vdots \\ \mathbf{A}^T_{:J:} \end{bmatrix}.
\end{aligned}\right\} \tag{10.2.13}$$

2. LMV Longitudinal Unfolding Method

The LMV longitudinal unfolding of a third-order tensor is represented as [282]

$$\left.\begin{aligned}
a^{(JK\times I)}_{(j-1)K+k,i} = a_{ijk} \quad &\Leftrightarrow \quad \mathbf{A}^{(JK\times I)} = \mathbf{A}^{(1)}_{\text{LMV}} = \begin{bmatrix} \mathbf{A}_{:1:} \\ \vdots \\ \mathbf{A}_{:J:} \end{bmatrix}, \\
a^{(KI\times J)}_{(k-1)I+i,j} = a_{ijk} \quad &\Leftrightarrow \quad \mathbf{A}^{(KI\times J)} = \mathbf{A}^{(2)}_{\text{LMV}} = \begin{bmatrix} \mathbf{A}_{::1} \\ \vdots \\ \mathbf{A}_{::K} \end{bmatrix}, \\
a^{(IJ\times K)}_{(i-1)J+j,k} = a_{ijk} \quad &\Leftrightarrow \quad \mathbf{A}^{(IJ\times K)} = \mathbf{A}^{(3)}_{\text{LMV}} = \begin{bmatrix} \mathbf{A}_{1::} \\ \vdots \\ \mathbf{A}_{I::} \end{bmatrix},
\end{aligned}\right\} \tag{10.2.14}$$

where the row subscript j is determined by (10.2.6).

3. Kolda Longitudinal Unfolding Method

The Kolda longitudinal unfolding of a third-order tensor is given by

$$\left.\begin{aligned}
a^{(JK\times I)}_{(k-1)J+j,i} = a_{ijk} \quad &\Leftrightarrow \quad \mathbf{A}^{(JK\times I)} = \mathbf{A}^{(1)}_{\text{Kolda}} = \begin{bmatrix} \mathbf{A}^T_{::1} \\ \vdots \\ \mathbf{A}^T_{::K} \end{bmatrix}, \\
a^{(KI\times J)}_{(k-1)I+i,j} = a_{ijk} \quad &\Leftrightarrow \quad \mathbf{A}^{(KI\times J)} = \mathbf{A}^{(2)}_{\text{Kolda}} = \begin{bmatrix} \mathbf{A}_{::1} \\ \vdots \\ \mathbf{A}_{::K} \end{bmatrix}, \\
a^{(IJ\times K)}_{(j-1)I+i,k} = a_{ijk} \quad &\Leftrightarrow \quad \mathbf{A}^{(IJ\times K)} = \mathbf{A}^{(3)}_{\text{Kolda}} = \begin{bmatrix} \mathbf{A}^T_{:1:} \\ \vdots \\ \mathbf{A}^T_{:J:} \end{bmatrix}.
\end{aligned}\right\} \tag{10.2.15}$$

The entries of the Kolda longitudinal matricization of an Nth-order tensor are also given by (10.2.12), but its row subscript j is determined by (10.2.7).

Regarding the horizontal and longitudinal unfoldings, there are the following relationships.

(1) The relationships between the longitudinal unfoldings of a third-order tensor are

$$\mathbf{A}_{\text{Kiers}}^{(1)} = \mathbf{A}_{\text{Kolda}}^{(1)}, \quad \mathbf{A}_{\text{Kiers}}^{(3)} = \mathbf{A}_{\text{Kolda}}^{(3)}, \quad \mathbf{A}_{\text{LMV}}^{(2)} = \mathbf{A}_{\text{Kolda}}^{(2)}. \tag{10.2.16}$$

(2) The relationships between the longitudinal unfolding and the horizontal unfolding of a third-order tensor are

$$\mathbf{A}_{\text{Kiers}}^{(n)} = \mathbf{A}_{\text{Kiers}(n)}^{T}, \quad \mathbf{A}_{\text{LMV}}^{(n)} = \mathbf{A}_{\text{LMV}(n)}^{T}, \quad \mathbf{A}_{\text{Kolda}}^{(n)} = \mathbf{A}_{\text{Kolda}(n)}^{T}. \tag{10.2.17}$$

(3) The LMV mode-n longitudinal unfolding matrix is a longitudinal arrangement of the Kiers mode-n horizontal unfolding matrix. Conversely, the Kiers mode-n longitudinal unfolding matrix is a longitudinal arrangement of the LMV mode-n horizontal unfolding matrix.

Some publications (e.g., [65], [245], [5], [7]) use the horizontal unfolding of tensors, while others (e.g., [302], [282], [257]) use longitudinal unfolding.

From the following example we can see the connection and difference between the Keirs, LMV and Kolda matricization results for the same tensor.

Example 10.3 Consider again the two frontal slice matrices of the tensor $\mathcal{A} \in \mathbb{R}^{3\times4\times2}$ given by [254]

$$\mathbf{A}_{::1} = \begin{bmatrix} 1 & 4 & 7 & 10 \\ 2 & 5 & 8 & 11 \\ 3 & 6 & 9 & 12 \end{bmatrix}, \quad \mathbf{A}_{::2} = \begin{bmatrix} 13 & 16 & 19 & 22 \\ 14 & 17 & 20 & 23 \\ 15 & 18 & 21 & 24 \end{bmatrix} \in \mathbb{R}^{3\times4}, \tag{10.2.18}$$

from Example 10.1. The horizontal and longitudinal unfolding matrices are as follows:

(1) The Keirs horizontal unfolding

$$\mathbf{A}_{\text{Keirs}(1)} = \begin{bmatrix} 1 & 4 & 7 & 10 & 13 & 16 & 19 & 22 \\ 2 & 5 & 8 & 11 & 14 & 17 & 20 & 23 \\ 3 & 6 & 9 & 12 & 15 & 18 & 21 & 24 \end{bmatrix},$$

$$\mathbf{A}_{\text{Keirs}(2)} = \begin{bmatrix} 1 & 13 & 2 & 14 & 3 & 15 \\ 4 & 16 & 5 & 17 & 6 & 18 \\ 7 & 19 & 8 & 20 & 9 & 21 \\ 10 & 22 & 11 & 23 & 12 & 24 \end{bmatrix},$$

$$\mathbf{A}_{\text{Keirs}(3)} = \begin{bmatrix} 1 & 2 & 3 & 4 & 5 & 6 & 7 & 8 & 9 & 10 & 11 & 12 \\ 13 & 14 & 15 & 16 & 17 & 18 & 19 & 20 & 21 & 22 & 23 & 24 \end{bmatrix}.$$

(2) The LMV horizontal unfolding

$$\mathbf{A}_{\mathrm{LMV}(1)} = \begin{bmatrix} 1 & 13 & 4 & 16 & 7 & 19 & 10 & 22 \\ 2 & 14 & 5 & 17 & 8 & 20 & 11 & 23 \\ 3 & 15 & 6 & 18 & 9 & 21 & 12 & 24 \end{bmatrix},$$

$$\mathbf{A}_{\mathrm{LMV}(2)} = \begin{bmatrix} 1 & 2 & 3 & 13 & 14 & 15 \\ 4 & 5 & 6 & 16 & 17 & 18 \\ 7 & 8 & 9 & 19 & 20 & 21 \\ 10 & 11 & 12 & 22 & 23 & 24 \end{bmatrix},$$

$$\mathbf{A}_{\mathrm{LMV}(3)} = \begin{bmatrix} 1 & 4 & 7 & 10 & 2 & 5 & 8 & 11 & 3 & 6 & 9 & 12 \\ 13 & 16 & 19 & 22 & 14 & 17 & 20 & 23 & 15 & 18 & 21 & 24 \end{bmatrix}.$$

(3) The Kolda horizontal unfolding

$$\mathbf{A}_{\mathrm{Kolda}(1)} = \begin{bmatrix} 1 & 4 & 7 & 10 & 13 & 16 & 19 & 22 \\ 2 & 5 & 8 & 11 & 14 & 17 & 20 & 23 \\ 3 & 6 & 9 & 12 & 15 & 18 & 21 & 24 \end{bmatrix},$$

$$\mathbf{A}_{\mathrm{Kolda}(2)} = \begin{bmatrix} 1 & 2 & 3 & 13 & 14 & 15 \\ 4 & 5 & 6 & 16 & 17 & 18 \\ 7 & 8 & 9 & 19 & 20 & 21 \\ 10 & 11 & 12 & 22 & 23 & 24 \end{bmatrix},$$

$$\mathbf{A}_{\mathrm{Kolda}(3)} = \begin{bmatrix} 1 & 2 & 3 & 4 & 5 & 6 & 7 & 8 & 9 & 10 & 11 & 12 \\ 13 & 14 & 15 & 16 & 17 & 18 & 19 & 20 & 21 & 22 & 23 & 24 \end{bmatrix}.$$

(4) The Kiers longitudinal unfolding

$$\mathbf{A}_{\mathrm{Keirs}}^{(1)} = \begin{bmatrix} 1 & 2 & 3 \\ 4 & 5 & 6 \\ 7 & 8 & 9 \\ 10 & 11 & 12 \\ 13 & 14 & 15 \\ 16 & 17 & 18 \\ 19 & 20 & 21 \\ 22 & 23 & 24 \end{bmatrix}, \quad \mathbf{A}_{\mathrm{Keirs}}^{(2)} = \begin{bmatrix} 1 & 4 & 7 & 10 \\ 13 & 16 & 19 & 22 \\ 2 & 5 & 8 & 11 \\ 14 & 17 & 20 & 23 \\ 3 & 6 & 9 & 12 \\ 15 & 18 & 21 & 24 \end{bmatrix}, \quad \mathbf{A}_{\mathrm{Keirs}}^{(3)} = \begin{bmatrix} 1 & 13 \\ 2 & 14 \\ 3 & 15 \\ \vdots & \vdots \\ 10 & 22 \\ 11 & 23 \\ 12 & 24 \end{bmatrix}.$$

(5) The LMV longitudinal unfolding

$$\mathbf{A}_{\mathrm{LMV}}^{(1)} = \begin{bmatrix} 1 & 2 & 3 \\ 13 & 14 & 15 \\ 4 & 5 & 6 \\ 16 & 17 & 18 \\ 7 & 8 & 9 \\ 19 & 20 & 21 \\ 10 & 11 & 12 \\ 22 & 23 & 24 \end{bmatrix}, \quad \mathbf{A}_{\mathrm{LMV}}^{(2)} = \begin{bmatrix} 1 & 4 & 7 & 10 \\ 2 & 5 & 8 & 11 \\ 3 & 6 & 9 & 12 \\ 13 & 16 & 19 & 22 \\ 14 & 17 & 20 & 23 \\ 15 & 18 & 21 & 24 \end{bmatrix},$$

and

$$\mathbf{A}_{\text{LMV}}^{(3)} = \begin{bmatrix} 1 & 4 & 7 & 10 & 2 & 5 & 8 & 11 & 3 & 6 & 9 & 12 \\ 13 & 16 & 19 & 22 & 14 & 17 & 20 & 23 & 15 & 18 & 21 & 24 \end{bmatrix}^T.$$

(6) The Kolda longitudinal unfolding

$$\mathbf{A}_{\text{Kolda}}^{(1)} = \begin{bmatrix} 1 & 2 & 3 \\ 4 & 5 & 6 \\ 7 & 8 & 9 \\ 10 & 11 & 12 \\ 13 & 14 & 15 \\ 16 & 17 & 18 \\ 19 & 20 & 21 \\ 22 & 23 & 24 \end{bmatrix}, \quad \mathbf{A}_{\text{Kolda}}^{(2)} = \begin{bmatrix} 1 & 4 & 7 & 10 \\ 2 & 5 & 8 & 11 \\ 3 & 6 & 9 & 12 \\ 13 & 16 & 19 & 22 \\ 14 & 17 & 20 & 23 \\ 15 & 18 & 21 & 24 \end{bmatrix}, \quad \mathbf{A}_{\text{Kolda}}^{(3)} = \begin{bmatrix} 1 & 13 \\ 2 & 14 \\ 3 & 15 \\ \vdots & \vdots \\ 10 & 22 \\ 11 & 23 \\ 12 & 24 \end{bmatrix}.$$

From the above results, it can be seen that

(a) the difference between the same mode-n horizontal unfoldings of the different methods is in the order in which the column vectors are arranged;
(b) the difference between the same mode-n longitudinal unfoldings of the different methods is in the order in which the row vectors are arranged.

The matricization of tensors brings convenience to tensor analysis, but it should be noted [65] that matricization may lead to the following problems:

(1) the mathematical stability of the model is slightly worse;
(2) the explanation of the model is slightly less clear;
(3) the predictability of the model is slightly worse;
(4) the number of the model parameters may be greater.

The matricization of tensors is an effective mathematical tool of tensor analysis. However, the ultimate goal of tensor analysis is sometimes to restore the original tensor from its matricization.

The operation of reconstructing the original tensor from its vectorization or matricization is called *tensorization* or *tensor reshaping* or *tensor reconstruction.*

Any horizontal unfolding matrices $\mathbf{A}^{(I\times JK)}$, $\mathbf{A}^{(J\times KI)}$ or $\mathbf{A}^{(K\times IJ)}$ or any longitudinal unfolding matrices $\mathbf{A}^{(KI\times J)}$, $\mathbf{A}^{(IJ\times K)}$ or $\mathbf{A}^{(JK\times I)}$ can be tensorized into the third-order tensor $\mathcal{A} \in \mathbb{K}^{I\times J\times K}$. For example, for the Kiers matricization matrix, its tensorization entry is defined as

$$\begin{aligned} a_{ijk} &= \mathbf{A}^{(I\times JK)}_{i,(k-1)J+j} = \mathbf{A}^{(J\times KI)}_{j,(i-1)K+k} = \mathbf{A}^{(K\times IJ)}_{k,(j-1)I+i} \\ &= \mathbf{A}^{(JK\times I)}_{(k-1)J+j,i} = \mathbf{A}^{(KI\times J)}_{(i-1)K+k,j} = \mathbf{A}^{(IJ\times K)}_{(j-1)I+i,k}, \end{aligned} \tag{10.2.19}$$

and, for the LMV matricization matrix,

$$\begin{aligned} a_{ijk} &= \mathbf{A}^{(I\times JK)}_{i,(j-1)K+k} = \mathbf{A}^{(J\times KI)}_{j,(k-1)I+i} = \mathbf{A}^{(K\times IJ)}_{k,(i-1)J+j} \\ &= \mathbf{A}^{(JK\times I)}_{(j-1)K+k,i} = \mathbf{A}^{(KI\times J)}_{(k-1)I+i,j} = \mathbf{A}^{(IJ\times K)}_{(i-1)J+j,k}. \end{aligned} \tag{10.2.20}$$

The tensorization of the column vector $\mathbf{a}^{(IJK\times 1)}$ or the row vector $\mathbf{a}^{(1\times IJK)}$ is defined as

$$a_{ijk} = \mathbf{a}^{(IJK\times 1)}_{(k-1)IJ+(j-1)I+i} = \mathbf{a}^{(1\times IJK)}_{(i-1)JK+(j-1)K+k}, \tag{10.2.21}$$

where $i = 1,\ldots,I$, $j = 1,\ldots,J$, $k = 1,\ldots,K$.

More generally, an Nth-order tensor can be reshaped or reconstructed using one of the mode-n horizontal unfolding formulas (10.2.4), (10.2.6) or (10.2.7).

10.3 Basic Algebraic Operations of Tensors

The basic algebraic operations of tensors involve mainly the products of tensors, the product of a tensor and a matrix and the rank of a tensor.

10.3.1 Inner Product, Norm and Outer Product

The inner product of two tensors is a generalization of the inner product of two vectors: first vectorize two tensor and then find the inner product of the two vectorized tensors to yield the inner product of the two tensors.

Definition 10.2 For two tensors $\mathcal{A},\mathcal{B} \in \mathcal{T}(I_1, I_2, \ldots, I_N)$, the *tensor inner product* $\langle \mathcal{A},\mathcal{B}\rangle$ is a scalar and is defined as the inner product of the column vectorizations of the two tensors, i.e.,

$$\begin{aligned} \langle \mathcal{A},\mathcal{B}\rangle &\stackrel{\text{def}}{=} \langle \text{vec}(\mathcal{A}), \text{vec}(\mathcal{B})\rangle = (\text{vec}(\mathcal{A}))^H \text{vec}(\mathcal{B}) \\ &= \sum_{i_1=1}^{I_1}\sum_{i_2=1}^{I_2}\cdots\sum_{i_n=1}^{I_N} a^*_{i_1 i_2\cdots i_n} b_{i_1 i_2\cdots i_n}. \end{aligned} \tag{10.3.1}$$

Using the concept of the inner product of tensors, we can directly introduce the definition of the Frobenius norm of a tensor.

Definition 10.3 The *Frobenius norm* of a tensor $\mathcal{A}$ is defined as

$$\|\mathcal{A}\|_F = \sqrt{\langle \mathcal{A},\mathcal{A}\rangle} \stackrel{\text{def}}{=} \left(\sum_{i_1=1}^{I_1}\sum_{i_2=1}^{I_2}\cdots\sum_{i_n=1}^{I_N} |a_{i_1 i_2\cdots i_n}|^2\right)^{1/2}. \tag{10.3.2}$$

The outer product of two vectors is a matrix, e.g., $\mathbf{X} = \mathbf{u}\mathbf{v}^T$. The outer product of more than two vectors yields a tensor.

Definition 10.4 The outer product of n vectors $\mathbf{a}^{(m)} \in \mathbb{K}^{m\times 1}, m = 1,\ldots,n$, denoted $\mathbf{a}^{(1)} \circ \cdots \circ \mathbf{a}^{(n)}$, is an nth-order tensor, namely

$$\mathcal{A} = \mathbf{a}^{(1)} \circ \mathbf{a}^{(2)} \circ \cdots \circ \mathbf{a}^{(n)}, \tag{10.3.3}$$

whose entries are

$$a_{i_1 i_2 \cdots i_n} = a_{i_1}^{(1)} a_{i_2}^{(2)} \cdots a_{i_n}^{(n)}, \tag{10.3.4}$$

where $a_j^{(m)}$ is the jth entry of the mode-m vector $\mathbf{a}^{(m)}$.

Example 10.4 The outer product of two vectors $\mathbf{u} \in \mathbb{K}^{m\times 1}$ and $\mathbf{v} \in \mathbb{K}^{n\times 1}$ is

$$\mathbf{X} = \mathbf{u} \circ \mathbf{v} = \begin{bmatrix} u_1 \\ \vdots \\ u_m \end{bmatrix} [v_1, \ldots, v_n] = \begin{bmatrix} u_1 v_1 & \cdots & u_1 v_n \\ \vdots & \ddots & \vdots \\ u_m v_1 & \cdots & u_m v_n \end{bmatrix} \quad \in \mathbb{K}^{m\times n}.$$

The outer product of three vectors $\mathbf{u} \in \mathbb{K}^{I\times 1}, \mathbf{v} \in \mathbb{K}^{J\times 1}, \mathbf{w} \in \mathbb{K}^{K\times 1}$ is

$$\mathcal{A} = \mathbf{u} \circ \mathbf{v} \circ \mathbf{w} = \begin{bmatrix} u_1 v_1 & \cdots & u_1 v_J \\ \vdots & \ddots & \vdots \\ u_I v_1 & \cdots & u_I v_J \end{bmatrix} \circ [w_1, \cdots, w_K] \quad \in \mathbb{K}^{I\times J\times K},$$

whose frontal slice matrix

$$\mathbf{A}_{::k} = \begin{bmatrix} u_1 v_1 w_k & \cdots & u_1 v_J w_k \\ \vdots & \ddots & \vdots \\ u_I v_1 w_k & \cdots & u_I v_J w_k \end{bmatrix}, \quad k = 1,\ldots,K.$$

That is, $a_{ijk} = u_i v_j w_k, i = 1,\ldots,I,\ j = 1,\ldots,J,\ k = 1,\ldots,K$.

The outer product of vectors is easily extended to the outer product of tensors.

Definition 10.5 Given tensors $\mathcal{A} \in \mathbb{K}^{I_1\times\cdots\times I_P}$ and $\mathcal{B} \in \mathbb{K}^{J_1\times\cdots\times J_Q}$, their *tensor outer product* is denoted $\mathcal{A} \circ \mathcal{B} \in \mathbb{K}^{I_1\times\cdots\times I_P\times J_1\times\cdots\times J_Q}$, and is defined as

$$(\mathcal{A} \circ \mathcal{B})_{i_1\cdots i_P j_1\cdots j_Q} = a_{i_1\cdots i_P} b_{j_1\cdots j_Q}, \quad \forall i_1,\ldots,i_P, j_1,\ldots,j_Q. \tag{10.3.5}$$

The inner product and the norm of tensors have the following properties [254].

1. The norm of a tensor can be transformed to the norm of its matricization function:
$$\|\mathcal{A}\| = \left\|\mathbf{A}^{(I_n\times I_1\cdots I_{n-1}I_{n+1}\cdots I_N)}\right\| = \left\|\mathbf{A}^{(I_1\cdots I_{n-1}I_{n+1}\cdots I_N\times I_n)}\right\|.$$
2. The norm of a tensor can be transformed to the norm of its vectorization function:
$$\|\mathcal{A}\| = \left\|\mathbf{a}^{(I_1 I_2\cdots I_N\times 1)}\right\| = \left\|\mathbf{a}^{(1\times I_1 I_2\cdots I_N)}\right\|.$$

3. The square of the norm of difference of two tensors satisfies

$$\|\mathcal{A} - \mathcal{B}\|^2 = \|\mathcal{A}\|^2 - 2\langle \mathcal{A}, \mathcal{B}\rangle + \|\mathcal{B}\|^2.$$

4. If $\mathbf{Q} \in \mathbb{K}^{J \times I_n}$ is an orthonormal matrix, i.e., $\mathbf{Q}\mathbf{Q}^H = \mathbf{I}_{J\times J}$ or $\mathbf{Q}^H\mathbf{Q} = \mathbf{I}_{I_n \times I_n}$, then the norm satisfies

$$\|\mathcal{A} \times_n \mathbf{Q}\| = \|\mathcal{A}\|,$$

where $\times_n$ is the mode-n product, to be defined below.

5. Let $\mathcal{A}, \mathcal{B} \in \mathbb{K}^{I_1 \times I_2 \times \cdots \times I_N}$, $\mathbf{a}_n, \mathbf{b}_n \in \mathbb{K}^{J \times I_n}$, such that $\mathcal{A} = \mathbf{a}_1 \circ \mathbf{a}_2 \circ \cdots \circ \mathbf{a}_N$ and $\mathcal{B} = \mathbf{b}_1 \circ \mathbf{b}_2 \circ \cdots \circ \mathbf{b}_N$; then

$$\langle \mathcal{A}, \mathcal{B}\rangle = \prod_{n=1}^{N} \langle \mathbf{a}_n, \mathbf{b}_n\rangle.$$

6. If $\mathcal{A} \in \mathbb{K}^{I_1 \times \cdots \times I_{n-1} \times J \times I_{n+1} \times \cdots \times I_N}$, $\mathcal{B} \in \mathbb{K}^{I_1 \times I_{n-1} \times K \times I_{n+1} \times \cdots \times I_N}$ and $\mathbf{C} \in \mathbb{K}^{J \times K}$ then

$$\langle \mathcal{A}, \mathcal{B} \times_n \mathbf{C}\rangle = \langle \mathcal{A} \times_n \mathbf{C}^H, \mathcal{B}\rangle.$$

10.3.2 Mode-n Product of Tensors

The product of a third-order tensor and a matrix was defined by Tucker [476], [477]. It is called the *Tucker product* of a third-order tensor.

Definition 10.6 Let $\mathcal{X} \in \mathbb{K}^{I_1 \times I_2 \times I_3}$, $\mathbf{A} \in \mathbb{K}^{J_1 \times I_1}$, $\mathbf{B} \in \mathbb{K}^{J_2 \times I_2}$ and $\mathbf{C} \in \mathbb{K}^{J_3 \times I_3}$. The Tucker mode-1 product $\mathcal{X} \times_1 \mathbf{A}$, the Tucker mode-2 product $\mathcal{X} \times_2 \mathbf{B}$ and the Tucker mode-3 product $\mathcal{X} \times_3 \mathbf{C}$ are respectively defined as [282]

$$(\mathcal{X} \times_1 \mathbf{A})_{j_1 i_2 i_3} = \sum_{i_1=1}^{I_1} x_{i_1 i_2 i_3} a_{j_1 i_1}, \quad \forall j_1, i_2, i_3, \tag{10.3.6}$$

$$(\mathcal{X} \times_2 \mathbf{B})_{i_1 j_2 i_3} = \sum_{i_2=1}^{I_2} x_{i_1 i_2 i_3} b_{j_2 i_2}, \quad \forall i_1, j_2, i_3, \tag{10.3.7}$$

$$(\mathcal{X} \times_3 \mathbf{C})_{i_1 i_2 j_3} = \sum_{i_3=1}^{I_3} x_{i_1 i_2 i_3} c_{j_3 i_3}, \quad \forall i_1, i_2, j_3. \tag{10.3.8}$$

The following is an interpretation of the Tucker mode-n tensor product. First, the mode-1 tensor product can be represented in tensor symbol as

$$\mathcal{Y} = \mathcal{X} \times_1 \mathbf{A}. \tag{10.3.9}$$

Second, from the entry formulas (10.2.8) for the mode-1 horizontal unfolding of the third-order tensor and from the longitudinal unfolding definition formula (10.2.14), we have respectively

$$\begin{aligned} y_{j_1 i_2 i_3} &= \sum_{i_1=1}^{I_1} x_{i_1 i_2 i_3} a_{j_1 i_1} = \sum_{i_1=1}^{I_1} \mathbf{X}^{(I_1 \times I_2 I_3)}_{i_1,(i_3-1)I_2+i_2} a_{j_1 i_1} \\ &= \left(\mathbf{A}\mathbf{X}^{(I_1 \times I_2 I_3)}\right)_{j_1,(i_3-1)I_2+i_2}, \end{aligned}$$

and

$$\begin{aligned} y_{j_1 i_2 i_3} &= \sum_{i_1=1}^{I_1} x_{i_1 i_2 i_3} a_{j_1 i_1} = \sum_{i_1=1}^{I_1} \mathbf{X}^{(I_2 I_3 \times I_1)}_{(i_2-1)I_3+i_3,i_1} a_{j_1 i_1} \\ &= \left(\mathbf{X}^{(I_2 I_3 \times I_1)}\mathbf{A}^T\right)_{(i_2-1)I_3+i_3,j_1}. \end{aligned}$$

Moreover, from (10.2.8) and (10.2.14) it is known that $\mathbf{Y}^{(J_1 \times I_2 I_3)}_{j_1,(i_3-1)I_2+i_2} = y_{j_1 i_2 i_3}$ and $\mathbf{Y}^{(I_2 I_3 \times J_1)}_{(i_2-1)I_3+i_3,j_1} = y_{j_1 i_2 i_3}$. Therefore, the mode-1 product of the third-order tensor can be represented in matrix form as

$$\begin{aligned} \mathbf{Y}^{(J_1 \times I_2 I_3)} &= (\mathcal{X} \times_1 \mathbf{A})^{(J_1 \times I_2 I_3)} = \mathbf{A}\mathbf{X}^{(I_1 \times I_2 I_3)}, \\ \mathbf{Y}^{(I_2 I_3 \times J_1)} &= (\mathcal{X} \times_1 \mathbf{A})^{(I_2 I_3 \times J_1)} = \mathbf{X}^{(I_2 I_3 \times I_1)}\mathbf{A}^T. \end{aligned}$$

Mimicking the mode-1 tensor product, we obtain matrix representations of the mode-2 and the mode-3 products below:

$$\mathcal{Y} = \mathcal{X} \times_1 \mathbf{A} \quad \Leftrightarrow \quad \begin{cases} \mathbf{Y}^{(J_1 \times I_2 I_3)} = \mathbf{A}\mathbf{X}^{(I_1 \times I_2 I_3)}, \\ \mathbf{Y}^{(I_2 I_3 \times J_1)} = \mathbf{X}^{(I_2 I_3 \times I_1)}\mathbf{A}^T. \end{cases} \tag{10.3.10}$$

$$\mathcal{Y} = \mathcal{X} \times_2 \mathbf{B} \quad \Leftrightarrow \quad \begin{cases} \mathbf{Y}^{(J_2 \times I_3 I_1)} = \mathbf{B}\mathbf{X}^{(I_2 \times I_3 I_1)}, \\ \mathbf{Y}^{(I_3 I_1 \times J_2)} = \mathbf{X}^{(I_3 I_1 \times I_2)}\mathbf{B}^T. \end{cases} \tag{10.3.11}$$

$$\mathcal{Y} = \mathcal{X} \times_3 \mathbf{C} \quad \Leftrightarrow \quad \begin{cases} \mathbf{Y}^{(J_3 \times I_1 I_2)} = \mathbf{C}\mathbf{X}^{(I_3 \times I_1 I_2)}, \\ \mathbf{Y}^{(I_1 I_2 \times J_3)} = \mathbf{X}^{(I_1 I_2 \times I_3)}\mathbf{C}^T. \end{cases} \tag{10.3.12}$$

Example 10.5 We revisit again the two frontal slices of the third-order tensor $\mathcal{X} \in \mathbb{R}^{3\times 4\times 2}$ given by

$$\mathbf{X}_{::1} = \begin{bmatrix} 1 & 2 & 3 & 4 \\ 5 & 6 & 7 & 8 \\ 9 & 10 & 11 & 12 \end{bmatrix}, \quad \mathbf{X}_{::2} = \begin{bmatrix} 13 & 14 & 15 & 16 \\ 17 & 18 & 19 & 20 \\ 21 & 22 & 23 & 24 \end{bmatrix}. \tag{10.3.13}$$

Now we will compute respectively the products of the tensor $\mathcal{X}$ and the following matrices:

$$\mathbf{A} = \begin{bmatrix} 1 & 2 & 3 \\ 4 & 5 & 6 \end{bmatrix}, \qquad \mathbf{B} = \begin{bmatrix} 1 & 2 \\ 3 & 4 \\ 5 & 6 \end{bmatrix}.$$

Solution From (10.3.6), the two frontal slices of $\mathcal{X} \times_1 \mathbf{A} \in \mathbb{R}^{2\times4\times2}$ are given by

$$(\mathcal{X} \times_1 \mathbf{A})_{::1} = \begin{bmatrix} 38 & 44 & 50 & 56 \\ 83 & 98 & 113 & 138 \end{bmatrix},$$

$$(\mathcal{X} \times_1 \mathbf{A})_{::2} = \begin{bmatrix} 110 & 116 & 122 & 128 \\ 263 & 278 & 293 & 308 \end{bmatrix}.$$

Moreover, from (10.3.8) the three frontal slices of $\mathcal{X} \times_3 \mathbf{B} \in \mathbb{R}^{3\times4\times3}$ are as follows:

$$(\mathcal{X} \times_3 \mathbf{B})_{::1} = \begin{bmatrix} 27 & 30 & 33 & 36 \\ 39 & 42 & 45 & 48 \\ 51 & 54 & 57 & 60 \end{bmatrix},$$

$$(\mathcal{X} \times_3 \mathbf{B})_{::2} = \begin{bmatrix} 55 & 62 & 69 & 76 \\ 83 & 90 & 97 & 104 \\ 111 & 118 & 125 & 132 \end{bmatrix},$$

$$(\mathcal{X} \times_3 \mathbf{B})_{::3} = \begin{bmatrix} 83 & 94 & 105 & 116 \\ 127 & 138 & 149 & 160 \\ 171 & 182 & 193 & 204 \end{bmatrix}.$$

Figure 10.9 shows the operational principle of the three-mode product $\mathcal{X} \times_1 \mathbf{u}_1 \times_2 \mathbf{u}_2 \times_3 \mathbf{u}_3$ for the third-order tensor $\mathcal{X} \in \mathbb{R}^{8\times7\times4}$, where three vectors are $\mathbf{u}_1 \in \mathbb{R}^{8\times1\times1}, \mathbf{u}_2 \in \mathbb{R}^{1\times7\times1}, \mathbf{u}_3 \in \mathbb{R}^{1\times1\times4}$.

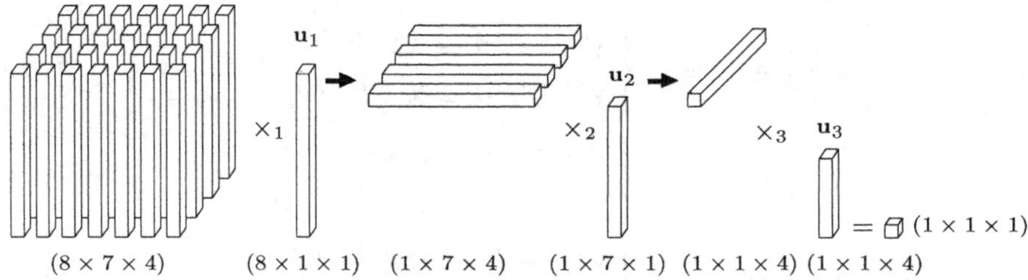

Figure 10.9 The three-mode product of a third-order tensor.

From Figure 10.9 we can conclude that

(a) each mode-vector-product reduces the tensor order by one, so that the tensor finally becomes a zero-order tensor or scalar;
(b) the orders of the three mode vectors of a third-order tensor can be exchanged in the three-mode product.

The Tucker product can be extended to the mode-n (matrix) product of an Nth-order tensor.

Definition 10.7 The *mode-n product* of an Nth-order tensor $\mathcal{X} \in \mathbb{K}^{I_1\times\cdots\times I_N}$

and a $J_n \times I_n$ matrix $\mathbf{U}^{(n)}$ is denoted $\mathcal{X} \times_n \mathbf{U}^{(n)}$. This is an $I_1 \times \cdots \times I_{n-1} \times J_n \times I_{n+1} \cdots \times I_N$ tensor whose entry is defined as [281]

$$(\mathcal{X} \times_n \mathbf{U}^{(n)})_{i_1 \cdots i_{n-1} j i_{n+1} \cdots i_N} \stackrel{\text{def}}{=} \sum_{i_n=1}^{I_n} x_{i_1 i_2 \cdots i_N} a_{j i_n}, \tag{10.3.14}$$

where $j = 1, \ldots, J_n$, $i_k = 1, \ldots, I_k$, $k = 1, \ldots, N$.

From the above definition it is easy to seen that the mode-n product of an Nth-order tensor $\mathcal{X} \in \mathbb{K}^{I_1 \times I_2 \times \cdots \times I_N}$ and the identity matrix $\mathbf{I}_{I_n \times I_n}$ is equal to the original tensor, namely

$$\mathcal{X} \times_n \mathbf{I}_{I_n \times I_n} = \mathcal{X}. \tag{10.3.15}$$

As for a third-order tensor, the mode-n product of an Nth-order tensor $\mathcal{X} \in \mathbb{R}^{I_1 \times I_N}$ and a matrix $\mathbf{U}^{(n)} \in \mathbb{R}^{J_n \times I_n}$ can be represented in the mode-n matricization of the tensor as follows:

$$\mathcal{Y} = \mathcal{X} \times_n \mathbf{U}^{(n)} \quad \Leftrightarrow \quad \mathbf{Y}_{(n)} = \mathbf{U}^{(n)} \mathbf{X}_{(n)} \quad \text{or} \quad \mathbf{Y}^{(n)} = \mathbf{X}^{(n)} (\mathbf{U}^{(n)})^T \tag{10.3.16}$$

in which $\mathbf{X}_{(n)} = \mathbf{X}^{(I_n \times I_1 \cdots I_{n-1} I_{n+1} \cdots I_N)}$ and $\mathbf{X}^{(n)} = \mathbf{X}^{(I_1 \cdots I_{n-1} I_{n+1} \cdots I_N \times I_n)}$ are the mode-n horizontal and longitudinal unfoldings of the tensor $\mathcal{X}$, respectively.

Let $\mathcal{X} \in \mathbb{K}^{I_1 \times I_2 \times \cdots \times I_N}$ be an Nth-order tensor; then its mode-n product has the following properties [281].

1. Given matrices $\mathbf{A} \in \mathbb{K}^{J_m \times I_m}, \mathbf{B} \in \mathbb{K}^{J_n \times I_n}$, if $m \neq n$, then

$$\mathcal{X} \times_m \mathbf{A} \times_n \mathbf{B} = (\mathcal{X} \times_m \mathbf{A}) \times_n \mathbf{B} = (\mathcal{X} \times_n \mathbf{B}) \times_m \mathbf{A} = \mathcal{X} \times_n \mathbf{B} \times_m \mathbf{A}.$$

2. Given matrices $\mathbf{A} \in \mathbb{K}^{J \times I_n}, \mathbf{B} \in \mathbb{K}^{I_n \times J}$, then

$$\mathcal{X} \times_n \mathbf{A} \times_n \mathbf{B} = \mathcal{X} \times_n (\mathbf{BA}).$$

3. If $\mathbf{A} \in \mathbb{K}^{J \times I_n}$ has a full column rank then

$$\mathcal{Y} = \mathcal{X} \times_n \mathbf{A} \Rightarrow \mathcal{X} = \mathcal{Y} \times_n \mathbf{A}^\dagger.$$

4. If $\mathbf{A} \in \mathbb{K}^{J \times I_n}$ is semi-orthonormal, i.e., $\mathbf{A}^H \mathbf{A} = \mathbf{I}_{I_n}$, then

$$\mathcal{Y} = \mathcal{X} \times_n \mathbf{A} \quad \Rightarrow \quad \mathcal{X} = \mathcal{Y} \times_n \mathbf{A}^H.$$

The above properties show that the tensor $\mathcal{X}$ can be reconstructed by the mode -n product $\mathcal{Y} = \mathcal{X} \times_n \mathbf{A}$ via $\mathcal{X} = \mathcal{Y} \times_n \mathbf{A}^\dagger$ or $\mathcal{X} = \mathcal{Y} \times_n \mathbf{A}^H$.

Example 10.6 Two frontal slices of a third-order tensor $\mathcal{X} \in \mathbb{R}^{I_1 \times I_2 \times I_3}$ are given by $\mathbf{X}_{::1} = \begin{bmatrix} 1 & 2 & 3 \\ 4 & 5 & 6 \end{bmatrix}$ and $\mathbf{X}_{::2} = \begin{bmatrix} 7 & 8 & 9 \\ 10 & 11 & 12 \end{bmatrix}$, and

$$\mathbf{A} = \begin{bmatrix} -0.7071 & 0.5774 \\ 0.0000 & 0.5774 \\ 0.7071 & 0.5774 \end{bmatrix}$$

is a semi-orthonormal matrix, i.e., $\mathbf{A}^T\mathbf{A} = \mathbf{I}_2$. Find the frontal slices of the tensor $\mathcal{X}$.

Solution Since $I_1 = 2, I_2 = 3, I_3 = 2, J = 3$, we obtain the mode-1 horizontal unfolding $\mathbf{Y}^{(J\times I_2I_3)} = \mathbf{A}\mathbf{X}^{(I_1\times I_2I_3)}$ of $\mathcal{Y} = \mathcal{X} \times_1 \mathbf{A} \in \mathbb{R}^{J\times I_2\times I_3}$ as follows:

$$\begin{aligned}\mathbf{Y}^{(J\times I_2I_3)} &\\ &= \begin{bmatrix} -0.70711 & 0.57735 \\ 0.00000 & 0.57735 \\ 0.70711 & 0.57735 \end{bmatrix}\begin{bmatrix} 1 & 2 & 3 & 7 & 8 & 9 \\ 4 & 5 & 6 & 10 & 11 & 12 \end{bmatrix} \\ &= \begin{bmatrix} 1.60229 & 1.47253 & 1.34277 & 0.82373 & 0.69397 & 0.56421 \\ 2.30940 & 2.88675 & 3.46410 & 5.77350 & 6.35085 & 6.92820 \\ 3.01651 & 4.30097 & 5.58543 & 10.72327 & 12.00773 & 13.29219 \end{bmatrix}.\end{aligned}$$

Therefore, the mode-1 horizontal unfolding of the tensor $\mathcal{X} = \mathcal{Y} \times_1 \mathbf{A}^T$ can be retrieved as $\mathbf{X}^{(I_1\times I_2I_3)} = \mathbf{A}^T\mathbf{Y}^{(J\times I_2I_3)}$, yielding the estimate

$$\begin{aligned}\hat{\mathbf{X}}^{(I_1\times I_2I_3)} &\\ &= \begin{bmatrix} -0.70711 & 0.00000 & 0.70711 \\ 0.57735 & 0.57735 & 0.57735 \end{bmatrix} \\ &\quad\times \begin{bmatrix} 1.60229 & 1.47253 & 1.34277 & 0.82373 & 0.69397 & 0.56421 \\ 2.30940 & 2.88675 & 3.46410 & 5.77350 & 6.35085 & 6.92820 \\ 3.01651 & 4.30097 & 5.58543 & 10.72327 & 12.00773 & 13.29219 \end{bmatrix} \\ &= \begin{bmatrix} 1.00000 & 2.00002 & 3.00003 & 7.00006 & 8.00007 & 9.00008 \\ 4.00000 & 5.00000 & 5.99999 & 9.99999 & 10.99999 & 11.99999 \end{bmatrix},\end{aligned}$$

and thus we have the following estimate for the frontal slices of the tensor as follows:

$$\hat{\mathbf{X}}_{::1} = \begin{bmatrix} 1.00000 & 2.00002 & 3.00003 \\ 4.00000 & 5.00000 & 5.99999 \end{bmatrix},$$

$$\hat{\mathbf{X}}_{::2} = \begin{bmatrix} 7.00006 & 8.00007 & 9.00008 \\ 9.99999 & 10.99999 & 11.99999 \end{bmatrix}.$$

10.3.3 Rank of Tensor

If the tensor $\mathcal{A} \in \mathcal{T}(I_1, I_2, \ldots, I_n)$ can be decomposed as

$$\mathcal{A} = \mathbf{a}^{(1)} \circ \mathbf{a}^{(2)} \circ \cdots \circ \mathbf{a}^{(n)}, \tag{10.3.17}$$

then it is said to be *decomposed*, and the vectors $\mathbf{a}^{(i)} \in \mathbb{K}^{I_i}, i = 1, \ldots, n$, are known as the components or *factors* of the decomposed tensor $\mathcal{A}$. Since all the factors are rank-1 vectors, the above decomposition is called a rank-1 decomposition of the tensor.

An entry of the decomposed tensor is defined as

$$a_{i_1 i_2 \cdots i_n} = a_{i_1}^{(1)} a_{i_2}^{(2)} \cdots a_{i_n}^{(n)}. \tag{10.3.18}$$

The set of all $I_1 \times I_2 \times \cdots \times I_n$ decomposed tensors is the *decomposed tensor set*, denoted $\mathcal{D}(I_1, I_2, \ldots, I_n)$ or simply $\mathcal{D}$.

LEMMA 10.1 [253] *For two decomposed tensors $\mathcal{A}, \mathcal{B} \in \mathcal{D}$, if*

$$\mathcal{A} = \mathbf{a}^{(1)} \circ \mathbf{a}^{(2)} \circ \cdots \circ \mathbf{a}^{(n)}, \quad \mathcal{B} = \mathbf{b}^{(1)} \circ \mathbf{b}^{(2)} \circ \cdots \circ \mathbf{b}^{(n)}, \tag{10.3.19}$$

then the following results are true:

- $\langle \mathcal{A}, \mathcal{B} \rangle = \prod_{i=1}^{n} \langle \mathbf{a}^{(i)}, \mathbf{b}^{(i)} \rangle$;
- $\|\mathcal{A}\|_F = \prod_{i=1}^{n} \|\mathbf{a}^{(i)}\|_2$ *and* $\|\mathcal{B}\|_F = \prod_{i=1}^{n} \|\mathbf{b}^{(i)}\|_2$;
- $\mathcal{A} + \mathcal{B} \in \mathcal{D}$ *if and only if $\mathcal{A}$ and $\mathcal{B}$ are different in only one component, all other components being the same.*

Let $\|\mathcal{A}\|_F = 1$ and $\|\mathcal{B}\|_F = 1$. Two decomposed tensors $\mathcal{A}$ and $\mathcal{B}$ are said to be orthogonal, denoted $\mathcal{A} \perp \mathcal{B}$, if their inner product is equal to zero, i.e.,

$$\langle \mathcal{A}, \mathcal{B} \rangle = \prod_{i=1}^{n} \langle \mathbf{a}^{(i)}, \mathbf{b}^{(i)} \rangle = 0.$$

In matrix algebra, the number of linearly independent rows (columns) of a matrix $\mathbf{A}$ is called the row (column) rank of $\mathbf{A}$, equivalently stated as: the column (row) rank of $\mathbf{A}$ is the dimension of the column (row) space of $\mathbf{A}$. The column rank and the row rank of a matrix are always equal. Therefore the rank, column rank and row rank are the same for a matrix. However, this important property of the matrix rank is no longer true for a higher-order tensor.

In analogy to the the column rank and row rank of a matrix (the mode-1 and mode-2 vectors), the rank of the mode-n vectors of a tensor is called the mode-n rank of the tensor.

DEFINITION 10.8 [280] Given an Nth-order tensor $\mathcal{A} \in \mathbb{K}^{I_1 \times \cdots \times I_N}$, among its $I_1 \cdots I_{n-1} I_{n+1} \cdots I_N$ mode-n vectors with dimension I_n, the maximum number of linearly independent vectors is called the *mode-n rank* of the tensor $\mathcal{A}$, denoted $r_n = \mathrm{rank}_n(\mathcal{A})$, equivalently stated as: the dimension of the subspace spanned by the mode-n vectors of a tensor is known as its mode-n rank.

For example, the mode-n rank for $n = 1, 2, 3$ of a third-order tensor $\mathcal{A} \in \mathbb{R}^{I \times J \times K}$ are denoted $r_n(\mathcal{A})$, and are defined as

$$r_1(\mathcal{A}) \stackrel{\text{def}}{=} \dim\big(\mathrm{span}_{\mathbb{R}}\{\mathbf{a}_{:jk} \,|\, j = 1, \ldots, J, k = 1, \ldots, K\}\big),$$
$$r_2(\mathcal{A}) \stackrel{\text{def}}{=} \dim\big(\mathrm{span}_{\mathbb{R}}\{\mathbf{a}_{i:k} \,|\, i = 1, \ldots, I, k = 1, \ldots, K\}\big),$$
$$r_3(\mathcal{A}) \stackrel{\text{def}}{=} \dim\big(\mathrm{span}_{\mathbb{R}}\{\mathbf{a}_{ij:} \,|\, i = 1, \ldots, I, j = 1, \ldots, J\}\big),$$

where $\mathbf{a}_{:jk}$, $\mathbf{a}_{i:k}$ and $\mathbf{a}_{ij:}$ are the mode-1, mode-2 and mode-3 vectors of the tensor, respectively.

The column rank and the row rank of a matrix are always the same, but the mode-i rank and the mode-j of a higher-order tensor are generally not the same if $i \neq j$.

The third-order tensor $\mathcal{A} \in \mathbb{K}^{I \times J \times K}$ is said to be of rank (r_1, r_2, r_3) if it has mode-1 rank r_1, mode-2 rank r_2 and mode-3 rank r_3. More generally, an Nth-order tensor $\mathcal{A}$ is of rank $(r_1, r_2, \ldots, r_N)$, if its model-n rank is equal to $r_n, n = 1, \ldots, N$. In particular, if the rank of every mode-n matricization is equal to 1 then the tensor is of the rank $(1, 1, \ldots, 1)$.

There are N mode-n ranks of an Nth-order tensor, which is inconvenient in use. For this purpose, it is necessary to define only one rank for a given tensor, the *tensor rank*.

Consider a tensor $\mathcal{A} \in \mathcal{T}$ that can be decomposed into the weighted sum of R decomposed tensors:

$$\mathcal{A} = \sum_{i=1}^{R} \sigma_i \mathcal{U}_i, \tag{10.3.20}$$

where $\sigma_i > 0, i = 1, \ldots, R$, $\mathcal{U}_i \in \mathcal{D}$ and $\|\mathcal{U}_i\|_F = 1, i = 1, \ldots, R$.

From [267], the tensor rank of $\mathcal{A}$ is denoted $\text{rank}(\mathcal{A})$, and is defined as the minimum value of R such that (10.3.20) holds. Then (10.3.20) is called the *rank decomposition* of the tensor.

In particular, if $R = 1$, i.e., $\mathcal{A} = \mathbf{a}^{(1)} \circ \mathbf{a}^{(2)} \circ \cdots \circ \mathbf{a}^{(n)}$, then $\mathcal{A}$ is said to be a rank-1 tensor. Therefore, a third-order tensor $\mathcal{A}$ is a rank-1 tensor, if it can be represented as the outer product of three vectors, i.e., $\mathcal{A} = \mathbf{a}^{(1)} \circ \mathbf{a}^{(2)} \circ \mathbf{a}^{(3)}$. Similarly, a third-order tensor $\mathcal{A}$ is a rank-2 tensor (i.e., the rank of the tensor is 2) if it can be represented as $\mathcal{A} = \mathbf{a}^{(1)} \circ \mathbf{a}^{(2)} \circ \mathbf{a}^{(3)} + \mathbf{b}^{(1)} \circ \mathbf{b}^{(2)} \circ \mathbf{b}^{(3)}$.

10.4 Tucker Decomposition of Tensors

In order to carry out the information mining of a tensor, tensor decomposition is necessary.

The concept of tensor decomposition stems from two papers of Hitchcock in 1927 [212], [213] who expressed a tensor as the sum of finite rank-1 tensors, and called it a canonical polyadic decomposition. The concept of multi-way models was proposed by Cattell in 1944 [88]. However, two aspects of tensor decomposition and multi-way models attracted people's attention in succession in the 1960s. Tucker published his three papers on factor decomposition of the tensor [475]–[477]. Later, Carroll and Chang [87] and Harshman [198] independently proposed canonical factor decomposition (CANDECOMP) and parallel factor decomposition (PARAFAC), respectively. These studies brought into being two major categories of tensor decomposition methods:

(1) Tucker decomposition, later known as higher-order SVD;
(2) canonical factor decomposition (CANDECOMP/PARAFAC), now simply called CP decomposition.

In this section we discuss the Tucker decomposition of tensors, while the CP decomposition of tensors will be considered in the next section.

10.4.1 Tucker Decomposition (Higher-Order SVD)

Tucker decomposition is related closely to the Tucker operator, while the Tucker operator is in turn an effective representation of the multimode product of a tensor and a matrix.

Definition 10.9 Given an Nth-order tensor $\mathcal{G} \in \mathbb{K}^{J_1 \times J_2 \times \cdots \times J_N}$ and matrices $\mathbf{U}^{(n)} \in \mathbb{K}^{I_n \times J_n}$, where $n \in \{1, \ldots, N\}$, the *Tucker operator* is defined as [254]

$$[\![\mathcal{G}; \mathbf{U}^{(1)}, \mathbf{U}^{(2)}, \ldots, \mathbf{U}^{(N)}]\!] \stackrel{\text{def}}{=} \mathcal{G} \times_1 \mathbf{U}^{(1)} \times_2 \mathbf{U}^{(2)} \times_3 \cdots \times_N \mathbf{U}^{(N)}. \tag{10.4.1}$$

The result is an Nth-order $I_1 \times I_2 \times \cdots \times I_N$ tensor.

Given an Nth-order tensor $\mathcal{G} \in \mathbb{K}^{J_1 \times \cdots \times J_N}$ and an index set $\mathcal{N} = \{1, \ldots, N\}$, then the Tucker operator has the following properties [254]:

1. If matrices $\mathbf{U}^{(n)} \in \mathbb{K}^{I_n \times J_n}, \mathbf{V} \in \mathbb{K}^{K_n \times I_n}$ for all $n \in \mathcal{N}$, then

$$[\![[\![\mathcal{G}; \mathbf{U}^{(1)}, \ldots, \mathbf{U}^{(N)}]\!]; \mathbf{V}^{(1)}, \ldots, \mathbf{V}^{(N)}]\!] = [\![\mathcal{G}; \mathbf{V}^{(1)}\mathbf{U}^{(1)}, \ldots, \mathbf{V}^{(N)}\mathbf{U}^{(N)}]\!].$$

2. If $\mathbf{U}^{(n)} \in \mathbb{K}^{I_n \times J_n}, n \in \mathcal{N}$ has full-column rank then

$$\mathcal{X} = [\![\mathcal{G}; \mathbf{U}^{(1)}, \ldots, \mathbf{U}^{(N)}]\!] \quad \Leftrightarrow \quad \mathcal{G} = [\![\mathcal{X}; \mathbf{U}^{(1)\dagger}, \ldots, \mathbf{U}^{(N)\dagger}]\!].$$

3. If $\mathbf{U}^{(n)} \in \mathbb{K}^{I_n \times J_n}$ is an orthonormal matrix such that $\mathbf{U}^{(n)T}\mathbf{U}^{(n)} = \mathbf{I}_{J_n}$, then

$$\mathcal{X} = [\![\mathcal{G}; \mathbf{U}^{(1)}, \ldots, \mathbf{U}^{(N)}]\!] \quad \Leftrightarrow \quad \mathcal{G} = [\![\mathcal{X}; \mathbf{U}^{(1)T}, \ldots, \mathbf{U}^{(N)T}]\!].$$

Proposition 10.1 [254] *Given a tensor $\mathcal{G} \in \mathbb{R}^{J_1 \times J_2 \times \cdots \times J_N}$, let $\mathcal{N} = \{1, \ldots, N\}$ and the matrices $\mathbf{U}^{(n)} \in \mathbb{R}^{I_n \times J_n}, n \in \mathcal{N}$. If every matrix has the QR decomposition $\mathbf{U}^{(n)} = \mathbf{Q}_n\mathbf{R}_n, \forall n \in \mathcal{N}$, where $\mathbf{Q}_n$ is an orthonormal matrix and $\mathbf{R}_n$ is an upper triangular matrix, then*

$$\big\|[\![\mathcal{G}; \mathbf{U}^{(1)}, \ldots, \mathbf{U}^{(n)}]\!]\big\| = \big\|[\![\mathcal{X}; \mathbf{R}_1, \ldots, \mathbf{R}_n]\!]\big\|.$$

The matrix $\mathbf{A} \in \mathbb{R}^{I_1 \times I_2}$ is a two-mode mathematical object: it has two companion vector spaces, column space and row space. The SVD makes these vector spaces orthogonal and decomposes $\mathbf{A}$ into the product of three matrices $\mathbf{A} = \mathbf{U}_1\mathbf{\Sigma}\mathbf{U}_2^T$, where the J_1 left singular vectors of $\mathbf{U}_1 \in \mathbb{R}^{I_1 \times J_1}$ span the column space of $\mathbf{A}$, and the J_2 right singular vectors of the $\mathbf{U}_2 \in \mathbb{R}^{I_2 \times J_2}$ span the row space of $\mathbf{A}$.

Because the role of the diagonal singular values is usually more important than the left and right singular vectors, the diagonal singular value matrix $\mathbf{\Sigma}$ is viewed as the core matrix of $\mathbf{A}$. If the diagonal singular value matrix $\mathbf{\Sigma}$ is regarded as a second-order tensor then $\mathbf{\Sigma}$ is naturally the core tensor of the second-order tensor $\mathcal{A}$, while the product of three matrices $\mathbf{A} = \mathbf{U}_1\mathbf{\Sigma}\mathbf{U}_2^T$ in the SVD of the matrix $\mathbf{A}$ can be rewritten as the mode-n product of the tensor $\mathbf{A} = \mathbf{\Sigma} \times_1 \mathbf{U}_1 \times_2 \mathbf{U}_2$.

This mode-n product of the SVD of a matrix is easily extended to the SVD of an Nth-order tensor or an N-dimensional *supermatrix* $\mathcal{A} \in \mathbb{C}^{I_1 \times \cdots \times I_N}$.

THEOREM 10.1 (Nth-order SVD) [281] *Every $I_1 \times I_2 \times \cdots \times I_N$ real tensor $\mathcal{X}$ can be decomposed into the mode-n product*

$$\mathcal{X} = \mathcal{G} \times_1 \mathbf{U}^{(1)} \times_2 \mathbf{U}^{(2)} \times_3 \cdots \times_N \mathbf{U}^{(N)} = [\![\mathcal{G}; \mathbf{U}^{(1)}, \mathbf{U}^{(2)}, \ldots, \mathbf{U}^{(N)}]\!] \tag{10.4.2}$$

with entries

$$x_{i_1 i_2 \cdots i_N} = \sum_{j_1=1}^{J_1} \sum_{j_2=1}^{J_2} \cdots \sum_{j_N=1}^{J_N} g_{i_1 i_2 \cdots i_N} u_{i_1 j_1}^{(1)} u_{i_2 j_2}^{(2)} \cdots u_{i_N j_N}^{(N)}, \tag{10.4.3}$$

where $\mathbf{U}^{(n)} = [\mathbf{u}_1^{(n)}, \ldots, \mathbf{u}_{J_n}^{(n)}]$ is an $I_n \times J_n$ semi-orthogonal matrix such that $(\mathbf{U}^{(n)})^T \mathbf{U}^{(n)} = \mathbf{I}_{J_n}$, with $J_n \leq I_n$, the core tensor $\mathcal{G}$ is a $J_1 \times J_2 \times \cdots \times J_N$ tensor and the subtensor $\mathcal{G}_{j_n=\alpha}$ is the tensor $\mathcal{X}$ with the fixed index $j_n = \alpha$. The subtensors have the following properties.

- *All-orthogonality: two subtensors $\mathcal{G}_{j_n=\alpha}$ and $\mathcal{G}_{j_n=\beta}$, for $\alpha \neq \beta$, are orthogonal:*

$$\langle \mathcal{G}_{j_n=\alpha}, \mathcal{G}_{j_n=\beta} \rangle = 0, \quad \forall \alpha \neq \beta,\ n = 1, \ldots, N. \tag{10.4.4}$$

- *Ordering:*

$$\|\mathcal{G}_{i_n=1}\|_F \geq \|\mathcal{G}_{i_n=2}\|_F \geq \cdots \geq \|\mathcal{G}_{i_n=N}\|_F. \tag{10.4.5}$$

Unlike the diagonal singular value matrix, the core tensor $\mathcal{G}$ does not have a diagonal construction, i.e., its off-diagonal entries are not equal to zero [253]. Thus the entries $g_{j_1 \cdots j_N}$ of the core tensor $\mathcal{G} = [\![g_{j_1 \cdots j_N}]\!]$ guarantee the interaction of the mode matrices $\mathbf{U}^{(n)}, n = 1, \ldots, N$.

The mode-n matrix $\mathbf{U}^{(n)}$ is required to have an orthogonal column construction, similarly to the left singular matrix $\mathbf{U}$ and the right singular matrix $\mathbf{V}$ of SVD, i.e., any two columns of $\mathbf{U}^{(n)}$ are mutually orthogonal.

Because Equation (10.4.2) is a decomposition form of SVD in a higher-order tensor, (10.4.2) is naturally called the *higher-order singular value decomposition* of tensor [281].

As early as the 1960s, Tucker proposed the factor decomposition of third-order tensors [475], [476], [477], now commonly called the Tucker decomposition. In comparison with the Tucker decomposition, higher-order SVD more closely reflects the connection to and generalization of the SVD of a matrix. However, Tucker decomposition and higher-order SVD are often used without any distinction being drawn between them.

Proposition 10.2 *There exists the following transform relation between the higher-order SVD of an Nth-order tensor $\mathcal{X} \in \mathbb{C}^{I_1 \times \cdots \times I_N}$ and its core tensor:*

$$\mathcal{X} = \mathcal{G} \times_1 \mathbf{U}^{(1)} \times_2 \mathbf{U}^{(2)} \cdots \times_N \mathbf{U}^{(N)}$$
$$\Rightarrow \quad \mathcal{G} = \mathcal{X} \times_1 \mathbf{U}^{(1)T} \times_2 \mathbf{U}^{(2)T} \cdots \times_N \mathbf{U}^{(N)T}, \tag{10.4.6}$$

where $\mathcal{G} \in \mathbb{C}^{J_1 \times \cdots J_N}$, $\mathbf{U}^{(n)} \in \mathbb{C}^{I_n \times J_n}$ and $J_n \leq I_n$.

Proof Continuously applying the property of the mode-n product of a tensor and a matrix $\mathcal{X} \times_m \mathbf{A} \times_n \mathbf{B} = \mathcal{X} \times_n \mathbf{B} \times_m \mathbf{A}$, it can be seen that Equation (10.4.2) can be equivalently written as $\mathcal{X} = \mathcal{G} \times_N \mathbf{U}^{(N)} \times_{N-1} \mathbf{U}^{(N-1)} \cdots \times_1 \mathbf{U}^{(1)}$. Use the property of the mode-n product $\mathcal{X} \times_n \mathbf{A} \times_n \mathbf{B} = \mathcal{X} \times_n (\mathbf{BA})$ and notice that $(\mathbf{U}^{(1)})^T \mathbf{U}^{(1)} = \mathbf{I}_{J_1}$; then

$$\begin{aligned}\mathcal{X} \times_1 (\mathbf{U}^{(1)})^T &= \mathcal{G} \times_N \mathbf{U}^{(N)} \times_{N-1} \mathbf{U}^{(N-1)} \cdots \times_1 \mathbf{U}^{(1)} \times_1 (\mathbf{U}^{(1)})^T \\ &= \mathcal{G} \times_N \mathbf{U}^{(N)} \times_{N-1} \mathbf{U}^{(N-1)} \cdots \times_2 \mathbf{U}^{(2)} \times_1 ((\mathbf{U}^{(1)})^T \mathbf{U}^{(1)}) \\ &= \mathcal{G} \times_N \mathbf{U}^{(N)} \times_{N-1} \mathbf{U}^{(N-1)} \cdots \times_2 \mathbf{U}^{(2)}.\end{aligned}$$

Mimicking this process, we have also $\mathcal{X} \times_1 (\mathbf{U}^{(1)})^T \times_2 (\mathbf{U}^{(2)})^T = \mathcal{G} \times_N \mathbf{U}^{(N)} \times_{N-1} \mathbf{U}^{(N-1)} \cdots \times_3 \mathbf{U}^{(3)}$. Finally, we have $\mathcal{G} = \mathcal{X} \times_1 (\mathbf{U}^{(1)})^T \times_2 (\mathbf{U}^{(2)})^T \cdots \times_N (\mathbf{U}^{(N)})^T$. □

Define

$$\mathbf{U}_{\otimes}^{(n)} = (\mathbf{U}^{(n-1)} \otimes \cdots \otimes \mathbf{U}^{(1)} \otimes \mathbf{U}^{(N)} \otimes \mathbf{U}^{(n+1)})^T \quad \text{(Kiers)}, \tag{10.4.7}$$
$$\mathbf{U}_{\otimes}^{(n)} = (\mathbf{U}^{(n+1)} \otimes \cdots \otimes \mathbf{U}^{(N)} \otimes \mathbf{U}^{(1)} \otimes \mathbf{U}^{(n-1)})^T \quad \text{(LMV)}, \tag{10.4.8}$$
$$\mathbf{U}_{\otimes}^{(n)} = (\mathbf{U}^{(N)} \otimes \cdots \otimes \mathbf{U}^{(n+1)} \otimes \mathbf{U}^{(n-1)} \otimes \mathbf{U}^{(1)})^T \quad \text{(Kolda)}. \tag{10.4.9}$$

Then the equivalent matrix representation of the higher-order SVD is given by

$$\mathbf{X}_{(n)} = \mathbf{U}^{(n)} \mathbf{G}_{(n)} \mathbf{U}_{\otimes}^{(n)}, \tag{10.4.10}$$
$$\mathbf{X}^{(n)} = (\mathbf{U}_{\otimes}^{(n)})^T \mathbf{G}^{(n)} (\mathbf{U}^{(n)})^T, \tag{10.4.11}$$

where $\mathbf{X}_{(n)}$ and $\mathbf{X}^{(n)}$ are horizontal and longitudinal unfoldings, respectively. When using the above matrix equivalent representation, two basic principles should be followed: there is no factor matrix $\mathbf{U}^{(n)}, n \leq 0$, and there are no factor matrices with the same subscript.

Table 10.1 summarizes a variety of mathematical representation of Tucker decompositions.

10.4.2 Third-Order SVD

The Tucker decomposition (third-order SVD) of a third-order tensor

$$\mathcal{X} = \mathcal{G} \times_1 \mathbf{A} \times_2 \mathbf{B} \times_3 \mathbf{C} \quad \in \mathbb{K}^{I \times J \times K}, \tag{10.4.12}$$

has the following properties [282].

Table 10.1 Mathematical representations of the Tucker decomposition

	Mathematical formula
Operator form	$\mathcal{X} = [\![\mathcal{G}; \mathbf{U}^{(1)}, \mathbf{U}^{(2)}, \ldots, \mathbf{U}^{(N)}]\!]$
Mode-N form	$\mathcal{X} = \mathcal{G} \times_1 \mathbf{U}^{(1)} \times_2 \mathbf{U}^{(2)} \cdots \times_N \mathbf{U}^{(N)}$
Entry form	$x_{i_1 \cdots i_N} = \sum_{j_1=1}^{J_1} \sum_{j_2=1}^{J_2} \cdots \sum_{j_N=1}^{J_N} g_{j_1 j_2 \cdots j_N} u_{i_1,j_1}^{(1)} u_{i_2,j_2}^{(2)} \cdots u_{i_N,j_N}^{(N)}$
Outer product	$\mathcal{X} = \sum_{j_1=1}^{J_1} \sum_{j_2=1}^{J_2} \cdots \sum_{j_N=1}^{J_N} g_{j_1 j_2 \cdots j_N} \mathbf{u}_{j_1}^{(1)} \circ \mathbf{u}_{j_2}^{(2)} \circ \cdots \circ \mathbf{u}_{j_N}^{(N)}$
Kiers	$\begin{cases} \mathbf{X}_{(n)} = \mathbf{U}^{(n)} \mathbf{G}_{(n)} (\mathbf{U}^{(n-1)} \otimes \cdots \otimes \mathbf{U}^{(1)} \otimes \mathbf{U}^{(N)} \otimes \cdots \otimes \mathbf{U}^{(n+1)})^T \\ \mathbf{X}^{(n)} = (\mathbf{U}^{(n-1)} \otimes \cdots \otimes \mathbf{U}^{(1)} \otimes \mathbf{U}^{(N)} \otimes \cdots \otimes \mathbf{U}^{(n+1)}) \mathbf{G}^{(n)} \mathbf{U}^{(n)T} \end{cases}$
LMV	$\begin{cases} \mathbf{X}_{(n)} = \mathbf{U}^{(n)} \mathbf{G}_{(n)} (\mathbf{U}^{(n+1)} \otimes \cdots \otimes \mathbf{U}^{(N)} \otimes \mathbf{U}^{(1)} \cdots \otimes \mathbf{U}^{(n-1)})^T \\ \mathbf{X}^{(n)} = (\mathbf{U}^{(n+1)} \otimes \cdots \otimes \mathbf{U}^{(N)} \otimes \mathbf{U}^{(1)} \cdots \otimes \mathbf{U}^{(n-1)}) \mathbf{G}^{(n)} \mathbf{U}^{(n)T} \end{cases}$
Kolda	$\begin{cases} \mathbf{X}_{(n)} = \mathbf{U}^{(n)} \mathbf{G}_{(n)} (\mathbf{U}^{(N)} \otimes \cdots \otimes \mathbf{U}^{(n+1)} \otimes \mathbf{U}^{(n-1)} \otimes \cdots \otimes \mathbf{U}^{(1)})^T \\ \mathbf{X}^{(n)} = (\mathbf{U}^{(N)} \otimes \cdots \otimes \mathbf{U}^{(n+1)} \otimes \mathbf{U}^{(n-1)} \otimes \cdots \otimes \mathbf{U}^{(1)}) \mathbf{G}^{(n)} \mathbf{U}^{(n)T} \end{cases}$

1. The mode-1 matrix $\mathbf{A} \in \mathbb{K}^{I \times P}$, the mode-2 matrix $\mathbf{B} \in \mathbb{K}^{J \times Q}$ and the mode-3 matrix $\mathbf{C} \in \mathbb{K}^{K \times R}$ are all columnwise orthonormal, i.e.,

$$\mathbf{A}^H \mathbf{A} = \mathbf{I}_P, \quad \mathbf{B}^H \mathbf{B} = \mathbf{I}_Q, \quad \mathbf{C}^H \mathbf{C} = \mathbf{I}_R.$$

2. The longitudinal unfoldings $\mathbf{G}^{(QR \times P)}$, $\mathbf{G}^{(RP \times Q)}$ and $\mathbf{G}^{(PQ \times R)}$ of the third-order core tensor $\mathcal{G} \in \mathbb{K}^{P \times Q \times R}$ are respectively columnwise orthogonal:

$$\begin{aligned} \langle \mathbf{G}_{:p_1}^{(QR \times P)}, \mathbf{G}_{:p_2}^{(QR \times P)} \rangle &= \sigma_1^2(p_1) \delta_{p_1,p_2}, \quad 1 \le p_1, p_2 \le P, \\ \langle \mathbf{G}_{:q_1}^{(RP \times Q)}, \mathbf{G}_{:q_2}^{(RP \times Q)} \rangle &= \sigma_2^2(q_1) \delta_{q_1,q_2}, \quad 1 \le q_1, q_2 \le Q, \\ \langle \mathbf{G}_{:r_1}^{(PQ \times R)}, \mathbf{G}_{:r_2}^{(PQ \times R)} \rangle &= \sigma_3^2(r_1) \delta_{r_1,r_2}, \quad 1 \le r_1, r_2 \le R. \end{aligned}$$

Here the singular values are arranged as follows:

$$\begin{aligned} &\sigma_1^2(1) \ge \sigma_1^2(2) \ge \cdots \ge \sigma_1^2(P), \\ &\sigma_2^2(1) \ge \sigma_2^2(2) \ge \cdots \ge \sigma_2^2(Q), \\ &\sigma_3^2(1) \ge \sigma_3^2(2) \ge \cdots \ge \sigma_3^2(R). \end{aligned}$$

The SVD of the third-order tensor $\mathcal{X} = [\![x_{ijk}]\!] \in \mathbb{R}^{I \times J \times K}$ has a total of three subscript variables i, j, k, and thus has $P_3 = 3! = 6$ possible horizontal unfolding matrices, which are given by

$$x^{(I\times JK)}_{i,(k-1)J+j} = x_{ijk} \quad \Rightarrow$$
$$\mathbf{X}^{(I\times JK)} = [\mathbf{X}_{::1},\ldots,\mathbf{X}_{::K}] = \mathbf{A}\mathbf{G}^{(P\times QR)}(\mathbf{C}\otimes\mathbf{B})^T, \tag{10.4.13}$$
$$x^{(J\times KI)}_{j,(i-1)K+k} = x_{ijk} \quad \Rightarrow$$
$$\mathbf{X}^{(J\times KI)} = [\mathbf{X}_{1::},\ldots,\mathbf{X}_{I::}] = \mathbf{B}\mathbf{G}^{(Q\times RP)}(\mathbf{A}\otimes\mathbf{C})^T, \tag{10.4.14}$$
$$x^{(K\times IJ)}_{k,(j-1)I+i} = x_{ijk} \quad \Rightarrow$$
$$\mathbf{X}^{(K\times IJ)} = [\mathbf{X}_{:1:},\ldots,\mathbf{X}_{:J:}] = \mathbf{C}\mathbf{G}^{(R\times PQ)}(\mathbf{B}\otimes\mathbf{A})^T, \tag{10.4.15}$$
$$x^{(I\times JK)}_{i,(j-1)K+k} = x_{ijk} \quad \Rightarrow$$
$$\mathbf{X}^{(I\times JK)} = [\mathbf{X}^T_{:1:},\ldots,\mathbf{X}^T_{:J:}] = \mathbf{A}\mathbf{G}^{(P\times QR)}(\mathbf{B}\otimes\mathbf{C})^T, \tag{10.4.16}$$
$$x^{(J\times KI)}_{j,(k-1)I+i} = x_{ijk} \quad \Rightarrow$$
$$\mathbf{X}^{(J\times KI)} = [\mathbf{X}^T_{::1},\ldots,\mathbf{X}^T_{::K}] = \mathbf{B}\mathbf{G}^{(Q\times RP)}(\mathbf{C}\otimes\mathbf{A})^T, \tag{10.4.17}$$
$$x^{(K\times IJ)}_{k,(i-1)J+j} = x_{ijk} \quad \Rightarrow$$
$$\mathbf{X}^{(K\times IJ)} = [\mathbf{X}^T_{1::},\ldots,\mathbf{X}^T_{I::}] = \mathbf{C}\mathbf{G}^{(R\times PQ)}(\mathbf{A}\otimes\mathbf{B})^T; \tag{10.4.18}$$

and the following six possible longitudinal unfolding matrix representations:

$$x^{(JK\times I)}_{(k-1)J+j,i} = x_{ijk} \quad \Rightarrow \quad \mathbf{X}^{(JK\times I)} = (\mathbf{C}\otimes\mathbf{B})\mathbf{G}^{(QR\times P)}\mathbf{A}^T, \tag{10.4.19}$$
$$x^{(KI\times J)}_{(i-1)K+k,j} = x_{ijk} \quad \Rightarrow \quad \mathbf{X}^{(KI\times J)} = (\mathbf{A}\otimes\mathbf{C})\mathbf{G}^{(RP\times Q)}\mathbf{B}^T, \tag{10.4.20}$$
$$x^{(IJ\times K)}_{(j-1)I+i,k} = x_{ijk} \quad \Rightarrow \quad \mathbf{X}^{(IJ\times K)} = (\mathbf{B}\otimes\mathbf{A})\mathbf{G}^{(PQ\times R)}\mathbf{C}^T, \tag{10.4.21}$$
$$x^{(JK\times I)}_{(j-1)K+k,i} = x_{ijk} \quad \Rightarrow \quad \mathbf{X}^{(JK\times I)} = (\mathbf{B}\otimes\mathbf{C})\mathbf{G}^{(QR\times P)}\mathbf{A}^T, \tag{10.4.22}$$
$$x^{(KI\times J)}_{(k-1)I+i,j} = x_{ijk} \quad \Rightarrow \quad \mathbf{X}^{(KI\times J)} = (\mathbf{C}\otimes\mathbf{A})\mathbf{G}^{(RP\times Q)}\mathbf{B}^T, \tag{10.4.23}$$
$$x^{(IJ\times K)}_{(i-1)J+j,k} = x_{ijk} \quad \Rightarrow \quad \mathbf{X}^{(IJ\times K)} = (\mathbf{A}\otimes\mathbf{B})\mathbf{G}^{(PQ\times R)}\mathbf{C}^T. \tag{10.4.24}$$

Here are three equivalent matricization representation methods for the third-order SVD.

(1) The third-order SVD matricization representations (Kiers method) [245], [491]:

$$\text{horizontal}\begin{cases} x^{(I\times JK)}_{i,(k-1)J+j} = x_{ijk} & \Rightarrow \quad \mathbf{X}^{(I\times JK)}_{\text{Kiers}} = \mathbf{A}\mathbf{G}^{(P\times QR)}_{\text{Kiers}}(\mathbf{C}\otimes\mathbf{B})^T, \\ x^{(J\times KI)}_{j,(i-1)K+k} = x_{ijk} & \Rightarrow \quad \mathbf{X}^{(J\times KI)}_{\text{Kiers}} = \mathbf{B}\mathbf{G}^{(Q\times RP)}_{\text{Kiers}}(\mathbf{A}\otimes\mathbf{C})^T, \\ x^{(K\times IJ)}_{k,(j-1)I+i} = x_{ijk} & \Rightarrow \quad \mathbf{X}^{(K\times IJ)}_{\text{Kiers}} = \mathbf{C}\mathbf{G}^{(R\times PQ)}_{\text{Kiers}}(\mathbf{B}\otimes\mathbf{A})^T. \end{cases}$$

$$\text{longitudinal}\begin{cases} x^{(JK\times I)}_{(k-1)J+j,i} = x_{ijk} & \Rightarrow \quad \mathbf{X}^{(JK\times I)}_{\text{Kiers}} = (\mathbf{C}\otimes\mathbf{B})\mathbf{G}^{(QR\times P)}_{\text{Kiers}}\mathbf{A}^T, \\ x^{(KI\times J)}_{(i-1)K+k,j} = x_{ijk} & \Rightarrow \quad \mathbf{X}^{(KI\times J)}_{\text{Kiers}} = (\mathbf{A}\otimes\mathbf{C})\mathbf{G}^{(RP\times Q)}_{\text{Kiers}}\mathbf{B}^T, \\ x^{(IJ\times K)}_{(j-1)I+i,k} = x_{ijk} & \Rightarrow \quad \mathbf{X}^{(IJ\times K)}_{\text{Kiers}} = (\mathbf{B}\otimes\mathbf{A})\mathbf{G}^{(PQ\times R)}_{\text{Kiers}}\mathbf{C}^T. \end{cases}$$

(2) The third-order SVD matricization representations (LMV method) [281], [282]:

$$\text{horizontal}\begin{cases} x^{(I\times JK)}_{i,(j-1)K+k} = x_{ijk} & \Rightarrow \quad \mathbf{X}^{(I\times JK)}_{\text{LMV}} = \mathbf{A}\mathbf{G}^{(P\times QR)}_{\text{LMV}}(\mathbf{B}\otimes\mathbf{C})^T,\\ x^{(J\times KI)}_{j,(k-1)I+i} = x_{ijk} & \Rightarrow \quad \mathbf{X}^{(J\times KI)}_{\text{LMV}} = \mathbf{B}\mathbf{G}^{(Q\times RP)}_{\text{LMV}}(\mathbf{C}\otimes\mathbf{A})^T,\\ x^{(K\times IJ)}_{k,(i-1)J+j} = x_{ijk} & \Rightarrow \quad \mathbf{X}^{(K\times IJ)}_{\text{LMV}} = \mathbf{C}\mathbf{G}^{(R\times PQ)}_{\text{LMV}}(\mathbf{A}\otimes\mathbf{B})^T. \end{cases}$$

$$\text{longitudinal}\begin{cases} x^{(JK\times I)}_{(j-1)K+k,i} = x_{ijk} & \Rightarrow \quad \mathbf{X}^{(JK\times I)}_{\text{LMV}} = (\mathbf{B}\otimes\mathbf{C})\mathbf{G}^{(QR\times P)}_{\text{LMV}}\mathbf{A}^T,\\ x^{(KI\times J)}_{(k-1)I+i,j} = x_{ijk} & \Rightarrow \quad \mathbf{X}^{(KI\times J)}_{\text{LMV}} = (\mathbf{C}\otimes\mathbf{A})\mathbf{G}^{(RP\times Q)}_{\text{LMV}}\mathbf{B}^T,\\ x^{(K\times IJ)}_{(i-1)J+j,k} = x_{ijk} & \Rightarrow \quad \mathbf{X}^{(IJ\times K)}_{\text{LMV}} = (\mathbf{A}\otimes\mathbf{B})\mathbf{G}^{(PQ\times R)}_{\text{LMV}}\mathbf{C}^T. \end{cases}$$

(3) The third-order SVD matricization representations (Kolda method) [65], [245], [164], [254], [7]:

$$\text{horizontal}\begin{cases} x^{(I\times JK)}_{i,(k-1)J+j} = x_{ijk} & \Rightarrow \quad \mathbf{X}^{(I\times JK)}_{\text{Kolda}} = \mathbf{A}\mathbf{G}^{(P\times QR)}_{\text{Kolda}}(\mathbf{C}\otimes\mathbf{B})^T,\\ x^{(J\times KI)}_{j,(k-1)I+i} = x_{ijk} & \Rightarrow \quad \mathbf{X}^{(J\times KI)}_{\text{Kolda}} = \mathbf{B}\mathbf{G}^{(Q\times RP)}_{\text{Kolda}}(\mathbf{C}\otimes\mathbf{A})^T,\\ x^{(K\times IJ)}_{k,(j-1)I+i} = x_{ijk} & \Rightarrow \quad \mathbf{X}^{(K\times IJ)}_{\text{Kolda}} = \mathbf{C}\mathbf{G}^{(R\times PQ)}_{\text{Kolda}}(\mathbf{B}\otimes\mathbf{A})^T. \end{cases}$$

$$\text{longitudinal}\begin{cases} x^{(JK\times I)}_{(k-1)J+j,i} = x_{ijk} & \Rightarrow \quad \mathbf{X}^{(JK\times I)}_{\text{Kolda}} = (\mathbf{C}\otimes\mathbf{B})\mathbf{G}^{(QR\times P)}_{\text{Kolda}}\mathbf{A}^T,\\ x^{(KI\times J)}_{(k-1)I+i,j} = x_{ijk} & \Rightarrow \quad \mathbf{X}^{(KI\times J)}_{\text{Kolda}} = (\mathbf{C}\otimes\mathbf{A})\mathbf{G}^{(RP\times Q)}_{\text{Kolda}}\mathbf{B}^T,\\ x^{(IJ\times K)}_{(j-1)I+i,k} = x_{ijk} & \Rightarrow \quad \mathbf{X}^{(IJ\times K)}_{\text{Kolda}} = (\mathbf{B}\otimes\mathbf{A})\mathbf{G}^{(PQ\times R)}_{\text{Kolda}}\mathbf{C}^T. \end{cases}$$

Below we give a proof of $\mathbf{X}^{(J\times KI)}_{\text{Kiers}} = \mathbf{B}\mathbf{G}^{(Q\times RP)}_{\text{Kiers}}(\mathbf{A}\otimes\mathbf{C})^T$. The other forms can be similarly proved.

Proof First, the Tucker decomposition can be written in the horizontal unfolding form

$$\begin{aligned}
\mathbf{X}_{i::} &= \sum_{p=1}^{P}\sum_{q=1}^{Q}\sum_{r=1}^{R} g_{pqr}a_{ip}\mathbf{B}_{:q}\mathbf{C}_{:r}^T\\
&= \sum_{p=1}^{P}\sum_{q=1}^{Q}\sum_{r=1}^{R} \mathbf{G}^{(Q\times RP)}_{q,(p-1)R+r}a_{ip}\mathbf{b}_q\mathbf{c}_r^T\\
&= \sum_{p=1}^{P}\sum_{r=1}^{R}[\mathbf{b}_1,\ldots,\mathbf{b}_Q]\begin{bmatrix}\mathbf{G}^{(Q\times RP)}_{1,(p-1)R+r}\\ \vdots\\ \mathbf{G}^{(Q\times RP)}_{Q,(p-1)R+r}\end{bmatrix}a_{ip}\mathbf{c}_r^T\\
&= \sum_{p=1}^{P}\sum_{r=1}^{R}\mathbf{B}\mathbf{G}^{(Q\times RP)}_{:(p-1)R+r}a_{ip}\mathbf{c}_r^T.
\end{aligned}$$

Expand the above equation to get

$$\mathbf{X}_{i::} = \mathbf{B}\Big[\mathbf{G}_{:1}^{(Q\times RP)},\ldots,\mathbf{G}_{:R}^{(Q\times RP)},\ldots,\mathbf{G}_{:(P-1)R+1}^{(Q\times RP)},\ldots,\mathbf{G}_{:RP}^{(Q\times RP)}\Big]\begin{bmatrix} a_{i1}\mathbf{c}_1^T \\ \vdots \\ a_{i1}\mathbf{c}_R^T \\ \vdots \\ a_{iP}\mathbf{c}_1^T \\ \vdots \\ a_{iP}\mathbf{c}_R^T \end{bmatrix},$$

where $\mathbf{a}$ is a row vector and $\mathbf{c}$ is a column vector. Hence, from $\mathbf{X}_{\text{Kiers}}^{(J\times KI)} = [\mathbf{X}_{1::},\ldots,\mathbf{X}_{I::}]$ we have

$$\mathbf{X}_{\text{Kiers}}^{(J\times KI)} = \mathbf{B}\mathbf{G}_{\text{Kiers}}^{(Q\times RP)}\begin{bmatrix} a_{11}\mathbf{c}_1^T & \cdots & a_{I1}\mathbf{c}_1^T \\ \vdots & \ddots & \vdots \\ a_{11}\mathbf{c}_R^T & \cdots & a_{I1}\mathbf{c}_R^T \\ \vdots & \ddots & \vdots \\ a_{1P}\mathbf{c}_1^T & \cdots & a_{IP}\mathbf{c}_1^T \\ \vdots & \ddots & \vdots \\ a_{1P}\mathbf{c}_R^T & \cdots & a_{IP}\mathbf{c}_R^T \end{bmatrix} = \mathbf{B}\mathbf{G}_{\text{Kiers}}^{(Q\times RP)}\begin{bmatrix} \mathbf{a}_1^T\mathbf{c}_1^T \\ \vdots \\ \mathbf{a}_1^T\mathbf{c}_R^T \\ \vdots \\ \mathbf{a}_P^T\mathbf{c}_1^T \\ \vdots \\ \mathbf{a}_P^T\mathbf{c}_R^T \end{bmatrix}$$

$$= \mathbf{B}\mathbf{G}_{\text{Kiers}}^{(Q\times RP)}(\mathbf{A}\otimes\mathbf{C})^T.$$ □

The above Tucker decomposition form is called the Tucker3 decomposition and has the following two simplified forms:

$$\text{Tucker2:}\quad x_{ijk} = \sum_{p=1}^{P}\sum_{q=1}^{Q} g_{pqk}a_{ip}b_{jq} + e_{ijk}, \tag{10.4.25}$$

$$\text{Tucker1:}\quad x_{ijk} = \sum_{p=1}^{P} g_{pjk}a_{ip} + e_{ijk}. \tag{10.4.26}$$

As compared with the Tucker3 decomposition, we have $\mathbf{C} = \mathbf{I}_K$ and $\mathcal{G} \in \mathbb{R}^{P\times Q\times K}$ in the Tucker2 decomposition; and $\mathbf{B} = \mathbf{I}_J, \mathbf{C} = \mathbf{I}_K$ and $\mathcal{G} \in \mathbb{R}^{P\times J\times K}$ in the Tucker1 decomposition.

10.4.3 Alternating Least Squares Algorithms

The Tucker or CP decomposition of an Nth-order tensor can be written as the unified mathematical model

$$\mathcal{X} = f(\mathbf{U}^{(1)},\mathbf{U}^{(2)},\ldots,\mathbf{U}^{(N)}) + \mathcal{E}, \tag{10.4.27}$$

where $\mathbf{U}^{(n)}, n = 1, \ldots, N$, are the decomposed factor or component matrices, and $\mathcal{E}$ is an Nth-order noise or error tensor. Hence, the factor matrices can be found by solving the following optimization problem:

$$(\hat{\mathbf{U}}^{(1)}, \ldots, \hat{\mathbf{U}}^{(N)}) = \underset{\mathbf{U}^{(1)},\ldots,\mathbf{U}^{(N)}}{\arg\min} \left\| \mathcal{X} - f(\mathbf{U}^{(1)}, \ldots, \mathbf{U}^{(N)}) \right\|_F^2. \tag{10.4.28}$$

This is an N-argument coupled optimization problem. As stated in Chapter 6, an effective method for solving this class of coupled optimization problems is the alternating least squares (ALS) algorithm.

The basic idea of the ALS algorithm for Tucker decomposition is: fixing the factor matrices $\mathbf{U}_{k+1}^{(1)}, \ldots, \mathbf{U}_{k+1}^{(i-1)}$ as they were updated at the $(k+1)$th iteration and $\mathbf{U}_k^{(i+1)}, \ldots, \mathbf{U}_k^{(N)}$ as they were updated at the kth iteration, find the LS solution of the factor matrix $\mathbf{U}^{(i)}$ at the $(k+1)$th iteration:

$$\hat{\mathbf{U}}_{k+1}^{(i)} = \underset{\mathbf{U}^{(i)}}{\arg\min} \left\| \mathcal{X} - f\big(\mathbf{U}_{k+1}^{(1)}, \ldots, \mathbf{U}_{k+1}^{(i-1)}, \mathbf{U}^{(i)}, \mathbf{U}_k^{(i+1)}, \ldots, \mathbf{U}_k^{(N)}\big) \right\|_F^2, \tag{10.4.29}$$

where $i = 1, \ldots, N$. For $k = 1, 2, \ldots$, apply LS methods alternately until all the factor matrices converge.

As an example, we discuss the solution of the following optimization problem for a Tucker decomposition:

$$\min_{\mathbf{A},\mathbf{B},\mathbf{C},\mathbf{G}^{(P\times QR)}} \left\| \mathbf{X}^{(I\times JK)} - \mathbf{A}\mathbf{G}^{(P\times QR)}(\mathbf{C}\otimes\mathbf{B})^T \right\|_F^2. \tag{10.4.30}$$

By the principle of the ALS, if we assume that the mode-2 matrix $\mathbf{B}$, the horizontal unfoldings of the mode-3 matrix $\mathbf{C}$ and the core tensor $\mathcal{G}$ are fixed then the above optimization problem can be decoupled as an optimization problem containing only the mode-1 matrix $\mathbf{A}$:

$$\min_{\mathbf{A}} \left\| \mathbf{X}^{(I\times JK)} - \mathbf{A}\mathbf{G}^{(P\times QR)}(\mathbf{C}\otimes\mathbf{B})^T \right\|_F^2.$$

This corresponds to finding an LS solution $\mathbf{A}$ for $\mathbf{X}^{(I\times JK)} = \mathbf{A}\mathbf{G}^{(P\times QR)}(\mathbf{C}\otimes\mathbf{B})^T$. Postmultiplying both sides of the matrix equation by $(\mathbf{C}\otimes\mathbf{B})$, we get

$$\mathbf{X}^{(I\times JK)}(\mathbf{C}\otimes\mathbf{B}) = \mathbf{A}\mathbf{G}^{(P\times QR)}(\mathbf{C}\otimes\mathbf{B})^T(\mathbf{C}\otimes\mathbf{B}). \tag{10.4.31}$$

Make the SVD $\mathbf{X}^{(I\times JK)}(\mathbf{C}\otimes\mathbf{B}) = \mathbf{U}_1\mathbf{S}_1\mathbf{V}_1^T$, and take the P principal left singular vectors as an estimate of the matrix $\mathbf{A}$, i.e., $\hat{\mathbf{A}} = \mathbf{U}_1(:, 1:P)$. This operation can be simply expressed as $[\mathbf{A}, \mathbf{S}, \mathbf{T}] = \text{SVD}[\mathbf{X}^{(I\times JK)}(\mathbf{C}\otimes\mathbf{B}), P]$.

Similarly, we can respectively find the estimates of $\mathbf{B}$ and $\mathbf{C}$. Then, fixing the estimated factor matrices $\hat{\mathbf{B}}$ and $\hat{\mathbf{C}}$, we can re-solve $\mathbf{A}$ and next recalculate $\mathbf{B}$ and $\mathbf{C}$ until all three factor matrices $\mathbf{A}, \mathbf{B}, \mathbf{C}$ converge.

When all the factor matrices converge and satisfy the orthogonality conditions $\mathbf{A}^T\mathbf{A} = \mathbf{I}_P, \mathbf{B}^T\mathbf{B} = \mathbf{I}_Q$ and $\mathbf{C}^T\mathbf{C} = \mathbf{I}_R$, we can premultiply both sides of (10.4.31) by the factor matrix $\mathbf{A}^T$:

$$\mathbf{G}^{(P\times QR)} = \mathbf{A}^T\mathbf{X}^{(I\times JK)}(\mathbf{C}\otimes\mathbf{B}), \tag{10.4.32}$$

since $\mathbf{A}^T\mathbf{A} = \mathbf{I}_P$ and $(\mathbf{C}^T \otimes \mathbf{B}^T)(\mathbf{C} \otimes \mathbf{B}) = (\mathbf{C}^T\mathbf{C}) \otimes (\mathbf{B}^T\mathbf{B}) = \mathbf{I}_R \otimes \mathbf{I}_Q = \mathbf{I}_{QR}$.

Similarly, we can find the other two horizontal unfolding matrices $\mathbf{G}^{(Q\times RP)}$ and $\mathbf{G}^{(R\times PQ)}$ of the core tensor, and hence get the core tensor itself. Thus one has an ALS algorithm for the Tucker decomposition, as shown in Algorithm 10.1.

Algorithm 10.1 ALS Algorithm for Tucker decomposition [7]

---

input: Third-order tensor $\mathcal{X}$.
initialize: Choose factor matrices $\mathbf{B}_0, \mathbf{C}_0$, and put $k = 0$.
repeat

1. Use $x^{(I\times JK)}_{i,(k-1)J+j} = x_{ijk}$ to form the horizontal unfolding matrix $\mathbf{X}^{(I\times JK)}$.
2. Compute the truncated SVD of $\mathbf{X}^{(I\times JK)}(\mathbf{C}_k \otimes \mathbf{B}_k)$
   $[\mathbf{A}_{k+1}, \mathbf{S}_1, \mathbf{T}_1] = \text{SVD}[\mathbf{X}^{(I\times JK)}(\mathbf{C}_k \otimes \mathbf{B}_k), P]$.
3. Compute the truncated SVD of $\mathbf{X}^{(J\times KI)}(\mathbf{A}_{k+1} \otimes \mathbf{C}_k)$
   $[\mathbf{B}_{k+1}, \mathbf{S}_2, \mathbf{T}_2] = \text{SVD}[\mathbf{X}^{(J\times KI)}(\mathbf{A}_{k+1} \otimes \mathbf{C}_k), Q]$.
4. Compute the truncated SVD of $\mathbf{X}^{(K\times IJ)}(\mathbf{B}_{k+1} \otimes \mathbf{A}_{k+1})$
   $[\mathbf{C}_{k+1}, \mathbf{S}_3, \mathbf{T}_3] = \text{SVD}[\mathbf{X}^{(K\times IJ)}(\mathbf{B}_{k+1} \otimes \mathbf{A}_{k+1}), R]$.
5. **exit if** $\|\mathbf{A}_{k+1} - \mathbf{A}_k\|_F \leq \epsilon, \|\mathbf{B}_{k+1} - \mathbf{B}_k\|_F \leq \epsilon, \|\mathbf{C}_{k+1} - \mathbf{C}_k\|_F \leq \epsilon$.

return $k \leftarrow k + 1$.

6. Compute $\mathbf{G}^{(P\times QR)} = \mathbf{A}^T\mathbf{X}^{(I\times JK)}(\mathbf{C} \otimes \mathbf{B})$.
7. Compute $\mathbf{G}^{(Q\times RP)} = \mathbf{B}^T\mathbf{X}^{(J\times KI)}(\mathbf{A} \otimes \mathbf{C})$.
8. Compute $\mathbf{G}^{(R\times PQ)} = \mathbf{C}^T\mathbf{X}^{(K\times IJ)}(\mathbf{B} \otimes \mathbf{A})$.

output: Factors $\mathbf{A} \in \mathbb{R}^{I\times P}, \mathbf{B} \in \mathbb{R}^{J\times Q}, \mathbf{C} \in \mathbb{R}^{K\times R}$; core tensor $\mathcal{G} \in \mathbb{R}^{P\times Q\times R}$.

Algorithm 10.1 is also available for the Kiers horizontal unfolding of a third-order tensor, and is easily extended to other matricization cases.

For example, if the Kiers horizontal unfolding matrices are replaced by the Kolda horizontal unfolding matrices

$$x^{(I\times JK)}_{i,(k-1)J+j} = x_{ijk}, \quad x^{(J\times KI)}_{j,(k-1)I+i} = x_{ijk}, \quad x^{(K\times IJ)}_{k,(j-1)I+i} = x_{ijk},$$

and correspondingly

$$\begin{aligned}
[\mathbf{A}, \mathbf{S}, \mathbf{V}] &= \text{SVD}(\mathbf{X}^{(I\times JK)}(\mathbf{C} \otimes \mathbf{B}), P),\\
[\mathbf{B}, \mathbf{S}, \mathbf{V}] &= \text{SVD}(\mathbf{X}^{(J\times KI)}(\mathbf{C} \otimes \mathbf{A}), Q),\\
[\mathbf{C}, \mathbf{S}, \mathbf{V}] &= \text{SVD}(\mathbf{X}^{(K\times IJ)}(\mathbf{B} \otimes \mathbf{A}), R),
\end{aligned}$$

then Algorithm 10.1 becomes the ALS algorithm for the Tucker decomposition proposed by Bro in 1998 [65], where P, Q, R are the number of principal singular values of the corresponding matrices. Therefore, the equivalent matrix forms of the higher-order SVD must correspond strictly to the matricization method of the tensor.

Consider the Tucker3 decomposition or the higher-order SVD of an Nth-order tensor $\mathcal{X}$:

$$\min_{\mathbf{U}^{(1)},\ldots,\mathbf{U}^{(N)},\mathbf{G}_{(n)}} \left\|\mathbf{X}_{(n)} - \mathbf{U}^{(n)}\mathbf{G}_{(n)}\mathbf{U}_{\otimes}^{(n)}\right\|_F^2, \tag{10.4.33}$$

where the Kronecker product $\mathbf{U}_{\otimes}^{(n)}$ is defined by (10.4.7) or (10.4.8) or (10.4.9).

In the higher-order SVD, the factor matrix $\mathbf{U}^{(n)} \in \mathbb{R}^{I_n \times J_n}$ satisfies the semi-orthogonality condition $(\mathbf{U}^{(n)})^T\mathbf{U}^{(n)} = \mathbf{I}_{J_n}$, so we have

$$\mathbf{U}_{\otimes}^{(n)}(\mathbf{U}_{\otimes}^{(n)})^T = \mathbf{I}_{J_1\cdots J_{n-1}J_{n+1}\cdots J_N}. \tag{10.4.34}$$

In order to solve the matrix equation $\mathbf{X}_{(n)} \approx \mathbf{U}^{(n)}\mathbf{G}_{(n)}\mathbf{U}_{\otimes}^{(n)}$, let SVD $\mathbf{X}_{(n)} = \mathbf{U}^{(n)}\mathbf{S}^{(n)}(\mathbf{V}^{(n)})^T$, then $\mathbf{U}^{(n)}\mathbf{S}^{(n)}(\mathbf{V}^{(n)})^T = \mathbf{U}^{(n)}\mathbf{G}_{(n)}\mathbf{U}_{\otimes}^{(n)}$. Premultiplying both sides by $(\mathbf{U}^{(n)})^T$ and postmultiplying by $(\mathbf{U}_{\otimes}^{(n)})^T$, then from (10.4.34) we have

$$\mathbf{G}_{(n)} = \mathbf{S}^{(n)}(\mathbf{V}^{(n)})^T(\mathbf{U}_{\otimes}^{(n)})^T. \tag{10.4.35}$$

Algorithm 10.2 is a higher-order SVD (HOSVD) algorithm, while Algorithm 10.3 is a higher-order orthogonal iteration (HOOI) algorithm.

Algorithm 10.2 HOSVD($\mathcal{X}, R_1, \ldots, R_N$) [257]

---

input: Nth-order tensor $\mathcal{X}$.
initialize: The horizontal unfolding matrices $\mathbf{X}_{(n)}, n = 1, \ldots, N$.
for $n = 1, \ldots, N$ **do**
  1. Compute the SVD $\mathbf{X}_{(n)} = \mathbf{U}_k\mathbf{\Sigma}_k\mathbf{V}_k^T$.
  2. Determine the effective rank R_n of $\mathbf{X}_{(n)}$.
  3. Update $\mathbf{U}^{(n)} \leftarrow \mathbf{U}_k(:, 1 : R_n)$.
end for
  4. Calculate $\mathcal{G} = \mathcal{X} \times_1 \mathbf{U}^{(1)T} \times_2 \mathbf{U}^{(2)T} \times_3 \cdots \times_N \mathbf{U}^{(N)T}$.
output: Core tensor $\mathcal{G}$ and factor matrices $\mathbf{U}^{(1)}, \ldots, \mathbf{U}^{(N)}$.

Algorithm 10.3 HOOI($\mathcal{X}, R_1, \ldots, R_N$) [257]

---

input: Nth-order tensor $\mathcal{X}$.
initialize: The horizontal unfolding matrices $\mathbf{X}_{(n)}, n = 1, \ldots, N$.
for $n = 1, \ldots, N$ **do**
  1. Compute the SVD $\mathbf{X}_{(n)} = \mathbf{U}_k\mathbf{\Sigma}_k\mathbf{V}_k^T$.
  2. Determine the effective rank R_n of $\mathbf{X}_{(n)}$.
  3. Update $\mathbf{U}^{(n)} \leftarrow \mathbf{U}_k(:, 1 : R_n)$ and let $k = 1$.
end for
repeat
  for $n = 1, \ldots, n$ **do**
    $\mathcal{Y} \leftarrow \mathcal{X} \times_1 \mathbf{U}^{(1)T} \cdots \times_{n-1} \mathbf{U}^{(n-1)} \times_{n+1} \mathbf{U}^{(n+1)T} \times_{n+2} \cdots \times_N \mathbf{U}^{(N)}$.
    $\mathbf{U}^{(n)} \leftarrow R_n$ leading left singular vectors of $\mathbf{Y}_{(n)}$.
  end for
  4. $\mathcal{G}_k = \mathcal{X} \times_1 \mathbf{U}^{(1)T} \times_2 \mathbf{U}^{(2)T} \times_3 \cdots \times_N \mathbf{U}^{(N)T}$.
  5. **exit if** $\|\mathcal{G}_{k+1} - \mathcal{G}_k\|_F \leq \epsilon, k > 1$ or maximum iterations exhausted.
return $k \leftarrow k + 1$.
output: Core tensor $\mathcal{G}$ and factor matrices $\mathbf{U}^{(1)}, \ldots, \mathbf{U}^{(N)}$.

The HOOI algorithm is the more effective technique for computing the factor matrices and the core tensor. The feature of HOOI is to compute the leading singular vectors of the horizontal unfolding matrix $\mathbf{X}_{(n)}$ even if they just form an orthonormal basis of its dominant subspace.

10.5 Parallel Factor Decomposition of Tensors

As mentioned early, canonical decomposition (CANDECOMP) and parallel factor (PARAFAC) decomposition are two data analysis methods for multi-way data, proposed independently by Carroll and Chang in 1970 [87] and Harshman in 1970 [198] on the basis of the principle of parallel proportional profiles (PP) proposed by Cattell in 1944 [88].

PARAFAC can be seen as a generalization of two-way factor analysis to multi-way data. The CANDECOMP/PARAFAC decomposition is usually simply called the CP decomposition.

10.5.1 Bilinear Model

Tensor analysis is essentially multilinear analysis, which in turn is a generalization of bilinear analysis; principal component analysis (PCA) and independent component analysis (ICA) are two typical bilinear analysis methods.

Via a bilinear or multilinear model, the factors (or components, or loadings) reflecting the linear combination of variables are extracted. These extracted factors are then used to explain the basic information content of the data.

In data analysis, given a two-way data matrix $\mathbf{X} \in \mathbb{K}^{I \times J}$, two-way bilinear analysis uses the model

$$x_{ij} = \sum_{r=1}^{R} a_{ir} b_{jr} + e_{ij} \tag{10.5.1}$$

for data fitting, where the meaning of each parameter is as follows:

x_{ij} is the entry of the ith row and the jth column of the data matrix $\mathbf{X}$;

R is the number of factors;

a_{ir} is the "factor loading";

b_{jr} is the "factor score";

e_{ij} is the measurement error of the data x_{ij} and is the entry of the ith row and jth column of the $I \times J$ error matrix $\mathbf{E}$.

If $b_{jr} = \beta_r$ is fixed at a constant value, then $x_{ij} = \sum_{r=1}^{R} \beta_r a_{ir}$ is a linear model for the factor loadings a_{ir}. Conversely, if $a_{ir} = \alpha_r$ is fixed at a constant value, then $x_{ij} = \sum_{r=1}^{R} \alpha_r b_{jr}$ is a linear model for the factor scores b_{jr}. Therefore, the two-way data model (10.5.1) is often called a two-way bilinear model.

Define respectively the *factor loading vector* and the *factor score vector* as

$$\mathbf{a}_r = [a_{1r}, \ldots, a_{Ir}]^T, \qquad \mathbf{b}_r = [b_{1r}, \ldots, b_{Jr}]^T;$$

then the two-way bilinear model can be rewritten in matrix form as

$$\mathbf{X} = \sum_{r=1}^{R} \mathbf{a}_r \circ \mathbf{b}_r = \sum_{r=1}^{R} \mathbf{a}_r \mathbf{b}_r^T \tag{10.5.2}$$

$$= \begin{bmatrix} a_{11}b_{11} + \cdots + a_{1R}b_{1R} & \cdots & a_{11}b_{J1} + \cdots + a_{1R}b_{JR} \\ \vdots & \ddots & \vdots \\ a_{I1}b_{11} + \cdots + a_{IR}b_{1R} & \cdots & a_{I1}b_{J1} + \cdots + a_{IR}b_{JR} \end{bmatrix}. \tag{10.5.3}$$

The above equation can be equivalently given as

$$\mathbf{X} = \mathbf{A}\mathbf{B}^T = \sum_{r=1}^{R} \mathbf{a}_r \circ \mathbf{b}_r, \tag{10.5.4}$$

where

$$\mathbf{A} = \begin{bmatrix} a_{11} & \cdots & a_{1R} \\ \vdots & \ddots & \vdots \\ a_{I1} & \cdots & a_{IR} \end{bmatrix} \in \mathbb{K}^{I\times R}, \tag{10.5.5}$$

$$\mathbf{B} = \begin{bmatrix} b_{11} & \cdots & b_{1R} \\ \vdots & \ddots & \vdots \\ b_{J1} & \cdots & b_{JR} \end{bmatrix} \in \mathbb{K}^{J\times R} \tag{10.5.6}$$

are the *factor loading matrix* and the *factor score matrix*, respectively.

Data analysis emphasizes the uniqueness of a structured model. That a structured model is unique means that the model identification requires no further constraint conditions. An unconstrained two-way bilinear model has no uniqueness: from (10.5.4) it is easy to see that in the two-way bilinear model there exists infinite rotational freedom. This is so because

$$\mathbf{X} = \mathbf{A}\mathbf{Q}(\mathbf{B}\mathbf{Q})^T = \mathbf{A}\mathbf{B}^T \quad \text{or} \quad \mathbf{X} = \mathbf{A}\mathbf{Q}^T(\mathbf{B}\mathbf{Q}^T)^T = \mathbf{A}\mathbf{B}^T.$$

That is, given only the two-way data matrix $\mathbf{X} \in \mathbb{R}^{I\times J}$, there are infinitely many sets of solutions $(\mathbf{A}, \mathbf{B})$ satisfying the two-way bilinear model. Hence, in order to keep the uniqueness of two-way bilinear model fitting, some constraints must be imposed on the factor matrices.

The PCA is a two-way bilinear analysis method with an orthogonality constraint on factor matrices. Assume that the data matrix $\mathbf{X} \in \mathbb{R}^{I\times J}$ has R leading singular values; then the PCA uses the truncated SVD $\mathbf{X} = \mathbf{U}_1\mathbf{\Sigma}_1\mathbf{V}_1^T$ as the two-way data model, where $\mathbf{\Sigma}_1$ is a $R \times R$ diagonal matrix whose diagonal entries are the R leading singular values while $\mathbf{U}_1 \in \mathbb{R}^{I\times R}$ and $\mathbf{V}_1 \in \mathbb{R}^{J\times R}$ respectively consist of the R left and right singular vectors corresponding to the R leading singular values. Letting

$$\mathbf{A} = \mathbf{U}_1\mathbf{\Sigma}_1, \quad \mathbf{B} = \mathbf{V}_1, \tag{10.5.7}$$

the PCA data model can be written as the orthogonally constrained bilinear model

$$\mathbf{X} = \mathbf{A}\mathbf{B}^T \quad \text{subject to} \quad \mathbf{A}^T\mathbf{A} = \mathbf{D},\ \mathbf{B}^T\mathbf{B} = \mathbf{I}, \tag{10.5.8}$$

where $\mathbf{D}$ is an $R \times R$ diagonal matrix and $\mathbf{I}$ is an $R \times R$ identity matrix. That is, the PCA requires the column vectors of the factor loading matrix $\mathbf{A}$ to be mutually orthogonal and each column vector of the factor score matrix $\mathbf{B}$ to be orthonormal.

Table 10.2 compares the PCA with an unconstrained two-way bilinear analysis.

Table 10.2 Comparison of PCA with two-way bilinear analysis

Method	Bilinear analysis	PCA analysis
Constructed model	$\mathbf{X} = \mathbf{A}\mathbf{B}^T$	$\mathbf{X} = \mathbf{A}\mathbf{B}^T$
Constraint	no	$\mathbf{A}^T\mathbf{A} = \mathbf{D}$, $\mathbf{B}^T\mathbf{B} = \mathbf{I}$
Cost function	$\lVert\mathbf{X} - \mathbf{A}\mathbf{B}^T\rVert_F^2$	$\lVert\mathbf{X} - \mathbf{A}\mathbf{B}^T\rVert_F^2$
Optimization	$\min_{\mathbf{A},\mathbf{B}} \lVert\mathbf{X} - \mathbf{A}\mathbf{B}^T\rVert_F^2$	$\min_{\mathbf{A},\mathbf{B}} \left\lVert\mathbf{X} - \mathbf{A}\mathbf{B}^T\right\rVert_F^2$ s.t. $\mathbf{A}^T\mathbf{A} = \mathbf{D}$, $\mathbf{B}^T\mathbf{B} = \mathbf{I}$

Two-way bilinear data analysis as such is not available for processing a multi-way data set, but it can be generalized, as we now discuss.

10.5.2 Parallel Factor Analysis

In order to generalize unconstrained two-way bilinear analysis to a multi-way data set, it is required to answer the following questions.

(1) How do we name the factors when there are three or more?
(2) How do we cope with the uncertainty caused by the degrees of rotational freedom?

In fact, in some modern applications (e.g., chemistry), it is already difficult to distinguish factor loadings and factor scores. However, if we want to promote the analysis to deal with three-way data, it is necessary to add a new factor, and this factor is all the more difficultly to name. A simple solution to this difficulty is to use different modes to distinguish the different factors. Specifically, we call the factor loading vector $\mathbf{a}_r$ of fitting data matrix $\mathbf{X}$ the mode-A vector, and the factor score vector $\mathbf{b}_r$ the mode-B vector. If we add a factor vector $\mathbf{c}_r$ then it is called the mode-C vector, and so on; there could be mode-D, mode-E vectors etc.

To cope with the uncertainty in parallel factor decomposition caused by the degrees of rotational freedom, Cattell in 1944 proposed the parallel proportional profiles principle [88]: a multi-way data model with no degrees of rotational freedom can be obtained only by assigning different proportions or weighting coefficients to the same profile or loading vectors of two-way or more than two-way data sets. Thus the two-way bilinear analysis representation (10.5.1) generalizes naturally to the parallel factor decomposition of a three-way tensor:

$$x_{ijk} = \sum_{r=1}^{R} a_{ir} b_{jr} c_{kr} + e_{ijk}, \tag{10.5.9}$$

where x_{ijk} are the entries of the tensor $\mathcal{X} = [\![x_{ijk}]\!] \in \mathbb{K}^{I\times J\times K}$, while e_{ijk} are the entries of the additive noise tensor $\mathcal{E} = [\![e_{ijk}]\!] \in \mathbb{K}^{I\times J\times K}$.

Define the mode-A, mode-B and mode-C vectors respectively as

$$\mathbf{a}_r = [a_{1r}, \ldots, a_{Ir}]^T \in \mathbb{K}^{I\times 1}, \tag{10.5.10}$$

$$\mathbf{b}_r = [b_{1r}, \ldots, b_{Jr}]^T \in \mathbb{K}^{J\times 1}, \tag{10.5.11}$$

$$\mathbf{c}_r = [c_{1r}, \ldots, c_{Kr}]^T \in \mathbb{K}^{K\times 1}; \tag{10.5.12}$$

then entry representation (10.5.9) of the factor decomposition of a three-way tensor can be equivalently expressed as

$$\mathcal{X} = \sum_{r=1}^{R} \mathbf{a}_r \circ \mathbf{b}_r \circ \mathbf{c}_r + \mathcal{E}. \tag{10.5.13}$$

This is the three-way generalization of the matrix representation (10.5.2) of two-way bilinear analysis.

Figure 10.10 shows, for comparison, two-way factor analysis and parallel factor analysis.

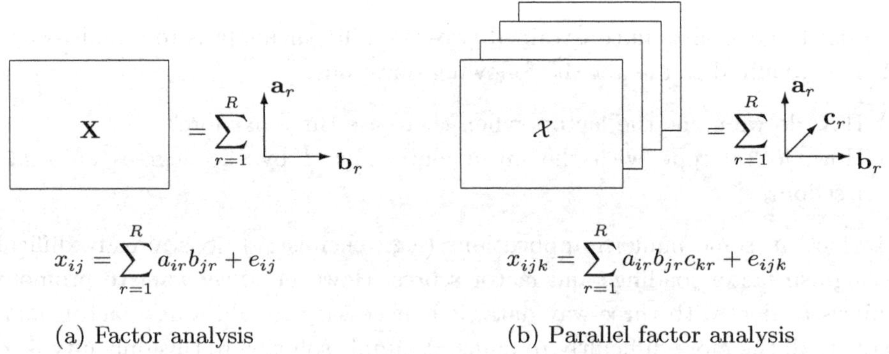

Figure 10.10 Factor analysis and parallel factor analysis.

Obviously, if only one mode factor, for example mode-A, can be varied, and the other two mode factors, mode-B and mode-C, are fixed, i.e., $b_{jr} = \alpha_r$ and

$c_{kr} = \beta_r$, then the three-way data model reduces to a linear combination of mode-A factor vectors. Similarly, the three-way data model can also be regarded as a linear combination of mode-B factors (mode-A and mode-C fixed) or the mode-C factors (mode-A and mode-B fixed), so the three-way data model (10.5.9) is a trilinear factor model.

Figure 10.11 shows a comparison of parallel factor decomposition with Tucker decomposition.

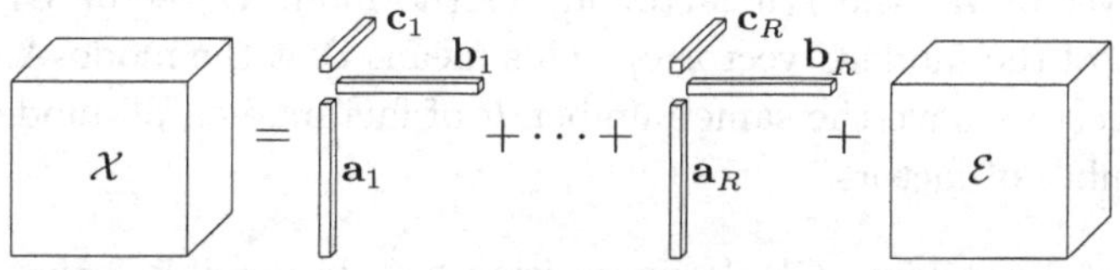

(a) Parallel factor decomposition

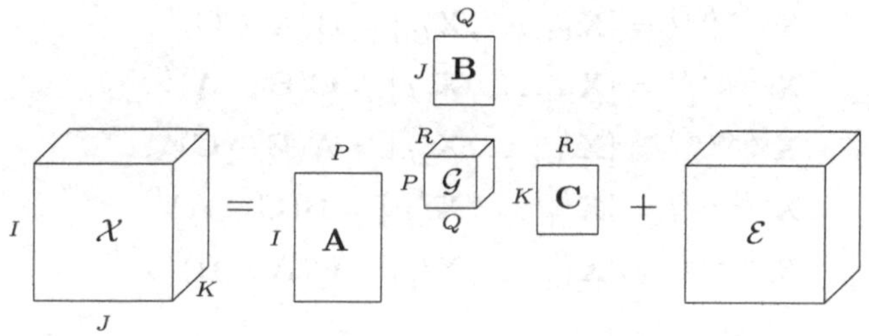

(b) Tucker decomposition

Figure 10.11 Parallel factor decomposition and Tucker decomposition.

Table 10.3 gives mathematical formulas for Tucker decomposition, CP decomposition and SVD.

Table 10.3 Three decomposition methods

Decomposition method	Mathematical formula
Tucker decomposition	$x_{ijk} = \sum_{p=1}^{P} \sum_{q=1}^{Q} \sum_{r=1}^{R} g_{pqr}\, a_{ip}\, b_{jq}\, c_{kr}$
CP decomposition	$x_{ijk} = \sum_{p=1}^{P} \sum_{q=1}^{Q} \sum_{r=1}^{R} a_{ip}\, b_{jq}\, c_{kr}$
SVD	$x_{ij} = \sum_{r=1}^{R} g_{rr}\, a_{ir}\, b_{jr}$

From Table 10.3 we have the following results.

(1) In the Tucker decomposition, the entry g_{pqr} of the core tensor $\mathcal{G}$ shows that there are interaction forms involving the entry a_{ip} of the mode-A vector $\mathbf{a}_i = [a_{i1}, \ldots, a_{iP}]^T$, the entry b_{jq} of the mode-B vector $\mathbf{b}_j = [b_{j1}, \ldots, b_{jQ}]^T$ and the entry c_{kr} of the mode-C vector $\mathbf{c}_k = [c_{k1}, \ldots, c_{kR}]^T$.

(2) In the CP decomposition, the core tensor is a unit tensor. Because the entries on the superdiagonal $p = q = r \in \{1, \ldots, R\}$ are equal to 1, and all the other entries are zero, there are interactions only between the rth factor a_{ir} of the mode-A vector $\mathbf{a}_i$, the rth factor b_{jr} of the mode-B vector $\mathbf{b}_j$ and the rth factor c_{kr} of the mode-C vector $\mathbf{c}_k$. This means that the mode-A, mode-B and mode-C vectors have the same number R of factors, i.e., all modes extract the same number of factors.

The horizontal unfolding CP decomposition has six possible forms:

$$\mathbf{X}^{(I \times JK)} = [\mathbf{X}_{::1}, \ldots, \mathbf{X}_{::K}] = \mathbf{A}(\mathbf{C} \odot \mathbf{B})^T, \tag{10.5.14}$$

$$\mathbf{X}^{(J \times KI)} = [\mathbf{X}_{1::}, \ldots, \mathbf{X}_{I::}] = \mathbf{B}(\mathbf{A} \odot \mathbf{C})^T, \tag{10.5.15}$$

$$\mathbf{X}^{(K \times IJ)} = [\mathbf{X}_{:1:}, \ldots, \mathbf{X}_{:J:}] = \mathbf{C}(\mathbf{B} \odot \mathbf{A})^T, \tag{10.5.16}$$

$$\mathbf{X}^{(I \times JK)} = [\mathbf{X}_{:1:}^T, \ldots, \mathbf{X}_{:J:}^T] = \mathbf{A}(\mathbf{B} \odot \mathbf{C})^T, \tag{10.5.17}$$

$$\mathbf{X}^{(J \times KI)} = [\mathbf{X}_{::1}^T, \ldots, \mathbf{X}_{::K}^T] = \mathbf{B}(\mathbf{C} \odot \mathbf{A})^T, \tag{10.5.18}$$

$$\mathbf{X}^{(K \times IJ)} = [\mathbf{X}_{1::}^T, \ldots, \mathbf{X}_{I::}^T] = \mathbf{C}(\mathbf{A} \odot \mathbf{B})^T, \tag{10.5.19}$$

where

$$\mathbf{A} = [\mathbf{a}_1, \ldots, \mathbf{a}_R] = \begin{bmatrix} a_{11} & \cdots & a_{1R} \\ \vdots & \ddots & \vdots \\ a_{I1} & \cdots & a_{IR} \end{bmatrix} \in \mathbb{R}^{I \times R}, \tag{10.5.20}$$

$$\mathbf{B} = [\mathbf{b}_1, \ldots, \mathbf{b}_R] = \begin{bmatrix} b_{11} & \cdots & b_{1R} \\ \vdots & \ddots & \vdots \\ b_{J1} & \cdots & b_{JR} \end{bmatrix} \in \mathbb{R}^{J \times R}, \tag{10.5.21}$$

and

$$\mathbf{C} = [\mathbf{c}_1, \ldots, \mathbf{c}_R] = \begin{bmatrix} c_{11} & \cdots & c_{1R} \\ \vdots & \ddots & \vdots \\ c_{K1} & \cdots & c_{KR} \end{bmatrix} \in \mathbb{R}^{K \times R}. \tag{10.5.22}$$

Here we demonstrate Equation (10.5.14), the other expressions can be similarly proved.

Proof By Cattell's principle of parallel proportional profiles, for the front slice matrix of a third-order tensor we have

$$\mathbf{X}_{::k} = \mathbf{a}_1 \mathbf{b}_1^T c_{k1} + \cdots + \mathbf{a}_R \mathbf{b}_R^T c_{kR}. \tag{10.5.23}$$

Thus $\mathbf{X}^{(I\times JK)} = [\mathbf{X}_{::1},\ldots,\mathbf{X}_{::K}]$ can be written as

$$\begin{aligned}\mathbf{X}^{(I\times JK)} &= \left[\mathbf{a}_1\mathbf{b}_1^T c_{11}+\cdots+\mathbf{a}_R\mathbf{b}_R^T c_{1R},\ldots,\mathbf{a}_1\mathbf{b}_1^T c_{K1}+\cdots+\mathbf{a}_R\mathbf{b}_R^T c_{KR}\right]\\ &= [\mathbf{a}_1,\ldots,\mathbf{a}_R]\begin{bmatrix} c_{11}\mathbf{b}_1^T & \cdots & c_{K1}\mathbf{b}_1^T\\ \vdots & \ddots & \vdots\\ c_{1R}\mathbf{b}_R^T & \cdots & c_{KR}\mathbf{b}_R^T\end{bmatrix}\\ &= \mathbf{A}(\mathbf{C}\odot\mathbf{B})^T. \qquad \square\end{aligned}$$

Table 10.4 summarizes the mathematical expressions for the horizontal unfolding matrices and for the CP decomposition of the third-order tensor.

Table 10.4 Horizontal unfolding and CP decomposition of third-order tensor

Matricization	Horizontal unfolding matrices and CP decompositions
Kiers method	$x^{(I\times JK)}_{i,(k-1)J+j} = x_{ijk} \Rightarrow \mathbf{X}^{(I\times JK)} = [\mathbf{X}_{::1},\ldots,\mathbf{X}_{::K}] = \mathbf{A}(\mathbf{C}\odot\mathbf{B})^T$
	$x^{(J\times KI)}_{j,(i-1)K+k} = x_{ijk} \Rightarrow \mathbf{X}^{(J\times KI)} = [\mathbf{X}_{1::},\ldots,\mathbf{X}_{I::}] = \mathbf{B}(\mathbf{A}\odot\mathbf{C})^T$
	$x^{(K\times IJ)}_{k,(j-1)I+i} = x_{ijk} \Rightarrow \mathbf{X}^{(K\times IJ)} = [\mathbf{X}_{:1:},\ldots,\mathbf{X}_{:J:}] = \mathbf{C}(\mathbf{B}\odot\mathbf{A})^T$
LMV method	$x^{(I\times JK)}_{i,(j-1)K+k} = x_{ijk} \Rightarrow \mathbf{X}^{(I\times JK)} = [\mathbf{X}^T_{:1:},\ldots,\mathbf{X}^T_{:J:}] = \mathbf{A}(\mathbf{B}\odot\mathbf{C})^T$
	$x^{(J\times KI)}_{j,(k-1)I+i} = x_{ijk} \Rightarrow \mathbf{X}^{(J\times KI)} = [\mathbf{X}^T_{::1},\ldots,\mathbf{X}^T_{::K}] = \mathbf{B}(\mathbf{C}\odot\mathbf{A})^T$
	$x^{(K\times IJ)}_{k,(i-1)J+j} = x_{ijk} \Rightarrow \mathbf{X}^{(K\times IJ)} = [\mathbf{X}^T_{1::},\ldots,\mathbf{X}^T_{I::}] = \mathbf{C}(\mathbf{A}\odot\mathbf{B})^T$
Kolda method	$x^{(I\times JK)}_{i,(k-1)J+j} = x_{ijk} \Rightarrow \mathbf{X}^{(I\times JK)} = [\mathbf{X}_{::1},\ldots,\mathbf{X}_{::K}] = \mathbf{A}(\mathbf{C}\odot\mathbf{B})^T$
	$x^{(J\times KI)}_{j,(k-1)I+i} = x_{ijk} \Rightarrow \mathbf{X}^{(J\times KI)} = [\mathbf{X}^T_{::1},\ldots,\mathbf{X}^T_{::K}] = \mathbf{B}(\mathbf{C}\odot\mathbf{A})^T$
	$x^{(K\times IJ)}_{k,(j-1)I+i} = x_{ijk} \Rightarrow \mathbf{X}^{(K\times IJ)} = [\mathbf{X}_{:1:},\ldots,\mathbf{X}_{:J:}] = \mathbf{C}(\mathbf{B}\odot\mathbf{A})^T$

If we apply the PCA or ICA to analyze (10.5.14), then it requires the use of a truncated SVD,

$$\mathbf{A}^{(I\times JK)} = \sum_{r=1}^{R}\sigma_r\mathbf{u}_r\mathbf{v}_r^H, \tag{10.5.24}$$

where R is the number of principal singular values. This involves $R(I+JK+1)$ parameters, because $\mathbf{u}_r\in\mathbb{K}^{I\times 1}, \mathbf{v}_r\in\mathbb{K}^{JK\times 1}$.

However, the CP method based on (10.5.13) requires only $R(I+J+K)$ parameters. Therefore, compared with the CP analysis method, the PCA or ICA methods requires a much larger number of free parameters, as $R(I+JK+1)\gg R(I+J+K)$. This is one of the outstanding advantages of the CP method.

Although most multi-way data analysis techniques can maintain the multi-way properties of the data, some simplified multi-way analysis technologies (such as Tucker1) are based on a matricization of a multi-way tensor and first transform a third-order or higher-order tensor into a two-way data set. Once a three-way tensor is unfolded to a two-way data set, two-way analysis methods (such as SVD, PCA, ICA) can be applied to extract the data construction.

However, it should be noted that when a multi-way array is rearranged to a two-way data set, it is possible that loss of information and error interpretation will result. If the data are contaminated by noise, this phenomenon will be more serious.

A typical example is a sensory data set, where eight judges evaluate 10 kinds of bread according to 11 class attributes [65]. When this data set is modeled using the PARAFAC model, a common evaluation guide is assumed, and each judge, in varying degrees, must comply with this evaluation guide. But, when the sensory data are unfolded to a two-way array, and the two-way factor model is used, there is no longer a common evaluation guide and each judge can make a completely independent evaluation. In this case, in order to explain the variations in the data, the two-way factor model needs to extract as many factors as possible.

Compared with the PARAFAC model, which explains the data variations just in forms of basic assumptions, the additional variations captured by the two-way factor model are in fact only a reflection of the role of noise rather than of the internal structure of the data. As a result, the multi-way model is superior to the two-way model in terms of interpretation and accuracy. An important fact is that the multilinear models (such as PARAFAC, Tucker decomposition and their variants) can capture the multi-variant linear structure in the data, but a bilinear model (such as SVD, PCA and ICA) cannot.

Similarly to the horizontal unfolding CP decomposition, the longitudinal unfolding CP decomposition also has six forms:

$$\mathbf{X}^{(JK\times I)} = \begin{bmatrix} \mathbf{X}_{:1:} \\ \vdots \\ \mathbf{X}_{:J:} \end{bmatrix} = (\mathbf{B}\odot\mathbf{C})\mathbf{A}^T, \quad \mathbf{X}^{(JK\times I)} = \begin{bmatrix} \mathbf{X}_{::1}^T \\ \vdots \\ \mathbf{X}_{::K}^T \end{bmatrix} = (\mathbf{C}\odot\mathbf{B})\mathbf{A}^T,$$

$$\mathbf{X}^{(KI\times J)} = \begin{bmatrix} \mathbf{X}_{::1} \\ \vdots \\ \mathbf{X}_{::K} \end{bmatrix} = (\mathbf{C}\odot\mathbf{A})\mathbf{B}^T, \quad \mathbf{X}^{(KI\times J)} = \begin{bmatrix} \mathbf{X}_{1::}^T \\ \vdots \\ \mathbf{X}_{I::}^T \end{bmatrix} = (\mathbf{A}\odot\mathbf{C})\mathbf{B}^T,$$

$$\mathbf{X}^{(IJ\times K)} = \begin{bmatrix} \mathbf{X}_{1::} \\ \vdots \\ \mathbf{X}_{I::} \end{bmatrix} = (\mathbf{A}\odot\mathbf{B})\mathbf{C}^T, \quad \mathbf{X}^{(IJ\times K)} = \begin{bmatrix} \mathbf{X}_{:1:}^T \\ \vdots \\ \mathbf{X}_{:J:}^T \end{bmatrix} = (\mathbf{B}\odot\mathbf{A})\mathbf{C}^T.$$

Here we give a proof that $\mathbf{X}^{(KI\times J)} = (\mathbf{C}\odot\mathbf{A})\mathbf{B}^T$:

$$\begin{aligned}\mathbf{X}^{(KI\times J)} &= \begin{bmatrix}\mathbf{X}_{::1}\\ \vdots \\ \mathbf{X}_{::K}\end{bmatrix} = \begin{bmatrix}\mathbf{a}_1\mathbf{b}_1^T c_{11} + \cdots + \mathbf{a}_R\mathbf{b}_R^T c_{1R}\\ \vdots \\ \mathbf{a}_1\mathbf{b}_1^T c_{K1} + \cdots + \mathbf{a}_R\mathbf{b}_R^T c_{KR}\end{bmatrix}\\ &= \begin{bmatrix} c_{11}\mathbf{a}_1 & \cdots & c_{1R}\mathbf{a}_R\\ \vdots & \ddots & \vdots\\ c_{K1}\mathbf{a}_1 & \cdots & c_{KR}\mathbf{a}_R\end{bmatrix}\begin{bmatrix}\mathbf{b}_1^T\\ \vdots\\ \mathbf{b}_R^T\end{bmatrix} = [\mathbf{c}_1\otimes\mathbf{a}_1,\ldots,\mathbf{c}_R\otimes\mathbf{a}_R]\mathbf{B}^T\\ &= (\mathbf{C}\odot\mathbf{A})\mathbf{B}^T.\end{aligned}$$

The other formulas can be similarly proved.

Table 10.5 summarizes the mathematical representations of the longitudinal unfolding and the CP decomposition of third-order tensors.

Table 10.5 Longitudinal unfolding and CP decomposition of third-order tensors

Matricizing method	Longitudinal unfolding and CP decomposition
Kiers	$x^{(JK\times I)}_{(k-1)J+j,i} = x_{ijk} \Leftrightarrow \mathbf{X}^{(JK\times I)} = [\mathbf{X}_{::1},\ldots,\mathbf{X}_{::K}]^T = (\mathbf{C}\odot\mathbf{B})\mathbf{A}^T$
	$x^{(KI\times J)}_{(i-1)K+k,j} = x_{ijk} \Leftrightarrow \mathbf{X}^{(KI\times J)} = [\mathbf{X}_{1::},\ldots,\mathbf{X}_{I::}]^T = (\mathbf{A}\odot\mathbf{C})\mathbf{B}^T$
	$x^{(IJ\times K)}_{(j-1)I+i,k} = x_{ijk} \Leftrightarrow \mathbf{X}^{(IJ\times K)} = [\mathbf{X}_{:1:},\ldots,\mathbf{X}_{:J:}]^T = (\mathbf{B}\odot\mathbf{A})\mathbf{C}^T$
LMV	$x^{(JK\times I)}_{(j-1)K+k,i} = x_{ijk} \Leftrightarrow \mathbf{X}^{(JK\times I)} = [\mathbf{X}^T_{:1:},\ldots,\mathbf{X}^T_{:J:}]^T = (\mathbf{B}\odot\mathbf{C})\mathbf{A}^T$
	$x^{(KI\times J)}_{(k-1)I+i,j} = x_{ijk} \Leftrightarrow \mathbf{X}^{(KI\times J)} = [\mathbf{X}^T_{::1},\ldots,\mathbf{X}^T_{::K}]^T = (\mathbf{C}\odot\mathbf{A})\mathbf{B}^T$
	$x^{(IJ\times K)}_{(i-1)J+j,k} = x_{ijk} \Leftrightarrow \mathbf{X}^{(IJ\times K)} = [\mathbf{X}^T_{1::},\ldots,\mathbf{X}^T_{I::}]^T = (\mathbf{A}\odot\mathbf{B})\mathbf{C}^T$
Kolda	$x^{(JK\times I)}_{(k-1)J+j,i} = x_{ijk} \Leftrightarrow \mathbf{X}^{(JK\times I)} = [\mathbf{X}_{::1},\ldots,\mathbf{X}_{::K}]^T = (\mathbf{C}\odot\mathbf{B})\mathbf{A}^T$
	$x^{(KI\times J)}_{(k-1)I+i,j} = x_{ijk} \Leftrightarrow \mathbf{X}^{(KI\times J)} = [\mathbf{X}^T_{::1},\ldots,\mathbf{X}^T_{::K}]^T = (\mathbf{C}\odot\mathbf{A})\mathbf{B}^T$
	$x^{(IJ\times K)}_{(j-1)I+i,k} = x_{ijk} \Leftrightarrow \mathbf{X}^{(IJ\times K)} = [\mathbf{X}^T_{:1:},\ldots,\mathbf{X}^T_{:J:}]^T = (\mathbf{B}\odot\mathbf{A})\mathbf{C}^T$

From Tables 10.4 and 10.5 it is easy to see that

$$\mathbf{X}^{(n)}_{\text{Kiers}} = \left(\mathbf{X}_{\text{Kiers}(n)}\right)^T,\quad \mathbf{X}^{(n)}_{\text{LMV}} = \left(\mathbf{X}_{\text{LMV}(n)}\right)^T,\quad \mathbf{X}^{(n)}_{\text{Kolda}} = \left(\mathbf{X}_{\text{Kolda}(n)}\right)^T.$$

Given an Nth-order tensor $\mathcal{X}\in\mathbb{R}^{I_1\times I_2\times\cdots\times I_N}$, its CP decomposition has the elementwise form

$$x_{i_1\cdots i_N} = \sum_{r=1}^{R} u^{(1)}_{i_1,r}u^{(2)}_{i_2,r}\cdots u^{(N)}_{i_N,r}, \tag{10.5.25}$$

which can be equivalently written as

$$\mathcal{X} = \sum_{r=1}^{R} \mathbf{u}_r^{(1)} \circ \mathbf{u}_r^{(2)} \circ \cdots \circ \mathbf{u}_r^{(N)}. \tag{10.5.26}$$

The CP decomposition can be also written as the *Kruskal operator form* [254]

$$\mathcal{X} = [\![\mathbf{U}^{(1)}, \ldots, \mathbf{U}^{(N)}]\!] = [\![\mathcal{I}; \mathbf{U}^{(1)}, \ldots, \mathbf{U}^{(N)}]\!]. \tag{10.5.27}$$

Clearly, CP decomposition is a special example of the Tucker decomposition when the core tensor $\mathcal{G} \in \mathbb{K}^{J_1 \times \cdots \times J_N}$ is an Nth-order unit tensor $\mathcal{I} \in \mathbb{R}^{R \times \cdots \times R}$.

Table 10.6 summarizes the mathematical representation of the CP decomposition, where $L = I_1 \cdots I_{n-1} I_{n+1} \cdots I_N$ and where $\mathbf{U}^{(n)} \in \mathbb{R}^{I_n \times R}, n = 1, \ldots, N$, are the factor matrices.

Table 10.6 Mathematical formulation of CP decomposition

Representation	Mathematical formulation
Operator form	$\mathcal{X} = [\![\mathbf{U}^{(1)}, \mathbf{U}^{(2)}, \ldots, \mathbf{U}^{(N)}]\!] = [\![\mathcal{I}; \mathbf{U}^{(1)}, \ldots, \mathbf{U}^{(N)}]\!]$
Mode-n form	$\mathcal{X} = \mathcal{I} \times_1 \mathbf{U}^{(1)} \times_2 \mathbf{U}^{(2)} \cdots \times_N \mathbf{U}^{(N)}$
Entry form	$x_{i_1 \cdots i_N} = \sum_{r=1}^{R} u_{i_1,r}^{(1)} u_{i_2,r}^{(2)} \cdots u_{i_N,r}^{(N)}$
Outer product	$\mathcal{X} = \sum_{r=1}^{R} \mathbf{u}_r^{(1)} \circ \mathbf{u}_r^{(2)} \circ \cdots \circ \mathbf{u}_r^{(N)}$
Kiers	$\begin{cases} x_{i_n j}^{(I_n \times L)} = x_{i_1 \cdots i_N} \quad (j \text{ determined by } (10.2.4)) \quad \Leftrightarrow \\ \mathbf{X}^{(I_n \times L)} = \mathbf{U}^{(n)} (\mathbf{U}^{(n-1)} \odot \cdots \odot \mathbf{U}^{(1)} \odot \mathbf{U}^{(N)} \odot \cdots \odot \mathbf{U}^{(n+1)})^T \end{cases}$
LMV	$\begin{cases} x_{i_n j}^{(I_n \times L)} = x_{i_1 \cdots i_N} \quad (j \text{ determined by } (10.2.6)) \quad \Leftrightarrow \\ \mathbf{X}^{(I_n \times L)} = \mathbf{U}^{(n)} (\mathbf{U}^{(n+1)} \odot \cdots \odot \mathbf{U}^{(N)} \odot \mathbf{U}^{(1)} \odot \cdots \odot \mathbf{U}^{(n-1)})^T \end{cases}$
Kolda	$\begin{cases} x_{i_n j}^{(I_n \times L)} = x_{i_1 \cdots i_N} \quad (j \text{ determined by } (10.2.7)) \quad \Leftrightarrow \\ \mathbf{X}^{(I_n \times L)} = \mathbf{U}^{(n)} (\mathbf{U}^{(N)} \odot \cdots \odot \mathbf{U}^{(n+1)} \odot \mathbf{U}^{(n-1)} \odot \cdots \odot \mathbf{U}^{(1)})^T \end{cases}$

Table 10.7 summarizes several variants of the CP decomposition of three-way data: PARAFAC2 is the relaxed form of the PARAFAC model, S-PARAFAC is shifted PARAFAC and cPARAFAC is convolution PARAFAC; PARALIND is a linearly dependent parallel factor analysis. All the methods listed in the table can be extended to N-way data.

In Table 10.7, the matrix $\mathbf{D}_k$ is the diagonal matrix whose diagonal entries are the entries of the kth row of the mode-C matrix $\mathbf{C}$, $\mathbf{H}$ is the interaction matrix of the factor matrices, while the matrix $\mathbf{A}_k$ is the mode-A matrix corresponding to

Table 10.7 Comparison of multi-way models [7]

Model	Mathematical formulation	Literature
PARAFAC	$x_{ijk} = \sum_{p=1}^{P}\sum_{q=1}^{Q}\sum_{r=1}^{R} g_{rrr}a_{ip}b_{jq}c_{kr}$	[87], [198]
PARAFAC2	$\mathbf{X}_{::k} = \mathbf{A}_k\mathbf{D}_k\mathbf{B}^T + \mathbf{E}_{::k}$	[199]
S-PARAFAC	$x_{ijk} = \sum_{r=1}^{R} a_{(i+s_{jr})r}b_{jr}c_{kr} + e_{ijk}$	[200]
PARALIND	$\mathbf{X}_{::k} = \mathbf{AHD}_k\mathbf{B}^T + \mathbf{E}_{::k}$	[64]
cPARAFAC	$x_{ijk} = \sum_{r=1}^{R} a_{ir}b_{(j-\theta)r}c_{kr}^{\theta} + e_{ijk}$	[334]
Tucker3	$x_{ijk} = \sum_{p=1}^{P}\sum_{q=1}^{Q}\sum_{r=1}^{R} g_{pqr}a_{ip}b_{jq}c_{kr}$	[477]
Tucker2	$x_{ijk} = \sum_{p=1}^{P}\sum_{q=1}^{Q} g_{pqk}a_{ip}b_{jq}$	[477], [479]
Tucker1	$x_{ijk} = \sum_{p=1}^{P} g_{pjk}a_{ip}$	[477], [479]
S-Tucker3	$x_{ijk} = \sum_{p=1}^{P}\sum_{q=1}^{Q}\sum_{r=1}^{R} g_{pqr}a_{(i+s_p)p}b_{jq}c_{kr}$	[200]

the front slice $\mathbf{X}_{::k}$; this mode-A matrix is required to obey the constraint $\mathbf{A}_k^T\mathbf{A}_k = \mathbf{\Phi}, k = 1,\ldots,K$, where $\mathbf{\Phi}$ is a matrix independent of $\mathbf{A}_k, k = 1,\ldots,K$.

10.5.3 Uniqueness Condition

The CP decomposition has two inherent indeterminacies: the permutation indeterminacy and the scaling indeterminacy of the factor vectors.

First, the factor vectors $(\mathbf{a}_r, \mathbf{b}_r, \mathbf{c}_r)$ and $(\mathbf{a}_p, \mathbf{b}_p, \mathbf{c}_p)$ can be exchanged; the result

$$\mathcal{X} = \sum_{r=1}^{R} \mathbf{a}_r \circ \mathbf{b}_r \circ \mathbf{c}_r$$

will remain unchanged. This permutation indeterminacy can be also represented as

$$\mathcal{X} = [\![\mathbf{A},\mathbf{B},\mathbf{C}]\!] = [\![\mathbf{AP},\mathbf{BP},\mathbf{CP}]\!], \quad \forall R\times R \text{ permutation matrices } \mathbf{P},$$

where $\mathbf{A} = [\mathbf{a}_1,\ldots,\mathbf{a}_R], \mathbf{B} = [\mathbf{b}_1,\ldots,\mathbf{b}_R]$ and $\mathbf{C} = [\mathbf{c}_1,\ldots,\mathbf{c}_R]$.

Moreover, if $\alpha_r\beta_r\gamma_r = 1$ holds for $r = 1,\ldots,R$ then

$$\mathcal{X} = \sum_{r=1}^{R}(\alpha_r\mathbf{a}_r)\circ(\beta_r\mathbf{b}_r)\circ(\gamma_r\mathbf{c}_r) = \sum_{r=1}^{R}\mathbf{a}_r\circ\mathbf{b}_r\circ\mathbf{c}_r.$$

In other words, the CP decomposition of a tensor is blind to the scale (i.e., the norm) of the factor vectors.

The permutation indeterminacy and the scaling indeterminacy of the CP decomposition do not affect the multilinear analysis of a tensor, and therefore are allowable. Apart from these two inherent indeterminacies, the uniqueness of the CP decomposition is that in the condition of the sum of all decomposition components is equal to the original tensor, the possible combination of the rank-1 tensors is unique.

The CP decomposition still has uniqueness under a more relaxed condition, in fact. This relaxed condition is related to the Kruskal rank.

Definition 10.10 [267] The *Kruskal rank* of a matrix $\mathbf{A} \in \mathbb{R}^{I\times J}$, simply called the k-rank, is denoted $k_{\mathbf{A}} = \text{rank}_k(\mathbf{A}) = r$ and is equal to r, where r is the maximum integer such that any r column vectors of $\mathbf{A}$ are linearly independent.

Because the matrix rank, $\text{rank}(\mathbf{A}) = r$, requires only that r is the maximum number of columns such that a group of column vectors is linearly independent, whereas the Kruskal rank requires that r is the maximum number of columns such that every group of r column vectors is linearly independent, the Kruskal rank is always less than or equal to the rank of $\mathbf{A}$, i.e., $\text{rank}_k(\mathbf{A}) \leq \text{rank}(\mathbf{A}) \leq \min\{I, J\}, \forall\, \mathbf{A} \in \mathbb{R}^{I\times J}$.

The sufficient condition for CP decomposition to have uniqueness, proposed by Kruskal in 1977 [267], is given by

$$k_{\mathbf{A}} + k_{\mathbf{B}} + k_{\mathbf{C}} \geq 2R + 2. \tag{10.5.28}$$

Berge and Sidiropoulos [37] showed, for $R = 2$ and $R = 3$, that Kruskal's above sufficient condition is not only a sufficient condition, but also a necessary condition of CP decomposition of a tensor. However, this conclusion is not true for the case $R > 3$.

Sidiropoulos and Bro [442] generalized Kruskal's sufficient condition, and showed that if the rank of an N-way tensor $\mathcal{X} \in \mathbb{R}^{I_1\times\cdots\times I_N}$ is R, and the CP decomposition of the tensor is

$$\mathcal{X} = \sum_{r=1}^{R} \mathbf{a}_r^{(1)} \circ \mathbf{a}_r^{(2)} \circ \cdots \circ \mathbf{a}_r^{(N)}, \tag{10.5.29}$$

then the sufficient condition for unique decomposition is given by

$$\sum_{n=1}^{N} \text{rank}_k(\mathbf{X}_{(n)}) \geq 2R + (N-1), \tag{10.5.30}$$

where $\text{rank}_k(\mathbf{X}_{(n)})$ is the Kruskal rank of the mode-n matricization $\mathbf{X}_{(n)} = \mathbf{X}^{(I_n\times L)}$ of the tensor $\mathcal{X}$.

From the longitudinal unfoldings $\mathbf{X}^{(JK\times I)} = (\mathbf{B}\odot\mathbf{C})\mathbf{A}^T, \mathbf{X}^{(KI\times J)} = (\mathbf{C}\odot\mathbf{A})\mathbf{B}^T$

and $\mathbf{X}^{(IJ\times K)} = (\mathbf{A} \odot \mathbf{B})\mathbf{C}^T$, Liu and Sidiropoulos [302] showed that the necessary condition for the CP decomposition of a third-order tensor is

$$\min\{\text{rank}(\mathbf{A} \odot \mathbf{B}), \text{rank}(\mathbf{B} \odot \mathbf{C}), \text{rank}(\mathbf{C} \odot \mathbf{A})\} = R, \tag{10.5.31}$$

and the necessary condition for the uniqueness for the CP decomposition of an Nth-order tensor is

$$\min_{n=1,\ldots,N} \text{rank}\,(\mathbf{A}_1 \odot \cdots \odot \mathbf{A}_{n-1} \odot \mathbf{A}_{n+1} \odot \cdots \odot \mathbf{A}_N) = R, \tag{10.5.32}$$

where $\mathbf{A}_n, n = 1, \ldots, N$ are the component matrices satisfying the longitudinal unfolding

$$\mathbf{X}^{((I_2\times\cdots\times I_N)\times I_1)} = (\mathbf{A}_N \odot \cdots \odot \mathbf{A}_3 \odot \mathbf{A}_2)\mathbf{A}_1^T. \tag{10.5.33}$$

Because $\text{rank}(\mathbf{A} \odot \mathbf{B}) \leq \text{rank}(\mathbf{A} \otimes \mathbf{B}) \leq \text{rank}(\mathbf{A})\text{rank}(\mathbf{B})$, a simpler necessary condition is [302]

$$\min_{n=1,\ldots,N} \left\{ \prod_{m=1, m\neq n}^{N} \text{rank}(\mathbf{A}_m) \right\} \geq R. \tag{10.5.34}$$

Lathauwer [279] showed that the CP decomposition of a third-order tensor $\mathcal{A} \in \mathbb{R}^{I\times J\times K}$ is generally unique if $R \leq K$ and

$$R(R-1) \leq I(I-1)J(J-1)/2. \tag{10.5.35}$$

By [257], the CP decomposition of a fourth-order tensor $\mathcal{A} \in \mathbb{R}^{I\times J\times K\times L}$ with rank R is generally unique if $R \leq L$ and

$$R(R-1) \leq IJK(3IJK - IJ - IK - JK - I - J - K + 3)/4. \tag{10.5.36}$$

10.5.4 Alternating Least Squares Algorithm

Given a third-order tensor $\mathcal{X}$, consider the separated optimization problems for the Kiers horizontal unfolding of the CP decomposition of $\mathcal{X}$:

$$\mathbf{A} = \arg\min_{\mathbf{A}} \left\|\mathbf{X}^{(I\times JK)} - \mathbf{A}(\mathbf{C} \odot \mathbf{B})^T\right\|_F^2, \tag{10.5.37}$$

$$\mathbf{B} = \arg\min_{\mathbf{B}} \left\|\mathbf{X}^{(J\times KI)} - \mathbf{B}(\mathbf{A} \odot \mathbf{C})^T\right\|_F^2, \tag{10.5.38}$$

$$\mathbf{C} = \arg\min_{\mathbf{C}} \left\|\mathbf{X}^{(K\times IJ)} - \mathbf{C}(\mathbf{B} \odot \mathbf{A})^T\right\|_F^2. \tag{10.5.39}$$

Their LS solutions are respectively given by

$$\mathbf{A} = \mathbf{X}^{(I\times JK)}\left((\mathbf{C} \odot \mathbf{B})^T\right)^\dagger, \tag{10.5.40}$$

$$\mathbf{B} = \mathbf{X}^{(J\times KI)}\left((\mathbf{A} \odot \mathbf{C})^T\right)^\dagger, \tag{10.5.41}$$

$$\mathbf{C} = \mathbf{X}^{(K\times IJ)}\left((\mathbf{B} \odot \mathbf{A})^T\right)^\dagger. \tag{10.5.42}$$

Using the property of the Moore–Penrose inverse of the Khatri–Rao product

$$(\mathbf{C} \odot \mathbf{B})^{\dagger} = (\mathbf{C}^T\mathbf{C} * \mathbf{B}^T\mathbf{B})^{\dagger}(\mathbf{C} \odot \mathbf{B})^T, \tag{10.5.43}$$

we obtain the solution of the factor matrix $\mathbf{A}$ as follows:

$$\mathbf{A} = \mathbf{X}^{(I\times JK)}(\mathbf{C} \odot \mathbf{B})(\mathbf{C}^T\mathbf{C} * \mathbf{B}^T\mathbf{B})^{\dagger}. \tag{10.5.44}$$

Mimicking this operation, we also obtain the LS solutions for the mode-B matrix and for mode-C matrix:

$$\mathbf{B} = \mathbf{X}^{(J\times KI)}(\mathbf{A} \odot \mathbf{C})(\mathbf{A}^T\mathbf{A} * \mathbf{C}^T\mathbf{C})^{\dagger}, \tag{10.5.45}$$

$$\mathbf{C} = \mathbf{X}^{(K\times IJ)}(\mathbf{B} \odot \mathbf{A})(\mathbf{B}^T\mathbf{B} * \mathbf{A}^T\mathbf{A})^{\dagger}. \tag{10.5.46}$$

Equations (10.5.44)–(10.5.46) give an ALS algorithm for CP decomposition shown in Algorithm 10.4 [65], [7].

Algorithm 10.4 CP–ALS algorithm via Kiers horizontal unfolding

---

input: Kiers unfolding matrices $\mathbf{X}^{(I\times JK)}, \mathbf{X}^{(J\times KI)}, \mathbf{X}^{(K\times IJ)}$ of $\mathcal{X}$.

initialize: $\mathbf{B}_0$,$\mathbf{C}_0$ and the number of factors R. Put $k = 0$.

repeat

1. $\mathbf{A}_{k+1} \leftarrow \mathbf{X}^{(I\times JK)}(\mathbf{C}_k \odot \mathbf{B}_k)\big(\mathbf{C}_k^T\mathbf{C}_k * \mathbf{B}_k^T\mathbf{B}_k\big)^{\dagger}$.
2. $\mathbf{B}_{k+1} \leftarrow \mathbf{X}^{(J\times KI)}(\mathbf{A}_{k+1} \odot \mathbf{C}_k)\big(\mathbf{A}_{k+1}^T\mathbf{A}_{k+1} * \mathbf{C}_k^T\mathbf{C}_k\big)^{\dagger}$.
3. $\mathbf{C}_{k+1} \leftarrow \mathbf{X}^{(K\times IJ)}(\mathbf{B}_{k+1} \odot \mathbf{A}_{k+1})\big(\mathbf{B}_{k+1}^T\mathbf{B}_{k+1} * \mathbf{A}_{k+1}^T\mathbf{A}_{k+1}\big)^{\dagger}$.
4. **exit if** $\big\|\mathbf{X}^{(I\times JK)} - \mathbf{A}_{k+1}(\mathbf{C}_{k+1} \odot \mathbf{B}_{k+1})^T\big\|_F^2 \leq \epsilon$.

return $k \leftarrow k+1$.

output: Factor matrices $\mathbf{A} \in \mathbb{R}^{I\times R}, \mathbf{B} \in \mathbb{R}^{J\times R}, \mathbf{C} \in \mathbb{R}^{K\times R}$.

Similarly, one has an ALS algorithm based on the LMV longitudinal unfolding for the CP decomposition of a third-order tensor, as given in Algorithm 10.5 [65], [282].

Algorithm 10.6 is an ALS algorithm for the CP decomposition of an Nth-order tensor via the Kolda horizontal unfolding.

The main advantage of ALS algorithms is that they are simple and easily implemented, but the main disadvantage is that most ALS algorithms converge slowly and may even be divergent. To cope with these disadvantages, it is necessary to regularize the cost function of an ALS algorithm. The following optimization problems give regularized ALS iterations for the CP decomposition of a third-order tensor.

Algorithm 10.5 CP–ALS algorithm via LMV longitudinal unfolding

input: LMV longitudinal unfolding matrices $\mathbf{X}^{(JK\times I)}, \mathbf{X}^{(KI\times J)}, \mathbf{X}^{(IJ\times K)}$ of $\mathcal{X}$.

initialize: $\mathbf{B}_0$, $\mathbf{C}_0$ and the number of factors R. Put $k = 0$.

repeat

1. $\mathbf{A}_{k+1} \leftarrow \big((\mathbf{B}_k \odot \mathbf{C}_k)^\dagger \mathbf{X}^{(JK\times I)}\big)^T$.
2. $\tilde{\mathbf{B}} \leftarrow \big((\mathbf{C}_k \odot \mathbf{A}_{k+1})^\dagger \mathbf{X}^{(KI\times J)}\big)^T$.
3. $[\mathbf{B}_{k+1}]_{:,r} = \tilde{\mathbf{B}}_{:,r}/\|\tilde{\mathbf{B}}_{:,r}\|_2,\ r = 1,\ldots,R$.
4. $\tilde{\mathbf{C}} \leftarrow \big((\mathbf{A}_{k+1} \odot \mathbf{B}_{k+1})^\dagger \mathbf{X}^{(IJ\times K)}\big)^T$.
5. $[\mathbf{C}_{k+1}]_{:,r} = \tilde{\mathbf{C}}_{:,r}/\|\tilde{\mathbf{C}}_{:,r}\|_2,\ r = 1,\ldots,R$.
6. **exit if** $\|\mathbf{X}^{(JK\times I)} - (\mathbf{B} \odot \mathbf{C})\mathbf{A}^T\|_F^2 \leq \epsilon$.

return $k \leftarrow k+1$.

output: The factor matrices $\mathbf{A} \in \mathbb{R}^{I\times R}, \mathbf{B} \in \mathbb{R}^{J\times R}, \mathbf{C} \in \mathbb{R}^{K\times R}$.

Algorithm 10.6 CP–ALS $(\mathcal{X}, R)$ [257]

input: Kolda horizontal unfolding matrices $\mathbf{X}_{(n)}, n = 1,\ldots,N$ and the number of factor R.

initialize: $\mathbf{A}_n \in \mathbb{R}^{I_n\times R}, n = 1,\ldots,N$. Put $k = 0$.

repeat

for $n = 1,\ldots,N$ **do**

1. $\mathbf{V} \leftarrow \mathbf{A}_1^T\mathbf{A}_1 * \cdots * \mathbf{A}_{n-1}^T\mathbf{A}_{n-1} * \mathbf{A}_{n+1}^T\mathbf{A}_{n+1} * \cdots * \mathbf{A}_N^T\mathbf{A}_N$.
2. $\mathbf{A}_n \leftarrow \mathbf{X}_{(n)}(\mathbf{A}_N \odot \cdots \odot \mathbf{A}_{n+1} \odot \mathbf{A}_{n-1} \odot \cdots \odot \mathbf{A}_1)\mathbf{V}^\dagger$.
3. $\lambda_n \leftarrow \|\mathbf{A}_n\|$.
4. $\mathbf{A}_n^{(k)} \leftarrow \mathbf{A}_n/\lambda_n$.

end for

exit if $\sum_{n=1}^N \|\mathbf{A}_n^{(k+1)} - \mathbf{A}_n^{(k)}\|_F \leq \epsilon$.

return $k \leftarrow k+1$.

output: $\boldsymbol{\lambda} = [\lambda_1,\ldots,\lambda_N]^T$ and $\mathbf{A}_n \in \mathbb{R}^{I_n\times R}, n = 1,\ldots,N$.

(1) Kolda horizontal unfolding [293]

$$\mathbf{A}_{k+1} = \arg\min_{\mathbf{A}} \left\{ \left\|\mathbf{X}^{(I\times JK)} - \mathbf{A}(\mathbf{C}_k \odot \mathbf{B}_k)^T\right\|_F^2 + \tau_k\|\mathbf{A} - \mathbf{A}_k\|_F^2 \right\},$$

$$\mathbf{B}_{k+1} = \arg\min_{\mathbf{B}} \left\{ \left\|\mathbf{X}^{(J\times KI)} - \mathbf{B}(\mathbf{C}_k \odot \mathbf{A}_{k+1})^T\right\|_F^2 + \tau_k\|\mathbf{B} - \mathbf{B}_k\|_F^2 \right\},$$

$$\mathbf{C}_{k+1} = \arg\min_{\mathbf{C}} \left\{ \left\|\mathbf{X}^{(K\times IJ)} - \mathbf{C}(\mathbf{B}_{k+1} \odot \mathbf{A}_{k+1})^T\right\|_F^2 + \tau_k\|\mathbf{C} - \mathbf{C}_k\|_F^2 \right\}.$$

(2) LMV longitudinal unfolding [338]

$$\mathbf{A}_{k+1}^T = \arg\min_{\mathbf{A}} \left\{ \left\| \mathbf{X}^{(JK\times I)} - (\mathbf{B}_k \odot \mathbf{C}_k)\mathbf{A}^T \right\|_F^2 + \tau_k \left\| \mathbf{A}^T - \mathbf{A}_k^T \right\|_F^2 \right\},$$

$$\mathbf{B}_{k+1}^T = \arg\min_{\mathbf{B}} \left\{ \left\| \mathbf{X}^{(KI\times J)} - (\mathbf{C}_k \odot \mathbf{A}_{k+1})\mathbf{B}^T \right\|_F^2 + \tau_k \left\| \mathbf{B}^T - \mathbf{B}_k^T \right\|_F^2 \right\},$$

$$\mathbf{C}_{k+1}^T = \arg\min_{\mathbf{C}} \left\{ \left\| \mathbf{X}^{(IJ\times K)} - (\mathbf{A}_{k+1} \odot \mathbf{B}_{k+1})\mathbf{C}^T \right\|_F^2 + \tau_k \left\| \mathbf{C}^T - \mathbf{C}_k^T \right\|_F^2 \right\}.$$

With the updating of the factor matrix $\mathbf{A}$ in the Kolda horizontal unfolding as an example, the regularization term $\|\mathbf{A} - \mathbf{A}_k\|_F^2$ can force the updated matrix $\mathbf{A}$ not to deviate from $\mathbf{A}_k$ too much, so as to avoid the divergence of the iterative process.

Finding the gradient matrix of the objective function with respect to the argument matrix $\mathbf{A}$ in the suboptimization problem, and then letting the gradient matrix equal the zero matrix, we obtain

$$\left((\mathbf{C}_k \odot \mathbf{B}_k)^T(\mathbf{C}_k \odot \mathbf{B}_k) + \tau_k \mathbf{I}\right)\mathbf{A}^T = (\mathbf{C}_k \odot \mathbf{B}_k)^T(\mathbf{X}^{I\times JK})^T + \tau_k(\mathbf{A}_k)^T.$$

From this one can deduce a regularized LS solution of the suboptimization problem as follows:

$$\begin{aligned}\mathbf{A} &= \left(\mathbf{X}^{I\times JK}(\mathbf{C}_k \odot \mathbf{B}_k) + \tau_k \mathbf{A}_k\right)\left((\mathbf{C}_k \odot \mathbf{B}_k)^T(\mathbf{C}_k \odot \mathbf{B}_k) + \tau_k \mathbf{I}\right)^{-1}\\ &= \left(\mathbf{X}^{I\times JK}(\mathbf{C}_k \odot \mathbf{B}_k) + \tau_k \mathbf{A}_k\right)\left(\mathbf{C}_k^T\mathbf{C}_k * \mathbf{B}_k^T\mathbf{B}_k + \tau_k \mathbf{I}\right)^{-1}.\end{aligned}$$

Similarly, one has also

$$\mathbf{B} = \left(\mathbf{X}^{(J\times KI)}(\mathbf{C}_k \odot \mathbf{A}_{k+1}) + \tau_k \mathbf{B}_k\right)\left(\mathbf{C}_k^T\mathbf{C}_k * \mathbf{A}_{k+1}^T\mathbf{A}_{k+1} + \tau_i \mathbf{I}\right)^{-1},$$

$$\mathbf{C} = \left(\mathbf{X}^{(K\times IJ)}(\mathbf{B}_{k+1} \odot \mathbf{A}_{k+1}) + \tau_k \mathbf{C}_k\right)\left(\mathbf{B}_{k+1}^T\mathbf{B}_{k+1} * \mathbf{A}_{k+1}^T\mathbf{A}_{k+1} + \tau_k \mathbf{I}\right)^{-1}.$$

Algorithm 10.7 gives a regularized ALS algorithm for the Kolda horizontal unfolding of the CP decomposition of a third-order tensor.

If $\tau_k \equiv 0, \forall\, k$, then Algorithm 10.7 reduces to the ordinary ALS algorithm. It can be seen that, when updating the factor matrix $\mathbf{A}$, the improvement of the regularized ALS algorithm over the ALS algorithm is mainly reflected in the following two aspects.

(1) The numerical stability problem caused by the possible singularity of $\mathbf{C}_k^T\mathbf{C}_k * \mathbf{B}_k^T\mathbf{B}_k$ is avoided by the use of $\left(\mathbf{C}_k^T\mathbf{C}_k * \mathbf{B}_k^T\mathbf{B}_k + \tau_k\mathbf{I}\right)^{-1}$ instead of $(\mathbf{C}_k^T\mathbf{C}_k * \mathbf{B}_k^T\mathbf{B}_k)^\dagger$.

(2) The sudden jump of $\mathbf{A}_{k+1}$ is avoided by the use of $\mathbf{X}^{I\times JK}(\mathbf{C}_k \odot \mathbf{B}_k) + \tau_k\mathbf{A}_k$ instead of $\mathbf{X}^{I\times JK}(\mathbf{C}_k \odot \mathbf{B}_k)$ because, through the weighting coefficient τ_k, the update of $\mathbf{A}_{k+1}$ is weakly correlated with $\mathbf{A}_k$.

It should be pointed out that if the regularization term is $\tau_k\|\mathbf{A}\|_F^2$ rather than

Algorithm 10.7 Regularized ALS algorithm CP–RALS $(\mathcal{X}, R, N, \lambda)$ [293]

input: Kolda horizontal unfolding matrices $\mathbf{X}^{(I\times JK)}, \mathbf{X}^{(J\times KI)}, \mathbf{X}^{(K\times IJ)}$ and the number R of factors.

initialize: $\mathbf{A}_0 \in \mathbb{R}^{I\times R}, \mathbf{B}_0 \in \mathbb{R}^{J\times R}, \mathbf{C}_0 \in \mathbb{R}^{K\times R}, \tau_0 = 1$. Put $k = 0$.

repeat

1. $\mathbf{W} \leftarrow \mathbf{X}^{I\times JK}(\mathbf{C}_k \odot \mathbf{B}_k) + \tau_k \mathbf{A}_k$.
2. $\mathbf{S} \leftarrow \mathbf{C}_k^T\mathbf{C}_k * \mathbf{B}_k^T\mathbf{B}_k + \tau_k\mathbf{I}$.
3. $\mathbf{A}_{k+1} \leftarrow \mathbf{W}\mathbf{S}^{-1}$.
4. $\mathbf{W} \leftarrow \mathbf{X}^{(J\times KI)}(\mathbf{C}_k \odot \mathbf{A}_{k+1}) + \tau_k \mathbf{B}_k$.
5. $\mathbf{S} \leftarrow \mathbf{C}_k^T\mathbf{C}_k * \mathbf{A}_{k+1}^T\mathbf{A}_{k+1} + \tau_i\mathbf{I}$.
6. $\mathbf{B}_{k+1} \leftarrow \mathbf{W}\mathbf{S}^{-1}$.
7. $\mathbf{W} \leftarrow \mathbf{X}^{(K\times IJ)}(\mathbf{B}_{k+1} \odot \mathbf{A}_{k+1}) + \tau_k \mathbf{C}_k$.
8. $\mathbf{S} \leftarrow \mathbf{B}_{k+1}^T\mathbf{B}_{k+1} * \mathbf{A}_{k+1}^T\mathbf{A}_{k+1} + \tau_k\mathbf{I}$.
9. $\mathbf{C}_{k+1} \leftarrow \mathbf{W}\mathbf{S}^{-1}$.
10. $\tau_{k+1} \leftarrow \delta \cdot \tau_k$.
11. **exit if** $\|\mathbf{A}_{k+1} - \mathbf{A}_k\|_F + \|\mathbf{B}_{k+1} - \mathbf{B}_k\|_F + \|\mathbf{C}_{k+1} - \mathbf{C}_k\|_F \leq \epsilon$.

return $k \leftarrow k + 1$.

output: $\mathbf{A} \in \mathbb{R}^{I\times R}, \mathbf{B} \in \mathbb{R}^{J\times R}, \mathbf{C} \in \mathbb{R}^{K\times R}$.

$\tau_k\|\mathbf{A}-\mathbf{A}_k\|_F^2$ then the regularized ALS algorithm only has advantage (1) and cannot have advantage (2).

Regarding CP decompositions, in practice, if a tensor has linear dependence among the components of factor matrices, or large differences in magnitude between the components [366], [468], then only a few algorithms can accurately retrieve the hidden components of the tensor [468], [389]. This involves the stability of the CP tensor decomposition.

By the stability of the CP tensor decomposition is meant the existence of a finite Cramer–Rao bound in a stochastic set-up where tensor entries are corrupted by additive Gaussian-distributed noises. On Cramer–Rao-induced bounds for CANDECOMP/PARAFAC tensor decompositions, the interested reader may refer to [463].

10.6 Applications of Low-Rank Tensor Decomposition

Without doubt this is the era of "big data". Technological advances in data acquisition are spurring on the generation of unprecedentedily massive data sets, thus posing a permanent challenge to data analysts [238]. Many such big data are expressed in tensor form. It is a key problem to extract useful information from big tensor data. Perhaps the most simple, yet useful, approach for solving this key problem is low-rank tensor decomposition. This section presents typical applications of low-rank tensor decomposition in various disciplines.

10.6.1 Multimodal Data Fusion

In various disciplines, information about the same phenomenon can be acquired from different types of detectors, at different conditions, in multiple experiments or from multiple subjects. Each such acquisition framework is known as a *modality* [272], and the existence of several such acquisition frameworks is called *multimodality*.

Definition 10.11 [193] *Data fusion* is the process integrating multiple data and knowledge representing the same real-world object into a consistent, accurate and useful representation.

The key to data fusion is diversity.

Definition 10.12 [272] *Diversity* (due to multimodality) is the property that allows one to enhance the uses, benefits and insights of data in a way that cannot be achieved with a single modality.

The following are some practical examples of diversity and multimodality [272].

1. *Audiovisual multimodality* A large number of audiovisual applications involve human speech and vision. In such applications, it is usually an audio channel that conveys the information of interest. Audiovisual multimodality is used for a broad range of applications, such as speech processing (including speech recognition, speech activity detection, speech enhancement, speaker extraction and separation), scene analysis (including tracking a speaker within a group), biometrics and monitoring for safety and security applications [28] and more.
2. *Human–machine interaction* (HMI) An important task of HMI is to design modalities that will make HMI as natural, efficient, and intuitive as possible [478]. The idea is to combine multiple interaction modes based on audiovision, touch, smell, movement (e.g., gesture detection and user tracking), interpretation of human language commands, and other multisensory functions [441], [478].
3. *Remote sensing and earth observations* Sensor technologies in remote sensing and earth observation are divided into passive and active sensors. Optical hyperspectral and multispectral imaging are passive, and they rely on natural illumination. Typical active sensors have light detection and ranging (LiDAR) and synthetic aperture radar (SAR), and they can operate at night and in shadowed areas [38]. Both technologies can provide information about elevation, the three-dimensional structure of the observed objects and their surface properties.

4. *Meteorological monitoring* Accurate measurements of atmospheric phenomena such as rain, water vapor, dew, fog and snow are required for meteorological analysis and forecasting, as well as for numerous applications in hydrology, agriculture and aeronautical services.
5. *Biomedicine and health* This includes understanding brain functionality, medical diagnosis and smart patient-monitoring.

Data fusion can be of two types: model-driven and data-driven. A *model-driven approach* relies on an explicit realistic model of the underlying processes [113], [49]. A *data-driven approach* relies only on the observations and an assumed model for them: it avoids external input [198]. Data-driven approaches are usually an extension of matrix-based approaches.

Given a point x in measurement space, consider an analytical model of x given by

$$x = f(\mathbf{z}), \tag{10.6.1}$$

where f is an intermediate transform with unknown elements $\mathbf{z} = \{z_1, \ldots, z_p\}$. The unknown elements of the model are known as *latent variables.*

The goal of data fusion is to estimate values of the latent variables $z_1, \ldots, z_p$ as precisely as possible by combining relevant information from two or more data sources.

A key property in any analytical model is the uniqueness of the estimated latent variables.

1. Diversity in Matrix Decomposition Models

When x in (10.6.1) is the (i, j)th entry of a data matrix $\mathbf{X} = [x_{ij}] \in \mathbb{K}^{I\times J}$, perhaps the simplest yet most useful implementation of (10.6.1) is

$$x_{ij} = \sum_{r=1}^{R} a_{ir} b_{jr}, \quad i = 1, \ldots, I,\ j = 1, \ldots, J, \tag{10.6.2}$$

which can be written in the matrix decomposition form

$$\mathbf{X} = \sum_{r=1}^{R} \mathbf{a}_i \mathbf{b}_j^T = \mathbf{A}\mathbf{B}^T, \tag{10.6.3}$$

where $\mathbf{A} = [\mathbf{a}_1, \ldots, \mathbf{a}_r] \in \mathbb{K}^{I\times R}$ with $\mathbf{a}_r = [a_{1r}, \ldots, a_{Ir}]^T$ and $\mathbf{B} = [\mathbf{b}_1, \ldots, \mathbf{b}_r] \in \mathbb{K}^{J\times R}$ with $\mathbf{b}_r = [b_{1r}, \ldots, b_{Jr}]^T$.

The model in (10.6.3) provides I linear combinations of the columns of $\mathbf{B}$ and J linear combinations of the columns of $\mathbf{A}$. By [443], $\mathbf{X}$ provides I-fold diversity for $\mathbf{B}$ and J-fold diversity for $\mathbf{A}$. However, the underlying factor matrices $\mathbf{A}$ and $\mathbf{B}$ are not uniquely determined, because for any invertible matrix $\mathbf{T} \in \mathbb{R}^{R\times R}$, it always holds that

$$\mathbf{X} = \mathbf{A}\mathbf{B}^T = (\mathbf{A}\mathbf{T})(\mathbf{T}^{-1}\mathbf{B}). \tag{10.6.4}$$

2. Diversity in Tensor Decomposition Models

When x in (10.6.1) is the (i, j, k)th entry of a data tensor $\mathcal{X} \in \mathbb{K}^{I\times J\times K}$, a simple yet efficient implementation of (10.6.1) is the CP decomposition given by

$$x_{ijk} = \sum_{r=1}^{R} a_{ir}b_{jr}c_{kr}, \quad i = 1,\ldots,I,\ j = 1,\ldots,J,\ k = 1,\ldots,K, \tag{10.6.5}$$

or written as the third-order tensor form

$$\mathcal{X} = \sum_{r=1}^{R} \mathbf{a}_r \circ \mathbf{b}_r \circ \mathbf{c}_r, \tag{10.6.6}$$

where $\mathbf{a}_r = [a_{1r},\ldots,a_{Ir}]^T$, $\mathbf{b}_r = [b_{1r},\ldots,b_{Jr}]^T$ and $\mathbf{c}_r = [c_{1r},\ldots,c_{Kr}]^T$.

The third-order tensor given by (10.6.6) provides three modes of linear diversity: JK linear combinations of the columns of $\mathbf{A}$, IK of the columns of $\mathbf{B}$ and IJ of the columns of $\mathbf{C}$ [443].

Equation (10.6.6) is easily extended to the Nth-order tensor model

$$\mathcal{X} = \sum_{r=1}^{R} \mathbf{u}_r^{(1)} \circ \mathbf{u}_r^{(2)} \circ \cdots \circ \mathbf{u}_r^{(N)}, \tag{10.6.7}$$

which provides N modes of linear diversity.

Unlike the matrix decomposition, which, as explained above, does not have uniqueness, the CP decomposition (10.6.6) is inherently "essentially unique" up to a scaled permutation matrix.

By [272], if a model has a unique decomposition, whether a structural constraint of factor matrices or an observational constraint (i.e., any nondegenerate mode of a higher-order tensor), it can be regarded as a node of "diversity" since it is "disjoint" from other such modes. In particular, each observational mode in the Nth-order tensor (10.6.7) is a mode of diversity.

10.6.2 Fusion of Multimodal Brain Images

The human brain is today perhaps the most challenging biological object under study. The reason is that the identification of the brain structures involved in cognitive processes would not only yield essential understanding about the human condition, but would also provide leverage to help deal with the staggering global burden of disease due to brain disorders [507], [238].

It is assumed for electroencephalography (EEG) and functional magnetic resonance imaging (fMRI) recorded concurrently that the time samples for the fMRI are obtained at integer multiples of those for EEG.

Consider PARAFAC in the context of time and frequency decompositions of EEG, which naturally lead to three-way tensors $\mathcal{S}$. Calculating the wavelet or Gabor spectrum for each channel $\mathbf{V}(i, :), i = 1,\ldots, I_E$, yields a matrix $\mathcal{S}_T(i, :, :)$ for given time frames and frequencies. The spectra for all channels may be shaped into a three-way tensor $\mathcal{S}_T \in \mathbb{R}^{I_E \times I_{T\delta} \times I_{F\delta}}$, where I_E is the number of sensors, $I_{T\delta}$ is the number of time samples and $I_{F\delta}$ is the number of frequency samples.

Define the following signature matrices of EEG:

$\mathbf{K} \in \mathbb{R}^{I_E \times I_{Cx}}$: EEG lead-field matrix,

$\mathbf{G} \in \mathbb{R}^{I_{Cx} \times I_T}$: primary current density matrix (generators) of EEG,

$\mathbf{M}_V \in \mathbb{R}^{I_E \times R}$: spectral signature of EEG atoms over electrodes,

$\mathbf{M}_G \in \mathbb{R}^{I_{Cx} \times R}$: spectral signature of EEG atoms over the cortical grid,

$\mathbf{T}_V \in \mathbb{R}^{I_{T\delta} \times R}$: total temporal signature of EEG atoms,

$\mathbf{F}_V \in \mathbb{R}^{I_{F\delta} \times R}$: total spectral signature of EEG atoms.

In PARAFAC, $\mathcal{S}_T$ is decomposed into R components, which can be expressed in tensor notation as [238]

$$\mathcal{S}_T = \mathbf{M}_V \times_R \mathbf{T}_V \times_R \mathbf{F}_V + \mathcal{E} = [\![\mathbf{M}_V, \mathbf{T}_V, \mathbf{F}_V]\!] + \mathcal{E} \tag{10.6.8}$$

or in scalar notation as

$$\mathcal{S}_T(i_E, i_{T\delta}, i_{F\delta}) = \sum_{r=1}^{R} \mathbf{M}_V(i_E, r)\mathbf{T}_V(i_{T\delta}, r)\mathbf{F}_V(i_{F\delta}, r) + \mathcal{E}(i_E, i_{T\delta}, i_{F\delta}), \tag{10.6.9}$$

where $i_E, i_{T\delta}$ and $i_{F\delta}$ are indices for space, time and frequency, respectively, and $\mathcal{E}$ denotes the noise. Moreover, $\mathbf{M}_V(:,r), \mathbf{T}_V(:,r), \mathbf{F}_V(:,r)$ are respectively the spatial, temporal and spectral signatures for atom r.

By the Lagrange multiplier method, one has

$$\begin{aligned}&(\hat{\mathbf{M}}_V, \hat{\mathbf{T}}_V, \hat{\mathbf{F}}_V)\\ &= \underset{\mathbf{M}_V, \mathbf{T}_V, \mathbf{F}_V}{\arg\min} \left\{ \frac{1}{2}\|\mathbf{S}_T - [\![\mathbf{M}_V, \mathbf{T}_V, \mathbf{F}_V]\!]\|_F^2 + \lambda_1 \|\mathbf{M}_V\|_F^2 + \lambda_2 \|\mathbf{L}\mathbf{M}_V\|_F^2 \right\}\end{aligned} \tag{10.6.10}$$

subject to

$$\mathbf{M}_V^T \mathbf{M}_V = \mathbf{I}, \quad \mathbf{M}_V \geq \mathbf{O}, \quad \mathbf{F}_V \geq \mathbf{O}, \tag{10.6.11}$$

with Laplacian matrix $\mathbf{L} \in \mathbb{R}^{I_E \times I_E}$.

Integrating PARAFAC and STONNICA (spatio-temporal orthogonal nonnegative independent component analysis), the model (10.6.8) is easily combined with EEG source estimation as follows [238]:

$$\mathcal{S}_T = [\![\mathbf{K}\mathbf{M}_G, \mathbf{T}_V, \mathbf{F}_V]\!] + \mathcal{E} \tag{10.6.12}$$

whose solution is given by

$$\begin{aligned}&(\hat{\mathbf{M}}_G, \hat{\mathbf{T}}_V, \hat{\mathbf{F}}_V)\\ &= \underset{\mathbf{M}_G, \mathbf{T}_V, \mathbf{F}_V}{\arg\min} \left\{ \frac{1}{2}\|\mathbf{S}_T - [\![\mathbf{K}\mathbf{M}_G, \mathbf{T}_V, \mathbf{F}_V]\!]\|_F^2 + \lambda_1 \|\mathbf{M}_G\|_F^2 + \lambda_2 \|\mathbf{L}_G\mathbf{M}_G\|_F^2 \right\}\end{aligned} \tag{10.6.13}$$

subject to

$$\mathbf{M}_G^T\mathbf{M}_G = \mathbf{I}, \quad \mathbf{M}_G \geq \mathbf{O}, \quad \mathbf{F}_V \geq \mathbf{O} \tag{10.6.14}$$

with the Laplacian matrix $\mathbf{L}_G \in \mathbb{R}^{I_{Cx}\times I_{Cx}}$; $\mathbf{F}_V$ is the spectral signature of the generator sources of EEG.

The coupled optimization problem (10.6.13) can be solved by using the alternating nonnegative least squares (ANLS) method; see Section 10.9.

10.6.3 Process Monitoring

Image data are being increasingly used for face recognition, video surveillance, monitoring of manufacturing processes etc. One of the main challenges in analyzing time-ordered image data is the high-dimensionality and complex correlation structure (i.e., temporal, spatial, and spectral correlations) of such data. Thus, effective dimension reduction and low-rank representation of images are essential for better face recognition, video surveillance and process monitoring [521].

As a low-rank decomposition method, the PCA has been traditionally used for feature extraction and the analysis of high-dimensional data such as nonlinear profiles and images. However, the regular PCA is available only for vectors and matrices and cannot be applied to color images, which are characterized by tensors. This subsection addresses extended PCA-based low-rank tensor recovery for image data.

The following are various extended PCA approaches.

1. *Unfold Principal Component Analysis* (UPCA) [521]

So-called *unfold principal component analysis* (UPCA) is an approach applying PCA to an unfolded tensor.

Let the tensor $\mathcal{X} \in \mathbb{R}^{I_1\times I_2\times I_3\times N}$ denote a sample of N images, where I_1 and I_2 are respectively the numbers of horizontal and vertical image pixels and I_3 is the number of color channels. For example, I_3 is 1 for grayscale and 3 for red, green and blue (RGB)-color images. In UPCA, the sample of N images ($\mathcal{X} \in \mathbb{R}^{I_1\times I_2\times I_3\times N}$) is represented as an matrix $\mathbf{X}_{(4)} \in \mathbb{R}^{N\times(I_1I_2I_3)}$ by unfolding the tensor along the 4-mode. Then, the PCA algorithm is applied to $\mathbf{X}_{(4)}$ to solve the eigendecomposition problem

$$\frac{1}{N-1}\tilde{\mathbf{X}}_{(4)}^T\tilde{\mathbf{X}}_{(4)} = \mathbf{U}\mathbf{D}\mathbf{U}^T. \tag{10.6.15}$$

Here $\tilde{\mathbf{X}}_{(4)} = \mathbf{X}_{(4)} - \bar{\mathbf{X}}_{(4)}$ is the centered matrix of $\mathbf{X}_{(4)}$, $\mathbf{U} \in \mathbb{R}^{(I_1I_2I_3)\times(I_1I_2I_3)}$ is the orthonormal matrix of the eigenvectors and $\mathbf{D}$ is a diagonal matrix consisting of the $I_1I_2I_3$ eigenvalues of $(N-1)^{-1}\tilde{\mathbf{X}}_{(4)}^T\tilde{\mathbf{X}}_{(4)}$.

Let $(N-1)^{-1}\tilde{\mathbf{X}}_{(4)}^T\tilde{\mathbf{X}}_{(4)}$ have p principal eigenvalues $\lambda_i, i = 1,\ldots,p$, and let their corresponding eigenvector matrix be $\mathbf{U}_p \in \mathbb{R}^{(I_1I_2I_3)\times p}$. Then, the low-rank unfolding matrix of the image tensor $\mathcal{X}$ can be obtained as

$$\hat{\mathbf{X}}_{(4)} = \mathbf{U}_p \mathbf{\Sigma}_p \mathbf{U}_p^T \in \mathbb{R}^{(I_1 I_2 I_3)\times(I_1 I_2 I_3)}, \tag{10.6.16}$$

where $\mathbf{\Sigma}_p = \mathbf{Diag}(\sqrt{\lambda_1}, \ldots, \sqrt{\lambda_p})$.

2. Multilinear Principal Component Analysis (MPCA)

Multilinear principal component analysis (MPCA), developed by Lu *et al.* [304], applies PCA directly to tensors rather than unfolded data.

Given N training samples of existing images, denoted $\mathcal{Y}_n \in \mathbb{R}^{I_1\times I_2\times I_3}$, $n = 1, \ldots, N$, the MPCA finds a set of semi-orthogonal factor matrices $\mathbf{U} = \{\mathbf{U}^{(k)} \in \mathbb{R}^{I_k\times P_k} | \mathbf{U}^{(k)T}\mathbf{U}^{(k)} = \mathbf{I}_{P_k}, P_k < I_k, k = 1, 2, 3\}$ such that the projected low-dimensional tensor captures the most of the variation observed in the original tensor.

Let

$$\mathcal{Z}_n = \tilde{\mathcal{Y}}_n \times_1 \mathbf{U}^{(1)} \times_2 \mathbf{U}^{(2)} \times_3 \mathbf{U}^{(3)}, \quad n = 1, \ldots, N, \tag{10.6.17}$$

where $\tilde{\mathcal{Y}}_n = \mathcal{Y}_n - \bar{\mathcal{Y}}$ is the centered tensor of $\mathcal{Y}_n$.

The MPCA model procedure is to find $\mathbf{U}^{(k)} \in \mathbb{R}^{I_k\times P_k}$ as follows:

$$\{\mathbf{U}^{(1)}, \mathbf{U}^{(2)}, \mathbf{U}^{(3)}\} = \underset{\mathbf{U}^{(1)},\mathbf{U}^{(2)},\mathbf{U}^{(3)}}{\arg\max} \sum_{n=1}^{N} \|\mathcal{Z}_n\|_F^2. \tag{10.6.18}$$

Proposition 10.3 *Let $\bar{\mathcal{X}} \in \mathbb{R}^{I_1\times I_2\times I_3\times N}$ denote a centered tensor of a sample of N images. The semi-orthogonal factor matrices obtained by the following Tucker decomposition of $\mathcal{X}$ are the same as the solutions to the MPCA model in (10.6.18):*

$$\{\mathcal{G}, \mathbf{U}^{(1)}, \mathbf{U}^{(2)}, \mathbf{U}^{(3)}\} = \underset{\mathcal{G},\mathbf{U}^{(1)},\mathbf{U}^{(2)},\mathbf{U}^{(3)}}{\arg\max} \left\|\bar{\mathcal{X}} - \mathcal{G} \times_1 \mathbf{U}^{(1)} \times_2 \mathbf{U}^{(2)} \times_3 \mathbf{U}^{(3)}\right\|_F^2. \tag{10.6.19}$$

Proof See [521]. □

This proposition indicates that one can use the MPCA to implement the Tucker decomposition.

For a new image $\mathbf{Y}_{\text{new}} \in \mathbb{R}^{I_1\times I_2\times I_3}$, the following give a comparison between UPCA and MPCA in process monitoring [521]:

(1) In UPCA, the feature vector and the residual vector are calculated from $\mathbf{z}_{\text{new}} = \text{vec}(\mathcal{Y}_{\text{new}})\mathbf{U}_p$ and $\mathbf{e}_{\text{new}} = \text{vec}(\mathcal{Y}_{\text{new}}) - \mathbf{z}_{\text{new}}\mathbf{U}_p^T$, respectively.
(2) In MPCA, the feature vector is calculated from $\mathcal{Z}_{\text{new}} = \mathcal{Y}_{\text{new}} \times_1 \mathbf{U}^{(1)} \times_2 \mathbf{U}^{(2)} \times_3 \mathbf{U}^{(3)}$, and the residual vector is obtained from $\mathcal{E}_{\text{new}} = \mathcal{Y}_{\text{new}} - \mathcal{Z}_{\text{new}} \times_1 \mathbf{U}^{(1)T} \times_2 \mathbf{U}^{(2)T} \times_3 \mathbf{U}^{(3)T}$. The monitoring features are defined by $\boldsymbol{\theta}_{\text{new}} = \text{vec}\,\mathcal{Z}_{\text{new}}$ and $\mathbf{e}_{\text{new}} = \text{vec}(\mathcal{E}_{\text{new}})$.

Since it breaks the tensor structure and has high computational complexity, UPCA is not an effective low-rank recovery technique for tensor data [203] although it has been extensively used in the analysis of image data.

10.6.4 Note on Other Applications

The high-order nonlinearities in a system are characterized by high-dimensional tensors such that these nonlinearities can be represented by just a few vectors obtained from the canonical tensor decomposition. A tensor-based nonlinear model order reduction (TNMOR) scheme was proposed in [299] for typical circuits with a few nonlinear components. The key feature of TNMOR is that it preserves the inherent sparsity through the use of low-rank tensors, thereby easing the memory requirement and speeding up computation via the exploitation of structures in the data.

Combining imaging technology and high spectral resolution spectroscopy in a unique system, the hyperspectral sensor provides a powerful means to discriminate the targets of interest in a scene [541]. A hyperspectral image consisting of two-dimensional spatial information and one-dimensional spectral information can be regarded as third-order tensor data. Recently, nonnegative CP decomposition has been used effectively to analyze hyperspectral large data [494], and an algorithm that combines tensor decomposition and PCA jointed algorithm was proposed in [322] for hyperspectral image denoising. On the other hand, in most practical situations it is difficult to specify the prior spectrum of the target of interest in advance. Hence, target detection without any target information, known as anomaly detection, is preferred and is more practicable (see [453], [314], [336] and the literature therein). A tensor-decomposition-based anomaly detection algorithm was recently developed in [541] for hyperspectral images.

Medical images are commonly used to visualize non-invasively the motion of the heart as it beats, using magnetic resonance imaging (MRI), echocardiography (ECHO) or computed tomography (CT). Such medical image sequences can be used to analyze the motion dynamics qualitatively. A number of nonrigid image registration techniques for cardiac motion tracking have been proposed in recent years; see [161], [500]. Quantifying cardiac motion at a population level is not a simple task, since images can vary widely in terms of image quality, size, resolution and pose. In order to study the components individually, McLeod *et al.* [320] proposed using tensor decomposition to decouple the spatial and temporal aspects.

The tensor clustering problem divides several Nth-order input tensors into groups by minimizing some criterion. A robust tensor clustering (RTC) was presented in [82] for robust face clustering; it first finds a lower-rank approximation to the original tensor data using a ℓ_1-norm optimization function and then computes a higher-order SVD of this approximate tensor to obtain the final clustering results.

In modern sensing, networking and communication, multidimensional data usually have high dimensionality and are represented as tensor data. These tensorial data have a large amount of redundancy, so they need to be carefully analyzed. In order to avoid the so-called "curse of dimensionality" (the various phenomena that arise in analyzing and organizing high-dimensional spaces and which do not occur in low-dimensional settings), it is desirable to have a mechanism to reduce such redundancy for the efficient application of learning algorithms. Recently, a heterogeneous Tucker decomposition model for tensor clustering was proposed in [459]. This is a novel subspace-clustering algorithm that exploits the tensorial structure of data.

Tensor decomposition relates to big data analytics. An excellent tutorial on state-of-the-art tensor decomposition, with original research on the parallel and distributed computation of low-rank decomposition for big tensors, can be found in [446].

10.7 Tensor Eigenvalue Decomposition

Eigenvalues of tensors, proposed by Qi [403] and Lim [296] independently in 2005, have attracted much attention in the literature and have found applications in science and engineering, such as in molecular conformation [125], blind source separation [252], signal processing [404], magnetic resonance imaging [405], [431], spectral hypergraph analysis [294], quantum entanglement [349] and so on.

10.7.1 Tensor–Vector Products

Eigenvalues of tensors are closely related to tensor–vector products. Suppose that $\mathcal{A} = [\![a_{i_1\cdots i_m}]\!] \in \mathbb{R}^{n\times n\times\cdots\times n}$ is a real-valued mth-order n-dimensional tensor with n^m entries.

A real-valued mth-order n-dimensional tensor $\mathcal{A}$ is said to be *supersymmetric* if its entries are invariant under permutation [259], i.e.,

$$a_{i_{p(1)}\cdots i_{p(m)}} = a_{i_1\ldots i_m} \quad \text{for all} \quad i_1,\ldots,i_m \in \{1,\ldots,n\} \quad \text{and} \quad p \in \Pi_m,$$

where Π_m denotes the space of all m-permutations. A supersymmetric tensor is usually referred to just as a *symmetric tensor*.

Let $\mathbb{R}^{[m,n]}$ and $\mathbb{S}^{[m,n]}$ denote the sets of all real-valued mth-order n-dimensional tensors and of all real symmetric tensors, respectively.

Putting $\mathcal{A} \in \mathbb{S}^{[m,n]}$ and $\mathbf{x} \in \mathbb{R}^n$, then we can define the following *tensor–vector products* [404], [259]:

$$\mathcal{A}\mathbf{x}^m = \sum_{i_1=1}^{n}\cdots\sum_{i_m=1}^{n} a_{i_1\cdots i_m} x_{i_1}\cdots x_{i_m}, \tag{10.7.1}$$

$$(\mathcal{A}\mathbf{x}^{m-1})_{i_1} = \sum_{i_2=1}^{n}\cdots\sum_{i_m=1}^{n} a_{i_1\cdots i_m} x_{i_2}\cdots x_{i_m} \quad \forall i_1 = 1,\ldots,n, \tag{10.7.2}$$

$$(\mathcal{A}\mathbf{x}^{m-2})_{i_1,i_2} = \sum_{i_3=1}^{n}\cdots\sum_{i_m=1}^{n} a_{i_1\cdots i_m} x_{i_3}\cdots x_{i_m} \quad \forall i_1, i_2 = 1,\ldots,n. \tag{10.7.3}$$

More generally, we have the following definition.

Definition 10.13 [258] Let $\mathcal{A} \in \mathbb{S}^{[m,n]}$ and $\mathbf{x} \in \mathbb{R}^n$. Then, for $0 \leq r \leq m-1$, the $(m-r)$-times product of a symmetric tensor $\mathcal{A}$ with the vector $\mathbf{x}$ is denoted by $\mathcal{A}\mathbf{x}^{m-r} \in \mathbb{R}^{[r,n]}$ and defined as

$$(\mathcal{A}\mathbf{x}^{m-r})_{i_1,\ldots,i_r} = \sum_{i_{r+1}=1}^{n} \cdots \sum_{i_m=1}^{n} a_{i_1\cdots i_m} x_{i_{r+1}} \cdots x_{i_m}$$

$$\text{for all } i_1,\ldots,i_r \in \{1,\ldots,n\}. \tag{10.7.4}$$

Regarding $(m-r)$-times tensor–vector products, we make the following remarks.

Remark 1 By the definition formulas (10.7.1)–(10.7.3), it easily follows that there exist the following relationships between the $(m-r)$-times tensor–vector products:

$$\mathcal{A}\mathbf{x}^m = \mathbf{x}^T \mathcal{A}\mathbf{x}^{m-1} \quad \text{and} \quad \mathcal{A}\mathbf{x}^{m-1} = (\mathcal{A}\mathbf{x}^{m-2})\mathbf{x}. \tag{10.7.5}$$

Remark 2 Since $\mathcal{A}\mathbf{x}^m = \mathbf{x}^T \mathcal{A}\mathbf{x}^{m-1}$, the scalar $\mathcal{A}\mathbf{x}^m$ can actually be regarded as the quadratic form of the symmetric tensor $\mathcal{A}$ and is suitable for using as a measure of the positive definiteness of tensors: the tensor $\mathcal{A} \in \mathbb{S}^{[m,n]}$ is called an mth-order n-dimensional *positive definite tensor* (*positive semi-definite tensor*), denoted $\mathcal{A} \in \mathbb{S}_{+}^{[m,n]}$ ($\mathcal{A} \in \mathbb{S}_{+0}^{[m,n]}$), if $\mathcal{A}\mathbf{x}^m > 0$ ($\mathcal{A}\mathbf{x}^m \geq 0$) for all $\mathbf{x} \neq \mathbf{0}$. Similarly, through $\mathcal{A}\mathbf{x}^m$, we can also define a negative definite tensor, negative semi-definite tensor and indefinite tensor, respectively.

Remark 3 The $(m-r)$-times tensor–vector product $\mathcal{A}\mathbf{x}^{m-r}$ is actually a tensor contraction [276]: $\mathcal{A}\mathbf{x}^{m-1}$ contracts the tensor $\mathcal{A} \in \mathbb{S}^{[m,n]}$ into a real-valued n-vector $\mathcal{A}\mathbf{x}^{m-1} \in \mathbb{R}^{[1,n]}$, i.e., an n-dimensional rank-1 tensor, and $\mathcal{A}\mathbf{x}^{m-2}$ contracts the tensor $\mathcal{A} \in \mathbb{S}^{[m,n]}$ into an $n \times n$ real-valued matrix whereas $\mathcal{A}\mathbf{x}^{m-r}$ contracts the tensor $\mathcal{A} \in \mathbb{S}^{[m,n]}$ into an rth-order n-dimensional tensor.

Remark 4 The derivatives of the tensor–vector product with respect to $\mathbf{x}$ are given by [259]

$$\nabla(\mathcal{A}\mathbf{x}^m) = m\mathcal{A}\mathbf{x}^{m-1} \quad \text{and} \quad \nabla^2(\mathcal{A}\mathbf{x}^m) = m(m-1)\mathcal{A}\mathbf{x}^{m-2}. \tag{10.7.6}$$

As we have seen, an $n \times n$ matrix with entries δ_{ij} is called a unit or identity matrix, denoted $\mathbf{I} \in \mathbb{R}^{n \times n}$, where δ is the standard Kronecker delta function, i.e.,

$$\delta_{ij} = \begin{cases} 1, & i = j, \\ 0, & i \neq j. \end{cases}$$

In contrast, the unit tensor has two different definitions.

Definition 10.14 A real-valued mth-order n-dimensional tensor $\mathcal{I} \in \mathbb{R}^{[m,n]}$ is called a unit tensor if its entries are defined by

$$i_{i_1\cdots i_m} = \delta_{i_1\cdots i_m} = \begin{cases} 1, & i_1 = \cdots = i_m, \\ 0, & \text{otherwise}. \end{cases} \tag{10.7.7}$$

From the above definition, we have

$$(\mathcal{I}\mathbf{x}^{m-1})_i = x_i^{m-1} \quad \text{or} \quad \mathcal{I}\mathbf{x}^{m-1} = \mathbf{x}^{[m-1]} = [x_1^{m-1}, \ldots, x_n^{m-1}]^T. \tag{10.7.8}$$

Definition 10.15 [258] For m even, if the entries of a real-valued mth-order n-dimensional tensor $\mathcal{E} \in \mathbb{R}^{[m,n]}$ are given by

$$e_{i_1\cdots i_m} = \frac{1}{m!} \sum_{p\in\Pi_m} \delta_{p(1)p(2)}\delta_{p(3)p(4)}\cdots\delta_{p(m-1)p(m)} \tag{10.7.9}$$

for $i_1, \ldots, i_m \in \{1, \ldots, n\}$, then it is known as a unit tensor.

From the definition (10.7.9), we have [258]

$$\mathcal{E}\mathbf{x}^{m-1} = \mathbf{x} \quad \text{for all} \quad \mathbf{x} \in \Sigma, \tag{10.7.10}$$

where Σ is the unit sphere defined by

$$\Sigma = \big\{\mathbf{x} \in \mathbb{R}^n \big| \, \|\mathbf{x}\| = 1\big\}. \tag{10.7.11}$$

10.7.2 Determinants and Eigenvalues of Tensors

The ith eigenvalue of the tensor $\mathcal{A}$ is denoted by $\lambda_i(\mathcal{A})$.

Definition 10.16 [403] Let $\mathcal{A} \in \mathbb{S}^{[m,n]}$ be an mth-order n-dimensional symmetric tensor. For some $\lambda \in \mathbb{C}$, if the polynomial system $\det(\mathcal{A} - \lambda\mathcal{I}) = 0$ (see below) has a solution $\mathbf{x} \in \mathbb{C}^n \setminus \{\mathbf{0}\}$ then λ is called a *tensor eigenvalue* of $\mathcal{A}$ and $\mathbf{x}$ is known as a *tensor eigenvector* of $\mathcal{A}$ associated with λ.

The two-tuple $(\lambda, \mathbf{x})$ is known as a *tensor eigenpair* of $\mathcal{A}$.

The symbol $\det(\mathcal{A})$ denotes the *hyperdeterminant* (or simply determinant) of the tensor $\mathcal{A}$. For the tensor $\mathcal{A} \in \mathbb{S}^{[m,n]}$, some properties of its determinants are summarized below.

Proposition 10.4 [403] *For an mth-order n-dimensional tensor $\mathcal{A}$, its determinant* $\det(\mathcal{A})$ *is the* resultant *of*

$$\mathcal{A}\mathbf{x}^{m-1} = \mathbf{0},$$

and is a homogeneous polynomial in the entries of $\mathcal{A}$, with degree $d = n(m-1)^{n-1}$. The degree of $[\mathcal{A}]_{i,\ldots,i}$ in $\det(\mathcal{A})$ *is not greater than $(m-1)^{n-1}$.*

The determinant of an mth-order n-dimensional tensor $\mathcal{A}$ has the following properties [403], [126], [221]:

1. $\det(\mathcal{A})$ is the resultant of $\mathcal{A}\mathbf{x}^{m-1} = \mathbf{0}$;
2. $\det(\mathcal{A}) = 0$ if and only if $\mathcal{A}\mathbf{x}^{m-1} = \mathbf{0}$ has a nonzero solution;
3. $\det(\mathcal{I}) = 1$, where $\mathcal{I}$ is the unit tensor;
4. the determinant of a nonsingular tensor is nonzero, i.e., $\det(\mathcal{A}) \neq 0$;

5. $\det(\mathcal{A})$ is an irreducible polynomial in the entries of $\mathcal{A}$;
6. if, for some i and a tensor $\mathcal{A}$, its entries $a_{ii_2\cdots i_m} = 0$ for all $i_2, \ldots, i_m = \{1, \ldots, n\}$ then $\det(\mathcal{A}) = 0$; in particular, the determinant of the zero tensor is zero;
7. $\det(\gamma\mathcal{A}) = \gamma^{n(m-1)^{n-1}} \det(\mathcal{A})$;
8. suppose that $\mathcal{T} \in \mathbb{C}^{[m,n]}$ is a complex-valued triangular tensor; then

$$\det(\mathcal{T}) = \prod_{i=1}^{n} (t_{i\cdots i})^{(m-1)^{(n-1)}}. \tag{10.7.12}$$

The set of eigenvalues of a tensor $\mathcal{A}$ is called the spectrum of $\mathcal{A}$, denoted $\sigma(\mathcal{A})$, and is defined as

$$\sigma(\mathcal{A}) = \{\lambda \in \mathbb{C} | \det(\mathcal{A} - \lambda\mathcal{I}) = 0\}. \tag{10.7.13}$$

Suppose that $\mathcal{T} \in \mathbb{C}^{[m,n]}$ is a triangular tensor. Then

$$\sigma(\mathcal{T}) = \{t_{i\cdots i} \in \mathbb{C} | i \in \{1, \ldots, n\}\}, \tag{10.7.14}$$

and the algebraic multiplicity of the eigenvalue $t_{i\cdots i}$ is $(m-1)^{n-1}$ for all $i \in \{1, \ldots, n\}$.

Definition 10.17 [222] Let $\mathcal{A}$ be an mth-order n-dimensional tensor of indeterminate variables and λ be an indeterminate variable. The determinant $\chi(\lambda) = \det(\mathcal{A} - \lambda\mathcal{I})$ is called the *characteristic polynomial* of the tensor $\mathcal{A}$.

Let $\mathcal{A}$ be an mth-order n-dimensional tensor of indeterminate variables. By [222], $\chi(\lambda)$ is homogeneous of degree $n(m-1)^{n-1}$.

There are various definitions of the eigenvalue of a symmetric tensor. Among them, H-eigenvalues and Z-eigenvalues are two particularly interesting definitions.

1. H-Eigenpair [403]

A real number λ is called an *H-eigenvalue* of the tensor $\mathcal{A}$ if λ and a nonzero real vector $\mathbf{x}$ are solutions of the homogeneous polynomial equation

$$\mathcal{A}\mathbf{x}^{m-1} = \lambda\mathbf{x}^{[m-1]}, \tag{10.7.15}$$

where $\mathbf{x}^{[m-1]}$ denotes a vector in $\mathbb{R}^n$ with ith entry $(x^{[m-1]})_i = x_i^{m-1}$, i.e.,

$$\mathbf{x}^{[m-1]} = [x_1^{m-1}, \ldots, x_n^{m-1}]^T \quad \text{and} \quad \mathcal{I}\mathbf{x}^{m-1} = \mathbf{x}^{[m-1]}. \tag{10.7.16}$$

Moreover, the solution $\mathbf{x}$ is known as an *H-eigenvector* of $\mathcal{A}$ associated with the H-eigenvalue λ, and $(\lambda, \mathbf{x})$ is called an *H-eigenpair* of $\mathcal{A}$.

The term "H-eigenvalue" is attributable to the fact that λ is a solution to the homogeneous polynomial equation (10.7.15).

A tensor all of whose entries are nonnegative is called a *nonnegative tensor.*

Theorem 10.2 [527] *An mth-order n-dimensional nonnegative tensor $\mathcal{A} = [\![a_{i_1\cdots i_m}]\!]$ has at least one H-eigenvalue, which has the largest modulus of all its eigenvalues.*

Theorem 10.3 [406] *Let* $\mathcal{A} = [\![a_{i_1 \cdots i_m}]\!]$ *be an* m*th-order* n*-dimensional symmetric nonnegative tensor with* $m \geq 2$*. Then the H-eigenvalues of* $\mathcal{A}$ *are always nonnegative.*

2. Z-Eigenpair [403]

A real number λ and a real vector $\mathbf{x} \in \mathbb{R}^n$ are called a *Z-eigenvalue* of $\mathcal{A}$ and a *Z-eigenvector* of $\mathcal{A}$ associated with the Z-eigenvalue λ, respectively, if they are solutions of the following system:

$$\mathcal{A}\mathbf{x}^{m-1} = \lambda \mathbf{x} \quad \text{and} \quad \mathbf{x}^T\mathbf{x} = 1. \tag{10.7.17}$$

A Z-eigenvalue and a corresponding Z-eigenvector form a *Z-eigenpair*.

The term "Z-eigenvalue" is attributed to Zhou, who suggested to the author of [403] the above definition in a private communication.

Since the definition (10.7.17) is associated with the Euclidean norm $\|\mathbf{x}\|_2 = 1$, the number $\lambda \in \mathbb{R}$ is also known as an *E-eigenvalue* of $\mathcal{A}$ if it and a nonzero vector $\mathbf{x} \in \mathbb{C}^n$ are solutions of the polynomial equation system (10.7.17), and the nonzero solution vector $\mathbf{x}$ is also called an *E-eigenvector* of $\mathbf{A}$ associated with the eigenvalue λ [403].

The Z-eigenpair defined by Qi [403] and the ℓ^2-eigenpair defined by Lim [296] are equivalent. Importantly, Lim [296] observed that any eigenpair $(\lambda, \mathbf{x})$ is a KKT point of the nonlinear optimization problem

$$\max_{\mathbf{x} \in \mathbb{R}^n} \mathcal{A}\mathbf{x}^m \quad \text{subject to} \quad \mathbf{x}^T\mathbf{x} = 1. \tag{10.7.18}$$

Define the *maximum Z-eigenvalue function* $\lambda_1^Z : \mathbb{S} \to \mathbb{R}$ as follows:

$$\lambda_1^Z = \{\lambda \in \mathbb{R} \,|\, \lambda \text{ is the largest Z-eigenvalue of } \mathcal{A}\}. \tag{10.7.19}$$

Lemma 10.2 [296], [403] *Let* $\mathcal{A} \in \mathbb{S}^{[m,n]}$*, where* m *is even. Then, we have*

$$\lambda_1^Z = \max_{\|\mathbf{x}\|_2 = 1} \mathcal{A}\mathbf{x}^m. \tag{10.7.20}$$

In spectral hypergraph analysis, by introducing the characteristic tensor of a hypergraph, Li *et al.* [294] recently showed that the maximum Z-eigenvalue function of the associated characteristic tensor provides a natural link between the combinatorial structure and the analytic structure of the underlying hypergraph.

Proposition 10.5 [406] *A real-valued* m*th-order* n*-dimensional symmetric tensor* $\mathcal{A}$ *always has Z-eigenvalues. If* m *is even then* $\mathcal{A}$ *always has at least one H-eigenvalue. When* m *is even, the following three statements are equivalent:*

- $\mathcal{A}$ *is positive semi-definite;*
- *all the H-eigenvalues of* $\mathcal{A}$ *are nonnegative;*
- *all the Z-eigenvalues of* $\mathcal{A}$ *are nonnegative.*

The relationship between H-eigenvalues and Z-eigenvalues is as follows.

(1) For an even-order symmetric tensor, both Z-eigenvalues and H-eigenvalues always exist [403], [92].
(2) Finding an H-eigenvalue of a symmetric tensor is equivalent to solving a homogeneous polynomial equation, whereas calculating a Z-eigenvalue is equivalent to solving a nonhomogeneous polynomial equation.
(3) In general, the properties of Z-eigenvalues and H-eigenvalues are different. For example, a diagonal symmetric tensor $\mathcal{A}$ has exactly n H-eigenvalues but may have more than n Z-eigenvalues [403].

3. D-Eigenpair [405]

Let $\mathbf{D}$ be a symmetric $n \times n$ matrix and assume that the order m of a tensor $\mathcal{A}$ is 4. Then λ and $\mathbf{x} \neq \mathbf{0}$ are called a *D-eigenvalue* and a *D-eigenvector* of the tensor $\mathcal{A}$, if

$$\mathcal{A}\mathbf{x}^{m-1} = \lambda \mathbf{D}\mathbf{x} \quad \text{and} \quad \mathbf{x}^T\mathbf{D}\mathbf{x} = 1; \tag{10.7.21}$$

$(\lambda, \mathbf{x})$ is said to be a *D-eigenpair* of the tensor $\mathcal{A}$.

Diffusion kurtosis imaging is a new model in medical engineering, in which a diffusion kurtosis tensor is involved. By introducing D-eigenvalues for a diffusion kurtosis tensor, the largest, the smallest and the average D-eigenvalues correspond to the largest, the smallest and the average apparent kurtosis coefficients of a water molecule in the space, respectively. See [405] for more details.

4. U-Eigenpair and US-Eigenpair [349]

Define the tensor–vector product of a complex-valued tensor $\mathcal{A} \in \mathbb{C}^{[m,n]}$ and a complex-valued vector $\mathbf{x} \in \mathbb{C}^n$ as

$$(\mathcal{A}^*\mathbf{x}^{m-1})_k = \sum_{i_2=1}^{n} \cdots \sum_{i_m=1}^{n} a^*_{ki_2\cdots i_m} x_{i_2} \cdots x_{i_m}, \quad k = 1, \ldots, n, \tag{10.7.22}$$

$$(\mathcal{A}\mathbf{x}^{(m-1)*})_k = \sum_{i_2=1}^{n} \cdots \sum_{i_m=1}^{n} a_{ki_2\cdots i_m} x^*_{i_2} \cdots x^*_{i_m}, \quad k = 1, \ldots, n. \tag{10.7.23}$$

A number $\lambda \in \mathbb{C}$ and a vector $\mathbf{x}$ are respectively called a *unitary eigenvalue (U-eigenvalue)* and a *U-eigenvector* of the tensor $\mathcal{A}$ if λ and $\mathbf{x}$ are solutions of the following equation system:

$$\left.\begin{aligned} \mathcal{A}^*\mathbf{x}^{m-1} &= \lambda \mathbf{x}^{[m-1]*}, \\ \mathcal{A}\mathbf{x}^{(m-1)*} &= \lambda \mathbf{x}^{[m-1]}, \\ \|\mathbf{x}\| &= 1. \end{aligned}\right\} \tag{10.7.24}$$

Let $\mathcal{S} \in \mathbb{S}^{[m,n]}$ be a complex-valued symmetric tensor and $\mathbf{x} = [x_1, \ldots, x_n]^T \in$

$\mathbb{C}^n$. Denote

$$(\mathcal{S}^*\mathbf{x}^{m-1})_k = \sum_{i_2=1}^{n} \cdots \sum_{i_m=1}^{n} s^*_{ki_2\cdots i_m} x_{i_2} \cdots x_{i_m}.$$

A number $\lambda \in \mathbb{C}$ and a vector $\mathbf{x} \neq \mathbf{0}$ are respectively called a *unitary symmetric eigenvalue (US-eigenvalue)* and *US-eigenvector* of $\mathcal{S}$ if λ and a nonzero vector $\mathbf{x} \in \mathbb{C}^n$ are solutions of the following equation system:

$$\left.\begin{aligned} \mathcal{S}^*\mathbf{x}^{m-1} &= \lambda\mathbf{x}^*, \\ \mathcal{S}\mathbf{x}^{(m-1)*} &= \lambda\mathbf{x}, \\ \|\mathbf{x}\| &= 1. \end{aligned}\right\} \tag{10.7.25}$$

Theorem 10.4 [349] *For complex-valued mth-order n-dimensional tensors $\mathcal{A}, \mathcal{B}, \mathcal{S} \in H = \mathbb{C}^{n\times\cdots\times n}$, the following results hold:*

- $\langle \mathcal{A}, \mathcal{B}\rangle = \langle \mathcal{B}, \mathcal{A}\rangle^*$;
- *all U-eigenvalues are real numbers;*
- *the US-eigenpair $(\lambda, \mathbf{x})$ of a complex-valued symmetric mth-order n-dimensional tensor $\mathcal{S}$ can also be defined by the following equation system:*

$$\left.\begin{aligned} \mathcal{S}^*\mathbf{x}^{m-1} &= \lambda\mathbf{x}^*, \\ \|\mathbf{x}\| = 1, &\quad \lambda \in \mathbb{R}, \end{aligned}\right\} \tag{10.7.26}$$

or

$$\left.\begin{aligned} \mathcal{S}\mathbf{x}^{(m-1)*} &= \lambda\mathbf{x}, \\ \|\mathbf{x}\| = 1, &\quad \lambda \in \mathbb{R}. \end{aligned}\right\} \tag{10.7.27}$$

5. *$\mathcal{B}$-Eigenpair* [92]

Let $\mathcal{A}$ be an mth-order n-dimensional tensor consisting of nm entries in $\mathbb{R}$:

$$\mathcal{A} = [\![a_{i_1\cdots i_m}]\!], \quad A_{i_1\cdots i_m} \in \mathbb{R},\ 1 \leq i_1, \ldots, i_m \leq n.$$

Definition 10.18 Let $\mathcal{A}$ and $\mathcal{B}$ be two mth-order n-dimensional tensors on $\mathbb{R}$. Assume that both $\mathcal{A}\mathbf{x}^{m-1}$ and $\mathcal{B}\mathbf{x}^{m-1}$ are not identically equal to zero. Then $(\lambda, \mathbf{x}) \in \mathbb{C} \times (\mathbb{C}^n \setminus \{\mathbf{0}\})$ is called an eigenpair of $\mathcal{A}$ relative to $\mathcal{B}$, if the n-system of equations

$$(\mathcal{A} - \lambda\mathcal{B})\mathbf{x}^{m-1} = \mathbf{0}, \tag{10.7.28}$$

i.e.,

$$\sum_{i_2=1}^{n} \cdots \sum_{i_m=1}^{n} (a_{ii_2\cdots i_m} - \lambda b_{ii_2\cdots i_m}) x_2 \cdots x_m = 0, \quad i = 1, \ldots, n \tag{10.7.29}$$

possesses a solution.

The scalar λ is known as a $\mathcal{B}$-*eigenvalue* of $\mathcal{A}$, and $\mathbf{x}$ is called a $\mathcal{B}$ *eigenvector* of $\mathcal{A}$.

In tensor analysis problems motivated by the geometric measurement of quantum entanglement, the largest U-eigenvalue $|\lambda|$ is the entanglement eigenvalue, and the corresponding U-eigenvector $\mathbf{x}$ is the closest separable state (for more details, see [349]).

10.7.3 Generalized Tensor Eigenvalues Problems

Let $\mathcal{A} = [\![a_{i_1\cdots i_m}]\!] \in \mathbb{C}^{n\times\cdots\times n}$ and $\mathcal{B} = [\![b_{i_1\cdots i_m}]\!] \in \mathbb{C}^{n\times\cdots\times n}$ be two complex-valued mth-order n-dimensional tensors. By [126], $(\mathcal{A}, \mathcal{B})$ is called a *regular tensor pair*, if

$$\det(\beta\mathcal{A} - \alpha\mathcal{B}) \neq 0 \quad \text{for some } (\alpha, \beta) \in \mathbb{C} \times \mathbb{C};$$

and $(\mathcal{A}, \mathcal{B})$ is known as a *singular tensor pair* if

$$\det(\beta\mathcal{A} - \alpha\mathcal{B}) = 0 \quad \text{for some } (\alpha, \beta) \in \mathbb{C} \times \mathbb{C}.$$

A *tensor pair* is also called a *tensor pencil*. In the following, we focus on the regular tensor pairs.

Definition 10.19 [126] For a regular tensor pair $(\mathcal{A}, \mathcal{B})$, if there exists $(\alpha, \beta) \in \mathbb{C} \times \mathbb{C}$ with a nonzero vector $\mathbf{x} \in \mathbb{C}^n$ such that

$$\beta\mathcal{A}\mathbf{x}^{m-1} = \alpha\mathcal{B}\mathbf{x}^{m-1} \tag{10.7.30}$$

then $\lambda = \alpha/\beta$ is called a *generalized eigenvalue* of the regular tensor pair $(\mathcal{A}, \mathcal{B})$ if $\det(\mathcal{B}) \neq 0$.

The set of eigenvalues (α, β) is called the spectrum of the tensor pair $(\mathcal{A}, \mathcal{B})$, and is denoted as

$$\sigma(\mathcal{A}, \mathcal{B}) = \big\{(\alpha, \beta) \in \mathbb{C} \times \mathbb{C} \big| \det(\beta\mathcal{A} - \alpha\mathcal{B}) = 0\big\}, \tag{10.7.31}$$

or, for a nonsingular $\mathcal{B}$, the spectrum is denoted as

$$\sigma(\mathcal{A}, \mathcal{B}) = \big\{\lambda \in \mathbb{C} \big| \det(\mathcal{A} - \lambda\mathcal{B}) = 0\big\}. \tag{10.7.32}$$

In particular, when $\beta = 1$, the generalized eigenvalue λ and the generalized eigenvector $\mathbf{x}$ reduce to the $\mathcal{B}$-eigenvalue and the $\mathcal{B}$-eigenvector of $\mathbf{A}$.

Several tensor eigenpair definitions described above can be unified under a generalized tensor eigenpair framework, as follows.

(1) When $\mathcal{B} = \mathcal{I}$ is a unit tensor defined by (10.7.7), the generalized tensor eigenvalue directly reduces to the H-eigenvalue of $\mathcal{A}$, because $\mathcal{I}\mathbf{x}^{m-1} = \mathbf{x}^{[m-1]}$ and thus $\mathcal{A}\mathbf{x}^{m-1} = \lambda\mathcal{I}\mathbf{x}^{m-1} = \lambda\mathbf{x}^{[m-1]}$.
(2) If $\mathcal{B} = \mathcal{E}$ is a unit tensor defined by (10.7.9) with $\|\mathbf{x}\| = 1$ then the generalized tensor eigenvalue reduces to the Z-eigenvalue of $\mathcal{A}$.

(3) Put $\mathcal{B}\mathbf{x}^{m-1} = \mathbf{D}\mathbf{x}$ and $\mathcal{B}\mathbf{x}^m = 1$ (i.e., $\mathbf{x}^T\mathcal{B}\mathbf{x}^{m-1} = \mathbf{x}^T\mathbf{D}\mathbf{x} = 1$). Then the generalized tensor eigenvalue becomes the D-eigenvalue of $\mathcal{A}$.
(4) When $\mathcal{A}$ and $\mathbf{x}$ are a complex-valued tensor and vector, respectively, and $\mathcal{B} = \mathcal{I}$, the definition of the generalized eigenvalue becomes the definition of the U-eigenvalue of $\mathcal{A}$.
(5) If $\mathcal{A}$ is a conjugate-symmetric tensor, $\mathbf{x} \in \mathbb{C}$ and $\mathcal{B} = \mathcal{E}$ with $\|\mathbf{x}\| = 1$ then the generalized tensor eigenvalue gives the US-eigenvalue of $\mathcal{A}$.

Premultiplying both sides of (10.7.30) by $\mathbf{x}^T$ and using (10.7.5), it can be seen that any generalized tensor eigenpair satisfies

$$\lambda = \frac{\mathcal{A}\mathbf{x}^m}{\mathcal{B}\mathbf{x}^m}. \tag{10.7.33}$$

Consider a computation method for generalized tensor eigenvalue decomposition. To this end, let $\mathbf{A} \in \mathbb{S}^{[m,n]}$ and $\mathcal{B} \in \mathbb{S}_+^{[m,n]}$ and define the nonlinear optimization problem [259]

$$\max_{\mathbf{x}} \left\{ f(\mathbf{x}) = \frac{\mathcal{A}\mathbf{x}^m}{\mathcal{B}\mathbf{x}^m}\|\mathbf{x}\|^m \right\} \quad \text{subject to} \quad \mathbf{x} \in \Sigma. \tag{10.7.34}$$

Here the constraint makes the term $\|\mathbf{x}\|^m$ in $f(\mathbf{x})$ superfluous.

Theorem 10.5 [259] *Let $f(\mathbf{x})$ be defined as in (10.7.34). For $\mathbf{x} \in \Sigma$, the gradient and the Hessian matrix are respectively given by*

$$g(\mathbf{x}) = \nabla f(\mathbf{x}) = \frac{m}{\mathcal{B}\mathbf{x}^m}\left((\mathcal{A}\mathbf{x}^m)\mathbf{x} + \mathcal{A}\mathbf{x}^{m-1} - \left(\frac{\mathcal{A}\mathbf{x}^m}{\mathcal{B}\mathbf{x}^m}\right)\mathcal{B}\mathbf{x}^{m-1} \right), \tag{10.7.35}$$

$$\begin{aligned} \mathbf{H}(\mathbf{x}) = \nabla^2 f(\mathbf{x}) = {} & \frac{m^2\mathcal{A}\mathbf{x}^m}{(\mathcal{B}\mathbf{x}^m)^3}(\mathcal{B}\mathbf{x}^{m-1} \circledcirc \mathcal{B}\mathbf{x}^{m-1}) + \frac{m}{\mathcal{B}\mathbf{x}^m}\Big((m-1)\mathcal{A}\mathbf{x}^{m-2} \\ & + \mathcal{A}\mathbf{x}^m\left(\mathbf{I} + (m-2)\mathbf{x}\mathbf{x}^T\right) + m(\mathcal{A}\mathbf{x}^{m-1} \circledcirc \mathbf{x}) \Big) \\ & - \frac{m}{(\mathcal{B}\mathbf{x}^m)^2}\Big((m-1)\mathcal{A}\mathbf{x}^m\mathcal{B}\mathbf{x}^{m-2} + m(\mathcal{A}\mathbf{x}^{m-1} \circledcirc \mathcal{B}\mathbf{x}^{m-1}) \\ & + m\mathcal{A}\mathbf{x}^m\left(\mathbf{x} \circledcirc \mathcal{B}\mathbf{x}^{m-1}\right) \Big), \end{aligned} \tag{10.7.36}$$

where the symbol $\circledcirc$ means the symmetrized outer product: $\mathbf{a} \circledcirc \mathbf{b} = \mathbf{a}\mathbf{b}^T + \mathbf{b}\mathbf{a}^T$.

Kolda and Mayo [259], in 2014, developed a *generalized eigenproblem adaptive power* (GEAP) *method*, as shown in Algorithm 10.8.

The GEAP method can be specialized to the Z-eigenpair adaptive power method, as shown in Algorithm 10.9.

In these algorithms, the choice $\beta = 1$ is for finding local maxima (when the function is convex); otherwise, the choice $\beta = -1$ is for finding local minima (when the function is concave). For the tolerance on being positive or negative definite, $\tau = 10^{-6}$ is taken and a convergence error $\epsilon = 10^{-15}$ is selected in [259].

Algorithm 10.8 Generalized eigenproblem adaptive power method [259]

given: $\mathcal{A} \in \mathbb{S}^{[m,n]}, \mathcal{B} \in \mathbb{S}_+^{[m,n]}$ and $\mathbf{x}_0 \leftarrow \hat{\mathbf{x}}_0/\|\hat{\mathbf{x}}_0\|$, where $\hat{\mathbf{x}}_0$ is an initial guess.
repeat
1. Compute $\mathcal{A}\mathbf{x}_k^{m-2}, \mathcal{B}\mathbf{x}_k^{m-2}, \mathcal{A}\mathbf{x}_k^{m-1}, \mathcal{B}\mathbf{x}_k^{m-1}, \mathcal{A}\mathbf{x}_k^m$ and $\mathcal{B}\mathbf{x}_k^m$.
2. $\lambda_k \leftarrow \mathcal{A}\mathbf{x}_k^m/(\mathcal{B}\mathbf{x}_k^m)$.
3. $\mathbf{H}_k \leftarrow \mathbf{H}(\mathbf{x}_k)$.
4. $\alpha_k \leftarrow \beta \max\{0, (\tau - \lambda_{\min}(\beta \mathbf{H}_k))/m\}$.
5. $\hat{\mathbf{x}}_{k+1} \leftarrow \beta \left(\mathcal{A}\mathbf{x}_k^{m-1} - \lambda_k \mathcal{B}\mathbf{x}_k^{m-1} + (\alpha_k + \lambda_k)\mathcal{B}\mathbf{x}_k^m \mathbf{x}_k\right)$.
6. $\mathbf{x}_{k+1} \leftarrow \hat{\mathbf{x}}_{k+1}/\|\hat{\mathbf{x}}_{k+1}\|$.
7. **exit if** $|\lambda_{k+1} - \lambda_k| \le \epsilon$.

return $k \leftarrow k+1$.
output: $\lambda \leftarrow \lambda_{k+1}$.

Algorithm 10.9 Z-eigenpair adaptive power method [259]

given: $\mathcal{A} \in \mathbb{S}^{[m,n]}$ and $\mathbf{x}_0 \leftarrow \hat{\mathbf{x}}_0/\|\hat{\mathbf{x}}_0\|$, where $\hat{\mathbf{x}}_0$ is an initial guess.
repeat
1. Compute $\mathcal{A}\mathbf{x}_k^{m-2}, \mathcal{A}\mathbf{x}_k^{m-1}$ and $\mathcal{A}\mathbf{x}_k^m$.
2. $\lambda_k \leftarrow \mathcal{A}\mathbf{x}_k^m$.
3. $\mathbf{H}_k \leftarrow m(m-1)\mathcal{A}\mathbf{x}_k^{(m-2)}$.
4. $\alpha_k \leftarrow \beta \max\{0, (\tau - \lambda_{\min}(\beta \mathbf{H}_k))/m\}$.
5. $\hat{\mathbf{x}}_{k+1} \leftarrow \beta \left(\mathcal{A}\mathbf{x}_k^{m-1} + \alpha_k \mathbf{x}_k\right)$.
6. $\mathbf{x}_{k+1} \leftarrow \hat{\mathbf{x}}_{k+1}/\|\hat{\mathbf{x}}_{k+1}\|$.
7. **exit if** $|\lambda_{k+1} - \lambda_k| \le \epsilon$.

return $k \leftarrow k+1$.
output: $\lambda \leftarrow \lambda_{k+1}$.

10.7.4 Orthogonal Decomposition of Symmetric Tensors

Denote the mth-order outer product of a vector $\mathbf{u} \in \mathbb{R}^n$ by $\mathbf{u}^{\otimes m} = \mathbf{u} \circ \mathbf{u} \circ \cdots \circ \mathbf{u}$ (m times). When $m = 2$, $\mathbf{u}^{\otimes 2} = \mathbf{u} \circ \mathbf{u} = \mathbf{u}\mathbf{u}^T$.

Using the eigenvalue decomposition, every real-valued symmetric matrix $\mathbf{M} \in \mathbb{S}^{[2,n]}$ can be written as

$$\mathbf{M} = \sum_{i=1}^{n} \lambda_i \mathbf{u}_i \mathbf{u}_i^T = \sum_{i=1}^{n} \lambda_i \mathbf{u}_i^{\otimes 2},$$

where $\mathbf{u}_1, \ldots, \mathbf{u}_n$ are orthonormal.

A real-valued mth-order n-dimensional symmetric tensor $\mathcal{A} \in \mathbb{S}^{[m,n]}$ is called *orthogonally decomposable* if

$$\mathcal{A} = \sum_{i=1}^{r} \lambda_i \mathbf{u}_i^{\otimes m}, \tag{10.7.37}$$

where $\mathbf{u}_1, \ldots, \mathbf{u}_r$ are orthonormal. The smallest r for which such a decomposition exists is known as the (symmetric) rank of $\mathcal{A}$.

If $\mathcal{A} \in \mathbb{S}^{[m,n]}$ is an orthogonally decomposable tensor, i.e., $\mathcal{A} = \sum_{i=1}^{r} \lambda_i \mathbf{u}_i^{\otimes m}$ and $\mathbf{u}_1, \ldots, \mathbf{u}_r \in \mathbb{R}^n$ are orthonormal then [415]

$$\mathcal{A}\mathbf{u}_j^{m-1} = \sum_{i=1}^{r} \lambda_i (\mathbf{u}_i \circ \mathbf{u}_j)^{m-1} \mathbf{u}_i = \lambda_j \mathbf{u}_j$$

for all $j = 1, \ldots, r$. Therefore, $\mathbf{u}_1, \ldots, \mathbf{u}_r$ are the Z-eigenvectors of $\mathcal{A}$, with corresponding Z-eigenvalues $\lambda_1, \ldots, \lambda_r$. Note that $\mathcal{A}$ and $\mathbf{u}_1, \ldots, \mathbf{u}_r$ are required to be real in order to force $\lambda_1, \ldots, \lambda_r$ to be real as well.

Definition 10.20 [415] A unit vector $\mathbf{u} = \mathbf{v}_k \in \mathbb{R}^n$ is a robust eigenvector of $\mathcal{A} \in \mathbb{S}^{[m,n]}$ if there exists $\epsilon > 0$ such that starting from one random vector $\mathbf{v}_0$, the repeated iteration

$$\mathbf{v}_{k+1} \leftarrow \frac{\mathcal{A}\mathbf{v}_k^{m-1}}{\|\mathcal{A}\mathbf{v}_k^{m-1}\|} \tag{10.7.38}$$

converges, i.e., $\|\mathbf{v}_{k+1} - \mathbf{v}_k\| < \epsilon$.

Algorithm 10.10 uses the repeated iteration (10.7.38) starting with random vectors $\mathbf{v}_0 \in \mathbb{R}^n$ to compute robust eigenvectors.

Algorithm 10.10 Tensor power method [259]

input: An orthogonally decomposable tensor $\mathcal{A}$.

initialization: $i = 1$.

repeat until $\mathcal{A} = 0$.

1. Choose one random vector $\mathbf{v}_0 \in \mathbb{R}^n$.
2. Let $\mathbf{u}_i = \mathbf{v}_k$ be the result of repeated iteration of (10.7.38) starting with $\mathbf{v}_0$.
3. Solve the equation $\mathcal{A}\mathbf{u}_i^{m-1} = \lambda_i \mathbf{u}_i$ to get the eigenvalue λ_i.
4. Set $\mathcal{A} = \mathcal{A} - \lambda_i \mathbf{u}_i^{\otimes m}$.
5. $i \leftarrow i + 1$.

output: $\mathbf{u}_1, \ldots, \mathbf{u}_r$ and $\lambda_1, \ldots, \lambda_r$.

10.8 Preprocessing and Postprocessing

In the previous sections, the theory and methods were introduced for multi-way data analysis. As with two-way data processing, multi-way data analysis also requires preprocessing and postprocessing [201], [36], [62], [245].

10.8.1 Centering and Scaling of Multi-Way Data

In matrix analysis, signal processing, pattern recognition, wireless communications etc., preprocessing most commonly involves the zero-mean normalization (i.e., centering) of data. Preprocessing in tensor analysis is more complicated than in matrix analysis, as it requires not only *data centering*, but also *data scaling*.

The main purpose of the centering is to remove the "direct current" component in the original data, which is the fixed component, keeping only the random components, which contain all the useful information.

For the third-order tensor $\mathcal{X} \in \mathbb{R}^{I\times J\times K}$, its *mode-n centering*, for $n = 1, 2, 3$, is respectively given by

$$x_{ijk}^{\text{cent }1} = x_{ijk} - \bar{x}_{:jk}, \quad \bar{x}_{:jk} = \frac{1}{I}\sum_{i=1}^{I} x_{ijk}, \tag{10.8.1}$$

$$x_{ijk}^{\text{cent }2} = x_{ijk} - \bar{x}_{i:k}, \quad \bar{x}_{i:k} = \frac{1}{J}\sum_{j=1}^{J} x_{ijk}, \tag{10.8.2}$$

$$x_{ijk}^{\text{cent }3} = x_{ijk} - \bar{x}_{ij:}, \quad \bar{x}_{ij:} = \frac{1}{K}\sum_{k=1}^{K} x_{ijk}. \tag{10.8.3}$$

If the centering is carried out on two modes then

$$x_{ijk}^{\text{cent }(1,2)} = x_{ijk} - \bar{x}_{::k}, \quad \bar{x}_{::k} = \frac{1}{IJ}\sum_{i=1}^{I}\sum_{j=1}^{J} x_{ijk}, \tag{10.8.4}$$

$$x_{ijk}^{\text{cent }(2,3)} = x_{ijk} - \bar{x}_{i::}, \quad \bar{x}_{i::} = \frac{1}{JK}\sum_{j=1}^{J}\sum_{k=1}^{K} x_{ijk}, \tag{10.8.5}$$

$$x_{ijk}^{\text{cent }(3,1)} = x_{ijk} - \bar{x}_{:j:}, \quad \bar{x}_{:j:} = \frac{1}{IK}\sum_{i=1}^{I}\sum_{k=1}^{K} x_{ijk}. \tag{10.8.6}$$

In addition, the tensor data must be scaled, i.e., every piece of data is divided by some fixed factor:

$$x_{ijk}^{\text{scal }1} = \frac{x_{ijk}}{s_i}, \quad s_i = \sqrt{\sum_{j=1}^{J}\sum_{k=1}^{K} x_{ijk}^2} \quad i = 1,\ldots,I, \tag{10.8.7}$$

$$x_{ijk}^{\text{scal }2} = \frac{x_{ijk}}{s_j}, \quad s_j = \sqrt{\sum_{i=1}^{I}\sum_{k=1}^{K} x_{ijk}^2} \quad j = 1,\ldots,J, \tag{10.8.8}$$

$$x_{ijk}^{\text{scal }3} = \frac{x_{ijk}}{s_k}, \quad s_k = \sqrt{\sum_{i=1}^{I}\sum_{j=1}^{J} x_{ijk}^2} \quad k = 1,\ldots,K. \tag{10.8.9}$$

It should be noted that it is usual only to center the columns of the unfolding matrices of a third-order tensor, and to scale the rows [62]. Therefore, we have

for Kiers matricization,

$$\begin{cases} \mathbf{X}^{I\times JK} = [\mathbf{X}_{::1},\ldots,\mathbf{X}_{::K}] & \text{centering } x_{ijk}^{\text{cent}\,2},\ \text{scaling } x_{ijk}^{\text{scal}\,1}, \\ \mathbf{X}^{J\times KI} = [\mathbf{X}_{1::},\ldots,\mathbf{X}_{I::}] & \text{centering } x_{ijk}^{\text{cent}\,3},\ \text{scaling } x_{ijk}^{\text{scal}\,2}, \\ \mathbf{X}^{K\times IJ} = [\mathbf{X}_{:1:},\ldots,\mathbf{X}_{:J:}] & \text{centering } x_{ijk}^{\text{cent}\,1},\ \text{scaling } x_{ijk}^{\text{scal}\,3}; \end{cases}$$

for LMV matricization,

$$\begin{cases} \mathbf{X}^{I\times JK} = [\mathbf{X}_{:1:}^T,\ldots,\mathbf{X}_{:J:}^T] & \text{centering } x_{ijk}^{\text{cent}\,3},\ \text{scaling } x_{ijk}^{\text{scal}\,1}, \\ \mathbf{X}^{J\times KI} = [\mathbf{X}_{::1}^T,\ldots,\mathbf{X}_{::K}^T] & \text{centering } x_{ijk}^{\text{cent}\,1},\ \text{scaling } x_{ijk}^{\text{scal}\,2}, \\ \mathbf{X}^{K\times IJ} = [\mathbf{X}_{1::}^T,\ldots,\mathbf{X}_{I::}^T] & \text{centering } x_{ijk}^{\text{cent}\,2},\ \text{scaling } x_{ijk}^{\text{scal}\,3}; \end{cases}$$

for Kolda matricization,

$$\begin{cases} \mathbf{X}^{I\times JK} = [\mathbf{X}_{::1},\ldots,\mathbf{X}_{::K}] & \text{centering } x_{ijk}^{\text{cent}\,2},\ \text{scaling } x_{ijk}^{\text{scal}\,1}, \\ \mathbf{X}^{J\times KI} = [\mathbf{X}_{::1}^T,\ldots,\mathbf{X}_{::K}^T] & \text{centering } x_{ijk}^{\text{cent}\,1},\ \text{scaling } x_{ijk}^{\text{scal}\,2}, \\ \mathbf{X}^{K\times IJ} = [\mathbf{X}_{:1:}^T,\ldots,\mathbf{X}_{:J:}^T] & \text{centering } x_{ijk}^{\text{cent}\,1},\ \text{scaling } x_{ijk}^{\text{scal}\,3}. \end{cases}$$

10.8.2 Compression of Data Array

When the dimension of a higher-order array is large (such as when I_n is in the hundreds or thousands), the low-convergence problem of the ALS for a CP decomposition is usually serious. In these cases, the data array has to be compressed: the original Nth-order tensor $\mathcal{X} \in \mathbb{R}^{I_1\times\cdots\times I_N}$ is compressed into an Nth-order core tensor $\mathcal{G} \in \mathbb{R}^{J_1\times\cdots\times J_N}$ with smaller dimensions, then the ALS algorithm is run on the CP decomposition of the core tensor and finally the mode matrices of the CP decomposition of the original tensor are reconstructed from the mode matrices of the CP decomposition of the core tensor.

As an example, consider the CP decomposition of a third-order tensor $\mathcal{X} \in \mathbb{R}^{I_1\times I_2\times I_3}$:

$$\mathbf{X}_{(1)} = \mathbf{A}(\mathbf{C}\odot\mathbf{B})^T, \quad \mathbf{X}_{(2)} = \mathbf{B}(\mathbf{A}\odot\mathbf{C})^T, \quad \mathbf{X}_{(3)} = \mathbf{C}(\mathbf{B}\odot\mathbf{A})^T.$$

Assume that some dimension (say I_3) is very large. Because the dimensions of the matricization $\mathbf{X}_{(n)}, n = 1,2,3$, are very large, an ALS algorithm for computing the factor matrices $\mathbf{A} \in \mathbb{R}^{I_1\times R}, \mathbf{B} \in \mathbb{R}^{I_2\times R}, \mathbf{C} \in \mathbb{R}^{I_3\times R}$ directly would face a troublesome amount of computation and slow convergence.

Choose three orthogonal matrices $\mathbf{U} \in \mathbb{R}^{I_1\times J_1}, \mathbf{V} \in \mathbb{R}^{I_2\times J_2}, \mathbf{W} \in \mathbb{R}^{I_3\times J_3}$ such that

$$\mathbf{A} = \mathbf{UP}, \quad \mathbf{B} = \mathbf{VQ}, \quad \mathbf{C} = \mathbf{WR}.$$

Then, the identification of the three mode matrices $\mathbf{A},\mathbf{B},\mathbf{C}$ of the CP decomposition of the third-order tensor is divided into two subproblems, that of identifying an orthogonal matrix triple $(\mathbf{U}, \mathbf{V}, \mathbf{W})$ and that of identifying a nonorthogonal matrix triple $(\mathbf{P}, \mathbf{Q}, \mathbf{R})$.

Using the relation between the Khatri–Rao product and the Kronecker product, $(\mathbf{WR}) \odot (\mathbf{VQ}) = (\mathbf{W} \otimes \mathbf{V})(\mathbf{R} \odot \mathbf{Q})$, it is easy to see that

$$\mathbf{X}_{(1)} = \mathbf{A}(\mathbf{C} \odot \mathbf{B})^T = \mathbf{UP}\left[(\mathbf{WR}) \odot (\mathbf{VQ})\right]^T = \mathbf{UP}(\mathbf{R} \odot \mathbf{Q})^T(\mathbf{W} \otimes \mathbf{V})^T.$$

If we let $\mathbf{G}_{(1)} = \mathbf{P}(\mathbf{R} \odot \mathbf{Q})^T$ then $\mathbf{X}_{(1)}$ can be rewritten as

$$\mathbf{X}_{(1)} = \mathbf{UG}_{(1)}(\mathbf{W} \odot \mathbf{V})^T. \tag{10.8.10}$$

Similarly, it can be shown that

$$\mathbf{X}_{(2)} = \mathbf{VG}_{(2)}(\mathbf{U} \odot \mathbf{W})^T, \tag{10.8.11}$$
$$\mathbf{X}_{(3)} = \mathbf{WG}_{(3)}(\mathbf{V} \odot \mathbf{U})^T, \tag{10.8.12}$$

where

$$\mathbf{G}_{(2)} = \mathbf{Q}(\mathbf{P} \odot \mathbf{R})^T, \quad \mathbf{G}_{(3)} = \mathbf{R}(\mathbf{Q} \odot \mathbf{P})^T.$$

Equations (10.8.10)–(10.8.12) show that:

(1) $\mathcal{G} \in \mathbb{R}^{J_1 \times J_2 \times J_3}$ is the core tensor of the Tucker decomposition of the third-order tensor $\mathcal{X}$, and $\mathbf{G}_{(n)}, n = 1, 2, 3$, are the matricization results for the core tensor;
(2) the matrix triple $(\mathbf{P}, \mathbf{Q}, \mathbf{R})$ is the factor matrix triple of the CP decomposition of the core tensor $\mathcal{G}$.

The above considerations result in the data array compression algorithm shown in Algorithm 10.11.

The following are some remarks on this data array compression algorithm.

Remark 1 The essence of data array compression is that the CP decomposition of the original tensor $\mathcal{X} \in \mathbb{R}^{I_1 \times I_2 \times I_3}$ with large dimension is compressed into the CP decomposition of a core tensor $\mathcal{G} \in \mathbb{R}^{J_1 \times J_2 \times J_3}$ with small dimension.

Remark 2 The compression algorithm needs no iteration; its computation is simple, and thus a convergence problem does not exist.

Remark 3 If the factor matrices $\mathbf{B}$ and $\mathbf{C}$ found by the compression algorithm are used as the initial matrices to run the ALS algorithm for the CP decomposition of the original third-order tensor $\mathcal{X} \in \mathbb{R}^{I_1 \times I_2 \times I_3}$, then the ALS algorithm requires only a few iterations to achieve convergence. Therefore, data array compression can effectively accelerate the CP decomposition of a higher-dimensional tensor.

Remark 4 The compression algorithm given in Algorithm 10.11 is based on the Kiers matricization. If in the unfolding matrices and the mode-n matrices appropriate replacements are made, Algorithm 10.11 can be used for the LMV or Kolda matricizations.

After finding the factor (i.e., loading) matrices $\mathbf{A}, \mathbf{B}, \mathbf{C}$ via the ALS algorithm, it

Algorithm 10.11 Data array compression algorithm [62]

input: $\mathcal{X} \in \mathbb{R}^{I_1 \times I_2 \times I_3}$.

initialize: $\mathbf{X}_{(1)}, \mathbf{X}_{(2)}, \mathbf{X}_{(3)}$.

1. Compute the SVD of the unfolding matrices

$$[\mathbf{U}, \mathbf{S}, \mathbf{T}] = \mathrm{SVD}(\mathbf{X}_{(1)}, J_1),$$
$$[\mathbf{V}, \mathbf{S}, \mathbf{T}] = \mathrm{SVD}(\mathbf{X}_{(2)}, J_2),$$
$$[\mathbf{W}, \mathbf{S}, \mathbf{T}] = \mathrm{SVD}(\mathbf{X}_{(3)}, J_3),$$

where J_i are the numbers of leading singular values of $\mathbf{X}_{(i)}, i = 1, 2, 3$.

2. Compute the mode-n matrices of the core tensor,

$$\mathbf{G}_{(1)} = \mathbf{U}^T \mathbf{X}_{(1)} (\mathbf{W} \otimes \mathbf{V}),$$
$$\mathbf{G}_{(2)} = \mathbf{V}^T \mathbf{X}_{(2)} (\mathbf{U} \otimes \mathbf{W}),$$
$$\mathbf{G}_{(3)} = \mathbf{W}^T \mathbf{X}_{(3)} (\mathbf{V} \otimes \mathbf{U}).$$

3. Apply Algorithm 10.5 to make the CP decompositions

$$\mathbf{G}_{(1)} = \mathbf{P}(\mathbf{R} \odot \mathbf{Q})^T,$$
$$\mathbf{G}_{(2)} = \mathbf{Q}(\mathbf{P} \odot \mathbf{R})^T,$$
$$\mathbf{G}_{(3)} = \mathbf{R}(\mathbf{Q} \odot \mathbf{P})^T,$$

where $\mathbf{P} \in \mathbb{R}^{J_1 \times R}, \mathbf{Q} \in \mathbb{R}^{J_2 \times R}, \mathbf{R} \in \mathbb{R}^{J_3 \times R}$.

4. Find the factor matrices of the CP decomposition of the original tensor:

$$\mathbf{A} = \mathbf{U}\mathbf{P},\ \mathbf{B} = \mathbf{V}\mathbf{Q},\ \mathbf{C} = \mathbf{W}\mathbf{R}.$$

output: $\mathbf{A} \in \mathbb{R}^{I_1 \times J_1}, \mathbf{B} \in \mathbb{R}^{I_2 \times J_2}, \mathbf{C} \in \mathbb{R}^{I_3 \times J_3}$.

---

is necessary to make a quality evaluation of these loading matrices. This evaluation is called *leverage analysis* [62].

Because the loading matrices in CP decomposition are nonorthogonal, one needs to compute a leverage vector for every loading matrix.

The *leverage vector* of loading matrix $\mathbf{U}$ is defined as the vector consisting of the diagonal entries of the projection matrix of $\mathbf{U}$, namely [62]

$$\mathbf{v} = \mathbf{diag}(\mathbf{U}(\mathbf{U}^T\mathbf{U})^{-1}\mathbf{U}^T). \tag{10.8.13}$$

Letting $\mathbf{U} = \mathbf{A}, \mathbf{B}, \mathbf{C}$ in sequence, one obtains the leverage vectors of three loading matrices. The values of the entries of a leverage vector are between 0 and 1, i.e., $0 \le v_i \le 1$. If the value of some leverage entry is larger then the leverage of the sample value used is larger; otherwise, the leverage is smaller. If some leverage entry is very large, then the sample data may perhaps contain some outlier; the model thus obtained would not be appropriate, and one would need to eliminate this outlier and restart the ALS algorithm to make a new tensor analysis.

10.9 Nonnegative Tensor Decomposition Algorithms

As stated in Chapter 6, nonnegative matrix factorization (NMF) is more useful than principal component analysis (PCA) or independent component analysis (ICA) in much modern data analysis. In this section, we consider the generalization of the NMF to a multi-way array (tensor).

The study of nonnegative tensor decomposition (NTD) was first proposed by researchers studying nonnegatively constrained PARAFAC in chemometrics, [63], [66], [366], [368].

The *nonnegative tensor decomposition* problem is: given an Nth-order nonnegative tensor $\mathcal{X} \in \mathbb{R}_+^{I_1\times\cdots\times I_N}$, decompose it into

$$\begin{aligned} \text{NTD 1:} \quad & \mathcal{X} \approx \mathcal{G} \times_1 \mathbf{A}^{(1)} \times_2 \cdots \times_N \mathbf{A}^{(N)} \\ & \text{with } \mathcal{G} \geq 0, \mathbf{A}^{(1)} \geq 0, \ldots, \mathbf{A}^{(N)} \geq 0, \end{aligned} \tag{10.9.1}$$

$$\begin{aligned} \text{NTD 2:} \quad & \mathcal{X} \approx \mathcal{I} \times_1 \mathbf{A}^{(1)} \times_2 \cdots \times_N \mathbf{A}^{(N)} \\ & \text{with } \mathbf{A}^{(1)} \geq 0, \ldots, \mathbf{A}^{(N)} \geq 0, \end{aligned} \tag{10.9.2}$$

where the expressions $\mathcal{G} \geq 0$ and $\mathbf{A} \geq 0$ denote that all the entries of the tensor $\mathcal{G}$ and of the matrix $\mathbf{A}$ respectively are nonnegative.

The NTD problem can be expressed as the minimization of the objective function:

$$\text{NTD 1:} \quad \min_{\mathcal{G},\mathbf{A}^{(1)},\ldots,\mathbf{A}^{(N)}} \frac{1}{2}\left\|\mathcal{X} - \mathcal{G} \times_1 \mathbf{A}^{(1)} \times_2 \cdots \times_N \mathbf{A}^{(N)}\right\|_F^2, \tag{10.9.3}$$

$$\text{NTD 2:} \quad \min_{\mathbf{A}^{(1)},\ldots,\mathbf{A}^{(N)}} \frac{1}{2}\left\|\mathcal{X} - \mathcal{I} \times_1 \mathbf{A}^{(1)} \times_2 \cdots \times_N \mathbf{A}^{(N)}\right\|_F^2. \tag{10.9.4}$$

The basic idea of NTD is that the NTD problem is rewritten in the basic form of NMF: $\mathbf{X} = \mathbf{AS}, \mathbf{A} \geq 0, \mathbf{S} \geq 0$. In NTD there are two classes of common algorithms: multiplication updating algorithms and ALS updating algorithms.

10.9.1 Multiplication Algorithm

The key to designing a multiplication algorithm for NTD is rewriting the Tucker decomposition or the CP decomposition of a nonnegative tensor in the standard NMF form $\mathbf{X} = \mathbf{AS}$.

1. Multiplication Algorithm for Nonnegative Tensor Decomposition

Let $\mathcal{X} = [\![x_{i_1\cdots i_N}]\!] \in \mathbb{R}_+^{I_1\times\cdots\times I_N}$ be a nonnegative tensor, and consider its approximate Tucker decomposition

$$\mathcal{X} \approx \mathcal{G} \times_1 \mathbf{A}^{(1)} \times_2 \mathbf{A}^{(2)} \times_3 \cdots \times_N \mathbf{A}^{(N)}, \tag{10.9.5}$$

where $x_{i_1\cdots i_N} \geq 0$ and $\mathbf{A}^{(n)} = [a_{ij}^{(n)}] \in \mathbb{R}_+^{I_n\times J_n}, n = 1,\ldots,N$ are N nonnegative factor matrices; $\mathcal{G} = [\![g_{j_1\cdots j_N}]\!] \in \mathbb{R}_+^{J_1\times\cdots\times J_N}$ is the nonnegative core tensor, i.e.,

$a_{ij}^{(n)} \geq 0, \forall i = 1,\ldots,I_n,\ j = 1,\ldots,J_n,\ n = 1,\ldots,N$ and $g_{j_1\cdots j_N} \geq 0, \forall j_n = 1,\ldots,J_n,\ n = 1,\ldots,N$. This Tucker decomposition with a nonnegative core tensor and all nonnegative factor matrices is known as a *nonnegative Tucker decomposition.*

In the Tucker decomposition of a general tensor, all the columns of every factor matrix are required to be orthogonal, but this requirement is no longer reasonable in a nonnegative Tucker decomposition. This is the essential difference between nonnegative Tucker decomposition and higher-order SVD.

Using the unfolding matrices of the nonnegative tensor $\mathcal{X}$, an NTD can be also expressed in the matricization form

$$\mathbf{X}_{(n)} = \mathbf{A}^{(n)}\mathbf{G}_{(n)}\mathbf{A}_{\otimes}^{(n)T}; \tag{10.9.6}$$

here $\mathbf{X}_{(n)} \in \mathbb{R}_{+}^{I_n\times L}$, $\mathbf{G}_{(n)} \in \mathbb{R}_{+}^{J_n\times(J_1\cdots J_{n-1}J_{n+1}\cdots J_N)}$, $\mathbf{A}_{\otimes}^{(n)} \in \mathbb{R}_{+}^{L\times(J_1\cdots J_{n-1}J_{n+1}\cdots J_N)}$, $\mathbf{A}^{(n)} \in \mathbb{R}_{+}^{I_n\times J_n}$ and

$$\mathbf{A}_{\otimes}^{(n)} = \begin{cases} \mathbf{A}^{(n-1)} \otimes \cdots \otimes \mathbf{A}^{(1)} \otimes \mathbf{A}^{(N)} \otimes \cdots \otimes \mathbf{A}^{(n+1)} & \text{(Kiers matricization)} \\ \mathbf{A}^{(n+1)} \otimes \cdots \otimes \mathbf{A}^{(N)} \otimes \mathbf{A}^{(1)} \otimes \cdots \otimes \mathbf{A}^{(n-1)} & \text{(LMV matricization)} \\ \mathbf{A}^{(N)} \otimes \cdots \otimes \mathbf{A}^{(n+1)} \otimes \mathbf{A}^{(n-1)} \otimes \cdots \otimes \mathbf{A}^{(1)} & \text{(Kolda matricization)} \end{cases} \tag{10.9.7}$$

denotes the Kronecker product of $n-1$ factor matrices, excepting for the mode-n matrix.

Defining the cost function

$$J(\mathbf{A}^{(n)},\mathbf{G}_{(n)}) = \frac{1}{2}\left\|\mathbf{X}_{(n)} - \mathbf{A}^{(n)}\mathbf{G}_{(n)}\mathbf{A}_{\otimes}^{(n)T}\right\|_F^2, \tag{10.9.8}$$

and writing its matrix differential as standard forms $\mathrm{d}J(\mathbf{A}^{(n)},\mathbf{G}_{(n)}) = \mathrm{tr}(\mathbf{W}\mathrm{d}\mathbf{A}^{(n)})$ and $\mathrm{d}J(\mathbf{A}^{(n)},\mathbf{G}_{(n)}) = \mathrm{tr}(\mathbf{V}\mathrm{d}\mathbf{G}_{(n)})$, respectively, then

$$\frac{\partial J(\mathbf{A}^{(n)},\mathbf{G}_{(n)})}{\partial \mathbf{A}^{(n)}} = -\mathbf{X}_{(n)}\mathbf{A}_{\otimes}^{(n)}\mathbf{G}_{(n)}^T + \mathbf{A}^{(n)}\mathbf{G}_{(n)}\mathbf{A}_{\otimes}^{(n)T}\mathbf{A}_{\otimes}^{(n)}\mathbf{G}_{(n)}^T, \tag{10.9.9}$$

$$\frac{\partial J(\mathbf{A}^{(n)},\mathbf{G}_{(n)})}{\partial \mathbf{G}_{(n)}} = -\mathbf{A}^{(n)T}\mathbf{X}_{(n)}\mathbf{A}_{\otimes}^{(n)} + \mathbf{A}^{(n)T}\mathbf{A}^{(n)}\mathbf{G}_{(n)}\mathbf{A}_{\otimes}^{(n)T}\mathbf{A}_{\otimes}^{(n)}. \tag{10.9.10}$$

Hence, we get the following gradient algorithms:

$$\mathbf{A}^{(n)} \leftarrow \mathbf{A}^{(n)} + \eta_A\left(\mathbf{X}_{(n)}\mathbf{M}_{(n)} - \mathbf{A}^{(n)}\mathbf{M}_{(n)}^T\mathbf{M}_{(n)}\right),$$

$$\mathbf{G}_{(n)} \leftarrow \mathbf{G}_{(n)} + \eta_G\left(\mathbf{A}^{(n)T}\mathbf{X}_{(n)}\mathbf{A}_{\otimes}^{(n)} - \mathbf{A}^{(n)T}\mathbf{A}^{(n)}\mathbf{G}_{(n)}\mathbf{A}_{\otimes}^{(n)T}\mathbf{A}_{\otimes}^{(n)}\right),$$

where $\mathbf{M}_{(n)} = \mathbf{A}_{\otimes}^{(n)}\mathbf{G}_{(n)}^T$.

Taking step sizes

$$\eta_{ij} = \frac{a^{(n)}(i,j)}{\left[\mathbf{A}^{(n)}\mathbf{M}_{(n)}^T\mathbf{M}_{(n)}\right]_{ij}}, \quad i=1,\ldots,I_n,\ j=1,\ldots,J_n$$

$$\eta_{jk} = \frac{g_{(n)}(j,k)}{\left[\mathbf{A}^{(n)T}\mathbf{A}^{(n)}\mathbf{G}_{(n)}\mathbf{A}_{\otimes}^{(n)T}\mathbf{A}_{\otimes}^{(n)}\right]_{jk}}, \quad j=1,\cdots,J_n,\ k=1,\ldots,\prod_{\substack{i=1\\ i\neq n}}^{N} J_i$$

then the above gradient algorithm becomes the *multiplication updating algorithm* for the nonnegative Tucker decomposition:

$$a^{(n)}(i,j) \leftarrow a^{(n)}(i,j)\frac{[\mathbf{X}_{(n)}\mathbf{M}_{(n)}]_{ij}}{\left[\mathbf{A}^{(n)}\mathbf{M}_{(n)}^T\mathbf{M}_{(n)}\right]_{ij}}, \tag{10.9.11}$$

$$g_{(n)}(j,k) \leftarrow g_{(n)}(j,k)\frac{\mathbf{A}^{(n)T}\mathbf{X}_{(n)}\mathbf{A}_{\otimes}^{(n)}}{\left[\mathbf{A}^{(n)T}\mathbf{A}^{(n)}\mathbf{G}_{(n)}\mathbf{A}_{\otimes}^{(n)T}\mathbf{A}_{\otimes}^{(n)}\right]_{jk}}, \tag{10.9.12}$$

which is written in matrix form as

$$\begin{aligned}
\mathbf{A}^{(n)} &\leftarrow \mathbf{A}^{(n)} * \Big(\big(\mathbf{X}_{(n)}\mathbf{M}_{(n)}\big) \oslash \big(\mathbf{A}^{(n)}\mathbf{M}_{(n)}^T\mathbf{M}_{(n)}\big)\Big),\\
\mathbf{G}_{(n)} &\leftarrow \mathbf{G}_{(n)} * \Big(\big(\mathbf{A}^{(n)T}\mathbf{X}_{(n)}\mathbf{A}_{\otimes}^{(n)}\big) \oslash \big(\mathbf{A}^{(n)T}\mathbf{A}^{(n)}\mathbf{G}_{(n)}\mathbf{A}_{\otimes}^{(n)T}\mathbf{A}_{\otimes}^{(n)}\big)\Big),
\end{aligned} \tag{10.9.13}$$

where $\mathbf{C} = \mathbf{A} * \mathbf{B}$ is the Hadamard Product $c_{ij} = a_{ij}b_{ij}$, and $\mathbf{F} = \mathbf{D} \oslash \mathbf{E}$ denotes $f_{ij} = d_{ij}/e_{ij}$.

2. Multiplication Algorithm for Nonnegative CP Decomposition

Consider the *nonnegative CP decomposition* of an Nth-order nonnegative tensor $\mathcal{X} \in \mathbb{R}_+^{I_1\times\cdots\times I_N}$:

$$x_{i_1\cdots i_N} = \sum_{r=1}^{R} a_{i_1,r}^{(1)} a_{i_2,r}^{(2)} \cdots a_{i_N,r}^{(N)} \quad \text{subject to} \quad a_{i_n,r}^{(n)} \geq 0,\ \forall\, i_n, r, \tag{10.9.14}$$

such that the reconstructed sum of squared errors (RE) is given by

$$\mathrm{RE}(a_{i_1,r}^{(1)},\ldots,a_{i_N,r}^{(N)}) = \sum_{i_1=1}^{I_1}\cdots\sum_{i_N=1}^{I_N}\left(x_{i_1\cdots i_N} - \sum_{r=1}^{R} a_{i_1,r}^{(1)} a_{i_2,r}^{(2)} \cdots a_{i_N,r}^{(N)}\right)^2,$$

which is minimized.

The nonnegative CP decomposition of the Nth-order nonnegative tensor can be equivalently written as

$$\mathbf{X}_{(n)} = \mathbf{A}^{(n)}\mathbf{S}_{(n)}^T \quad \in \mathbb{R}_+^{I_n\times I_1\cdots I_{n-1}I_{n+1}\cdots I_N}, \tag{10.9.15}$$

where $\mathbf{A}^{(n)} \in \mathbb{R}_+^{I_n \times R}, \mathbf{S}_{(n)} \in \mathbb{R}_+^{(I_1 \cdots I_{n-1} I_{n+1} \cdots I_N) \times R}$ and

$$\mathbf{S}_{(n)} = \begin{cases} \mathbf{A}^{(n-1)} \odot \cdots \odot \mathbf{A}^{(1)} \odot \mathbf{A}^{(N)} \odot \cdots \odot \mathbf{A}^{(n+1)} & \text{(Kiers matricization)}, \\ \mathbf{A}^{(n+1)} \odot \cdots \odot \mathbf{A}^{(N)} \odot \mathbf{A}^{(1)} \odot \cdots \odot \mathbf{A}^{(n-1)} & \text{(LMV matricization)}, \\ \mathbf{A}^{(N)} \odot \cdots \odot \mathbf{A}^{(n+1)} \odot \mathbf{A}^{(n-1)} \odot \cdots \odot \mathbf{A}^{(1)} & \text{(Kolda matricization)}. \end{cases} \tag{10.9.16}$$

Because $\mathbf{X}_{(n)} = \mathbf{A}^{(n)} \mathbf{S}_{(n)}^T$ has the standard NMF form $\mathbf{X} = \mathbf{AS}$, the multiplication algorithm for NMF can be directly extended as the multiplication algorithm for nonnegative CP decomposition of an Nth-order tensor:

$$a^{(n)}(i_n, r) \leftarrow a^{(n)}(i_n, r) \frac{[\mathbf{X}_{(n)} \mathbf{S}_{(n)}]_{i_n, r}}{[\mathbf{A}^{(n)} \mathbf{S}_{(n)}^T \mathbf{S}_{(n)}]_{i_n, r}}, \quad i_n = 1, \ldots, I_n,$$

$$s_{(n)}^T(i, r) \leftarrow s_{(n)}^T(i, r) \frac{[\mathbf{A}^{(n)T} \mathbf{X}_{(n)}]_{i,r}}{[\mathbf{A}^{(n)T} \mathbf{A}^{(n)} \mathbf{S}_{(n)}^T]_{i,r}}, \quad i = 1, \ldots, I_1 \cdots I_{n-1} I_{n+1} \cdots I_N.$$

Here $s_{(n)}^T(i, r)$ denotes the (i, r)th-entry of $\mathbf{S}_{(n)}^T$, and $r = 1, \ldots, R$.

The above entrywise multiplication algorithm can be also written in the following matrix forms:

$$\mathbf{A}^{(n)} \leftarrow \mathbf{A}^{(n)} * \left((\mathbf{X}_{(n)} \mathbf{S}_{(n)}) \oslash (\mathbf{A}^{(n)} \mathbf{S}_{(n)}^T \mathbf{S}_{(n)}) \right), \tag{10.9.17}$$

$$\mathbf{S}_{(n)}^T \leftarrow \mathbf{S}_{(n)}^T * \left((\mathbf{A}^{(n)T} \mathbf{X}_{(n)}) \oslash (\mathbf{A}^{(n)T} \mathbf{A}_{(n)} \mathbf{S}_{(n)}^T) \right). \tag{10.9.18}$$

The above discussion can be summarized as a multiplication algorithm for nonnegative CP (NCP) decomposition; see Algorithm 10.12.

Regarding the positive-tensor decomposition algorithms, interested readers can refer to [506].

10.9.2 ALS Algorithms

In the following, we discuss ALS methods for the Tucker decomposition and CP decomposition of nonnegative tensors.

1. ALS Method for Nonnegative Tucker Decomposition

Consider the nonnegative Tucker decomposition

$$\min_{\mathcal{G}, \mathbf{A}^{(1)}, \ldots, \mathbf{A}^{(N)}} \frac{1}{2} \left\| \mathcal{G} \times_1 \mathbf{A}^{(1)} \times_2 \cdots \times_N \mathbf{A}^{(N)} - \mathcal{X} \right\|_F^2, \tag{10.9.19}$$

which can be written in matrix form as

$$\min_{\mathcal{G}, \mathbf{A}^{(1)}, \ldots, \mathbf{A}^{(N)}} \frac{1}{2} \sum_{n=1}^{N} \left\| \mathbf{A}^{(n)} \mathbf{G}_{(n)} \mathbf{A}_{\otimes}^{n} - \mathbf{X}_{(n)} \right\|_F^2, \tag{10.9.20}$$

where $\mathbf{A}_{\otimes}^{n}$ are given by (10.9.7).

Algorithm 10.12 Multiplication algorithm for NCP decomposition [254]

input: $\mathcal{X}$.

initialize: $\mathbf{A}_0^{(n)} \in \mathbb{R}^{I_n \times J_n}, n = 1, \ldots, N$.

repeat

1. **for** $n = 1, \ldots, N$ **do**

   1.1. Denote $I_n = \{i_1, \ldots, i_{n-1}, i_{n+1}, \ldots, i_N\}$, and define $y_{i_n, I_n} = x_{i_1 \cdots i_N}$.

   1.2. $m_{I_n, r} = a_{i_1, r}^{(1)} \cdots a_{i_{n-1}, r}^{(n-1)} a_{i_{n+1}, r}^{(n+1)} \cdots a_{i_N, r}^{(N)}, r = 1, \ldots, R$.

   1.3. $\mathbf{A}^{(n)} \leftarrow \mathbf{A}^{(n)} * ((\mathbf{YM}) \oslash (\mathbf{A}^{(n)}\mathbf{M}^T\mathbf{M}))$.

   1.4. $K_r^{(n)} = \sqrt{\sum_{n=1}^N \left(a_{i_n, r}^{(n)}\right)^2}, r = 1, \ldots, R$.

   1.5. $a_{i_n, r}^{(n)} \leftarrow a_{i_n, r}^{(n)} \dfrac{\left(\prod_{i=1}^N K_r^{(i)}\right)^{1/N}}{K_r^{(n)}}, r = 1, \ldots, R$.

2. **end for**

3. **exit if** the stopping criterion is satisfied.

return

output: $\mathbf{A}^{(n)} \in \mathbb{R}^{I_n \times R}, n = 1, \ldots, N$.

At the $(k+1)$th iteration, if $\mathbf{A}_{k+1}^{(1)}, \ldots, \mathbf{A}_{k+1}^{(n-1)}, \mathbf{A}_k^{(n+1)}, \ldots, \mathbf{A}_k^{(N-1)}$ are fixed, and the horizontal unfolding matrices $\mathbf{G}_{(n)}^k$ of the core tensor are known, then, from (10.9.20), it is easy to see that finding $\mathbf{A}_{k+1}^{(n)}$ corresponds to finding the LS solution of the matrix equation $\mathbf{A}^{(n)}\mathbf{G}_{(n)}\mathbf{A}_\otimes^n \approx \mathbf{X}_{(n)}$, namely

$$\mathbf{A}_{k+1}^{(n)} = \mathcal{P}_+ \left(\mathbf{X}_{(n)}(\mathbf{G}_{(n)}^k \mathbf{S}_{k+1}^{(n)})^\dagger\right), \quad n = 1, \ldots, N, \tag{10.9.21}$$

where $\mathbf{B}^\dagger$ is the Moore–Penrose inverse matrix of $\mathbf{B}$, $[\mathcal{P}_+(\mathbf{C})]_{ij} = \max\{0, C_{ij}\}$ denotes the nonnegative constraint of the matrix entry, and

$$\mathbf{S}_{k+1}^{(n)} = \begin{cases} \mathbf{A}_{k+1}^{(n-1)} \otimes \cdots \otimes \mathbf{A}_{k+1}^{(1)} \otimes \mathbf{A}_k^{(N)} \otimes \cdots \otimes \mathbf{A}_k^{(n+1)} & \text{(Kiers matricization)}, \\ \mathbf{A}_k^{(n+1)} \otimes \cdots \otimes \mathbf{A}_k^{(N)} \otimes \mathbf{A}_{k+1}^{(1)} \otimes \cdots \otimes \mathbf{A}_{k+1}^{(n-1)} & \text{(LMV matricization)}, \\ \mathbf{A}_k^{(N)} \otimes \cdots \otimes \mathbf{A}_k^{(n+1)} \otimes \mathbf{A}_{k+1}^{(n-1)} \otimes \cdots \otimes \mathbf{A}_{k+1}^{(1)} & \text{(Kolda matricization)}. \end{cases} \tag{10.9.22}$$

In order to update $\mathbf{G}_{(n)}^k$, using the vectorization formula $\text{vec}(\mathbf{ABC}) = (\mathbf{C}^T \otimes \mathbf{A})\text{vec}\,\mathbf{B}$, then the equivalent matrix equation $\mathbf{A}^{(n)}\mathbf{G}_{(n)}\mathbf{A}_\otimes^n \approx \mathbf{X}_{(n)}$ in (10.9.20) can be written in vectorization form as

$$\text{vec}(\mathbf{G}_{(n)}^{k+1}) = \mathcal{P}_+ \left((\mathbf{S}_{k+1}^{(n)T} \otimes \mathbf{A}_{k+1}^{(n)})^\dagger \text{vec}\,\mathbf{X}_{(n)}\right). \tag{10.9.23}$$

The above discussions can be summarized as an ALS algorithm for the nonnegative Tucker decomposition of an Nth-order tensor, as shown in Algorithm 10.13 [164].

Algorithm 10.13 ALS algorithm for nonnegative Tucker decomposition

---

input: Nth-order tensor $\mathcal{X}$.

initialize: $\mathbf{A}_0^{(n)} \in \mathbb{R}^{I_n \times J_n}, n = 1, \ldots, N$. Put $k = 0$.

repeat

1. Horizontally unfold $\mathcal{X}$ into the matrices $\mathbf{X}_{(n)}, n = 1, \ldots, N$.
2. Use (10.9.22) to update $\mathbf{S}_{k+1}^{(n)}, n = 1, \ldots, N$.
3. Use (10.9.21) to update $\mathbf{A}_{k+1}^{(n)}, n = 1, \ldots, N$.
4. Use (10.9.23) to update $\text{vec}(\mathbf{G}_{(n)}^{k+1}), n = 1, \ldots, N$.
5. **exit if** the convergence criterion is met or the maximum number of iteration is exhausted.

return $k \leftarrow k + 1$.

output: $\mathbf{A}^{(n)}$ and $\text{vec}\, \mathbf{G}_{(n)}$, $n = 1, \ldots, N$.

---

2. ALS Algorithm for Nonnegative CP Decomposition

The ALS method and the regularized ALS method for the CP decomposition of a tensor, described in Section 10.5, are easily generalized to an ALS method and a regularized ALS method for nonnegative tensor CP decomposition, respectively. The only operation added is to constrain every updated factor matrix $\mathbf{A}_{k+1}^{(n)}$ to meet the nonnegative condition $\mathcal{P}_+(\mathbf{A}_{k+1}^{(n)})$, where $[\mathcal{P}_+(\mathbf{A}_{k+1}^{(n)})]_{ij} = \max\{0, \mathbf{A}_{k+1}^{(n)}(i,j)\}$.

An Nth-order nonnegative CP decomposition can be written in the standard form of the NMF,

$$\mathbf{X}_{(n)} = \mathbf{A}^{(n)}\mathbf{S}^{(n)T}, \tag{10.9.24}$$

where

$$\mathbf{S}^{(n)} = \begin{cases} \mathbf{A}^{(n-1)} \odot \cdots \odot \mathbf{A}^{(1)} \odot \mathbf{A}^{(N)} \odot \cdots \odot \mathbf{A}^{(n+1)} & \text{(Kiers matricization)}, \\ \mathbf{A}^{(n+1)} \odot \cdots \odot \mathbf{A}_k^{(N)} \odot \mathbf{A}^{(1)} \odot \cdots \odot \mathbf{A}^{(n-1)} & \text{(LMV matricization)}, \\ \mathbf{A}_k^{(N)} \odot \cdots \odot \mathbf{A}^{(n+1)} \odot \mathbf{A}^{(n-1)} \odot \cdots \odot \mathbf{A}^{(1)} & \text{(Kolda matricization)}. \end{cases} \tag{10.9.25}$$

From (10.9.24) we obtain directly an LS solution of the factor matrix,

$$\mathbf{A}^{(n)} = \mathbf{X}_{(n)}(\mathbf{S}^{(n)T})^\dagger = \mathbf{X}_{(n)}\mathbf{S}^{(n)}(\mathbf{S}^{(n)}\mathbf{S}^{(n)T})^\dagger \tag{10.9.26}$$

which can be written as

$$\mathbf{A}^{(n)} = \mathbf{X}_{(n)}\mathbf{S}^{(n)}\mathbf{W}^\dagger, \tag{10.9.27}$$

where $\mathbf{W} = \mathbf{S}^{(n)}\mathbf{S}^{(n)T}$.

Algorithm 10.14 shows an ALS algorithm for nonnegative CP decomposition (CP-NALS) [164].

Algorithm 10.14 CP-NALS $(\mathcal{X}, R)$

---

input: Nth-order tensor $\mathcal{X}$ and the number R of factors.

initialize: $\mathbf{A}_0^{(n)} \in \mathbb{R}^{I_n \times R}, n = 1, \ldots, N$. Put $k = 0$.

repeat

1. Horizontally unfold $\mathcal{X}$ into the matrices $\mathbf{X}_{(n)}, n = 1, \ldots, N$.
2. Use (10.9.25) to compute $\mathbf{S}_k^{(n)}$, where $\mathbf{A}^{(i)} = \mathbf{A}_{k+1}^{(i)}, i = 1, \ldots, n-1$, and $\mathbf{A}^{(i)} = \mathbf{A}_k^{(i)}, i = n+1, \ldots, N$.
3. Calculate $\mathbf{W}_k = \mathbf{S}_k^{(n)} \mathbf{S}_k^{(n)T}$.
4. Update the factor matrices $\mathbf{A}_{k+1}^{(n)} = \mathbf{X}_{(n)} \mathbf{S}_k^{(n)} \mathbf{W}_k^{\dagger}$.
5. **exit if** the convergence criterion is satisfied or the maximum number of iterations is exhausted.

return $k \leftarrow k + 1$.

output: $\mathbf{A}^{(n)} \in \mathbb{R}^{I_n \times R}, n = 1, \ldots, N$.

Using the relation between the ALS method and the regularized ALS method for CP decomposition described in Section 10.5, as long as the factor matrix updating formula in step 4 of Algorithm 10.14 is amended to

$$\mathbf{A}_{k+1}^{(n)} = (\mathbf{X}_{(n)} \mathbf{S}_k^{(n)} + \tau_k \mathbf{A}_k^{(n)})(\mathbf{W}_k + \tau_k \mathbf{I})^{-1}, \tag{10.9.28}$$

one obtains the regularized nonnegative ALS algorithm CP-RNALS $(\mathcal{X}, R)$ for CP decomposition, where τ_k is the regularization parameter.

10.10 Tensor Completion

Tensor decomposition based on low-rank approximation is a powerful technique to extract useful information from multi-way data, because many real-world multi-way data are usually high-dimensional but have an intrinsically low-dimensional structure.

On the other hand, in many engineering problems the multi-way data are incompletely observed. Therefore, tensor low-rank decomposition with missing entries (i.e., requiring tensor completion) has applications in many data analysis problems such as recommender systems, scan completion, appearance acquisition completion, image inpainting, video decoding and video inpainting. In contrast with matrix completion, tensor completion can capture the intrinsic multidimensional structure of the multi-way data.

The approaches for completing tensors can be divided roughly into local and global approaches [170].

(1) *Local approach* This looks at neighboring pixels or voxels of a missing element and locally estimates the unknown values on the basis of some difference measure between the adjacent entries.
(2) *Global approach* This takes advantage of a global property of the data.

10.10.1 Simultaneous Tensor Decomposition and Completion

In an Nth-order tensor, each order represents one factor. It is usually assumed that the *within-factor variation* and *joint-factor variation* (i.e., variation between factors) are known *a priori* and can be regarded as auxiliary information. For example, for a video object described by a third-order tensor with variations spanned by the rows, columns and time axis, adjacent rows, columns and frames are respectively highly correlated. This auxiliary information comprises the *factor priors* for tensor analysis [99].

Given an incomplete Nth-order tensor $\mathcal{X}_0 \in \mathbb{R}^{I_1\times\cdots\times I_N}$ with available entries in a subset Ω in $\mathbb{R}^{I_1\times I_2\times\cdots\times I_N}$, consider finding a tensor $\mathcal{X}$ with components $\mathcal{Z} \in \mathbb{R}^{I_1\times\cdots\times I_N}$ and $\mathbf{V}_n \in \mathbb{R}^{I_n\times I_n}, n = 1,\ldots,N$, such that $\mathcal{X}_0$ and $\mathcal{X}$ have the same observed entries:

$$\mathcal{X} = \mathcal{Z} \times_1 \mathbf{V}_1^T \times_2 \cdots \times_N \mathbf{V}_N^T, \quad \Omega(\mathcal{X}) = \Omega(\mathcal{X}_0). \tag{10.10.1}$$

This problem is known as the *simultaneous tensor decomposition and completion* (STDC) problem [99]. Clearly, if $\Omega = \mathbb{R}^{I_1\times I_2\times\cdots\times I_N}$ then the STDC problem reduces to a tensor decomposition problem.

If $\mathcal{X}$ is a low-rank tensor then either the core tensor $\mathcal{Z}$ is also a low-rank tensor or the factor matrices $\mathbf{V}_1,\ldots,\mathbf{V}_N$ are a set of low-rank matrices. Such a low-rank property is usually regarded as a *global prior* in tensor completion [99].

Equation (10.10.1) has infinite solutions if any priors are not included in the model components $\mathcal{Z},\mathbf{V}_1,\ldots,\mathbf{V}_N$.

Define $\|\mathcal{Z}\|_F^2$ and $\|\mathbf{V}_n\|_*, n = 1,\ldots,N$, as the factor priors for the within-factor relation and

$$\mathrm{tr}\big((\mathbf{V}_1 \otimes \cdots \otimes \mathbf{V}_N)\mathbf{L}(\mathbf{V}_1 \otimes \cdots \otimes \mathbf{V}_N)^T\big) \tag{10.10.2}$$

as the factor prior for the joint-factor relations, where $\mathbf{L}$ is a positive semi-definite Laplacian matrix.

Another example of factor priors is found in the Vandermonde factor matrix. For the tensor decomposition $\mathcal{Z} = \sum_{r=1}^R g_r \mathbf{u}_r^{(1)} \circ \cdots \circ \mathbf{u}_r^{(N)}$, the factor matrices

$$\mathbf{U}^{(n)} = \big[\mathbf{u}_1^{(n)},\ldots,\mathbf{u}_R^{(n)}\big], \quad n = 1,\ldots,N, \tag{10.10.3}$$

are said to be *Vandermonde factor matrices* [451] if

$$\mathbf{u}_r^{(n)} = \left[1, z_{n,r}, z_{n,r}^2, \ldots, z_{n,r}^{I_n-1}\right]^T. \tag{10.10.4}$$

The Vandermonde structure may result from the following uses [451]:

(1) sinusoidal carrier in telecommunication applications;

(2) uniform linear arrays (ULAs) in array processing [444], [493];

(3) Doppler effects, carrier frequency offset or small delay spreads in array processing and in wireless communication [351], [445].

By using factor priors, Equation (10.10.1) can be rewritten as the following STDC problem based on factor priors [99]:

$$(\hat{\mathcal{X}}, \hat{\mathcal{Z}}, \hat{\mathbf{V}}_1, \ldots, \hat{\mathbf{V}}_N) = \operatorname{argmin}\Bigg\{ \gamma\|\mathcal{Z}\|_F^2 + \sum_{n=1}^N \alpha_n \|\mathbf{V}_n\|_* + \beta\,\mathrm{tr}\left((\mathbf{V}_1 \otimes \cdots \otimes \mathbf{V}_N)\mathbf{L}(\mathbf{V}_1 \otimes \cdots \otimes \mathbf{V}_N)^T\right) \Bigg\}, \tag{10.10.5}$$

subject to Equation (10.10.1).

Define an augmented Lagrangian function for the above constrained optimization problem as follows:

$$\begin{aligned} L(\mathcal{X}, \mathcal{Z}, \mathbf{V}_1, \ldots, \mathbf{V}_N, \mathcal{Y}, \mu) &= \sum_{n=1}^N \alpha_n \|\mathbf{V}_n\|_* + \beta\,\mathrm{tr}((\mathbf{V}_1 \otimes \cdots \otimes \mathbf{V}_N)\mathbf{L}(\mathbf{V}_1 \otimes \cdots \otimes \mathbf{V}_N)^T) + \gamma\|\mathcal{Z}\|_F^2 \\ &\quad + \langle \mathcal{Y}, \mathcal{X} - \mathcal{Z} \times_1 \mathbf{V}_1^T \cdots \times_N \mathbf{V}_N^T \rangle + \frac{\mu}{2}\|\mathcal{X} - \mathcal{Z} \times_1 \mathbf{V}_1^T \cdots \times_N \mathbf{V}_N^T\|_F^2. \end{aligned} \tag{10.10.6}$$

Owing to the operations of the Kronecker product and the mode-n product, Equation (10.10.6) is a nonconvex function that contains many local minimizers. Using alternating least squares, if one variable at a time is optimized with the others fixed, then every subproblem can be simplified to a convex optimization problem with a global optimum.

1. Optimization of Factor Matrices

When one factor matrix $\mathbf{V}_n$ is optimized with the others fixed, the objective function in (10.10.6) becomes

$$\begin{aligned} L(\mathbf{V}_n) = \alpha_n \|\mathbf{V}_n\|_* + \beta\,\mathrm{tr}\left(\mathbf{V}_n \mathbf{H}_n^T \mathbf{H}_n \mathbf{U}_n^T\right) + \left\langle \mathbf{Y}^{(n)}, -\mathbf{V}_n^T \mathbf{Z}^{(n)} \mathbf{U}_n \right\rangle \\ + \frac{\mu}{2}\left\|\mathbf{X}^{(n)} - \mathbf{V}_n^T \mathbf{Z}^{(n)} \mathbf{U}_n\right\|_F^2 \end{aligned} \tag{10.10.7}$$

with

$$\mathbf{U}_n = \mathbf{V}_1 \otimes \cdots \otimes \mathbf{V}_{n-1} \otimes \mathbf{V}_{n+1} \otimes \cdots \otimes \mathbf{V}_N. \tag{10.10.8}$$

The nuclear norm minimization of the objective function $L(\mathbf{V}_n)$ can be solved via the linearized augmented Lagrangian and alternating direction methods, as follows [524]:

$$\hat{\mathbf{V}}_n \approx \arg\min_{\mathbf{V}_n} \left\{ \frac{\alpha_n}{\gamma} \|\mathbf{V}_n\|_* + \frac{1}{2} \left\| \mathbf{V}_n - \left(\mathbf{V}_n^t - \frac{1}{\tau} \nabla f_2(\mathbf{V}_n^t) \right) \right\|_F^2 \right\}, \tag{10.10.9}$$

where $\mathbf{V}_n^t$ is the approximation from the previous iteration,

$$\begin{aligned} f_2(\mathbf{V}_n^t) = & 2\beta \mathbf{V}_n^t \mathbf{H}_n^T \mathbf{H}_n + \mu \mathbf{Z}^{(n)} \mathbf{U}_n \mathbf{U} N^T \mathbf{Z}_n^{(n)T} \mathbf{V}_n \\ & - \mathbf{Z}^{(n)} \mathbf{U}_n (\mu \mathbf{X}^{(n)} + \mathbf{Y}^{(n)})^T, \end{aligned} \tag{10.10.10}$$

and the Lipschitz constant τ is defined by

$$\tau = \sigma_1 (2\beta \mathbf{H}_k^T \mathbf{H}_k) + \sigma_1 (\mu \mathbf{Z}^{(k)} \mathbf{U}_k \mathbf{U}_k^T \mathbf{Z}^{(k)T}). \tag{10.10.11}$$

2. Optimization of Core Tensor $\mathcal{Z}$

Minimizing the objective function in (10.10.6) with respect to $\mathcal{Z}$ gives the result

$$\hat{\mathcal{Z}} = \arg\min_{\mathcal{Z}} \left\{ \gamma \|\mathcal{Z}\|_F^2 + \frac{\mu}{2} \left\| \left(\mathcal{X} + \frac{1}{\mu} \mathcal{Y} \right) - \mathcal{Z} \times_1 \mathbf{V}_1^T \times_2 \cdots \times_N \mathbf{V}_N^T \right\|_F^2 \right\} \tag{10.10.12}$$

or

$$\hat{\mathbf{z}} = \arg\min_{\mathbf{z}} \left\{ \gamma \|\mathbf{z}\|_F^2 + \frac{\mu}{2} \left\| \left(\mathbf{x} + \frac{1}{\mu} \mathbf{y} \right) - \mathbf{V}^T \mathbf{z} \right\|_F^2 \right\}, \tag{10.10.13}$$

where $\mathbf{x} = \text{vec}(\mathcal{X})$, $\mathbf{y} = \text{vec}(\mathcal{Y})$, $\mathbf{z} = \text{vec}(\mathcal{Z})$, and $\mathbf{V} = \mathbf{V}_1 \otimes \cdots \otimes \mathbf{V}_N$.

Although a closed solution to Equation (10.10.13) is given by $\hat{\mathbf{z}} = (\mu \mathbf{V}\mathbf{V}^T + 2\gamma \mathbf{I})^{-1} \mathbf{V}(\mathbf{x} + \mu^{-1}\mathbf{y})$, this solution is difficult to compute when the size of $\mathbf{V}$ is very large. A better way is to use the conjugate gradient (CG) method to approximate the global minimizer [99].

3. Optimization of $\mathcal{X}$

Minimizing the objective function in (10.10.6) with respect to $\mathcal{X}$ yields

$$\hat{\mathcal{X}} = \arg\min_{\mathcal{X}} \left\| \mathcal{X} - \left(\mathcal{Z} \times_1 \mathbf{V}_1^T \times_2 \cdots \times_N \mathbf{V}_N^T - \frac{1}{\mu} \mathcal{Y} \right) \right\|_F^2 \tag{10.10.14}$$

with the constraints $\Omega(\mathcal{X}) = \Omega(\mathcal{X}_0)$ and

$$\bar{\Omega}(\mathcal{X}) = \bar{\Omega} \left(\mathcal{Z} \times_1 \mathbf{V}_1^T \times_2 \cdots \times_N \mathbf{V}_N^T - \frac{1}{\mu} \mathcal{Y} \right). \tag{10.10.15}$$

Algorithm 10.15 summarizes the steps of the simultaneous tensor decomposition and completion method.

Algorithm 10.15 Simultaneous tensor decomposition and completion [99]

---

input: Incomplete tensor $\mathcal{X}_0 \in \mathbb{R}^{I_1 \times \cdots \times I_N}$, a positive semi-definite matrix $\mathbf{L}$ and the parameters $\alpha_1, \ldots, \alpha_N, \beta, \gamma, \mu$.

initialization: $\mathbf{V}_k = \mathbf{I}_{I_k}, k = 1, \ldots, N, \mathcal{Z} = \mathcal{X} = \mathcal{X}_0$, $\mathcal{Y}_{i_1, \ldots, i_N} \equiv 0$, $t = 1, \mu^1 = \mu$.

repeat

1. **for** $k = 1, \ldots, N$ **do**

   Use the alternating LS method to solve (10.10.9) for $\mathbf{V}_k$.

   end for

2. Use the conjugate gradient (CG) method to solve (10.10.13) for $\mathbf{z}$.
3. Use (10.2.1) to tensorize $\mathbf{z}$ for $\mathcal{Z}$.
4. Solve the optimization problem (10.10.14) for $\mathcal{X}$.
5. Update $\mathcal{Y} \leftarrow \mathcal{Y} + \mu^t(\mathcal{X} - \mathcal{Z} \times_1 \mathbf{V}_1^T \times_2 \cdots \times_N \mathbf{V}_N^T)$.
6. Compute $\mu^{t+1} = \rho\mu^t, \rho \in [1.1, 1.2]$.

exit if $t > 100$ and $\|\mathcal{X} - \mathcal{Z} \times_1 \mathbf{V}_1^T \times_2 \cdots \times_N \mathbf{V}_N^T\|_F^2 < 10^{-4}\|\mathcal{X}_0\|_F^2$.

return $t \leftarrow t + 1$.

output: factor matrices $\mathbf{V}_1, \ldots, \mathbf{V}_N$, core tensor $\mathcal{Z}$, the completed tensor $\mathcal{X}$.

10.10.2 Smooth PARAFAC Tensor Completion

The state-of-the art tensor completion methods can be categorized into three types: low-rank nuclear norm, low-rank Tucker decomposition and low-rank CP decomposition with optional smoothness constraints [533]. It has recently been realized (see e.g., [408], [196]) that smoothness is quite an important factor for visual data completion.

Given an incomplete input tensor $\mathcal{D} \in \mathbb{R}^{I_1 \times I_2 \times \cdots \times I_N}$, let Ω denote the indices of the available elements of $\mathcal{D}$.

Consider a *smooth PARAFAC tensor completion* (SPC), formulated as the optimization problem [533]:

$$\min_{\mathbf{g}, \mathbf{U}^{(1)}, \ldots, \mathbf{U}^{(N)}} \left\{ \frac{1}{2}\|\mathcal{X} - \mathcal{Z}\|_F^2 + \sum_{r=1}^{R} \frac{g_r^2}{2} \sum_{n=1}^{N} \rho^{(n)} \|\mathbf{L}^{(n)} \mathbf{u}_r^{(n)}\|_p^p \right\} \tag{10.10.16}$$

subject to

$$\mathcal{Z} = \sum_{r=1}^{R} g_r \mathbf{u}_r^{(1)} \circ \mathbf{u}_r^{(2)} \circ \cdots \circ \mathbf{u}_r^{(N)},$$
$$\mathcal{X}_\Omega = \mathcal{D}_\Omega, \mathcal{X}_\Omega = \mathcal{Z}_\Omega, \|\mathbf{u}_r^{(n)}\|_2 = 1, \forall r \in \{1,\ldots,R\}, \forall n \in \{1,\ldots,N\},$$

where $\mathcal{Z}$ is a completed output tensor, $\mathbf{U}^{(n)} = [\mathbf{u}_1^{(n)},\ldots,\mathbf{u}_R^{(n)}]$, $n = 1,\ldots,N$, are the factor matrices, $\boldsymbol{\rho} = [\rho_1,\ldots,\rho_R]^T$ is the smoothness parameter vector and the constrained conditions $\mathcal{X}_\Omega = \mathcal{D}_\Omega$ and $\mathcal{X}_\Omega = \mathcal{Z}_\Omega$ stand for

$$\mathcal{X}(i_1,\ldots,i_N) = \begin{cases} \mathcal{D}(i_1,\ldots,i_N), & (i_1,\ldots,i_N) \in \Omega, \\ \mathcal{Z}(i_1,\ldots,i_N), & \text{otherwise.} \end{cases} \tag{10.10.17}$$

The parameters g_r^2 in (10.10.16) are adaptively updated, depending on the penalty levels of the smoothness constraint terms. This adaptively enforces the different levels of smoothness for the different components (factors). For image completion, the SPC model (10.10.16) decomposes an image adaptively into a strong smooth background and a weaker smooth foreground. Because the low-frequency components are generally stronger than the high-frequency components in the Fourier domain of the image data, different penalty levels are used for individual rank-1 components, depending on their scales g_r [533].

Let $\mathbf{L}^{(n)} \in \mathbb{R}^{(I_n-1)\times I_n}$ be a smoothness constraint matrix defined typically as

$$\mathbf{L}^{(n)} = \begin{bmatrix} 1 & -1 & & & 0 \\ & 1 & -1 & & \\ & & \ddots & \ddots & \\ 0 & & & 1 & -1 \end{bmatrix}. \tag{10.10.18}$$

Then, the second term in (10.10.16) is seen to be a penalty term which enforces the smoothness of the individual feature vectors $\mathbf{u}_r^{(n)}$ via

$$\|\mathbf{L}^{(n)}\mathbf{u}_r^{(n)}\|_p^p = \sum_{i=1}^{I_n-1} |\mathbf{u}_r^{(n)}(i) - \mathbf{u}_r^{(n)}(i+1)|^p. \tag{10.10.19}$$

Two common choices of the parameter p in (10.10.19) correspond to two different smoothness constraints [533]:

- *Total variation* (TV) Here $p = 1$ and is thus constrained to be the sum of the differences between neighboring entries of the individual feature vectors.
- *Quadratic variation* (QV) Here $p = 2$ and is thus constrained to equal the sum of the squared differences between neighboring entries of the individual feature vectors.

In image recovery problems, such as denoising and restoration, it is generally considered that the TV constraint is better than QV.

Define the function

$$F_p^{(r,n)}(\mathbf{u}_r^{(n)}) = \frac{1}{2}\left\|\mathcal{Y}_r - g_r\mathbf{u}_r^{(1)} \circ \cdots \circ \mathbf{u}_r^{(N)}\right\|_F^2 + \frac{g_r^2}{2}\sum_{m=1}^{N} \rho^{(m)}\left\|\mathbf{L}^{(m)}\mathbf{u}_r^{(m)}\right\|_p^2, \tag{10.10.20}$$

for $r = 1, \ldots, R$, $n = 1, \ldots, N$.

Let SDR denote the signal-distortion ratio. Algorithm 10.16 summarizes a low-rank smooth PARAFAC tensor completion (SPC) algorithm.

Algorithm 10.16 has the following advantages and characteristics.

(1) The factor number R can be adaptively determined.
(2) When $p = 1$, the SPC algorithm is a tensor completion algorithm with the TV constraint. When $p = 2$ is used, the SPC algorithm is a tensor completion algorithm with the QV constraint.
(3) The SPC algorithm focuses upon "good approximation" rather than "minimum rank decomposition", since estimation of the number of factors for the smooth PARAFAC decomposition model is difficult, and its accuracy does not need to be very high provided that the reconstructed tensor is sufficiently well approximated in practice.

It should be noted that the SPC algorithm does not guarantee global convergence since tensor decomposition is in general a nonconvex optimization problem.

10.11 Software

The available software sources for tensor decompositions and tensor operations include the following (from [103], [238]).

(1) The MATLAB *Tensor Toolbox* version 2.5 [21] offers classes for tensor operations and has several tensor decomposition algorithms.
(2) The *TDALAB* [546] and *TENSORBOX* [388] provide a user friendly interface and advanced algorithms for CPD, nonnegative TKD and MWCA.
(3) The *Tensorlab toolbox* [450] builds upon a complex optimization framework and offers numerical algorithms for computing the CPD, BTD, and TKD and includes the structured data fusion of tensors and complex optimization tools.
(4) The *N-way toolbox* [15] includes (constrained) CPD, TKD and PLS in the context of chemometrics applications; many of these methods can handle constraints (e.g., nonnegativity and orthogonality) and missing elements.
(5) The *TT toolbox* [361] includes functions for basic operations that are low-parametric representations of high-dimensional tensors. A Python implementation of the TT Toolbox is also available [362].

Algorithm 10.16 SPC algorithm [533]

1. **input:** $\mathcal{T}, \Omega, p, \rho$, SDR, and ν.
2. $\epsilon \leftarrow 10^{(-\text{SDR}/10)}\|\mathcal{T}_\Omega\|_F^2$.
3. $\mathcal{X}_\Omega \leftarrow \mathbf{T}_\Omega$, $\mathbf{X}_{\bar{\Omega}} \leftarrow$ average of $\mathcal{T}_\Omega$.
4. Construct matrix $\mathbf{L}^{(n)}, n = 1, \ldots, N$ by (10.10.19).
5. $R \leftarrow R + 1$.
6. **initialize:** $\mathbf{u}_r^{(n)} \leftarrow \mathbf{u}_r^{(n)}/\|\mathbf{u}_r^{(n)}\|_2, r = 1, \ldots, R$, $n = 1, \ldots, N$.
7. $g_R \leftarrow \langle \mathcal{X}_R, \mathbf{u}_R^{(1)} \circ \mathbf{u}_R^{(2)} \circ \cdots \circ \mathbf{u}_R^{(N)} \rangle$.
8. $\mathcal{E} = \mathcal{X} - \sum_{r=1}^R g_r \mathbf{u}_r^{(1)} \circ \mathbf{u}_r^{(2)} \circ \cdots \circ \mathbf{u}_r^{(N)}$.
9. $\mathcal{E}_{\bar{\Omega}} \leftarrow 0$.
10. $t \leftarrow 0$.
11. $\mu_t \leftarrow \|\mathcal{E}\|_F^2$.
12. **repeat**
13. **for** $r = 1, \ldots, R$ **do**

    $\mathcal{Y}_r \leftarrow \mathcal{E} + g_r \mathbf{u}_r^{(1)} \circ \mathbf{u}_r^{(2)} \circ \cdots \circ \mathbf{u}_r^{(N)}$.

    for $n = 1, \ldots, N$ **do**

    Use the ALS method to solve $\mathbf{u}_r^{(n)} \leftarrow \arg\min_{\mathbf{u}_r^{(n)}} F_p^{(r,n)}(\mathbf{u}_r^{(n)})$.

    $\mathbf{u}_r^{(n)} = \mathbf{u}_r^{(n)}/\|\mathbf{u}_r^{(n)}\|_2$.

    end for

    $g_r \leftarrow \dfrac{\langle \mathcal{Y}_r, \mathbf{u}_r^{(1)} \circ \mathbf{u}_r^{(2)} \circ \cdots \circ \mathbf{u}_r^{(N)} \rangle}{1 + \sum_{n=1}^N \rho^{(n)} \|\mathbf{L}^{(n)} \mathbf{u}_r^{(n)}\|_p^p}$.

    $\mathcal{E} \leftarrow \mathcal{Y}_r - g_r \mathbf{u}_r^{(1)} \circ \mathbf{u}_r^{(2)} \circ \cdots \circ \mathbf{u}_r^{(N)}$ and $\mathcal{E}_{\bar{\Omega}} \leftarrow 0$.

    end for
14. $\mu_{t+1} \leftarrow \|\mathcal{E}\|_F^2$.
15. **if** $\dfrac{|\mu_t - \mu_{t+1}|}{|\mu_{t+1} - \epsilon|} \leq \nu$ **then**

    $R \leftarrow R + 1$.

    initialize $\mathbf{u}_r^{(n)} \leftarrow \mathbf{u}_r^{(n)}/\|\mathbf{u}_r^{(n)}\|_2, r = 1, \ldots, R$, $n = 1, \ldots, N$.

    $g_R \leftarrow \langle \mathcal{E}, \mathbf{u}_R^{(1)} \circ \mathbf{u}_R^{(2)} \circ \cdots \circ \mathbf{u}_R^{(N)} \rangle$.

    $\mathcal{E} \leftarrow \mathcal{E} - g_R \mathbf{u}_R^{(1)} \circ \mathbf{u}_R^{(2)} \circ \cdots \circ \mathbf{u}_R^{(N)}$.

    $\mathcal{E}_{\bar{\Omega}} \leftarrow 0$.
16. **end if**
17. $t \leftarrow t + 1$.
18. **exist if** $\mu_t \leq \epsilon$.
19. $\mathcal{Z} \leftarrow \sum_{r=1}^R g_r \mathbf{u}_r^{(1)} \circ \mathbf{u}_r^{(2)} \circ \cdots \circ \mathbf{u}_r^{(N)}$.
20. $\mathcal{X}_{\bar{\Omega}} \leftarrow \mathcal{Z}_{\bar{\Omega}}$.
21. **return**
22. **output:** $\mathcal{X}, \mathcal{Z}$.

(6) The *Hierarchical Tucker toolbox* [261] provides the construction and manipulation of tensors.

(7) The *Tensor calculus library* [147] provides tensor tools for scientific computing.

(8) The *CuBatch* [183] is a MATLAB-based software for both tensor and classical data analysis, and includes validation tools providing graphical outputs.

(9) The *MATLAB CMTF Toolbox* [4] provides constrained and unconstrained algorithms for the joint decomposition of tensors with different orders, using PARAFAC.

(10) The *ERPWAVELAB* [333] is a MATLAB-based toolbox for the multi-channel time/frequency analysis of EEG and MEG by various tensor decompositions, including PARAFAC, shifted PARAFAC and Tucker.

(11) The *Tensor Toolbox in Python* [50] includes the decomposition of tensors in tensor-train format[1] and spectral tensor-train format.

(12) The *TensorReg Toolbox* for MATLAB [546] provides sparse PARAFAC and Tucker regression functions.

(13) The code developed for multiway analysis is also available from the Three-Mode Company [266].

Many algorithms developed by the authors in [238] can be found at

http://www.research.cneuro.cu/categories/tensor,

http://neurosignal.boun.edu.tr/software/tensor.

Exercises

10.1 Given that the front slice matrices of the third-order tensor $\mathcal{X} \in \mathbb{R}^{3\times 4\times 2}$ are

$$\mathbf{X}_1 = \begin{bmatrix} 1 & 4 & 7 & 10 \\ 2 & 5 & 8 & 11 \\ 3 & 6 & 9 & 12 \end{bmatrix}, \quad \mathbf{X}_2 = \begin{bmatrix} 15 & 18 & 21 & 24 \\ 16 & 19 & 22 & 25 \\ 17 & 20 & 23 & 26 \end{bmatrix},$$

let $\mathbf{U} = \begin{bmatrix} 1 & 3 & 5 \\ 2 & 4 & 6 \end{bmatrix}$ and find the front slice matrices $\mathbf{Y}_1$ and $\mathbf{Y}_2$ of $\mathcal{Y} = \mathcal{X} \times_1 \mathbf{U}$.

10.2 Given a third-order tensor $\mathcal{X}$ with entries

$$X_{1,1,1} = 1, \quad X_{1,2,1} = 3, \quad X_{1,1,2} = 5, \quad X_{1,2,2} = 7,$$
$$X_{2,1,1} = 2, \quad X_{2,2,1} = 4, \quad X_{2,1,2} = 6, \quad X_{2,2,2} = 8,$$

[1] Consider a given tensor $\mathcal{B}$ approximated by a tensor $\mathcal{A}$ with entries

$$a_{i_1\dots i_d} = \sum_{\alpha_0,\dots,\alpha_{d-1},\alpha_d} g^{(1)}_{\alpha_0,i_1,\alpha_1} g^{(2)}_{\alpha_1,i_2,\alpha_2} \cdots g^{(d)}_{\alpha_{d-1},i_d,\alpha_d}.$$

This tensor decomposition can be represented graphically by a linear tensor network. This picture looks like a train with carriages and links between them, and that justifies the name *tensor-train decomposition*. See Oseledest O. V. Tensor-train decomposition. SIAM J. Sci. Comput., 33(5): 2295–2317, 2011.

find the LMV and Kolda horizontal unfolding matrices of the tensor $\mathcal{X}$.

10.3 Given that the frontal slices of $\mathcal{X} \in \mathbb{R}^{3\times 4\times 2}$ are

$$\mathbf{X}_1 = \begin{bmatrix} 1 & -4 & 7 & -10 \\ 2 & -5 & 8 & -11 \\ 3 & -6 & 9 & -12 \end{bmatrix}, \quad \mathbf{X}_2 = \begin{bmatrix} 13 & -16 & 19 & -22 \\ 14 & -17 & 20 & -23 \\ 15 & -18 & 21 & -24 \end{bmatrix},$$

find the three mode-n unfoldings of $\mathcal{X}$ and its vectorized version.

10.4 Given that the frontal slices of $\mathcal{X} \in \mathbb{R}^{3\times 4\times 2}$ are

$$\mathbf{X}_1 = \begin{bmatrix} 1 & 4 & 7 & 10 \\ 2 & 5 & 8 & 11 \\ 3 & 6 & 9 & 12 \end{bmatrix}, \quad \mathbf{X}_2 = \begin{bmatrix} 13 & 16 & 19 & 22 \\ 14 & 17 & 20 & 23 \\ 15 & 18 & 21 & 24 \end{bmatrix},$$

for a 2×3 matrix

$$\mathbf{U} = \begin{bmatrix} 1 & 3 & 5 \\ 2 & 4 & 6 \end{bmatrix}$$

find the frontal slices of the mode-1 product $\mathcal{Y} = \mathcal{X} \times_1 \mathbf{U} \in \mathbb{R}^{2\times 4\times 2}$.

10.5 Prove that

$$x^{(I\times JK)}_{i,(k-1)J+j} = x_{ijk}$$
$$\Rightarrow \quad \mathbf{X}^{(I\times JK)} = [\mathbf{X}_{::1}, \ldots, \mathbf{X}_{::K}] = \mathbf{A}\mathbf{G}^{(P\times QR)}(\mathbf{C} \otimes \mathbf{B})^T.$$

10.6 Let $\mathbf{X}_{i::} = \mathbf{b}_1\mathbf{c}_1^T a_{i1} + \cdots + \mathbf{b}_R\mathbf{c}_R^T a_{iR}$. Show that the CP decomposition of the horizonal unfolding matrix is given by

$$\mathbf{X}^{(J\times KI)} = [\mathbf{X}_{1::}, \ldots, \mathbf{X}_{I::}] = \mathbf{B}(\mathbf{A} \odot \mathbf{C})^T$$

and the CP decomposition of the longitudinal unfolding matrix is given by

$$\mathbf{X}^{(IJ\times K)} = \begin{bmatrix} \mathbf{X}_{1::} \\ \vdots \\ \mathbf{X}_{I::} \end{bmatrix} = (\mathbf{A} \odot \mathbf{B})\mathbf{C}^T.$$

10.7 Let $\mathbf{X}_{:j:} = \mathbf{a}_1\mathbf{c}_1^T b_{j1} + \cdots + \mathbf{a}_R\mathbf{c}_R^T b_{jR}$. Show that the CP decomposition of the horizonal unfolding matrix is

$$\mathbf{X}^{(K\times IJ)} = [\mathbf{X}_{:1:}, \ldots, \mathbf{X}_{:J:}] = \mathbf{C}(\mathbf{B} \odot \mathbf{A})^T$$

and the CP decomposition of the longitudinal unfolding matrix is

$$\mathbf{X}^{(JK\times I)} = \begin{bmatrix} \mathbf{X}_{:1:} \\ \vdots \\ \mathbf{X}_{:J:} \end{bmatrix} = (\mathbf{B} \odot \mathbf{C})\mathbf{A}^T,$$

where $\mathbf{A} = [\mathbf{a}_1, \ldots, \mathbf{a}_R]^T, \mathbf{B} = [b_{jk}]_{j=1,k}^{J,R}$ and $\mathbf{C} = [\mathbf{c}_1, \ldots, \mathbf{c}_R]^T$.

10.8 Let $\mathbf{X}_{::k} = \mathbf{a}_1\mathbf{b}_1^T c_{k1} + \cdots + \mathbf{a}_R\mathbf{b}_R^T c_{kR}$. Show that the CP decomposition of the horizonal unfolding matrix is

$$\mathbf{X}^{(J\times KI)} = [\mathbf{X}_{::1}^T, \ldots, \mathbf{X}_{::K}^T] = \mathbf{B}(\mathbf{C}\odot\mathbf{A})^T$$

and the CP decomposition of the longitudinal unfolding matrix is

$$\mathbf{X}^{(KI\times J)} = \begin{bmatrix} \mathbf{X}_{::1} \\ \vdots \\ \mathbf{X}_{::K} \end{bmatrix} = (\mathbf{C}\odot\mathbf{A})\mathbf{B}^T.$$

10.9 Use the definition of the $(m-r)$th-order tensor–vector products to prove the following results:

(a) $\mathcal{A}\mathbf{x}^m = \mathbf{x}^T\mathcal{A}\mathbf{x}^{m-1}$;

(b) $\mathcal{A}\mathbf{x}^{m-1} = (\mathcal{A}\mathbf{x}^{m-2})\mathbf{x}$.

10.10 Let $\mathbf{x} = [x_1, x_2, x_3]^T$ be a direction vector and $\mathcal{W}$ be a fourth-order three-dimensional fully symmetric tensor, called the diffusion kurtosis (DK) tensor. Prove that the critical points of the maximization problem

$$\max\ \mathcal{W}\mathbf{x}^4 \quad \text{subject to} \quad \mathbf{D}\mathbf{x}^2 = 1$$

satisfy the following equations:

$$\begin{cases} \mathcal{W}\mathbf{x}^3 = \lambda\mathbf{D}\mathbf{x}, \\ \mathbf{D}\mathbf{x}^2 = 1. \end{cases}$$

Here $\mathbf{D}$ is a positive definite matrix.

10.11 Let $f_1(\mathbf{x}) = \mathcal{A}\mathbf{x}^m$ and $f_2(\mathbf{x}) = \|\mathbf{x}\|_2^m$. Find the gradient vectors and the Hessian matrices of $f_1(\mathbf{x})$ and $f_2(\mathbf{x})$, respectively.

10.12 Find the gradient vector and the Hessian matrix of the function

$$f(\mathbf{x}) = \frac{\mathcal{A}\mathbf{x}^m}{\mathcal{B}\mathbf{x}^m}.$$

10.13 Prove that $\mathcal{I}\mathbf{x}^{m-1} = \mathbf{x}^{[m-1]}$, where $\mathcal{I}$ is an identity tensor.

10.14 Create a table for comparison of second-order matrix analysis and the higher-order matrix analysis (i.e., tensor analysis).

References

[1] Abatzoglou T. J., Mendel J. M., Harada G. A. The constrained total least squares technique and its applications to harmonic superresolution. IEEE Trans. Signal Processing, 39: 1070–1087, 1991.

[2] Abbott D. The Biographical Dictionary of Sciences: Mathematicians. New York: P. Bedrick Books, 1986.

[3] Abraham R., Marsden J. E., Ratiu T. Manifolds, Tensor Analysis, and Applications. New York: Addison-Wesley, 1983.

[4] Acar E. The MATLAB CMTF Toolbox. 2014.

[5] Acar E., Aykut-Bingo C., Bingo H., Bro R., Yener B. Multiway analysis of epilepsy tensors. Bioinformatics, 23: i10–i18, 2007.

[6] Acar E., Camtepe S. A., Krishnamoorthy M., Yener B. Modeling and multiway analysis of chatroom tensors. In Proc. IEEE International Conference on Intelligence and Security Informatics. Springer, 256–268, 2005.

[7] Acar E., Yener B. Unsupervised multiway data analysis: A literature survey. IEEE Trans. Knowledge and Data Engineering, 21(1): 6–20, 2009.

[8] Acar R., Vogel C. R. Analysis of bounded variation penalty methods for ill-posed problems. Inverse Problems, 10: 1217–1229, 1994.

[9] Adamyan V. M., Arov D. Z. A general solution of a problem in linear prediction of stationary processes. Theory Probab. Appl., 13: 294–407, 1968.

[10] Adib A., Moreau E., Aboutajdine D. Source separation contrasts using a reference signal. IEEE Signal Processing Lett., 11(3): 312–315, 2004.

[11] Afriat S. N. Orthogonal and oblique projectors and the characteristics of pairs of vector spaces. Math. Proc. Cambridge Philos. Soc., 53: 800–816, 1957.

[12] Alexander S. T. Adaptive Signal Processing: Theory and Applications. New York: Springer, 1986.

[13] Alter O., Brown P. O., Botstein D. Generalized singular value decomposition for comparative analysis of genome-scale expression data sets of two different organisms. Proc. Nat. Aca. Sci., USA, 100(6): 3351–3356, 2003.

[14] Amari S., Nagaoka H. Methods of Information Geometry. New York: Oxford University Press, 2000.

[15] Andersson C., Bro R. The N-way toolbox for MATLAB. [Online]. Chemomet. Intell. Lab. Syst., 52(1): 1–4, 2000. Available at http://www.models.life.ku.dk/nwaytoolbox/ (2000).

[16] Andrews H., Hunt B. Digital Image Restoration. Engliwood Cliffs, NJ: Prentice-Hall, 1977.

[17] Aronszajn N. Theory of reproducing kernels. Trans. Amer. Math. Soc., 68: 800–816, 1950.

[18] Auslender A. Asymptotic properties of the Fenchel dual functional and applications to decomposition problems. J. Optim. Theory Appl., 73(3): 427–449, 1992.

[19] Autonne L. Sur les groupes lineaires, reelles et orthogonaus. Bull Soc. Math., France, 30: 121–133, 1902.

[20] Aybat N. S., Iyengar G. A first-order augmented Lagrangian method for compressed sensing. SIAM J. Optim., 22(2): 429–459, 2012.

[21] Bader B. W., Kolda T. G., *et al.* MATLAB Tensor Toolbox version 2.5. Available at http://www.sandia.gov/tgkolda/TensorToolbox/ (2012).

[22] Banachiewicz T. Zur Berechungung der Determinanten, wie auch der Inverse, und zur darauf basierten Auflösung der Systeme linearer Gleichungen. Acta Astronomica, Sér. C, 3: 41–67, 1937.

[23] Bapat R. Nonnegative Matrices and Applications. Cambridge University Press, 1997.

[24] Barbarossa S., Daddio E., Galati G. Comparison of optimum and linear prediction technique for clutter cancellation. Proc. IEE, Part F, 134: 277–282, 1987.

[25] Barnett S. Matrices: Methods and Applications. Oxford: Clarendon Press, 1990.

[26] Barzilai J., Borwein J. M. Two-point step size gradient methods. IMA J. Numer. Anal. 8: 141–148, 1988.

[27] Basri R., Jacobs D. Lambertian reflectance and linear subspaces. IEEE Trans. Pattern Anal. Machine Intell., 25(2): 218–233, 2003.

[28] Beal M., Jojic N., Attias H. A graphical model for audiovisual object tracking. IEEE Trans. Pattern Anal. Machine Intell., 25(7): 828–836, 2003.

[29] Beck A., Teboulle M. A fast iterative shrinkage-thresholding algorithm for linear inverse problems. SIAM J. Imaging Sci., 2(1): 183–202, 2009.

[30] Beck A., Elder Y. C. Structured total maximum likelihood: an alternative to structured total squares. SIAM J. Matrix Anal. Appl., 31(5): 2623–2649, 2010.

[31] Behrens R. T., Scharf L. L. Signal processing applications of oblique projection operators. IEEE Trans. Signal Processing, 42(6): 1413–1424, 1994.

[32] Bellman R. Introduction to Matrix Analysis, 2nd edn. McGraw-Hill, 1970.

[33] Belochrani A., Abed-Merain K., Cardoso J. F., Moulines E. A blind source separation technique using second-order statistics. IEEE Trans. Signal Processing, 45(2): 434–444 1997.

[34] Beltrami E. Sulle funzioni bilineari, Giomale di Mathematiche ad Uso Studenti Delle Universita. 11: 98–106, 1873. An English translation by D Boley is available as Technical Report 90–37, University of Minnesota, Department of Computer Science, 1990.

[35] Berberian S. K. Linear Algebra. New York: Oxford University Press, 1992.

[36] Berge J. M. F. T. Convergence of PARAFAC preprocessing procedures and the Deming–Stephan method of iterative proportional fitting. In Multiway Data Analysis (Coppi R., Bolasco S. eds.). Amsterdam: Elsevier, 53–63, 1989.

[37] Berge J. M. F. T., Sidiropolous N. D. On uniqueness in CANDECOMP/ PARAFAC. Psychometrika, 67: 399–409, 2002.

[38] Berger C., Voltersen M., Eckardt R., Eberle J. Multi-modal and multi-temporal data fusion: outcome of the 2012 GRSS data fusion contest. IEEE J. Selected Topics Appl. Earth Observat. Remote Sensing, 6(3): 1324–1340, 2013.

[39] Berman A., Plemmons R. Nonnegative Matrices in the Mathematical Sciences. Philadephia: SIAM, 1994.

[40] Berry M. W., Browne M., Langville A. N., Pauca V. P., Plemmons R. J. Algorithms and applications for approximate nonnegative matrix factorization. Computational Statistics & Data Analysis, 52: 155–173, 2007.

[41] Bertsekas D. P. Multiplier methods: a survey. Automatica, 12: 133–145, 1976.

[42] Bertsekas D. P. Nonlinear Programming, 2nd edn. Belmont: Athena Scientific, 1999.

[43] Bertsekas D. P., Nedich A., Ozdaglar A. Convex Analysis and Optimization. Belmont: Athena Scientific, 2003.

[44] Bertsekas D. P., Tseng P. Partial proximal minimization algorithms for convex programming. SIAM J. Optimization, 4(3): 551–572, 1994.

[45] Bi H., Jiang C., Zhang B., Wang Z., Hong W. Radar change imaging with undersampled data based on matrix completion and Bayesian compressive sensing. IEEE Geosci. Remote Sensing Lett., 12(7): 1546–1550, 2015.

[46] Bi H., Zhang B., Hong W., Zhou S. Matrix-completion-based airborne tomographic SAR inversion under missing data. IEEE Geosci. Remote Sensing Lett. 12(11): 2346–2350, 2015.

[47] Bienvenu G., Kopp L. Principède la goniomgraveetre passive adaptive. Proc. 7th Colloqucum GRESIT, Nice, France, 106/1–106/10, 1979.

[48] Bien J., Tibshirani R. J. Sparse estimation of a covariance matrix. Biometrika, 98(4): 807–820, 2011.

[49] Biessmann F., Plis S., Meinecke F. C., Eichele T., Mller K. R. Analysis of multimodal neuroimaging data. IEEE Reviews in Biomedical Engineering, 4: 26–58, 2011.

[50] Bigoni D. Tensor toolbox, 2015.

[51] Boot J. Computation of the generalized inverse of singular or rectangular matrices. Amer. Math. Monthly, 70: 302–303, 1963.

[52] Boyd S., Parikh N., Chu E., Peleato B., Eckstein J. Distributed optimization and statistical learning via the alternating direction method of multipliers. Foundations and Trends in Machine Learning, 3(1): 1–122, 2010.

[53] Boyd S., Vandenberghe L. Convex Optimization. Cambridge, UK: Cambridge University Press, 2004.

[54] Boyd S., Vandenberghe L. Subgradients. Notes for EE364b, Stanford University, Winter 2006–2007, April 13, 2008.

[55] Brachat J., Common P., Mourrain B., Tsigaridas E. Symmetric tensor decomposition. Linear Algebra Appl., 433(11-12): 1851–1872, 2010.

[56] Bramble J., Pasciak J. A preconditioning technique for indefinite systems resulting from mixed approximations of elliptic problems. Math. Comput., 50(181): 1–17, 1988.

[57] Brandwood D. H. A complex gradient operator and its application in adaptive array theory. Proc. IEE, 130: 11–16, 1983.

[58] Branham R. L. Total least squares in astronomy. In Recent Advances in Total Least Squares Techniques and Error-in-Variables Modeling (Van Huffel S. ed.). Philadelphia, PA: SIAM, 1997.

[59] Bregman L. M. The method of successive projection for finding a common point of convex sets. Soviet Math. Dokl., 6: 688–692, 1965.

[60] Brewer J. W. Kronecker products and matrix calculus in system theory. IEEE Trans. Circuits Syst., 25: 772–781, 1978.

[61] Bridges T. J., Morris P. J. Differential eigenvalue problems in which the parameters appear nonlinearly. J. Comput. Phys., 55: 437–460, 1984.

[62] Bro R. PARAFAC: tutorial and applications. Chemometrics and Intelligent Laboratory Systems, 38: 149–171, 1997.

[63] Bro R., de Jong S. A fast non-negatively constrained least squares algorithm. J. Chemometrics, 11(5): 393–401, 1997.

[64] Bro R., Harshman R. A., Sidiropoulos N. D. Modeling multi-way data with linearly dependent loadings. Technical Report 2005-176, KVL, 2005.

[65] Bro R. Multiway analysis in the food industry: models, algorithms and applications, Doctoral dissertation, University of Amsterdam, 1998.

[66] Bro R., Sidiropoulos N. D. Least squares algorithms under unimodality and non-negativity constraints. J. Chemometrics, 12(4): 223–247, 1998.

[67] Brockwell P. J., Davis R. A. Time Series: Theory and Methods. New York: Springer-Verlag, 1987.

[68] Brookes M. Matrix Reference Manual 2004. Available at http://www.ee.ic.ac.uk/hp/staff/dmb/matrix/intro.html (2005).

[69] Bruckstein A. M., Donoho D. L., Elad M. From sparse solutions of systems of equations to sparse modeling of signals and images. SIAM Review, 51(1): 34–81, 2009.

[70] Brunet J. P., Tamayo P., Golub T. R., Mesirov J. P. Metagenes and molecular pattern discovery using matrix factorization. Proc. Nat. Acad. Sci. USA, 101(12): 4164–4169, 2004.

[71] Bunse-Gerstner A. An analysis of the HR algorithm for computing the eigenvalues of a matrix. Linear Alg. Applic., 35: 155–173, 1981.

[72] Byrd R. H., Hribar M. E., Nocedal J. An interior point algorithm for large scale nonlinear programming. SIAM J. Optimization, 1999, 9(4): 877–900.

[73] Byrne C., Censor Y. Proximity function minimization using multiple Bregman projections, with applications to split feasibility and Kullback–Leibler distance minimization. Ann. Oper. Rese., 105: 77–98, 2001.

[74] Cadzow J. A. Spectral estimation: an overdetermined rational model equation approach. Proc. IEEE, 70: 907–938, 1982.

[75] Cai J. F., Candès E. J., Shen Z. A singular value thresholding algorithm for matrix completion. SIAM J. Optim., 20(4): 1956–1982, 2010.

[76] Cai T. T., Wang L., Xu G. New bounds for restricted isometry constants. IEEE Trans. Inform. Theory, 56(9): 4388–4394, 2010.

[77] Candès E. J., Li X., Ma Y., Wright J. Robust principal component analysis? J. ACM, 58(3): 1–37, 2011.

[78] Candès E. J., Plan Y. Matrix completion with noise. Proc. IEEE, 98(6): 925–936, 2010.

[79] Candès E. J., Recht B. Exact matrix completion via convex optimization. Found. Comput. Math., 9: 717–772, 2009.

[80] Candès E. J., Romberg J., Tao T. Stable signal recovery from incomplete and inaccurate information. Commun. Pure Appl. Math., 59: 1207–1233, 2005.

[81] Candès E. J., Tao T. Near optimal signal recovery from random projections: universal encoding strategies. IEEE Trans. Inform. Theory, 52(12): 5406–5425, 2006.

[82] Cao X., Wei X., Han Y., Lin D. Robust face clustering via tensor decomposition. IEEE Trans. Cybernetics, 45(11): 2546–2557, 2015.

[83] Cardoso J. F. On the performance of orthogonal source separation algorithms. In Proc. Europ. Assoc. Signal Processing 94, VII, Edinburgh, U.K., 776C779, 1994.

[84] Cardoso J. F., Souloumiac A. Blind beamforming for non-Gaussian signals. Proc. IEE, F, 40(6): 362–370, 1993.

[85] Cardoso J. F., Souloumiac A. Jacobi angles for simultaneous diagonalization. SIAM J. Matrix Anal. Appl., 17(1): 161–164, 1996.

[86] Carroll C. W. The created response surface technique for optimizing nonlinear restrained systems. Oper. Res., 9: 169–184, 1961.

[87] Carroll J. D., Chang J. Analysis of individual differences in multidimensional scaling via an N-way generalization of Eckart–Young decomposition. Psychometrika, 35: 283–319, 1970.

[88] Cattell R. B. Parallel proportional profiles and other principles for determining the choice of factors by rotation. Psychometrika, 9: 267–283, 1944.

[89] Chan R. H., Ng M. K. Conjugate gradient methods for Toepilitz systems. SIAM Review, 38(3): 427–482, 1996.

[90] Chandrasekaran V., Sanghavi S., Parrilo P. A., Wilisky A S. Rank-sparsity incoherence for matrix decomposition. SIAM J. Optim. 21(2): 572–596, 2011.

[91] Chang C. I., Du Q. Estimation of number of spectrally distinct signal sources in hyperspectral imagery. IEEE Trans. Geosci. Remote Sensing, 42(3): 608–619, 2004.

[92] Chang K. C., Pearson K., Zhang T. On eigenvalue problems of real symmetric tensors. J. Math. Anal. Appl., 290(350): 416–422, 2009.

[93] Chatelin F. Eigenvalues of Matrices. New York: Wiley, 1993.

[94] Chen B., Petropulu A. P. Frequency domain blind MIMO system identification based on second- and higher order statistics. IEEE Trans. Signal Processing, 49(8): 1677–1688, 2001.

[95] Chen S. S., Donoho D. L., Saunders M. A. Atomic decomposition by basis pursuit. SIAM J. Sci. Comput., 20(1): 33–61, 1998.

[96] Chen S. S., Donoho D. L., Saunders M. A. Atomic decomposition by basis pursuit. SIAM Review, 43(1): 129–159, 2001.

[97] Chen W., Chen M., Zhou J. Adaptively regularized constrained total least-squares image restoration. IEEE Trans. Image Processing, 9(4): 588–596, 2000.

[98] Chen Y. Incoherence-optimal matrix completion. IEEE Trans. Inform. Theory, 61(5): 2909–2913, 2015.

[99] Chen Y. L., Hsu C. T., Liao H. Y. M. Simultaneous tensor decomposition and completion using factor priors. IEEE Trans. Pattern Anal. Machine Intell., 36(3): 577–591, 2014.

[100] Chua L. O. Dynamic nonlinear networks: state-of-the-art. IEEE Trans. Circuits Syst., 27: 1024–1044, 1980.

[101] Cichocki A., Cruces S., Amari S. I. Generalized alpha-beta divergences and their application to robust nonnegative matrix factorization. Entropy, 13: 134–170, 2011.

[102] Cichocki A., Lee H., Kim Y. D., Choi S. Non-negative matrix factorization with α-divergence. Pattern Recog. Lett., 29: 1433–1440, 2008.

[103] Cichocki A., Mandic D. P., Phan A. H., Caiafa C. F., Zhou G., Zhao Q., Lathauwer L. D. Tensor decompositions for signal processing applications. IEEE Signal Processing Mag., 34(3): 145–163, 2015.

[104] Cichocki A., Zdunek R., Amari S. I. Nonnegative matrix and tensor factorization. IEEE Signal Processing Mag., 25(1): 142–145, 2008.

[105] Cichocki A., Zdunek R., Amari S. I. Csiszar's divergences for nonnegative matrix factorization: family of new algorithms. In Lecture Notes in Computer Science; Springer: Charleston, SC, USA, 3889: 32–39, 2006.

[106] Cichocki A., Zdunek R., Choi S., Plemmons R., Amari S. I. Novel multi-layer nonnegative tensor factorization with sparsity constraints. Springer LNCS, 4432: 271–280, 2007.

[107] Cirrincione G., Cirrincione M., Herault J., *et al.* The MCA EXIN neuron for the minor component analysis. IEEE Trans. Neural Networks, 13(1): 160–187, 2002.

[108] Clark J. V., Zhou N., Pister K. S. J. Modified nodal analysis for MEMS with multi-energy domains. In Proc. Inter. Conf. on Modeling and Simulation of Microsystems, Semiconductors, Sensors and Actuators. San Diego, USA, 2000. Available at www-bsac.EECS.Berkely.EDU/-cfm/publication.html

[109] Combettes P. L., Pesquet J. C. Proximal splitting methods in signal processing. In Fixed-Point Algorithms for Inverse Problems in Science and Engineering, New York: Springer, 185–212, 2011.

[110] Comon P., Golub G., Lim L. H., Mourrain B. Symmetric tensors and symmetric tensor rank. SM Technical Report 06–02, Stanford University, 2006.

[111] Comon P., Golub G., Lim L. H., Mourrain B. Symmetric tensora and symmetric tensor rank. SIAM J. Matrix Anal. Appl., 30(3): 1254–1279, 2008.

[112] Comon P., Moreau E. Blind MIMO equalization and joint-diagonalization criteria. Proc. 2001 IEEE Inter. Conf. on Acoustics, Speech, and Signal Processing (ICASSP '01), 5: 2749–2752, 2001.

[113] Correa N. M., Adall T., Li Y. O., Calhoun V. D. Canonical correlation analysis for data fusion and group inferences. IEEE Signal Processing Mag., 27(4): 39–50, 2010.

[114] Dai Y. H., Fletcher R. Projected Barzilai–Borwein methods for large-scale box-constrained quadratic programming. Num. Math., 100: 21–47, 2005.

[115] Dai W., Milenkovic O. Subspace pursuit for compressive sensing signal reconstruction. IEEE Trans. Inform. Theory, 55(5): 2230–2249, 2009.

[116] Davila C. E. A subspace approach to estimation of autoregressive parameters from noisy measurements, IEEE Trans. Signal Processing, 46: 531–534, 1998.

[117] Davis G. A fast algorithm for inversion of block Toeplitz matrix. Signal Processing, 43: 3022–3025, 1995.

[118] Davis G., Mallat S., Avellaneda M. Adaptive greedy approximation. J. Constr. Approx., 13(1): 57–98, 1997.

[119] Decell Jr. H. P. An application of the Cayley–Hamilton theorem to generalized matrix inversion. SIAM Review, 7(4): 526–528, 1965.

[120] Delsarte P., Genin Y. The split Levinson algorithm. IEEE Trans. Acoust., Speech, Signal Processing, 34: 471–478, 1986.

[121] Delsarte P., Genin Y. On the splitting of classical algorithms in linear prediction theory. IEEE Trans. Acoust., Speech, Signal Processing, 35: 645–653, 1987.

[122] Dembo R. S., Steihaug T. Truncated-Newton algorithms for large-scale unconstrainted optimization. Math Progr., 26: 190–212, 1983.

[123] Demoment G. Image reconstruction and restoration: overview of common estimation problems. IEEE Trans. Acoust., Speech, Signal Processing, 37(12): 2024–2036, 1989.

[124] Dewester S., Dumains S., Landauer T., Furnas G., Harshman R. Indexing by latent semantic analysis. J. Soc. Inf. Sci., 41(6): 391–407, 1990.

[125] Diamond R. A note on the rotational superposition problem. Acta Crystallogr. Sect. A, 44: 211–216, 1988.

[126] Ding W., Wei Y. Generalized tensor eigenvalue problems. SIAM J. Matrix Anal. Appl., 36(3): 1073–1099, 2015.

[127] Doclo S., Moonen M. GSVD-based optimal filtering for single and multimicrophone speech enhancement. IEEE Trans. Signal Processing, 50(9): 2230–2244, 2002.

[128] Donoho D. L. Compressed sensing. IEEE Trans. Inform. Theory, 52(4): 1289–1306, 2006.

[129] Donoho D. L. For most large underdetermined systems of linear equations, the minimal ℓ^1 solution is also the sparsest solution. Commun. Pure Appl. Math., LIX: 797–829, 2006.

[130] Donoho D. L., Elad M. Optimally sparse representations in general (nonorthogonal) dictionaries via ℓ^1 minimization. Proc. Nat. Acad. Sci. USA, 100(5): 2197–2202, 2003.

[131] Donoho D. L., Huo X. Uncertainty principles and ideal atomic decomposition. IEEE Trans. Inform. Theory, 47(7): 2845–2862, 2001.

[132] Donoho D. L., Tsaig Y. Fast solution of l_1-norm minimization problems when the solution may be sparse. IEEE Trans. Inform. Theory, 54(11): 4789–4812, 2008.

[133] Donoho D. L., Tsaig T., Drori T., Starck J. -L. Sparse solution of underdetermined linear equations by stagewise orthogonal matching pursuit (StOMP). Stanford Univ., Palo Alto, CA, Stat. Dept. Tech. Rep. 2006-02, Mar. 2006.

[134] Drakakis K., Rickard S., de Frein R., Cichocki A. Analysis of financial data using non-negative matrix factorization. International Mathematical Forum 3(38): 1853–1870, 2008.

[135] Drmac Z. Accurate computation of the product-induced singular value decomposition with applications. SIAM J. Numer. Anal., 35(5): 1969–1994, 1998.

[136] Drmac Z. A tangent algorithm for computing the generalized singular value decomposition. SIAM J. Numer. Anal., 35(5): 1804–1832, 1998.

[137] Drmac Z. New accurate algorithms for singular value decomposition of matrix triplets. SIAM J. Matrix Anal. Appl., 21(3): 1026–1050, 2000.

[138] Duda R. O., Hart P E. Pattern Classification and Scene Analysis. New York: Wiley, 1973.

[139] Duncan W. J. Some devices for the solution of large sets of simultaneous linear equations. The London, Edinburgh, anf Dublin Philosophical Magazine and J. Science, Seventh Series, 35: 660–670, 1944.

[140] Eckart C., Young G. The approximation of one matrix by another of lower rank. Psychometrica, 1: 211–218, 1936.

[141] Eckart C., Young G. A Principal axis transformation for non-Hermitian matrices. Null Amer. Math. Soc., 45: 118–121, 1939.

[142] Edelman A., Arias T. A., Smith S. T. The geometry of algorithms with orthogonality constraints. SIAM J. Matrix Anal. Appl., 20(2): 303–353, 1998.

[143] Efron B., Hastie T., Johnstone I., Tibshirani R. Least angle regression. Ann. Statist., 32: 407–499, 2004.

[144] Efroymson G., Steger A., Steeinberg S. A matrix eigenvalue problem. SIAM Review, 22(1): 99–100, 1980.

[145] Eldar Y. C., Oppenheim A. V. MMSE whitening and subspace whitening. IEEE Trans. Inform. Theory, 49(7): 1846–1851, 2003.

[146] Eldar Y. C., Opeenheim A. V. Orthogonal and projected orthogonal matched filter detection. Signal Processing, 84: 677–693, 2004.

[147] Espig M., Schuster M., Killaitis A., Waldren N., *et al.* Tensor calculus library. Available at http://gitorious.org/tensorcalculus/ (2012).

[148] Estienne F., Matthijs N., Massart D. L., Ricoux P., Leibovici D. Multi-way modelling of high-dimensionality electroencephalographic data. Chemometrics Intell. Lab. Syst., 58(1): 59–72, 2001.

[149] Faddeev D. K., Faddeeva V. N. Computational Methods of Linear Algebra. San Francisco: W H Freedman Co., 1963.

[150] Farina A., Golino G., Timmoneri L. Comparison between LS and TLS in adaptive processing for radar systems. IEE P-Radar Sonar Nav., 150(1): 2–6, 2003.

[151] Fernando K. V., Hammarling S. J. A product induced singular value decomposition (PSVD) for two matrices and balanced relation. In Proc. Conf. on Linear Algebra in Signals, Systems and Controls, Society for Industrial and Applied Mathematics (SIAM). PA: Philadelphia, 128–140, 1988.

[152] Fiacco A. V., McCormick G. P. Nonlinear Programming: Sequential Unconstrained Minimization Techniques. New York: Wiley, 1968; or Classics Appl. Math. 4, SIAM, PA: Philadelphia, Reprint of the 1968 original, 1990.

[153] Field D. J. Relation between the statistics of natural images and the response properties of cortical cells. J. Opt. Soc. Amer. A, 4: 2370–2393, 1984.

[154] Figueiredo M. A. T., Nowak R. D., Wright S. J. Gradient projection for sparse reconstruction: application to compressed sensing and other inverse problems. IEEE J. Sel. Top. Signal Process., 1: 586–597, 2007.

[155] Flanigan F. Complex Variables: Harmonic and Analytic Functions, 2nd edn. New York: Dover Publications, 1983.

[156] Fletcher R. Conjugate gradient methods for indefinite systems. In Proc. Dundee Conf. on Num. Anal. New York: Springer-Verlag, 73–89, 1975.

[157] Fletcher R. Practical Methods of Optimization, 2nd edn. New York: John Wiley & Sons, 1987.

[158] Forsgren A., Gill P. E., Wright M. H. Interior methods for nonlinear optimization. SIAM Review, 44: 525–597, 2002.

[159] Forsythe G. E., Golub G. H. On the stationary values of a second-degree polynomial on the unit sphere. J. Soc. Ind. Appl. Math., 13(4): 1050–1068, 1965.

[160] Foucart S., Lai M. J. Sparsest solutions of underdetermined linear systems via l_q-minimization for $0 < q \leq 1$. Appl. Comput. Harmonic Anal., 26(3): 395–407, 2009.

[161] Frangi A. F., Niessen W. J., Viergever M. A. Three-dimensional modeling for functional analysis of cardiac images: a review. IEEE Trans. Medical Imaging, 20(1): 2–5, 2001.

[162] Frankel T. The Geometry of Physics: An Introduction (with corrections and additions). Cambridge: Cambridge University Press, 2001.

[163] Friedman J., Hastie T., Höfling H., Tibshirani R. Pathwise coordinate optimization. Ann. Appl. Statist., 1: 302–332, 2007.

[164] Friedlander M. P., Hatz K. Computing nonnegative tensor factorizations. Available at http://www.optimization-online.org/DBHTML/2006/10/1494.html (2006).

[165] Frobenius G. Über Matrizen aus nicht negativen Elementen. Berlin: S.-B. Preuss Acad. Wiss. 456–477, 1912.

[166] Fuhrmann D. R. An algorithm for subspace computation with applications in signal processing. SIAM J. Matrix Anal. Appl., 9: 213–220, 1988.

[167] Gabay D., Mercier B. A dual algorithm for the solution of nonlinear variational problems via finite element approximations. Comput. Math. Appl., 2: 17–40, 1976.

[168] Galatsanou N. P., Katsaggelos A. K. Methods for choosing the regularization parameter and estimating the noise variance in image restoration and their relation. IEEE Trans. Image Processing, 1(3): 322–336, 1992.

[169] Gander W., Golub G. H., Von Matt U. A constrained eigenvalue problem. Linear Algebra Appl., 114–115: 815–839, 1989.

[170] Gandy S., Recht B., Yamada I. Tensor completion and low-n-rank tensor recovery via convex optimization. Inverse Prob., 27(2): 25 010–25 028, 2011.

[171] Gantmacher F. R. The Theory of Matrices. London: Chelsea Publishing, 1977.

[172] Gersho A., Gray R. M. Vector Quantization and Signal Compression. Boston: Kluwer Acad. Press, 1992.

[173] Gilbert A. C., Muthukrishnan M., Strauss M. J. Approximation of functions over redundant dictionaries using coherence. Proc. 14th Annu. ACM-SIAM Symp. Discrete Algorithms, Jan. 2003.

[174] Gillies A. W. On the classfication of matrix generalized inverse. SIAM Review, 12(4): 573–576, 1970.

[175] Gleser L. J. Estimation in a multivariate "errors in variables" regression model: large sample results. Ann. Statist., 9: 24-44, 1981.

[176] Glowinski R., Marrocco A. Sur l'approximation, par elements finis d'ordre un, et la resolution, par penalisation-dualité, d'une classe de problems de Dirichlet non lineares. Revue Française d'Automatique, Informatique, et Recherche Opérationelle, 9: 41–76, 1975.

[177] Goldstein T., Osher S. The split Bregman method for L1-regularized problems. SIAM J. Imaging Sci., 2(2): 323–343, 2009.

[178] Golub G. H., Van Loan C. F. An analysis of the total least squares problem. SIAM J. Numer. Anal., 17: 883–893, 1980.

[179] Golub G. H., Van Loan C. F. Matrix Computation, 2nd edn. Baltimore: The John Hopkins University Press, 1989.

[180] Gonzaga C. C., Karas E. W. Fine tuning Nesterovs steepest descent algorithm for differentiable convex programming. Math. Program., Ser.A, 138: 141–166, 2013.

[181] Gonzales E. F., Zhang Y. Accelerating the Lee–Seung algorithm for non-negative matrix factorization. Technical report. Department of Computational and Applied Mathematics, Rice University, 2005.

[182] Gorodnitsky I. F., Rao B. D. Sparse signal reconstruction from limited data using FOCUSS: a re-weighted norm minimization algorithm, IEEE Trans. Signal Process., 1997, 45: 600–616.

[183] Gourvenec S., *et al.* CuBatch, 2005. Available at http://www.mathworks.com.

[184] Grassmann H. G. Die Ausdehnungslehre. Berlin: Enslin, 1862.

[185] Gray R. M. On the asymptotic eigenvalue distribution of Toeplitz matrices. IEEE Trans Inform. Theory, 18(6): 267–271, 1972.

[186] Graybill F. A. Matrices with Applications in Statistics. Balmont CA: Wadsworth International Group, 1983.

[187] Graybill F. A., Meyer C. D., Painter R. J. Note on the computation of the generalized inverse of a matrix. SIAM Review, 8(4): 522–524, 1966.

[188] Greville T. N. E. Some applications of the pseudoinverse of a matrix. SIAM Review, 2: 15–22, 1960.

[189] Griffiths J. W. Adaptive array processing: a tutorial. Proc. IEE, Part F, 130: 137–142, 1983.

[190] Grippo L., Sciandrone M. On the convergence of the block nonlinear Gauss–Seidel method under convex constraints. Operations Research Letter, 26: 127–136, 1999.

[191] Guan N., Tao D., Lou Z., Yu B. NeNMF: an optimal gradient method for non-negative matrix factorization. IEEE Trans. Signal Processing, 60(6): 2082–2098, 2012.

[192] Guttman L. Enlargement methods for computing the inverse matrix. Ann Math Statist, 17: 336–343, 1946.

[193] Haghighat M., Abdel-Mottaleb M., Alhalabi W. Discriminant correlation analysis: real-time feature level fusion for multimodal biometric recognition. IEEE Trans. Inform. Forensics Security, 11(9): 1984–1996, 2016.

[194] Hale E. T., Yin W., Zhang Y. Fixed-point continuation for ℓ_1-minimization: methodology and convergence. SIAM J. Optim., 19(3): 1107–1130, 2008.

[195] Halmos P. R. Finite Dimensional Vector Spaces. New York: Springer-Verlag, 1974.

[196] Han X., Wu J., Wang L., Chen Y., Senhadji L., Shu H. Linear total variation approximate regularized nuclear norm optimization for matrix completion. Abstract Appl. Anal., ID 765782, 2014.

[197] Hanchez Y., Dooren P. V. Elliptic and hyperbolic quadratic eigenvalue problems and associated distance problems. Linear Algebra Appl., 371: 31–44, 2003.

[198] Harshman R. A. Foundation of the PARAFAC procedure: models and conditions for an "explanatory" multi-modal factor analysis. UCLA Working papers in Phonetics, 16: 1–84, 1970.

[199] Harshman R. A. Parafac2: mathematical and technical notes. UCLA Working papers in Phonetics, 22: 30–44, 1972.

[200] Harshman R. A., Hong S., Lundy M. E. Shifted factor analysis – Part i: models and properties. J. Chemometrics, 17(7): 363–378, 2003.

[201] Harshman R. A., Lundy M. E. Data preprocessing and the extended PARAFAC model. In Research Methods for Multimode Data Analysis (Law

H. G., Snyder C. W., Hattie J. A., McDonald R. P. eds.). New York: Praeger, pp. 216–284, 1984.

[202] Hazan T., Polak T., Shashua A. Sparse image coding using a 3d nonnegative tensor factorization. Technical Report, The Hebrew University, 2005.

[203] He X., Cai D., Niyogi P. Tensor subspace analysis. Adv. Neural Informat. Process. Syst., 18: 499, 2006.

[204] Heeg R. S., Geurts B. J. Spatial instabilities of the incompressible attachment-line flow using sparse matrix Jacobi–Davidson techniques. Appl. Sci. Res., 59: 315–329, 1998.

[205] Helmke U., Moore J. B. Optimization and Dynamical Systems. London, UK: Springer-Verlag, 1994.

[206] Hendeson H. V., Searle S. R. On deriving the inverse of a sum of matrices. SIAM Review, 23: 53–60, 1981.

[207] Henderson H. V., Searle S. R. The vec-permutation matrix, the vec operator and Kronecker products: a review. Linear Multilinear Alg., 9: 271–288, 1981.

[208] Herzog R., Sachs E. Preconditioned conjugate gradient method for optimal control problems with control and state constraints. SIAM. J. Matrix Anal. and Appl., 31(5): 2291–2317, 2010.

[209] Hestenes M. R. Multiplier and gradient methods, J. Optim. Theory Appl., 4: 303–320, 1969.

[210] Hestenes M. R., Stiefel E. Methods of conjugate gradients for solving linear systems. J. Res. Nat. Bureau of Standards, 49: 409–436, 1952.

[211] Hindi H. A tutorial on convex optimization, in Proc. 2004 American Control Conf., Boston, Masachusetts June 30-July 2, 3252–3265, 2004.

[212] Hitchcock F. L. The expression of a tensor or a polyadic as a sum of products. J. Math. Phys., 6: 164–189, 1927.

[213] Hitchcock F. L. Multilple invariants and generalized rank of a p-way matrix or tensor. J. Math. Phys., 7: 39–79, 1927.

[214] Horn R. A., Johnson C. R. Matrix Analysis. Cambridge: Cambridge University Press, 1985.

[215] Horn R. A., Johnson C. R. Topics in Matrix Analysis. Cambridge: Cambridge University Press, 1991.

[216] Hotelling H. Some new methods in matrix calculation. Ann. Math. Statist, 14: 1–34, 1943.

[217] Hotelling H. Further points on matrix calculation and simultaneous equations. Ann. Math. Statist, 14: 440–441, 1943.

[218] Howland P., Jeon M., Park H. Structure preserving dimension reduction for clustered text data based on the generalized singular value decomposition. SIAM J. Matrix Anal. Appl., 25(1): 165–179, 2003.

[219] Howland P., Park H. Generalizing discriminant analysis using the generalized singular value decomposition. IEEE Trans. Pattern Anal. Machine Intell., 26(8): 995–1006, 2004.

[220] Hoyer P. O. Non-negative matrix factorization with sparseness constraints. J. Machine Learning Res., 5: 1457–1469, 2004.

[221] Hu Y., Zhang D., Ye J., Li X., He X. Fast and accurate matrix completion via truncated nuclear norm regularization. IEEE Trans. Pattern Anal. Machine Intell., 35(9): 2117–2130, 2013.

[222] Hu S., Huang Z. H., Ling C., Qi L. On determinants and eigenvalue theory of tensors. J. Symbolic Comput., 50: 508–531, 2013.

[223] Huang B. Detection of abrupt changes of total least squares models and application in fault detection. IEEE Trans. Control Syst. Technol., 9(2): 357–367, 2001.

[224] Huang L. Linear Algebra in System and Control Theory (in Chinese). Beijing: Science Press, 1984.

[225] Huber P. J. Robust estimation of a location parameter. Ann. Statist., 53: 73–101, 1964.

[226] Huffel S. V., Vandewalle J. The Total Least Squares Problems: Computational Aspects and Analysis. Frontiers in Applied Mathemamatics, vol. 9, Philadelphia: SIAM, 1991.

[227] Hyland D. C., Bernstein D. S. The optimal projection equations for model reduction and the relationships among the methods of Wilson, Skelton and Moore. IEEE Trans. Automatic Control, 30: 1201–1211, 1985.

[228] Jain P. K., Ahmad K. Functional analysis, 2nd edn. New Age International, 1995.

[229] Jain S. K., Gunawardena A. D. Linear Algebra: An Interactive Approach. New York: Thomson Learning, 2003.

[230] Jennings A., McKeown J. J. Matrix Computations. New York: John Wiley & Sons, 1992.

[231] Johnson D. H., Dudgeon D. E. Array Signal Processing: Concepts and Techniques. Englewood Cliffs, NJ: PTR Prentice Hall, 1993.

[232] Johnson L. W., Riess R. D., Arnold J. T. Introduction to Linear Algebra, 5th edn. New York: Prentice-Hall, 2000.

[233] Joho M., Mathis H. Joint diagonalization of correlation matrices by using Gradient methods with application to blind signal separation. In IEEE Proc. SAM, 273–277, 2002.

[234] Joho M., Rahbar K. Joint diagonalization of correlation matrices by using Newton methods with application to blind signal separation. In IEEE Proc. SAM, 403–407, 2002.

[235] Jolliffe I. Principal Component Analysis. New York: Springer-Verlag, 1986.

[236] Jordan C. Memoire sur les formes bilineaires. J. Math. Pures Appl., Deuxieme Serie, 19: 35–54, 1874.

[237] Kalogerias D. S., Petropulu A. P. Matrix completion in colocated MIMO radar: recoverability, bounds and theoretical guarantees. IEEE Trans. Signal Processing, 62(2): 309–321, 2014.

[238] Karahan E., Rojas-Lopez P. A., Bringas-Vega M. L., Valdes-Hernandez P. A., and Valdes-Sosa P. A. Tensor analysis and fusion of multimodal brain images. Proc. IEEE, 103(9): 1531–1559, 2015.

[239] Karmarkar N. A new polynomial-time algorithm for linear programming. Combinatorica, 4(4): 373–395, 1984.

[240] Kato T. A Short Introduction to Perturbation Theory for Linear Operators. New York: Springer-Verlag, 1982.

[241] Kay S. M. Modern Spectral Estimation: Theory and Applications. Englewood Cliffs, NJ: Prentice-Hall, 1988.

[242] Kayalar S., Weinert H. L. Oblique projections: formulas, algorithms, error bounds. Math. Control, Signals, Syst., 2(1): 33–45, 1989.

[243] Kelley C. T. Iterative methods for linear and nonlinear equations. Frontiers in Applied Mathematics, vol. 16, Philadelphia, PA: SIAM, 1995.

[244] Khatri C. G., Rao C. R. Solutions to some functional equations and their applications to characterization of probability distributions. Sankhya: The Indian J. Stat., Series A, 30: 167–180, 1968.

[245] Kiers H. A. L. Towards a standardized notation and terminology in multiway analysis. J. Chemometrics, 14: 105–122, 2000.

[246] Kim J., Park H. Fast nonnegative matrix factorization: an active-set-like method and comparisons. SIAM J. Sci. Comput., 33(6): 3261–3281, 2011.

[247] Kim H., Park H. Sparse non-negative matrix factorizations via alternating non-negativity-constrained least squares for microarray data analysis. Bioinformatics, 23(12): 1495–1502, 2007.

[248] Kim S. J., Koh K., Lustig M., Boyd S., Gorinevsky D. An interior-point method for large-scale ℓ_1-regularized least squares. IEEE J. Selec. Topics Signal Processing, 1(4): 606–617, 2007.

[249] Klein J. D., Dickinson B. W. A normalized ladder form of residual energy ratio algorithm for PARCOR estimation via projections. IEEE Trans. Automatic Control, 28: 943–952, 1983.

[250] Klema V. C., Laub A. J. The singular value decomposition: its computation and some applications. IEEE Trans. Automatic Control, 25: 164–176, 1980.

[251] Klemm R. Adaptive airborne MTI: an auxiliary channel approach. Proc. IEE, Part F, 134: 269–276, 1987.

[252] Kofidis E., Regalia Ph. On the best rank-1 approximation of higher-order supersymmetric tensors. SIAM J. Matrix Anal. Appl. 23: 863–884, 2002.

[253] Kolda T. G. Orthogonal tensor decompositions. SIAM J. Matrix Anal. Appl., 23(1): 243–255, 2001.

[254] Kolda T. G. Multilinear operators for higher-order decompositions. Sandia Report SAND2006-2081, California, Apr. 2006.

[255] Kolda T. G., Bader B. W., Kenny J. P. Higher-order web link analysis using multilinear algebra. Proc. 5th IEEE Inter. Conf. on Data Mining, 242–249, 2005.

[256] Kolda T. G., Bader B. W. The tophits model for higher-order web link analysis. In Workshop on Link Analysis, Counterterrorism and Security, 2006.

[257] Kolda T. G., Bader B. W. Tensor decompositions and applications. SIAM Review, 51(3): 455–500, 2009.

[258] Kolda T. G., Mayo J. R. Shifted power method for computing tensor eigenpairs. SIAM J. Matrix Anal. Appl., 32(4): 1095–1124, 2011.

[259] Kolda T. G., Mayo J. R. An adaptive shifted power method for computing generalized tensor eigenpairs. SIAM J. Matrix Anal. Appl., 35(4): 1563–1581, 2014.

[260] Komzsik L. Implicit computational solution of generalized quadratic eigenvalue problems. Finite Elemnets in Analysis and Design, 37: 799–810, 2001.

[261] Kressner D., Tobler C. htucker-A MATLAB toolbox for tensors in hierarchical Tucker format. MATHICSE, EPF Lausanne. Available at http://anchp.epfl.ch/htucker/ (2012).

[262] Kreutz-Delgado K. Real vector derivatives and gradients. Available at http://dsp.ucsd.edu/ kreutz/PEI05.html (2005).

[263] Kreutz-Delgado K. The complex gradient operator and the calculus. 2005. Available at http://dsp.ucsd.edu/ kreutz/PEI05.html (2005).

[264] Kreyszig E. Advanced Engineering Mathematics, 7th edn. New York: John Wiley & Sons, Inc., 1993.

[265] Krishna H., Morgera S. D. The Levinson recurrence and fast algorithms for solving Toeplitz systems of linear equations. IEEE Trans. Acoust., Speech, Signal Processing, 35: 839–847, 1987.

[266] Kroonenberg P. The three-mode company: a company devoted to creating three-mode software and promoting three-mode data analysis. Available at http://three-mode.leidenuniv.nl/ (2000).

[267] Kruskal J. B. Three-way arrays: rank and uniqueness of trilinear decompositions, with application to arithmetic complexity and statistics. Linear Algebra Appl., 18: 95–138, 1977.

[268] Kruskal J. B. Rank, decomposition, uniqueness for 3-way and N-way arrays, in Multiway Data Analysis (Coppi and Bolasco eds.), North-Holland: Elsevier Science Publishers B.V. 7–18, 1989.

[269] Kumar R. A fast algorithm for solving a Toeplitz system of equations. IEEE Trans Acoust., Speech, Signal Processing, 33: 254–267, 1985.

[270] Kumaresan R. Rank reduction techniques and burst error-correction decoding in real/complex fields. In Proc. 19h Asilomar Conf. Circuits Syst. Comput., CA: Pacific Grove, 1985.

[271] Kumaresan R., Tufts D. W. Estimating the angle of arrival of multiple plane waves. IEEE Trans. Aerospace Electron Syst., 19: 134–139, 1983.

[272] Lahat D., Adali T., Jutten C. Multimodal data fusion: an overview of methods, challenges, prospects. Proc. IEEE, 103(9): 1449–1477, 2015.

[273] Lancaster P. Lambda-Matrices and Vibrating Systems. Oxford: Pergamon Press, 1966.

[274] Lancaster P. Quadratic eigenvalue problems. Linear Alg. Appl., 150: 499–506, 1991.

[275] Lancaster P., Tismenetsky M. The Theory of Matrices with Applications, 2nd edn. New York: Academic, 1985.

[276] Landsberg J. M. Tensors: Geometry and Applications. Graduate Studres Mathematics, vol. 128, Providence RI: American Mathematical Society, 2011.

[277] Langville A. N., Meyer C. D., Albright R., Cox J., Duling D. Algorithms, initializations, convergence for the nonnegative matrix factorization. SAS Technical Report, ArXiv: 1407.7299, 2014.

[278] Lasdon L. Optimization Theory for Large Systems. New York: Macmillan, 1970.

[279] Lathauwer L. D. A link between the canonical decomposition in multilinear algebra and simultaneous matrix diagonalization. SIAM J. Matrix Anal. Appl., 28: 642–666, 2006.

[280] Lathauwer L. D. Decompositions of a higher-order tensor in block terms – Part II: definitions and uniqueness. SIAM J. Matrix Anal. Appl., 30(3): 1033–1066, 2008.

[281] Lathauwer L. D., Moor B. D., Vandewalle J. A multilinear singular value decomposition. SIAM J. Matrix Anal. Appl., 21: 1253–1278, 2000.

[282] Lathauwer L. D., Nion D. Decompositions of a higher-order tensor in block terms – Part III: alternating least squares algorithms. SIAM J. Matrix Anal. Appl., 30(3): 1067–1083, 2008.

[283] Laub A. J., Heath M. T., Paige C. C., Ward R. C. Computation of system balancing transformations and other applications of simultaneous diagonalization algorithms. IEEE Trans. Automatic Control, 32: 115–122, 1987.

[284] Lauritzen S. Graphical Models. London: Oxford University Press, 1996.

[285] Lay D. C. Linear Algebra and Its Applications, 2nd edn. New York: Addison-Wesley, 2000.

[286] Lee D. D., Seung H. S. Learning the parts of objects by non-negative matrix factorization. Nature, 401: 788–791, 1999.

[287] Lee D.D., Seung H. S. Algorithms for non-negative matrix factorization. Advances in Neural Information Proc. 13 (Proc. of NIPS 2000), MIT Press, 13: 556–562, 2001.

[288] Lee K., Bresler Y. Admira: atomic decomposition for minimum rank approximation. IEEE Trans. Inform. Theory, 56(9): 4402–4416, 2010.

[289] Leonard I. E. The matrix exponential. SIAM Review, 38(3): 507–512, 1996.

[290] Letexier D., Bourennane S., Blanc-Talon J. Nonorthogonal tensor matricization for hyperspectral image filtering. IEEE Geosci. Remote Sensing. Lett., 5(1): 3-7, 2008.

[291] Levinson N. The Wiener RMS (root-mean-square) error criterion in filter design and prediction, J. Math. Phys., 25: 261–278, 1947.

[292] Lewis A. S. The mathematics of eigenvalue optimization. Math. Program., 97(1-2): 155–176, 2003.

[293] Li N., Kindermannb S., Navasca C. Some convergence results on the regularized alternating least-squares method for tensor decomposition. Linear Algebra Appl., 438(2): 796–812, 2013.

[294] Li G., Qi L., Yu G. The Z-eigenvalues of a symmetric tensor and its application to spectral hypergraph theory. Numer. Linear Algebra Appl., 20: 1001–1029, 2013.

[295] Li X. L., Zhang X. D. Non-orthogonal approximate joint diagonalization free of degenerate solution. IEEE Trans. Signal Processing, 55(5): 1803–1814, 2007.

[296] Lim L. H. Singular values and eigenvalues of tensors: a variational approach. In Proc. IEEE Inter. Workshop on Computational Advances in Multi-Sensor Adaptive Processing, vol. 1: 129–132, 2005.

[297] Lin C. J. Projected gradient methods for nonnegative matrix factorization. Neural Comput., 19(10): 2756–2779, 2007.

[298] Lin Z., Chen M., Ma Y. The augmented Lagrange multiplier method for exact recovery of corrupted low-rank matrices. Technical Report UILU-ENG-09-2215, Nov. 2009.

[299] Liu H., Daniel L., Wong N. Model reduction and simulation of nonlinear circuits via tensor decomposition. IEEE Trans. Computer-Aided Design of Integ. Circuits Syst., 34(7): 1059–1069, 2015.

[300] Liu J., Musialski P., Wonka P., Ye J. Tensor completion for estimating missing values in visual data. IEEE Trans. Pattern Anal. Machine Intell., 35(1): 208–220, 2013.

[301] Liu S., Trenklerz G. Hadamard, Khatri–Rao, Kronecker and other matrix products. Inter. J. Inform. Syst., 4(1): 160–177, 2008.

[302] Liu X., Sidiropoulos N. D. Cramer–Rao lower bounds for low-rank decomposition of multidimensional arrays, IEEE Trans. Signal Processing, 49(9): 2074–2086, 2001.

[303] Lorch E. R. On a calculus of operators in reflexive vector spaces. Trans. Amer. Math. Soc., 45: 217–234, 1939.

[304] Lu H., Plataniotis K. N., Venetsanopoulos A. N. MPCA: multilinear principal component analysis of tensor objects. IEEE Trans. Neural Networks, 19(1): 18–39, 2008.

[305] Luenberger D. An Introduction to Linear and Nonlinear Programming, 2nd edn. MA: Addison-Wesley, 1989.

[306] Luo Y., Liu T., Tao D., Xu C. Multiview matrix completion for multilabel image classification. IEEE Trans. Image Processing, 24(8): 2355–2369, 2015.

[307] Lütkepohl H. Handbook of Matrices. New York: John Wiley & Sons, 1996.

[308] Lyantse V. E. Some properties of idempotent operators. Troret, i Prikl. Mat, 1: 16–22, 1958.

[309] MacDuffee C. C. The Theory of Matrices. Berlin: Springer-Verlag, 1933.

[310] Magnus J. R, Neudecker H. The commutation matrix: some properties and applications. Ann. Statist., 7: 381–394, 1979.

[311] Magnus J. R., Neudecker H. Matrix Differential Calculus with Applications in Statistics and Econometrics, revised edn. Chichester: Wiley, 1999.

[312] Mahalanobis P. C. On the generalised distance in statistics. Proc. of the Natl. Inst. Sci. India, 2(1): 49–55, 1936.

[313] Mallat S. G., Zhang Z. Matching pursuits with time-frequency dictionaries. IEEE Trans. Signal Processing, 41(12): 3397–3415, 1993.

[314] Manolakis D., Truslow E., Pieper M., Cooley T., Brueggeman M. Detection algorithms in hyperspectral imaging systems: an overview of practical algorithms. IEEE Signal Processing. Mag., 31(1): 24–33, 2014.

[315] Marshall Jr. T. G. Coding of real-number sequences for error correction: a digital signal processing problem. IEEE J. Select. Areas Commun., 2(2): 381–392, 1984.

[316] Mateos G., Bazerque J. A., Giannakis G. B. Distributed sparse linear regression. IEEE Trans. Signal Processing, 58(10): 5262–5276, 2010.

[317] Mathew G., Reddy V. Development and analysis of a neural network approach to Pisarenko's harmonic retrieval method. IEEE Trans. Signal Processing, 42: 663–667, 1994.

[318] Mathew G., Reddy V. Orthogonal eigensubspaces estimation using neural networks. IEEE Trans. Signal Processing, 42: 1803–1811, 1994.

[319] Matthews K. R. Linear algebra notes. Department of Mathematics, University of Queensland, MP274, 1991.

[320] McLeod K., Sermesant M., Beerbaum P., Pennec X. Spatio-temporal tensor decomposition of a polyaffine motion model for a better analysis of pathological left ventricular dynamics. IEEE Trans. Medical Imaging, 34(7): 1562–1575, 2015.

[321] Meerbergen K. Locking and restarting quadratic eigenvalue solvers. SIAM J. Sci. Comput., 22(5): 1814–1839, 2001.

[322] Meng S., Huang L. T., Wang W. Q. Tensor decomposition and PCA jointed algorithm for hyperspectral image denoising. IEEE Signal Processing Lett., 13(7): 897–901, 2016.

[323] Mesarovic V. Z., Galatsanos N. P., Katsaggelos K. Regularized constrained total least squares image restoration. IEEE Trans. Image Processing, 4(8): 1096–1108, 1995.

[324] Meyer C. Matrix Analysis and Applied Linear Algebra. Philadephia: SIAM, 2000.

[325] Michalewicz Z., Dasgupta D., Le Riche R., Schoenauer M. Evolutionary algorithms for constrained engineering problems. Computers & Industrial Eng. J., 30: 851–870, 1996.

[326] Micka O. J., Weiss A. J. Estimating frequencies of exponentials in noise using joint diagonalization. IEEE Trans. Signal Processing, 47(2): 341–348, 1999.

[327] Million E. The Hadamard product. Available at http://buzzard.ups.edu/courses/2007spring/projects/million-paper.pdf (2007).

[328] Mirsky L. Symmetric gauge functions and unitarily invariant norms. Quart J. Math. Oxford, 11: 50–59, 1960.

[329] Mohanty N. Random Signal Estimation and Identification. New York: Van Nostrand Reinhold, 1986.

[330] Moore E. H. General analysis, Part 1. Mem. Amer. Philos. Soc., 1: 1, 1935.

[331] Moreau J. J. Fonctions convexes duales et points proximaux dans un espace Hilbertien. Rep. Paris Acad. Sci., Series A, 255: 2897–2899, 1962.

[332] Moreau E. A generalization of joint-diagonalization criteria for source separation. IEEE Trans. Signal Processing, 49(3): 530–541, 2001.

[333] Morup M. ERPWAVELAB, 2006.

[334] Morup M., Hansen L. K., Herrmann C. S., Parnas J., Arnfred S. M. Parallel factor analysis as an exploratory tool for wavelet transformed event-related EEG. NeuroImage, 29: 938–947, 2006.

[335] Murray F. J. On complementary manifofolds and projections in L_p and l_p. Trans. Amer. Math. Soc, 43: 138–152, 1937.

[336] Nasrabadi N. M. Hyperspectral target detection: an overview of current and future challenges. IEEE Signal Processing Magazine, 31(1); 34–44, 2014.

[337] Natarajan B K. Sparse approximate solutions to linear systems. SIAM J. Comput., 24: 227–234, 1995.

[338] Navasca C., Lathauwer L. D., Kindermann S. Swamp reducing technique for tensor decomposition. In Proc. 16th European Signal Processing Conf., Lausanne, Switzerland, August 25–29, 2008.

[339] Neagoe V. E. Inversion of the Van der Monde Matrix. IEEE Signal Processing Letters, 3: 119–120, 1996.

[340] Needell D., Vershynin R. Uniform uncertainty principle and signal recovery via regularized orthogonal matching pursuit. Found. Comput. Math., 9(3): 317–334, 2009.

[341] Needell D., Vershynin R. Signal recovery from incomplete and inaccurate measurements via regularized orthogonal matching pursuit. IEEE J. Sel. Topics Signal Processing, 4(2): 310–316, 2009.

[342] Needell D., Tropp J. A. CoSaMP: iterative signal recovery from incomplete and inaccurate samples. Appl. Comput. Harmonic Anal., 26(3): 301–321, 2009.

[343] Nesterov Y. A method for solving a convex programming problem with rate of convergence $O(1/k^2)$. Soviet Math. Doklady, 269(3): 543–547, 1983.

[344] Nesterov Y. Introductory Lectures on Convex Optimization: A Basic Course. Boston MA: Kluwer Academic, 2004.

[345] Nesterov Y. Smooth minimization of nonsmooth functions. Math. Program. 103(1): 127–152, 2005.

[346] Nesterov Y., Nemirovsky A. A general approach to polynomial-time algorithms design for convex programming. Report of the Central Economical and Mathematical Institute, USSR Academy of Sciences, Moscow, 1988.

[347] Neumaier A. Solving ill-conditioned and singular linear systems: a tutorial on regularization. SIAM Review, 40(3): 636–666, 1998.

[348] Ng L., Solo V. Error-in-variables modeling in optical fow estimation. IEEE Trans. Image Processing, 10(10): 1528–1540, 2001.

[349] Ni G., Qi L., Bai M. Geometric measure of entanglement and U-eigenvalues of tensors. SIAM J. Matrix Anal. Appl., 35(1): 73–87, 2014.

[350] Nievergelt Y. Total least squares: state-of-the-art regression in numerical analysis. SIAM Review, 36(2): 258–264, 1994.

[351] Nionand D., Sidiropoulos N. D. Tensor algebra and multidimensional harmonic retrieval in signal processing for MIMO radar. IEEE Trans. Signal Processing, 58(11): 5693–5705, 2010.

[352] Noble B., Danniel J. W. Applied Linear Algebra, 3rd edn. Englewood Cliffs, NJ: Prentice-Hall, 1988.

[353] Nocedal J., Wright S. J. Numerical Optimization. New York: Springer-Verlag, 1999.

[354] Nour-Omid B., Parlett B. N., Ericsson T., Jensen P. S. How to implement the spectral transformation. Math. Comput., 48: 663–673, 1987.

[355] Noutsos D. Perron Frobenius theory and some extensions. Department of Mathematics, University of Ioannina, May 2008.

[356] Ohlsson H., Ljung L., Boyd S. Segmentation of ARX-models using sum-of-norms regularization. Automatica, 46: 1107–1111, 2010.

[357] Ohsmann M. Fast cosine transform of Toeplitz matrices, algorithm and applications. IEEE Trans. Signal Processing, 41: 3057–3061, 1993.

[358] Oja E. The nonlinear PCA learning rule in independent component analysis. Neurocomputing, 17: 25–45, 1997.

[359] Olson L., Vandini T. Eigenproblems from finite element analysis of fluid–structure interactions. Comput. Struct., 33: 679–687, 1989.

[360] Ortega J. M., Rheinboldt W. C. Iterative Solution of Nonlinear Equations in Several Variables. New York/London: Academic Press, 1970.

[361] Oseledets I. TT-toolbox 2.2. Available at http://github.com/oseledets/TT-Toolbox/ (2012).

[362] Oseledets I., Saluev T., Savostyanov D. V., Dolgov S. V. ttpy: python implementation of the tensor train, 2014.

[363] Osborne M., Presnell B., Turlach B. A new approach to variable selection in least squares problems. IMA J. Numer. Anal., 20: 389–403, 2000.

[364] Osher S., Burger M., Goldfarb D., Xu J., Yin W. An iterative regularization method for total variation-based image restoration. Multiscale Model. Simul., 4(2): 460–489, 2005.

[365] Ottersten B., Asztely D., Kristensson M., Parkvall S. A statistical approach to subspace based estimation with applications in telecommunications. In Recent Advances in Total Least Squares Techniques and Error-in-Variables Modeling (Van Huffel S. ed.), Philadelphia, PA: SIAM, 1997.

[366] Paatero P. A weighted non-negative least squares algorithm for three-way PARAFAC factor analysis. Chemometr. Intell. Lab. Syst., 38: 223–242, 1997.

[367] Paatero P., Tapper U. Positive matrix factorization: a non-negative factor model with optimal utilization of error estimates of data values. Environmetrics, 5: 111–126, 1994.

[368] Paatero P., Tapper U. Least squares formulation of robust non-negative factor analysis. Chemometrics Intell. Lab, 37: 23–35, 1997.

[369] Paige C. C. Computing the generalized singular value decomposition. SIAM J. Sci. Statist. Comput., 7: 1126–1146, 1986.

[370] Paige C. C., Saunders N. A. Towards a generalized singular value decomposition. SIAM J. Numer. Anal., 18: 269–284, 1981.

[371] Pajunnen P., Karhunen J. Least-squares methods for blind source Separation based on nonlinear PCA. Int. J. of Neural Syst., 8: 601–612, 1998.

[372] Papoulis A. Probability, Random Variables and Stochastic Processes. New York: McGraw-Hill, 1991.

[373] Parikh N., Bord S. Proximal algorithms. Found. Trends Optim., 1(3): 123–231, 2013.

[374] Parlett B. N. The Rayleigh quotient iteration and some generalizations for nonormal matrices. Math. Comput., 28(127): 679–693, 1974.

[375] Parlett B. N. The Symmetric Eigenvalue Problem. Englewood Cliffs, NJ: Prentice-Hall, 1980.

[376] Parra L., Spence C., Sajda P., Ziehe A., Muller K. Unmixing hyperspectral data. In Advances in Neural Information Processing Systems, vol. 12. Cambridge, MA: MIT Press, 942–948, 2000.

[377] Pati Y. C., Rezaiifar R., Krishnaprasad P. S. Orthogonal matching pursuit: recursive function approximation with applications to wavelet decomposition. In Proc. 27th Ann. Asilomar Conf. Signals Syst. Comput., vol. 1, 40–44, 1993.

[378] Pauca V. P., Piper J., Plemmons R. Nonnegative matrix factorization for spectral data analysis. Linear Algebra Appl., 416(1): 29–47, 2006.

[379] Pauca V. P., Shahnaz F., Berry M. W., Plemmons R. J. Text mining using non- hnegative matrix factorizations. In Proc. 4th SIAM Inter. Conf. on Data Mining, Lake Buena Vista, Florida, April 22-24, 2004.

[380] Pavon M. New results on the interpolation problem for continue-time stationary-increments processes. SIAM J. Control Optim., 22: 133–142, 1984.

[381] Pearson K. On lines and planes of closest fit to points in space. Phil. Mag, 559–572, 1901.

[382] Pease M. C. Methods of Matrix Algebra. New York: Academic Press, 1965.

[383] Peng C. Y., Zhang X. D. On recursive oblique projectors. IEEE Signal Processing Lett., 12(6): 433–436, 2005.

[384] Penrose R. A. A generalized inverse for matrices. Proc. Cambridge Philos. Soc., 51: 406–413, 1955.

[385] Perron O. Zur Theorie der Matrices. Math. Ann., 64(2): 248–263, 1907.

[386] Petersen K. B., Petersen M. S. The Matrix Cookbook. Available at http://matrixcookbook.com (2008).

[387] Pham D. T. Joint approximate diagonalization of positive defnite matrices. SIAM J. Matrix Anal. Appl., 22(4): 1136–1152, 2001.

[388] Phan A. H., Tichavsky P., Cichocki A. TENSORBOX: a MATLAB package for tensor decomposition, LABSP, RIKEN, Japan. Available at http://www.bsp.brain.riken.jp/phan/tensorbox.php/ (2012).

[389] Phan A. H., Tichavsky P., Cichocki A. Low complexity damped Gauss–Newton algorithms for CANDECOMP/PARAFAC. SIAM J. Matrix Anal. Appl., 34: 126–147, 2013.

[390] Phillip A., Regalia P. A., Mitra S. Kronecker propucts, unitary matrices and signal processing applications. SIAM Review, 31(4): 586–613, 1989.

[391] Piegorsch W. W., Casella G. The early use of matrix diagonal increments in statistical problems. SIAM Review, 31: 428–434, 1989.

[392] Piegorsch W. W., Casella G. Erratum: inverting a sum of matrices. SIAM Review, 32: 470, 1990.

[393] Pintelon R., Guillaume P., Vandersteen G., Rolain Y. Analyzes, development and applications of TLS algorithms in frequency domain system identification. In Recent Advances in Total Least Squares Techniques and Error-in-Variable Modeling (Van Huffel S. ed.), Philadelphia, PA: SIAM, 1997.

[394] Pisarenko V. F. The retrieval of harmonics from a covariance function. Geophysics, J. Roy. Astron. Soc., 33: 347–366, 1973.

[395] Piziak R., Odell P. L. Full rank factorization of matrices. Math. Mag., 72(3): 193–202, 1999.

[396] Polyak B. T. Introduction to Optimization. New York: Optimization Software Inc., 1987.

[397] Poularikas A. D. The Handbook of Formulas and Tables for Signal Processing. New York: CRC Press, Springer, IEEE Press, 1999.

[398] Powell M. J. D. A method for nonlinear constraints in minimization problems. In Optimization (Fletcher R. ed.), New York: Academic Press, 283–298, 1969.

[399] Powell M. J. D. On search directions for minimization algorithms. Math. Progr., 4: 193–201, 1973.

[400] Price C. The matrix pseudoinverse and minimal variance estimates. SIAM Review, 6: 115–120, 1964.

[401] Pringle R. M., Rayner A. A. Generalized Inverse of Matrices with Applications to Statistics. London: Griffin 1971.

[402] Puig A. T., Wiesel A., Fleury G., Hero A. O. Multidimensional shrinkage-thresholding operator and group LASSO penalties. IEEE Signal Processing Lett., 18(6): 343–346, 2011.

[403] Qi L. Eigenvalues of a real supersymmetric tensor. J. Symb. Comput. 40: 1302–1324, 2005.

[404] Qi L., Teo K. L. Multivariate polynomial minimization and its application in signal processing. J. Global Optim., 26: 419–433, 2003.

[405] Qi L., Wang Y., Wu E. X. D-eigenvalues of diffusion kurtosis tensors, J. Comput. Appl. Math., 221: 150–157, 2008.

[406] Qi L., Xu C., Xu Y. Nonnegative tensor factorization, completely positive tensors, hierarchical elimination algorithm. SIAM J. Matrix Anal. Appl., 35(4): 1227–1241, 2014.

[407] Rado R. Note on generalized inverse of matrices. Proc. Cambridge Philos. Soc., 52: 600–601, 1956.

[408] Ram I., Elad M., Cohen I. Image processing using smooth ordering of its patches. IEEE Trans. Image Process., 22(7): 2764–2774, 2013.

[409] Rao C. R. Estimation of heteroscedastic variances in linear models. J. Amer. Statist. Assoc., 65: 161–172, 1970.

[410] Rao C. R., Mitra S. K. Generalized Inverse of Matrices. New York: John Wiley & Sons, 1971.

[411] Rayleigh L. The Theory of Sound, 2nd edn. New York: Macmillian, 1937.

[412] Recht B., Fazel M., Parrilo P. A. Guaranteed minimum-rank solutions of linear matrix equations via nuclear norm minimization. SIAM Review, 52(3): 471–501, 2010.

[413] Regalia P. A., Mitra S. K. Kronecker products, unitary matrices and signal processing applications. SIAM Review, 31(4): 586–613, 1989.

[414] Riba J., Goldberg J., Vazquez G. Robust beamforming for interference rejection in mobile communications. IEEE Trans. Signal Processing, 45(1): 271–275, 1997.

[415] Robeva E. Orthogonal decomposition of symmetric tensors. SIAM J. Matrix Anal. Appl., 37(1): 86–102, 2016.

[416] Roos C. A full-Newton step $O(n)$ infeasible interior-point algorithm for linear optimization. SIAM J. Optim., 16(1): 1110–1136, 2006.

[417] Roos C., Terlaky T., Vial J. Ph. Theory and Algorithms for Linear Optimization: An Interior-Point Approach. Chichester, UK: John Wiley & Sons, 1997.

[418] Roth V. The Generalized LASSO. IEEE Trans. Neural Networks, 15(1): 16–28, 2004.

[419] Roy R., Kailath T. ESPRIT – Estimation of signal parameters via rotational invariance techniques. IEEE Trans. Acoust., Speech, Signal Processing, 37: 297–301, 1989.

[420] Rudin L., Osher S., Fatemi E. Nonlinear total variation based noise removal algorithms. Physica D, 60: 259–268, 1992.

[421] Saad Y. The Lanczos biothogonalization algorithm and other oblique projection methods for solving large unsymmetric systems. SIAM J. Numer. Anal., 19: 485-506, 1982.

[422] Saad Y. Numerical Methods for Large Eigenvalue Problems. New York: Manchester University Press, 1992.

[423] Salehi H. On the alternating projections theorem and bivariate stationary stichastic processes. Trans. Amer. Math. Soc., 128: 121–134, 1967.

[424] Samson C. A unified treatment of fast algorithms for identification. Inter. J. Control, 35: 909–934, 1982.

[425] Scales L. E. Introduction to Non-linear Optimization. London: Macmillan, 1985.

[426] Schmidt R. O. Multiple emitter location and signal parameter estimation. Proc RADC Spectral Estimation Workshop, NY: Rome, 243–258, 1979.

[427] Schmidt R. O. Multiple emitter location and signal parameter estimation. IEEE Trans. Antenna Propagat., 34: 276–280, 1986.

[428] Schott J. R. Matrix Analysis for Statistics. Wiley: New York, 1997.

[429] Schoukens J., Pintelon R., Vandersteen G., Guillaume P. Frequency-domain system identification using nonparametric noise models estimated from a small number of data sets. Automatica, 33(6): 1073–1086, 1997.

[430] Schutz B. Geometrical Methods of Mathematical Physics. Cambridge: Cambridge University Press, 1980.

[431] Schultz T., Seidel H. P. Estimating crossing fibers: a tensor decomposition approach. IEEE Trans. Visualization Comput. Graphics, 14: 1635–1642, 2008.

[432] Scutari G., Palomar D. P., Facchinei F., Pang J. S. Convex optimization, game theory, variational inequality theory. IEEE Signal Processing Mag., 27(3): 35–49, 2010.

[433] Searle S. R. Matrix Algebra Useful for Statististics. New York: John Wiley & Sons, 1982.

[434] Shahnaz F., Berry M. W., Pauca V. P., Plemmons R. J. Document clustering using nonnegative matrix factorization. Inf. Processing & Management, 42(2): 373–386, 2006.

[435] Shashua A., Levin A. Linear image coding for regression and classification using the tensor-rank principle. In Proc. IEEE Conf. on Computer Vision and Pattern Recognition, 2001.

[436] Shashua A., Zass R., Hazan T. Multi-way clustering using super-symmetric nonnegative tensor factorization. In Proc. European Conf. on Computer Vision, Graz, Austria, May 2006.

[437] Shavitt I., Bender C. F., Pipano A., Hosteny R. P. The iterative calculation of several of the lowest or highest eigenvalues and corresponding eigenvectors of very large symmetric matrices. J. Comput. Phys., 11: 90–108, 1973.

[438] Sherman J., Morrison W. J. Adjustment of an inverse matrix corresponding to changes in the elements of a given column or a given row of the original matrix (abstract). Ann. Math. Statist., 20: 621, 1949.

[439] Sherman J., Morrison W. J. Adjustment of an inverse matrix corresponding to a change in one element of a given matrix. Ann. Math. Statist., 21: 124–127, 1950.

[440] Shewchuk J. R. An introduction to the conjugate gradient method without the agonizing pain. Available at http://quake-papers/painless-conjugate-gradient-pics.ps (1994).

[441] Shivappa S. T., Trivedi M. M., Rao B. D. Audiovisual information fusion in human-computer interfaces and intelligent environments: a survey. Proc. IEEE, 98(10): 1692–1715, 2010.

[442] Sidiropoulos N. D., Bro R. On the uniqueness of multilinear decomposition of N-way arrays. J. Chemometrics, 14: 229–239, 2000.

[443] Sidiropoulos N. D., Bro R. On communication diversity for blind identifiability and the uniqueness of low-rank decomposition of N-way arrays. In Proc. Int. Conf. Acoust. Speech Signal Processing, vol. 5: 2449–2452, 2000.

[444] Sidiropoulos N. D., Bro R., Giannakis G. B. Parallel factor analysis in sensor array processing. IEEE Trans. Signal Processing, 48(8): 2377–2388, 2000.

[445] Sidiropoulos N. D., Liu H. Identifiability results for blind beamforming in incoherent multipath with small delay spread. IEEE Trans. Signal Processing, 49(1): 228–236, 2001.

[446] Sidiropoulos N. D., Papalexakis E. E., Faloutsos C. Parallel randomly compressed cubes: a scalable distributed architecture for big tensor decomposition. IEEE Signal Processing Mag., 33(9): 57–70, 2014.

[447] Silva V., Lim L. H. Tensor rank and the ill-posedness of the best low-rank approximation problem, SIAM J. Matrix Anal. Appl., 30(3): 1084–1127, 2008.

[448] Simon J. C. Patterns and Operators: The Foundations and Data Representation. North Oxford Academic Publishers, 1986.

[449] Solomentsev E. D. Euclidean space. Encyclopedia of Mathematics. Springer, 2011.

[450] Sorber L., Barel M. V., Lathauwer D. L. Tensorlab v2.0. Available at http://www.tensorlab.net/ (2014).

[451] Sorensen M., Lathauwer L. D. Blind signal separation via tensor decomposition with Vandermonde factor: canonical polyadic decomposition. IEEE Trans. Signal Processing, 61(22): 5507–5519, 2013.

[452] Spivak M. A Comprehensive Introduction to Differential Geometry, 2nd edn. (five volumes). New York: Publish or Perish Press, 1979.

[453] Stein D. W. J., Beaven S. G., Hoff L. E., Winter E. M., Schaum A. P., Stocker A. D. Anomaly detection from hyperspectral imagery. IEEE Signal Processing Mag., 19(1): 58–69, 2002.

[454] Stewart G. W. On the early history of the singular value decomposition. SIAM Review, 35(4): 551–566, 1993.

[455] Stewart G. W., Sun J. G. Matrix Perturbation Theory. New York: Academic Press, 1990.

[456] Stiefel E. Richtungsfelder und ferparallelismus in n-dimensionalem mannig faltigkeiten. Commentarii Math Helvetici, 8: 305–353, 1935.

[457] Stoica P., Sorelius J., Cedervall M., Söderström T. Error-in-variables modeling: an instrumental variable approach. In Recent Advances in Total Least Squares Techniques and Error-in-Variables Modeling (Van Huffel S. ed.). Philadelphia, PA: SIAM, 1997.

[458] Sun J., Zeng H., Liu H., Lu Y., Chen Z. Cubesvd: a novel approach to personalized web search. In Proc. 14th Inter. Conf. on World Wide Web, 652–662, 2005.

[459] Sun Y., Gao J., Hong X., Mishra B., Yin B. Heterogeneous tensor decomposition for clustering via manifold optimization. IEEE Trans. Pattern Anal. Machine Intell., 38(3): 476–489, 2016.

[460] Syau Y. R. A note on convex functions. Inter. J. Math. & Math. Sci. 22(3): 525–534, 1999.

[461] Takeuchi K., Yanai H., Mukherjee B. N. The Foundations of Multivariante Analysis. New York: Wiley, 1982.

[462] Tibshirani R. Regression shrinkage and selection via the lasso. J. R. Statist. Soc. B, 58: 267–288, 1996.

[463] Tichavsky P., Phan A. H., Koldovsky Z. Cramer–Rao-induced bounds for CANDECOMP/PARAFAC tensor decomposition. IEEE Trans. Signal Processing, 61(8): 1986–1997, 2013.

[464] Tikhonov A. Solution of incorrectly formulated problems and the regularization method, Soviet Math. Dokl., 4: 1035–1038, 1963.

[465] Tikhonov A., Arsenin V. Solution of Ill-Posed Problems. New York: Wiley, 1977.

[466] Tisseur F., Meerbergen K. Quadratic eigenvalue problem. SIAM Review, 43(2): 235–286, 2001.

[467] Toeplitz O. Zur Theorie der quadratischen und bilinearen Formen von unendlichvielen Veränderlichen. I Teil: Theorie der L-Formen, Math Annal, 70: 351–376, 1911.

[468] Tomasi G., Bro R. A comparison of algorithms for fitting the PARAFAC model. Computat. Statist. Data Anal., 50(7): 1700–1734, 2006.

[469] Tou J. T., Gonzalez R. C. Pattern Recognition Principles. London: Addison-Wesley Publishing Comp, 1974.

[470] Tropp J. A. Greed is good: algorithmic results for sparse approximation. IEEE Trans. Inform. Theory, 50(10): 2231–2242, 2004.

[471] Tropp J. A. Just relax: convex programming methods for identifying sparse signals in noise. IEEE Trans. Inform. Theory, 52(3): 1030–1051, 2006.

[472] Tropp J. A., Wright S. J. Computational methods for sparse solution of linear inverse problems. Proc. IEEE, 98(6): 948–958, 2010.

[473] Tsallis C. Possible generalization of Boltzmann–Gibbs statistics. J. Statist. Phys., 52: 479–487, 1988.

[474] Tsallis C. The nonadditive entropy S_q and its applications in physics and elsewhere: some remarks. Entropy, 13: 1765–1804, 2011.

[475] Tucker L. R. Implications of factor analysis of three-way matrices for measurement of change. In Problems in Measuring Change (Harris C. W. ed.), Wisconsin: University of Wisconsin Press, 122–137, 1963.

[476] Tucker L. R. The extension of factor analysis to three-dimensional matrices. In Contributions to Mathematical Psychology (Gulliksen H., Frederiksen N. eds.), New York: Holt, Rinehart & Winston, 109–127, 1964.

[477] Tucker L. R. Some mathematical notes on three-mode factor analysis. Psychometrika, 31: 279–311, 1966.

[478] Turk M. Multimodal interaction: a review. Pattern Recog. Lett., 36(1): 189–195, 2014.

[479] van der Kloot W. A., Kroonenberg P. M. External analysis with three-mode principal component models, Psychometrika, 50: 479-494, 1985.

[480] van der Veen A. J. Joint diagonalization via subspace fitting techniques. Proc. 2001 IEEE Inter. Conf. on Acoust., Speech, Signal Processing, vol. 5: 2773–2776, 2001.

[481] Van Huffel S. (ed.). Recent Advances in Total Least Squares Techniques and Error-in-Variables Modeling. Philadelphia, PA: SIAM, 1997.

[482] Van Huffel S. TLS applications in biomedical signal processing. In Recent Advances in Total Least Squares Techniques and Error-in-Variables Modeling (Van Huffel S. ed.). Philadelphia, PA: SIAM, 1997.

[483] Van Loan C. F. Generalizing the singular value decomposition. SIAM J. Numer. Anal., 13: 76–83, 1976.

[484] Van Loan C. F. Matrix computations and signal processing. In Selected Topics in Signal Processing (Haykin S. ed.). Englewood Cliffs: Prentice-Hall, 1989.

[485] van Overschee P., De Moor B. Subspace Identification for Linear Systems. Boston, MA: Kluwer, 1996.

[486] Van Huffel S., Vandewalle J. Analysis and properties of the generalized total least squares problem $Ax = b$ when some or all columns in A are subject to error. SIAM J. Matrix Anal. Appl., 10: 294–315, 1989.

[487] Vandaele P., Moonen M. Two deterministic blind channel estimation algorithms based on oblique projections. Signal Processing, 80: 481–495, 2000.

[488] Vandenberghe L. Lecture Notes for EE236C (Spring 2011-12), UCLA.

[489] Vanderbei R. J. An interior-point algorithm for nonconvex nonlinear programming. Available at http://orfe.princeton.edu/rvdb/pdf/talks/level3/nl.pdf (2000)

[490] Vanderbei R. J., Shanno D. F. An interior-point algorithm for nonconvex nonlinear programming. Comput. Optim. Appl., 13(1-3): 231–252, 1999.

[491] Vasilescu M. A. O., Terzopoulos D. Multilinear analysis of image ensembles: TensorFaces. In Proc. European Conf. on Computer Vision, Copenhagen, Denmark, 447–460, 2002.

[492] Vasilescu M. A. O., Terzopoulos D. Multilinear image analysis for facial recognition. In Proc. Inter. Conf. on Pattern Recognition, Quebec City, Canada, 2002.

[493] Veen A. V. D. Algebraic methods for deterministic blind beamforming. Proc. IEEE, 86: 1987–2008, 1998.

[494] Veganzones M. A., Cohen J. E., Farias R. C., Chanussot J., Comon P. Non-negative tensor CP decomposition of hyperspectral data. IEEE Trans. Geosci. Remote Sensing, 54(5): 2577–2588, 2016.

[495] Viberg M., Ottersten B. Sensor array processing based on subspace fitting. IEEE Trans. Signal Processing, 39: 1110–1121, 1991.

[496] Vollgraf R., Obermayer K. Quadratic optimization for simultaneous matrix diagonalization. IEEE Trans. Signal Processing, 54(9): 3270–3278, 2006.

[497] Wächter A., Biegler L. T. On the implementation of an interior-point filter line-search algorithm for large-scale nonlinear programming. Math. Progr. Ser. A, 106(1): 25–57, 2006.

[498] Wang H. Coordinate descent algorithm for covariance graphical lasso. Stat. Comput., 24: 521–529, 2014.

[499] Wang H., Ahuja N. Compact representation of multidimensional data using tensor rank-one decomposition. In Proc. Inter. Conf. on Pattern Recognition, Vol. 1, 44–47, 2004.

[500] Wang H., Amini A. Cardiac motion and deformation recovery from MRI: a review. IEEE Trans. Medical Imaging, 31(2): 487–503, 2012.

[501] Wang W., Sanei S., Chambers J. A. Penalty function-based joint diagonalization approach for convolutive blind separation of nonstationary sources. IEEE Trans. Signal Processing, 53(5): 1654–1669, 2005.

[502] Watkins D. S. Understanding the QR algorithm. SIAM Review, 24(4): 427–440, 1982.

[503] Watson G. A. Characterization of the subdifferential of some matrix norms, Linear Algebra Appl., 170: 33–45, 1992.

[504] Wax M., Sheinvald J. A least squares approach to joint diagonalization. IEEE Signal Processing Lett., 4(2): 52–53, 1997.

[505] Weiss A. J., Friedlander B. Array processing using joint diagonalization. Signal Processing, 50(3): 205–222, 1996.

[506] Welling M., Weber M. Positive tensor factorization. Pattern Recognition Lett., 22: 1255–1261, 2001.

[507] Whiteford H. A., Degenhardt L.., *et al.* Global burden of disease attributable to mental and substance use disorders: findings from the Global Burden of Disease Study 2010. Lancet, 382(9904): 1575–1586, 2013.

[508] Wiesel A., Eldar Y. C., Beck A. Maximum likelihood estimation in linear models with Gaussian model matrix, IEEE Signal Processing Lett., 13: 292–295, 2006.

[509] Wiesel A., Eldar Y. C., Yeredor A. Linear regression with Gaussian model uncertainty: algorithms and bounds, IEEE Trans. Signal Processing, 56: 2194–2205, 2008.

[510] Wilkinson J. H. The Algerbaic Eigenvalue Problem. Oxford, UK: Clarendon Press, 1965.

[511] Wirtinger W. Zur formalen Theorie der Funktionen von mehr Komplexen *vromnderlichen*. Math. Ann., 97: 357–375, 1927.

[512] Wolf J. K. Redundancy, the discrete Fourier transform, and impulse noise cancellation. IEEE Trans. Commun, 31: 458–461, 1983.

[513] Woodbury M. A. Inverting modified matrices. Memorandum Report 42, Statistical Research Group, Princeton, 1950.

[514] Wright J., Ganesh A., Rao S., Peng Y., Ma Y. Robust principal component analysis: exact recovery of corrupted low-rank matrices via convex optimization. In Proc. Adv. Neural Inf. Processing Syst., 87(4): 20:3–20:56, 2009.

[515] Xu H., Caramanis C., Mannor S. Robust regression and Lasso. IEEE Trans. Inform. Theory, 56(7): 3561–3574, 2010.

[516] Xu G., Cho Y., Kailath T. Application of fast subspace decomposition to signal processing and communication problems. IEEE Trans. Signal Processing, 42: 1453–1461, 1994.

[517] Xu G., Kailath T. A fast algorithm for signal subspace decomposition and its performance analysis. In Proc. Inter. Conf. of Acoust., Speech and Signal Processing, Toronto, Canada, 3069–3072, 1991.

[518] Xu G., Kailath T. Fast subspace decomposition. IEEE Trans. Signal Processing, 42: 539–551, 1994.

[519] Xu G., Kailath T. Fast estimation of principal eigenspace using Lanczos algorithm. SIAM J. Matrix Anal. Appl. 15(3): 974–994, 1994.

[520] Xu L., Oja E., Suen C. Modified Hebbian learning for curve and surface fitting. Neural Networks, 5: 441–457, 1992.

[521] Yan H., Paynabar K., Shi J. Image-based process monitoring using low-rank tensor decomposition. IEEE Trans. Automation Sci. Eng., 12(1): 216–227, 2015.

[522] Yang B. Projection approximation subspace tracking. IEEE Trans. Signal Processing, 43: 95–107, 1995.

[523] Yang B. An extension of the PASTd algorithm to both rank and subspace tracking. IEEE Signal Processing Lett., 2(9): 179–182, 1995.

[524] Yang J., Yuan X. Linearized augmented Lagrangian and alternating direction methods for nuclear norm minimization. Math. Comput., 82: 301–329, 2013.

[525] Yang J. F., Kaveh M. Adaptive eigensubspace algorithms for direction or frequency estimation and tracking. IEEE Trans. Acoust., Speech, Signal Processing, 36: 241–251, 1988.

[526] Yang X., Sarkar T. K., Arvas E. A survey of conjugate gradient algorithms for solution of extreme eigen-problems of a symmetric matrix. IEEE Trans. Acoust., Speech, Signal Processing, 37: 1550–1556, 1989.

[527] Yang Y., Yang Q. Further results for Perron–Frobenius theorem for nonnegative tensors. SIAM J. Matrix Anal. Appl., 31: 2517–2530, 2010.

[528] Yang W., Gao Y., Shi Y., Cao L. MRM-Lasso: a sparse multiview feature selection method via low-rank analysis. IEEE Trans. Neural Networks Learning Syst., 26(11): 2801-2815, 2015.

[529] Yeniay O., Ankara B. Penalty function methods for constrained optimization with genetic algorithms. Math. Comput. Appl., 10(1): 45–56, 2005.

[530] Yeredor A. Non-orthogonal joint diagonalization in the least squares sense with application in blind source separation. IEEE Trans. Signal Processing, 50(7): 1545–1553, 2002.

[531] Yeredor A. Time-delay estimation in mixtures. In Proc. 2003 IEEE Inter. Conf. on Acoustics, Speech, and Signal Processing, 5: 237–240, 2003.

[532] Yin W., Osher S., Goldfarb D., Darbon J. Bregman iterative algorithms for l_1-minimization with applications to compressed sensing. SIAM J. Imaging Sci., 1(1): 143–168, 2008.

[533] Yokota T., Zhao Q., Cichocki A. Smooth PARAFAC decomposition for tensor completion. IEEE Trans. Signal Processing, 64(20): 5423–5436, 2016.

[534] Youla D. C. Generalized image restoration by the method of altenating projections. IEEE Trans. Circuits Syst., 25: 694–702, 1978.

[535] Yu X., Tong L. Joint channel and symbol estimation by oblique projections. IEEE Trans. Signal Processing, 49(12): 3074–3083, 2001.

[536] Yu Y. L. Nesterov's optimal gradient method. 2009. Available at http://www.webdocs.cs. ualberta.ca/yaoliang/mytalks/NS.pdf (2009).

[537] Yuan M., Lin Y. Model selection and estimation in regression with grouped variables. J. Royal Statist. Soc., Series B, 68(1): 49–67, 2006.

[538] Zass R., Shashua A. Nonnegative sparse PCA. In Proc. Conf. on Neural Information Processing Systems, Vancuver, Canada, 2006.

[539] Zdunek R., Cichocki A. Nonnegative matrix factorization with constrained second-order optimization. Signal Processing, 87(8): 1904–1916, 2007.

[540] Zha H. The restricted singular value decomposition of matrix triplets. SIAM J. Matrix Anal. Appl., 12: 172–194, 1991.

[541] Zhang X., Wen G., Dai W. A tensor decomposition-based anomaly detection algorithm for hyperspectral image. IEEE Trans. Geosci. Remote Sensing, 54(10): 5801–5820, 2016.

[542] Zhang X. D. Numerical computations of left and right pseudo inverse matrices (in Chinese). Kexue Tongbao, 7(2): 126, 1982.

[543] Zhang X. D., Liang Y. C. Prefiltering-based ESPRIT for estimating parameters of sinusoids in non-Gaussian ARMA noise. IEEE Trans. Signal Processing, 43: 349–353, 1995.

[544] Zhang Z. Y. Nonnegative matrix factorization: models, algorithms and applications. Intell. Syst. Ref. Library, 24(6): 99–134, 2012.

[545] Zhou G., Cichocki A. TDALAB: tensor decomposition laboratory, LABSP, Wako-shi, Japan. Available at http://bsp.brain.riken.jp/TDALAB/ (2013).

[546] Zhou H. TensorReg toolbox for Matlab, 2013.

Index